Hepar-, Hepato- liver   Hepatitis (hep-a-TĪT-is), inflammation of the liver.

Hist-, Histio- tissue   Histology (his-TOL-ō-jē), the study of tissues.

Hydr- water   Hydrocele (HI-drō-sēl), accumulation of fluid in a saclike cavity.

Hyster- uterus   Hysterectomy (his′-te-REK-tō-mē), surgical removal of the uterus.

Ileo- ileum   Ileocecal (il′-ē-ō-SĒ-kal) valve, folds at the opening between ileum and cecum.

Ilio- ilium   Iliosacral (il′-ē-ō-SĀ-kral), pertaining to ilium and sacrum.

Kines- motion   Kinesiology (ki-nē-sē-OL-ō-jē), study of movement of body parts.

Labi- lip   Labial (LĀ-bē-al), pertaining to a lip.

Lachry-, Lacri- tears   Nasolacrimal (nā-zō-LAK-rim-al), pertaining to the nose and lacrimal apparatus.

Laparo- loin, flank, abdomen   Laparoscopy (lap′-a-ROS-kōpē), examination of the interior of the abdomen by means of a laparoscope.

Leuco-, Leuko- white   Leukocyte (LYOO-kō-sīt), white blood cell.

Lingua- tongue   Lingual (LIN-gwal), pertaining to the tongue.

Lip-, Lipo fat   Lipoma (li-PŌ-ma), a fatty tumor.

Lith- stone   Lithiasis (li-THĒ-a-sis), the formation of stones.

Lumbo- lower back, loin   Lumbar (LUM-bar), pertaining to the loin.

Macul- spot, blotch   Macula (MAK-yoo-la), spot or blotch.

Malign- bad, harmful   Malignant (ma-LIG-nant), condition that gets worse and results in death.

Mamm- breast   Mammography (ma-MOG-ra-fē), x-ray of the mammary gland.

Mast- breast   Mastitis (ma-STĪT-is), inflammation of the mammary gland.

Meningo- membrane   Meningitis (men-in-JĪT-is), inflammation of the membranes of spinal cord and brain.

Metro- uterus   Endometrium (en′-dō-MĒ-trē-um), lining of the uterus.

Morpho- form, shape   Morphology (mor-FOL-o-jē), the study of form and structure of things.

Myelo- marrow, spinal cord   Poliomyelitis (pō-lē-ō-mī′-a-LĪT-is), inflammation of the gray matter of the spinal cord.

Myo- muscle   Myocardium (mi-ō-KARD-ē-um), heart muscle.

Necro- corpse, dead   Necrosis (ne-KRŌ-sis), death of areas of tissue surrounded by healthy tissue.

Nephro- kidney   Nephrosis (ne-FRŌ-sis), degeneration of kidney tissue.

Neuro- nerve   Neuroblastoma (nyoor′-ō-blas-TŌ-ma), malignant tumor of the nervous system composed of embryonic nerve cells.

Oculo- eye   Binocular (bī-NOK-yoo-lar), pertaining to the two eyes.

Odont- tooth   Orthodontic (or-thō-DONT-ik), pertaining to the proper positioning and relationship of the teeth.

Onco- mass, tumor   Oncology (ong-KOL-ō-jē), study of tumors.

Oo- egg   Oocyte (Ō-ō-sīt), original egg cell.

Oophor- ovary, egg carrier   Oophorectomy (ō′-öf-o-REK-tō-mē), surgical removal of ovaries.

Ophthalm- eye   Ophthalmology (of-thal-MOL-ō-jē), the study of the eye and its diseases.

Or- mouth   Oral (Ō-ral), pertaining to the mouth.

Orchido- testicle   Orchidectomy (or′-ki-DEK-tō-mē), surgical removal of a testicle.

Osmo- odor, sense of smell   Anosmia (an-OZ-mē-a), absence of sense of smell.

Oss-, Osseo-, Osteo- bone   Osteoma (os-tē-Ō-ma), bone tumor.

Oto- ear   Otosclerosis (ō′-tō-skle-RŌ-sis), formation of bone in the labyrinth of the ear.

Palpebr- eyelid   Palpebra (PAL-pe-bra), eyelid.

Part- birth, delivery, labor   Parturition (par′-too-RISH-un), act of giving birth.

Patho- disease   Pathogenic (path′-ō-JEN-ik), causing disease.

Ped- children   Pediatrician (pēd-ē-a-TRISH-an), medical specialist in the treatment of children.

Peps- digest   Peptic (PEP-tik), pertaining to digestion.

Phag-, Phago- to eat   Phagocytosis (fag′-ō-sī-TŌ-sis), the process by which cells ingest particulate matter.

Philic-, Philo- to like, have an affinity for   Hydrophilic (hī-drō-FIL-ik), having an affinity for water.

Phleb- vein   Phlebitis (fle-BĪT-is), inflammation of the veins.

Phon- voice, sound   Phonogram (FŌ-nō-gram), record made of sound.

Phren- diaphragm   Phrenic (FREN-ik), pertaining to the diaphragm.

Pilo- hair   Depilatory (de-PIL-a-tō-rē), hair remover.

Pneumo- lung, air   Pneumothorax (nyoo-mō-THOR-aks), air in the thoracic cavity.

Pod- foot   Podiatry (po-DĪ-a-trē), the diagnosis and treatment of foot disorders.

Procto- anus, rectum   Proctoscopy (prok-TOS-kō-pē), instrumental examination of the rectum.

Psycho- soul, mind   Psychiatry (sī-KĪ-a-trē), treatment of mental disorders.

Pulmon- lung   Pulmonary (PUL-mō-ner′-ē), pertaining to the lungs.

Pyle-, Pyloro opening, passage   Pyloric (pi-LOR-ik), pertaining to the pylorus of the stomach.

Pyo- pus   Pyuria (pī-YOOR-ē-a), pus in the urine.

Ren- kidneys   Renal (RĒ-nal), pertaining to the kidney.

Rhin- nose   Rhinitis (ri-NĪT-is), inflammation of nasal mucosa.

Salpingo- uterine (Fallopian) tube   Salpingitis (sal′-pin-JĪ-tis), inflammation of the uterine (Fallopian) tubes.

Scler-, Sclero- hard   Atherosclerosis (ath′-er-ō-skle-RŌ-sis), hardening of the arteries.

Sep-, Septic- toxic condition due to microorganisms   Septicemia (sep′-ti-SĒ-mē-a), presence of bacterial toxins in the blood (blood poisoning).

Soma-, Somato- body   Somatotropic (sō-mat-ō-TRŌ-pik), having a stimulating effect on body growth.

Somni- sleep   Insomnia (in-SOM-nē-a), inability to sleep.

Stasis-, Stat- stand still   Homeostasis (hō′-mē-ō-STĀ-sis), achievement of a steady state.

Sten- narrow   Stenosis (ste-NŌ-sis), narrowing of a duct or canal.

Tegument- skin, covering   Integumentary (in-teg-yoo-MEN-ta-rē), pertaining to the skin.

Therm- heat   Thermometer (ther-MOM-et-er), instrument used to measure and record heat.

Thromb- clot, lump   Thrombus (THROM-bus), clot in a blood vessel or heart.

Tox-, Toxic- poison   Toxemia (tok-SĒ-mē-a), poisonous substances in the blood.

Trich- hair   Trichosis (trik-Ō-sis), disease of the hair.

Tympan- eardrum   Tympanic (tim-PAN-ik) membrane, eardrum.

Vas- vessel, duct   Cerebrovascular (se-rē-brō-VAS-kyoo-lar), pertaining to the blood vessels of the cerebrum of the brain.

Viscer- organ   Visceral (VIS-e-ral), pertaining to the abdominal organs.

Zoo- animal   Zoology (zō-OL-ō-jē), the study of animals.

Zyg(o)- joined   Zygote (ZĪ-gōt), cell resulting from fertilization of an ovum by a sperm cell.

Prefixes

A-, An- without, lack of, deficient   Anesthesia (an′-es-THĒ-zha), without sensation.

Ab- away from, from   Abnormal (ab-NOR-mal), away from normal.

Ad- to, near, toward   Adduction (a-DUK-shun), movement of a limb toward the axis of the body.

Alb- white   Albino (al-BĪ-no), person whose skin, hair, and eyes lack the pigment melanin.

Alveol- cavity, socket   Alveolus (al-VĒ-o-lus), air sac in the lung.

(continued on back endpapers)

Introduction to the *Human Body*

Aug. 12th - Sat.
Cadaver lab

Food - Brain Chemistry
& Behavior
Jeffery Fortuna
1990's

Gerard J. Tortora

Jerry Tortora is a professor of biology; he teaches human anatomy and physiology and microbiology at Bergen Community College in Paramus, New Jersey, where he is the Biology Coordinator. He has just completed 34 years as a teacher, the past 28 at Bergen CC. He received a B.S. in biology from Fairleigh Dickinson University in 1962 and an M.A. in biology from Montclair State College in 1965. He has also taken graduate courses in education and science at Columbia University and Rutgers University. Professor Tortora belongs to numerous scientific organizations, such as the Human Anatomy and Physiology Society (HAPS), the American Association for the Advancement of Science (AAAS), the American Association of Microbiology (ASM), and the Metropolitan Association of College and University Biologists (MACUB). He is the author of a number of best-selling anatomy and physiology, anatomy, and microbiology textbooks and several laboratory manuals.

Introduction to the

Human Body

the essentials of anatomy and physiology

FOURTH EDITION

Gerard J. Tortora

Biology Coordinator
Bergen Community College

An imprint of Addison Wesley Longman, Inc.

Menlo Park, California • Reading, Massachusetts • New York • Harlow, England
Don Mills, Ontario • Sydney • Mexico City • Madrid • Amsterdam

Executive Editor: Bonnie Roesch
Senior Developmental Editor: Thom Moore
Project Coordination and Text Design: Electronic Publishing Services Inc.
Cover Designer: Yvo Riezebos
Cover Photo: © Corvis-Bettman
Art Coordinator: Claudia Durrell
Photo Researcher: Mira Schachne
Electronic Production Manager: Valerie L. Zaborski
Manufacturing Manager: Helene G. Landers
Electronic Page Makeup: Electronic Publishing Services Inc.
Printer and Binder: RR Donnelley & Sons Company
Cover Printer: Phoenix Color Corp.

For permission to use copyrighted material, grateful acknowledgment is made to the copyright holders on pp. C-1–C-3, which are hereby made part of this copyright page.

Library of Congress Cataloging-in-Publication Data

Tortora, Gerard J.
    Introduction to the human body: the essentials of anatomy and physiology / Gerard J. Tortora.—4th ed.
        p. cm.
    Includes index.
    ISBN 0-673-98222-X
    1. Human physiology.   2. Body, Human.   I. Title
QP36.T67 1996
612–dc20   96-27182
                                                                CIP

0-673-98222-X

5678910—DOW—99

To Angelina M. Tortora,
my mother,
whose love, support, guidance, and patience
have been such an important part of my
personal and professional life

Contents in brief

Contents in detail

14

THE CARDIOVASCULAR SYSTEM: BLOOD 321

15

THE CARDIOVASCULAR SYSTEM: HEART 341

16

THE CARDIOVASCULAR SYSTEM: BLOOD VESSELS 362

Introduction to the Human Body: The Essentials of Anatomy and Physiology, Fourth Edition, is designed for courses in human anatomy and physiology or in human biology. It assumes no previous study of the human body. The successful approach of the previous editions—to provide students with a basic understanding of the structure and functions of the human body with an emphasis on homeostasis—has been retained. In the development of the fourth edition, we focused on improving the acknowledged strengths of the text as well as introducing several new and innovative features.

New to This Edition

A fine illustration program is one of the signature features of *Introduction to the Human Body: The Essentials of Anatomy and Physiology.* Once again, artwork and photographs have been carefully reviewed, revised, and updated to continue the standard of excellence that instructors and students alike have come to expect from this book. Extensive analysis of the previous edition's art program has resulted in the following new features for the fourth edition:

- **New illustrations** have been added throughout the book to amplify both anatomical and physiological concepts. A good example is the new Figure 18.8 (above right), which details the structure of an alveolus (an air sac in the lungs). Notice how the caption, Key Concept Statement (described immediately following), and figure question provide a powerful learning and study framework for the artwork. In addition, many illustrations from the previous edition have been revised for greater clarity. These additions and enhancements are in keeping with the standard of excellence that *Introduction to the Human Body* always strives to exceed.

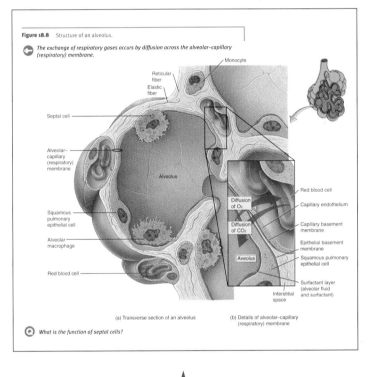

Figure 18.8 Structure of an alveolus.

The exchange of respiratory gases occurs by diffusion across the alveolar–capillary (respiratory) membrane.

(a) Transverse section of an alveolus

(b) Details of alveolar–capillary (respiratory) membrane

What is the function of septal cells?

- **Key Concept Statements** are new and unique to anatomy and physiology textbooks. They are concise statements incorporated into most illustrations, symbolized by a 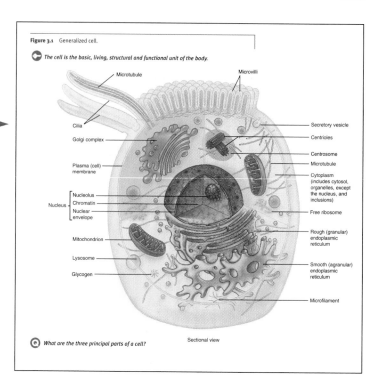 designed to capture the essence of a key concept that has been discussed in the text and then amplified in the illustration. ────────→

Figure 3.1 Generalized cell.

The cell is the basic, living, structural and functional unit of the body.

What are the three principal parts of a cell?

Sectional view

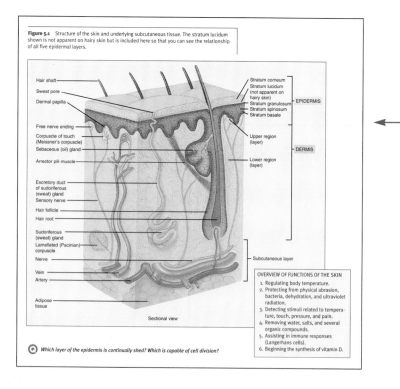

Figure 5.1 Structure of the skin and underlying subcutaneous tissue. The stratum lucidum shown is not apparent on hairy skin but is included here so that you can see the relationship of all five epidermal layers.

Which layer of the epidermis is continually shed? Which is capable of cell division?

OVERVIEW OF FUNCTIONS OF THE SKIN
1. Regulating body temperature.
2. Protecting from physical abrasion, bacteria, dehydration, and ultraviolet radiation.
3. Detecting stimuli related to temperature, touch, pressure, and pain.
4. Removing water, salts, and several organic compounds.
5. Assisting in immune responses (Langerhans cells).
6. Beginning the synthesis of vitamin D.

Sectional view

- **Overview of Functions** is another unique new feature which juxtaposes the anatomical components and a brief functional overview for each body system. These function "boxes" accompany the first figure of chapters dealing with body systems. They permit students to integrate visually the anatomy and physiology of a body system at the outset. ←────────

- **Orientation insets,** introduced in the previous edition, have been greatly modified and expanded for this edition. In one type of inset, not only are planes used to indicate where certain sections are made, but the planes are now labeled so that the reader can more easily relate the planes to the sections that result when a part of the body is cut (see Figure 10.6a). ———————→

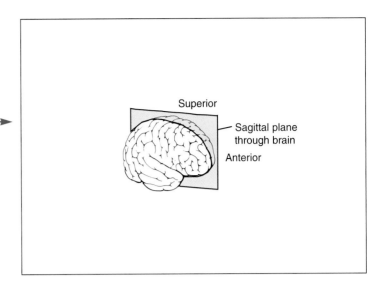

Other insets contain a directional arrow to indicate the direction from which the body is viewed—superior, inferior, posterior, anterior, and so on (see Figure 12.5a). Still other insets have arrows leading from or to them to direct attention to enlarged and detailed parts of illustrations (see Figure 19.8a).

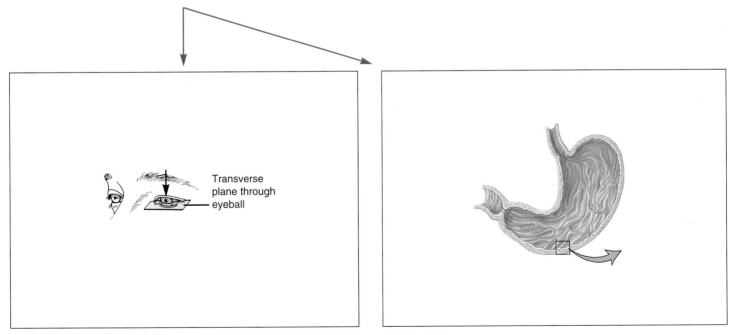

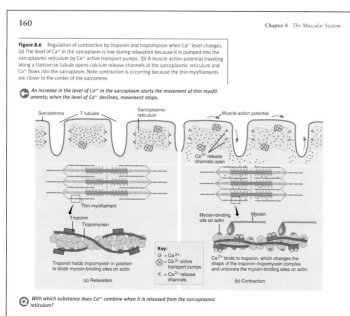

160 — Chapter 8 The Muscular System

Figure 8.6 Regulation of contraction by troponin and tropomyosin when Ca²⁺ level changes. (a) The level of Ca²⁺ in the sarcoplasm is low during relaxation because it is pumped into the sarcoplasmic reticulum by Ca²⁺ active transport pumps. (b) A muscle action potential traveling along a transverse tubule opens calcium release channels in the sarcoplasmic reticulum and Ca²⁺ flows into the sarcoplasm. Note contraction is occurring because the thin myofilaments are closer to the center of the sarcomere.

An increase in the level of Ca²⁺ in the sarcoplasm starts the movement of thin myofilaments; when the level of Ca²⁺ declines, movement stops.

Key:
⊕ = Ca²⁺
⊗ = Ca²⁺ active transport pumps
< = Ca²⁺ release channels

(a) Relaxation — Troponin holds tropomyosin in position to block myosin-binding sites on actin.

(b) Contraction — Ca²⁺ binds to troponin, which changes the shape of the troponin–tropomyosin complex and uncovers the myosin-binding sites on actin.

With which substance does Ca²⁺ combine when it is released from the sarcoplasmic reticulum?

The following sequence occurs during sliding of the filaments (Figure 8.7):

1. When a muscle is relaxed, ATP attaches to ATP-binding sites on the myosin heads (cross bridges). A portion of each myosin head acts as an ATPase, an enzyme that splits the ATP into ADP + ℗. This reaction transfers energy from ATP to the myosin head, even before contraction begins. The myosin heads are thus in an activated (energized) state.

2. When the sarcoplasmic reticulum releases Ca²⁺ and Ca²⁺ level rises in the sarcoplasm, tropomyosin slides away from its blocking position.

3. The activated myosin heads simultaneously bind to the myosin-binding sites on actin.

4. The shape change that occurs when myosin binds to actin produces the **power stroke** of contraction. The power stroke is the force that causes the thin actin myofilaments to slide past the thick myosin myofila-

ments. In other words, during the power stroke, the myosin heads swivel toward the center of the sarcomere, like the oars of a boat. This action draws the thin filaments past the thick filaments of a sarcomere. As the myosin heads swivel, they release ADP.

5. Once the power stroke is complete, ATP again combines with the ATP-binding sites on the myosin heads. As ATP binds, the myosin head detaches from actin.

6. Again, ATP is split, giving its energy to the myosin head, which returns to its original upright position.

7. The myosin head is then ready to combine with another myosin-binding site further along the thin filament.

The cycle of steps 3 through 7 repeats over and over as long as ATP is available and the Ca²⁺ level near the thin myofilaments is high. The myosin heads keep rotating back and forth with each power stroke, pulling the thin myofilaments toward the H zone. At any one instant, about half of the myosin heads are bound to actin and are swiveling. The other

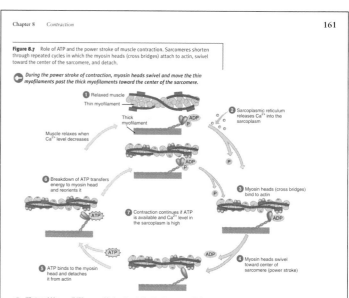

Chapter 8 Contraction — 161

Figure 8.7 Role of ATP and the power stroke of muscle contraction. Sarcomeres shorten through repeated cycles in which the myosin heads (cross bridges) attach to actin, swivel toward the center of the sarcomere, and detach.

During the power stroke of contraction, myosin heads swivel and move the thin myofilaments past the thick myofilaments toward the center of the sarcomere.

1. Relaxed muscle
2. Sarcoplasmic reticulum releases Ca²⁺ into the sarcoplasm
3. Myosin heads (cross bridges) bind to actin
4. Myosin heads swivel toward center of sarcomere (power stroke)
5. ATP binds to the myosin head and detaches it from actin
6. Breakdown of ATP transfers energy to myosin head and reorients it
7. Contraction continues if ATP is available and Ca²⁺ level in the sarcoplasm is high

Muscle relaxes when Ca²⁺ level decreases

What would happen if ATP were suddenly not available after the sarcomere had started to shorten?

half are detached and preparing to swivel again. Contraction is analogous to running on a nonmotorized treadmill. One foot (myosin head) strikes the belt (thin myofilament) and pushes it backward (toward the H zone). Then the other foot comes down and imparts a second push. The belt soon moves smoothly while the runner (thick myofilament) remains stationary. And like the legs of a runner, the myosin heads need a constant supply of energy to keep going!

This continual movement of myosin heads applies the force that draws the Z discs toward each other, and the sarcomere shortens. The myofibrils thus contract, and the whole muscle fiber shortens. During a maximal muscle contraction, the distance between Z discs can decrease to half the resting length.

Relaxation

Two changes permit a muscle fiber to relax after it has contracted. First, acetylcholine is rapidly broken down by the enzyme acetylcholinesterase (AChE). When action potentials cease in the motor neuron, release of ACh stops, and AChE rapidly breaks down the ACh already present in the synaptic cleft. This ends the generation of muscle action potentials, and the Ca²⁺ release channels in the sarcoplasmic reticulum membrane close.

Second, Ca²⁺ is rapidly removed from the sarcoplasm into the sarcoplasmic reticulum. As the Ca²⁺ level drops in the sarcoplasm, the tropomyosin–troponin complex slides back over the myosin-binding sites on actin. This prevents further myosin head binding to actin, and the thin myofilaments slip back to their relaxed positions. Figure 8.8 summarizes the events associated with contraction and relaxation of a muscle fiber.

After death, autolysis begins in muscle fibers and Ca²⁺ leaks out of the sarcoplasmic reticulum. The Ca²⁺ binds to troponin and triggers sliding of the thin myofilaments. ATP production has ceased, however, so the myosin heads cannot detach from actin. The resulting condition, in which muscles are in a state of rigidity (cannot contract or stretch), is called **rigor mortis** (rigidity of death). Rigor mortis lasts about 24 hours but disappears as tissues begin to disintegrate.

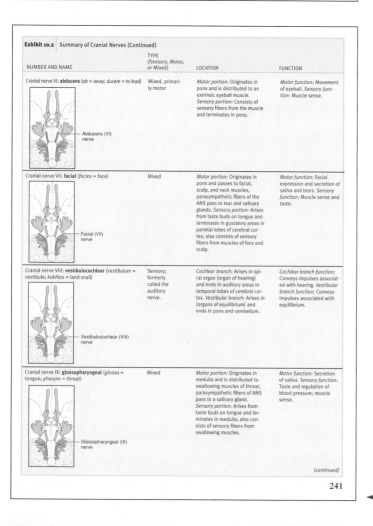

Exhibit 10.2 Summary of Cranial Nerves (Continued)

NUMBER AND NAME	TYPE (Sensory, Motor, or Mixed)	LOCATION	FUNCTION
Cranial nerve VI: **abducens** (ab = away; ducere = to lead)	Mixed, primarily motor	Motor portion: Originates in pons and is distributed to an extrinsic eyeball muscle. Sensory portion: Consists of sensory fibers from the muscle and terminates in pons.	Motor function: Movement of eyeball. Sensory function: Muscle sense.
Cranial nerve VII: **facial** (facies = face)	Mixed	Motor portion: Originates in pons and passes to facial, scalp, and neck muscles, parasympathetic fibers of the ANS pass to tear and salivary glands. Sensory portion: Arises from taste buds on tongue and terminates in gustatory areas in parietal lobes of cerebral cortex; also consists of sensory fibers from muscles of face and scalp.	Motor function: Facial expression and secretion of saliva and tears. Sensory function: Muscle sense and taste.
Cranial nerve VIII: **vestibulocochlear** (vestibulum = vestibule; kokhlos = land snail)	Sensory; formerly called the auditory nerve.	Cochlear branch: Arises in spiral organ (organ of hearing) and ends in auditory areas in temporal lobes of cerebral cortex. Vestibular branch: Arises in (organs of equilibrium) and ends in pons and cerebellum.	Cochlear branch function: Conveys impulses associated with hearing. Vestibular branch function: Conveys impulses associated with equilibrium.
Cranial nerve IX: **glossopharyngeal** (glossa = tongue; pharynx = throat)	Mixed	Motor portion: Originates in medulla and is distributed to swallowing muscles of throat; parasympathetic fibers of ANS pass to a salivary gland. Sensory portion: Arises from taste buds on tongue and terminates in medulla; also consists of sensory fibers from swallowing muscles.	Motor function: Secretion of saliva. Sensory function: Taste and regulation of blood pressure; muscle sense.

(continued)

241

- **Correlation of sequential processes** in text and art is achieved through the use of numbered lists in the text that correspond to numbered segments in the accompanying art. This is done extensively throughout the book and permits the reader to connect more easily the text description with the illustration under consideration.

- **Artwork in Exhibits.** Exhibits summarize important, detailed information in tabular form, and have been made more effective by the addition of artwork to selected exhibits throughout the book.

In addition to changes in the art program, several other new features have been added:

- **Greater emphasis on Critical Thinking** through (1) the addition of new **"Think it over"** questions with the Wellness Focus essays and (2) a chapter-end section containing six **Critical Thinking Applications.** These applications, created by Joan Barber of Delaware Technical and Community College, are essay-style problems that encourage students to think about and apply the concepts they have studied in each chapter.

- **Learning Objectives** now appear at the beginning of each chapter, as well as in the body of the text, where they act as "checkpoints" for student reading. Chapter-opening objectives are also page-referenced.

- **A Look Ahead,** chapter-opening outlines of chapter content, are now page-referenced for greater ease of use.

- **New icons are used throughout the book:**

   Key concept statements are identified by a key.

  Ⓠ Questions with figures by a circle with a Q.

  Common Disorders by a stethoscope.

Organization and Content

Like the previous edition of *Introduction to the Human Body: The Essentials of Anatomy and Physiology,* the fourth edition of the book is divided into 24 chapters, and follows a systemic approach to the study of the human body. Following is a brief summary of some of the significant changes and new features in selected chapters.

In Chapter 3, "Cells," we completely revised the material on the discussion of cancer.

In response to students and instructors, we added new line drawings to accompany the connective tissue photos in Chapter 4, "Tissues."

Our new version of Chapter 6, "The Skeletal System," concentrates on revising the discussions of intramembranous ossification, bone and mineral homeostasis, and osteoporosis.

Chapter 7, "Articulations," now features a new section on arthroscopy.

For Chapter 8, "The Muscular System," we thoroughly updated the discussion of the physiology of muscular contraction. You will also find here a new exhibit on naming skeletal muscles, as well as revisions of several overview sections dealing with skeletal muscles.

Chapter 9, "Nervous Tissue," features revised sections on the organs of the nervous system and grouping of neural tissue.

In Chapter 10, "Central and Somatic Nervous Systems," the discussion of sensory and motor pathways has been revised, and the sections dealing with the cerebellum and spinal cord injury have been expanded. There is also a new exhibit on the functions of the parts of the brain.

In Chapter 13, "The Endocrine System," we have revised the discussions of the comparison of the endocrine and nervous systems and the chemistry of hormones.

The three chapters on the cardiovascular system have been thoroughly revised and updated, too. Chapter 14, "Blood," features a new exhibit on blood cells and a new section on clot-dissolving chemicals. Chapter 15, "Heart," includes a revision of the section on the cardiac cycle. Chapter 16, "Blood Vessels," has a revised section on hypertension.

For Chapter 17, "The Lymphatic System, Nonspecific Resistance, and Immunity," the coverage of immunity and AIDS, and types of T cells have been carefully revised and updated. There are also new sections dealing with functions of antibodies and systemic lupus erythematosus (SLE).

In Chapter 18, "The Respiratory System," the discussion of SIDS has been modified.

Chapter 19, "The Digestive System," now contains new material on hepatitis D and E and a revised discussion of the histology of the stomach and small intestine.

Chapter 20, "Nutrition and Metabolism," has a new section on guidelines for healthy eating.

In Chapter 21, "The Urinary System," the discussions of hormonal regulation of glomerular filtration rate and hemodialysis have been revised.

Chapter 23, "The Reproductive Systems," contains revised sections on testosterone production, histology of the ovaries, oogenesis, and the female reproductive cycle.

In the last chapter, Chapter 24, "Development and Inheritance," we have revised the discussions of embryonic membranes, birth control methods, and variations on dominant–recessive inheritance.

Learning Aids

In response to users of the previous edition of *Introduction to the Human Body,* we have retained the learning aids that students and instructors find most useful, and have tried to improve them wherever possible.

A Look Ahead

These comprehensive outlines provide a quick overview of chapter topics, concepts, and organization. As mentioned before, page references are now provided.

Learning Objectives

In the previous edition, learning objectives were interspersed at key points throughout the text, rather than placed in a list at the beginning of each chapter. However, we discovered that students and instructors wanted it both ways—so now it is, with page references added to the chapter-opening list, too!

Exhibits

Health-science students are generally expected to learn a great deal about the anatomy of certain organ systems. In order to avoid interrupting discussion of concepts, anatomical details have been presented in tabular form in exhibits, many of which are accompanied by illustrations. There are also summary exhibits featuring information about physiological principles and even clinical applications of what the student is learning. Students have told us that they find the exhibits helpful for review, for mental organization of key details and concepts, and for continual self-checking.

Color Coding

As in the previous edition of this book, colors are used in a consistent and meaningful manner throughout the text in order to emphasize structural and functional relations. For example, sensory structures, sensory neurons, and sensory regions of the brain are shades of blue ■, whereas motor structures are light red ■. Membrane phospholipids are gray ■ and orange ■, the cytosol is sand ■, and extracellular fluid is blue ■. Negative and positive feedback loops also use color cues to aid the students in recognizing and understanding the concept. Stimulus and response are both orange ■ since they both alter the controlled condition. The controlled condition is green ■, the receptor is blue ■, the control center is purple ■, and the effector is red ■. Such color cues provide additional help for students who are trying to learn complex anatomical and physiological concepts.

Questions with Figures and Answers to Questions with Figures

We received a very enthusiastic response to the questions with figures in the last edition of this book. Students and instructors have told us that they find these questions really useful for self-quizzing, concept checking, and critical thinking. As in the last edition, the answers to the questions with figures are located at the end of each chapter.

Cross-Referencing

This new edition features more cross-references than ever before, both for specific pages and for chapters. Most cross-references are intended to help students relate new concepts to previously learned material. However, we acknowledge the really ambitious student by including some cross-references to material that has yet to be considered.

Phonetic Pronunciations

We have carefully revised all phonetic pronunciations, which appear in parentheses after many anatomical and physiological terms. The pronunciations are given both in the text, where the term is introduced, and in the Glossary at the end of the book. Even more pronunciations have been added to this edition.

Word Roots

These are derivations designed to provide an understanding of the meaning of new terms. They appear in parentheses when a term is introduced. More than 200 new word roots have been added to this edition.

Wellness Focus Essays

A successful feature of the last two editions has been the "Wellness Focus" essays, written by Barbara Brehm Curtis of Smith College. These essays are intended to increase students' appreciation of the relevancy of the concepts presented in the text to good health. The wellness philosophy supports the notion that life-style choices that individuals make throughout the years have an important influence on their mental and physical well-being. An understanding of the anatomy and physiology of the human body increases your ability to understand how life-style factors such as diet, exercise, and stress management affect the maintenance of health.

Many brand-new Wellness Focus essays appear in this edition, along with revised versions of some of the more popular essays from the previous edition. Also, as mentioned, each essay now has a "think it over" concept application exercise.

We believe that the information contained in these Wellness Focus essays is timely and interesting; we hope students and instructors continue to feel this way, too.

Common Disorders

Located in a special box at the end of appropriate chapters, Common Disorders provides a review of normal body processes, and demonstrates the importance of the study of anatomy and physiology to a career in any of the health fields. These sections answer many questions students ask about medical disorders and diseases. As previously noted, the Common Disorders sections are now marked with a special "stethoscope" icon.

Medical Terminology and Conditions

Vocabulary-building glossaries of selected medical terms and conditions appear at the end of appropriate chapters.

Study Outline

The Study Outline at the end of each chapter summarizes major topics and includes specific page references so that students can easily turn to full text discussions.

Self-Quizzes

In response to feedback from instructors and students, Self-Quizzes have been retained at the end of every chapter. These quizzes are meant not only for students to test their ability to memorize the facts presented in each chapter, but also to sharpen their critical thinking skills by applying the concepts and processes that are part of the way the human body is structured and how it functions. Answers to the Self-Quiz items are presented at the end of the book.

Glossary

A full glossary of terms with phonetic pronunciations appears at the end of the book.

Inside Cover Materials

A "value-added" feature of this book is contained in the endpapers, where the student will find useful information about prefixes, suffixes, word roots, and combining forms of the terminology used in the study of anatomy and physiology.

Supplements

The following supplementary items are available to accompany *Introduction to the Human Body: The Essentials of Anatomy and Physiology,* Fourth Edition. This extensive support system for the text includes printed materials, visual aids, software, and a laserdisk, all carefully produced to enhance teaching and learning environments. For complete information on any of the following items, please call either your local Benjamin Cummings representative or the Benjamin Cummings customer service line at 1-800-447-2226. Or write to:

Marketing Manager—Allied Health Sciences
Benjamin Cummings
2725 Sand Hill Road
Menlo Park, CA 94025

Instructor's Manual

Each chapter in the Instructor's Manual contains a chapter overview, a list of chapter objectives, a complete lecture outline, teaching tips, and a list of additional resources related to the chapter topics.

Testbank

The testbank contains numerous questions in a variety of forms (multiple choice, essay, short answer, true–false, and matching) for each of the 24 chapters in the book. It is available in printed form as well as on *Testmaster* for use with IBM or Macintosh computers.

Quizmaster

Quizmaster is a new program that coordinates with the *Testmaster* test generator program. *Quizmaster* allows students to take timed or untimed computerized tests created with *Testmaster.* Upon completing a test, a student can see his or her score and view or print a diagnostic report that lists topics or objectives that have either been mastered or that need to be restudied. When *Quizmaster* is installed on a network, student scores are saved on disk, and instructors can use the *Quizmaster* utility program to view records and print reports for individual students, class sections, and entire courses.

Transparencies

A set of over 150 transparencies is available to adopters. Illustrations match those in the text. Special care has been taken to enlarge labels and to make the transparencies clear and usable for overhead projection, even in a large lecture hall.

Student Learning Guide

Newly written for the fourth edition, this valuable aid for students includes numerous and varied exercises, labeling and coloring diagrams, and mastery tests. Each chapter begins with a framework that helps students visualize the relationships among key concepts and terms.

Laboratory Manual

A laboratory manual that follows the organization of the text and uses similar illustrations is available for those courses with a lab coordinated to the lecture. Written specifically to accompany this text, the manual focuses on laboratory experiences that will enhance what is being taught in lecture.

Coloring Books

Many students benefit from the interactive and enjoyable activity of studying with one or more of the very popular Benjamin Cummings Coloring Books, including a newly revised *Anatomy Coloring Book,* Second Edition, by Kapit and Elson; *The Physiology Coloring Book,* by Kapit, Macey, and Meisami; and *The Human Brain Coloring Book,* by Diamond, Scheibel, and Elson.

Software

Several software packages are available, supporting both anatomical and physiological study. They can be operated on a variety of electronic platforms, including IBM compatibles and Macintosh, on disks and CD-ROM. Demonstration disks are available for review. Please contact your Benjamin Cummings representative for details on the various packages available for instructor or student use and to receive a review disk.

Videodisc

The Anatomy and Physiology Videodisc, developed and produced in cooperation with Videodiscovery, Inc., is available to qualified adopters. This outstanding work features numerous still illustrations that match those found in the text, both with and without labels, as well as with pointers for quizzing in class. Animations of physiological topics are included. Numerous short movie clips, short filmed segments on applied and clinical situations, extensive histological slides, and the complete *Bassett Atlas of the Human Body* round out the coverage. Please contact your Benjamin Cummings representative for further information or a demonstration.

Acknowledgments

I wish to thank the following people for their helpful contributions to this edition.

Joan Barber of Delaware Technical and Community College thoroughly revised and updated the chapter-end Self-Quizzes, providing many new and challenging items. Joan is also responsible for the excellent—and often wonderfully humorous—chapter-end Critical Thinking Applications; to her I convey a hearty thank you. Special thanks are also due Barbara Brehm Curtis of Smith College, who contributed the Wellness Focus essays and whose work considerably enhances the practical aspect of the text.

I must convey yet another thanks to Joan Barber for her painstaking, close reading of the entire final draft of the manuscript of this edition. Shirley Mulcahy of San Diego Mesa College, like Joan, also read the entire final draft and made many incisive, helpful comments. These two individuals, 3,000 miles apart, arrived at many of the same conclusions about how to make this book better for students. I owe them both an enormous debt of gratitude!

Thom Moore, senior developmental editor, read the entire manuscript and recommended numerous modifications throughout the text. His attention to detail and constructive input have guided the book to its successful completion. Thom's impact on my Benjamin Cummings publications is obvious and significant. Claudia Durrell, the art coordinator, exhibited her usual care and expertise in overseeing the all-important visual elements of the book. It is hard to imagine doing this text without her input. Mira Schachne's work as photo researcher has been of the highest quality. Over the years, I have learned that nothing is impossible for Mira. Her tenacity is a model for us all. Bonnie Roesch, the executive editor, kept a keen eye on the development process and provided continuous support and encouragement. Bonnie's professionalism, pleasant manner, editorial judgment, and abiding wisdom have meant so much to me for so many years.

The following reviewers commented on the entire manuscript and made many helpful suggestions:

Joan Barber, Delaware Technical and Community College

Susan Newton Bruch, De Anza College

William Kleinelp, Jr., Middlesex Community College

Carey Miller, Brookdale Community College

Shirley Mulcahy, San Diego Mesa College

Kevin Petti, San Diego Miramar College

Carl Pratt, College of New Rochelle

Caryl Tickner, Stark Technical College

James Timmons, Lee College

Alexander Varkey, Liberty University

I would like to invite all readers and users of the book to send their reactions and suggestions to me so that plans can be made for subsequent editions.

Gerard J. Tortora
Biology Coordinator
Science and Health, S229
Bergen Community College
400 Paramus Road
Paramus, NJ 07652

Note to the student

Please read carefully the **learning objectives** at the beginning of each chapter, and again as you encounter them in the body of the chapter. Each objective is a statement of a skill or knowledge that you should acquire. To meet these objectives, you will have to perform several activities. Obviously, you must read the section of the chapter that follows the objective carefully. If there is a section that you do not understand after one reading, you should reread it before continuing.

In conjunction with your reading, pay particular attention to the figures and exhibits; they have been carefully coordinated with the textual narrative. **A Look Ahead,** also found at the beginning of each chapter, provides a brief, page-referenced overview of the material to be covered.

Studying the **figures** (illustrations that include artwork and photographs) in this book is as important as reading the text. We have designed the figures so that they will be as effective as possible in your study of the topic. In order to get the most out of the visual parts of this book, you should learn to integrate all the tools we have added to the figures to help you understand the concepts being presented.

Each figure starts with a clearly written **legend** (caption), which describes what you are looking at in the drawing or photograph.

Following the legend, for most figures you will see a **Key Concept Statement** marked with a "key" icon. These are concise statements that capture the essence of a key concept as it was stated in the text and is now reflected in the artwork or photograph.

Added to many figures you will also find an **Orientation Inset** to help you understand the perspective from which you are viewing a particular piece of anatomical art.

The first piece of artwork in the chapters devoted to body systems also features a special boxed area off to the side. This box is called **Overview of Functions** and contains a brief, listed summary of the functions of the body system being presented. The overview of functions is intended to help you visually integrate the anatomy and physiology being presented.

Finally, at the bottom of all figures (except occasional "summary" figures) you will find **Questions with Figures.** If you try to answer these questions as you go along, they will serve as self-checks to help you understand the material. You will find that there are three basic ways to respond to the various questions: (1) Often it will be possible to answer a question by examining the figure itself. Such questions reinforce a visual message by putting what you are looking at in the figure into words. (2) Other questions will encourage you to integrate the knowledge you've gained by carefully reading the text associated with the figure. (3) Still other questions may prompt you to think critically about the topic at hand or predict a consequence in advance of its description in the text. If, after a reasonable amount of reading and thought, you find that you are still stumped for an answer, you may refer to the **Answers to Questions with Figures** at the end of each chapter—but we hope you will need to use the answers only to confirm that you are on the right track!

At the end of each chapter are three—sometimes four—learning guides that you may find useful. The first, **Study Outline,** is a concise summary of important topics discussed in the chapter. This section is designed to consolidate the essential points covered in the chapter, so that you may recall and relate them to one another. For convenience, page numbers are listed next to key concepts so you can easily refer to specific passages in the text for clarification or amplification. A second aid, **Medical Terminology and Conditions,** appears in some chapters. This is a listing of terms or conditions designed to build your medical vocabulary. A third guide, **Self-Quizzes,** is designed to help you evaluate your understanding of the chapter contents. Finally, you will find six **Critical Thinking Applications** at the end of each chapter. These are word problems presenting you with specific situations that allow you to apply the concepts you have studied in the chapter.

As a further aid, we have included **Pronunciations** and, sometimes, **Word Roots,** for many terms that may be new to you. These appear in parentheses immediately following the new words, and the pronunciations are repeated in the glossary of terms at the back of the book. (Of course, because there will always be some disagreement among medical personnel and dictionaries about pronunciation, you will come across variations in different sources.) Look at the words carefully and say them out loud several times. Learning to pronounce a new word will help you remember it and make it a useful part of your medical vocabulary. Take a few minutes now to read the following pronunciation key, so it will be familiar as you encounter new words. The key is repeated at the beginning of the Glossary of Terms, page G-1.

Pronunciation Key

1. The most strongly accented syllable appears in capital letters, for example, bilateral (bī-LAT-er-al) and diagnosis (dī-ag-NŌ-sis).

2. If there is a secondary accent, it is noted by a prime symbol ('), for example, constitution (kon'-sti-TOO-shun) and physiology (fiz'-ē-OL-ō-jē). Any additional secondary accents are also noted by a prime, for example, decarboxylation (dē'-kar-bok'-si-LĀ-shun).

3. Vowels marked by a line above the letter are pronounced with the long sound as in the following common words:

   ā as in *māke*

   ē as in *bē*

   ī as in *īvy*

   ō as in *pōle*

4. Vowels not so marked are pronounced with the short sound as in the following words:

   e as in *bet*

   i as in *sip*

   o as in *not*

   u as in *bud*

5. Other phonetic symbols are used to indicate the following sounds:

   a as in *above*

   oo as in *sue*

   yoo as in *cute*

   oy as in *oil*

student learning objectives

1. Describe the levels of structural organization that compose the human body.   *2*

2. Briefly explain how body systems relate to each other.   *2*

3. Define the life processes of humans.   *5*

4. Define homeostasis and describe its importance in health and disease.   *5*

5. Describe several planes that may be passed through the human body and explain how sections are made.   *8*

ORGANIZATION OF THE HUMAN BODY

a look ahead

You are beginning a study of the human body so that you can learn how it is organized and how it functions. In order to understand what happens to the body when it is injured, diseased, or placed under extreme stress, you must first have a basic understanding of how the body is organized, how its different parts normally work, and the various conditions that will affect the operation of its different parts in order to maintain health and life.

In this chapter, you will be introduced to the various systems that compose the human body. You will also learn how the various systems generally cooperate with each other to maintain the health of the body as a whole. In later chapters, when you have studied body systems in some detail, it will be pointed out how body systems interact to keep you healthy.

Anatomy and Physiology Defined

To understand the structures and functions of the human body, we study the sciences of anatomy and physiology. *Anatomy* (a-NAT-ō-mē; *anatome* = to cut up) refers to the study of *structure* and the relationships among structures. *Physiology* (fiz'-ē-OL-ō-jē) deals with *functions* of the body parts, that is, how the body parts work. Because function can never be separated completely from structure, you will learn about the human body by studying anatomy and physiology together. You will see how each structure of the body is designed to carry out a particular function and how the structure of a part often determines the functions it performs. For example, the hairs lining the nose filter air that we inhale. The bones of the skull are tightly joined to protect the brain. The bones of the fingers, by contrast, are more loosely joined to permit various types of movements.

Levels of Structural Organization

objective: *Describe the levels of structural organization that compose the human body.*

The human body consists of several levels of structural organization that are associated with one another (Figure 1.1). The *chemical level* includes all chemicals needed to maintain life. Chemicals are made up of atoms, the smallest units of matter, and certain of these, such as carbon (C), hydrogen (H), oxygen (O), nitrogen (N), calcium (Ca), potassium (K), and sodium (Na), are essential for maintaining life. Atoms combine to form molecules, two or more atoms joined together. Familiar examples of molecules are proteins, carbohydrates, fats, and vitamins.

Molecules, in turn, combine to form the next higher level of organization: the *cellular level. Cells* are the basic structural and functional units of an organism. Among the many kinds of cells in your body are muscle cells, nerve cells, and blood

cells. Figure 1.1 shows four different types of cells from the lining of the stomach. Each has a different structure, and each performs a different function.

The third higher level of organization is the *tissue level. Tissues* are groups of similar cells that, together, perform a particular function. The four basic types of tissue in the body are *epithelial tissue, connective tissue, muscle tissue,* and *nervous tissue.* The cells in Figure 1.1 form an epithelial tissue that lines the stomach. Each cell has a specific function in digestion.

When different kinds of tissues join together, they form the next higher level of organization: the *organ level. Organs* are composed of two or more different tissues, have specific functions, and usually have a shape that can be recognized. Examples of organs are the heart, liver, lungs, brain, and stomach. Figure 1.1 shows several tissues that make up the stomach. The *serosa* is a layer of connective tissue and epithelial tissue around the outside of the stomach that protects it and reduces friction when the stomach moves and rubs against other organs around it. The *muscle tissue layers* of the stomach are under the serosa and contract to mix food and pass it on to the next digestive organ (small intestine). The *epithelial tissue layer* lining the stomach produces mucus, acid, and enzymes that aid in digestion.

The fifth highest level of organization is the *system level.* A *system* consists of related organs that have a common function. The digestive system, which functions in the breakdown and absorption of food, is composed of these organs: mouth, salivary glands, pharynx (throat), esophagus, stomach, small intestine, large intestine, rectum, liver, gallbladder, and pancreas.

The highest level of organization is the *organismic level.* All the systems of the body functioning with one another make up the total *organism*—one living individual.

The chapters that follow take up the anatomy and physiology of each of the body systems. Exhibit 1.1 on page 4 lists these systems.

How Body Systems Work Together

objective: *Briefly explain how body systems relate to each other.*

As the body systems are studied in more detail in later chapters, you will see how they work together to maintain health, protect you from disease, and allow for reproduction of the species. For now, however, we will consider how just two body systems—the integumentary and skeletal systems—cooperate with each other.

The integumentary system (skin, hair, and nails) protects all body systems, including the skeletal system, by serving as a barrier between the outside environment and internal tissues and organs. The skin is also involved in the production of vitamin D, which the body needs in order to use calcium properly.

Figure 1.1 Levels of structural organization in the human body.

🗝️ *The levels of structural organization are chemical, cellular, tissue, organ, system, and organismic.*

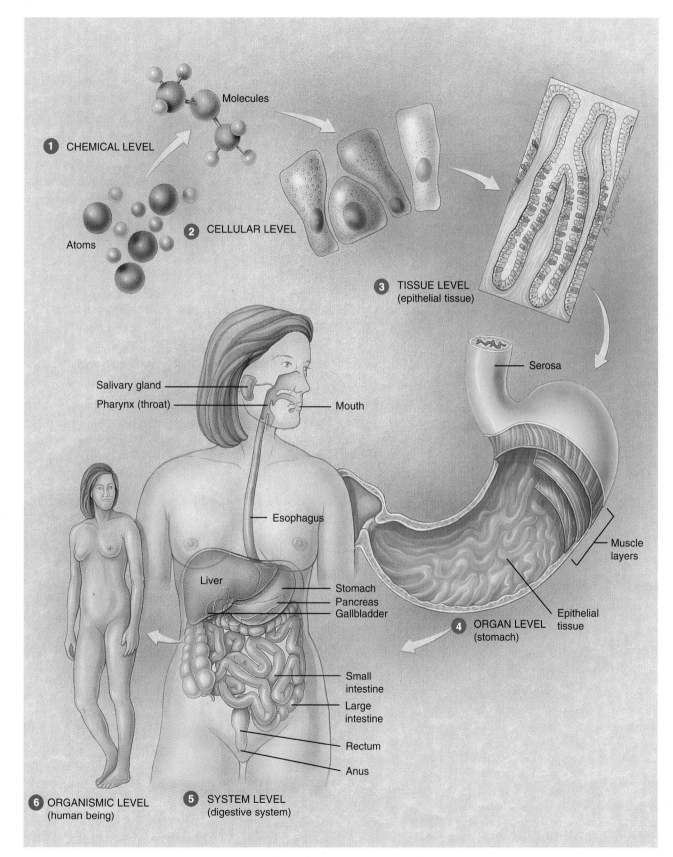

1 CHEMICAL LEVEL

Molecules

Atoms

2 CELLULAR LEVEL

3 TISSUE LEVEL
(epithelial tissue)

Serosa

Muscle layers

Epithelial tissue

4 ORGAN LEVEL
(stomach)

Salivary gland

Pharynx (throat)

Mouth

Esophagus

Liver

Stomach

Pancreas

Gallbladder

Small intestine

Large intestine

Rectum

Anus

6 ORGANISMIC LEVEL
(human being)

5 SYSTEM LEVEL
(digestive system)

Q *Which level of structural organization is composed of two or more different types of tissues and has a usually recognizable shape?*

| **Exhibit 1.1** | Principal Systems of the Human Body, Representative Organs, and Functions |

[handwritten: a structural system]
[handwritten: more of a functional system]

1. Integumentary

Definition: Skin, and structures derived from it, such as hair, nails, and sweat and oil glands.

Function: Helps regulate body temperature, protects the body, eliminates wastes, helps make vitamin D, receives certain stimuli such as temperature, pressure, and pain.

Reference: See Figure 5.1.

2. Skeletal

Definition: All the bones of the body, their associated cartilages, and joints.

Function: Supports and protects the body, assists with body movements, houses cells that produce blood cells, stores minerals.

Reference: See Figure 6.6.

3. Muscular

Definition: Specifically refers to skeletal muscle tissue, which is muscle usually attached to bones (other muscle tissues include smooth and cardiac).

Function: Participates in bringing about movement, maintains posture, produces heat.

Reference: See Figure 8.13.

4. Nervous *[handwritten: Control system of the body]*

Definition: Brain, spinal cord, nerves, and sense organs, such as the eye and ear.

Function: Regulates body activities through nerve impulses by detecting changes in the environment, interpreting the changes, and responding to the changes by bringing about muscular contractions or glandular secretions.

Reference: See Figures 10.2a and 10.6.

5. Endocrine *[handwritten: regulates cellular processes]*

Definition: All glands and tissues that produce chemical regulators of body functions called hormones.

Function: Regulates body activities through hormones transported by the blood of the cardiovascular system to various target organs.

Reference: See Figure 13.1.

6. Cardiovascular

Definition: Blood, heart, and blood vessels.

Function: Distributes oxygen and nutrients to cells, carries carbon dioxide and wastes from cells, helps maintain the acid–base balance of the body, protects against disease, prevents hemorrhage by forming blood clots, helps regulate body temperature.

Reference: See Figures 15.1 and 16.7.

7. Lymphatic and Immune

Definition: Lymph, lymphatic vessels, and structures or organs containing lymphatic tissue (large numbers of white blood cells called lymphocytes), such as the spleen, thymus gland, lymph nodes, and tonsils.

Function: Returns proteins and plasma (liquid portion of blood) to the cardiovascular system, transports fats from the gastrointestinal tract to the cardiovascular system, serves as a site of maturation and proliferation of certain white blood cells, and helps protect against disease by the production of antibodies, as well as other responses.

Reference: See Figure 17.1.

8. Respiratory

Definition: Lungs and associated passageways such as the pharynx (throat), larynx (voice box), trachea (windpipe), and bronchial tubes leading into and out of them.

Function: Supplies oxygen, eliminates carbon dioxide, helps regulate the acid–base balance of the body, helps produce vocal sounds.

Reference: See Figure 18.1.

9. Digestive

Definition: A long tube called the gastrointestinal tract and associated organs such as the salivary glands, liver, gallbladder, and pancreas. *[handwritten: into molecules]*

Function: Breaks down and absorbs food for use by cells, eliminates solid and other wastes.

Reference: See Figure 19.1.

10. Urinary

Definition: Kidneys, ureters, urinary bladder, and urethra that, together, produce, store, and eliminate urine.

Function: Regulates the volume and chemical composition of blood, eliminates wastes *[handwritten: nitrogenous]*, regulates fluid and electrolyte balance and volume, helps maintain the acid–base balance of the body, secretes a hormone that helps regulate red blood cell production.

Reference: See Figure 21.1.

11. Reproductive

Definition: Organs (testes and ovaries) that produce reproductive cells (sperm and ova) and other organs that transport, store, and nourish reproductive cells (vagina, uterine tubes, uterus, vas deferens, urethra, penis).

Function: Reproduces the organism and produces hormones that regulate metabolism.

Reference: See Figures 23.1 and 23.8.

(Calcium is a mineral required for the growth and development of bones.) The skeletal system, in turn, provides support for the integumentary system.

Life Processes

objective: *Define the life processes of humans.*

All living organisms have certain characteristics that set them apart from nonliving things. Following are the important life processes of humans:

1. **Metabolism** (me-TAB-ō-lizm; *metabole* = change) is the sum of all the chemical processes that occur in the body. One phase of metabolism, called **catabolism** (ca-TAB-ō-lizm; *cata* = downward), involves breaking down large, complex molecules into smaller, simpler ones. An example is the splitting of proteins in food into amino acids, the building blocks of proteins. The other phase, called **anabolism** (a-NAB-ō-lizm; *ana* = upward), uses the energy from catabolism to build the body's structural and functional components. An example of anabolism is the synthesis of proteins that make up muscles and bones.

2. **Responsiveness** is the ability to detect and respond to changes in the external environment (the environment outside the body) or internal environment (the environment inside the body). Different cells detect different changes and respond in characteristic ways. For example, neurons (nerve cells) respond by generating electrical signals, known as nerve impulses, and sometimes carrying them over long distances, such as between your big toe and your brain.

3. **Movement** includes motion of the whole body, individual organs, single cells, or even structures inside cells. For example, the coordinated contraction of several leg muscles moves your whole body from place to place when you walk or run. During digestion, food moves out of the stomach into the small intestine.

4. **Growth** refers to an increase in size. It may be due to an increase in the size of existing cells, the number of cells, or the amount of substance surrounding cells.

5. **Differentiation** is the process whereby unspecialized cells become specialized cells. Specialized cells differ in structure and function from the cells from which they originated. For example, following the union of a sperm and ovum, the fertilized egg undergoes tremendous differentiation and progresses through various stages into a unique individual who is similar to, yet quite different from, either of the parents (see Chapter 24).

6. **Reproduction** refers to either the formation of new cells for growth, repair, or replacement or the production of a new individual.

Homeostasis: Maintaining Physiological Limits

objective: *Define homeostasis and describe its importance in health and disease.*

As we have seen, the human body is composed of various systems and organs, each of which consists of millions of cells. These cells need relatively stable conditions in order to function effectively and contribute to the survival of the body as a whole. The maintenance of stable conditions for its cells is an essential function of the human body, which physiologists call homeostasis, one of the major themes of this textbook.

Homeostasis (hō′-mē-ō-STĀ-sis; *homeo* = same; *stasis* = standing still) is a condition in which the body's internal environment remains within certain physiological limits. The internal environment refers to the fluid around body cells, called interstitial (intercellular) fluid, discussed in detail in Chapter 3 (see Figure 3.3). An organism is said to be in homeostasis when its internal environment contains the appropriate concentration of chemicals, is at an appropriate temperature, and is at an appropriate pressure. When homeostasis is disturbed, ill health may result. If body fluids are not eventually brought back into homeostasis, death may occur.

Stress and Homeostasis

Homeostasis may be disturbed by *stress*, which is any stimulus that creates an imbalance in the internal environment. The stress may come from the external environment in the form of stimuli such as heat, cold, or lack of oxygen. Or the stress may originate within the body in the form of stimuli such as high blood pressure, tumors, or unpleasant thoughts. Most stresses are mild and routine. Extreme stress might be caused by poisoning, overexposure to temperature extremes, and surgical operations.

Fortunately, the body has many regulating (homeostatic) devices that may bring the internal environment back into balance. Every body structure, from the cellular to the system level, attempts to keep the internal environment within normal physiological limits.

The homeostatic mechanisms of the body are under the control of the nervous system and the endocrine system. The nervous system regulates homeostasis by detecting when the body deviates from its balanced state and then sending messages (nerve impulses) to the proper organs to counteract the stress. The endocrine system is a group of glands that secrete chemical messengers, called hormones, into the blood. Whereas nerve impulses coordinate homeostasis rapidly, hormones work more slowly. Following is an example of how the nervous system regulates homeostasis.

Homeostasis of Blood Pressure (BP)

Blood pressure (BP) is the force of blood as it passes through blood vessels, especially arteries. In order to sustain life, blood must not only be kept circulating, it must also circulate at an appropriate pressure. For example, if blood pressure is too low, organs of the body, such as the brain, will not receive adequate oxygen and nutrients to function properly. High blood pressure, on the other hand, has adverse effects on organs such as the heart, kidneys, and brain. High blood pressure contributes to the development of heart attacks and stroke. Among other factors, blood pressure depends on the rate and strength of the heartbeat. If some stress causes the heartbeat to speed up, the following sequence occurs (Figure 1.2). As the heart pumps faster, it pushes more blood into the arteries, increasing blood pressure. The higher pressure is detected by pressure-sensitive nerve cells in the walls of certain arteries, which respond by sending nerve impulses to the brain. The brain, in turn, responds by sending impulses to the heart and certain blood vessels to slow the heart rate, thus decreasing blood pressure. The continual monitoring of blood pressure by the nervous system is an attempt to maintain a normal blood pressure and involves what is called a feedback system.

A *feedback system* involves a cycle of events in which information about the status of body conditions is continually monitored and fed back (reported) to a central control region. A feedback system consists of three basic components—control center, receptor, and effector (Figure 1.2).

1. The *control center* determines the point at which some aspect of the body, called a *controlled condition,* should be maintained. In the body, there are hundreds of controlled conditions. The one considered here is blood pressure. Other examples are heart rate, acidity of the blood, blood sugar level, body temperature, and breathing rate. The control center receives information about the status of a controlled condition from a receptor and then determines an appropriate course of action.

2. The *receptor* monitors changes in the controlled condition and then sends the information, called the *input,* to the control center. Any stress that changes a controlled condition is called a *stimulus.* For example, a stimulus such as avoiding hitting someone with your car makes your heart beat faster and this raises blood pressure (the controlled condition). Pressure-sensitive nerve cells in arteries (receptors) send nerve impulses (input) to the control center, which in this case is the brain.

3. The *effector* is the part of the body that receives information, called the *output,* from the control center and produces a *response* (effect). In this example, the brain sends nerve impulses (output) to the heart (effector). Heart rate decreases and blood pressure drops (response). This helps return blood pressure (controlled condition) to normal, and homeostasis is restored.

The response that occurs is continually monitored by the receptor and fed back to the control center. If the response reverses the original stimulus, as in the example just described, the system is a *negative feedback system.* If the response enhances the original stimulus, the system is a *positive feedback system.*

Negative feedback systems, such as the one shown in Figure 1.2, require frequent monitoring and adjustment within physiological limits. Such systems include blood pressure, body temperature, and blood sugar level. Positive feedback systems, on the other hand, are important for conditions that do not occur often and do not require continual fine-tuning. Unlike

Figure 1.2 Homeostasis of blood pressure by a negative feedback system. The response is fed back into the system, and the system continues to lower blood pressure until there is a return to homeostasis. **Note:** Wherever feedback cycle diagrams are used throughout the book, they will be similar to this illustration in terms of style and color.

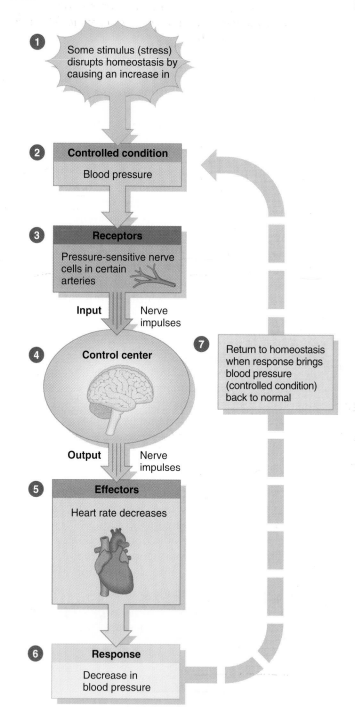

Q *What are the two ways that negative and positive feedback systems differ?*

negative feedback systems, positive feedback systems tend to intensify a controlled condition. For example, in the system shown in Figure 1.2, if the brain had signaled the heart to beat even faster and the blood pressure had continued to rise, the system would be a positive feedback system.

Most of the feedback systems of the body are negative. Although many positive feedback systems can be destructive and result in various disorders, some are normal and beneficial, such as blood clotting and labor contractions in childbirth. Blood clotting helps stop the loss of blood from a wound. When labor contractions begin in childbirth, a certain hormone is released into the blood. This hormone intensifies the contractions, which, in turn, stimulate the release of more of this hormone. The cycle is broken with the birth of the infant.

another example is allergic reaction

Anatomical Position
the Basic position from which every point is named & reffered to

In anatomy, there is universal agreement that descriptions of the human body assume the body is in a specific position, called the **anatomical position.** In the anatomical position, the subject is standing upright, facing the observer, with the upper limbs (extremities) placed at the sides, the palms turned forward, and the feet flat on the floor (Figure 1.3). The common names and anatomical terms for several body regions are also presented in Figure 1.3.

Figure 1.3   Anatomical position. The common names and anatomical terms, in parentheses, are indicated for many regions of the body. For example, the chest is the thoracic region.

Descriptions of the human body assume that it is in a specific position called the anatomical position.

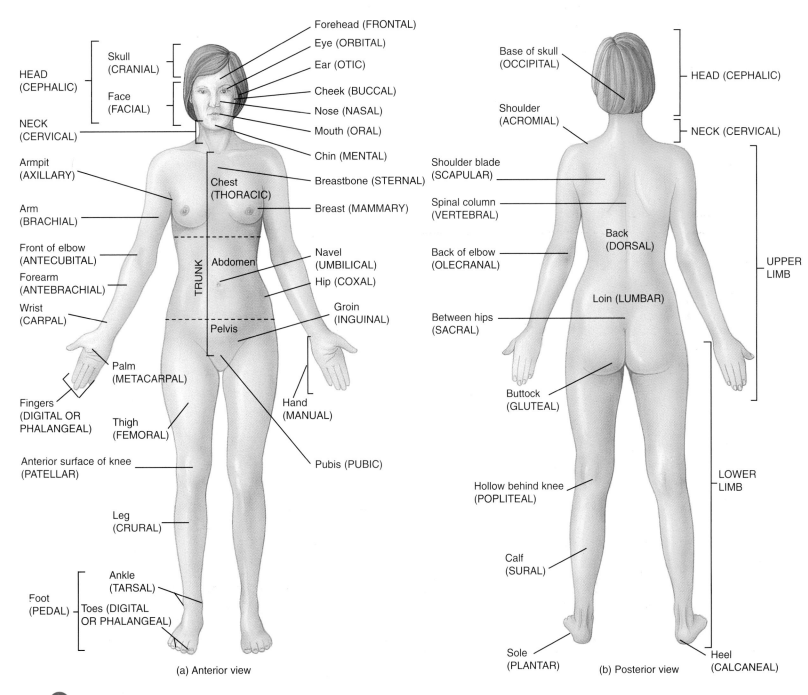

(a) Anterior view

(b) Posterior view

What are the characteristics of the anatomical position?

Directional Terms

To locate various body structures in relationship to each other, anatomists use certain *directional terms*. Directional terms are defined in Exhibit 1.2; the examples given there are shown in Figures 1.4 and 1.8. Study the exhibit and figures together.

Planes and Sections

objective: *Describe several planes that may be passed through the human body and explain how sections are made.*

The human body may also be described in terms of the *planes* (imaginary flat surfaces) that pass through it (Figure 1.5 on page 10). A *sagittal* (SAJ-i-tal; *sagittalis* = arrow) *plane* is a plane that divides the body into right and left sides. A *midsagittal plane* passes through the midline of the body and divides the body into *equal* right and left sides. A *parasagittal* (*para* = near) *plane* does not pass through the midline of the body and divides the body into *unequal* left and right portions. A *frontal* (*coronal*; kō-RO-nal) *plane* is a plane that divides the body into anterior (front) and posterior (back) portions. A *transverse* (*cross-sectional* or *horizontal*) *plane* divides the body into superior (upper) and inferior (lower) portions. An *oblique* (ō-BLĒK) *plane* passes through the body or an organ at an angle between the transverse plane and either the midsagittal, parasagittal, or frontal plane.

When you study a body structure, you will often view it in section, meaning that you look at only one surface of the three-dimensional structure. Figure 1.6 on page 11 indicates how three different sections—a *transverse* (cross) *section*, a *frontal section*, and a *midsagittal section*—provide different views of the brain.

Exhibit 1.2 Directional Terms[a]		
TERM	**DEFINITION**	**EXAMPLE**
Superior (soo'-PEER-ē-or) (**cephalic** or **cranial**)	Toward the head or the upper part of a structure.	The heart is superior to the liver.
Inferior (in'-FEER-ē-or) (caudal)	Away from the head or toward the lower part of a structure.	The stomach is inferior to the lungs.
Anterior (an-TEER-ē-or) (ventral)	Nearer to or at the front of the body.	The heart is anterior to the backbone.
Posterior (pos-TEER-ē-or) (dorsal)	Nearer to or at the back of the body.	The esophagus (food tube) is posterior to the trachea (windpipe).
Medial (MĒ-dē-al)	Nearer to the midline of the body or a structure. The midline is an imaginary vertical line that divides the body into equal left and right sides.	The ulna is on the medial side of the forearm.
Lateral (LAT-er-al)	Farther from the midline of the body or a structure.	The radius is on the lateral side of the forearm.
Intermediate (in'-ter-MĒ-dē-at)	Between two structures.	The ring finger is intermediate between the little and middle fingers.
Proximal (PROK-si-mal)	Nearer to the attachment of a limb to the trunk or a structure; nearer to the point of origin.	The humerus is proximal to the radius.
Distal (DIS-tal)	Farther from the attachment of a limb to the trunk or a structure; farther from the point of origin.	The phalanges (bones of the fingers) are distal to the carpals (wrist bones).
Superficial (soo'-per-FISH-al)	Toward or on the surface of the body.	The sternum (breastbone) is superficial to the heart (see Figure 1.8).
Deep (DĒP)	Away from the surface of the body.	The ribs are deep to the skin of the chest (see Figure 1.8).

[a]Study this exhibit with Figures 1.4 and 1.8. Locate the structures mentioned in each example.

Figure 1.4   Directional terms. Study Exhibit 1.2 with this figure to understand the directional terms: *superior, inferior, anterior, posterior, medial, lateral, intermediate, proximal,* and *distal.*

🔑 *Directional terms are very precise terms that locate various parts of the body in relation to one another.*

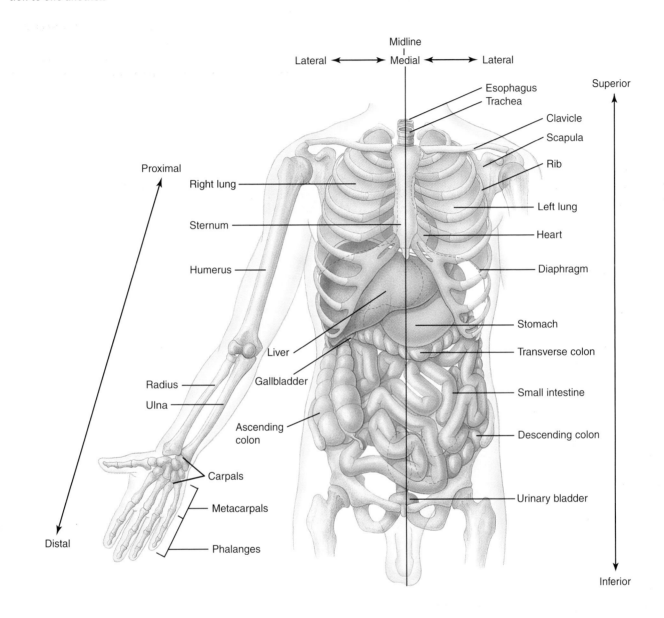

🅠 *Is the radius proximal to the humerus? Is the esophagus anterior to the trachea? Are the ribs superficial to the lungs? Is the urinary bladder medial to the transverse colon? Is the sternum lateral to the descending colon?*

Figure 1.5   Planes of the human body.

🔑 *The frontal, transverse, sagittal, and oblique planes divide the body in specific ways.*

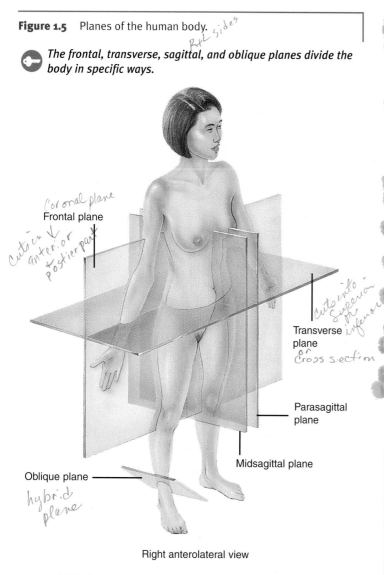

Coronal plane
Frontal plane

Cuts in anterior & posterior part

Transverse plane

Cuts into superior or inferior

Cross section

Parasagittal plane

Midsagittal plane

Oblique plane
hybrid plane

Right anterolateral view

❓ *Which plane divides the heart into anterior and posterior portions?*

Body Cavities

Spaces within the body that contain internal organs are called *body cavities.* The cavities help protect, separate, and support internal organs. Figure 1.7 on page 12 shows the two principal body cavities: dorsal and ventral. The *dorsal body cavity* is located near the posterior or dorsal (back) surface of the body. It is composed of a *cranial cavity,* which is formed by the cranial (skull) bones and contains the brain and its coverings (called *meninges;* me-NIN-jēz), and a *vertebral (spinal) canal,* which is formed by the vertebrae (individual bones) of the vertebral column (backbone) and contains the spinal cord and its coverings (also *meninges*) as well as the beginnings (*roots*) of spinal nerves.

The *ventral body cavity* is located on the anterior or ventral (front) aspect of the body and contains organs collectively called *viscera* (VIS-er-a). Like the dorsal body cavity, the ventral body cavity has two principal subdivisions—an upper portion, called the *thoracic* (thō-RAS-ik) *cavity* (or chest cavity), and a lower portion, called the *abdominopelvic* (ab-dom′-i-nō-

PEL-vik) *cavity.* The diaphragm (DĪ-a-fram; *diaphragma* = partition or wall), a dome-shaped sheet of muscle, an important muscle for breathing, divides the ventral body cavity into the thoracic and abdominopelvic cavities.

The thoracic cavity contains two *pleural cavities* around each lung and a *pericardial* (per′-i-KAR-dē-al; *peri* = around; *cardi* = heart) *cavity,* a space around the heart (Figure 1.8 on page 13).

The *mediastinum* (mē′-dē-as-TĪ-num; *medias* = middle; *stare* = stand in), in the thoracic cavity, is the mass of tissues between the lungs that extends from the sternum (breastbone) to the vertebral column (backbone) (Figure 1.8). The mediastinum includes all the structures in the thoracic cavity except the lungs themselves. Among the structures in the mediastinum are the heart, thymus gland, esophagus, trachea, and many large blood vessels, such as the aorta.

The abdominopelvic cavity, as the name suggests, is divided into two portions, although no specific structure separates them (see Figure 1.7). The upper portion, the *abdominal cavity,* contains the stomach, spleen, liver, gallbladder, pancreas, small intestine, and most of the large intestine. The lower portion, the *pelvic cavity,* contains the urinary bladder, portions of the large intestine, and the internal reproductive organs. The pelvic cavity is located between two imaginary planes, which are indicated by dashed lines in Figure 1.7a.

A summary of the body cavities is presented in Exhibit 1.3.

Exhibit 1.3	Summary of Body Cavities
CAVITY	**COMMENTS**
Dorsal	
Cranial	Formed by cranial bones and contains brain and its coverings.
Vertebral	Formed by vertebral column and contains spinal cord and beginnings of spinal nerves.
Ventral	
Thoracic	Chest cavity; separated from abdominal cavity by diaphragm.
Pleural	Contains lungs.
Pericardial	Contains heart.
Mediastinum	Region between the lungs from the breastbone to backbone that contains heart, thymus gland, esophagus, trachea, bronchi, and many large blood and lymphatic vessels.
Abdominopelvic	Subdivided into abdominal and pelvic cavities.
Abdominal	Contains stomach, spleen, liver, gallbladder, pancreas, small intestine, and most of large intestine.
Pelvic	Contains urinary bladder, portions of the large intestine, and internal female and male reproductive organs.

Figure 1.6   Planes through different parts of the brain. The planes are shown in the diagram on the left and the resulting sections are shown in the photographs on the right.

 Planes divide the body in various ways to produce sections.

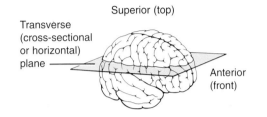

(a)

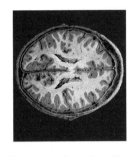

Transverse (cross) section

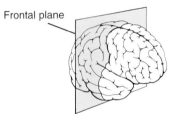

(b)

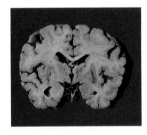

Frontal section

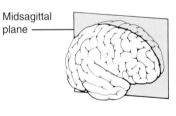

(c)

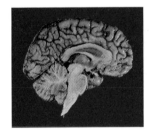

Midsagittal section

 Which plane divides the brain into equal right and left sides?

Figure 1.7   Body cavities.

🔑 *The two principal cavities are the dorsal and ventral body cavities.*

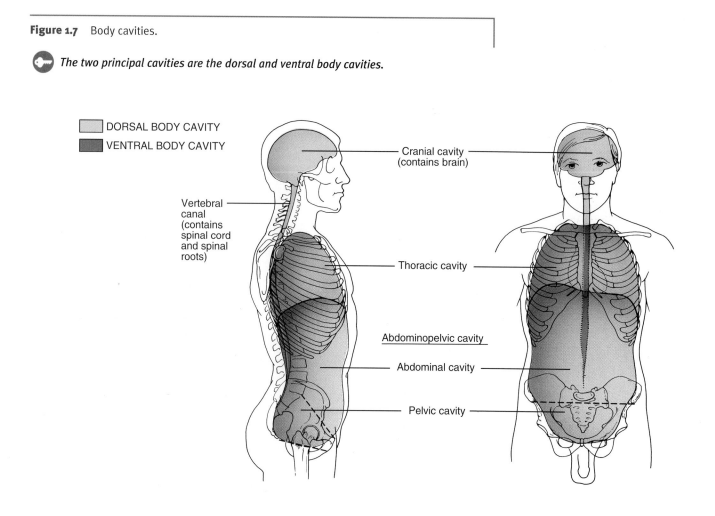

DORSAL BODY CAVITY
VENTRAL BODY CAVITY

Cranial cavity
(contains brain)

Vertebral
canal
(contains
spinal cord
and spinal
roots)

Thoracic cavity

Abdominopelvic cavity

Abdominal cavity

Pelvic cavity

❓ *In which cavities are the following structures located: urinary bladder, stomach, heart, pancreas, small intestine, lungs, internal female reproductive organs, mediastinum, spleen, rectum, liver? Use the following symbols for your response: T = thoracic, A = abdominal, or P = pelvic.*

Figure 1.8   Thoracic cavity. The two pleural cavities surround the right and left lungs and the pericardial cavity surrounds the heart. The mediastinum is found between the lungs and extends from the sternum to the vertebral column. The arrow in the inset indicates the direction from which the thoracic (chest) cavity is viewed (superior).

🔑 *The diaphragm separates the thoracic cavity from the abdominal cavity.*

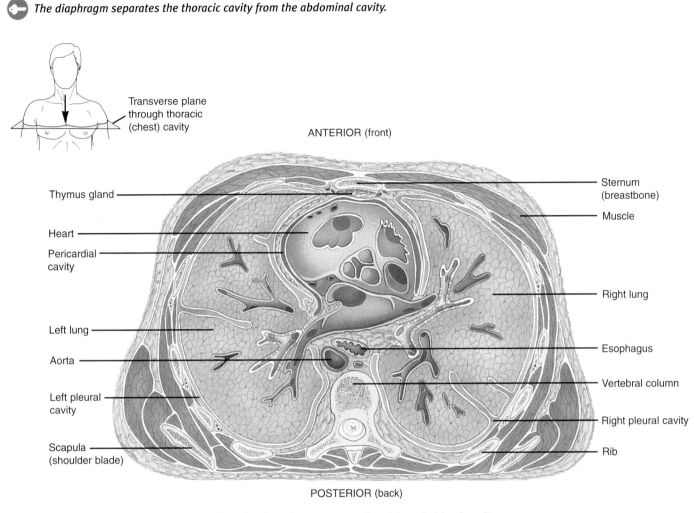

Superior view of transverse section of thoracic (chest) cavity

❓ *Which of the following structures is contained in the mediastinum: thymus gland, right lung, heart, esophagus, aorta, rib, left pleural cavity?*

Abdominopelvic Regions and Quadrants

To locate organs easily, the abdominopelvic cavity is divided into nine *abdominopelvic regions* shown in Figure 1.9 on page 14. The abdominopelvic cavity may also be divided into *quadrants* (*quad* = a one-fourth part). These are shown in Figure 1.10 on page 14. A horizontal line and a vertical line are passed through the umbilicus (navel). These two lines divide the abdomen into a *right upper quadrant (RUQ)*, *left upper quadrant (LUQ)*, *right lower quadrant (RLQ)*, *and left lower quadrant (LLQ)*. Whereas the nine-region division is more widely used for anatomical studies, clinicians find the quadrant division is better suited for locating the site of an abdominopelvic pain, tumor, or other abnormality.

Figure 1.9 Abdominopelvic cavity. The nine regions. The internal reproductive organs in the pelvic cavity are shown in Figures 23.1 and 23.8.

The nine-region designation is used for anatomical studies.

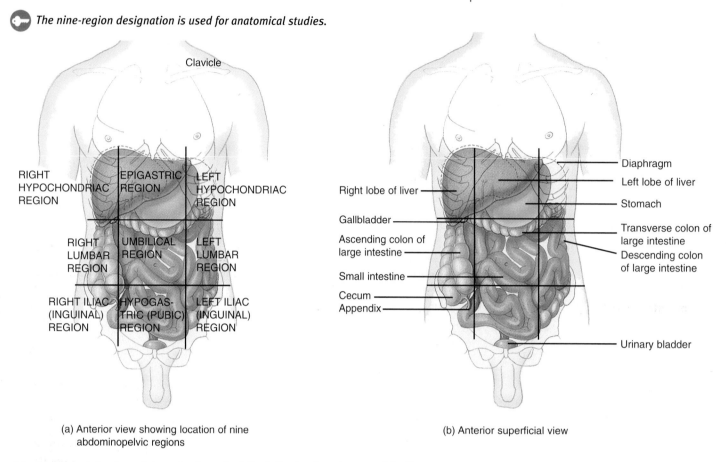

(a) Anterior view showing location of nine abdominopelvic regions

(b) Anterior superficial view

In which abdominopelvic region is each of the following found: most of the liver, transverse colon, urinary bladder, and appendix?

Figure 1.10 Quadrants of the abdominopelvic cavity. The two lines cross at right angles at the umbilicus (navel).

The quadrant designation is used to locate the site of pain, tumor, or some other abnormality.

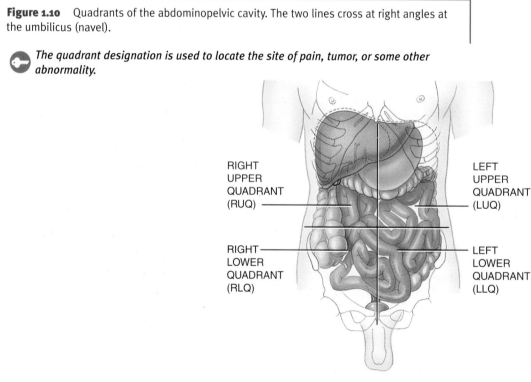

Anterior view

In which quadrant would the pain from appendicitis (inflammation of the appendix) be felt?

wellness focus
Homeostasis Is the Basis

You've seen "homeostasis" defined as a condition in which the body's internal environment—the fluid around body cells—remains within certain physiological limits. Okay, now what does this mean to you in your everyday life? Well, it can mean a lot, depending on where you are and what you do.

Homeostasis: The Power to Heal

The body's ability to maintain homeostasis gives it tremendous healing power and a remarkable resistance to abuse. But for most people, lifelong good health is not something that just happens. Two important factors in this balance called health are the environment and your own behavior. Your homeostasis is affected by the air you breathe, the food you eat, and even the thoughts you think.

The way you live can either support or interfere with your body's ability to maintain homeostasis and recover from the inevitable stresses life throws your way. Let's consider the stress imposed by the common cold. You support your natural healing processes when you take care of yourself. Plenty of rest and fluids allow the immune system to do its job. The cold runs its course and you are soon back on your feet.

If, instead of taking care of yourself, you continue to smoke two packs of cigarettes a day, skip meals, and pull several all-nighters studying for an anatomy and physiology exam, you interfere with the immune system's ability to fend off attacking microorganism, and bring the body back to homeostasis and good health.

Other infections take advantage of your weakened state, and pretty soon the cold has "turned into" bronchitis or pneumonia.

Homeostasis and Prevention

Many diseases are the result of years of poor health behavior that interferes with the body's natural drive to maintain homeostasis. An obvious example is smoking-related illness. Smoking tobacco exposes sensitive lung tissue to a multitude of chemicals that cause cancer and damage the lung's ability to repair itself. Because diseases such as emphysema and lung cancer are difficult to treat and very rarely cured, it is much wiser to quit smoking (or never start) than to hope the doctor can fix you up once you are diagnosed with the disease.

This is not to say that all illness is the result of bad habits. We do not yet understand what causes many illnesses, and we have little or no control over their development. And even when lifestyle habits may have contributed to the development of an illness, we must acknowledge that changing habits is difficult, and so must be aware of "blaming the victim" for bringing on a disease.

The Wellness Lifestyle

For many years, the word "health" was used to mean simply an absence of disease. Many people felt that there was more to health than simply not being sick. The term "wellness" came into use to indicate a kind of high-level health that represents more than the absence of disease and includes well-being in all areas of your life, not just the physical. The wellness concept is based on the notion that lifestyle choices you make throughout the years have an important impact on your mental, physical, and spiritual well-being. Wellness means taking responsibility for your health, preventing accidents and illness, and working with health care providers when necessary. Wellness encourages consumer awareness and promotes the establishment of social systems and environments conducive to health-promoting behavior. While disease prevention is an important goal of the wellness lifestyle, disease and disability do not prevent a wellness lifestyle, since wellness means doing what you can to maximize your personal potential for optimal well-being and to construct a meaningful and rewarding life.

critical thinking

What health habits have you developed over the past several years to prevent disease or enhance your body's ability to maintain health and homeostasis?

Cardiologist George Sheehan once said that when he thought about life expectancy, he preferred to think about what he expected from life. How does his statement illustrate the wellness concept?

| Study Outline |

Anatomy and Physiology Defined (p. 2)

1. Anatomy is the study of structure and the relationship among structures.
2. Physiology is the study of how body structures function.

Levels of Structural Organization (p. 2)

1. The human body consists of several levels of organization: chemical, cellular, tissue, organ, system, and organismic.
2. Cells are the basic structural and functional units of an organism.
3. Tissues consist of groups of similarly specialized cells.
4. Organs are structures that are composed of two or more different tissues, have specific functions, and usually have a recognizable shape.
5. Systems consist of associations of organs that have a common function.
6. The human organism is a collection of structurally and functionally integrated systems.
7. The systems of the human body are the integumentary, skeletal, muscular, nervous, endocrine, cardiovascular, lymphatic and immune, respiratory, digestive, urinary, and reproductive (see Exhibit 1.1).

How Body Systems Work Together (p. 2)

1. All body systems work together to maintain health and protect against disease.
2. Each system interacts with the others to help the entire organism function as a unit.

Life Processes (p. 5)

1. All living forms have certain characteristics that distinguish them from nonliving things.
2. Among the life processes in humans are metabolism, responsiveness, movement, growth, differentiation, and reproduction.

Homeostasis: Maintaining Physiological Limits (p. 5)

1. Homeostasis is a condition in which the internal environment (interstitial fluid) of the body remains within certain physiological limits in terms of chemical composition, temperature, and pressure.
2. All body systems attempt to maintain homeostasis.
3. Homeostasis is controlled mainly by the nervous and endocrine systems.

Stress and Homeostasis (p. 5)

1. Stress is an external or internal stimulus that creates a change in the internal environment.
2. If a stress acts on the body, homeostatic mechanisms attempt to counteract the effects of the stress and bring the condition back to normal.

Homeostasis of Blood Pressure (BP) (p. 5)

1. Blood pressure (BP) is the force exerted by blood as it passes through blood vessels, especially arteries.

2. If a stress causes the heartbeat to increase, blood pressure also increases. Pressure-sensitive nerve cells in certain arteries inform the brain, and the brain responds by sending impulses that decrease heartbeat, thus decreasing blood pressure back to normal.
3. Any cycle of events in which information about the status of something is continually fed back to a control region is called a feedback system.
4. A feedback system consists of (1) a control center that determines the point at which a controlled condition should be maintained, (2) receptors that monitor changes in the controlled condition and send the information (input) to the control center, and (3) effectors that receive information (output) from the control center and produce a response (effect).
5. If a response reverses the original stimulus, the system is a negative feedback system.
6. If a response enhances the original stimulus, the system is a positive feedback system.

Anatomical Position (p. 7)

1. When in the anatomical position, the subject stands erect facing the observer, with upper limbs at the sides, palms turned forward, and feet flat on the floor.
2. Parts of the body have both anatomical and common names for different regions. Examples include cranial (skull), thoracic (chest), brachial (arm), patellar (knee), cephalic (head), and gluteal (buttock).

Directional Terms (p. 8)

1. Directional terms indicate the relationship of one part of the body to another.
2. Commonly used directional terms are superior (toward the head or upper part of a structure), inferior (away from the head or toward the lower part of a structure), anterior (near or at the front of the body), posterior (near or at the back of the body), medial (nearer the midline of the body or a structure), intermediate (between a medial and lateral structure), proximal (nearer the attachment of an extremity to the trunk or a structure), distal (farther from the attachment of an extremity to the trunk or a structure), superficial (toward or on the surface of the body), and deep (away from the surface of the body).

Planes and Sections (p. 8)

1. Planes are imaginary flat surfaces that are used to divide the body or organs into definite areas. A midsagittal plane is a plane through the midline of the body that divides the body (or an organ) into equal right and left sides. A parasagittal plane is a plane that does not pass through the midline of the body and divides the body or organs into unequal right and left sides. A frontal (coronal) plane divides the body into anterior and posterior portions. A transverse (cross-sectional or horizontal) plane divides the body into superior and inferior portions. An oblique plane passes through the body or organs at an angle between a transverse plane and either a midsagittal, parasagittal, or frontal plane.

2. Sections result from cuts through body structures. They are named according to the plane on which the cut is made and include transverse (cross) sections, frontal sections, and midsagittal sections.

Body Cavities (p. 10)

1. Spaces in the body that contain internal organs are called cavities.

2. The dorsal and ventral cavities are the two principal body cavities.

3. The dorsal cavity is subdivided into the cranial cavity, which contains the brain, and the vertebral (spinal) canal, which contains the spinal cord and beginnings of spinal nerves.

4. The ventral body cavity is subdivided by the diaphragm into an upper thoracic cavity and a lower abdominopelvic cavity.

5. The thoracic cavity contains two pleural cavities, a pericardial cavity, and the mediastinum.

6. The mediastinum is a mass of tissues between the lungs that extends from the sternum to the vertebral column. It contains all the structures of the thoracic cavity, except the lungs.

7. The abdominopelvic cavity is divided into a superior abdominal and an inferior pelvic cavity. No specific structure divides them.

8. Viscera (organs) of the abdominal cavity include the stomach, spleen, pancreas, liver, gallbladder, small intestine, and most of the large intestine.

9. Viscera of the pelvic cavity include the urinary bladder, sigmoid colon, rectum, and internal female and male reproductive structures.

Abdominopelvic Regions and Quadrants (p. 13)

1. To describe the location of organs easily, the abdominopelvic cavity may be divided into nine regions.

2. The names of the nine abdominopelvic regions are epigastric, right hypochondriac, left hypochondriac, umbilical, right lumbar, left lumbar, hypogastric (pubic), right iliac (inguinal), and left iliac (inguinal).

3. The abdominopelvic cavity may also be divided into quadrants by passing an imaginary horizontal and vertical line through the umbilicus.

4. The names of the abdominopelvic quadrants are right upper quadrant (RUQ), left upper quadrant (LUQ), right lower quadrant (RLQ), and left lower quadrant (LLQ).

Self-Quiz

1. Which of the following is NOT true:

   a. The structure of a body part determines its function.
   b. When unspecialized cells become specialized, the process is known as differentiation.   c. Growth, repair, or replacement of cells is known as responsiveness.   d. Digesting foods in order to obtain the energy contained within them is part of your body's metabolism.   e. Movement is considered one of the important life processes.

2. Match the following:

   _A_ a. Carries oxygen, nutrients, and carbon dioxide
   _B_ b. Breaks down and absorbs food
   _E_ c. Body movement, posture, and heat production
   _C_ d. Regulates activities through hormones
   _F_ e. Supports and protects body
   _G_ f. Regulates chemical composition of blood
   _D_ g. Protects body and regulates body temperature

   A. cardiovascular system
   B. digestive system
   C. endocrine system
   D. integumentary system
   E. muscular system
   F. skeletal system
   G. urinary system

3. Which of the following lists best illustrates the idea of increasing levels of organizational complexity?

   a. chemical, tissue, cellular, organ, organismic, system
   b. chemical, cellular, tissue, organ, system, organismic
   c. cellular, chemical, tissue, organismic, organ, system

   d. chemical, cellular, tissue, system, organ, organismic
   e. tissue, cellular, chemical, organ, system, organismic

4. Fill in the missing blanks in the following table:

SYSTEM	MAJOR ORGANS	FUNCTIONS
Integumentary	a. Skin + assoc. struc.	b. reg. Temp, elim. waste, protects, vit. D
c. LYMPHATIC IMMUNE	Lymphatic vessels, spleen, thymus gland, tonsils, lymph nodes	d. protect against disease, filters body fluid, transports fats, returns protien + plasma to Cardio system
e. Respiratory	f. nose Throat Larynx Trachea lunge	Supplies oxygen to cells; eliminates carbon dioxide; regulates acid–base balance
Reproductive	g. penis, testes, ovaries vagina, uterine tubes	h. reproduction
i. URINARY	Kidneys, urinary bladder, ureters, urethra	j. elim. wastes, reg. Blood comp, reg. fluid + electolyte bal, help control RBC prod.

5. Homeostasis is

   a. the sum of all the chemical reactions in the body   b. a form of reproduction involving only one sex   c. the combination of growth, repair, and energy release that is basic to life   d. the internal environment remaining within certain physiological limits   e. caused by stress

6. Which of the following is normally controlled by a positive feedback system?

   a. blood pressure   b. heart rate   c. blood sugar   d. labor of childbirth   e. body temperature

7. An itch in your axillary region would cause you to scratch
   a. your armpit   b. in front of your elbow   c. your neck   d. the top of your head   e. your calf

8. Your nose is _____ to your lips.
   a. inferior   b. lateral   c. superior   d. deep   e. posterior

9. Your skull is _____ in relation to your brain.
   a. intermediate   b. superior   c. deep   d. superficial   e. proximal

10. The hand is _____ to the wrist.
    a. distal.   b. posterior.   c. medial   d. proximal   e. anterior

11. A magician is about to separate his assistant's body into top and bottom portions. The plane through which he will pass his magic wand is the
    a. midsagittal   b. frontal   c. transverse   d. parasagittal   e. oblique

12. The sweat glands are included in the _____ system.
    a. urinary   b. digestive   c. endocrine   d. integumentary   e. lymphatic and immune

13. The gallbladder is part of the _____ system.
    a. urinary   b. digestive   c. endocrine   d. integumentary   e. lymphatic and immune

14. Match the following:

    D a. contains urinary bladder and reproductive organs
    A b. contains brain
    H c. cavity that contains heart
    F d. region between the lungs, from the breastbone to the backbone
    G e. separates thoracic and abdominal cavities
    E f. contains lungs
    C g. contains spinal cord
    B h. contains stomach, pancreas, liver

    A. cranial cavity
    B. abdominal cavity
    C. vertebral cavity
    D. pelvic cavity
    E. pleural cavity
    F. mediastinum
    G. diaphragm
    H. pericardial cavity

15. To find the urinary bladder, you would look in the
    a. hypochondriac region   b. umbilical region   c. epigastric region   d. iliac region   e. hypogastric region

16. If a person is having his/her appendix removed, the surgeon would prepare which area for surgery?
    a. right upper quadrant   b. right lower quadrant   c. left upper quadrant   d. left lower quadrant   e. left hypochondriac region

Critical Thinking Applications

1. A woman jogs past you at the park. List as many body systems as you can think of that are affected by exercise. Describe how these systems help maintain homeostasis.

2. The term hypochondriac means below (hypo-) cartilage (chondro). Explain why this region of the abdominopelvic cavity was given this name. (Refer to the chapter figures if necessary.) Can you think of a reason why a person with many imaginary ailments is called a "hypochondriac"?

3. Taylor was going for the record for the longest upside down hang from the monkey bars. She didn't make it and she may have broken her arm. The emergency room technician would like an x-ray film of her arm in the anatomical position. Use the proper anatomical terms to describe the position of Taylor's arm in the x-ray.

4. Breakfast this morning was a large chocolate candy bar and a cup of coffee. Design a negative feedback system showing how your body maintains homeostasis following the stress of this breakfast to your internal environment.

5. Your friend is rushed to the hospital suffering from acute abdominal pain. A decision is made to remove the appendix. Using anatomical terminology, describe the location of the appendix within the abdominopelvic cavity.

6. Imagine that a manned space flight lands on Mars. The astronaut life-specialist observes lumpy shapes that may be life-forms or mud balls. What are some characteristics of living organisms that may help the astronaut determine if these are life-forms or mud balls?

Answers to Figure Questions

1.1 Organ level.

1.2 In negative feedback systems, the response reverses the original stimulus; in positive feedback systems, the response enhances the original stimulus.
     Negative feedback systems tend to maintain conditions that require frequent monitoring and adjustment within physiological limits; positive feedback systems are involved with conditions that do not occur often and do not require continual fine-tuning.

1.3 The subject stands erect facing the observer, the feet are flat on the floor, the upper limbs are at the sides, and the palms are turned forward.

1.4 No, No, Yes, Yes, No.

1.5 Frontal plane.

1.6 Midsagittal plane.

1.7 P, A, T, A, A, T, P, T, A, P, A.

1.8 Thymus gland, heart, esophagus, aorta.

1.9 Epigastric, umbilical, hypogastric, right iliac.

1.10 Right lower quadrant.

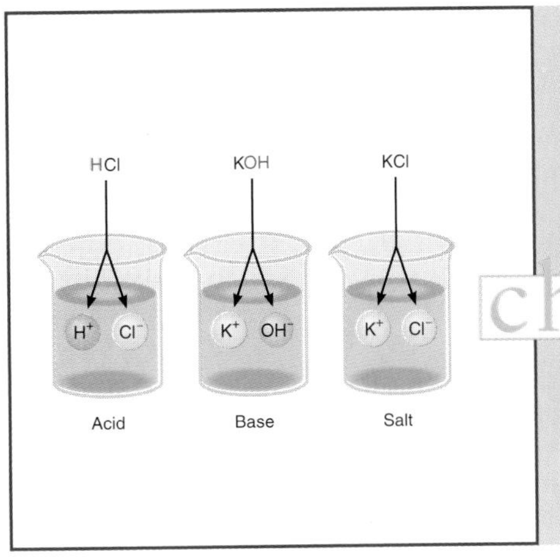

HCl   KOH   KCl

H^+ Cl^- K^+ OH^- K^+ Cl^-

Acid   Base   Salt

chapter 2

INTRODUCTORY CHEMISTRY

a look ahead

any common substances we eat and drink—water, sugar, table salt, cooking oil—play vital roles in keeping us alive. In this chapter, you will learn how these substances function in your body. Because your body is composed of chemicals and all body activities are chemical in nature, it is important to become familiar with the language and basic ideas of chemistry.

Introduction to Basic Chemistry

Chemical Elements

All living and nonliving things consist of *matter,* which is anything that occupies space and has mass (the amount of matter that a substance contains). Matter may exist as a solid, liquid, or gas. All forms of matter are made up of a limited number of building units called *chemical elements,* substances that cannot be broken down into simpler substances by ordinary chemical reactions. At present, scientists recognize 109 different elements. Elements are designated by letter abbreviations called *chemical symbols:* H (hydrogen), C (carbon), O (oxygen), N (nitrogen), Na (sodium), K (potassium), Fe (iron), and Ca (calcium).

Twenty-six of the elements are found in the human organism. Oxygen, carbon, hydrogen, and nitrogen make up about 96 percent of the body's mass. Other elements such as calcium, phosphorus, potassium, sulfur, sodium, chlorine, magnesium, iodine, and iron make up about 3.9 percent of the body's mass. Thirteen other chemical elements found in the human body are called *trace elements* because they are found in low concentrations and compose the remaining 0.1 percent (Exhibit 2.1). *metals - for the most part are trace elements*

Structure of Atoms

objective: *Describe the structure of an atom.*

Each element is made up of *atoms,* the smallest units of matter that enter into chemical reactions. An element is composed of all the same type of atoms. A sample of the element carbon, such as pure coal, contains only carbon atoms. A tank of oxygen contains only oxygen atoms.

An atom consists of two basic parts: the nucleus and the electrons (Figure 2.1). The centrally located *nucleus* contains positively charged particles called *protons* (p^+) and uncharged (neutral) particles called *neutrons* (n^0). Because each proton has one positive charge, the nucleus itself is positively charged. *Electrons* (e^-) are negatively charged particles that move around the nucleus. The number of electrons in an atom always equals the number of protons. Because each electron carries one negative charge, the negatively charged electrons and the positively charged protons balance each other, and the atom is electrically neutral. As you will see later, certain atoms can gain or lose electrons when they participate in chemical reactions.

What makes the atoms of one element different from those of another? The answer is the number of protons. Figure 2.2 on page 22 shows that a hydrogen atom contains one pro-

Figure 2.1 Structure of an atom. In this highly simplified version of a carbon atom, note the centrally located nucleus. The nucleus contains six neutrons and six protons, although all are not visible in this view. The six electrons move about the nucleus in regions called electron shells, shown here as circles.

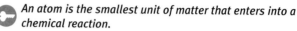

An atom is the smallest unit of matter that enters into a chemical reaction.

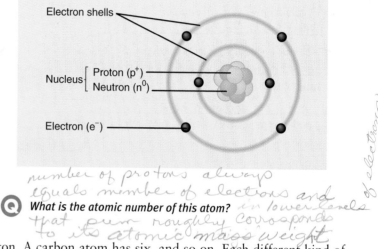

Q *What is the atomic number of this atom?*

number of protons always equals number of electrons and that sum roughly corresponds to its atomic mass weight *in lower levels of electrons*

ton. A carbon atom has six, and so on. Each different kind of atom has a different number of protons in its nucleus. The number of protons in an atom is called the atom's *atomic number.* Therefore each kind of atom, or element, has a different atomic number. The total number of protons and neutrons in an atom is its *mass number.* Thus an atom of sodium has a mass number of 23 because there are 11 protons and 12 neutrons in its nucleus.

Atoms and Molecules

When atoms combine with or break apart from other atoms, a *chemical reaction* occurs. In the process, new products with different properties are formed. Chemical reactions are the foundation of all life processes.

It is the electrons of an atom that participate in chemical reactions. The electrons move around the nucleus in regions (shown in Figure 2.2 as circles) called *electron shells.* Each electron shell has a maximum number of electrons it can hold. For instance, the electron shell nearest the nucleus never holds more than two electrons, no matter what the element. This electron shell is referred to as the first electron shell. The second electron shell never holds more than eight electrons. The third electron shell can hold up to 18 electrons. Higher electron shells (there are as many as seven shells) can contain many more electrons.

An atom always attempts to fill its outermost electron shell with the maximum number of electrons it can hold. To do this, the atom may give up an electron, take on an electron, or share an electron with another atom—whichever is easier. The *valence* (combining capacity) is the number of extra or deficient electrons in the outermost electron shell. Look at the

CHEMICAL ELEMENT (SYMBOL)	PERCENTAGE OF TOTAL BODY MASS	COMMENT
Oxygen (O)	65.0	Constituent of water and organic molecules (carbon- and hydrogen-containing, made by a living system); needed for cellular respiration, which produces adenosine triphosphate (ATP), an energy-rich chemical in cells.
Carbon (C)	18.5	Found in every organic molecule.
Hydrogen (H)	9.5	Constituent of water, all foods, and most organic molecules; when it is a cation (positively charged ion, H^+), it is an acid.
Nitrogen (N)	3.2	Component of all proteins and nucleic acids. The nucleic acids are deoxyribonucleic acid (DNA) and ribonucleic acid (RNA).
Calcium (Ca)	1.5	Contributes to hardness of bone and teeth; needed for many body processes, for example, blood clotting and contraction of muscle.
Phosphorus (P)	1.0	Component of many proteins, nucleic acids, and adenosine triphosphate (ATP); required for normal bone and tooth structure.
Potassium (K)	0.4	Most abundant cation (K^+) inside cells; important in conduction of nerve impulses and muscle contraction.
Sulfur (S)	0.3	Component of many proteins.
Sodium (Na)	0.2	Most plentiful cation (Na^+) outside cells; essential in blood to maintain water balance; needed for conduction of nerve impulses and muscle contraction.
Chlorine (Cl)	0.2	Most plentiful anion (negatively charged particle, Cl^-) outside cells; essential in blood and interstitial fluid to maintain water balance.
Magnesium (Mg)	0.1	Needed for many enzymes to function properly.
Iodine (I)	0.1	Vital to production of hormones by the thyroid gland.
Iron (Fe)	0.1	Cations (Fe^{2+} and Fe^{3+}) are components of hemoglobin (oxygen-carrying protein in blood) and some enzymes needed for ATP production.
Aluminum (Al), Boron (B), Chromium (Cr), Cobalt (Co), Copper (Cu), Fluorine (F), Manganese (Mn), Molybdenum (Mo), Selenium (Se), Silicon (Si), Tin (Sn), Vanadium (V), Zinc (Zn)		These elements are called **trace elements** because they are present in minute concentrations.

Exhibit 2.1 Chemical Elements Found in the Body

Compose about 96% of total body mass.

Compose about 3.9% of total body mass.

Compose about 0.1% of total body mass.

chlorine atom in Figure 2.2. Its outermost electron shell, which happens to be the third electron shell, has seven electrons. Because the third level of an atom can hold up to 18 electrons, one stable form is reached at 8 electrons. Thus, chlorine having 7 electrons can be described as having a shortage of one electron. In fact, chlorine usually does try to pick up an extra electron (see Figure 2.3b). Sodium, by contrast, has only one electron in its outermost electron shell (Figure 2.2). This again happens to be the third electron shell. It is much easier for sodium to get rid of the one electron than to fill the third level by taking on seven more electrons (see Figure 2.3a). Atoms of a few elements, like helium, have completely filled outer electron shells and do not need to gain or lose electrons. These are called *inert elements* and do not enter into chemical reactions.

Figure 2.2 Atomic structures of some representative atoms that have important roles in the human body.

 When atoms take part in chemical reactions, they lose, gain, or share electrons in their outermost electron shell.

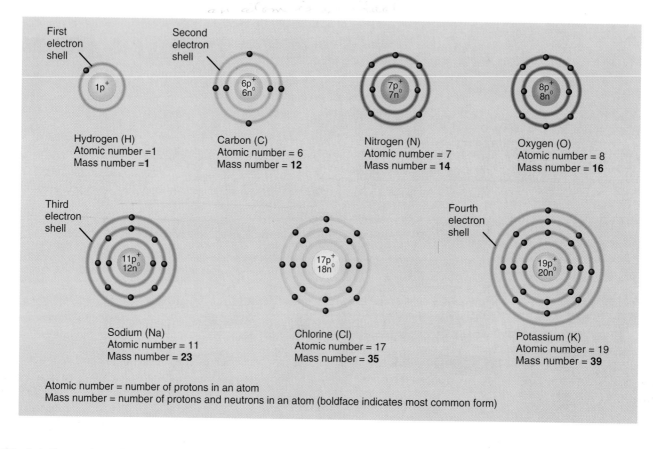

Atomic number = number of protons in an atom

Mass number = number of protons and neutrons in an atom (boldface indicates most common form)

Q *Why is helium an inert chemical element?*

Atoms with incompletely filled outer electron shells, like sodium and chlorine, tend to combine with other atoms in a chemical reaction. During the reaction, the atoms can trade off or share electrons and thereby fill their outer electron shells.

When two or more atoms combine in a chemical reaction, the resulting combination is called a ***molecule*** (MOL-e-kyool). A molecule may contain two atoms of the same kind, as in the hydrogen molecule: H_2. The subscript 2 indicates that there are two hydrogen atoms in the molecule. Molecules may also contain two or more different kinds of atoms, as in the hydrochloric acid molecule: HCl. Here an atom of hydrogen is attached to an atom of chlorine.

A ***compound*** is a chemical substance composed of two or more *different* elements. Compounds can be broken down into their constituent elements by chemical means. Whereas hydrochloric acid (HCl) is a compound, a molecule of hydrogen (H_2) is not. The atoms in molecules and compounds are held together by electrical forces of attraction called ***chemical bonds***.

Ionic Bonds

objective: *Explain how chemical bonds form.*

Atoms are electrically neutral because the number of positively charged protons equals the number of negatively charged electrons. But when an atom gains or loses electrons, this balance is upset. If the atom gains electrons, it acquires an overall negative charge. If the atom loses electrons, it acquires an overall positive charge. A particle with a negative or positive charge is called an ***ion*** (I-on). An ion is always symbolized by writing the chemical abbreviation followed by the number of positive (+) or negative (−) charges the ion acquires.

Consider a sodium ion (Figure 2.3a). A sodium atom (Na) has 11 protons and 11 electrons, with 1 electron in its outer electron shell. When sodium gives up the single electron in its outer electron shell, it is left with 11 protons and only 10 electrons. It is an ***electron donor*** because it gives up electrons. The atom now has an overall positive charge of one (+1) and

Figure 2.3   Formation of an ionic bond. (a) An atom of sodium attains stability by passing a single electron to an electron acceptor. The loss of this single electron results in the formation of a sodium ion (Na^+). (b) An atom of chlorine attains stability by accepting a single electron from an electron donor. The gain of this single electron results in the formation of a chloride ion (Cl^-). (c) When Na^+ and Cl^- ions are combined, they are held together by the attraction of opposite charges, which is known as an ionic bond, and a molecule of sodium chloride (NaCl) is formed.

 An ionic bond is an attraction that holds together ions with different charges.

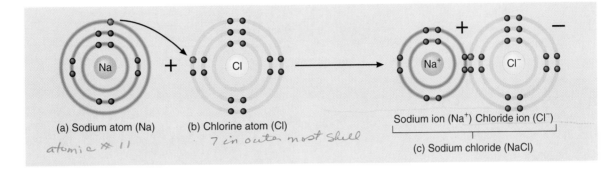

(a) Sodium atom (Na)   (b) Chlorine atom (Cl)

atomic # 11   7 in outer most shell

Sodium ion (Na^+) Chloride ion (Cl^-)

(c) Sodium chloride (NaCl)

Will potassium (K) be more likely to be an electron donor or an electron acceptor? (Look back to Figure 2.2 for the atomic structure of K.)

is called a sodium ion (written Na^+). Generally, atoms whose outer electron shell is less than half-filled lose electrons and form positively charged ions called *cations* (KAT-i-ons). Other examples of cations are the potassium ion (K^+), calcium ion (Ca^{2+}), and iron ion (Fe^{2+}).

Another example is the formation of a chloride ion (Figure 2.3b). Chlorine has a total of 17 electrons, 7 of them in the outer electron shell. Because this energy level can hold 8 electrons, chlorine tends to pick up an electron that has been lost by another atom. Chlorine is an *electron acceptor* because it picks up electrons. By accepting an electron, chlorine acquires a total of 18 electrons. However, it still has only 17 protons in its nucleus. The chloride ion therefore has a negative charge of one (-1) and is written Cl^-. Atoms whose outer electron shell is more than half-filled tend to gain electrons and form negatively charged ions called *anions* (AN-i-ons). Other examples of anions include the iodine ion (I^-) and sulfur ion (S^{2-}).

The positively charged sodium ion (Na^+) and the negatively charged chloride ion (Cl^-) attract each other—opposite charges attract each other. The attraction, called an *ionic bond*, holds the two ions together, and a molecule is formed (Figure 2.3c). The formation of this molecule is a solid substance, called sodium chloride (NaCl) or table salt, and is one of the most common examples of ionic bonding. Thus an ionic bond is an attraction between ions in which one atom loses electrons and another atom gains electrons.

Covalent Bonds

The *covalent bond* is a more common bond in the human body and is more stable than an ionic bond. When a covalent bond

is formed, neither of the combining atoms loses or gains an electron. Instead, the two atoms *share* one, two, or three electron pairs. Look at the hydrogen atom again. One way a hydrogen atom can fill its outer electron shell is to combine with another hydrogen atom to form the molecule H_2 (Figure 2.4a). In the H_2 molecule, the two atoms share a pair of electrons. Each hydrogen atom has its own electron plus one electron from the other atom. When one pair of electrons is shared between atoms, as in the H_2 molecule, a *single covalent bond* is formed. When two or three pairs of electrons are shared between two atoms, a *double covalent bond* (Figure 2.4b) or a *triple covalent bond* (Figure 2.4c) is formed.

The same principles that apply to covalent bonding between atoms of the same element also apply to atoms of different elements. Methane (CH_4), also known as marsh gas, is an example of covalent bonding between atoms of different elements (Figure 2.4d). The outer electron shell of the carbon atom can hold eight electrons but has only four of its own. Each hydrogen atom can hold two electrons but has only one of its own. In the methane molecule, the carbon atom shares four pairs of electrons. One pair is shared with each hydrogen atom.

Hydrogen Bonds

A *hydrogen bond* consists of a hydrogen atom covalently bonded to one oxygen atom or one nitrogen atom but also attracted to another oxygen or nitrogen atom. Because hydrogen bonds are weak, only about 5 percent as strong as covalent bonds, they do not bind atoms into molecules. However, they do serve to hold together different molecules or various parts of the same molecule. As weak bonds, they may be formed and

Figure 2.4   Covalent bond formation. To the right are simpler ways to represent these molecules. In structural formulas, each covalent bond is written as a straight line between the symbols for two atoms. In molecular formulas, the number of atoms in each molecule is noted by subscripts.

🔑   *In a covalent bond, two atoms share one, two, or three pairs of valence electrons.*

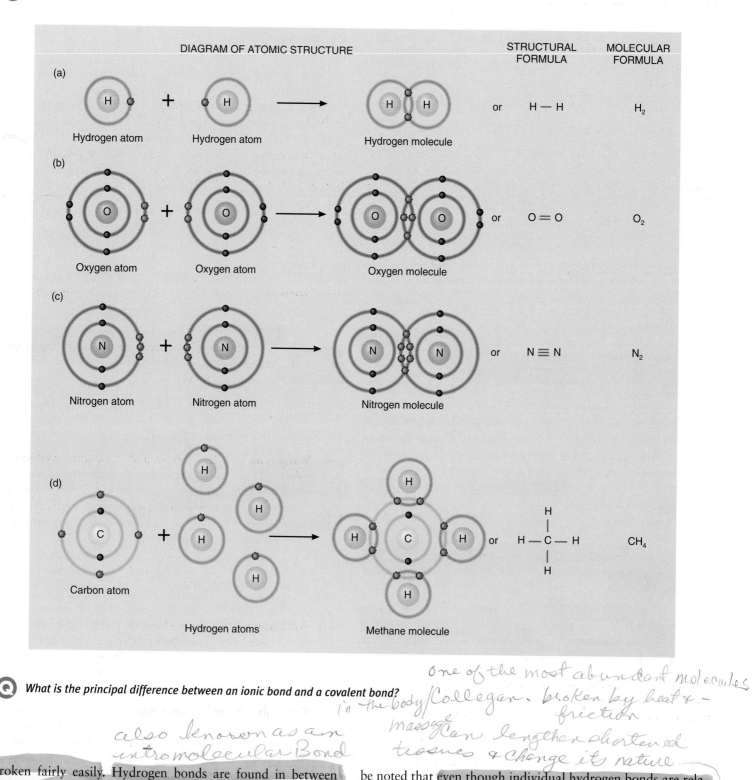

❓   **What is the principal difference between an ionic bond and a covalent bond?**

[handwritten annotations]

one of the most abundant molecules in the body/Collegan. broken by heat & friction

massage can lengthen shorten ed tissues & change its nature

also known as an intromolecular Bond

broken fairly easily. Hydrogen bonds are found in between water molecules and in large complex molecules such as proteins and nucleic acids (see Figure 2.11 on page 32). It should

be noted that even though individual hydrogen bonds are relatively weak, large molecules may contain several hundred of these bonds, resulting in considerable strength and stability.

sol/gel continuum

massage causes physiotropic effect   rearranges

Chemical Reactions

objective: *Define a chemical reaction and explain why it is important to the human body.*

Chemical reactions involve the making or breaking of bonds between atoms. These reactions occur continually in all cells of your body. Chemical reactions are the means by which body structures are built and body functions are carried out. After a chemical reaction, the total number of atoms remains the same, but because they are rearranged, there are new molecules with new properties. This section looks at the basic chemical reactions common to all living cells.

Synthesis Reactions

anabolic reaction

When two or more atoms, ions, or molecules combine to form new and larger molecules, the process is called a **synthesis reaction.** The word *synthesis* means "combination," and synthesis reactions involve the *forming of new bonds.* Synthesis reactions may be expressed as follows:

An example of a synthesis reaction is:

$$2H \quad + \quad O \quad \longrightarrow \quad H_2O$$

Two hydrogen atoms   One oxygen atom   One water molecule

Decomposition Reactions

Catabolic reaction

The reverse of a synthesis reaction is a **decomposition reaction.** The word *decompose* means to break down into smaller parts. In a decomposition reaction, the *bonds are broken.* Large molecules are broken down into smaller molecules, atoms, or ions. A decomposition reaction occurs in this way:

Under the proper conditions, methane can decompose into carbon and hydrogen:

$$CH_4 \quad \longrightarrow \quad C \quad + \quad 2H_2$$

One methane molecule   One carbon atom   Four hydrogen atoms

Energy and Chemical Reactions

Combination of anabolic + catabolic

Chemical energy is the energy released or absorbed in the breaking or forming of chemical bonds. When a chemical bond is formed, energy is required. When a bond is broken, energy is released. In other words, synthesis reactions need energy and decomposition reactions give off energy. The building processes of the body—the construction of bones, the growth of hair and nails, the replacement of injured cells—occur basically through synthesis reactions. The breakdown of foods, on the other hand, occurs through decomposition reactions that release energy that

can be used in the body's building processes. As you will see later, the energy from decomposition reactions is stored in a molecule called ATP (adenosine triphosphate) until needed by the body at some future point. Then, when energy is needed for synthesis reactions, it is provided by the breakdown of ATP. These reactions are covered in detail later in the chapter.

Chemical Compounds and Life Processes

Most of the chemicals in the body exist in the form of compounds that are divided into two principal classes: inorganic and organic. **Inorganic compounds** are usually small, usually lack carbon, and many contain ionic bonds. They include water, oxygen, carbon dioxide, and many acids, bases, and salts. **Organic compounds** always contain carbon and are held together mostly or entirely by covalent bonds. Organic compounds in the body include carbohydrates, lipids (such as triglycerides or fats), proteins, nucleic acids, and adenosine triphosphate (ATP). Both organic and inorganic compounds are vital to life functions.

Inorganic Compounds

The body is a container that water invented to carry itself around in.

objective: *Discuss the functions of water and inorganic acids, bases, and salts.*

Water

One of the most abundant inorganic substances in the human organism is *water.* In an average adult male, water makes up about 60 percent of body weight. With few exceptions, water makes up most *to 80 percent* of the volume of cells and body fluids. The following properties of water explain why it is such a vital compound:

1. **Water is an excellent solvent and suspending medium.** A **solvent** is a liquid or gas in which some other material, called a **solute,** has been dissolved. The combination of solvent plus solute is called a **solution.** Water is the solvent that carries nutrients, oxygen, and wastes throughout the body. As a suspending medium, water is also vital to survival. Many large organic molecules are suspended in the water of body cells, which allows them to come in contact with other chemicals, leading to essential chemical reactions.

2. **Water can participate in chemical reactions.** In digestion, water helps break down large nutrient molecules. Water molecules are also part of the synthesis reactions that produce hormones and enzymes and nutrients such as carbohydrates, lipids, and proteins.

3. **Water absorbs and releases heat very slowly.** In comparison with other substances, water requires a large input of heat to increase its temperature and a great loss of heat to decrease its temperature. Thus the large amount of water in the body moderates the effects of fluctuating environmental temperatures, thereby helping maintain the homeostasis of body temperature.

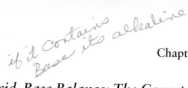
[handwritten: if it contains Base its alkaline]

4. **Water requires a large amount of heat to change from a liquid to a gas.** When water (perspiration) evaporates from the skin, it takes with it large quantities of heat, providing an excellent body cooling mechanism.

5. **Water serves as a lubricant.** It is a major part of saliva, mucus, and other lubricating fluids. Lubrication is especially necessary in the chest and abdomen where internal organs touch and slide over each other, and at joints where bones, ligaments, and tendons rub against each other.

Inorganic Acids, Bases, and Salts

When molecules of inorganic acids, bases, or salts dissolve in the water of the body cells, they undergo *ionization* (ī-on-i-ZĀ-shun) or *dissociation* (dis′-sō-sē-Ā-shun); that is, they break apart into ions. An *acid* ionizes into one or more *hydrogen ions* (H^+) and one or more *anions* (negative ions). A *base*, by contrast, ionizes into one or more *hydroxide ions* (OH^-) and one or more cations (positive ions). A *salt*, when dissolved in water, ionizes into cations and anions, neither of which is H^+ or OH^- (Figure 2.5). Acids and bases react with one another to form salts. For example, the combination of hydrochloric acid (HCl), an acid, and potassium hydroxide (KOH), a base, produces potassium chloride (KCl), a salt, and water (H_2O). This reaction occurs as follows:

$$HCl + KOH \longrightarrow KCl + H_2O$$

Acid   Base   Salt   Water

Figure 2.5   Acids, bases, and salts. (a) When placed in water, hydrochloric acid (HCl) ionizes into H^+ and Cl^-. (b) When the base potassium hydroxide (KOH) is placed in water, it ionizes into OH^- and K^+. (c) When the salt potassium chloride (KCl) is placed in water, it ionizes into positive and negative ions (K^+ and Cl^-), neither of which is H^+ or OH^-.

🔑 *Ionization is the separation of inorganic acids, bases, and salts into ions in a solution.*

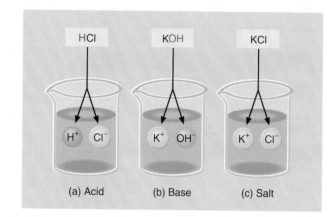

(a) Acid   (b) Base   (c) Salt

❓ *The compound $CaCO_3$ (calcium carbonate) ionizes into a calcium ion (Ca^{2+}) and a carbonate ion (CO_3^{2-}). Is it an acid, a base, or a salt? What about H_2SO_4, which ionizes into two H^+ and one SO_4^{2-}?*

Acid–Base Balance: The Concept of pH

objective: *Define pH and explain how the body attempts to keep pH within the limits of homeostasis.*

Because acids ionize into hydrogen ions (H^+) and bases ionize into hydroxide ions (OH^-), it follows that the more hydrogen ions in a solution, the more acidic it is. Conversely, the more hydroxide ions, the more basic (alkaline) the solution. The term *pH* is used to describe the degree of *acidity* or *alkalinity* (*basicity*) of a solution.

A solution's acidity or alkalinity is expressed on a *pH scale* that runs from 0 to 14 (Figure 2.6). The pH scale is based on the number of H^+ in a solution (expressed in certain chemical units called moles per liter). A solution that is 0 on the pH scale has many H^+ and few OH^-. A solution that rates 14, by contrast, has many OH^- and few H^+. The midpoint is 7, where the concentrations of H^+ and OH^- are equal. A solution that has a pH of 7, for example, pure water, is neutral. A solution with more H^+ than OH^- is *acidic* and has a pH below 7. A solution with more OH^- than H^+ is *basic* (*alkaline*) and has a pH above 7. A change of one whole number on the pH scale

[handwritten: distilled water is neutral; tap water is slightly acidic — partly because of chlorine]

Figure 2.6   pH scale. At pH 7 (neutrality), the concentrations of H^+ and OH^- are equal. A pH value below 7 indicates an acidic solution. A pH value above 7 indicates an alkaline (basic) solution. A change in one whole number on the pH scale represents a 10-fold change from the previous concentration.

🔑 *At pH 7 (neutrality), the concentrations of H^+ and OH^- are equal.*

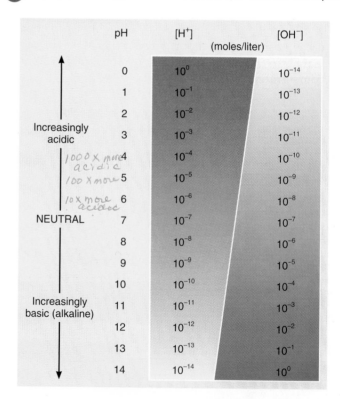

pH	[H^+]	[OH^-]
	(moles/liter)	
0	10^0	10^{-14}
1	10^{-1}	10^{-13}
2	10^{-2}	10^{-12}
3	10^{-3}	10^{-11}
4	10^{-4}	10^{-10}
5	10^{-5}	10^{-9}
6	10^{-6}	10^{-8}
7	10^{-7}	10^{-7}
8	10^{-8}	10^{-6}
9	10^{-9}	10^{-5}
10	10^{-10}	10^{-4}
11	10^{-11}	10^{-3}
12	10^{-12}	10^{-2}
13	10^{-13}	10^{-1}
14	10^{-14}	10^0

Increasingly acidic *[handwritten: 1000 X more acidic (4), 100 X more (5), 10 X more acidic (6)]*

NEUTRAL   7

Increasingly basic (alkaline)

❓ *Which pH is more acid: 6.82 or 6.91? Which pH is closer to neutrality: 8.41 or 5.59?*

represents a 10-fold change from the previous concentration. That is, a pH of 2 is 10 times more acidic than a pH of 3 and a pH of 1 is 100 times more acidic than a pH of 3.

Your body fluids must maintain a fairly constant balance of acids and bases because the biochemical reactions that occur in living systems are extremely sensitive to even small changes in the acidity or alkalinity of the environment. Any departure from normal H^+ and OH^- concentrations can seriously affect a cell's functions.

Maintaining pH: Buffer Systems

Although the pH of various body fluids may differ, the normal limits for each are quite specific and narrow. Exhibit 2.2 shows the pH values for certain body fluids compared with common substances. Even though acids and bases are continually taken into the body as foods and beverages, the pH levels of body

wine is acidic or basic depending upon what you are eliminating

Exhibit 2.2	pH Values of Selected Substances	
SUBSTANCE		**pH VALUE**
Gastric juice (digestive juice of the stomach)[a]		1.2–3.0
Lemon juice		2.2–2.4
Grapefruit juice, vinegar, wine		3.0
Carbonated soft drink		3.0–3.5
Vaginal fluid[a] *acidic – immune system – helps prevent infection*		3.5–4.5
Pineapple juice, orange juice		3.5
Tomato juice		4.2
Coffee		5.0
Urine[a]		4.6–8.0
Saliva[a]		6.35–6.85
Milk		6.6–6.9
Distilled (pure) water		7.0
Blood[a]		7.35–7.45
Semen (fluid containing sperm)[a] *nutralizes acidity of vagina*		7.20–7.60
Cerebrospinal fluid (fluid associated with nervous system)[a]		7.4
Pancreatic juice (digestive juice of the pancreas)[a]		7.1–8.2
Eggs		7.6–8.0
Bile (liver secretion that aids fat digestion)[a]		7.6–8.6
Milk of magnesia		10.0–11.0
Lye		14.0

[a] Substances in the human body.

fluids remain relatively constant because of the body's **buffer systems.** Buffers are found in the body's fluids. They prevent drastic changes in pH and help maintain homeostasis. (More will be said about buffers in Chapter 22.)

Organic Compounds

objective: *Discuss the functions of carbohydrates, lipids, and proteins.*

Cells need Glucose for Energy

Carbohydrates

Carbohydrates are sugars and starches. They are the body's most readily available source of energy. They also provide energy reserves. For example, glycogen (animal starch), which is stored glucose, is found in the liver and skeletal muscles for emergency energy needs. Some carbohydrates are used in building cell structures. Also, some carbohydrates are components of deoxyribonucleic acid (DNA), the molecule that carries hereditary information, and ribonucleic acid (RNA), the molecule involved in protein synthesis. DNA and RNA are discussed a little later.

Carbohydrates are composed of carbon, hydrogen, and oxygen. *(only covalently Bonds)* The ratio of hydrogen to oxygen atoms is typically 2:1, for example, glucose ($C_6H_{12}O_6$). Carbohydrates are divided into three major groups on the basis of size: monosaccharides, disaccharides, and polysaccharides. Monosaccharides and disaccharides are referred to as **simple sugars** and polysaccharides are referred to as **complex sugars** (starches).

1. **Monosaccharides** (mon-ō-SAK-a-rīds; *mono* = one; *sakcharon* = sugar) are simple sugars and are the building blocks of carbohydrates. Monosaccharides contain from three to seven carbon atoms. Deoxyribose is a component of genes, and glucose is the main energy-supplying molecule of the body. *too little causes loss of cognitive function*

2. **Disaccharides** (dī-SAK-a-rīds; *di* = two) are simple sugars that consist of two small monosaccharides joined chemically into a large, more complex molecule. When two monosaccharides combine to form a disaccharide, a molecule of water is always lost. This reaction is known as **dehydration synthesis** (*dehydration* = loss of water) and is shown in Figure 2.7. Glucose and fructose are two monosaccharides that form the disaccharide sucrose (table sugar). Disaccharides can be broken down into their smaller, simpler molecules by the addition of water. This reverse chemical reaction is called **hydrolysis (digestion),** which means to split by using water. Sucrose, for example, may be hydrolyzed (digested) into glucose and fructose by the addition of water. See Figure 2.7.

3. **Polysaccharides** (pol'-ē-SAK-a-rīds; *poly* = many) are complex sugars (starches) that consist of tens or hundreds of monosaccharides joined together through dehydration synthesis. Like disaccharides, polysaccharides can be broken down into their constituent sugars through hydrolysis reactions. Glycogen is a polysaccharide.

Figure 2.7 Dehydration synthesis and hydrolysis of a molecule of sucrose. In the dehydration synthesis reaction (read from left to right), the two smaller molecules, glucose and fructose, are joined to form a larger molecule of sucrose. Note the loss of a water molecule. In hydrolysis (read from right to left), the sucrose molecule is broken down into the two smaller molecules, glucose and fructose. Here, a molecule of water is added to sucrose for the reaction to occur.

Monosaccharides are the building blocks of carbohydrates. *Know*

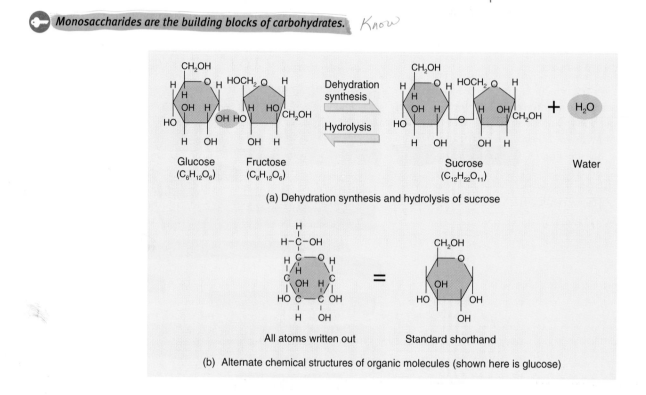

(a) Dehydration synthesis and hydrolysis of sucrose

(b) Alternate chemical structures of organic molecules (shown here is glucose)

How many carbons can you count in fructose? In sucrose?

Lipids

Like carbohydrates, *lipids* (*lipose* = fat) are also composed of carbon, hydrogen, and oxygen, but they do not have a 2:1 ratio of hydrogen to oxygen. There is less oxygen in lipids than in carbohydrates. Most lipids do not dissolve in water, but they readily dissolve in solvents such as chloroform and ether. Functionally, lipids, such as triglycerides (fats), protect, insulate, and serve as a source of energy. Triglycerides represent the body's most highly concentrated source of energy. They provide more than twice as much energy per weight as either carbohydrates or proteins. However, triglycerides are less efficient as body fuels than are carbohydrates, because they are more difficult to break down. Other lipids, such as phospholipids, make up parts of plasma (cell) membranes. Still other lipids are components of bile salts, hormones, some vitamins, and cholesterol, which also makes up part of plasma membranes and is used to produce certain sex hormones. Among the classes of lipids are triglycerides (fats and oils), phospholipids (lipids that contain phosphorus), steroids (such as cholesterol, vitamin D,

and sex hormones like estrogens and testosterone), carotenes (chemicals used to make vitamin A, which is necessary for proper vision), vitamins E and K, and eicosanoids (ī-KŌ-sanoids). This last class of lipids includes prostaglandins, which contribute to inflammation, regulate body temperature, and help form blood clots; and leukotrienes, which participate in allergic and inflammatory reactions. Because lipids are a large and diverse group of compounds, we discuss only triglycerides in detail at this point.

A molecule of *triglyceride* consists of two building blocks: *glycerol* and *fatty acids* (Figure 2.8a,b). A single molecule of triglyceride is formed when a molecule of glycerol combines with three molecules of fatty acids. This reaction, like the one described for disaccharide formation, is a dehydration synthesis reaction. During digestion, a single molecule of triglyceride is broken down (hydrolyzed) into fatty acids and glycerol.

Because of their relationship to cardiovascular disease, triglycerides are very important in our daily lives. At this point, it will be helpful to compare three types of dietary triglycerides

Figure 2.8   Triglycerides. Structure and reactions of (a) glycerol and (b) fatty acids. Each time a glycerol and a fatty acid are joined in dehydration synthesis, a molecule of water is lost. (c) Triglycerides consist of one molecule of glycerol joined to three molecules of fatty acids. Shown here is a molecule of a triglyceride that contains two saturated fatty acids and one monounsaturated fatty acid.

 Glycerol and fatty acids are the building blocks of triglycerides.

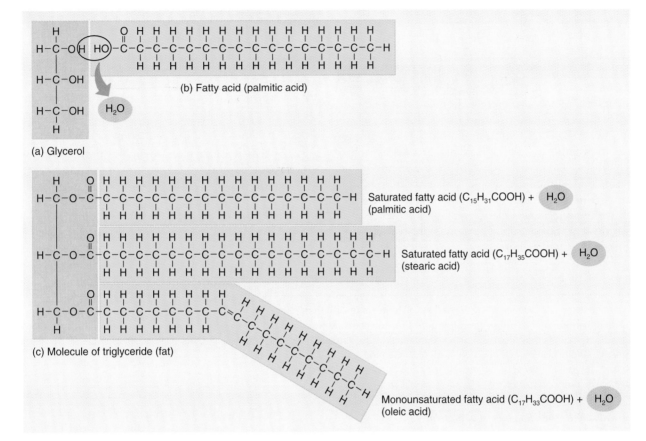

(a) Glycerol

(b) Fatty acid (palmitic acid)

(c) Molecule of triglyceride (fat)

Saturated fatty acid ($C_{15}H_{31}COOH$) + H_2O (palmitic acid)

Saturated fatty acid ($C_{17}H_{35}COOH$) + H_2O (stearic acid)

Monounsaturated fatty acid ($C_{17}H_{33}COOH$) + H_2O (oleic acid)

Q *Which type of fat is associated with raising blood cholesterol level?*

and how they relate to cholesterol. A *saturated fat* is one in which all the carbon atoms are bonded to the maximum number of hydrogen (saturated) (Figure 2.8c). Saturated fats (and some cholesterol) occur mostly in animal foods such as meats, milk products, and eggs. They also occur in some plant products such as cocoa butter, palm oil, and coconut oil. Because the liver produces cholesterol from some breakdown products of saturated fats, eating these fats is discouraged for individuals with high cholesterol levels. *Monounsaturated* and *polyunsaturated fats* are fats that are not completely saturated with hydrogen atoms. Examples of monounsaturated are olive oil, canola oil, and peanut oil, which are believed to help reduce cholesterol levels. Corn oil, safflower oil, sunflower oil, cottonseed oil, sesame oil, and soybean oil are examples of polyunsaturated fats, which researchers believe also help to reduce cholesterol in the blood.

Proteins

Proteins are much more complex in structure than carbohydrates or lipids and are involved in numerous physiological activities. Proteins are largely responsible for the structure of body cells. Some proteins in the form of enzymes function as catalysts to speed up certain chemical reactions (discussed shortly). Other proteins assume an important role in muscular contraction. Antibodies are proteins that defend the body against invading microbes. Some types of hormones are proteins.

Chemically, proteins always contain carbon, hydrogen, oxygen, and nitrogen, and sometimes sulfur. *Amino acids* are the building blocks of proteins. Each amino acid consists of a basic (alkaline) *amino group* (—NH$_2$), an acid *carboxyl group* (—COOH), and a *side chain* (R group) that is different for *[handwritten: Enzymes designated by suffix (ase)]*

each of the 20 different amino acids (Figure 2.9a). In protein formation, amino acids combine to form more complex molecules; the covalent bonds formed between amino acids are called *peptide bonds* (Figure 2.9b).

When two amino acids combine, a *dipeptide* results (Figure 2.9b). Adding another amino acid to a dipeptide produces a *tripeptide.* Further additions of amino acids result in the formation of *peptides* (4–10 amino acids) or *polypeptides* (10–2000 or more amino acids). All have the same basic composition, but each also has additional atoms arranged in a specific way. Because each variation in the number or sequence of amino acids produces a different protein, a great variety of proteins is possible. The situation is similar to using an alphabet of 20 letters to form words. Each letter would be equivalent to an amino acid, and each word would be a different protein.

If a protein encounters a hostile environment in which temperature, pH, or ion concentration is altered, it may unravel and lose its characteristic shape. This process is called *denaturation* (dē-nā′-chur-Ā-shun). Denatured proteins are no longer functional. A common example of denaturation is seen in frying an egg. In a raw egg the protein (albumin) is soluble and the egg white appears as a clear, viscous fluid. When heat is applied to the egg, however, the protein changes shape, becomes insoluble, and looks white.

Enzymes

[handwritten: is a type of or a class of a protein]

As we have seen, chemical reactions occur when chemical bonds are made or broken as atoms, ions, or molecules collide with one another. Normal body temperature and pressure are too low for chemical reactions to occur at a rate rapid enough to maintain life. *Enzymes* are the living cell's solution to this problem. They speed up chemical reactions by increasing the frequency of collisions and properly orienting the colliding molecules. And they do this without increasing temperature or pressure—in other words, without disrupting or killing the cell. Substances that can speed up chemical reactions by increasing the frequency of collisions, *without themselves being altered*, are called *catalysts* (KAT-a-lists). In living cells, enzymes function as catalysts.

Enzymes catalyze selected reactions with great specificity, efficiency, and control.

1. **Specificity.** Enzymes are highly specific catalysts. Each particular enzyme affects only specific *substrates* (molecules on which the enzyme acts). In some cases, a portion of the enzyme, called the *active site,* is thought to "fit" the substrate like a key fits a lock (see Figure 2.10). In other cases, the active site changes its shape to fit snugly around the substrate once the substrate enters

Figure 2.9 Amino acids and peptide bond formation. (a) In keeping with their name, amino acids have an amino group and a carboxyl (acid) group. The side chain (R group) is different in each amino acid. (b) When two or more amino acids are chemically united, the resulting covalent bond between them is called a peptide bond. Here, the amino acids glycine and alanine are joined to form the dipeptide glycylalanine. The peptide bond is formed at the point where water is lost.

🔑 *Amino acids are the building blocks of proteins.*

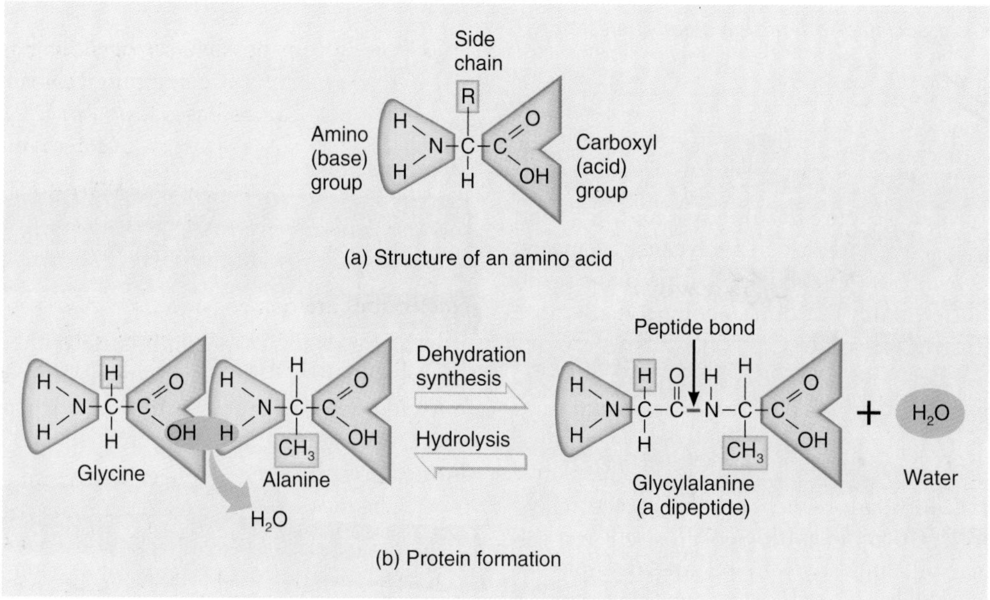

(a) Structure of an amino acid

(b) Protein formation

❓ *What type of reaction is involved in protein catabolism?*

the active site. Of the more than 1000 known enzymes, each has a characteristic three-dimensional shape with a specific surface shape, which allows it to recognize and bond to certain substrates. When an enzyme is denatured, the active site loses its unique shape and can no longer fit together with its substrate.

2. **Efficiency.** Under optimal conditions, enzymes can catalyze reactions at rates that are millions to billions of times more rapid than those of similar reactions occurring without enzymes. The *turnover number* (number of substrate molecules converted to product per enzyme molecule in 1 second) is generally between 1 and 10,000 and can be as high as 600,000.

3. **Control.** Enzymes are subject to a variety of cellular controls. Their rate of synthesis and their concentration at any given time are under the control of a cell's genes. Substances within the cell may either enhance or inhibit activity of a given enzyme. Many enzymes occur in both active and inactive forms in cells.

An enzyme is thought to work as shown in Figure 2.10.

Figure 2.10   How an enzyme works.

🔑 *An enzyme speeds up a chemical reaction without being altered or consumed.*

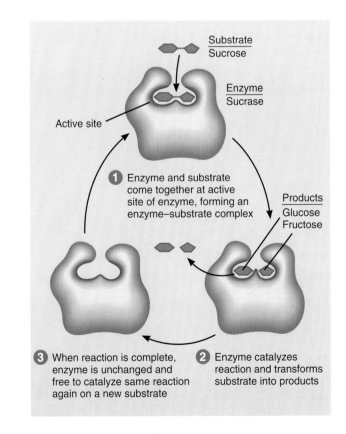

Substrate
Sucrose

Enzyme
Sucrase

Active site

❶ Enzyme and substrate come together at active site of enzyme, forming an enzyme–substrate complex

Products
Glucose
Fructose

❸ When reaction is complete, enzyme is unchanged and free to catalyze same reaction again on a new substrate

❷ Enzyme catalyzes reaction and transforms substrate into products

❓ *What part of an enzyme combines with its substrate?*

---

❶ The substrate makes contact with the active site on the enzyme molecule, forming a temporary compound called the **enzyme–substrate complex.**

❷ The substrate molecule is transformed by rearrangement of existing atoms, breakdown of the substrate molecule, or combination of several substrate molecules. The transformed substrate molecules are called the **products** of the reaction.

❸ After the reaction is completed and the products of the reaction move away from the enzyme, the unchanged enzyme is free to attach to another substrate molecule.

The names of enzymes usually end in the suffix *-ase*. All enzymes can be grouped according to the types of chemical reactions they catalyze. For example, *oxidases* add oxygen, *dehydrogenases* remove hydrogen, *hydrolases* add water, and *transferases* transfer groups of atoms.

Nucleic Acids: Deoxyribonucleic Acid (DNA) and Ribonucleic Acid (RNA)   *Know*

objective: *Explain the importance of deoxyribonucleic acid (DNA), ribonucleic acid (RNA), and adenosine triphosphate (ATP).*

Nucleic (noo-KLĒ-ic) *acids* are exceedingly large organic molecules containing carbon, hydrogen, oxygen, nitrogen, and phosphorus. There are two principal kinds: *deoxyribonucleic* (dē-ok′-sē-rī-bō-noo-KLĒ-ik) *acid (DNA)* and *ribonucleic acid (RNA)*.

The building blocks of nucleic acids are *nucleotides.* A molecule of DNA is a chain composed of repeating nucleotide units. Each nucleotide of DNA consists of three basic parts (Figure 2.11a):

1. One of four possible *nitrogenous* (*nitrogen*) bases, which are ring-shaped structures containing atoms of C, H, O, and N. The nitrogenous bases found in DNA are adenine (A), thymine (T), cytosine (C), and guanine (G).

2. A pentose sugar called *deoxyribose.*

3. *Phosphate groups* (PO_4^{3-}).

Nucleotides are named for their nitrogenous bases; for example, a nucleotide containing thymine is called a *thymine nucleotide.*

Figure 2.11b shows the following structural characteristics of the DNA molecule:

1. The molecule consists of two strands with crossbars. The strands twist about each other in the form of a *double helix* so that the shape resembles a twisted ladder.

2. The uprights of the DNA ladder consist of alternating phosphate groups and the deoxyribose portions of the nucleotides.

Figure 2.11   DNA molecule.

🔑 *Nucleotides are the building blocks of nucleic acids.*

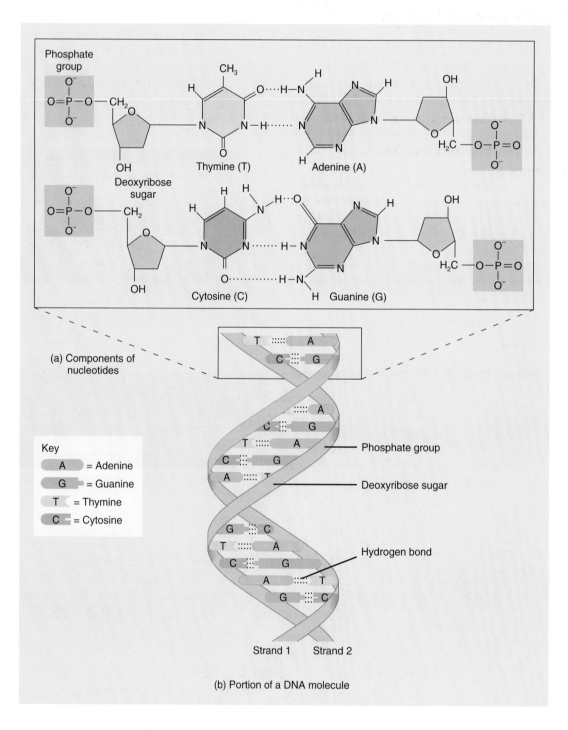

(a) Components of nucleotides

Key

A = Adenine
G = Guanine
T = Thymine
C = Cytosine

Phosphate group
Deoxyribose sugar
Hydrogen bond

Strand 1   Strand 2

(b) Portion of a DNA molecule

🅠 *Which nitrogenous base is not present in RNA?*

3. The rungs of the ladder contain paired nitrogenous bases. Adenine always pairs with thymine and cytosine always pairs with guanine. About 1000 rungs of DNA comprise a *gene,* a portion of a DNA strand that composes the hereditary material in cells.

Humans have about 100,000 functional genes that direct the production of body proteins (Chapter 3). Genes determine which traits we inherit, and they control all the activities that take place in our cells throughout a lifetime.

Chapter 2   *Chemical Compounds and Life Processes*   DNA is the molecule that builds our Genes   **33**

RNA, the other kind of nucleic acid, differs from DNA in several respects. RNA is single stranded; DNA is double stranded. The sugar in the RNA nucleotide is ribose. And RNA does not contain the nitrogenous base thymine. Instead of thymine, RNA has the nitrogenous base uracil. Three different kinds of RNA have been identified in cells. Each type has a specific role to perform with DNA in protein synthesis reactions (see Chapter 3).

Adenosine Triphosphate (ATP)

A molecule that is indispensable to the life of a cell is *adenosine* (a-DEN-ō-sēn) *triphosphate (ATP)*. ATP is found universally in living systems. Its essential function is to store energy for the cell's basic life activities.

Structurally, ATP consists of three phosphate groups (PO_4^{3-}) and an adenosine unit composed of adenine and the sugar ribose (Figure 2.12). ATP releases a great amount of usable energy when it is broken down by the addition of a water molecule (hydrolysis).

When the terminal phosphate group (here symbolized as Ⓟ) is hydrolyzed and therefore removed, the changed molecule is called *adenosine diphosphate (ADP)*. This reaction liberates energy and is represented as follows:

$$ATP \longrightarrow ADP + Ⓟ + E$$

The energy supplied by the catabolism of ATP into ADP is constantly being used by the cell. Because the supply of ATP at any given time is limited, a mechanism exists to replenish it: a phosphate group is added to ADP to manufacture more ATP. The reaction is represented as follows:

$$ADP + Ⓟ + E \longrightarrow ATP$$

Energy is required to make ATP. This energy is required to attach a phosphate group to ADP and is supplied primarily by the breakdown of the monosaccharide glucose in the cell in a process called cellular respiration (Chapter 20).

Figure 2.12   Structure of ATP and ADP. The two phosphate bonds that can be used to transfer energy are indicated by "squiggles" (~). Most often energy transfer involves hydrolysis of the terminal phosphate bond.

ATP stores chemical energy for various cellular activities.

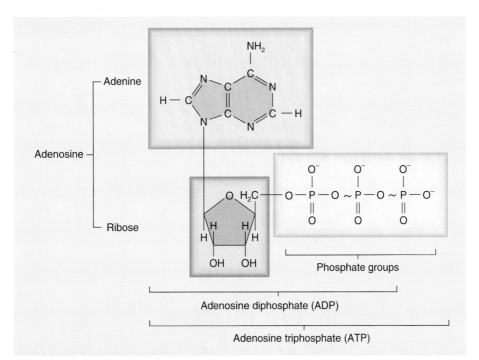

Is there more energy in ATP or ADP?

wellness focus

Fat-Burning Exercise: Fat-Burning Fantasies?

One of the most interesting chemistry topics for many people is fat metabolism. They yearn to answer the million-dollar question: how can fat loss be facilitated? Most readers have heard that exercise is an essential part of any weight control program. But what kind of exercise is best?

The ABC's of Fuel Mobilization

You will learn about how your body makes energy in Chapter 20. In order to understand what makes your body decide to burn fat, you need only understand a few simple concepts for now. In order to release energy, your body catabolizes certain molecules that are stored specifically to provide energy when needed. The two primary sources of energy the body uses are carbohydrates (in the form of glycogen, stored in the liver and in skeletal muscles) and fats (primarily triglycerides, stored in adipose tissue).

How does your body decide which to use during exercise? Fuel utilization is regulated by an interplay of many factors. One of these is exercise intensity. During high-intensity exercise, your body prefers to metabolize carbohydrates because carbohydrate metabolism is more efficient and yields more energy per unit of oxygen (O_2).

This preference for carbohydrates at high intensity makes intuitive sense when you consider the molecular structure of carbohydrates and lipids.

Remember from high school biology that animals use oxygen plus food (some mixture of carbon, hydrogen, oxygen, and sometimes nitrogen) to produce energy, in the process creating carbon dioxide (CO_2) plus water (H_2O). Because lipids have less oxygen than carbohydrates, more oxygen is required to get lipids to the CO_2 plus H_2O stage, and it takes more "effort" in terms of the chemical pathways involved.

Does Low Intensity Mean High Fat Loss?

The percentage of energy expenditure due to fat combustion is directly related to exercise intensity. At very low intensities, your body relies predominantly on fat stores. At moderate intensities you use some fat and some carbohydrates. However, carbohydrates are the sole source of energy during high-intensity exercise. This has led some people to conclude that in order to burn fat, *low-intensity* exercise is preferable.

This conclusion takes people in the wrong direction because the percentage of calories that come from fat is only one small part of the energy balance picture. Studies holding caloric expenditure equivalent for people exercising at different exercise intensities have failed to show any difference in weight or fat loss. In order to lose weight, you must burn more calories than you eat. Calories do count. You will burn a similar number of fat calories in a leisurely or brisk walk. It really doesn't matter whether those calories come from glycogen or fat. Using up glycogen is fine; after your next meal, calories will go to replace any depleted carbohydrate stores rather than to make more fat.

The Best Kind of Exercise

Both high- and low-intensity exercise have advantages and disadvantages. People who are very overweight will find low-intensity exercise safer and more comfortable, even though it takes longer to expend a given number of calories. Physical activity, even if low-intensity, helps normalize many metabolic disorders commonly present in obese people, such as problems with blood sugar and cholesterol regulation. People with high fitness levels can take advantage of high-intensity exercise to burn more calories in a shorter period of time and achieve even greater improvements in physical fitness.

critical thinking

Why do you think abdominal exercises such as sit-ups are ineffective for reducing abdominal fat stores?

A friend asks you whether running or walking is better for weight loss. How would you respond?

| Study Outline |

Introduction to Basic Chemistry (p. 20)

Chemical Elements (p. 20)

1. Matter is anything that occupies space and has mass. It is made up of building units called chemical elements.
2. Oxygen, carbon, hydrogen, and nitrogen make up 96 percent of the body's weight.

Structure of Atoms (p. 20)

1. Units of chemical elements are called atoms.
2. Atoms consist of a nucleus, which contains protons and neutrons, and electrons that move about the nucleus in electron shells.
3. The total number of protons in an atom is its atomic number. This number is equal to the number of electrons in the atom.
4. The combined total of protons and neutrons in an atom is its mass number.

Atoms and Molecules (p. 20)

1. Electrons are the parts of an atom that participate in chemical reactions.
2. A molecule is two or more chemically combined atoms. A molecule containing different kinds of atoms is a compound.
3. In an ionic bond, outer electron shell electrons are transferred from one atom to another. The transfer forms ions, whose unlike charges attract each other and form ionic bonds.
4. In a covalent bond, there is a sharing of pairs of outer electron shell electrons.
5. Hydrogen bonding provides temporary bonding between certain atoms within large complex molecules such as proteins and nucleic acids.

Chemical Reactions (p. 25)

1. Synthesis reactions produce a new molecule.
2. In decomposition reactions, a substance breaks down into other substances.
3. When chemical bonds are formed, energy is usually needed. When bonds are broken, energy is released.

Chemical Compounds and Life Processes (p. 25)

1. Inorganic substances usually lack carbon and many contain ionic bonds.
2. Organic substances always contain carbon. Most organic substances are held together mostly or entirely by covalent bonds.

Inorganic Compounds (p. 25)

1. Water is the most abundant substance in the body. It is an excellent solvent and suspending medium, participates in chemical reactions, absorbs and releases heat slowly, and lubricates.
2. Inorganic acids, bases, and salts dissociate into ions in water. Cations are positively charged ions; anions are negatively charged ions. An acid ionizes into H^+; a base ionizes into OH^-. A salt ionizes into neither H^+ nor OH^- ions.
3. The pH of different parts of the body must remain fairly constant for the body to remain healthy. On the pH scale, 7 represents neutrality. Values below 7 indicate acid solutions, and values above 7 indicate alkaline solutions.
4. The pH values of different parts of the body are maintained by buffer systems that eliminate excess H^+ and excess OH^- ions in order to maintain pH homeostasis.

Organic Compounds (p. 27)

1. Carbohydrates are sugars or starches and are the most common sources of the energy that is needed for life. They may be monosaccharides, disaccharides, or polysaccharides. Carbohydrates, and other organic molecules, are joined together to form larger molecules with the loss of water (dehydration synthesis). In the reverse process, called hydrolysis (digestion), large molecules are broken down into smaller ones upon the addition of water.
2. Lipids are a diverse group of compounds that includes triglycerides (fats), phospholipids, steroids, carotenes, vitamins E and K, and eicosanoids. Triglycerides protect, insulate, and provide energy.
3. Proteins are constructed from amino acids. They give structure to the body, regulate processes as enzymes, provide protection, and help muscles to contract.
4. Enzymes are catalysts that are highly specific in terms of substrates with which they react, efficient in terms of the number of substrate molecules with which they react, and subject to a variety of cellular controls.
5. Deoxyribonucleic acid (DNA) and ribonucleic acid (RNA) are nucleic acids consisting of nitrogenous bases, sugar, and phosphate groups. DNA is a double helix and is the primary chemical in genes. RNA differs in structure and chemical composition from DNA and is mainly concerned with protein synthesis reactions.
6. The principal energy-storing molecule in the body is adenosine triphosphate (ATP). When an ATP molecule is decomposed, energy is liberated and the changed molecule is known as adenosine diphosphate (ADP). ATP is manufactured from ADP and Ⓟ primarily by using the energy supplied by the decomposition of glucose in a process called cellular respiration.

Self-Quiz

1. A substance that dissociates in water to form H^+ and one or more anions is called a(n)

   a. base   b. salt   c. buffer   d. acid   e. nitrogenous base

2. Ionic bonds are characterized by

   a. sharing electrons between atoms   b. their ability to form strong, stable bonds   c. atoms transferring electrons between atoms   d. the type of bonding formed in most organic compounds   e. an attraction between water molecules

3. Amino acids are the basic structural units of

   a. carbohydrates   b. proteins   c. complex polysaccharides   d. triglycerides   e. nucleic acids

4. The kind of chemical reaction by which a disaccharide is formed from two monosaccharides is known as a(n)

   a. decomposition reaction   b. dehydration synthesis   c. hydrolysis   d. exchange reaction   e. dissociation

5. If an atom has two electrons in its second electron shell and its first electron shell is filled, it will tend to

   a. lose two electrons from its second electron shell   b. lose the electrons from its first electron shell   c. lose all the electrons from its first and second electron shells   d. gain six electrons in its second electron shell   e. share two electrons in its second electron shell

6. Which of the following is a salt?

   a. H_2O b. HCl   c. NaOH   d. NaCl   e. CO_2

7. Which solution is most basic (alkaline)?

   a. pH 0   b. pH 4   c. pH 7   d. pH 10   e. pH 14

8. The concentration of H^+ is equal to that of OH^- at

   a. pH 0   b. pH 4   c. pH 7   d. pH 10   e. pH 14

9. Chlorine (Cl) has an atomic number of 17. An atom of chlorine may become a chloride ion (Cl^-) by

   a. losing one electron   b. losing one neutron   c. gaining one proton   d. gaining one electron   e. gaining two electrons

10. An organic compound that consists of C, H, O and that may be broken down into glycerol and fatty acids is a(n)

    a. triglyceride   b. nucleic acid   c. glycogen   d. carbohydrate   e. protein

11. Matter that cannot be broken down into simpler substances by chemical reactions is known as a(n)

    a. molecule   b. anion   c. compound   d. neutron   e. chemical element

12. Which of the following are incorrectly matched?

    a. synthesis reaction—anabolism   b. dehydration reaction—synthesis reaction   c. hydrolysis—anabolism

    d. decomposition reaction—catabolism   e. buffer reaction—exchange reaction

13. The difference in H^+ concentration between solutions with a pH of 3 and a pH of 5 is that the solution with the pH of 3 has _____ H^+.

    a. two times more   b. two times less   c. 10 times more   d. 100 times more   e. 100 times less

14. What is the principal energy storage molecule in the body?

    a. DNA   b. ATP   c. ADP   d. RNA   e. protein

15. Which of the following statements about water is false?

    a. It is involved in many chemical reactions in the body.   b. It is an important solvent in the human body.   c. It helps lubricate a variety of structures in the body.   d. It can absorb a large amount of heat without changing its temperature.   e. It requires very little heat to change from a liquid to a gas.

16. Which of the following is NOT a true statement about enzyme activity?

    a. Enzymes form a temporary complex with their substrates.   b. Enzymes are not permanently altered by the chemical reaction   c. Most enzymes work with a wide variety of substrates.   d. Enzymes are considered organic catalysts.   e. Enzymes are subject to cellular control.

17. For each item in the following list, place an R if it applies to RNA and a D if it refers to DNA; use R and D if it applies to both RNA and DNA.

    R+D   a. composed of nucleotides

    D   b. forms a double helix

    D   c. contains thymine

    R   d. contains the sugar ribose

    R   e. contains the nitrogenous base uracil

    D   f. is the hereditary material of cells

    D   g. contains the sugar deoxyribose

    R   h. single stranded

    R+D   i. contains adenosine

    R+D   j. contains phosphate groups

18. Match the following:

    F   a. inorganic compound   A. amino acid

    D   b. monosaccharide   B. glycerol

    A   c. building block of protein   C. glycogen

    E   d. building block of DNA   D. glucose

    C   e. polysaccharide   E. nucleic acid

    B   f. component of triglycerides   F. water

Critical Thinking Applications

1. An MRI (magnetic resonance image) detects the response of protons to the magnetizing effect of radio waves. What are protons? Where are they found in the human body?

2. A new health food product is advertised as "all natural—no chemicals." Discuss the validity of this claim.

3. After your first A & P exam, you're suffering from heartburn (caused by excess secretion of stomach acid). What food or drink could you take to buffer the stomach acid? Explain the role of buffers in the body.

4. Your lab partner Nat is taking A & P for the third time. He just spilled a bottle labeled NaCl all over both of you and he's headed for the emergency shower. Should you join him? Why or why not?

5. Your cousin Edith adds milk and lemon juice to her tea. The tea now has strange white lumps floating in it. What caused the milk to curdle?

6. The cafeteria serves a lunch special: hamburger, french fries, and a soft drink. List the organic compounds that you will be eating, their basic structural units, and their uses in the body.

Answers to Figure Questions

2.1 Six.

2.2 Helium (He), is an inert chemical element because it has a completely filled outermost electron shell.

2.3 K is an electron donor; when it ionizes, it becomes a cation, K^+.

2.4 An ionic bond involves the loss and gain of electrons; a covalent bond involves the sharing of pairs of electrons.

2.5 $CaCO_3$ is a salt and H_2SO_4 is an acid.

2.6 A pH of 6.82 is more acidic than a pH of 6.91. Both pH = 8.41 and pH = 5.59 are 1.41 pH units from neutral (pH = 7).

2.7 There are 6 carbons in fructose, 12 in sucrose.

2.8 Saturated fat.

2.9 Hydrolysis.

2.10 Active site.

2.11 Thymine.

2.12 ATP.

3 chapter

CELLS = little protein factories

student learning objectives

*I*t is at the cellular level of organization that activities essential to life occur and disease processes originate. A *cell* is the basic, living, structural and functional unit of the body. *Cytology* (sī-TOL-ō-jē; *cyto* = cell; *logos* = study of) is the branch of science concerned with the study of cells. This chapter examines the structure, functions, and reproduction of cells.

put together so that we can study the common features of many cells at one time. As you will see later, different cells of the body have different types and numbers of various structures, depending on their functions. Examine the generalized cell in Figure 3.1.

For convenience, we divide the generalized cell into three principal parts:

Generalized Cell

objective: *List the parts of a generalized cell.*

A *generalized cell* is a composite of many different cells in the body. Although no such single cell actually exists, it has been

1. **Plasma (cell) membrane.** The outer, limiting membrane separating the cell's internal parts from the extracellular materials and external environment. (Extracellular materials—substances outside cells—will be examined in Chapter 4.) *Composed of phospholipids + proteins*

prototypical cell)

Figure 3.1 Generalized cell. *a composite of all cells*

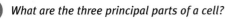 The cell is the basic, living, structural and functional unit of the body.

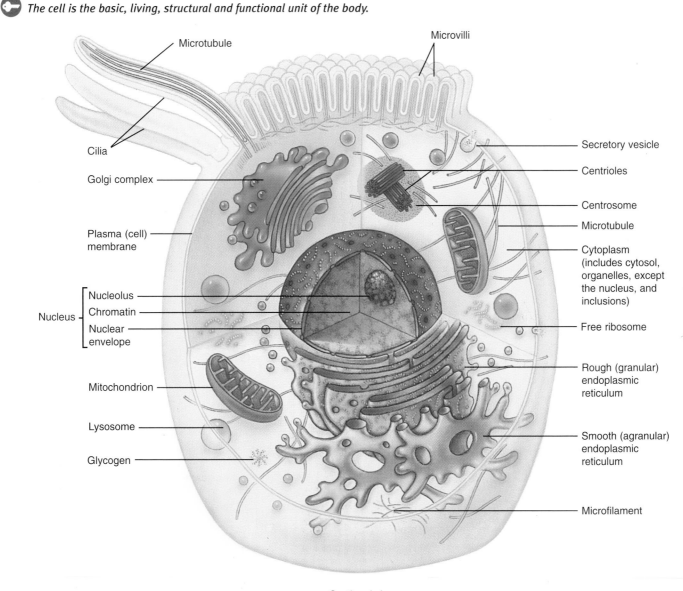

- Microtubule
- Microvilli
- Cilia
- Golgi complex
- Plasma (cell) membrane
- Nucleolus
- Chromatin
- Nuclear envelope
- Nucleus
- Mitochondrion
- Lysosome
- Glycogen
- Secretory vesicle
- Centrioles
- Centrosome
- Microtubule
- Cytoplasm (includes cytosol, organelles, except the nucleus, and inclusions)
- Free ribosome
- Rough (granular) endoplasmic reticulum
- Smooth (agranular) endoplasmic reticulum
- Microfilament

Sectional view

❓ *What are the three principal parts of a cell?*

liquid environment

2. **Cytosol.** The term *cytoplasm* (SĪ-tō-plazm) refers to all cellular contents between the plasma membrane and nucleus. The thick, semifluid portion of cytoplasm is called *cytosol* (SĪ-tō-sol).

3. **Organelles.** Highly organized structures with characteristic shapes that are specialized for specific cellular activities.
include organelles

Plasma (Cell) Membrane

objective: *Explain the structure and functions of the plasma membrane.*

The exceedingly thin structure that separates one cell from other cells and from the external environment is called the *plasma (cell) membrane.* It is a gatekeeper that regulates the passage of substances into and out of cells.

Chemistry and Structure

The plasma membrane consists mostly of phospholipids (lipids that contain phosphorus) and proteins. Other chemicals in lesser amounts include cholesterol (a lipid), glycolipids (combinations of carbohydrates and lipids), and glycoproteins (combinations of sugars and proteins) (Figure 3.2).

The phospholipid molecules are arranged in two parallel rows, forming a *phospholipid bilayer.* There are two kinds of membrane proteins: *integral proteins,* which penetrate through the phospholipid bilayer, and *peripheral proteins,* which are loosely attached to the exterior or interior surface of the membrane (Figure 3.2).

Some integral proteins form *channels* that have a *pore* (hole) through which certain substances move into and out of cells. Others act as *transporters* (carriers) to move a substance from one side of the membrane to another. Some serve as recognition sites called *receptors,* which identify and attach to specific molecules such as hormones, nutrients, antibodies, and other chemicals. Still others function as enzymes. Membrane glycoproteins and glycolipids are often *cell identity markers,* which enable a cell to recognize other similar cells or to recognize and respond to potentially dangerous foreign cells.

Functions

Following are the functions of the plasma membrane:

1. **Communication.** The plasma membrane functions in cellular communication through its receptors, glycoproteins, and glycolipids.

2. **Electrochemical gradient.** As you will see later, the plasma membrane sets up an electrical and chemical gradi-

---

Figure 3.2   Chemistry and structure of the plasma membrane.

The plasma membrane consists mostly of phospholipids and proteins.

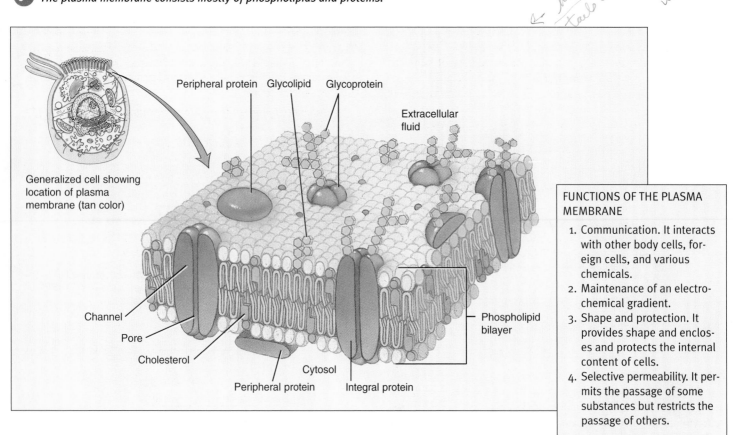

FUNCTIONS OF THE PLASMA MEMBRANE

1. Communication. It interacts with other body cells, foreign cells, and various chemicals.
2. Maintenance of an electrochemical gradient.
3. Shape and protection. It provides shape and encloses and protects the internal content of cells.
4. Selective permeability. It permits the passage of some substances but restricts the passage of others.

What are four factors that affect the permeability of plasma membranes?

ent (difference) between the inside and outside of the cell. This gradient is important for the proper functioning of most cells.

3. **Shape and protection.** The plasma membrane provides a flexible boundary that gives shape to cells and encloses and protects the internal contents of cells. Some membrane proteins can also change membrane shape during processes such as cell division, movement, and ingestion.

4. **Selective permeability.** The plasma membrane regulates entry and exit of materials. It permits passage of certain substances and restricts the passage of others. This property of membranes is called *selective permeability* (per′-mē-a-BIL-i-tē). The selective permeability of a plasma membrane to different substances depends on several factors that relate to the structure of the membrane:

   - **Lipid solubility.** Substances that dissolve in lipids, such as steroids and fat-soluble vitamins (A, E, D, and K) pass easily through the phospholipid bilayer of the plasma membrane.

   - **Size.** Most large molecules, including most proteins, cannot pass through the plasma membrane. A few very small molecules, such as water, oxygen, and carbon dioxide, can pass through the phospholipid bilayer.

   - **Charge.** The phospholipid bilayer portion of the plasma membrane is impermeable to all ions. Some charged ions do pass through the membrane, however, by moving through a pore in an integral protein or by being carried from one side to the other by integral proteins that function as transporters (Figure 3.2).

   - **Presence of channels and transporters.** Plasma membranes are permeable to a variety of substances that cannot cross the phospholipid bilayer. Proteins in the membrane help several substances to cross the membrane, much as a ferryboat helps people and cars cross a river. These integral proteins increase membrane permeability in two ways. Some proteins form water-filled pores through the membrane. Other proteins act as transporters, picking up a substance and shuttling it through to the other side before releasing it. Most proteins that form pores and function as transporters are very selective, allowing only a specific substance to cross the membrane. This will be described shortly.

Movement of Materials Across Plasma Membranes

objective: *Describe how materials move across plasma membranes.*

Before actually discussing how materials move into and out of a cell, we will examine the various fluids through which they move. Most body fluid is inside body cells and is called *intracellular* (*intra* = within) *fluid (ICF)* or *cytosol.* Fluid outside body cells is called *extracellular* (*extra* = outside) *fluid (ECF).* The ECF filling the microscopic spaces between the cells of tis-

sues is *interstitial* (in′-ter-STISH-al) *fluid* (*inter* = between) or *intercellular fluid.* The ECF in blood vessels is called *plasma* (Figure 3.3) and in lymphatic vessels is called *lymph.* Interstitial fluid contains gases, nutrients, and ions, all needed for maintaining life. Interstitial fluid circulates in the spaces between tissue cells. Essentially, all body cells are surrounded by the same fluid environment. For this reason, interstitial fluid is often called the body's *internal environment.*

The processes involved in the movement of substances across plasma membranes are classified as either passive or active. In *passive processes,* substances move across plasma membranes without the use of some of the cell's own energy (from the splitting of ATP). Instead, these substances rely on a type of energy called kinetic energy. *Kinetic energy* is the energy of motion of molecules. Because of it, molecules in solutions are constantly moving about, colliding with one another, and moving in various directions. Another feature of passive processes is the movement of substances from regions of higher concentration to regions of lower concentration. They move from an area where there are more of them to an area where there are fewer of them. The movement from high to low continues until the molecules are evenly distributed, a condition called *equilibrium* (ē′-kwi-LIB-rē-um). The difference between high and low concentrations is called the *concentration gradient.* Molecules moving from the high-concentration area to the low-concentration area are said to move *down* (with) the concentration gradient. The movement involves only the kinetic energy of individual molecules. The substances move on their own down the concentration gradient. In *active processes,* the cell uses some of its own energy (from the splitting of ATP). Let's take a closer look at passive and active processes.

[handwritten annotation: passive does not require energy have kinetic energy]

[handwritten annotation: active processes require energy]

Figure 3.3   Body fluids. Intracellular (ICF) and extracellular fluid (ECF). Intracellular fluid is the fluid within cells. Extracellular fluid is found outside cells, in blood vessels as plasma, and between tissue cells as interstitial fluid.

🔑 *Interstitial fluid is referred to as the internal environment of the body.*

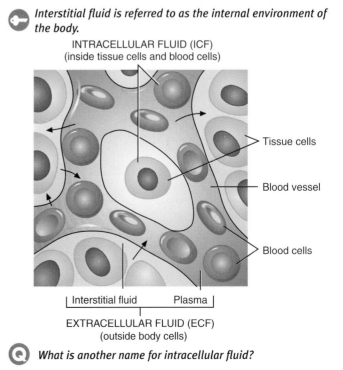

INTRACELLULAR FLUID (ICF)
(inside tissue cells and blood cells)

Tissue cells

Blood vessel

Blood cells

Interstitial fluid   Plasma

EXTRACELLULAR FLUID (ECF)
(outside body cells)

❓ *What is another name for intracellular fluid?*

Passive Processes

Do not require energy
HAVE KINETIC ENERGY.

In passive processes, the motion of molecules is affected by several factors, including (1) concentration of the molecules of solute and solvent, (2) temperature—as temperature increases, the rate of molecular motion increases, and (3) pressures exerted by gravity and other forces.

SIMPLE DIFFUSION   **Simple diffusion** occurs where there is a net (greater) movement of molecules or ions from a region of their higher concentration to a region of their lower concentration (Figure 3.4) until the molecules are evenly distributed (equilibrium). Remember, the movement involves only the kinetic energy of individual molecules and the substances move on their own down the concentration gradient.

For example, if a dye pellet is placed in a beaker filled with water, the color of the dye is seen immediately around the pellet. At increasing distances from the pellet, the color becomes lighter. Later, however, the water solution will be a uniform color. The dye molecules possess kinetic energy and move about at random and they move down the concentration gradient from an area of high dye concentration to an area of low dye concentration. The water molecules also move from their high-concentration to their low-concentration area. When dye molecules and water molecules are evenly distributed among themselves, equilibrium is reached and diffusion ceases, even though molecular movements continue. If the water in the beaker were boiled, the rate of diffusion of the dye molecules would increase tremendously because an increase in temperature speeds up diffusion.

Diffusion across selectively permeable membranes follows the same principles as seen in the example of the dye pellet in water. In the human body, diffusion is important in activities such as the movement of oxygen and carbon dioxide between blood and body cells and blood and lungs during respiration, the absorption of nutrients by body cells, the excretion of wastes by body cells, and electrolyte balance—all activities that contribute to homeostasis.

FACILITATED DIFFUSION   In a type of diffusion called **facilitated diffusion,** large and lipid-insoluble substances move across plasma membranes with the assistance of integral proteins that function as transporters (carriers). Most sugars, including glucose, are transported via facilitated diffusion. As an example of facilitated diffusion, we will use the transport of glucose across a plasma membrane (Figure 3.5).

1 First, glucose attaches to a transporter on the outside of the membrane.

2 Then the transporter changes shape.

3 Glucose passes through the membrane and is released inside the cell.

Figure 3.4   Simple diffusion. Small lipid-insoluble substances may diffuse through pores in channels formed by integral proteins, whereas larger, lipid-soluble substances may diffuse through the phospholipid bilayer.

In simple diffusion, there is a net (greater) movement of molecules or ions from a region of their higher concentration to a region of their lower concentration.

Rely on Kinetic energy.

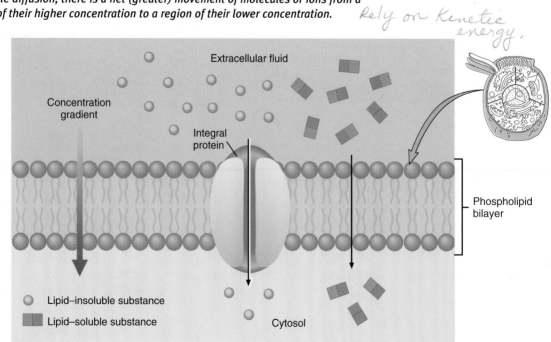

How would fever affect body processes that involve diffusion?

Figure 3.5   Facilitated diffusion. Large and lipid-insoluble substances move across plasma membranes by facilitated diffusion.

In facilitated diffusion, integral proteins function as transporters (carriers).

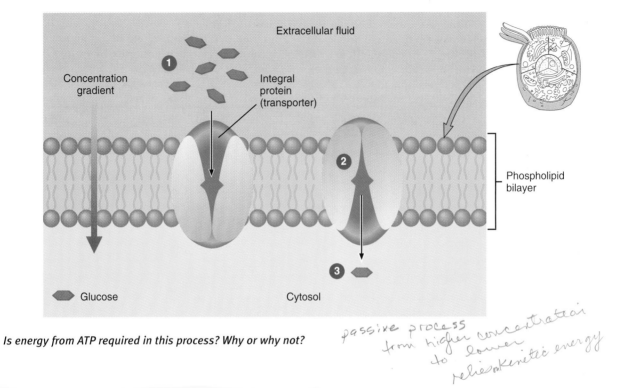

Glucose

Cytosol

Is energy from ATP required in this process? Why or why not?

passive process from higher concentration to lower relies on kinetic energy

In facilitated diffusion, the cell does not use energy from the splitting of ATP, and the movement of relatively large and lipid-insoluble substances is from a region of their higher concentration to a region of their lower concentration.

OSMOSIS   **Osmosis** (oz-MŌ-sis) is the net movement of water molecules through a selectively permeable membrane from an area of higher water concentration to an area of lower water concentration. Osmosis refers only to water. The water molecules pass through pores in integral proteins in the plasma membrane. Here again, the only energy involved is the kinetic energy of motion.

The apparatus in Figure 3.6 demonstrates osmosis. It consists of a cellophane sac, a selectively permeable membrane. The cellophane sac is filled with a colored, 20 percent sugar (sucrose) solution (20 parts sugar and 80 parts water) and placed into a beaker containing distilled (pure) water (0 parts sugar and 100 parts water). There is a lower concentration of water inside the cellophane sac than outside. Because of this difference, water begins to move from the beaker into the cellophane sac. This passage of water through a selectively permeable membrane produces a pressure called osmotic pressure. **Osmotic pressure** is the pressure needed to stop the flow of water across the membrane. The greater the solute concentration of the solution, the greater its osmotic pressure. There is no movement of sugar from the cellophane sac inside the beaker, because the cellophane is impermeable to molecules of sugar: sugar molecules are too large to go through the pores of the

membrane. As water moves into the cellophane sac, the sugar solution becomes increasingly diluted and the increased volume and pressure force the mixture up the glass tubing. In time, the weight of the water that has accumulated in the cellophane sac and the glass tube will be equal to the osmotic pressure, and this keeps the water volume from increasing in the tube. When water molecules leave and enter the cellophane sac at the same rate, equilibrium is reached. Osmotic pressure is an important force in the movement of water between various compartments of the body in order to maintain homeostasis.

ISOTONIC, HYPOTONIC, AND HYPERTONIC SOLUTIONS   Osmosis may also be understood by considering the effects of different water concentrations on red blood cells. If the normal shape of a red blood cell is to be maintained, the cell must be in an **isotonic** (iso = same) **solution,** one where the concentrations of water molecules (solvent) and solute (solid) molecules are the same on both sides of the selectively permeable cell membrane. Under ordinary circumstances, a 0.90 percent NaCl (salt) solution is isotonic for red blood cells. In this condition, water molecules enter and exit the cell at the same rate, allowing the cell to maintain its normal shape (Figure 3.7a).

A different situation results if red blood cells are placed in a solution that has a lower concentration of solutes and therefore a higher concentration of water. This is called a **hypotonic** (hypo = less than) **solution.** Water molecules then enter the cells faster than they can leave, causing the red blood cells to swell and eventually burst. The rupture of red blood

Figure 3.6 Principle of osmosis. In (a) the cellophane sac (a selectively permeable membrane) contains a 20 percent sugar solution and is immersed in a beaker of distilled (pure) water. The arrows indicate that water molecules can pass freely into the sac. Sugar molecules, however, cannot pass out of the sac. As water moves into the sac by osmosis, the sugar solution becomes more dilute and its volume increases. At equilibrium (b), the sugar solution has moved part of the way up the glass tubing. Now the number of water molecules entering and the number leaving the cellophane sac are equal.

🔑 *Osmosis is the net diffusion of water molecules through a selectively permeable membrane.*

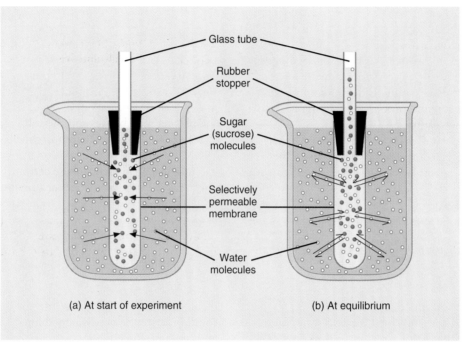

Glass tube

Rubber stopper

Sugar (sucrose) molecules

Selectively permeable membrane

Water molecules

(a) At start of experiment

(b) At equilibrium

❓ *What is osmotic pressure?*

cells is called *hemolysis* (hē-MOL-i-sis) (Figure 3.7b). Distilled water is a strongly hypotonic solution.

A *hypertonic* (*hyper* = greater than) *solution* has a higher concentration of solutes and a lower concentration of water than the red blood cells. A hypertonic solution might be 10 percent NaCl. In such a solution, water molecules move out of the cells faster than they can enter, causing the cells to shrink. This shrinkage is called *crenation* (krē-NĀ-shun) (Figure 3.7c).

FILTRATION *Filtration* involves the movement of solvents such as water and dissolved substances such as sugar across a selectively permeable membrane by gravity or mechanical pressure, usually hydrostatic (water) pressure. Such movement is always from an area of higher pressure to an area of lower pressure and continues as long as a pressure difference exists. Most small to medium-sized molecules can be pushed through a cell membrane by filtration pressure.

Filtration is a very important process in which water and nutrients in the blood are pushed into interstitial fluid for use by body cells. It is also the primary force that begins the process of urine formation.

Figure 3.7 Principle of osmosis applied to red blood cells. Shown here are the effects on red blood cells when placed in an isotonic, hypotonic, and hypertonic solution.

🔑 *An isotonic solution is one in which cell shape stays the same because there is no net water movement into or out of the cell.*

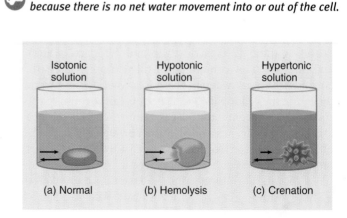

Isotonic solution

Hypotonic solution

Hypertonic solution

(a) Normal

(b) Hemolysis

(c) Crenation

❓ *Is a 2 percent NaCl solution hypotonic, hypertonic, or isotonic?*

Active Processes *Do require energy (ATP)*

When cells actively participate in moving substances across membranes, they must expend some of their own energy by splitting ATP. Such processes, as we have said, are called active processes. The active processes we will consider in some detail here are active transport and endocytosis (phagocytosis and pinocytosis).

ACTIVE TRANSPORT   The process by which substances are transported across plasma membranes, from an area of low concentration to an area of high concentration, is called **active transport.** Movement is against the concentration gradient (Figure 3.8). A typical body cell is estimated to use up to 40 percent of its ATP for active transport.

In active transport, the substance being moved enters a pore in an integral protein and makes contact with a region in the channel. At contact, ATP splits and releases energy that changes the shape of the integral protein, which, in turn, moves the substance (into or out of the cell).

Active transport is vitally important in maintaining the concentrations of some ions inside body cells and other ions outside body cells. For example, before a nerve cell can conduct a nerve impulse the concentration of potassium ions (K⁺) must be considerably higher inside the nerve cell than outside (see Figure 9.4a). At the same time, the nerve cells must have a higher concentration of sodium ions (Na⁺) outside than inside (see Figure 9.4a).

ENDOCYTOSIS   Large molecules and particles pass through plasma membranes by **endocytosis** (*endo* = into; *cyt* = cell; *osis* = process), in which a portion of the plasma membrane surrounds the substance, encloses it, and brings it into the cell. The reverse process of endocytosis is **exocytosis** (*exo* = out of), in which substances are discharged from cells. Endocytosis and exocytosis are important because molecules and particles of material that would normally be restricted from crossing the plasma membrane because of their large size can be brought in or removed from the cell.

We will now take a closer look at two types of endocytosis: phagocytosis and pinocytosis. In **phagocytosis** (fag′-ō-sī-TŌ-sis), or "cell eating," projections of the plasma membrane called **pseudopods** (SOO-dō-pods) surround a large solid particle outside the cell and bring it into the cell (Figure 3.9 on page 46). Once the particle is surrounded, the membrane folds inward, forming a membrane sac around the particle. This newly formed sac, called a **phagocytic vesicle,** breaks off from the plasma membrane, and the solid material inside the vesicle is digested by enzymes from organelles called lysosomes (discussed shortly). Phagocytosis is a vital body defense mechanism. Through phagocytosis, certain white blood cells engulf and destroy bacteria and other foreign substances.

In **pinocytosis** (pi′-nō-sī-TŌ-sis), or "cell drinking," the engulfed material is an extracellular liquid rather than a solid. Moreover, no pseudopods are formed. Instead, a minute droplet of liquid is attracted to the surface of the membrane.

Figure 3.8   Active transport. The substance transported moves from lower to higher concentration and the cell uses energy from the splitting of ATP.

🔑 *Active transport is necessary to maintain proper ion concentrations inside and outside cells for nerve impulse conduction.*

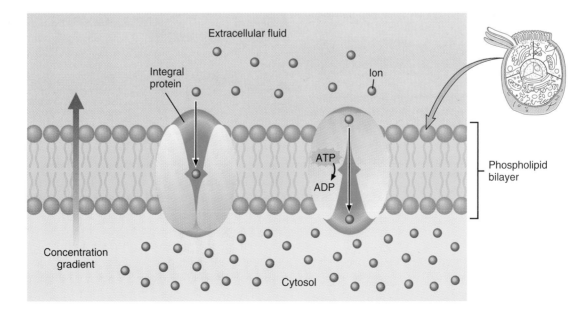

Q *Why is this considered an active process?*

Figure 3.9   Phagocytosis.

🔑 *Phagocytic cells destroy microbes and other foreign substances, an activity that helps protect against disease.*

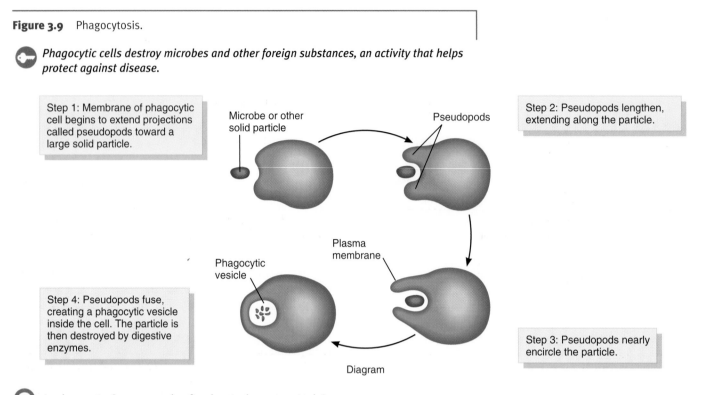

Step 1: Membrane of phagocytic cell begins to extend projections called pseudopods toward a large solid particle.

Microbe or other solid particle

Pseudopods

Step 2: Pseudopods lengthen, extending along the particle.

Phagocytic vesicle

Plasma membrane

Step 4: Pseudopods fuse, creating a phagocytic vesicle inside the cell. The particle is then destroyed by digestive enzymes.

Step 3: Pseudopods nearly encircle the particle.

Diagram

Ⓠ *Is phagocytosis an example of endocytosis or exocytosis?*

The membrane folds inward, forms a *pinocytic vesicle* that surrounds the liquid, and detaches from the membrane. Indigestible particles and cell products are removed from the cell by exocytosis. Whereas only a few cells are capable of phagocytosis, most carry out pinocytosis.

The various passive and active processes by which substances move across plasma membranes are summarized in Exhibit 3.1.

Cytosol *Chemical stew within a cell*

All the cellular contents between the plasma membrane and nucleus make up the *cytoplasm.* The thick, semifluid portion of cytoplasm is the *cytosol* (see Figure 3.1).

Physically, cytosol is a thick, transparent, gel-like fluid. Chemically, cytosol is 75–90 percent water plus various organic and inorganic particles dissolved or suspended in the water. Some chemical reactions, such as energy-releasing decomposition reactions, occur in the cytosol. Cytosol is also the site where new substances are synthesized for cellular use.

Organelles

objective: *Describe the structure and functions of organelles.*

Despite the numerous chemical activities occurring at the same time in the cell, there is little interference of one reaction with another. This is because the cell has many different compartments provided by its *organelles* ("little organs"). These spe-

cialized structures are usually surrounded by one or two membranes. Organelles have a characteristic appearance and specific roles in growth, maintenance, repair, and control. The numbers and types of organelles vary in different kinds of cells, depending on their functions.

Nucleus *membranous*

The *nucleus* (NOO-klē-us; *nucleus* = kernel) is generally spherical or oval. It is the largest structure in the cell (Figure 3.10). The nucleus is the control center of the cell because it contains the body's genes, which carry the inherited directions for cell structure and cellular activities. These genes are arranged in single file along DNA-containing structures called *chromosomes.* Most body cells contain a single nucleus, although some, such as mature red blood cells, do not have a nucleus. Skeletal muscle cells and a few other cells contain several nuclei.

The nucleus is bounded by a membranous sac called the *nuclear envelope* (*membrane*). The surface of the nuclear envelope is studded with ribosomes (described next) and is continuous at certain points with rough (granular) endoplasmic reticulum (also described shortly). Throughout the nuclear membrane are openings called *nuclear pores* that provide a means of communication between the nucleus and cytosol, specifically for the passage of large molecules such as RNA and various proteins.

Inside the nucleus are one or more spherical bodies called *nucleoli* (noo-KLĒ-ō-lī; singular is *nucleolus*). They are composed of protein, DNA, and RNA, and they are not enclosed by a membrane. They can disperse and disappear, which they do during cell division (reproduction of cells), reforming once new cells are formed. A type of RNA called *ribosomal RNA (rRNA)*, which is very important in protein synthesis, is produced in

PROCESS	DESCRIPTION
Exhibit 3.1	Summary of Processes by Which Substances Move Across Membranes
Passive processes	Substances move down a concentration gradient from an area of higher to lower concentration or pressure; cell does not expend energy (ATP).
Simple diffusion	Net movement of molecules or ions due to their kinetic energy from an area of higher to lower concentration until an equilibrium is reached.
Facilitated diffusion	Diffusion of larger molecules across a selectively permeable membrane with the assistance of membrane proteins in the membrane that serve as carriers.
Osmosis	Net movement of *water* molecules due to kinetic energy across a selectively permeable membrane from an area of higher to lower concentration of water until an equilibrium is reached. The solute concentration produces osmotic pressure.
Filtration	Movement of solvents (such as water) and solutes (such as glucose) across a selectively permeable membrane as a result of gravity or hydrostatic (water) pressure from an area of higher to lower pressure.
Active processes	Substances move against a concentration gradient from an area of lower to higher concentration; cell must expend energy (ATP). *require cellular energy*
Active transport	Movement of substances across a selectively permeable membrane from a region of lower to higher concentration with membrane proteins as carriers; the process requires energy expenditure. *ions*
Endocytosis	Movement of large molecules and particles through plasma membranes in which the membrane surrounds the substance, encloses it, and brings it into the cell. Examples include phagocytosis ("cell eating") and pinocytosis ("cell drinking"). *receptor mediated –*
Exocytosis	Export of substances from the cell by reverse endocytosis. *cells expell things... reverse of phagocytosis –*

Figure 3.10   Nucleus with one nucleolus.

The nucleus contains most of the genes, which are located on chromosomes.

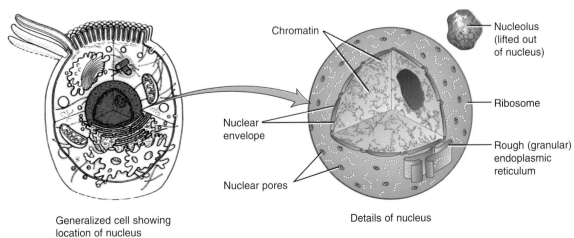

Generalized cell showing location of nucleus

Details of nucleus

Why is the nucleus called the control center of the cell?

nucleoli. Finally, inside the nucleus are DNA (the hereditary material) and protein, which form a loose mass known as *chromatin* (this coils into rod-shaped chromosomes during cell division).

Endoplasmic Reticulum (ER) *membranous*

Throughout the cytoplasm winds a system of membranous channels (*cisterns*) of varying shapes called the **endoplasmic reticulum** (en-dō-PLAS-mik re-TIK-yoo-lum; *endo* = within; *plasmic* = cytoplasm; *reticulum* = network) or **ER** (Figure 3.11a). The channels are continuous with the nuclear envelope. There are two types of ER. **Rough (granular) ER** is studded with ribosomes; **smooth (agranular) ER** has no ribosomes.

Ribosomes associated with rough ER synthesize proteins. The rough ER also serves as a temporary storage area for newly synthesized molecules and may add sugar groups to certain proteins, thus forming glycoproteins. Together, the rough ER and the Golgi complex (another organelle, described next) synthesize and package molecules that will be secreted from the cell.

Smooth ER is the site of fatty acid, phospholipid, and steroid synthesis. Also, in certain cells, enzymes within smooth ER can inactivate or detoxify a variety of chemicals, including alcohol, pesticides, and carcinogens (cancer-causing agents).

Ribosomes *non-membranous*

Ribosomes (RĪ-bō-sōms) are tiny granules composed of ribosomal RNA (rRNA) and protein. Structurally, ribosomes consist of two subunits, one about half the size of the other (Figure 3.11b). Ribosomes are the sites of protein synthesis (which is discussed later in the chapter).

Some ribosomes are scattered in the cytosol; others are attached to endoplasmic reticulum (ER).

Golgi Complex *membranous*

The **Golgi** (GOL-jē) **complex** is generally near the nucleus. It consists of flattened sacs (*cisterns*) that are stacked on each other like a pile of dishes with expanded areas at their ends (Figure 3.12). Associated with the cisterns are small **Golgi vesicles,** which cluster along the expanded edges of the cisterns.

The Golgi complex receives newly synthesized proteins and lipids from the endoplasmic reticulum and then sorts,

---

Figure 3.11 Endoplasmic reticulum (ER) and ribosomes.

Endoplasmic reticulum provides a surface area for chemical reactions and transports, stores, synthesizes, packages, and detoxifies molecules; ribosomes are the sites of protein synthesis.

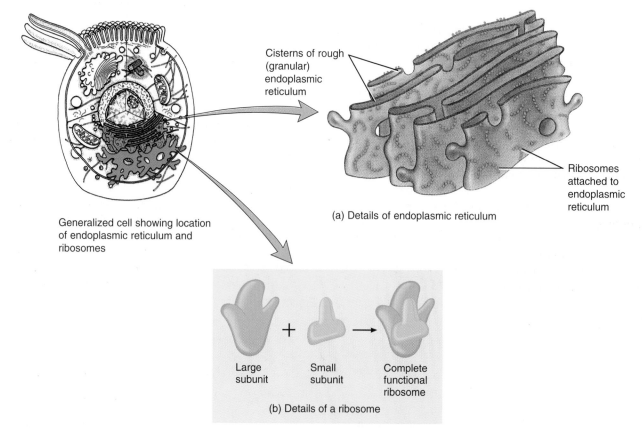

Cisterns of rough (granular) endoplasmic reticulum

Ribosomes attached to endoplasmic reticulum

(a) Details of endoplasmic reticulum

Generalized cell showing location of endoplasmic reticulum and ribosomes

Large subunit + Small subunit → Complete functional ribosome

(b) Details of a ribosome

Q *What is the basic structural and functional difference between smooth ER and rough ER?*

Figure 3.12   Golgi complex.

🔑 *All proteins destined for export from the cell pass into secretory vesicles.*

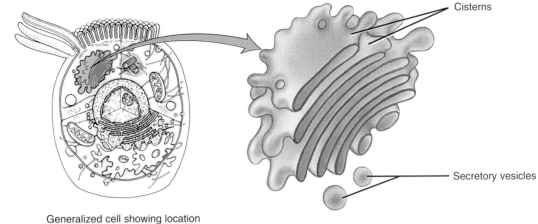

Generalized cell showing location
of Golgi complex

Details of a Golgi complex

Cisterns

Secretory vesicles

❓ *What is the main function of the Golgi complex?*

packages, and delivers them for inclusion in the plasma membrane and lysosomes (described next). All proteins destined for export from the cell pass into **secretory vesicles,** structures that release the proteins to the exterior of the cell by exocytosis.

Lysosomes *membranous   i.e, stomach or pancreas*

Lysosomes (LĪ-sō-sōms; *lysis* = dissolution; *soma* = body) are membrane-enclosed sacs that form from the Golgi complex and contain powerful digestive enzymes capable of breaking down many kinds of molecules (see Figure 3.1). These enzymes are also capable of digesting bacteria and other substances that enter the cell in phagocytic vesicles (see Figure 17.5). White blood cells, which ingest bacteria by phagocytosis, contain large numbers of lysosomes.

Lysosomes also use their enzymes to recycle the cell's own molecules. A lysosome can engulf another organelle, digest it, and return the digested components to the cytosol for reuse. In this way, worn-out cellular structures are continually renewed. The process by which worn-out organelles are digested is called *autophagy* (aw-TOF-a-jē; *auto* = self; *phagio* = to eat).

Lysosomal enzymes may also destroy their host cell, a process known as *autolysis* (aw-TOL-i-sis; *lysis* = dissolution). Autolysis occurs after death and in some pathological conditions.

Lysosomes also function in extracellular digestion. Lysosomal enzymes released at sites of injury help digest away cellular debris, which prepares the injured area for effective repair.

membranous   Some think these are extracellular parasites

Mitochondria *has its own DNA interesting but not founded*

Small, rod-shaped structures called *mitochondria* (mī-tō-KON-drē-a; *mitos* = thread; *chondros* = granule) appear throughout the cytoplasm. Because of their function in generating energy, they are referred to as "powerhouses" of the cell (Figure 3.13). A mitochondrion consists of two membranes, each of which is similar in structure to the plasma membrane. The outer mitochondr-

ial membrane is smooth, but the inner membrane is arranged in a series of folds called *cristae* (KRIS-tē; *crista* = ridge). The center of a mitochondrion is called the *matrix.*

The cristae provide an enormous surface area for chemical reactions, and it is here that ATP is produced. Cells that expend a lot of energy, such as muscle, liver, and kidney tubule cells, have a large number of mitochondria.

Cytoskeleton *non membranous microfilaments*

Cytosol has an internal structure consisting of <u>microfilaments,</u> microtubules, and intermediate filaments, together referred to as the *cytoskeleton* (see Figure 3.1). All consist of different proteins.

Microfilaments are rodlike structures that are involved in the contraction of muscle cells. In nonmuscle cells, microfilaments provide support and shape and assist in locomotion of entire cells (as in the movement of phagocytes) and movements within cells.

Microtubules are straight, slender tubes that provide support and shape for cells and serve as channels for substances to move through the cytosol. As you will see shortly, other cell structures are also composed of microtubules. These include flagella and cilia, centrioles, and the mitotic spindle.

Intermediate filaments appear to provide structural reinforcement in some cells and assist in contraction in others.

Flagella and Cilia *the male sperm cell   non-membranous*

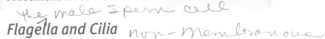

Some body cells possess projections for moving the entire cell or for moving substances along the surface of the cell. If the projections are few (usually occurring singly or in pairs) and long, they are called *flagella* (fla-GEL-a; singular is *flagellum*). Flagellum means whip. The only example of a flagellum in the human body is the tail of a sperm cell, used for locomotion (see Figure 23.4). If the projections are numerous and short, resembling many hairs, they are called *cilia* (*cilia* = eyelashes).

Figure 3.13   Mitochondrion.

🔑 *Chemical reactions within mitochondria generate ATP.*

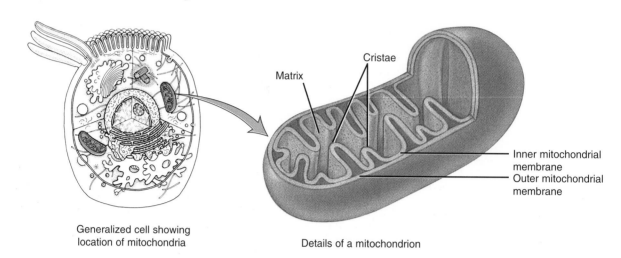

Generalized cell showing
location of mitochondria

Details of a mitochondrion

❓ *How do the cristae of a mitochondrion contribute to its energy-producing function?*

Ciliated cells of the respiratory tract move mucus that has trapped foreign particles (see Figure 18.4). Both cilia and flagella consist of microtubules surrounded by cytosol and enclosed by plasma membrane.

Centrosome and Centrioles

A **centrosome** is a dense area of cytosol near the nucleus. Centrosomes are involved in cell reproduction, to be described shortly. In the centrosome is a pair of cylindrical structures called **centrioles** (see Figure 3.1). Centrioles are composed of microtubules and they are at right angles to each other. Centrioles play a role in the formation and regeneration of flagella and cilia.

The major parts of the cell and their functions are summarized in Exhibit 3.2.

Gene Action

objective: *Define a gene and explain the sequence of events involved in protein synthesis.*

Protein Synthesis

Cells are basically protein factories that constantly produce large numbers of different proteins. Some proteins are structural, helping to form plasma membranes, microfilaments, microtubules, centrioles, flagella, cilia, the mitotic spindle, and other parts of cells. Others serve as hormones, antibodies, and contractile elements in muscle tissue. Still others serve as enzymes that regulate the chemical reactions that occur in cells.

The instructions for making proteins are contained in DNA. Cells make proteins by transferring the genetic infor-

mation encoded in DNA into specific proteins in a two-step process. In the first step, the genetic information in DNA is *transcribed* (copied) to produce a molecule of RNA. In the second step, the information in RNA is *translated* into a new protein molecule. Let us take a look at how DNA directs protein synthesis through transcription and translation.

Transcription

Transcription is the process by which genetic information is transferred from DNA to a type of RNA called **messenger RNA** (**mRNA**). It is called transcription because it resembles the transcription, or copying, of a sequence of words from one tape to another. In transcription, a segment of DNA uncoils and free nitrogenous bases attach to one-half of the segment, forming a strand of RNA, which now has the blueprint or pattern for a particular type of protein. Thus, the genetic information stored in the nitrogenous bases of one side of that segment of DNA is copied to the nitrogenous bases of the mRNA strand (Figure 3.14 on page 52). Specifically, the copying works this way: cytosine (C) in the DNA template or pattern dictates a guanine (G) in the mRNA strand being made; a G in the DNA template dictates a C in the mRNA strand; and a thymine (T) in the DNA template dictates an adenine (A) in the mRNA. Because RNA contains uracil (U) instead of T, an A in the DNA template dictates a U in the mRNA. For example,

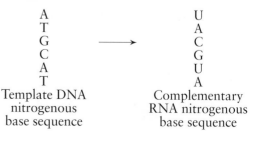

A		U
T		A
G	→	C
C		G
A		G
T		U

Template DNA
nitrogenous
base sequence

Complementary
RNA nitrogenous
base sequence

Exhibit 3.2 | Cell Parts and Their Functions

PART	FUNCTIONS
Plasma membrane	Provides communication with other cells and various chemicals, establishes an electrochemical gradient, provides shape and protection, and regulates entrance and exit of materials (selective permeability).
Cytosol	Serves as the substance in which chemical reactions occur.
Organelles	
Nucleus	Contains genes and therefore controls cellular activities.
Ribosomes	Sites of protein synthesis.
Endoplasmic reticulum (ER)	Provides surface area for many types of chemical reactions; ribosomes attached to rough ER synthesize proteins that will be secreted; smooth ER synthesizes lipids and detoxifies certain molecules.
Golgi complex	Processes, sorts, packages, and delivers proteins and lipids for inclusion in the plasma membrane and lysosomes, or for export from the cell; forms lysosomes.
Lysosomes	Digest cell structures and foreign microbes.
Mitochondria	Sites for production of ATP.
Cytoskeleton	
Microfilaments	Involved in muscle cell contraction; provide support and shape; assist in cellular and intracellular movement.
Microtubules	Provide support and shape; form intracellular conducting channels; assist in cellular movement; form flagella, cilia, centrioles, and mitotic spindle.
Intermediate filaments	Provide structural reinforcement in some cells.
Flagella and cilia	Allow movement of entire cell (flagella) or movement of particles along surface of cell (cilia).
Centrosome and centrioles	Centrosome helps organize mitotic spindle during cell division; centrioles form and regenerate flagella and cilia.

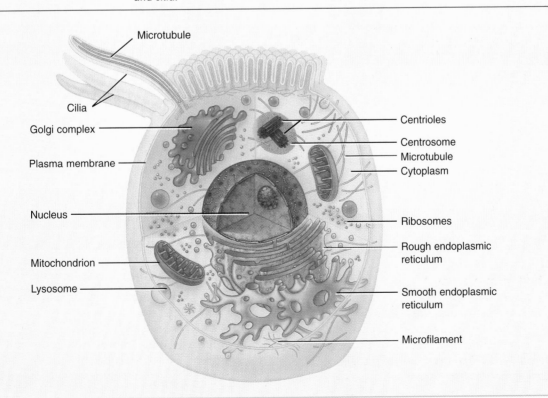

Figure 3.14 Transcription. When RNA synthesis is complete, mRNA leaves the nucleus and enters the cytoplasm, where translation occurs.

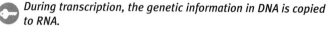

During transcription, the genetic information in DNA is copied to RNA.

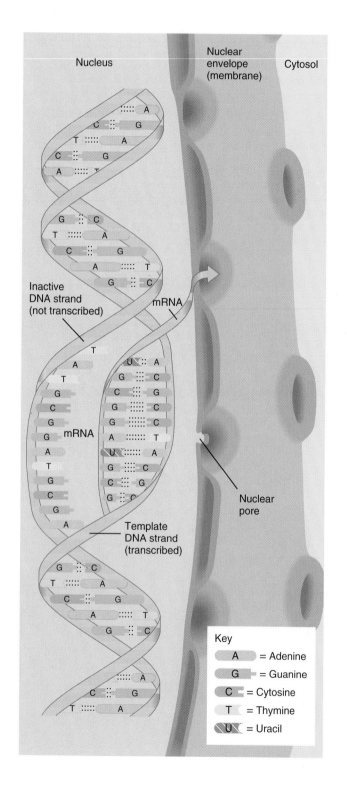

If the DNA template had the nitrogenous base sequence AGCT, what would the mRNA base sequence be?

DNA synthesizes mRNA through transcription. DNA also synthesizes two other kinds of RNA. One is *ribosomal RNA (rRNA)* that, with protein, makes up ribosomes. The other is called *transfer RNA (tRNA)*, to be explained next. Once synthesized, mRNA, rRNA, and tRNA leave the nucleus of the cell. In the cytosol, they participate in the next step in protein synthesis—translation.

Translation

Just as DNA provides the template for mRNA to be made, so mRNA provides a template for a protein to be made. *Translation* is the process by which the information in the nitrogenous bases of mRNA "directs" the arrangement of the amino acids of a protein. It is the arrangement of the amino acids that defines a protein, that is, whether it is an enzyme, or plasma membrane protein, or contractile element in a muscle, and so on. Translation then is the key part of protein synthesis.

Translation and all of protein synthesis take place on a ribosome. One protein molecule is made per ribosome.

1 The process begins when an mRNA strand attaches to a ribosome (Figure 3.15). Now amino acids must be brought to the ribosome so the actual protein can be formed, and that is the job of tRNA.

2 The tRNA picks up an amino acid in the cytosol that will participate in protein synthesis. For each different amino acid, there is a different tRNA. One end of a tRNA molecule attaches to an amino acid. The other end has three nitrogenous bases (triplet) known as an *anticodon.*

3 The tRNA, with the attached amino acid, travels to the ribosome. There, the anticodon recognizes a corresponding triplet, or *codon,* on the mRNA and attaches to it. If the tRNA anticodon is UAC, the mRNA codon would be AUG.

4 Once the tRNA attaches to mRNA, the ribosome moves along the mRNA strand and the next tRNA with its amino acid attaches.

5 These two amino acids are joined by a peptide bond through dehydration synthesis and the first tRNA is released and can pick up another molecule of the same amino acid if necessary.

6 Amino acids are attached one by one until the protein is complete.

7 When a specified protein is complete, synthesis is stopped by a special *stop codon.* The newly formed protein molecule is then released from the ribosome.

8 After protein synthesis, the large and small ribosomal subunits separate.

As the ribosome moves along the mRNA strand, it "reads" the information coded in mRNA and synthesizes a protein according to that information; the ribosome synthe-

Figure 3.15   Translation.

During translation, information in mRNA specifies the amino acid sequence of a protein.

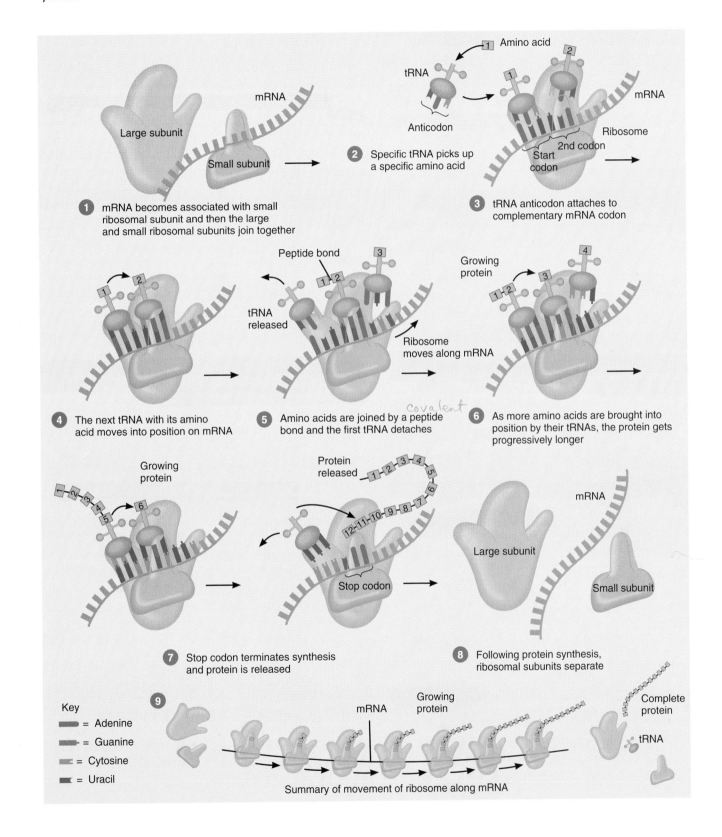

1 mRNA becomes associated with small ribosomal subunit and then the large and small ribosomal subunits join together

2 Specific tRNA picks up a specific amino acid

3 tRNA anticodon attaches to complementary mRNA codon

4 The next tRNA with its amino acid moves into position on mRNA

5 Amino acids are joined by a peptide bond and the first tRNA detaches

6 As more amino acids are brought into position by their tRNAs, the protein gets progressively longer

7 Stop codon terminates synthesis and protein is released

8 Following protein synthesis, ribosomal subunits separate

9

Key
= Adenine
= Guanine
= Cytosine
= Uracil

Summary of movement of ribosome along mRNA

What is a gene relative to protein synthesis?

sizes the protein by translating the codon sequences into an amino acid sequence.

Based on this description of protein synthesis, we can now define a *gene* as a group of nucleotides on a DNA molecule that serves as the master pattern for synthesizing a specific protein. Genes average about 1000 pairs of nucleotides, which appear in a specific sequence on the DNA molecule.

Remember that the nitrogenous base sequence of the gene determines the sequence of the bases in the mRNA. The sequence of the bases in the mRNA then determines the sequence of amino acids that will form the protein. Thus each gene is responsible for making a particular protein as follows:

$$DNA \xrightarrow{\text{Transcription}} mRNA \xrightarrow{\text{Translation}} Protein$$

Normal Cell Division

objective: | *Discuss the stages, events, and significance of cell division.*

Most of the cell activities discussed thus far maintain the cell on a day-to-day basis. However, cells must be replaced when they become damaged, diseased, or wear out and die. In a 24-hour period, the average adult loses billions of cells from different parts of the body. Cells that have a short life span, like those of the outer layer of skin, are continually replaced. Cells with a long life span, such as certain muscle cells, are replaced less frequently. In addition to replacement, new cells must also be produced as part of the growth process. Also, sperm and egg cells must be produced.

Cell division is the process by which cells reproduce themselves. There are two kinds of cell division.

In the first kind of division, *somatic cell division,* a single starting cell called a *parent cell* divides to produce two identical cells called *daughter cells.* This process consists of a nuclear division called *mitosis* plus a cytoplasmic division called *cytokinesis.* Somatic cell division ensures that each daughter cell has the same *number* and *kind* of chromosomes as the original parent cell and thus the same hereditary material and genetic potential as the parent cell. Somatic cell division results in an increase in the number of body cells and is the means by which dead or injured cells are replaced and new ones are added for body growth.

The second type of cell division is called *reproductive cell division.* It is the mechanism by which sperm and egg cells are produced, the cells required to form a new organism. This process consists of a nuclear division called *meiosis* plus *cytokinesis.* Here we will discuss somatic cell division; reproductive cell division will be discussed in Chapter 23.

Somatic Cell Division

Human cells, except for egg and sperm cells, contain 23 pairs of chromosomes, *for a total of 46 chromosomes.* A *chromosome* is a DNA molecule that stores hereditary information in genes. Following somatic cell division, both daughter cells must also contain 23 pairs of chromosomes identical to those of the parent cell DNA. Therefore, prior to mitosis, the parent cell must produce a duplicate set of chromosomes by replicating (producing duplicates of) its DNA.

DNA replication takes place when a cell is between divisions, when it is said to be in *interphase.* Also during this stage, the RNA and proteins needed to produce all the structures doubled in cell division are manufactured. When DNA replicates, its helix partially uncoils (Figure 3.16), and the DNA appears as a granular mass called *chromatin* (see Figure 3.17a on page 56). To uncoil, the DNA separates where its nitrogenous bases are connected. Each exposed nitrogenous base then picks up a complementary nitrogenous base. Uncoiling and complementary base pairing continue until each of the two original DNA strands is matched and joined with two newly formed DNA strands. The original DNA molecule has become two DNA molecules. Also during interphase, the paired centrioles in the cell replicate so that two pairs are present. Now mitosis (nuclear division) begins.

Mitosis

Mitosis is the distribution of the two sets of chromosomes into two separate and equal nuclei following the replication of the chromosomes of the parent nucleus. It results in the *exact* duplication of genetic information. Biologists divide the process into four stages: prophase, metaphase, anaphase, and telophase. These are arbitrary classifications, but mitosis is actually a continuous process, with one stage merging into the next.

PROPHASE   The first stage of mitosis is called *prophase* (pro = before) (Figure 3.17b). During early prophase, the chromatin condenses and shortens into visible chromosomes. Because DNA replication took place during interphase, each prophase chromosome consists of a pair of identical double-stranded DNA molecules called *chromatids.* Each pair of chromatids is held together by a small spherical body called a *centromere.*

Later in prophase, the nucleoli disappear, during which time synthesis of RNA is temporarily halted, and the nuclear envelope breaks down and is absorbed in the cytosol. In addition, a centrosome and its centrioles each move to opposite poles (ends) of the cell. As they do so, the centrosomes start to form the *mitotic spindle,* a football-shaped assembly of microtubules that are responsible for the movement of chromosomes. The lengthening of microtubules between centrosomes pushes the centrosomes to the poles of the cell so that the spindle extends from pole to pole. The spindle is an attachment site for chromosomes, and it also distributes chromosomes to opposite poles of the cell.

Figure 3.16   Replication of DNA. The two strands of the double helix separate by breaking the bonds between nucleotides. New nucleotides attach at the proper sites, and a new strand of DNA is paired off with each of the original strands. After replication, the two DNA molecules, each consisting of a new and an old strand, return to their helical structure.

🔑 *Replication doubles the amount of DNA. Thus, each daughter cell contains the normal amount after a somatic cell division.*

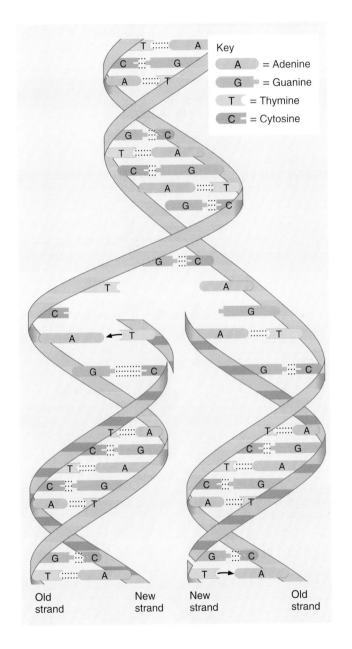

Key
A = Adenine
G = Guanine
T = Thymine
C = Cytosine

Old strand   New strand   New strand   Old strand

❓ *Why is DNA replication a key step that occurs prior to somatic cell division?*

METAPHASE   During **metaphase** (*meta* = after), the second stage of mitosis, the centromeres of the chromatid pairs line up at the exact center of the mitotic spindle. This midpoint region is called the *metaphase plate* or *equatorial plane region* (Figure 3.17c).

ANAPHASE   The third stage of mitosis, **anaphase** (*ana* = upward), is characterized by the splitting and separation of centromeres and the movement of the two sister chromatids of each pair toward opposite poles of the cell (Figure 3.17d). Once separated, the sister chromatids are referred to as daughter chromosomes. As the chromosomes move during anaphase, they appear V-shaped as the centromeres lead the way and seem to drag the trailing parts of the chromosomes toward opposite poles of the cell.

TELOPHASE   The final stage of mitosis, **telophase** (*telos* = far or end), is essentially the opposite of prophase and it begins as soon as chromosomal movement stops. During telophase, the identical sets of chromosomes at opposite poles of the cell uncoil and revert to their threadlike chromatin form. Microtubules disappear and a new nuclear envelope reforms around each chromatin mass; new nucleoli reappear in the daughter nuclei; and eventually the mitotic spindle breaks up (Figure 3.17e).

Cytokinesis

Division of the parent cell's cytoplasm is called **cytokinesis** (sī-tō-ki-NĒ-sis; *cyto* = cell; *kinesis* = motion). It begins in late anaphase or early telophase with formation of a **cleavage furrow**, a slight indentation of the plasma membrane that extends around the center of the cell (see Figure 3.17d–f). The furrow gradually deepens until the opposite surfaces of the cell make contact and the cell is split in two. When cytokinesis is complete, interphase begins. The result of cytokinesis is two separated daughter cells, each with separate portions of cytoplasm and its own set of identical chromosomes.

Abnormal Cell Division: Cancer *uncontrolled Cell Division*

objective: *Describe cancer as a homeostatic imbalance of cells.*

Definition

When cells in some area of the body duplicate without control, the excess tissue that develops is called a **tumor, growth,** or **neoplasm.** The study of tumors is called **oncology** (on-KOL-ō-gē; *onco* = swelling or mass; *logos* = study of) and a physician who specializes in this field is called an **oncologist.** Tumors may be cancerous and sometimes fatal or they may

process of DNA Replication
happens during enterphase

Figure 3.17   Cell division: mitosis and cytokinesis. Photomicrographs and diagrammatic representations of the various stages of cell division in whitefish eggs. Read the sequence starting at (a) and move clockwise until you complete the cycle.

 In somatic cell division, a single diploid parent cell divides to produce two identical diploid daughter cells.

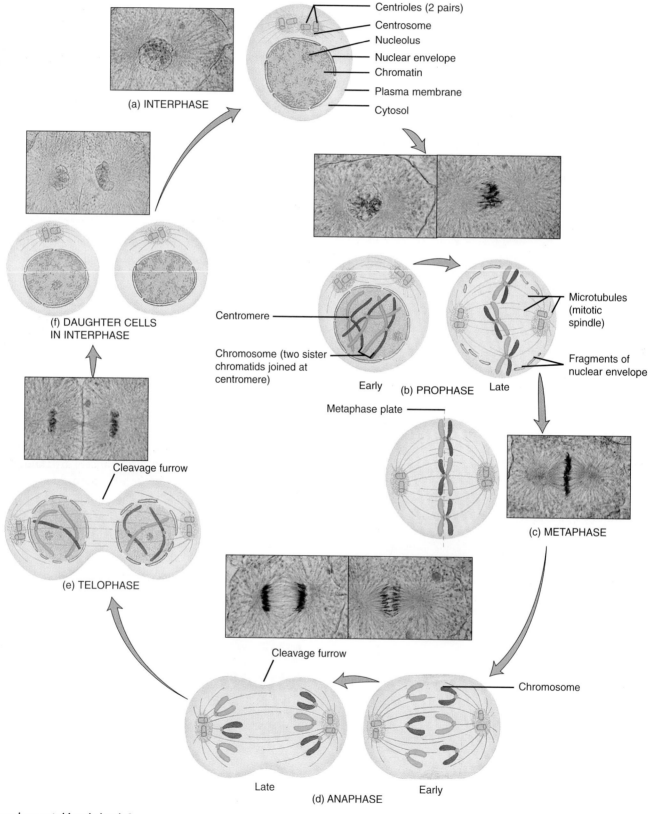

(a) INTERPHASE

Centrioles (2 pairs)
Centrosome
Nucleolus
Nuclear envelope
Chromatin
Plasma membrane
Cytosol

(f) DAUGHTER CELLS
IN INTERPHASE

Centromere

Chromosome (two sister
chromatids joined at
centromere)

Microtubules
(mitotic
spindle)

Fragments of
nuclear envelope

Early   (b) PROPHASE   Late

Metaphase plate

(c) METAPHASE

Cleavage furrow

(e) TELOPHASE

Cleavage furrow

Chromosome

Late   (d) ANAPHASE   Early

Q   *When does cytokinesis begin?*

[handwritten notes in top margin: Prostate cancer tends to metastasis to the sacrum and leads to chronic low back pain. become undifferentiated cell growth]

be quite harmless. A cancerous growth is called a *malignant tumor* or *malignancy* and may spread to other parts of the body. A noncancerous growth is called a *benign tumor.* Benign tumors do not spread to other parts of the body, but they may be removed if they interfere with a normal body function or are disfiguring.

Growth and Spread

Cells of malignant tumors duplicate continuously and very often quickly and without control. Such an increase in the number of cells due to an increase in the frequency of cell division is called *hyperplasia* (hī-per-PLĀ-zē-a). They are also capable of *metastasis* (me-TAS-ta-sis), the spread of cancerous cells to other parts of the body. Initially, malignant cells invade surrounding tissues. As the cancer grows, it expands and begins to compete with normal tissues for space and nutrients. Eventually, the normal tissue decreases in size (atrophies) and dies.

Following the invasion, some of the malignant cells may detach from the initial (primary) tumor. They may invade a body cavity or enter the blood or lymph, which can lead to widespread metastasis. Next, those malignant cells that survive in the blood or lymph invade other body tissues and establish secondary tumors. Finally, the secondary tumors become vascularized (develop networks of blood vessels).

Causes

What triggers a perfectly normal cell to lose control and become abnormal? Several factors contribute. First, there are environmental agents: substances in our air, water, and food. A chemical or other environmental agent that produces cancer is called a *carcinogen* (car-SIN-ō-jen). The World Health Organization estimates that carcinogens may be associated with 60 to 90 percent of all human cancers. The hydrocarbons found in cigarette tar are carcinogens. Ninety percent of all lung cancer patients are smokers. Another environmental factor is radiation. Ultraviolet (UV) light from the sun, for example, may cause genetic mutations in exposed skin cells and lead to skin cancer, especially among light-skinned people.

Viruses are a second cause of cancer, at least in animals. These agents are tiny packages of nucleic acids, either DNA or RNA, that are capable of infecting cells and converting

them to virus producers. With over 100 separate viruses identified as carcinogens in many species and tissues of animals, it is also probable that at least some cancers in humans are due to viruses.

A great deal of cancer research is now directed toward studying *oncogenes* (ONG-kō-jēnz)—genes found in every human cell that have the ability to transform a normal cell into a cancerous cell when they are *inappropriately activated.* Oncogenes develop from normal genes that regulate growth and development, called *proto-oncogenes.* Every cell contains proto-oncogenes, which carry out normal cellular functions until a malignant change occurs. It appears that some proto-oncogenes are activated to oncogenes by mutations in which the DNA of the proto-oncogene is altered. A *mutation* is a permanent structural change in a gene. Carcinogens may induce mutations. Viruses are believed to cause cancer by inserting their own oncogenes or proto-oncogenes into the host cell's DNA.

Researchers have also determined that some cancers are not caused by oncogenes but may be caused by a loss of genes called *tumor-suppressor genes.* These genes produce proteins that normally inhibit cell division.

The National Research Council has issued a series of guidelines related to diet and cancer. The main recommendations are to (1) reduce fat intake; (2) increase consumption of fiber, fruits, and vegetables; (3) increase intake of complex carbohydrates (potatoes, pasta); and (4) reduce consumption of salted, smoked, and pickled foods and simple carbohydrates (refined sugars).

Treatment

Treating cancer is difficult because it is not a single disease and because all the cells in a single tumor population do not behave in the same way. Although most cancers are thought to derive from a single abnormal cell, by the time a tumor reaches a clinically detectable size, the cancer may contain a diverse population of cells. For example, some cells metastasize and others do not. Some cells divide and others do not. Some are sensitive to drugs and some are resistant.

Besides chemotherapy, radiation therapy, surgery, cryothermia (freezing), hyperthermia (abnormally high temperature), and immunotherapy (bolstering the body's own defenses) may be used alone or in combination.

*C*ancer. Few words conjure up more terrifying images, despite the fact that many cancers are curable. In a quest to prevent the seemingly uncontrollable, many people look to lifestyle factors that they can control, especially the foods they eat.

Does Diet Really Matter?

Many lifestyle factors are much more important than diet when it comes to preventing cancer. Don't let a healthful diet be a substitute for quitting smoking, exercising, practicing safer sex, limiting alcohol intake, and working for a cleaner environment.

Dietary factors seem to matter most with cancers of the digestive tract and, to a limited extent, other cancers, such as those of the lung, breast, and prostate. Conveniently, a cancer-prevention diet also appears to reduce risk of heart disease, high blood pressure, and obesity.

Eat Your Fruits and Vegetables

Several studies have associated a diet rich in fruits and vegetables with lower cancer risk. Researchers believe that some carcinogens cause cancer by producing highly reactive oxygen atoms called free radicals, which may cause damage to cell components such as DNA. Fruits and vegetables contain a number of nutrients and chemicals, known as phytochemicals, that appear to interfere with carcinogenesis (development of cancer) caused or promoted by free radicals. They help either by detoxifying carcinogens (cancer-causing substances) or keeping them from penetrating cells or suppressing malignant

changes in cells that have been exposed to carcinogens. These cancer-prevention compounds include familiar nutrients like vitamins C and E, carotinoids (vitamin A precursors, including beta-carotene), and selenium. Others include indoles, isothiocyanates, flavonoids, and isoflavones.

Increase your intake of these cancer-prevention compounds by consuming 5 to 9 servings of a variety of fruits and vegetables each day. Try to have a serving from at least three of the following categories: cruciferous vegetables (for example, broccoli, brussels sprouts, cabbage, and cauliflower); citrus fruits; dark green leafy vegetables; and dark yellow/orange/red vegetables.

Eat Lower on the Food Chain

Grains, nuts, seed, and legumes (dried beans and their products) also contain a variety of cancer-prevention compounds. Soy products have provoked a great deal of interest because a high intake of these is associated with lower risk of breast and prostate cancers. Phytoestrogens ("plant estrogens") appear to be the ingredients responsible for this effect. Phytoestrogens alter the behavior of sex hormones in both men and women, and the role these hormones play in the process of carcinogenesis.

Avoid Fat and Carcinogens

Carcinogens are found in foods that are smoked, cured, and pickled. Consumption of pesticide residues on fruits and vegetables and in meats may increase cancer risk. Some experts believe that fat intake may be

the major and most controllable dietary carcinogen in North America. Red meat consumption has also been associated with increased cancer risk.

Reducing fat intake also helps prevent obesity, another cancer risk factor. Statistics indicate that people who are more than 20 percent above their recommended weight have a higher than average risk of many types of cancer. Cancer risk increases with amount of extra weight.

Pass It On

Studies suggest that fiber may help prevent cancers of the colon and rectum. Fiber is plant material that people can't digest and is found in plant foods. Fiber may help prevent cancer in several ways. First, it speeds the passage of stools on their journey through the colon, thus reducing the exposure of the colon lining to carcinogens present in these wastes. Second, fiber increases stool bulk and water content, thus diluting carcinogen concentration. And lastly, fat consumption, which is positively associated with cancer risk, often drops when people increase their intake of foods high in fiber.

---

critical thinking

What are some dietary changes you could make that would turn your current eating habits into a cancer-prevention diet? What foods would you eat more of? Less of?

Medical Terminology and Conditions

Note to the Student

Each chapter in this text that discusses a major system of the body is followed by a glossary of *medical terminology and conditions.* Both normal and pathological conditions of the system are included in these glossaries.

Some of these disorders, as well as disorders discussed in the text, are referred to as local or systemic. A *local disease* is one that affects one part of a limited area of the body. A *systemic disease* affects either the entire body or several parts.

The science that deals with why, when, and where diseases occur and how they are transmitted in a human community is known as epidemiology (ep′-i-dē′-mē-OL-ō-jē; *epidemios* = prevalent; *logos* = study of). The science that deals with the effects and uses of drugs in the treatment of disease is called *pharmacology* (far′-ma-KOL-ō-jē; *pharmakon* = medicine; *logos* = study of).

Atrophy (AT-rō-fē; *a* = without; *tropho* = nourish) A decrease in the size of cells with subsequent decrease in the size of the affected tissue or organ; wasting away.

Biopsy (BĪ-op-sē; *bio* = life; *opsis* = vision) The removal and microscopic examination of tissue from the living body for diagnosis.

Dysplasia (dis-PLĀ-zē-a; *dys* = abnormal; *plas* = to grow) Alteration in the size, shape, and organization of cells due to chronic irritation or inflammation; may progress to neoplasia (tumor formation, usually malignant) or revert to normal if the stress is removed.

Hyperplasia (hī′-per-PLĀ-zē-a; *hyper* = over) Increase in the number of cells due to an increase in the rate of cell division.

Hypertrophy (hī-PER-trō-fē) Increase in the size of cells without cell division.

Metaplasia (met′-a-PLĀ-zē-a; *meta* = change) The transformation of one type of cell into another.

Metastasis (me-TAS-ta-sis; *stasis* = standing still) The spread of cancer to surrounding tissues (local metastasis) or to other body tissues (distant metastasis).

Necrosis (ne-KRŌ-sis; *necros* = death; *osis* = condition) Death of a group of cells.

Neoplasm (NĒ-ō-plazm; *neo* = new) Any abnormal formation or growth, usually a malignant tumor.

Study Outline

Generalized Cell (p. 39)

1. A cell is the basic, living, structural and functional unit of the body.
2. A generalized cell is a composite that represents various cells of the body.
3. Cytology is the science concerned with the study of cells.
4. The principal parts of a cell are the plasma (cell) membrane, cytosol, and organelles.

Plasma (Cell) Membrane (p. 40)

Chemistry and Structure (p. 40)

1. The plasma (cell) membrane surrounds the cell and separates it from other cells and the external environment.
2. It is composed primarily of phospholipids and proteins.

Functions (p. 40)

1. Functionally, the plasma membrane provides communication between cells and between cells and chemicals, establishes an electrochemical gradient, and provides shape and protection.
2. The membrane's selectively permeable nature restricts the passage of certain substances. Substances can pass through the membrane depending on their lipid solubility, size, electrical charges, and the presence of channels and transporters.

Movement of Materials Across Plasma Membranes (p. 41)

1. Fluid outside cells is called extracellular fluid (interstitial fluid, plasma, lymph); fluid inside cells is called intracellu-

lar fluid. Interstitial fluid is the body's internal environment.

2. Substances move across plasma membranes between extracellular and intracellular fluids.
3. Passive processes involve the kinetic energy of individual molecules; no ATP is required and substances move from areas of higher to areas of lower concentration until they reach equilibrium.
4. Simple diffusion is the net movement of molecules or ions from an area of higher concentration to an area of lower concentration until equilibrium is reached.
5. In facilitated diffusion, certain molecules, such as glucose, combine with a carrier to become soluble in the phospholipid portion of the cell membrane.
6. Osmosis is the movement of water through a selectively permeable membrane from an area of higher water concentration to an area of lower water concentration until equilibrium is reached.
7. In an isotonic solution, red blood cells maintain their normal shape; in a hypotonic solution, they undergo hemolysis; in a hypertonic solution, they undergo crenation.
8. Filtration is the movement of water and dissolved substances across a selectively permeable membrane by pressure.
9. Active processes involve the use of ATP by the cell.
10. Active transport is the movement of ions across a cell membrane from lower to higher concentration.

11. Endocytosis is the movement of substances through plasma membranes in which the membrane surrounds the substance, encloses it, and brings it into the cell.

12. Phagocytosis is the ingestion of solid particles by pseudopods. It is the process used by white blood cells to destroy bacteria.

13. Pinocytosis is the ingestion of a liquid by the plasma membrane. In this process, the liquid becomes surrounded by a vacuole.

Cytosol (p. 46)

1. Cytosol is the thick, semifluid portion of cytoplasm between the plasma membrane and nucleus.

2. It is composed of mostly water plus proteins, carbohydrates, lipids, and inorganic substances.

3. Functionally, cytosol is the medium in which some chemical reactions occur.

Organelles (p. 46)

1. Organelles are specialized structures in the cell that are usually surrounded by one or two membranes that have a characteristic shape and carry on specific activities.

2. They assume specific roles in cellular growth, maintenance, repair, and control.

Nucleus (p. 46)

1. Usually the largest organelle, the nucleus controls cellular activities and contains the genetic information.

2. Most body cells have a single nucleus; some (red blood cells) have none; others (skeletal muscle cells) have several.

3. The parts of the nucleus include the nuclear envelope, nucleoli, and genetic material (DNA), which comprises the chromosomes.

Endoplasmic Reticulum (ER) (p. 48)

1. The ER is a network of parallel membranes continuous with the plasma membrane and nuclear membrane.

2. Rough (granular) ER has ribosomes attached to it. Smooth (agranular) ER does not contain ribosomes.

3. The ER stores newly synthesized molecules, synthesizes and packages molecules, detoxifies chemicals, and releases calcium ions involved in muscle contraction.

Ribosomes (p. 48)

1. Ribosomes are granular structures consisting of ribosomal RNA and proteins.

2. They occur free or attached to endoplasmic reticulum.

3. Functionally, ribosomes are the sites of protein synthesis.

Golgi Complex (p. 48)

1. The Golgi complex consists of stacked, flattened membranous sacs (cisterns).

2. The principal function of the Golgi complex is to process, sort, and deliver proteins to the plasma membrane, lysosomes, and secretory vesicles.

Lysosomes (p. 49)

1. Lysosomes are spherical structures that contain digestive enzymes. They are formed from Golgi complexes.

2. They are found in large numbers in white blood cells, which carry on phagocytosis.

3. Digestion by lysosomes of worn-out cell parts is called autophagy.

4. Lysosomes also function in extracellular digestion.

Mitochondria (p. 49)

1. Mitochondria consist of a smooth outer membrane and a folded inner membrane surrounding the interior matrix. The inner folds are called cristae.

2. The mitochondria are called "powerhouses of the cell" because ATP is produced in them.

Cytoskeleton (p. 49)

1. Microfilaments, microtubules, and intermediate filaments form the cytoskeleton.

2. Microfilaments are rodlike structures that are involved in muscular contraction, support, and movement.

3. Microtubules are slender tubes that support, provide movement, and form flagella, cilia, centrioles, and the mitotic spindle.

4. Intermediate filaments appear to provide structural reinforcement in some cells.

Flagella and Cilia (p. 49)

1. These cellular projections have the same basic structure and are used in movement.

2. If projections are few (usually occurring singly or in pairs) and long, they are called flagella. If they are numerous and hairlike, they are called cilia.

3. Flagella move an entire cell (e.g., the flagellum on a sperm cell). Cilia move objects along a cell surface (e.g., the cilia move particles in mucus toward the throat for elimination).

Centrosome and Centrioles (p. 50)

1. The dense area of cytosol containing the centrioles is called a centrosome. It is important in cell reproduction.

2. Centrioles are paired cylinders arranged at right angles to one another. They play a role in the formation and regeneration of flagella and cilia.

Gene Action (p. 50)

1. Cells are basically protein factories.

2. Cells make proteins by transcription and translation.

3. In transcription, genetic information encoded in DNA is passed to a strand of messenger RNA (mRNA).

4. DNA also synthesizes ribosomal RNA (rRNA) and transfer RNA (tRNA).

5. In translation, the information in the nitrogenous base sequence of mRNA dictates the amino acid sequence of a protein.

6. Protein synthesis occurs on ribosomes where a strand of mRNA is attached.

7. tRNA transports specific amino acids to the mRNA at the ribosome. A portion of the tRNA has a triplet of bases called an anticodon; it matches a codon of three bases on the mRNA.

8. The ribosome moves along an mRNA strand as amino acids are joined to form the protein molecule.

9. Peptide bonds hold the amino acids together.

Normal Cell Division (p. 54)

1. Cell division is the process by which cells reproduce themselves.

2. Cell division that results in an increase in body cells is called somatic cell division and involves a nuclear division called mitosis and cytokinesis.

3. Cell division that results in the production of sperm and eggs is called reproductive cell division and consists of a nuclear division called meiosis and cytokinesis.

Somatic Cell Division (p. 54)

1. Prior to mitosis and cytokinesis, the DNA molecules, or chromosomes, replicate themselves so the same hereditary traits are passed on to future generations of cells.

2. DNA replication occurs during interphase.

3. Mitosis is the distribution of two sets of chromosomes into separate and equal nuclei following their replication.

4. It consists of prophase, metaphase, anaphase, and telophase.

5. Cytokinesis usually begins in late anaphase and terminates in telophase.

6. A cleavage furrow forms at the cell's equatorial plane and progresses inward, cutting through the cell to form two separate portions of cytosol.

Abnormal Cell Division: Cancer (p. 55)

1. Cancerous tumors are referred to as malignant; non-cancerous tumors are called benign; the study of tumors is called oncology.

2. The spread of cancer from its primary site is called metastasis.

3. Carcinogens include environmental agents and viruses.

4. Treating cancer is difficult because all the cells in a single population do not behave the same way.

Self-Quiz

1. If the extracellular fluid contains a greater concentration of dissolved particles than the cytosol of the cell, the extracellular fluid is said to be

   a. isotonic   b. hypertonic   c. hypotonic   d. hyperplasic   e. osmotic

2. A red blood cell placed in a hypotonic solution will undergo

   a. hemolysis   b. diffusion of salt into the cell   c. crenation   d. a decrease in osmotic pressure   e. equilibrium

3. Sea water is about 3.5 percent NaCl. This solution _____ a cell in the body.

   a. is in equilibrium with   b. is isotonic to   c. is hypertonic to   d. is hypotonic to   e. moves by osmosis into

4. Which of the following would NOT normally pass through a selectively permeable plasma membrane?

   a. water   b. sodium ions   c. large proteins   d. oxygen   e. steroids

5. Which of the following is a passive transport process?

   a. phagocytosis   b. active transport   c. facilitated diffusion   d. exocytosis   e. pinocytosis

6. Which of the following processes requires cellular ATP?

   a. diffusion   b. active transport   c. osmosis   d. facilitated diffusion   e. filtration

7. Match the following:

   B a. cellular movement
   F b. selective permeability
   G c. protein synthesis
   H d. lipid synthesis, detoxification
   C e. packages proteins and lipids
   E f. ATP production
   D g. digest bacteria, worn-out organelles
   A h. forms mitotic spindle

   A. centrosome
   B. cytoskeleton
   C. Golgi complex
   D. lysosomes
   E. mitochondria
   F. plasma membrane
   G. ribosomes
   H. smooth ER

8. If a DNA strand has a nitrogenous base sequence T, A, C, G, A, the sequence of bases on the corresponding mRNA would be

   a. A, U, G, C, U   b. A, T, G, C, T   c. T, A, C, G, A   d. U, A, C, G, A   e. U, C, G, U, A

9. Transcription involves

   a. transferring information from the mRNA to tRNA   b. codon binding with anticodons   c. joining amino acids by peptide bonds   d. copying information contained in the DNA to the mRNA   e. synthesizing the protein on the ribosome

10. Place the following events in the proper order.

    1. DNA uncoils and mRNA is transcribed
    2. tRNA with an attached amino acid pairs with mRNA
    3. mRNA passes from the nucleus into the cytoplasm and attaches to a ribosome
    4. protein is formed
    5. two amino acids are linked by a peptide bond
    a. 1, 2, 3, 4, 5   b. 1, 3, 2, 5, 4   c. 1, 2, 3, 5, 4   d. 1, 5, 3, 2, 4,   e. 2, 1, 3, 4, 5

11. The codon CAA will pair with the anticodon

    a. CUU   b. GUU   c. GAA   d. GTT   e. CTT

12. Match the following descriptions with the phases of mitosis:

    C a. nuclear envelope (membrane) and nucleoli reappear.
    E b. centromeres of the chromatid pairs line up in the center of the mitotic spindle.
    F c. DNA and centrioles duplicate.
    B d. cleavage furrow splits cell into two daughter cells.
    D e. chromosomes move toward opposite poles of cell.
    A f. chromatids are attached at centromeres; mitotic spindle forms.

    A. prophase
    B. cytokinesis
    C. telophase
    D. anaphase
    E. metaphase
    F. interphase

13. A benign tumor is also called a

    a. malignancy   b. metastasis   c. biopsy   d. neoplasm   e. cancer

14. Which of the following is NOT true concerning the plasma membrane?

    a. Its surface contains folds known as cristae.   b. It is composed primarily of phospholipids and proteins.   c. It is selectively permeable.   d. It is responsible for maintaining an electrochemical gradient between the inside and the outside of the cell.   e. It enables the cell to communicate with other cells.

15. If a virus were to enter a cell and destroy its ribosomes, how would the cell be affected?

    a. It would be unable to undergo mitosis.   b. It would no longer be able to produce ATP.   c. Movement of the cell would cease.   d. It would undergo autophagy   e. It would be unable to synthesize proteins.

Critical Thinking Applications

1. Your 65-year-old aunt has spent every sunny day at the beach for as long as you can remember. Her skin looks a lot like a comfortable lounge chair—brown and wrinkled. Recently, her dermatologist removed a suspicious growth from the skin of her face. What would you suspect is the problem and its likely cause?

2. Mucin is a protein present in saliva. When mixed with water, mucin becomes the slippery substance known as mucus. Trace the route taken by mucin through the cells of the salivary glands from its synthesis to its secretion (release), listing all the organelles and processes involved.

3. Nicotine and other compounds present in cigarette smoke can inhibit the action of the cilia within the respiratory system. What problems could result from this and other effects of smoking?

4. You want to dilute a sample of red blood cells to observe under the microscope. Your lab partner Bart hasn't read the chapter on cells yet. Explain to Bart the type of saline (NaCl) solution needed for your experiment.

5. A child was brought to the emergency room after eating rat poison containing arsenic. Arsenic kills rats by blocking the function of the mitochondria. What effect does arsenic have on the child's body functions?

6. Imagine that a new chemotherapy agent has been discovered that disrupts microtubules in cancerous cells but leaves normal cells unaffected. What effect would this agent have on the cancer cells?

Answers to Figure Questions

3.1 Plasma membrane, cytosol, and organelles.

3.2 Lipid solubility, size of molecules, membrane charge, presence of channels and transporters.

3.3 Cytosol.

3.4 Because a fever represents an increase in body temperature, all diffusion processes would be increased.

3.5 No. Facilitated diffusion is a passive process, so no energy is expended to transport glucose across the membrane.

3.6 The pressure required to prevent the movement across a selectively permeable membrane.

3.7 Hypertonic.

3.8 Energy is used from the splitting of ATP.

3.9 Endocytosis.

3.10 It controls cellular activity by regulating which proteins become synthesized.

3.11 Rough ER has attached ribosomes whereas smooth ER does not. Rough ER synthesizes proteins that will be exported from the cell, while smooth ER is associated with lipid synthesis and other metabolic reactions.

3.12 It processes, sorts, packages, and delivers proteins and lipids.

3.13 They increase surface area for chemical reactions.

3.14 UCGA.

3.15 A group of nucleotides on a DNA molecule that serves as the pattern for synthesizing a specific protein.

3.16 Because each of the new daughter cells must have a complete set of genes.

3.17 Usually starts in late anaphase or early telophase.

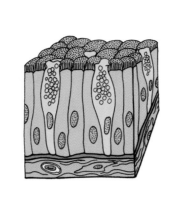

chapter 4

TISSUES

a look ahead

muscle cells
+ nerve cells
conduct electricity
the others Do not

Cells are highly organized units, but they do not function by themselves. They work together in groups of similar cells called tissues.

Types of Tissues

A *tissue* is a group of similar cells and their intercellular substance that function together to perform a specialized activity. The science that deals with the study of tissues is called *histology* (his´-TOL-ō-jē; *histio* = tissue; *logos* = study of). There are four types of tissue in the body:

Know

1. **Epithelial** (ep´-i-THĒ-lē-al) **tissue,** which covers body surfaces; lines body cavities, hollow organs, and ducts (tubes); and forms glands.
2. **Connective tissue,** which protects and supports the body and its organs, binds organs together, stores energy reserves as fat, and provides immunity.
3. **Muscle tissue,** which is responsible for movement.
4. **Nervous tissue,** which initiates and transmits nerve impulses that coordinate body activities.

Epithelial tissue and connective tissue will be discussed in detail in this chapter. Detailed discussions of bone tissue, blood, muscle tissue, and nervous tissue are taken up later in the book.

Epithelial Tissue

Epithelial tissue, or more simply *epithelium* (plural is *epithelia*), may be divided into two types: (1) *covering and lining epithelium* and (2) *glandular epithelium.* Covering and lining epithelium forms the outer covering of the skin and the outer covering of some internal organs. It also lines body cavities, blood vessels, ducts, and the interiors of the respiratory, digestive, urinary, and reproductive systems. It makes up, along with nervous tissue, the parts of the sense organs for hearing, vision, and touch. Glandular epithelium constitutes the secreting portion of glands, such as sweat glands.

General Features of Epithelial Tissue

objective: *Describe the general features of epithelial tissue.*

outermost layer of skin

Following are the general features of epithelial tissue:

1. Epithelium consists largely or entirely of closely packed cells with little extracellular material between cells.
2. Epithelial cells are arranged in continuous sheets, in either single or multiple layers.
3. Epithelial cells have a free surface, which is exposed to a body cavity, lining of an internal organ, or the exterior of the body, and a basal surface, which is attached to a basement membrane. The basement membrane is lo-

1 layer of protein molecules that glue together 2 types of tissue

cated between the epithelium and the underlying connective tissue layer.

4. Epithelia are *avascular* (*a* = without; *vascular* = blood vessels). The vessels that supply nutrients and remove wastes are located in adjacent connective tissues. The exchange of materials between epithelium and connective tissue is by diffusion.
5. Epithelia have a nerve supply.
6. Because epithelium is subject to a certain amount of wear and tear and injury, it has a high capacity for renewal (high mitotic rate).
7. Functions of epithelia include protection, secretion, absorption, excretion, sensory reception, and reproduction.

mitotic

nerves penetrate into epithelial tissue

Have extensive Supply of nerve endings in them

Covering and Lining Epithelium

objective: *Explain how covering and lining epithelium is classified.*

Classification by Cell Shape

Epithelial tissue, which covers or lines various parts of the body, contains cells with four basic shapes:

1. *Squamous* (SKWĀ-mus; *squama* = flat) cells are flat and attach to each other like tiles. Their thinness allows for the active and passive movement of substances through them (see Chapter 3).
2. *Cuboidal* cells are thicker, being cube or hexagon shaped. They produce several important body *secretions* (fluids that are produced and released by cells, such as sweat and enzymes). They may also be used for *absorption* (intake) of fluids and other substances, such as digested foods in the intestines.
3. *Columnar* cells are tall and cylindrical, thereby protecting underlying tissues. They may also be specialized for secretion and absorption. Some may also have cilia.
4. *Transitional* cells range in shape from flat to columnar and often change shape due to distention (stretching), expansion, or movement of body parts.

Classification by Arrangement of Layers

Epithelial cells tend to be arranged in one or more layers depending on the function of the particular body part. The terms used to refer to their classification by arrangement of layers include the following:

1. *Simple epithelium* is a single layer of cells found in areas where passive and active movement of molecules is needed.
2. *Stratified epithelium* contains two or more layers of cells used for protection of underlying tissues in areas where there is considerable wear and tear.
3. *Pseudostratified columnar epithelium* contains one layer of cells. The tissue looks multilayered but not all

SUBTYPES

See page 65

cells reach the surface; those that do are either ciliated or secrete mucus (see Exhibit 4.1 on p. 70, Pseudo-stratified columnar).

In summary, epithelium may be classified by both shape and arrangement of layers of cells as indicated as follows:

Simple

1. Squamous
2. Cuboidal
3. Columnar

Stratified

1. Squamous*
2. Cuboidal*
3. Columnar*
4. Transitional

Pseudostratified columnar

Types and Functions

objective: *Describe the various types and functions of covering and lining epithelium.*

Each of the epithelial tissues described in the following section is illustrated in Exhibit 4.1. Along with the illustrations are descriptions, locations, and functions of the tissues.

Simple Epithelium

SIMPLE SQUAMOUS EPITHELIUM   This tissue looks like a tiled floor when viewed from above. The flat, single layer of cells allows for the diffusion of respiratory gases (oxygen and carbon dioxide) in the lungs and the movement of fluid and dissolved substances between the blood and tissue cells by osmosis and filtration.

SIMPLE CUBOIDAL EPITHELIUM   The cuboidal shape of these cells is obvious only when the tissue is viewed from the side. Its single layer of cells is important in the secretion of substances, such as tears and saliva, and in absorption, such as the reabsorption of water by kidney cells.

SIMPLE COLUMNAR EPITHELIUM   In side view, the cells appear rectangular. Simple columnar epithelium exists in two forms: *nonciliated simple columnar epithelium* and *ciliated simple columnar epithelium.* The nonciliated type contains microvilli and goblet cells. *Microvilli* (*micro* = small; *villus* = tuft of hair) (see Figure 3.1) are microscopic fingerlike cytoplasmic projections that serve to increase the surface area of the plasma membrane, which allows larger amounts of digested nutrients and fluids to be absorbed into the body. *Goblet cells* are modified columnar cells that secrete mucus, which accumulates in the upper half of the cell, causing that area to bulge out. The whole cell then resembles a goblet or wine glass. The secreted mucus provides lubrication and protection.

Another modification of columnar epithelium is found in cells with hairlike processes called *cilia* (*ciliaris* = resembling an eyelash). In a few parts of the upper respiratory tract, ciliated columnar cells are interspersed with goblet cells. Mucus secreted by the goblet cells forms a film over the respiratory surface that traps foreign particles that are inhaled. The cilia wave in unison and move the mucus, with any foreign particles, toward the throat, where it can be swallowed or spit out.

Stratified Epithelium

In contrast to simple epithelium, stratified epithelium has at least two layers of cells. Thus it is more durable and can protect underlying tissues from the external environment and from wear and tear. The name of the specific kind of stratified epithelium depends on the shape of the surface cells.

STRATIFIED SQUAMOUS EPITHELIUM   In the superficial layers of this type of epithelium, the cells are flat, whereas in the deep layers, cells vary in shape from cuboidal to columnar. The bottom cells continually replicate by cell division. As new cells grow, the cells of the bottom layer continually shift upward and outward and push the surface cells outward. As they move farther from the deep layer and their blood supply (in the underlying connective tissue), they become dehydrated, shrunken, and harder. At the surface, the cells lose their attachments to other cells and are rubbed off. Old cells are sloughed off and replaced as new cells continually emerge. The Pap smear, a screening test for precancer and cancer of the uterus, cervix, and vagina, consists of examining cells that are sloughed off.

There are two forms of stratified squamous epithelium. In *keratinized stratified squamous epithelium,* a tough layer of keratin is deposited in the surface cells. *Keratin* (*keras* = hornlike) is a protein that protects the skin from injury and microbial invasion and makes it waterproof. *Nonkeratinized stratified squamous epithelium* does not contain keratin and remains moist.

STRATIFIED CUBOIDAL EPITHELIUM   This fairly rare type of epithelium sometimes consists of more than two layers of cells. Its function is mainly protective. *Surface of skin*

STRATIFIED COLUMNAR EPITHELIUM   This type of tissue is also uncommon in the body. It functions in protection and secretion.

TRANSITIONAL EPITHELIUM   This kind of epithelium is variable in appearance, depending on whether it is relaxed or distended (stretched). In the relaxed state, it looks similar to stratified cuboidal epithelium except that the upper cells tend to be large and rounded. This allows the tissue to be stretched without the outer cells breaking apart from one another. When stretched, they are drawn out into squamous epithelium. Because of this capability, transitional epithelium lines hollow structures that are subjected to expansion from within, such as the urinary bladder. Its function is to help prevent a rupture of these organs.

Pseudostratified Columnar Epithelium

The third category of covering and lining epithelium is called pseudostratified columnar epithelium. The nuclei of the cells are at varying depths. Even though all the cells are attached to the basement membrane in a single layer, some cells do not reach the free surface. These features give the false impression of a multilayered tissue, the reason for the name pseudostratified epitheli-

*This classification is based on the shape of the outermost layer of cells.

Exhibit 4.1 | Epithelial Tissues

COVERING AND LINING EPITHELIUM   *(single)*
                                      one cell layer—flat

Simple squamous epithelium

Description: Single layer of flat cells; centrally located nuclei.

Location: Lines air sacs of lungs, heart, blood and lymphatic vessels, and abdominal cavity and makes up walls of blood capillaries throughout the body, including the kidney (glomeruli).

Function: Diffusion, osmosis filtration, and secretion in serous membranes.

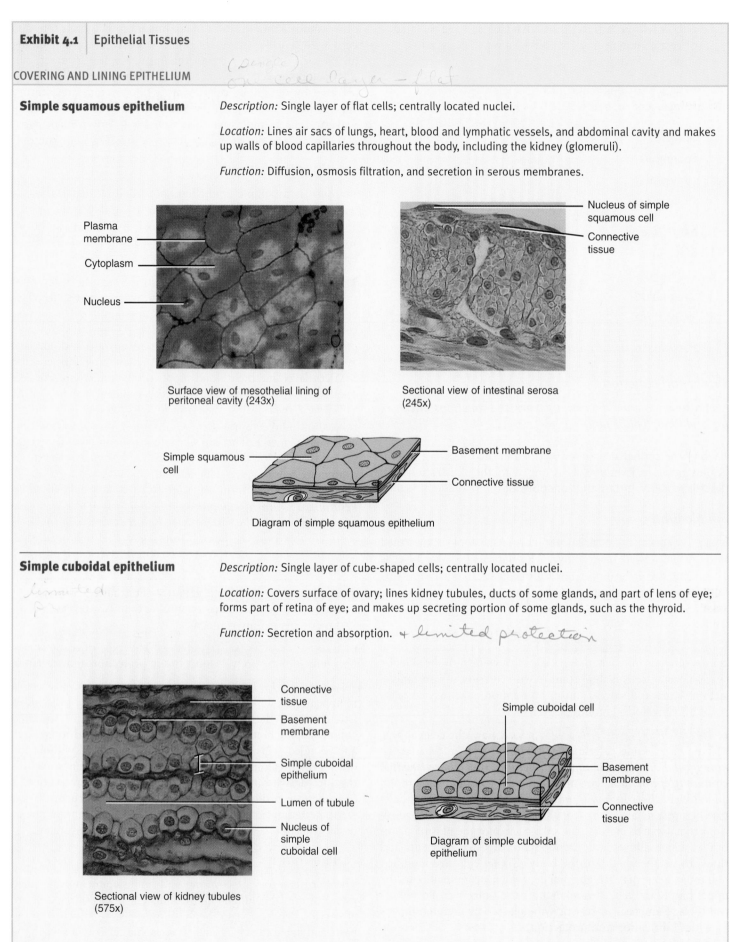

Surface view of mesothelial lining of peritoneal cavity (243x)

Sectional view of intestinal serosa (245x)

Diagram of simple squamous epithelium

Simple cuboidal epithelium

Description: Single layer of cube-shaped cells; centrally located nuclei.

Location: Covers surface of ovary; lines kidney tubules, ducts of some glands, and part of lens of eye; forms part of retina of eye; and makes up secreting portion of some glands, such as the thyroid.

Function: Secretion and absorption. *+ limited protection*

Sectional view of kidney tubules (575x)

Diagram of simple cuboidal epithelium

(continued)

Exhibit 4.1 | Epithelial Tissues (Continued)

Nonciliated simple columnar epithelium

Description: Single layer of nonciliated rectangular cells; nuclei at base of cells; contains goblet cells and microvilli in some locations.

Location: Lines the gastrointestinal tract from the stomach to the anus, excretory ducts of many glands, and gallbladder.

Function: Secretion and absorption.

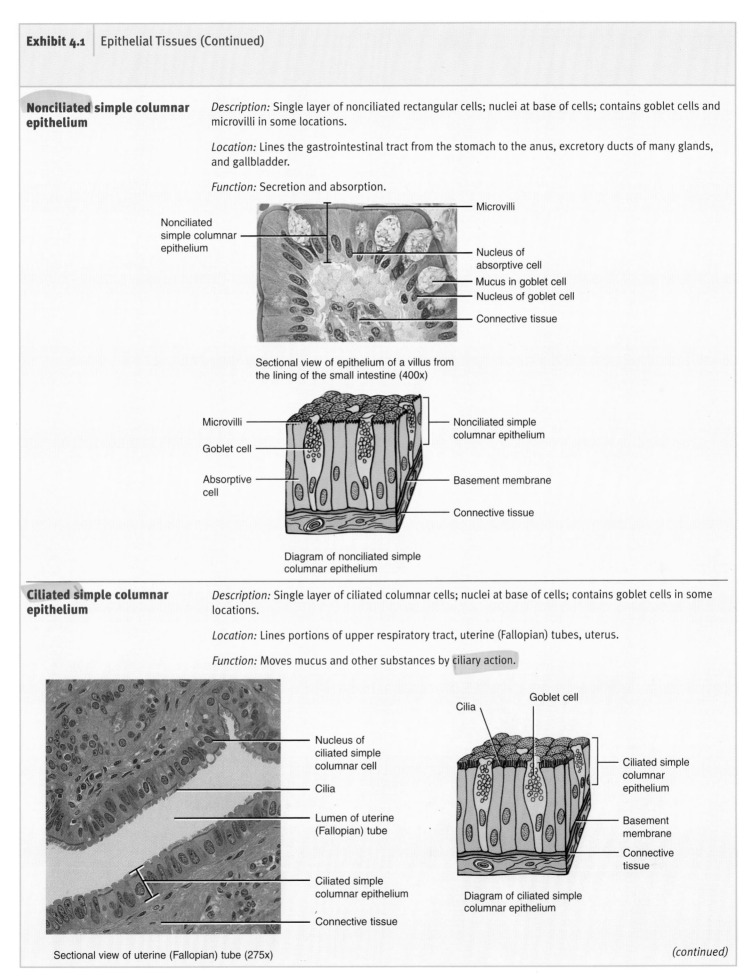

Nonciliated simple columnar epithelium

Microvilli

Nucleus of absorptive cell

Mucus in goblet cell

Nucleus of goblet cell

Connective tissue

Sectional view of epithelium of a villus from the lining of the small intestine (400x)

Microvilli

Goblet cell

Absorptive cell

Nonciliated simple columnar epithelium

Basement membrane

Connective tissue

Diagram of nonciliated simple columnar epithelium

Ciliated simple columnar epithelium

Description: Single layer of ciliated columnar cells; nuclei at base of cells; contains goblet cells in some locations.

Location: Lines portions of upper respiratory tract, uterine (Fallopian) tubes, uterus.

Function: Moves mucus and other substances by ciliary action.

Nucleus of ciliated simple columnar cell

Cilia

Lumen of uterine (Fallopian) tube

Ciliated simple columnar epithelium

Connective tissue

Cilia

Goblet cell

Ciliated simple columnar epithelium

Basement membrane

Connective tissue

Diagram of ciliated simple columnar epithelium

Sectional view of uterine (Fallopian) tube (275x)

(continued)

Exhibit 4.1 | Epithelial Tissues (Continued)

function

found in areas of mechanical stress

Stratified squamous epithelium

Description: Several layers of cells; squamous cells in superficial layers; cuboidal to columnar cells in deep layers; bottom cells replace surface cells as they are lost.

Location: Keratinized variety forms outer layer of skin; nonkeratinized variety lines wet surfaces, such as lining of the mouth, esophagus, and vagina, and covers the tongue.

Function: Protection.

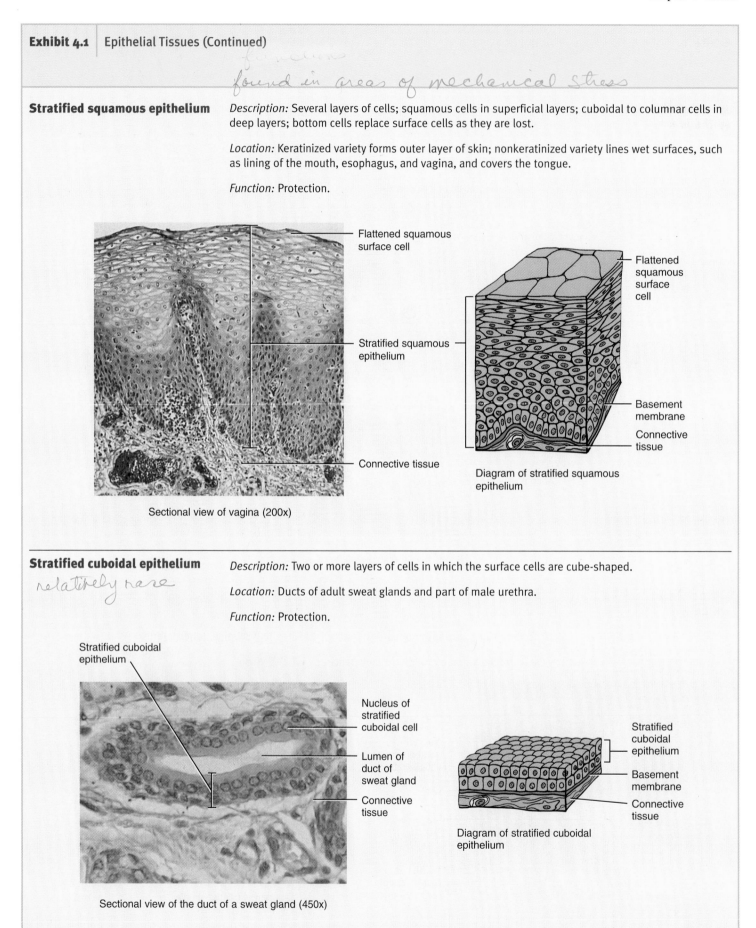

Flattened squamous surface cell

Stratified squamous epithelium

Connective tissue

Sectional view of vagina (200x)

Flattened squamous surface cell

Basement membrane

Connective tissue

Diagram of stratified squamous epithelium

Stratified cuboidal epithelium

relatively rare

Description: Two or more layers of cells in which the surface cells are cube-shaped.

Location: Ducts of adult sweat glands and part of male urethra.

Function: Protection.

Stratified cuboidal epithelium

Nucleus of stratified cuboidal cell

Lumen of duct of sweat gland

Connective tissue

Sectional view of the duct of a sweat gland (450x)

Stratified cuboidal epithelium

Basement membrane

Connective tissue

Diagram of stratified cuboidal epithelium

(continued)

Exhibit 4.1 | Epithelial Tissues (Continued)

Stratified columnar epithelium

Description: Several layers of many-sided cells; columnar cells only in superficial layer.

Location: Lines part of urethra, large excretory ducts of some glands, small areas in anal mucous membrane.

Function: Protection and secretion.

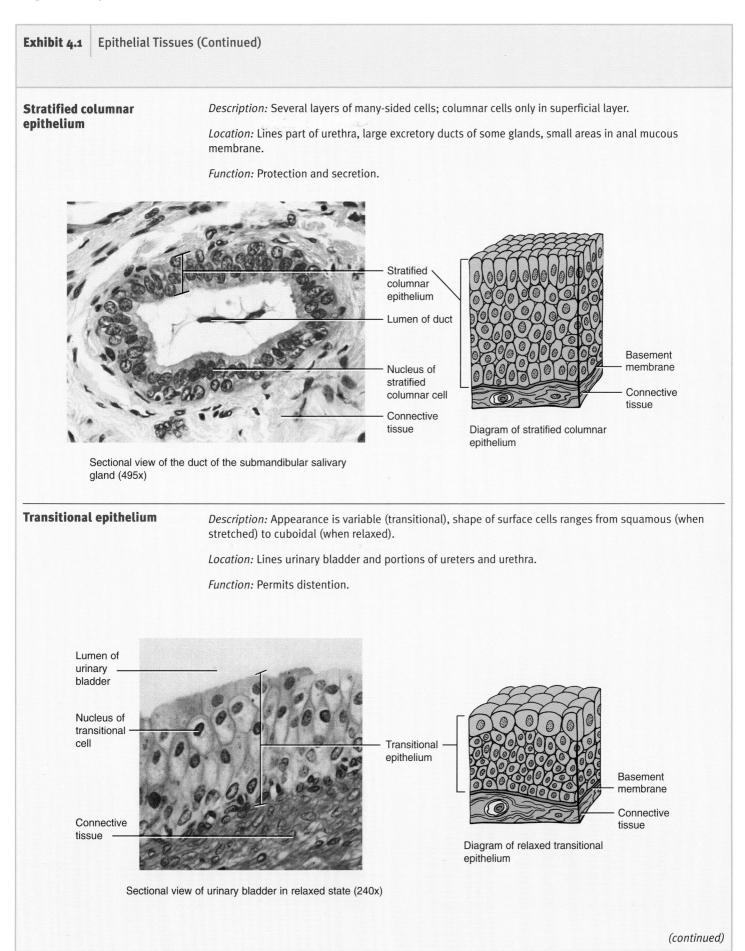

Stratified columnar epithelium

Lumen of duct

Nucleus of stratified columnar cell

Connective tissue

Basement membrane

Connective tissue

Diagram of stratified columnar epithelium

Sectional view of the duct of the submandibular salivary gland (495x)

Transitional epithelium

Description: Appearance is variable (transitional), shape of surface cells ranges from squamous (when stretched) to cuboidal (when relaxed).

Location: Lines urinary bladder and portions of ureters and urethra.

Function: Permits distention.

Lumen of urinary bladder

Nucleus of transitional cell

Connective tissue

Transitional epithelium

Basement membrane

Connective tissue

Diagram of relaxed transitional epithelium

Sectional view of urinary bladder in relaxed state (240x)

(continued)

Exhibit 4.1 | Epithelial Tissues (Continued)

Pseudostratified columnar epithelium

Description: Not a true stratified tissue; nuclei of cells at different levels; all cells attached to basement membrane, but not all reach surface.

Location: Pseudostratified ciliated columnar epithelium lines most of upper respiratory tract, pseudostratified nonciliated columnar epithelium lines larger ducts of many glands, epididymis, and part of male urethra.

Function: Secretion and movement of mucus by ciliary action.

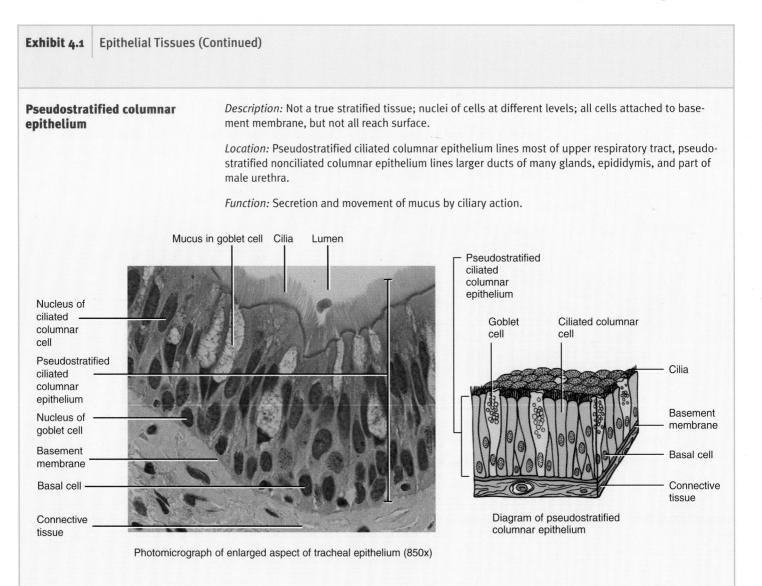

Photomicrograph of enlarged aspect of tracheal epithelium (850x)

Diagram of pseudostratified columnar epithelium

GLANDULAR EPITHELIUM

Exocrine glands

Description: Secrete products into ducts or directly onto external or internal body surfaces.

Location: Sweat, oil, wax, and mammary glands of the skin; digestive glands such as salivary glands that secrete into mouth cavity, and pancreas that secretes into the small intestine.

Function: Produce mucus, perspiration, oil, wax, milk, or digestive enzymes.

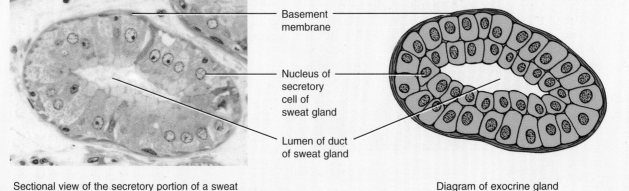

Sectional view of the secretory portion of a sweat gland (1300x)

Diagram of exocrine gland

(continued)

Exhibit 4.1	Epithelial Tissues (Continued)

Endocrine glands

Description: Secrete hormones into blood.

Location: Examples are pituitary gland at base of brain, thyroid and parathyroid glands near voice box, adrenal glands above kidneys, pancreas near the stomach, ovaries in pelvic cavity, testes in scrotum, pineal gland at base of brain, and thymus gland in thoracic cavity.

Function: Produce hormones that regulate various body activities.

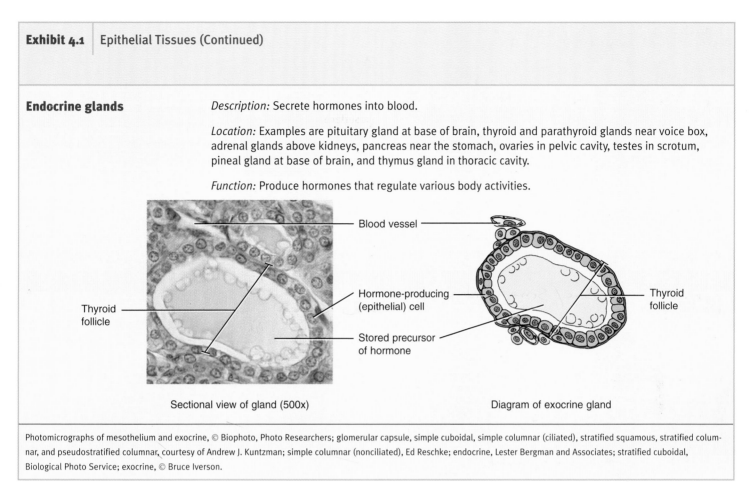

Sectional view of gland (500x)   Diagram of exocrine gland

Photomicrographs of mesothelium and exocrine, © Biophoto, Photo Researchers; glomerular capsule, simple cuboidal, simple columnar (ciliated), stratified squamous, stratified columnar, and pseudostratified columnar, courtesy of Andrew J. Kuntzman; simple columnar (nonciliated), Ed Reschke; endocrine, Lester Bergman and Associates; stratified cuboidal, Biological Photo Service; exocrine, © Bruce Iverson.

um (*pseudo* = false). In pseudostratified ciliated columnar epithelium, the cells that reach the surface either secrete mucus or bear cilia that sweep away mucus and trapped foreign particles for eventual elimination from the body. Pseudostratified nonciliated columnar epithelium contains no cilia or goblet cells.

Glandular Epithelium

objective: *Define a gland and distinguish between exocrine and endocrine glands.*

The function of glandular epithelium is secretion, which is accomplished by cells that lie in clusters deep below the covering and lining epithelium. A *gland* may consist of one cell or a group of highly specialized epithelial cells that secrete substances into ducts, onto a surface, or into the blood. The production of such substances always requires active work by the cells and results in an expenditure of energy. All glands of the body are classified as exocrine or endocrine.

Exocrine (*exo* = outside; *krin* = to secrete) *glands* secrete their products into ducts (tubes) that empty at the surface of covering and lining epithelium or directly onto a free surface. The product of an exocrine gland may be released at the skin surface or into the lumen (cavity) of a hollow organ. The secretions of exocrine glands include mucus, perspiration, oil, wax, and digestive enzymes. Examples of exocrine glands include sweat glands and salivary glands.

Endocrine (*endo* = within) *glands* are ductless; their secretory products enter the extracellular fluid and diffuse into the blood.

The secretions of endocrine glands are always hormones, chemicals that regulate various physiological activities. The pituitary, thyroid, and adrenal glands are examples of endocrine glands.

Connective Tissue

The most abundant tissue in the body is *connective tissue*. It binds together, supports, and strengthens other body tissues, protects and insulates internal organs, and compartmentalizes structures such as skeletal muscles.

General Features of Connective Tissue

Following are the general features of connective tissue:

1. Connective tissue consists of three basic elements: (1) cells, (2) ground substance, and (3) fibers. Together, the ground substance and fibers, which are outside the cells, form the matrix (discussed in detail below). Connective tissue cells rarely touch one another; they are separated by a considerable amount of matrix. *Sol-gel properties*

2. Connective tissue usually does not occur on free surfaces, such as the coverings or linings of internal organs or the external surface of the body.

3. Except for cartilage, connective tissue, like epithelium, has a nerve supply.

pancreas - secretes both endocrine & exocrine

embryonic

4. Connective tissue usually is highly vascular (has a rich blood supply). Exceptions include cartilage, which is avascular, and tendons, which have a scanty blood supply.

5. The *matrix* of a connective tissue, which may be fluid, gelatinous, fibrous, or calcified, is usually secreted by the connective tissue cells and adjacent cells and determines the tissue's qualities.

Connective Tissue Cells

Following are some of the cells contained in various types of connective tissue. The specific tissues to which they belong will be described shortly.

Fixed *Fixed*

1. *Fibroblasts* (FĪ-brō-blasts; *fibro* = fiber) are the most numerous cells. They secrete the molecules that form the ground substance and fibers.

Fixed in tissue/ some wander through

2. *Macrophages* (MAK-rō-fā-jez; *macro* = large; *phagein* = *+issue* to eat) are capable of engulfing bacteria and cellular debris by phagocytosis (see Chapter 3). They thus provide a vital defense for the body.

3. *Plasma cells* secrete antibodies and, accordingly, provide another defense mechanism through immunity.

4. *Mast cells* are abundant alongside blood vessels. They produce histamine, a chemical that dilates (widens) small blood vessels during inflammation.

melanocytes are also fixed

Connective Tissue Matrix

Each type of connective tissue has unique properties due to the matrix between the cells. Matrix contains protein *fibers* embedded in a fluid, gel, or solid *ground substance*. Connective tissue cells usually produce the ground substance and deposit it in the space between the cells.

Besides chemicals normally found in extracellular fluid, ground substance also contains several other materials. Two examples of these materials are: (1) *Hyaluronic* (hī-a-loo-RON-ik) *acid,* a viscous, slippery substance that binds cells together, lubricates joints, and helps maintain the shape of the eyeballs. It also appears to play a role in helping phagocytes (cells capable of phagocytosis) migrate through connective tissue during development and repair of a wound. (2) *Chondroitin* (kon-DROY-tin) *sulfate,* a jellylike substance that provides support and adhesiveness in cartilage, bone, the skin, and blood vessels.

The ground substance supports cells, binds them together, and provides a medium through which substances are exchanged between blood and cells. The ground substance is quite active in tissue development, migration, and metabolic functions.

Fixed

Fibers in the matrix provide strength and support for tissues. Three types of fibers are embedded in the matrix between the cells of connective tissue: collagen, elastic, and reticular fibers.

Collagen (*kolla* = glue) *fibers* are very tough and resistant to a pulling force, yet allow some flexibility. These fibers often occur in bundles, an arrangement that affords great strength. Chemically, collagen fibers consist of the protein *collagen*. This is the most abundant protein in your body, representing about 25 percent of the total protein. Collagen fibers are found in most types of connective tissues, especially bone, cartilage, tendons, and ligaments.

Elastic fibers are smaller than collagen fibers and freely branch and rejoin one another. They consist of a protein called *elastin*. Like collagen fibers, elastic fibers provide strength. In addition, they can be stretched considerably without breaking. Elastic fibers are plentiful in the skin, blood vessels, and lungs.

Reticular (*rete* = net) *fibers,* consisting of *collagen* and a coating of *glycoprotein,* provide support in the walls of blood vessels and form a network around cells, nerve fibers, and skeletal and smooth muscle cells. Produced by fibroblasts, they are much thinner than collagen fibers and form branching networks. Like collagen fibers, reticular fibers provide support and strength.

Classification of Connective Tissue

| objective: | *Explain how connective tissue is classified.*

We will classify connective tissue as follows:

I. Embryonic connective tissue
   A. Mesenchyme *from which all other connective tissues are produced*
   B. Mucous connective tissue *jelly like/umbilical cord*
II. Mature connective tissue
   A. Loose connective tissue
     1. Areolar connective tissue
     2. Adipose tissue
     3. Reticular connective tissue
   B. Dense connective tissue
     1. Dense regular connective tissue
     2. Dense irregular connective tissue
     3. Elastic connective tissue
   C. Cartilage
     1. Hyaline cartilage
     2. Fibrocartilage
     3. Elastic cartilage
   D. Bone (osseous) tissue
   E. Blood (vascular) tissue

Most of the connective tissues described in the following sections are illustrated in Exhibit 4.2. Along with the illustrations are the descriptions, locations, and functions of the tissues.

Embryonic Connective Tissue

Connective tissue that is present primarily in the embryo or fetus is called *embryonic connective tissue.*

An embryonic connective tissue found almost exclusively in the embryo is *mesenchyme* (MEZ-en-kīm)—the tissue from which all other connective tissues are produced. It is composed of irregularly shaped mesenchymal cells, a semifluid ground substance, and delicate reticular fibers.

Mucous connective tissue is a jellylike embryonic connective tissue found primarily in the fetus in the umbilical cord, where it supports the wall of the cord. It consists of widely scattered fibroblasts, a jellylike ground substance, and collagen fibers.

Mature Connective Tissue

Mature connective tissue exists in the newborn, has cells produced from mesenchyme, and does not change after birth. It is subdivided into (1) loose connective tissue, (2) dense connective tissue, (3) cartilage, (4) bone, and (5) blood.

Loose Connective Tissue *[Proper]*

In *loose connective tissue* the fibers are *loosely* arranged and there are many cells. The types are areolar connective tissue, adipose tissue, and reticular tissue.

AREOLAR (a-RĒ-o-lar) CONNECTIVE TISSUE   This tissue is one of the most widely distributed connective tissues in the body. It contains several kinds of cells, including fibroblasts, macrophages, plasma cells, and mast cells. All three types of fibers—collagen, elastic, and reticular—are present. The fluid, or gelatinous ground substance, contains mostly hyaluronic acid. Combined with adipose tissue, areolar connective tissue forms the *subcutaneous* (sub′-kyoo-TĀ-ne-us; *sub* = under; *cutis* = skin) *layer*—the layer of tissue that attaches the skin to underlying tissues and organs.

ADIPOSE TISSUE   This tissue is a loose connective tissue in which the cells, called *adipocytes* (*adeps* = fat), are specialized for triglyceride (fat) storage. Adipocytes are derived from fibroblasts. Because of the accumulation of a large triglyceride droplet, the cytoplasm and nucleus are pushed to the edge of the cell. Adipose tissue is a good insulator and can therefore reduce heat loss through the skin. It is also a major energy reserve and generally supports and protects various organs. *[most predominantly breasts - hips - heart]*

RETICULAR CONNECTIVE TISSUE   Reticular connective tissue consists of fine interlacing fibers and cells. It forms the *stroma* (framework) of certain organs and helps to bind together the cells of smooth muscle tissue. *[liver - kidney - spleen, Bone marrow]*

Dense Connective Tissue *[PROPER]*

Dense connective tissue contains more numerous, thicker and *denser* fibers but considerably fewer cells than loose connective tissue. The types are dense regular connective tissue, dense irregular connective tissue, and elastic connective tissue.

DENSE REGULAR CONNECTIVE TISSUE   *[muscle - ligaments - tendons]* In this tissue, bundles of collagen fibers have a *regular* (orderly), parallel arrangement that gives the tissue great strength. The tissue structure withstands pulling in one direction. Fibroblasts, which produce the fibers and ground substance, appear in rows between the fibers. The tissue is silvery white, and tough yet somewhat pliable.

DENSE IRREGULAR CONNECTIVE TISSUE   This tissue contains collagen fibers that are usually *irregularly* arranged. It is found in parts of the body where tensions are exerted in various directions. The tissue usually occurs in sheets. *[Dermis, Nerve & muscle sheaths]*

ELASTIC CONNECTIVE TISSUE   This tissue has a predominance of freely branching elastic fibers. Fibroblasts are present in the spaces between the fibers. Elastic connective tissue can be stretched and will snap back into shape (elasticity). The tissue provides stretch and strength, allowing structures to perform their functions efficiently. *[between Vertebrae of spinal column]*

Cartilage *[Supporting tissue]*

Cartilage is capable of enduring considerably more stress than the tissues just discussed. Unlike other connective tissues, cartilage has no blood vessels or nerves, except for those in the *perichondrium* (per′-i-KON-dre-um; *peri* = around; *chondros* = cartilage), the dense irregular connective tissue membrane around the surface of cartilage. Cartilage consists of a dense network of collagen and elastic fibers embedded in a firm matrix of chondroitin sulfate, a jellylike substance. Whereas the strength of cartilage is due to its collagen fibers, its resilience (ability to assume its original shape after deformation) is due to chondroitin sulfate. The cells of mature cartilage, called *chondrocytes* (KON-dro-sīts), occur singly or in groups within spaces called *lacunae* (la-KOO-ne; *lacuna* = little lake) in the matrix. There are three kinds of cartilage: hyaline, fibrocartilage, and elastic. (See Exhibit 4.2.)

HYALINE CARTILAGE   This cartilage, also called *gristle*, appears in the body as a bluish-white, shiny substance. The fine collagen fibers, although present, are not visible with ordinary staining techniques, and the prominent chondrocytes are found in lacunae. Hyaline cartilage is the most abundant kind of cartilage in the body. It affords flexibility and support and reduces friction and absorbs shock at joints. Hyaline cartilage is the weakest of the three types of cartilage. The role of hyaline cartilage in bone formation is discussed in Chapter 6. *[Larynx nasal septum rib]*

FIBROCARTILAGE   Chondrocytes are scattered among clearly visible bundles of collagen fibers within the matrix of this type of cartilage. This tissue combines strength and rigidity and is the strongest of the three types of cartilage. *[pads in knee joint, public bones (between) intervertebral Disks]*

ELASTIC CARTILAGE   In this tissue, chondrocytes are located in a threadlike network of elastic fibers within the matrix. Elastic cartilage provides strength and maintains the shape of certain organs. *[outer-ear tip of nose epiglottis]*

Bone (Osseous) Tissue *[Supporting]*

Together, cartilage, joints, and *bone* or *osseous* (OS-e-us) *tissue* comprise the skeletal system. The details of bone tissue are considered in Chapter 6. *[matrix is solid/calcium Bone is 1/3 collagen fibers + 2/3 solids]*

Blood (Vascular) Tissue *[Liqued Types]*

Blood (*vascular*) *tissue* is a connective tissue with a liquid matrix called plasma. Suspended in the plasma are formed elements—cells and cell fragments called red blood cells, white blood cells, and platelets.

The details of blood are considered in Chapter 14. *[protein fibers are dissolved in blood]*

[LYMPH]

Membranes

objective: *Describe the various types of membranes in the body.*

The combination of an epithelial layer and an underlying connective tissue layer forms an *epithelial membrane.* The three types of epithelial membranes are mucous, serous, and cutaneous. The fourth kind of membrane in the body, synovial, does not contain epithelium.

Exhibit 4.2 | Connective Tissues

EMBRYONIC CONNECTIVE TISSUE

Mesenchyme

Description: Consists of irregularly shaped, mesenchymal cells embedded in a semifluid ground substance that contains reticular fibers.

Location: Under skin and along developing bones of an embryo and in adults along blood vessels.

Function: Forms all other kinds of connective tissue.

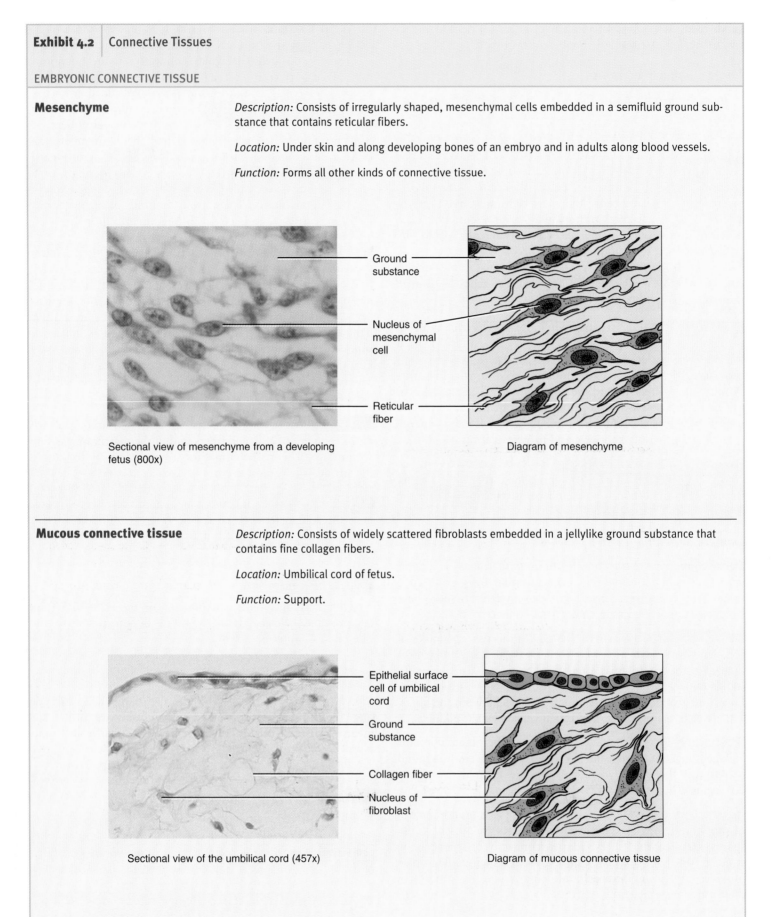

Ground substance

Nucleus of mesenchymal cell

Reticular fiber

Sectional view of mesenchyme from a developing fetus (800x)

Diagram of mesenchyme

Mucous connective tissue

Description: Consists of widely scattered fibroblasts embedded in a jellylike ground substance that contains fine collagen fibers.

Location: Umbilical cord of fetus.

Function: Support.

Epithelial surface cell of umbilical cord

Ground substance

Collagen fiber

Nucleus of fibroblast

Sectional view of the umbilical cord (457x)

Diagram of mucous connective tissue

(continued)

Exhibit 4.2 | Connective Tissues (Continued)

LOOSE CONNECTIVE TISSUE

Areolar connective tissue

Description: Consists of fibers (collagen, elastic, and reticular) and several kinds of cells (fibroblasts, macrophages, plasma cells, and mast cells) embedded in a semifluid ground substance.

Location: Subcutaneous layer of skin; upper part of dermis of skin, mucous membranes, blood vessels, nerves, and around body organs.

Function: Strength, elasticity, and support.

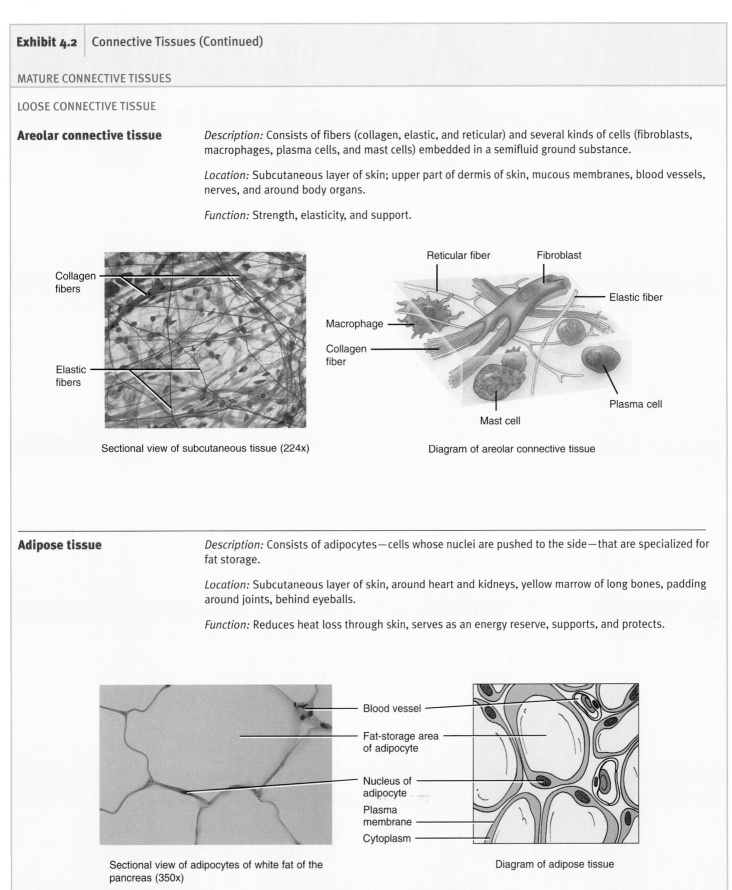

Sectional view of subcutaneous tissue (224x)

Diagram of areolar connective tissue

Adipose tissue

Description: Consists of adipocytes—cells whose nuclei are pushed to the side—that are specialized for fat storage.

Location: Subcutaneous layer of skin, around heart and kidneys, yellow marrow of long bones, padding around joints, behind eyeballs.

Function: Reduces heat loss through skin, serves as an energy reserve, supports, and protects.

Sectional view of adipocytes of white fat of the pancreas (350x)

Diagram of adipose tissue

(continued)

Exhibit 4.2 | Connective Tissues (Continued)

Reticular connective tissue

Description: Consists of network of interlacing reticular fibers and reticular cells.

Location: Stroma (framework) of liver, spleen, lymph nodes; portion of bone marrow that gives rise to blood cells; around blood vessels and muscle.

Function: Forms stroma of organs; binds together smooth muscle tissue cells.

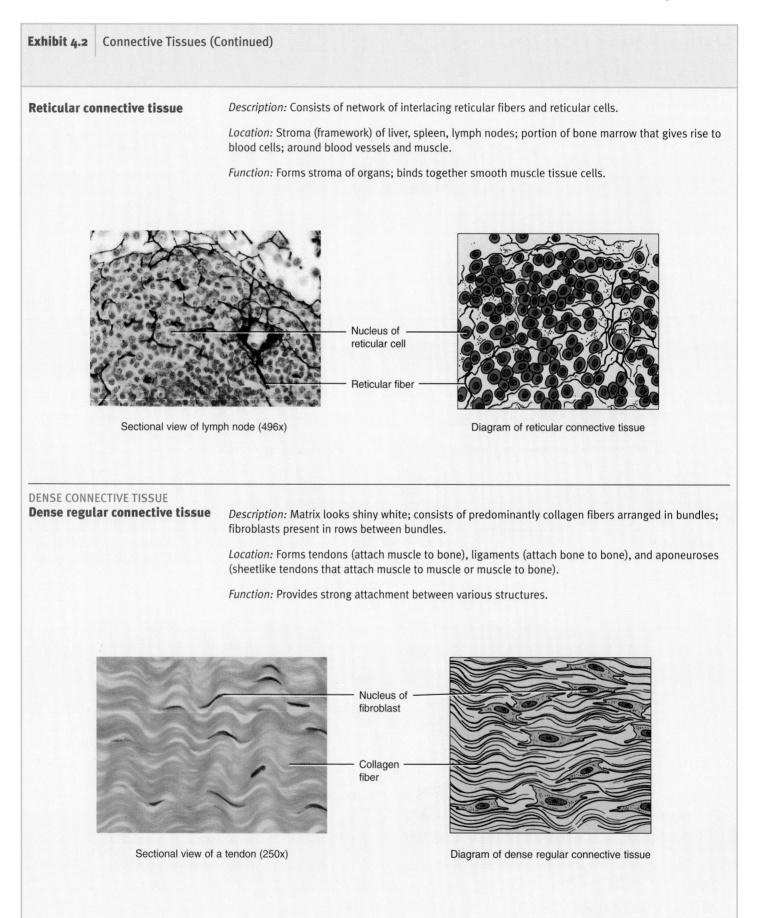

Nucleus of reticular cell

Reticular fiber

Sectional view of lymph node (496x)

Diagram of reticular connective tissue

DENSE CONNECTIVE TISSUE
Dense regular connective tissue

Description: Matrix looks shiny white; consists of predominantly collagen fibers arranged in bundles; fibroblasts present in rows between bundles.

Location: Forms tendons (attach muscle to bone), ligaments (attach bone to bone), and aponeuroses (sheetlike tendons that attach muscle to muscle or muscle to bone).

Function: Provides strong attachment between various structures.

Nucleus of fibroblast

Collagen fiber

Sectional view of a tendon (250x)

Diagram of dense regular connective tissue

(continued)

Exhibit 4.2 | Connective Tissues (Continued)

DENSE CONNECTIVE TISSUE

Dense irregular connective tissue

Description: Consists of predominantly collagen fibers, randomly arranged, and a few fibroblasts.

Location: Fasciae (tissue beneath skin and around muscles and other organs), deeper region of dermis of skin, periosteum (membrane) around joint capsules, membrane capsules around various organs (kidneys, liver, testes, lymph nodes), heart valves.

Function: Provides strength.

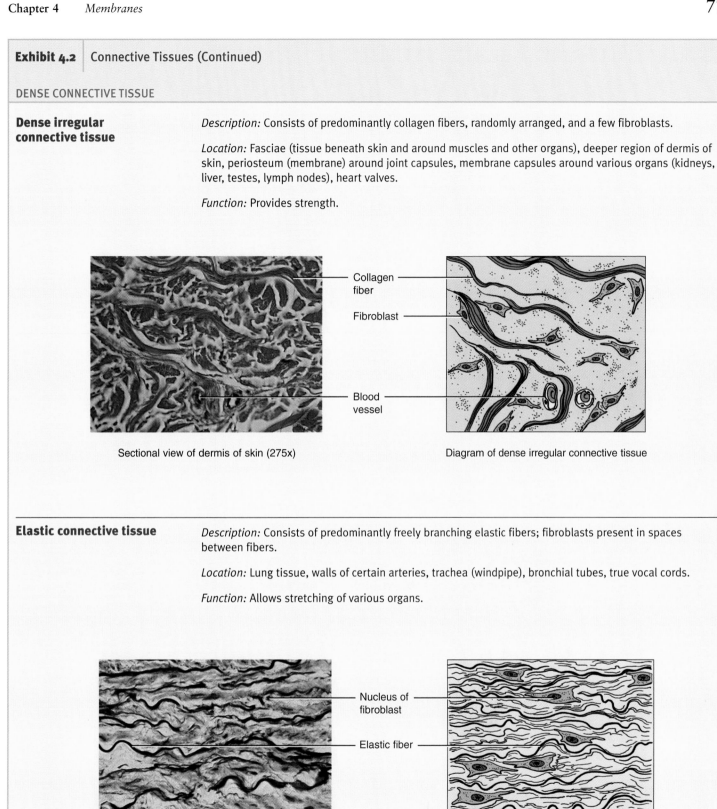

Collagen fiber

Fibroblast

Blood vessel

Sectional view of dermis of skin (275x)   Diagram of dense irregular connective tissue

Elastic connective tissue

Description: Consists of predominantly freely branching elastic fibers; fibroblasts present in spaces between fibers.

Location: Lung tissue, walls of certain arteries, trachea (windpipe), bronchial tubes, true vocal cords.

Function: Allows stretching of various organs.

Nucleus of fibroblast

Elastic fiber

Sectional view of aorta (largest artery in the body) (335x)   Diagram of elastic connective tissue

(continued)

Exhibit 4.2 | Connective Tissues (Continued)

CARTILAGE

Hyaline cartilage

Description: Also called gristle; appears as a bluish white, glossy mass; contains numerous chondrocytes; is the most abundant type of cartilage.

Location: Ends of long bones (articular cartilage) and ribs (costal cartilage), nose, parts of larynx (voice box), trachea (windpipe), bronchial tubes, embryonic skeleton.

Function: Provides movement at joints, flexibility, and support.

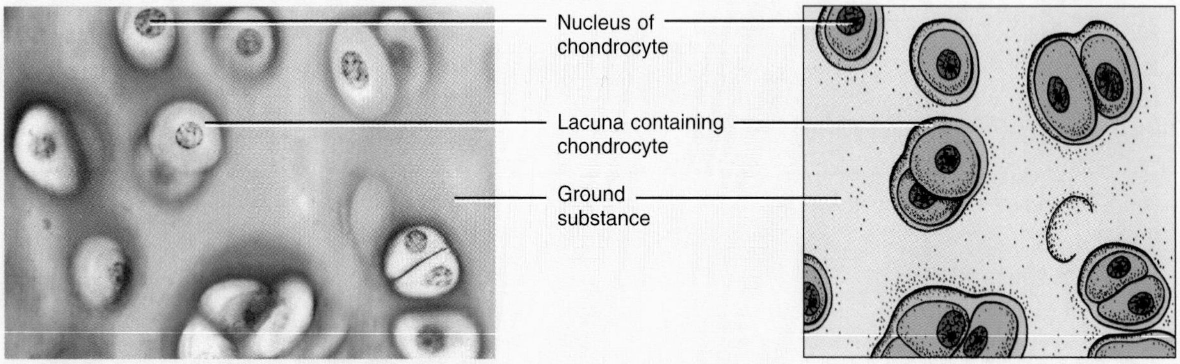

Nucleus of chondrocyte

Lacuna containing chondrocyte

Ground substance

Sectional view of hyaline cartilage from trachea (512x)

Diagram of hyaline cartilage

Fibrocartilage

Description: Consists of chondrocytes scattered among bundles of collagen fibers.

Location: Joint between hipbones (pubic symphysis), discs between backbones (intervertebral discs), knees.

Function: Support and fusion.

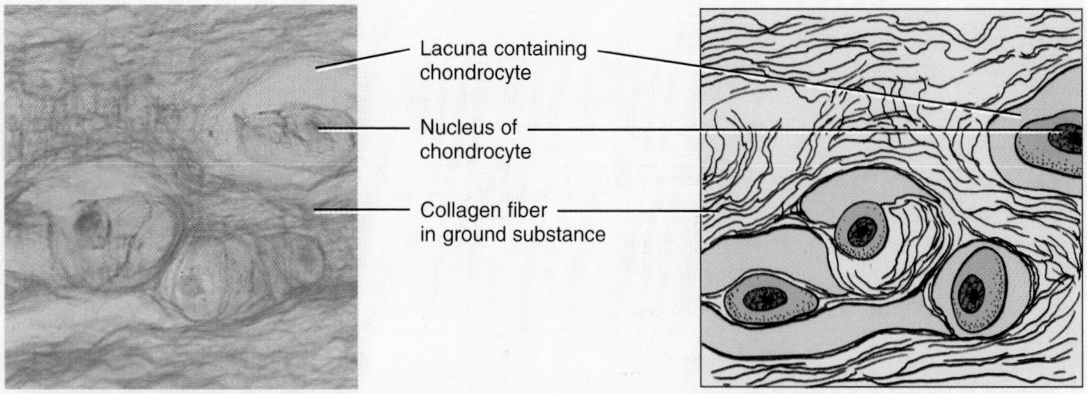

Lacuna containing chondrocyte

Nucleus of chondrocyte

Collagen fiber in ground substance

Sectional view of fibrocartilage from patellar tendon insertion (742x)

Diagram of fibrocartilage

(continued)

Exhibit 4.2	Connective Tissues (Continued)

Elastic cartilage

Description: Consists of chondrocytes located in a threadlike network of elastic fibers.

Location: Lid on top of larynx (epiglottis), external ear, and auditory (Eustachian) tubes.

Function: Gives support and maintains shape.

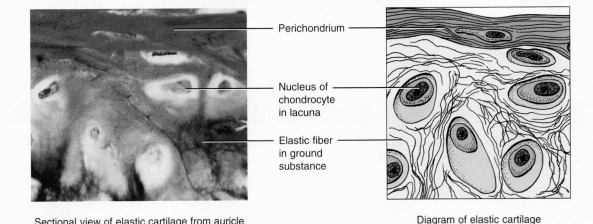

Perichondrium

Nucleus of chondrocyte in lacuna

Elastic fiber in ground substance

Sectional view of elastic cartilage from auricle of ear (742x)

Diagram of elastic cartilage

Mucous Membranes

A ***mucous membrane*** lines a body cavity that opens directly to the exterior. Mucous membranes line the entire gastrointestinal, respiratory, excretory, and reproductive tracts.

The epithelial layer of a mucous membrane secretes mucus, which prevents the cavities from drying out. It also traps particles in the respiratory passageways, lubricates and absorbs food as it moves through the gastrointestinal tract, and secretes digestive enzymes.

The connective tissue layer binds the epithelium to the underlying structures. It also provides the epithelium with oxygen and nutrients and removes wastes since the blood vessels are located here (see Figure 19.2).

Serous Membranes

A ***serous membrane*** lines a body cavity that does not open directly to the exterior, and it covers the organs that lie within the cavity. Serous membranes consist of two portions. The part attached to the cavity wall is called the ***parietal*** (pa-RĪ-e-tal; *paries* = wall) ***portion***; the part that covers the organs is the ***visceral*** (*viscus* = body organ) ***portion.*** The serous membrane lining the thoracic cavity and covering the lungs is called the ***pleura*** (see Figure 18.5). The one lining the heart cavity and covering the heart is the ***pericardium*** (*cardio* = heart). The one lining the abdominal cavity and covering the abdominal organs and some pelvic organs is called the ***peritoneum.***

The epithelial layer, called ***mesothelium,*** secretes a lubricating serous fluid that allows the organs to glide against one another or against the cavity walls. The connective tissue layer is a thin layer of areolar connective tissue.

Cutaneous Membrane

The ***cutaneous*** (kyoo-TĀ-nē-us) ***membrane,*** or skin, constitutes an organ of the integumentary system and is discussed in the next chapter.

Synovial Membranes *contains no epithelial membranes*

Synovial (sin-Ō-vē-al) ***membranes*** line the cavities of some joints (see Figure 7.1). They are composed of areolar connective tissue with elastic fibers and varying amounts of fat; they do not have an epithelial layer. Synovial membranes secrete ***synovial fluid,*** which lubricates the ends of bones as they move at joints and nourishes the articular cartilage covering the bones.

Muscle Tissue

Muscle tissue consists of fibers (cells) that are highly specialized for contraction. The results of contraction are motion, maintenance of posture, and production of heat. Muscle tissue is described in detail in Chapter 8.

Nervous Tissue *Brain + Spinal Cord*

The nervous system consists of two kinds of cells: neurons and neuroglia. *Neurons* (*neuron* = nerve), or nerve cells, are highly specialized cells capable of picking up stimuli, converting the stimuli to nerve impulses, and conducting the nerve impulses to other neurons, muscle fibers, or glands. *Neuroglia* (*glia* = glue) protect and support neurons. Nervous tissue is described in detail in Chapter 9.

muscle - skeletal - cardiac - smooth
voluntary   involuntary
control

Tissue Adaptation: Health Benefits of Strength Training

The phrase "use it or lose it" refers to the adaptation that occurs in muscle and associated connective tissue structures to environmental stimuli. A lifestyle whose highest level of exertion consists of walking from the couch to the refrigerator does not demand much musculoskeletal strength. Muscles accustomed to such a lifestyle atrophy and weaken. On the other hand, regular strength training sends a message to the musculoskeletal system that a challenging environment requires muscular strength. Tissues adapt to stress—this is known as the training effect. Stress muscles and they become stronger; stress bones and they become denser.

Strength training refers to any kind of exercise in which the muscles exert force against a resistance. Weight training, for example, uses weight machines and free weights to apply resistance. Resistance can also be applied with calisthenics such as push ups, where the body's weight is the resistance against which muscles (in this case, the muscles in your arms) work.

Strength to Carry on

Certainly the most well-known benefit of weight training is an increase in muscle size and strength. But the benefits of weight training go beyond the acquisition of muscular prowess. In our sedentary society, many orthopedic problems are the result of weakness and inflexibility, which are often shrugged off as "natural" attributes of the aging process. Studies have shown that between the ages of 20 and 70, people lose about 30 percent of their muscle mass, a loss that averages several pounds a decade. Some of this loss is inevitable and is due to the fact that we lose muscle cells as we age. But much of this loss results from low activity levels rather than the aging process. In "sedentary" muscles, the remaining cells shrink and become weaker. Regular strength training helps prevent this muscle cell atrophy.

Another important result of weight training is that other connective tissue structures, such as tendons, ligaments, and joint capsules, also increase in strength. Stronger joints are less prone to injury and strength improvements and healthier joints increase sports skill level and enjoyment. Weight training apparently also strengthens bones and may thus help maximize deposition of bone mineral in young adults and prevent, or at least slow, the loss of bone mineral in later life.

Preliminary research suggests that strength-training exercise offers some of the heart disease prevention benefits traditionally associated with aerobic exercise, including the control of cholesterol, blood sugar, and blood pressure.

Metabolic Frolic

People trying to lose weight are usually told to participate in aerobic exercise because it burns more calories than weight training. While aerobic exercise requires a greater expenditure of energy per minute, weight training can still be a helpful supplement to aerobic exercise for two reasons. First, weight training may help prevent injuries, especially for people who have a low level of fitness. Many injuries that occur in new exercisers are due to pushing a deconditioned body to do too much, too soon. A new exerciser can benefit from "getting into shape" before beginning an exercise program. Second, weight training can play an important part in a weight control program because it will help preserve or increase the amount of one's muscle tissue. Because metabolic processes occur at a much faster rate—even at rest—in nonfat body tissues, the greater your fat-free mass (a large proportion of which is muscle tissue), the higher your metabolic rate. The higher your metabolic rate, the more calories you can eat without gaining weight.

critical thinking

Why do you think muscle tissue has a higher metabolic rate than fat tissue?

What do you say to your 65-year-old uncle who thinks strength training is just for athletes?

Know for Boards

Study Outline

Types of Tissues (p. 64)

1. A tissue is a group of similar cells and their intercellular substance that are specialized for a particular function.
2. The tissues of the body are classified into four principal types: epithelial, connective, muscular, and nervous.

Know

Epithelial Tissue (p. 64)

General Features of Epithelial Tissue (p. 64)

1. The general types of epithelium include covering and lining epithelium and glandular epithelium.
2. Some general characteristics of epithelium are: consists mostly of cells with little extracellular material, arranged in sheets, attached to connective tissue by a basement membrane, avascular (no blood vessels), has a nerve supply, and can replace itself.

Covering and Lining Epithelium (p. 64)

1. Cell shapes include squamous (flat), cuboidal (cubelike), columnar (rectangular), and transitional (variable); layers are arranged as simple (one layer), stratified (several layers), and pseudostratified (one layer that appears to be several).
2. Simple squamous epithelium consists of a single layer of flat cells. It is adapted for diffusion and filtration and is found in lungs and kidneys.
3. Simple cuboidal epithelium consists of a single layer of cube-shaped cells. It is adapted for secretion and absorption. It is found covering ovaries, in kidneys and eyes, and lining some glandular ducts.
4. Nonciliated simple columnar epithelium consists of a single layer of nonciliated rectangular cells. It lines most of the gastrointestinal tract. Specialized cells containing microvilli perform absorption. Goblet cells secrete mucus. Ciliated simple columnar epithelium consists of a single layer of ciliated rectangular cells. It is found in a few portions of the upper respiratory tract where it moves foreign particles trapped in the mucus out of the body.
5. Stratified squamous epithelium consists of several layers of cells in which the top layer is flat. It is protective. It lines the upper gastrointestinal tract and vagina (nonkeratinized) and forms the outer layer of skin (keratinized).
6. Stratified cuboidal epithelium consists of several layers of cells in which the top layer is cube-shaped. It is found in sweat glands and a portion of the male urethra.
7. Stratified columnar epithelium consists of several layers of cells in which the top layer is rectangular. It protects and secretes. It is found in the male urethra and large excretory ducts.
8. Transitional epithelium consists of several layers of cells whose appearance is variable. It lines the urinary bladder and is capable of stretching.
9. Pseudostratified columnar epithelium has only one layer but gives the appearance of many. It lines larger excretory ducts, parts of the urethra, auditory (Eustachian) tubes, and most upper respiratory structures, where it protects and secretes.

Glandular Epithelium (p. 71)

1. A gland is a single cell or a mass of epithelial cells adapted for secretion.
2. Exocrine glands (sweat, oil, and digestive glands) secrete into ducts or directly onto a free surface.
3. Endocrine glands secrete hormones into the blood.

Connective Tissue (p. 71)

1. Connective tissue is the most abundant body tissue.
2. Some general characteristics of connective tissue are: consists of cells, ground substance, and fibers; has abundant matrix with relatively few cells; does not occur on free surfaces; has a nerve supply (except for cartilage); and has a rich blood supply (except for cartilage and tendons).

Connective Tissue Cells (p. 72)

1. Cells in connective tissue include fibroblasts (secrete matrix and fibers), macrophages (phagocytes), plasma cells (secrete antibodies), and mast cells (produce histamine).

Connective Tissue Matrix (p. 72)

1. The ground substance and fibers comprise the matrix.
2. Substances found in the ground substance include hyaluronic acid and chondroitin sulfate.
3. The ground substance supports, binds, provides a medium for the exchange of materials, and is active in influencing cell functions.
4. The fibers provide strength and support and are of three types.
5. Collagen fibers (composed of collagen) are found in bone, tendons, and ligaments; elastic fibers (composed of elastin) are found in skin, blood vessels, and lungs; and reticular fibers (composed of collagen and glycoprotein) are found around fat cells, nerve fibers, and skeletal and smooth muscle cells.

Embryonic Connective Tissue (p. 72)

1. Mesenchyme forms all other connective tissues.
2. Mucous connective tissue is found in the umbilical cord of the fetus, where it gives support.

Mature Connective Tissue (p. 72)

1. Mature connective tissue is connective tissue that exists in the newborn and does not change after birth. It is subdivided into several kinds: loose connective tissue, dense connective tissue, cartilage, bone, and blood.
2. Loose connective tissue includes areolar connective tissue, adipose tissue, and reticular connective tissue.
3. Areolar connective tissue consists of three types of fibers, several kinds of cells, and a semifluid ground substance. It is found in the subcutaneous layer and mucous membranes and around blood vessels, nerves, and body organs.
4. Adipose tissue consists of adipocytes that store fat. It is found in the subcutaneous layer, around various organs, and in the yellow bone marrow of long bones.

5. Reticular connective tissue consists of reticular fibers and reticular cells and is found in the liver, spleen, and lymph nodes.

6. Dense connective tissue includes dense regular connective tissue, dense irregular connective tissue, and elastic connective tissue.

7. Dense regular connective tissue consists of bundles of collagen fibers and fibroblasts. It forms tendons, ligaments, and aponeuroses.

8. Dense irregular connective tissue consists of randomly arranged collagen fibers and a few fibroblasts. It is found in the dermis of skin and in membrane capsules.

9. Elastic connective tissue consists of elastic fibers and fibroblasts. It is found in the lungs, walls of arteries, and bronchial tubes.

10. Cartilage has a firm jellylike matrix (chondroitin sulfate) containing collagen and elastic fibers and chondrocytes.

11. Hyaline cartilage is found in the embryonic skeleton, at the ends of bones, in the nose, and in respiratory structures. It is flexible, allows movement, and provides support.

12. Fibrocartilage is found in the pubic symphysis, intervertebral discs, and knees.

13. Elastic cartilage maintains the shape of organs such as the epiglottis of the larynx, auditory (Eustachian) tubes, and external ear.

14. Bone (osseous) tissue, cartilage, and joints make up the skeletal system.

15. Blood (vascular) tissue consists of plasma and formed elements (red blood cells, white blood cells, and platelets).

Membranes (p. 73)

1. An epithelial membrane is an epithelial layer overlying a connective tissue layer. Examples are mucous, serous, and cutaneous membranes.

2. Mucous membranes line cavities that open to the exterior, such as the gastrointestinal tract.

3. Serous membranes (pleura, pericardium, peritoneum) line closed cavities and cover the organs in the cavities. These membranes consist of parietal and visceral portions.

4. Synovial membranes line joint cavities and do not contain epithelium.

Muscle Tissue (p. 79)

1. Muscle tissue is specialized for contraction.

2. Through contraction, muscle tissue provides motion, maintenance of posture, and production of heat.

Nervous Tissue (p. 79)

1. The nervous system is composed of neurons (nerve cells) and neuroglia.

2. Neurons generate and conduct nerve impulses; neuroglia protect and support neurons.

Self-Quiz

1. Epithelial tissue functions in

   a. conducting nerve impulses   b. storing fat   c. covering and lining the body and its parts   d. movement   e. storing minerals

2. Which of the following is NOT considered a type of connective tissue?

   a. blood (vascular)   b. adipose   c. reticular   d. neuroglia   e. cartilage

3. The epithelial tissue type that is composed of many cell layers, the most superficial layers being flat, is _____ epithelium.

   a. transitional   b. simple squamous   c. pseudostratified   d. stratified cuboidal   e. stratified squamous

4. Stratified cuboidal epithelium functions in

   a. protection   b. secretion   c. diffusion   d. movement of cilia   e. absorption

5. Which of the following structures would NOT be found in the matrix of connective tissue?

   a. collagen fibers   b. elastic fibers   c. keratin fibers   d. reticular fibers   e. chondroitin sulfate

6. Mast cells produce

   a. histamine   b. antibodies   c. fibers of the matrix   d. hyaluronic acid   e. mucus

7. Match the following tissue types with their description:

   C a. fat storage
   G b. distensible (stretchy)
   E c. forms the stroma (framework) of many soft organs
   D d. composes the intervertebral discs
   H e. stores red bone marrow, protects, supports
   F f. transports O_2 and CO_2
   B g. forms tendons and ligaments
   A h. involved in diffusion
   I i. covers ends of bones; in nose

   A. simple squamous epithelium
   B. dense regular connective tissue
   C. adipose tissue
   D. fibrocartilage
   E. reticular connective tissue
   F. blood (vascular) tissue
   G. transitional epithelium
   H. bone (osseous) tissue
   I. hyaline cartilage

8. All of the following are types of epithelial membranes except

   a. cutaneous membrane b. synovial membrane c. mucous membrane d. serous membrane e. peritoneum

9. The tissue type characterized by a gelatinous ground substance embedded with reticular, collagen, and elastic fibers and several cell types, including fibroblasts and macrophages, is

   a. dense irregular connective tissue b. mucous connective tissue c. hyaline cartilage d. areolar connective tissue e. mesenchyme

10. The four principal tissue types are:

    a. blood, connective, muscle, nervous b. connective, epithelial, muscle, nervous c. epithelial, embryonic, blood, nervous d. epithelial, muscle, loose, nervous e. epithelial, connective, muscle, membranous

Critical Thinking Applications

1. Two college students return home at the end of their freshman year. The two students had weighed exactly the same when they played on their high school soccer team. Now one student is 35 pounds heavier than the other. Discuss the possible reasons for their weight difference and the possible consequences if the weight difference continues to grow.

2. A child starts preschool with her arm in a cast. During the school year she has several more broken bones following minor accidents. The concerned teacher suggests to the parents that they speak to the family doctor about the child's connective tissue disorder. Discuss the possible cause of the disorder.

3. How does your respiratory system trap and remove airborne substances? Trace the route taken by pollen, dust, and cigarette smoke particles trapped by the respiratory system. Include the anatomical terms for the membrane and the various types of cells and their functions.

4. You've gone out to eat at your favorite fast food joint: General Bellisimo's Fried Chicken. A health food zealot waves a chicken leg and declares "This is all fat!" Using your knowledge of tissues, defend the contents of the chicken leg.

5. The neighborhood kids are walking around with common pins and sewing needles stuck through their fingertips. There is no visible bleeding. What tissue layer have they pierced? How do you know?

6. Many cosmetic skin cremes include collagen or elastin in their ingredients. Some individuals have had collagen preparations injected into their lips and wrinkles. Compare the effects on your skin structure of topical cosmetics versus a collagen injection.

5 chapter

THE INTEGUMENTARY SYSTEM

student learning objectives

1. Describe the structure and functions of the skin. *85*

2. Explain the pigments involved in skin color. *88*

3. Describe the structure and functions of the accessory organs of the skin. *88*

4. Describe the effects of aging on the integumentary system. *90*

5. Explain how the skin helps regulate body temperature. *91*

a look ahead

A group of tissues that performs a specific function is an *organ.* The next higher level of organization is a *system*—a group of organs operating together to perform specialized functions. The skin and its accessory organs, such as hair, nails, glands, and several specialized receptors, constitute the *integumentary* (in′-teg-yoo-MEN-tar-ē; *integumentum* = covering) *system* of the body.

Of all the body's organs, none is more easily inspected or exposed to infection, disease, and injury than the skin. Because of its visibility, skin reflects our emotions and some aspects of normal physiology, as evidenced by frowning, blushing, and sweating. Changes in skin color may indicate homeostatic imbalances in the body; for example, abnormal skin eruptions or rashes such as chickenpox, cold sores, or measles may reveal systemic infections or diseases of internal organs. Other disorders may involve just the skin itself, such as warts, age spots, or pimples. The skin's location makes it vulnerable to damage from trauma, sunlight, microbes, or pollutants in the environment.

Many interrelated factors may affect both the appearance and health of the skin, including nutrition, hygiene, circulation, age, immunity, genetic traits, psychological state, and drugs. So important is the skin to body image that people spend much time and money to restore it to a more youthful appearance.

Skin

objective: *Describe the structure and functions of the skin.*

The *skin* is an organ because it consists of different tissues that are joined to perform specific activities. It is one of the largest organs in surface area and weight. In adults, the skin covers an area of about 2 square meters (22 square feet). The skin is not just a simple, thin covering that keeps the body together and provides protection. It performs several essential functions that will be described shortly. *Dermatology* (der′-ma-TOL-ō-jē; *dermato* = skin; *logos* = study of) is the medical specialty that deals with the diagnosis and treatment of skin disorders.

Structure

Structurally, the skin consists of two principal parts (Figure 5.1). The outer, thinner portion, which is composed of *epithelium,* is called the *epidermis.* The epidermis is attached to the inner, thicker, *connective tissue* part, which is called the *dermis.* Beneath the dermis is a *subcutaneous (subQ) layer,* also called the *hypodermis,* which attaches the skin to underlying structures. Recall that the subcutaneous layer consists of adipose tissue and areolar connective tissue.

Functions

Among the numerous functions of the skin are the following:

1. **Regulation of body temperature.** In response to high environmental temperature or strenuous exercise, the evaporation of sweat from the skin surface helps lower an elevated body temperature to normal. Changes in the flow of blood in the skin also help regulate body temperature (described later in the chapter).

2. **Protection.** The skin covers the body and provides a physical barrier that protects underlying tissues from physical abrasion, bacterial invasion, dehydration, and ultraviolet (UV) radiation. Hair and nails also have protective functions, as described shortly.

3. **Sensation.** The skin contains abundant nerve endings and receptors that detect stimuli related to temperature, touch, pressure, and pain (see Chapter 12).

4. **Excretion.** Small amounts of water, salts, and several organic compounds (components of perspiration) are excreted by sweat glands.

5. **Immunity.** Certain cells of the epidermis (Langerhans cells) are important components of the immune system, which fends off foreign invaders to the body (see Chapter 17).

6. **Synthesis of vitamin D.** Exposure of the skin to ultraviolet (UV) radiation helps with the production of vitamin D, a substance that aids in the absorption of calcium and phosphorus from the digestive system into the blood.

Epidermis

The *epidermis* (*epi* = above) is composed of keratinized stratified squamous epithelium and contains four types of cells (Figure 5.2 on page 87). The most numerous (about 90 percent) is known as a *keratinocyte* (ker-a-TIN-ō-sīt; *kerato* = horny), a cell that undergoes keratinization. In the process of *keratinization,* cells formed in the basal layers are pushed to the surface. As the cells move upward they accumulate *keratin,* a protein that helps protect the skin and underlying tissue. At the same time, the cytoplasm, nucleus, and other organelles disappear, and the cells die. Eventually, the keratinized cells slough off and are replaced by underlying cells that, in turn, become keratinized. The whole process takes two to four weeks.

The second type of cell is called a *melanocyte* (MEL-a-nō-sīt; *melan* = black), which can also be found in the dermis. It produces *melanin,* one of the pigments responsible for skin color, and absorbs ultraviolet (UV) radiation.

The third type of cell in the epidermis is known as a *Langerhans* (LANG-er-hans) *cell.* These cells function in immune responses and are easily damaged by UV radiation.

A fourth type of cell found in the epidermis is called a *Merkel cell.* These cells are located in the deepest layer of the epidermis of hairless skin. Merkel cells are thought to function in the sensation of touch.

Four or five distinct layers of cells form the epidermis. In most regions of the body the epidermis has four recognizable layers. Where exposure to friction is greatest, such as in the

Know for Boards

Figure 5.1   Structure of the skin and underlying subcutaneous tissue. The stratum lucidum shown is not apparent on hairy skin but is included here so that you can see the relationship of all five epidermal layers.

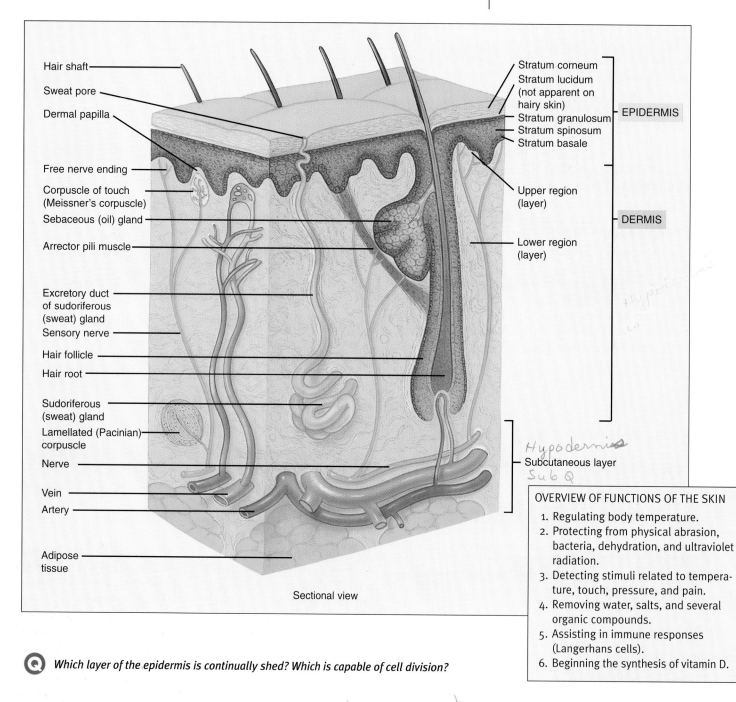

Hair shaft

Sweat pore

Dermal papilla

Free nerve ending

Corpuscle of touch
(Meissner's corpuscle)

Sebaceous (oil) gland

Arrector pili muscle

Excretory duct
of sudoriferous
(sweat) gland

Sensory nerve

Hair follicle

Hair root

Sudoriferous
(sweat) gland

Lamellated (Pacinian)
corpuscle

Nerve

Vein

Artery

Adipose
tissue

Stratum corneum

Stratum lucidum
(not apparent on
hairy skin)

Stratum granulosum

Stratum spinosum

Stratum basale

EPIDERMIS

Upper region
(layer)

DERMIS

Lower region
(layer)

Hypodermis

Subcutaneous layer

Sub Q

Sectional view

OVERVIEW OF FUNCTIONS OF THE SKIN

1. Regulating body temperature.
2. Protecting from physical abrasion, bacteria, dehydration, and ultraviolet radiation.
3. Detecting stimuli related to temperature, touch, pressure, and pain.
4. Removing water, salts, and several organic compounds.
5. Assisting in immune responses (Langerhans cells).
6. Beginning the synthesis of vitamin D.

Q *Which layer of the epidermis is continually shed? Which is capable of cell division?*

palms and the soles, the epidermis is thicker (1 to 2 mm) and five layers are recognizable (Figure 5.2). Constant exposure of thin or thick skin to friction or pressure stimulates formation of a callus, an abnormal thickening of the epidermis.

The names of the five layers (strata), from the deepest to the most superficial, are:

Germ layer

1. **Stratum basale** (ba-SA-lē; *basale* = base). This single layer of cuboidal to columnar cells contains cells that are capable of continued cell division. It also contains

Stratum germavatum

melanocytes. The cells multiply, producing keratinocytes, which push up toward the surface and become part of the more superficial layers (keratinization). The stratum basale also contains Merkel cells that are sensitive to touch.

Fixed

2. **Stratum spinosum** (spi-NŌ-sum; *spinosum* = thornlike or prickly). This layer of the epidermis has about ten rows (sheets) of polyhedral (many-sided) cells with spinelike projections. Melanin is also found in this layer.

Figure 5.2   Structure of the epidermis.

🔑 *The superficial layer of the skin—the epidermis—is keratinized stratified squamous epithelium.*

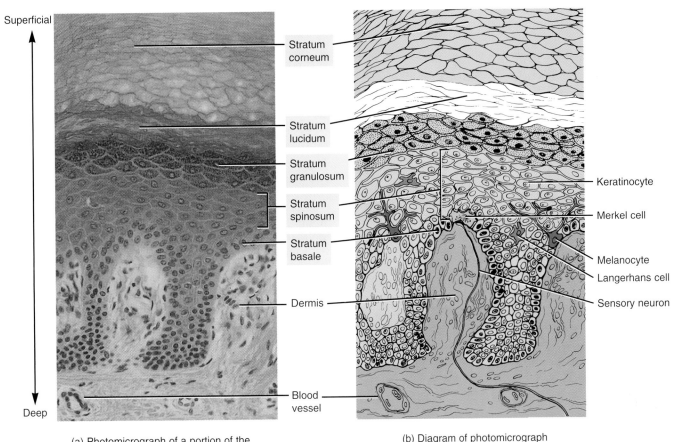

(a) Photomicrograph of a portion of the skin (496x)

(b) Diagram of photomicrograph

Ⓠ *Which epidermal cell functions in immunity?*

3. **Stratum granulosum** (gran-yoo-LŌ-sum; *granulum* = little grain). The third layer of the epidermis consists of about five rows of flattened cells with darkly staining granules.

4. **Stratum lucidum** (LOO-si-dum; *lucidum* = clear). Normally, only the thick skin of the palms and soles has this layer. It consists of about five rows of clear, flat, dead cells.

5. **Stratum corneum** (COR-nē-um; *corneum* = horny). This layer consists of about 30 rows of flat, dead cells completely filled with keratin.

[handwritten annotation: Extra layer only in thick skin]

[handwritten annotation: (protein) Thickens skin, also in hair & nails (Keratin)]

[handwritten annotation: Cornfied means filled with Keratin.]

Dermis

The second principal part of the skin, the ***dermis*** (*derma* = skin), is composed of connective tissue containing collagen and elastic fibers (see Figure 5.1). The combination of collagen and elastic fibers gives the skin its strength, ***extensibility*** (the ability to stretch), and ***elasticity*** (the ability to return to original shape after extension.) The ability of the skin to stretch can readily be seen during pregnancy, obesity, and tissue swelling

(edema). Small tears in the skin due to extensive stretching that remain visible as silvery white streaks are called ***striae*** (STRĪ-ē; *stria* = streak). The few cells in the dermis include fibroblasts, macrophages, and adipocytes. The dermis is very thick in the palms and soles and very thin in the eyelids and scrotum. The upper region of the dermis consists of areolar connective tissue and its surface area is greatly increased by small, fingerlike projections called ***dermal papillae*** (pa-PIL-ē; *papilla* = nipple). Dermal papillae cause ridges in the epidermis, which produce fingerprints and help us to grip objects. Some dermal papillae contain ***corpuscles of touch*** (***Meissner's*** (MĪS-nerz) ***corpuscles***), nerve endings sensitive to touch. Others contain blood capillaries. *[handwritten: ← reticular]*

The lower region of the dermis consists of dense, irregular connective tissue, adipose tissue, hair follicles, nerves, oil glands, and the ducts of sweat glands. It is attached to underlying bone and muscle by the subcutaneous layer. The subcutaneous layer contains nerve endings called ***lamellated*** (***Pacinian***) ***corpuscles,*** which are sensitive to pressure.

Cold and warmth receptors are found in the upper and middle dermis.

Skin Color

objective: *Explain the pigments involved in skin color.*

Skin color is due to melanin, carotene, and hemoglobin. The amount of *melanin,* a pigment found in the epidermis, varies the skin color from pale yellow to black. Because the number of melanocytes is about the same in all races, differences in skin color are due to the amount of pigment the melanocytes produce and distribute. An inherited inability of an individual in any race to produce melanin results in *albinism* (AL-bin-izm; *albus* = white); albinism may be noticed by the absence of pigment in the hair and eyes as well as the skin. In some people, melanin tends to form in patches called *freckles.*

When the skin is repeatedly exposed to ultraviolet radiation, the amount and darkness of melanin increase, which tans and protects the body against radiation. Overexposure to ultraviolet light, however, may lead to skin cancer. Among the most serious skin cancers is *malignant melanoma* (*melano* = dark colored; *oma* = tumor), cancer of the melanocytes. Fortunately, most skin cancers involve basal and squamous cells that can be removed surgically.

The pigment *carotene* (KAR-o-tēn; *keraton* = carrot) has a yellow-orange color and is found in the stratum corneum and fatty areas of the dermis and subcutaneous layer. Together, carotene and melanin account for the yellowish color of skin.

Caucasian skin will appear pale to pink to red depending on the amount and quality of the *hemoglobin* in the red blood cells moving through blood vessels in the dermis.

Erythema (er-e-THĒ-ma; *erythros* = red), redness of the skin, is caused by engorgement of capillaries in the dermis with blood. Exercise and embarrassment may cause noticeable erythema, especially of the face. Erythema also occurs with skin injury, infection, inflammation, or allergic reactions.

Accessory Organs of the Skin

objective: *Describe the structure and functions of the accessory organs of the skin.*

Accessory organs of the skin that develop from the epidermis of an embryo—hair, glands, nails—perform vital functions. Hair and nails protect the body. Sweat glands help regulate body temperature.

Hair

Hairs or *pili* (PI-lē) are growths of the epidermis that are variously distributed over the body. Their primary function is protection. Hair on the head guards the scalp from injury and the sun's rays; eyebrows and eyelashes protect the eyes from foreign particles; hair in the nostrils protects against inhaling insects and foreign particles.

Each hair is a thread of fused, dead, keratinized cells that consists of a shaft and a root (Figure 5.3). The *shaft* is the superficial portion, most of which projects above the surface of the skin. The *root* is the portion below the surface that penetrates into the dermis and even into the subcutaneous layer.

Surrounding the root is the *hair follicle,* which is composed of two layers of epidermal cells: *external* and *internal root sheaths* surrounded by a connective tissue sheath.

The base of each follicle is enlarged into an onion-shaped structure, the *bulb.* This structure contains an indentation, the *papilla of the hair,* which contains many blood vessels and provides nourishment for the growing hair. The bulb also contains a region of cells called the *matrix,* which produces new hairs by cell division when older hairs are shed.

Each hair follicle goes through a *growth cycle,* which consists of a *growth stage* and a *resting stage.* During the growth stage, a hair is formed by cells of the matrix that differentiate, become keratinized, and die. As new cells are added at the base of the hair root, the hair grows longer. In time, the growth of the hair stops and the resting stage begins. After the resting stage, a new growth cycle begins in which a new hair replaces the old hair and the old hair is pushed out of the hair follicle. In general, scalp hair grows for about three years and rests for about one to two years. At any given time, most hair is in the growth stage.

Normal hair loss (shedding) in the adult scalp is about 100 hairs per day. Both the rate of growth and the replacement cycle may be altered by any of the following factors: illness, radiation therapy and chemotherapy, diet, age, genetics, gender, and severe emotional stress. The rate of shedding also increases for three to four months after childbirth.

Minoxidil (Rogaine®) is a potent vasodilator, that is, a drug that widens blood vessels and increases circulation. When applied topically to the skin, it does stimulate some hair regrowth in some persons with thinning hair. For many, however, the hair growth is meager, and it does not help people who already are bald.

The color of hair is due to melanin. It is synthesized by melanocytes in the matrix of the bulb and passes into cells of the root and shaft. Dark-colored hair contains mostly true melanin. Blond and red hair contain variants of melanin in which there is iron and more sulfur. Gray hair occurs with a decline in the synthesis of melanin. White hair results from accumulation of air bubbles in the hair shaft.

Associated with hairs is a bundle of smooth muscle called *arrector* (*arrector* = to raise) *pili.* It is located along the side of the hair follicle (Figure 5.3). These muscles contract under stresses of fright and cold, pulling the hairs into a vertical position and resulting in "goose-bumps" or "gooseflesh."

Glands

Three kinds of glands associated with the skin are sebaceous, sudoriferous, and ceruminous.

Sebaceous (Oil) Glands [Secrete into hair follicle which open to skin]

Sebaceous (se-BĀ-shus; *sebaceus* = oily) or *oil glands,* with few exceptions, are connected to hair follicles (Figure 5.3). The secreting portions of the glands lie in the dermis and open into the necks of hair follicles or directly onto a skin surface. There are no sebaceous glands in the palms and soles.

Sebaceous glands secrete an oily substance called *sebum* (SĒ-bum). Sebum keeps hair from drying out, prevents exces-

Figure 5.3 Principal parts of a hair root and associated structures.

Hairs are growths of the epidermis composed of dead, keratinized cells.

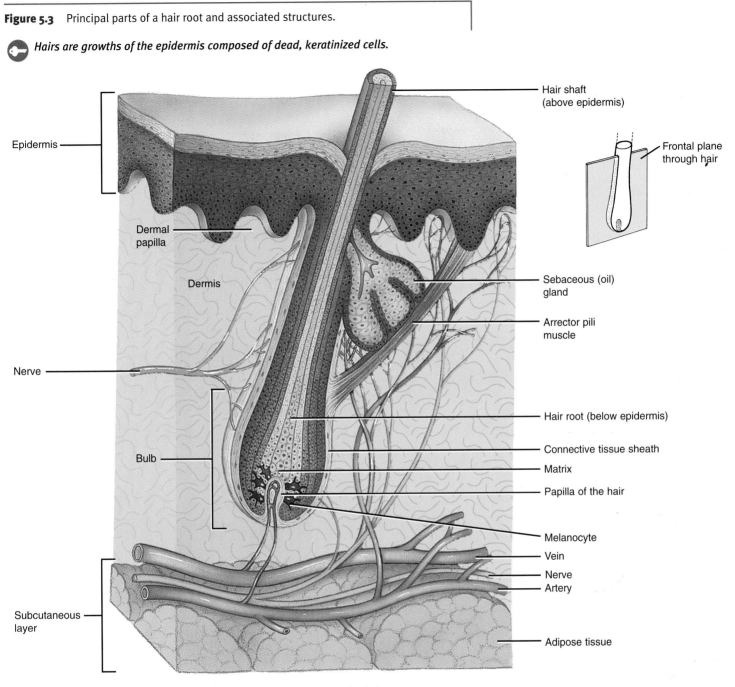

Frontal section of a hair

Why does the papilla of the hair contain blood vessels?

sive evaporation of water from the skin, keeps the skin soft, and inhibits the growth of certain bacteria.

When sebaceous glands of the face become enlarged because of accumulated sebum, *blackheads* develop. Because sebum is nutritive to certain bacteria, *pimples* or *boils* often result. The color of blackheads is due to melanin and oxidized oil, not dirt. Sebaceous gland activity increases during adolescence.

Sudoriferous (Sweat) Glands

Sudoriferous (soo′-dor-IF-er-us; *sudor* = sweat; *ferre* = to bear) or *sweat glands* are divided into two types. **Apocrine** (*apo* =

from) *sweat glands* are found in the skin of the axilla (armpit), pubic region, and pigmented areas (areolae) of the breasts. Apocrine sweat gland ducts open into hair follicles. They begin to function at puberty and produce a viscous (sticky) secretion. They are stimulated during emotional stresses and sexual excitement and the secretions are commonly known as "cold sweat."

Eccrine (*ekkrinein* = to secrete) *sweat glands* are distributed throughout the skin except for areas such as the margins of the lips, nail beds of the fingers and toes, and eardrums. Eccrine sweat glands are most numerous in the skin of the palms and the soles. Eccrine sweat gland ducts terminate at a pore at the surface of the epidermis (see Figure 5.1) and function throughout life and produce a more watery secretion than that of apocrine sweat glands.

Perspiration, or **sweat,** is the substance produced by sudoriferous glands. Its principal function is to help regulate body temperature. It also helps eliminate wastes.

Because the mammary glands are actually modified sudoriferous glands, they could be discussed here. However, because of their relationship to the reproductive system, they will be considered in Chapter 23.

Ceruminous Glands   *Subset of Apocrine*

Ceruminous (se-ROO-mi-nus; *cera* = wax) **glands** are present in the external auditory meatus (canal), the outer ear canal. Their ducts open either directly onto the surface of the external auditory meatus or into ducts of sebaceous glands. The combined secretion of the ceruminous and sebaceous glands is called **cerumen.** Cerumen and the hairs in the external auditory meatus provide a sticky barrier against foreign bodies.

Nails

Plates of tightly packed, hard, keratinized cells of the epidermis are referred to as **nails.** Each nail (Figure 5.4) consists of a nail body, a free edge, and a nail root. The **nail body** is the portion of the nail that is visible; the **free edge** is the part that extends past the end of the finger or toe; the **nail root** is the portion that is not visible. Most of the nail body is pink because of the underlying blood capillaries. The whitish semilunar area near the nail root is called the **lunula** (LOO-nyoo-la; *lunula* = little moon). It appears whitish because the vascular tissue underneath does not show through due to the thickened stratum basale in the area.

Nail growth occurs by the transformation of superficial cells of the **nail matrix** into nail cells. The average growth of

fingernails is about 1 mm (0.04 inch) per week. The **cuticle** consists of stratum corneum.

Functionally, nails help us to grasp and manipulate small objects, provide protection to the ends of the digits, and allow us to scratch various parts of the body.

Aging and the Integumentary System

objective: *Describe the effects of aging on the integumentary system.*

Although skin is constantly aging, pronounced effects do not become noticeable until a person reaches the late forties. Around that time, collagen fibers decrease in number, stiffen, break apart, and form into a shapeless, matted tangle. Elastic fibers lose some of their elasticity, thicken into clumps, and fray. As a result, the skin forms indentations known as wrinkles. Fibroblasts, which produce both collagen and elastic fibers, decrease in number, and macrophages become less efficient phagocytes. With increased age, the hair and nails grow more slowly. Langerhans cells dwindle in number, thus decreasing the immune response of older skin. Decreased size of sebaceous (oil) glands leads to dry and broken skin that is more susceptible to infection and contributes to wrinkles. Production of sweat diminishes, which probably contributes to the increased incidence of heat stroke in the elderly. There is a decrease in the number of functioning melanocytes, resulting in gray hair and atypical skin pigmentation. An increase in the size of some melanocytes produces blotching (liver spots). Aged skin also heals poorly and becomes more susceptible to pathological conditions such as skin cancer, itching, and pressure sores.

Figure 5.4   Structure of nails. Shown is a fingernail.

🔑 *Nail cells arise by transformation of superficial cells of the nail matrix into nail cells.*

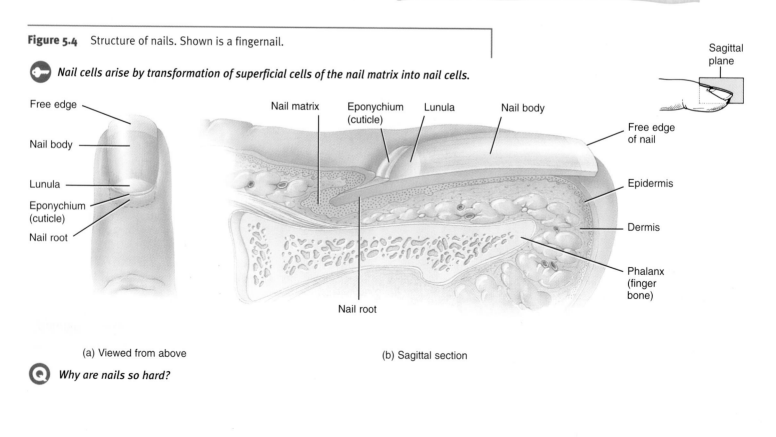

(a) Viewed from above   (b) Sagittal section

❓ *Why are nails so hard?*

Homeostasis of Body Temperature

objective: *Explain how the skin helps regulate body temperature.*

One of the best examples of homeostasis in humans is the regulation of body temperature by the skin. As warm-blooded animals, we are able to maintain a remarkably constant body temperature of 37°C (98.6°F) even though the environmental temperature varies greatly.

Suppose you are in an environment where the temperature is 38°C (101°F). Heat (the stimulus) continually flows from the environment to your body, raising body temperature. To counteract these changes in a controlled condition, a sequence of events is set into operation (Figure 5.5). Temperature-sensitive receptors (nerve endings) in the skin called *thermoreceptors* detect the stimulus and send nerve impulses (input) to your brain (control center). A temperature control region of the brain (called the hypothalamus, see Figure 10.6) then sends nerve impulses (output) to the sweat glands (effectors), which produce perspiration more rapidly. As the sweat evaporates from the surface of your skin, heat is lost and your body temperature drops to normal (return to homeostasis). When environmental temperature is low, sweat glands produce less perspiration.

Your brain also sends output to blood vessels (a second set of effectors), dilating (widening) those in the dermis so that skin blood flow increases. As more warm blood flows through capillaries close to the body surface, more heat can be lost to the environment, which lowers body temperature. Thus heat is lost from the body, and body temperature falls to the normal value to restore homeostasis.

Note that this temperature regulation involves a *negative feedback system* because the response (cooling) is opposite to the stimulus (heating) that started the cycle. Also, the thermoreceptors continually monitor body temperature and feed this information back to the brain. The brain, in turn, continues to send impulses to the sweat glands and blood vessels until the temperature returns to 37°C (98.6°F).

Regulating the rate of sweating and changing dermal blood flow are only two mechanisms by which body temperature can be adjusted. Other mechanisms include regulating metabolic rate (a slower metabolic rate reduces heat production) and regulating skeletal muscle contractions (decreased muscle tone results in less heat production). These mechanisms are discussed in detail in Chapter 20.

If environmental temperature is low or some factor causes body temperature to decrease, a series of responses takes place in order to raise body temperature to normal. For example, blood vessels in the dermis constrict, thus reducing heat loss to the environment. Also, certain hormones are released that increase metabolic rate and therefore increase heat production. Finally, skeletal muscles contract involuntarily (shivering) and this too increases heat production. All of these responses attempt to raise body temperature back to normal.

Figure 5.5   One role of the skin in regulating the homeostasis of body temperature.

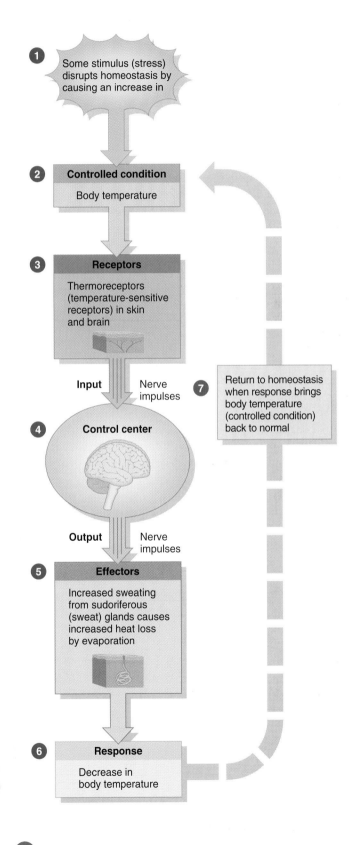

1. Some stimulus (stress) disrupts homeostasis by causing an increase in

2. **Controlled condition**
   Body temperature

3. **Receptors**
   Thermoreceptors (temperature-sensitive receptors) in skin and brain

Input   Nerve impulses

7. Return to homeostasis when response brings body temperature (controlled condition) back to normal

4. **Control center**

Output   Nerve impulses

5. **Effectors**
   Increased sweating from sudoriferous (sweat) glands causes increased heat loss by evaporation

6. **Response**
   Decrease in body temperature

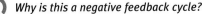 *Why is this a negative feedback cycle?*

Scan the ads in almost any women's or men's "health" magazines and you will get the impression that the cosmetics companies have succeeded where the Spanish explorer Ponce de Leon (1460–1521) long ago failed: they seem to have discovered the fountain of youth. Buy the right products and your skin will escape the aging process.

What Causes the Skin to Age?

Dermatologists estimate that approximately 90 percent of the skin changes attributed to aging are in fact the result of photodamage, which means exposure to too much sun. Skin that has experienced the effects of time without the effects of the sun appears thinner than young skin, but with fewer deep wrinkles, "age spots," or other areas of irregular pigmentation. Skin that has sustained years of photodamage has more and deeper wrinkles, and appears coarser and yellower, with many areas of irregular pigmentation. The elastic fibers in the dermis are thickened and tangled, with significantly less collagen (a fibrous protein) than skin that has aged without photodamage.

To everyone except sun worshipers, this is good news. While you can do very little (even with cosmetics) to turn back the biological clock, you can do a great deal to prevent sun exposure and photodamage.

What Can Cosmetics Do?

A number of products claim to reverse epidermal damage associated with too much sun, especially the coarser appearance, irregular pigmentation, dryness, and fine lines seen in aging skin. But don't believe everything you read in the ad or on the label: skin-care products have few labeling or advertising restrictions. Some of these products do contain ingredients that research has shown to be somewhat helpful for some people. Many brands, however, contain too little of the ingredient to be effective. A number of these products are extremely expensive. And once use of the product is discontinued, the benefits usually vanish.

A few substances have been found to improve the skin's appearance, at least when it examined under a microscope. Tretinoin (Retin-A) appears to be the most potent product for reversing photodamage. Microscopic examination of skin biopsies suggests that tretinoin helps normalize the epidermis, restore the collagen and elastic fibers in the dermis, and improve the blood flow between skin layers. Because tretinoin has a number of side-effects, it is available only by prescription. Over-the-counter products contain milder substances, such as alphahydroxy acids and other acids, which work by irritating the epidermis. This speeds the exfoliation of excess dead epidermal cells, giving the skin a smoother, moister appearance. Some products contain antioxidants, which are intended to interfere with the production of harmful chemicals known as free radicals, produced by photodamage and other natural processes. While clinical tests of over-the-counter antioxidant products show some promise, it is not yet clear whether they are effective. Moisturizers do not reverse the aging process, but they temporarily trap water in the skin, which improves the skin's appearance.

Prevent Premature Aging and Photodamage

Because the restorative effects of cosmetics appear to be limited, the most important actions you can take to reverse photodamage and slow the aging of your skin are to protect it from the sun and, if you smoke, to quit (smoking accelerates skin aging). Research suggests that protecting skin from the sun can heal some of the photodamage that has occurred over the years. Staying out of the sun and covering the body when outdoors remain the best protection against UV exposure. A topical sunscreen that shields you from both UVB and UVA exposure helps protect uncovered body parts.

critical thinking

Sunscreens may not protect against melanoma and, in fact, may lead to a false sense of security by protecting against sunburn and this exposing skin to the sun's rays for extended periods of time. How can you protect skin without relying on sunscreens?

List other recommendations you have heard for enjoying the outdoors while minimizing photodamage.

Common Disorders

Burns

Tissue damage from excessive heat, electricity, radioactivity, or corrosive chemicals that destroy the protein in cells is called a *burn.* Burns disrupt homeostasis because they destroy the protection afforded by the skin. They permit microbial invasion and infection, loss of fluid, and loss of regulation of body temperature.

A *first-degree burn* involves only the surface epidermis. It is characterized by mild pain and redness, but no blisters. A *second-degree burn* involves the entire epidermis and possibly part of the dermis. Some skin functions are lost. Redness, blister formation, edema, and pain are characteristic of such a burn. *A third-degree burn* destroys the epidermis, dermis, and the epidermal organs, and skin functions are lost.

Acne

Acne is an inflammation of sebaceous (oil) glands that usually begins at puberty when the glands grow in size and increase production of sebum.

Pressure Sores

Pressure sores are also called *decubitus* (dē-KYOO-bi-tus) *ulcers* and *bedsores.* They are caused by constant deficiency of blood to tissues over a bony projection that has been subjected to prolonged pressure against an object like a bed, cast, or splint.

Skin Cancer

[handwritten: massage therapists may be the 1st to notice this on clients back in particular]

Excessive sun exposure can result in *skin cancer,* the most common cancer in Caucasian people. There are three common forms of skin cancer. *Basal cell carcinomas (BCCs)* account for over 75 percent of all skin cancers. The tumors arise from the epidermis and rarely spread. *Squamous cell carcinomas (SCCs)* also arise from the epidermis but are less common than BCCs and have a variable tendency to spread. Most SCCs arise from preexisting lesions on sun-exposed skin. *Malignant melanomas* arise from melanocytes and are the leading cause of death from all skin diseases because they spread rapidly. Malignant melanomas account for only about 3 percent of all skin cancers. *[handwritten: least common + most deadly]*

Sunburn

Sunburn is injury to the skin as a result of acute, prolonged exposure to the ultraviolet (UV) rays of sunlight. ∎

Medical Terminology and Conditions

Abrasion (a-BRĀ-shun; *ab* = away; *rasion* = scraped) A portion of the skin that has been scraped away.

Athlete's (ATH-lēts) *foot* A superficial fungus infection of the skin of the foot.

Cold sore (KŌLD sor) A lesion, usually in the oral mucous membrane, caused by type 1 herpes simplex virus (HSV), transmitted by oral or respiratory routes. Triggering factors include ultraviolet (UV) radiation, hormonal changes, and emotional stress. Also called a *fever blister.*

Corn (KORN) A painful thickening of the skin that may be hard or soft, depending on location. Hard corns are usually found over toe joints, and soft corns are usually found between the fourth and fifth toes.

Cyst (SIST; *cyst* = sac containing fluid) A sac with a distinct connective tissue wall, containing a fluid or other material.

Dermabrasion (der-ma-BRĀ-shun; *derm* = skin) Removal of acne, scars, tattoos, or moles by sandpaper or a high-speed brush.

Dermatitis (der-ma-TĪ-tis) Inflammation of the skin.

Hemangioma (hē-man′-jē-Ō-ma; *hemo* = blood; *angio* = blood vessel; *oma* = tumor) Localized tumor of the skin and subcutaneous layer that results from an abnormal increase in blood vessels; one type is a *port-wine stain,* a flat, pink, red, or purple lesion present at birth, usually at the nape of the neck.

Hives (HĪVZ) Condition of the skin marked by reddened elevated patches that are often itchy. Most commonly caused by infections, physical trauma, medications, emotional stress, food additives, and certain foods.

Hypodermic (hī-pō-DER-mik; *hypo* = under) Relating to the area beneath the skin. Also called *subcutaneous.*

Impetigo (im′-pe-TĪ-go) Superficial skin infection caused by staphylococci or streptococci; most common in children.

Intradermal (in′-tra-DER-mal; *intra* = within) Within the skin. Also called *intracutaneous.*

Keratosis (ker′-a-TŌ-sis; *kera* = horn) Formation of a hardened growth of tissue.

Laceration (las′-er-Ā-shun; *lacerare* = to tear) Wound or irregular tear of the skin.

Nevus (NE-vus) A round, pigmented, flat, or raised skin area that may be present at birth or develop later. Varying in color from yellow-brown to black. Also called a *mole* or *birthmark.*

Pruritus (proo-RĪ-tus; *pruire* = to itch) Itching, one of the most common dermatological disorders. It may be caused by skin disorders (infections), systemic disorders (cancer, kidney failure), or psychogenic factors (emotional stress).

Topical (TOP-i-kal) Pertaining to a definite area; local. Also in reference to a medication, applied to the surface rather than ingested or injected.

Vitiligo (vit-i-LĪ-gō) Partial or complete loss of melanocytes from patches of skin that results in irregular white spots.

Wart (WORT) Mass produced by uncontrolled growth of epithelial skin cells; caused by a virus (papillomavirus). Most warts are noncancerous.

Study Outline   *Know for Boards*

Skin (p. 85)

1. The skin and its accessory organs (hair, glands, and nails) constitute the integumentary system.

2. The principal parts of the skin are the outer epidermis and inner dermis. The dermis overlies the subcutaneous layer.

3. Following are the functions of the skin: regulation of body temperature, protection, sensation, excretion, immunity, and synthesis of vitamin D.

4. Epidermal cells include keratinocytes, melanocytes, Langerhans cells, and Merkel cells. The epidermal layers, from deepest to most superficial, are the strata basale, spinosum, granulosum, lucidum, and corneum. The basale undergoes continuous cell division and produces all other layers.

5. The dermis consists of two regions. The upper region is areolar connective tissue containing blood vessels, nerves, hair follicles, dermal papillae, and corpuscles of touch (Meissner's corpuscles). The lower region is dense, irregularly arranged connective tissue containing adipose tissue, hair follicles, nerves, sebaceous (oil) glands, and ducts of sudoriferous (sweat) glands.

6. Skin color is due to melanin, carotene, and hemoglobin.

Accessory Organs of the Skin (p. 88)

1. Accessory organs of the skin are structures developed from the epidermis of an embryo.

2. They include hair, skin glands (sebaceous, sudoriferous, and ceruminous), and nails.

Hair (p. 88)

1. Hairs are threads of fused, dead keratinized cells that function in protection.

2. Hairs consist of a shaft above the surface, a root that penetrates the dermis and subcutaneous layer, and a hair follicle.

3. Associated with hairs is a bundle of smooth muscle called arrector pili.

4. New hairs develop from cell division of the matrix in the bulb; hair replacement and growth occur in a cyclic pattern that consists of a growth stage and a resting stage.

Glands (p. 88)

1. Sebaceous (oil) glands are usually connected to hair follicles; they are absent in the palms and soles. Sebaceous glands produce sebum, which moistens hairs and waterproofs the skin.

2. Sudoriferous (sweat) glands are divided into apocrine and eccrine. They produce perspiration, which carries small amounts of wastes to the surface and assists in maintaining body temperature.

3. Ceruminous glands are modified sudoriferous glands that secrete cerumen. They are found in the external auditory meatus.

Nails (p. 90)

1. Nails are hard, keratinized epidermal cells covering the terminal portions of the fingers and toes.

2. The principal parts of a nail are the body, free edge, root, lunula, cuticle, and matrix. Cell division of the matrix cells produces new nails.

Aging and the Integumentary System (p. 90)

1. Most effects of aging occur when an individual reaches the late forties.

2. Among the effects of aging are wrinkling, loss of subcutaneous fat, atrophy of sebaceous glands, and decrease in the number of melanocytes and Langerhans cells.

Homeostasis of Body Temperature (p. 91)

1. One of the functions of the skin is the regulation of normal body temperature of 37°C (98.6°F).

2. If environmental temperature is high, skin receptors sense the stimulus (heat) and generate impulses (input) that are transmitted to the brain (control center). The brain then sends impulses (output) to sweat glands and blood vessels (effectors) to produce perspiration and vasodilation. As the perspiration evaporates, the skin is cooled and the body temperature returns to normal.

3. The skin-cooling response is a negative feedback mechanism.

Self-Quiz

1. Skin coloration

   a. is due to melanin found in the subcutaneous layer   b. in Caucasians is due primarily to carotene   c. is related to apocrine glands   d. is stimulated by exposure to the sun   e. is produced by Merkel cells

2. Which layer of the epidermis is most superficial?

   a. stratum basale   b. stratum spinosum   c. stratum granulosum   d. stratum lucidum   e. stratum corneum

3. Which layer of the epidermis contains cells that multiply and cells that produce melanin?

   a. stratum basale   b. stratum spinosum   c. stratum granulosum   d. stratum lucidum   e. stratum corneum

4. When you have your hair cut, scissors are cutting through the hair

   a. follicle   b. root   c. bulb   d. papilla   e. shaft

5. A person with albinism has a defect in the production of

   a. carotene   b. keratin   c. collagen   d. cerumen   e. melanin

6. The epidermis is composed of _____ tissue.

   a. keratinized simple squamous epithelium   b. keratinized stratified squamous epithelium   c. dense irregular connective   d. areolar connective   e. stratified transitional epithelium

7. Which of the following statements about eccrine sweat glands is false?

   a. They are most numerous on the palms and the soles.   b. They help regulate body temperature.   c. They produce a viscous (sticky) secretion.   d. They function throughout life.   e. They terminate at pores at the skin's surface.

8. What tissue is the main type found in the lower dermis?

   a. dense, irregular connective   b. stratified squamous epithelium   c. smooth muscle   d. nervous   e. cartilage

9. Which of the following is NOT a function of skin?

   a. calcium production   b. vitamin D synthesis   c. protection   d. immunity   e. temperature regulation

10. Which of the following structures is NOT an accessory organ of the skin?

    a. dermal papillae   b. sudoriferous glands   c. sebaceous glands   d. ceruminous glands   e. nails

11. Which of the following is NOT true concerning hair?

    a. Hair growth cycles include a growth stage and a resting stage.   b. Normally, adults lose about 1000 hairs per day.   c. Hair color is due to melanin.   d. Sebaceous glands are associated with hair.   e. Contraction of the arrector pili muscles makes hair stand erect.

12. Sebaceous glands

    a. secrete an oily substance   b. are located on the palms and soles   c. are responsible for breaking out in a "cold sweat"   d. are involved in body temperature regulation   e. are found in the external auditory meatus

13. As keratinocytes in the epidermis are pushed toward the skin's surface, they

    a. begin to divide more rapidly   b. become more elastic   c. begin to die   d. lose their melanin   e. begin to assume a columnar shape

14. To produce vitamin D, the skin cells need to be exposed to

    a. calcium and phosphorus   b. ultraviolet light   c. heat   d. pressure   e. keratin

15. In order to prevent an unwanted hair from growing back, you must make sure which structure is destroyed?

    a. shaft   b. cuticle   c. lunula   d. matrix   e. arrector pili

16. The portion of the nail that is responsible for nail growth is the

    a. cuticle   b. nail matrix   c. lunula   d. nail body   e. nail root

17. Match the following:

    D a. Langerhans cell
    F b. Merkel cell
    E c. keratin
    G d. melanin
    H e. lamellated (Pacinian) corpuscles
    A f. cerumen
    C g. carotene
    B h. striae
    I i. corpuscles of touch (Meissner's) corpuscles

    A. ear wax
    B. silvery white streaks
    C. yellow-orange pigment
    D. function in immune responses
    E. waterproofing pigment of skin, hair
    F. touch receptor found in epidermis
    G. yellow to black pigment
    H. nerve endings sensitive to pressure
    I. touch receptors found in dermal papillae

Critical Thinking Applications

1. It's 92°F as you sit under the shade of a tree having some iced tea and a sandwich. What mechanisms does your body use to keep you from overheating? After lunch, it's time for your A & P class. The instructor (who's wearing a suit) sets the thermostat for 65°F "'cause it's hot out there." You're wearing shorts, a tank top, and sandals. Explain how your body will respond to the chilly environment.

2. Fair-skinned Brad has a new job as a lifeguard. He wants a gorgeous tan NOW so he does not wear any sun protection during his first 8-hour shift. Predict the result of Brad's action and explain what he should have done.

3. A new mother complains of excessive hair loss and white streaks on her abdomen. What would you tell her about the cause and possible duration of these problems?

4. A neighbor gave birth to a child who was diagnosed with albinism. How would you explain this condition to your other neighbors? What type of precautions should be taken when exposing this child to the sun? Why?

5. Fifteen-year-old Carole has a bad case of "blackheads." According to her Aunt Freida, Carole's skin problems are from too much late night TV, chicken nuggets, and ched-dar popcorn. Explain the real cause of blackheads to Aunt Freida.

6. An accident while slicing a bagel lands you in the emergency room. After examining the injury, the nurse states that you've cut through to the subQ layer. Discuss the structures and functions of the skin layers that you've exposed.

Answers to Figure Questions

5.1   Stratum corneum; stratum basale.
5.2   Langerhans cell.
5.3   To provide nourishment for growing hair.
5.4   They are composed of tightly packed, hard, keratinized epidermal cells.
5.5   The result of the effectors (lowering body temperature) is opposite to the initial stimulus (rising body temperature).

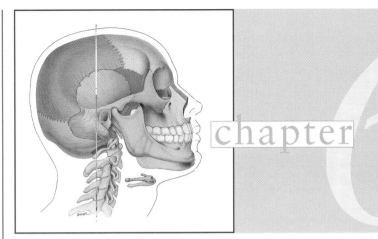

chapter 6

THE SKELETAL SYSTEM

a look ahead

The framework of bones and cartilage that protects our organs and allows us to move is called the *skeletal system.* Each bone in the skeletal system is an individual organ. Among the tissues associated with bones are bone tissue, cartilage, dense connective tissue, epithelium, blood, adipose tissue, and nervous tissue. The specialized branch of medicine that deals with the preservation and restoration of the skeletal system, articulations (joints), and associated structures is called *orthopedics* (or′-thō-PĒ-diks; *ortho* = correct or straighten; *pais* = child).

Functions

objective: *Discuss the functions of the skeletal system.*

The skeletal system performs the following functions:

1. **Support.** The skeleton provides a framework for the body and, as such, supports soft tissues and provides points of attachment for skeletal muscles.

2. **Protection.** Internal organs are protected from injury by the skeleton. For example, the brain is protected by the cranial bones while the heart and lungs are protected by the ribs and sternum (breastbone).

3. **Movement.** Skeletal muscles are attached to bones. When muscles contract, they pull on bones and together they produce movement. This function is discussed in detail in Chapter 8.

4. **Mineral storage and homeostasis.** Bones store several minerals, especially calcium and phosphorus, that can be distributed to other parts of the body on demand. The homeostatic mechanism that deposits and removes calcium and phosphorus is discussed in detail on pp. 104.

5. **Site of blood cell production.** In certain bones, a connective tissue called red bone marrow produces blood cells, a process called *hemopoiesis* (hēm′-ō-poy-Ē-sis; *haimatos* = blood; *poiein* = to make). *Red bone marrow,* one type of bone marrow, consists of immature blood cells, adipose cells, and macrophages. It is found in developing bones and adult bones such as the pelvis, ribs, breastbone, backbones, skull, and ends of the arm bones and thigh bones. Hemopoiesis is discussed in detail in Chapter 14.

6. **Storage of energy.** Lipids stored in cells of another type of bone marrow called *yellow bone marrow* are an important chemical energy reserve. Yellow bone marrow is composed mostly of adipose tissue and a few blood cells.

Types of Bones

The bones of the body may be classified into four principal types on the basis of shape: long, short, flat, and irregular. *Long bones* have greater length than width and consist of (1) a main, cylindrical, central portion called a shaft or body and (2) extremities (ends). Long bones consist mostly of compact bone tissue (dense bone with few spaces) but also have considerable amounts of spongy bone tissue (bone with large spaces). The details of compact and spongy bone are discussed shortly. Long bones include bones of the thighs, legs, toes, arms, forearms, and fingers. Figure 6.1 shows the parts of a long bone.

Short bones are somewhat cube-shaped and nearly equal in length and width. They are spongy except at the surface, where there is a thin layer of compact bone. Short bones include the wrist and ankle bones.

Flat bones are generally thin and are composed of two more or less parallel plates of compact bone enclosing a layer of spongy bone. Flat bones afford considerable protection and provide extensive areas for muscle attachment. Flat bones include the cranial bones, sternum (breastbone), ribs, and scapulas (shoulder blades).

Irregular bones have complex shapes and cannot be grouped into any of the three categories just described. They also vary in the amount of spongy and compact bone present. Such bones include the vertebrae (backbones) and certain facial bones.

There are two additional types of bones not included in this classification by shape; they are classified by location. *Sutural* (SOO-chur-al) *bones,* also called *Wormian bones,* are small bones between the joints of certain cranial bones. Their number varies greatly from person to person. *Sesamoid bones* are small bones in tendons where considerable pressure develops, for instance, in the wrist. These, like sutural bones, also vary in number. Two sesamoid bones, the patellas (kneecaps), are present in all individuals.

Parts of a Long Bone

The gross structure of a long bone—in this case, the humerus (arm bone)—will be examined before considering the microscopic structure of bone. A typical long bone consists of the following parts (Figure 6.1):

1. *Diaphysis* (dī-AF-i-sis; *dia* = through; *physis* = growth). The shaft or body, the long, main, cylindrical portion of the bone.

Figure 6.1 Parts of a long bone. *Tend to be compact bones*
growth until 25 years of age

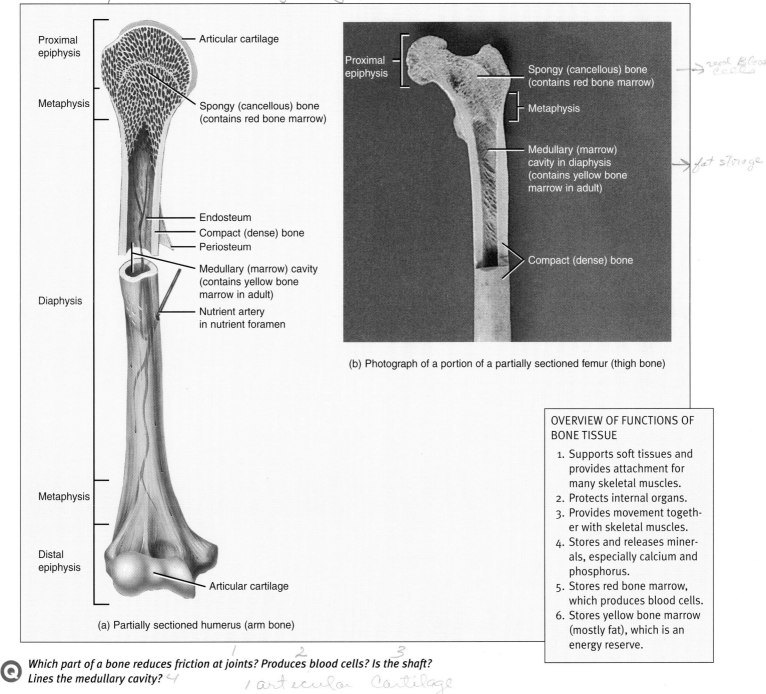

Proximal epiphysis

Articular cartilage

Metaphysis

Spongy (cancellous) bone (contains red bone marrow)

Endosteum
Compact (dense) bone
Periosteum

Diaphysis

Medullary (marrow) cavity (contains yellow bone marrow in adult)

Nutrient artery in nutrient foramen

Metaphysis

Distal epiphysis

Articular cartilage

(a) Partially sectioned humerus (arm bone)

Proximal epiphysis

Spongy (cancellous) bone (contains red bone marrow)   → *red blood cells*

Metaphysis

Medullary (marrow) cavity in diaphysis (contains yellow bone marrow in adult)   → *fat storage*

Compact (dense) bone

(b) Photograph of a portion of a partially sectioned femur (thigh bone)

OVERVIEW OF FUNCTIONS OF BONE TISSUE

1. Supports soft tissues and provides attachment for many skeletal muscles.
2. Protects internal organs.
3. Provides movement together with skeletal muscles.
4. Stores and releases minerals, especially calcium and phosphorus.
5. Stores red bone marrow, which produces blood cells.
6. Stores yellow bone marrow (mostly fat), which is an energy reserve.

Q *Which part of a bone reduces friction at joints? Produces blood cells? Is the shaft?*
Lines the medullary cavity?

1 articular cartilage
2 red bone marrow
3 Diaphysis
4 endosteum

2. *Epiphyses* (ē-PIF-i-sēz; *epi* = above). The extremities or ends of the bone. (Singular is *epiphysis*.)

3. *Metaphysis* (me-TAF-i-sis; *meta* = after). In mature bone, the region where the diaphysis joins the epiphysis. In a growing bone, the region that contains a layer of hyaline cartilage called the *epiphyseal* (*growth*) *plate* (site where bone growth in length occurs).

4. *Articular cartilage*. A thin layer of hyaline cartilage covering the epiphysis where the bone forms an articulation (joint) with another bone. The cartilage reduces friction and absorbs shock at freely movable joints.

5. *Periosteum* (per'-ē-OS-tē-um; *peri* = around; *osteon* = bone). The periosteum is a tough white fibrous membrane around the surface of the bone not covered by articular cartilage. It consists of dense, irregular connective tissue, blood vessels, and nerves that pass into the bone, and various types of bone cells. The periosteum is necessary for the protection, nutrition, growth in diameter, and repair of bones and is the site of attachment for ligaments and tendons.

6. *Medullary* (MED-yoo-lar'-ē; *medulla* = central part of a structure) or *marrow cavity*. The space within the diaphysis that contains the fatty yellow bone marrow in adults.

7. *Endosteum* (end-OS-tē-um; *endo* = within). The lining of the medullary cavity that consists of osteoprogenitor cells and osteoclasts (described shortly).

Histology

| objective: | *Describe the microscopic structure of compact and spongy bone tissue.*

The skeletal system consists of four types of connective tissue: cartilage, bone, bone marrow, and the periosteum. We described the microscopic structure of cartilage in Chapter 4. Here, we will discuss the microscopic structure of bone tissue.

Like other connective tissues, *bone* or *osseous* (OS-ē-us) *tissue* contains a great deal of matrix (intercellular substance) surrounding widely separated cells. The matrix consists of an inorganic component (mineral salts), which makes bone hard, and an organic component (mostly collagen fibers), which gives bone its strength. There are four types of cells in bone tissue (Figure 6.2a): osteoprogenitor (osteogenic) cells, osteoblasts, osteocytes, and osteoclasts. *Osteoprogenitor* (os'-tē-ō-prō-JEN-i-tor; *osteo* = bone; *pro* = precursor; *gen* = to produce) *cells* undergo mitosis to become osteoblasts. They are found in the periosteum, endosteum, and canals in bone that contain blood vessels. *Osteoblasts* (OS-tē-ō-blasts'; *blast* = germ or bud) are the cells that form bone, but they do not have the ability to divide by mitosis. They are found on the surfaces of bone.

Osteoblasts initially form collagen and other organic compounds needed to build bone, but once they become isolated in the bony matrix, they are called *osteocytes* (OS-tē-ō-sīts'; *cyte* = cell), or mature bone cells. They are the principal cells of bone tissue. Like osteoblasts, osteocytes have no mitotic potential. Osteocytes maintain daily cellular activities of bone tissue. *Osteoclasts* (OS-tē-ō-clasts'; *clast* = to break) are found on the surfaces of bone and function in bone resorption (destruction of matrix), which is important in the development, growth, maintenance, and repair of bone.

Unlike other connective tissues, the matrix of bone contains abundant mineral salts, primarily calcium phosphate and some calcium carbonate. As these salts are deposited by osteoblasts around the collagen fibers of the matrix, the tissue hardens. This hardening process is called *calcification*.

Bone is not completely solid. In fact, all bone has some spaces (sometimes only microscopic) between its hard components. The spaces provide channels for blood vessels that supply bone cells with nutrients. The spaces also make bones lighter. Depending on the size and location of the spaces, bone may be classified as compact or spongy (see Figure 6.1).

Compact Bone Tissue

Compact (*dense*) *bone tissue* contains few spaces. It forms the external layer of all bones and the bulk of the shaft of long bones. Compact bone tissue provides protection and support and helps long bones resist the stress of weight placed on them.

Note in Figure 6.2b that compact bone has a concentric-ring structure. Nutrient arteries and nerves from the periosteum penetrate compact bone through *perforating* (*Volkmann's*) *canals*. These blood vessels connect with blood vessels and nerves of the medullary cavity and those of the *central* (*Haversian*) *canals*. The central canals run lengthwise through the bone. Around the canals are *concentric lamellae* (la-MEL-ē)—rings of hard, calcified matrix. Between the lamellae are small spaces called *lacunae* (la-KOO-nē; *lacuna* = little lake), which contain osteocytes. Projecting outward in all directions from the lacunae are minute canals called *canaliculi* (kan'-a-LIK-yoo-lī), which contain slender processes of osteocytes. The canaliculi connect with those of other lacunae and, eventually, with the central canals. Thus an intricate branching network of canaliculi is formed throughout the bone to provide numerous interconnected routes for nutrients and oxygen to reach osteocytes and remove wastes. Each central canal, with its surrounding lamellae, lacunae, osteocytes, and canaliculi, is called an *osteon* (*Haversian system*).

Spongy Bone Tissue

In contrast to compact bone, *spongy* (*cancellous*) *bone tissue* usually does not contain true osteons. It consists of an irregular latticework of thin plates of bone called *trabeculae* (tra-BEK-yoo-lē). See Figure 6.2b. The macroscopic spaces between

Figure 6.2 Histology of bone. A photomicrograph of compact bone tissue is provided in Exhibit 4.2 on p. 75.

🔑 *Osteocytes lie in lacunae arranged in concentric circles around a central canal in compact bone tissue and in lacunae arranged irregularly in trabeculae of spongy bone tissue.*

[handwritten: Matrix of Bone 1/3 protein fibers (Collagen) 2/3 Calcium phosphate crystals Ca₃(Po4)₂ 2% of mass of a bone are the cells themselves]

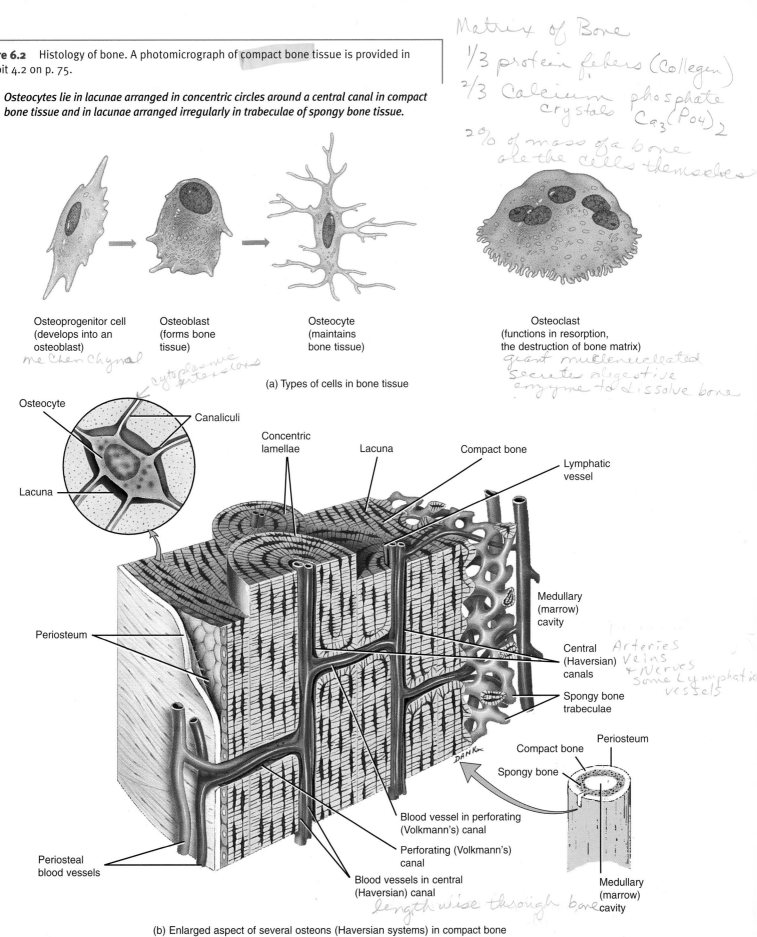

Osteoprogenitor cell (develops into an osteoblast) *[handwritten: mesenchymal]*

Osteoblast (forms bone tissue)

Osteocyte (maintains bone tissue) *[handwritten: cytoplasmic extensions]*

Osteoclast (functions in resorption, the destruction of bone matrix) *[handwritten: giant multinucleated secrete digestive enzyme to dissolve bone]*

(a) Types of cells in bone tissue

Osteocyte

Canaliculi

Lacuna

Concentric lamellae

Lacuna

Compact bone

Lymphatic vessel

Medullary (marrow) cavity

Central (Haversian) canals *[handwritten: Arteries Veins & Nerves some Lymphatic vessels]*

Spongy bone trabeculae

Periosteum

Periosteal blood vessels

Blood vessel in perforating (Volkmann's) canal

Perforating (Volkmann's) canal

Blood vessels in central (Haversian) canal

Compact bone

Periosteum

Spongy bone

Medullary (marrow) cavity

[handwritten: lengthwise through bone]

(b) Enlarged aspect of several osteons (Haversian systems) in compact bone

Ⓠ *As people age, some central (Haversian) canals may become blocked. What effect would this have on an osteocyte?*

the trabeculae of some bones are filled with red bone marrow. Within the trabeculae lie lacunae, which contain osteocytes. Blood vessels from the periosteum penetrate through to the spongy bone, and osteocytes in the trabeculae are nourished directly from the blood circulating through the marrow cavities.

Spongy bone tissue makes up most of the bone tissue of short, flat, and irregularly shaped bones and most of the epiphyses of long bones. Spongy bone tissue in the hipbones, ribs, breastbone (sternum), backbones (vertebrae), skull, and ends of some long bones is the only site of red bone marrow storage and thus hemopoiesis in adults.

Most people think of all bone as a very hard, rigid material. Yet the bones of an infant are quite soft and become rigid only after growth stops during late adolescence. Even then, bone is constantly broken down and rebuilt. It is a dynamic, living tissue. Let us now see how bones are formed and how they grow.

Ossification: Bone Formation

objective: *Explain the steps involved in ossification (bone formation).*

The process by which bone forms is called *ossification* (os′-i-fi-KĀ-shun; *facere* = to make). The "skeleton" of a human embryo is composed of fibrous connective tissue membranes or hyaline cartilage. Both are shaped like bones and provide the sites for ossification. Ossification begins around the sixth or seventh week of embryonic life and continues throughout adulthood.

Two methods of bone formation occur. The first is called *intramembranous* (in′-tra-MEM-bra-nus; *intra* = within; *membranous* = membrane) *ossification.* This refers to the formation of bone directly on or within fibrous connective tissue membranes. The second kind, *endochondral* (en′-dō-KON-dral; *endo* = within; *chondro* = cartilage) *ossification,* refers to the formation of bone within a cartilage model. These two methods of ossification do *not* lead to differences in the structure of mature bones. They are simply different methods of bone development. Both mechanisms involve the replacement of a preexisting connective tissue with bone. The "soft spots" that help the fetal skull pass through the birth canal are replaced by bone through intramembranous ossification.

The first stage in the development of bone is the appearance of osteoprogenitor cells that undergo mitosis to produce osteoblasts, which will produce bony matrix by either intramembranous or endochondral ossification.

Intramembranous Ossification

Of the two methods of bone formation, *intramembranous ossification* is easier to understand. The flat bones of the skull, the mandible (lower jawbone), and the clavicles (collarbones) are formed in this way. The process occurs as follows (Figure 6.3):

1. At the site where bone will develop, cells in mesenchyme (recall that *mesenchyme* is the tissue from which all other connective tissues eventually arise) come together and develop into osteoprogenitor cells and then into osteoblasts. This site is called a *center of ossification.* Osteoblasts secrete the organic matrix of bone until they are completely surrounded by it.

2. Then secretion of matrix stops and the cells, now called osteocytes, are in lacunae and extend narrow cytoplasmic processes into canaliculi. Within a few days, calcium and other mineral salts are deposited and the matrix hardens (calcification).

3. As the bone matrix forms, it develops into *trabeculae* that fuse with one another to form spongy bone. The spaces between trabeculae fill with red bone marrow. On the outside of the bone, the mesenchyme condenses.

4. The mesenchyme develops into the periosteum. Eventually, most surface layers of the spongy bone are replaced by compact bone, but spongy bone remains in the center of the bone. Much of this newly formed bone will be remodeled (destroyed and reformed) so the bone may reach its final adult size and shape.

Endochondral Ossification

The replacement of cartilage by bone is called *endochondral ossification.* Most bones of the body are formed this way, but the process is best observed in a long bone. It occurs as follows (Figure 6.4 on page 105):

red bone marrow forms white & red blood cells
yellow marrow form adipose (fat)

Figure 6.3   Intramembranous ossification.

Intramembranous ossification involves the formation of bone directly on or within loose fibrous connective tissue membranes.

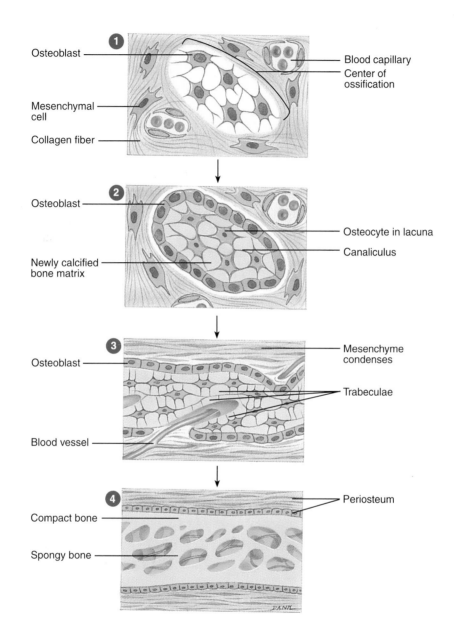

Which bones of the body develop by intramembranous ossification?

1 **Development of the cartilage model.** At the site where the bone will develop, cells in mesenchyme come together in the shape of the future bone. The cells then develop into cartilage producing cells that change the model into hyaline cartilage. In addition, a membrane called the *perichondrium* (per-i-KON-drē-um) develops around the cartilage.

2 **Growth of the cartilage model.** The cartilage model grows in length and thickness. As the cartilage model continues to grow, cartilage cells in its midregion bring about chemical changes that trigger calcification. Once the cartilage becomes calcified, other cartilage cells die because nutrients no longer diffuse quickly enough through the matrix. The lacunae of the cells that have died are now empty, and the thin partitions between them break down forming small cavities. In the meantime, a nutrient artery penetrates the bone through a hole (nutrient foramen) in the midregion of the model. This stimulates osteoprogenitor cells in the perichondrium to develop into osteoblasts. The cells produce a thin layer of compact bone under the perichondrium. Once the perichondrium starts to form bone, it is known as the *periosteum.*

3 **Development of the primary ossification center.** Near the middle of the model, capillaries of the periosteum grow into the disintegrating calcified cartilage. The capillaries stimulate growth of a *primary ossification center,* a region where bone tissue will replace most of the cartilage. Osteoblasts then begin to deposit bone matrix over the remnants of calcified cartilage, forming spongy bone trabeculae. As the ossification center enlarges toward the ends of the bone, osteoclasts break down the newly formed spongy bone trabeculae. This activity leaves a cavity, the medullary (marrow) cavity, in the core of the model. The cavity then fills with red bone marrow.

4 **Development of the diaphysis and epiphysis.** The diaphysis (shaft), which was once a solid mass of hyaline cartilage, is replaced by compact bone, the core of which contains a red bone marrow-filled medullary cavity. When blood vessels (epiphyseal arteries) enter the epiphyses, *secondary ossification centers* develop, usually around the time of birth.

5 In the secondary ossification centers, bone formation is similar to that in the primary ossification centers. One difference, however, is that spongy bone remains in the interior of the epiphyses (no medullary cavities are formed in the epiphyses). Also, hyaline cartilage remains covering the epiphyses as the articular cartilage and between the diaphysis and epiphysis as the *epiphyseal plate,* which is responsible for the lengthwise growth of long bones.

The epiphyseal plate allows the diaphysis of the bone to increase in length until early adulthood. The rate of growth is controlled by hormones such as human growth hormone (hGH). The epiphyseal plate cartilage cells stop dividing and the cartilage is eventually replaced by bone. The new structure is called the *epiphyseal line.* With the appearance of the epiphyseal line, bone growth in length stops.

Growth in diameter occurs along with growth in length. In this process, the bone lining the medullary cavity is destroyed by osteoclasts in the endosteum so that the cavity increases in diameter. At the same time, osteoblasts from the periosteum add new bone tissue around the outer surface of the bone.

Homeostasis

| objective: | *Describe the factors involved in bone growth and maintenance.*

Bone Growth and Maintenance

Bone, like skin, continually replaces itself throughout adult life. *Remodeling* is the replacement of old bone tissue by new bone tissue. Compact bone is formed by the transformation of spongy bone. The diameter of a long bone is increased by the destruction of bone internally and the construction of new bone externally. Even after bones have reached their adult shapes and sizes, old bone is continually destroyed and new bone is formed in its place. Bone is never metabolically at rest; it constantly remodels. Remodeling removes worn and injured bone, replacing it with new tissue. It also allows bone to serve as the body's storage area for calcium (to be discussed shortly). Many other tissues in the body need calcium to perform their functions. The blood continually trades off calcium with the bones, removing calcium when it and other tissues are not receiving enough and resupplying the bones to keep them from losing mass.

Osteoclasts are responsible for the resorption (destruction of matrix) of bone tissue. A delicate homeostasis is maintained between the action of the osteoclasts in removing calcium and collagen and the action of the bone-making osteoblasts in depositing calcium and collagen. Should too much new tissue be formed, the bones become abnormally thick and heavy. If too much calcium is deposited, the surplus may form thick bumps (spurs) that interfere with movement at joints. A loss of too much tissue or calcium makes bones breakable or too flexible.

Normal bone growth in the young, bone remodeling in the adult, and repair of fractured bone depend on (1) adequate minerals, most importantly calcium, phosphorus, and magnesium; (2) vitamins A, C, and D; (3) several hormones, most importantly human growth hormone, sex hormones (estrogens and testosterone), insulin, insulin-like growth factors, thyroid hormones, calcitonin, and parathyroid hormone; and (4) exercise that places stress on bones (weight-bearing activities). See Exhibit 6.1 on p. 107.

Bone and Mineral Homeostasis

Bones contain more calcium than any other organs, but calcium is also needed for muscle and nerve function and blood clotting. Adequate levels of calcium are needed in the blood but excessive amounts may also be harmful, and the bones serve as useful storage sites when blood calcium increases beyond its normal homeostatic level. Bones also store more phosphate (phosphorus) than any other organs. Appropriate levels of phosphate are required to produce nucleic acids (DNA and RNA) and ATP.

Figure 6.4 Endochondral ossification of the tibia (shinbone).

🔑 *Endochondral ossification involves the replacement of cartilage by bone.*

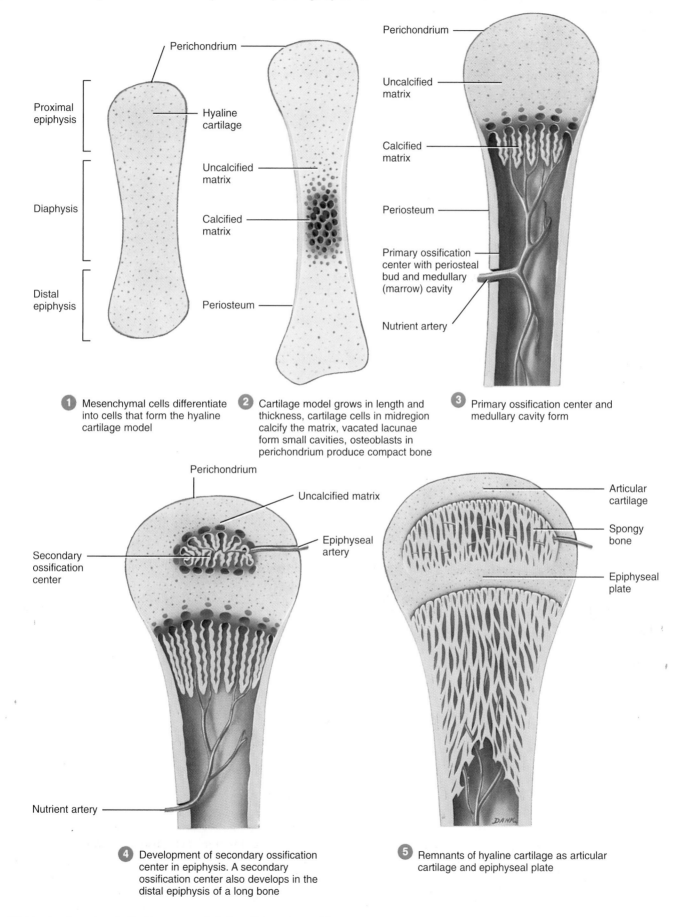

Proximal epiphysis

Diaphysis

Distal epiphysis

Perichondrium

Hyaline cartilage

Uncalcified matrix

Calcified matrix

Periosteum

Perichondrium

Uncalcified matrix

Calcified matrix

Periosteum

Primary ossification center with periosteal bud and medullary (marrow) cavity

Nutrient artery

1 Mesenchymal cells differentiate into cells that form the hyaline cartilage model

2 Cartilage model grows in length and thickness, cartilage cells in midregion calcify the matrix, vacated lacunae form small cavities, osteoblasts in perichondrium produce compact bone

3 Primary ossification center and medullary cavity form

Perichondrium

Uncalcified matrix

Secondary ossification center

Epiphyseal artery

Nutrient artery

Articular cartilage

Spongy bone

Epiphyseal plate

4 Development of secondary ossification center in epiphysis. A secondary ossification center also develops in the distal epiphysis of a long bone

5 Remnants of hyaline cartilage as articular cartilage and epiphyseal plate

❓ *Which structure signals that bone growth in length has stopped?*

inhibits osteoblasts
stemulates osteoclasts
Causes bone to be dissolved

The most important hormone that regulates Ca²⁺ exchange between bone and blood is *parathyroid hormone* (*PTH*), secreted by the parathyroid glands. Negative feedback systems adjust blood Ca²⁺ concentration (controlled condition). Look at the negative feedback cycle in Figure 6.5. If some stimulus causes blood Ca²⁺ level to fall, parathyroid gland cells (receptors) detect this change. The control center is the gene for PTH within the nucleus of a parathyroid gland cell. One input signal to the control center is an increased level of a molecule known as cyclic AMP (adenosine monophosphate) in the cytosol. Cyclic AMP accelerates reactions that "turn on" the PTH gene, PTH synthesis speeds up, and more PTH (output) is released into the blood. PTH increases the number and activity of osteoclasts (effectors), which step up the pace of bone resorption. The resulting release of Ca²⁺ (and phosphate ions) from bone into blood (response) returns the blood Ca²⁺ level to normal.

PTH also affects the kidneys. It promotes (1) recovery of Ca²⁺ so it is not lost in the urine, (2) elimination of phosphate in the urine, and (3) formation of vitamin D. These kidney effects of PTH all elevate blood Ca²⁺ concentration.

Another hormone also contributes to the homeostasis of blood Ca²⁺. *Calcitonin* (*CT*) is secreted by thyroid gland cells when blood Ca²⁺ rises above normal. It inhibits osteoclastic activity, speeds Ca²⁺ uptake by bone from blood, and accelerates Ca²⁺ deposit into bones. The net result is that calcitonin promotes bone formation and decreases blood Ca²⁺ level.

Figure 6.5   Negative feedback system for the regulation of blood calcium ion (Ca²⁺) concentration.

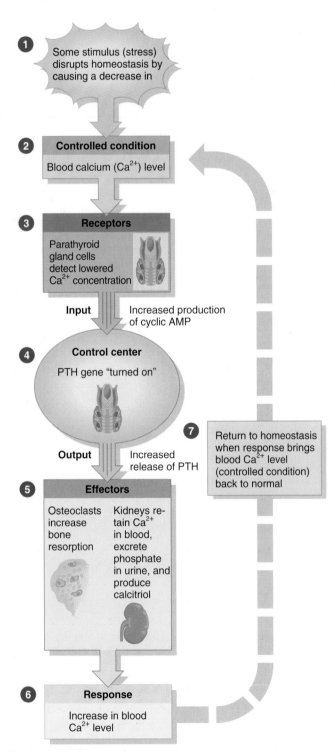

1. Some stimulus (stress) disrupts homeostasis by causing a decrease in

2. **Controlled condition**
   Blood calcium (Ca²⁺) level

3. **Receptors**
   Parathyroid gland cells detect lowered Ca²⁺ concentration

   Input   Increased production of cyclic AMP

4. **Control center**
   PTH gene "turned on"

   Output   Increased release of PTH

5. **Effectors**
   Osteoclasts increase bone resorption   Kidneys retain Ca²⁺ in blood, excrete phosphate in urine, and produce calcitriol

6. **Response**
   Increase in blood Ca²⁺ level

7. Return to homeostasis when response brings blood Ca²⁺ level (controlled condition) back to normal

Ⓠ *What body functions depend on proper levels of Ca²⁺?*

Exercise and Bone

Within limits, bone has the ability to alter its strength in response to mechanical stress. When placed under mechanical stress, bone tissue becomes stronger through increased deposition of mineral salts and production of collagen fibers. Another effect of stress is to increase the production of calcitonin, which inhibits bone reabsorption. Without mechanical stress, bone does not remodel normally because reabsorption outstrips bone formation. Removal of mechanical stress weakens the bone through *demineralization* (loss of bone minerals) and collagen reduction. The main mechanical stresses on bone are those that result from the pull of skeletal muscles caused by weight-bearing and exercise and the pull of gravity. If a person is bedridden or has a fractured bone in a cast, the strength of the unstressed bones diminishes. Astronauts subjected to the weightlessness of space also lose bone mass. In both cases, the bone loss can be dramatic, as much as 1 percent per week. Bones of athletes, which are repetitively and highly stressed, become notably thicker than those of nonathletes. Weight-bearing activities, such as walking or moderate weight lifting, help build and retain bone mass.

Exhibit 6.1 summarizes the factors that influence bone growth, remodeling, and repair.

Aging and Bone

With aging comes a loss of calcium from bones. In females, it begins after age 30, accelerates greatly around age 40 to 45 as levels of estrogens decrease, and continues until as much as 30 percent of bone calcium is lost by age 70. In males, calcium loss typically does not begin until after age 60. The loss of calcium from bones causes osteoporosis, which will be described shortly.

The second principal effect of aging on the skeletal system is a decrease in the rate of protein formation, which results in a decreased ability to produce the organic portion of bone matrix. Consequently, bone accumulates less organic matrix and more inorganic matrix. In some elderly individuals, this process causes their bones to become brittle and more susceptible to fracture.

Exhibit 6.1	Summary of Factors that Influence Growth, Remodeling, and Repair of Bone
FACTOR	**COMMENT**
Minerals	
Calcium and phosphorus	Make bone matrix hard.
Magnesium	Deficiency inhibits osteoblasts.
Vitamins	
Vitamin A	Controls activity, distribution, and coordination of osteoblasts and osteoclasts; toxic in high doses.
Vitamin C	Helps maintain bone matrix; deficiency leads to decreased collagen production, which inhibits bone growth and delays fracture repair.
Vitamin D	Active form is formed in the skin and kidneys from dietary precursor; helps build bone by increasing absorption of calcium from intestine into blood; may reduce the risk of osteoporosis but is toxic in high doses.
Hormones	
Human growth hormone (hGH)	Secreted by the anterior pituitary gland at base of brain; promotes general growth of all body tissues, including bone.
Sex hormones (several estrogens and testosterone)	Estrogens secreted by ovaries and testosterone secreted by testes increase bone-building activity of osteoblasts to promote bone growth; responsible for characteristic feminine and masculine skeletal differences.
Thyroid hormones (thyroxine and triiodothyronine)	Secreted by thyroid gland just below voice box; promote normal bone growth and maturity.
Calcitonin (CT)	Secreted by thyroid gland; promotes bone formation by inhibiting activity of osteoclasts, speeding up Ca^{2+} absorption from blood, and accelerating Ca^{2+} deposit in bones.
Parathyroid hormone (PTH)	Secreted by parathyroid glands attached to back surface of thyroid gland; promotes bone resorption by increasing the number and activity of osteoclasts; enhances recovery of Ca^{2+} from urine; promotes formation of the active form of vitamin D.
Exercise	Weight-bearing activities help build thicker, stronger bones and retard the loss of bone mass that occurs as people age.

Bone Surface Markings ✓

The surfaces of bones have various structural features adapted to specific functions. These features are called *bone surface markings.* Long bones that bear a great deal of weight have large, rounded ends that can form sturdy joints, for example. Other bones have depressions that receive the rounded ends.

Following are examples of bone surface markings:

I. *Depressions and Openings*

    A. *Foramen* (fō-RĀ-men; *foramen* = hole) An opening through which blood vessels, nerves, or ligaments pass, such as the foramen magnum of the occipital bone (see Figure 6.8).

    B. *Meatus* (mē-Ā-tus; *meatus* = canal) A tubelike passageway running within a bone, such as the external auditory meatus of the temporal bone (see Figure 6.7b).

    C. *Paranasal sinus* (*sinus* = cavity) An air-filled cavity within a bone connected to the nasal cavity, such as the frontal sinus in the frontal bone (see Figure 6.7c).

    D. *Fossa* (*fossa* = ditch, trench) A depression in or on a bone, such as the mandibular fossa of the temporal bone (see Figure 6.8).

II. *Processes that Form Joints*

    A. *Condyle* (KON-dīl; *condylus* = knucklelike process) A large, rounded prominence that forms a joint, such as the medial condyle of the femur (see Figure 6.28).

    B. *Head* A rounded projection that forms a joint and is supported on the constricted portion (neck) of a bone, such as the head of the femur (see Figure 6.28).

    C. *Facet* A smooth, flat surface such as the facet on a vertebra (see Figure 6.16).

III. *Processes to Which Tendons, Ligaments, and Other Connective Tissues Attach*

   A. *Tuberosity*   A large, rounded, usually roughened process, such as the deltoid tuberosity of the humerus (see Figure 6.22).

   B. *Spinous process* (*spine*)   A sharp, slender projection, such as the spinous process of a vertebra (see Figure 6.17).

   C. *Trochanter* (trō-KAN-ter)   A large, blunt projection found only on the femur, such as the greater trochanter (see Figure 6.28).

   D. *Crest*   A prominent border or ridge, such as the iliac crest of the hipbone (see Figure 6.25).

Divisions of the Skeletal System

objective: *Classify the bones of the body into axial and appendicular divisions.*

The adult human skeleton consists of 206 bones grouped as the *axial skeleton* and the *appendicular skeleton.* The axial skeleton consists of the bones of the skull, auditory ossicles (ear bones), hyoid bone, ribs, breastbone, and backbone. The appendicular skeleton consists of the bones of the upper and lower limbs (extremities), plus the bones called girdles, which connect the limbs to the axial skeleton. There are 80 bones in the axial skeleton and 126 in the appendicular (Exhibit 6.2). Refer to Figure 6.6 to see how the two divisions are joined to form the complete skeleton.

Skull

The *skull* rests on top of the vertebral column and is composed of two sets of bones: cranial bones and facial bones. The eight *cranial bones* enclose and protect the brain. They are the frontal bone (1), parietal bones (2), temporal bones (2), occipital bone (1), sphenoid bone (1), and ethmoid bone (1). There are 14 *facial bones:* nasal bones (2), maxillae (2), zygomatic bones (2), mandible (1), lacrimal bones (2), palatine bones (2), inferior nasal conchae (2), and vomer (1). Locate the skull bones in the various views of the skull (Figure 6.7 on pp. 110–111).

Sutures

A *suture* (SOO-chur; *sutura* = seam) is an immovable joint found only between skull bones (Figure 6.7). Sutures hold skull bones together. Examples are the following:

1. **Coronal** (kō-RŌ-nal; *corona* = crown) **suture.** Unites the frontal bone and the two parietal bones.

2. **Sagittal** (SAJ-i-tal; *sagitta* = arrow) **suture.** Unites the two parietal bones.

3. **Lambdoid** (LAM-doyd) **suture.** Unites the parietal bones and the occipital bone.

4. **Squamous** (SKWĀ-mus; *squama* = flat) **suture.** Unites the parietal bones and the temporal bones.

Exhibit 6.2	Divisions of the Adult Skeletal System

REGIONS OF THE SKELETON	NUMBER OF BONES
Axial Skeleton	
Skull	
Cranium	8
Face	14
Hyoid	1
Auditory ossicles 3 pair = 6	6
Vertebral column	26
Thorax	
Sternum	1
Ribs	24
Subtotal = 80	
Appendicular Skeleton	
Pectoral (shoulder) girdles	
Clavicle	2
Scapula	2
Upper limbs (extremities)	
Humerus	2
Ulna	2
Radius	2
Carpals	16
Metacarpals	10
Phalanges 14 ea. hand	28
Pelvic (hip) girdle = 3 but fused	
Hip, pelvic, or coxal bone	2
Lower limbs (extremities)	
Femur	2
Fibula	2
Tibia	2
Patella	2
Tarsals	14
Metatarsals	10
Phalanges	28
Subtotal = 126	
Total = 206	

total does not include pearmaid Bones

Board? 206 Bones in Body

Figure 6.6 Divisions of the skeletal system. The axial skeleton is indicated in blue. Note the position of the hyoid bone in Figure 6.7b.

The adult human skeleton consists of 206 bones grouped into axial and appendicular divisions.

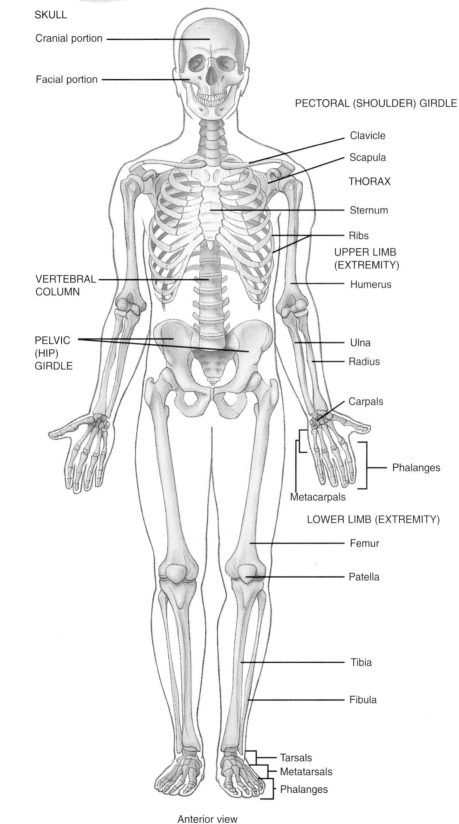

SKULL

Cranial portion

Facial portion

PECTORAL (SHOULDER) GIRDLE

Clavicle

Scapula

THORAX

Sternum

Ribs

UPPER LIMB (EXTREMITY)

Humerus

VERTEBRAL COLUMN

Ulna

Radius

PELVIC (HIP) GIRDLE

Carpals

Phalanges

Metacarpals

LOWER LIMB (EXTREMITY)

Femur

Patella

Tibia

Fibula

Tarsals

Metatarsals

Phalanges

Anterior view

Is each of the following bones part of the axial or the appendicular skeleton: skull, clavicle, vertebral column, pectoral girdle, humerus, pelvic girdle, and femur?

Figure 6.7 Skull. Although the hyoid bone is not part of the skull, it is indicated in (b) and (c) for reference.

🔑 *The skull consists of two sets of bones: cranial and facial.*

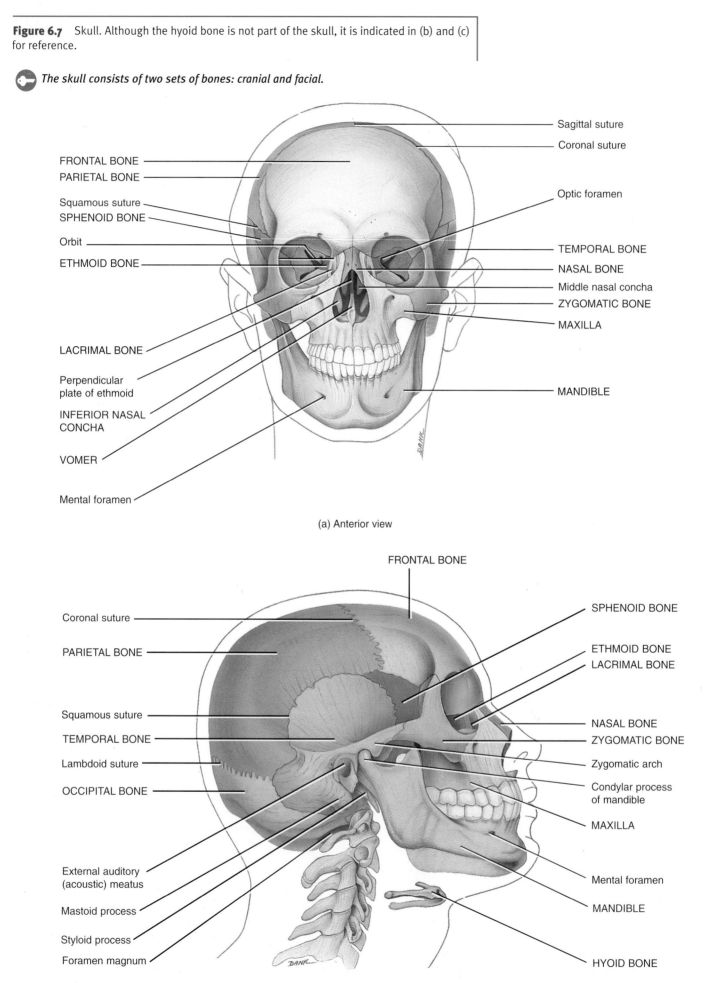

FRONTAL BONE
PARIETAL BONE
Squamous suture
SPHENOID BONE
Orbit
ETHMOID BONE

LACRIMAL BONE

Perpendicular plate of ethmoid

INFERIOR NASAL CONCHA

VOMER

Mental foramen

Sagittal suture
Coronal suture

Optic foramen

TEMPORAL BONE
NASAL BONE
Middle nasal concha
ZYGOMATIC BONE
MAXILLA

MANDIBLE

(a) Anterior view

FRONTAL BONE

Coronal suture

PARIETAL BONE

Squamous suture
TEMPORAL BONE
Lambdoid suture
OCCIPITAL BONE

External auditory (acoustic) meatus

Mastoid process

Styloid process
Foramen magnum

SPHENOID BONE

ETHMOID BONE
LACRIMAL BONE

NASAL BONE
ZYGOMATIC BONE
Zygomatic arch
Condylar process of mandible

MAXILLA

Mental foramen

MANDIBLE

HYOID BONE

(b) Right lateral view

Figure 6.7   (Continued)

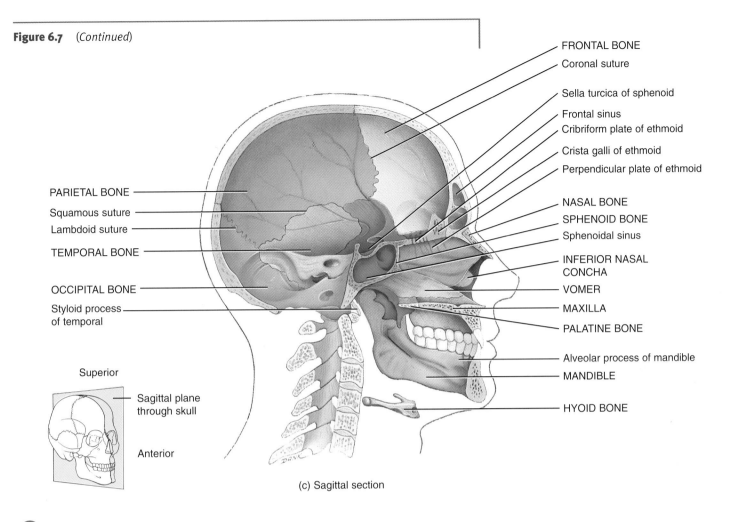

FRONTAL BONE
Coronal suture

Sella turcica of sphenoid
Frontal sinus
Cribriform plate of ethmoid
Crista galli of ethmoid
Perpendicular plate of ethmoid

NASAL BONE
SPHENOID BONE
Sphenoidal sinus

INFERIOR NASAL CONCHA
VOMER
MAXILLA
PALATINE BONE

Alveolar process of mandible
MANDIBLE

HYOID BONE

PARIETAL BONE
Squamous suture
Lambdoid suture
TEMPORAL BONE

OCCIPITAL BONE
Styloid process of temporal

Superior

Sagittal plane through skull

Anterior

(c) Sagittal section

Q *What are the names of the cranial bones?*

Cranial Bones

Frontal Bone

The *frontal bone* forms the forehead (the anterior part of the cranium), part of the *orbits* (eye sockets), and most of the anterior (front) part of the *cranial floor*. The *frontal sinuses* lie deep within the frontal bone (see Figure 6.7c). These mucous-membrane-lined cavities act as sound chambers that give the voice resonance. Other functions of the sinuses are given on p. 116.

Parietal Bones

The two *parietal bones* (pa-RĪ-e-tal; *paries* = wall) form the greater portion of the sides and roof of the cranial cavity (Figure 6.7).

Temporal Bones

The two *temporal bones* (*tempora* = temples) form the inferior (lower) sides of the cranium and part of the cranial floor.

In the lateral (side) view of the skull in Figure 6.7b, note that the temporal and zygomatic bones join to form the *zygomatic arch*. The *mandibular fossa* forms a joint with a

projection on the mandible (lower jawbone) called the condylar process to form the temporomandibular joint (TMJ). The mandibular fossa is seen best in Figure 6.8.

The *external auditory* (*acoustic*) *meatus* is the canal in the temporal bone that leads to the middle ear. The *mastoid* (*mastoid* = breast-shaped) *process* is a rounded projection of the temporal bone posterior to (behind) the external auditory meatus. It serves as a point of attachment for several neck muscles. The *styloid* (*stylos* = pillar) *process* projects downward from the undersurface of the temporal bone and serves as a point of attachment for muscles and ligaments of the tongue and neck. The *carotid foramen* is a hole through which the internal carotid artery passes.

Occipital Bone

The occipital (ok-SIP-i-tal; *occipital* = back of head) *bone* forms the posterior (back) part and a prominent portion of the base of the cranium (Figure 6.8).

The *foramen magnum* is the largest foramen in the skull. It is a large hole in the occipital bone through which the medulla oblongata (part of the brain that is continuous with the spinal cord) and the vertebral arteries and nerves pass.

Figure 6.8   Skull. The arrow in the inset indicates the direction from which the skull is viewed (inferior).

🔑 *A suture is an immovable joint between skull bones that holds them together.*

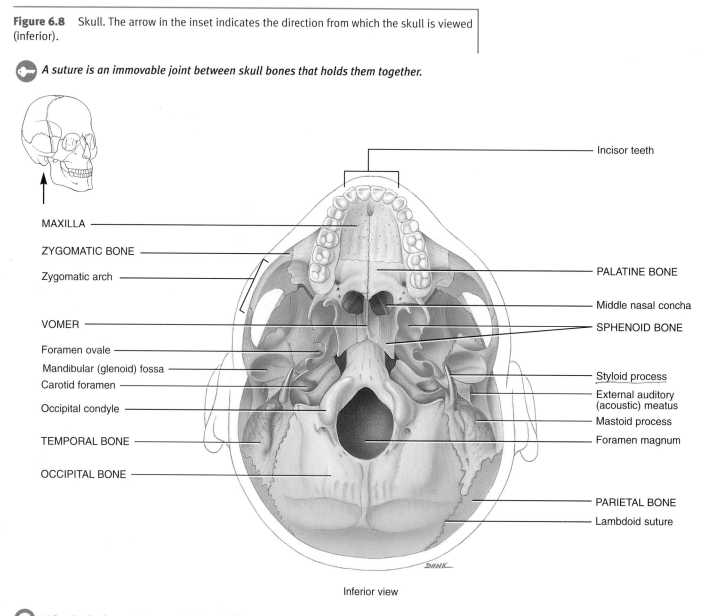

Inferior view

Q *What is the largest foramen in the skull?*

The *occipital condyles* are oval projections, one on either side of the foramen magnum, that articulate (form a joint) with the first cervical vertebra.

Sphenoid Bone

The *sphenoid* (SFĒ-noyd; *spheno* = wedge-shaped) *bone* is situated at the middle part of the base of the skull (Figure 6.9). This bone is referred to as the keystone of the cranial floor because it attaches to all the other cranial bones. The shape of the sphenoid is frequently described as a bat with outstretched wings.

The cubelike central portion of the sphenoid bone contains the *sphenoidal* (sfē-NOY-dal) *sinuses,* which drain into the nasal cavity (see Figure 6.12). On the superior surface of the sphenoid body is a depression called the *sella turcica* (SEL-a TUR-si-ka = Turk's saddle). This depression contains the pituitary gland. A branch of the trigeminal nerve (mandibular) passes through the

foramen ovale. The sphenoid bone also contains the *optic foramen,* through which the optic nerve passes.

Ethmoid Bone

The *ethmoid* (ETH-moid; *ethmoid* = sievelike) *bone* is a light, spongy bone located in the front part of the floor of the cranium between the orbits (Figure 6.10 on p. 114).

The ethmoid contains 3 to 18 air spaces, or "cells." It is from these "cells" that the bone derives its name. The ethmoid "cells" together form the *ethmoidal sinuses.* The sinuses are shown in Figure 6.12. The *perpendicular plate* forms the upper portion of the nasal septum (see Figure 6.11). The *cribriform* (KRIB-ri-form) *plate* forms the roof of the nasal cavity. It contains *olfactory foramina* through which the olfactory nerve passes. Projecting upward from the cribriform plate is a triangular process called the *crista galli* (= cock's comb), which

Test?   distinguish cranial Bones
        from facial Bones

Figure 6.9   Sphenoid bone. The arrow in the inset indicates the direction from which the skull
section is viewed (superior).

 *The sphenoid bone is called the keystone of the cranial floor because it articulates
with all other cranial bones, holding them together.*

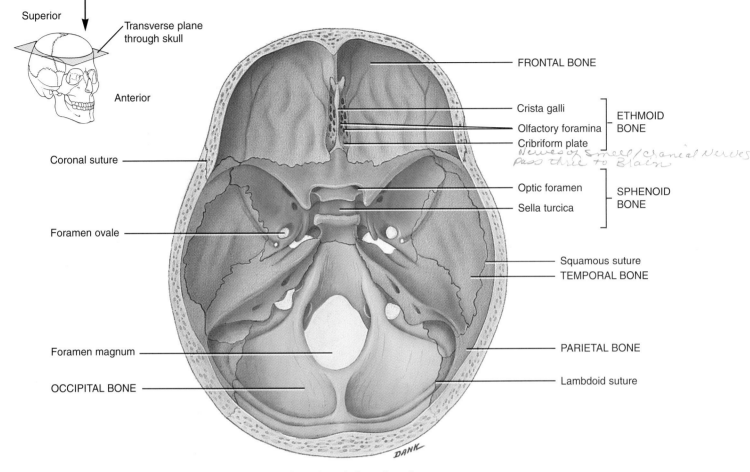

Viewed from above in floor of cranium

*Starting at the crista galli of the ethmoid bone and going in a clockwise direction,
name the bones that make contact with the sphenoid bone.*

serves as a point of attachment for the membranes (meninges) that cover the brain.

The ethmoid bone also contains two thin, scroll-shaped bones on either side of the nasal septum. These are called the *superior nasal concha* (KONG-ka; *concha* = shell) and the *middle nasal concha. Conchae* (KONG-kē) is the plural. The conchae filter air before it passes into the trachea, bronchi, and lungs.

Facial Bones

Nasal Bones

The paired *nasal bones* form part of the bridge of the nose (see Figure 6.7a). The lower portion of the nose, indeed the major portion, consists of cartilage.

Maxillae

The *maxillae* (mak-SIL-ē; *macera* = to chew) unite to form the upper jawbone and articulate with every bone of the face except the mandible, or lower jawbone (Figure 6.11 on p. 115).

Each maxillary bone contains a *maxillary sinus* that empties into the nasal cavity (see Figure 6.12). The *alveolar* (al-Vē-ō-lar; *alveolus* = hollow) *process* contains the *alveoli* (bony sockets) into which the maxillary (upper) teeth are set. The maxilla forms the anterior three-fourths of the hard palate. The left and right sides of the maxillary bones fuse before birth. If the fusion does not occur, a condition called *cleft palate* results. Incomplete fusion of the sides of the palatine bones may also occur (see Figure 6.8). A *cleft lip,* which involves a split upper lip, may also be associated with cleft palate. Depending on the extent

Figure 6.10   Ethmoid bone.

🔑 *The ethmoid bone is the major supporting structure of the nasal cavity.*

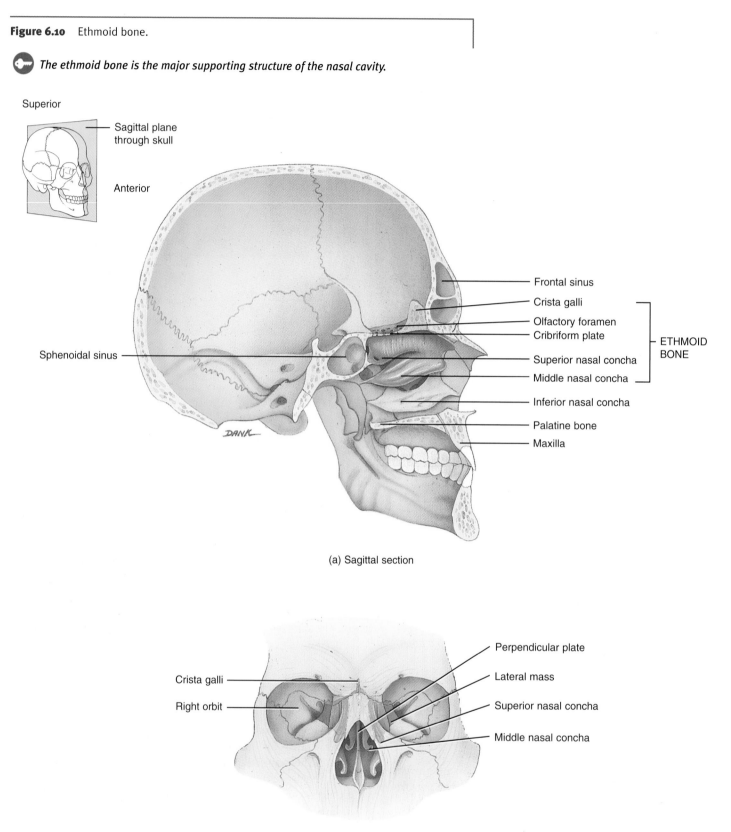

Superior

Sagittal plane through skull

Anterior

Frontal sinus

Crista galli

Olfactory foramen

Cribriform plate

ETHMOID BONE

Superior nasal concha

Middle nasal concha

Sphenoidal sinus

Inferior nasal concha

Palatine bone

Maxilla

DANK

(a) Sagittal section

Perpendicular plate

Crista galli

Lateral mass

Right orbit

Superior nasal concha

Middle nasal concha

(b) Anterior view of position of ethmoid bone in skull

❓ *What part of the ethmoid bone forms the top part of the nasal septum?*   perp. plate

Figure 6.11   Maxillae.

🔑 *The maxillae articulate with every bone of the face, except the mandible.*

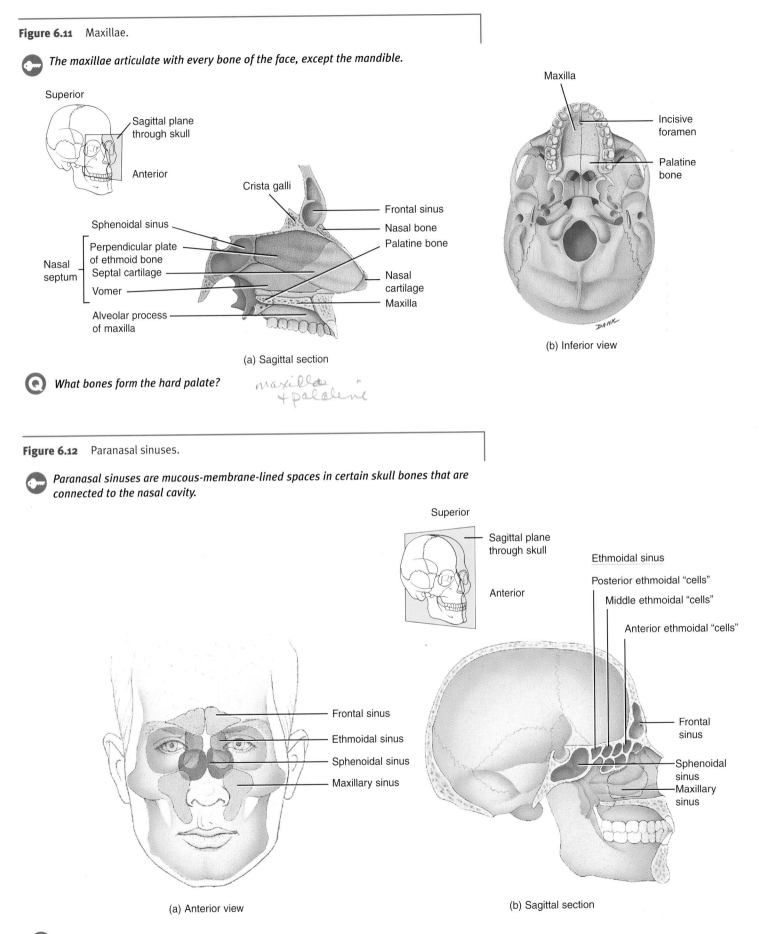

(a) Sagittal section

(b) Inferior view

Q *What bones form the hard palate?*   maxilla + palatine

Figure 6.12   Paranasal sinuses.

🔑 *Paranasal sinuses are mucous-membrane-lined spaces in certain skull bones that are connected to the nasal cavity.*

(a) Anterior view

(b) Sagittal section

Q *Name the two main functions of the paranasal sinuses.*

and position of the cleft, speech and swallowing may be affected. A surgical procedure can sometimes improve the condition.

Paranasal Sinuses

Paired cavities, called *paranasal* (*para* = beside) *sinuses,* are located in certain bones near the nasal cavity (Figure 6.12 on p. 115). The paranasal sinuses are lined with mucous membranes that are continuous with the lining of the nasal cavity. Cranial bones containing paranasal sinuses are the frontal, sphenoid, ethmoid, and maxillae. (The paranasal sinuses were described in the discussion of each of these bones.) Besides producing mucus, the paranasal sinuses serve as resonating chambers for sound as we speak.

Secretions of the mucous membrane of a paranasal sinus drain into the nasal cavity. An inflammation of the membrane due to an allergy or infection is called *sinusitis.* If the membrane swells enough to block drainage into the nasal cavity, fluid pressure builds up in the paranasal sinus and a sinus headache results.

Zygomatic Bones

The two *zygomatic bones,* commonly referred to as the cheekbones, form the prominences of the cheeks. Recall that the temporal and zygomatic bones unite to form the zygomatic arch (see Figure 6.7b).

Mandible

The *mandible* (*mandere* = to chew), or lower jawbone, is the largest, strongest facial bone. It is the only movable skull bone. The mandible has a *condylar* (KON-di-lar) *process* that forms a joint with the mandibular fossa of the temporal bone to form the temporomandibular joint (TMJ). See Figure 6.7b. The *alveolar process* is an arch containing the *alveoli* (sockets) for the mandibular (lower) teeth (see Figure 6.7c). The *mental* (*mentum* = chin) *foramen* is a hole in the mandible used as a landmark for dentists to inject anesthetics (see Figure 6.7a).

Lacrimal Bones

The paired *lacrimal* (LAK-ri-mal; *lacrima* = teardrop) *bones* are thin bones roughly resembling a fingernail in size and shape. They are the smallest bones of the face. They can be seen in the anterior and lateral views of the skull in Figure 6.7.

Palatine Bones

The two *palatine* (PAL-a-tīn) *bones* are L-shaped bones that form the posterior portion of the hard palate. These can be seen in Figure 6.8.

Inferior Nasal Conchae

Refer to the views of the skull in Figures 6.7a and 6.10. The two *inferior nasal conchae* (KONG-kē) are scroll-like bones that project into the nasal cavity inferior to the superior and middle nasal conchae of the ethmoid bone.

Vomer

The *vomer* (= plowshare) is a roughly triangular bone that forms the lower and back part of the nasal septum. It is clearly seen in the anterior view of the skull in Figure 6.7a and the inferior view in Figure 6.8.

The lower border of the vomer articulates with the cartilage septum that divides the nose into a right and left nostril. Its upper border articulates with the perpendicular plate of the ethmoid bone. Thus the structures that form the *nasal septum,* or partition, are the perpendicular plate of the ethmoid, the septal cartilage, and the vomer (see Figure 6.11a). A *deviated nasal septum* projects sideways from the midline of the nose. If the deviation is severe, it may entirely block the nasal passageway. Even an incomplete blockage may lead to nasal congestion, blocked paranasal sinuses, chronic sinusitis, headaches, or nosebleeds.

Fontanels

The "skeleton" of a newly formed embryo consists of cartilage or fibrous connective tissue membranes shaped like bones. Gradually, the cartilage or fibrous connective tissue membranes are replaced by bone, but at birth, *fontanels* (fon'-ta-NELZ; = little fountains) are still found between cranial bones (Figure 6.13). These "soft spots" are areas of fibrous connective tissue membranes that will eventually be replaced by bone through intramembranous ossification. They enable the fetal skull to compress as it passes through the birth canal and permit rapid growth of the brain during infancy. Although an infant may have many fontanels at birth, the form and location of six are fairly constant (Exhibit 6.3).

Foramina

The names and locations of certain *foramina* (singular is *foramen*) are included in the illustrations of the bones of the skull. In later chapters, when reference is made to the foramina, you can refer back to the illustrations to locate them.

Hyoid Bone

The single *hyoid* (*hyoedes* = U-shaped) *bone* is a unique component of the axial skeleton because it does not articulate with (attach to) any other bone. Rather, it is suspended from the styloid process of the temporal bone by ligaments and muscles. The hyoid bone is located in the neck between the mandible and larynx. Refer to Figure 6.7b to see the position of the hyoid bone. It supports the tongue and provides attachment for some of its muscles and muscles of the neck and pharynx (throat).

The hyoid bone is frequently fractured during strangulation. As a result, it is carefully examined in an autopsy when strangulation is suspected.

Figure 6.13   Fontanels of the skull at birth.

🔑 *Fontanels are membrane-filled spaces between cranial bones that are present at birth.*

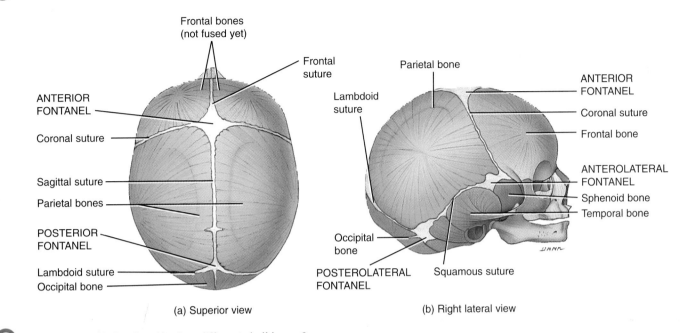

(a) Superior view

(b) Right lateral view

❓ *Which fontanel is bordered by four different skull bones?*

FONTANEL		LOCATION	DESCRIPTION
Anterior	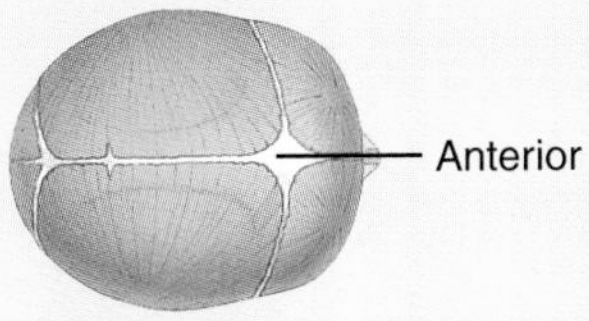 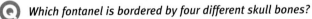Anterior	Between the two parietal bones and the frontal bone.	The largest of the six fontanels; usually closes 18–24 months after birth.
Posterior	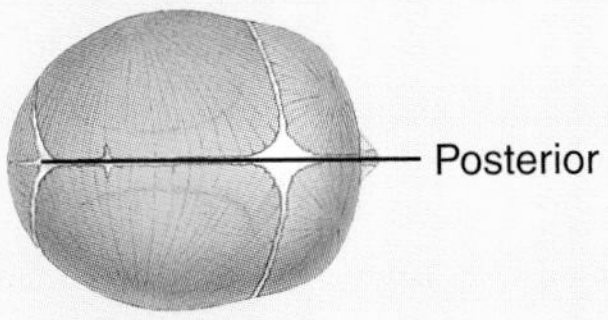 Posterior	Between the two parietal bones and the occipital bone.	Considerably smaller than the anterior fontanel; generally closes about 2 months after birth.
Anterolateral	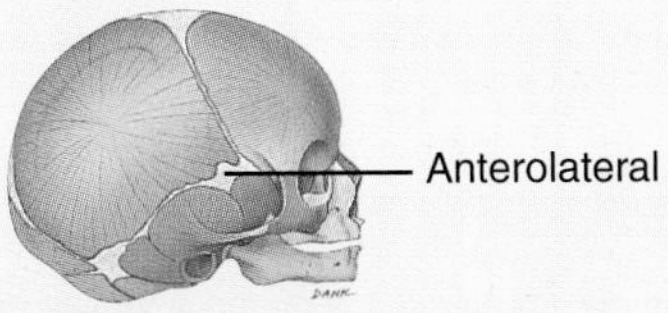 Anterolateral	One on each side of the skull between the frontal, parietal, temporal, and sphenoid bones.	Normally close about 3 months after birth.
Posterolateral	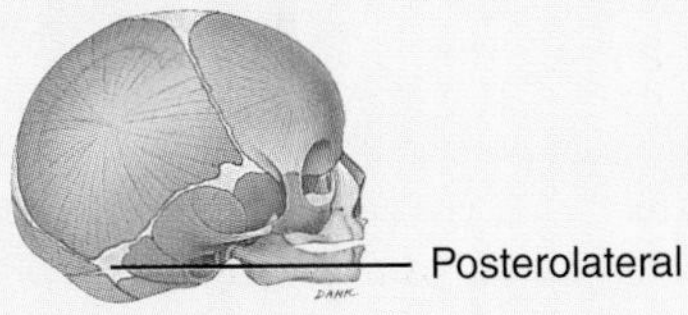 Posterolateral	One lies on each side of the skull between the parietal, occipital, and temporal bones.	Begin to close 1–2 months after birth, closure generally complete by 12 months.

Exhibit 6.3   Fontanels

Vertebral Column

objective: *Describe the structural and functional features of the vertebral column.*

Divisions

The **vertebral column** (**spine**) or **backbone** is composed of a series of bones called **vertebrae** (VER-te-brē; singular is **vertebra**). The vertebral column is a strong, flexible rod that moves forward, backward, and sideways. It encloses and protects the spinal cord, supports the head, and serves as a point of attachment for the ribs and the muscles of the back. Between vertebrae are openings called **intervertebral foramina.** The nerves that connect the spinal cord to various parts of the body pass through these openings.

The adult vertebral column contains 26 vertebrae (Figure 6.14): 7 **cervical** (*cervix* = neck) **vertebrae** in the neck region; 12

thoracic (*thorax* = chest) **vertebrae** posterior to the thoracic cavity; 5 **lumbar** (*lumbus* = loin) **vertebrae** supporting the lower back; 5 **sacral vertebrae** fused into one bone called the **sacrum** (SĀ-krum); and 4 **coccygeal** (kok-SIJ-ē-al) **vertebrae** fused into one bone called the **coccyx** (KOK-six). Prior to the fusion of the sacral and coccygeal vertebrae, the total number of vertebrae is 33.

Between adjacent vertebrae from the second vertebra (axis) to the sacrum are **intervertebral discs.** Each disc is composed of an outer layer consisting of fibrocartilage and an inner soft, pulpy, highly elastic structure. The discs form strong joints, permit various movements of the vertebral column, and absorb vertical shock.

Normal Curves

When viewed from the side, the vertebral column shows four **curves** (Figure 6.14a). Two of them, called cervical and lumbar curves, are convex (bulging out), relative to the front of the

Figure 6.14   Vertebral column.

🔑 *The adult vertebral column typically contains 26 vertebrae.*

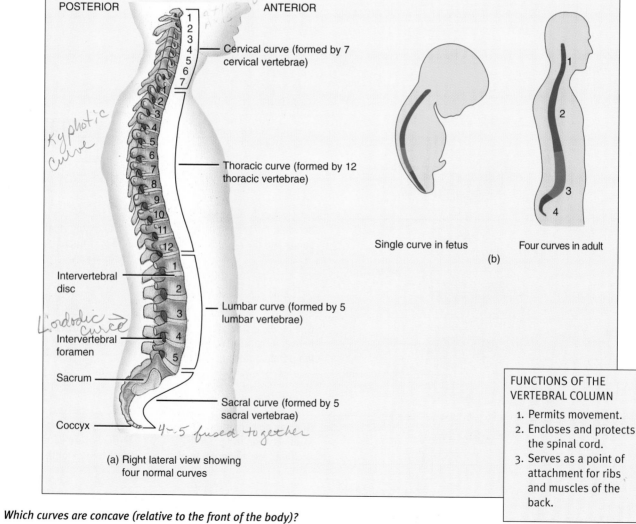

POSTERIOR   ANTERIOR

Cervical curve (formed by 7 cervical vertebrae)

Thoracic curve (formed by 12 thoracic vertebrae)

Intervertebral disc

Lumbar curve (formed by 5 lumbar vertebrae)

Intervertebral foramen

Sacrum

Lumbar curve (formed by 5 lumbar vertebrae)

Coccyx

Sacral curve (formed by 5 sacral vertebrae)

(a) Right lateral view showing four normal curves

Single curve in fetus   Four curves in adult

(b)

FUNCTIONS OF THE VERTEBRAL COLUMN

1. Permits movement.
2. Encloses and protects the spinal cord.
3. Serves as a point of attachment for ribs and muscles of the back.

Q   *Which curves are concave (relative to the front of the body)?*

body, and two, called thoracic and sacral, are concave (cupping in). The curves of the vertebral column increase its strength, help maintain balance in the upright position, absorb shocks from walking, and help protect the column from fracture.

In the fetus, there is only a single anteriorly concave curve (Figure 6.14b). At approximately the third postnatal month, when an infant begins to hold its head erect, the *cervical curve* develops. Later, when the child stands and walks, the *lumbar curve* develops.

Typical Vertebra

Vertebrae in different regions of the column vary in size, shape, and detail, but they are similar enough that we can discuss a typical vertebra (Figure 6.15).

1. The *body* is the thick, disc-shaped front portion; it is the weight-bearing part of a vertebra.

2. The *vertebral arch* extends behind the body. It is formed by two short, thick processes, the *pedicles* (PED-i-kuls; *pediculus* = little foot), which unite with the *laminae* (LAM-i-nē; *lamina* = thin layer), flat parts that end in a single spinous process, a sharp slender projection. The space between the vertebral arch and body contains the spinal cord and is known as the *vertebral foramen.* The vertebral foramina of all vertebrae together form the *vertebral* (*spinal*) *canal.* The *intervertebral foramen* permits the passage of a single spinal nerve.

3. Seven *processes* arise from the vertebral arch. At the point where a lamina and pedicle join, a *transverse process* extends laterally. There is the single *spinous process* (*spine*) behind and below the junction of the laminae. These processes serve as points of muscular attachment. The remaining processes form joints with other vertebrae. The two *superior articular processes* of a vertebra articulate with the vertebra immediately above them, and the two *inferior articular processes* articulate with the vertebra below them. The articulating surfaces of the articular processes are referred to as *facets* (*facet* = little face).

Figure 6.15   Typical vertebra as illustrated by a thoracic vertebra.

🔑 *A vertebra consists of a body, vertebral arch, and several processes.*

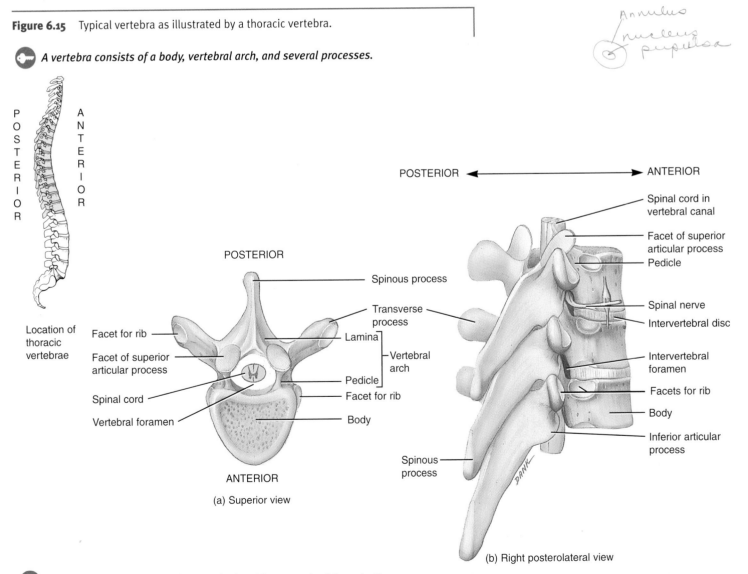

(a) Superior view

(b) Right posterolateral view

❓ *What are the functions of the vertebral and intervertebral foramina?*

Cervical Region

Vertebrae are numbered by region from top to bottom. There are seven cervical vertebrae (C1–C7). The spinous processes of the second through sixth *cervical vertebrae* are often *bifid,* that is, with a cleft (Figure 6.16). All cervical vertebrae have three foramina: one vertebral foramen and two transverse foramina.

Each cervical transverse process contains a *transverse foramen* through which blood vessels and nerves pass.

The first two cervical vertebrae differ considerably from the others. The first cervical vertebra (C1), the *atlas,* is named for its support of the head. It lacks a body and a spinous process. The upper surface contains *superior articular facets* that articulate with the occipital bone. This articulation per-

Figure 6.16   Cervical vertebrae.

🔑 *The cervical vertebrae are found in the neck region.*

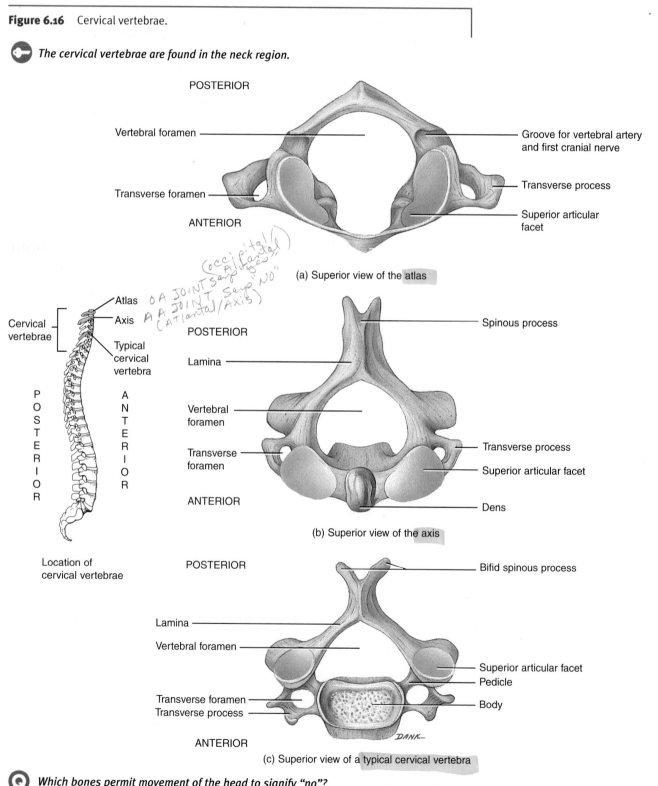

(a) Superior view of the atlas

(b) Superior view of the axis

(c) Superior view of a typical cervical vertebra

Ⓠ **Which bones permit movement of the head to signify "no"?**

mits you to nod your head. The inferior surface contains **inferior articular facets** that articulate with the second cervical vertebra.

The second cervical vertebra (C2), the **axis,** does have a body. A peglike process called the **dens** (*dens* = tooth) or **odontoid process** projects up through the ring of the atlas. The dens makes a pivot on which the atlas and head rotate. This arrangement permits side-to-side movement of the head.

The third through sixth cervical vertebrae (C3–C6) are similar to the typical cervical vertebra described previously.

The seventh cervical vertebra (C7), called the **vertebra prominens,** is somewhat different. It is marked by a large, spinous process that may be seen and felt at the base of the neck.

Thoracic Region

Thoracic vertebrae (T1–T12) are considerably larger and stronger than cervical vertebrae. Thoracic vertebrae have surfaces called *facets,* for articulating with the ribs (see Figure 6.15).

Lumbar Region

The **lumbar vertebrae** (L1–L5) are the largest and strongest in the column (Figure 6.17). The spinous processes are well adapted for the attachment of the large back muscles.

Sacrum and Coccyx   ? *fusion dates*

The **sacrum** is a triangular bone formed by the union of five sacral vertebrae. These are indicated in Figure 6.18 as S1

through S5. Fusion begins between 16 and 18 years of age and is usually completed by the mid-twenties. The sacrum serves as a strong foundation for the pelvic girdle. It is positioned at the back portion of the pelvic cavity between the two hipbones.

The anterior and posterior sides of the sacrum contain four pairs of **sacral foramina.** Nerves and blood vessels pass through the foramina. The **sacral canal** is a continuation of the vertebral canal. The lower entrance is called the **sacral hiatus** (hī-Ā-tus; *hiatus* = opening). Anesthetic agents are sometimes injected through the hiatus during childbirth. The anterior top border of the sacrum has a projection, the **sacral promontory** (PROM-on-tō′-rē), which is used as a landmark for measuring the pelvis prior to childbirth.

The **coccyx** is also triangular in shape and is formed by the fusion (usually between 20 and 30 years of age) of the coccygeal vertebrae, usually the last four. These are indicated in Figure 6.18 as Co1 through Co4. The coccyx articulates superiorly with the sacrum.

Thorax

Anatomically, the term **thorax** refers to the entire chest. The skeletal portion of the thorax is a bony cage formed by the sternum, costal cartilage, ribs, and the bodies of the thoracic vertebrae (Figure 6.19).

The thoracic cage encloses and protects the organs in the thoracic cavity and supports the bones of the shoulder girdle and upper limbs.

Figure 6.17   Lumbar vertebra.   *large lima Bean shaped*

🔑 *Lumbar vertebrae are found in the lower back.*

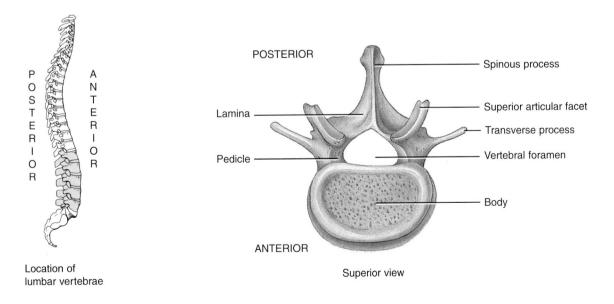

Location of lumbar vertebrae

POSTERIOR

Lamina

Pedicle

ANTERIOR

Spinous process

Superior articular facet

Transverse process

Vertebral foramen

Body

Superior view

🔍 *Why are the lumbar vertebrae the largest and strongest in the vertebral column?*

Figure 6.18 Sacrum and coccyx.

🔑 *The sacrum is formed by the union of five sacral vertebrae and the coccyx is formed by the union of usually four coccygeal vertebrae.*

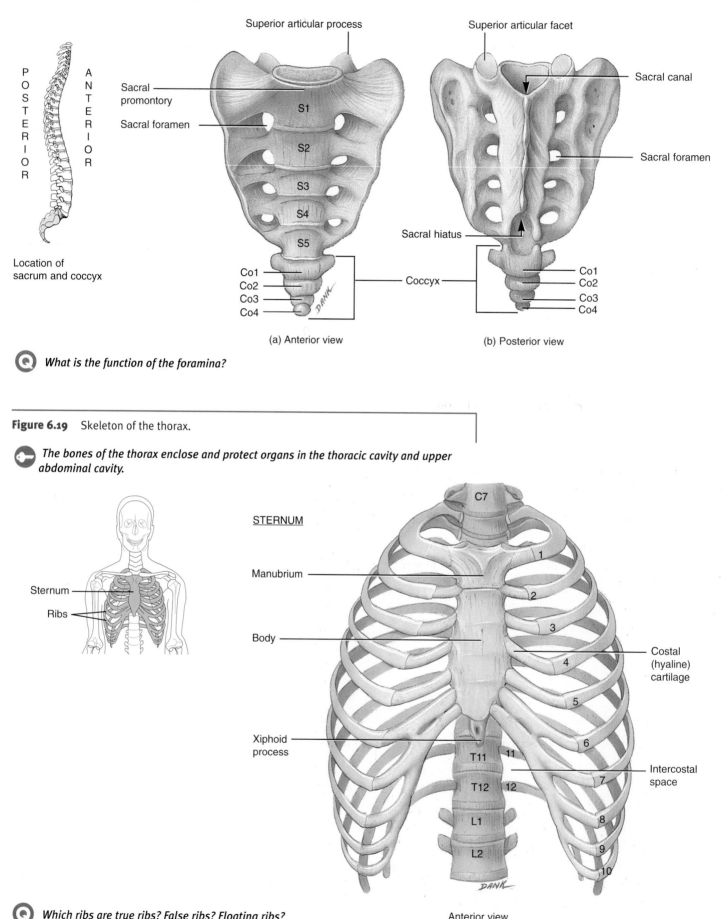

Superior articular process

Superior articular facet

Sacral promontory

Sacral foramen

S1
S2
S3
S4
S5

Co1
Co2
Co3
Co4

POSTERIOR   ANTERIOR

Location of sacrum and coccyx

Sacral canal

Sacral foramen

Sacral hiatus

Coccyx

Co1
Co2
Co3
Co4

(a) Anterior view

(b) Posterior view

❓ *What is the function of the foramina?*

Figure 6.19 Skeleton of the thorax.

🔑 *The bones of the thorax enclose and protect organs in the thoracic cavity and upper abdominal cavity.*

Sternum

Ribs

STERNUM

Manubrium

Body

Xiphoid process

C7

1
2
3
4
5
6
7
8
9
10
11
12

T11
T12
L1
L2

Costal (hyaline) cartilage

Intercostal space

❓ *Which ribs are true ribs? False ribs? Floating ribs?*

Anterior view

manubrium — sternal notch (head)

Sternum

The ***sternum,*** or breastbone, is a flat, narrow bone located in the center of the anterior thoracic wall.

The sternum (see Figure 6.19) is divided into the ***manubrium*** (ma-NOO-brē-um; *manubrium* = a handle), the upper portion; the ***body,*** the middle, largest portion; and the ***xiphoid*** (ZĪ-foyd; *xiphos* = a sword) ***process,*** the lower, smallest portion. The manubrium articulates with the clavicles and the first and second ribs. The body of the sternum articulates directly or indirectly with the second through tenth ribs. The xiphoid process has no ribs attached to it but provides attachment for some abdominal muscles. If the hands of a rescuer are mispositioned during cardiopulmonary resuscitation (CPR), there is the danger of fracturing the xiphoid process and driving it into the liver.

Since the sternum houses the red bone marrow throughout life and because it is readily accessible and has thin compact bone, it is a common site for withdrawal of marrow for biopsy. Under a local anesthetic, a wide-bore needle is introduced into the marrow cavity of the sternum for aspiration of a sample of red bone marrow. This procedure is called ***sternal puncture.***

Ribs

Twelve pairs of ***ribs*** make up the sides of the thoracic cavity (see Figure 6.19). The ribs increase in length from the first through seventh. Then they decrease in length to the twelfth rib. Each rib articulates posteriorly with its corresponding thoracic vertebra.

The first through seventh ribs have a direct anterior attachment to the sternum by a strip of hyaline cartilage, called ***costal*** (*costa* = rib) ***cartilage.*** These ribs are called ***true ribs.*** The remaining five pairs of ribs are referred to as ***false ribs*** because their costal cartilages do not attach directly to the sternum or do not attach at all. The cartilages of the eighth, ninth, and tenth ribs attach to each other and then to the cartilage of the seventh rib. The eleventh and twelfth false ribs are also called ***floating ribs*** because their anterior ends do not attach at all; they attach only posteriorly to the thoracic vertebrae. Spaces between ribs, called ***intercostal*** (*inter* = between) ***spaces,*** are occupied by intercostal muscles, blood vessels, and nerves.

Pectoral (Shoulder) Girdle

The ***pectoral*** (PEK-tō-ral) or ***shoulder girdles*** attach the bones of the upper limbs to the axial skeleton (Figure 6.20). Each of the two pectoral girdles consists of two bones: a clavicle and a scapula. The clavicle is the anterior component and articulates with the sternum at the sternoclavicular joint. The posterior component, the scapula, articulates with the clavicle and humerus. The pectoral girdles do not articulate with the vertebral column.

Figure 6.20   Right pectoral (shoulder) girdle and upper limb.

🔑 *The shoulder girdle attaches the bones of the upper limb to the axial skeleton.*

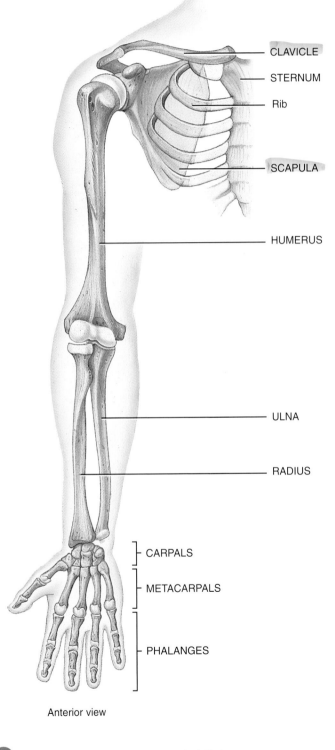

CLAVICLE
STERNUM
Rib
SCAPULA
HUMERUS
ULNA
RADIUS
CARPALS
METACARPALS
PHALANGES

Anterior view

❓ *What bones make up the pectoral girdle?*

Although the shoulder joints are not very stable, they are freely movable and thus allow movement in many directions.

Clavicle

The **clavicles** (KLAV-i-kuls; *clavis* = key), or collarbones, are long, slender bones with a double curvature. They lie horizontally and are above the first rib. The medial end articulates with the sternum. The lateral end articulates with the acromion of the scapula (see Figure 6.21).

Scapula

The **scapulae** (SCAP-yoo-lē), or shoulder blades, are large, triangular, flat bones situated in the posterior part of the thorax (Figure 6.21).

A sharp ridge, the **spine,** runs diagonally across the posterior surface of the **body.** The end of the spine is called the **acromion** (a-KRŌ-mē-on; *acro* = top). This process articulates with the clavicle. Inferior to the acromion is a depression called the **glenoid cavity.** This cavity articulates with the head of the humerus to form the shoulder joint. Also present on the scapula is a projection called the **coracoid** (KOR-a-koyd) *process,* to which muscles attach.

Upper Limb

| objective: | *Identify the bones of the upper limb.* |

The **upper limbs** consist of 60 bones. The skeleton of the right upper limb is shown in Figure 6.20. Each upper limb includes a humerus in the arm, ulna and radius in the forearm, carpals (wrist bones), metacarpals (palm bones), and phalanges (finger bones) of the hand.

Humerus

The **humerus** (HYOO-mer-us) is the longest and largest bone of the upper limb (Figure 6.22). It articulates with the scapula and at the elbow with both ulna and radius.

The proximal end of the humerus consists of a **head** that articulates with the glenoid cavity of the scapula. It also has an **anatomical neck,** which is a groove just below the head.

The **body** (**shaft**) of the humerus contains a roughened area called the **deltoid tuberosity,** which serves as a point of attachment for the deltoid muscle.

The following parts are found at the distal end of the humerus. The **capitulum** (ka-PIT-yoo-lum), meaning a small

---

Figure 6.21   Right scapula.

🔑 *The glenoid cavity of the scapula articulates with the head of the humerus to form the shoulder joint.*

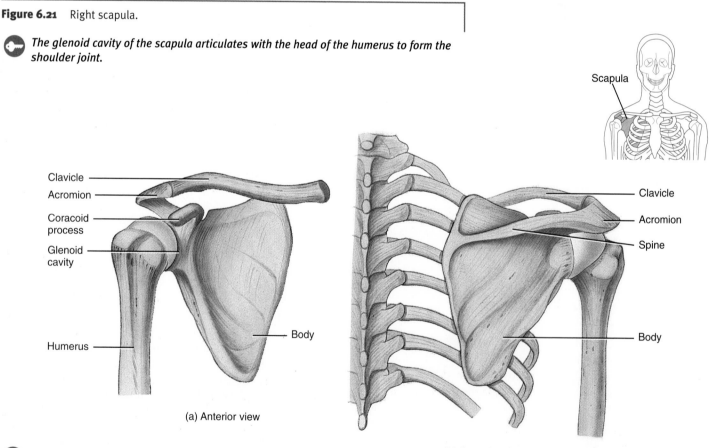

(a) Anterior view

(b) Posterior view

🜨 *What is the common term we use for the scapula?*

funny bone is actually ulnar nerve

Figure 6.22   Right humerus in relation to the scapula, ulna, and radius.

🔑 *The humerus is the longest and largest bone of the upper limb.*

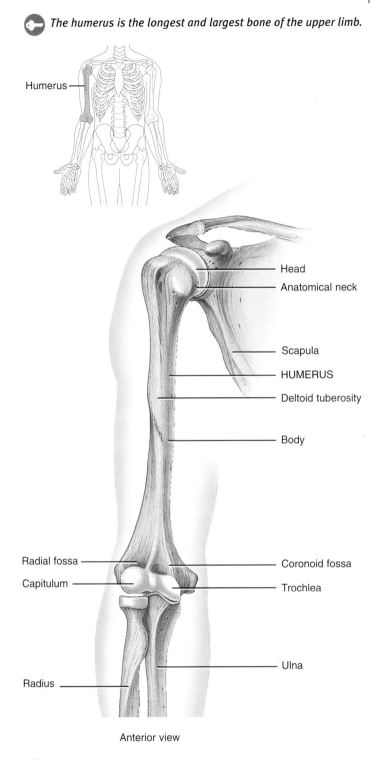

Humerus

Head
Anatomical neck

Scapula

HUMERUS

Deltoid tuberosity

Body

Radial fossa — — Coronoid fossa

Capitulum — — Trochlea

— Ulna

Radius

Anterior view

❓ *With which part of the scapula does the humerus articulate?*

head, is a rounded knob that articulates with the head of the radius. The *radial fossa* is a depression that receives the head of the radius when the forearm is flexed (bent). The *trochlea*

(TRŌK-lē-a) is a surface that articulates with the ulna. The *coronoid* (*korone* = crown-shaped) *fossa* is a depression that receives part of the ulna when the forearm is flexed. The *olecranon* (ō-LEK-ra-non) *fossa* is a depression on the back of the bone that receives the olecranon of the ulna when the forearm is extended (straightened).

Ulna and Radius

The *ulna* is the medial bone of the forearm (Figure 6.23). In other words, it is located at the little finger side. The proximal end of the ulna has an *olecranon,* which forms the elbow. The *coronoid process,* together with the olecranon, receives the trochlea of the humerus. The *trochlear notch* is a curved area between the olecranon and the coronoid process. The trochlea of the humerus fits into this notch. The *radial notch* is a depression for the head of the radius. A *styloid* (*stilus* = pointed instrument) *process* is at the most distal end.

The *radius* is the lateral bone of the forearm, that is, situated on the thumb side. The proximal end of the radius articulates with the capitulum of the humerus and radial notch of the ulna. It has a raised, roughened area called the *radial tuberosity* that provides a point of attachment for the biceps brachii muscle. The body (shaft) of the radius articulates with two bones (carpals) of the wrist. Also at the distal end is a *styloid process.*

Carpals, Metacarpals, and Phalanges

The *carpus* (*wrist*) of the hand consists of eight small bones, the *carpals,* connected to each other by ligaments (Figure 6.24 on p. 127). The bones are arranged in two transverse rows, with four bones in each row. In the anatomical position, the carpals in the top row, from the lateral to medial position, are the *scaphoid* or *navicular, lunate, triquetrum,* and *pisiform.* In about 70 percent of carpal fractures, only the scaphoid is broken. The carpals in the bottom row, from the lateral to medial position, are the *trapezium, trapezoid, capitate,* and *hamate.*

The five bones of the *metacarpus* (*meta* = after) constitute the palm of the hand. Each metacarpal bone consists of a proximal *base,* a middle *shaft,* and a distal *head.* The metacarpal bones are numbered I to V, starting with the lateral bone in the thumb. The heads of the metacarpals are commonly called the "knuckles" and are readily visible when the fist is clenched.

The *phalanges* (fa-LAN-jēz; *phalanx* = closely knit), or bones of the fingers, number 14 in each hand. A single bone of the finger (or toe) is referred to as a *phalanx* (FĀ-lanks). Each phalanx consists of a proximal *base,* a middle *shaft,* and a distal *head.* There are two phalanges (proximal and distal) in the first digit, called the thumb (*pollex*), and three phalanges (proximal, middle, and distal) in each of the remaining four digits. These digits are commonly referred to as the index finger, middle finger, ring finger, and little finger.

Figure 6.23   Right ulna and radius in relation to the humerus and carpals. The arrow in the inset in (b) indicates the direction from which the ulna is viewed (lateral).

🗝 *The ulna is the medial bone of the forearm and the radius is the lateral bone.*

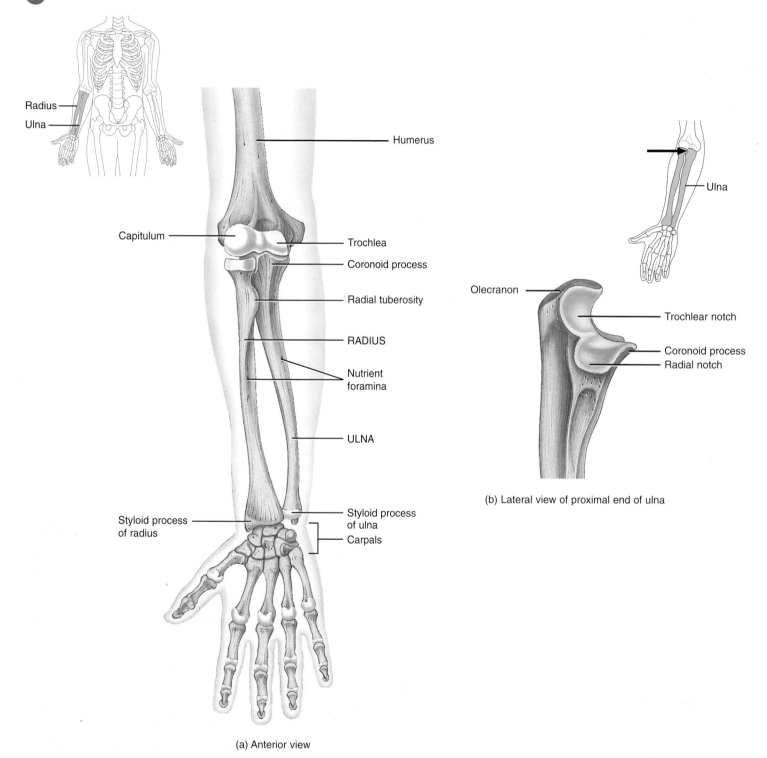

(a) Anterior view

(b) Lateral view of proximal end of ulna

Ⓠ *What part of the ulna is called the elbow?*

Figure 6.24   Right wrist and hand in relation to the ulna and radius.

🔑 *The skeleton of the hand consists of the carpals, metacarpals, and phalanges.*

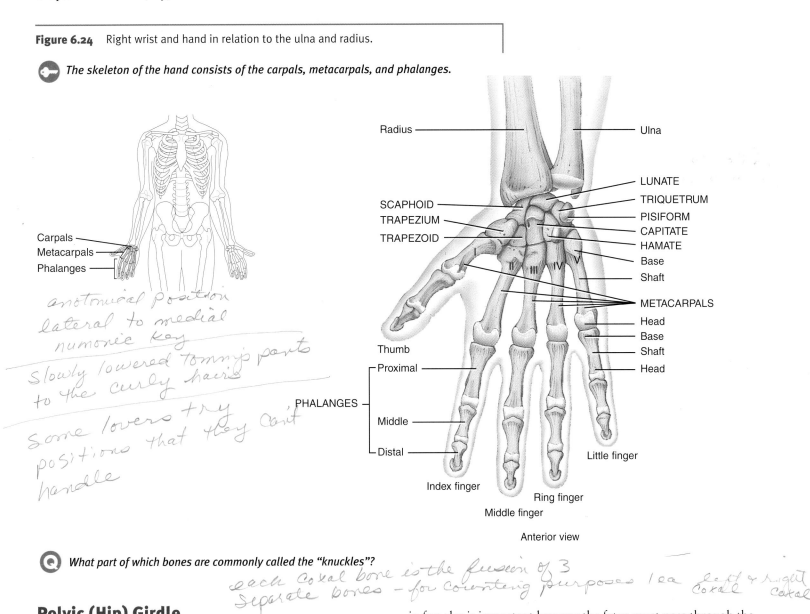

anatomical position
lateral to medial
numonic key
slowly lowered Tommy's pants
to the curly hairs

some lovers try
positions that they can't
handle

Anterior view

❓ *What part of which bones are commonly called the "knuckles"?*

each coxal bone is the fusion of 3
separate bones — for counting purposes 1 ea left + right
coxal   coxal

Pelvic (Hip) Girdle

The *pelvic (hip) girdle* consists of the two *hipbones,* also called *coxal bones* (Figure 6.25). The pelvic girdle provides a strong and stable support for the vertebral column and viscera. The hipbones are united to each other anteriorly at a fibrocartilaginous joint called the *pubic symphysis* (PYOO-bik SIM-fi-sis). They unite posteriorly to the sacrum.

Together with the sacrum and coccyx, the two hipbones of the pelvic girdle form the basinlike structure called the *pelvis.* The pelvis is divided into a greater pelvis and a lesser pelvis. The *greater (false) pelvis* is the portion situated superior to the pelvic brim. The *lesser (true) pelvis* is inferior to the pelvic brim. The lesser pelvis contains a superior opening called the *pelvic inlet* and an inferior opening called the *pelvic outlet.*

Pelvimetry is the measurement of the size of the pelvic inlet and outlet of the birth canal. The size of the pelvic cavity

in females is important because the fetus must pass through the narrower opening of the lesser pelvis at birth.

Each of the two *hipbones* of a newborn consists of a superior *ilium* (meaning flank), an inferior and anterior *pubis* (meaning pubic hair), and an inferior and posterior *ischium* (IS-kē-um), meaning hip (Figure 6.26 on p. 129). Eventually, the three separate bones fuse into one. The area of fusion is a deep, lateral fossa (depression) called the *acetabulum* (as′-e-TAB-yoo-lum; *acetabulum* = vinegar cup), which serves as the socket for the head of the femur. Although the adult hipbones are single bones, it is common to discuss them as if they still consisted of three portions.

The ilium is the largest of the three subdivisions. Its upper border is called the *iliac crest* (border or ridge). On the lower surface is the *greater sciatic* (sī-AT-ik) *notch.* The ischium joins with the pubis and together they surround the *obturator* (OB-too-rā′-tor) *foramen.*

Figure 6.25 Female pelvic (hip) girdle.

🗝 *The hipbones are united in front at the pubic symphysis and in back at the sacrum.*

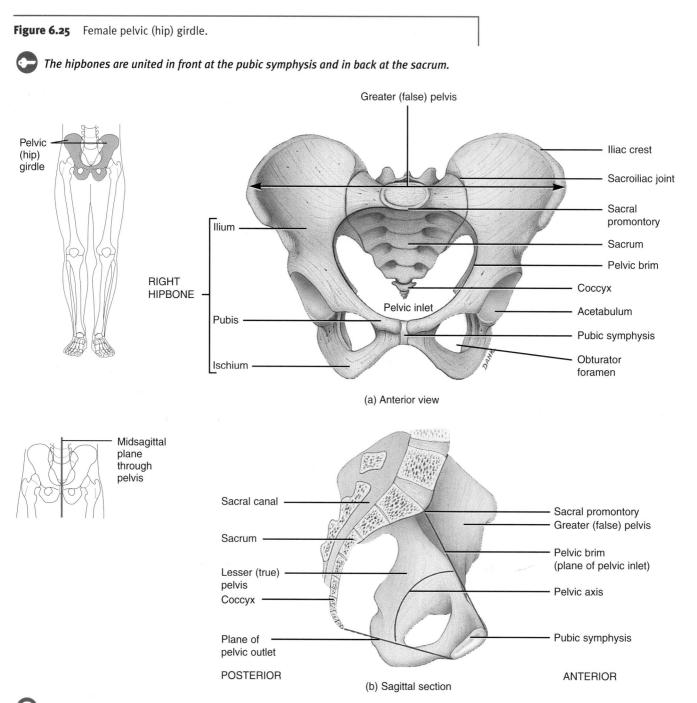

(a) Anterior view

(b) Sagittal section

Ⓠ *What part of the pelvis surrounds the pelvic organs in the pelvic cavity?*

Lower Limb

objective: *Identify the bones of the lower limb.*

The *lower limbs* are composed of 60 bones (Figure 6.27). Each limb includes the femur in the thigh, patella (kneecap), fibula and tibia in the leg, tarsals (ankle bones), metatarsals, and phalanges (toes) in the foot.

Femur

The *femur,* or thighbone, is the longest and heaviest bone in the body (Figure 6.28 on p. 130). Its proximal end articulates

with the hipbone. Its distal end articulates with the tibia. The *body* (*shaft*) of the femur bows medially so that the knee joints are brought nearer to the body's line of gravity. The convergence is greater in the female because the female pelvis is broader.

The *head* of the femur articulates with the acetabulum of the hipbone. The *neck* of the femur is a constricted region below the head. A fairly common fracture in the elderly occurs at the neck of the femur. The neck may become so weak it fails to support the body. The *greater trochanter* (trō-KAN-ter) is a projection felt and seen in front of the hollow on the side of the hip. It is used as a landmark for an intramuscular injection.

Figure 6.26   Right hipbone. The lines of fusion of the ilium, ischium, and pubis are not always visible in an adult bone.

🔑 *The two hipbones form the pelvic girdle, which attaches the lower limbs to the axial skeleton and supports the vertebral column and viscera.*

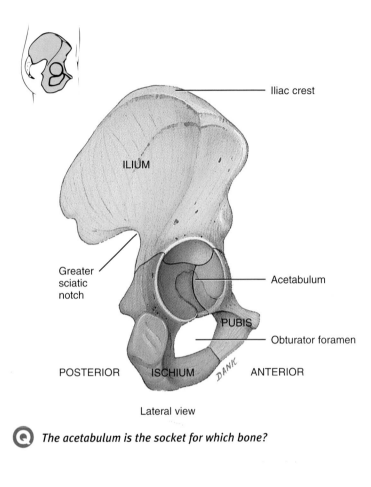

Lateral view

🅠 *The acetabulum is the socket for which bone?*

Figure 6.27   Right pelvic (hip) girdle and lower limb.

🔑 *The pelvic girdle attaches the lower limb to the axial skeleton.*

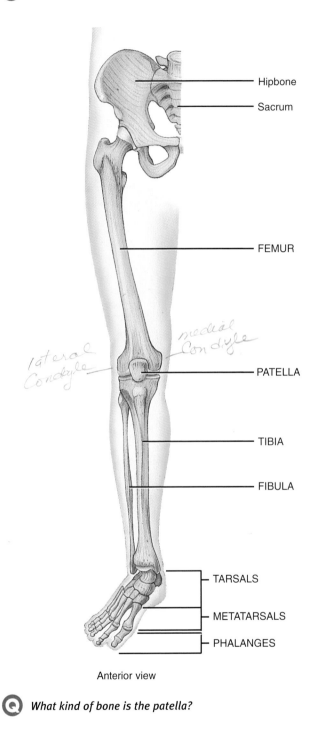

Anterior view

🅠 *What kind of bone is the patella?*

The body of the femur contains a rough vertical ridge called the ***linea aspera*** (LIN-ē-a AS-per-a), which serves to attach several thigh muscles.

The distal end of the femur expands into the ***medial condyle*** and ***lateral condyle,*** which articulate with the tibia. The ***patellar surface*** is located between the condyles on the anterior surface.

Patella

The ***patella*** (*patella* = small plate), or kneecap, is a small, triangular bone in front of the knee joint (see Figure 6.27). It is a sesamoid bone that develops in the tendon of the quadriceps femoris muscle.

Tibia and Fibula

The ***tibia,*** or shinbone, is the larger medial bone of the leg (Figure 6.29). It bears the weight of the body. The tibia articu-

lates at its proximal end with the femur and fibula and at its distal end with the fibula and talus bone of the ankle.

The proximal end of the tibia expands into a ***lateral condyle*** and a ***medial condyle,*** which articulate with the

Figure 6.28   Right femur in relation to the hipbone, patella, tibia, and fibula.

The acetabulum of the hipbone and head of the femur articulate to form the hip joint.

Figure 6.29   Right tibia and fibula in relation to the femur, patella, and talus.

The tibia articulates with the femur and fibula proximally and fibula and talus distally.

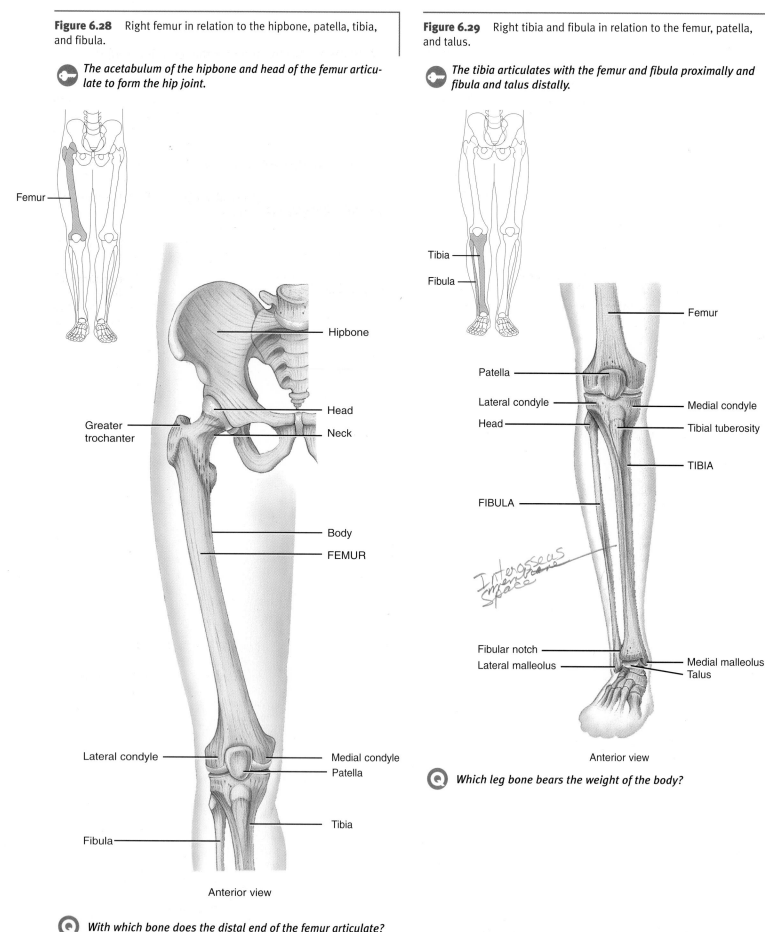

Anterior view

Q *With which bone does the distal end of the femur articulate?*

Anterior view

Q *Which leg bone bears the weight of the body?*

condyles of the femur. The ***tibial tuberosity*** is a point of attachment for the patellar ligament.

The medial surface of the distal end of the tibia forms the ***medial malleolus*** (mal-LĒ-ō-lus; *malleus* = little hammer), which articulates with the talus bone of the ankle and forms the bump that can be felt on the medial surface of your ankle. The ***fibular notch*** articulates with the fibula.

Shinsplints is the name given to soreness or pain along the tibia. It is probably caused by inflammation of the periosteum brought on by repeated tugging of the muscles and tendons attached to the periosteum, often the result of walking or running up and down hills or vigorous activity followed by relative inactivity.

The ***fibula*** is parallel to the tibia (Figure 6.29) and considerably smaller. The proximal end of the fibula articulates with the lateral condyle of the tibia just below the knee joint. The distal end has a projection called the ***lateral malleolus*** that articulates with the talus bone of the ankle. This forms the prominence on the lateral surface of the ankle. The inferior portion of the fibula also articulates with the tibia at the fibular notch.

Tarsals, Metatarsals, and Phalanges

The ***tarsus*** is a collective term for the seven bones of the ankle called ***tarsals*** (Figure 6.30). The ***talus*** (TĀ-lus; *talas* = ankle bone) and ***calcaneus*** (kal-KĀ-nē-us) are located on the posterior part of the foot. The anterior part contains the ***cuboid***, ***navicular*** (boat-shaped), and three ***cuneiform bones*** called the ***first*** (***medial***), ***second*** (***intermediate***), and ***third*** (***lateral***) ***cuneiform*** (*cuneiform* = wedge-shaped). The talus is the only bone of the foot that articulates with the fibula and tibia. It is surrounded by the medial malleolus of the tibia and the lateral malleolus of the fibula. The calcaneus, or heel bone, is the largest and strongest tarsal bone.

The ***metatarsus*** consists of five metatarsal bones numbered I to V from the medial to lateral position. Like the metacarpals of the palm, each metatarsal consists of a proximal ***base***, a middle ***shaft***, and a distal ***head***. The first metatarsal is thicker than the others because it bears more weight.

The ***phalanges*** of the foot resemble those of the hand both in number and arrangement. Each also consists of a prox-

---

Figure 6.30   Right foot. The arrow in the inset indicates the direction from which the foot is viewed (superior).

🔑 *The skeleton of the foot consists of the tarsals, metatarsals, and phalanges.*

Superior view

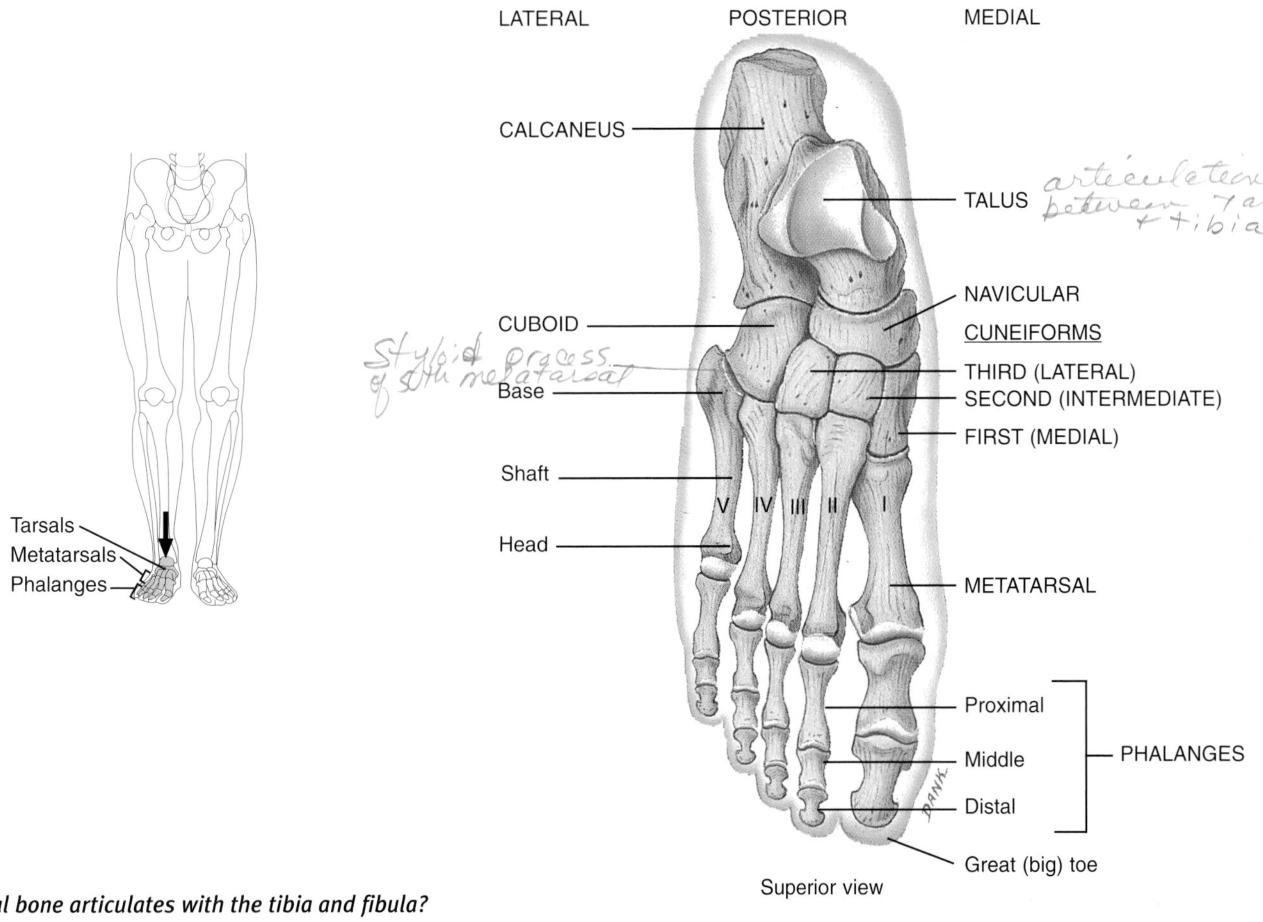

❓ *What tarsal bone articulates with the tibia and fibula?*

imal *base,* a middle *shaft,* and a distal *head.* The great (big) toe, or *hallux,* has two large, heavy phalanges (proximal and distal) while the other four toes each have three phalanges (proximal, middle, and distal).

Arches of the Foot

The bones of the foot are arranged in two *arches* (Figure 6.31). These arches enable the foot to support the weight of the body and provide leverage while walking. The arches are not rigid. They yield as weight is applied and spring back when the weight is lifted, thus helping to absorb shocks.

The *longitudinal arch* runs from the front to the back of the foot, from the calcaneus to the distal ends of the metatarsals. It consists of a medial (inner) and lateral (outer) part. The *transverse arch* is formed by the navicular, three cuneiforms, and the proximal ends of the five metatarsals.

The bones composing the arches are held in position by ligaments and tendons. If these ligaments and tendons are weakened, the height of the longitudinal arch may "fall." The result is *flatfoot.*

plantar fascia between is tissue between arches

Comparison of Female and Male Skeletons

objective: *Compare the principal structural and functional differences between female and male skeletons.*

The bones of the male are generally larger and heavier than those of the female. The articular ends are thicker in relation to the shafts. In addition, because certain muscles of the male are larger than those of the female, the points of attachment—tuberosities, lines, ridges—are larger in the male skeleton.

Many significant structural differences between female and male skeletons occur in the pelvis. Most are structural adaptations for pregnancy and childbirth. The female pelvis provides more room in the lesser (true) pelvis, especially in the pelvic inlet and pelvic outlet, to accommodate the passage of the infant's head at birth. Typical differences are listed in Exhibit 6.4 and illustrated in Figure 6.32.

Exhibit 6.4	Comparison of a Typical Female and Male Pelvis (see Figure 6.32)	
POINT OF COMPARISON	FEMALE	MALE
General structure	Light and thin	Heavy and thick
Greater (false) pelvis	Shallow	Deep
Pelvic inlet	Larger and more oval	Heart-shaped
Sacrum	Short, wide, flat, curving forward in lower part	Long, narrow, with smooth concavity
Pubic arch	Great than a 90° angle	Less than a 90° angle
Ilium	Less vertical	More vertical
Iliac crest	Less curved	More curved
Acetabulum	Small	Large
Obturator foramen	Oval	Round

Figure 6.31   Arches of the right foot.

🔑 *Arches help the foot support and distribute the weight of the body and provide leverage while walking.*

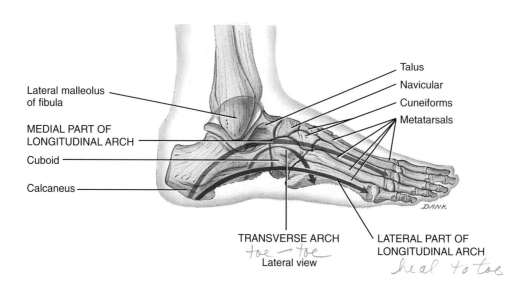

toe → toe
Lateral view

heal to toe

 What structural aspect of the arches allow them to absorb shocks?

Figure 6.32   Comparison of female and male pelvises.

Many structural adaptations of the female pelvis are related to pregnancy and child-birth.

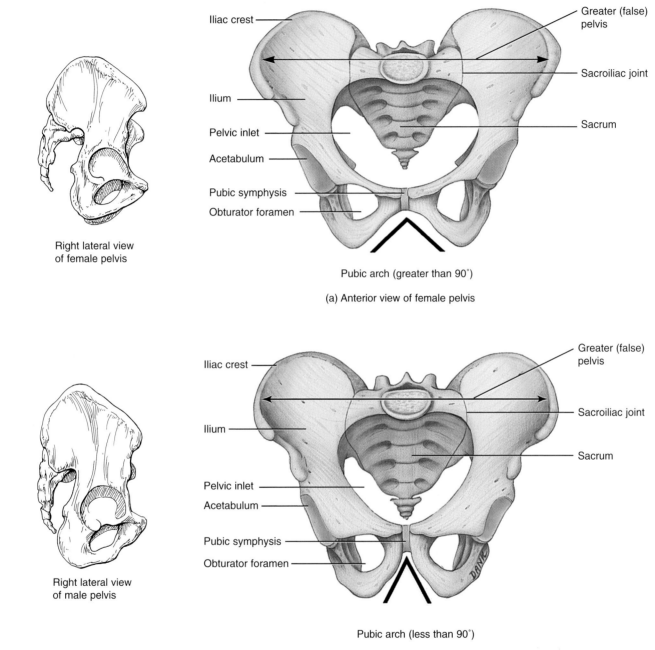

Right lateral view
of female pelvis

Iliac crest

Ilium

Pelvic inlet

Acetabulum

Pubic symphysis

Obturator foramen

Greater (false)
pelvis

Sacroiliac joint

Sacrum

Pubic arch (greater than 90˚)

(a) Anterior view of female pelvis

Right lateral view
of male pelvis

Iliac crest

Ilium

Pelvic inlet

Acetabulum

Pubic symphysis

Obturator foramen

Greater (false)
pelvis

Sacroiliac joint

Sacrum

Pubic arch (less than 90˚)

(b) Anterior view of male pelvis

What is the main difference between the pelvic inlet in females and males?

wellness focus
Steps to Healthy Feet

We take the structure and function of our feet for granted—until they start to hurt. And even then we often continue to mistreat them, cramming them into shoes that are too tight, and then asking them to walk on concrete sidewalks and take us on long shopping expeditions. No wonder foot problems are such a common complaint! Fortunately, most foot problems are preventable by understanding the foot's structure and function, and then using good footwear to support them in their work.

These Feet Were Made for Walking

Each time you take a step, your heel strikes the ground first. Then you roll through the arches, over the ball of your foot, and onto your toes. Your arches flatten slightly as they absorb the weight of your body. One foot continues to bear your weight until the heel of the other foot touches the ground. As you walk, your big toe maintains your balance while the other toes give your foot some resiliency. The two outer metatarsals move to accommodate for uneven surfaces, while the inner three stay rigid for support.

The most common cause of foot problems is ill-fitting shoes that interfere with the mechanical actions described above. The high heel is a case in point, and explains why 80 percent of those suffering from foot problems are women. While many people think high heels look good and are fun to wear, they should not be used for walking, because they make the body's weight fall onto the forefoot. Thus the arches of the foot are not allowed to absorb the force of the body's weight. This unnatural stress can injure soft-tissue structures, joints, and bones.

Sensible Shoes

Sensible shoes prevent many foot problems; they are especially important if you are doing any amount of walking. A good shoe has a sole that is strong and flexible and provides a good gripping surface. Cushioned insoles help protect feet from hard surfaces. Arch supports help distribute weight over a broader area, just like the arches in your foot.

Many people spend a great deal of time researching which brand of shoes to buy yet do not spend adequate time evaluating whether or not the shoes suit their feet. A high-quality shoe is only worth buying if it fits! It's better to buy a lesser-quality shoe that fits well than a poorly fitting high-quality shoe. As one avid shopper put it, "If the shoe fits, buy it."

Shop for shoes in the late afternoon when your feet are at their largest. One foot is often bigger than the other; always buy for the bigger foot. The shoes you try on should feel comfortable immediately—don't plan on shoes stretching with wear. The heel should fit snugly, the instep should not gape open, and the toe box should be wide enough for you to wiggle all your toes.

If you participate in some kind of sport or work in a physically demanding occupation, choose well-fitting shoes that will accommodate the extra demands placed on your feet. Walking shoes, provide support and shock absorption and are built to accommodate the walking biomechanics of your foot. Good walking shoes include heel cushioning, arch support, and flexible soles, and prevent the foot from rolling sideways in or out.

critical thinking

Why do you think excess body weight is associated with an increased risk of foot problems?

Consider walking on rough, rocky, uneven terrain. What kind of attributes should a good hiking boot have?

Common Disorders

Osteoporosis

Osteoporosis (os′-tē-ō-pō-RŌ-sis) is a disorder characterized by decreased bone mass and increased susceptibility to fractures as a result of decreased levels of estrogens. Estrogens maintain osseous tissue by stimulating osteoblasts to form new bone.

The disorder primarily affects middle-aged and elderly people—especially women (whites more than blacks). Women produce smaller amounts of sex hormones, especially estrogens, after menopause, and both men and women produce smaller amounts during old age. As a result, the osteoblasts become less active, and there is a decrease in bone mass.

Osteoporosis can also occur in male marathoners whose caloric intake is inadequate, teenagers on junk-food diets, young women suffering from eating disorders, people allergic to dairy products, nursing mothers, and those exposed to prolonged treatment with cortisone. Often, the first symptom of osteoporosis is a pathological fracture. Bone mass becomes so depleted that the skeleton can no longer withstand the mechanical stresses of everyday living. For example, a hip fracture might result from sitting down too quickly. Osteoporosis causes more than 250,000 hip fractures a year, and complications from osteoporosis are the 12th leading cause of death in the U.S. It is responsible for shrinkage of the backbone (vertebrae) and height loss, hunched backs, bone fractures, and considerable pain. Osteoporosis afflicts the entire skeletal system.

Risk factors for developing osteoporosis, besides race and gender, are (1) body build (short females are at greater risk because they have less total bone mass), (2) weight (thin females, anorectic females, and those who overdo exercise are at greater risk because adipose tissue is a source of an estrogen that retards bone loss), (3) smoking (smoking decreases blood levels of estrogens), (4) calcium deficiency or malabsorption, (5) vitamin D deficiency, (6) exercise (sedentary people are more likely to develop bone loss), (7) certain drugs (alcohol, some diuretics, cortisone, and tetracycline promote bone loss), (8) a family history of osteoporosis (daughters of women with osteoporosis have reduced bone mass), and (9) menopause (which may be premature due to excessive exercise).

Adequate diet and exercise are the mainstays for preventing osteoporosis. In postmenopausal women, estrogen replacement therapy (ERT; low doses of estrogen and sometimes progesterone, another sex hormone), calcium supplements, and weight-bearing exercise help prevent or retard the development of osteoporosis. The most important aspect of treatment is prevention. Adequate calcium intake and exercise in her early years may be more beneficial to a woman than ERT and calcium supplements when she is older.

Rickets

Rickets is a vitamin D deficiency in children in which the body does not absorb calcium and phosphorus. As a result, when the child walks, the weight of the body causes the bones in the legs to bow.

Paget's Disease

Paget's disease is an irregular thickening and softening of the bones, especially among bones of the skull, pelvis, and limbs.

Fractures

A *fracture* is any break in a bone. Several types of fractures include:

1. **Partial** (**incomplete**). The break across the bone is incomplete.

2. **Complete.** The break across the bone is complete.

3. **Closed** (**simple**). The bone does not break through the skin.

4. **Open** (**compound**). The broken ends of the bone protrude through the skin.

The healing process for bone varies with several factors such as the extent of the fracture, age, nutrition, the general health of the individual, and the type of treatment received.

Herniated (Slipped) Disc

A *herniated* (*slipped*) disc is characterized by a protrusion of the inner, soft elastic material of an intervertebral disc.

Abnormal Curves

Scoliosis (skō′-lē-Ō-sis; *scolio* = bent) is a lateral bending of the vertebral column, usually in the thoracic region. *Kyphosis* (kī-FŌ- sis; *kypho* = hunchback) is an exaggeration of the thoracic curve of the vertebral column. Mild kyphosis is termed "round-shouldered." *Lordosis* (lor-DŌ-sis; *lordo* = swayback) is an exaggeration of the lumbar curve of the vertebral column.

Spina Bifida

Spina bifida (SPĪ-na BIF-i-da) is a congenital defect of the vertebral column in which laminae fail to unite at the midline. ■

Medical Terminology and Conditions

Bunion (BUN-yun) A deformity of the great toe. Although the condition may be inherited, it is typically caused by wearing tightly fitting shoes. The condition produces inflammation of bursae (fluid-filled sacs at the joint), bone spurs, and calluses.

Clawfoot A condition in which the longitudinal arch is abnormally elevated. It is frequently caused by muscle imbalance, such as that which results from poliomyelitis.

Craniotomy (krā-nē-OT-ō-mē; *cranium* = skull; *tome* = a cutting) Any surgery that requires cutting through the bones surrounding the brain.

Osteogenic (os′-tē-ō-JEN-ik) *sarcoma* (sar-KŌ-ma; *sarcoma* = connective tissue tumor) Bone cancer that primarily affects osteoblasts and occurs most often in the bones of teenagers during their growth spurt; most common sites are the metaphyses of the thighbone (femur), shinbone (tibia), and armbone (humerus); metastases occur most often in lungs; treatment consists of multidrug chemotherapy and removal of the malignant growth or amputation of the limb.

Osteomyelitis (os′-tē-ō-mī-el-Ī-tis); *myelos* = marrow) Inflammation of a bone, especially the marrow, caused by a pathogenic organism, especially *Staphylococcus aureus*.

Osteopenia (os′-tē-ō-PĒ-nē-a; *penia* = poverty) Reduced bone mass due to a decrease in the rate of bone synthesis to a level insufficient to compensate for normal bone breakdown; any decrease in bone mass below normal. An example is osteoporosis.

Study Outline

Functions (p. 98)

1. The skeletal system consists of all bones attached at joints and cartilage between joints.
2. The functions of the skeletal system include support, protection, movement, mineral storage, housing blood-forming tissue, and storage of energy.

Types of Bones (p. 98)

1. On the basis of shape, bones are classified as long, short, flat, or irregular.
2. Sutural bones are found between the sutures of certain cranial bones. Sesamoid bones develop in tendons or ligaments. The patellas are sesamoid bones.

Parts of Long Bone (p. 98)

1. Parts of a long bone include the diaphysis (shaft), epiphyses (ends), metaphysis, articular cartilage, periosteum, medullary (marrow) cavity, and endosteum.
2. The diaphysis is covered by periosteum.

Histology (p. 100)

1. Bone tissue consists of widely separated cells surrounded by large amounts of matrix (intercellular substance). The four principal types of cells are osteoprogenitor cells, osteoblasts, osteocytes, and osteoclasts. The matrix contains collagen fibers (organic) and mineral salts that consist mainly of calcium phosphate (inorganic).
2. Compact (dense) bone tissue consists of osteons (Haversian systems) with little space between them. Compact bone lies over spongy bone and composes most of the bone tissue of the diaphysis. Functionally, compact bone protects, supports, and resists stress.
3. Spongy (cancellous) bone tissue consists of trabeculae surrounding many red bone marrow-filled spaces. It forms most of the structure of short, flat, and irregular bones and the epiphyses of long bones. Functionally, spongy bone stores some red and yellow bone marrow and provides some support.

Ossification: Bone Formation (p. 102)

1. Bone forms by a process called ossification.
2. The two types of ossification, intramembranous and endochondral, involve the replacement of a preexisting connective tissue with bone.
3. Intramembranous ossification occurs within fibrous connective tissue membranes of the embryo and the adult.
4. Endochondral ossification occurs within a cartilage model. The primary ossification center of a long bone is in the diaphysis. Cartilage degenerates, leaving cavities that merge to form the medullary (marrow) cavity. Osteoblasts lay down bone. Next, ossification occurs in the epiphyses, where bone replaces cartilage, except for articular cartilage and the epiphyseal plate.
5. Because of the activity of the epiphyseal plate, the diaphysis of a bone increases in length.
6. Bone grows in diameter as a result of the addition of new bone tissue around the outer surface of the bone.

Homeostasis (p. 104)

1. The homeostasis of bone growth and development depends on a balance between bone formation and resorption.
2. Old bone is constantly destroyed by osteoclasts, while new bone is constructed by osteoblasts. This process is called remodeling.
3. Normal growth depends on minerals (calcium, phosphorus, magnesium), vitamins (D, C, A), and hormones (human growth hormone, sex hormones, thyroid hormones, calcitonin, and parathyroid hormone).
4. Under mechanical stress, bone increases in mass.
5. Bones store and release calcium and phosphate. This is controlled by calcitonin (CT) and parathyroid hormone (PTH).
6. When placed under mechanical stress, bone tissue becomes stronger.
7. The important mechanical stresses result from the pull of skeletal muscles and the pull of gravity.
8. With aging, bone loses calcium and organic matrix.

Bone Surface Markings (p. 107)

1. Bone surface markings are structural features visible on the surfaces of bones.
2. Each marking has a specific function—joint formation, muscle attachment, or passage of nerves and blood vessels.

Divisions of the Skeletal System (p. 108)

1. The axial skeleton consists of bones arranged along the longitudinal axis. The parts of the axial skeleton are the skull, hyoid bone, auditory ossicles, vertebral column, sternum, and ribs.
2. The appendicular skeleton consists of the bones of the girdles and the upper and lower limbs (extremities). The parts of the appendicular skeleton are the pectoral (shoulder) girdles, bones of the upper limbs, pelvic (hip) girdle, and bones of the lower limbs.

Skull (p. 108)

1. The skull consists of the cranium and the face.
2. Sutures are immovable joints between bones of the skull. Examples are the coronal, sagittal, lambdoid, and squamous sutures.
3. The eight cranial bones include the frontal (1), parietal (2), temporal (2), occipital (1), sphenoid (1), and ethmoid (1).
4. The 14 facial bones are the nasal (2), maxillae (2), zygomatic (2), mandible (1), lacrimal (2), palatine (2), inferior nasal conchae (2), and vomer (1).
5. Paranasal sinuses are cavities in bones of the skull that communicate with the nasal cavity. They are lined by mucous membranes. Cranial bones containing paranasal sinuses are the frontal, sphenoid, ethmoid, and maxillae.
6. Fontanels are membrane-filled spaces between the cranial bones of fetuses and infants. The major fontanels are the anterior, posterior, anterolaterals, and posterolaterals.
7. The foramina of the skull bones provide passages for nerves and blood vessels.

Hyoid Bone (p. 116)

1. The hyoid bone is a U-shaped bone that does not articulate with any other bone.
2. It supports the tongue and provides attachment for some of its muscles as well as some neck muscles.

Vertebral Column (p. 118)

1. The bones of the adult vertebral column are the cervical vertebrae (7), thoracic vertebrae (12), lumbar vertebrae (5), the sacrum (5, fused), and the coccyx (4, fused).
2. The vertebral column contains normal curves that give strength, support, and balance.
3. The vertebrae are similar in structure, each consisting of a body, vertebral arch, and seven processes. Vertebrae in the different regions of the column vary in size, shape, and detail.

Thorax (p. 121)

1. The thoracic skeleton consists of the sternum, ribs, costal cartilages, and thoracic vertebrae.
2. The thoracic cage protects vital organs in the chest area.

Pectoral (Shoulder) Girdle (p. 123)

1. Each pectoral (shoulder) girdle consists of a clavicle and scapula.
2. Each attaches an upper limb to the trunk.

Upper Limb (p. 124)

1. The bones of each upper limb include the humerus, ulna, radius, carpals, metacarpals, and phalanges.

Pelvic (Hip) Girdle (p. 127)

1. The pelvic (hip) girdle consists of two hipbones.
2. It attaches the lower limbs to the trunk at the sacrum.
3. Each hipbone consists of three fused components—ilium, pubis, and ischium.

Lower Limb (p. 128)

1. The bones of each lower limb include the femur, patella, tibia, fibula, tarsals, metatarsals, and phalanges.
2. The bones of the foot are arranged in two arches, the longitudinal arch and the transverse arch, to provide support and leverage.

Comparison of Female and Male Skeletons (p. 132)

1. The female pelvis is adapted for pregnancy and childbirth. Differences in pelvic structure are listed in Exhibit 6.4.
2. Male bones are generally larger and heavier than female bones and have more prominent markings for muscle attachment.

Self-Quiz

1. Which of the following statements about the skeletal system is NOT true?

   a. Bones store calcium and phosphorus. b. Yellow bone marrow is the site of hemopoiesis. c. Skeletal muscles pull on bones to produce movement. d. The skeleton is the body's framework. e. Red bone marrow contains immature blood cells.

2. Bones whose length is almost equal to their width, so that they resemble a cube in their shape, are called

   a. sutural bones b. flat bones c. irregular bones d. short bones e. sesamoid bones

3. A membrane composed of dense irregular connective tissue that covers a bone is called the

   a. endosteum b. osteon c. perichondrium d. periosteum e. lamella

4. The cell that functions in bone reabsorption is the

   a. osteoblast b. osteoclast c. osteocyte d. osteoprogenitor e. osteoporocyte

5. Which of the following do NOT go together?

   a. spongy bone—trabeculae b. spongy bone—red bone marrow c. compact bone—diaphysis d. cancellous bone—osteon e. spongy bone—osteocytes

6. An opening through which blood vessels, nerves, or ligaments pass is called a

   a. facet   b. fossa   ⓒ foramen   d. meatus
   e. condyle

7. Place the stages of endochondral ossification in the proper order:

   1. The primary ossification center appears in the diaphysis.
   2. Cartilage degenerates, creating a cavity.
   3. Ossification occurs in the epiphyses.
   4. Osteoblasts lay down bone in the primary ossification center.
   5. A cartilage model is formed.
      ⓐ 5,1,2,4,3   b. 1,5,4,3,2   c. 5,2,4,1,3
      d. 4,3,1,5,2   e. 3,1,4,5,2

8. Bone strength and bone formation are encouraged by all of the following except

   ⓐ parathyroid hormone   b. vitamin D   c. calcium and phosphorus   d. human growth hormone   e. mechanical stress

9. The ribs articulate with the

   ⓐ thoracic vertebrae   b. sacrum   c. cervical vertebrae
   d. lumbar vertebrae   e. atlas and axis

10. Match the following:

    __E__ a. run lengthwise through bone   A. lamellae
    __D__ b. connect central canals with   B. lacunae
            lacunae   C. perforating
    __A__ c. concentric rings of matrix   (Volkmann's)
    __C__ d. connect nutrient arteries and   canal
            nerves from the periosteum   D. canaliculi
            to the central (Haversian)   E. central
            canals   (Haversian)
    __B__ e. spaces that contain   canal
            osteocytes

11. Some bones have distinctive regions or markings. Which of the following bones is incorrectly paired with its region or marking.

    a. occipital bone—foramen magnum   b. axis (C2)—dens   c. hipbone (coxal bone)—acetabulum   d. sternum—xiphoid process   ⓔ humerus—olecranon

12. For each of the following bones, place an AX in the blank if it belongs to the axial skeleton, and an AP in the blank if it is part of the appendicular skeleton.

    __AX__ a. lacrimal bones   __AP__ p. fibula
    __AP__ b. clavicle   __AX__ q. palatine
    __AP__ c. radius   __AX__ r. hyoid
    __AX__ d. mandible   __AP__ s. tibia
    __AP__ e. patella   __AX__ t. sphenoid
    __AP__ f. carpals   __AX__ u. vertebrae
    __AP__ g. scapula   __AP__ v. hipbones
    __AX__ h. sternum   __AX__ w. maxilla
    __AP__ i. phalanges   __AX__ x. frontal
    __AP__ j. tarsals   __AX__ y. inferior nasal concha
    __AX__ k. ethmoid   __AP__ z. humerus
    __AP__ l. metatarsals   __AP__ aa. ulna
    __AX__ m. temporal   __AP__ bb. femur
    __AP__ n. metacarpals   __AX__ cc. ribs
    __AX__ o. vomer   __AX__ dd. occipital

13. Match the bone to its shape:

    __C__ a. humerus   A. flat
    __E__ b. carpus   B. irregular
    __B__ c. vertebrae   C. long
    __D__ d. patella   D. sesamoid
    __A__ e. sternum   E. short

Critical Thinking Applications

1. While investigating her new baby brother, a 4-year-old girl discovers a soft spot on the baby's skull and announces that the baby needs to go back because "it's not finished yet." Explain the presence of soft spots in the infant and their later ossification.

2. Brutus missed the punching bag and hit a brick wall with his fist. The doctor says Brutus has fractured the heads of metacarpals II through IV. Describe the location of the breaks in a layperson's terms and briefly explain the healing process.

3. A young boy is the shortest child in his class. The pediatrician determines that his height is at less than the 5th percentile on the chart for his age group. What laboratory tests and other information may the pediatrician need to diagnose his condition?

4. The clavicle is often fractured in falls. Trace the path of the force from a fall from the initial impact on the hand to the clavicle naming all the bones involved.

5. A very overweight person may have trouble walking due to the effects of the added weight on the vertebral column and on the joints of the lower extremities. Discuss some locations where problems may develop.

6. Lynda, a 55-year-old couch potato who smokes heavily, wants to lose 50 pounds before her next class reunion. Her diet consists mostly of diet soda and crackers. Explain the effects of her age and lifestyle on her bone composition.

Answers to Figure Questions

6.1   Articular cartilage; red bone marrow; diaphysis; endosteum.

6.2   Because the central canals are the main blood supply to the osteocytes of an osteon, their blockage would lead to death of the osteocytes.

6.3   Flat bones of the skull, mandible (lower jawbone), and clavicles (collarbones).

6.4   Epiphyseal line.

6.5   Heartbeat, respiration, nerve cell functioning, enzyme functioning, and blood clotting, to name just a few.

6.6   Axial skeleton: skull, vertebral column; appendicular skeleton: clavicle, pectoral girdle, humerus, pelvic girdle, femur.

6.7   Frontal, parietal, temporal, occipital, sphenoid, and ethmoid.

6.8   Foramen magnum.

6.9   Crista galli, frontal, parietal, temporal, occipital, temporal, parietal, frontal, crista galli.

6.10 Perpendicular plate.

6.11 Maxillae and palatine bones.

6.12 They produce mucus and serve as resonating chambers for vocalization.

6.13 Anterolateral (sphenoidal) fontanel.

6.14 Thoracic and sacral curves.

6.15 The vertebral foramen encloses the spinal cord while the intervertebral foramina provide spaces for spinal nerves to exit the vertebral column.

6.16 Atlas and axis.

6.17 The body weight supported by vertebrae increases toward the lower end of the backbone.

6.18 They are passageways for blood vessels and nerves.

6.19 True ribs (pairs 1–7), false ribs (pairs 8–12), and floating ribs (pairs 11 and 12).

6.20 Clavicle and scapula.

6.21 Shoulder blade.

6.22 Glenoid cavity.

6.23 Olecranon.

6.24 Heads of the metacarpals.

6.25 Lesser pelvis.

6.26 Femur.

6.27 Sesamoid bone.

6.28 Tibia.

6.29 Tibia.

6.30 Talus.

6.31 They are not rigid, yielding when weight is applied and springing back when weight is lifted.

6.32 Larger and more oval in females and heart-shaped in males.

7 chapter

ARTICULATIONS

student learning objectives

*B*ones are too rigid to bend without being damaged. Fortunately, flexible connective tissues form joints that hold bones together while still permitting some degree of movement, in most cases. Because most body movements occur at joints, you can appreciate their importance if you imagine how a cast over the knee joint makes walking difficult or how a splint on a finger limits the ability to manipulate small objects. A few joints do not permit any movement at all but do provide a great degree of protection.

Articulations

objective: *Define an articulation (joint) and describe how the structure of an articulation determines its function.*

An *articulation (joint)* is a point of contact between bones, between cartilage and bones, or between teeth and bones. When we say one bone *articulates* with another bone, we mean that one bone forms a joint with another bone. The scientific study of joints is referred to as *arthrology* (ar-THROL-ō-jē; *arthro* = joint; *logos* = study of).

The joint's structure determines how it functions. Some joints permit no movement, others permit slight movement, and still others afford fairly free movement. In general, the closer the fit at the point of contact, the stronger the joint. At tightly fitted joints, however, movement is restricted. The looser the fit, the greater the movement, but loosely fitted joints are prone to dislocation (displacement). Movement at joints is also determined by (1) the structure (shape) of articulating bones, (2) the flexibility (tension or tautness) of the connective tissue that binds the bones together, and (3) the position of associated ligaments, muscles, and tendons. Joint flexibility may also be affected by hormones. For example, relaxin, a hormone produced by the placenta and ovaries, relaxes the pubic symphysis and ligaments between the sacrum, hipbone, and coccyx toward the end of pregnancy, which assists in delivery.

Classification of Joints

Joints may be categorized into structural classes, based on anatomical characteristics, or into functional classes, based on the type of movement they permit.

Structural Classification

The structural classification of joints is based on the presence or absence of a space between the articulating bones called a *synovial (joint) cavity* (described shortly) and the type of connective tissue that binds the bones together. Structurally, joints are classified as (1) *fibrous,* if there is no synovial cavity and the bones are held together by fibrous connective tissue; (2) *cartilaginous,* if there is no synovial cavity and the bones are held together by cartilage; or (3) *synovial,* if there is a synovial cavity and the bones forming the joint are united by a surrounding articular capsule and frequently accessory ligaments (described in detail later).

Functional Classification

The functional classification of joints takes into account the degree of movement they permit. Functionally, a joint is classified as follows:

1. A *synarthrosis* (sin'-ar-THRŌ-sis; *syn* = together; *arthros* = joint) is an immovable joint; *synarthroses* (sēz) is plural.

2. An *amphiarthrosis* (am'-fē-ar-THRŌ-sis; *amphi* = both sides) is a slightly movable joint; *amphiarthroses* (sēz) is plural.

3. A *diarthrosis* (dī-ar-THRŌ-sis) is a freely movable joint; *diarthroses* (sēz) is plural.

We will discuss the joints of the body based on their functional classification, referring to their structural characteristics as well.

Synarthrosis (Immovable Joint)

A synarthrosis, or immovable joint, may be one of three types: suture, gomphosis, and synchondrosis.

Suture

A *suture* (SOO-cher; *sutura* = seam) is a fibrous joint found between bones of the skull. In a suture, the bones are united by dense fibrous connective tissue. The suture's irregular structure gives it added strength and decreases the chance of fractures. An example is the coronal suture between the parietal and frontal bones (see Figure 6.7b).

Gomphosis

A *gomphosis* (gom-FŌ-sis; *gomphosis* = to bolt together) is a type of fibrous joint in which a cone-shaped peg fits into a socket. The only examples are the articulations of the roots of the teeth in the alveoli (sockets) of the maxillae and mandible.

Synchondrosis

A *synchondrosis* (sin'-kon-DRŌ-sis; *syn* = together; *chondro* = cartilage) is a cartilaginous joint in which the connecting material is hyaline cartilage. The joint is eventually replaced by bone. Examples include the epiphyseal plate (see Figure 6.4) and the joint between the first rib and the sternum (see Figure 6.19).

Amphiarthrosis (Slightly Movable Joint)

An amphiarthrosis, or slightly movable joint, may be of two types: syndesmosis and symphysis.

Syndesmosis   *[handwritten: fibrous]*

In a **syndesmosis** (sin′-dez-MŌ-sis; *syndesmo* = band or ligament) joint, there is a much more dense fibrous connective tissue than in a suture, but the fit between bones is not quite as tight, which permits some flexibility. An example is the distal articulation of the tibia and fibula (see Figure 6.29).

Symphysis   *[handwritten: Cartilaginous]*

A **symphysis** (SIM-fi-sis; *symphysis* = growing together) is a cartilaginous joint in which the connecting material is a broad, flat disc of fibrocartilage. Examples include the intervertebral joints (see Figure 6.14) and the pubic symphysis (see Figure 6.25).

[handwritten: most central joint in body — true midline]

Diarthrosis (Freely Movable Joint)

A diarthrosis, or freely movable joint, has a variety of shapes and permits several different types of movements. At this point, we will discuss the general structure of a diarthrosis and later consider the various types.

Structure of a Diarthrosis

objective: *Describe the structure of a typical diarthrosis.*

A distinguishing anatomical feature of a diarthrosis is the space, called a **synovial** (si-NŌ-vē-al) (*joint*) **cavity** (Figure 7.1), that separates the articulating bones. For this reason, diarthroses are also called **synovial joints.** Another characteristic of such joints is the presence of **articular cartilage.** Articular cartilage (which is hyaline cartilage) covers the surfaces of the articulating bones, where it reduces friction when the bones move and helps absorb shock.

A sleevelike **articular** (*joint*) **capsule** surrounds a diarthrosis, encloses the synovial cavity, and unites the articulating bones. The articular capsule is composed of two layers. The outer layer, the **fibrous capsule,** usually consists of dense, irregular connective tissue. It attaches to the periosteum of the articulating bones. The flexibility of the fibrous capsule permits movement at a joint, whereas its great strength resists dislocation. The fibers of some capsules are arranged in parallel bundles of fibers called **ligaments** (*ligare* = to bind) and are given specific names (see Figure 7.5). The strength of the ligaments is one of the principal factors in holding bone to bone. Diarthroses are freely movable joints because of the synovial cavity and the arrangement of the articular capsule and ligaments.

[handwritten margin note: Blends & becomes fibrous capsule]

The inner layer of the articular capsule is a **synovial membrane,** which is composed of areolar connective tissue. It secretes **synovial fluid,** which looks and feels like uncooked egg white. The fluid fills the synovial cavity, where it lubricates and reduces friction in the joint and supplies nutrients to and removes wastes from cartilage cells of the articular cartilage.

Many diarthroses also contain additional ligaments called **accessory ligaments.** Some of these lie outside the articular capsule; others occur within the articular capsule (see Figure 7.5).

Inside some diarthroses, such as the knee joint, are pads of fibrocartilage called **articular discs** (**menisci**). They lie between two bones with different shapes and help stabilize the joint by providing a tighter fit (see Figure 7.5d). A tearing of

Figure 7.1   Generalized structure of a diarthrosis (synovial joint).

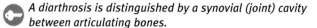

A diarthrosis is distinguished by a synovial (joint) cavity between articulating bones.

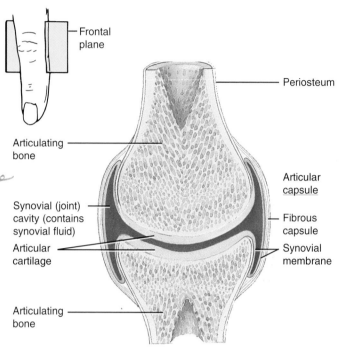

Frontal plane

Periosteum

Articulating bone

Articular capsule

Synovial (joint) cavity (contains synovial fluid)

Fibrous capsule

Articular cartilage

Synovial membrane

Articulating bone

Frontal section of a generalized diarthrotic (synovial) joint

In addition to the structures shown, what other structures may be found in diarthroses?

articular discs in the knee, commonly called **torn cartilage,** occurs frequently among athletes.

The various movements of the body create friction between moving parts. To reduce this friction, saclike structures called **bursae** (*bursa* = pouch) are situated in body tissues. They are filled with a fluid similar to synovial fluid. Bursae are located where bone and softer tissues meet, such as between the skin and bone, between tendons and bones, muscles and bones, and ligaments and bones. As fluid-filled sacs, they cushion the movement of one part of the body over another. An inflammation of a bursa is called **bursitis.**

Types of Diarthroses

objective: *Describe the types of diarthroses and the movements that occur at each.*

Though all diarthroses are similar in structure, variations exist in the shape of the articulating surfaces. Accordingly, diarthroses are divided into six subtypes: gliding joint, hinge joint, pivot joint, condyloid joint, saddle joint, and ball-and-socket joint.

Gliding Joint   *[handwritten: Arthroidal joint]*

The articulating surfaces of bones in a **gliding joint** are usually flat. These joints permit a **gliding movement** in which one surface moves back-and-forth and from side-to-side over another surface without any angular or rotary motion (Figure 7.2a).

Figure 7.2   Types of diarthroses (synovial joints). For each subtype shown, there is a drawing of the actual joint and a simplified diagram.

🔑 *Diarthroses are classified into subtypes on the basis of the shapes of the articulating bone surfaces.*

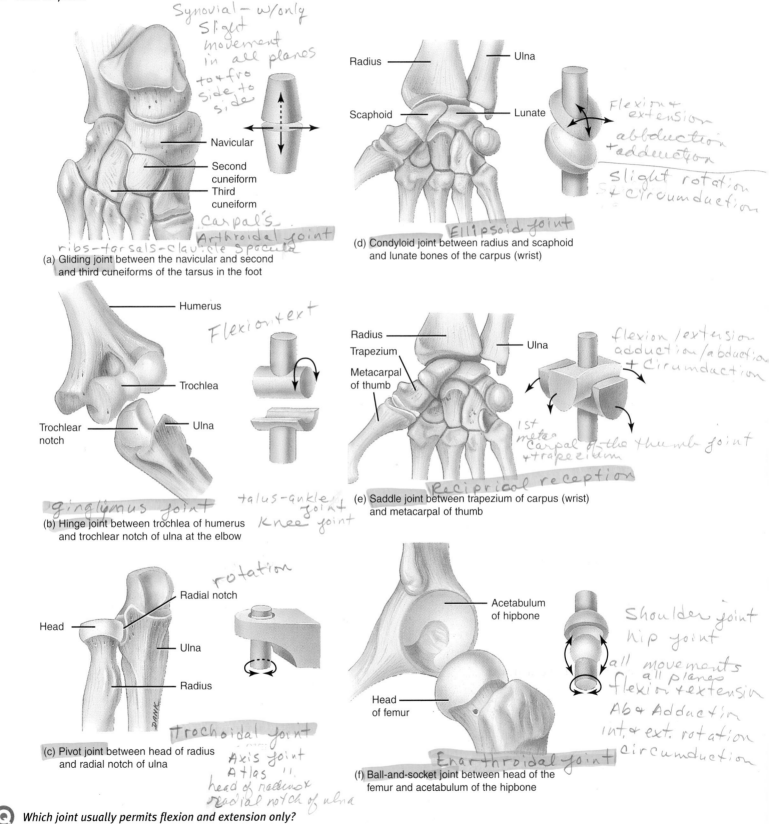

(a) Gliding joint between the navicular and second and third cuneiforms of the tarsus in the foot

(b) Hinge joint between trochlea of humerus and trochlear notch of ulna at the elbow

(c) Pivot joint between head of radius and radial notch of ulna

(d) Condyloid joint between radius and scaphoid and lunate bones of the carpus (wrist)

(e) Saddle joint between trapezium of carpus (wrist) and metacarpal of thumb

(f) Ball-and-socket joint between head of the femur and acetabulum of the hipbone

❓ *Which joint usually permits flexion and extension only?*

Examples of gliding joints are those between the carpals, tarsals, sternum and clavicle, and scapula and clavicle.

Hinge Joint — *Ginglymus*

In a *hinge joint,* the convex (()) surface of one bone fits into the concave (()) surface of another bone (Figure 7.2b). Movement is primarily in a single plane, similar to that of a hinged door. Movement is usually flexion and extension. *Flexion* decreases the angle between articulating bones and occurs when you bend your knee or elbow (Figure 7.3c,f). *Extension* increases the angle between articulating bones to restore a body part to its anatomical position after it has been flexed (Figure 7.3c,f). Other hinge joints are the ankle joints, interphalangeal joints, and the joint between the occipital bone and atlas. Some hinge joints are capable of *hyperextension,* continuation of extension beyond the anatomical position, such as when the head bends backward (Figure 7.3a).

Pivot Joint — *Trochoidal joint*

In a *pivot joint,* a rounded or pointed surface of one bone articulates within a ring formed partly by bone and partly by a ligament (see Figure 7.2c). The primary movement is *rotation,* movement of a bone around its own axis. We rotate the atlas around the axis when we turn the head from side-to-side (Figure 7.3j). Another pivot joint is found between the proximal ends of the ulna and radius and it allows us to turn the palms forward (or upward as in supination) and downward (or backward as in pronation).

Condyloid Joint — *Ellipsoid joint*

In a *condyloid joint,* the oval-shaped articulating surface of one bone fits into a depression of another bone (see Figure 7.2d). The joint at the wrist between the radius and carpals is a condyloid joint. The movement permitted by such a joint is

allows movement in 2 planes
flexion/extension
abduction/abduction

Figure 7.3 Representative movements at diarthroses (synovial joints).

🔑 *Whereas flexion decreases the angle at a joint, extension increases the angle.*

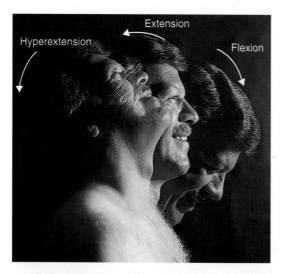

(a) Joints between atlas and occipital bone and between backbones

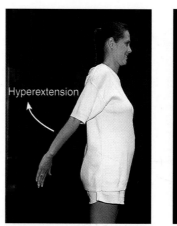 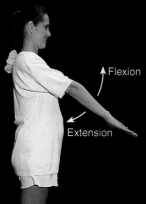

(b) Shoulder joint

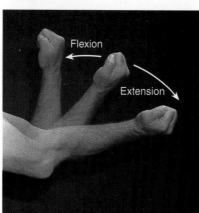

(c) Elbow joint

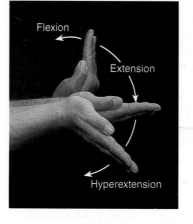

(d) Wrist joint

(e) Hip joint

Figure 7.3   *(Continued)*

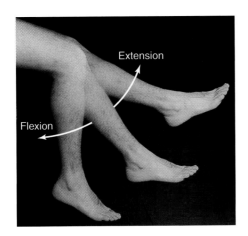

(f) Knee joint

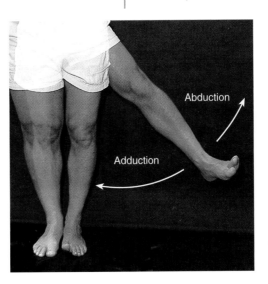

(g) Hip joint

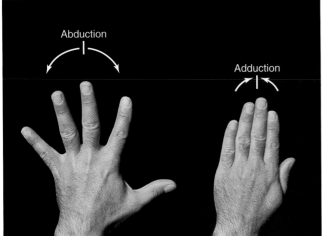

(h) Joints between metacarpals and phalanges

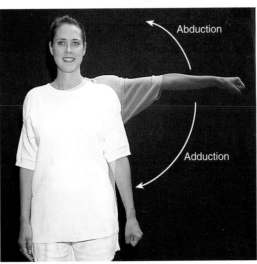

(i) Shoulder joint

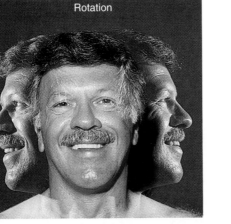

(j) Joint between atlas and axis

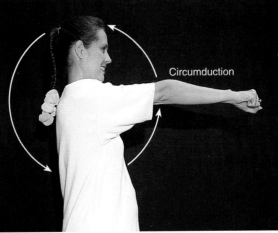

(k) Shoulder joint

movement
Composit of all
planes

Q   *Which movement moves the distal end of a bone in a circle? Moves a bone toward the midline? Moves a bone around its axis?*

side-to-side and back-and-forth, as when you flex and extend, and abduct and adduct the wrist. *Abduction* refers to movement away from the midline of the body (see Figure 7.3g–i). *Adduction* refers to movement toward the midline of the body (see Figure 7.3g–i). At a condyloid joint, it is possible to combine flexion–extension and abduction–adduction in sequence to produce a movement called *circumduction.* In circumduction, the distal end of a bone moves in a circle. For example, you can circumduct your hand by turning your hand in a circle (Figure 7.3k).

Saddle Joint

In a *saddle joint,* the articular surface of one bone is saddle-shaped and the articular surface of the other is shaped like the legs of a rider sitting in a saddle. Movements at a saddle joint are side-to-side and back-and-forth (see Figure 7.2e). The joint between the trapezium of the carpus and metacarpal of the thumb is a saddle joint. Such joints also permit circumduction, as in moving the thumb in a circle.

Ball-and-Socket Joint

A *ball-and-socket joint* consists of a ball-like surface of one bone fitted into a cuplike depression of another bone (see Figure 7.2f). Such a joint permits movement in three planes: flexion–extension, abduction–adduction, and rotation–circumduction (see Figure 7.2f). Examples are the shoulder joint and hip joint. The shoulder joint also permits circumduction, as when winding up to pitch a baseball (Figure 7.3k).

Special Movements

| **objective:** | *Describe several special movements that occur at diarthroses.* |

In addition to gliding movements, flexion, extension, hyperextension, rotation, abduction, adduction, and circumduction, several other movements also occur at diarthroses. These are called *special movements* and occur only at particular joints (Figure 7.4).

   Elevation is an upward movement of part of the body, and *depression* is a downward movement of part of the body (Figure 7.4a,b). You elevate your mandible when you close your mouth and depress your mandible when you open your mouth. You can also elevate and depress your shoulders. *Protraction* is the movement of the mandible or shoulder girdle forward. Thrusting the jaw outward is protraction of the mandible (Figure 7.4c). *Retraction* is the movement of a protracted part of the body backward. Pulling the lower jaw back in line with the upper jaw is retraction of the mandible (Figure 7.4d).

   Six special movements relate to the foot and hand (for all, refer to Figure 7.4e–h). *Inversion* is the movement of the sole inward (medially) so that the soles face toward each other. *Eversion* is the movement of the sole outward so the soles face

away from each other. *Dorsiflexion* involves bending the foot upward, and *plantar flexion* involves bending the foot downward. *Supination* is a movement of the forearm in which the palm of the hand is turned forward or upward. *Pronation* is a movement of the forearm in which the palm is turned backward or downward.

Knee Joint

In order to give you an idea of the complexity of diarthroses, we will examine some of the structural features of the knee joint. Among the main structures of the knee are the following (Figure 7.5 on page 148):

1. **Articular capsule.** In the knee, the capsule is strengthened by muscle tendons surrounding the joint.
2. **Patellar ligament.** *or Patellar* Tendon of insertion of the quadriceps femoris muscle that extends from the patella to the tibia. This ligament strengthens the anterior surface of the joint.
3. **Oblique popliteal** (pop-LIT-ē-al) **ligament.** Ligament that extends from the femur to the tibia. The ligament affords strength for the posterior surface of the joint.
4. **Arcuate popliteal ligament.** Ligament that extends from the femur to the fibula. It strengthens the lower lateral part of the posterior surface of the joint.
5. **Tibial (medial) collateral ligament.** Ligament on the medial surface of the joint that extends from the femur to the tibia. The ligament strengthens the medial aspect of the joint.
6. **Fibular (lateral) collateral ligament.** Ligament on the lateral surface of the joint that extends from the femur to the fibula. The ligament strengthens the lateral aspect of the joint.
7. **Anterior cruciate** (KROO-shē-āt) **ligament (ACL).** Extends posteriorly and laterally from the tibia to the femur. This ligament is stretched or torn in about 70 percent of all serious knee injuries.
8. **Posterior cruciate ligament (PCL).** Extends anteriorly and medially from the tibia to the femur.
9. **Articular discs (menisci).** Fibrocartilage discs between the tibial and femoral condyles. They help to compensate for the irregular shapes of the articulating bones. There are two types of articular discs:
   a. **Medial meniscus.** Semicircular piece of fibrocartilage (C-shaped) on the lateral aspect of the knee.
   b. **Lateral meniscus.** Nearly circular piece of fibrocartilage (approaches a complete O in shape) on the medial aspect of the knee.
10. **Bursae.** Saclike structures filled with fluid that help to reduce friction.

Figure 7.4   Special movements.

🔑 *Special movements occur only at specific joints.*

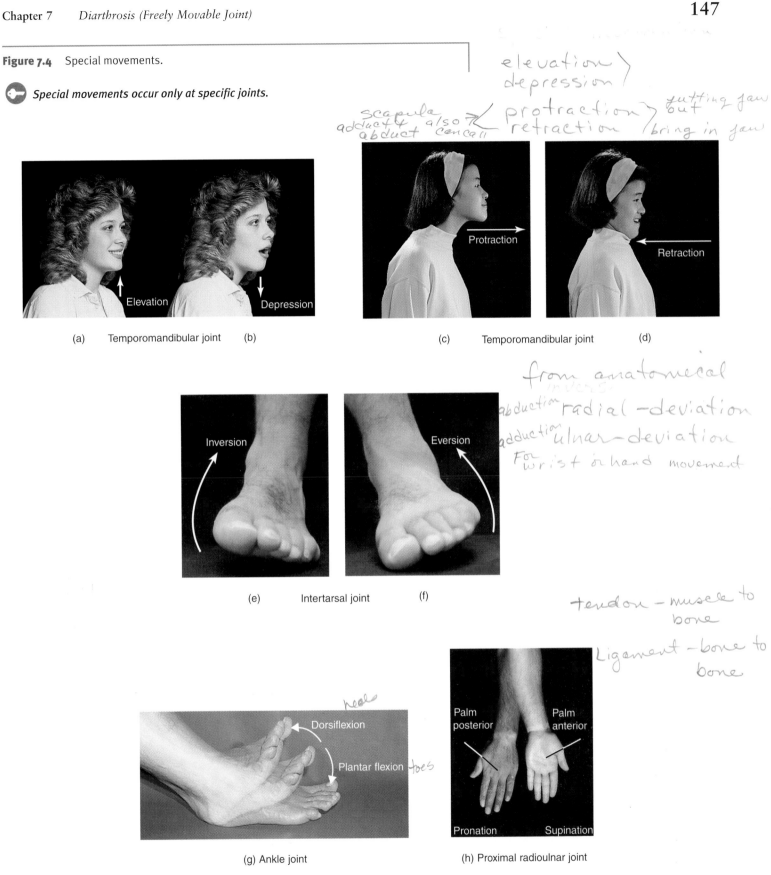

(a)   Temporomandibular joint   (b)

(c)   Temporomandibular joint   (d)

(e)   Intertarsal joint   (f)

(g) Ankle joint   (h) Proximal radioulnar joint

Handwritten notes:

elevation ⟩
depression ⟩

scapula adduct + abduct also ⟩ protraction ⟩ jutting jaw out
cancell   retraction ⟩ bring in jaw

from anatomical
abduction radial-deviation
adduction ulnar-deviation
For wrist or hand movement

tendon - muscle to bone
Ligament - bone to bone

heel
toes

Q *Which special movement is used when shrugging the shoulders?*

Figure 7.5   Structure of the knee joint.

Synovial joint at distal end
articulation between femur & Tibia

 The knee joint is the largest and most complex joint in the body.

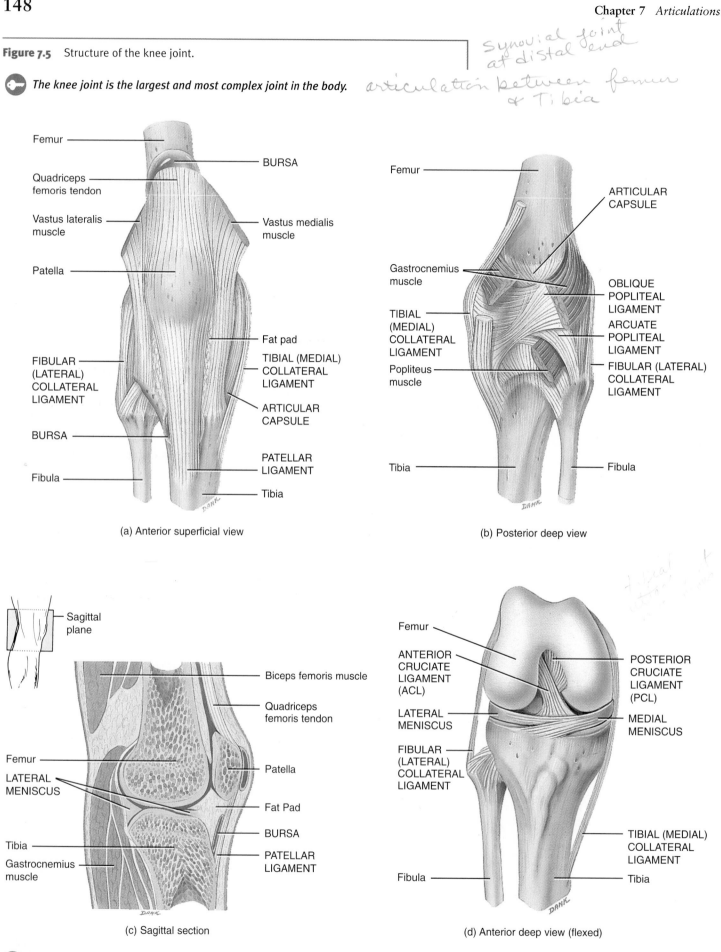

(a) Anterior superficial view

(b) Posterior deep view

(c) Sagittal section

(d) Anterior deep view (flexed)

 *Which structures are damaged in "torn cartilage"?*

massage therapists are subject CTD
cumulative TRAUMA Disorder
avoid by - good body mechanics
& doing stretches

wellness focus

Repetitive Motion Injuries

diarthroses (freely movable joints) were made for movement. But the human body is not a machine, and diarthroses were not designed to withstand the repetition of a given motion over and over and over again, all day long. When you repeat the same motion for extended periods of time, you may overstress the joint or joints responsible for that motion and associated soft-tissue structures, such as the articular capsule, ligaments, bursae, muscles, tendons, and nerves. Repeated episodes of mechanical stress can lead to the development of *repetitive motion injuries*.

Repeat That?

Repetitive motion injuries are a type of *cumulative trauma disorder* (CTD), defined by OSHA (Occupational Health and Safety Administration) as a group of disorders characterized by ongoing damage to soft tissues. Repetitive motion injuries are the most common type of CTD, but CTDs may also involve trauma due to exposure to cold or hot temperatures, certain types of lighting, vibration, and so forth. Repetitive motion injuries are similar in many ways to the *overuse injuries* that athletes frequently experience. Just as tennis players may develop epicondylitis (tennis elbow), so too may building maintenance personnel and construction workers who perform repeated elbow flexion and extension in their work.

Many readers are familiar with carpal tunnel syndrome, in which pressure develops on the median nerve as it passes through the carpal tunnel, a narrow tunnel of bone and ligament at the wrist. Pressure on this nerve causes numbness, tingling, and pain in some or all of the fingers. While carpal tunnel syndrome can be caused by many factors, it is also a common repetitive motion injury in people who perform repeated motions with their hands.

Repetitive motions alone may cause repetitive motion injuries. Risk increases when repetitive motions are coupled with poor posture and biomechanics, which put excess strain on joints. Joint stress also increases when a person must apply force with the motion, such as when gripping or lifting heavy materials. The joints at highest risk are those that are the weakest: wrists, backs, elbows, shoulders, and necks, the most common sites of repetitive motion injury.

Repetitive motion injuries usually develop slowly over a long period of time. They typically begin with mild to moderate discomfort in the affected joints, especially at night. Other symptoms include swelling in the joint, muscle fatigue, numbness, and tingling. Early symptoms may come and go at first, but then they become constant. Symptoms of advanced damage include more intense pain, muscle weakness, and nerve problems. If left untreated, repetitive motion injuries can be extremely painful and severely limit a joint's range of motion. Fortunately, because they develop slowly, most repetitive motion injuries are discovered early enough to be treated successfully.

Ergonomics: Avoid a Repeat Performance

Treatment begins with diagnosis of the cause of the repetitive motion stress. People trained in biomechanics have developed a field known as *ergonomics*, an applied science concerned with designing and arranging things people use so that people and things interact more efficiently. Specialists in ergonomics analyze how to set up computer workstations, industrial assembly lines, and even the interiors of automobiles in order to produce the least amount of mechanical stress on the bodies of the people operating the equipment. Just as coaches help athletes use correct biomechanics to avoid injury, ergonomics specialists also train workers to perform their jobs in ways that avoid injury. However, even improved equipment design, correct posture, and appropriate biomechanics will not prevent or correct all repetitive motion injuries. Sometimes job requirements must be adjusted to include a wider variety of tasks in order to reduce the need for repetitive motions. Medical treatment is often required for repetitive motion injuries, and usually includes the use of pain relievers, ice, rest, and physical therapy to increase the strength and flexibility of the injured area.

critical thinking

Why do you think dental hygienists, musicians, and people who type all day are at increased risk for the development of carpal tunnel syndrome? What kinds of things could these people do to lessen their risk of developing carpal tunnel syndrome?

Why does strength training help prevent job-related repetitive motion injuries?

149

Common Disorders

Rheumatism

Rheumatism (*rheumat* = subject to flux) refers to any painful state of the supporting structures of the body—its bones, ligaments, joints, tendons, or muscles. Arthritis is a form of rheumatism in which the joints are inflamed.

Osteoarthritis *1st most common*

Osteoarthritis (os'-tē-ō-ar-THRĪ-tis) is a degenerative joint disease characterized by deterioration of articular cartilage and spur (new bone) formation. *localized to Over used or injured joints*

Gouty Arthritis

In *gouty* (GOW-tē) *arthritis,* sodium urate crystals are deposited in the soft tissues such as the kidneys and cartilage of joints. The crystals eventually destroy all the joint tissues.

Rheumatoid Arthritis (RA) *2nd most common*
Systemic

Rheumatoid (ROO-ma-toyd) *arthritis* (**RA**) is an autoimmune disease in which an individual's antibodies attack the joint tissues. Although RA is a less common type of arthritis, it is a more severe and potentially crippling disease. The primary symptom is inflammation of the synovial membrane.

Lyme Disease

Lyme disease is caused by a bacterium (*Borrelia burgdorferi*) transmitted to humans by deer ticks and other ticks. Within a few weeks of the tick bite, a rash may appear at the site and may be accompanied by joint stiffness, fever and chills, headache, stiff neck, nausea, and low back pain.

Dislocation

A *dislocation or luxation* (luks-Ā-shun) is the displacement of a bone from a joint with tearing of ligaments, tendons, and articular capsules. A partial or incomplete dislocation is called a *subluxation.*

Sprain and Strain

A *sprain* is the forcible wrenching or twisting of a joint with partial rupture or other injury to its attachments without dislocation. A sprain is more serious than a *strain,* which is the overstretching of a muscle. ∎

Medical Terminology and Conditions

Arthralgia (ar-THRAL-jē-a; *arth* = joint; *algia* = pain) Pain in a joint.

Arthroscopy (ar-THROS-kō-pē; *arthro* = joint; *skopein* = to view) A procedure that involves examination of the interior of a joint, usually the knee, using an arthroscope, a lighted instrument the diameter of a pencil. It is used to determine the nature and extent of damage following knee injury; to remove torn cartilage and repair cruciate ligaments in the knee; to obtain tissue samples for analysis and to perform surgery on other joints, such as the shoulder, elbow, ankle, and wrist; and to monitor the progression of disease and the effects of therapy.

Arthrosis (ar-THRŌ-sis) Refers to an articulation; also a disease of a joint.

Chondritis (kon-DRĪ-tis; *chondro* = cartilage) Inflammation of cartilage.

Rheumatology (roo'-ma-TOL-ō-jē; *rheumat* = subject to flux) The study of joints; the field of medicine devoted to joint diseases and related conditions.

Synovitis (sin'-ō-VĪ-tis; *synov* = joint) Inflammation of a synovial membrane in a joint.

Study Outline

Classification of Joints (p. 141)

1. An articulation (joint) is a point of contact between two or more bones.
2. Structural classification is based on the presence of a synovial (joint) cavity and type of connecting tissue. Structurally, joints are classified as fibrous, cartilaginous, and synovial.
3. Functional classification of joints is based on the degree of movement permitted. Joints may be synarthroses, amphiarthroses, or diarthroses.

Synarthrosis (Immovable Joint) (p. 141)

1. A synarthrosis is an immovable joint.
2. Types of synarthroses include a suture (found between skull bones), a gomphosis (roots of teeth in alveoli of mandible and maxilla), and a synchondrosis (temporary cartilage between diaphyses and epiphyses).

Amphiarthrosis (Slightly Movable Joint) (p. 141)

1. An amphiarthrosis is a slightly movable joint.

2. Types of amphiarthroses include a syndesmosis (such as the tibiofibular articulation) and a symphysis (the pubic symphysis).

Diarthrosis (Freely Movable Joint) (p. 142)

1. A diarthrosis is a freely movable joint (synovial joint).

2. Diarthroses contain a synovial (joint) cavity, articular cartilage, and a synovial membrane; some also contain ligaments, articular discs (menisci), and bursae.

Types of Diarthroses (p. 142)

1. Diarthroses (synovial joints) can be classified according to the shape of the articulating surfaces.

2. Types of diarthroses include gliding joints (wrist bones), hinge joints (elbow), pivot joints (between radius and ulna), condyloid joints (between radius and wrist), saddle joints (between wrist and thumb), and ball-and-socket joints (shoulder).

Special Movements (p. 146)

1. In addition to gliding movements, flexion, extension, hyperextension, rotation, abduction, adduction, and circumduction, special movements also occur at synovial joints.

2. Special movements include inversion, eversion, dorsiflexion, plantar flexion, protraction, retraction, supination, pronation, elevation, and depression.

Knee Joint (p. 146)

1. The knee joint is a diarthrosis that illustrates the complexity of this type of joint.

2. It contains an articular capsule, several ligaments within and around the outside of the joint, articular discs (menisci), and bursae.

Self-Quiz

1. The point of contact between two or more bones is known as a(n)

   a. synapse   b. conjunction   c. diarthrosis   d. articulation   e. fusion

2. The roots of the teeth in the alveolar sockets of the mandible and maxilla are what type of joint?

   a. diarthrosis   b. gomphosis   c. symphysis   d. synchondrosis   e. suture

3. Structural classification of joints is based on which of the following?

   a. degree of movement   b. presence of a synovial cavity   c. type of connective tissue at the joint   d. both a and b are correct   e. both b and c are correct

4. Articular cartilage and bursae would most likely be found in which of the following?

   a. gomphosis   b. a suture   c. the pubic symphysis   d. the knee   e. a synchondrosis

5. Which of the following structures provides flexibility to a joint while at the same time resisting dislocation?

   a. bursae   b. articular cartilage   c. synovial fluid   d. muscles   e. articular capsule

6. A joint in which there is no cavity and the bones are held together by fibrous connective tissue would be structurally classified as a _____ joint.

   a. synovial   b. symphysis   c. fibrous   d. synchondrosis   e. amphiarthrosis

7. The joints between the vertebrae and the joint between the pubic bones are examples of which joint type?

   a. synovial   b. symphysis   c. fibrous   d. synchondrosis   e. suture

8. A fibrocartilage pad located within a synovial joint is called a(n)

   a. articular cartilage   b. articular capsule   c. articular disc (meniscus)   d. bursa   e. synovial membrane

9. The joint of the humerus and ulna at the elbow is a _____ joint.

   a. ball-and-socket   b. hinge   c. gliding   d. condyloid   e. saddle

10. The joint type seen between carpals and tarsals is the _____ joint.

    a. ball-and-socket   b. hinge   c. gliding   d. pivot   e. saddle

11. A pivot joint can be found between which of the following bones?

    a. talus and tibia   b. clavicle and sternum   c. radius and ulna   d. humerus and ulna   e. scapula and humerus

12. Which of the following diarthrotic joints allows for the greatest degree of movement?

    a. ball-and-socket   b. hinge   c. gliding   d. pivot   e. saddle

13. Match the following:

    ____ a. movement of a bone around its own axis

    ____ b. movement away from the midline of the body

    ____ c. palm faces upward or forward

    ____ d. downward movement of a body part

    ____ e. movement toward the body midline

    ____ f. movement of the mandible or shoulder backward

    ____ g. turning the palm so it faces downward or backward

    ____ h. upward movement of a body part

    ____ i. movement of a distal end of a body part in a circle

    ____ j. beyond the plane of extension

    A. rotation
    B. supination
    C. depression
    D. adduction
    E. retraction
    F. pronation
    G. abduction
    H. hyperextension
    I. circumduction
    J. elevation

Critical Thinking Applications

1. Periodontal disease is a major cause of dental problems and tooth loss in adults. One area affected is the joint between the teeth and the mandible and maxillae. Name this type of joint and describe its structure. How will structural damage affect the joint function?

2. John tripped over a kitchen chair leg. He twisted his knee joint during his fall and then hit the tiled floor directly on his patella. John now has a bad case of "water on the knee." What is the scientific name for the fluid in the joint? What knee structures may have been damaged in the fall?

3. An elderly aunt complains of rheumatism (arthritis) in her neck and arms. She is a plump woman who has always enjoyed baking for her nieces and nephews. You've always enjoyed eating her creations. What can you suggest to ease her rheumatism?

4. Ehlers–Danlos syndrome, an inherited connective tissue disorder, causes changes in the structure of collagen (the strong protein of fibrous connective tissues). What effects would weakened collagen have on joint structure and function?

5. A father quickly jerks back on his toddler's arm to prevent a fall down the stairs. The child holds his arm to his body and screams in pain. Which joint was probably dislocated or injured during the rescue? What structural features of this joint make it susceptible to such injury?

6. After your second A & P exam, you dropped to one knee, raised one arm over your head with hand clenched in a fist, pumped your arm up and down, bent your head back to look straight up, and let out a loud "YES!" Use the proper terms to describe the movements undertaken by the various joints.

Answers to Figure Questions

7.1   Accessory ligaments, articular discs (menisci), and bursae.
7.2   Hinge joint.
7.3   Circumduction; adduction; rotation.

7.4   Elevation.
7.5   Articular discs (menisci).

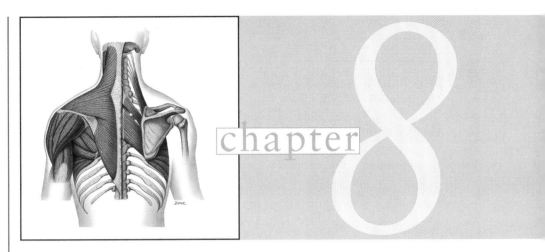

chapter 8

THE MUSCULAR SYSTEM

a look ahead

*some say the liver/some the heart/ but actually
muscle system consumes the most energy
skeletal muscle is an organ (4 types of tissue)*

Although bones and joints form the framework of the body, they cannot move the body by themselves. Motion results from the contraction and relaxation of muscles. Muscle tissue constitutes about 40 to 50 percent of the total body weight and is composed of highly specialized cells. The scientific study of muscles is known as *myology* (mī-OL-ō-jē; *myo* = muscle; *logos* = study of).

Types of Muscle Tissue

There are three types of muscle tissue.

Skeletal muscle tissue, which is named for its location, is attached mostly to bones and it moves parts of the skeleton. It is *striated*; that is, striations, or alternating light and dark bands, are visible under a microscope. It is *voluntary* because it can be made to contract and relax by conscious control.

Cardiac muscle tissue forms the bulk of the wall of the heart. It is *striated* and *involuntary*; that is, its contractions are not under conscious control. *pushes Blood thru ♡*

Smooth muscle tissue is involved with internal processes. It is located in the walls of hollow internal structures, such as blood vessels, the stomach, and the intestines. It is *nonstriated,* because it lacks striations, and *involuntary.* *regulates diameter of vessels*

Functions of Muscle Tissue

Through sustained contraction or alternating contraction and relaxation, muscle tissue has four key functions: producing motion, moving substances within the body, providing stabilization, and generating heat.

1. *Motion.* Motion is obvious in movements such as walking and running, and in localized movements, such as grasping a pencil or nodding the head. These movements rely on the integrated functioning of bones, joints, and skeletal muscles.

2. *Movement of substances within the body.* Cardiac muscle produces contractions that move blood through the heart and blood vessels. Smooth muscle contractions move food through the gastrointestinal tract, sperm and ova through the reproductive systems, and urine through the urinary system.

3. *Stabilizing body positions and regulating organ volume.* Skeletal muscle contractions maintain the body in stable positions, such as standing or sitting. Postural muscles display sustained contractions when a person is awake; for example, partially contracted neck muscles hold the head upright. In a similar manner, sustained contractions of smooth muscles may prevent outflow of the contents of a hollow organ. Temporary storage of food in the stomach or urine in the urinary bladder is possible because smooth muscles close off the exit route.

4. *Heat production.* As skeletal muscle contracts to perform work, a by-product is heat. Much of the heat released by muscle is used to maintain normal body temperature. Muscle contractions are thought to generate as much as 85 percent of all body heat. This is why active cheering helps warm you up during a cold weather football game.

Characteristics of Muscle Tissue

Muscle tissue has four principal characteristics that are important in understanding its functions:

1. *Excitability* is the ability of muscle tissue to receive and respond to stimuli.

2. *Contractility* is the ability to shorten and thicken (contract).

3. *Extensibility* is the ability of muscle tissue to stretch (extend).

4. *Elasticity* is the ability of muscle tissue to return to its original shape after contraction or extension.

Skeletal Muscle Tissue

objective: *Describe the connective tissue components, blood and nerve supply, and histology of skeletal muscle tissue.*

To understand how muscles move, you need some knowledge of their connective tissue coverings, nerve and blood supply, and the structure of individual muscle fibers (cells).

Connective Tissue Components

The term *fascia* (FASH-ē-a; *fascia* = bandage) is applied to a sheet or broad band of fibrous connective tissue beneath the skin or around muscles and other organs of the body. There are two types of fascia: (1) *superficial fascia* (*subcutaneous layer*), which is composed of areolar connective tissue and adipose tissue and is immediately under the skin; and, more important to the study of muscles, (2) *deep fascia,* which is composed of dense, irregular connective tissue and holds muscles together, separating them into functional groups.

Several connective tissue coverings extend from the deep fascia (Figure 8.1). The entire muscle is wrapped in *epimysium* (ep′-i-MĪZ-ē-um; *epi* = upon). Bundles of muscle fibers (cells) called *fascicles* (FAS-i-kuls) are covered by *perimysium* (per′-i-MĪZ-ē-um; *peri* = around). Finally, *endomysium* (en′-dō-MĪZ-ē-um; *endo* = within) wraps each individual muscle fiber.

Epimysium, perimysium, and endomysium extend beyond the muscle as a *tendon* (*tendere* = to stretch out)—a cord of connective tissue that attaches a muscle to a bone. An

✓(3) Sub-serous fascia — loose connective tissue lies between deep fascia + serous membranes lining cavities

blast = bud/or germinate a mother cell
a muscle cell is a myofiber originate from myoblast cells are multinucleate contain many mitochondria

Figure 8.1 Relationships of connective tissue to skeletal muscle showing the relative positions of the epimysium, perimysium, and endomysium.

sarco - muscle
sarcolemma - plasma membrane (outer layer) of cell

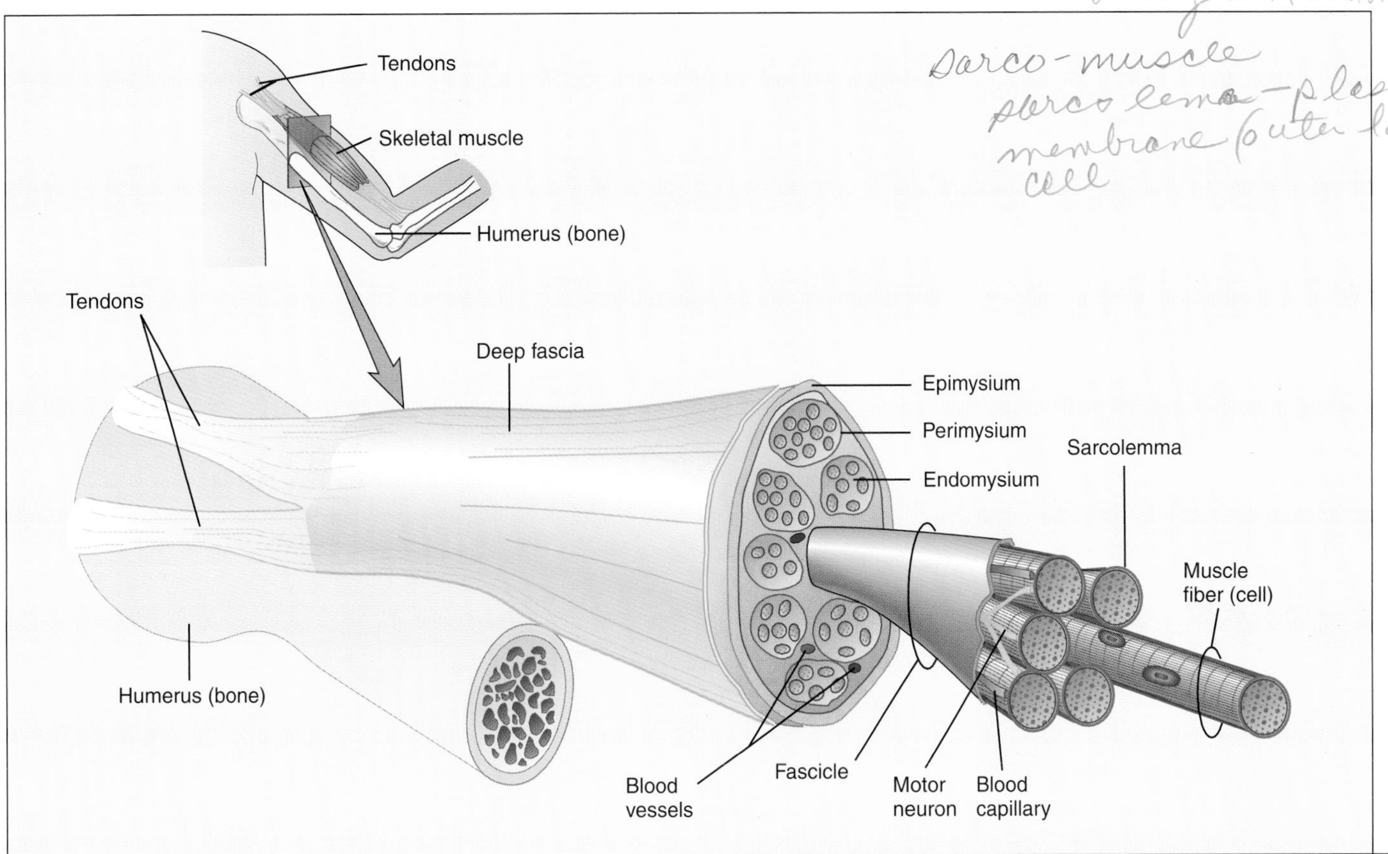

OVERVIEW OF FUNCTIONS OF MUSCLE TISSUE

1. Motion. Movements such as walking and running.
2. Movement of substances within the body. All three types of muscle tissue help move substances such as blood, food, sperm and ova, and urine.
3. Stabilizing body positions and regulating organ volume. Skeletal muscle contractions maintain stable body positions and posture. Sustained contractions of smooth muscles block exit of food from the stomach and urine from the urinary bladder for temporary storage.
4. Heat production. Skeletal muscle contractions may generate 85 percent of all body heat to help maintain normal body temperature.

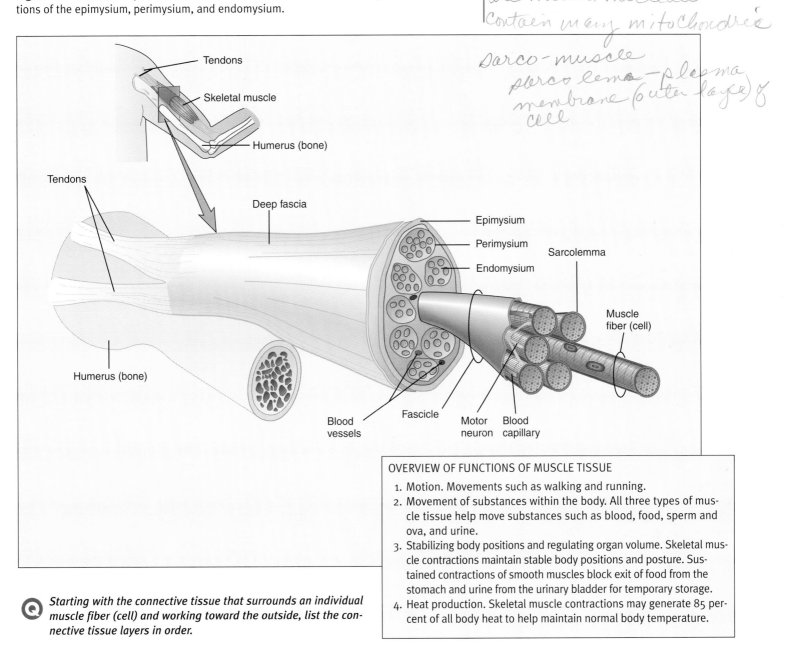

Starting with the connective tissue that surrounds an individual muscle fiber (cell) and working toward the outside, list the connective tissue layers in order.

example is the calcaneal (Achilles) tendon of the gastrocnemius muscle (see Figure 8.25a).

Nerve and Blood Supply

Skeletal muscles are well supplied with nerves and blood vessels, both of which are directly related to contraction, the chief characteristic of muscle (see Figure 8.1). For a skeletal muscle fiber to contract, it must first be stimulated by an electric current called a ***muscle action potential.*** Muscle contraction also requires a good deal of energy and therefore large amounts of nutrients and oxygen. Moreover, the waste products of these energy-producing reactions must be eliminated. Thus prolonged muscle action depends on a rich blood supply to deliver nutrients and oxygen and remove wastes.

Generally, an artery and one or two veins accompany each nerve that penetrates a skeletal muscle. Microscopic blood vessels called capillaries are distributed within the endomysium. Each muscle fiber is thus in close contact with one or more capillaries. Each skeletal muscle fiber also makes contact with a portion of a motor neuron (nerve cell) called a synaptic end bulb (see Figure 8.5a).

Histology

muscle as a tissue (3 types)

Microscopic examination of a skeletal muscle reveals that it consists of thousands of elongated, cylindrical cells arranged parallel to each other and are called ***muscle fibers*** or ***myofibers*** (Figure 8.2c). Each fiber is covered by a plasma membrane called the ***sarcolemma*** (*sarco* = flesh; *lemma* = sheath). The cytoplasm,

a muscle cell is a muscle fiber

Figure 8.2   Organization of skeletal muscle from gross to molecular levels.

The structural organization of a skeletal muscle from macroscopic to microscopic is: skeletal muscle, fascicle (bundle of muscle fibers), muscle fiber, myofibril, thin and thick myofilaments.

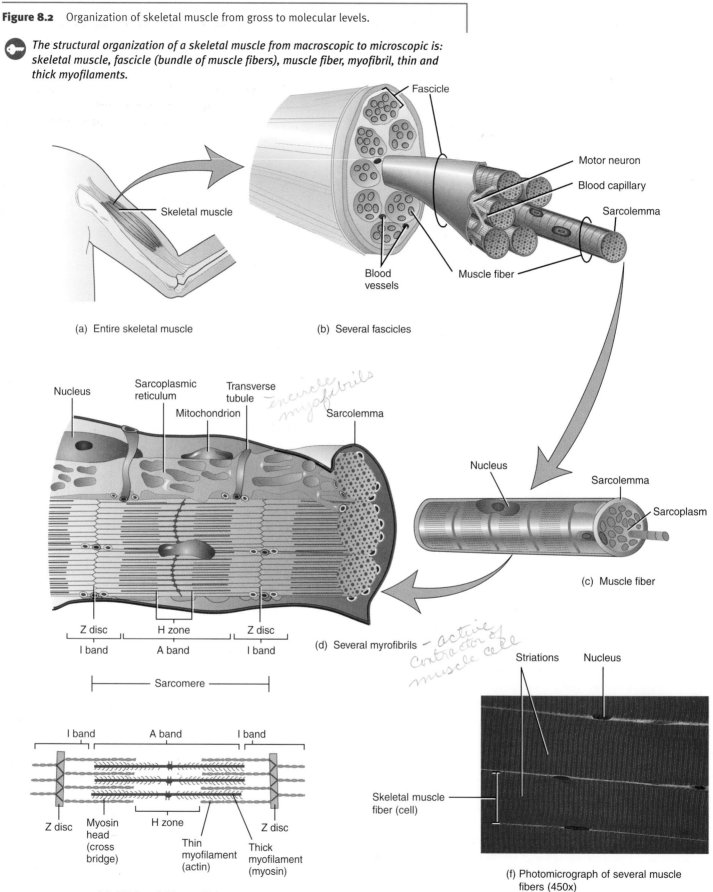

(a) Entire skeletal muscle

(b) Several fascicles

(c) Muscle fiber

(d) Several myofibrils

(e) Thick and thin myofilaments

(f) Photomicrograph of several muscle fibers (450x)

Of what does the A band consist? Of what does the I band consist? Where are calcium ions (Ca²⁺) stored?

called *sarcoplasm,* of a muscle fiber contains many nuclei (multinucleate). In the sarcoplasm are numerous mitochondria. The large numbers are related to the large amounts of energy (ATP) that muscle tissue must generate in order to contract. The sarcoplasm also contains myofibrils (described shortly), special high-energy molecules (also to be described shortly), enzymes, and *sarcoplasmic reticulum* (sar'-kō-PLAZ-mik re-TIK-yoo-lum), a network of membrane-enclosed tubules similar to smooth endoplasmic reticulum (Figure 8.2d). Perpendicular to the sarcoplasmic reticulum are *transverse tubules (T tubules).* The tubules are tunnel-like extensions of the sarcolemma that pass through the muscle fiber and also open to the exterior of the muscle fiber. Skeletal muscle fibers also contain *myoglobin,* a reddish pigment similar to hemoglobin in blood. Myoglobin stores oxygen until needed by mitochondria to generate ATP.

Each skeletal muscle fiber is composed of cylindrical structures called *myofibrils,* which run lengthwise through the muscle fiber. They, in turn, consist of two kinds of even smaller structures called *thin myofilaments* and *thick myofilaments.*

Myofilaments do not extend the entire length of a muscle fiber; they are arranged in compartments called *sarcomeres* (*meros* = part), which are the basic functional units of striated muscle fibers (Figure 8.2d). Sarcomeres are separated from one another by narrow zones of dense material called *Z discs (lines).* Within a sarcomere is a dark area, called the *A band,* composed of mostly thick myofilaments. A narrow *H zone* in the center of each A band contains thick, but no thin, myofilaments. A light-colored area called the *I band* is composed of thin myofilaments.

This combination of alternating dark A bands and light I bands gives the muscle fiber its striated (striped) appearance.

Thick myofilaments are composed mostly of the protein *myosin,* which is shaped like two golf clubs twisted together. The tails (handles of the golf club) are arranged parallel to each other, forming the shaft of the thick myofilament. The heads of the golf clubs project outward on the surface of the shaft. These projecting heads are referred to as *myosin heads* or *cross bridges* (Figure 8.3b).

Thin myofilaments are anchored to the Z discs. They are composed mostly of the protein *actin* arranged in two single strands that resemble two strands of pearls twisted together (Figure 8.3c). Each actin molecule contains a *myosin-binding site* for a myosin head. Besides actin, the thin myofilaments contain two other protein molecules, *tropomyosin* and *troponin,* that help regulate muscle contraction. In a relaxed muscle fiber, they block the myosin-binding sites.

Contraction

objective: *Explain the factors involved in the contraction of skeletal muscle fibers.*

Sliding-Filament Mechanism

During muscle contraction, myosin heads (cross bridges) pull on the thin myofilaments, causing them to slide inward toward

Figure 8.3 Detailed structure of myofilaments. (a) The relation of thick (myosin) and thin (actin) myofilaments in a sarcomere. (b) About 200 myosin molecules comprise a thick myofilament. The myosin tails all point toward the center of the sarcomere. (c) Thin myofilaments contain actin, troponin, and tropomyosin.

Myofibrils contain thick and thin myofilaments.

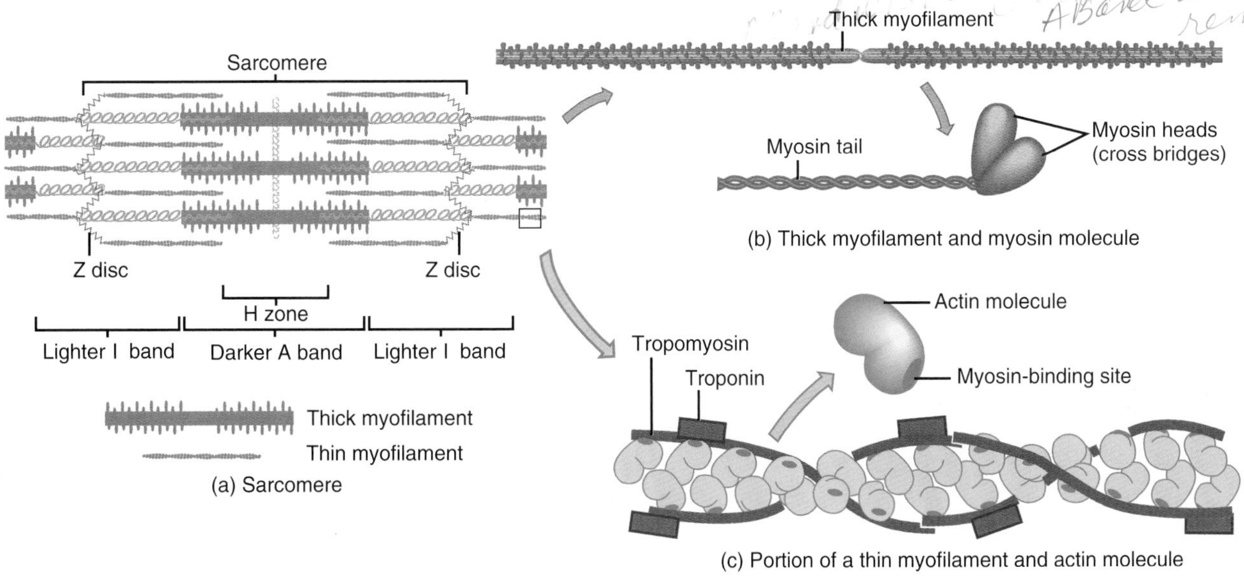

(a) Sarcomere

(b) Thick myofilament and myosin molecule

(c) Portion of a thin myofilament and actin molecule

What proteins are present in the A band? In the I band?

the center of a sarcomere (Figure 8.4). The sarcomere shortens, but the lengths of the thin and thick myofilaments themselves do not change. The myosin heads of the thick myofilaments connect with the actin of the thin myofilaments. The myosin heads move like the oars of a boat on the surface of the thin myofilaments, and the thin and thick myofilaments slide past each other. The myosin heads may pull the thin myofilaments of each sarcomere so far inward toward the center of a sarcomere that they overlap. As the thin myofilaments slide inward, the Z discs are drawn toward each other and the sarcomere is shortened. The sliding of myofilaments and shortening of sarcomeres cause the shortening of the muscle fibers. This is the **sliding-filament mechanism** of muscle contraction. This process occurs only when there are sufficient calcium ions (Ca²⁺) and an adequate supply of energy.

Neuromuscular Junction

For a skeletal muscle fiber to contract, it must be stimulated by a nerve cell, or **neuron**. The particular type of neuron that stimulates muscle tissue is called a **motor neuron**. A **motor unit** is composed of a motor neuron and all the muscle fibers it stimulates. A single motor neuron connects to many muscle fibers. Stimulation of one motor neuron causes all the muscle fibers in that motor unit to contract simultaneously. Muscles that control precise movements, such as the external eye muscles, have fewer than 10 muscle fibers in each motor unit but many motor units. Muscles of the body that are responsible for gross (large) movements, such as the biceps brachii in the arm and gastrocnemius in the leg, may have as many as 2000 muscle fibers in each motor unit but few motor units.

When the axon of a motor neuron enters a skeletal muscle, it branches into axon terminals that approach—but do not touch—the sarcolemma of a muscle fiber. *NERVE TERMINAL* (handwritten) This region of the sarcolemma near the axon terminal is known as the **motor end plate**. The term **neuromuscular junction** refers to the axon terminal of a motor neuron together with the motor end plate (Figure 8.5). The ends of the axon terminals enlarge into swellings known as **synaptic end bulbs.** They contain sacs called **synaptic vesicles** filled with chemicals called **neurotransmitters.** The space between the axon terminal and sarcolemma is known as a **synaptic cleft.**

When a nerve impulse (nerve action potential) reaches an axon terminal, Ca²⁺ ions from interstitial fluid enter the synaptic end bulb, causing synaptic vesicles to release a neurotransmitter (Figure 8.5c). The neurotransmitter released at neuromuscular junctions is **acetylcholine** (as′-ē-til-KŌ-lēn), or **ACh.** The ACh diffuses across the synaptic cleft to combine with ACh receptors on the sarcolemma. This produces a muscle action potential that travels along the sarcolemma and results in contraction. The details of nerve impulse generation and muscle action potential are discussed in Chapter 9.

As long as ACh is present in the neuromuscular junction, it will stimulate the muscle fiber. Continuous stimulation by ACh is prevented by an enzyme called **acetylcholinesterase** (**AChE**) (as′-ē-til-kō′-lin-ES-ter-ās). AChE is found in the synaptic cleft. AChE inactivates ACh within 1/500 second. This permits the muscle fiber to prepare itself so another muscle action potential may be generated. When the next nerve impulse comes through, the synaptic vesicles release more ACh, another muscle action potential is generated, and AChE again inactivates ACh. This cycle is repeated over and over again.

Physiology of Contraction

In order for a muscle to contract, both calcium ions (Ca²⁺) and energy, in the form of ATP, are required.

When a muscle fiber is relaxed (not contracting), there is a low concentration of Ca²⁺ in the sarcoplasm (Figure 8.6a on page 160). This is because the sarcoplasmic reticulum membrane contains Ca²⁺ active transport pumps that move Ca²⁺ from the sarcoplasm into the sarcoplasmic reticulum. However, when a muscle action potential travels along the sarcolemma and into the transverse tubule system, Ca²⁺ release channels open in the sarcoplasmic reticulum membrane (Figure 8.6b). The result is a flood of Ca²⁺ into the sarcoplasm around the thick and thin myofilaments. The Ca²⁺ combines with troponin, causing the troponin to change shape. This shape change slides the troponin–tropomyosin complex away from the myosin-binding sites on actin (Figure 8.6b).

Figure 8.4   Sliding-filament mechanism of muscle contraction.

🔑 *During muscle contraction, thin myofilaments move inward toward the H zone.*

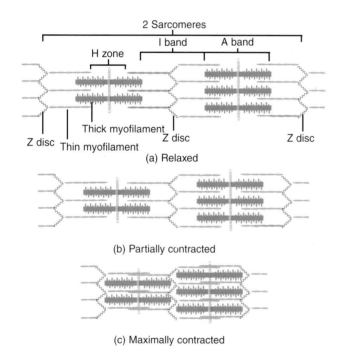

(a) Relaxed

(b) Partially contracted

(c) Maximally contracted

❓ *What happens to the I band as muscle contracts? Do the lengths of the thick and thin myofilaments change?*

Figure 8.5 Neuromuscular junction.

A neuromuscular junction refers to the contact between the axon terminal of a motor neuron and its motor end plate.

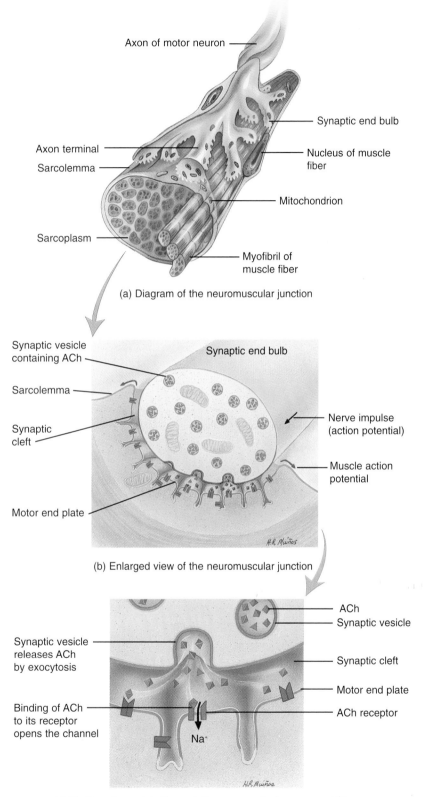

Axon of motor neuron

Synaptic end bulb

Axon terminal

Sarcolemma

Nucleus of muscle fiber

Mitochondrion

Sarcoplasm

Myofibril of muscle fiber

(a) Diagram of the neuromuscular junction

Synaptic vesicle containing ACh

Synaptic end bulb

Sarcolemma

Nerve impulse (action potential)

Synaptic cleft

Muscle action potential

Motor end plate

(b) Enlarged view of the neuromuscular junction

ACh
Synaptic vesicle

Synaptic vesicle releases ACh by exocytosis

Synaptic cleft

Motor end plate

Binding of ACh to its receptor opens the channel

ACh receptor

Na⁺

(c) Binding of acetylcholine to ACh receptors in the motor end plate

What is the motor end plate?

Figure 8.6 Regulation of contraction by troponin and tropomyosin when Ca²⁺ level changes. (a) The level of Ca²⁺ in the sarcoplasm is low during relaxation because it is pumped into the sarcoplasmic reticulum by Ca²⁺ active transport pumps. (b) A muscle action potential traveling along a transverse tubule opens calcium release channels in the sarcoplasmic reticulum and Ca²⁺ flows into the sarcoplasm. Note contraction is occurring because the thin myofilaments are closer to the center of the sarcomere.

 An increase in the level of Ca²⁺ in the sarcoplasm starts the movement of thin myofilaments; when the level of Ca²⁺ declines, movement stops.

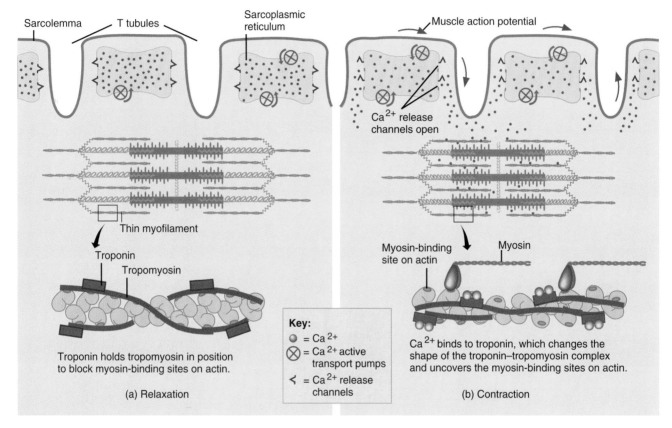

Troponin holds tropomyosin in position to block myosin-binding sites on actin.

Key:
⊙ = Ca²⁺
⊗ = Ca²⁺ active transport pumps
< = Ca²⁺ release channels

Ca²⁺ binds to troponin, which changes the shape of the troponin–tropomyosin complex and uncovers the myosin-binding sites on actin.

(a) Relaxation

(b) Contraction

With which substance does Ca²⁺ combine when it is released from the sarcoplasmic reticulum?

The following sequence occurs during sliding of the filaments (Figure 8.7):

1 When a muscle is relaxed, ATP attaches to ATP-binding sites on the myosin heads (cross bridges). A portion of each myosin head acts as an ATPase, an enzyme that splits the ATP into ADP + Ⓟ. This reaction transfers energy from ATP to the myosin head, even before contraction begins. The myosin heads are thus in an activated (energized) state.

2 When the sarcoplasmic reticulum releases Ca²⁺ and Ca²⁺ level rises in the sarcoplasm, tropomyosin slides away from its blocking position.

3 The activated myosin heads simultaneously bind to the myosin-binding sites on actin.

4 The shape change that occurs when myosin binds to actin produces the *power stroke* of contraction. The power stroke is the force that causes the thin actin myofilaments to slide past the thick myosin myofila-

ments. In other words, during the power stroke, the myosin heads swivel toward the center of the sarcomere, like the oars of a boat. This action draws the thin filaments past the thick filaments of a sarcomere. As the myosin heads swivel, they release ADP.

5 Once the power stroke is complete, ATP again combines with the ATP-binding sites on the myosin heads. As ATP binds, the myosin head detaches from actin.

6 Again, ATP is split, giving its energy to the myosin head, which returns to its original upright position.

7 The myosin head is then ready to combine with another myosin-binding site further along the thin filament.

The cycle of steps **3** through **7** repeats over and over as long as ATP is available and the Ca²⁺ level near the thin myofilaments is high. The myosin heads keep rotating back and forth with each power stroke, pulling the thin myofilaments toward the H zone. At any one instant, about half of the myosin heads are bound to actin and are swiveling. The other

Figure 8.7   Role of ATP and the power stroke of muscle contraction. Sarcomeres shorten through repeated cycles in which the myosin heads (cross bridges) attach to actin, swivel toward the center of the sarcomere, and detach.

🔑 *During the power stroke of contraction, myosin heads swivel and move the thin myofilaments past the thick myofilaments toward the center of the sarcomere.*

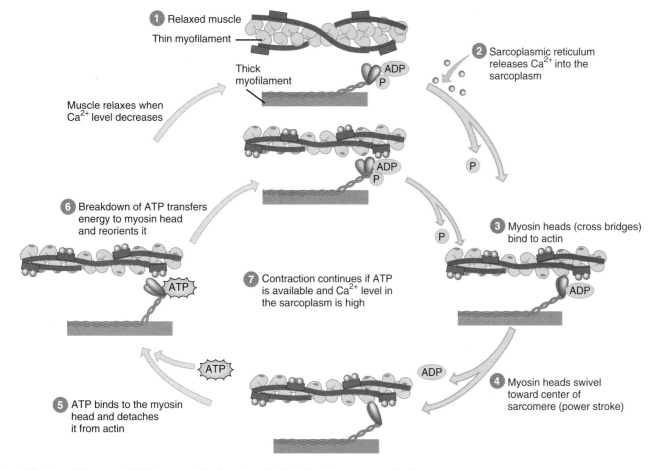

1 Relaxed muscle

Thin myofilament

2 Sarcoplasmic reticulum releases Ca²⁺ into the sarcoplasm

Thick myofilament

Muscle relaxes when Ca²⁺ level decreases

6 Breakdown of ATP transfers energy to myosin head and reorients it

3 Myosin heads (cross bridges) bind to actin

7 Contraction continues if ATP is available and Ca²⁺ level in the sarcoplasm is high

5 ATP binds to the myosin head and detaches it from actin

4 Myosin heads swivel toward center of sarcomere (power stroke)

❓ *What would happen if ATP were suddenly not available after the sarcomere had started to shorten?*

half are detached and preparing to swivel again. Contraction is analogous to running on a nonmotorized treadmill. One foot (myosin head) strikes the belt (thin myofilament) and pushes it backward (toward the H zone). Then the other foot comes down and imparts a second push. The belt soon moves smoothly while the runner (thick myofilament) remains stationary. And like the legs of a runner, the myosin heads need a constant supply of energy to keep going!

This continual movement of myosin heads applies the force that draws the Z discs toward each other, and the sarcomere shortens. The myofibrils thus contract, and the whole muscle fiber shortens. During a maximal muscle contraction, the distance between Z discs can decrease to half the resting length.

Relaxation

Two changes permit a muscle fiber to relax after it has contracted. First, acetylcholine is rapidly broken down by the enzyme acetylcholinesterase (AChE). When action potentials cease in the

motor neuron, release of ACh stops, and AChE rapidly breaks down the ACh already present in the synaptic cleft. This ends the generation of muscle action potentials, and the Ca²⁺ release channels in the sarcoplasmic reticulum membrane close.

Second, Ca²⁺ is rapidly removed from the sarcoplasm into the sarcoplasmic reticulum. As the Ca²⁺ level drops in the sarcoplasm, the tropomyosin–troponin complex slides back over the myosin-binding sites on actin. This prevents further myosin head binding to actin, and the thin myofilaments slip back to their relaxed positions. Figure 8.8 summarizes the events associated with contraction and relaxation of a muscle fiber.

After death, autolysis begins in muscle fibers and Ca²⁺ leaks out of the sarcoplasmic reticulum. The Ca²⁺ binds to troponin and triggers sliding of the thin myofilaments. ATP production has ceased, however, so the myosin heads cannot detach from actin. The resulting condition, in which muscles are in a state of rigidity (cannot contract or stretch), is called ***rigor mortis*** (rigidity of death). Rigor mortis lasts about 24 hours but disappears as tissues begin to disintegrate.

Figure 8.8   Summary of events of contraction and relaxation.

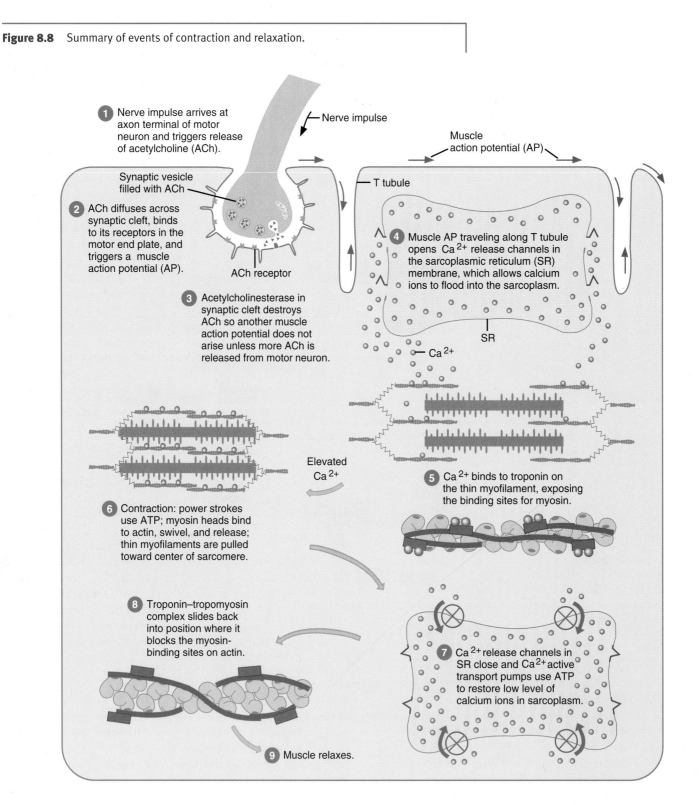

1 Nerve impulse arrives at axon terminal of motor neuron and triggers release of acetylcholine (ACh).

Nerve impulse

Muscle action potential (AP)

Synaptic vesicle filled with ACh

2 ACh diffuses across synaptic cleft, binds to its receptors in the motor end plate, and triggers a muscle action potential (AP).

ACh receptor

T tubule

4 Muscle AP traveling along T tubule opens Ca²⁺ release channels in the sarcoplasmic reticulum (SR) membrane, which allows calcium ions to flood into the sarcoplasm.

3 Acetylcholinesterase in synaptic cleft destroys ACh so another muscle action potential does not arise unless more ACh is released from motor neuron.

SR

Ca²⁺

Elevated Ca²⁺

5 Ca²⁺ binds to troponin on the thin myofilament, exposing the binding sites for myosin.

6 Contraction: power strokes use ATP; myosin heads bind to actin, swivel, and release; thin myofilaments are pulled toward center of sarcomere.

8 Troponin–tropomyosin complex slides back into position where it blocks the myosin-binding sites on actin.

7 Ca²⁺ release channels in SR close and Ca²⁺ active transport pumps use ATP to restore low level of calcium ions in sarcoplasm.

9 Muscle relaxes.

Energy for Contraction

Skeletal muscle fibers, unlike other body cells, alternate between virtual inactivity and continuous activity. Although ATP is the immediate source of energy for muscular contraction, muscle fibers usually contain only enough ATP to sustain activity for about 5 seconds. Skeletal muscle fibers contain a high-energy molecule called ***creatine phosphate*** (KRĒ-a-tin fos′-FĀT), which can be used to produce more ATP quickly during prolonged exercise.

ATP is broken down into ADP + ℗ and energy is released as follows:

$$\text{ATP} \longrightarrow \text{ADP} + ℗ + \text{Energy}$$

Creatine phosphate breaks down into creatine and phosphate, and in the process large amounts of energy are released as follows:

$$\text{Creatine phosphate} \longrightarrow \text{Creatine} + \text{Phosphate} + \text{Energy}$$

The released energy is used to convert ADP to ATP and the ATP is then used as a source of energy for contraction. This mechanism provides enough energy for muscles to contract maximally for about 15 seconds, as in a 100-meter dash. In other words, it is good for maximal short bursts of activity.

When muscle activity continues so that even the supply of creatine phosphate is depleted, glucose must be broken down. Skeletal muscles store glucose in the form of glycogen. During exercise, glycogen is converted back to glucose. Once this occurs, glucose is split into two molecules of pyruvic acid in the sarcoplasm, a process called *glycolysis.* In the process, energy is released and used to form ATP. Glycolysis does not require oxygen, so it is an *anaerobic process.* For this reason, glycolysis is referred to as *anaerobic respiration.* It may be summarized as follows:

$$1 \text{ Glucose} \longrightarrow 2 \text{ Pyruvic acid} + 2 \text{ ATP (energy)}$$

Normally, muscle fibers contain oxygen (O_2). This allows the pyruvic acid formed by glycolysis to enter mitochondria, where it is completely catabolized to carbon dioxide and water. Because this breakdown requires O_2, it is referred to as *aerobic respiration* or *cellular respiration.* The complete catabolism of pyruvic acid also yields energy that is used to generate most of a muscle fiber's ATP:

$$2 \text{ Pyruvic acid} + O_2 \longrightarrow CO_2 + H_2O + 36 \text{ ATP (energy)} + \text{Heat}$$

If there is insufficient oxygen for the complete catabolism of pyruvic acid, most of the pyruvic acid is converted to lactic acid, some of which diffuses out of the muscle fibers and into blood. Anaerobic respiration (glycolysis) and aerobic respiration (cellular respiration), together, allow for prolonged muscular activity, such as jogging, and will continue as long as nutrients and adequate oxygen last. Anaerobic respiration can provide sufficient energy for about 30 seconds of maximal muscle activity, such as a 400-meter dash.

All-or-None Principle

The weakest stimulus from a neuron that can still initiate a contraction is called a *threshold stimulus.* According to the *all-or-none principle,* when a threshold, or greater, stimulus is applied, individual muscle fibers of a motor unit will contract to their fullest extent or will not contract at all. In other words, *individual muscle fibers* do not partly contract. This does not mean the entire muscle must be either fully relaxed or fully contracted because, of the many motor units that comprise the entire muscle, some are contracting and some are relaxing. Thus the muscle as a whole can contract to a greater or lesser degree. Strength of contraction may be decreased by fatigue, lack of nutrients, or lack of oxygen.

Homeostasis

objective: *Explain the relationship of muscle tissue to homeostasis.*

Muscle tissue plays a vital role in maintaining the body's homeostasis. Three examples are the relationship of muscle tissue to oxygen, to fatigue, and to heat production.

Oxygen Debt

During exercise, blood vessels in muscles dilate, blood flow is increased, and oxygen delivery increases. But when muscular exertion is very great, oxygen cannot be supplied to muscle fibers fast enough, and the aerobic breakdown of pyruvic acid cannot produce all the ATP required for further muscle contraction. Additional ATP is then generated by anaerobic glycolysis. In the process, however, most of the pyruvic acid is converted to lactic acid. About 80 percent of this lactic acid diffuses from the skeletal muscles and is transported to the liver for conversion back to glucose or glycogen, but some lactic acid accumulates in muscle tissue.

Ultimately, this lactic acid must be broken down completely into carbon dioxide and water. After exercise has stopped, extra oxygen is required to metabolize the lactic acid; replenish ATP, creatine phosphate, and glycogen; and pay back any oxygen that has been borrowed from hemoglobin, myoglobin, air in the lungs, and body fluids. The additional oxygen that must be taken into the body after vigorous exercise to restore all systems to their normal states is called *oxygen debt.* The debt is paid back by labored breathing that continues after exercise has stopped. Thus accumulated lactic acid causes hard breathing and sufficient discomfort to stop muscle activity until homeostasis is restored.

Muscle Fatigue

If a skeletal muscle or group of skeletal muscles is continuously stimulated for an extended period of time, the contraction becomes progressively weaker until the muscles no longer respond. The inability of a muscle to maintain its strength of contraction is called *muscle fatigue.* It is related to an inability of muscle to produce sufficient energy to meet its needs. Although its exact mechanism is not completely understood, it may be related to insufficient oxygen, depletion of glycogen, and/or lactic acid buildup. Increased lactic acid would cause a decrease in the pH of the cells' environment. Muscle fatigue may therefore be viewed as a homeostatic mechanism that prevents pH levels from dropping below the normal acceptable range for the homeostasis of cells.

Heat Production

Homeostatic mechanisms are used to regulate body temperature (as described in Chapter 5). Of the total energy released during muscular contraction, only a small amount is used for mechanical work (contraction). As much as 85 percent is released as heat to help maintain normal body temperature. Excessive heat loss by the body results in shivering, an increase in muscle contraction, which increases the rate of heat production by several hundred percent as an effort to raise body temperature back to normal.

Kinds of Contractions

objective: *Describe the various kinds of muscle contractions.*

Skeletal muscles produce different kinds of contractions, depending on how frequently they are stimulated.

Twitch

The ***twitch contraction*** is a brief contraction of all the muscle fibers in a motor unit in response to a single action potential in its motor neuron. Figure 8.9a is a graph (***myogram***) of a twitch contraction. Note that a brief period exists between application of the stimulus and the beginning of contraction, the ***latent period***. During this time, Ca^{2+} is released from the sarcoplasmic reticulum and myosin head activity begins. The second phase is the ***contraction period*** (the upward tracing), and the third phase, the ***relaxation period*** (the downward tracing). All periods are very short for muscles that move the eyes, but longer for large leg muscles.

If two stimuli are applied one immediately after the other, the muscle will respond to the first stimulus but not to the second. When a muscle fiber receives enough stimulation to contract, it temporarily loses its excitability and cannot contract again until its responsiveness is regained. This period of lost excitability is the ***refractory period***. Skeletal muscle has a short refractory period. Cardiac muscle has a long refractory period.

Tetanus

When two stimuli are applied but the second is delayed until the refractory period is over, the skeletal muscle will respond to both stimuli. In fact, if the second stimulus is applied after the refractory period, but before the muscle has finished relaxing, the second contraction will be stronger than the first. This phenomenon, in which stimuli arrive at different times, is called ***wave summation*** (Figure 8.9b).

If a human muscle is accidentally shocked (electrocuted) or a frog muscle is electrically stimulated in a laboratory at a rate of 20 to 30 stimuli per second, the muscle can only partly relax between stimuli. As a result, the muscle maintains a sustained contraction called ***incomplete tetanus*** (Figure 8.9b). Stimulation at an increased rate (80 to 100 stimuli per second) results in ***complete tetanus***, a sustained contraction that lacks even partial relaxation between stimuli (Figure 8.9b).

Both kinds of tetanus result from the additional Ca^{2+} ions released at the second stimulus while Ca^{2+} ions are still in the sarcoplasm from the first stimulus. This causes the rapid succession of separate twitches. Relaxation is either partial or does not occur at all. Voluntary contractions, such as contraction of the biceps brachii muscle in order to flex the forearm, are tetanic contractions. In fact, most of our muscular contractions are short-term tetanic contractions and are thus smooth sustained contractions.

Isotonic and Isometric

In an ***isotonic*** (iso = equal; tonos = tension) ***contraction***, the muscle shortens and pulls on another structure, such as a bone, to produce movement. During such a contraction, the tension remains constant and energy is expended. Picking up a book from a desk is an example of an isotonic contraction.

In an ***isometric contraction***, there is a minimal shortening of the muscle, but the *tension* on the muscle increases greatly. Holding a book in your hand at your side is an example of

Figure 8.9 Myograms of various contractions. (a) Twitch contraction. The arrow at 0 indicates the point at which the stimulus is applied. (b) Comparison between twitch, wave summation, incomplete tetanus, and complete tetanus. The small arrows indicate the points at which stimuli are applied.

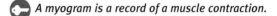

A myogram is a record of a muscle contraction.

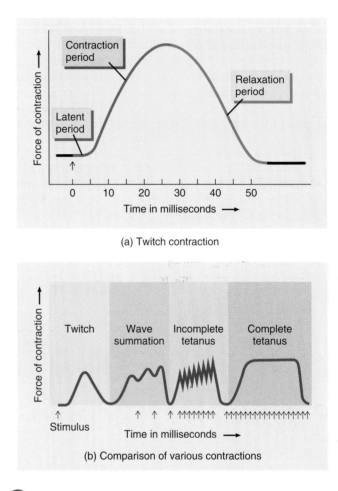

(a) Twitch contraction

(b) Comparison of various contractions

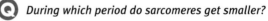

During which period do sarcomeres get smaller?

an isometric contraction. Although isometric contractions do not result in body movement, energy is still expended. Isometric contractions are important in body movements because they stabilize some joints as others are moved.

Muscle Tone

objective: *Define muscle tone and describe abnormalities related to it.*

A muscle may be in a state of partial contraction even though the muscle fibers operate on an all-or-none basis. Under normal conditions, at any given time, some fibers in a muscle are contracted while others are relaxed. This contraction tightens a muscle, but there may not be enough fibers contracting at the time to produce movement. Fibers contracting at different times (recruitment) allow the contraction to be sustained for long periods.

A sustained partial contraction of portions of a skeletal muscle results in ***muscle tone*** (*tonos* = tension). Tone is essential for maintaining posture. For example, when the muscles in the back of the neck are in tonic contraction, they keep the head from slumping forward, but they do not apply enough force to pull the head all the way back.

Abnormalities of muscle tone are referred to as hypotonia or hypertonia. *Hypotonia* means decreased or lost muscle tone. Such muscles are said to be *flaccid* (FLAK-sid or FLAS-sid). Flaccid muscles are loose and their normal rounded contour is replaced by a flattened appearance.

Hypertonia means increased muscle tone and is characterized by increased muscle stiffness and sometimes associated with a change in normal reflexes.

Muscular Atrophy and Hypertrophy

massage therapists work often on Hypertonicity

Muscular atrophy (A-trō-fē) refers to a wasting away of muscles. Individual muscle fibers decrease in size owing to a progressive loss of myofibrils. Muscles atrophy if they are not used (***disuse atrophy***). Bedridden individuals and people with casts may experience atrophy because the flow of impulses to the inactive muscle is greatly reduced. If the nerve supply to a muscle is cut, it will undergo complete atrophy (***denervation atrophy***). In about six months to two years, the muscle will be one-quarter its original size and the muscle fibers will be replaced by fibrous tissue. The transition to fibrous tissue, when complete, cannot be reversed.

Muscular hypertrophy (hī-PER-trō-fē) is the reverse of atrophy. It is an increase in the diameters of muscle fibers owing to the production of more myofibrils, mitochondria, sarcoplasmic reticulum, nutrients, and energy-supplying molecules (ATP and creatine phosphate). Hypertrophic muscles result from very forceful muscular activity or repetitive muscular activity at moderate levels. It is believed that the number of muscle fibers does not increase after birth. During childhood, the increase in the *size* of muscle fibers appears to be at least partially under the control of human growth hormone (hGH), which is produced by the anterior pituitary gland. A further increase in the *size* of muscle fibers appears to be due to the hormone testosterone, produced by the testes. The influence of testosterone probably accounts for the generally larger muscles in males than females. More forceful muscular contractions, as in weight lifting, also contribute to larger muscles.

Cardiac Muscle Tissue

one nucleus, more sarcoplasm & mitochondria, larger T-tubules, less developed sarcoplasmic reticulum

objective: *Describe the structure and function of cardiac muscle tissue.*

The heart wall is chiefly composed of ***cardiac muscle tissue.*** Although it is striated in appearance like skeletal muscle, it is involuntary (Figure 8.10). Unlike skeletal muscle fibers, which contain several peripherally located nuclei, cardiac muscle cells

have only a single centrally located nucleus. Cardiac muscle fibers are also shorter in length, larger in diameter, and squarish rather than circular in transverse section. Cardiac muscle fibers also exhibit branching.

Cardiac muscle fibers form two separate networks. The muscular walls and partition of the upper chambers (atria) of the heart compose one network. The muscular walls and partition of the lower chambers (ventricles) of the heart compose the other network. Within each network, the cardiac muscle fibers branch and interconnect with one another. At the same time, each fiber in a network is connected to the next fiber by an irregular transverse thickening of the sarcolemma called an ***intercalated*** (in-TER-ka-lāt-ed; *intercalere* = to insert between) ***disc*** (Figure 8.10). These discs contain ***gap junctions,*** which help conduct muscle action potentials from one muscle fiber to another. When a single fiber in a network is stimulated, all the fibers in the network become stimulated. Thus each network contracts as a functional unit.

Under normal resting conditions, cardiac muscle tissue contracts and relaxes rapidly, continuously, and rhythmically about 75 times a minute without stopping. This is a major physiological difference between cardiac and skeletal muscle tissue. Accordingly, cardiac muscle tissue requires a constant supply of oxygen. Energy generation occurs in numerous large mitochondria. Another difference is the source of stimulation. While skeletal muscle tissue ordinarily contracts only when stimulated by a nerve impulse, cardiac muscle tissue can contract without extrinsic (external) nerve stimulation. It is stimulated by the pacemaker* and by a specialized, intrinsic (internal) conducting tissue within the heart. Nerve stimulation merely causes the conducting tissue to increase or decrease its rate of discharge. This spontaneous, rhythmical self-excitation is called ***autorhythmicity*** (aw′-tō-rith-MIS-i-tē). Although all cardiac muscle fibers have autorhythmicity, the pacemaker fires at a greater rate of frequency than other cardiac muscle fibers and thus establishes the rate of contraction for the entire heart.

Cardiac muscle tissue also has a long refractory period, which allows time for the heart to relax between beats. The long refractory period permits heart rate to be increased significantly but prevents the heart itself from undergoing incomplete or complete tetanus. Tetanus of heart muscle would stop blood flow through the body and result in death.

Smooth Muscle Tissue

most able to regenerate

objective: *Describe the structure and function of smooth muscle tissue.*

non-striated

Like cardiac muscle tissue, ***smooth muscle tissue*** is involuntary. Smooth muscle fibers are considerably smaller than skeletal

---

*The "pacemaker" referred to here, also called the SA node, is composed of specialized muscle fibers in the right atrium of the heart and controls the rhythm of the electrical impulses that cause the heartbeat. Don't confuse this "natural" pacemaker with the electronic device that is sometimes implanted in heart patients in order to restore normal heartbeat.

Figure 8.10 Histology of cardiac muscle tissue.

🔑 *Cardiac muscle tissue is striated and involuntary.*

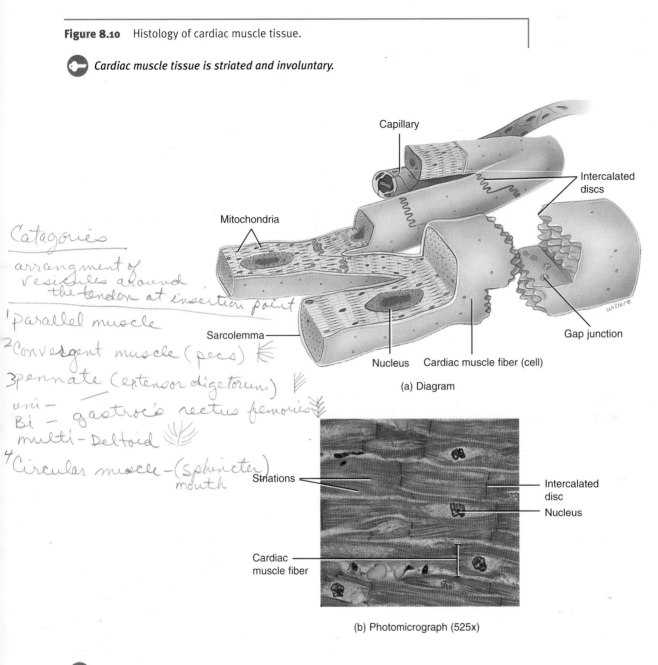

(a) Diagram

(b) Photomicrograph (525x)

Handwritten margin notes:

Catagories

arrangment of
vesicles around
the tendon at insertion point

1. parallel muscle
2. Convergent muscle (pecs)
3. pennate (extensor digetorum)
   uni — gastroc's rectus femoris
   Bi —
   multi — Deltoid
4. Circular muscle — (sphincter)
   mouth

❓ *What characteristic of cardiac muscle tissue prevents it from undergoing tetanus?*

muscle fibers. The fibers are widest in their midportion and tapered at both ends, and within each fiber is a single, oval, centrally located nucleus (Figure 8.11). In addition to thick and thin myofilaments, smooth muscle fibers also contain **intermediate filaments.** None of these various myofilaments has a regular pattern, and thus it is nonstriated, or smooth.

The intermediate filaments stretch between structures called **dense bodies,** which are similar to Z discs. Some are attached to the sarcolemma; others are dispersed in the sarcoplasm. In contraction, those attached to the sarcolemma cause a lengthwise shortening of the fiber. Note in Figure 8.11 how this causes a bubblelike appearance.

There are two kinds of smooth muscle tissue, visceral and multiunit. The more common type is **visceral (single-unit) muscle tissue.** It is found in wraparound sheets that form part of the walls of small arteries and veins and hollow viscera such as the stomach, intestines, uterus, and urinary bladder. The fibers in visceral muscle tissue are tightly bound together to form a continuous network. They contain gap junctions to facilitate muscle action potential conduction between fibers. When a neuron stimulates one fiber, the muscle action potential travels over the other fibers so that contraction occurs in a wave over many adjacent fibers. Whereas skeletal muscle fibers contract as individual units, visceral muscle cells con-

Figure 8.11   Histology of smooth muscle tissue. Depicted in (a) is a smooth muscle fiber before contraction (left) and after contraction (right).

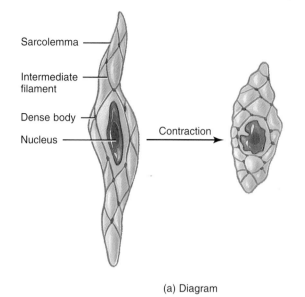

Sarcolemma

Intermediate filament

Dense body

Nucleus

Contraction

(a) Diagram

Smooth muscle fiber

Nucleus

(b) Photomicrograph (840x)

Q *Which type of smooth muscle tissue is found in the walls of hollow organs?*

tract in sequence as the action potential spreads from one cell to another.

The second kind of smooth muscle tissue, ***multiunit smooth muscle tissue,*** consists of individual fibers. Each has its own motor-nerve endings. Whereas stimulation of a single visceral muscle fiber causes contraction of many adjacent fibers, stimulation of a single multiunit fiber causes contraction of only that fiber, as with skeletal muscle. Multiunit smooth muscle tissue is found in the walls of large arteries, in large airways to the lungs, in the arrector pili muscles attached to hair follicles, and in the internal eye muscles.

Smooth muscle tissue exhibits several important physiological differences from striated muscle tissue. First, the duration of contraction of smooth muscle fibers is 5 to 500 times longer than in skeletal muscle fibers. Second, smooth muscle tissue can undergo sustained, long-term tone, which is important in the gastrointestinal tract (where the wall of the tract maintains a steady pressure on its contents), in the walls of the blood vessels called arterioles that maintain a steady pressure on blood, and in the wall of the urinary bladder that maintains a steady pressure on urine.

Smooth muscle is not under voluntary control. Some smooth muscle fibers contract in response to nerve impulses from the autonomic (involuntary) nervous system. Others contract in response to hormones or local factors such as pH, oxygen and carbon dioxide levels, temperature, and ion concentrations.

Finally, unlike skeletal muscle fibers, smooth muscle fibers can stretch considerably without becoming tense or taut. This is important because it permits smooth muscle to accommodate great changes in size while still retaining the ability to contract. Smooth muscle in the wall of hollow organs such as the uterus, stomach, intestines, and urinary bladder can stretch as the viscera enlarge, while the pressure within them remains the same.

Exhibit 8.1 presents a summary of the major characteristics of the three types of muscle tissue.

How Skeletal Muscles Produce Movement

objective: *Describe how skeletal muscles produce movement.*

Origin and Insertion

Based on the description of muscle tissue, we can define a ***skeletal muscle*** as an organ composed of several different types of tissues. These include skeletal muscle tissue, vascular tissue (blood vessels and blood), nervous tissue (motor neurons), and several connective tissues.

Skeletal muscles produce movements by pulling on tendons, which, in turn, pull on bones. Most muscles cross at least one joint and are attached to the articulating bones that form the joint (Figure 8.12 on page 169). When the muscle contracts, it draws one bone toward the other. The two bones do not move equally. One is held nearly in its original position; the attachment of a muscle tendon to the stationary bone is called the ***origin.*** The attachment of the other muscle tendon to the movable bone is the ***insertion.*** A good analogy is a spring on a door. The part of the spring attached to the door represents the insertion; the part attached to the frame is the origin. The fleshy portion of the muscle between the tendons of the origin and insertion is called the ***belly*** (***gaster***).

Group Actions

Most movements require several skeletal muscles acting in groups rather than individually. Also, most skeletal muscles are arranged in opposing pairs at joints, that is, flexors–extensors, abductors–adductors, and so on. Bending the elbow is an example. A muscle that causes a desired action is referred to as the ***prime mover*** or ***agonist*** (*agogos* = leader). In this instance, the biceps brachii is the prime mover (see Figure 8.21). While the biceps brachii is contracting, another muscle, called the ***antagonist*** (*anti-*

Exhibit 8.1	Summary of the Principal Features of Muscle Tissue		
CHARACTERISTICS	SKELETAL MUSCLE	CARDIAC MUSCLE	SMOOTH MUSCLE
Cell appearance and features	Long cylindrical fiber with many peripherally located nuclei; striated; unbranched.	Branched cylinder usually with one centrally located nucleus; striated; intercalated discs join neighboring fibers.	Spindle-shaped fiber with one, centrally positioned nucleus; no striations.
Location	Attached primarily to bones.	Heart.	Walls of hollow viscera, airways, blood vessels, iris and ciliary body of eye, arrector pili or hair follicles.
Fiber diameter	Very large.	Large.	Small.
Fiber length	Very large.	Small.	Intermediate.
Sarcomeres	Yes.	Yes.	No.
Sarcoplasmic reticulum	Yes.	Yes.	Scanty.
Transverse tubules	Yes, aligned with each A–I band junction.	Yes, aligned with each Z disc.	No.
Speed of contraction	Fast.	Moderate.	Slow.
Nervous control	Voluntary.	Involuntary.	Involuntary.
Capacity for regeneration	Limited.	None.	Considerable compared with other muscle tissues but limited compared with tissues such as epithelium.

agonistes = opponent), is relaxing, here, the triceps brachii (see Figure 8.21). The antagonist has an effect opposite to that of the prime mover; that is, the antagonist relaxes and yields to the movement of the prime mover. Do not assume, however, that the biceps brachii is always the prime mover and the triceps brachii is always the antagonist. For example, when straightening the elbow, the triceps brachii serves as the prime mover and the biceps brachii functions as the antagonist. If the prime mover and antagonist contracted together with equal force, there would be no movement, as in an isometric contraction.

Most movements also involve muscles called *synergists* (SIN-er-gists; *syn* = together; *ergon* = work), which help the prime mover function more efficiently by reducing unnecessary movement.

Some muscles in a group also act as *fixators,* which stabilize the origin of the prime mover so that the prime mover can act more efficiently. Under different conditions and depending on the movement, many muscles act at various times as prime movers, antagonists, synergists, or fixators.

Naming Skeletal Muscles

objective: *List and describe several ways that skeletal muscles are named.*

The names of most of the nearly 700 skeletal muscles are based on specific characteristics. Learning the terms used to indicate specific characteristics will help you remember the names of the muscles (Exhibit 8.2 on page 170).

Principal Skeletal Muscles

objective: *For various regions of the body, describe the location of skeletal muscles and identify their functions.*

Exhibits 8.3 through 8.15 list the principal muscles of the body with their origins, insertions, and actions. (By no means have

Figure 8.12   Relationship of skeletal muscles to bones. (a) Skeletal muscles produce movements by pulling on bones. (b) Bones serve as levers, and joints act as fulcrums for the levers. Here the lever–fulcrum principle is illustrated by the movement of the forearm lifting a weight. Note where the resistance and effort are applied in this example.

 In the limbs, the origin of a muscle is proximal and the insertion is distal.

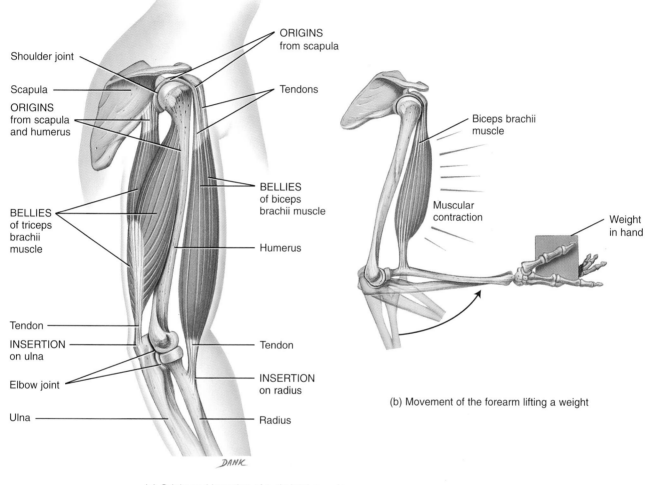

(a) Origin and insertion of a skeletal muscle

(b) Movement of the forearm lifting a weight

Q *Which muscle produces the desired action? Which muscle performs the opposite action of the one producing the desired action? Which muscle prevents unwanted movements?*

all the muscles of the body been included.) For each exhibit, an *overview* section provides a general orientation to the muscles and their functions or unique characteristics. Refer to Chapter 6 to review bone markings, since they serve as points of origin and insertion for muscles. Students often have difficulty in pronouncing names of skeletal muscles and understanding how they are named. To make this task easier, we have provided phonetic pronunciations and derivations that indicate how the

muscles are named. If you have mastered the naming of the muscles, their actions will have more meaning and be easier to remember.

The muscles are divided into groups according to the part of the body on which they act. Figure 8.13 on pages 170-171 shows general anterior and posterior views of the muscular system. As you study groups of muscles in the following exhibits, refer to Figure 8.13 to see how each group is related to all others.

Exhibit 8.2 | Naming Skeletal Muscles

CHARACTERISTIC	DESCRIPTION	EXAMPLE
Direction of muscle fibers	Direction of muscle fibers relative to the midline of the body.	
	Rectus means the fibers run parallel to the midline.	Rectus abdominis (see Fig. 8.17b)
	Transverse means the fibers run perpendicular to the midline.	Transversus abdominis (see Fig. 8.17b)
	Oblique means the fibers run diagonally to the midline.	External oblique (see Fig. 8.17a)
Location	Structure near which a muscle is found.	
	A muscle near the frontal bone.	Frontalis (see Fig. 8.14)
	A muscle near the tibia.	Tibialis anterior (see Fig. 8.25d)
Size	Relative size of the muscle.	
	Maximus means the largest.	Gluteus maximus (see Fig. 8.13b)
	Minimus means the smallest.	Gluteus minimus (see Fig. 8.24b)
	Longus means the longest.	Adductor longus (see Fig. 8.24a)
	Brevis means short.	Peroneus brevis
Number of origins	Number of tendons of origin.	
	Biceps means two origins.	Biceps brachii (see Fig. 8.20)
	Triceps means three origins.	Triceps brachii (see Fig. 8.21)
	Quadriceps means four origins.	Quadriceps femoris (see Fig. 8.24a)
Shape	Relative shape of the muscle.	
	Deltoid means having a triangular shape.	Deltoid (see Fig. 8.17a)
	Trapezius means having a trapezoid shape.	Trapezius (see Fig. 8.19c)
	Serratus means having a saw-toothed shape.	Serratus anterior (see Fig. 8.17b)
	Rhomboideus means having a rhomboid (diamond) shape.	Rhomboideus major (see Fig. 8.19d)
Origin and insertion	Sites where muscle originates and inserts.	Sternocleidomastoid originates on sternum and clavicle and inserts on mastoid process of temporal bone (see Fig. 8.13a)
Action	Principal action of the muscle.	
	Flexor (FLEK-sor): decreases the angle at a joint.	Flexor carpi radialis (see Fig. 8.22a)
	Extensor (eks-TEN-sor): increases the angle at a joint.	Extensor carpi ulnaris (see Fig. 8.22b)
	Abductor (ab-DUK-tor): moves a bone away from the midline.	Abductor longus
	Adductor (ad-DUK-tor): moves a bone closer to the midline.	Adductor longus (see Fig. 8.24a)
	Levator (le-VĀ-tor): produces an upward movement.	Levator scapulae (see Fig. 8.19d)
	Depressor (de-PRES-or): produces a downward movement.	Depressor labii inferioris (see Fig. 8.14b)
	Supinator (soo'pi-NĀ-tor): turns the palm upward or anteriorly.	Supinator (not illustrated)
	Pronator (pro-NĀ-tor): turns the palm downward or posteriorly.	Pronator teres (see Fig. 8.22a)
	Sphincter (SFINGK-ter): decreases the size of an opening.	External anal sphincter (not illustrated)
	Tensor (TEN-sor): makes a body part more rigid.	Tensor fasciae latae (see Fig. 8.24a)
	Rotator (RŌ-tāt-or): moves a bone around its longitudinal axis.	Rotatores (not illustrated)

Figure 8.13   Principal superficial skeletal muscles.

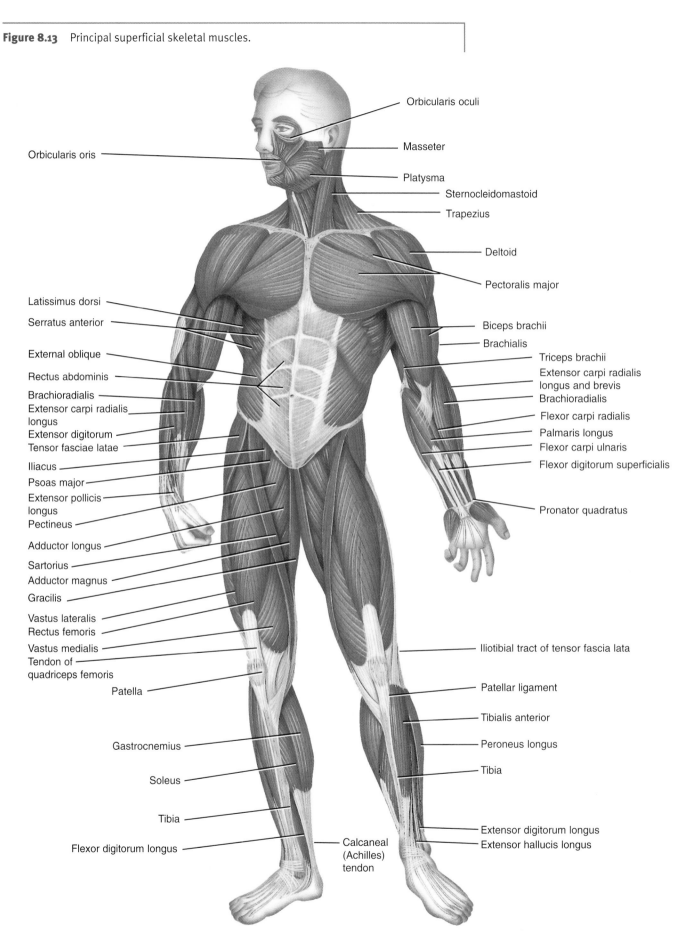

(a) Anterior view

Figure 8.13 *(Continued)*

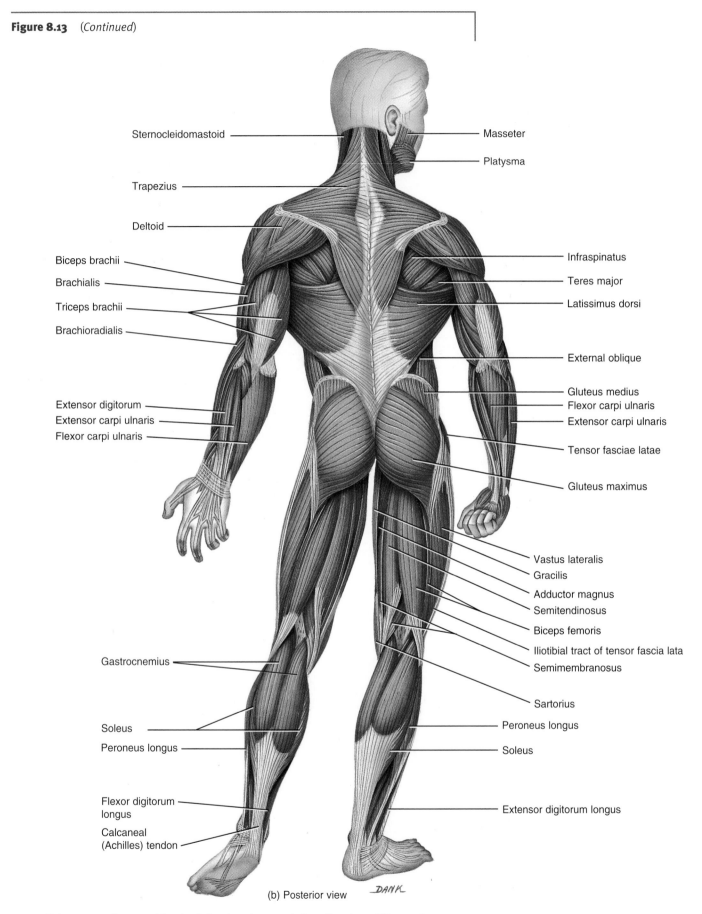

Sternocleidomastoid

Masseter

Platysma

Trapezius

Deltoid

Biceps brachii

Brachialis

Triceps brachii

Brachioradialis

Infraspinatus

Teres major

Latissimus dorsi

External oblique

Gluteus medius

Flexor carpi ulnaris

Extensor carpi ulnaris

Tensor fasciae latae

Gluteus maximus

Extensor digitorum

Extensor carpi ulnaris

Flexor carpi ulnaris

Vastus lateralis

Gracilis

Adductor magnus

Semitendinosus

Biceps femoris

Iliotibial tract of tensor fascia lata

Semimembranosus

Gastrocnemius

Sartorius

Peroneus longus

Soleus

Soleus

Peroneus longus

Extensor digitorum longus

Flexor digitorum longus

Calcaneal (Achilles) tendon

(b) Posterior view

DANK

Q *Select a muscle named for the following characteristics: direction of fibers, shape, action, size, origin and insertion, location, and number of origins.*

Exhibit 8.3 | Muscles of Facial Expression (Figure 8.14)

Overview: The muscles in this group provide humans with the ability to express a wide variety of emotions, including frowning, surprise, fear, and happiness. The muscles themselves lie within the layers of superficial fascia. As a rule, they arise from the fascia or bones of the skull and insert into the skin. Because of their insertions, the muscles of facial expression move the skin rather than a joint when they contract.

MUSCLE	ORIGIN	INSERTION	ACTION
Frontalis (fron-TA-lis; *front* = forehead)	Galea aponeurotica. (flat tendon that attaches to the frontalis and occipitalis muscles)	Skin superior to orbit.	Draws scalp forward, raises eyebrows, and wrinkles skin of forehead horizontally.
Occipitalis (ok-si'-pi-TA-lis; *occipito* = base of skull)	Occipital and temporal bones.	Galea aponeurotica.	Draws scalp backward.
Orbicularis oris (*orb* = circular; *or* = mouth)	Muscle fibers surrounding opening of mouth.	Skin at corner of mouth.	Closes lips, compresses lips against teeth, protrudes lips, and shapes lips during speech.
Zygomaticus (zī-gō-MA-ti-kus) major (*zygomatic* = cheekbone; *major* = greater)	Zygomatic bone.	Skin at angle of mouth and orbicularis oris.	Draws angle of mouth upward and outward as in smiling or laughing.
Buccinator (BUK-si-NĀ-tor; *bucc* = cheek)	Maxilla and mandible.	Orbicularis oris.	Major cheek muscle; compresses cheek as in blowing air out of mouth and causes cheeks to cave in, producing the action of sucking.
Platysma (pla-TIZ-ma; *platy* = flat, broad)	Fascia over deltoid and pectoralis major muscles.	Mandible, muscles around angle of mouth, and skin of lower face.	Draws outer part of lower lip downward and backward as in pouting; depresses mandible.
Oribcularis oculi (or-bi'-kyoo-LAR-is- O-kyoo-lī; *ocul* = eye)	Medial wall of orbit.	Circular path around orbit.	Closes eye.
Levator palpebrae superioris (le-VĀ-tor PAL-pe-brē soo-per'-e-OR-is; *palpebrae* = eyelids) (see also Figure 8.16)	Roof of orbit.	Skin of upper eyelid.	Elevates upper eyelid.

Figure 8.14 Muscles of facial expression. In this and subsequent figures in the chapter, the muscles indicated in all uppercase letters are the ones specifically referred to in the corresponding exhibit.

🔑 *When they contract, muscles of facial expression move the skin rather than a joint.*

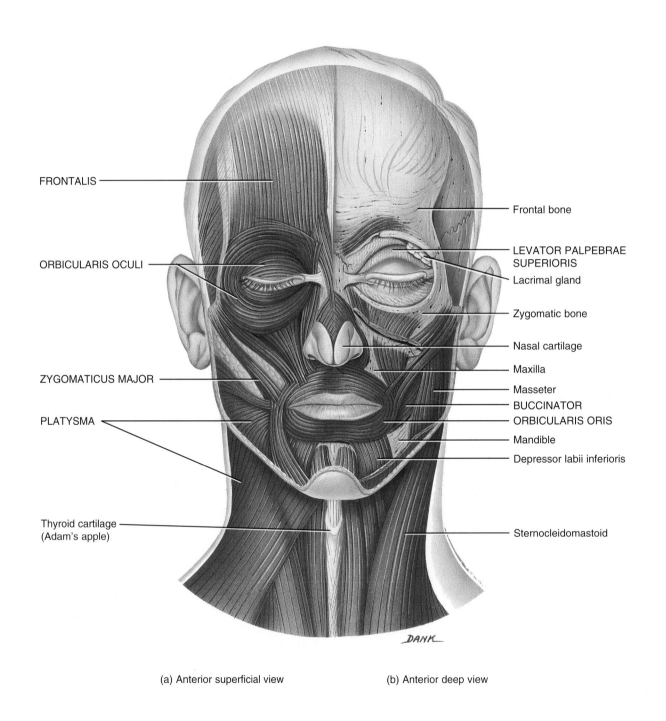

FRONTALIS

ORBICULARIS OCULI

ZYGOMATICUS MAJOR

PLATYSMA

Thyroid cartilage
(Adam's apple)

Frontal bone

LEVATOR PALPEBRAE
SUPERIORIS

Lacrimal gland

Zygomatic bone

Nasal cartilage

Maxilla

Masseter

BUCCINATOR

ORBICULARIS ORIS

Mandible

Depressor labii inferioris

Sternocleidomastoid

DANK

(a) Anterior superficial view   (b) Anterior deep view

❓ *What major muscle causes frowning? Smiling? Pouting? Squinting?*

Exhibit 8.4	Muscles that Move the Mandible (Lower Jaw) (Figure 8.15)

Overview: Muscles that move the mandible (lower jaw) are also known as muscles of mastication because they are involved in biting and chewing. These muscles also assist in speech.

MUSCLE	ORIGIN	INSERTION	ACTION
Masseter (MA-se-ter; *master* = chewer)	Maxilla and zygomatic arch.	Mandible.	Elevates mandible as in closing mouth, assists in side-to-side movement of mandible and protracts (protrudes) mandible.
Temporalis (tem'-por-A-lis; *tempora* = temples)	Temporal bone.	Mandible.	Elevates and retracts mandible, and assists in side-to-side movement of mandible.

Figure 8.15   Muscles that move the mandible (lower jaw).

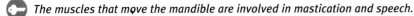

🔑 *The muscles that move the mandible are involved in mastication and speech.*

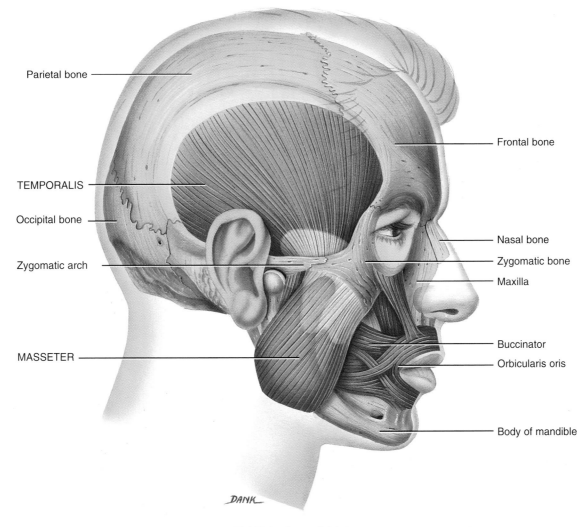

Right lateral superficial view

Ⓠ *What is the attachment to the movable bone called? What is the attachment to the stationary bone called?*

Exhibit 8.5 | Muscles That Move the Eyelids: Extrinsic Muscles (Figure 8.16)

Overview: Two types of muscles are associated with the eyeball, extrinsic and intrinsic. *Extrinsic muscles* originate outside the eyeball and are inserted on its outer surface (sclera). They move the eyeballs in various directions. *Intrinsic muscles* originate and insert entirely within the eyeball. They move structures within the eyeballs.

Movements of the eyeballs are controlled by three pairs of extrinsic muscles. Two pairs of rectus muscles move the eyeball in the direction indicated by their respective names—superior, inferior, lateral, and medial. One pair of muscles, the oblique muscles—superior and inferior— rotate the eyeball on its axis. The extrinsic muscles of the eyeballs are among the fastest contracting and most precisely controlled skeletal muscles of the body.

MUSCLE	ORIGIN	INSERTION	ACTION
Superior rectus (REK-tus; *superior* = above; *rectus* = in this case, muscle fibers running parallel to long axis of eyeball)	Tendinous ring attached to bony orbit around optic foramen.	Superior and central part of eyeball.	Rolls eyeball upward.
Inferior rectus (REK-tus; *inferior* = below)	Same as above.	Inferior and central part of eyeball.	Rolls eyeball downward.
Lateral rectus (REK-tus)	Same as above.	Lateral side of eyeball.	Rolls eyeball laterally.
Medial rectus (REK-tus)	Same as above.	Medial side of eyeball.	Rolls eyeball medially.
Superior oblique (ō-BLĒK; *oblique* = in this case, muscle fibers running diagonal to long axis of eyeball)	Same as above.	Eyeball between superior and lateral recti.	Rotates eyeball on its axis; directs cornea downward and laterally; the muscle moves through a ring of fibrocartilaginous tissue called the trochlea (*trochlea* = pulley).
Inferior oblique (ō-BLĒK)	Maxilla.	Eyeball between inferior and lateral recti.	Rotates eyeball on its axis; directs cornea upward and laterally.

Figure 8.16   Extrinsic muscles of the eyeballs.

🔑 *The extrinsic muscles of the eyeballs are among the fastest contracting and most precisely controlled skeletal muscles in the body.*

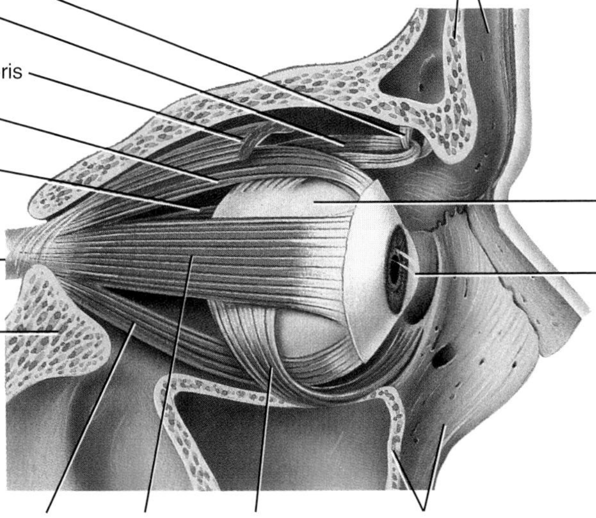

Lateral view of right eyeball

❓ *Which muscles roll the eyeballs laterally?*

Exhibit 8.6	Muscles That Act on the Anterior Abdominal Wall (Figure 8.17)

Overview: The anterior and lateral abdominal wall is composed of skin, fascia, and four pairs of flat, sheetlike muscles: rectus abdominis, external oblique, internal oblique, and transversus abdominis. The anterior surfaces of the rectus abdominis muscles are interrupted by three transverse fibrous bands of tissue called *tendinous intersections*. The aponeuroses of the external oblique, internal oblique, and transversus abdominis muscles meet at the midline to form the *linea alba* (white line), a tough fibrous band that extends from the xiphoid process of the sternum to the pubic symphysis. The inferior free border of the external oblique aponeurosis, plus some collagen fibers, forms the *inguinal ligament*.

MUSCLE	ORIGIN	INSERTION	ACTION
Rectus abdominis (REK-tus ab-DOM-in-is; *rectus* = fibers parallel to midline; *abdomino* = abdomen)	Pubis and pubic symphysis.	Cartilage of fifth to seventh ribs and xiphoid process.	Compresses abdomen to aid in defecation, urination, forced expiration, and childbirth and flexes vertebral column.
External oblique (ō-BLĒK; *external* = closer to surface; *oblique* = fibers diagonal to midline)	Lower eight ribs.	Ilium and linea alba (midline aponeurosis).	Contraction of both compresses abdomen; contraction of one side alone bends vertebral column laterally.
Internal oblique (ō-BLĒK; *internal* = farther from surface)	Ilium, inguinal ligament, and thoracolumbar fascia.	Cartilage of last three or four ribs.	Compresses abdomen; contraction of one side alone bends vertebral column laterally.
Transversus abdominis (trans-VER-sus ab-DOM-in-is; *transverse* = fibers perpendicular to midline)	Ilium, inguinal ligament, lumbar fascia, and cartilages of last six ribs.	Sternum, linea alba, and pubis.	Compresses abdomen.

Figure 8.17   Muscles of the anterior abdominal wall.

The inguinal ligament separates the thigh from the body wall.

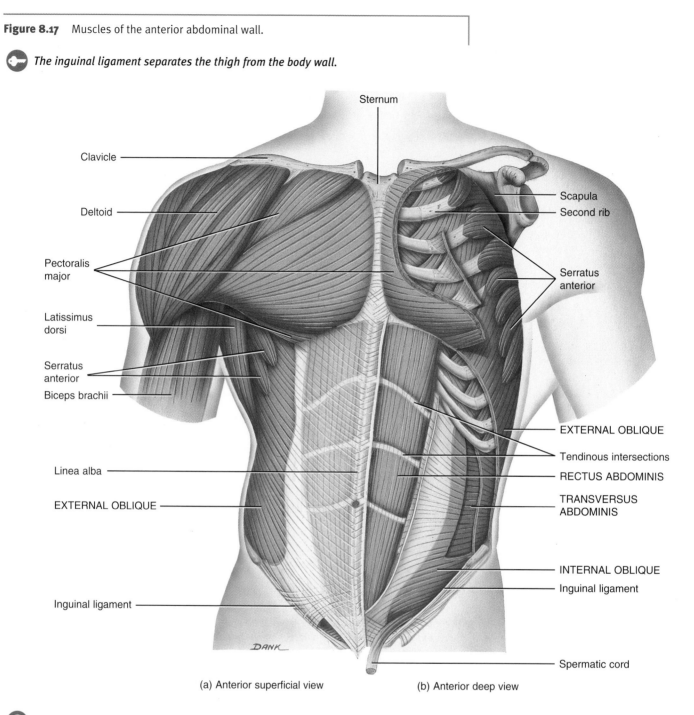

(a) Anterior superficial view   (b) Anterior deep view

Which abdominal muscle aids in urination?

Exhibit 8.7 | Muscles Used in Breathing (Figure 8.18)

Overview: The muscles described here are attached to the ribs and by their contraction and relaxation alter the size of the thoracic cavity during breathing. Essentially, inspiration occurs when the thoracic cavity increases in size. Expiration occurs when the thoracic cavity decreases in size. The principal muscles of **inspiration** during normal breathing are the diaphragm and external intercostals. During forced inspiration, accessory muscles, such as the sternocleidomastoid, scalenes, and pectoralis minor are also used. The principal muscles of **expiration** during normal breathing are also the diaphragm and external intercostals. During forced expiration, accessory muscles, such as the internal intercostals and abdominal muscles (external oblique, internal oblique, transversus abdominis, and rectus abdominis) are also used.

The diaphragm, one of the muscles used in breathing, is dome-shaped and has three major openings through which various structures pass between the thorax and abdomen. These structures include the aorta along with the thoracic duct and azygos vein, which pass through the **aortic hiatus**, the esophagus with accompanying vagus (X) nerves, which pass through the **esophageal hiatus**, and the inferior vena cava, which passes through the **foramen for the vena cava**. In a condition called a hiatus hernia, the stomach protrudes superiorly through the esophageal hiatus.

MUSCLE	ORIGIN	INSERTION	ACTION
Diaphragm (DĪ a-fram; *dia* = across; *phragma* = wall)	Sternum, costal cartilages of last six ribs, and lumbar vertebrae.	Central tendon.	Forms floor of thoracic cavity, pulls central tendon downward during inspiration and thus increases vertical length of thorax.
External intercostals (in'-ter-KOS-tals; *external* = closer to surface; *inter* = between; *costa* = rib)	Inferior border of rib above.	Superior border of rib below.	May elevate ribs during inspiration and thus increase lateral, anterior, and posterior dimensions of thorax.
Internal intercostals (in'-ter-KOS-tals; *internal* = farther from surface)	Superior border of rib below.	Inferior border of rib above.	May draw adjacent ribs together during forced expiration and thus decrease lateral, anterior, and posterior dimensions of thorax.

Figure 8.18   Muscles used in breathing.

Openings in the diaphragm permit the passage of the aorta, esophagus, and inferior vena cava.

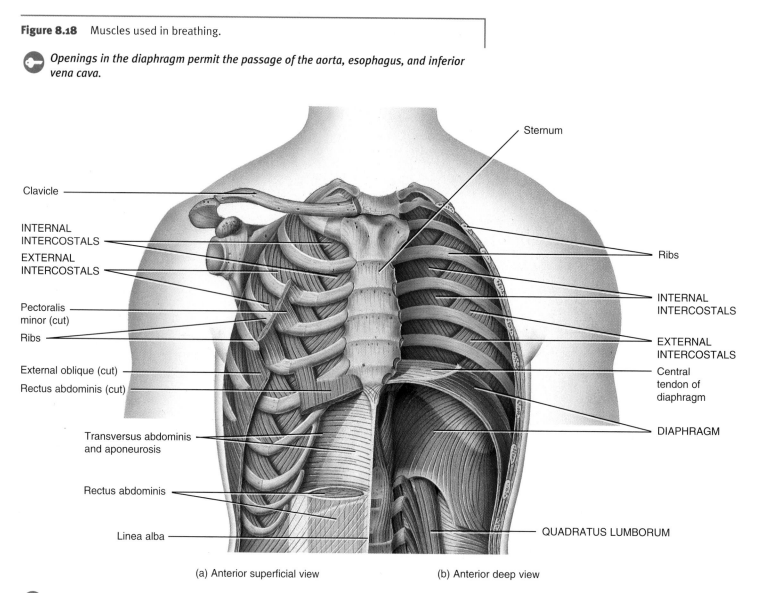

Sternum

Clavicle

INTERNAL INTERCOSTALS

EXTERNAL INTERCOSTALS

Pectoralis minor (cut)

Ribs

External oblique (cut)

Rectus abdominis (cut)

Transversus abdominis and aponeurosis

Rectus abdominis

Linea alba

Ribs

INTERNAL INTERCOSTALS

EXTERNAL INTERCOSTALS

Central tendon of diaphragm

DIAPHRAGM

QUADRATUS LUMBORUM

(a) Anterior superficial view   (b) Anterior deep view

Which respiratory muscles are used in normal inspiration?

Exhibit 8.8 | Muscles That Move the Pectoral (Shoulder) Girdle (Figure 8.19)

Overview: Muscles that move the pectoral (shoulder) girdle originate on the axial skeleton and insert on the clavicle or scapula. The muscles can be distinguished into *anterior* and *posterior*. The principal action of the muscles is to hold the scapula in place so that it can function as a stable point of origin for most of the muscles that move the humerus (arm).

MUSCLE	ORIGIN	INSERTION	ACTION
Anterior			
Subclavius (sub-KLĀ-vē-us; *sub* = under; *clavius* = clavicle)	First rib.	Clavicle.	Depresses clavicle.
Pectoralis minor (pek'-tor-A-lis; *pectus* = breast, chest, thorax; *minor* = lesser)	Third through fifth ribs.	Scapula.	Depresses and moves scapula and elevates third through fifth ribs during forced inspiration when scapula is fixed.
Serratus (ser-Ā-tus) anterior *serratus* = saw-toothed; *anterior* = front)	Upper eight or nine ribs.	Scapula.	Rotates scapula upward and laterally and elevates ribs when scapula is fixed.
Posterior			
Trapezius (tra-PĒ-zē-us; *trapezoides* = trapezoid-shaped)	Occipital bone and spines of seventh cervical and all thoracic vertebrae.	Clavicle and scapula.	Elevates clavicle, adducts scapula, rotates scapula upward, elevates or depresses scapula, and extends head.
Levator scapulae (le-VĀ-tor SKA-pyoo-lē; *levator* = raises; *scapulae* = scapula)	Upper four or five cervical vertebrae.	Scapula.	Elevates scapula and slightly rotates it downward.
Rhomboideus (rom-BOID-ē-us) major (*rhomboides* = rhomboid or diamond-shaped)	Spines of second to fifth thoracic vertebrae.	Scapula.	Adducts scapula and slightly rotates it downward.

Figure 8.19 Muscles that move the pectoral (shoulder) girdle.

Muscles that move the pectoral girdle originate on the axial skeleton and insert on the clavicle or scapula.

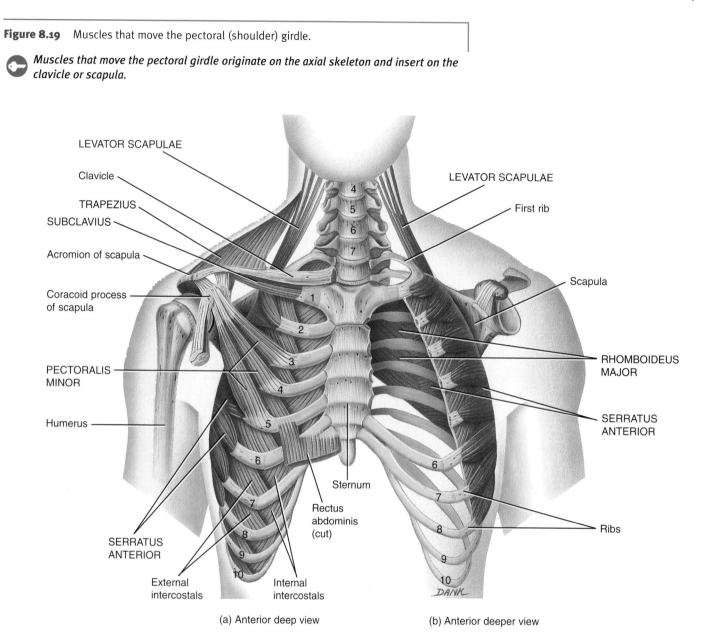

LEVATOR SCAPULAE

Clavicle

TRAPEZIUS

SUBCLAVIUS

Acromion of scapula

Coracoid process
of scapula

PECTORALIS
MINOR

Humerus

SERRATUS
ANTERIOR

External
intercostals

Internal
intercostals

Rectus
abdominis
(cut)

Sternum

LEVATOR SCAPULAE

First rib

Scapula

RHOMBOIDEUS
MAJOR

SERRATUS
ANTERIOR

Ribs

(a) Anterior deep view

(b) Anterior deeper view

Figure 8.19   (*Continued*)

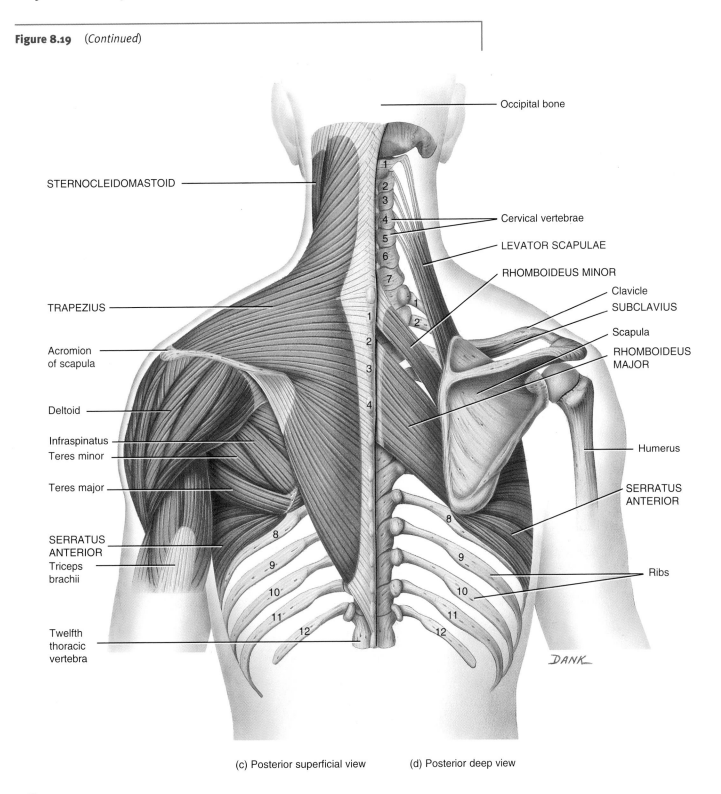

Occipital bone

STERNOCLEIDOMASTOID

Cervical vertebrae

LEVATOR SCAPULAE

RHOMBOIDEUS MINOR

Clavicle

SUBCLAVIUS

TRAPEZIUS

Scapula

RHOMBOIDEUS
MAJOR

Acromion
of scapula

Deltoid

Infraspinatus

Teres minor

Humerus

Teres major

SERRATUS
ANTERIOR

SERRATUS
ANTERIOR

Triceps
brachii

Ribs

Twelfth
thoracic
vertebra

DANK

(c) Posterior superficial view   (d) Posterior deep view

Q   *Which muscles originate on the ribs? The vertebrae?*

Exhibit 8.9   Muscles That Move the Humerus (Arm) (Figure 8.20)

Overview: Of the muscles that cross the shoulder joint, only two of them (pectoralis major and latissimus dorsi) do not originate on the scapula. These two muscles are thus designated as **axial muscles**, since they originate on the axial skeleton. The remaining muscles, the **scapular muscles**, arise from the scapula.

The strength and stability of the shoulder joint are not provided by the shape of the articulating bones or its ligaments. Instead, four deep muscles of the shoulder and their tendons—subscapularis, supraspinatus, infraspinatus, and teres minor—strengthen and stabilize the shoulder joint. The tendons are so arranged as to form a nearly complete circle around the joint. This arrangement is referred to as the **rotator musculotendinous) cuff** and is a common site of injury to baseball pitchers, especially tearing of the supraspinatus muscle.

MUSCLE	ORIGIN	INSERTION	ACTION
Axial			
Pectoralis (pek'-tor-A-lis) major (see also Figure 8.17a)	Clavicle, sternum, cartilages of second to sixth ribs.	Humerus.	Flexes, adducts, and rotates arm medially.
Latissimus dorsi (la-TIS-i-mus DOR-sī; *latissimus* = widest; *dorsum* = back)	Spines of lower six thoracic vertebrae, lumbar vertebrae, sacrum and ilium, lower four ribs.	Humerus.	Extends, adducts, and rotates arm medially; draws arm downward and backward.
Scapular			
Deltoid (DEL-toyd; *delta* = triangular) (see also Figure 8.17a)	Clavicle and scapula.	Humerus.	Abducts, flexes, extends, and rotates arm.
Subscapularis (sub-scap'-yoo-LA-ris; *sub* = below; *scapularis* = scapula)	Scapula.	Humerus.	Rotates arm medially.
Supraspinatus (soo'-pra-spi-NĀ-tus; *supra* = above; *spinatus* = spine of scapula)	Scapula.	Humerus.	Assists deltoid muscle in abducting arm.
Infraspinatus (in'-fra-spi-NĀ-tus; *infra* = below) (see also Figure 8.19c)	Scapula.	Humerus.	Rotates arm laterally; adducts arm.
Teres (TE-res) major (*teres* = long and round) (see also Figure 8.19c)	Scapula.	Humerus.	Extends arm; assists in adduction and medial rotation of arm.
Teres minor	Scapula.	Humerus.	Rotates arm laterally; extends and adducts arm.
Coracobrachialis (kor'-a-kō-BRĀ-kē-a'-lis; *coraco* = coracoid process)	Scapula.	Humerus.	Flexes and adducts arm.

Figure 8.20   Muscles that move the humerus (arm).

The strength and stability of the shoulder joint are provided by the tendons of the muscles that form the rotator cuff.

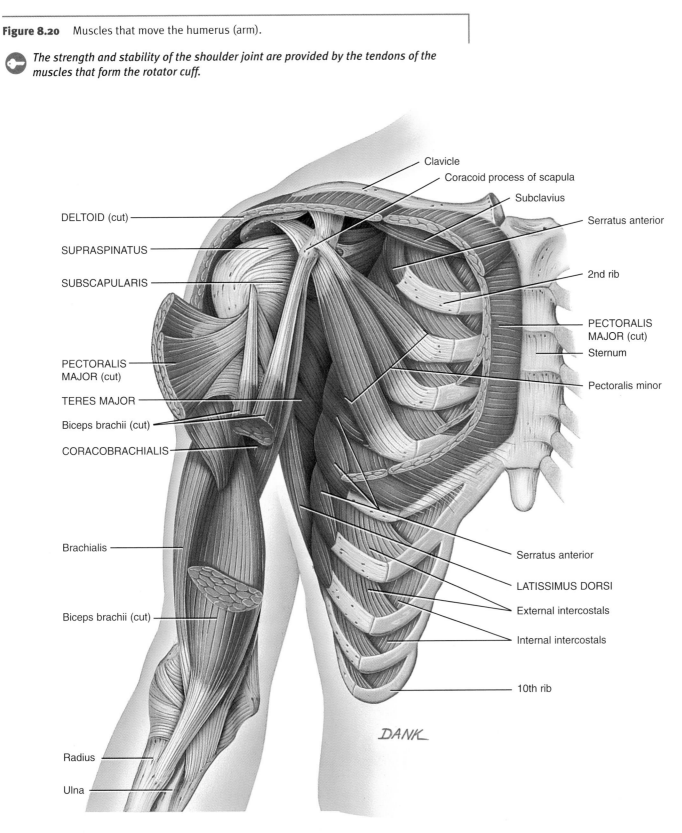

Clavicle
Coracoid process of scapula
Subclavius
Serratus anterior
2nd rib
PECTORALIS MAJOR (cut)
Sternum
Pectoralis minor
Serratus anterior
LATISSIMUS DORSI
External intercostals
Internal intercostals
10th rib

DELTOID (cut)
SUPRASPINATUS
SUBSCAPULARIS
PECTORALIS MAJOR (cut)
TERES MAJOR
Biceps brachii (cut)
CORACOBRACHIALIS
Brachialis
Biceps brachii (cut)
Radius
Ulna

DANK

Anterior deep view

 *Which muscles are axial muscles?*

Exhibit 8.10 | Muscles that Move the Radius and Ulna (Forearm) (Figure 8.21)

Overview: Most of the muscles that move the radius and ulna (forearm) are divided into *flexors* and *extensors*. Recall that the elbow joint is a hinge joint, capable only of flexion and extension. Whereas the biceps brachii, brachialis, and brachioradialis are flexors of the elbow joint, the triceps brachii is an extensor. Other muscles that move the radius and ulna are concerned with supination and pronation.

MUSCLE	ORIGIN	INSERTION	ACTION
Flexors			
Biceps brachii (BĪ-seps BRĀ-kē-ī; *biceps* = two heads of origin; *brachion* = arm)	Scapula.	Radius.	Flexes and supinates forearm; flexes arm.
Brachialis (brā-kē-A-lis)	Humerus.	Ulna.	Flexes forearm.
Brachioradialis (brā′-kē-ō-rā′-dē-A-lis; *radialis* = radius) (see also Figure 8.22a)	Humerus.	Radius.	Flexes forearm.
Extensor			
Triceps brachii (TRĪ-seps BRĀ-kē-ī; *triceps* = three heads of origin)	Scapula and humerus.	Ulna.	Extends forearm; extends arm.
Supinator			
Supinator (SOO-pi-nā-tor; *supination* = turning palm upward or anteriorly). Not illustrated.	Humerus and ulna.	Radius.	Supinates forearm and hand.
Pronator			
Pronator teres (PRŌ-nā′-ter TE-rez; *pronation* = turning palm downward or posteriorly) (see Figure 8.22a)	Humerus and ulna.	Radius.	Pronates forearm and hand

Figure 8.21   Muscles that move the radius and ulna (forearm).

🔑 *Whereas anterior arm muscles flex the forearm, posterior arm muscles extend it.*

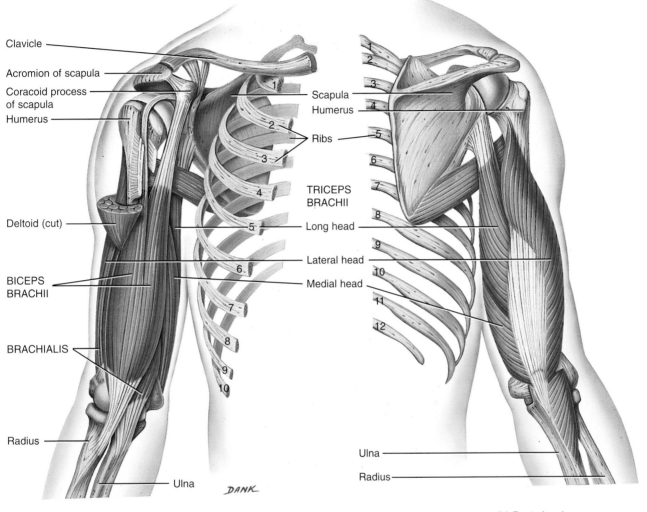

(a) Anterior view

(b) Posterior view

❓ *Identify the three flexors of the forearm.*

Exhibit 8.11 | Muscles That Move the Wrist and Fingers (Figure 8.22)

Overview: Muscles that move the wrist, hand, and fingers are located on the forearm and are many and varied. However, as you will see, their names for the most part give some indication of their origin, insertion, or action. On the basis of location and function, the muscles are divided into two groups—anterior and posterior. The **anterior compartment muscles** function as flexors. They originate on the humerus and typically insert on the carpals, metacarpals, and phalanges. The bellies of these muscles form the bulk of the proximal forearm. The **posterior compartment muscles** function as extensors. These muscles arise on the humerus and insert on the metacarpals and phalanges. Each of the two principal groups is also divided into superficial and deep muscles.

The tendons of the muscles of the forearm that attach to the wrist or continue into the hand, along with blood vessels and nerves, are held close to bones by fascia. The tendons are also surrounded by tendon sheaths. At the wrist, the deep fascia is thickened into fibrous bands called **retinacula** (*retinere* = retain). The **flexor retinaculum** (**transverse carpal ligament**) is located over the palmar surface of the carpal bones. Through it pass the long flexor tendons of the digits and wrist and the median nerve. The **extensor retinaculum** (**dorsal carpal ligament**) is located over the dorsal surface of the carpal bones. Through it pass the extensor tendons of the wrist and digits.

MUSCLE	ORIGIN	INSERTION	ACTION
Anterior Compartment (flexors)			
Flexor carpi radialis (FLEK-sor KAR-pē rā'-dē-A-lis; *flexor* = decreases angle at joint; *carpus* = writs; *radialis* = radius)	Humerus.	Second and third metacarpals.	Flexes and abducts wrist.
Flexor carpi ulnaris (FLEK-sor KAR-pē ul-NAR-is; *ulnaris* = ulna)	Humerus and ulna.	Pisiform, hamate, and fifth metacarpal.	Flexes and adducts wrist.
Palmaris longus (pal-MA-ris LON-gus; *palma* = palm)	Humerus.	Flexor retinaculum.	Flexes wrist.
Flexor digitorum profundus (FLEK-sor di'-ji-TOR-um pro-FUN-dus; *digit* = finger or toe; *profundus* = deep). Not illustrated.	Ulna.	Bases of distal phalanges.	Flexes distal phalanges of each finger.
Flexor digitorum superficialis (FLEK-sor di'-ji-TOR-um soo'-per-fish'-ē-A-lis; *superficialis* = closer to surface)	Humerus, ulna, and radius.	Middle phalanges.	Flexes middle phalanges of each finger.
Posterior Compartment (extensors)			
Extensor carpi radialis longus (eks-TEN-sor KAR-pē rā'-dē-A-lis LON-gus; *extensor* = increases angle at joint; *longus* = long)	Humerus.	Second metacarpal.	Extends and abducts wrist.
Extensor carpi ulnaris (eks-TEN-sor KAR-pē ul-NAR-is)	Humerus and ulna.	Fifth metacarpal.	Extends and adducts wrist.
Extensor digitorum (eks-TEN-sor di'-ji-TOR-um)	Humerus.	Second through fifth phalanges.	Extends phalanges.

Figure 8.22   Muscles that move the wrist, hand, and fingers.

The anterior compartment muscles function as flexors and the posterior compartment muscles function as extensors.

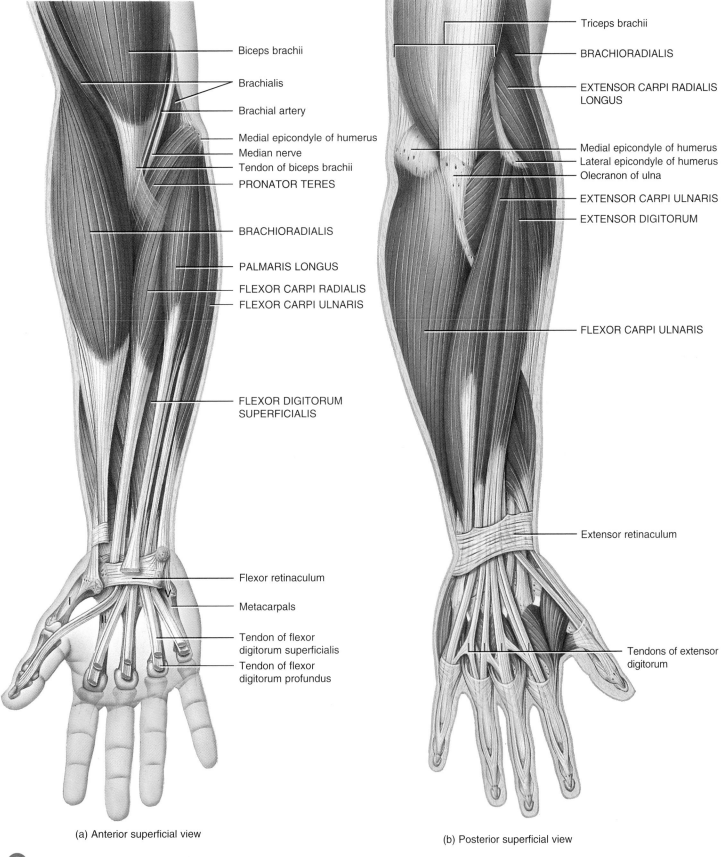

Biceps brachii

Brachialis

Brachial artery

Medial epicondyle of humerus
Median nerve
Tendon of biceps brachii
PRONATOR TERES

BRACHIORADIALIS

PALMARIS LONGUS

FLEXOR CARPI RADIALIS
FLEXOR CARPI ULNARIS

FLEXOR DIGITORUM
SUPERFICIALIS

Flexor retinaculum

Metacarpals

Tendon of flexor
digitorum superficialis
Tendon of flexor
digitorum profundus

Triceps brachii

BRACHIORADIALIS

EXTENSOR CARPI RADIALIS
LONGUS

Medial epicondyle of humerus
Lateral epicondyle of humerus
Olecranon of ulna

EXTENSOR CARPI ULNARIS

EXTENSOR DIGITORUM

FLEXOR CARPI ULNARIS

Extensor retinaculum

Tendons of extensor
digitorum

(a) Anterior superficial view

(b) Posterior superficial view

Circle three muscles that flex the wrist. Underline three muscles that extend the wrist.

Exhibit 8.12 | Muscles That Move the Vertebral Column (Backbone) (Figure 8.23)

Overview: The muscles that move the vertebral column (backbone) are quite complex because they have multiple origins and insertions and there is considerable overlapping among them.

MUSCLE	ORIGIN	INSERTION	ACTION
Sternocleidomastoid (ster'-nō-klī -dō-MAS-toid; *sternum* = breastbone; *cleido* = clavicle; *mastoid* = mastoid process of temporal bone) (see Figure 8.19c)	Sternum and clavicle.	Temporal bone.	Contractions of both muscles flex the cervical part of the vertebral column, draw the head forward, and elevate chin; contraction of one muscle rotates face toward side opposite contracting muscle.
Rectus abdominis (REK-tus ab-DOM-in-is) (see Figure 8.17)	Pubis and pubic symphysis.	Cartilages of fifth through seventh ribs and sternum.	Flexes vertebral column at lumbar spine and compresses abdomen.
Quadratus lumborum (kwod-RĀ-tus lum-BOR-um; *quadratus* = squared, four-sided; *lumb* = lumbar region) (see Figure 8.24a)	Ilium	Twelfth rib and upper four lumbar vertebrae.	Flexes vertebral column laterally.
Erector spinae (e-REK-tor SPI-nē)	This is the largest muscular mass of the back and consists of three groupings: iliocostalis, longissimus, and spinalis. These groups, in turn, consist of a series of overlapping muscles. The iliocostalis group is laterally placed, the longissimus group is intermediate in placement, and the spinalis is medially placed. All three groups of muscles extend various parts of the vertebral column.		

Figure 8.23   Muscles that move the vertebral column.

The erector spinae group is the largest muscular mass of the body.

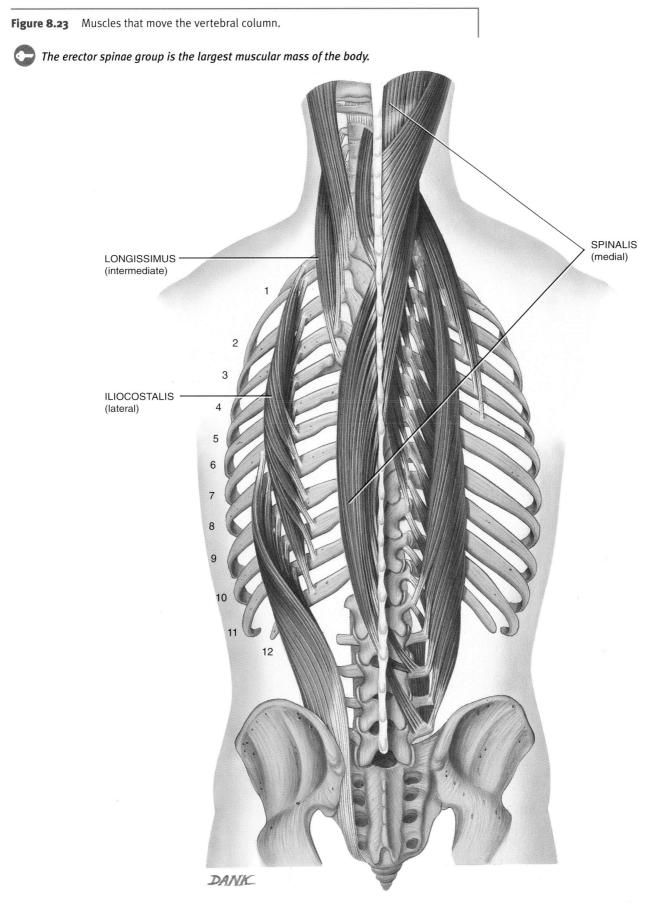

LONGISSIMUS
(intermediate)

SPINALIS
(medial)

ILIOCOSTALIS
(lateral)

1
2
3
4
5
6
7
8
9
10
11
12

DANK

Posterior view

Which muscles extend the spinal column? Which muscles flex it?

Exhibit 8.13 | Muscles That Move the Femur (Thigh) (Figure 8.24)

Overview: Muscles of the lower limbs are larger and more powerful than those of the upper limbs since lower limb muscles function in stability, locomotion, and maintenance of posture. Upper limb muscles are characterized by versatility of movement. In addition, muscles of the lower limbs frequently cross two joints and act equally on both. The majority of muscles that act on the femur (thigh) originate on the pelvic (hip) girdle and insert on the femur. The anterior muscles are the psoas major and iliacus, together referred to as the iliopsoas muscle. The remaining muscles (except for the pectineus, adductors, and tensor fasciae latae) are posterior muscles. Technically, the pectineus and adductors are components of the medial compartment of the thigh, but they are included in this exhibit because they act on the thigh. The tensor fasciae latae muscle is laterally placed. The ***fascia lata*** is a deep fascia of the thigh that encircles the entire thigh. It is well developed laterally, where together with the tendons of the gluteus maximus and tensor fasciae latae muscles, it forms a structure called the ***iliotibial tract***. The tract inserts into the lateral condyle of the tibia.

MUSCLE	ORIGIN	INSERTION	ACTION
Psoas (SŌ-as) major (*psoa* = muscle of loin)	Lumbar vertebrae.	Femur.	Flexes and rotates thigh laterally; flexes vertebral column.
Iliacus (il'-ē-AK-us; *iliacus* = ilium)	Ilium.	Tendon of psoas major.	Flexes and rotates thigh laterally.
Gluteus maximus (GLOO-tē-us MAK-si-mus; *glutos* = buttock; *maximus* = largest)	Ilium, sacrum, coccyx, and aponeurosis of sacrospinalis.	Iliotibial tract of fascia lata and femur.	Extends and rotates thigh laterally.
Gluteus medius (GLOO-tē-us MĒ-dē-us; *media* = middle)	Ilium.	Femur.	Abducts and rotates thigh medially.
Gluteus minimus (GLOO-tē-us MIN-i-mus; *minimus* = smallest)	Ilium.	Femur.	Abducts and rotates thigh medially.
Tensor fasciae latae (TEN-sor FA-shē-ē LĀ-tē; *tensor* = makes tense; *fascia* = band; *latus* = wide)	Ilium.	Tibia by way of the iliotibial tract.	Flexes and abducts thigh.
Adductor longus (LONG-us; *adductor* = moves part closer to midline; *longus* = long)	Pubis and pubic symphysis.	Femur.	Adducts, laterally rotates, and flexes thigh.
Adductor magnus (MAG-nus; *magnus* = large)	Pubis and ischium.	Femur.	Adducts, flexes, laterally rotates and extends thigh (anterior part flexes, posterior part extends).
Piriformis (pir-i-FOR-mis; *pirum* = pear; *forma* = shape)	Sacrum.	Femur.	Rotates thigh laterally and abducts it.
Pectineus (pek-TIN-ē-us; *pecten* = comb-shaped)	Pubis.	Femur.	Flexes and adducts thigh.

Figure 8.24   Muscles that move the femur (thigh).

Most muscles of the thigh originate on the pelvic (hip) girdle and insert on the femur.

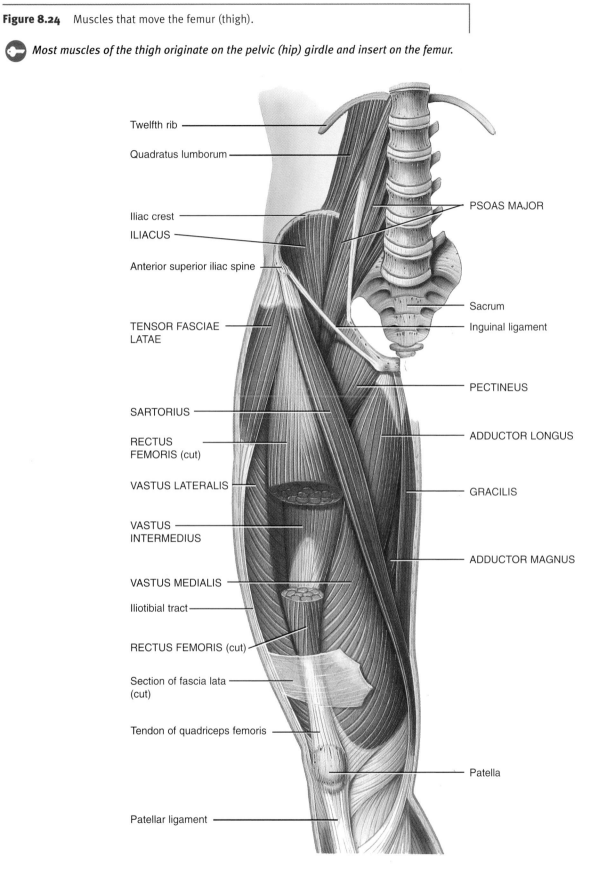

Twelfth rib

Quadratus lumborum

PSOAS MAJOR

Iliac crest

ILIACUS

Anterior superior iliac spine

Sacrum

Inguinal ligament

TENSOR FASCIAE
LATAE

PECTINEUS

SARTORIUS

RECTUS
FEMORIS (cut)

ADDUCTOR LONGUS

VASTUS LATERALIS

GRACILIS

VASTUS
INTERMEDIUS

ADDUCTOR MAGNUS

VASTUS MEDIALIS

Iliotibial tract

RECTUS FEMORIS (cut)

Section of fascia lata
(cut)

Tendon of quadriceps femoris

Patella

Patellar ligament

(a) Anterior superficial view

Figure 8.24   *(Continued)*

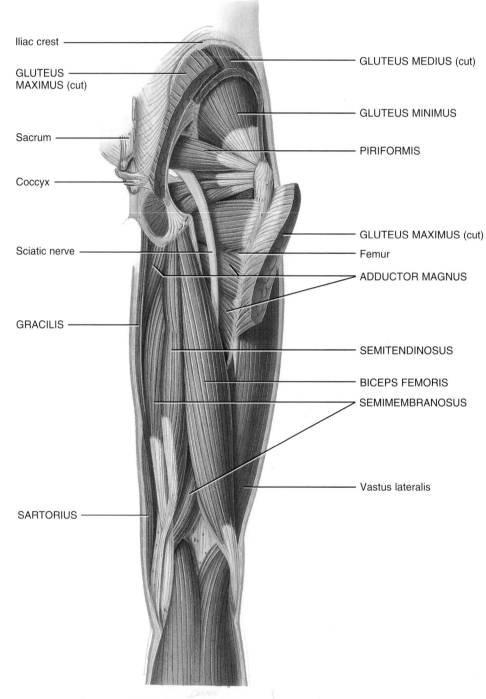

Iliac crest

GLUTEUS
MAXIMUS (cut)

Sacrum

Coccyx

Sciatic nerve

GRACILIS

SARTORIUS

GLUTEUS MEDIUS (cut)

GLUTEUS MINIMUS

PIRIFORMIS

GLUTEUS MAXIMUS (cut)

Femur

ADDUCTOR MAGNUS

SEMITENDINOSUS

BICEPS FEMORIS

SEMIMEMBRANOSUS

Vastus lateralis

(b) Posterior superficial view

Ⓠ   *Put an x next to the muscles that are part of the quadriceps femoris. Now also put an*
x next to the hamstring muscles.

Exhibit 8.14 | Muscles That Act on the Tibia and Fibula (Leg) (Figure 8.24)

Overview: The muscles that act on the tibia and fibula (leg) originate in the hip and thigh and are separated into compartments by deep fascia. The *medial (adductor) compartment* is so named because its muscles adduct the thigh. The adductor magnus, adductor longus, and pectineus muscles, components of the medial compartment, are included in Exhibit 8.13 because they act on the femur. The gracilis, the other muscle in the medial compartment, not only adducts the thigh but also flexes the leg. For this reason, it is included in this exhibit.

The *anterior (extensor) compartment* is so designated because its muscles act to extend the leg, and some also flex the thigh. It is composed of the quadriceps femoris and sartorius muscles. The quadriceps femoris muscle is a composite muscle that includes four distinct parts, usually described as four separate muscles (rectus femoris, vastus lateralis, vastus medialis, and vastus intermedius). The common tendon for the four muscles is the *quadriceps tendon*, which attaches to the patella. The tendon continues below the patella as the *patellar ligament* and attaches to the tibial tuberosity. The rectus femoris and sartorius muscles are also flexors of the thigh.

The *posterior (flexor) compartment* is so named because its muscles flex the leg (but also extend the thigh). Included are the hamstrings (biceps femoris, semitendinosus, and semimembranosus). The hamstrings are so named because their tendons are long and stringlike in the popliteal area. The *popliteal fossa* is a diamond-shaped space on the posterior aspect of the knee bordered laterally by the tendons of the biceps femoris and medially by the semitendinosus and semimembranosus muscles.

MUSCLE	ORIGIN	INSERTION	ACTION
Medial (adductor) compartment			
Adductor magnus (MAG-nus)	See Exhibit 8.13.		
Adductor longus (LONG-us)			
Pectineus (pek-TIN-ē-us)			
Gracilis (gra-SIL-is; *gracilis* = slender)	Pubic symphysis.	Tibia.	Adducts thigh and flexes leg.
Anterior (extensor) compartment			
Quadriceps femoris (KWOD-ri-seps FEM-or-is; *quadriceps* = four heads of origin; *femoris* = femur)			
Rectus femoris (REK-tus FEM-or-is; *rectus* = fibers parallel to midline)	Ilium.	Patella and tibial tuberosity through patellar ligament (tendon of quadriceps).	All four heads extend leg; rectus portion alone also flexes thigh.
Vastus lateralis (VAS-tus lat'-er-A-lis; *vastus* = large; *lateralis* = lateral)	Femur.		
Vastus medialis (VAS-tus mē'-dē-A-lis; *medialis* = medial)	Femur.		
Vastus intermedius (VAS-tus in'-ter-MĒ-dē-us; *intermedius* = middle)	Femur.		
Sartorius (sar-TOR-ē-us; *sartor* = tailor; refers to cross-legged position of tailors)	Ilium.	Tibia.	Flexes leg; flexes thigh and rotates it laterally, thus crossing leg.
Posterior (flexor) compartment			
Hamstrings			
Biceps femoris (BĪ-seps FEM-or-is; *biceps* = two heads of origin)	Ischium and femur.	Fibula and tibia.	Flexes legs and extends thigh.
Semitendinosus (sem'-ē-TEN-di-nō-sus; *semi* = half; *tendo* = tendon)	Ischium.	Tibia.	Flexes legs and extends thigh.
Semimembranosus (sem'-ē-MEM-bra-nō-sus; *membran* = membrane)	Ischium.	Tibia.	Flexes legs and extends thigh.

Exhibit 8.15 | Muscles That Move the Foot and Toes (Figure 8.25)

Overview: Muscles that move the foot and toes are located in the leg. The musculature of the leg, like that of the thigh, is divided into three compartments by deep fascia. The *anterior compartment* consists of muscles that dorsiflex the foot. In a situation analogous to the wrist, the tendons of the muscles of the anterior compartment are held firmly to the ankle by thickenings of deep fascia called the *superior extensor retinaculum (transverse ligament of the ankle)* and *inferior extensor retinaculum (cruciate ligament of the ankle)*.

The *lateral compartment* contains muscles that plantar flex and evert the foot. The *posterior compartment* consists of muscles that are divisible into superficial and deep groups. The superficial muscles share a common tendon of insertion, the calcaneal (Achilles) tendon, the strongest tendon of the body, that inserts into the calcaneus bone of the ankle. The superficial and most deep muscles plantar flex the foot.

MUSCLE	ORIGIN	INSERTION	ACTION
Anterior compartment			
Tibialis (tib-ē-A-lis) anterior (*tribialis* = tibia; *anterior* = front)	Tibia.	First metatarsal and first cuneiform.	Dorsiflexes and inverts foot.
Extensor digitorum longus (eks-TEN-sor di'-ji-TOR-um LON-gus; *extensor* = increases angle at joint)	Tibia and fibula.	Middle and distal phalanges of four outer toes.	Dorsiflexes and everts foot and extends toes.
Lateral compartment			
Peroneus longus (per'-ō-NĒ-us LON-gus; *perone* = fibula; *longus* = leg)	Fibula and tibia.	First metatarsal and first cuneiform.	Plantar flexes and everts foot.
Posterior compartment			
Gastrocnemius (gas'-trok-NĒ-mē-us; *gaster* = belly; *kneme* = leg)	Femur.	Calcaneus by way of calcaneal (Achilles) tendon.	Plantar flexes foot and flexes leg.
Soleus (SŌ-lē-us; *soleus* = sole of foot)	Fibula and tibia.	Calcaneus by way of calcaneal (Achilles) tendon.	Plantar flexes foot.
Tibialis (TIB-ē-A-lis) posterior (*posterior* = back)	Tibia and fibula.	Second, third, and fourth metatarsals; navicular; all three cuneiforms, and cuboid.	Plantar flexes and inverts foot.
Flexor digitorum longus (FLEK-sor di'-ji-TOR-um LON-gus; *flexor* = decreases angle at joint; *digitorum* = finger or toe)	Tibia.	Distal phalanges of four outer toes.	Plantar flexes and inverts foot; flexes toes.

Figure 8.25   Muscles that move the foot and toes.

The superficial muscles of the posterior compartment share a common tendon of insertion, the calcaneal (Achilles) tendon, that inserts into the calcaneal bone of the ankle.

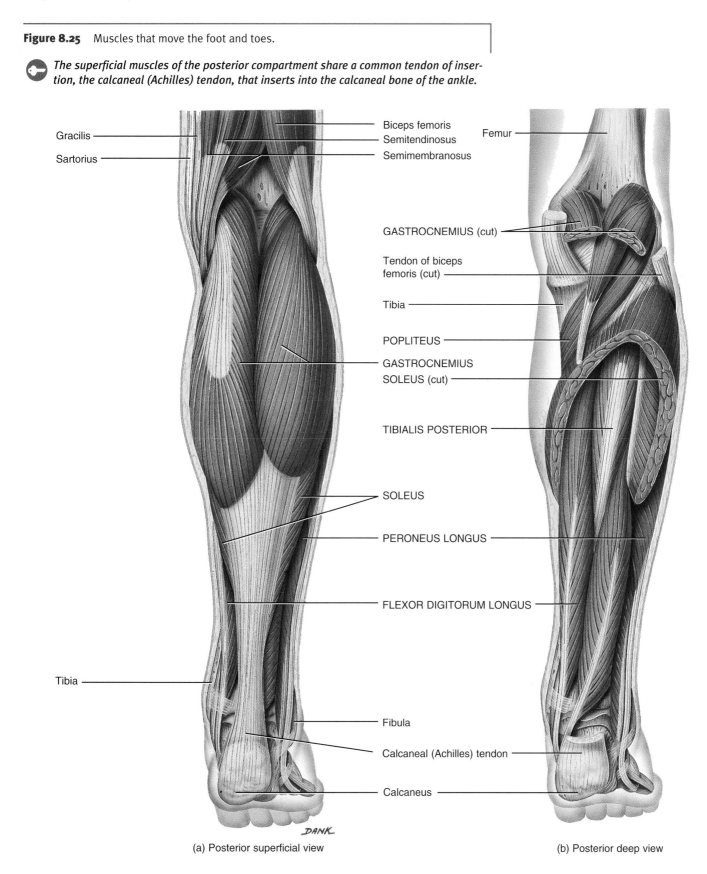

(a) Posterior superficial view

(b) Posterior deep view

Figure 8.25 (Continued)

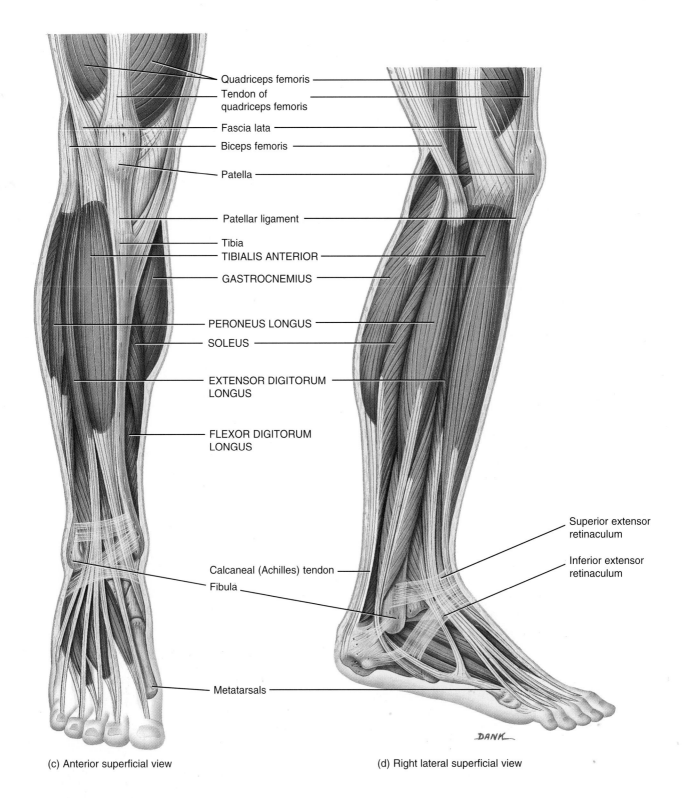

Quadriceps femoris

Tendon of
quadriceps femoris

Fascia lata

Biceps femoris

Patella

Patellar ligament

Tibia

TIBIALIS ANTERIOR

GASTROCNEMIUS

PERONEUS LONGUS

SOLEUS

EXTENSOR DIGITORUM
LONGUS

FLEXOR DIGITORUM
LONGUS

Calcaneal (Achilles) tendon

Fibula

Metatarsals

Superior extensor
retinaculum

Inferior extensor
retinaculum

DANK

(c) Anterior superficial view

(d) Right lateral superficial view

Circle the muscles that dorsiflex the foot.

wellness focus

Flexibility: A Question of Degree

a certain degree of elasticity is an important attribute of muscle tissue, because elasticity contributes to a characteristic known as flexibility. Flexibility refers to a joint's range of motion (ROM). Range of motion is the maximum ability to move the bones of a joint through an arc. For example, if an injured knee can go from a 50° angle fully flexed to a 120° angle fully extended, the knee's ROM is 70°. Physical therapists measure improvements in joint mobility by increases in ROM.

Most athletes know the importance of flexibility. Sports such as dance, gymnastics, figure skating, and diving require a great degree of flexibility. But even athletes in sports not noted for their flexibility requirements include stretching exercises in their workouts to increase or maintain flexibility. Flexibility improves performance and prevents injury. For example, runners use stretching exercises to prevent muscle tightness in the lower back and hamstring muscles, and to improve stride length and overall running form.

Some degree of flexibility is necessary for everyone, if only for the performance of daily simple tasks. Limitation in your shoulder's ROM can make it difficult to pull a shirt on over your head, lift a dish from a shelf, or even brush your hair. As we age, muscles tend to shorten, especially if we do not do stretching exercises regularly. Shorter muscles decrease a joint's ROM. This may be part of the reason age is associated with stiff joints. It's important to remember, however, that this stiffness is due more to lack of exercise than to merely getting older.

Stretching It

When a relaxed muscle is physically stretched, its ability to elongate is limited by connective tissue structures, such as fasciae. Regular stretching gradually lengthens these structures, but the process occurs very slowly. In order to see an improvement in flexibility, stretching exercises must be performed regularly—daily, if possible—for many weeks.

Collagenous tissue stretches best when slow, gentle force is applied at elevated tissue temperatures. An external source of heat, such as hot packs or ultrasound, can be used. But ten or more minutes of muscular contraction, if possible, is the best way to raise muscle temperature, because this heats the muscle more deeply and thoroughly. That's where the name "warm-up" comes from. It's important to warm-up *before* stretching, not vice versa. Stretching cold muscles does not increase flexibility and may even cause injury.

Contraction Is Contraindicated

The easiest and safest way to increase flexibility is with static stretching. A good static stretch is slow and gentle. After warming up, you get into a comfortable stretching position and then relax. Continuing to relax and breathe deeply, you reach just a little farther . . . and a little farther, holding the stretch for at least 30 seconds. If you have difficulty relaxing, you know you have stretched too far. Ease up until you feel a stretch but no strain.

When stretching, it is important to relax. Sounds simple, right? But if you ever visit an exercise class, you'll notice some people who are tense, rigid, and hunched up, because the stretching positions are uncomfortable; their muscles tighten up in protest. These people figure they'd better push a little harder, and they tense up even more. They are unintentionally activating the motor nerves that initiate muscular contraction in the very muscles they are supposed to be relaxing, which of course interferes with the muscle's ability to elongate and stretch!

critical thinking

Using the information presented in this chapter, figure out which muscles are being stretched when you reach over and touch your toes.

What stretching positions stretch the pectoralis major?

Common Disorders

Fibrosis

The formation of fibrous connective tissue in locations where it normally does not exist is called *fibrosis.*

Fibromyalgia

Fibromyalgia (*algia* = painful condition) refers to a group of common nonarticular rheumatic disorders characterized by pain, tenderness, and stiffness of muscles, tendons, and surrounding soft tissues.

Muscular Dystrophies

Muscular dystrophies (*dystrophy* = degeneration) are hereditary muscle-destroying diseases characterized by degeneration of individual muscle fibers (cells), which leads to a progressive atrophy of the skeletal muscle.

Myasthenia Gravis (MG)

Myasthenia (mī-as-THĒ-nē-a) *gravis* (MG) is a weakness of skeletal muscles caused by an abnormality at the neuromuscular junction that prevents muscle fibers from contracting. Myasthenia gravis is an autoimmune disorder caused by antibodies that bind to the receptors and hinder the attachment of ACh.

Abnormal Contractions

One kind of abnormal muscular contraction is a *spasm,* a sudden involuntary contraction of large groups of muscles. *Tremor* is a rhythmic, involuntary, purposeless contraction of opposing muscle groups. A *tic* is a spasmodic twitching made involuntarily by muscles that are ordinarily under voluntary control.

Effects of Muscle-Building Anabolic Steroids

These steroids, a derivative of the hormone testosterone, are used to build muscle proteins and therefore increase strength and endurance. However, anabolic steroids have a number of side effects, including liver cancer, kidney damage, increased risk of heart disease, irritability and aggressive behavior, psychotic symptoms (including hallucinations), and wide mood swings. In females, additional side effects include sterility, the development of facial hair, deepening of the voice, decrease in size of the breasts and uterus, enlargement of the clitoris, and irregularities of menstruation. In males, additional side effects include decrease in size of the testes, baldness, excessive development of breast glands, and diminished hormone secretion and sperm production by the testes.   ■

Medical Terminology and Conditions

Electromyography or EMG (e-lek′-trō-mī-OG-ra-fē; *electro* = electricity; *myo* = muscle; *graph* = to write) The recording and study of electrical changes that occur in muscle tissue.

Gangrene (GANG-rēn; *gangraena* = an eating sore) Death of a soft tissue, such as muscle, that results from interruption of its blood supply.

Myalgia (mī-AL-jē-a; *algia* = painful condition) Pain in or associated with muscles.

Myoma (mī-Ō-ma; *oma* = tumor) A tumor consisting of muscle tissue.

Myomalacia (mī′-ō-ma-LĀ-shē-a; *malaco* = soft) Softening of a muscle.

Myopathy (mī-OP-a-thē; *pathos* = disease) Any disease of muscle tissue.

Myositis (mī′-ō-SĪ-tis; *itis* = inflammation of) Inflammation of muscle fibers (cells).

Myospasm (MĪ-ō-spazm) Spasm of a muscle.

Myotonia (mī-ōTŌ-nē-a; *tonia* = tension) Increased muscular excitability and contractility with decreased power of relaxation; tonic spasm of the muscle.

Paralysis (pa-RAL-a-sis; *para* = beyond; *lyein* = to loosen) Loss or impairment of motor (muscular) function resulting from a lesion of nervous or muscular origin.

Wryneck or torticollis (RĪ-neck; *tortus* = twisted; *collum* = neck) Spasmodic contraction of several superficial and deep muscles of the neck that produces twisting of the neck and an unnatural position of the head.

Study Outline

Types of Muscle Tissue (p. 154)

1. Skeletal muscle tissue is mostly attached to bones. It is striated and voluntary.

2. Cardiac muscle tissue forms most of the wall of the heart. It is striated and involuntary.

3. Smooth muscle tissue is located in viscera. It is nonstriated and involuntary.

Functions of Muscle Tissue (p. 154)

1. Through contraction, muscle tissue performs four important functions.

2. These functions are motion, movement of substances within the body, stabilization of body positions, and heat production.

Characteristics of Muscle Tissue (p. 154)

1. Excitability is the property of receiving and responding to stimuli.

2. Contractility is the ability to shorten and thicken (contract).

3. Extensibility is the ability to be stretched (extend).

4. Elasticity is the ability to return to original shape after contraction or extension.

Skeletal Muscle Tissue (p. 154)

Connective Tissue Components (p. 154)

1. The term fascia is applied to a sheet or broad band of fibrous connective tissue underneath the skin or around muscles and organs of the body.

2. Other connective tissue components are epimysium, covering the entire muscle; perimysium, covering fasciculi; and endomysium, covering fibers.

3. Tendons and aponeuroses are extensions of connective tissue beyond muscle fibers that attach the muscle to bone or other muscle.

Nerve and Blood Supply (p. 155)

1. Nerves convey impulses (action potentials) for muscular contraction.

2. Blood provides nutrients and oxygen for contraction.

Histology (p. 155)

1. Skeletal muscle consists of fibers covered by a sarcolemma. The fibers contain sarcoplasm, nuclei, mitochondria, myoglobin, high-energy molecules, sarcoplasmic reticulum, and transverse tubules.

2. Each fiber also contains myofibrils that consist of thin and thick myofilaments. The myofilaments are compartmentalized into sarcomeres.

3. Thin myofilaments are composed of actin, tropomyosin, and troponin; thick myofilaments consist of myosin.

4. Projecting myosin heads are called cross bridges.

Contraction (p. 157)

Sliding-Filament Mechanism (p. 157)

1. A muscle action potential travels over the sarcolemma and enters the transverse tubules and sarcoplasmic reticulum.

2. The action potential leads to the release of calcium ions from the sarcoplasmic reticulum.

3. Actual contraction is brought about when the thin myofilaments of a sarcomere slide toward each other as myosin heads pull on actin myofilaments.

Neuromuscular Junction (p. 158)

1. A motor neuron transmits a nerve impulse (nerve action potential) to a skeletal muscle, which stimulates contraction.

2. A motor neuron and the muscle fibers it stimulates form a motor unit.

3. A single motor unit may innervate as few as 10 or as many as 2000 muscle fibers.

4. A neuromuscular junction refers to an axon terminal of a motor neuron and the portion of the muscle fiber sarcolemma close to, but not touching, it (motor end plate).

Physiology of Contraction (p. 158)

1. When a nerve impulse (nerve action potential) reaches an axon terminal, the synaptic vesicles of the terminal release acetylcholine (ACh), which initiates a muscle action potential in the sarcolemma. The muscle action potential then travels into the transverse tubules and sarcoplasmic reticulum to release some of its stored Ca^{2+} into the sarcoplasm.

2. The released Ca^{2+} combines with troponin, causing it to pull on tropomyosin to change its position, thus exposing myosin-binding sites on actin.

3. The immediate, direct source of energy for muscle contraction is ATP. ATPase splits ATP into ADP + Ⓟ and the released energy activates (energizes) myosin heads (cross bridges).

4. Activated myosin heads attach to actin and a change in the orientation of the myosin heads occurs (power stroke); their movement results in the sliding of thin myofilaments.

5. Relaxation is brought about when ACh is broken down and Ca^{2+} is pumped from the sarcoplasm into the sarcoplasmic reticulum.

Energy for Contraction (p. 162)

1. The immediate, direct source of energy for muscle contraction is ATP.

2. Muscle fibers generate ATP continuously. This involves creatine phosphate and the metabolism of nutrients such as glycogen.

All-or-None Principle (p. 163)

1. The weakest stimulus capable of causing contraction is a threshold stimulus.

2. Muscle fibers of a motor unit contract to their fullest extent or not at all.

Homeostasis (p. 163)

1. Oxygen debt is the amount of O_2 needed to convert accumulated lactic acid into CO_2 and H_2O. It occurs during strenuous exercise and is paid back by rapid breathing after exercising until homeostasis between muscular activity and oxygen requirements is restored.

2. Muscle fatigue results from diminished availability of oxygen, toxic effects of carbon dioxide, and lactic acid built up during exercise.

3. The heat given off during muscular contraction maintains the homeostasis of body temperature.

Kinds of Contractions (p. 163)

1. The various kinds of contractions are twitch, tetanus, isotonic, and isometric.

2. A record of a contraction is called a myogram. The refractory period is the time when a muscle has temporarily lost excitability. Skeletal muscles have a short refractory period. Cardiac muscle has a long refractory period.

3. Wave summation is the increased strength of a contraction resulting from the application of a second stimulus before the muscle has completely relaxed after a previous stimulus.

Muscle Tone (p. 164)

1. A sustained partial contraction of portions of a skeletal muscle results in muscle tone.

2. Tone is essential for maintaining posture.

Muscular Atrophy and Hypertrophy (p. 165)

1. Muscular atrophy refers to a state of wasting away of muscles.

2. Muscular hypertrophy refers to an increase in the diameter of muscle fibers.

Cardiac Muscle Tissue (p. 165)

1. This muscle tissue is found only in the heart. It is striated and involuntary.

2. The fibers are squarish in transverse section and usually contain a single centrally placed nucleus.

3. The fibers branch freely and are connected via gap junctions.

4. Intercalated discs provide strength and aid action potential conduction.

5. Unlike skeletal muscle tissue, cardiac muscle tissue contracts and relaxes rapidly, continuously, and rhythmically. Energy is supplied by numerous, large mitochondria.

6. Cardiac muscle tissue can contract without extrinsic stimulation and can remain contracted longer than skeletal muscle tissue.

7. Cardiac muscle tissue has a long refractory period, which prevents tetanus.

Smooth Muscle Tissue (p. 165)

1. Smooth muscle tissue is nonstriated and involuntary.

2. Smooth muscle fibers contain intermediate filaments, in addition to thin and thick myofilaments, and dense bodies.

3. Visceral (single-unit) smooth muscle is found in the walls of viscera. The fibers are arranged in a network so that contraction occurs in a wave over many adjacent fibers.

4. Multiunit smooth muscle is found in blood vessels and the eyes. The fibers operate singly rather than as a unit.

5. The duration of contraction and relaxation of smooth muscle is longer than in skeletal muscle.

6. Smooth muscle fibers contract in response to nerve impulses, hormones, and local factors.

7. Smooth muscle fibers can stretch considerably without developing tension.

How Skeletal Muscles Produce Movement (p. 167)

1. Skeletal muscles produce movement by pulling on bones.

2. The attachment to the stationary bone is the origin. The attachment to the movable bone is the insertion.

3. The prime mover produces the desired action. The antagonist produces an opposite action. The synergist assists the prime mover by reducing unnecessary movement. The fixator stabilizes the origin of the prime mover so that it can act more efficiently.

Naming Skeletal Muscles (p. 168)

1. The names of most skeletal muscles indicate specific characteristics.

2. The major descriptive categories are direction of fibers, location, size, number of origins (or heads), shape, origin and insertion, and action.

Principal Skeletal Muscles (p. 168)

1. The principal skeletal muscles of the body are grouped according to region in Exhibits 8.3 through 8.15.

2. In studying groups, refer to Figure 8.13 to see how each group is related to all others.

Self-Quiz

1. The connective tissue sheath that surrounds a skeletal muscle is known as the
   a. perimysium b. myofilament c. epimysium d. fascicle e. tendon

2. Which of the following is NOT a function of muscle tissue?
   a. motion b. production of heat c. production of hemoglobin d. maintenance of posture e. regulation of organ volume

3. Bundles of muscle cells (fibers) wrapped in perimysium are called
   a. myofilaments b. myofibers c. sarcomeres d. sarcolemmas e. fasciculi

4. Each individual muscle fiber is wrapped by
   a. epimysium b. endomysium c. superficial fascia d. perimysium e. sarcomeres

5. After 5 to 6 seconds, the ATP on the myosin heads is exhausted and the muscle must rely on _____ to quickly produce more ATP for contraction.
   a. acetylcholine b. creatine phosphate c. lactic acid d. pyruvic acid e. acetylcholinesterase

6. Glycolysis involves
   a. the breakdown of glycerol for energy b. the use of oxygen to break down glucose c. pyruvic acid and oxygen d. the breakdown of glucose to pyruvic acid e. creatine phosphate and ADP

7. Following muscle contraction in response to a stimulus, the muscle temporarily cannot respond to anther stimulus. This is known as

   a. the refractory period   b. the summation period   c. the latent period   d. oxygen debt   e. complete tetanus

8. When a threshold stimulus is applied to a muscle fiber, the muscle fiber will contract fully or it will not contract at all. This is known as

   a. the power stroke   b. the sliding-filament theory   c. the "all-or-none principle"   d. tetanus   e. the contraction period

9. The ability of muscle tissue to return to its original shape after contraction is called

   a. excitability   b. extensibility   c. elasticity   d. contractibility   e. entropy

10. The network of membrane-enclosed tubules that stores calcium ions in the muscle fiber is the

    a. sarcoplasmic reticulum   b. sarcolemma   c. sarcomere   d. sarcoplasm   e. transverse tubules

11. The striations of skeletal muscle are due to the

    a. transverse tubules   b. arrangement of the nuclei in the multinucleated muscle fiber   c. cross section of myofibrils   d. alternating A bands and I bands   e. myosin heads (cross bridges)

12. A motor unit consists of

    a. a transverse tubule and its associated sarcomeres   b. a motor neuron and all the muscle fibers it stimulates   c. a muscle and all its motor neurons   d. all the myofilaments enclosed within the sarcomere   e. the motor end plate and the transverse tubules

13. Thick myofilaments

    a. are composed of actin, troponin, and tropomyosin   b. make up the I band   c. stretch the entire length of the sarcomere   d. have binding sites for Ca^{2+} e. have myosin heads (cross bridges) used for the power stroke

14. While the biceps brachii is contracting, the triceps brachii acts as a(n)

    a. agonist   b. antagonist   c. isometric contraction   d. isotonic contraction   e. synergist

15. Skeletal muscles are named using several characteristics. Which characteristic is NOT used to name skeletal muscles.

    a. direction of fibers   b. size   c. speed of contraction   d. location   e. shape

16. Arrange the following in the correct order for skeletal muscle contraction:

    1. sarcoplasmic reticulum releases Ca^{2+}

    2. Ca^{2+} combines with troponin

    3. acetylcholine is released from the axon terminal

    4. action potential travels from sarcolemma into transverse tubules

    5. activated myosin heads (cross bridges) attach to actin

    6. thin filaments slide

       a. 3,4,1,2,5,6   b. 4,3,2,1,5,6   c. 1,2,3,4,5,6   d. 4,1,3,5,2,6   e. 3,1,4,5,2,6

17. All of the following may result in an increase in muscle size EXCEPT

    a. weight training   b. denervation atrophy   c. human growth hormone   d. testosterone   e. isotonic contraction

18. Smooth muscle

    a. fibers branch freely and contain intercalated discs   b. fibers are long, cylindrical, and multinucleated   c. develops considerable tension when stretched   d. contraction may occur in a wave   e. contracts rapidly, continuously, and rhythmically

19. Match the following:

    ____ a. thick myofilaments containing myosin heads (cross bridges)

    ____ b. reddish pigment in muscle cells

    ____ c. dense area that separates sarcomeres

    ____ d. thin myofilament containing troponin

    ____ e. striated zone of the sarcomere composed of thin myofilaments

    ____ f. striated zone of the sarcomere composed of thick and thin myofilaments

    ____ g. organelle of the muscle that contains calcium

    ____ h. region of sarcolemma near the adjoining axon terminal

    ____ i. sac containing neurotransmitter

    ____ j. space between axon terminal and the sarcolemma of the muscle

    ____ k. extends and laterally rotates the hip (thigh)

    ____ l. adducts and laterally rotates the hip (thigh)

    ____ m. compresses abdomen and flexes vertebral column

    ____ n. flexes the neck

    ____ o. flexes and abducts wrist

    ____ p. flexes wrist

    ____ q. flexes, adducts, rotates shoulder (arm) medially

    ____ r. adducts, flexes, extends, rotates shoulder (arm) medially and laterally

    ____ s. elevates, depresses, adducts scapula

    ____ t. elevates mandible, closes mouth

    ____ u. draws scalp forward

    ____ v. extends, adducts, and rotates arm medially

    ____ w. dorsiflexes and inverts foot

    ____ x. plantar flexes foot and flexes leg

    ____ y. extends the hip (leg)

    A. trapezius
    B. myosin
    C. myoglobin
    D. adductor longus
    E. Z discs
    F. quadriceps group
    G. actin
    H. A band
    I. synaptic cleft
    J. gastrocnemius
    K. deltoid
    L. masseter
    M. palmaris longus
    N. I band
    O. sarcoplasmic reticulum
    P. motor end plate
    Q. rectus abdominis
    R. tibialis anterior
    S. frontalis
    T. synaptic vesicle
    U. latissimus dorsi
    V. pectoralis major
    W. sternocleidomastoid
    X. flexor carpi radialis
    Y. gluteus maximus

Critical Thinking Applications

1. Breast meat from domesticated turkeys (which do not fly) is "white" meat. Leg meat from domesticated turkeys (which do run) is "dark" meat. Skeletal muscle color is determined by the supply of blood, myoglobin, and mitochondria. Explain the structural and functional differences between white meat and dark meat.

2. Irene was late for her tennis match and did not warm-up before the game. While making a quick move, she felt a sudden, sharp pain in her calf. Later, Irene's leg had a large black-and-blue area diagnosed as intramuscular bleeding. Can you explain the cause of the pain and bleeding?

3. Bill tore some ligaments in his knee while skiing. He was in a toe-to-thigh cast for 6 weeks. When the cast was removed, the "healed" leg was noticeably thinner than the uncasted leg. What happened to his leg and what should he do about it?

4. The newspaper reported several cases of botulism poisoning following a fund-raiser potluck dinner for the local clinic. The cause appeared to be three-bean salad "flavored" with the bacterium *Clostridium botulinum*. The *C. botulinum* produces a toxin that blocks the release of acetylcholine from the motor neuron. What would be the result of botulism poisoning on muscle function?

5. The wife of a body-building competitor has been trying to get pregnant for over a year. The wife blames the husband's use of anabolic steroids. The husband thinks his steroids are "all natural" so that can't be the problem. What do you think?

6. Imagine that the uterus was composed of skeletal muscle instead of smooth muscle. How would this change affect uterine function? How would substituting skeletal muscle for cardiac muscle affect heart function?

Answers to Figure Questions

8.1 Endomysium, perimysium, epimysium, deep fascia, and superficial fascia.

8.2 A band is mostly thick myofilaments; I band is thin myofilaments; calcium ions (Ca^{2+}) are stored in sarcoplasmic reticulum.

8.3 A band: myosin, actin, troponin, and tropomyosin; I band: actin, troponin, and tropomyosin.

8.4 They disappear; no.

8.5 The region of the sarcolemma near the axon terminal.

8.6 Troponin.

8.7 The myosin heads (cross bridges) would not be able to detach from actin.

8.9 Contraction period.

8.10 Long refractory period.

8.11 Visceral (single-unit).

8.12 Prime mover (agonist); antagonist; synergist.

8.13 Possible responses: direction of fibers—external oblique; shape—deltoid; action—extensor digitorum; size—gluteus maximus; origin and insertion—sternocleidomastoid; location—tibialis anterior; number of origins—biceps brachii.

8.14 Frowning—frontalis; smiling—zygomaticus major; pouting—platysma; squinting—orbicularis oculi.

8.15 Origin; insertion.

8.16 Lateral rectus, superior oblique, and inferior oblique.

8.17 Rectus abdominis.

8.18 Diaphragm and external intercostals.

8.19 Ribs—subclavius, pectoralis major, serratus anterior; vertebrae—trapezius, levator scapulae, rhomboideus major.

8.20 Pectoralis major and latissimus dorsi.

8.21 Biceps brachii, brachialis, and brachioradialis.

8.22 Flexor carpi radialis, flexor carpi ulnaris, and palmaris longus flex wrist; exterior carpi radialis longus, extensor carpi ulnaris, and extensor digitorum extend wrist.

8.23 Erector spinae extends; sternocleidomastoid, rectus abdominis, and quadratus lumborum flex.

8.24 Quadriceps femoris—rectus femoris, vastus lateralis, vastus medialis, and vastus intermedius; hamstrings—biceps femoris, semitendinosus, semimembranosus.

8.25 Tibialis anterior and extensor digitorum longus.

chapter 9

NERVOUS TISSUE

a look ahead

The body has two control centers, one for rapid response—the nervous system—and one that makes slower but no less important adjustments for maintaining homeostasis—the endocrine system. The nervous system has three basic functions. First, it *senses* changes (stimuli) within the body (internal environment) and in the outside environment (external environment); this is its sensory function. Second, it analyzes the sensory information, stores some of it, and makes decisions regarding appropriate behaviors; this is its *integrative* function. Third, it *responds* to stimuli by initiating action in the form of muscular contractions or glandular secretions; this is its motor function.

The branch of medical science that deals with the normal functioning and disorders of the nervous system is called **neurology** (noo-ROL-ō-jē; *neuro* = nerve or nervous system; *logos* = study of).

Organization

objective: *Describe the organization of the nervous system.*

The nervous system has two principal divisions, the **central nervous system (CNS)** and the **peripheral** (pe-RIF-er-al) **nervous system (PNS)** (Figure 9.1).

The CNS consists of the **brain** and **spinal cord.** Within the CNS, many different kinds of incoming sensory information are integrated and correlated, thoughts and emotions are generated, and memories are formed and stored. Most nerve impulses that stimulate muscles to contract and glands to secrete originate in the CNS.

The CNS is connected to sensory receptors, muscles, and glands in peripheral parts of the body by the PNS. The PNS consists of **cranial nerves** that arise from the brain and **spinal**

Figure 9.1 Organization of the nervous system.

🔑 *The two principal subdivisions of the nervous system are (1) the central nervous system (CNS), consisting of the brain and spinal cord, and (2) the peripheral nervous system (PNS), consisting of cranial and spinal nerves; the PNS has somatic and autonomic components.*

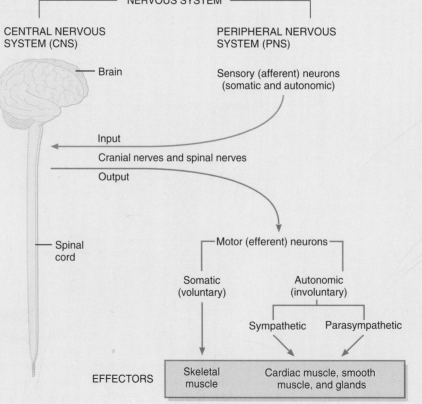

🔁 *Which division of the nervous system is voluntary?*

nerves that emerge from the spinal cord. These nerves carry nerve impulses into and out of the CNS.

The input component of the PNS consists of nerve cells called *sensory* or *afferent* (AF-er-ent; *ad* = toward; *ferre* = to carry) *neurons* (NU-rons). They conduct nerve impulses from sensory receptors in various parts of the body to the CNS and end within the CNS. The output component consists of nerve cells called *motor* or *efferent* (EF-er-ent; *ex* = away from; *ferre* = to carry) *neurons*. They originate within the CNS and conduct nerve impulses from the CNS to muscles and glands. Neurons called *association* (*interneuron*) *neurons* carry nerve impulses from sensory to motor neurons and are located in the CNS. Most neurons are association neurons.

The PNS may be subdivided further into a *somatic* (*soma* = body) *nervous system (SNS)* and an *autonomic* (*auto* = self; *nomos* = law) *nervous system (ANS)*. The SNS consists of sensory neurons that convey information from cutaneous and special sense receptors primarily in the head, body wall, and limbs to the CNS and motor neurons from the CNS that conduct impulses to *skeletal muscles* only. Because these motor responses can be consciously controlled, this portion of the SNS is *voluntary*.

The ANS (Chapter 11) consists of mostly motor neurons from the CNS that conduct impulses to *smooth muscle, cardiac muscle,* and *glands*. Since its motor responses are not normally under conscious control, the ANS is *involuntary*.

The motor portion of the ANS consists of two branches, the *sympathetic division* and the *parasympathetic division*. With few exceptions, the viscera receive instructions from both. Usually, the two divisions have opposing actions. For example, sympathetic neurons speed the heartbeat while parasympathetic neurons slow it down. Processes promoted by sympathetic neurons often involve expenditure of energy while those promoted by parasympathetic neurons restore and conserve body energy.

Histology

objective: *Compare the structure and functions of neuroglia and neurons.*

Despite the complexity of the nervous system, it consists of only two kinds of cells: neuroglia and neurons. Neuroglia serve to support and protect the neurons. Neurons are specialized for nerve impulse conduction and for all special functions, such as thinking, controlling muscle activity, and regulating glands.

Neuroglia

The *neuroglia* (noo-ROG-lē-a; *neuro* = nerve; *glia* = glue) or *glia* serve numerous support and protective functions for the neurons. Neuroglia are generally smaller than neurons and outnumber them by 5 to 50 times. There are six types of neuroglia: astrocytes, oligodendrocytes, microglia, ependymal cells, neurolemmocytes (Schwann cells), and satellite cells. Exhibit 9.1 summarizes their functions. Neuroglia are a common source of gliomas (tumors), and it is estimated that gliomas account for almost half of brain tumors. Such tumors are highly malignant and grow rapidly.

Neurons

Neurons conduct impulses from one part of the body to another. They are the basic information-processing units of the nervous system. The terms nerve cell and neuron mean the same thing.

Structure

Neurons have three distinct parts: (1) cell body, (2) dendrites, and (3) axon (Figure 9.2a on page 209). The *cell body* contains a well-defined nucleus and nucleolus surrounded by a granular cytoplasm along with typical organelles such as lysosomes, mitochondria, and a Golgi complex. It does not contain a mitotic apparatus. The inability of neurons to regenerate in adults is related to the absence of a mitotic apparatus.

Neurons have two kinds of cytoplasmic processes: dendrites and axons. *Dendrites* (*dendro* = tree) are usually short, thick, highly branched extensions of the cytoplasm. A neuron usually has several main dendrites. They function to receive impulses and conduct them toward the cell body.

The second type is the *axon* (*axon* = axis), a single long, thin extension that sends impulses to another neuron or tissue. Axons vary in length from a few millimeters (1 mm = 0.04 inch) in the brain to a meter (3.28 ft) or more between the spinal cord and toes. Along the length of an axon, there may be side branches called *axon collaterals*. The axon and its collaterals terminate by branching into many fine filaments called *axon terminals*. The ends of axon terminals contain bulblike structures called *synaptic end bulbs,* which contain sacs called *synaptic vesicles* that store chemicals called neurotransmitters. Neurotransmitters determine whether an impulse passes from one neuron to another or from a neuron to another tissue (see Chapter 8).

Formation of Myelin

The axons of most neurons are surrounded by a many-layered white, lipid (fatty) and protein covering produced by neuroglia that is called the *myelin sheath.* The sheath electrically insulates the axon of a neuron and increases the speed of nerve impulse conduction. Axons with such a covering are said to be *myelinated,* whereas those without it are said to be *unmyelinated*.

Two types of neuroglia produce myelin sheaths (see Exhibit 9.1): neurolemmocytes and oligodendrocytes. In the PNS, *neurolemmocytes* (*Schwann cells*) form myelin sheaths around the axons during fetal development. In forming a sheath, a neurolemmocyte wraps around the axon in a spiral fashion in such a way that its nucleus and cytoplasm end up in the outside layer. The inner portion, up to 100 layers of neurolemmocyte membrane, is the myelin sheath.

The outer nucleated cytoplasmic layer of the neurolemmocyte, which encloses the myelin sheath, is called the *neurolemma* (*sheath of Schwann*). A neurolemma is found only around axons in the PNS. When an axon is injured, the neurolemma aids regeneration by forming a regeneration tube that guides and stimulates regrowth of the axon. At intervals along an axon, the myelin sheath has gaps called *neurofibral nodes* (*nodes of Ranvier*; RON-vē-ā). See Figure 9.2.

In the CNS, an *oligodendrocyte* myelinates many axons in somewhat the same manner that neurolemmocytes myelinate PNS axons. Oligodendrocytes merely deposit a myelin sheath, however, without forming a neurolemma. Neurofibral

Exhibit 9.1 | Neurologia

TYPE, MICROSCOPIC APPEARANCE AND DESCRIPTION	FUNCTIONS	TYPE, MICROSCOPIC APPEARANCE, AND DESCRIPTION	FUNCTIONS

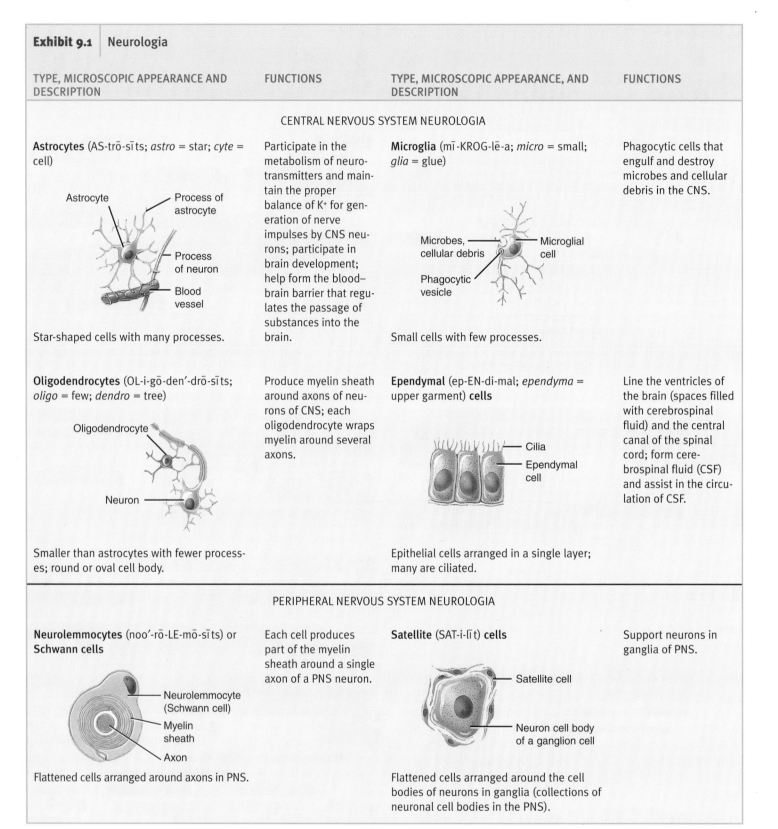

CENTRAL NERVOUS SYSTEM NEUROLOGIA

Astrocytes (AS-trō-sīts; *astro* = star; *cyte* = cell)

Astrocyte — Process of astrocyte — Process of neuron — Blood vessel

Star-shaped cells with many processes.

Participate in the metabolism of neuro-transmitters and maintain the proper balance of K+ for generation of nerve impulses by CNS neurons; participate in brain development; help form the blood–brain barrier that regulates the passage of substances into the brain.

Microglia (mī-KROG-lē-a; *micro* = small; *glia* = glue)

Microbes, cellular debris — Microglial cell — Phagocytic vesicle

Small cells with few processes.

Phagocytic cells that engulf and destroy microbes and cellular debris in the CNS.

Oligodendrocytes (OL-i-gō-den'-drō-sīts; *oligo* = few; *dendro* = tree)

Oligodendrocyte — Neuron

Smaller than astrocytes with fewer processes; round or oval cell body.

Produce myelin sheath around axons of neurons of CNS; each oligodendrocyte wraps myelin around several axons.

Ependymal (ep-EN-di-mal; *ependyma* = upper garment) **cells**

Cilia — Ependymal cell

Epithelial cells arranged in a single layer; many are ciliated.

Line the ventricles of the brain (spaces filled with cerebrospinal fluid) and the central canal of the spinal cord; form cerebrospinal fluid (CSF) and assist in the circulation of CSF.

PERIPHERAL NERVOUS SYSTEM NEUROLOGIA

Neurolemmocytes (noo'-rō-LE-mō-sīts) or **Schwann cells**

Neurolemmocyte (Schwann cell) — Myelin sheath — Axon

Flattened cells arranged around axons in PNS.

Each cell produces part of the myelin sheath around a single axon of a PNS neuron.

Satellite (SAT-i-līt) **cells**

Satellite cell — Neuron cell body of a ganglion cell

Flattened cells arranged around the cell bodies of neurons in ganglia (collections of neuronal cell bodies in the PNS).

Support neurons in ganglia of PNS.

Figure 9.2   Structure of a typical neuron. Arrows indicate the direction of information flow. The break indicates that the axon is actually longer than shown.

The basic parts of a neuron are (1) dendrites, (2) cell body, and (3) axon.

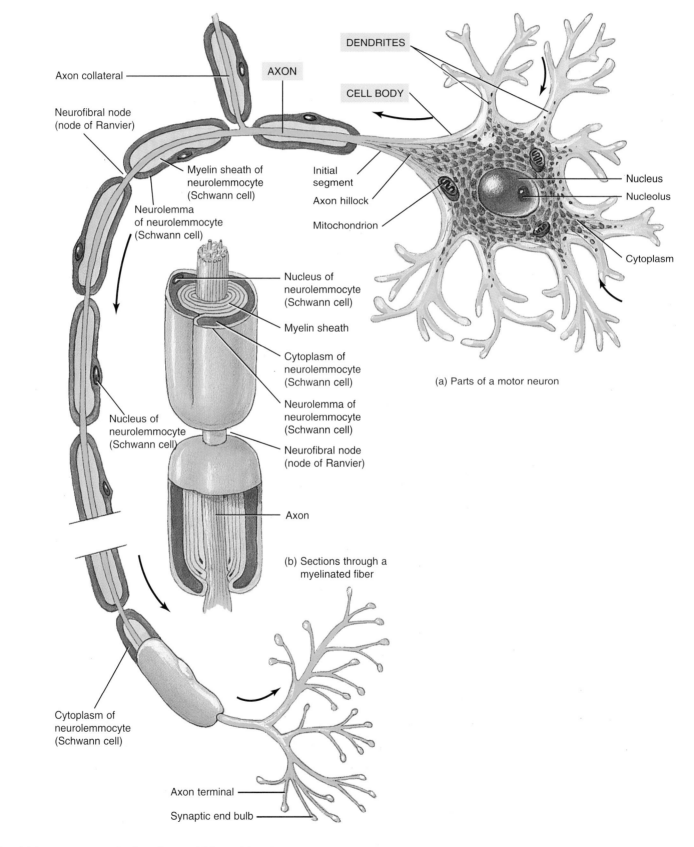

Axon collateral

AXON

Neurofibral node
(node of Ranvier)

Myelin sheath of
neurolemmocyte
(Schwann cell)

Neurolemma
of neurolemmocyte
(Schwann cell)

DENDRITES

CELL BODY

Initial
segment

Axon hillock

Mitochondrion

Nucleus

Nucleolus

Cytoplasm

Nucleus of
neurolemmocyte
(Schwann cell)

Myelin sheath

Cytoplasm of
neurolemmocyte
(Schwann cell)

Neurolemma of
neurolemmocyte
(Schwann cell)

Neurofibral node
(node of Ranvier)

Axon

Nucleus of
neurolemmocyte
(Schwann cell)

(a) Parts of a motor neuron

(b) Sections through a
myelinated fiber

Cytoplasm of
neurolemmocyte
(Schwann cell)

Axon terminal

Synaptic end bulb

Which processes receive impulses? Which send impulses?

nodes are present, but they are fewer in number. Axons in the CNS display little regrowth after injury. This is thought to be due, in part, to the absence of a neurolemma.

The amount of myelin increases from birth to maturity, and its presence greatly increases the speed of nerve impulse conduction. Because myelination is still in progress during infancy, an infant's responses to stimuli are not as rapid or coordinated as those of an older child or adult. Certain diseases such as multiple sclerosis (see p. 244) or Tay–Sachs disease (see p. 244) involve destruction of myelin sheaths.

Grouping of Neural Tissue

A **nerve fiber** is a general term for any process projecting from the cell body, in other words, a dendrite or an axon. Most commonly though, it refers to an axon and its sheaths. Figure 9.2 shows a typical nerve fiber.

The term **white matter** refers to groups of myelinated axons from many neurons. Myelin has a whitish color that gives white matter its name. A **nerve** consists of a group of myelinated nerve fibers in the PNS. An example of a nerve is the sciatic nerve in the thigh. Most nerves contain both sensory and motor fibers. The term **tract** is used to indicate a bundle of fibers located in the CNS. Tracts may run long distances up or down the spinal cord or connect parts of the brain with each other and with the spinal cord. Spinal tracts that conduct impulses up the cord and carry sensory impulses are called *ascending tracts*. Spinal tracts that carry impulses down the cord carry motor impulses and are called *descending tracts*.

The **gray matter** of the nervous system contains either neuron cell bodies and dendrites or bundles of unmyelinated axons. The absence of myelin in these neuron parts accounts for the gray color. Groups of neuron cell bodies located in the PNS are known as **ganglia** (*ganglion* = singular). However, clusters of neuron cell bodies and dendrites in the CNS are known as **nuclei**.

In the brain, gray matter is found covering its outer surface and the nuclei lie in deeper regions. In the spinal cord, gray matter is located internally in regions called **horns**, which are surrounded by groups of white matter tracts called **columns** (see Figure 10.3, p. 223).

Functions

Two striking features of neurons are (1) their highly developed ability to produce and conduct electrical messages called nerve impulses and (2) their limited ability to regenerate.

Nerve Impulses

objective: *Describe how a nerve impulse is generated and conducted.*

Nerve impulses are like tiny electric currents that pass along neurons. These impulses result from the movement of ions (electrically charged particles) in and out through the plasma membranes of neurons. Before discussing changes in electrical charge, it is necessary to understand the kinds of channels in plasma membranes that permit movement of ions between the external and internal surfaces of the neuron's membrane.

Ion Channels in Plasma Membranes

Plasma membranes contain a variety of **ion channels** formed by membrane proteins. These channels are usually highly selective with respect to which ions pass through. Although some ionic channels are always open, most are subject to regulation in which they spend some time open (conducting) and some time closed (nonconducting). In such channels, the passage of ions is controlled by protein molecules that form a gate, which can change its shape to open or close the channel in response to various signals. For example, some of these channels open and close in response to changes in voltage of the plasma membrane or differences in concentration of ions on either side of the plasma membrane of the cell. Others open and close in response to chemicals, such as neurotransmitters and hormones, which bind to the channels.

Membrane Potentials

In a resting neuron (one that is not conducting an impulse), there is a difference in electrical charges on the outside and inside of the plasma membrane. The outside has a positive charge and the inside has a negative charge. What causes this difference? There are two factors. One is the different numbers of potassium ions (K^+) and sodium ions (Na^+) on either side of the membrane. There is 30 times more K^+ inside the cell than outside. At the same time, there is about 15 times more Na^+ outside than inside. The other factor is the presence of large negatively charged ions (for example, proteins) trapped inside the cell. Let us examine how these factors work to cause the difference in charge.

Even when a nerve cell is not conducting an impulse, it is *actively* transporting Na^+ out of the cell and K^+ in, at the same time, by means of the **sodium–potassium pump** (Figure 9.3a–d). (The mechanism of active transport may be reviewed in Figure 3.8.)

Because Na^+ is positive and is actively transported out, a positive charge develops outside the membrane. Even though K^+ is also positive and is actively transported into the cell, there is not enough K^+ to balance the even larger number of negative ions trapped in the cell.

In addition, the operation of the sodium–potassium pump creates a concentration and electrical gradient (difference) for Na^+ and K^+, which means that K^+ tends to diffuse (leak) out of the cell, and Na^+ tends to diffuse in. Moreover, membranes tend to be much more permeable to K^+, so K^+ diffuses out along its concentration gradient easily and quickly while Na^+ enters along its gradient much more slowly. The final result is a net positive charge outside and net negative charge inside. The difference in charge on either side of the membrane of a resting neuron is the **resting membrane potential**. Such a membrane is said to be **polarized** (Figure 9.3e). Said another way, a polarized membrane is one that is positive on the outside and negative on the inside. The electrical charge on the inside of a polarized membrane is −70 *millivolts* (abbreviated as mV). One millivolt equals one-thousandth of a volt; a 1.5-volt battery will power an ordinary flashlight.

Excitability

The ability of nerve cells to respond to stimuli and convert them into nerve impulses is called **excitability**. A **stimulus** is anything in the environment capable of changing the resting

Figure 9.3   Sodium–potassium pump and development of the resting membrane potential.

When a nerve cell is not conducting an impulse, it is actively transporting Na⁺ out of the cell and K⁺ into the cell at the same time by means of the sodium–potassium pump.

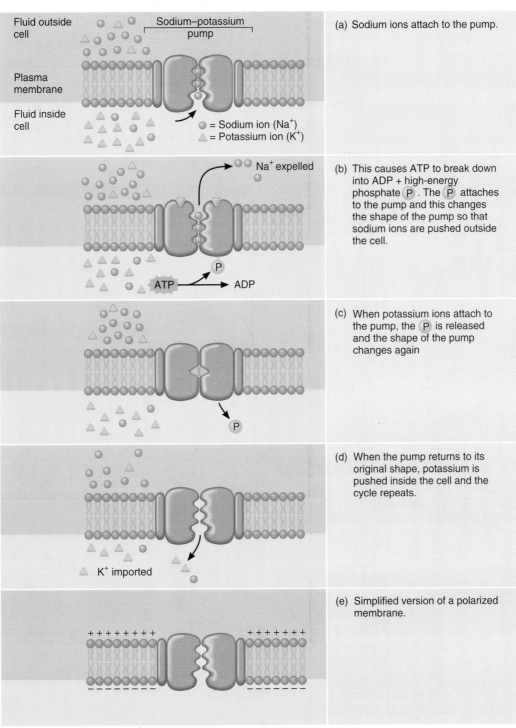

(a) Sodium ions attach to the pump.

(b) This causes ATP to break down into ADP + high-energy phosphate P . The P attaches to the pump and this changes the shape of the pump so that sodium ions are pushed outside the cell.

(c) When potassium ions attach to the pump, the P is released and the shape of the pump changes again

(d) When the pump returns to its original shape, potassium is pushed inside the cell and the cycle repeats.

(e) Simplified version of a polarized membrane.

Q *What ions contribute to the negative charge inside the membrane?*

membrane potential. We will now look at how a nerve impulse, or nerve action potential, is generated.

Before we examine the details of nerve impulse generation, it will be helpful to understand the meaning of two terms: depolarization and repolarization. *Depolarization* is the loss and then the reversal of polarization due to the rapid opening of sodium ion channels. *Repolarization* is the recovery of the resting membrane potential due to the slower opening of potassium ion channels and the closing of sodium ion channels. Together, depolarization and repolarization comprise a nerve impulse and take only about 1 *millisecond* (abbreviated as msec). One millisecond is equal to 1/1000 of a second.

If a stimulus of adequate strength is applied to a polarized membrane (Figure 9.4a,e), the membrane becomes less negative and reaches a critical level called *threshold* (about −55 mV). During this time, sodium ion channels open to permit Na⁺ to enter the membrane (Figure 9.4b,e). Sodium ions also move in because they are attracted to the negative ions on the inside of the membrane. As more of the positive Na⁺ enter the cell, the negative value (−70 mV) on the inside begins to shift toward 0 and then to positive (+30 mV). In relation to the inside of the membrane, the outside is now relatively negatively charged. This change in polarization is called *depolarization* and the membrane is said to be *depolarized* (Figure 9.4b). Throughout depolarization, the Na⁺ continues to rush inside until the membrane potential is reversed: the inside of the membrane becomes positive and the outside negative.

Depolarization continues until the membrane potential reaches +30 mV. Once depolarization has taken place at a specific point on the membrane (Figure 9.4e), that point immediately becomes *repolarized*; that is, its resting potential is restored (Figure 9.4d,e). Just as depolarization results from changes in membrane permeability, so does repolarization. Repolarization, however, involves potassium ion channels. When the membrane is polarized (resting), the potassium ion channel gate is nearly closed and K⁺ remains inside (Figure 9.4a). After the membrane becomes depolarized, the channel opens and K⁺ rapidly diffuses out. At the same time, the sodium channels are closing. The loss of positive ions leaves the inner membrane negative again, that is, repolarized. During repolarization, the outflow of K⁺ may be so great that *hyperpolarization* occurs (Figure 9.4e). In this state, polarization is more negative than the resting level (−70 mV). Hyperpolarization is the basis for inhibitory actions of the nervous system. As potassium ion channels close, the membrane potential returns to the resting level.

Once the events of depolarization and repolarization have taken place, we say that a *nerve impulse* (*nerve action potential*) has occurred. An *action potential* is a rapid change in membrane potential that involves a depolarization followed by a repolarization (restoration of the resting potential, that is, a return to negative inside, positive outside). The impulse self-propagates along the outside surface of the membrane of a neuron. Of all the cells of the body, only muscle fibers and nerve cells produce action potentials.

For the nerve impulse to communicate information to another part of the body, it must be propagated (transmitted) along the neuron. A nerve impulse generated at any one point on the membrane excites (depolarizes) adjacent portions of the membrane, causing depolarization of the adjacent areas, opening of new sodium ion channels, inward movement of Na⁺, and development of new nerve impulses at successive points along the membrane (Figure 9.4b–d).

Following depolarization, repolarization returns the cell to its resting membrane potential and the neuron is now prepared to receive another stimulus and conduct it in the same manner. In fact, until repolarization occurs, the neuron cannot conduct another nerve impulse. The period of time during which the neuron cannot generate another nerve action potential is called the *refractory period.*

Generally, a nerve impulse travels in only one direction along a neuron. A sensory neuron is stimulated at its dendrite by a receptor, a structure sensitive to changes in the environment. Association and motor neurons are stimulated at their dendrites or cell bodies by another neuron.

The initiation and conduction of a *muscle action potential* are basically similar to a nerve action potential (nerve impulse), although the duration of a muscle action potential is considerably longer, while the velocity of conduction of a nerve action potential is about 18 times faster.

All-or-None Principle

Any stimulus strong enough to initiate a nerve impulse is referred to as a *threshold stimulus*. A single nerve cell, just like a single muscle fiber, transmits an action potential according to the *all-or-none principle:* if a stimulus is strong enough to generate a nerve action potential, the impulse is conducted along the entire neuron at maximum strength, unless conduction is altered by conditions such as toxic materials in cells or fatigue.

An analogy helps in understanding this principle. If a long trail of gunpowder were spilled along the ground and ignited at one end, it would send a blazing signal down the entire length of the trail. It would not matter how tiny or great the triggering spark or flame or explosion was; the blazing signal moving along the trail of gunpowder would be just the same, a maximal one (or none at all if it never started).

Any stimulus weaker than a threshold stimulus is a *subthreshold stimulus*. If it occurs only once, it is incapable of initiating a nerve impulse. If, however, a second stimulus or a series of subthreshold stimuli are quickly applied to the neuron, the cumulative effect may be sufficient to initiate an impulse. This phenomenon is called summation and is discussed shortly.

Continuous and Saltatory Conduction

The nerve impulse conduction we have been considering so far applied to unmyelinated fibers. This step-by-step depolarization of each adjacent area of the axon or dendrite membrane is called *continuous conduction*. In myelinated fibers, conduction is somewhat different. The myelin in a myelin sheath does not conduct an electric current. It is a fatty insulating layer that virtually inhibits the movement of ions. However, as you recall, a myelin sheath is interrupted at various intervals called neurofibril nodes (nodes of Ranvier). Depolarization can occur at the nodes and nerve impulses can be generated and conducted. In a myelinated fiber, the nerve impulse jumps from node to node. This type of impulse conduction is called *saltatory conduction* (SAL-ta-tō-rē; *saltare* = leaping).

Saltatory conduction is important in maintaining homeostasis. Because an impulse jumps long intervals as it moves from one node to the next, it travels much faster than in the step-by-step depolarization. This is especially important when split-second responses are necessary.

Speed of Nerve Impulses

When a neuron receives a threshold stimulation, the speed of the nerve impulse is determined by temperature, the diameter of the fiber, and the presence or absence of myelin. The speed of a nerve impulse is not dependent on stimulus strength.

When warmed, nerve fibers conduct impulses at higher speeds; when cooled, at lower speeds. Consequently, pain

Figure 9.4   Initiation and propagation of a nerve impulse. (a–d) The red area containing the straight arrow represents the region of the membrane that is propagating the nerve impulse. The curved arrows represent local currents. (e) Record of potential changes of a nerve impulse.

 Inflow of sodium ions (Na⁺) causes depolarization and outflow of potassium ions (K⁺) causes repolarization.

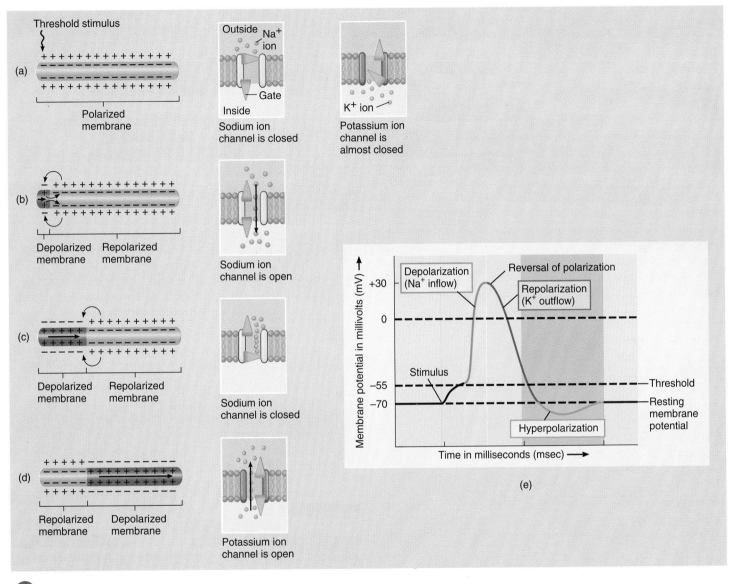

In a polarized membrane, what is the charge just inside the membrane?

resulting from injured tissue can be reduced by the application of cold because the nerve fibers carrying the pain sensation are partially blocked.

Fibers with large diameters conduct impulses faster than those with small ones. Fibers with the largest diameter are all myelinated and therefore capable of saltatory conduction. Fibers with the smallest diameter are unmyelinated, so their conduction is continuous.

Conduction Across Synapses

A nerve impulse is conducted not only along the length of a neuron but also from one neuron to another or to a structure such as a muscle or gland. Impulses from a neuron to a muscle fiber are conducted across a *neuromuscular junction*, which was discussed in Chapter 8 (see Figure 8.5). Between a neuron and glandular cell, the impulse crosses a *neuroglandular junction*.

Impulses are conducted from one neuron to another or from a neuron to another cell such as a muscle fiber or glandular cell across a *synapse*—a junction between the cells. Within a synapse, the involved cells approach, but do not quite touch, one another. The synapse is essential for homeostasis because of its ability to transmit certain impulses and inhibit others. Most diseases of the brain and many psychiatric disorders result from a disruption of synaptic communication.

Synapses are also the sites of action for most drugs that affect the brain, both therapeutic and addictive substances.

Figure 9.5 shows the three parts of a chemical synapse between neurons. In such a synapse, the presynaptic electrical signal (nerve impulse) is converted into a chemical signal (liberated neurotransmitter). The **presynaptic neuron** is a neuron located before a synapse. It is the neuron through which an impulse travels. The **postsynaptic neuron** is located after a synapse. The space between, filled with extracellular fluid, is the **synaptic cleft**. Axon terminals of neurons end in bulblike structures called **synaptic end bulbs**.

In the synapse shown in Figure 9.5, a neuron secretes a chemical substance called a **neurotransmitter** that acts on receptors of the next neuron (postsynaptic neuron), a muscle fiber at a neuromuscular junction, or a glandular cell at a neuroglandular junction. Neurotransmitters are made by the neuron, usually from amino acids, and are stored in the synaptic end bulbs in small sacs called **synaptic vesicles**. (The various kinds of neurotransmitters will be studied in Chapter 10.)

How are neurotransmitters released? When a nerve impulse arrives at the synaptic end bulb of a presynaptic neuron and depolarization occurs, calcium ion channels in the bulb open. Calcium ions flood in from interstitial fluid, attract synaptic vesicles to the plasma membrane, and help liberate the neurotransmitter molecules from the vesicles into the synaptic cleft (Figure 9.5). (In general, each neuron liberates only one type of neurotransmitter.)

Figure 9.5 Parts of a chemical synapse.

🔑 *A synapse is a junction between cells, such as between neurons or between neurons and muscle fibers or glandular cells.*

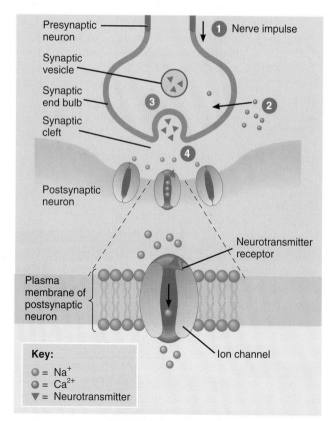

Presynaptic neuron

1 Nerve impulse

Synaptic vesicle

Synaptic end bulb

Synaptic cleft

3

2

4

Postsynaptic neuron

Neurotransmitter receptor

Plasma membrane of postsynaptic neuron

Ion channel

Key:
- 🔵 = Na⁺
- 🔵 = Ca²⁺
- ▼ = Neurotransmitter

Q *Which neuron is responsible for one-way impulse conduction?*

What happens once the neurotransmitter enters the synaptic cleft depends on the chemical nature of the neurotransmitter and how it interacts with the receptors on the postsynaptic plasma membrane. A neurotransmitter may produce an **excitatory transmission**, which creates a new nerve impulse, or an **inhibitory transmission**, which prevents further nerve impulses.

As you can see in the synapse shown in Figure 9.5, there is only **one-way impulse conduction**—from a presynaptic axon to a postsynaptic cell. This is because only synaptic end bulbs of presynaptic neurons can release neurotransmitter. As a result, nerve impulses must move forward over their pathways. They cannot back up into another presynaptic neuron, a situation that would seriously disrupt homeostasis.

A single postsynaptic neuron receives signals from many presynaptic neurons. Some neurotransmitters produce excitation (small depolarizations) and some produce inhibition (small hyperpolarizations). The sum of all the effects, excitatory and inhibitory, determines the final effect on the postsynaptic neuron. Thus the postsynaptic neuron is an **integrator**. It receives signals, integrates them, and then responds accordingly. The postsynaptic neuron may respond in the following ways:

1. If the excitatory (depolarization) effect is greater than the inhibitory (hyperpolarization) effect, but less than the threshold level of stimulation, the result is **facilitation,** that is, near-threshold excitation so that subsequent stimuli can more easily generate a nerve impulse.

2. If the excitatory effect is greater than the inhibitory effect, but equal to or higher than the threshold level of stimulation, the result is *generation of a nerve impulse.*

3. If the inhibitory effect is greater than the excitatory effect, the membrane hyperpolarizes, and the result is inhibition of the postsynaptic neuron and thus an *inability to generate a nerve impulse.*

Regeneration of Nervous Tissue

Unlike the cells of epithelial tissue, neurons have only limited powers of **regeneration,** that is, the capacity to replicate or repair themselves. Around six months of age, virtually all nerve cells lose their ability to reproduce. Thus when a neuron is damaged or destroyed, it cannot be replaced by other neurons. A neuron destroyed is permanently lost, and only some types of damage may be repaired.

In the peripheral nervous system (PNS), damage to some types of myelinated axons and dendrites can be repaired if the cell body remains intact and if the cell that performs the myelination remains active. If the myelinating cell is a neurolemmocyte (Schwann cell), it helps regeneration. These cells proliferate following axonal damage, and their neurolemmas form a regeneration tube that assists in regeneration. Axons in the central nervous system (CNS) are myelinated by oligodendrocytes, which do not form neurolemmas to assist in regeneration and do not survive following axonal damage. An added complication in the CNS is that, following axonal damage, astrocytes appear to stop axons from regenerating by forming a type of scar tissue. This scar tissue acts as a physical barrier to regeneration. Thus an injury to the brain or spinal cord has permanent effects. An injury to a nerve in the PNS may repair itself before scar tissue forms. As a result, some nerve function may be restored.

*i*t's normal to feel sad when you experience a loss, or when things just aren't going right. At such times, you might say to a friend, "I feel depressed." The mildest form of depression, often called "the blues," consists of a low mood that lasts for less than two weeks. These feelings of depression can usually be relieved, at least temporarily, by engaging in activities that help you feel better.

But when someone experiences feelings of sadness for a longer period than seems reasonable, given the objective circumstances of his or her life, that person may be diagnosed with a more serious depressive disorder. There are several categories of depressive disorders.

Major depression is marked by a severely depressed mood or loss of interest in almost all activities. Someone with major depression has several of the following symptoms:

- An inability to feel pleasure in anything.
- Changes in appetite and weight, sometimes with significant weight loss or gain.
- Problems with sleep, including either insomnia or sleeping much more than usual.
- Feeling nervous and agitated.
- Feeling lethargic and tired all the time.
- Low self-esteem, feeling worthless, guilty, dejected, helpless, and hopeless.
- Loss of interest in sex.
- Difficulty concentrating and making decisions.

- Withdrawal from social contacts.
- A negative outlook; an ability to see only the bad things in life.
- Thoughts of death or suicide.

Dysthymia is a chronic, mild depression distinguished by being in a depressed mood most of the time for the past two years. People with dysthymia are functioning, but not at an optimal level. They have some of the symptoms of major depression (listed above) which may go away for a month or two but then return. Dysthymia predisposes people to developing major depression or other problems, such as addictive disorders, including alcoholism and eating disorders.

Bipolar (manic) depression consists of periods of severe depression that alternate with episodes of mania, or elation. People with bipolar depression feel wonderful during manic periods, but have a distorted perception of reality and often behave inappropriately and make poor decisions.

Seasonal affective disorder (SAD) is a form of depression associated with the short, gray days of winter. SAD usually lasts from October to March. Symptoms typically include lethargy, food cravings, and weight gain.

Depression is called "the common cold of mental illness" because it occurs so frequently, affecting about one in five Americans at some point during their lifetimes.

What Causes Depression?

Depression appears to be caused by a number of physiological and psychological variables that interact in a complex manner. Although risk factors for depression have been identified, it's important to note that depression can develop in anyone for no apparent reason. Risk factors for depression include a family history of depression, loss and severe stress, physical illness, and substance abuse. Major depression occurs twice as often in women as in men.

Depression is a physiological illness, marked by changes in nervous system function. In particular, depression is associated with neurotransmitter imbalances. Psychologists do not yet know whether the feelings of depression cause or are caused by these physiological changes.

Catch It While You Can

One of the most important characteristics of depression is its tendency to worsen over time if left untreated. Repeated episodes of the blues can lead to dysthymia, which in turn can develop into major depression. The more severe the depression, the more the nervous system, is physiologically out of balance, and the more difficult the disorder is to treat. Treating depression at its earliest stages may short-circuit the tendency of depression to worsen over the years.

critical thinking

Therapists recommend that when medication is prescribed for depression, the patient should also receive some form of psychotherapy. If the medication seems to correct the feelings of depression, why do you think therapy is recommended?

Study Outline

Organization (p. 206)

1. The nervous system helps control and integrate all body activities by sensing changes (sensory), analyzing them (integrative), and responding to them (motor).
2. The nervous system has two principal divisions: central nervous system (CNS) and peripheral nervous system (PNS).
3. The central nervous system (CNS) consists of the brain and spinal cord.
4. The peripheral nervous system (PNS) consists of cranial and spinal nerves. It has sensory (afferent) and motor (efferent) components.
5. The PNS also is subdivided into somatic (voluntary) and autonomic (involuntary) nervous systems.
6. The somatic nervous system (SNS) consists of neurons that conduct impulses from cutaneous and special sense receptors to the CNS and motor neurons from the CNS to skeletal muscle tissue. The SNS is voluntary.
7. The autonomic nervous system (ANS) contains sensory neurons from visceral organs and motor neurons that convey impulses from the CNS to smooth muscle tissue, cardiac muscle tissue, and glands. The ANS is involuntary.

Histology (p. 207)

Neuroglia (p. 207)

1. Neuroglia are specialized tissue cells that support neurons, attach neurons to blood vessels, produce the myelin sheath around axons of the CNS, and carry out phagocytosis.
2. Neuroglia include astrocytes, oligodendrocytes, microglia, ependymal cells, neurolemmocytes (Schwann cells), and satellite cells.

Neurons (p. 207)

1. Neurons (nerve cells) consist of a cell body, dendrites that receive stimuli, and a single axon that sends impulses to another neuron or to a muscle or gland.
2. Myelin is formed by neurolemmocytes in the PNS and oligodendrocytes in the CNS.
3. Neurons are grouped in various ways to form nerves, tracts, ganglia, nuclei, horns, and columns.

Functions (p. 210)

Nerve Impulses (p. 210)

1. The nerve impulse (nerve action potential) is the body's quickest way of controlling and maintaining homeostasis.
2. Plasma membranes contain ion channels. A channel is a membrane protein.

3. The membrane of a nonconducting neuron is positive outside and negative inside, owing to the differing numbers of K^+ and Na^+, large negatively charged proteins, and the operation of the sodium–potassium pump. Such a membrane is said to be polarized.
4. When a stimulus causes the inside of the cell membrane to become positive and the outside negative, the membrane is said to have an action potential, which travels along the membrane. The traveling action potential is a nerve impulse. The ability of a neuron to respond to a stimulus and convert it into a nerve impulse is called excitability.
5. Restoration of the resting membrane potential is called repolarization. The period of time during which the membrane recovers and cannot initiate another action potential is called the refractory period.
6. According to the all-or-none principle, if a stimulus is strong enough to generate an action potential, the impulse travels at a constant and maximum strength. A stronger stimulus will not cause a larger impulse.
7. Nerve impulse conduction that occurs as a step-by-step process is called continuous conduction.
8. Conduction in which the impulse jumps from node to node is called saltatory conduction.
9. Fibers with larger diameters conduct impulses faster than those with smaller diameters; myelinated fibers conduct impulses faster than unmyelinated.

Conduction Across Synapses (p. 213)

1. Nerve impulse conduction can occur from one neuron to another or from a neuron to an effector.
2. The junction between neurons, neurons and muscle fibers, or neurons and glandular cells is called a synapse.
3. At a synapse, there is only one-way nerve impulse conduction from a presynaptic axon to a postsynaptic dendrite, cell body, or axon.
4. The postsynaptic neuron is an integrator. It receives signals, integrates them, and then responds accordingly.
5. At a synapse, there may be facilitation (a state of near-threshold excitation, so that additional stimuli can generate an impulse more easily), impulse generation, or impulse inhibition.

Regeneration of Nervous Tissue (p. 214)

1. At about six months of age, the neuron loses its ability to divide.
2. Nerve fibers in the PNS have a neurolemma and are thus capable of regeneration.
3. Axons and dendrites in the CNS do not have a neurolemma, which means injury to the brain or spinal cord has permanent effects.

Self-Quiz

1. Which of the following are incorrectly paired?

   a. central nervous system–brain and spinal cord
   b. somatic nervous system–motor neurons to skeletal muscles   c. sympathetic nervous system–motor neurons to skeletal, smooth, and cardiac muscles   d. peripheral nervous system–cranial and spinal nerves   e. autonomic nervous system–parasympathetic and sympathetic neurons

2. The division of the peripheral nervous system that receives information from special receptors in the head and limbs and conducts impulses to skeletal muscle is the

   a. parasympathetic nervous system   b. somatic nervous system   c. sympathetic nervous system   d. autonomic nervous system   e. central nervous system

3. The division of the nervous system that integrates and correlates information is the

   a. parasympathetic nervous system   b. somatic nervous system   c. sympathetic nervous system   d. autonomic nervous system   e. central nervous system

4. The type of cell that produces the myelin sheath around axons in the CNS is the

   a. astrocyte   b. ependymal cell   c. neurolemmocyte (Schwann cell)   d. oligodendrocyte   e. satellite cell

5. Repolarization is caused by a

   a. rush of Na^+ into the nerve cell   b. rush of Na^+ out of the nerve cell   c. rush of K^+ into the nerve cell   d. rush of K^+ out of the nerve cell   e. pumping of K^+ into the cell

6. The speed of nerve impulse conduction is increased

   a. by cold   b. by a very strong stimulus   c. during saltatory conduction   d. by small fiber size   e. in the absence of myelin

7. Continuous conduction requires the presence of a

   a. myelin sheath   b. neurofibral nodes (nodes of Ranvier)   c. white matter   d. membrane potential   e. subthreshold stimulus

8. Which of the following is NOT true regarding saltatory conduction?

   a. The nerve impulse jumps from node to node.   b. It is more energy efficient.   c. It transmits an action potential at a slower rate.   d. It only occurs along myelinated neurons.   e. It is important in maintaining homeostasis.

9. True/False.

   _F_ a. The speed of a nerve impulse is dependent on the strength of the stimulus.

   _T_ b. Fibers that are larger in diameter conduct impulses faster.

   _F_ c. Excitatory transmission causes hyperpolarization of a nerve cell membrane.

   _T_ d. An impulse may only be conducted one way across a synapse.

10. Match the following:

    _H_ a. the portion of a neuron containing the nucleus

    _A_ b. rounded structures at the distal end of an axon terminal

    _D_ c. receives impulses and conducts them to the cell body

    _E_ d. sac in which neurotransmitter is stored

    _I_ e. transmits impulses away from the cell body

    _G_ f. produces the myelin sheath

    _J_ g. unmyelinated gaps between the myelin sheaths

    _F_ h. substance that increases the speed of conduction while insulating a neuron

    _C_ i. neuron that transmits information to the CNS

    _B_ j. neuron that conveys information from the CNS back to an effector

    _L_ k. bundle of many nerve fibers in the PNS

    _N_ l. bundle of many nerve fibers in the CNS

    _K_ m. group of cell bodies in the PNS

    _M_ n. substance that conducts impulses across synapse

    A. synaptic end bulb
    B. motor neuron
    C. sensory neuron
    D. dendrite
    E. synaptic vesicle
    F. myelin sheath
    G. neurolemmocyte (Schwann cell)
    H. cell body
    I. axon
    J. neurofibral node (node of Ranvier)
    K. ganglion
    L. nerve
    M. neurotransmitter
    N. tract

11. Anything that is capable of reducing the resting membrane potential of a neuron is known as a(n)

    a. pump   b. threshold   c. action potential   d. stimulus   e. ion

12. Place the following events in the correct order of occurrence:

    _B_ 1. Sodium channels open and permit Na^+ to rush inside the neuron.

    _E_ 2. The sodium–potassium pump restores the ions to their original site.

    _A_ 3. A stimulus of threshold strength is applied to the neuron.

    _C_ 4. The charge inside the membrane goes from negative (-55 mV) to positive ($+30$ mV).

    _D_ 5. Potassium channels open and K^+ flows to the outside of the neuron.

    a. 4, 1, 2, 3, 5   b. 4, 3, 1, 2, 5   c. 3, 1, 4, 2, 5   d. 5, 3, 1, 4, 2   e. 3, 1, 4, 5, 2

Critical Thinking Applications

1. Kayla fell asleep while watching television. She woke up with "pins-and-needles" in her leg. Explain what happened to her leg during her nap.

2. A gunpowder analogy was used for the "all-or-none principle" in this chapter. Create your own analogy to illustrate continuous versus saltatory conduction.

3. Julie was cramming for her second A & P exam. She was on her second pot of coffee and second pack of nicotine gum (she was trying to quit smoking) when her legs started to twitch and she just couldn't sit still. What's happened to Julie?

4. A high school senior dove head first into a murky pond. Unfortunately, he hit a submerged log with his head and broke his neck. He is now paralyzed. Can you deduce the location of the damage to his nervous system? What is the likelihood of recovery from his injury?

5. An oral surgeon extracts your friend's wisdom tooth under local anesthesia. She sends him home with directions to hold an icepack against the side of his face. How does anesthesia affect nerve function? What effect does cold have on nerve function?

6. Julie just figured out that her A & P class actually starts at 10 AM and not at 10:15 AM, which has been her arrival time since the beginning of the term. One of the other students remarks that Julie's "gray matter is pretty thin." Should Julie thank him?

Answers to Figure Questions

9.1 Somatic nervous system.
9.2 Dendrites; axons.
9.3 Proteins.
9.4 Negative.
9.5 Presynaptic neuron.

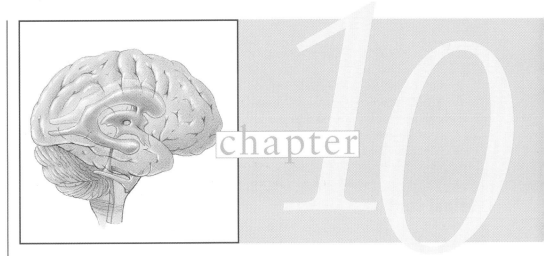
CENTRAL AND SOMATIC NERVOUS SYSTEMS

a look ahead

*Brain &
Spinal cord*   *Body &
mind*

In this chapter we will examine the structure and functions of the central nervous system—the brain and spinal cord. We will also take a look at spinal nerves and cranial nerves, which are part of the somatic nervous system of the peripheral nervous system (see Figure 9.1).

*primary function is to
maintain homeostasis*

Spinal Cord

objective: *Describe how the spinal cord is protected.*

Protection and Coverings

Vertebral Canal

The spinal cord is located in the vertebral canal of the vertebral column (see Figure 6.14). Because the wall of the vertebral canal is essentially a ring of bone, the cord is well protected. Additional protection is provided by the meninges, cerebrospinal fluid, and vertebral ligaments. *and fat*

Meninges

The **meninges** (me-NIN-jēz; singular, **meninx,** MĒ-ninks) are connective tissue coverings that run continuously around the spinal cord and brain. They are called, respectively, the **spinal meninges** (Figure 10.1) and the **cranial meninges** (see Figure 10.7). The outermost of the three layers of the meninges is called the **dura mater** (DYOO-ra MĀ-ter; *dura* = tough; *mater* = mother). Its tough, dense, irregular connective tissue helps to protect the delicate structures of the CNS. The spinal dura mater continues as the cranial dura mater of the brain. The tube of spinal dura mater ends just below the spinal cord around the second lumbar vertebra. The spinal cord is also protected by a cushion of fat and connective tissue located in the **epidural space,** a space between the dura mater and vertebral canal. Injection of an anesthetic into the epidural space results in a temporary sensory and motor paralysis called an **epidural block.** Such injections in the lower lumbar region are used to control pain during childbirth.

The middle layer is called the **arachnoid** (a-RAK-noyd; *arachne* = spider) because of its delicate spider's web arrangement of collagen and elastic fibers. It is also continuous with the arachnoid of the brain.

The inner layer is the **pia mater** (PĪ-a MĀ-ter; *pia* = delicate), or "delicate mother," a transparent layer of collagen and elastic fibers that adheres to the surface of the spinal cord and brain. It contains numerous blood vessels. Between the arachnoid and the pia mater is the **subarachnoid space,** where cerebrospinal fluid circulates. Inflammation of the meninges is known as **meningitis.**

Cerebrospinal fluid may be removed from the subarachnoid space between the third and fourth or fourth and fifth lumbar vertebrae by a **spinal tap** (**lumbar puncture**). A spinal tap may also be performed to introduce antibiotics and anesthetics and to administer chemotherapy.

know

Figure 10.1 Spinal meninges.

🔑 *Meninges are connective tissue coverings that surround the brain and spinal cord.*

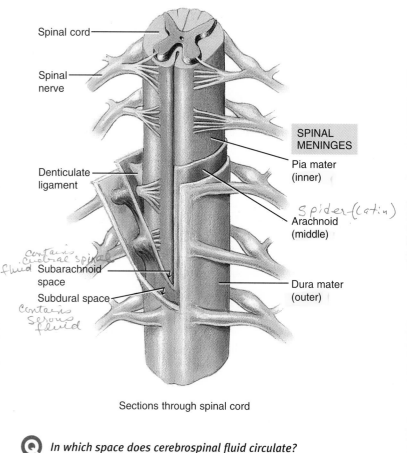

Spinal cord

Spinal nerve

Denticulate ligament

SPINAL MENINGES

Pia mater (inner)

Spider (Latin)

Arachnoid (middle)

Contains cerebral spinal fluid Subarachnoid space

Contains serous fluid Subdural space

Dura mater (outer)

Sections through spinal cord

❓ *In which space does cerebrospinal fluid circulate?*

General Features

objective: *Describe the structure and functions of the spinal cord.*

The length of the adult **spinal cord** ranges from 42 to 45 cm (16 to 18 inches). It extends from the foramen magnum of the occipital bone to the top of the second lumbar vertebra (Figure 10.2). It does not run the entire length of the vertebral column. Consequently, nerves arising from the lowest portion of the cord angle down the vertebral canal like wisps of flowing hair. They are appropriately named the **cauda equina** (KAW-da ē-KWĪ-na), meaning horse's tail.

The cord is not a straight, narrow structure. It has two conspicuous enlargements. The **cervical enlargement** contains nerves that supply the upper limbs; the **lumbar enlargement** contains nerves supplying the lower limbs.

The cord consists of 31 **spinal segments,** each giving rise to a pair of spinal nerves (see Figure 10.3).

Figure 10.2 Spinal cord and spinal nerves. In (a), selected nerves are labeled on the left side of the figure. The arrow in the inset in (b) indicates the direction from which the spinal cord is viewed (superiorly).

The spinal cord extends from the foramen magnum of the occipital bone to the top of the second lumbar vertebra.

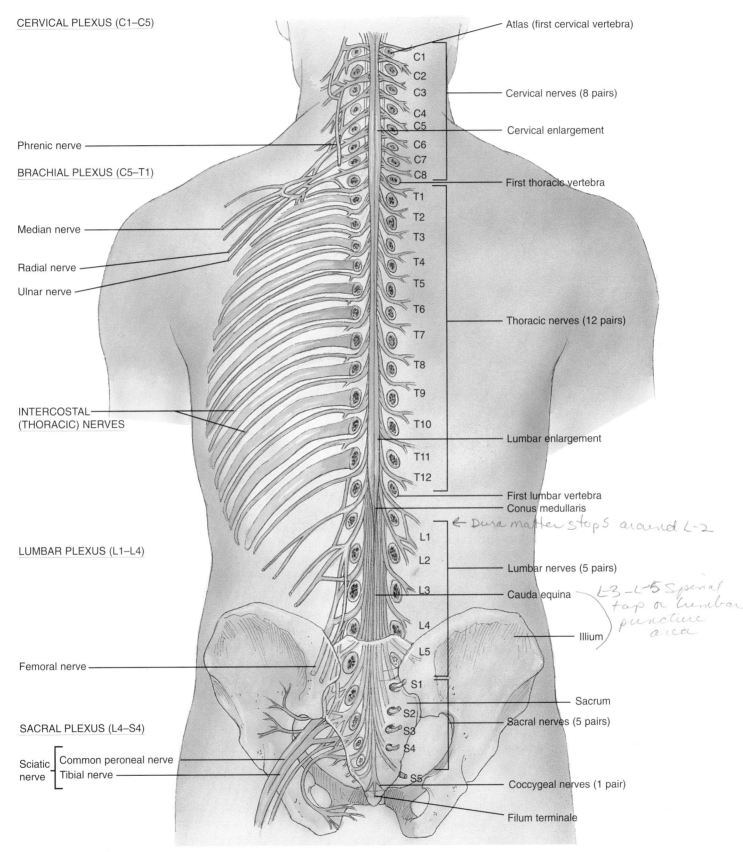

30 vertebrae nerve pairs
31 spinal nerve pairs

CERVICAL PLEXUS (C1–C5)

Phrenic nerve

BRACHIAL PLEXUS (C5–T1)

Median nerve

Radial nerve

Ulnar nerve

INTERCOSTAL (THORACIC) NERVES

LUMBAR PLEXUS (L1–L4)

Femoral nerve

SACRAL PLEXUS (L4–S4)

Sciatic nerve { Common peroneal nerve / Tibial nerve

Atlas (first cervical vertebra)

Cervical nerves (8 pairs)

Cervical enlargement

First thoracic vertebra

Thoracic nerves (12 pairs)

Lumbar enlargement

First lumbar vertebra

Conus medullaris

← Dura matter stops around L-2

Lumbar nerves (5 pairs)

Cauda equina

L3–L5 Spinal tap or lumbar puncture area

Illium

Sacrum

Sacral nerves (5 pairs)

Coccygeal nerves (1 pair)

Filum terminale

(a) Posterior view of entire spinal cord and portions of the spinal nerves

223 + 224 introduce you to what happens in cord + terminology to understand reflex arc

Figure 10.2 *(Continued)*

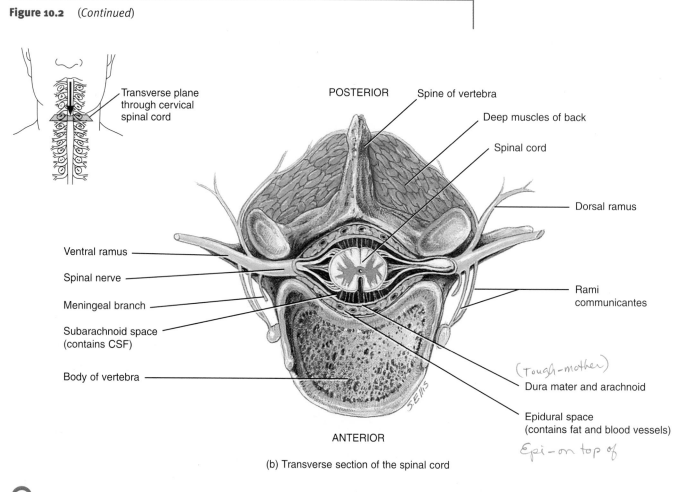

Transverse plane through cervical spinal cord

POSTERIOR

Spine of vertebra

Deep muscles of back

Spinal cord

Dorsal ramus

Ventral ramus

Spinal nerve

Meningeal branch

Rami communicantes

Subarachnoid space (contains CSF)

Body of vertebra

(Tough-mother)

Dura mater and arachnoid

Epidural space (contains fat and blood vessels)

Epi — on top of

ANTERIOR

(b) Transverse section of the spinal cord

Q *Spinal nerves are part of which division of the nervous system?*

Structure in Cross Section

The cord is divided into right and left halves by two grooves, the deep *anterior median fissure* and the shallow *posterior median sulcus* (Figure 10.3). As previously described, the spinal cord contains a centrally located H-shaped mass of gray matter surrounded by white matter. In the center of the gray matter is the *central canal,* which runs the length of the cord and contains cerebrospinal fluid. The sides of the H are divided into regions called *horns,* named relative to their location: anterior, lateral, and posterior. The gray matter consists mainly of association and motor neurons that serve as relay stations for impulses (more on this later). The white matter is also organized into regions called anterior, lateral, and posterior *columns.* The columns consist of myelinated axons organized into sensory (ascending) and motor (descending) *tracts* that convey impulses between the brain and spinal cord.

Functions

The spinal cord has two principal functions. (1) The *white matter tracts* in the spinal cord are highways for nerve impulse conduction. Along these highways sensory impulses flow from the periphery to the brain and motor impulses flow from the

non-myelinated

brain to the periphery. (2) The *gray matter* of the spinal cord receives and integrates incoming and outgoing information. Both functions of the spinal cord are essential to maintaining homeostasis.

Impulse Conduction Along Tracts

Often, the name of a tract indicates its position in the white matter, where it begins and ends, and the direction of nerve impulse propagation. For example, the anterior spinothalamic tract is located in the *anterior* white column, it begins in the *spinal cord,* and it ends in the *thalamus* (a region of the brain). Since it conveys nerve impulses from the cord toward the brain, it is a sensory (ascending) tract.

Sensory information transmitted from receptors up the spinal cord to the brain is conducted along two general pathways on each side of the cord: the anterolateral pathways and the posterior column–medial lemniscus pathway. The *anterolateral (spinothalamic) pathways* consist of two tracts: anterior spinothalamic and lateral spinothalamic that convey impulses for sensing pain, temperature (cold and warmth), crude (poorly localized) touch, pressure, tickle, and itch. The *posterior column–medial lemniscus pathway* (the fasciculus gracilis and fasciculus cuneatus) carries nerve impulses for

myelinated

will cover later in book this section is confusing written

reflex action processed here & things travel faster in white matter - example patellar Reflex

Figure 10.3   Spinal cord. The organization of gray and white matter in the spinal cord as seen in transverse section. The right side of the figure has been sectioned at a lower level than the left so that you can see what is inside the posterior root ganglion, posterior root of the spinal nerve, anterior root of the spinal nerve, and the spinal nerve.

🔑 *The spinal cord conducts nerve impulses along tracts between the brain and periphery and serves as an integrating center for spinal reflexes.*

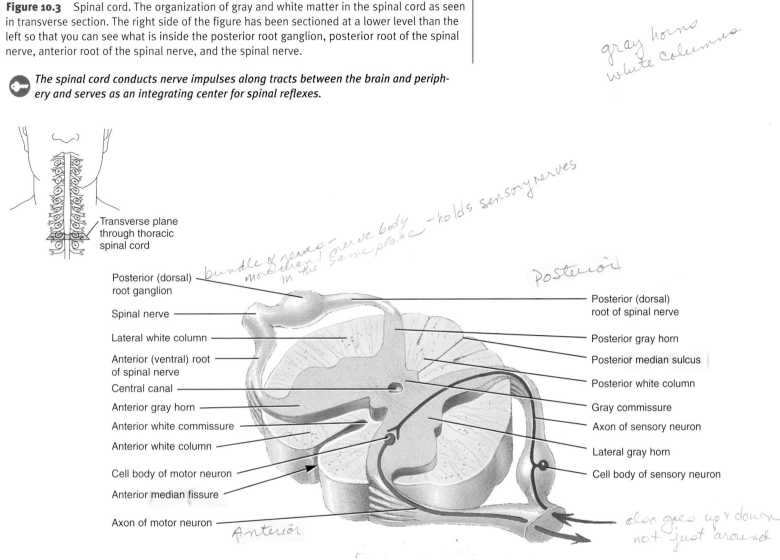

Transverse plane through thoracic spinal cord

Posterior (dorsal) root ganglion

Spinal nerve

Lateral white column

Anterior (ventral) root of spinal nerve

Central canal

Anterior gray horn

Anterior white commissure

Anterior white column

Cell body of motor neuron

Anterior median fissure

Axon of motor neuron

Posterior (dorsal) root of spinal nerve

Posterior gray horn

Posterior median sulcus

Posterior white column

Gray commissure

Axon of sensory neuron

Lateral gray horn

Cell body of sensory neuron

Transverse section of the thoracic spinal cord

❓ *The posterior root contains which type of nerve fibers? The anterior root contains which type of nerve fibers?*

sensing (1) ***proprioception*** (awareness of the position of muscles, tendons, joints, and body movements), (2) ***discriminative touch*** (the ability to feel exactly what part of the body is touched), (3) ***stereognosis*** (the ability to recognize by feel, the size, shape, and texture of an object), (4) ***weight discrimination*** (the ability to determine the weight of an object), and (5) ***vibratory sensations*** (the ability to detect vibration).

Figure 10.4a illustrates the anterolateral (spinothalamic) pathways as an example of a sensory (ascending) tract.

The sensory systems keep the central nervous system aware of the external and internal environments. Responses to this information are brought about by the motor systems, which enable us to move about and change our relationship to the world around us. As sensory information is conveyed to the central nervous system, it becomes part of a large pool of sensory input. We do not respond actively to every bit of input the central nervous system receives. Rather, each piece of incoming

information is integrated with all the other information arriving from activated sensory receptors. The integration process occurs not just once but at many stations along the pathways of the central nervous system with the help of association neurons. A part of the brain called the cerebral motor cortex assumes the major role for controlling precise, discrete, muscular movements. Other brain regions largely integrate semivoluntary movements like walking, swimming, and laughing. Still other parts of the brain help make body movements smooth and coordinated.

When the input reaches the highest center, sensory–motor integration occurs. This involves not only using information contained within that center but also information coming to it from other centers in the central nervous system. After integration occurs, the output from the center is sent down the spinal cord in two major descending motor pathways: the direct (pyramidal) pathways and indirect (extrapyramidal) pathways.

Figure 10.4　Sensory and motor pathways.

Sensory (ascending) tracts convey nerve impulses from the periphery to the brain and motor (descending) tracts convey nerve impulses from the brain to the periphery.

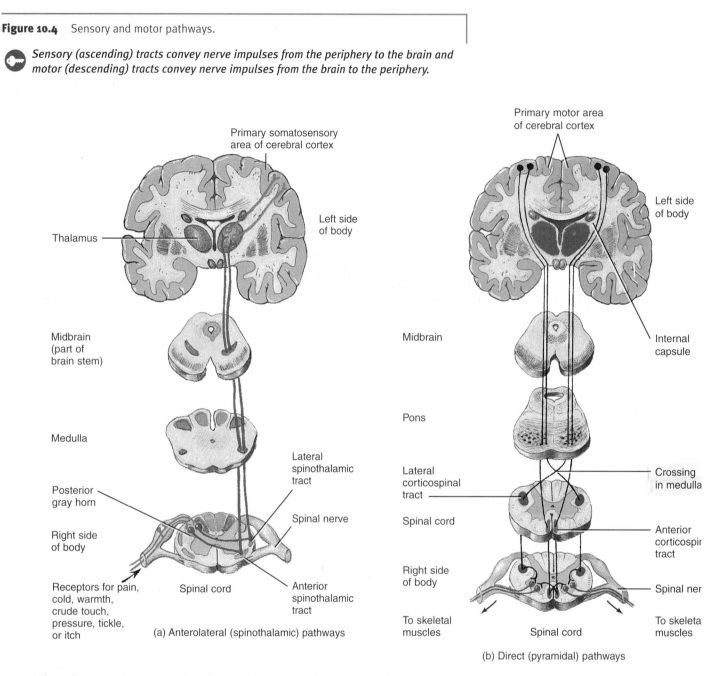

(a) Anterolateral (spinothalamic) pathways

(b) Direct (pyramidal) pathways

Where does conscious perception of sensations occur? Where do motor tracts originate?

Direct (pyramidal) pathways (the lateral corticospinal, anterior corticospinal, and corticobulbar tracts) convey nerve impulses destined to cause precise, voluntary movements of skeletal muscles (Figure 10.4b). ***Indirect (extrapyramidal) pathways*** (rubrospinal, tectospinal, and vestibulospinal tracts) convey nerve impulses that program automatic movements, help coordinate body movements with visual stimuli, maintain skeletal muscle tone and posture, and play a major role in equilibrium by regulating muscle tone in response to movements of the head. Motor pathways conduct impulses from the primary motor area of the cerebral cortex to skeletal muscles.

Reflex Center

The second principal function of the spinal cord is to serve as an integrating center for ***spinal reflexes.*** Reflexes are fast, predictable, automatic responses to changes in the environment. Spinal nerves are the paths of communication between the spinal cord tracts and the periphery. Each pair of spinal nerves is connected to the cord at two points called roots (see Figure 10.3). The *posterior* or *dorsal (sensory) root* contains sensory nerve fibers only and conducts impulses from the periphery to the spinal cord, specifically, the posterior (dorsal) gray horn. Each posterior root also has a swelling, the *posterior* or *dorsal (sensory) root ganglion,* which contains the cell bodies of the sensory neurons from the periphery. The other point of attachment of a spinal nerve to the cord is the *anterior* or *ventral (motor) root.* It contains motor nerve axons only and conducts impulses from the spinal cord to the periphery. The cell bodies of the motor neurons are located in the gray matter of the cord.

Reflex Arc and Homeostasis

The path an impulse follows from its origin in the dendrites or cell body of a neuron in one part of the body to its termination elsewhere in the body is called a *pathway.* All pathways consist of several connected neurons. One kind of pathway is known as a *reflex arc.* The basic components of a reflex arc are as follows (Figure 10.5):

1 *Receptor.* The dendrite of a sensory neuron or an associated sensory structure serves as a receptor. It responds

to a specific stimulus, a change in the internal or external environment, by producing one or more nerve impulses.

2 *Sensory neuron.* The nerve impulses pass from the receptor to the axon terminals of the sensory neuron, which are located in the gray matter of the spinal cord or brain stem.

3 *Integrating center.* This is one or more regions of gray matter within the CNS. The integrating center for the simplest type of reflex is a single synapse, between a sensory neuron and a motor neuron. More often, however, the integrating center consists of one or more association neurons, which may relay the impulse to other association neurons as well as to a motor neuron.

4 *Motor neuron.* Impulses triggered by the integrating center pass out of the CNS along a motor neuron to the part of the body that will respond.

5 *Effector.* The part of the body that responds to the motor nerve impulse, such as a muscle or gland, is the effector. Its action is a reflex.

Figure 10.5   General components of a reflex arc. The arrows show the direction of nerve impulse conduction.

🔑 *Reflexes are fast, predictable, automatic responses to changes in the environment.*

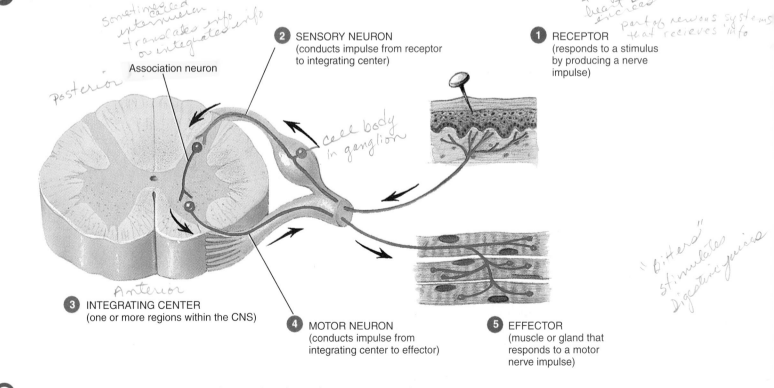

2 SENSORY NEURON (conducts impulse from receptor to integrating center)

Association neuron

1 RECEPTOR (responds to a stimulus by producing a nerve impulse)

3 INTEGRATING CENTER (one or more regions within the CNS)

4 MOTOR NEURON (conducts impulse from integrating center to effector)

5 EFFECTOR (muscle or gland that responds to a motor nerve impulse)

❓ *Which attachment of a spinal nerve to the spinal cord contains motor neurons?*

When we discussed skeletal muscle physiology, we described skeletal muscle as voluntary. However, if we had to think about every action we do, we might suffer severe injuries by the time we decided on the right action and we would be so preoccupied with our body movements we would have little time left for other more serious considerations. Reflexes that result in the contraction of skeletal muscles are *somatic reflexes*. One example is the *patellar reflex* (knee jerk). This reflex is an extension of the knee by contraction of the quadriceps femoris muscle in response to tapping the patellar ligament. The reflex helps us to remain standing erect despite the effects of gravity. Another is the *withdrawal* reflex. This reflex protects us from serious cuts or burns by causing immediate withdrawal from a source of injury, usually before we are even aware of any pain. During a physical examination when a physician tests somatic reflexes, the responses are a reflection of the health of the nervous system.

The functions of smooth and cardiac muscles and many glands are also regulated by reflexes called *autonomic* (*visceral*) *reflexes.* Later in the book we will examine the reflex nature of swallowing, coughing, sneezing, digestion, urination, and defecation.

how they look as they branched out.

Spinal Nerves

objective: *Describe the composition, coverings, and branches of a spinal nerve.*

Names

Spinal nerves, which connect the CNS to sensory receptors, muscles, and glands, are components of the somatic nervous system (SNS) of the peripheral nervous system (PNS). The 31 pairs of spinal nerves are named and numbered according to the region and level of the spinal cord from which they emerge (see Figure 10.2). There are 8 pairs of cervical nerves, 12 pairs of thoracic, 5 pairs of lumbar, 5 pairs of sacral, and 1 pair of coccygeal nerves. The first cervical pair emerges between the atlas and the occipital bone. All other spinal nerves leave the vertebral column from the intervertebral foramina (holes) between vertebrae.

Composition and Coverings

a lot of stress

A *spinal nerve* has two separate points of attachment to the cord, a posterior (dorsal) root and an anterior (ventral) root (see Figure 10.3). These roots unite to form a spinal nerve at the intervertebral foramen. Because the posterior root contains sensory fibers and the anterior root contains motor fibers, a spinal nerve is composed of both and therefore is a *mixed nerve.* Each spinal nerve consists of bundles of fibers wrapped in connective tissue and supplied with blood vessels, very sim-

ilar to the structure of skeletal muscles but, of course, even longer and thinner.

Distribution

Branches

After a spinal nerve leaves its intervertebral foramen, it divides into several branches, known as *rami* (RĀ-mī). *Ramus* is the singular. See Figure 10.2b. The *dorsal ramus* (RĀ-mus) innervates (supplies) the deep muscles and skin of the back. The *ventral ramus* innervates the superficial back muscles, all the structures of the limbs, and the lateral and ventral trunk. In addition, the *meningeal branch* supplies the vertebrae, vertebral ligaments, blood vessels of the spinal cord, and the meninges. The *rami communicantes* (ko-myoo-nī-KAN-tēz) are components of the autonomic nervous system that are discussed in the next chapter.

Plexuses

The ventral rami of spinal nerves, except for thoracic nerves T2 to T11, do not go directly to the body structures they supply. Instead, they form networks on either side of the body by joining with adjacent nerves. Such a network is called a *plexus* (*plexus* = braid). The principal plexuses are the cervical plexus, brachial plexus, lumbar plexus, and sacral plexus (see Figure 10.2). Emerging from the plexuses are nerves bearing names that are often descriptive of the general regions they supply or the course they take. Each of the nerves, in turn, may have several branches named for the specific structures they innervate.

The *cervical plexus* supplies the skin and muscles of the head, neck, upper part of the shoulders, and the diaphragm. The phrenic nerves, which send motor impulses to the diaphragm, arise from the cervical plexus. Damage to the spinal cord above the origin of the phrenic nerves may cause respiratory failure. The *brachial plexus* constitutes the nerve supply for the upper limbs and a number of neck and shoulder muscles. The median nerve, which arises from this plexus, may be compressed by trauma, edema, and repeated flexion of the wrist. This condition is called *carpal tunnel syndrome.* The *lumbar plexus* supplies the abdominal wall, external genitals, and part of the lower limbs. The *sacral plexus* supplies the buttocks, perineum, and lower limbs. The sciatic nerve, the longest nerve in the body, arises from the sacral plexus.

basically lower back problems

Intercostal (Thoracic) Nerves   *Do not intermingle & braid*

Spinal nerves T2 to T11 do not form plexuses. They are known as *intercostal* (*thoracic*) *nerves* and go directly to the structures they supply, such as muscles between ribs, abdominal muscles, and skin of the chest and back (see Figure 10.2). *between ribs*

Now we will consider the principal parts of the brain, how the brain is protected, and how it is related to the spinal cord and cranial nerves.

Brain

objective: *Discuss how the brain is protected and supplied with blood, and name the principal parts of the brain and explain the function of each part.*

Principal Parts

The **brain** is made up of about 100 billion neurons and 1000 billion neuroglia. It is one of the largest organs of the body, weighing about 1300g (3 lb). It is mushroom shaped and divided into four principal parts: brain stem, diencephalon, cerebrum, and cerebellum (Figure 10.6). The **brain stem,** the stalk of the mushroom, is continuous with the spinal cord and con-

sists of the medulla oblongata, pons, and midbrain. Above the brain stem is the **diencephalon** (dī-en-SEF-a-lon; *dia* = between; *enkephalos* = brain), consisting mostly of the thalamus and hypothalamus. The **cerebrum** (se-RĒ-brum; *cerebrum* = brain) spreads over the diencephalon. It has two sides called cerebral hemispheres and occupies most of the cranium. Below the cerebrum and behind the brain stem is the **cerebellum** (ser'-e-BEL-um; *cerebellum* = little brain).

Protection and Coverings

The brain is protected by the cranium and meninges. The **cranial meninges,** as extensions of the spinal meninges, have the same names: the outermost **dura mater,** middle **arachnoid,** and innermost **pia mater** (Figure 10.7 on page 229).

Figure 10.6   Brain. The infundibulum and pituitary gland are discussed in conjunction with the endocrine system in Chapter 13.

 The four principal parts of the brain are the (1) brain stem, (2) cerebellum, (3) diencephalon, and (4) cerebrum.

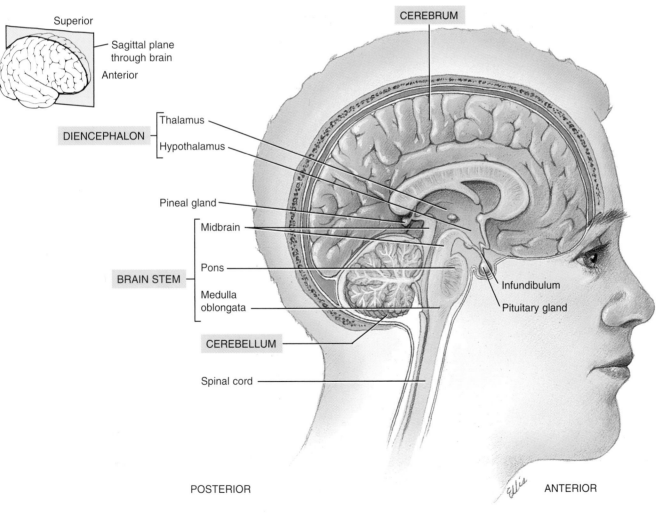

(a) Diagram of medial view of sagittal section

Figure 10.6   *(Continued)*

POSTERIOR   ANTERIOR

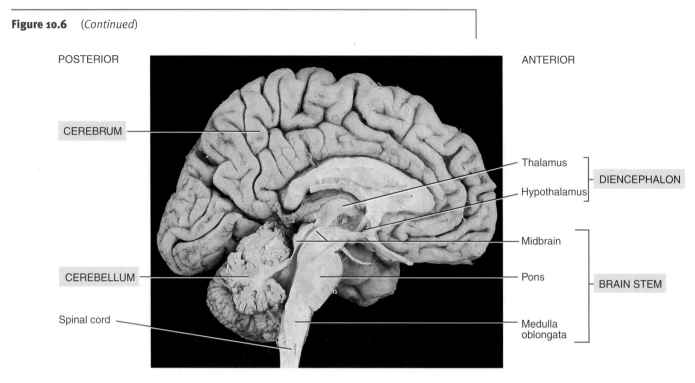

(b) Photograph of medial view of sagittal section

Q *Which part of the brain is attached to the spinal cord?*

[handwritten: Brain does not need insulin]

[handwritten: nutrients — most important, glucose + carries away waste]

Cerebrospinal Fluid (CSF)

[handwritten margin: hold — cerebral spinal fluid]

The brain, as well as the rest of the CNS, is further protected against injury by *cerebrospinal fluid* (**CSF**). CSF circulates through the subarachnoid space, around the brain and spinal cord, and through the ventricles of the brain.

The *ventricles* (VEN-tri-kuls; *ventriculus* = little belly or cavity) are cavities in the brain that are connected to each other, with the central canal of the spinal cord, and with the subarachnoid space (Figure 10.7). There are four ventricles, two *lateral ventricles,* one *third ventricle,* and one *fourth ventricle.*

The entire CNS contains between 80 and 150 ml (3 to 5 oz) of cerebrospinal fluid. It is a clear, colorless liquid. It contains proteins, glucose, urea, and salts. CSF has two principal functions related to homeostasis: protection and circulation. It serves as a shock-absorbing medium to protect the brain and spinal cord from jolts that would otherwise be traumatic. The fluid also buoys the brain so that it "floats" in the cranial cavity. In its circulatory function, CSF delivers nutritive substances filtered from blood and removes wastes and toxic substances produced by brain and spinal cord cells.

Cerebrospinal fluid is formed by filtration and secretion from *choroid* (KŌ-royd; *chorion* = membrane) *plexuses,* specialized capillaries in the ventricles (Figure 10.7). CSF circulates continually. From the ventricles, it flows into the subarachnoid space around the back of the brain and downward around the posterior surface of the spinal cord, up the anterior surface of the spinal cord, and around the anterior part of the brain. From there it is gradually reabsorbed into veins, mostly into a vein called the *superior sagittal sinus.* The absorption actually occurs through *arachnoid villi,* which are fingerlike projections of the arachnoid that push into the superior sagittal sinus (Figure 10.7). Normally, cerebrospinal fluid is absorbed as rapidly as it is formed.

If an obstruction or inflammation causes CSF to accumulate in the ventricles or the subarachnoid space, the condition is called *hydrocephalus* (*hydro* = water; *enkephalos* = brain). This accumulation of fluid can place pressure on the brain and cause brain damage. Hydrocephalus is treated by inserting a shunt in the ventricles to drain off the excess fluid into a vein in the neck.

Blood Supply

The brain is well supplied with oxygen and nutrients from a special circulatory route at the base of the brain called the *cerebral arterial circle* (*circle of Willis*) (see Figure 16.9 on p. 375). Although the brain comprises only about 2 percent of total body weight, it requires about 20 percent of the body's oxygen supply. The brain is one of the most metabolically active organs of the body, and the amount of oxygen it uses varies with the degree of mental activity.

[handwritten margin: won't let brain starve — glucose + oxy keep flowing even if one part is shut off]

Figure 10.7 Meninges and ventricles of the brain. In (a), arrows indicate the direction of flow of cerebrospinal fluid.

Cerebrospinal fluid protects the brain and spinal cord and delivers nutrients from the blood to the brain and spinal cord and removes wastes from the brain and spinal cord to the blood.

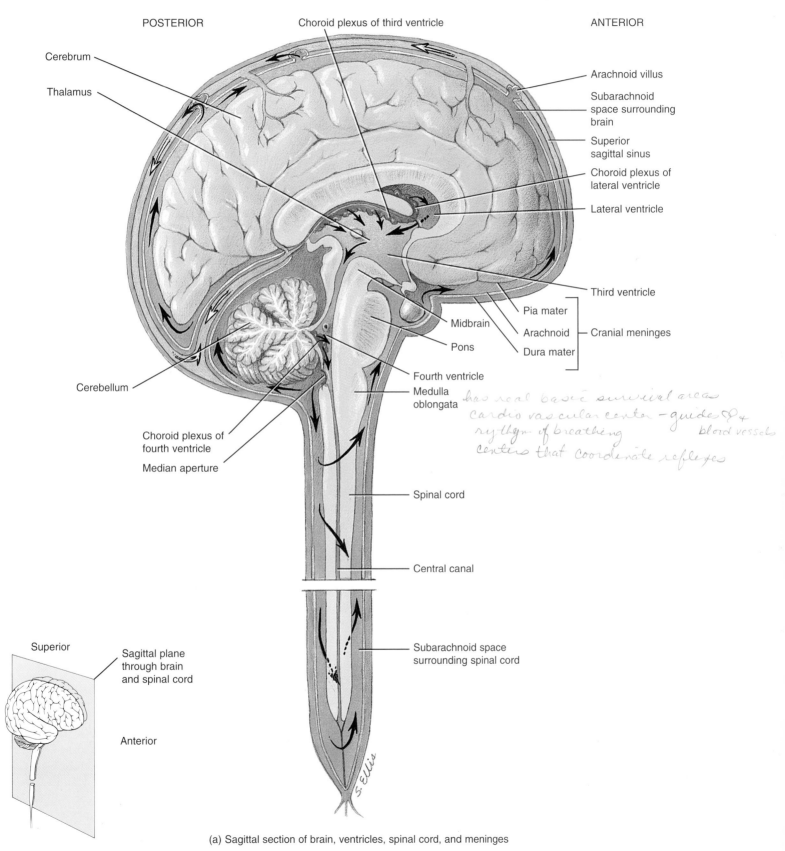

POSTERIOR

Choroid plexus of third ventricle

ANTERIOR

Cerebrum

Thalamus

Arachnoid villus

Subarachnoid space surrounding brain

Superior sagittal sinus

Choroid plexus of lateral ventricle

Lateral ventricle

Third ventricle

Pia mater
Arachnoid — Cranial meninges
Dura mater

Midbrain

Pons

Cerebellum

Fourth ventricle

Medulla oblongata

Choroid plexus of fourth ventricle

Median aperture

Spinal cord

Central canal

Superior

Sagittal plane through brain and spinal cord

Anterior

Subarachnoid space surrounding spinal cord

S. Ellis

[handwritten: has real basic survival areas cardio vascular center — guide of + blood vessels rythm of breathing centers that coordinate reflexes]

(a) Sagittal section of brain, ventricles, spinal cord, and meninges

Figure 10.7   *(Continued)*

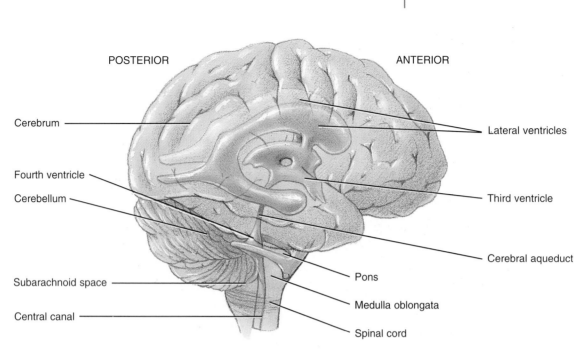

POSTERIOR   ANTERIOR

Cerebrum ———————————————————— Lateral ventricles

Fourth ventricle

Cerebellum ——————————————————— Third ventricle

——————————————————— Cerebral aqueduct

Subarachnoid space ——————————————————— Pons

Central canal ——————————————————— Medulla oblongata

——————————————————— Spinal cord

(b) Lateral view from right side of brain

Q *Where is cerebrospinal fluid formed and absorbed?*

If the blood flow to the brain is interrupted even briefly, unconsciousness may result. If the cells are totally deprived of oxygen for 4 or more minutes, many are permanently injured because lysosomes of brain cells are extremely sensitive to decreased oxygen. If the condition persists, they break open and release enzymes that bring about self-destruction of brain cells.

Blood supplying the brain also contains glucose, the principal source of energy for brain cells. Because carbohydrate storage in the brain is limited, the supply of glucose must be continuous. If blood entering the brain has a low glucose level, mental confusion, dizziness, convulsions, and loss of consciousness may occur.

Glucose, oxygen, and certain ions pass rapidly from the circulating blood into brain cells. Other substances enter slowly or not at all. The different rates of passage of certain materials from the blood into most parts of the brain are due to the **blood–brain barrier** (**BBB**). Brain capillary walls have more tightly connected cells with a thicker basement membrane, plus they are surrounded by astrocytes (a type of neuroglial cell), thus forming a barrier to all but the smallest molecules or those selectively admitted through active transport. Brain cells are protected from harmful substances this way. Unfortunately, most antibiotics cannot enter either. Trauma, inflammation, and toxins can cause a breakdown of the blood–brain barrier.

Brain Stem

Medulla Oblongata

The **medulla oblongata** (me-DULL-la ob′-long-GA-ta), or just simply the **medulla,** is a continuation of the spinal cord. It forms the inferior part of the brain stem (Figures 10.8 and 10.6).

The medulla contains all sensory (ascending) and motor (descending) tracts running between the spinal cord and other parts of the brain. These tracts constitute the white matter of the medulla. Some tracts cross as they pass through the medulla. Let us see how this crossing occurs and what it means.

In the medulla are two roughly triangular structures called **pyramids** (Figures 10.8 and 10.9 on page 232). The pyramids contain the largest motor tracts that pass from the outer region of the cerebrum (cerebral cortex) to the spinal cord. Most of the fibers in the left pyramid cross to the right side, and most of the fibers in the right pyramid cross to the left. This crossing is called the **decussation** (dē′-ku-SĀ-shun) *of pyramids.* Decussation explains why one side of the cerebral cortex controls the opposite side of the body. Motor fibers that originate in the left cerebral cortex activate muscles on the right side of the body, and vice versa. Similarly, most sensory fibers also cross over in the medulla so that nearly all sensory impulses received on one side of the body are perceived in the opposite side of the cerebral cortex.

Figure 10.8 Brain stem. The arrow in the inset indicates the direction from which the brain is viewed (inferior).

🔑 *The brain stem consists of the medulla oblongata, pons, and midbrain.*

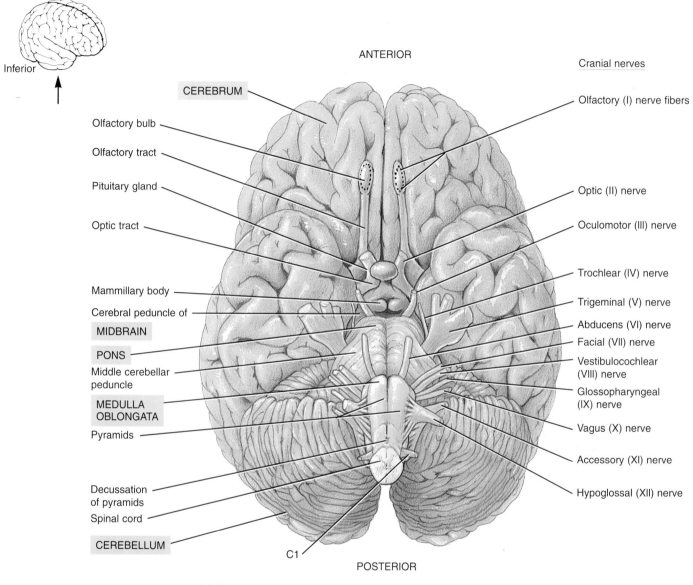

Inferior aspect of brain showing brain stem in relation to cranial nerves and associated structures

❓ *What part of the brain stem contains pyramids? The cerebral peduncles? Literally means "bridge"?*

Also within the medulla are the ***cardiovascular center,*** which regulates the rate and force of heartbeat and the diameter of blood vessels (p. 354 and Figure 15.8) and the ***medullary rhythmicity area,*** which regulates the basic rhythm of breathing (p. 422 and Figure 18.15). Other centers in the medulla coordinate swallowing, vomiting, coughing, sneezing, and hiccuping. Finally, the medulla contains the nuclei of origin for several pairs of cranial nerves: vestibulocochlear (VIII), accessory (IX), vagus (X), and hypoglossal nerves (XII) (Figures 10.8 and 10.9 and Exhibit 10.1).

In view of the many vital activities controlled by the medulla, it is not surprising that a hard blow to the base of the skull can be fatal.

Figure 10.9   Medulla. Shown are the internal anatomy and the decussation of pyramids.

The pyramids of the medulla contain the largest motor tracts from the cerebrum to the spinal cord.

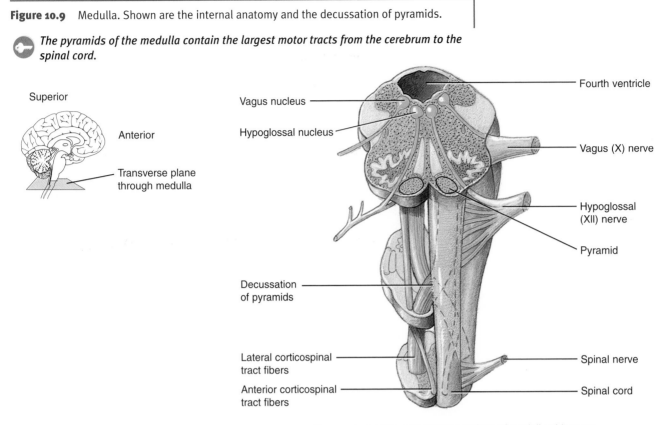

Transverse section and anterior surface of medulla oblongata

What does decussation mean? Functionally, what is the consequence of decussation of pyramids?

Pons

regulatory center. how much air lungs should take in before they over inflate

The *pons*, which means bridge, lies directly above the medulla and in front of the cerebellum (Figures 10.5 and 10.8). Like the medulla, the pons consists of nuclei and scattered white fibers. The pons connects the spinal cord with the brain and parts of the brain with each other. The nuclei for certain pairs of cranial nerves originate in the pons: the trigeminal (V), abducens (VI), facial (VII), and the vestibular branch of the vestibulocochlear nerves (VIII) (Figure 10.8 and Exhibit 10.1). The pons also contains nuclei that help regulate breathing (see Chapter 18). These are called the *pneumotaxic area* and *apneustic area*.

Midbrain

one of the old parts of brain

The *midbrain* extends from the pons to the lower portion of the diencephalon (Figures 10.6 and 10.8). It contains a pair of tracts referred to as *cerebral peduncles* (pe-DUNG-kulz). The cerebral peduncles contain motor fibers that connect the cerebral cortex to the pons and spinal cord, and sensory fibers that connect the spinal cord to the thalamus. The cerebral peduncles constitute the main connection for tracts between upper and lower parts of the brain and spinal cord. The nuclei for the oculomotor (III) and trochlear (IV) cranial nerves originate in the midbrain (see Figure 10.8 and Exhibit 10.1).

Throughout the brain stem—medulla, pons, and midbrain, and diencephalon—is a group of widely scattered neurons that give a netlike (reticular) appearance and is referred to as the *reticular formation*. Neurons of the reticular formation receive and integrate input from the cerebral cortex, hypothalamus, thalamus, cerebellum, and spinal cord. The neurons also send impulses to all levels of the central nervous system.

A severe blow to the mandible twists and distorts the brain stem and overwhelms the reticular formation by sending a sudden volley of nerve impulses to the brain, resulting in unconsciousness.

Diencephalon

relay station a little more evolved than pons. very central spot in brain

The *diencephalon* (*dia* = through; *enkephalos* = brain) consists principally of the thalamus and hypothalamus. See Figure 10.6.

Thalamus

The *thalamus* (THAL-a-mus; *thalamos* = inner chamber) is an oval structure above the midbrain that consists mostly of paired masses of gray matter organized into nuclei (Figure 10.10). Some nuclei in the thalamus serve as relay stations for sensory impulses from other parts of the CNS to the cerebral cortex.

Figure 10.10   Diencephalon: thalamus and hypothalamus. Also shown are the basal ganglia (discussed later).

🔑 *The thalamus is the principal relay station for sensory impulses that reach the cerebral cortex from other parts of the brain and the spinal cord.*

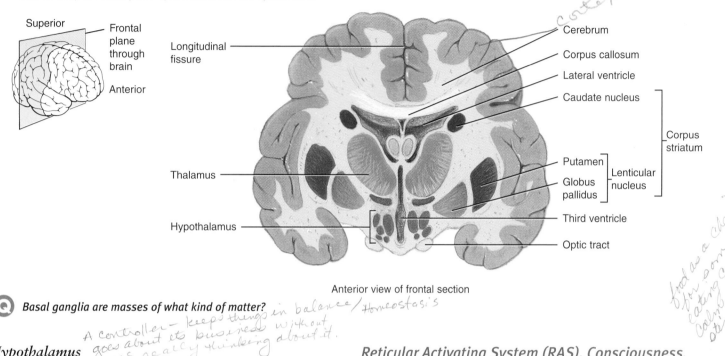

Anterior view of frontal section

Ⓠ *Basal ganglia are masses of what kind of matter?*

Hypothalamus

The **hypothalamus** (*hypo* = under) is the small portion of the diencephalon that lies below the thalamus and above the pituitary gland, the "master gland" of the body (Figure 10.10).

Despite its small size, nuclei in the hypothalamus control many body activities, most of them related to homeostasis.

1. The hypothalamus controls and integrates activities of the autonomic nervous system, which regulates activities such as heart rate, movement of food through the gastrointestinal tract, and contraction of the urinary bladder.

2. It controls the release of many hormones from the pituitary gland and thus serves as a primary connection between the nervous system and endocrine system. These systems are considered the major control systems of the body.

3. It controls normal body temperature.

4. It is associated with feelings of rage, aggression, pain, and pleasure.

5. It regulates food intake through two centers. The **feeding** (*hunger*) **center** is responsible for hunger sensations. When sufficient food has been ingested, the **satiety** (sa-TĪ-e-tē; *satis* = full, satisfied) **center** is stimulated and sends out nerve impulses that inhibit the feeding center.

6. It contains a **thirst center** that regulates the intake of liquids.

7. It is one of the areas of the brain that maintains consciousness and sleep patterns.

Reticular Activating System (RAS), Consciousness, and Sleep

Humans awaken and sleep in a fairly constant 24-hour rhythm called a **circadian** (ser-KĀ-dē-an) **rhythm.** When the brain is aroused or awake, it is in a state of readiness and able to react consciously to various stimuli. The aroused state depends largely on the reticular formation. Because stimulating a portion of the reticular formation increases cortical activity, this area is also known as the **reticular activating system** (**RAS**). When stimulated, many nerve impulses pass upward to widespread areas of the cerebral cortex. The RAS produces consciousness and **arousal,** that is, awakening from deep sleep.

Following arousal, the RAS and cerebral cortex continue to activate each other through a feedback system consisting of many circuits. The result is a state of wakefulness called **consciousness.** The RAS is the physical basis of consciousness. It continuously sifts and selects, forwarding only the essential, unusual, or dangerous to the conscious mind. Because humans experience different levels of consciousness (alertness, attentiveness, relaxation, inattentiveness), it is assumed that the level of consciousness depends on the number of feedback circuits operating at the time.

Consciousness may be altered by cocaine and amphetamines (which produce extreme alertness), meditation (which causes relaxed, focused consciousness), and alcohol and anesthetics (which cause various levels of unconsciousness called anesthesia). Damage or disease can produce a lack of consciousness called **coma.**

Inactivation of the RAS produces **sleep,** a state of partial unconsciousness from which an individual can be aroused. Just as there are different levels of consciousness, there are different levels of sleep.

— 1st half Quiz to here
minus impulse conduction center

Cerebrum

Supported on the diencephalon and brain stem and forming the bulk of the brain is the *cerebrum* (see Figure 10.6). The surface is composed of a thin area of gray matter and is called the ***cerebral cortex*** (*cortex* = rind or bark). The cortex consists of six layers of nerve cell bodies, beneath which lies the cerebral white matter.

During embryonic development, when there is a rapid increase in brain size, the gray matter of the cerebral cortex enlarges much faster than the underlying white matter. As a result, the cerebral cortex rolls and folds upon itself so that it can fit into the cranial cavity. The folds are called ***gyri*** (JĪ-ri) or ***convolutions*** (Figure 10.11). *Gyrus* (*gyros* = circle) is the singular.

The deep grooves between folds are ***fissures;*** the shallow grooves are ***sulci*** (SUL-sī). *Sulcus* (*sulcus* = groove) is singular. The ***longitudinal fissure*** separates the cerebrum into right and left halves, or ***cerebral hemispheres.*** Internally, however, the

---

Figure 10.11 Cerebrum. The inset in (a) indicates the relative differences among a gyrus, sulcus, and fissure.

ability to think — have thought & learning processes

🔑 **The cerebrum is the seat of intelligence and provides us with the ability to read, write, and speak; make calculations and compose music; remember the past and plan for the future; and create works.**

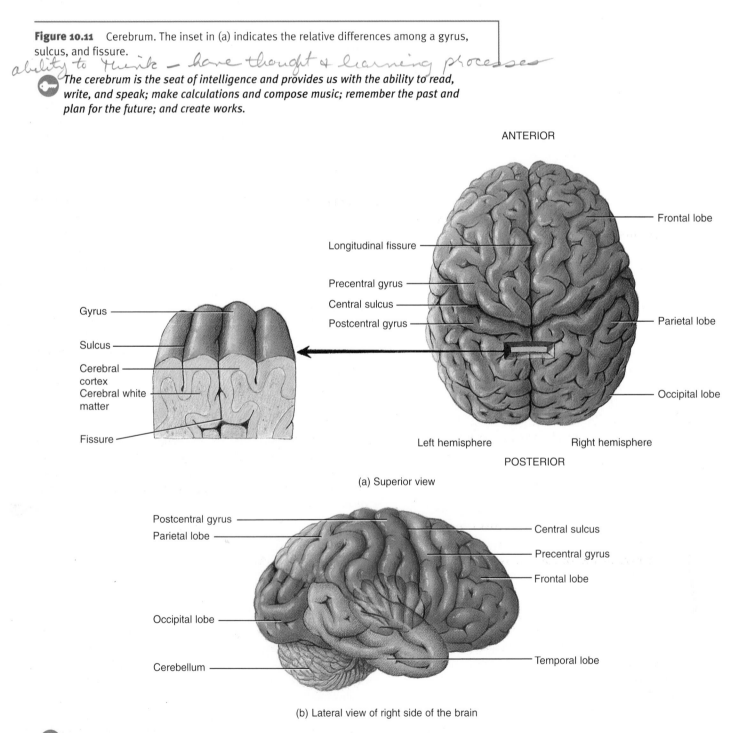

(a) Superior view

(b) Lateral view of right side of the brain

ⓠ *What structure holds the cerebral hemispheres together?*

left side more logical
R side more creative
√ language
R arming music
Patellar area

hemispheres are connected by a large bundle of transverse fibers composed of white matter called the **corpus callosum** (kal-LŌ-sum; *corpus* = body; *callosus* = hard). This structure is larger in females than in males, which may account for differences in emotional responses in males and females. It appears that the male brain has emotional capacities in just one hemisphere (right side), whereas the female brain has emotional capacities in both hemispheres. Because of the smaller corpus callosum, information flow is slower between the emotional (right) side and the verbal (left) side of the male brain. Thus it is assumed that males typically express their emotions less effectively than females.

Lobes

Each cerebral hemisphere is subdivided into four lobes by sulci or fissures. The lobes are named frontal, parietal, temporal, and occipital (Figure 10.11). The **central sulcus** separates the frontal and parietal lobes. A major gyrus, the **precentral gyrus,** is located immediately anterior to the central sulcus. The gyrus contains the primary motor area of the cerebral cortex. Another major gyrus, the **postcentral gyrus,** is located immediately posterior to the central sulcus. This gyrus contains the primary somatosensory area of the cerebral cortex, which is discussed shortly.

As you will see later, the olfactory and optic nerves are associated with specific lobes of the cerebrum.

Brain Lateralization (*Split-Brain Concept*)

Detailed examination of the brain reveals anatomical differences between the two cerebral hemispheres. For example, in left-handed people the parietal and occipital lobes of the right hemisphere and the frontal lobe of the left hemisphere are typically narrower.

In addition to structural differences, there are also functional differences. In most people, the left hemisphere is more important for right-handed control, language, numerical and scientific skills, and reasoning. The right hemisphere is more important for left-handed control, artistic awareness, and imagination.

White Matter

The white matter underlying the cerebral cortex consists of myelinated axons that transmit impulses between gyri in the same hemisphere, from the gyri in one cerebral hemisphere to the corresponding gyri in the opposite cerebral hemisphere, and from the cerebrum to other parts of the brain and spinal cord.

Basal Ganglia (*Cerebral Nuclei*)

The **basal ganglia** (**cerebral nuclei**) are paired masses of gray matter within the white matter of each cerebral hemisphere (see Figure 10.10).

The largest basal ganglion is the **corpus striatum** (strī-Ā-tum; *corpus* = body; *striatus* = striped). It consists of the **caudate** (*cauda* = tail) **nucleus** and the **lenticular** (*lenticula* = shaped like a lentil or lens) **nucleus.** The lenticular nucleus, in

turn, is subdivided into the **putamen** (pu-TĀ-men; *putamen* = shell) and the **globus pallidus** (*globus* = ball; *pallid* = pale).

The basal ganglia control large subconscious movements of skeletal muscles, such as swinging the arms while walking (such gross movements are also consciously controlled by the cerebral cortex), and regulate muscle tone required for specific body movements.

Damage to the basal ganglia results in uncontrollable shaking (tremor) or involuntary muscle movements, such as in Parkinson's disease. Destruction of a substantial portion of the basal ganglia results in total **paralysis** (loss or impairment of motor and sensory functions) of the side of the body opposite to the damage.

Limbic System

reptilian part of brain = emotions

Certain parts of the cerebral hemispheres and diencephalon constitute the **limbic** (*limbus* = border) **system.** It is a wishbone-shaped group of structures that encircles the brain stem and assumes a primary function in emotions such as pain, pleasure, anger, rage, fear, sorrow, sexual feelings, docility, and affection. It is therefore sometimes called the "visceral" or "emotional" brain. Although behavior is a function of the entire nervous system, the limbic system controls most of its involuntary aspects, the aspects related to survival, and animal experiments suggest that it has a major role in controlling the overall pattern of behavior. Together with portions of the cerebrum, the limbic system also functions in memory; memory impairment results from damage to the limbic system.

Functional Areas of Cerebral Cortex

The functions of the cerebrum are numerous and complex. In a general way, the cerebral cortex is divided into three areas. The **sensory areas** receive and interpret sensory impulses, the **motor areas** control muscular movement, and the **association areas** are concerned with integrative functions such as memory, emotions, reasoning, will, judgment, personality traits, and intelligence. These areas are shown in Figure 10.12 on page 236.

SENSORY AREAS   Sensory input to the cerebral cortex flows mainly to the posterior half of the cerebral hemispheres, to regions behind the central sulci. In the cerebral cortex, primary sensory areas have the most direct connections with peripheral sensory receptors.

Somato-touch

1. **Primary somatosensory** (sō′-mat-ō-SEN-sō-rē; *soma* = body; *sensus* = sense) **area** or **general somatosensory area.** Located directly posterior to the central sulcus of each cerebral hemisphere in the postcentral gyrus of the parietal lobe. In Figure 10.12, the primary somatosensory area is designated by the areas numbered 1, 2, and 3.*
The primary somatosensory area receives sensations from cutaneous and muscular receptors in various parts of the body. Some parts of the body are represented by large areas in the primary somatosensory area. These include

---

*These numbers, as well as most of the others shown, are based on K. Brodman's cytoarchitectural map of the cerebral cortex. His map, first published in 1909, attempts to correlate structure and function.

Figure 10.12   Functional areas of the cerebrum.

🔑 *Particular areas of the cerebral cortex process sensory, motor, and integrative signals.*

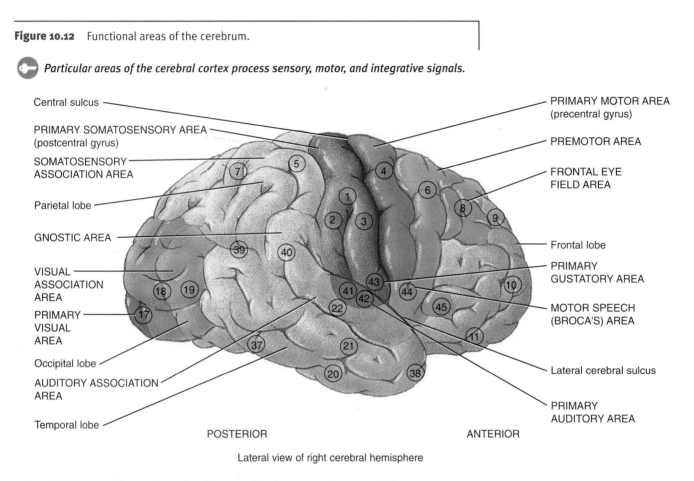

Lateral view of right cerebral hemisphere

❓ *Which part of the cerebrum localizes exactly where sensations occur? Translates thoughts into speech? Interprets sensations related to taste? Integrates and interprets sensations? Controls specific muscles or groups of muscles? Interprets sensations related to smell? Controls voluntary scanning movements of the eyes?*

the lips, face, and thumb. Other body parts, such as the trunk and lower limbs, are represented by relatively small areas. The amount of space given a particular part of the body is determined by the functional importance of the part and its need for sensitivity. The major function of the primary somatosensory area is to localize exactly the points of the body where the sensations originate.

2. *Primary visual area* (area 17). Located in the occipital lobe, it receives sensory impulses from the eyes and interprets shape, color, and movement.

3. *Primary auditory area* (areas 41 and 42). Located in the temporal lobe, it interprets the basic characteristics of sound such as pitch and rhythm.

4. *Primary gustatory area* (area 43). Located at the base of the postcentral gyrus, it interprets sensations related to taste.

5. *Primary olfactory area.* Located in the temporal lobe, it interprets sensations related to smell.

MOTOR AREAS   Just as the primary somatosensory area of the cortex has been mapped to reflect the amounts of sensory information coming from different body parts, the *motor cortex* has been mapped to indicate which specific areas control

particular muscle groups. There is a motor cortex in both the left and right cerebral hemispheres. The degree of representation is proportional to the precision of movement required of a particular body part. For example, the thumb, fingers, lips, tongue, and vocal cords have large representations. The trunk has a relatively small representation.

1. *Primary motor area* (area 4). Located in the precentral gyrus of the frontal lobe (Figure 10.12). Like the primary somatosensory area, the primary motor area consists of regions that control specific muscles or groups of muscles. Stimulation of a specific point of the primary motor area results in a muscular contraction, usually on the opposite side of the body.

2. *Motor speech areas* (area 44). The production of speech occurs in the *motor speech area,* also called *Broca's* (BRŌ-kaz) *area,* located in the frontal lobe (usually the left one) just superior to the lateral cerebral sulcus.

   Broca's area and other language areas are located in the left cerebral hemisphere of most individuals regardless of whether they are left- or right-handed. Injury to the sensory or motor speech areas results in *aphasia* (a-FĀ-zē-a; *a* = without; *phasis* = speech), an inability to speak; *agraphia* (*a* = without; *graph* = write), an inability to write; *word deafness,* an inability to understand spoken words; or *word blindness,* an inability to understand written words.

ASSOCIATION AREAS   The *association areas* of the cerebrum consist of association tracts that connect motor and sensory areas and large parts of the cerebral cortex of the occipital, parietal, and temporal lobes and the frontal lobes anterior to the motor areas (Figure 10.12). Association areas include:

1. *Somatosensory association area* (areas 5 and 7). Just posterior to the primary somatosensory area, its role is to integrate and interpret sensations. This area permits you to determine the exact shape and texture of an object without looking at it, to determine the orientation of one object to another as they are felt, and to sense the relationship of one body part to another.

2. *Visual association area* (areas 18 and 19). Located in the occipital lobe, it relates present to past visual experiences with recognition and evaluation of what is seen.

3. *Auditory association area* (area 22). Located below the primary auditory area in the temporal cortex, it determines if a sound is speech, music, or noise. It also translates words into thoughts.

4. *Gnostic* (NOS-tik; *gnosis* = knowledge) *area* (areas 5, 7, 39, and 40). This area is located among the somatosensory, visual, and auditory association areas. It integrates sensory interpretations from the association areas and impulses from other areas so that a common thought can be formed from the various sensory inputs.

5. *Premotor area* (area 6). Immediately in front of the primary motor area, the premotor area deals with learned motor activities of a complex and sequential nature. It generates nerve impulses that cause a specific group of muscles to contract in a specific sequence, for example, to write a word.

6. *Frontal eye field area* (area 8). This area in the frontal cortex controls voluntary scanning movements of the eyes—searching for a word in a dictionary, for instance.

7. *Language areas.* From Broca's area, nerve impulses pass to the premotor regions that control the muscles of the larynx, pharynx, and mouth. The impulses from the premotor area to the muscles result in specific, coordinated contractions that enable you to speak. Simultaneously, impulses are sent from the motor speech area to the primary motor area. From here, impulses reach your breathing muscles to regulate the proper flow of air past the vocal cords. The coordinated contractions of your speech and breathing muscles enable you to speak your thoughts.

Memory

At this point, we will take a closer look at memory. Although memory has been studied by scientists for decades, there is still no satisfactory explanation for how we remember. *Memory* is the ability to recall thoughts. For an experience to become part of memory, it must produce changes in the central nervous system that represent the experience.

Memory is classified into two categories, short-term and long-term. *Short-term memory* lasts only seconds or hours and is the ability to recall bits of information. One example is looking up a telephone number and remembering it long enough to dial. *Long-term memory,* on the other hand, lasts from days to years. If you frequently use a telephone number, it becomes part of long-term memory and can be retrieved for use for quite a long period. Such reinforcement use is called *memory consolidation.*

The portions of the brain known to be associated with memory include the association cortex of the frontal, parietal, occipital, and temporal lobes; parts of the limbic system; and the diencephalon.

Electroencephalogram (EEG)

Brain cells generate electrical activity as a result of literally millions of action potentials (nerve impulses) of individual neurons. The electrical potentials are called *brain waves* and indicate activity of the cerebral cortex. Brain waves pass easily through the skull and can be detected by sensors called electrodes. A record of such waves is called an *electroencephalogram* (EEG). The EEG is used clinically in the diagnosis of epilepsy, infectious diseases, tumors, trauma, and blood clots. In cases of doubt, a flat EEG is increasingly being taken as one criterion of *brain death.*

Cerebellum

The *cerebellum* is the second-largest portion of the brain. It is behind the medulla and pons and below the occipital lobes of the cerebrum (see Figure 10.6).

The cerebellum consists of two *cerebellar hemispheres.* The surface of the cerebellum, called the *cerebellar cortex,* consists of gray matter. Beneath the cortex are *white matter tracts* that resemble branches of a tree. Deep within the white matter are masses of gray matter, the *cerebellar nuclei.* The cerebellum is attached to the brain stem by three paired bundles of fibers called *cerebellar peduncles* (see Figure 10.8).

The cerebellum compares the intended movement programmed by motor areas in the cerebrum with what is actually happening. It constantly receives sensory input from muscles, tendons, and joints, receptors for equilibrium, and visual receptors of the eyes. The cerebellum helps to smooth and coordinate complex sequences of skeletal muscle contractions. The cerebellum is the main brain region that regulates posture and balance and makes possible all skilled motor activities, from catching a baseball to dancing.

Damage to the cerebellum through trauma or disease creates symptoms involving skeletal muscles. The effects are on the same side of the body as the damaged side of the cerebellum because of a double crossing of tracts within the cerebellum. There may be lack of muscle coordination, called *ataxia* (*a* = without; *taxis* = order). Blindfolded people with ataxia cannot touch the tip of their nose with a finger because they cannot co-ordinate movement with their sense of where a body part is located. Another sign of ataxia is a changed speech pattern due to uncoordinated speech muscles. Cerebellar damage may also result in **disturbances of gait** (staggering or abnormal walking

movements) and **severe dizziness.** Alcohol inhibits the cerebellum and individuals who consume too much alcohol show signs of ataxia.

A summary of the principal parts of the brain and their functions is presented in Exhibit 10.1.

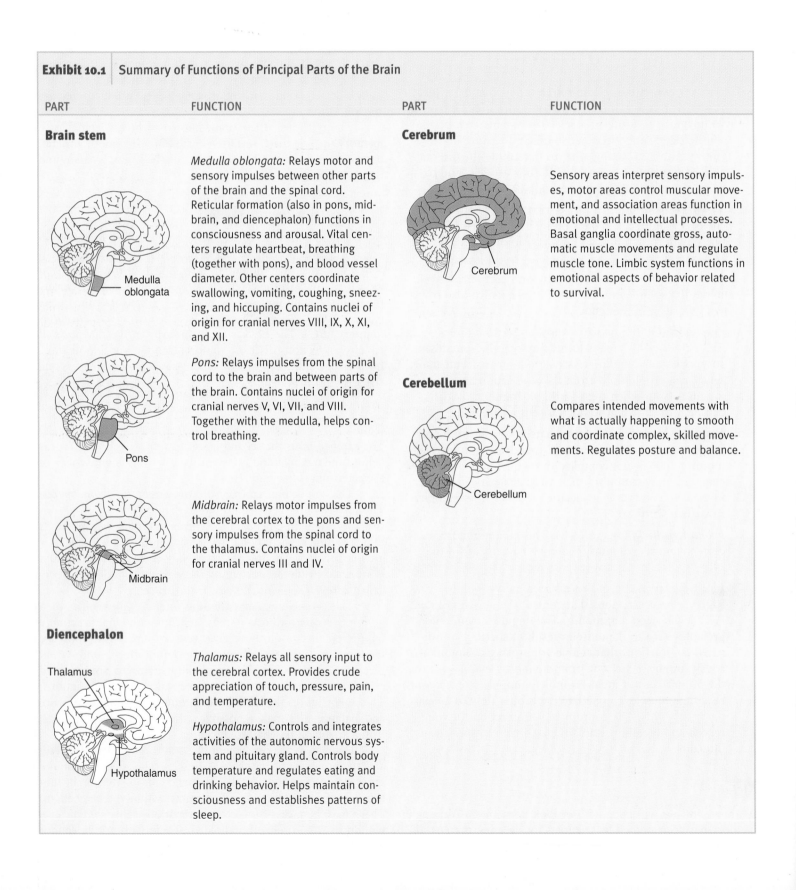

Exhibit 10.1 | **Summary of Functions of Principal Parts of the Brain**

PART	FUNCTION	PART	FUNCTION
Brain stem	*Medulla oblongata:* Relays motor and sensory impulses between other parts of the brain and the spinal cord. Reticular formation (also in pons, midbrain, and diencephalon) functions in consciousness and arousal. Vital centers regulate heartbeat, breathing (together with pons), and blood vessel diameter. Other centers coordinate swallowing, vomiting, coughing, sneezing, and hiccuping. Contains nuclei of origin for cranial nerves VIII, IX, X, XI, and XII.	**Cerebrum**	Sensory areas interpret sensory impulses, motor areas control muscular movement, and association areas function in emotional and intellectual processes. Basal ganglia coordinate gross, automatic muscle movements and regulate muscle tone. Limbic system functions in emotional aspects of behavior related to survival.
	Pons: Relays impulses from the spinal cord to the brain and between parts of the brain. Contains nuclei of origin for cranial nerves V, VI, VII, and VIII. Together with the medulla, helps control breathing.	**Cerebellum**	Compares intended movements with what is actually happening to smooth and coordinate complex, skilled movements. Regulates posture and balance.
	Midbrain: Relays motor impulses from the cerebral cortex to the pons and sensory impulses from the spinal cord to the thalamus. Contains nuclei of origin for cranial nerves III and IV.		
Diencephalon	*Thalamus:* Relays all sensory input to the cerebral cortex. Provides crude appreciation of touch, pressure, pain, and temperature.		
	Hypothalamus: Controls and integrates activities of the autonomic nervous system and pituitary gland. Controls body temperature and regulates eating and drinking behavior. Helps maintain consciousness and establishes patterns of sleep.		

Neurotransmitters

| objective: | *Explain the functions of selected neurotransmitters.* |

About 60 substances are either known or suspected *neurotransmitters*. These substances are released from synaptic vesicles in axon terminals and establish lines of communication between nerve cells; they facilitate, excite, or inhibit postsynaptic neurons in the CNS. Nervous system activities depend on proper levels and regulation of the neurotransmitters. Many nervous system disorders are due to neurotransmitter problems.

Following are some examples of neurotransmitters:

1. **Acetylcholine.** Found in cerebral cortex, all skeletal neuromuscular junctions, and autonomic nervous system; usually excitatory; blockage of acetylcholine receptors in skeletal muscles leads to myasthenia gravis.

2. **Dopamine.** Concentrated in brain; generally excitatory; involved in emotional responses and subconscious movements of skeletal muscles; decreased levels associated with Parkinson's disease.

3. **Norepinephrine.** Released at some neuromuscular and neuroglandular junctions; concentrated in the brain stem; also found in CNS; usually excitatory; may be related to arousal, dreaming, and regulation of mood.

4. **Serotonin.** Found in CNS; generally inhibitory; may be involved in inducing sleep, sensory perception, temperature regulation, and control of mood.

5. **Gamma aminobutyric acid (GABA).** Concentrated in brain; inhibitory; probably a target for antianxiety drugs such as valium, which enhances the action of gamma aminobutyric acid.

6. **Substance P.** Found in sensory nerves and CNS; stimulates perception of pain.

7. **Enkephalins** (en-KEF-a-lins). Concentrated in CNS; inhibit pain impulses by suppressing substance P.

8. **Endorphins** (en-DOR-fins). Concentrated in pituitary gland and brain; inhibit pain by inhibiting substance P; may have a role in memory and learning, sexual activity, control of body temperature; have been linked to depression and schizophrenia.

Cranial Nerves

| objective: | *Identify the 12 pairs of cranial nerves by name, number, type, location, and function.* |

Cranial nerves, like spinal nerves, are part of the somatic nervous system (SNS) of the peripheral nervous system. Of the 12 pairs of *cranial nerves*, 10 originate from the brain stem (see Figure 10.8), but all pass through foramina (holes) of the skull. The cranial nerves are designated with roman numerals and with names. The roman numerals indicate the order in which the nerves arise from the brain (front to back). The names indicate the distribution or function.

Some cranial nerves contain only sensory fibers and thus are called *sensory nerves.* The remainder contain both sensory and motor fibers and are referred to as *mixed nerves.* The cell bodies of sensory fibers are found outside the brain, whereas the cell bodies of motor fibers lie in nuclei within the brain.

Exhibit 10.2 summarizes cranial nerves and clinical applications.

Exhibit 10.2	Summary of Cranial Nerves[a]			
NUMBER AND NAME	**TYPE** (Sensory, Motor, or Mixed)	**LOCATION**		**FUNCTION**
Cranial nerve I: **olfactory** (*olfacere* = to smell)	Sensory	Arises in lining of nose and terminates in primary olfactory areas in temporal lobes of cerebral cortex.		*Function:* Smell.

Olfactory (I) nerve
Olfactory tract

(continued)

Exhibit 10.2 | Summary of Cranial Nerves*ᵃ* (Continued)

NUMBER AND NAME	TYPE (Sensory, Motor, or Mixed)	LOCATION	FUNCTION
Cranial nerve II: **optic** (*optikos* = vision, eye, or optics) 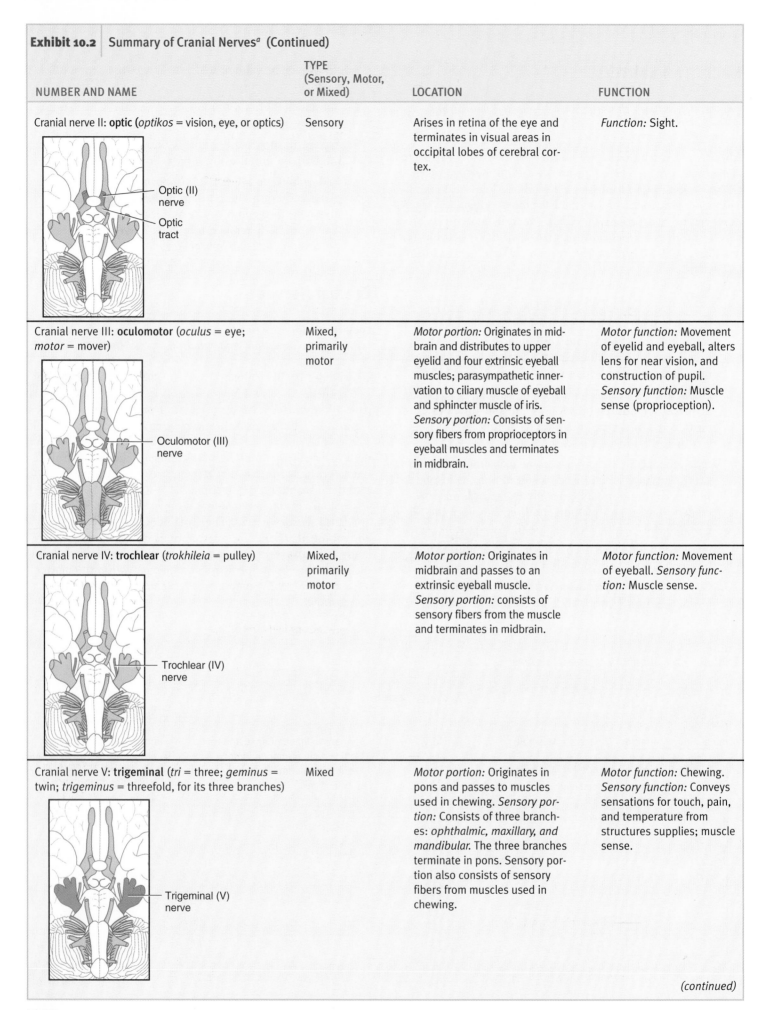 Optic (II) nerve — Optic tract	Sensory	Arises in retina of the eye and terminates in visual areas in occipital lobes of cerebral cortex.	*Function:* Sight.
Cranial nerve III: **oculomotor** (*oculus* = eye; *motor* = mover) Oculomotor (III) nerve	Mixed, primarily motor	*Motor portion:* Originates in midbrain and distributes to upper eyelid and four extrinsic eyeball muscles; parasympathetic innervation to ciliary muscle of eyeball and sphincter muscle of iris. *Sensory portion:* Consists of sensory fibers from proprioceptors in eyeball muscles and terminates in midbrain.	*Motor function:* Movement of eyelid and eyeball, alters lens for near vision, and construction of pupil. *Sensory function:* Muscle sense (proprioception).
Cranial nerve IV: **trochlear** (*trokhileia* = pulley) Trochlear (IV) nerve	Mixed, primarily motor	*Motor portion:* Originates in midbrain and passes to an extrinsic eyeball muscle. *Sensory portion:* consists of sensory fibers from the muscle and terminates in midbrain.	*Motor function:* Movement of eyeball. *Sensory function:* Muscle sense.
Cranial nerve V: **trigeminal** (*tri* = three; *geminus* = twin; *trigeminus* = threefold, for its three branches) Trigeminal (V) nerve	Mixed	*Motor portion:* Originates in pons and passes to muscles used in chewing. *Sensory portion:* Consists of three branches: *ophthalmic, maxillary, and mandibular.* The three branches terminate in pons. Sensory portion also consists of sensory fibers from muscles used in chewing.	*Motor function:* Chewing. *Sensory function:* Conveys sensations for touch, pain, and temperature from structures supplies; muscle sense.

(continued)

Exhibit 10.2 | Summary of Cranial Nerves (Continued)

NUMBER AND NAME	TYPE (Sensory, Motor, or Mixed)	LOCATION	FUNCTION
Cranial nerve VI: **abducens** (*ab* = away; *ducere* = to lead) Abducens (VI) nerve	Mixed, primarily motor	*Motor portion:* Originates in pons and is distributed to an extrinsic eyeball muscle. *Sensory portion:* Consists of sensory fibers from the muscle and terminates in pons.	*Motor function:* Movement of eyeball. *Sensory function:* Muscle sense.
Cranial nerve VII: **facial** (*facies* = face) Facial (VII) nerve	Mixed	*Motor portion:* Originates in pons and passes to facial, scalp, and neck muscles, parasympathetic fibers of the ANS pass to tear and salivary glands. *Sensory portion:* Arises from taste buds on tongue and terminates in gustatory areas in parietal lobes of cerebral cortex; also consists of sensory fibers from muscles of face and scalp.	*Motor function:* Facial expression and secretion of saliva and tears. *Sensory function:* Muscle sense and taste.
Cranial nerve VIII: **vestibulocochlear** (*vestibulum* = vestibule; *kokhlos* = land snail) Vestibulocochlear (VIII) nerve	Sensory; formerly called the auditory nerve.	*Cochlear branch:* Arises in spiral organ (organ of hearing) and ends in auditory areas in temporal lobes of cerebral cortex. *Vestibular branch:* Arises in (organs of equilibrium) and ends in pons and cerebellum.	*Cochlear branch function:* Conveys impulses associated with hearing. *Vestibular branch function:* Conveys impulses associated with equilibrium.
Cranial nerve IX: **glossopharyngeal** (*glossa* = tongue; *pharynx* = throat) Glossopharyngeal (IX) nerve	Mixed	*Motor portion:* Originates in medulla and is distributed to swallowing muscles of throat; parasympathetic fibers of ANS pass to a salivary gland. *Sensory portion:* Arises from taste buds on tongue and terminates in medulla; also consists of sensory fibers from swallowing muscles.	*Motor function:* Secretion of saliva. *Sensory function:* Taste and regulation of blood pressure; muscle sense.

(continued)

Exhibit 10.2 | Summary of Cranial Nerves (Continued)

NUMBER AND NAME	TYPE (Sensory, Motor, or Mixed)	LOCATION	FUNCTION
Cranial nerve X: **vagus** (*vagus* = vagrant or wandering) Vagus (X) nerve	Mixed	*Major portion:* Originates in medulla and terminates in muscles of throat, voice box, respiratory passageways, lungs, esophagus, heart, stomach, small intestine, most of large intestine, and gallbladder; parasympathetic fibers of the ANS pass to innervate involuntary muscles and glands of the gastrointestinal tract. *Sensory portion:* Arises from essentially same structures supplied by motor fibers and terminates in medulla and pons; also consists of sensory fibers from muscles supplied and taste buds.	*Motor function:* Visceral muscle movement. *Sensory function:* Sensations from organs supplied; muscle sense; taste.
Cranial nerve XI: **accessory** (*accessorius* = assisting) Accessory (XI) nerve	Mixed, primarily motor	*Motor portion:* From medulla and supplies voluntary muscles of throat and voice box and originates from cervical spinal cord, which supplies sternocleidomastoid and trapezius muscles. *Sensory portion:* Consists of sensory fibers from muscles supplied.	*Motor function:* Movement of tongue during speech, swallowing, and head movements. *Sensory function:* Muscle sense.
Cranial nerve XII: **hypoglossal** (*hypo* = below; *glossa* = tongue) Hypoglossal (XII) nerve	Mixed, primarily motor	*Motor portion:* From medulla and supplies voluntary muscles of throat and voice box and originates from cervical spinal cord, which supplies sternocleidomastoid and trapezius muscles. *Sensory portion:* Consists of sensory fibers from muscles supplied.	*Motor function:* Movement of tongue during speech, swallowing, and head movements. *Sensory function:* Muscle sense.

[a]A mnemonic device used to remember the names of the nerves is: "Oh, Oh, Oh, to touch and feel very green vegetables—AH!" The initial letter of each word corresponds to the initial letter of each pair of cranial nerves.

*W*hile those who get it usually take it for granted, people who don't would do almost anything for it. Although it's extremely valuable, you can't buy it. You can't borrow it, steal it, or give it away. And the harder you try to get it, the less likely you are to succeed. What is it? A good night's sleep.

Insomnia may include any or all of the following problems:

1. Taking a long time to fall asleep.
2. Awakening frequently during the night.
3. Awakening too early in the morning.
4. Feeling tired and dissatisfied with one's sleep upon awakening.

Insomnia that continues for more than a few weeks is a problem requiring medical attention, because it may signal a more serious health problem, such as depression.

Why Not Take a Sleeping Pill?

Sleeping pills are among the most commonly prescribed drugs in North America. An occasional dose of sleep medication may be helpful when taken as directed, but in general, medication only worsens the problem. Sleeping pills disturb the sleep cycle, so sleep is less satisfying even though you may get more of it. They often leave the user with a hangover that results in daytime fatigue. The user quickly builds up a tolerance to sleep medication so that it becomes less effective within a week. One of the biggest dangers of sleeping pills is addiction, which can be very difficult to overcome. Some sleeping pills can even cause insomnia by suppressing the brain's production of dopamine, a neurotransmitter that helps you go to sleep. The biggest problem with sleeping pills is that they do not address the real causes of sleeping disorders: bad habits and stress.

Sleep Therapy

While it is normal for some insomnia to occur during times of stress, it should disappear once the stressful situation is resolved. When insomnia continues, it is often because the person has developed a poor sleep environment or poor sleep habits.

The sleep environment should be comfortable, restful, and associated with relaxation and sleep. Sleep and work areas should be separate. If traffic or neighbor noises are a problem, "white noise" machines or tapes can provide a soothing tone that helps cover the disturbing sounds. Some people even resort to earplugs, though these can be problematic if you need to awaken to an alarm clock. Shades that block light can help darken rooms with windows that face bright outside light sources.

Regular use of stimulants can interfere with one's ability to fall and stay asleep. A decrease in the consumption of caffeine (found in coffee, tea, and cola drinks) often improves sleep quality. The nicotine in cigarettes is also a stimulant and should be avoided. Many people believe alcohol will help them relax and go to sleep, but like sleeping pills, alcohol disrupts the sleep cycle. While alcohol may help you fall asleep, it usually produces light, restless sleep, and the sleeper often wakes suddenly during the night and is unable to go back to sleep.

A large meal before bed can inhibit sleep. A light snack, however, can help one sleep better. Exercise helps decrease muscle tension and improve sleep quality. Exercise can also improve one's ability to manage stress, to feel less worried and more in control. But beware: exercise too close to bedtime can wind you up instead of down.

Sleep comes more easily to those who go to bed with a peaceful mental attitude. It is helpful to relax for at least an hour before bed. Read, listen to music, knit, take a warm shower. Avoid activities that wake you up. A pre-bed routine helps get you ready for sleep and for a visit from Morpheus, the god of dreams in Greek mythology.

critical thinking

What are some factors common to students' lifestyles that might contribute to insomnia?

How could each of these factors be addressed to reduce sleep problems?

Common Disorders

Cerebrovascular Accident (CVA)

The most common brain disorder is a *cerebrovascular accident* (CVA), also called a *stroke*. A CVA is characterized by a relatively abrupt onset of persisting neurological symptoms due to the destruction of brain tissue resulting from disorders in the blood vessels that supply the brain. CVAs may be due to a decreased blood supply or to a blood vessel that bursts.

Transient Ischemic Attack (TIA)

A *transient ischemic attack* (TIA) is a period of temporary cerebral dysfunction caused by an interference of the blood supply to the brain.

Spinal Cord Injury

The spinal cord may be damaged in various ways. Depending on the location and extent of the injury, paralysis may occur. *Paralysis* refers to total loss of motor function resulting from damage to nervous tissue or a muscle. Paralysis may be classified as follows: *monoplegia* (*mono* = one; *plege* = stroke) is paralysis of one limb only and is usually a result of a **stroke** (brain tissue damage due to an interrupted blood supply or ruptured blood vessel); *paraplegia* (*para* = beyond) is paralysis of both lower limbs and is due to a severed spinal cord in the thoracic and upper lumbar regions; *hemiplegia* (*hemi* = half) is paralysis of the upper limb, trunk, and lower limb on one side of the body and is usually due to a stroke; and *quadriplegia* (*quad* = four) is paralysis of the two upper and two lower limbs and is associated with a severed spinal cord in the cervical region.

Neuritis

Neuritis is inflammation of a single nerve, two or more nerves in separate areas, or many nerves simultaneously.

Sciatica

Sciatica (sī-AT-i-ka) is a type of neuritis characterized by severe pain along the path of the sciatic nerve or its branches. It may be caused by a slipped disc, pelvic injury, osteoarthritis of the backbone, and pressure from an expanding uterus during pregnancy.

Shingles

Shingles is an acute infection of the peripheral nervous system. It is caused by *herpes zoster* (HER-pēz ZOS-ter), the chickenpox virus.

Brain Tumors

A *brain tumor* refers to any benign or malignant growth within the cranium.

Poliomyelitis

Poliomyelitis (*infantile paralysis*), or simply *polio* (*polios* = gray), is caused by a virus called poliovirus. The virus destroys motor nerve cell bodies, specifically in the anterior horns of the spinal cord and in the nuclei of the cranial nerves and produces paralysis and atrophy of skeletal muscles.

Cerebral Palsy (CP)

The term *cerebral palsy* (CP) refers to a group of motor disorders resulting in loss of muscle control. It is caused by damage to the motor areas of the brain.

Parkinson's Disease (PD)

Parkinson's disease (PD) is a progressive disorder of the central nervous system that typically affects its victims around age 60. The cause is unknown. In PD, there is a degeneration of dopamine-producing neurons in the substantia nigra (nucleus in the midbrain), and the severe reduction of dopamine in the basal ganglia brings about most of the symptoms: tremor, impaired motor performance, rigid facial muscles, impaired walking, poor posture, autonomic dysfunction, and sensory complaints.

Multiple Sclerosis (MS)

Multiple sclerosis (MS) is the progressive destruction of the myelin sheaths of neurons in the central nervous system that interferes with the transmission of nerve impulses from one neuron to another. Although the cause of MS is unclear, there is some evidence that it results from a virus that triggers an autoimmune response in which the body's myelin-producing oligodendrocytes are destroyed by a defensive cell called a cytotoxic (killer) T cell, a type of white blood cell known as a lymphocyte (refer to Chapter 17).

Epilepsy

Epilepsy is characterized by short, recurrent, periodic attacks of motor, sensory, or psychological malfunction. The attacks, called *epileptic seizures*, are initiated by abnormal and irregular discharges of electricity from millions of neurons in the brain.

Dyslexia

In *dyslexia* (dis-LEK-sē-a; *dys* = difficulty; *lexis* = words), the brain's ability to translate images received from the eyes or ears into understandable language is impaired. The condition is a genetic disorder and is unrelated to basic intellectual capacity. Some peculiarity in the brain's organizational pattern distorts the ability to read, write, and count.

Tay–Sachs Disease

Tay–Sachs disease is an inherited disease in which neurons of the brain degenerate because of excessive amounts of a lipid called ganglioside. There is no known cure, and children who are born with the disorder usually die before age 5.

Headache

One of the most common human afflictions is *headache* or *cephalgia* (*enkephalos* = brain; *algia* = painful condition). Based on origin, two general types are distinguished: intracranial (within the brain) and extracranial (outside the brain). Serious headaches of intracranial origin are caused by brain tumors, blood vessel abnormalities, inflammation of the brain or meninges, decrease in oxygen supply to the brain, or damage to brain cells. Extracranial headaches are related to infections of the eyes, ears, nose, and sinuses and are commonly felt as headaches because of the location of these structures. Tension headaches are extracranial headaches associated with stress, fatigue, and anxiety and usually occur in the occipital and temporal muscles. A *migraine* is a usually generalized headache but may affect only one side. It is accompanied by nausea, anorexia, and vomiting. Evidence suggests that a migraine is a genetic disorder that is related to regional alterations in cerebral blood flow.

Trigeminal Neuralgia (Tic Douloureux)

Irritation of the trigeminal (V) nerve causes *trigeminal neuralgia* or *tic douloureux* (doo-loo-ROO), characterized by brief but extreme pain in the face and forehead on the affected side.

Reye's Syndrome (RS)

Reye's (RĪZ) *syndrome* (*RS*) seems to occur following a viral infection, particularly chickenpox or influenza. Aspirin at normal doses is believed to be a risk factor, and most often, children and teenagers are affected. The disease is characterized by vomiting and brain dysfunction (disorientation, lethargy, and personality changes) and may progress to coma and death.

Alzheimer's Disease (AD)

Alzheimer's (ALTZ-hī-merz) *disease,* or *AD,* is a disabling neurological disorder that afflicts about 11 percent of the population over age 65. AD patients initially have trouble remembering recent events. Next, they become more confused and forgetful, often repeating questions or getting lost while traveling to previously familiar places. Disorientation grows and memories of past events disappear. As their minds continue to deteriorate, they lose their ability to read, write, talk, eat, or walk. The disease culminates in dementia, which is the loss of reason and ability to care for oneself. A person with AD usually dies of some complication that afflicts bedridden patients, such as pneumonia. Its causes are unknown, its effects are irreversible, and it has no cure. Among the conditions linked to AD are a genetic defect, a slow-acting virus, and environmental toxins. It is the fourth leading cause of death among the elderly following heart disease, cancer, and stroke.

Delirium

Delirium (de-LIR-ē-um; *deliria* = off the tract) is a temporary disorder of abnormal cognition (perception, thinking, and memory) accompanied by fever, disturbances of the sleep–wake cycle and speech, and hyperactive or hypoactive movements. ∎

Medical Terminology and Conditions

Agnosia (ag-NŌ-zē-a; *a* = without; *gnosis* = knowledge) Inability to recognize the significance of sensory stimuli.

Analgesia (an-al-JĒ- zē-a; *an* = without; *algia* = painful condition) Pain relief.

Anesthesia (an'-es-THĒ-zē-a; *esthesia* = feeling) Loss of feeling.

Dementia (de-MEN-shē-a; *de* = away from; *mens* = mind) An organic mental disorder that results in permanent or progressive loss of intellectual abilities (memory, judgment, abstract thinking) and changes in personality.

Electroconvulsive therapy (*ECT*) (e-lek-trō-con-VUL-siv THER-a-pē) A form of shock therapy in which convulsions are induced by passing a brief electric current through the brain. It is a treatment option for severe depression and acute mania in adults.

Huntington's chorea (HUNT-ing-tunz kō-RĒ-a; *choreia* = dance) A rare hereditary disease characterized by involuntary jerky movements and mental deterioration that terminates in dementia.

Nerve block Loss of sensation in a region, such as in local dental anesthesia, from injection of a local anesthetic.

Neuralgia (noo-RAL-jē-a; *neur* = nerve) Attacks of pain along the entire course or branch of a peripheral sensory nerve.

Stupor (STOO-por) Unresponsiveness from which a patient can be aroused only briefly and by vigorous and repeated stimulation.

Viral encephalitis (VĪ-ral en'-sef-a-LĪ-tis) An acute inflammation of the brain caused directly by various viruses or by an allergic reaction to one of the many viruses that are normally harmless to the central nervous system. If the virus affects the spinal cord as well, it is called *encephalomyelitis.*

Study Outline

Spinal Cord (p. 220)

Protection and Coverings (p. 220)

1. The spinal cord is protected by the vertebral canal, meninges, cerebrospinal fluid, and vertebral ligaments.

2. The meninges are three connective tissue coverings that run continuously around the spinal cord and brain: dura mater, arachnoid, and pia mater.

3. Removal of cerebrospinal fluid from the subarachnoid space is called a spinal puncture. The procedure is used to

diagnose pathologies and to introduce antibiotics or contrast media.

General Features (p. 220)

1. The spinal cord extends from the foramen magnum of the occipital bone to the top of the second lumbar vertebra.
2. It contains cervical and lumbar enlargements that serve as points of origin for nerves to the limbs.
3. Nerves arising from the lowest portion of the cord are called the cauda equina.
4. The gray matter in the spinal cord is divided into horns and the white matter into columns.
5. In the center of the spinal cord is the central canal, which runs the length of the spinal cord and contains cerebrospinal fluid.
6. The spinal cord contains sensory (ascending) and motor (descending) tracts.

Structure in Cross Section (p. 222)

1. Parts of the spinal cord observed in cross section are the central canal; anterior, posterior, and lateral gray horns; anterior, posterior, and lateral white columns; and ascending and descending tracts.
2. The spinal cord conveys sensory and motor information by way of the sensory (ascending) and motor (descending) tracts, respectively.

Functions (p. 222)

1. One major function of the spinal cord is to convey sensory impulses from the periphery to the brain and to conduct motor impulses from the brain to the periphery.
2. The other major function is to serve as a reflex center. The posterior root, posterior root ganglion, and anterior root are involved in conveying an impulse.
3. A reflex arc is the shortest route that can be taken by an impulse from a receptor to an effector. Its basic components are a receptor, a sensory neuron, an integrating center, a motor neuron, and an effector.
4. A reflex is a quick, predictable, automatic response to a stimulus that passes along a reflex arc. Reflexes represent the body's principal mechanisms for responding to changes (stimuli) in the internal and external environments.

Spinal Nerves (p. 226)

Names (p. 226)

1. The 31 pairs of spinal nerves are named and numbered according to the region and level of the spinal cord from which they emerge.
2. There are 8 pairs of cervical, 12 pairs of thoracic, 5 pairs of lumbar, 5 pairs of sacral, and 1 pair of coccygeal nerves.

Composition and Coverings (p. 226)

1. Spinal nerves are attached to the spinal cord by means of a posterior root and an anterior root. All spinal nerves are mixed.
2. Individual fibers, bundles of nerves, and the entire nerve are wrapped in connective tissue.

Distribution (p. 226)

1. Branches of a spinal nerve include the dorsal ramus, ventral ramus, meningeal branch, and rami communicantes.

2. The ventral rami of spinal nerves, except for T2 to T11, form networks of nerves called plexuses.
3. The principal plexuses are called the cervical, brachial, lumbar, and sacral plexuses. The plexuses branch and rebranch.
4. Nerves T2 to T11 do not form plexuses and are called intercostal (thoracic) nerves. They are distributed directly to the structures they supply in intercostal spaces.

Brain (p. 227)

Principal Parts (p. 227)

1. The principal parts of the brain are the brain stem, diencephalon, cerebrum, and cerebellum.
2. The brain stem consists of the medulla oblongata, pons, and midbrain. The diencephalon consists of the thalamus and hypothalamus.

Protection and Coverings (p. 227)

1. The brain is protected by cranial bones, meninges, and cerebrospinal fluid.
2. The cranial meninges are continuous with the spinal meninges and are named dura mater, arachnoid, and pia mater.

Cerebrospinal Fluid (CSF) (p. 228)

1. Cerebrospinal fluid is formed in the choroid plexuses and circulates continually through the subarachnoid space, ventricles, and central canal.
2. Cerebrospinal fluid protects by serving as a shock absorber. It also delivers nutritive substances from the blood and removes wastes.

Blood Supply (p. 228)

1. Any interruption of the oxygen supply to the brain can weaken, permanently damage, or kill brain cells.
2. Glucose deficiency may produce dizziness, convulsions, and unconsciousness.
3. The blood–brain barrier (BBB) explains the different rates of passage of certain material from the blood into the brain.

Brain Stem (p. 230)

1. The medulla oblongata is continuous with the upper part of the spinal cord. It contains regions for regulating heart rate, diameter of blood vessels, respiratory rate, swallowing, coughing, vomiting, sneezing, and hiccuping. The vestibulocochlear, accessory, vagus, and hypoglossal nerves also originate there.
2. The pons is superior to the medulla. It connects the spinal cord with the brain and links parts of the brain with one another. It relays impulses related to voluntary skeletal movements from the cerebral cortex to the cerebellum. The trigeminal, abducens, facial, and vestibular branches of the vestibulocochlear nerves also originate there. The pons contains two regions that control respiration.
3. The midbrain connects the pons and diencephalon. It conveys motor impulses from the cerebrum to the cerebellum and cord and sensory impulses from cord to thalamus.

Diencephalon (p. 232)

1. The diencephalon consists mainly of the thalamus and hypothalamus.
2. The thalamus is superior to the midbrain and contains nuclei that serve as relay stations for sensory impulses to

the cerebral cortex. It also provides crude recognition of pain, temperature, touch, pressure, and vibration.

3. The hypothalamus is below the thalamus. It controls and integrates the autonomic nervous system and pituitary gland, functions in rage and aggression, controls body temperature, regulates food and fluid intake, and maintains consciousness and sleep patterns.

Reticular Activating System (RAS), Consciousness, and Sleep (p. 233)

1. The RAS functions in arousal (awakening from deep sleep) and consciousness (wakefulness).

2. Sleep is a state of partial unconsciousness from which an individual can be aroused.

Cerebrum (p. 234)

1. The cerebrum is the largest part of the brain. Its cortex contains convolutions, fissures, and sulci.

2. The cerebral lobes are named the frontal, parietal, temporal, and occipital.

3. The left cerebral hemisphere is more important for right-handed control, spoken and written language, numerical and scientific skills, and reasoning; the right hemisphere is more important for left-handed control, musical and artistic awareness, space and pattern perception, insight, imagination, and generating mental images of sight, sound, touch, taste, and smell.

4. The white matter is under the cerebral cortex and consists of myelinated axons running in three principal directions.

5. The basal ganglia are paired masses of gray matter in the cerebral hemispheres. They help to control muscular movements.

6. The limbic system is found in the cerebral hemispheres and diencephalon. It functions in emotional aspects of behavior and memory.

7. The sensory areas of the cerebral cortex receive and interpret sensory impulses. The motor areas govern muscular movement. The association areas are concerned with emotional and intellectual processes.

8. Memory is the ability to recall thoughts and is generally classified into two kinds: short-term and long-term memory.

9. Brain waves generated by the cerebral cortex are recorded as an electroencephalogram (EEG). It may be used to diagnose epilepsy, infections, and tumors.

Cerebellum (p. 237)

1. The cerebellum occupies the inferior and posterior aspects of the cranial cavity. It consists of two cerebellar hemispheres with a cerebellar cortex of gray matter and an interior of white matter tracts.

2. It is attached to the brain stem by three pairs of cerebellar peduncles.

3. The cerebellum coordinates skeletal muscles and maintains normal muscle tone and body equilibrium.

Neurotransmitters (p. 239)

1. Numerous substances are either known or suspected neurotransmitters that function to facilitate, excite, or inhibit postsynaptic neurons.

2. Examples of neurotransmitters include acetylcholine, dopamine, norepinephrine, serotonin, gamma aminobutyric acid, substance P, enkephalins, and endorphins.

Cranial Nerves (p. 239)

1. Twelve pairs of cranial nerves originate from the brain.

2. Like spinal nerves, they are part of the PNS. See Exhibit 10.2 for a summary of cranial nerves.

Self-Quiz

1. A bundle of fibers located outside the central nervous system is known as a
   a. nucleus   b. tract   c. nerve   d. dendrite   e. column

2. The term gray matter refers to
   a. neurons that are unmyelinated   b. neurons that are myelinated   c. protective connective tissue coverings over the brain and spinal cord   d. ascending and descending tracts   e. the lining of the ventricles

3. Fast responses to changes in the environment are referred to as
   a. plexuses   b. spinal pathways   c. pyramidal pathways   d. conduction pathways   e. reflexes

4. A sensory neuron enters the spinal cord from the
   a. anterior root   b. center root   c. lateral horn   d. posterior root   e. superior ramus

5. Impulses sent from the central nervous system to an effector are conveyed by way of the
   a. motor neuron   b. sensory neuron   c. association neuron   d. ascending tracts   e. gray horns

6. Which of the following is NOT a function of cerebrospinal fluid?
   a. protection   b. circulation   c. nerve impulse transmission   d. nutrition   e. shock absorber.

7. A needle used during a spinal tap would penetrate (in order)
   1. arachnoid
   2. dura mater
   3. epidural space
   4. pia mater
   5. subarachnoid
   a. 1,2,3,5   b. 2,3,1,5,4   c. 3,2,1,5   d. 3,1,5,2   e. 5,1,3,4,2

8. The spinal tract that conducts impulses to skeletal muscles involved in precise movements is the _____ pathway.
   a. anterolateral (spinothalamic)   b. direct (pyramidal)   c. indirect (extrapyramidal)   d. posterior column–medial lemniscus   e. ventral ramus

9. The spinal tract that conducts impulses for sensing pain, temperature, crude touch, pressure, tickle and itch is the _____ pathway.

    (a.) anterolateral (spinothalamic)   b. direct (pyramidal)   c. indirect (extrapyramidal)   d. posterior column–medial lemniscus   e. ventral ramus

10. Trace a reflex arc starting with the stimulus and ending with the response (reflex).

    E 1. effector
    C 2. integrating center
    D 3. motor neuron
    A 4. receptor
    B 5. sensory neuron

    (a.) 4, 5, 2, 3, 1   b. 1, 5, 2, 3, 4   c. 4, 3, 2, 5, 1
    d. 5, 2, 3, 4, 1   e. 3, 1, 4, 5, 2

11. The correct arrangement of spinal nerves is ____ pairs of cervical nerves, ____ pairs of thoracic nerves, ____ pairs of lumbar nerves, ____ pairs of thoracic nerves, and ____ pair(s) of coccygeal nerves

    a. 12, 8, 5, 5, 1   b. 12, 5, 8, 5, 1   c. 8, 5, 12, 5, 3
    d. 5, 5, 8, 12, 3   (e.) 8, 12, 5, 5, 1

12. The _____ plexus supplies the upper limbs and some neck and shoulder muscles.

    (a.) brachial   b. cervical   c. lumbar   d. thoracic   e. sacral

13. The diencephalon is composed of the

    a. medulla, pons, and hypothalamus   b. midbrain, hypothalamus, and thalamus   c. cerebellum and midbrain   d. medulla, pons, and midbrain   (e.) hypothalamus and thalamus

14. The region of the cerebral cortex involved in control of muscle movement is the

    a. primary somatosensory area   (b.) primary motor area   c. gnostic area   d. auditory association area   e. primary gustatory area

15. Which two of the following cranial nerves are NOT involved in controlling movement of the eyeball?

    a. II   b. III   c. IV   d. VI   e. X

16. The cranial nerve associated with hearing and equilibrium is

    a. V   b. VII   c. VIII   d. X   e. XII

17. The cranial nerve that functions in secretion of saliva and taste is

    a. V   b. VII   c. VIII   d. X   e. XII

Critical Thinking Applications

1. Dennis' first visit to the dentist after a 10- year absence resulted in extensive dental work. The dentist injected a "numbing" anesthetic in several locations during the session. While having lunch right after the visit, soup drips out of Dennis' mouth because he has no feeling around his left upper lip, right lower lip, and tip of his tongue. What did the dentist do to Dennis? *anesthetic prohibits release of acetylcholine*

2. A pediatrician suspects an infection is present in the protective coverings over a child's brain. She orders a spinal tap performed. List the protective coverings and their functions. How can a spinal tap give the doctor information about the brain?

3. Compare the effects of damage to cranial nerve II and damage to the occipital lobe.

4. While checking cranial nerve function on an emergency room patient, a nurse finds that the patient can see a light but cannot track the light with her eyes. What do these results tell you about cranial nerve function in this patient?

5. An elderly relative suffered a stroke and now has difficulty controlling the movement of her right arm and also has some speech problems. What areas of the brain were damaged by the stroke?

6. A drunken driver had a car accident that caused severe damage to the brain stem, crushing his medulla oblongata. Predict the outcome of his injuries.

Answers to Figure Questions

10.1 Subarachnoid space.
10.2 Peripheral nervous system.
10.3 Sensory; motor.
10.4 Primary somatosensory area of cerebral cortex; primary motor area of cerebral cortex.
10.5 Anterior root.
10.6 Medulla oblongata.
10.7 Formed in the choroid plexuses and reabsorbed into arachnoid villi.
10.8 Medulla oblongata; midbrain; pons.
10.9 Decussation means crossing to the opposite side. The pyramids contain motor tracts that extend from the cortex into the spinal cord. Because they convey impulses for contraction of skeletal muscle, the functional consequence of decussation of the pyramids is that one side of the cerebrum controls muscles on the opposite side of the body.
10.10 Gray matter.
10.11 Corpus callosum.
10.12 Primary somatosensory; motor speech area; primary gustatory area; somatosensory association area; primary motor area; primary olfactory area; frontal eye field area.

chapter *11*

AUTONOMIC NERVOUS SYSTEM

a look ahead

Respiratory System
Voluntary & involuntary
lung action

The part of the nervous system that regulates smooth muscle, cardiac muscle, and glands is the ***autonomic nervous system (ANS).*** (Recall that together the ANS and somatic nervous system comprise the peripheral nervous system; see Figure 9.1.) Functionally, it usually operates without conscious control. The system was originally named *autonomic* because physiologists thought it was autonomous or self-governing; that is, it functioned with no control from the central nervous system. It is now recognized that many parts of the CNS are connected to the ANS and exert considerable influence over it. Autonomic centers in the cerebral cortex are connected to autonomic centers of the thalamus, which, in turn, connect with the hypothalamus. It is at the level of the hypothalamus that the major control and integration of the ANS occur. (This connection between the CNS and ANS is the reason why anxiety, which can result from either conscious or subconscious stimulation in the cerebral cortex, results in sweaty palms and palpitations and why a shocking or horrifying sight can result in lowered blood pressure and fainting.) The hypothalamus receives input from areas of the nervous system concerned with emotions and receptors associated with visceral functions, olfaction (smell), gustation (taste), and changes in temperature and the electrolyte composition of the blood.

Comparison of Somatic and Autonomic Nervous Systems

objective: *Compare the main structural and functional differences between the somatic and autonomic nervous systems.*

The somatic nervous system consists of sensory neurons that convey information from receptors primarily in the head, body wall, and limbs and motor neurons that produce conscious movement in skeletal muscles. The autonomic nervous system consists mostly of motor neurons that regulate visceral activities, usually involuntarily and automatically. For example, the ANS regulates the size of the pupils, the shape of the lens for near vision, dilation of blood vessels, rate and force of the heartbeat, movements of the gastrointestinal tract, and secretion by most glands.

The ANS is generally considered to be entirely motor. All its axons are motor fibers, which transmit impulses from the central nervous system to visceral effectors. Autonomic fibers are thus called ***visceral motor fibers.*** *Visceral effectors* include cardiac muscle, smooth muscle, and glandular epithelium. Impulses that arise from receptors in viscera travel along sensory neurons of spinal nerves to the spinal cord or cranial nerves to the lower portions of the brain. The sensory impulses are delivered to various autonomic centers and the returning motor impulse usually produces an adjustment in a visceral effector without conscious recognition. Some visceral sensations, however, such as hunger, nausea, fullness of the urinary bladder, and pain from viscera, produce conscious recognition.

The autonomic nervous system consists of two divisions: *sympathetic* and *parasympathetic.* Many organs innervated by the ANS receive visceral motor fibers from both divisions. In general, impulses from one division stimulate the organ to start or increase activity, whereas impulses from the other decrease the organ's activity. Organs that receive impulses from both sympathetic and parasympathetic fibers are said to have ***dual innervation.*** Thus autonomic innervation may be excitatory or inhibitory. In the somatic nervous system, stimulation of a skeletal muscle is always excitatory. When the neuron ceases to stimulate a muscle, contraction stops altogether.

In the ANS, there are two motor neurons and a ganglion between them. The first motor neuron runs from the CNS to a ganglion, where it synapses with the second motor neuron. This neuron ultimately synapses on a visceral effector. The second motor neuron may release either acetylcholine (ACh) or norepinephrine (NE). In the somatic nervous system, all motor pathways involve only one motor neuron from the CNS, which synapses directly on a skeletal muscle and secretes only ACh.

Figure 11.1 and Exhibit 11.1 present a summary of the principal differences between the somatic and autonomic nervous systems.

Structure of the Autonomic Nervous System

objective: *Identify the structural features of the autonomic nervous system.*

Autonomic Motor Pathways

Autonomic motor pathways involve two motor neurons. One extends from the CNS to a ganglion; the other from the ganglion to the effector (muscle or gland).

The first motor neuron in an autonomic pathway is called a ***preganglionic neuron*** (Figure 11.1b). Its cell body is in the brain or spinal cord. Its myelinated axon, called a ***preganglionic fiber,*** passes out of the CNS as part of a cranial or spinal nerve. The fiber travels to an autonomic ganglion, where it synapses with the second motor neuron in the pathway, called a postganglionic neuron.

The ***postganglionic neuron*** lies entirely outside the CNS. Its cell body and dendrites are located in the autonomic ganglion, where the synapse with one or more preganglionic fibers occurs. The axon of a postganglionic neuron, called a ***postganglionic fiber,*** is unmyelinated and ends in a visceral effector.

Thus preganglionic neurons convey motor impulses from the CNS to autonomic ganglia. Postganglionic neurons relay motor impulses from autonomic ganglia to visceral effectors.

Preganglionic Neurons

The sympathetic division is also called the ***thoracolumbar*** (thō-ra-kō-LUM-bar) ***division*** because the preganglionic neurons have their cell bodies in the *thoracic* and *lumbar* segments of the spinal cord (Figure 11.2 on page 252). The parasympathetic division is also called the ***craniosacral*** (krā-nē-ō-SĀ-kral) ***division*** because the cell bodies of the preganglionic neurons are located in the nuclei of *cranial nerves* III, VII, IX, and X in the brain stem and in the *sacral* segments of the spinal cord.

Important feature →

also has sensory fibers

1 motor neuron + no ganglion

2 motor neurons has ganglion

Figure 11.1   Comparison of somatic and autonomic nervous systems.

🔑 *In the somatic nervous system, stimulation of a skeletal muscle is always excitatory; in the autonomic nervous system, visceral effectors may be excited or inhibited.*

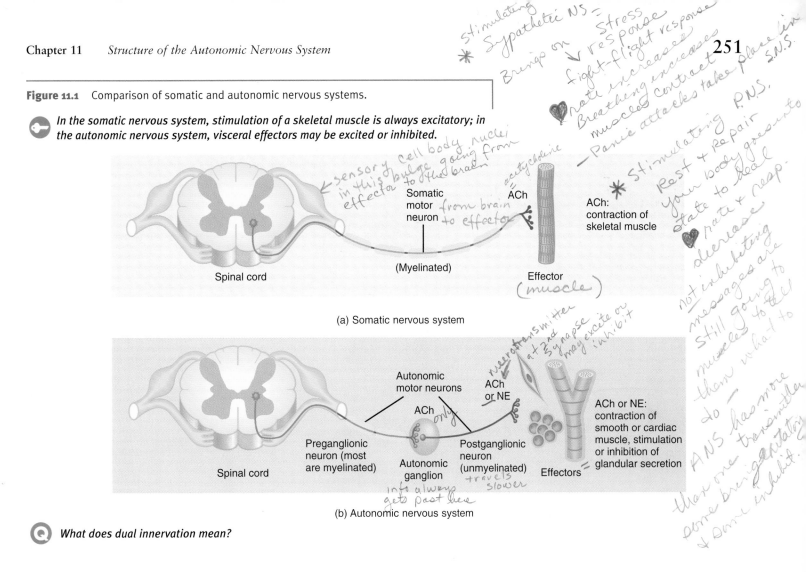

(a) Somatic nervous system

(b) Autonomic nervous system

❓ *What does dual innervation mean?*

Autonomic Ganglia

Autonomic pathways always include *autonomic ganglia,* where synapses between visceral motor fibers occur. Autonomic ganglia may be divided into three groups:

Sympathetic trunk ganglia lie on either side of the backbone, from the base of the skull to the coccyx (Figure 11.3 on page 253). They receive preganglionic fibers only from the *sympathetic division* (see Figure 11.2). Because of this, sympathetic preganglionic fibers tend to be short.

Prevertebral ganglia (Figure 11.3) lie anterior to the backbone and close to the large abdominal arteries from which the names of the ganglia are derived. Examples are the celiac ganglion, the superior mesenteric ganglion, and the inferior

mesenteric ganglion (see Figure 11.2). Prevertebral ganglia also receive preganglionic fibers from the *sympathetic division.*

Preganglionic fibers from the *parasympathetic division* make synapses in the third kind of autonomic ganglia, *terminal ganglia* (see Figure 11.2). The ganglia of this group are located at the end of a visceral motor pathway very close to visceral effectors or actually within the walls of visceral effectors. Because of this, parasympathetic preganglionic fibers tend to be long (see Figure 11.2).

Postganglionic Neurons

Axons from preganglionic neurons of the sympathetic division can (1) synapse with postganglionic neurons in the sympathetic

Exhibit 11.1	Comparison of Somatic and Autonomic Nervous Systems	
	SOMATIC	AUTONOMIC
Effectors	Skeletal muscles.	Cardiac muscle, smooth muscle, glandular epithelium.
Type of control	Voluntary.	Involuntary.
Neural pathway	One motor neuron extends from CNS and synapses directly on a skeletal muscle.	One motor neuron extends from the CNS and synapses with another motor neuron in a ganglion; the second motor neuron synapses on a visceral effector.
Action on effector	Always excitatory.	May be excitatory or inhibitory.
Neurotransmitters	Acetylcholine.	Acetylcholine or norepinephrine.

Figure 11.2   Structure of the autonomic nervous system.

In an autonomic motor pathway, a preganglionic neuron synapses with a postganglionic neuron in an autonomic ganglion.

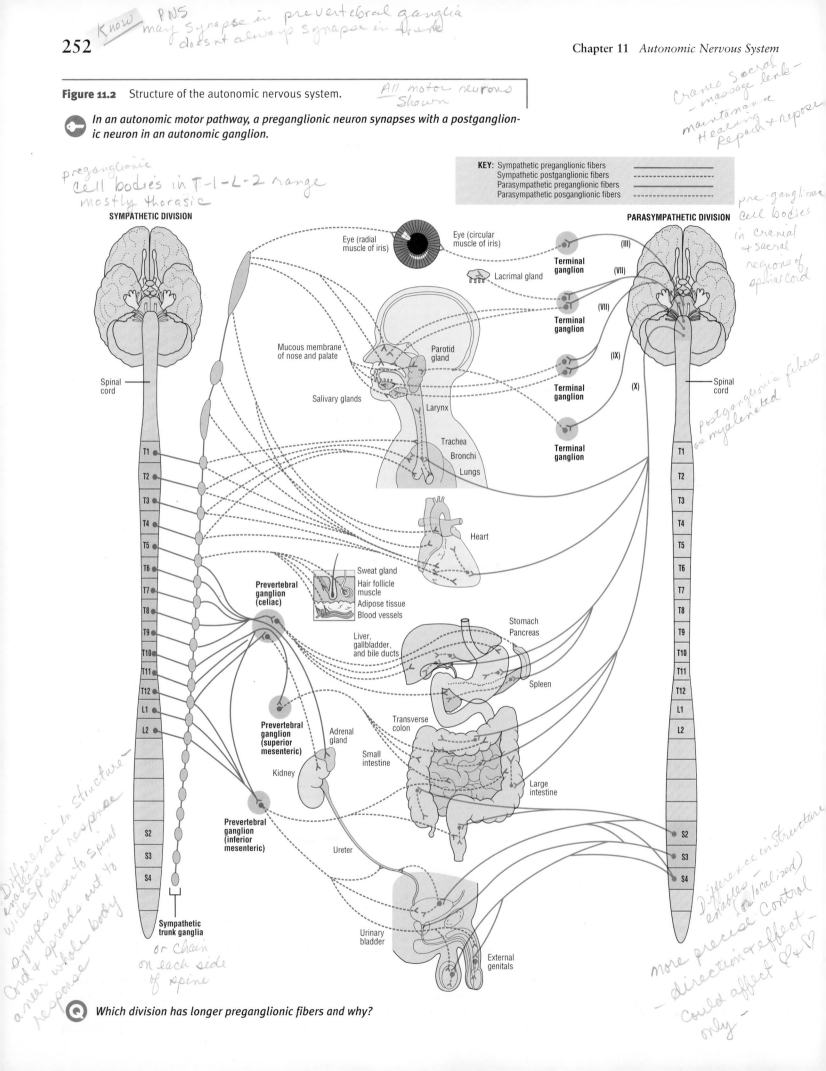

KEY: Sympathetic preganglionic fibers
        Sympathetic postganglionic fibers
        Parasympathetic preganglionic fibers
        Parasympathetic posganglionic fibers

SYMPATHETIC DIVISION

PARASYMPATHETIC DIVISION

Eye (radial muscle of iris)

Eye (circular muscle of iris)

Lacrimal gland

Mucous membrane of nose and palate

Parotid gland

Salivary glands

Larynx

Trachea
Bronchi
Lungs

Spinal cord

Spinal cord

Terminal ganglion   (III)

Terminal ganglion   (VII)

(VII)

(IX)

Terminal ganglion   (X)

Terminal ganglion

Heart

Sweat gland
Hair follicle muscle
Adipose tissue
Blood vessels

Prevertebral ganglion (celiac)

Liver, gallbladder, and bile ducts

Stomach
Pancreas

Spleen

Prevertebral ganglion (superior mesenteric)

Adrenal gland

Kidney

Transverse colon

Small intestine

Large intestine

Prevertebral ganglion (inferior mesenteric)

Ureter

Sympathetic trunk ganglia

Urinary bladder

External genitals

T1
T2
T3
T4
T5
T6
T7
T8
T9
T10
T11
T12
L1
L2
S2
S3
S4

T1
T2
T3
T4
T5
T6
T7
T8
T9
T10
T11
T12
L1
L2
S2
S3
S4

Q *Which division has longer preganglionic fibers and why?*

Figure 11.3   Ganglia of the sympathetic division of the autonomic nervous system.

Whereas sympathetic trunk and prevertebral ganglia are associated with the sympathetic division, terminal ganglia are associated with the parasympathetic division.

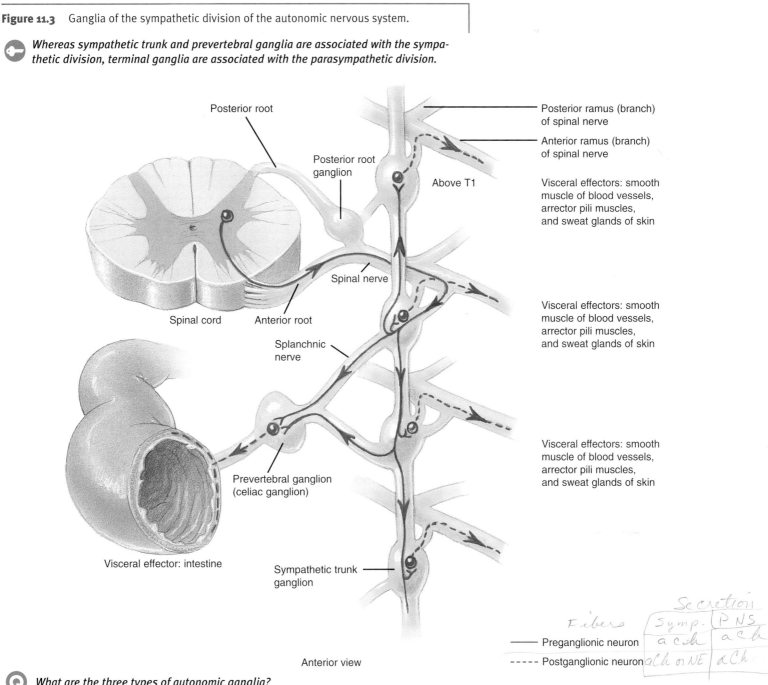

Posterior root

Posterior root ganglion

Posterior ramus (branch) of spinal nerve

Anterior ramus (branch) of spinal nerve

Above T1

Visceral effectors: smooth muscle of blood vessels, arrector pili muscles, and sweat glands of skin

Spinal nerve

Visceral effectors: smooth muscle of blood vessels, arrector pili muscles, and sweat glands of skin

Spinal cord   Anterior root

Splanchnic nerve

Visceral effectors: smooth muscle of blood vessels, arrector pili muscles, and sweat glands of skin

Prevertebral ganglion (celiac ganglion)

Visceral effector: intestine

Sympathetic trunk ganglion

——— Preganglionic neuron

- - - - Postganglionic neuron

Anterior view

What are the three types of autonomic ganglia?

trunk ganglia or (2) continue, without synapsing, through the sympathetic trunk ganglia to a prevertebral ganglion and synapse with the postganglionic neurons there (see Figure 11.3). Each sympathetic preganglionic fiber synapses with many postganglionic fibers in the ganglion, and the postganglionic fibers pass to several visceral effectors. After exiting their ganglia, the post-synaptic fibers innervate their visceral effectors.

Axons from preganglionic neurons of the parasympathetic division pass to terminal ganglia near or within a visceral effector. In the ganglion, the presynaptic neuron usually synapses with four or five postsynaptic neurons, which terminate in a single visceral effector. After exiting their ganglia, the postsynaptic fibers supply their visceral effectors.

With this background in mind, we can now examine some specific structural features of the sympathetic and parasympathetic divisions.

Sympathetic Division

Preganglionic fibers are myelinated and leave the spinal cord through the anterior root of a spinal nerve. After exiting, they pass to the nearest sympathetic trunk ganglion on the same side. A preganglionic fiber may terminate (synapse) in several ways, but in most cases, the synapse will be with usually 20 or more postganglionic cell bodies in a ganglion. Often the postganglionic fibers then terminate in widely separated organs of the body. Thus an impulse that starts in a single preganglionic neuron may reach several visceral effectors. For this reason, most sympathetic responses have widespread effects on the body.

Parasympathetic Division   *Know*

Preganglionic cell bodies of the parasympathetic division are found in nuclei in the brain stem and the sacral segments of the spinal cord (see Figure 11.2). Their fibers emerge as part of a cranial or spinal nerve. Preganglionic fibers of both the cranial and sacral outflows end in terminal ganglia, where they synapse with postganglionic neurons. The arrangement of neurons in the parasympathetic division allows an individual to more precisely control the body's functioning by stimulating only the specific organs needed; it allows a more localized body response.

The sympathetic and parasympathetic divisions are compared in Exhibit 11.2.

Functions of the Autonomic Nervous System

objective: *Describe the functions of the autonomic nervous system.*

ANS Neurotransmitters

Autonomic fibers, like other axons of the nervous system, release neurotransmitters at synapses as well as at points of contact with visceral effectors (smooth and cardiac muscle and glands). These points are called **neuroeffector junctions.** Neuroeffector junctions may be either neuromuscular or neuroglandular junctions. On the basis of the neurotransmitter produced, autonomic fibers may be classified as either cholinergic or adrenergic.

Cholinergic (kō′-lin-ER-jik) **fibers** release **acetylcholine** (**ACh**) and include the following: (1) all sympathetic and parasympathetic preganglionic axons, (2) all parasympathetic postganglionic axons, and (3) a few sympathetic postganglionic axons. Because acetylcholine is quickly inactivated by the enzyme **acetylcholinesterase** (**AChE**), the effects produced by cholinergic fibers are short-lived and local. The fact that postganglionic parasympathetic neurons release the short-lived ACh is important to maintain the precise control needed by the parasympathetic division.

Adrenergic (ad′-ren-ER-jik) **fibers** produce **norepinephrine** (**NE**). Most sympathetic postganglionic axons are adrenergic. Because norepinephrine is inactivated much more slowly than acetylcholine and because norepinephrine may enter the blood-

stream, the effects of sympathetic stimulation are longer lasting and more widespread than parasympathetic stimulation. It is important for postganglionic sympathetic neurons to release NE so that the organs are stimulated long enough for the individual to deal with the emergency.

Activities

As noted earlier, impulses from one division generally stimulate the organ's activities, whereas impulses from the other inhibit the organ's activities. The stimulating division may be either the sympathetic or the parasympathetic, depending on the organ. For example, sympathetic impulses increase heart activity, whereas parasympathetic impulses decrease it. On the other hand, parasympathetic impulses increase digestive activities, whereas sympathetic impulses inhibit them. The actions of the two systems are integrated to help maintain homeostasis. A summary of ANS functions is presented in Exhibit 11.3.

The parasympathetic division is primarily concerned with activities that *conserve and restore body energy.* For instance, parasympathetic impulses to the digestive glands and smooth muscle of the gastrointestinal tract normally dominate over sympathetic impulses. Thus energy-supplying food can be digested and absorbed by the body.

The sympathetic division, in contrast, is primarily concerned with processes involving the *expenditure of energy.* When the body is in homeostasis, the main function of the sympathetic division is to counteract the parasympathetic effects just enough to carry out normal processes requiring energy. During extreme stress, however, the sympathetic dominates the parasympathetic. Confronted with stress, the body becomes alert and capable of unusual feats. Fear stimulates the sympathetic division as do a variety of other emotions and physical activities.

Activation of the sympathetic division sets into operation a series of physiological responses collectively called the **fight-or-flight response.** Among the effects it produces are the following (see Exhibit 11.3 for additional sympathetic responses):

1. The pupils of the eyes dilate.

2. Heart rate and force of contraction and blood pressure increase.

3. The blood vessels of nonessential organs such as the skin and viscera constrict.

4. Blood vessels of organs involved in fighting off danger—skeletal muscles, cardiac muscles, brain, and

Exhibit 11.2	**Structural Features of Sympathetic and Parasympathetic Divisions**
SYMPATHETIC	**PARASYMPATHETIC**
Contains sympathetic trunk and prevertebral ganglia.	Contains terminal ganglia.
Ganglia are close to the CNS and distant from visceral effectors.	Ganglia are near or within visceral effectors.
Each preganglionic fiber is short and synapses with many postganglionic neurons that pass to many visceral effectors.	Each preganglionic fiber is long and usually synapses with very few postganglionic neurons that pass to a single visceral effector.
Distributed throughout the body, including the skin.	Distribution limited primarily to head and viscera of thorax, abdomen, and pelvis.

Exhibit 11.3 | Activities of Autonomic Nervous System

[handwritten: stress (flight or fright) Survival response evolutionary process] *[handwritten: Rest + Restore.]*

VISCERAL EFFECTOR	EFFECT OF SYMPATHETIC STIMULATION	EFFECT OF PARASYMPATHETIC STIMULATION
Glands	*[handwritten: to maintain homeostasis we get rid of heat from muscle contraction]*	
Sweat	Increases secretion.	No known functional innervation.
Lacrimal (tear)	No known functional innervation.	Stimulates secretion. *[handwritten: specks of dust in eye + emotional grief or sadness]*
Adrenal	Promotes epinephrine and norepinephrine secretion. *[handwritten: helps to speed up heart + lungs]*	No known functional innervation.
Liver	Promotes the breakdown of glycogen in the liver into glucose, the conversion of noncarbohydrates in the liver to glucose, and decreases bile secretion. *[handwritten: expending energy stored in glycogen]*	Promotes glycogen synthesis; increases bile secretion.
Kidney	Stimulates secretion of renin (enzyme), which helps raise blood pressure, decreases urine volume. *[handwritten: save energy + keeps more fluid in]* *[handwritten: more blood to brain + extremities]*	No known functional innervation.
Pancreas	Inhibits secretion of enzymes and insulin (hormone that lowers blood sugar level); promotes secretion of glucagon (hormone that raises blood sugar level). *[handwritten: helps increase glucose]*	Promotes secretion of enzymes and insulin. *[handwritten: puts blood sugar into cells]*
Smooth muscle		
Radial muscle of iris of eye	Contraction that results in dilation of pupil. *[handwritten: lets more light in + see more]*	No known functional innervation.
Circular muscle of iris of eye	No known functional innervation.	Contraction that results in constriction of pupil.
Ciliary muscle of eye	Relaxation that results in far vision.	Contraction that results in near vision.
Salivary glands	Decreases secretion. *[handwritten: dry mouth / saving fluid + energy]*	Stimulates secretion.
Gastric glands	Inhibits secretion. *[handwritten: about same thing save energy]*	Promotes secretion.
Intestinal glands	Inhibits secretion.	Promotes secretion.
Gallbladder and ducts	Relaxation.	Contraction ⟶ increases release of bile into small intestine.
Stomach	Decreases motility (movement) and tone; contracts sphincters.	Increases motility and tone; relaxes sphincters.
Intestines	Decreases motility and tone; contracts sphincters. *[handwritten: (movement)]*	Increases motility and tone; relaxes sphincters.
Lungs (smooth muscle of bronchi)	Relaxation ⟶ airway widening. *[handwritten: dilates Bronchi] [handwritten: take in more oxygen to keep everything going]*	Contraction ⟶ airway narrowing.
Urinary bladder	Relaxation of muscular wall; contraction of internal sphincter. *[handwritten: more in kids more in adults]*	Contraction of muscular wall; relaxation of internal sphincter.
Spleen	Contraction and discharge of stored blood into general circulation.	No known functional innervation.
Arrector pili of hair follicles	Contraction that results in erection of hairs. *[handwritten: goose bumps]*	No known functional innervation.
Uterus	Inhibits contraction if nonpregnant; stimulates contraction if pregnant.	Minimal effect.
Sex organs	In male, produces ejaculation. *[handwritten: after erection]*	Vasodilation and erection in both sexes.
Cardiac muscle	*[handwritten: Point + Shoot PNS ANS]*	
Heart	Increases rate and strength of contraction.	Decreases rate and strength of contraction.

lungs—dilate to allow faster flow of blood.

5. Bronchioles (small air tubes leading into and out of the lungs) dilate to provide more oxygen to muscles for energy production.

6. Blood sugar level rises as liver glycogen converts to glucose for extra energy.

7. The adrenal glands produce epinephrine and norepinephrine to intensify and prolong the sympathetic effects just described.

8. Processes not essential for meeting the stress are inhibited; for example, muscular movements of the gastrointestinal tract and digestive secretions slow down or stop.

Stress and Health: Personality and Perception

*t*oo much stress. We've all had that feeling. Sometimes it's experienced as tightness in the throat, a churning stomach, or clenched teeth. Breathing becomes shallow and more rapid. Muscle tension increases, and we develop headaches and back, neck, and shoulder pain. We feel distracted and can't concentrate. Our thoughts race from one thing to the next and then back again without resolution. We run faster and faster just to stay in place on the treadmill of life. Because stress *feels* so bad, people often wonder whether stress is in fact harmful to their health.

Can Stress Cause Illness?

Concern about the effects of chronic stress on physical and mental health is nothing new. In the fifth century B.C., Hippocrates, who has been called the father of medicine, counseled medical students to consider emotional factors in their diagnosis and treatment of disease. Aristotle, the Greek philosopher and scientist, believed that body and soul were inseparable, and that emotions play an important role in health and illness. Throughout history, people have observed that hard times and ill health often go hand in hand. We hear about people who become ill and die shortly after the loss of a spouse. We see middle-aged relatives and friends living high-pressure lifestyles who develop heart disease. And when ominous happenings loom on the horizon, we say we are "worried sick."

Stress can cause illness. But the relationship between stress and health is not a simple one. A certain amount or type of stress does not automatically cause a given health condition. The impact of stress on health is mediated by a number of important genetic, environmental, and personality variables.

People respond to stress in many different ways. Some people have a very high tolerance and even a fondness for life in the fast lane; they seek situations packed full of challenging demands. Others are more comfortable with a slower pace. But each of us seems to have our own "Achilles heel," our own special area of vulnerability in which excess stress manifests itself physically and psychologically. As we observe ourselves over time, we come to understand our own personal tolerances. We can then use our physical and psychological vulnerabilities as "stress barometers" to warn us when we need to address sources of stress in order to cope with demands and regain our equilibrium. Otherwise minor symptoms can worsen over time and turn into a major health problem.

It's the Thought That Counts

Research shows that *perception* is one of the most important mediators of the stress–illness relationship, because our emotional and physical responses to stress vary with our perceptions. For example, if we perceive that a stressful situation is totally outside our control, we feel helpless and hopeless. These feelings are strongly associated with negative health effects. However, if we feel we at least have some options to choose from in addressing a stressful situation, we experience a ray of hope, a less harmful stress response, and fewer negative health effects.

Researchers have found that emotions such as anger and hostility are especially harmful to our health. Feelings of isolation and alienation have also shown strong associations with negative health consequences. People who manage to avoid getting sick despite a lot of stress are more apt to see sources of stress as challenges, and to approach the tasks at hand with an optimistic outlook and a feeling of control. They reach out to others for help and emotional support. And best of all, they have more fun.

critical thinking

Consider the sympathetic activation known as the fight-or-flight response, described in this chapter. Discuss how chronic overactivation of this response could be harmful to one's health.

Can merely thinking about something stressful activate the fight-or-flight response? Why?

Study Outline

Comparison of Somatic and Autonomic Nervous Systems (p. 250)

1. The somatic nervous system produces conscious movement in skeletal muscles.

2. The autonomic nervous system regulates visceral activities, that is, activities of smooth muscle, cardiac muscle, and glands, and it usually operates without conscious control.

   Know

3. The ANS is regulated by centers in the brain, in particular, by the cerebral cortex and hypothalamus.

4. A single somatic motor neuron synapses on skeletal muscles; in the ANS, there are two motor neurons—one from the CNS to a ganglion and one from a ganglion to a visceral effector.

5. Somatic motor neurons release acetylcholine (ACh) and autonomic motor neurons release either acetylcholine or norepinephrine (NE).

Structure of the Autonomic Nervous System (p. 250)

1. The ANS consists of visceral motor neurons organized into nerves, ganglia, and plexuses.

2. It is entirely motor. All autonomic axons are motor fibers.

3. Motor neurons are preganglionic (with myelinated axons) and postganglionic (with unmyelinated axons).

4. The autonomic nervous system consists of two principal divisions: sympathetic and parasympathetic.

5. Autonomic ganglia are classified as sympathetic trunk ganglia (on sides of backbone), prevertebral ganglia (anterior to backbone), and terminal ganglia (near or inside visceral effectors).

Functions of the Autonomic Nervous System (p. 254)

1. Autonomic fibers release neurotransmitters at synapses. On the basis of the neurotransmitter produced, these fibers may be classified as cholinergic or adrenergic.

2. Cholinergic fibers release acetylcholine (ACh). Adrenergic fibers produce norepinephrine (NE).

3. Sympathetic responses are widespread and, in general, concerned with energy expenditure. Parasympathetic *Know* responses are restricted and are typically concerned with energy restoration and conservation.

Self-Quiz

1. Comparing the somatic nervous system with the autonomic nervous system, which of the following statements is true?

   a. The autonomic nervous system controls involuntary movement in skeletal muscle.   b. The somatic nervous system controls voluntary movement in glands and smooth muscle.   c. The autonomic nervous system produces voluntary movement in smooth muscle and glands.   d. The autonomic nervous system controls involuntary movement in smooth muscle, viscera, and glands.   e. The somatic nervous system controls involuntary movement in smooth muscle, viscera, and glands.

2. Major control and integration of the autonomic nervous system occur in what portion of the central nervous system?

   a. spinal cord   b. brain stem   c. basal ganglia   d. cerebellum   e. hypothalamus

3. Neurons in the autonomic nervous system include

   a. two motor neurons and one ganglion   b. one motor neuron and two ganglia   c. two motor neurons and two ganglia   d. one motor and one sensory neuron and no ganglia   e. one motor neuron, one sensory neuron, and one ganglion

4. Which of the following would be considered an effect of parasympathetic stimulation?

   a. increased blood pressure   b. increased heart rate   c. increased synthesis of glycogen from glucose   d. increased constriction of some blood vessels   e. dilation of the pupils of the eye

5. Which of these statements about the parasympathetic division of the autonomic nervous system is NOT true? The parasympathetic division

   a. arises from the cranial and sacral spinal cord segments   b. is concerned with conserving and restoring energy

   c. uses acetylcholine as its neurotransmitter   d. has ganglia near or within visceral effectors   e. results in "fight-or-flight" response

6. Which of the following choices is NOT true concerning the autonomic nervous system?

   a. Sympathetic responses are generally widespread throughout the body.   b. The autonomic division is entirely motor.   c. Parasympathetic responses are generally local effects.   d. Sensory neurons include pre- and postganglionic fibers.   e. Most visceral effectors have dual innervation.

7. True/False. *with digestive system (trigger responses may be sensory)*   *Book is wishy washy on this*

   T or F a. The ANS includes both sensory and motor fibers.

   __T__ b. Sympathetic nervous system effects are generally concerned with energy expenditure.

   __T__ c. "Fight-or-flight" responses are associated with the sympathetic division of the autonomic *Know* nervous system.

   __F__ d. The two principal divisions of the autonomic nervous system include the sympathetic and the unsympathetic.

8. Which of the following are NOT paired correctly?

   a. adrenergic—norepinephrine   b. cholinergic—acetylcholine   c. adrenergic—sympathetic postganglionic fiber   d. cholinergic—parasympathetic postganglionic fiber   e. adrenergic—all preganglionic fibers

9. The autonomic ganglia associated with the parasympathetic division are the

   a. sympathetic trunk ganglia   b. prevertebral ganglia   c. dorsal root ganglia   d. terminal ganglia   e. basal ganglia

Critical Thinking Applications

1. Sky-diving, hang-gliding, and ski-jumping can all get you killed or give you a great adrenaline rush. How do these activities cause this type of physiological effect?

2. It's Thanksgiving and you've just eaten a huge turkey dinner with all the trimmings. Now you're heading for the football game—on the television. Which division of the nervous system will be handling your body's post-dinner activities? List several organs and the effects of nervous system stimulation on their functions.

3. The condition "white coat hypertension" is characterized by a significant increase in a patient's blood pressure whenever the patient sees a health professional in his/her white coat. Which division of the nervous system is responsible for the increase in blood pressure? How does this happen?

4. Jimmy wanted a toy on the top of the wood bookcase, so he climbed up the shelves. His mother ran in when she heard the crash and lifted the heavy bookcase with one arm while pulling her son out with the other. Later that day, she could not lift the bookcase back into position by herself. How do you explain the temporary "supermom" effect?

5. Ron's car hit a deer on the way to class. The animal was badly injured and the accident scene was a very messy one. After the accident, Ron felt shaky and a bit faint, his palms were sweaty and he felt sick to his stomach. Explain his physical response.

6. Your friend the art major suggests that you try yoga or meditation as a way to improve your attitude and your grades. Yoga and meditation have been shown to influence the autonomic nervous system. Use your knowledge of the nervous system to explain yoga's calming effect on body functions.

Answers to Figure Questions

11.1. An organ receives impulses from both the sympathetic and parasympathetic divisions of the ANS.

11.2. Most parasympathetic preganglionic fibers are longer than most sympathetic preganglionic fibers because parasympathetic ganglia are in the walls of visceral organs.

11.3. Sympathetic trunk, prevertebral, and terminal.

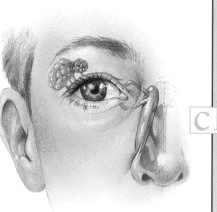

chapter 12

SENSATIONS

a look ahead

259

Having examined the structure of the nervous system and its activities, we will now see how its different parts cooperate in receiving sensory information and transmitting motor impulses that result in movement or secretion.

Sensations

[handwritten: Quiz 5 will have emphasis on this section]

| objective: | *Define a sensation and describe the conditions necessary for a sensation to occur.* |

Consider what would happen if you could not feel the pain of a hot pot handle or an inflamed appendix, or if you could not see, hear, smell, taste, or maintain your balance. In short, if you could not "sense" your environment and make the necessary homeostatic adjustments, you could not survive very well on your own.

Definition

Sensation refers to the awareness of external or internal conditions of the body. For a sensation to occur, four conditions must be satisfied.

[handwritten: - Really bright - sudden light / - chemical burn / - Stress or extreme heat]

1. A *stimulus,* or change in the environment, which is capable of activating certain sensory neurons must occur. *[handwritten: stimulus then actuates]*

2. A *receptor* or *sense organ* must pick up the stimulus and convert it to a nerve impulse (nerve action potential) by way of a *generator potential* (described shortly). A receptor or sense organ is specialized nervous tissue that is extremely sensitive to internal or external stimuli. *[handwritten: Sensory / Motor side]*

3. The nerve impulse must be *conducted* along a neural pathway from the receptor or sense organ to the brain.

4. A region of the brain must receive and *integrate* the nerve impulse into a sensation.

[handwritten left margin: Changes / Change — / depolarization / afferent / sensory / must reach / brain or its / not a sensation]

A stimulus received by a receptor may be light, heat, pressure, mechanical energy, or chemical energy. When an adequate stimulus is applied to a receptor, it responds by altering its membrane's permeability to small ions. This results in a change in the resting membrane potential called a *generator potential.* When the generator potential reaches the threshold level, it initiates a nerve impulse that is transmitted along the nerve fiber. The function of a generator potential is to convert a stimulus into a nerve impulse. *[handwritten: electromagnetic waves / magnetic / vibration / smell + taste]*

Receptors vary in their complexity. The simplest are free dendrite endings in the skin (for example, pain receptors). Others are housed in complex sense organs such as the eye. Regardless of complexity, all sense receptors contain the dendrites of sensory neurons, either alone or in close association with specialized cells of other tissues. *[handwritten: Nerve / energy]*

A receptor converts a stimulus into a nerve impulse, and only after that impulse has been conducted to a region of the spinal cord or brain can it be integrated into a sensation. The nature of the sensation and the type of reaction generated vary with the level of the central nervous system at which the sensation is integrated.

Sensory fibers terminating in the spinal cord generate spinal reflexes without action by the brain. Sensory fibers terminating in the lower brain stem bring about more complex motor reactions. At the lower brain stem, they cause subconscious motor reactions. Sensory impulses that reach the thalamus are localized crudely in the body and sorted by specific sensations such as touch, pressure, pain, position, hearing, or taste. When sensory information reaches the cerebral cortex, we experience precise localization. It is at this level that memories of previous sensory information are stored and the perception of sensation occurs on the basis of past experience.

Characteristics

Conscious sensations and perceptions occur in the cerebral cortex. In other words, you see, hear, and feel in the brain. You seem to see with your eyes and feel pain in an injured part of your body, but that is because the cortex interprets the sensation as coming from the stimulated sense receptor. *Projection* is the name of the process by which the brain refers sensations to their point of stimulation. *[handwritten: Back]*

[handwritten: Know] A second characteristic of many sensations is *adaptation,* which is a decrease in sensitivity to continued stimuli. In fact, the perception of a sensation may actually disappear, even though the stimulus is still being applied. For example, when you first get into a tub of hot water, you feel a burning sensation. But soon the sensation becomes one of comfortable warmth, even though the stimulus (hot water) is still present, and in time, even the sensation of warmth disappears. *[handwritten: constantly wearing a watch / noises / smell]*

Sensations are also characterized by *afterimages*; that is, some sensations persist even though the stimulus has been removed. This is the reverse of adaptation. When you look at a bright light, then look away or close your eyes and still see the light for several seconds afterward, you are experiencing afterimage. *[handwritten: Sound / flash bulb / sea legs]*

Although all nerve impulses are the same, one sensation can be distinguished from another, such as sights from sounds. *Modality* refers to that specific characteristic of each sensation which allows it to be distinguished from other types. Accordingly, nerve impulses generated in the eyes are interpreted by the occipital lobe as sight, whereas those from the ears are interpreted by the temporal lobes as hearing.

Classification of Receptors

Receptors are classified on the basis of location, stimulus detected, and simplicity or complexity. Exhibit 12.1 presents a straightforward classification system; please refer to this exhibit now.

see handout

Exhibit 12.1   Classification of Receptors

A. Location

1. **Exteroceptors** (eks′-ter-ō-SEP-tors). Located at or near surface of body; provide information about *external* environment; transmit sensations of hearing, sight, smell, taste, touch, pressure, temperature, and pain.

2. **Interoceptors** (in-ter-ō-SEP-tors). Located in blood vessels and viscera; provide information about *internal* environment; sensations such as pain, pressure, fatigue, hunger, thirst, and nausea arise from within the body.

3. **Proprioceptors** (prō′-prē-ō-SEP-tors). Located in muscles, tendons, joints, and the internal ear; provide information about body position and movement; transmit information related to muscle tension, position and tension of joints, and equilibrium.

B. Stimulus detected

1. **Mechanoreceptors.** Detect pressure or stretching; stimuli so detected are related to touch, pressure, proprioception, hearing, equilibrium, and blood pressure.

2. **Thermoreceptors.** Detect changes in temperature.

3. **Nociceptors** (nō′-sē-SEP-tors). Detect pain, usually as a result of physical or chemical damage to tissues.

4. **Photoreceptors.** Detect light in retina of eye.

5. **Chemoreceptors.** Detect taste in mouth, smell in nose, and chemicals such as oxygen, carbon dioxide, water, and glucose in body fluids.

C. Simplicity or complexity

1. **Simple receptors.** Simple structures and neural pathways that are associated with general senses (touch, pressure, heat, cold, and pain).

2. **Complex receptors.** Complex structures and neural pathways that are associated with special senses (smell, taste, sight, hearing, and equilibrium).

Know 4 Quizs

General Senses

also known as

Cutaneous Sensations = Skin

objective: *List and describe the cutaneous sensations.*

Cutaneous (kyoo-TĀ-nē-us; cuta = skin) *sensations* include tactile sensations (touch and pressure), thermal sensations (cold and heat), and pain. Receptors for these sensations are located in the skin, connective tissues under the skin, and mucous membranes in the mouth and anus.

Cutaneous receptors are distributed over the body surface in such a way that some areas are densely populated with receptors and are consequently very sensitive (tip of the

tongue, lips, fingertips, palms, soles, nipples, and external genitals), while other areas contain only a few and are less sensitive (back of the hand and back of the neck).

Cutaneous receptors consist of the dendrites of sensory neurons. Recall that cutaneous sensations are interpreted by the parietal lobes of the cerebral cortex.

Tactile Sensations

Even though the *tactile* (TAK-tĭl; *tact* = touch) *sensations* are divided into touch and pressure, they are both detected by mechanoreceptors.

TOUCH   *Touch sensations* come from stimulation of tactile receptors in the skin or tissues immediately beneath the skin. *Crude touch* refers to the perception that something has touched the skin, though its exact location, shape, size, or texture cannot be determined. *Discriminative touch* refers to the ability to recognize exactly what point of the body is touched.

Tactile receptors for touch include hair root plexuses, type I cutaneous mechanoreceptors, and corpuscles of touch (Figure 12.1). *Hair root plexuses* are around the roots of hairs. They detect movements mainly on the surface of the body when hairs are disturbed. *more crude*

Type I cutaneous mechanoreceptors, also called *tactile* or *Merkel* (MER-kel) *discs*, make contact with the stratum basale. They function in discriminative touch. *Corpuscles of touch*, or *Meissner's* (MĪS-nerz) *corpuscles*, are egg-shaped receptors that are located in the dermal papillae of the skin, especially in the fingertips, palms, and soles. They are also abundant in the eyelids, tip of the tongue, lips, nipples, clitoris, and tip of the penis. Corpuscles of touch function in discriminative touch. *most superficial* / *more discrim*

PRESSURE   *Pressure sensations* generally occur in deeper tissues. In comparison with touch, pressure lasts longer, varies less in intensity, and is felt over a larger area. Pressure is really sustained touch. *adaptation takes longer slower response*

Pressure receptors are type II cutaneous mechanoreceptors and lamellated corpuscles. *Type II cutaneous mechanoreceptors,* or *end organs of Ruffini*, are embedded deep in the dermis and in deeper tissues of the body. They detect heavy and continuous touch. *Lamellated,* or *Pacinian* (pa-SIN-ē-an), *corpuscles* (Figure 12.1) appear layered like an onion. They are located in subcutaneous tissues and around joints, tendons, and muscles. *actually subcutaneous*

Thermal Sensations = also are free nerve endings — serve both temp + nerve impulses

The *thermal* (*therm* = heat) *sensations* are heat and cold. *Thermoreceptors* are free nerve endings. Thermoreceptors can detect temperatures from as low as 10°C (50°F) to as high as 45°C (113°F). Below 10°C, nerve impulses cannot be generated; for this reason cold is an excellent anesthetic. Also, below 10°C, pain receptors rather than cold receptors are stimulated, which produces the sensation of freezing. Likewise, above 45°C, pain receptors rather than thermoreceptors are stimulated, producing the sensation of burning. Because freezing and burning feel almost alike, very cold stimuli may be reported as hot and vice versa. Thermoreceptors tend to adapt to continu- *why use hydro therapy*

→ depolarization is not happening

Figure 12.1   Structure and location of cutaneous receptors.

Cutaneous sensations include tactile sensations (touch and pressure), thermal sensations (hot and cold), and pain.

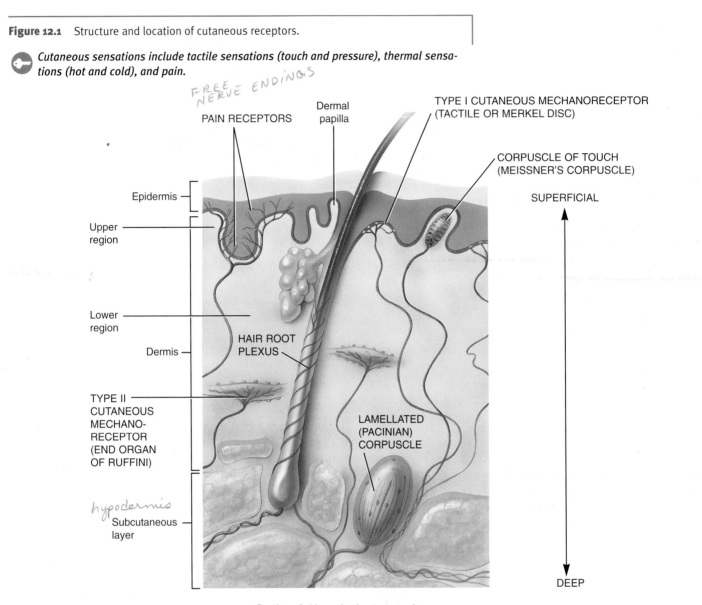

Section of skin and subcutaneous layer

Which receptors are especially abundant in the fingertips, palms, and soles?

ous stimulation, which can easily lead to tissue injuries due to burns or frostbite.

Pain Sensations

The ability to perceive pain is indispensable for a normal life, providing us with information about tissue-damaging stimuli and thus often enabling us to protect ourselves from greater damage. Pain initiates our search for medical assistance, and our description and indication of the location of the pain may help pinpoint the underlying cause of disease.

The receptors for *pain*, called *nociceptors* (nō-sē-SEP-tors; *noci* = harmful), are free nerve endings (Figure 12.1). Pain receptors are found in practically every tissue of the body and they respond to any type of stimulus. When stimuli for other sensations reach a certain threshold, they stimulate pain receptors as well. Excessive stimulation of a sense organ causes pain.

Other stimuli include excessive stretching of a structure, prolonged muscular contractions, inadequate blood flow to an organ, or the presence of certain chemical substances.

During tissue irritation or injury, chemicals such as prostaglandins are released, which stimulate nociceptors. This explains why pain persists even after the initial trauma occurs because nociceptors adapt only slightly or not at all to the presence of these substances, which are only slowly removed from the tissues following an injury. If there were adaptation to pain, it would cease to be sensed and irreparable damage could result.

Recognition of the kind and intensity of most pain occurs in the cerebral cortex. In most instances of somatic pain, the cortex projects the pain back to the stimulated area. If you burn your finger, you feel the pain in your finger. In most instances of visceral pain, however, the sensation is not projected back to the point of stimulation. Rather, the pain is felt in the skin overlying the stimulated organ or in a surface area

See pg 150?
Anatomy coloring Book on Dermatone

Triggerpoints
explained
Know

far from the stimulated organ. This phenomenon is called **referred pain.** It occurs because the area to which the pain is referred and the visceral organ involved are innervated by the same segment of the spinal cord. For example, sensory fibers from the heart as well as from the skin over the heart and left upper limb enter spinal cord segments T1 to T5. Thus the pain of a heart attack is typically felt in the skin over the heart and along the left arm.

A kind of pain frequently experienced by patients who have had a limb amputated is called **phantom pain.** They still experience sensations such as itching, pressure, tingling, or pain in the limb as if the limb were still there. An explanation for this phenomenon is that the remaining proximal portions of the sensory nerves that previously received impulses from the limb are being stimulated by the trauma of the amputation. Stimuli from these nerves are interpreted by the brain as coming from the nonexistent (phantom) limb.

Pain may be controlled by interfering with nerve impulse transmission to the cerebrum. This may be done by a variety of drugs, surgery, acupuncture, hypnosis, relaxation, massage, biofeedback, and electrical stimulation directly to the affected nerves or via the skin over those nerve pathways.

w/ touch you overload pain message so pain transmission is inhibited by touch message to the brain & releasing endorphins to inhibit Substance P

Proprioceptive Sensations

objective: *Define proprioception and describe the structure of proprioceptive receptors.*

An awareness of the activities of muscles, tendons, and joints and of equilibrium (balance) is provided by the **proprioceptive** (*proprio* = one's own), or **kinesthetic** (kin′-es-THET-ik; *kinesis* = motion), *sense*. It informs us of the degree to which muscles are contracted, the amount of tension in tendons, the change of position of joints, and the position of the head relative to the ground and in response to movements. Proprioception tells us the location and rate of movement of one body part in relation to others, so we can walk, type, or dress without using our eyes. It also allows us to estimate weight and determine the muscular work necessary to perform a task. For example, it takes less muscular work to lift a bag of feathers than it does to lift a bag of the same size that contains stones.

Most proprioceptors adapt only slightly. This feature is advantageous because the brain must be apprised of the status of different parts of the body at all times so that adjustments can be made to ensure coordination.

The sensory pathway for muscle sense consists of impulses generated by proprioceptors via cranial and spinal nerves to the central nervous system. Impulses for conscious proprioception pass along sensory tracts in the cord and are relayed to the primary somatosensory area in the parietal lobe of the cerebral cortex posterior (postcentral gyrus). Proprioceptive impulses also pass to the cerebellum along spinocerebellar tracts and contribute to subconscious proprioception.

Receptors = *mechanisms —* KNOW!

Proprioceptive receptors are located in skeletal muscles, tendons in and around synovial joints, and the internal ear.

Sense stretch & tension

MUSCLE SPINDLES *Muscle spindles* are delicate proprioceptive receptors between skeletal muscle fibers. When a muscle is stretched, the spindle is stretched and it sends an impulse to the CNS, indicating how much and how fast the muscle is changing its length. Within the CNS, the information is integrated to coordinate muscle activity.

receptors

TENDON ORGANS *Tendon organs* (*Golgi tendon organs*) are found at the junction of a tendon with a muscle. They protect tendons and their associated muscles from damage due to excessive tension. When tension is applied to a tendon, tendon organs relay the information to the central nervous system.

JOINT KINESTHETIC RECEPTORS *Joint kinesthetic receptors* are found in and around synovial joints. They respond to pressure, acceleration and deceleration, and excessive strain on a joint. *+ speed or stretch*

MACULAE AND CRISTAE The proprioceptors in the internal ear are the maculae and the cristae. Their function in equilibrium is discussed later in the chapter.

Special Senses

The special senses—smell, taste, sight, hearing, and equilibrium—have receptor organs that are structurally more complex than receptors for general sensations. The sense of smell is the least specialized, and the sense of sight, the most. Like the general senses, the special senses allow us to detect changes in our environment.

Olfactory Sensations

Chemoreceptor *cerebral cortex area*

objective: *Describe the receptors for olfaction and the olfactory pathway to the brain.*

Structure of Receptors

The receptors for the **olfactory** (ol-FAK-tō-rē; *olfact* = smell) *sense* or sense of smell are located in the superior portion of the nasal cavity (Figure 12.2). The **olfactory receptors** are neurons that contain a knob-shaped dendrite at one end. Several cilia called **olfactory hairs** project from the dendrite. These hairs react to odors in the air and then stimulate the olfactory receptors. The opposite end of each olfactory receptor contains an axon. Within the connective tissue beneath the olfactory epithelium are glands that produce mucus. This mucus moistens the surface of the olfactory epithelium and serves as a solvent for odoriferous substances. The continuous secretion of mucus also serves to freshen the surface film of fluid and prevents continuous stimulation of olfactory hairs by the same odor.

doesn't smell —(no gas) — must have 8 molecules to "smell" it. *Cold food sometimes*

Stimulation of Receptors

For a substance to be smelled, it must be capable of becoming a gas so that the gaseous particles can enter the nostrils. Also, the substance must be water-soluble so that it can dissolve in the nasal mucus to make contact with olfactory receptors.

Figure 12.2 Olfactory receptors. (a) Location in nasal cavity. (b) Details.

🔑 *Olfactory receptors are neurons whose one end contains olfactory hairs that project from a knob-shaped dendrite.*

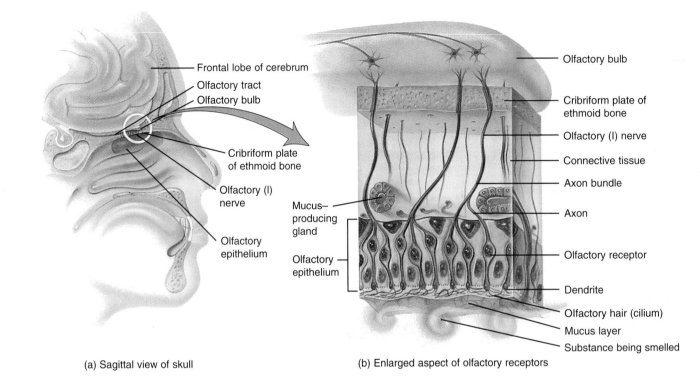

(a) Sagittal view of skull

(b) Enlarged aspect of olfactory receptors

🔍 *Impulses from olfactory bulbs pass to which structures?*

Finally, the substance must be lipid-soluble to pass through the plasma membranes of olfactory hairs and initiate an impulse. Because adaptation to odors is rapid, the situation could be potentially dangerous. *also receptors — if damaged — can regenerate (sometimes not fully) & do so about every month*

Olfactory Pathway

The axons of the olfactory receptors unite to form the *olfactory (I) nerves,* which pass through holes in the cribriform plate of the ethmoid bone (Figure 12.2). The olfactory (I) nerves convey the impulses to *olfactory bulbs,* which lie beneath the frontal lobes of the cerebrum, and then to the *olfactory tract,* which conveys the impulses to the primary olfactory area of the cerebral cortex in the temporal lobe of the cerebrum. In the cerebral cortex, the impulses are interpreted as odor and give rise to the sensation of smell.

Don't need to know

Gustatory Sensations

objective: *Describe the receptors for gustation and the gustatory pathway to the brain.*

Chemoreceptor

Structure of Receptors

The receptors for *gustatory* (GUS-ta-tō′-rē; *gust* = taste) *sensations,* or sensations of taste, are located in taste buds

(Figure 12.3). *Taste buds* contain *gustatory receptors.* Each receptor has a *gustatory hair* (microvillus) that projects to the external surface through an opening in the taste bud called the *taste pore.*

Taste buds are found in elevations on the tongue called *papillae* (pa-PIL-ē). The papillae give the upper surface of the tongue its rough appearance (Figure 12.3a,b). *Circumvallate* (ser-kum-VAL-āt) *papillae* are at the posterior portion of the tongue. *Fungiform* (FUN-ji-form; meaning mushroom-shaped) *papillae* are scattered over the entire surface of the tongue. All circumvallate and most fungiform papillae contain taste buds. *Filiform* (FIL-i-form) *papillae* are also distributed over the entire surface of the tongue. They rarely contain taste buds.

Stimulation of Receptors

For gustatory receptors to be stimulated, substances must be dissolved in saliva so they can enter taste pores. Once the taste substance makes contact with plasma membranes of the gustatory hairs, a generator potential is developed. Then the generator potential initiates a nerve impulse.

There are four primary taste sensations: sour, salt, bitter, and sweet. All other "tastes," such as chocolate, pepper, and coffee, are combinations of these four that are modified by accompanying olfactory sensations.

Persons with colds or allergies sometimes complain that they cannot taste their food. Actually, their taste sensations are

almost no food has "one" taste

Figure 12.3   Gustatory receptors.

🔑 *Gustatory receptors are found in taste buds.*

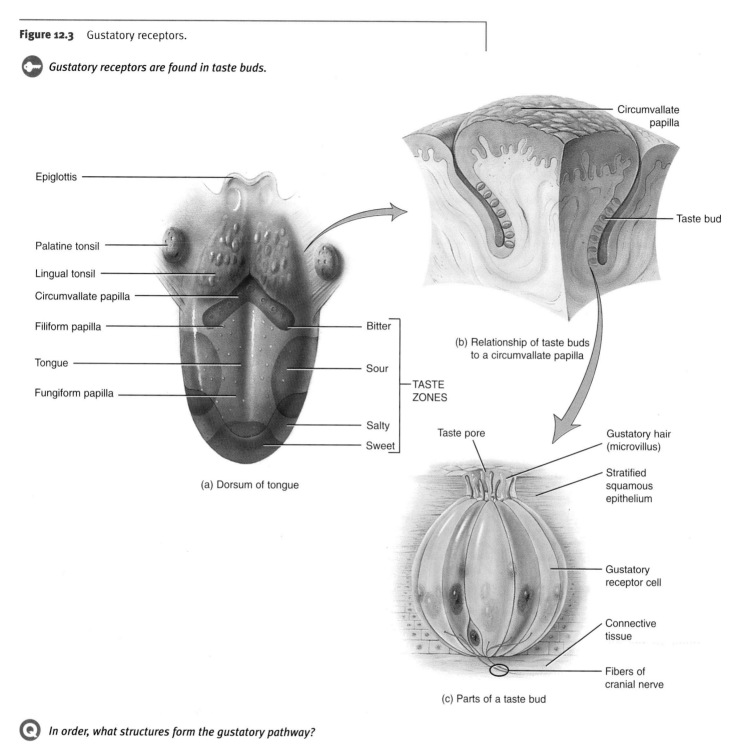

Circumvallate papilla

Epiglottis

Palatine tonsil

Lingual tonsil

Circumvallate papilla

Filiform papilla — Bitter

Tongue — Sour

Fungiform papilla — TASTE ZONES

Salty

Sweet

Taste bud

(b) Relationship of taste buds to a circumvallate papilla

(a) Dorsum of tongue

Taste pore

Gustatory hair (microvillus)

Stratified squamous epithelium

Gustatory receptor cell

Connective tissue

Fibers of cranial nerve

(c) Parts of a taste bud

❓ *In order, what structures form the gustatory pathway?*

probably operating normally, but their olfactory sensations are not. Much of what we think of as taste is actually smell because odors from foods pass upward to stimulate the olfactory system. In fact, a given concentration of a substance stimulates the olfactory system thousands of times more than it stimulates the gustatory system.

Certain regions of the tongue react more strongly to particular primary tastes than others. The tip of the tongue reacts to all four primary taste sensations but is highly sensitive to sweet and salty substances. The back portion is highly sensitive to bitter substances, and the lateral edges are predominantly sensitive to sour substances (Figure 12.3a).

Gustatory Pathway

Taste impulses are conveyed from the gustatory receptors in taste buds along the facial (VII), glossopharyngeal (IX), and vagus (X) cranial nerves to the medulla oblongata and then to the thalamus. They terminate in the primary gustatory area in the parietal lobe of the cerebral cortex.

Visual Sensations *most specialized of all special senses*

objective: *Describe the receptors for visual sensations and the visual pathway to the brain.*

The study of the structure, function, and diseases of the eye is known as **ophthalmology** (of'-thal-MOL-ō-jē; *opthalmo* = eye; *logos* = study of). A physician who specializes in the diagnosis and treatment of eye disorders with drugs, surgery, and corrective lenses is known as an **ophthalmologist,** whereas an **optometrist** has a doctorate in optometry and is licensed to test the eyes and treat visual defects by prescribing corrective lenses. An **optician** is a technician who fits, adjusts, and dispenses corrective lenses on prescription of an ophthalmologist or optometrist.

Vision involves the eyeballs, the optic (II) nerves, the occipital lobes of the cerebrum, and a number of accessory structures.

Accessory Structures of the Eye

The **accessory structures** of the eye are the eyebrows, eyelids, eyelashes, and the lacrimal (tearing) apparatus (Figure 12.4). The **eyebrows** help protect the eyeballs from foreign objects, perspiration, and direct rays of the sun. The upper and lower **eyelids** shade the eyes during sleep, protect the eyes from excessive light and foreign objects, and spread lubricating secretions over the eyeballs (by blinking).

The **lacrimal** (*lacrima* = tear) **apparatus** refers to the glands, ducts, canals, and sacs that manufacture and drain tears. A **lacrimal gland** produces tears. Each is about the size and shape of an almond and has ducts that empty tears onto the surface of the upper lid. Tears then pass over the surface of the eyeball toward the nose into two **lacrimal canals** and a **nasolacrimal duct,** which causes the tears to drain into the lower portion of the nasal cavity.

Tears are a watery solution containing salts, some mucus, and a bactericidal enzyme called **lysozyme.** Tears clean, lubricate, and moisten the surface of the eyeball exposed to air to prevent drying out. Normally, tears are cleared away by evaporation or by passing into the nasal cavities as fast as they are produced. If, however, an irritating substance makes contact with the eye, the lacrimal glands are stimulated to oversecrete and tears accumulate. This is a protective mechanism because the tears dilute and wash away the irritant. Humans are unique in their ability to cry to express certain emotions such as happiness and sadness. In response to parasympathetic stimulation, the lacrimal glands produce excessive tears that may spill over the edges of the eyelids and even fill the nasal cavity with fluid.

Structure of the Eyeball

The adult **eyeball** measures about 2.5 cm (1 inch) in diameter and is divided into three layers: fibrous tunic, vascular tunic, and retina or nervous tunic (Figure 12.5a).

Figure 12.4   Accessory structures of the eye.

Accessory structures of the eye are the eyelids, eyelashes, eyebrows, and lacrimal apparatus.

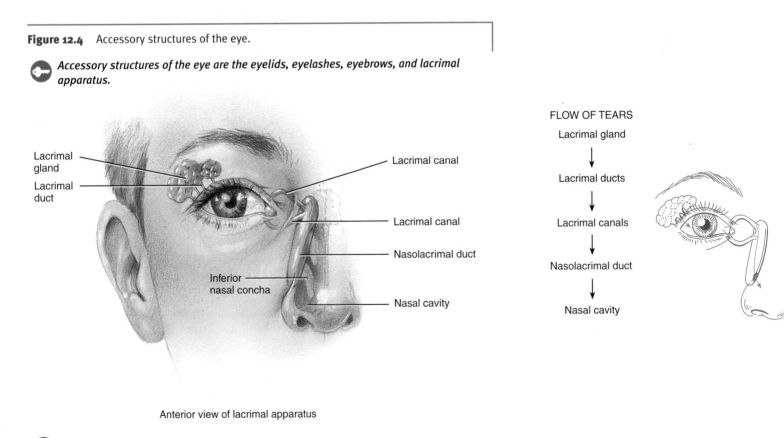

Lacrimal gland
Lacrimal duct
Lacrimal canal
Lacrimal canal
Nasolacrimal duct
Inferior nasal concha
Nasal cavity

Anterior view of lacrimal apparatus

FLOW OF TEARS

Lacrimal gland
↓
Lacrimal ducts
↓
Lacrimal canals
↓
Nasolacrimal duct
↓
Nasal cavity

Q *What substances are in tears and what are their functions?*

Fibrous Tunic

The *fibrous tunic* is the outer coat of the eyeball consisting of an anterior cornea and a posterior sclera. The ***cornea*** (KOR-nē-a) is a nonvascular, transparent fibrous coat that covers the colored part of the eyeball, the iris. The cornea's outer surface is covered by an epithelial layer called the ***conjunctiva,*** which also lines the eyelid. The ***sclera*** (SKLE-ra; *skleros* = hard), the "white" of the eye, is a coat of dense connective tissue that covers all of the eyeball except the cornea. The sclera gives shape to the eyeball, makes it more rigid, and protects its inner parts.

The cornea bends light rays entering the eyeball in order to produce a clear image. If the cornea is not curved properly, the image is not focused on the area of sharpest vision on the retina and blurred vision occurs. A defective cornea can be removed and replaced with a donor cornea of similar diameter, a procedure called a ***corneal transplant.*** Corneal transplants are considered the most successful type of transplantation because corneas do not contain blood vessels and the body is less likely to reject them.

Vascular Tunic

The ***vascular tunic*** is the middle layer of the eyeball and is composed of the choroid, ciliary body, and iris. The ***choroid*** (KŌ-royd) is a thin, dark brown membrane that lines most of

---

Figure 12.5   Structure of the eyeball and responses of the pupil to light. The arrow in the inset in (a) indicates the direction from which the eyeball is viewed (superior).

🔑 *The wall of the eyeball consists of the fibrous tunic, vascular tunic, and retina (nervous tunic).*

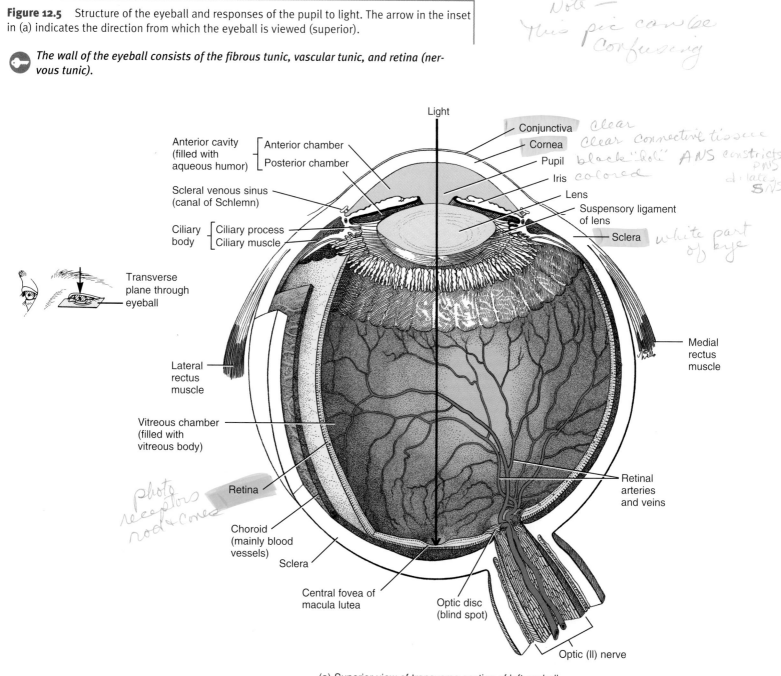

(a) Superior view of transverse section of left eyeball

Figure 12.5 (*Continued*)

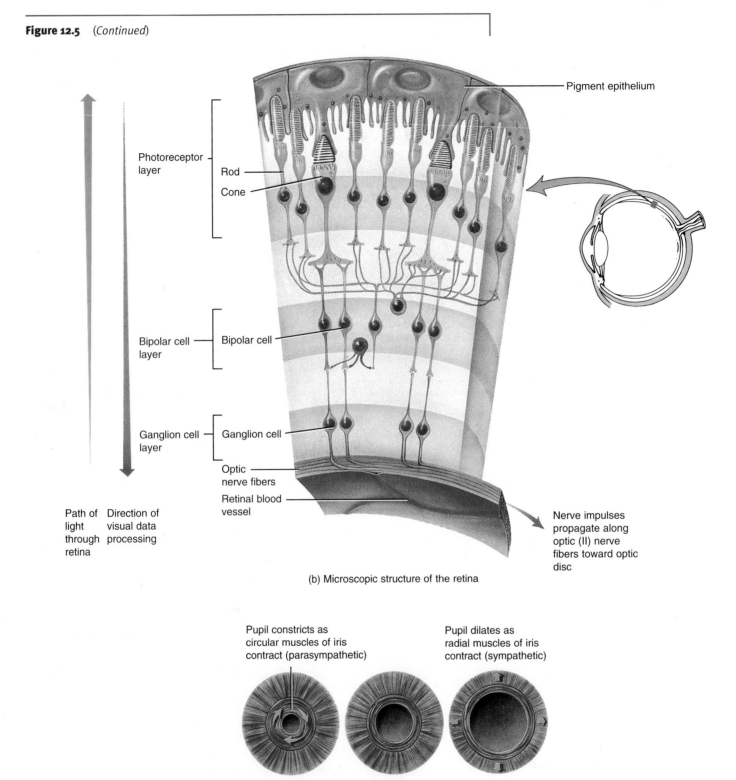

Pigment epithelium

Photoreceptor layer

Rod

Cone

Bipolar cell layer — Bipolar cell

Ganglion cell layer — Ganglion cell

Optic nerve fibers

Retinal blood vessel

Path of light through retina

Direction of visual data processing

Nerve impulses propagate along optic (II) nerve fibers toward optic disc

(b) Microscopic structure of the retina

Pupil constricts as circular muscles of iris contract (parasympathetic)

Pupil dilates as radial muscles of iris contract (sympathetic)

Bright light

Normal light

Dim light

(c) Anterior view of responses of the pupil to varying brightness of light

Q *Which photoreceptor is specialized for vision in dim light and allows us to see shapes and movement? Which photoreceptor is specialized for color and visual acuity?*

the internal surface of the sclera. It contains numerous blood vessels and a large amount of pigment. The choroid absorbs light rays so they are not reflected within the eyeball and its blood supply nourishes the retina.

At the front of the eye, the choroid becomes the *ciliary* (SIL-ē-ar′-ē) *body.* The ciliary body consists of the *ciliary processes,* whose capillaries secrete a watery fluid called aqueous humor, and the *ciliary muscle,* a smooth muscle that alters the shape of the lens for near or far vision.

The *iris* (*irid* = colored circle) is the doughnut-shaped colored portion of the eyeball. It consists of circular and radial smooth muscle fibers. The hole in the center of the iris is the *pupil,* through which light enters the eyeball. The iris regulates the amount of light entering the eyeball. When the eye is stimulated by bright light, the circular muscles of the iris contract and decrease the size of the pupil, called constriction. When the eye must adjust to dim light, the radial muscles contract and the pupil increases in size (dilates). See Figure 12.5c. These muscles are controlled by the autonomic nervous system (see Exhibit 11.3).

Retina (Nervous Tunic)

The third and inner coat of the eye, the *retina* (*nervous tunic*), lies only in the posterior three-quarters portion of the eyeball. Its primary function is image formation. The retina is one of the few places in the body where blood vessels can be seen directly. With a special reflecting light called an *ophthalmoscope,* it is possible to examine the retina and detect vascular changes associated with hypertension, atherosclerosis, and diabetes. The retina consists of an inner nervous tissue layer (visual portion) and an outer pigmented layer (nonvisual portion).

The nervous tissue layer of the retina contains three layers of neurons, a *photoreceptor cell layer,* a *bipolar cell layer,* and a *ganglion cell layer* (Figure 12.5b). The photoreceptors are of two types called rods and cones because of their shapes. They are visual receptors highly specialized for stimulation by light rays. Functionally, rods and cones develop generator potentials. *Rods* are specialized for vision in dim light. They also allow us to discriminate between different shades of dark and light and to see shapes and movement. *Cones* are specialized for color vision and sharpness of vision (*visual acuity*). They are stimulated only by bright light, which is why we cannot see color by moonlight.

There are about 6 million cones and 120 million rods. Cones are most densely concentrated in the *central fovea,* a small depression in the center of the macula lutea. The *macula lutea* (MAK-yoo-la LOO-tē-a), or yellow spot, is in the exact center of the retina. The fovea is the area of sharpest vision because of its high concentration of cones. Rods are absent from the fovea and macula and increase in density toward the periphery of the retina.

When information has passed through the photoreceptor neurons, it is conducted to the bipolar cells, then to the ganglion cells. The axons of the ganglion cells extend posteriorly to a small area of the retina called the *optic disc* (*blind spot*), where they all exit as the optic (II) nerve. The optic disc is called the blind spot because it contains no photoreceptors. The optic nerve transmits visual impulses to the occipital lobe of the cerebral cortex for interpretation as sight (see Chapter 10).

A frequently encountered problem related to the retina is a *detached retina,* which may occur in trauma, such as a blow to the head. The actual detachment occurs between the nervous part of the retina and the underlying pigmented layer. Fluid accumulates between these layers, resulting in distorted vision and blindness. Often, the retina may be surgically reattached.

Lens

The *lens* is a transparent structure that normally focuses light rays onto the retina. It is constructed of numerous layers of protein fibers. *Suspensory ligaments* attached to the lens hold it in position behind the pupil. A loss of transparency of the lens is known as a *cataract.*

Interior

The interior of the eyeball is a large space divided into two cavities by the lens, the anterior cavity and the vitreous chamber. The *anterior cavity* lies in front of the lens and is filled with *aqueous* (*aqua* = water) *humor,* a watery fluid, similar to cerebrospinal fluid. The fluid is secreted into the anterior cavity by blood capillaries of the ciliary processes. It then is drained off into the *scleral venous sinus* (*canal of Schlemm*), an opening where the sclera and cornea meet, and reenters the blood.

The pressure in the eye, called *intraocular pressure* (*IOP*), is produced mainly by the aqueous humor. The intraocular pressure, along with the vitreous body (described shortly), maintains the shape of the eyeball and keeps the retina smoothly attached to the choroid so the retina will be well nourished and form clear images. Normal intraocular pressure (about 16 mm Hg) is maintained by drainage of the aqueous humor as described above. Besides maintaining intraocular pressure, the aqueous humor also helps nourish the lens and cornea because neither has blood vessels.

The second, and larger, cavity of the eyeball is the *vitreous chamber.* It lies behind the lens and contains a clear jellylike substance called the *vitreous body.* This substance contributes to intraocular pressure, helps prevent the eyeball from collapsing, and holds the retina flush against the eyeball. The vitreous body, unlike the aqueous humor, does not undergo constant replacement. It is formed during embryonic life and is not replaced thereafter.

A summary of structures associated with the eyeball is presented in Exhibit 12.2.

Physiology of Vision

Read — but don't need to know all of it @ for Olery

| **objective:** | *Describe the mechanism involved in vision.* |

Before light can reach the rods and cones of the retina to result in image formation, it must pass through the cornea, aqueous humor, pupil, lens, and vitreous body. For vision to

Exhibit 12.2 | Summary of Structures Associated with the Eyeball

STRUCTURE	FUNCTION	STRUCTURE	FUNCTION
Fibrous tunic	*Cornea:* Admits and refracts (bends) light. *Sclera:* Provides shape and protects inner parts.	**Lens**	Refracts light.
Vascular tunic	*Iris:* Regulates amount of light that enters eyeball. ***CILIARY BODY: SECRETES AQUEOUS HUMOR AND ALTERS SHAPE OF LENS FOR NEAR OR FAR VISION (ACCOMMODATION).*** *Choroid:* Provides blood supply and absorbs scattered light.	**Anterior cavity**	Contains aqueous humor that helps maintain shape of eyeball.
Retina (nervous tunic)	Receives light and converts light into receptor potentials and nerve impulses. Output to brain is via the optic (II) nerve.	**Vitreous chamber**	Contains vitreous body that helps maintain shape of eyeball and keeps retina attached to choroid.

occur, light reaching the rods and cones must form an image on the retina.

Retinal Image Formation

The formation of an image on the retina requires four basic processes, all concerned with focusing light rays: (1) refraction of light rays, (2) accommodation (increase in curvature) of the lens, (3) constriction of the pupil, and (4) convergence of the eyes (movement of the eyeballs inward). Accommodation and pupil size are functions of the ciliary muscle and the muscles of the iris. They are termed *intrinsic eye muscles,* because they are inside the eyeball. Convergence is a function of the voluntary muscles attached to the outside of the eyeball called the *extrinsic eye muscles* (see Figure 8.16).

REFRACTION OF LIGHT RAYS   When light rays traveling through a transparent medium (such as air) pass into a second transpar-ent medium with a different density (such as water), they bend at the surface of the two media. This bending is called *refraction* (Figure 12.6a) and occurs at the cornea and the lens.

ACCOMMODATION OF THE LENS   If the surface of a lens curves outward, as in a convex lens, the lens will refract incoming rays toward each other so they eventually cross. The greater the curve, the more the light rays bend toward each other. The lens of the eye is convex on both its front and back surfaces (bicon-vex). It can change focusing power by becoming moderately curved at one moment and greatly curved the next. This increase in the curvature of the lens is called *accommodation* (Figure 12.6c).

In near vision, the ciliary muscle contracts, pulling the cil-iary process toward the lens, releasing the tension on the sus-pensory ligament and the lens. Because of its elasticity, the lens shortens, thickens, and bulges, becoming more convex and refracting more acutely. In far vision, the ciliary muscle is

Figure 12.6   Refraction of light rays and accommodation.

🔑 *Refraction is the bending of light rays.*

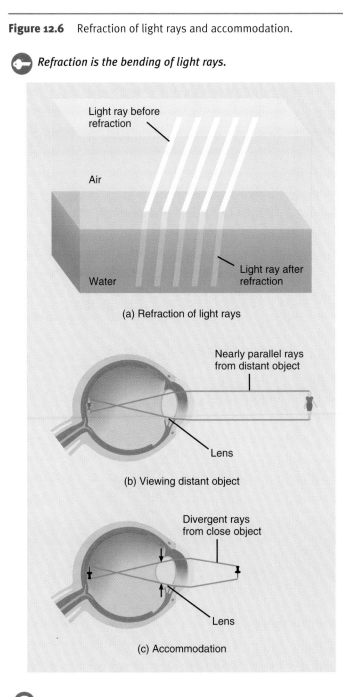

Light ray before refraction

Air

Water

Light ray after refraction

(a) Refraction of light rays

Nearly parallel rays from distant object

Lens

(b) Viewing distant object

Divergent rays from close object

Lens

(c) Accommodation

❓ *What changes occur during accommodation for near vision?*

relaxed and the lens becomes flatter. With aging, the lens loses elasticity and thus its ability to accommodate.

The normal eye, known as an *emmetropic* (em′-e-TROP-ik) *eye,* can sufficiently refract light rays from an object to focus a clear image on the retina (Figure 12.7a). Many individuals, however, do not have normal vision because of improper refraction. They may have *myopia* (mī-Ō-pē-a) or nearsightedness (an inability to clearly see distant objects), *hypermetropia* (hī′-per-mē-TRŌ-pē-a) or farsightedness (an inability to clearly see near objects), or *astigmatism* (a-STIG-ma-tizm), irregularities in the surface of the lens or cornea. Myopia and hypermetropia are illustrated and explained in

Figure 12.7b–e. The inability to focus on nearby objects due to loss of elasticity of the lens with age is called *presbyopia* (prez-bē-ŌP-ē-a).

CONSTRICTION OF PUPIL   Part of the accommodation mechanism involves the circular muscle fibers of the iris, which constrict the pupil. *Constriction of the pupil* means narrowing the diameter of the hole through which light enters the eye. This action occurs simultaneously with accommodation so that light rays do not enter the eye at the edge of the lens. Rays entering at the edge would not be brought to focus on the retina and would result in blurred vision.

CONVERGENCE   In humans, both eyes normally focus on only one set of objects—a characteristic called *single binocular vision.* With single binocular vision, when we stare straight ahead at a distant object, the incoming light rays are aimed directly at both pupils and are refracted to identical spots on the retinas of both eyes. But as we move closer to the object, our eyes must move toward the nose for the light rays to hit the same points on both retinas. The term *convergence* refers to this movement of the two eyeballs toward the nose so they are both directed to the object being viewed. The nearer the object, the greater the convergence required. Convergence is brought about automatically by the extrinsic eye muscles. Convergence and binocular vision are necessary for *depth perception* and for perceiving objects that have a three-dimensional appearance.

INVERTED IMAGE   Images are focused upside down on the retina. They also undergo mirror reversal; that is, light received from the right side of an object hits the left side of the retina and vice versa. Note in Figure 12.6 how the light from the top of the object hits the bottom of the central fovea and vice versa. We do not see an inverted world, however, because the brain learns early in life to coordinate visual images with the location of objects. The brain automatically turns visual images right-side-up and right-side-around.

Stimulation of Photoreceptors

EXCITATION OF RODS   After an image is formed on the retina by refraction, accommodation, constriction of the pupil, and convergence, light impulses must be converted into nerve impulses. The initial step is the development of generator potentials by rods and cones. To understand how this occurs, we first examine the role of photopigments.

A *photopigment* is a substance that can absorb light and undergo a change in structure to produce a generator potential. The photopigment in rods is called *rhodopsin* (*rhodo* = rose; *opsis* = vision) and is composed of a protein called *opsin* and a derivative of vitamin A called *retinal.*

Rhodopsin is a highly unstable compound in the presence of even very small amounts of light. Any amount of light in a darkened room will cause rhodopsin to split into opsin and retinal. This will trigger a generator potential that ultimately causes the depolarization of rods, thereby initiating visual impulses. In low-light situations, opsin and retinal will quickly recombine back into rhodopsin and its production will be able to keep pace with its rate of breakdown, thereby providing night vision. Rods usually are nonfunctional in daylight, because rhodopsin is bro-

Read ✓

Figure 12.7 Normal and abnormal refraction in the eyeball. (a) In the normal (emmetropic) eye, light rays from an object are bent sufficiently by the refracting media and converge on the central fovea. A clear image is formed. (b) In the nearsighted (myopic) eye, the image is focused in front of the retina. The condition may result from an elongated eyeball or thickened lens. (c) Correction is by use of a concave lens that causes entering light rays to diverge so that they have to travel further through the eyeball and are focused directly on the retina. (d) In the farsighted (hypermetropic) eye, the image is focused behind the retina. The condition results from a shortened eyeball or a thin lens. (e) Correction is by a convex lens that causes entering light rays to converge so that they focus directly on the retina.

🔑 *In myopia (nearsightedness), distant objects can't be seen clearly; in hypermetropia (farsightedness), near objects can't be seen clearly.*

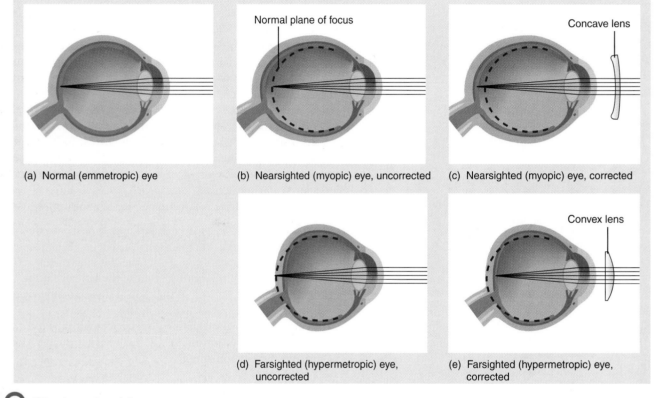

(a) Normal (emmetropic) eye

(b) Nearsighted (myopic) eye, uncorrected

(c) Nearsighted (myopic) eye, corrected

(d) Farsighted (hypermetropic) eye, uncorrected

(e) Farsighted (hypermetropic) eye, corrected

Normal plane of focus

Concave lens

Convex lens

Ⓠ *What is presbyopia?*

ken down faster than it can be re-formed. After going from bright sunlight into a dark room, it will take several minutes before the rods will function again. The period of adjustment is the time needed for the completely dissociated rhodopsin to re-form.

Night blindness is the lack of normal night vision following the adjustment period. It is most often caused by vitamin A deficiency. *rods are for Dim light*

EXCITATION OF CONES Cones are the receptors for brighter light and color. As in rods, photopigment breakdown produces the generator potential. The photopigment in cones also contains retinal, but the protein is different. Unlike rhodopsin, the photopigment of the cones requires bright light for breakdown and it re-forms quickly. There are three types of cones, each containing a different combination of retinal and protein. One type of cone responds best to red light, the second to green, and the third to blue. Just as an artist can obtain almost any color

by mixing colors, the cones can perceive any color by differential stimulation. When looking at an object, if all three types of cones are stimulated, the object will be perceived to be white in color; if none is stimulated, the object looks black.

If one of the types of cones is missing from the retina, an individual cannot distinguish some colors from others and is said to be *color-blind*. The most common type is *red–green color blindness* in which usually one photopigment is missing. Color blindness is an inherited condition that affects males far more frequently than females. The inheritance of the condition is discussed in Chapter 24 and illustrated in Figure 24.13.

Visual Pathway

Figure 12.5b shows that the retina (nervous tunic) is composed of three layers of neurons: photoreceptor (rods and cones), bipolar, and ganglion. For light to reach rods and cones and stimulate them to produce generator potentials, it must first pass through the ganglion and bipolar layers.

Once generator potentials are developed by rods and cones, the potentials produce signals in bipolar neurons. In response, bipolar neurons become excited and then transmit the excitatory visual signal on to ganglion cells. The ganglion cells become depolarized and initiate nerve impulses. The cell bodies of the ganglion cells lie in the retina, and their axons leave the eyeball as the *optic (II) nerve* (Figure 12.8) and pass through the *optic chiasm* (kī-AZM), a point where some fibers cross to the opposite side and others remain uncrossed. On passing through the optic chiasm, the fibers, now part of the *optic tract,* terminate in the thalamus. Here the fibers synapse with neurons whose axons pass to the primary visual areas in the occipital lobes of the cerebral cortex. Because of the crossing at the optic chiasm, the right side of the cortex interprets visual sensations from the left side of an object and the left side interprets visual sensations from the right side of an object.

Figure 12.8 Visual pathway.

🔑 *The optic chiasm is the crossing point of the optic nerves.*

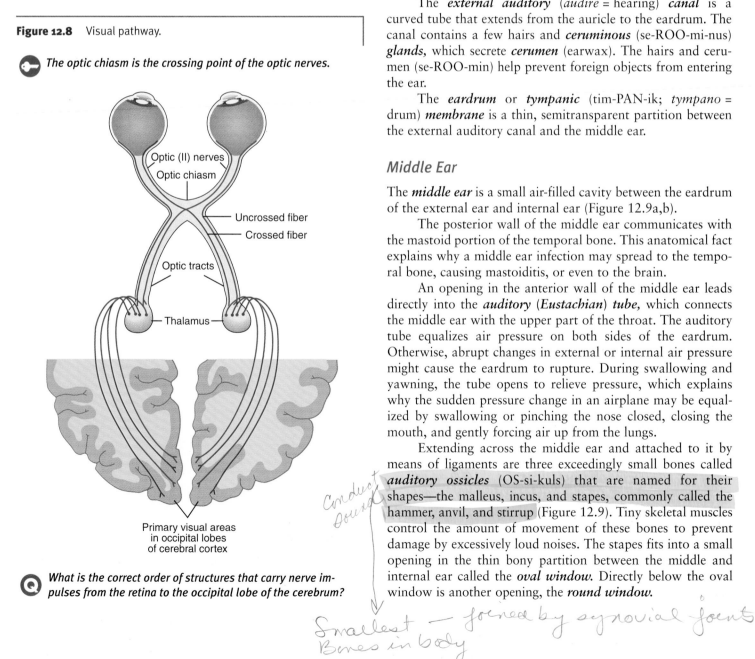

What is the correct order of structures that carry nerve impulses from the retina to the occipital lobe of the cerebrum?

Auditory Sensations and Equilibrium

| **objective:** | *Describe the receptors for hearing and equilibrium and their pathways to the brain.* |

In addition to containing receptors for sound waves, the ear also contains receptors for equilibrium (balance). The ear is divided into three principal regions: external ear, middle ear, and internal ear.

External Ear

The *external ear* is designed to collect sound waves and pass them inward (Figure 12.9a). It consists of an auricle, external auditory canal, and eardrum.

The *auricle* is a flap of elastic cartilage shaped like the flared end of a trumpet and attached to the head by ligaments and muscles. It directs sound waves into the external auditory canal.

The *external auditory* (*audire* = hearing) *canal* is a curved tube that extends from the auricle to the eardrum. The canal contains a few hairs and *ceruminous* (se-ROO-mi-nus) *glands,* which secrete *cerumen* (earwax). The hairs and cerumen (se-ROO-min) help prevent foreign objects from entering the ear.

The *eardrum* or *tympanic* (tim-PAN-ik; *tympano* = drum) *membrane* is a thin, semitransparent partition between the external auditory canal and the middle ear.

Middle Ear

The *middle ear* is a small air-filled cavity between the eardrum of the external ear and internal ear (Figure 12.9a,b).

The posterior wall of the middle ear communicates with the mastoid portion of the temporal bone. This anatomical fact explains why a middle ear infection may spread to the temporal bone, causing mastoiditis, or even to the brain.

An opening in the anterior wall of the middle ear leads directly into the *auditory* (*Eustachian*) *tube,* which connects the middle ear with the upper part of the throat. The auditory tube equalizes air pressure on both sides of the eardrum. Otherwise, abrupt changes in external or internal air pressure might cause the eardrum to rupture. During swallowing and yawning, the tube opens to relieve pressure, which explains why the sudden pressure change in an airplane may be equalized by swallowing or pinching the nose closed, closing the mouth, and gently forcing air up from the lungs.

Extending across the middle ear and attached to it by means of ligaments are three exceedingly small bones called *auditory ossicles* (OS-si-kuls) that are named for their shapes—the malleus, incus, and stapes, commonly called the hammer, anvil, and stirrup (Figure 12.9). Tiny skeletal muscles control the amount of movement of these bones to prevent damage by excessively loud noises. The stapes fits into a small opening in the thin bony partition between the middle and internal ear called the *oval window.* Directly below the oval window is another opening, the *round window.*

Figure 12.9   Structure of the auditory apparatus, showing the division of the right ear into external, middle, and internal portions.

🔑 *The ear is divided into three principal regions: external (outer), middle, and internal (inner).*

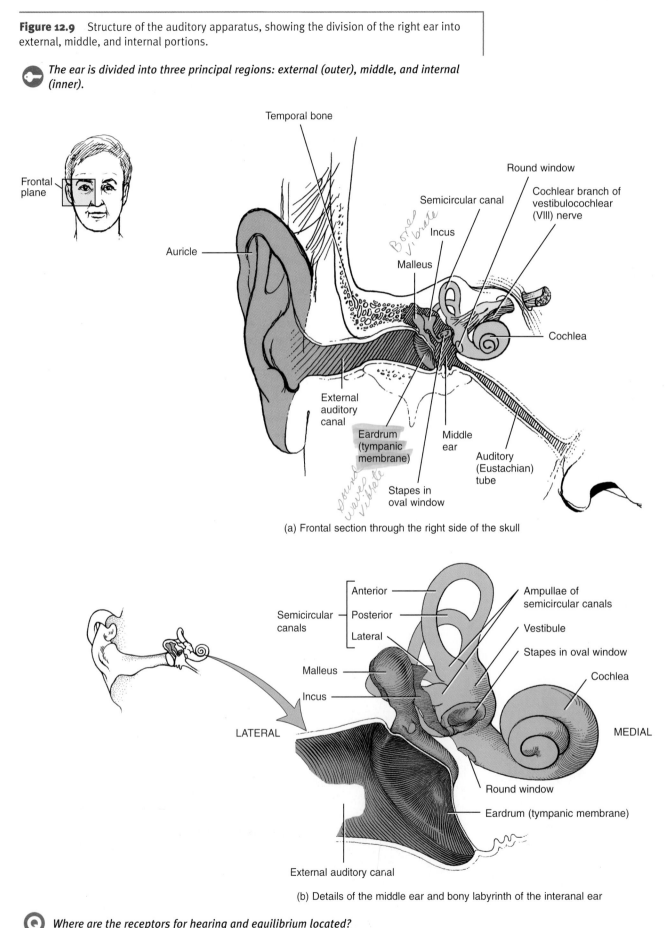

(a) Frontal section through the right side of the skull

(b) Details of the middle ear and bony labyrinth of the interanal ear

Ⓠ *Where are the receptors for hearing and equilibrium located?*

Internal Ear

The ***internal ear*** is divided into the outer bony labyrinth and inner membranous labyrinth, which lies within the bony labyrinth (Figure 12.10). The ***bony labyrinth*** (LAB-i-rinth) is a series of cavities in the temporal bone. It can be divided into three areas named on the basis of shape—vestibule, cochlea, and semicircular canals. The bony labyrinth contains a fluid called ***perilymph.*** This fluid surrounds the inner ***membranous labyrinth,*** a series of sacs and tubes having the same general form as the bony labyrinth. The membranous labyrinth contains a fluid called ***endolymph.***

The ***vestibule*** is the middle part of the bony labyrinth. The membranous labyrinth in the vestibule consists of two sacs called the ***utricle*** (YOO-tri-kul = little bag) and ***saccule*** (SAK-yool = little sac). Behind the vestibule are the three bony semicircular canals. The anterior and posterior semicircular

canals are both vertical but in different planes, whereas the lateral one is horizontal. One end of each canal enlarges into a swelling called the ***ampulla*** (am-POOL-la; *ampulla* = little jar). Inside the bony semicircular canals are the ***semicircular ducts,*** which communicate with the utricle of the vestibule.

In front of the vestibule is the ***cochlea*** (KOK-lē-a = snail's shell), a bony spiral canal that resembles a snail's shell. A cross section through the cochlea shows that it is divided into three channels, the ***scala vestibuli*** that ends at the oval window, the ***scala tympani*** that ends at the round window, and the ***cochlear duct.*** Between the cochlear duct and the scala vestibuli is the ***vestibular membrane.*** Between the cochlear duct and scala tympani is the ***basilar membrane.***

Resting on the basilar membrane is the ***spiral organ (organ of Corti),*** the organ of hearing. The spiral organ consists of supporting cells and hair cells, which are the receptors for auditory sensations. The hair cells have long hairlike

---

Figure 12.10   Details of the internal ear of the right ear. (a) Relationship of the scala tympani, cochlear duct, and scala vestibuli. The arrows indicate the transmission of sound waves, which is discussed shortly. (b) The origins of the vestibular and cochlear branches of the vestibulo-cochlear (VIII) nerve. (c) Enlargement of the spiral organ (organ of Corti).

The three channels in the cochlea are the (1) scala vestibuli, (2) scala tympani, and (3) cochlear duct.

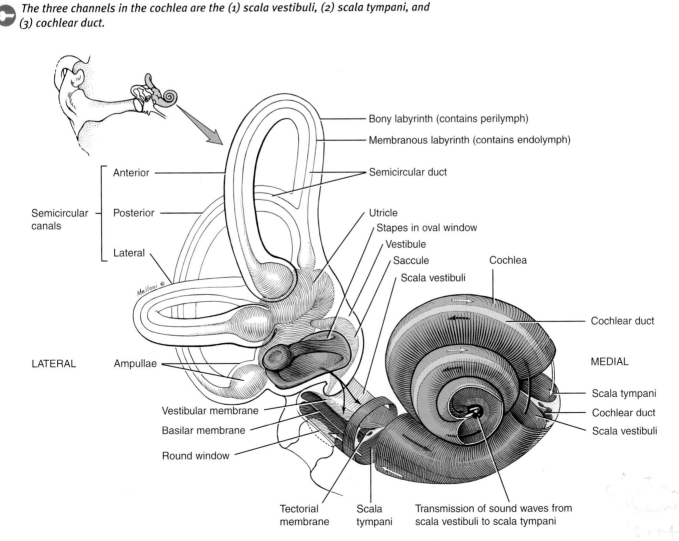

(a) Sections through the cochlea of the right ear

Figure 12.10   (*Continued*)

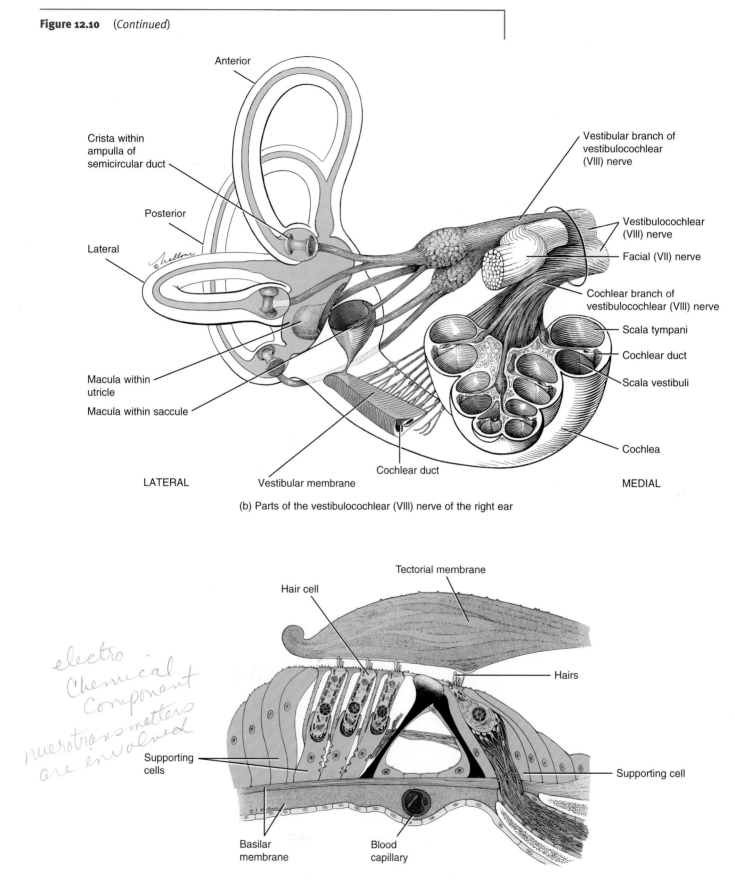

(b) Parts of the vestibulocochlear (VIII) nerve of the right ear

(c) Details of the spiral organ (organ of Corti) of the right ear

electro Chemical Componant nuerotransmatters are envolved

Q *What structures separate the external ear from the middle ear and the middle ear from the inner ear?*

processes at their free ends that extend into the endolymph of the cochlear duct. The hair cells are in contact with the cochlear branch of the vestibulocochlear (VIII) nerve. Over the hair cells and in contact with them is the **tectorial** (*tectum* = cover) **membrane,** a delicate and flexible membrane.

Hair cells of the spiral organ are easily damaged by exposure to high-intensity noises such as those produced by jet planes and loud music. They rearrange into disorganized patterns or they and their supporting cells degenerate.

Sound Waves

Sound waves originate from a vibrating object, much the same way that waves travel over the surface of water. The sounds heard most acutely by human ears are from sources that vibrate at frequencies between 500 and 5000 cycles per second (Hz). The entire range of hearing extends from 20 to 20,000 Hz.

The frequency of vibration is its **pitch**. The greater the frequency, the higher the pitch. The greater the force of the vibration, the louder the sound. Intensity or loudness is measured in **decibels** (**dB**). The point at which a person can just detect sound from silence is 0 dB. Rustling leaves have a deci-

bel rating of 15, normal conversation 45, crowd noise 60, a vacuum cleaner 75, and a pneumatic drill 90. Hearing loss may result from prolonged exposure to sounds over 90 dB. Between 115 and 120 dB is the pain threshold.

Physiology of Hearing

The events involved in the physiology of hearing sound waves are as follows (Figure 12.11):

1 Sound waves that reach the ear are directed by the auricle into the external auditory canal.

2 When the waves strike the eardrum, the eardrum vibrates. The distance the eardrum moves is relative to the intensity and frequency of the sound waves. It vibrates slowly in response to low-frequency sounds and rapidly in response to high-frequency sounds.

3 The center of the eardrum is connected to the malleus, which also starts to vibrate. The vibration is then picked up by the incus, which transmits the vibration to the stapes.

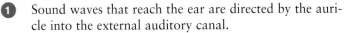

Figure 12.11 Physiology of hearing in the right ear. The numbers correspond to the events listed in the text. The cochlea has been uncoiled in order to more easily visualize the transmission of sound waves and their subsequent distortion of the vestibular or basilar membranes of the cochlear duct.

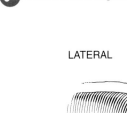 *Sound waves originate from a vibrating object.*

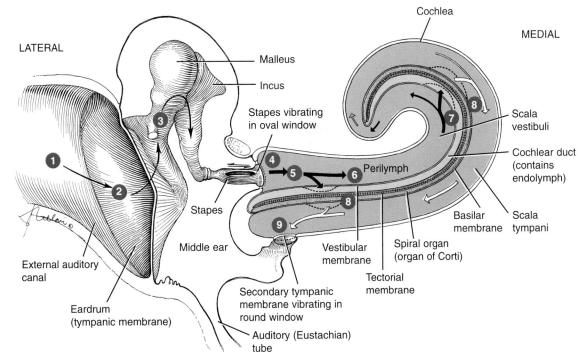

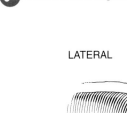 *What is the function of the hair cells?*

4 As the stapes moves back and forth, it pushes the oval window in and out.

5 The movement of the oval window sets up waves in the perilymph of the cochlea.

6 As the oval window bulges inward, it pushes on the perilymph of the scala vestibuli. Pressure waves are transmitted from the scala vestibuli to the scala tympani and eventually to the round window, causing it to bulge outward into the middle ear. (See number nine Figure 12.11.)

7 As the pressure waves push on the walls of the scala vestibuli and scala tympani, they also push the vestibular membrane back and forth. As a result, the pressure of the endolymph inside the cochlear duct changes.

8 The pressure changes in the endolymph move the basilar membrane slightly. When the basilar membrane vibrates, the hair cells of the spiral organ move against the tectorial membrane. The bending of the hairs ultimately leads to the generation of nerve impulses.

The function of hair cells is to convert a mechanical force (stimulus) into an electrical signal (nerve impulse). It is believed to occur this way: When the hairs at the top of the cell are moved, the hair cell membrane depolarizes, producing the generator potential. Depolarization spreads through the cell and causes the release of a neurotransmitter from the hair cell, which excites a sensory nerve fiber at the base of the hair cell.

The impulses are then passed on to the cochlear branch of the vestibulocochlear (VIII) nerve (Figure 12.11) and then to the medulla oblongata. Here most impulses cross to the opposite side and then travel to the midbrain, thalamus, and finally to the auditory area of the temporal lobe of the cerebral cortex.

Differences in pitch result when sound waves of various frequencies cause different regions of the basilar membrane to vibrate more intensely than others. Loudness is determined by the intensity of sound waves. High-intensity waves cause greater vibration of the basilar membrane. Thus more hair cells are stimulated and more impulses reach the brain.

Physiology of Equilibrium

Equilibrium (ē'-kwi-LIB-rē-um) refers to a state of balance between opposing forces. There are two kinds of equilibrium. *Static equilibrium* refers to the balance and posture of the body (mainly the head) when the body is not moving. *Dynamic equilibrium* refers to the balance and posture of the body (mainly the head) in response to rotational movements. Collectively, the receptor organs for equilibrium are in the internal ear and are referred to as the *vestibular* (ves-TIB-

yoo-lar) *apparatus.* It includes the saccule, utricle, and semicircular ducts.

Static Equilibrium

The walls of the utricle and saccule contain a small, flat region called a *macula* (see Figure 12.10b). The maculae (plural) are the receptors for static equilibrium. The maculae resemble the spiral organ (organ of Corti). They consist of two kinds of cells: *hair (receptor) cells* and *supporting cells* (Figure 12.12). Hair cells contain long hairlike extensions of the cell membrane. Floating over the hair cells is a thick, jellylike substance called the *otolithic membrane.* A layer of calcium carbonate crystals, called *otoliths* (*oto* = ear; *lithos* = stone), extends over the entire surface of the otolithic membrane.

The otolithic membrane sits on top of the macula. If you tilt your head forward, the membrane (and the otoliths) is pulled by gravity and slides downhill over the hair cells in the direction of the tilt. This stimulates the hair cells and ultimately results in a nerve impulse that is then transmitted to the vestibular branch of the vestibulocochlear (VIII) nerve (see Figure 12.10b).

Most of the vestibular branch fibers enter the brain stem and terminate in the medulla oblongata. The remaining fibers enter the cerebellum. Fibers from the medulla send impulses to the cranial nerves that control eye movements and head and neck movements. The cerebellum continuously receives updated sensory information from the utricle and saccule concerning static equilibrium. Using this information, the cerebellum sends continuous impulses to the motor areas of the cerebrum, causing the motor system to increase or decrease its impulses to specific skeletal muscles to maintain static equilibrium.

Dynamic Equilibrium

The three semicircular ducts maintain dynamic equilibrium (see Figure 12.10a). The ducts are positioned at right angles to one another in three planes. This positioning permits detection of an imbalance in three planes. In the ampulla, the dilated portion of each duct, there is a small elevation called the *crista* (Figure 12.13, p. 280). Each crista is composed of a group of *hair (receptor) cells* and *supporting cells* covered by a jellylike material called the *cupula.* When the head moves, the endolymph in the semicircular ducts flows over the hairs and bends them. The movement of the hairs stimulates sensory neurons, and the resulting impulses pass over the vestibular branch of the vestibulocochlear (VIII) nerve. The impulses follow the same pathways as those for static equilibrium and are eventually sent to the muscles that must contract to maintain body balance and posture.

A summary of the structures of the ear related to hearing and equilibrium is presented in Exhibit 12.3, p. 281.

Figure 12.12   Location and structure of receptors in the maculae of the right ear.

The movement of stereocilia ultimately results in the development of a nerve impulse.

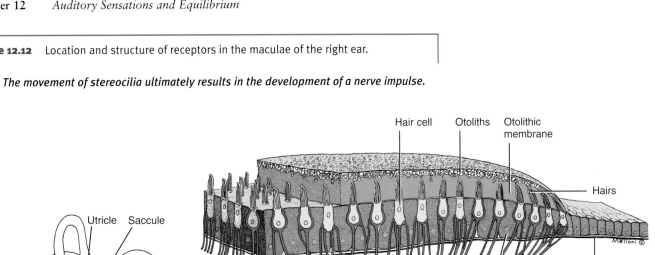

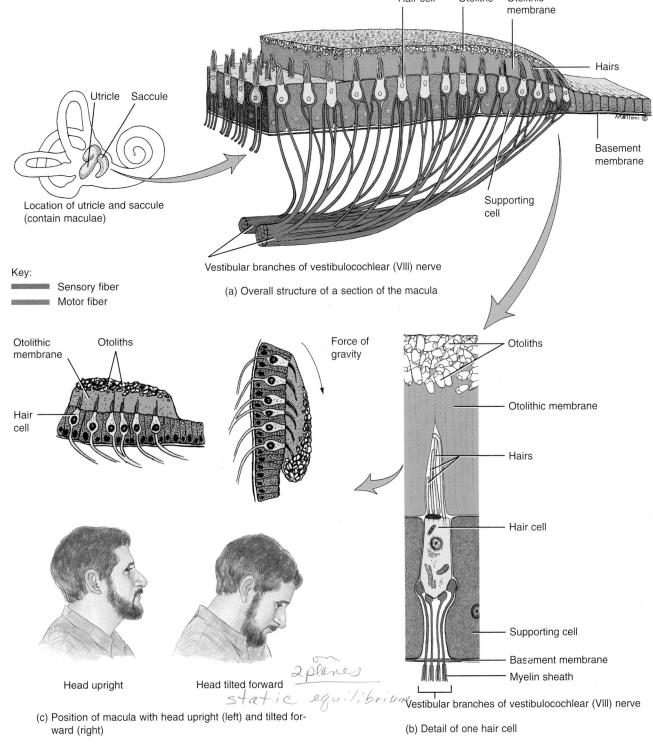

(a) Overall structure of a section of the macula

Key:
Sensory fiber
Motor fiber

Location of utricle and saccule
(contain maculae)

Vestibular branches of vestibulocochlear (VIII) nerve

(c) Position of macula with head upright (left) and tilted forward (right)

Head upright   Head tilted forward

(b) Detail of one hair cell

With which type of equilibrium are the maculae mainly concerned?

Figure 12.13   Semicircular ducts of the right ear. The ampullary nerves are branches of the
vestibular division of the vestibulocochlear (VIII) nerve.

🔑 *The positions of the semicircular ducts permit detection of rotational movements.*

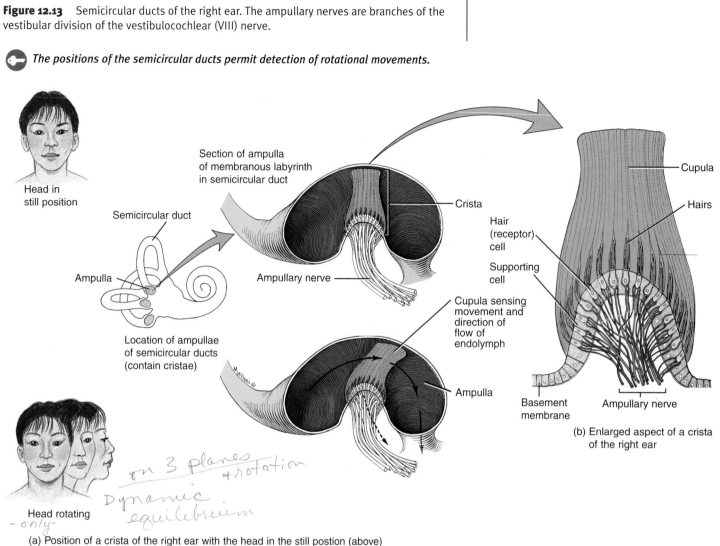

Head in
still position

Semicircular duct

Ampulla

Location of ampullae
of semicircular ducts
(contain cristae)

Section of ampulla
of membranous labyrinth
in semicircular duct

Crista

Ampullary nerve

Cupula sensing
movement and
direction of
flow of
endolymph

Ampulla

Cupula

Hairs

Hair
(receptor)
cell

Supporting
cell

Basement
membrane

Ampullary nerve

(b) Enlarged aspect of a crista
of the right ear

Head rotating

on 3 planes
+rotation

Dynamic
equilibrium

—only—

(a) Position of a crista of the right ear with the head in the still postion (above)
and when the head rotates (below)

❓ *With which type of equilibrium are the semicircular ducts associated?*

Exhibit 12.3	Summary of Structures of the Ear Related to Hearing and Equilibrium

REGIONS OF THE EAR AND KEY STRUCTURES	FUNCTIONS
External (outer) ear	*Auricle*: Collects sound waves.
	External auditory canal (meatus): Directs sound waves to eardrum.
	Eardrum (tympanic membrane): Sound waves cause it to vibrate, which, in turn, causes the malleus to vibrate.
Middle ear	*Auditory ossicles:* Transmit and amplify vibrations from tympanic membrane to oval window.
	Auditory (Eustachian) tube: Equalizes air pressure on both sides of the tympanic membrane.
Internal (Inner) ear	*Cochlea:* Contains a series of fluids, channels, and membranes that transmit vibrations to the spiral organ (organ of Corti), the organ of hearing; hair cells in the spiral organ ultimately produce nerve impulses in the cochlear branch of the vestibulocochlear (VIII) nerve.
	Semicircular ducts: Contain cristae, site of hair cells for dynamic equilibrium.
	Utricle: Contains macula, site of hair cells for static equilibrium.
	Saccule: Contains macula, site of hair cells for static equilibrium.

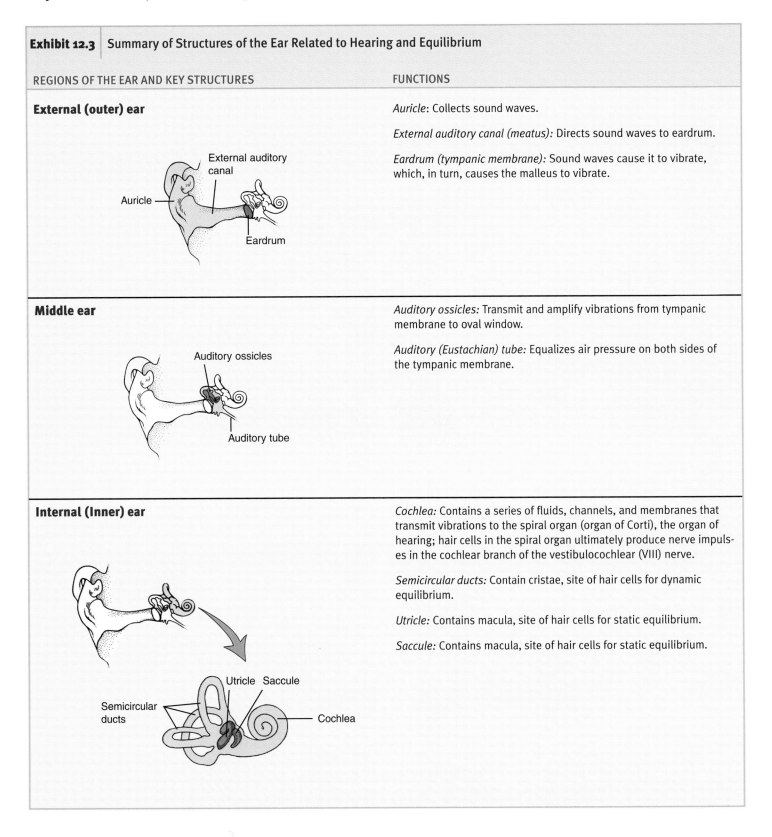

Pain Management: Sensation Modulation

Pain is a useful sensation when it alerts us to an injury that needs attention. We pull our finger away from a hot stove, we take off shoes that are too tight, and we rest an ankle that has been sprained. We do what we can to help the injury heal and meanwhile take over-the-counter or prescription painkillers until the pain goes away.

When the Pain Won't Go Away

Pain that persists for longer than two or three months despite appropriate medical or surgical treatment is known as *chronic pain*. The most common forms of chronic pain are lower back pain and headaches. Cancer, arthritis, fibromyalgia, and many other disorders are associated with chronic pain. People experiencing chronic pain often experience chronic frustration, as they are sent from one specialist to another in an attempt to diagnose accurately the cause of their pain. Their lives may turn into nightmares of fear and worry. The American Chronic Pain Association estimates that about 86 million people are affected by chronic pain.

Pain management programs have been developed to help people with chronic pain. Their goal is to decrease pain as much as possible, and then help patients learn to cope with whatever pain is left. Because no single treatment works for everyone, pain management programs typically offer a wide variety of treatments, from surgery and nerve blocks to acupuncture and exercise therapy. Following are some of the therapies that complement medical and surgical treatment for the management of chronic pain.

Counseling

Pain used to be regarded as a purely physical response to physical injury. Psychological factors are now understood to serve as important mediators in the perception of pain. Feelings such as fear and anxiety strengthen the perception of pain. Pain may be used to avoid certain situations, or to gain attention. Depression and associated symptoms such as sleep disturbances can contribute to chronic pain.

The perception and interpretation of one's chronic pain can affect pain intensity. Interesting research comparing the pain experiences of World War II soldiers who were homebound because of their wounds and civilians with comparable wounds due to surgery found that the civilians experienced significantly greater pain.

Counseling techniques can help people with chronic pain confront psychological issues that may be exacerbating their pain. They can provide psychological support and enhance a person's sense of control over the pain. Counselors also help people with chronic pain and their families cope with the many difficulties that may accompany chronic pain, such as marital stress, limited employment opportunities, and financial problems.

Relaxation and Meditation

Relaxation and meditation techniques may reduce pain by decreasing anxiety and giving patients a sense of personal control over the pain. These techniques often involve deep breathing, visualization of positive images, and muscular relaxation. They may help people become more aware of thoughts and situations that exacerbate or decrease pain. Some techniques provide a mental distraction from the sensations of pain. For example, patients learn to reinterpret pain into a sensation of pressure that doesn't hurt so much, or to imagine that the painful body part belongs to someone else.

Exercise

People with chronic pain tend to avoid movement, since it hurts. They may spend their days moving from bed to chair and back again. Muscles and joint structures atrophy, which may eventually cause the pain to worsen. Regular exercise and improved fitness help to relieve pain. Why? Exercise may stimulate the production of endorphins, chemicals produced by the body to relieve pain. Exercise may improve self-confidence, or serve as a distraction from pain. Exercise also improves sleep quality, which is often a problem for people with chronic pain.

critical thinking

In what part of the nervous system do relaxation techniques have their effect? How do you think they achieve this effect?

Why might *ergonomics* be part of a pain management program?

Common Disorders

Cataract

A *cataract,* meaning "waterfall," is a clouding of the lens or its capsule so that it becomes opaque or milk white.

Glaucoma

Glaucoma is an abnormally high intraocular pressure (IOP), owing to a buildup of aqueous humor inside the anterior chamber of the eyeball. Glaucoma can progress from mild visual impairment to a point where neurons of the retina are destroyed, resulting in blindness.

Conjunctivitis (Pinkeye)

Conjunctivitis (*pinkeye*) is an inflammation of the conjunctiva, which is the membrane that lines the insides of the eyelids and covers the cornea. Conjunctivitis caused by bacteria is very contagious; conjunctivitis caused by dust, smoke, and pollutants in the air is not contagious.

Trachoma

Trachoma (tra-KŌ-ma) is a serious form of chronic contagious conjunctivitis, which is caused by a bacterium called *Chlamydia trachomatis.*

Deafness

Deafness is significant or total hearing loss. It may be caused by an impaired cochlea or cochlear branch of the vestibulocochlear (VIII) nerve or by impaired external and middle ear mechanisms for transmitting sounds to the cochlea. Among the factors that contribute to deafness are atherosclerosis, which reduces blood supply to the ears; repeated exposure to loud noise, which destroys hair cells of the spiral organ (organ of Corti); certain drugs, such as aspirin and streptomycin; impacted cerumen; injury to the eardrum; and aging, which results in thickening of the eardrum, stiffened auditory ossicle joints, and decreased numbers of hair cells due to diminished cell division.

Labyrinthine Disease

Labyrinthine (lab′-i-RIN-thēn) *disease* refers to a malfunction of the internal ear that is characterized by deafness, tinnitus (ringing in the ears), vertigo (sensation of spinning), nausea, and vomiting. There may also be blurred vision, nystagmus (rapid, involuntary movement of the eyeballs), and a tendency to fall in a certain direction.

Ménière's Syndrome

Ménière's (men-YAIRZ) *syndrome* is characterized by an increased amount of endolymph that enlarges the internal ear. Among the symptoms are hearing loss, attacks of vertigo, and roaring tinnitus (ringing in the ears). The cause of Ménière's syndrome is unknown.

Vertigo

Vertigo (VER-ti-gō; *vertex* = whorl) is a sensation of spinning or movement in which the world is revolving or the person is revolving in space.

Otitis Media

Otitis media is an acute bacterial infection of the middle ear. It is characterized by pain, malaise, fever, and a reddening and outward bulging of the eardrum, which may rupture without prompt treatment.

Motion Sickness

Motion sickness is a functional disorder brought on by repetitive motion and characterized by various symptoms, primarily nausea and vomiting. The cause is excessive stimulation of the vestibular apparatus. Nerve impulses pass from the internal ear to the vomiting center in the medulla oblongata and this causes the nausea and vomiting. ∎

Medical Terminology and Conditions

Ametropia (am′-e-TRŌ-pē-a; *ametro* = disproportionate; *ops* = eye) Refractive defect of the eye resulting in an inability to focus images properly on the retina.

Audiometer (aw-dē-OM-e-ter; *audire* = to hear; *metron* = to measure) An instrument used to measure hearing by producing acoustic stimuli of known frequency and intensity.

Blepharitis (blef-a-RĪ-tis; *blepharo* = eyelid; *itis* = inflammation of) An inflammation of the eyelid.

Dyskinesia (dis′-ki-NĒ-zē-a; *dys* = difficult; *kinesis* = movement) Abnormality of motor function characterized by involuntary, purposeless movements.

Eustachitis (yoo′-stā-KĪ-tis) An inflammation or infection of the auditory (Eustachian) tube.

Exotropia (ek′-sō-TRŌ-pē-a; *ex* = out; *tropia* = turning) Turning outward of the eyes.

Kinesthesis (kin′-es-THĒ-sis; *aisthesis* = sensation) The sense of perception of movement.

Labyrinthitis (lab′-i-rin-THĪ-tis) An inflammation of the internal ear.

Mydriasis (mi-DRĒ-a-sis) Dilated pupil.

Myringitis (mir′-in-JĪ-tis; *myringa* = eardrum) An inflammation of the eardrum; also called *tympanitis.*

Nystagmus (nis-TAG-mus; *nystazein* = to nod) A rapid involuntary movement of the eyeballs.

Otalgia (o-TAL-jē-a; *oto* = ear; *algia* = pain) Earache.

Otosclerosis (ō′-tō-skle-RŌ-sis; *oto* = ear; *sclerosis* = hardening) Pathological process that may be hereditary in which new bone is deposited around the oval window. The result may be immobilization of the stapes, leading to deafness.

Photophobia (fō′-tō-FŌ-bē-a; *photo* = light; *phobia* = fear) Abnormal visual intolerance to light.

Ptosis (TŌ-sis; *ptosis* = fall) Falling or drooping of the eyelid. (This term is also used for the slipping of any organ below its normal position.)

Retinoblastoma (ret′-i-nō-blas-TŌ-ma; *blast* = bud; *oma* = tumor) A tumor arising from immature retinal cells and accounting for 2 percent of childhood malignancies.

Scotoma (skō-TŌ-ma; *scotoma* = darkness) An area of reduced or lost vision in the visual field. Also called a **blind spot** (other than the normal blind spot or optic disc).

Strabismus (stra-BIZ-mus) An imbalance in the extrinsic eye muscles that a person cannot overcome. In **convergent strabismus (cross-eye)**, the visual axes converge. In **divergent strabismus (walleye)**, the visual axes diverge. **Amblyopia** is the term used to describe the loss of vision in an otherwise normal eye that, because of muscle imbalance, cannot focus in sync with the other eye.

Tinnitus (ti-NĪ-tus) A ringing, roaring, or clicking in the ears.

Study Outline

Sensations (p. 260)
Definition (p. 260)

1. Sensation is a state of conscious or unconscious awareness of external and internal conditions of the body.
2. The prerequisites for a sensation to occur are reception of a stimulus, conversion of the stimulus into a nerve impulse by a receptor, conduction of the impulse to the brain, and integration of the impulse into a sensation by a region of the brain.
3. Each stimulus is capable of causing the membrane of a receptor to depolarize. This is called the generator potential.

Characteristics (p. 260)

1. Projection occurs when the brain refers a sensation to the point of stimulation.
2. Adaptation is the loss of sensation even though the stimulus is still applied.
3. An afterimage is the persistence of the sensation even though the stimulus is removed.
4. Modality is the property by which one sensation is distinguished from another.

Classification of Receptors (p. 260)

1. According to location, receptors are classified as exteroceptors, interoceptors, and proprioceptors.
2. On the basis of type of stimulus detected, receptors are classified as mechanoreceptors, thermoreceptors, nociceptors, photoreceptors, and chemoreceptors.
3. In terms of simplicity or complexity, simple receptors are associated with general senses and complex receptors are associated with special senses.

General Senses (p. 261)
Cutaneous Sensations (p. 261)

1. Cutaneous sensations include tactile sensations (touch and pressure), thermal sensations (heat and cold), and pain. Receptors for these sensations are located in the skin, connective tissues under the skin, and mucous membranes of the mouth and anus.
2. Receptors for touch are hair root plexuses, type I cutaneous mechanoreceptors (tactile discs), and corpuscles of touch (Meissner's corpuscles). Receptors for pressure are type II cutaneous mechanoreceptors and lamellated (Pacinian) corpuscles.
3. Thermoreceptors, which are free nerve endings, adapt to continuous stimulation.
4. Pain receptors (nociceptors) are located in nearly every body tissue.
5. Referred pain is felt in the skin near or away from the organ sending pain impulses.
6. Phantom pain is the sensation of pain in a limb that has been amputated.

Proprioceptive Sensations (p. 263)

1. Receptors located in skeletal muscles, in tendons, in and around joints, and in the internal ear convey impulses related to muscle tone, movement of body parts, and body position.
2. The receptors include muscle spindles, tendon organs (Golgi tendon organs), joint kinesthetic receptors, and the maculae and cristae.

Olfactory Sensations (p. 263)

1. The receptors for olfaction, the olfactory receptors, are in the nasal epithelium.
2. Substances to be smelled must be gaseous, water-soluble, and lipid-soluble.
3. Olfactory receptors convey impulses to olfactory (I) nerves, olfactory bulbs, olfactory tracts, and the olfactory area in the temporal lobe of the cerebral cortex.

Gustatory Sensations (p. 264)

1. The receptors for gustation, the gustatory receptors, are located in taste buds.

2. Substances to be tasted must be in solution in saliva.

3. The four primary tastes are salt, sweet, sour, and bitter.

4. Gustatory cells convey impulses to cranial nerves VII, IX, and X, the medulla oblongata, thalamus, and the primary gustatory area in the parietal lobe of the cerebral cortex.

Visual Sensations (p. 266)

1. Accessory structures of the eyes include the eyebrows, eyelids, eyelashes, and the lacrimal apparatus.

2. The eye is constructed of three coats: (a) fibrous tunic (sclera and cornea), (b) vascular tunic (choroid, ciliary body, and iris), and (c) retina (nervous tunic), which contains rods and cones.

3. The anterior cavity contains aqueous humor; the vitreous chamber contains the vitreous body.

4. The refractive media of the eye are the cornea and lens.

5. Retinal image formation involves refraction of light, accommodation of the lens, constriction of the pupil, convergence, and inverted image formation.

6. Improper refraction may result from myopia (nearsightedness), hypermetropia (farsightedness), and astigmatism (corneal or lens abnormalities).

7. Rods and cones develop generator potentials and ganglion cells initiate nerve impulses.

8. Impulses from ganglion cells are conveyed through the retina to the optic (II) nerve, the primary optic chiasm, the optic tract, the thalamus, and the visual area in the occipital lobe of the cerebral cortex.

Auditory Sensations and Equilibrium (p. 273)

1. The ear consists of three anatomical subdivisions: (a) the external ear (auricle, external auditory canal, and eardrum), (b) the middle ear (auditory or Eustachian tube, ossicles, oval window, and round window), and (c) the internal ear (bony labyrinth and membranous labyrinth). The internal ear contains the spiral organ (organ of Corti), the organ of hearing.

2. Sound waves enter the external auditory canal, strike the eardrum, pass through the ossicles, strike the oval window, set up waves in the perilymph, strike the vestibular membrane and scala tympani, increase pressure in the endolymph, strike the basilar membrane, and stimulate hairs on the spiral organ. A sound impulse is then initiated.

3. Static equilibrium is the orientation of the body relative to the pull of gravity. The maculae of the utricle and saccule are the sense organs of static equilibrium.

4. Dynamic equilibrium is the maintenance of body position in response to movement. The cristae in the semicircular ducts are the sense organs of dynamic equilibrium.

Self-Quiz

1. Match the receptors with their functions.

   ___J___ a. color vision
   ___E___ b. taste
   ___F___ c. smell
   ___I___ d. dynamic equilibrium
   ___C___ e. vision in dim light
   ___G___ f. stretch in a muscle
   ___H___ g. static equilibrium
   ___A___ h. pressure
   ___B___ i. discriminative touch
   ___D___ j. detects pain

   A. lamellated (pacinian) corpuscle
   B. type I cutaneous mechanoreceptor
   C. rods
   D. nociceptors
   E. gustatory receptors
   F. olfactory receptors
   G. muscle spindles
   H. maculae
   I. cristae
   J. cones

   answers of 24 7 may be switched in back of book

2. Which of the following receptor types would be considered interoreceptors?

   a. photoreceptors   b. taste buds   c. pressure receptors in blood vessels   d. tactile receptors   e. proprioceptors in muscles and joints   *should be?*

3. Which of the following is NOT a required condition for a sensation to occur?

   a. the presence of a stimulus   b. the persistence of afterimages   c. a receptor specialized to detect a stimulus   d. a sensory neuron to conduct an impulse   e. a region of the brain for integration of the nerve impulse

4. Which of the following sensory receptors respond to virtually any type of excessive stimuli?

   a. lamellated (Pacinian) corpuscles   b. nociceptors   c. thermoreceptors   d. interoreceptors   e. proprioceptors

5. Which of the following choices is NOT a tactile sensation?

   a. heat   b. crude touch   c. pressure   d. vibration   e. discriminative touch

6. Equilibrium and activities of the muscles and joints are monitored by

   a. proprioceptors   b. olfactory receptors   c. nociceptors   d. tactile receptors   e. gustatory receptors

7. Which of the following pairs is NOT matched correctly?   *all matched*

   a. exteroreceptors—external environment   b. proprioceptors—body position   c. nociceptors—pain   d. thermoreceptors—heat and cold   e. interoreceptors—internal ear   *proprioceptor is internal ear*

8. Place the following events concerning the visual pathway in the correct order:

   1. The action potentials exit the eye via the optic (II) nerve.
   2. Neurons terminate in the thalamus.
   3. Light reaches the retina.
   4. Rods and cones are stimulated.
   5. Synapses occur in the thalamus and continue along to the primary visual area of the cerebral cortex in the occipital lobe.
   6. Bipolar neurons and ganglion cells develop action potentials.

   a. 4, 1, 2, 5, 6, 3   b. 5, 4, 1, 3, 2, 6   c. 5, 4, 3, 1, 2, 6   d. 3, 4, 6, 1, 2, 5   e. 3, 4, 5, 6, 1, 2

9. Of the following choices, which choice is true regarding the auditory (Eustachian) tube?

a. It equalizes pressure on both sides of the eardrum.
b. It is filled with a fluid called endolymph.   c. It is concerned with equilibrium and balance.   d. It is the partition between the outer and middle ear.   e. It is where sound waves are directed through to the cochlea.

10. The bony labyrinth of the internal ear is divided into three main areas. They are the

    a. cochlea, saccule, utricle   b. cochlear duct, scala vestibuli, scala tympani   c. vestibule, semicircular canals, tympanic membrane   d. vestibule, cochlea, semicircular canals   e. vestibule, cochlea, tympanic membrane

11. You are seated at your desk and you drop your pencil. As you lean over to retrieve your pencil, what is occurring in your inner ear?

    a. The hair cells on the macula are responding to changes in static equilibrium.   b. The cristae of each semicircular duct are responding to changes in dynamic equilibrium.   c. The hair cells in the cochlea are responding to changes in dynamic equilibrium.   d. Nothing occurs because gravity is constant.   e. Nothing occurs until you fall out of your chair.

12. The feeling of pain in the left arm in response to a heart attack is called

a. phantom pain   b. referred pain   c. afterimage   d. projection   e. adaptation

13. The nonvascular, transparent portion of the fibrous tunic is the

    a. conjunctiva   b. sclera   c. pupil   d. lens   e. cornea

14. The vascular tunic does NOT include which of the following structures?

    a. scleral venous sinus (canal of Schlemm)   b. choroid   c. ciliary muscle   d. ciliary process   e. iris

15. Intraocular pressure is produced mainly by the

    a. suspensory ligaments   b. ciliary body   c. aqueous humor   d. vitreous chamber   e. lacrimal apparatus

16. The inability to focus properly on near objects due to the loss of lens elasticity with age is called

    a. astigmatism   b. blind spot   c. presbyopia   d. emmetropia   e. cataract

17. The organ of hearing

    a. is located in the middle ear   b. is also known as the organ of Corti   c. is stimulated by the movement of otoliths   d. is the tympanic membrane   e. passes impulses to the occipital lobe of the cerebral cortex

Critical Thinking Applications

1. When someone hits the back of the head during a fall, s/he will sometimes "see stars." If this is not caused by the sight of stars in the sky, what is the cause of this effect?

2. Discuss how the two types of photoreceptors in the retina differ in location and function. Use your explanation to explain what occurs when you "see out of the corner of your eye."

3. When you first enter a chemistry lab the odors are quite strong. After several minutes, the odor in the lab is barely noticeable. What has happened to the odors or has something happened to you?

4. Bet you've never seen a rabbit with glasses! Carrots contain the pigment carotene, which the body converts into vitamin A. What role does vitamin A play in vision?

5. Cliff works the night shift and sometimes falls asleep in A & P class. What is the effect on the structures of his internal ear when his head falls back and he slumps in his seat?

6. When Bruce listens to an A & P lecture it goes in one ear and out the other. Trace the pathway of sound waves from the external ear to the cranial nerve exiting the internal ear.

Answers to Figure Questions

12.1   Corpuscles of touch (Meissner's corpuscles).
12.2   Olfactory tracts.
12.3   Gustatory receptors; cranial nerves VII, IX, and X; medulla oblongata; thalamus; primary gustatory area in the parietal lobe of the cerebral cortex.
12.4   Tears contain water, salts, some mucus, and lysozyme. Tears clean, lubricate, and moisten the eyeball.
12.5   Rod; cone.
12.6   The ciliary muscle contracts ⟶ suspensory ligaments slacken ⟶ lens shortens, thickens, and bulges, and refracts light more acutely.
12.7   The loss of elasticity in the lens that occurs with aging.

12.8   Retina ⟶ optic nerve (cranial nerve II) ⟶ optic chiasm ⟶ optic tract ⟶ thalamus ⟶ primary visual area in occipital lobe of the cerebral cortex.
12.9   Inner ear: cochlea (hearing) and semicircular ducts (equilibrium).
12.10 The eardrum (tympanic membrane) separates the external ear from the middle ear. The oval and round windows separate the middle ear from the internal ear.
12.11 To convert a mechanical force (stimulus) into an electrical signal (nerve impulse).
12.12 Static equilibrium.
12.13 Dynamic equilibrium.

THE ENDOCRINE SYSTEM

a look ahead

student learning objectives

Around age 12, as they enter puberty, boys and girls start to develop striking differences in physical appearance and behavior. During this time, the impact of the nervous and endocrine systems in directing development and regulating body functions is clearly evident. Changes in the brain and pituitary gland markedly increase the synthesis of sex hormones from the gonads (ovaries in the female and testes in the male). In girls, fatty tissue starts to accumulate in the breasts and hips. At the same time, or a little later in boys, protein synthesis increases; muscle mass builds; and the longer, larger vocal cords produce a lower-pitched voice. These changes provide just a few examples of the powerful influence of secretions from endocrine glands.

Endocrine Glands   *Ductless*

The body contains two kinds of glands: exocrine and endocrine. *Exocrine* (*ex* = out; *krinein* = to secrete) *glands,* such as sweat (sudoriferous), oil (sebaceous), mucous, and digestive glands, secrete their products into ducts that carry the secretions into body cavities or on to body surfaces. *Endocrine* (*endo* = within) *glands,* by contrast, are ductless. They secrete their products—called hormones—into the extracellular space around the secretory cells. From there, the hormones diffuse into capillaries and are carried away by the blood. The endocrine glands of the body (Figure 13.1) constitute the *endocrine system* and include the pituitary gland, thyroid gland, parathyroid glands, adrenal glands, pineal gland, and thymus gland. In addition, several organs of the body contain endocrine tissue but are not endocrine glands exclusively. These include the hypothalamus, pancreas, ovaries, testes, kidneys, stomach, liver, small intestine, skin, heart, and placenta (during pregnancy). *Secrete hormones picked up by blood stream*

The science concerned with the structure and functions of the endocrine glands and endocrine tissue and the diagnosis and treatment of disorders of the endocrine system is called *endocrinology* (en'-dō-kri-NOL-ō-jē; *logos* = study of).

Comparison of Nervous and Endocrine Systems

objective: *Compare the functions of the nervous and endocrine systems in maintaining homeostasis.*

Together, the nervous and endocrine systems coordinate functions of all body systems. The nervous system controls homeostasis through nerve impulses that trigger the release of neurotransmitter molecules. The result is either excitation or inhibition of other neurons, muscle fibers (cells), or gland *1st step* cells. In contrast, the endocrine system releases its messenger molecules, called *hormones* (*hormon* = to urge on), into the bloodstream. A hormone is a secretion of endocrine tissue that alters the physiological activity of other body tissues. The cardiovascular system then delivers hormones to cells throughout the body. *2nd step*

The nervous and endocrine systems are coordinated as an interlocking supersystem. Certain parts of the nervous system stimulate or inhibit the release of hormones. Hormones, in turn, may promote or inhibit the generation of nerve impulses. And certain molecules, for example, norepinephrine, act as hormones in some locations and as neurotransmitters in others.

The nervous system causes muscles to contract and glands to secrete either more or less of their product. The endocrine system alters metabolic activities, regulates growth and development, and guides reproductive processes. Thus it not only helps regulate the activity of smooth and cardiac muscle and some glands, it affects virtually all other tissues as well.

Nerve impulses most often produce their effects within a few milliseconds. While some hormones can act within seconds, others can take several hours or more to bring about their responses. Also, the effects of activating the nervous system are generally briefer than effects produced by the endocrine system.

A comparison between the nervous and endocrine regulation of homeostasis is presented in Exhibit 13.1 on page 290.

Overview of Hormonal Effects

Although the *effects of hormones* are many and varied, their actions can be categorized into seven broad areas:

1. Hormones regulate the chemical composition and volume of the internal environment (extracellular fluid).

2. Hormones help regulate metabolism and energy balance.

3. Hormones help regulate contraction of smooth and cardiac muscle fibers and secretion by glands.

4. Hormones help maintain homeostasis despite emergency environmental disruptions such as infection, trauma, emotional stress, and dehydration.

5. Hormones regulate certain activities of the immune system.

6. Hormones play a role in the smooth, sequential integration of growth and development.

7. Hormones contribute to the basic processes of reproduction, including gamete (oocyte and sperm) production, nourishment of the embryo and fetus, and delivery.

Figure 13.1   Location of many endocrine glands, other organs containing endocrine tissue, and associated structures.

Endocrine glands secrete hormones that diffuse into the blood for transport to target tissues.

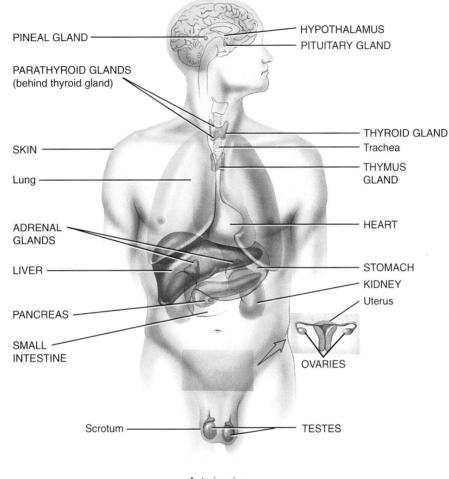

PINEAL GLAND

PARATHYROID GLANDS
(behind thyroid gland)

SKIN

Lung

ADRENAL
GLANDS

LIVER

PANCREAS

SMALL
INTESTINE

Scrotum

HYPOTHALAMUS
PITUITARY GLAND

THYROID GLAND
Trachea
THYMUS
GLAND

HEART

STOMACH
KIDNEY
Uterus

OVARIES

TESTES

Anterior view

What is the basic difference between an endocrine gland and an exocrine gland?

Chemistry of Hormones

objective: *Distinguish hormones on the basis of their chemistry.*

Chemically, there are three general classes of hormones (see Exhibit 13.2): (1) lipid derivatives, (2) amino acid derivatives, and (3) peptides and proteins.

1. **Lipid derivatives.** There are two groups of hormones that are derived from lipids: steroid hormones and eicosanoids. *Steroid hormones* are derived from cholesterol. *Eicosanoids* (ī-KŌ-sa-noids) are derived from a fatty acid. Two families of eicosanoids, *prostaglandins* (pros′-ta-GLAN-dins) and *leukotrienes* (loo-kō-TRĪ-ēns) act as local hormones in most tissues of the body. This means that their site of action is the immediate area in

Exhibit 13.1	Comparison of Nervous System and Endocrine System Regulation of Homeostasis	
CHARACTERISTIC	NERVOUS SYSTEM	ENDOCRINE SYSTEM
Mechanism of control	Neurotransmitters released in response to nerve impulses.	Hormones delivered to tissues throughout the body by the blood.
Cells affected	Muscle cells, gland cells, other neurons.	Virtually all body cells.
Type of action that results	Muscular contraction or glandular secretion.	Changes in metabolic activities.
Time to onset of action	Typically within milliseconds.	Seconds to hours or days.
Duration of action	Generally briefer.	Generally longer.

which they are produced. Leukotrienes are important in tissue inflammation. Prostaglandins are important in the normal physiology of smooth muscle (such as uterine contraction), blood flow, reproduction, platelet function, nerve impulse transmission, and the immune response. Prostaglandins also help induce inflammation, promote fever, and intensify pain. Drugs such as aspirin and ibuprofen (Motrin) inhibit prostaglandin synthesis and thus reduce fever and decrease pain.

↓ wide ranges—a few decrease fever

2. **Amino acid derivatives.** These are the simplest hormones and are derived from amino acids.
3. **Peptides and proteins.** These hormones consist of chains of amino acids, anywhere from 3 to 200.

Exhibit 13.2 contains a summary of the chemical classes of hormones, examples of each, and sites of production.

The one function all hormones have in common is maintaining homeostasis by changing the physiological activities of cells.

Exhibit 13.2	Chemical Classes of Hormones, Examples, and Sites of Production	
CHEMICAL CLASS	EXAMPLES	WHERE PRODUCED
1. Lipid derivatives		
Steroid hormones *all made from Cholesterol*	Aldosterone, cortisol, and androgens (male sex hormones)	Adrenal cortex
	Testosterone	Testes
	Estrogens and progesterone (female sex hormones)	Ovaries
Eicosanoids	*Lipid derived* Prostaglandins and leukotrienes *Ibuprofin Blocks prostaglandin*	All cells except red blood cells
2. Amino acid derivatives	T_3 and T_4 (thyroid hormones)	Thyroid (follicular cells)
	Epinephrine and norepinephrine	Adrenal medulla
3. Peptides and proteins *larger molecules*	All hypothalamic releasing and inhibiting hormones	Hypothalamus
	Oxytocin, antidiuretic hormone	Hypothalamus
	All anterior pituitary gland hormones	Anterior pituitary gland
	Insulin, glucagon	Pancreas

go thru membrane

Bind to receptors

Mechanism of Hormonal Action

| objective: | *Explain how hormones act on body cells.*

Overview

The amount of hormone released by an endocrine gland or tissue is determined by the body's *requirement* for the hormone at any given time (Figure 13.2). Hormone-producing cells are sent information from sensing and signaling systems that permit them to regulate the amount and duration of hormone release. Various characteristics of the internal environment, for example, the blood levels of glucose, Na^+, and O_2, are detected by these sensors and the information is received by the endocrine cells responsible for regulating that specific characteristic or sub-

---

Figure 13.2 Overview of physiology of the endocrine system.

 Depending on the body's need at a particular time, a hormone may be released or inhibited.

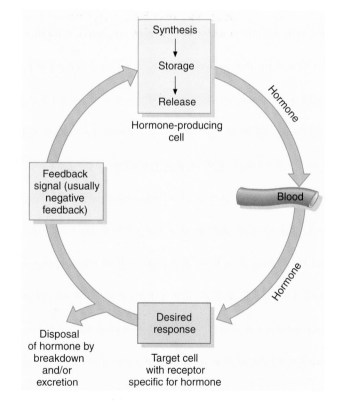

Which organs eliminate hormones that are broken down once they have accomplished their goal?

stance. Feedback systems (described later) regulate endocrine glands to maintain normal hormone production so that there is no overproduction or underproduction of a hormone.

Receptors

Although a given hormone travels throughout the body in the blood, it will only affect certain cells called **target cells.** Most body cells have **receptors** (proteins) that bind to hormones. Only target cells for a certain hormone have receptors that bind and recognize that hormone. Therefore, each hormone will influence only its target cells but not other cells in the body. Once a hormone binds to a target cell, it causes that cell to respond in a manner that will maintain some characteristic of the internal environment in homeostasis. After a hormone performs its function, it is broken down by the target cell and eliminated from the body by the liver or kidneys.

Generally, hormones cause target cells to alter their rate of function. The specific way in which hormones produce this effect depends on whether they are lipid-soluble or water-soluble and how they interact with target cell receptors.

Lipid-Soluble Hormones

Lipid-soluble hormones can diffuse through the phospholipid bilayer of the plasma membrane to get inside cells. These hormones bind to receptors *within* target cells. Their mechanism of action is as follows (Figure 13.3a):

1 A lipid-soluble hormone diffuses from the blood, through interstitial fluid, and through the phospholipid bilayer of the plasma membrane into a cell.

2 If the cell is a target cell, the hormone will bind to and activate receptors located within the cell. An activated receptor then alters cell function by turning specific genes on or off.

3 Newly formed messenger RNA (mRNA) leaves the nucleus and enters the cytosol. There, it directs synthesis of new proteins.

4 The new proteins alter the cell's activity and cause the physiological responses of that hormone.

Water-Soluble Hormones

Certain hormones, such as peptide and protein hormones, are not lipid-soluble and thus cannot diffuse through the phospholipid bilayer of the plasma membrane. The receptors for these water-soluble hormones are plasma membrane integral proteins. Since the hormone delivers its message to the plasma membrane, it is called the **first messenger.** A *second messenger* is needed to relay the message inside the cell, where hormone-stimulated responses can take place.

One second messenger is *cyclic AMP* (*cAMP*). It is synthesized from ATP, the main energy-providing molecule in cells. Cyclic AMP and other second messengers then alter cell function in specific ways.

A typical mechanism of action of a water-soluble hormone is as follows (Figure 13.3b):

1 A water-soluble hormone diffuses from the blood and binds to its receptor in a target cell's plasma membrane. This binding starts a reaction that ultimately converts ATP into cyclic AMP in the cytosol of the cell.

2 Cyclic AMP (the second messenger) causes the activation of several enzymes.

3 Activated enzymes catalyze reactions that produce physiological responses.

4 After a brief period of time, cyclic AMP is inactivated. Thus the cell's response is turned off unless new hormone molecules continue to bind to their receptors in the plasma membrane.

Control of Hormonal Secretions: Feedback Control

objective: *Explain how levels of hormones in the blood are regulated.*

The secretion of hormones is normally regulated so that there is no overproduction or underproduction of a particular hormone. This regulation is one of the very important mechanisms the body uses to maintain homeostasis. If the regulating mechanism does not operate properly and hormonal levels are

+ feedback — continues in direction its going
Labor pains
Blood clotting

— feedback — changes direction negates
Rising blood pressure — baro
falling blood pressure
thyroid imbalance — Chemoreceptor
Thermo-receptor

Figure 13.3   Proposed mechanisms of hormonal action.

Lipid-soluble hormones bind to receptors inside target cells. Water-soluble hormones bind to receptors embedded in the plasma membrane of target cells.

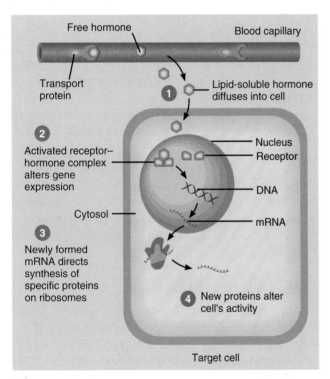

(a) Activation of intracellular receptor by lipid-soluble hormone

(b) Interaction between first messenger (water-soluable hormone) and plasma membrane receptor in which there is an increase in the production of cAMP

Q *Why is cAMP called a second messenger?*

excessive or deficient, disorders result, several of which are discussed at the end of the chapter.

Hormonal secretions are most often regulated by **negative feedback control** (see Figures 13.2 and 1.2). Information regarding the hormone level or its effect is fed back to the gland, which then responds accordingly. In a few cases, positive feedback controls regulate hormonal secretions. Also, some hormones of the pituitary gland are released in response to nerve impulses from the hypothalamus. Here we will describe the three negative feedback systems for controlling hormonal secretion.

Levels of Chemicals in Blood

In one type of negative feedback system, control of the hormone is triggered by levels of certain chemicals in blood. For example, blood calcium level is controlled by parathyroid hormone (PTH), produced by the parathyroid glands, and calcitonin (CT), produced by the thyroid gland. If blood calcium level is low, this serves as a stimulus for the parathyroids to release more PTH (see Figure 13.12). PTH then exerts its effects in various parts of the body until the blood calcium level is raised to normal. A high blood calcium level serves as a stimulus for the parathyroids to cease their production of PTH. However, the thyroid gland increases its production of CT to lower blood calcium level to normal. Note that in negative feedback control the body's response (increased or decreased calcium level) is opposite (negative) to the stimulus (low or high calcium level).

Nerve Impulses

In another type of negative feedback system, the hormone is released as a direct result of nerve impulses that stimulate the endocrine gland. For example, epinephrine and norepinephrine (NE) are released from the adrenal glands in response to sympathetic nerve impulses during stress.

Chemical Secretions from the Hypothalamus

In still another type of negative feedback system, the hormone is controlled through several different chemical secretions from the hypothalamus (see Figure 13.6). Hypothalamic secretions that stimulate the release of the hormone into the blood are called **releasing hormones**. Those that prevent the release of the hormone are called **inhibiting hormones**.

Most often, negative feedback systems maintain homeostasis of hormonal secretions. As we discuss the effects of various hormones in this chapter, we will also describe how the secretions of the hormones are controlled. At that time you will be able to see which type of negative feedback system is operating. As noted earlier, a positive feedback system contributes to regulation of hormone secretion. In a positive feedback, the response intensifies the initiating stimulus, which intensifies the response, and so on. One example of this occurs during childbirth. The hormone oxytocin (discussed in detail a little later) stimulates the contraction of the uterus. Uterine contractions, in turn, stimulate more oxytocin release (see Figure 13.8).

Pituitary Gland

objective: *Describe the location, histology, and functions of the pituitary gland.*

The hormones of the **pituitary gland** regulate so many body activities that the pituitary has been nicknamed the "master gland." It is a small, round structure that is attached to the hypothalamus of the brain by a stalklike structure, the **infundibulum** (see Figure 13.4). It is composed of a larger anterior pituitary gland and a smaller posterior pituitary gland. The **anterior pituitary gland** forms the glandular part of the pituitary. Blood vessels connect the anterior pituitary gland with the hypothalamus. The **posterior pituitary gland** contains axon terminals of neurons whose cell bodies are in the hypothalamus. Nerve fibers connect the posterior pituitary gland directly with the hypothalamus.

Anterior Pituitary Gland

The anterior pituitary gland releases hormones that regulate a whole range of body activities from growth to reproduction. Some of these hormones influence other endocrine glands and are called **tropic** (TRŌ-pik) **hormones.** The release of these hormones is controlled by hypothalamic releasing and inhibiting hormones that are produced by secretory neurons in the hypothalamus called **neurosecretory cells.**

These hormones are delivered to the anterior pituitary gland directly from the hypothalamus through blood vessels without first circulating through the heart (Figure 13.4). The short route allows the hormones to act quickly on the anterior pituitary gland and prevents their dilution or destruction.

The anterior pituitary gland secretes seven hormones (Figure 13.5 on page 295):

1. **Human growth hormone (hGH),** which controls general body growth and regulates metabolism.

2. **Thyroid-stimulating hormone (TSH),** which controls secretion of the thyroid gland.

3. **Follicle-stimulating hormone (FSH),** which stimulates sperm production by the testes and the production of oocytes (potential ova) and estrogens (female sex hormones) by the ovaries.

4. **Luteinizing hormone (LH),** which stimulates secretion of testosterone (male sex hormones) by the testes, secretion of estrogens and progesterone (female sex hormone) by the ovaries, and ovulation (release of an oocyte into the pelvic cavity).

5. **Prolactin (PRL),** which, together with other hormones, initiates and maintains milk production in the mammary glands.

6. **Adrenocorticotropic hormone (ACTH),** which stimulates the adrenal cortex (outer region) to secrete its hormones.

Figure 13.4 Pituitary gland and its blood supply. Note in the small figure to the right that releasing and inhibiting hormones synthesized by hypothalamic neurosecretory cells are transported into capillaries and are carried by veins to the anterior pituitary gland.

🔑 *Hypothalamic hormones are an important link between the nervous and endocrine systems.*

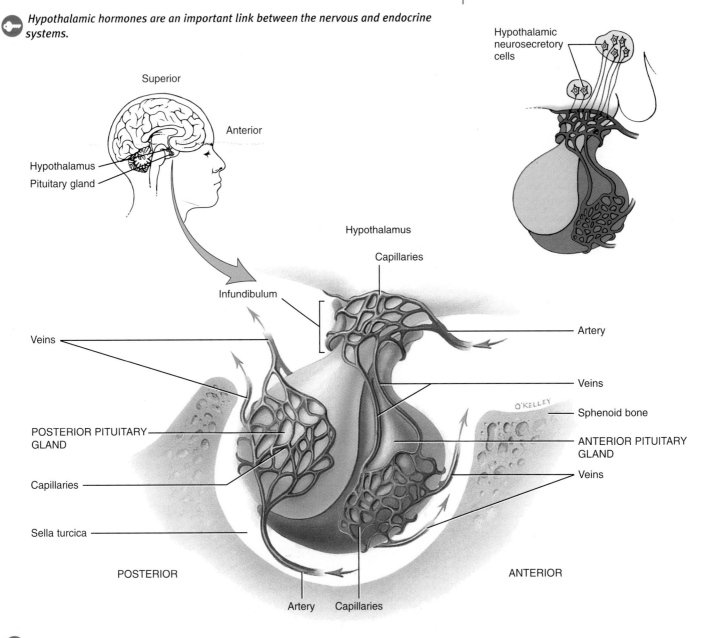

Q *Which portion of the pituitary gland is the glandular portion?*

7. **Melanocyte-stimulating hormone (MSH),** which affects skin pigmentation.

Human Growth Hormone (hGH)

Human growth hormone (hGH or **GH)** causes body cells to grow. It acts on the skeleton and skeletal muscles, in particular, to increase their rate of growth and maintain their size once growth is attained. The hormone causes cells to grow and multiply by increasing the rate at which amino acids enter cells and are built up into proteins. It also promotes fat break-

down and inhibits the use of glucose (blood sugar) for ATP production. The release of hGH from the anterior pituitary gland is controlled by two hormones from the hypothalamus: a releasing hormone and an inhibiting hormone. The releasing hormone circulates to the anterior pituitary gland and stimulates the release of hGH. The inhibiting hormone prevents the release of hGH.

Among the stimuli that promote hGH secretion is *hypoglycemia,* that is, low blood sugar level. When blood sugar level is low, the hypothalamus is stimulated to secrete the releasing hormone, which causes the release of hGH. This hor-

Figure 13.5   Hormones produced by the anterior pituitary gland and their functions.   *Know*

🔑 *Hormones that influence other endocrine glands are called tropic hormones.*

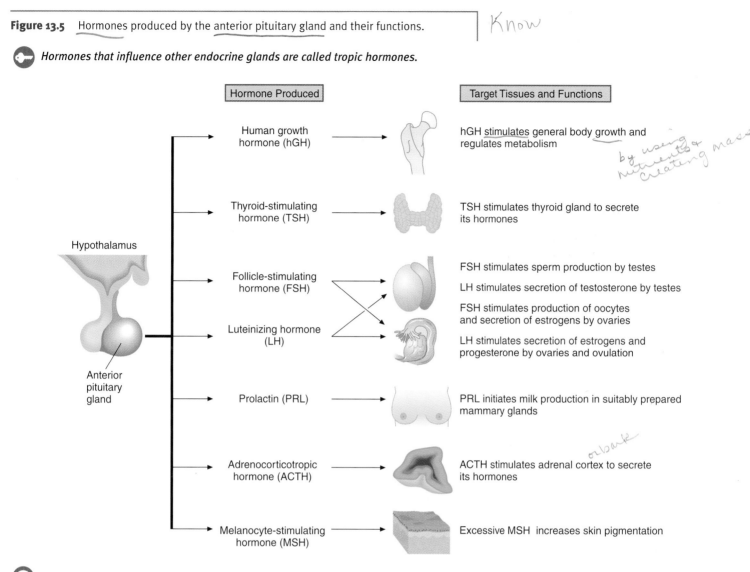

Hormone Produced	Target Tissues and Functions

Human growth hormone (hGH) — hGH stimulates general body growth and regulates metabolism *by using nutrients & creating mass*

Thyroid-stimulating hormone (TSH) — TSH stimulates thyroid gland to secrete its hormones

Follicle-stimulating hormone (FSH) — FSH stimulates sperm production by testes

Luteinizing hormone (LH) — LH stimulates secretion of testosterone by testes

FSH stimulates production of oocytes and secretion of estrogens by ovaries

LH stimulates secretion of estrogens and progesterone by ovaries and ovulation

Prolactin (PRL) — PRL initiates milk production in suitably prepared mammary glands

Adrenocorticotropic hormone (ACTH) — ACTH stimulates adrenal cortex to secrete its hormones *or bark*

Melanocyte-stimulating hormone (MSH) — Excessive MSH increases skin pigmentation

Hypothalamus

Anterior pituitary gland

❓ *What other endocrine glands are regulated by anterior pituitary gland hormones?*

mone, in turn, raises blood sugar level by converting glycogen into glucose and releasing it into the blood. As soon as blood sugar level returns to normal, releasing hormone secretion shuts off (Figure 13.6). This is a negative feedback system. Hyperglycemia (high blood sugar level) has the opposite effect as shown in Figure 13.6.

The secretion of hGH peaks at the end of the adolescent growth spurt. Genetic differences control when this will occur and thus account for differences in height.

Thyroid-Stimulating Hormone (TSH)

Thyroid-stimulating hormone (*TSH*) stimulates the production and secretion of hormones from the thyroid gland. Secretion is controlled by a hypothalamic releasing hormone. Release of the hypothalamic releasing hormone, in turn, depends on blood levels of thyroid hormones and the body's metabolic rate and operates according to a negative feedback system.

Adrenocorticotropic Hormone (ACTH)

Adrenocorticotropic hormone (*ACTH*) controls the production and secretion of certain adrenal cortex hormones. Secretion is controlled by a hypothalamic releasing hormone. Release of the hypothalamic releasing hormone, in turn, depends on a number of stimuli such as low blood glucose and physical stress and operates as a negative feedback system.

Follicle-Stimulating Hormone (FSH)

In the female, *follicle-stimulating hormone* (*FSH*) is transported from the anterior pituitary gland by the blood to the ovaries, where it stimulates the development of follicles (secretory cells around an oocyte) each month. FSH also stimulates cells in the ovaries to secrete estrogens, or female sex hormones. In the male, FSH stimulates the testes to produce sperm. Secretion of FSH is controlled by a hypothalamic releasing hormone. Release of a hypothalamic releasing hormone depends on blood levels of estrogens and testosterone

Figure 13.6 Regulation of human growth hormone (hGH) secretion.

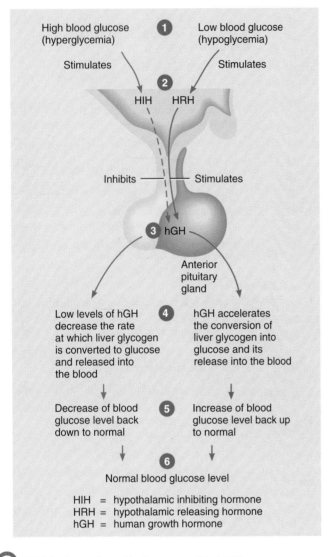

🔑 *Secretion of hGH is stimulated by a releasing hormone in response to hypoglycemia and is inhibited by an inhibiting hormone in response to hyperglycemia.*

HIH = hypothalamic inhibiting hormone
HRH = hypothalamic releasing hormone
hGH = human growth hormone

❓ *Which tissues in particular respond to hGH?*

(male sex hormone) and operates as a negative feedback system. High levels of estrogens and testosterone inhibit the release of a hypothalamic releasing hormone and thus the release of FSH.

Luteinizing Hormone (LH)

In the female, *luteinizing* (LOO-tē-in′-ĭz-ing) *hormone* (**LH**), together with FSH, stimulates secretion of estrogens by the ovaries and brings about the release of an oocyte by the ovary, a process called ovulation. LH also stimulates formation of the corpus luteum (ovarian structure formed after ovulation) in the ovary and secretion of progesterone (another female sex hormone) by the corpus luteum. In the male, LH stimulates the

testes to develop and secrete large amounts of testosterone. Secretion of LH, like that of FSH, is controlled by a hypothalamic releasing hormone.

Prolactin (PRL)

Prolactin (PRL), together with other hormones, initiates and maintains milk production by the mammary glands. The actual release of milk by the mammary glands depends on the hormone oxytocin, which is released from the posterior pituitary gland (discussed shortly). Together, milk secretion and ejection are referred to as lactation. By itself, PRL has only a weak effect; the mammary glands require preparation by other hormones, such as estrogens and progesterone. When the mammary glands have been primed by these hormones, PRL brings about milk secretion. The function of PRL in males is not known, but its oversecretion causes impotence (inability to have an erection of the penis). In females, oversecretion of PRL causes absence of menstrual cycles.

The hypothalamus secretes both an inhibiting and a releasing hormone that regulate PRL secretion. PRL level rises during pregnancy, apparently stimulated by a hypothalamic releasing hormone. Nursing an infant causes a reduction in secretion of the hypothalamic hormone.

Melanocyte-Stimulating Hormone (MSH)

The exact role of *melanocyte-stimulating hormone* (**MSH**) in humans is unknown. However, continued administration of MSH for several days does produce a darkening of the skin, and without MSH, the skin may be pale. A hypothalamic releasing hormone promotes MSH release, whereas a hypothalamic inhibiting hormone suppresses release.

Posterior Pituitary Gland

In a strict sense, the posterior pituitary gland is not an endocrine gland because it does not *make* hormones. Instead, it *stores* and, later, releases hormones. The posterior pituitary gland contains axon terminals of secretory neurons located in the hypothalamus called *neurosecretory cells* (Figure 13.7). The cell bodies of the neurons produce two hormones, *oxytocin* (**OT**) and *antidiuretic hormone* (**ADH**), which are transported through the axons to the posterior pituitary gland for storage and release.

Oxytocin (OT)

Oxytocin (ok′-sē-TŌ-sin; *oxytocia* = rapid childbirth) or **OT** stimulates the contraction of the smooth muscle cells in the pregnant uterus and the contractile cells of the mammary glands. It is released in large quantities just prior to giving birth (Figure 13.8 on page 298). When labor begins, the cervix (narrow, lower portion) of the uterus is distended (stretched) by the baby's body or head. Stretch receptors in the cervix send nerve impulses to the neurosecretory cells in the hypothalamus that stimulate the synthesis of OT. The OT is transported to the posterior pituitary gland and, from there, released into the blood and carried to the uterus to reinforce uterine contractions. As the contractions become more forceful, more OT is synthesized. As the baby's

Figure 13.7   Posterior pituitary gland connection to the hypothalamus. Note in the small figure to the right that hormones synthesized by neurosecretory cells in the hypothalamus pass in their axons down to the axon terminals in the posterior pituitary gland. Nerve impulses discharge the hormones, which diffuse into veins for distribution to target cells.

The posterior pituitary gland is not actually an endocrine gland because it stores and releases hormones rather than producing them.

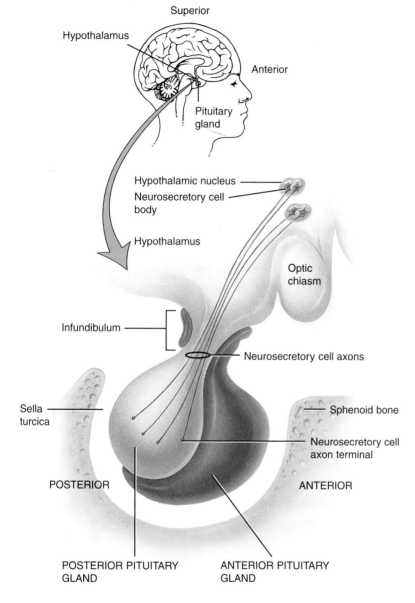

Which hormones are produced in the hypothalamus and stored in the posterior pituitary gland?

head passes through the uterine cervix, further distension of the cervix triggers the release of still more OT. Thus a positive feedback cycle is established. The cycle is broken by the birth of the infant because distension of the cervix suddenly lessens. Synthetic OT is used clinically to induce labor (the trade name is Syntocinon or Pitocin). In addition, Pitocin is used immediately after birth to increase uterine tone and control hemorrhage.

OT affects milk ejection. Milk formed by the glandular cells of the breasts is stored until the baby begins active sucking. This initiates a mechanism similar to that involved in forming and releasing OT for uterine muscle contractions. OT is transported from the posterior pituitary gland via the blood to the mammary glands, where it stimulates smooth muscle cells around the glandular cells and ducts to contract and eject milk. This response is called *milk ejection* (*let-down*).

Antidiuretic Hormone (ADH)

An *antidiuretic* is any chemical substance that decreases urine production. The principal physiological activity of *antidiuretic*

Figure 13.8   Regulation of the secretion of oxytocin (OT) during labor and delivery of a baby.

🔑 *OT secretion is regulated by a positive feedback cycle.*

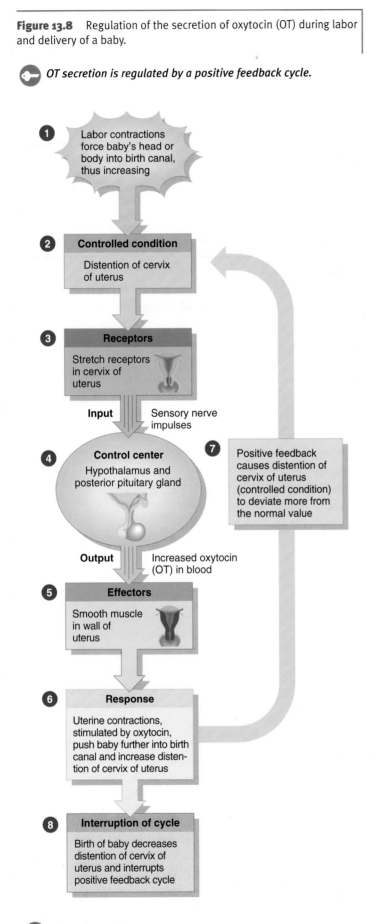

❓ *What are the two functions of OT?*

hormone (ADH) is its effect on urine volume. ADH causes the kidneys to remove water from fluid that will become urine and return it to the bloodstream, thus decreasing urine volume (antidiuresis).

ADH can also raise blood pressure by bringing about constriction of arterioles. For this reason, ADH is also referred to as *vasopressin.* If there is a severe loss of blood volume due to hemorrhage, ADH output increases.

The amount of ADH normally secreted varies with the body's needs (Figure 13.9). When the body is dehydrated, receptors in the hypothalamus called *osmoreceptors* (osmotic pressure detectors) detect the low water concentration in the blood and stimulate the neurosecretory cells in the hypothalamus to synthesize ADH, which is then transported to the posterior pituitary gland, released into the bloodstream, and transported to the kidneys. The kidneys respond by reabsorbing water back into the blood and decreasing urine output. ADH also decreases the rate at which perspiration is produced during dehydration.

Figure 13.9   Effects of antidiuretic hormone (ADH) and regulation of its secretion.

🔑 *ADH raises blood pressure.*

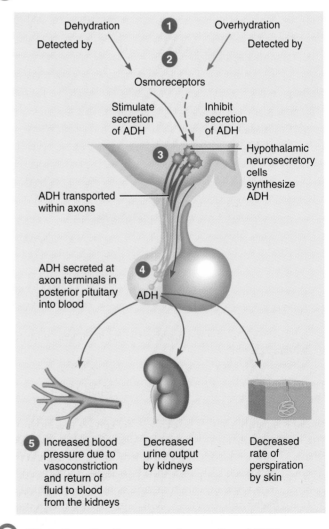

❓ *What other stimuli may promote secretion of ADH?*

Secretion of ADH can also be altered by a number of other conditions. Pain, stress, trauma, anxiety, nicotine, high blood levels of sodium ions (Na⁺), and drugs such as morphine and some anesthetics stimulate secretion of the hormone. Alcohol inhibits secretion and thereby increases urine output. This may be why thirst is one symptom of a hangover.

Exhibit 13.3 summarizes the actions and controls of secretion of pituitary gland hormones.

Thyroid Gland

objective: *Describe the location, histology, and functions of the thyroid gland.*

The *thyroid gland* is located just below the larynx (voice box) and in front of the trachea (windpipe). It consists of two lobes connected by a mass of tissue called an *isthmus* (IS-mus) (Figure 13.10a). It has a rich blood supply and thus can deliver high levels of hormone in a short period of time if necessary.

The thyroid gland is filled with *thyroid follicles*, which consist of *follicular cells* and *parafollicular cells* (Figure 13.10b). The follicular cells manufacture *thyroxine* (thi-ROK-sēn) or *T4,* because it contains four atoms of iodine, and *triiodothyronine* (trī-ī'-ōd-ō-THĪ-rō-nēn) or *T3,* because it contains three atoms of iodine. Together, these hormones are referred to as the *thyroid hormones*. They both contain iodine and both have the same effect. The parafollicular cells produce *calcitonin* (kal-si-TŌ-nin) or *CT.*

Function and Control of Thyroid Hormones

The thyroid hormones regulate (1) metabolism, (2) growth and development, and (3) the activity of the nervous system. In the regulation of metabolism, the thyroid hormones stimulate protein synthesis; increase lipolysis (fat breakdown); enhance cholesterol excretion; and increase the use of glucose for ATP production.

Exhibit 13.3	Summary of Pituitary Gland Hormones, Principal Actions, and Control of Secretion	
HORMONE	PRINCIPAL ACTIONS	CONTROL OF SECRETION
Anterior pituitary gland		
Human growth hormone (hGH)	Growth of body cells; promotes protein buildup; promotes fat breakdown; inhibits use of glucose for ATP production.	Releasing hormone and inhibiting hormone.
Thyroid-stimulating hormone (TSH)	Controls secretion of thyroid hormones by thyroid glands.	Releasing hormone.
Adrenocorticotropic hormone (ACTH)	Controls secretion of some hormones by adrenal cortex.	Releasing hormone.
Follicle-stimulating hormone (FSH)	In females, initiates development of oocytes and induces ovarian secretion of estrogens. In males, stimulates testes to produce sperm.	Releasing hormone.
Luteinizing hormone (LH)	In females, stimulates ovulation and formulation of corpus luteum, which secretes estrogens and progesterone; in males, stimulates testes to produce testosterone.	Releasing hormone.
Prolactin (PRL)	In females, initiates and maintains milk secretion by the mammary glands.	Releasing hormone and inhibiting hormone.
Melanocyte-stimulating hormone (MSH)	Produces darkening of the skin.	Releasing hormone and inhibiting hormone.
Posterior pituitary gland		
Oxytocin (OT)	Stimulates contraction of smooth muscle cells of pregnant uterus during labor and stimulates milk ejection from mammary glands.	Neurosecretory cells of hypothalamus secrete OT in response to uterine distension and stimulation of nipples.
Antidiuretic hormone (ADH)	Principal effect is to decrease urine volume; also raises blood pressure by constricting arterioles.	Neurosecretory cells of hypothalamus secrete ADH in response to low water concentration of the blood, stress, trauma, anxiety, nicotine, some anesthetics, and tranquilizers; alcohol inhibits secretion.

Figure 13.10   Thyroid gland.

Thyroid hormones regulate (1) metabolism, (2) growth and development, and (3) the activity of the nervous system.

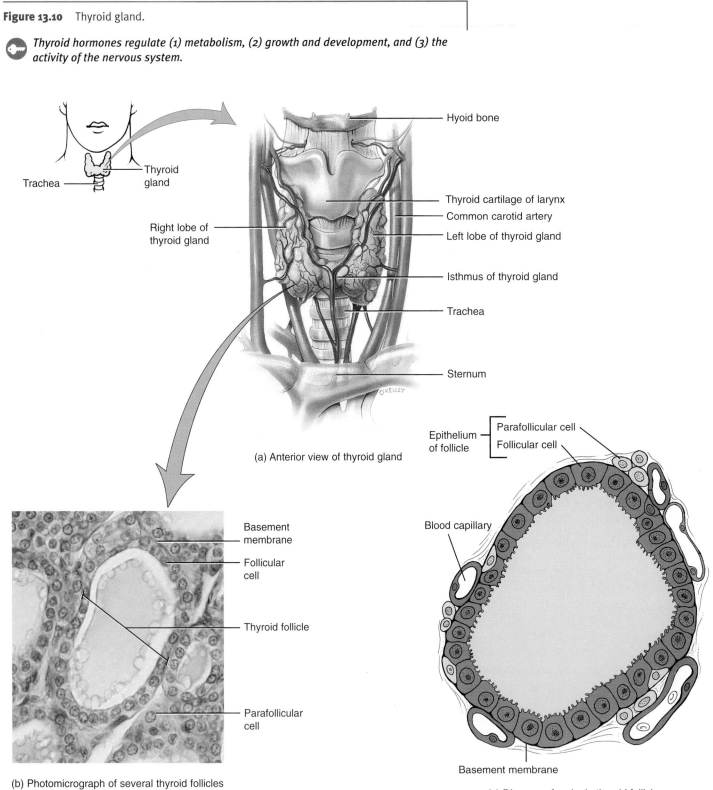

Trachea — Thyroid gland

Hyoid bone

Thyroid cartilage of larynx
Common carotid artery

Right lobe of thyroid gland

Left lobe of thyroid gland

Isthmus of thyroid gland

Trachea

Sternum

O'KELLEY

(a) Anterior view of thyroid gland

Epithelium of follicle — Parafollicular cell / Follicular cell

Blood capillary

Basement membrane

Follicular cell

Thyroid follicle

Parafollicular cell

Basement membrane

(b) Photomicrograph of several thyroid follicles (500x)

(c) Diagram of a single thyroid follicle

 Which cells secrete T3 and T4? Calcitonin? Which of these hormones are also called thyroid hormones?

Together with hGH and insulin, thyroid hormones accelerate body growth, particularly the growth of nervous tissue. Deficiency of thyroid hormones during fetal development can result in fewer and smaller neurons, defective myelination of axons, and mental retardation. During the early years of life, deficiency of thyroid hormones results in small stature and poor development of certain organs such as the brain and reproductive organs.

The secretion of thyroid hormones is stimulated by several factors (Figure 13.11).

1 If thyroid hormone levels in the blood fall below normal or the metabolic rate decreases, chemical sensors in the hypothalamus detect the change in blood chemistry and stimulate the hypothalamus to secrete a releasing hormone.

2 This hormone stimulates the anterior pituitary gland to secrete thyroid-stimulating hormone (TSH).

3 TSH stimulates the thyroid gland.

4 The thyroid gland releases thyroid hormones.

5 Blood levels of thyroid hormones increase until the metabolic rate returns to normal.

6 Thyroid hormones inhibit the release of releasing hormone and TSH. *from ant. pituitary*

Conditions that increase the body's need for energy—a cold environment, high altitude, pregnancy—increase the secretions of thyroid hormones.

Thyroid activity can be inhibited by a number of factors. When large amounts of certain sex hormones (estrogens and androgens) are circulating in the blood, for example, TSH secretion diminishes. Aging slows down the activities of most glands, and thyroid production may decrease. This is one of the factors that often accounts for weight gain as people age.

Calcitonin (CT)

The hormone produced by the parafollicular cells of the thyroid gland is *calcitonin* or *CT*. It is involved in the homeostasis of blood calcium (Ca^{2+}) and phosphate ($HPO4^{2-}$) levels. CT lowers the amount of calcium and phosphate in the blood by (1) inhibiting bone breakdown (specifically, by inhibiting osteoclasts—bone-destroying cells), (2) accelerating the uptake of calcium and phosphate by the bones, and (3) increasing movement of calcium and phosphate from urine into blood. *the output of blood into bones*

Blood calcium level directly controls the secretion of CT by a negative feedback system (Figure 13.12). *Decreases blood calcium*

Parathyroid Glands *= Breaking Bone Down / Increases blood Calcium*

objective: *Describe the location, histology, and functions of the parathyroid glands.* *Calcium back into blood from bone*

Attached to the posterior surfaces of the thyroid gland are small, round masses of tissue called the *parathyroid (para =*

Figure 13.11   Regulation of thyroid secretion.

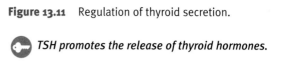

TSH promotes the release of thyroid hormones.

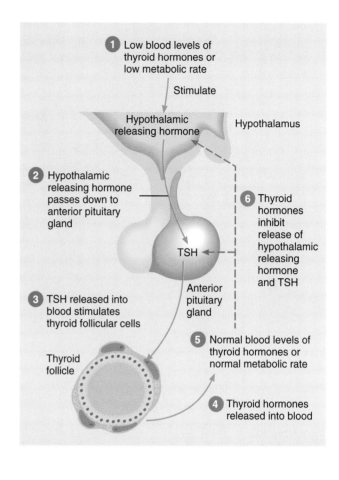

What is the effect of thyroid hormones on metabolic rate?

beside) *glands.* Usually, two parathyroids, superior and inferior, are attached to each thyroid lobe (Figure 13.13a, p. 303).

Microscopically, the parathyroids contain two kinds of epithelial cells (Figure 13.13b, p. 303). The more numerous cells, called *principal (chief) cells,* are believed to be the major producer of *parathyroid hormone* (PTH). The function of the other cells, called *oxyphil cells,* is unknown.

Parathyroid Hormone (PTH)

Parathyroid hormone (PTH) helps to control the level of calcium (Ca^{2+}) and phosphate (HPO_4^{2-}) ions in the blood, but in a different way than calcitonin. PTH helps activate vitamin D and increases the rate of calcium and phosphate absorption from the gastrointestinal tract into the blood. PTH also increases the number and activity of osteoclasts (bone-destroying cells), which cause bone tissue to break down and additional calcium and phosphate to be released into the blood.

Figure 13.12　Regulation of the secretion of the parathyroid hormone (PTH) and calcitonin (CT).

🔑 *As far as blood levels of calcium are concerned, PTH and CT have opposite functions.*

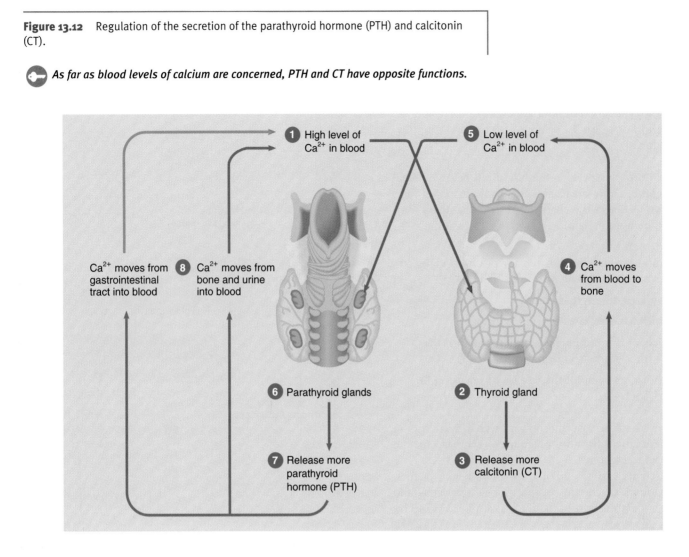

❓ *Will a high blood level of calcium cause increased secretion by the thyroid gland or parathyroid glands?*

Finally, PTH increases the rate at which the kidneys remove calcium from urine that is being formed and returns it to the blood, and increases phosphate loss in urine. More phosphate is lost through the urine than is gained from the bones.

The overall effect of PTH with respect to ions, then, is to decrease blood phosphate level and increase blood calcium level. As far as blood calcium level is concerned, PTH and CT have opposite functions.

When the calcium level of the blood falls, more PTH is released (see Figure 13.12). Conversely, when the calcium level of the blood rises, less PTH (and more CT) is secreted. This is another example of a negative feedback control system.

Adrenal Glands

objective:　*Describe the location, histology, and functions of the adrenal glands.*

The body has two *adrenal glands,* one of which is located superior to each kidney (Figure 13.14 on page 304). Each adrenal gland is composed of two regions: the outer *adrenal cortex,* which makes up the bulk of the gland, and the inner *adrenal medulla.* Each region produces different hormones. Like the thyroid gland, the adrenal glands have a very rich blood supply.

Figure 13.13   Parathyroid glands.

🔑 *The parathyroid glands are attached to the posterior surface of the thyroid gland.*

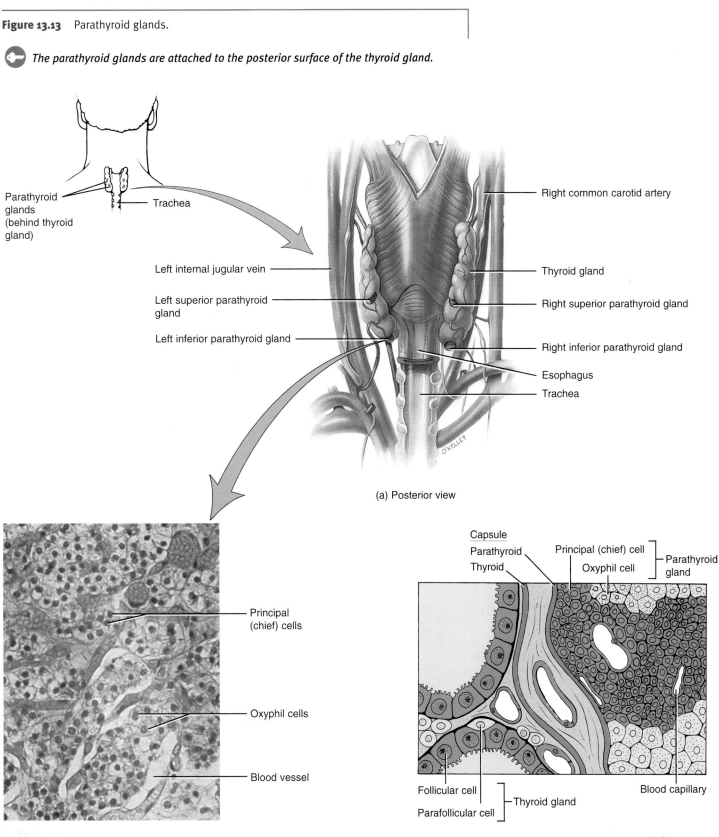

Parathyroid glands (behind thyroid gland)

Trachea

Left internal jugular vein

Left superior parathyroid gland

Left inferior parathyroid gland

Right common carotid artery

Thyroid gland

Right superior parathyroid gland

Right inferior parathyroid gland

Esophagus

Trachea

O'KELLEY

(a) Posterior view

Principal (chief) cells

Oxyphil cells

Blood vessel

b) Photomicrograph of parathyroid gland (340x)

Capsule
Parathyroid
Thyroid

Principal (chief) cell
Oxyphil cell

Parathyroid gland

Follicular cell
Parafollicular cell

Thyroid gland

Blood capillary

(c) Diagram of a portion of the thyroid gland (left) and parathyroid gland (right)

❓ *What effect does PTH have on osteoclasts?*

Figure 13.14   Adrenal glands.

The adrenal cortex secretes steroid hormones that are essential for life.

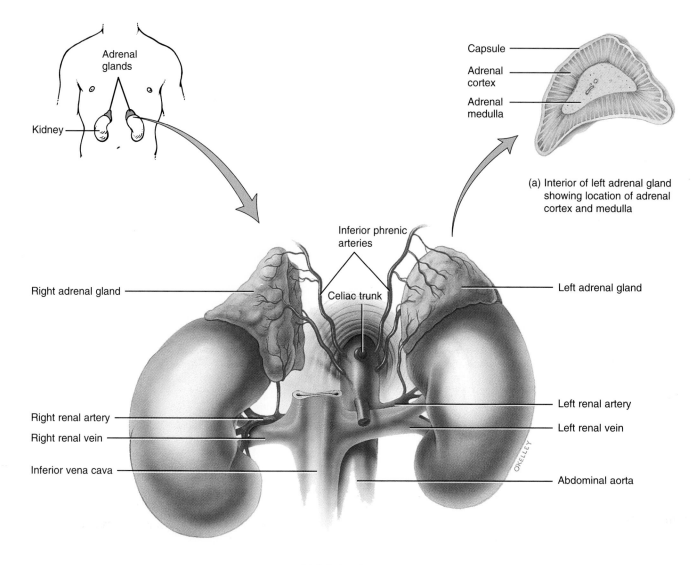

Adrenal
glands

Kidney

Capsule

Adrenal
cortex

Adrenal
medulla

(a) Interior of left adrenal gland
showing location of adrenal
cortex and medulla

Inferior phrenic
arteries

Right adrenal gland

Celiac trunk

Left adrenal gland

Right renal artery

Right renal vein

Inferior vena cava

Left renal artery

Left renal vein

Abdominal aorta

(b) Anterior view

Figure 13.14   (Continued)

SUPERFICIAL

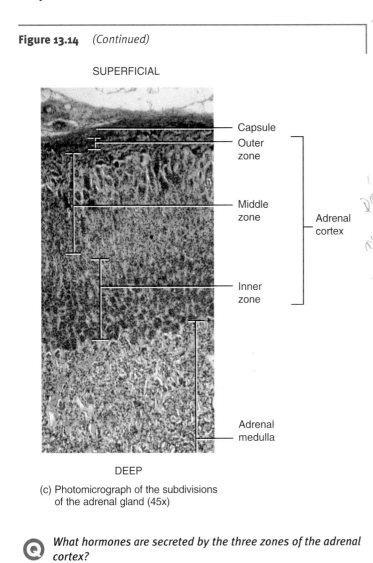

DEEP

(c) Photomicrograph of the subdivisions
of the adrenal gland (45x)

What hormones are secreted by the three zones of the adrenal cortex?

Adrenal Cortex

The adrenal cortex is subdivided into three zones, and each zone secretes different groups of steroid hormones. The outer zone secretes hormones called mineralocorticoids (min'-er-al-ō-KOR-ti-koyds) because they affect mineral homeostasis. The middle zone secretes hormones called glucocorticoids (gloo'-kō-KOR-ti-koyds) because they affect glucose homeostasis. The inner zone synthesizes small amounts of androgens (male sex hormones).

Mineralocorticoids

Mineralocorticoids help control the homeostasis of water, sodium ions (Na⁺), and potassium ions (K⁺). Although the adrenal cortex secretes several mineralocorticoids, the one

responsible for about 95 percent of mineralocorticoid activity is *aldosterone* (al-do-STER-ōn). Aldosterone acts on certain cells in the kidneys to increase their reabsorption of sodium ions (Na⁺) from the urine and thus return them to the blood. At the same time, aldosterone stimulates excretion of potassium ions (K⁺), so that large amounts of potassium are lost in the urine.

Control of aldosterone secretion involves several mechanisms operating simultaneously:

1 The most important mechanism of control involves the *renin–angiotensin* (an'jē-ō-TEN-sin) *pathway* (Figure 13.15). Conditions such as dehydration, Na⁺ deficiency, or hemorrhage may trigger the renin–angiotensin pathway.

2 These conditions cause a decrease in blood volume.

3 Decreased blood volume causes a decrease in blood pressure.

4 Lowered blood pressure stimulates kidney cells called juxtaglomerular cells to secrete an enzyme called *renin* (RĒ-nin).

5 This results in increased levels of renin in the blood.

6 Renin converts *angiotensinogen*, a plasma protein produced by the liver, into *angiotensin I*.

7 Increased levels of angiotensin I circulate to the lungs.

8 As blood flows through the lungs, angiotensin I is converted into *angiotensin II*.

9 This results in increased levels of angiotensin II.

10 Angiotensin II is a hormone that stimulates the adrenal cortex to secrete aldosterone.

11 Increased levels of aldosterone circulate to the kidneys.

12 In the kidneys, aldosterone increases Na⁺ reabsorption, and water follows by osmosis. Aldosterone also leads to increased K⁺ secretion by the kidneys into urine.

13 As a result, there is an increase in blood volume.

14 Increased blood volume raises blood volume to normal.

15 As blood volume increases, blood pressure increases to normal.

16 Angiotensin II also acts on the smooth muscle in the walls of arterioles, which responds by contracting to produce vasoconstriction.

17 Vasoconstriction of arterioles also increases blood pressure and thus helps raise blood pressure to normal.

Figure 13.15   Regulation of the secretion of aldosterone by the renin–angiotensin pathway.

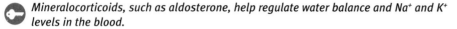

Mineralocorticoids, such as aldosterone, help regulate water balance and Na⁺ and K⁺ levels in the blood.

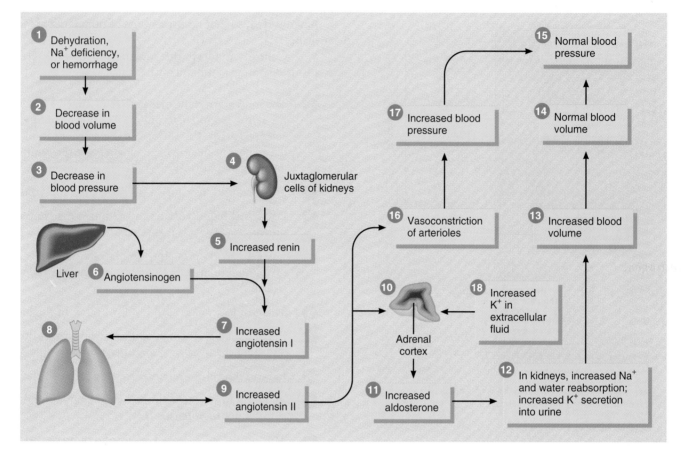

Q   *How does angiotensin II increase blood pressure?*

---

18   A second mechanism for the control of aldosterone secretion is the blood K⁺ level. An increase in the K⁺ concentration of blood and thus of interstitial fluid directly stimulates aldosterone secretion by the adrenal cortex and causes the kidneys to eliminate excess K⁺. A decline in the blood K⁺ level has the opposite effect.

Glucocorticoids

Glucocorticoids are concerned with metabolism and resistance to stress. *Cortisol* is the most abundant and is responsible for most glucocorticoid activity. The glucocorticoids have the following effects on the body:

1. Glucocorticoids work with other hormones to promote normal metabolism by making sure enough ATP (energy) is available. They increase the rate at which proteins are catabolized and amino acids are transported to the liver to be synthesized into new proteins. Or the liver may convert the amino acids to glucose if the body's reserves of glycogen and fat are low. Glucocorticoids also stimulate the breakdown of fat and the release of fatty acids from adipose tissue as an additional energy source.

2. Glucocorticoids help provide resistance to stress. A sudden increase in available glucose makes the body more alert and supplies the ATP needed to combat stresses ranging from fright and temperature extremes to high

altitude and surgery. Glucocorticoids also help raise blood pressure, which is especially advantageous if the stress happens to be blood loss, which causes a drop in blood pressure.

3. Glucocorticoids are anti-inflammatory compounds; that is, they inhibit the release of chemicals that cause inflammation. Unfortunately, they also slow wound healing. Although high doses of glucocorticoid drugs cause atrophy of immune system organs, thus depressing the body's ability to fight disease, they may be useful in the treatment of chronic inflammation.

The control of glucocorticoid secretion is a typical negative feedback mechanism (Figure 13.16). Low blood levels of glucocorticoids stimulate the hypothalamus to secrete a releasing hormone, which initiates the release of ACTH from the anterior pituitary gland, which, in turn, stimulates glucocorticoid secretion.

Androgens   gonado corticoids

The adrenal cortex secretes small amounts of androgens, male sex hormones. The concentration of sex hormones normally secreted by adult male adrenal glands is so low as to be insignificant in comparison to the levels of testosterone. In females, androgens contribute to sex drive (libido). Androgens help in the prepubertal growth spurt and early development of axillary and pubic hair in boys and girls.

Adrenal Medulla

The adrenal medulla consists of sympathetic autonomic nervous system (ANS) postganglionic cells that are specialized to secrete hormones. Hormone secretion is directly controlled by the ANS and the gland responds rapidly to a stimulus.

Epinephrine and Norepinephrine (NE)

The two principal hormones synthesized by the adrenal medulla are *epinephrine* and *norepinephrine* (*NE*), also called adrenaline and noradrenaline, respectively. Epinephrine constitutes about 80 percent of the gland's total secretion and is more potent than norepinephrine. Both hormones produce effects that mimic those brought about by the sympathetic division of the autonomic nervous system. To a large extent, they are responsible for the fight-or-flight response. Like the glucocorticoids of the adrenal cortices, they help the body resist stress.

Under stress, impulses received by the hypothalamus are conveyed to sympathetic preganglionic neurons and then to

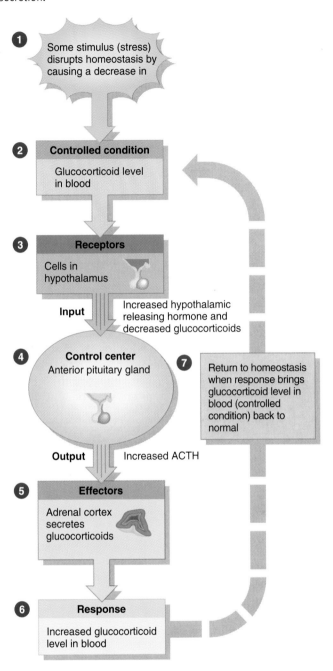

Figure 13.16   Negative feedback regulation of glucocorticoid secretion.

① Some stimulus (stress) disrupts homeostasis by causing a decrease in

② **Controlled condition**
Glucocorticoid level in blood

③ **Receptors**
Cells in hypothalamus

Input   Increased hypothalamic releasing hormone and decreased glucocorticoids

④ **Control center**
Anterior pituitary gland

⑦ Return to homeostasis when response brings glucocorticoid level in blood (controlled condition) back to normal

Output   Increased ACTH

⑤ **Effectors**
Adrenal cortex secretes glucocorticoids

⑥ **Response**
Increased glucocorticoid level in blood

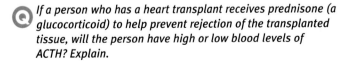 *If a person who has a heart transplant receives prednisone (a glucocorticoid) to help prevent rejection of the transplanted tissue, will the person have high or low blood levels of ACTH? Explain.*

the adrenal medulla, which increases the output of epinephrine and norepinephrine. Both hormones increase blood pressure by increasing heart rate and constricting blood vessels, accelerating the rate of respiration, dilating respiratory passageways, decreasing the rate of digestion, increasing the efficiency of muscular contractions, increasing blood sugar level, and stimulating cellular metabolism. Hypoglycemia (low blood sugar) may also stimulate secretion of epinephrine and norepinephrine.

Pancreas *Know hormonal aspect*

objective: *Describe the location, histology, and functions of the pancreas.*

The *pancreas* (*pan* = all; *kreas* = flesh) can be classified as both an endocrine and an exocrine gland. We will treat its endocrine functions here and discuss its exocrine functions with the digestive system (see Chapter 19). The pancreas is a flattened organ located behind and slightly below the stomach (Figure 13.17a).

The endocrine portion of the pancreas consists of clusters of cells called *pancreatic islets* or *islets of Langerhans* (LAHNG-er-hanz) (Figure 13.17b). The islets contain numerous blood capillaries and are surrounded by cells that form the exocrine part of the gland. At least four major kinds of cells are found in these clusters: (1) *alpha cells* that secrete the hormone glucagon; (2) *beta cells* that secrete the hormone insulin; (3) *delta cells* that secrete somatostatin, which inhibits secretion of insulin and glucagon; and (4) *F-cells* that secrete pancreatic polypeptide, which regulates the release of digestive enzymes from the pancreas. Glucagon and insulin are the chief regulators of blood sugar level.

Glucagon

Glucagon (GLOO-ka-gon) increases the blood glucose level (Figure 13.18 on page 310). It does this by (1) accelerating the conversion of glycogen in the liver into glucose; (2) promoting the conversion in the liver of other nutrients, such as amino acids and lactic acid, into glucose; and (3) stimulating the release of glucose from the liver into the blood. As a result, blood sugar level rises.

Secretion of glucagon is directly controlled by the level of blood sugar via a negative feedback system. When the blood sugar level falls below normal, chemical sensors in the alpha cells of the islets stimulate the cells to secrete glucagon. When blood sugar rises, the cells are no longer stimulated and production decreases. If for some reason the self-regulating device fails and the alpha cells secrete glucagon continuously, hyperglycemia (high blood sugar level) may result. High-protein meals, which raise the amino acid level of the blood, stimulate glucagon secretion, whereas somatostatin inhibits it.

Insulin

Insulin's action is opposite that of glucagon. It decreases blood sugar level when it is above normal. It does this in several ways (Figure 13.18):

1. Insulin accelerates the transport of glucose from blood into cells, especially skeletal muscle fibers.

2. Insulin accelerates the conversion of glucose into glycogen and synthesis of fatty acids.

3. Insulin accelerates the movement of amino acids into body cells, thus speeding up protein synthesis within cells.

4. Insulin decreases the conversion of glycogen in the liver into glucose.

5. Insulin slows down glucose formation from lactic acid and certain amino acids.

The regulation of insulin secretion, like that of glucagon secretion, is directly determined by the level of sugar in the blood and is based on a negative feedback system. Increased blood levels of certain amino acids also stimulate insulin release, and several other hormones also stimulate insulin secretion.

The role of insulin along with other factors that are used to control diabetes is discussed in the Wellness Focus box for this chapter.

Ovaries and Testes *Skip this section*

objective: *Describe the location, histology, and functions of the ovaries and testes.*

The female gonads, called the *ovaries,* are paired oval bodies located in the pelvic cavity. They produce *estrogens* and *progesterone,* which are responsible for the development and maintenance of the female sexual characteristics. Along with FSH and LH of the anterior pituitary gland, the sex hormones regulate the menstrual cycle, maintain pregnancy, and prepare the mammary glands for lactation. The ovaries (and placenta) also produce a hormone called *relaxin,* which relaxes the pubic symphysis and helps dilate the uterine cervix toward the end of pregnancy. *Inhibin,* a hormone that inhibits secretion of FSH, is also produced by the ovaries.

The male has two oval glands, called *testes,* that lie in the scrotum. The testes produce *testosterone,* the primary male sex hormone, which stimulates the development of male sexual characteristics. The testes also produce inhibin that inhibits secretion of FSH. The detailed structure of the ovaries and testes and the specific roles of sex hormones will be discussed in Chapter 23.

Figure 13.17 Pancreas.

🔑 *The pancreas is both an endocrine gland and an exocrine gland.*

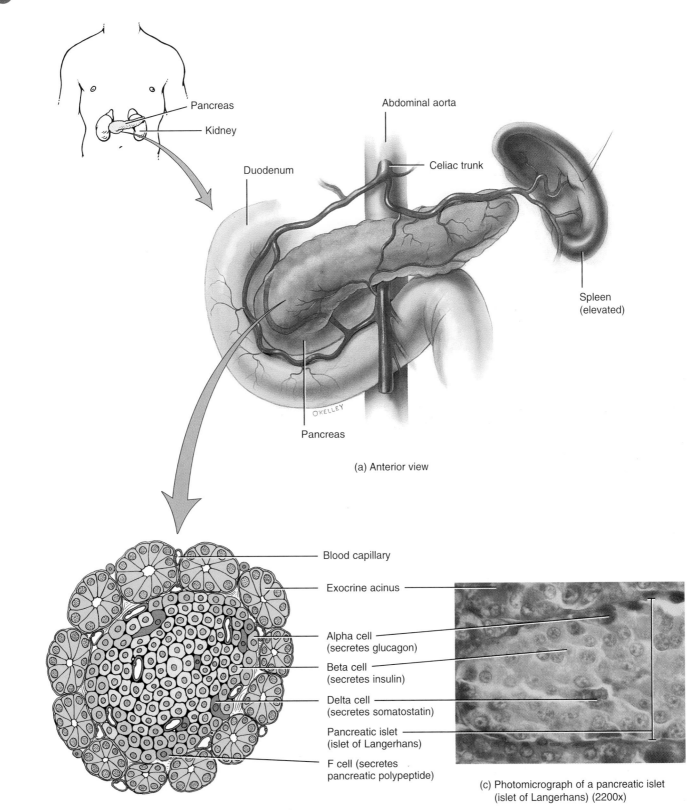

Pancreas

Kidney

Abdominal aorta

Duodenum

Celiac trunk

Spleen (elevated)

OKELLEY

Pancreas

(a) Anterior view

Blood capillary

Exocrine acinus

Alpha cell (secretes glucagon)

Beta cell (secretes insulin)

Delta cell (secretes somatostatin)

Pancreatic islet (islet of Langerhans)

F cell (secretes pancreatic polypeptide)

(b) Diagram of a pancreatic islet (islet of Langerhans)

(c) Photomicrograph of a pancreatic islet (islet of Langerhans) (2200x)

❓ *Which pancreatic hormones are the chief regulators of blood sugar level?*

Figure 13.18 Regulation of the secretion of glucagon and insulin.

Low blood glucose level stimulates release of glucagon whereas high blood glucose level stimulates secretion of insulin.

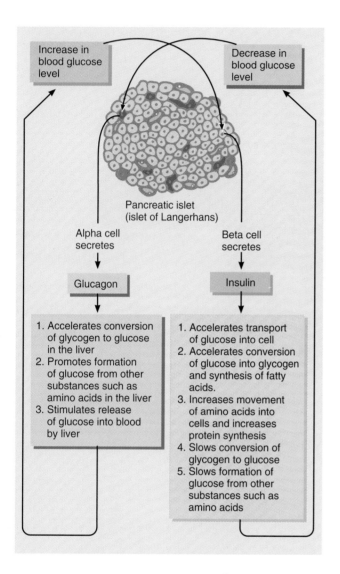

Why is glucagon sometimes called an anti-insulin hormone?

Pineal Gland

objective: *Describe the location, histology, and the functions of the pineal gland.*

The *pineal* (PĪN-ē-al; *pinealis* = shaped like a pine cone) *gland* is located near the thalamus in the brain (see Figure

13.1). Although the anatomy of the pineal gland is well known, its physiology is still obscure. One hormone secreted by the pineal gland is *melatonin.* More melatonin is produced during darkness. Its formation is interrupted when light entering the eyes inhibits melatonin secretion. In this way, the release of melatonin is governed by the daily dark–light cycle. In humans, melatonin may inhibit the secretion of FSH and LH by the anterior pituitary gland and may help to regulate the menstrual cycle and control the onset of puberty.

Thymus Gland

The *thymus gland* is behind the sternum between the lungs. It plays two roles in the body's immune system. The larger role, which involves inducing certain cells to produce antibodies, will be discussed in Chapter 17. The other role is hormonal and has to do with the maturation of T cells, a type of white blood cell. T cells destroy foreign microbes and substances. The thymic hormones called *thymosin, thymic humoral factor* (*THF*), *thymic factor* (*TF*), and *thymopoietin* promote the development of T cells. There is some evidence that thymic hormones may retard aging.

Other Endocrine Tissues

There are body tissues other than those normally classified as endocrine glands that contain endocrine tissue and thus secrete hormones. These are summarized in Exhibit 13.4.

Eicosanoids

Two families of eicosanoids, *prostaglandins* (pros′-ta-GLAN-dins), or *PGs,* and *leukotrienes* (loo-kō-TRĪ-ēns), or *LTs,* act as local hormones in most tissues of the body. This means that their site of action is the immediate area in which they are produced. They do not exert their effects on distant tissues like most other hormones. Leukotrienes are important in tissue inflammation. Prostaglandins are important in the normal physiology of smooth muscle (such as uterine contraction), secretion, blood flow, reproduction, platelet function, respiration, nerve impulse transmission, fat metabolism, and the immune response. Prostaglandins are also involved in certain pathologies; they help induce inflammation, promote fever, and intensify pain. Drugs such as aspirin and ibuprofen (Motrin) inhibit prostaglandin synthesis and thus reduce fever and decrease pain.

Exhibit 13.4	Summary of Hormones Produced by Organs That Contain Endocrine Cells
SITE OF PRODUCTION AND HORMONES	**ACTION**
Gastrointestinal tract	
Gastrin	Promotes secretion of gastric juice and increases movements of the gastrointestinal tract.
Gastric inhibitory peptide	Inhibits secretion of gastric juice and decreases movements of the gastrointestinal tract.
Secretin	Stimulates secretion of pancreatic juice and bile.
Cholecystokinin	Stimulates secretion of pancreatic juice, regulates release of bile from the gallbladder, and brings about a feeling of fullness following eating.
Placenta	
Human chorionic gonadotropin	Stimulates the production of estrogens and progesterone by the ovaries to maintain pregnancy.
Estrogens and progesterone	Maintain pregnancy and prepare mammary glands to secrete milk.
Human chorionic somatomammotropin	Stimulates the development of the mammary glands for lactation.
Kidney	
Erythropoietin	Stimulates red blood cell production.
Skin	
Vitamin D	Aids in the absorption of calcium and phosphorus.
Heart	
Atrial natriuretic peptide	Decreases blood pressure.

Stress and the General Adaptation Syndrome (GAS)

Theory — not fact of science

objective: *Describe how the body responds to stress and how stress and disease are related.*

Homeostatic mechanisms attempt to counteract the everyday stresses of living. If the mechanisms are successful, the internal environment maintains normal physiological limits of its chemicals, temperature, and pressure. If a stress is extreme, unusual, or long-lasting, however, the normal mechanisms may not be sufficient. In this case, the stress triggers a wide-ranging set of bodily changes called the *general adaptation syndrome* (GAS). It should be pointed out that it is impossible to remove all stress from our everyday lives. In fact, there is "positive," productive stress, referred to as *eustress* (*eu* = true), as well as "negative," harmful stress, referred to as *distress*.

Stressors

The hypothalamus is the body's watchdog. It has sensors that detect changes in the chemistry, temperature, and pressure of the blood and it is informed of emotions through tracts that connect it with the emotional centers of the cerebral cortex. When the hypothalamus senses stress, it initiates the general adaptation syndrome.

A *stressor* may be almost any disturbance—heat or cold, environmental poisons, surgery, or a strong emotional reaction. Stressors vary among different people and even in the same person at different times.

When a stressor appears, it stimulates the hypothalamus to initiate the syndrome through two pathways. The first pathway produces an immediate set of responses called the alarm reaction. The second pathway, called the resistance reaction, is slower to start, but its effects last longer.

Alarm Reaction   *Stage 1*

The *alarm reaction,* also called the *fight-or-flight response* or *sympathetic response,* is actually a complex of reactions initiated by hypothalamic stimulation of the sympathetic nervous system and the adrenal medulla (Figure 13.19a). The reactions are immediate and short-lived, mobilizing the body's resources for immediate physical activity. In essence, the alarm reaction brings large amounts of glucose and oxygen to the organs that are most active in warding off danger. These are the brain, which must become highly alert; the skeletal muscles, which may have to contract very forcefully to fight off an attacker; and the heart, which must work furiously to pump enough materials to the brain and muscles. See Chapter 11 on activities of the autonomic nervous system to review the specific fight-or-flight responses.

Sympathetic impulses to the adrenal medulla increase its secretion of epinephrine and norepinephrine (NE). These hormones supplement and prolong many sympathetic responses—increasing heart rate and strength, increasing blood pressure, constricting blood vessels, accelerating the rate of breathing, widening respiratory passageways, increasing the rate of glycogen catabolism to increase the blood sugar level, and decreasing the rate of digestion.

If you group the stress responses of the alarm reaction by function, you will note that they are designed to increase circulation rapidly, promote catabolism for energy production, and decrease nonessential activities. If the stress is great enough, the body mechanisms may not be able to cope and death can result. During the alarm reaction, digestive, urinary, and reproductive activities are inhibited.

Resistance Reaction   *Stage 2*

The second stage of the GAS is the *resistance reaction* (Figure 13.19b). Unlike the short-lived alarm reaction that is initiated by nerve impulses from the hypothalamus, the resistance reaction is initiated by releasing hormones secreted by the hypothalamus and is a long-term reaction.

One releasing hormone stimulates the anterior pituitary gland to increase secretion of ACTH. ACTH stimulates the release of mineralocorticoids and glucocorticoids. The mineralocorticoids conserve sodium ions. Sodium retention also leads to water retention, thus maintaining the high blood pressure that is typical of the alarm reaction. This would make up for fluid lost through severe bleeding. The glucocorticoids bring about a number of stress reactions; these were detailed earlier in the chapter.

Two other releasing hormones are secreted by the hypothalamus in response to stress. They cause the anterior pitu-

itary gland to secrete thyroid-stimulating hormone (TSH) and human growth hormone (hGH). TSH stimulates the thyroid to secrete thyroxine and triiodothyronine, which increase the catabolism of carbohydrates. hGH stimulates the catabolism of fats and the conversion of glycogen to glucose. The combined actions of TSH and hGH increase catabolism and thereby supply additional energy for the body.

The resistance reaction of the GAS allows the body to continue fighting a stressor long after the alarm reaction dissipates. It provides the energy, functional proteins, and circulatory changes required for meeting emotional crises, performing strenuous tasks, fighting infection, or resisting the threat of bleeding to death. However, cells use glucose at the same rate it enters the bloodstream, which means blood sugar level returns to normal. Blood pressure remains abnormally high, though, because the retention of water increases the volume of blood.

Generally, the resistance reaction is successful in seeing us through a stressful situation, and our bodies then return to normal. Occasionally, the resistance reaction fails to combat the stressor. In this case, the general adaptation syndrome moves into the stage of exhaustion.

Exhaustion   *Stage 3*

A major cause of exhaustion is loss of potassium ions. When the mineralocorticoids stimulate the kidneys to retain sodium ions, potassium and hydrogen ions are exchanged for sodium ions and secreted in the fluid that will become urine. As the chief positive ion in cells, potassium is partly responsible for controlling the water concentration of the cytosol. As the cells lose more and more potassium, they function less and less effectively. Finally, they start to die. This condition is called the *stage of exhaustion.* Unless it is rapidly reversed, vital organs cease functioning and the person dies.

Another cause of exhaustion is depletion of the adrenal glucocorticoids. In this case, blood glucose level suddenly falls, and the cells do not receive enough nutrients. A final cause of exhaustion is weakened organs. A long-term or strong resistance reaction puts heavy demands on the body, particularly on the heart, blood vessels, and adrenal cortex. They may not be up to handling the demands, or they may fail under the strain. In this respect, ability to handle stressors is determined to a large degree by general health.

Stress and Disease

Although the exact role of stress in human diseases is not known, it is becoming quite clear that stress can lead to certain diseases by temporarily inhibiting certain components of the immune system. Stress-related conditions include gastritis, ulcerative colitis, irritable bowel syndrome, peptic ulcers, hypertension, asthma, migraine headaches, anxiety, and depression. Individuals under stress are also at greater risk of developing chronic disease or dying prematurely.

Figure 13.19   Responses to stressors during the general adaptation syndrome (GAS). Colored arrows indicate immediate reactions. Black arrows indicate long-term reactions.

🔑 *Stressors stimulate the hypothalamus to initiate the GAS through the alarm reaction and the resistance reaction.*

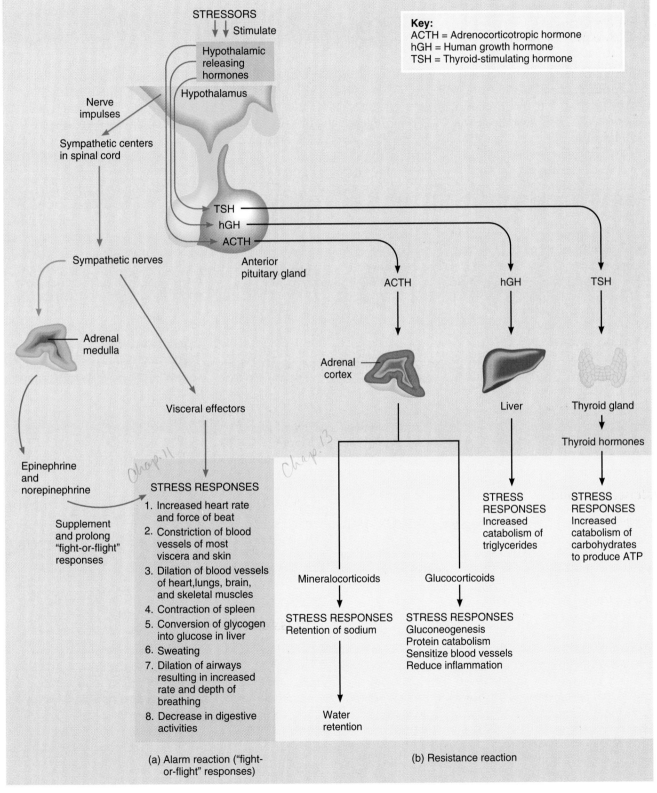

(a) Alarm reaction ("fight-or-flight" responses)

(b) Resistance reaction

Ⓠ *What is the basic difference between the GAS and homeostasis?*

wellness focus

One of the most common endocrine disorders is Type II diabetes mellitus, also known as non-insulin-dependent diabetes mellitus (NIDDM). NIDDM is characterized by high levels of insulin in the blood. Insulin levels appear to be high because of *insulin resistance,* which means the cells' receptors are not responding appropriately to insulin. For some reason, in people with NIDDM these receptors gradually become less sensitive, and it takes more insulin than it should to accomplish the important task of moving glucose from the bloodstream into the cell. Despite a more than adequate amount of insulin, blood sugar levels remain high, since the receptors are not letting insulin help the glucose across the membrane and into the cells. Another term often used in conjunction with NIDDM is *glucose intolerance,* which means poor blood sugar regulation.

In recent years, it has been observed that a large majority of people who develop NIDDM also develop hypertension (high blood pressure) and high blood cholesterol levels. They also tend to be somewhat overweight and sedentary. This cluster of disorders: NIDDM, hypertension, high serum cholesterol, and abdominal obesity is referred toby various names, including metabolic syndrome and insulin resistance syndrome. The fat pattern distribution of people with this syndrome may be the variable that explains this cluster of disorders.

A Little Riddle About Fat in the Middle

There are many ways for the body to store extra fat. Recently, people have been classified as "apples" or "pears," depending on their fat pattern distribution. Apples—people with extra abdominal fat—have a tenfold greater risk of developing NIDDM than people without excess fat. Pears—people with extra fat in the lower body (hips and thighs)—are also at increased risk, but the risk is only three to four times greater than that of people with no extra fat.

Why is abdominal fat riskier? All fat stores are not created equal. Adipocytes in different regions appear to differ in metabolic activity. Adipocytes in the abdominal region are metabolically "more active" than lower body fat cells; they are more responsive to hormones such as epinephrine. This means they release fat more readily into the bloodstream, which in the abdominal area flows to the liver. The liver takes up this fat and produces very low-density lipoprotein triglycerides, which are later converted into low-density lipoprotein cholesterol, the kind of cholesterol associated with the progression of arterial disease. This would help explain the elevation in cholesterol associated with metabolic syndrome.

The elevation in triglycerides may disrupt blood sugar regulation and trigger a rise in insulin. Elevated insulin levels in turn stimulate the sympathetic nervous system, which increases blood pressure. And there you have it, all in one package: high blood sugar, high blood lipids, hypertension, and abdominal obesity, a package that significantly increases arterial disease risk.

A Wellness Lifestyle: Prevention and Control

No one knows exactly what causes metabolic syndrome, but it appears to be related to both genetic and behavioral factors. A family history of NIDDM is a risk factor. Smoking, alcohol consumption, poor diet, and a sedentary lifestyle also predispose a person to the development of this disorder.

Both exercise and weight loss (in people who are overweight) increase the sensitivity of insulin receptors and improve glucose tolerance. This is often enough to correct the problem, or at least to delay its progression significantly.

critical thinking

Why do you think extra fat tends to be lost more easily in the abdominal region than in the hips and thighs?

Are you at risk? Use a tape measure to take the circumferences of your hip and waist. A waist-hip ratio greater than 1.0 for men or 0.8 for women is considered risky.

Common Disorders

Disorders of the endocrine system, in general, involve underproduction or overproduction of hormones.

Pituitary Gland Disorders

Several disorders of the anterior pituitary gland involve human growth hormone (hGH). If hGH is undersecreted during the growth years, bone growth is slow and the epiphyseal plates close before normal height is reached. This condition is called *pituitary dwarfism.* Other organs of the body also fail to grow, and the individual is childlike physically in many respects.

Oversecretion of hGH during childhood results in *giantism (gigantism),* an abnormal increase in the length of long bones. The person grows to be very large, but body proportions are about normal.

Oversecretion of hGH during adulthood is called *acromegaly* (ak'-rō-MEG-a-lē), which is shown in Figure 13.20a. The hormone cannot produce further lengthening of the long bones because the epiphyseal plates are already closed. Instead, the bones of the hands, feet, cheeks, and jaws thicken. Other tissues also grow; the eyelids, lips, tongue, and nose enlarge, and the skin thickens and furrows, especially on the forehead and soles.

The principal abnormality associated with dysfunction of the posterior pituitary gland is *diabetes insipidus* (in-SIP-i-dus; *diabetes* = overflow; *inspididus* = tasteless). Diabetes insipidus is the result of undersecretion of ADH, usually caused by damage to the posterior pituitary gland or hypothalamus.

Thyroid Gland Disorders

Undersecretion of thyroid hormones during fetal life or infancy results in *cretinism* (KRĒ-tin-izm), shown in Figure 13.20b. Two outstanding clinical symptoms of the cretin are dwarfism and mental retardation.

Hypothyroidism (*hypo* = under) during the adult years produces *myxedema* (mix-e-DĒ-ma) characterized by an edema that causes the facial tissues to swell and look puffy. Myxedema is more common in females.

Oversecretion of thyroid hormones increases metabolic rate, elevates heat production, and increases food intake. Symptoms include heat intolerance, increased sweating, weight loss despite good appetite, insomnia, and nervousness. The most common form of oversecretion of thyroid hormones is *Graves' disease,* an autoimmune disorder. A primary sign is an enlarged thyroid gland called *goiter* (GOY-ter; *guttur* = throat). Goiter also occurs in other thyroid diseases and if dietary intake of iodine is inadequate (Figure 13.20c). Graves' patients also have an edema behind the eye, which causes the eye to protrude (*exophthalmos*), which is shown in Figure 13.20d.

Parathyroid Gland Disorders

A deficiency of calcium owing to *hypoparathyroidism* causes neurons to depolarize without the usual stimulus. As a result, nervous impulses increase and cause muscle twitches, spasms, and convulsions. This condition is called *tetany* (*tetanos* = stretched). Hypoparathyroidism results from surgical removal of the parathyroids or from damage caused by parathyroid disease, infection, or injury.

Adrenal Gland Disorders

Oversecretion of the mineralocorticoid aldosterone results in *aldosteronism,* characterized by an increase in sodium and decrease in potassium in the blood. If potassium depletion is great, neurons cannot depolarize and muscular paralysis results. Excessive retention of sodium, and therefore of water, increases the volume of the blood and also causes hypertension.

Undersecretion of glucocorticoids (and aldosterone) results in *Addison's disease.* Increased potassium and decreased sodium lead to low blood pressure, dehydration, hypoglycemia, excessive skin pigmentation, arrhythmias, and potential cardiac arrest.

Cushing's syndrome is an oversecretion of glucocorticoids, especially cortisol and cortisone (Figure 13.20e). It is characterized by a redistribution of fat resulting in spindly legs accompanied by a "moon face," "buffalo hump" on the back, and pendulous abdomen.

Congenital adrenal hyperplasia (CAH) is a genetic disorder characterized by enlarged adrenal glands. Affected individuals are unable to produce cortisol. Because certain steps leading to synthesis of cortisol are blocked, precursor molecules of cortisol accumulate, and some of these are converted to androgens. The result is *virilism,* or masculinization. In a female, virile characteristics include growth of a beard (Figure 13.20f), development of a much deeper voice, occasionally the development of baldness, development of a masculine distribution of body hair, atrophy of the breasts, infrequent or absent menstruation, and increased muscularity that produces a male-like physique.

Pancreatic Endocrine Disorders

Diabetes mellitus (MEL-i-tus) refers to a group of hereditary diseases, all of which ultimately lead to elevated blood glucose (hyperglycemia) and excretion of glucose in the urine as hyperglycemia increases. Diabetes mellitus is also characterized by the three "polys": an inability to reabsorb water, resulting in increased urine production (*polyuria*), excessive thirst (*polydipsia*), and excessive eating (*polyphagia*).

Two major types of diabetes mellitus have been distinguished, type I and type II. *Type I diabetes,* which occurs abruptly, is characterized by a deficiency of insulin due to a decline in the number of insulin-producing beta cells. This is called *insulin-dependent diabetes* because periodic administration of insulin is required to treat it. It is also called *juvenile-onset diabetes* because it most commonly develops in people younger than age 20, though it persists throughout life. People who develop type I diabetes appear to have certain genes that make them more susceptible, but some triggering factor, usually a viral infection, is required. Insulin deficiency accelerates the breakdown of the body's fat reserves, which produces organic acids called ketones. Excess ketones cause a form of acidosis called *ketosis,* which lowers the pH of the blood and can result in death.

Type II diabetes is much more common than type I, representing more than 90 percent of all cases. Type II most often occurs in people who are over 40 and overweight. It is also called *maturity-onset diabetes.* Clinical symptoms are mild, and high glucose levels in the blood can usually be controlled by diet, exercise, and/or antidiabetic drugs such as glyburide (DiaBeta). Many type II diabetics have a sufficient amount or even a surplus of insulin in the blood. For these individuals, diabetes arises not from a shortage of insulin but probably because cells lose insulin receptors and become less sensitive to insulin. Type II diabetes is therefore also called *noninsulin-dependent diabetes.*

Oversecretion results in *hypoglycemia* and is much rarer than undersecretion and is generally the result of a malignant tumor in an islet. The principal symptom is a decreased blood glucose level, which stimulates the secretion of epinephrine, glucagon, and human growth hormone. As a consequence, anxiety, sweating, tremor, increased heart rate, and weakness occur. Moreover, brain cells do not have enough glucose to function efficiently. This condition leads to mental disorientation, convulsions, unconsciousness, and shock. ■

Figure 13.20 Photographs of various endocrine disorders.

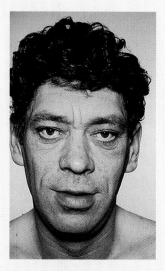

(a) Acromegaly (hypersecretion of hGH during adulthood)

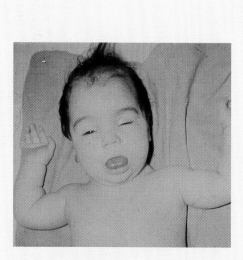

(b) Cretinism (hypersecretion of thyroid hormones during fetal life or infancy)

(c) Goiter (hypersecretion of thyroid hormones as in Graves' disease, or inadequate iodine intake)

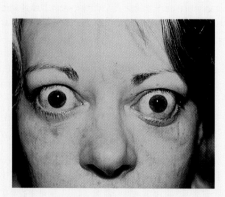

(d) Exophthalmos (hypersecretion of thyroid hormones, as in Graves' disease)

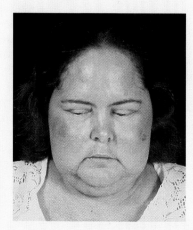

(e) Cushing's syndrome (hypersecretion of glucocorticoids)

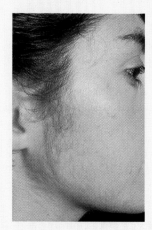

(f) Virilism (hypersecretion of androgens)

Study Outline

Endocrine Glands (p. 288)

1. Exocrine glands (sweat, oil, mucous, digestive) secrete their products through ducts into body cavities or onto body surfaces.
2. Endocrine glands secrete hormones into the blood.
3. The endocrine system consists of endocrine glands and several organs that contain endocrine tissue.

Comparison of Nervous and Endocrine Systems (p. 288)

1. The nervous system controls homeostasis through nerve impulses; the endocrine system uses hormones.
2. The nervous system causes muscles to contract and glands to secrete; the endocrine system affects virtually all body tissues.

Overview of Hormonal Effects (p. 288)

1. Hormones regulate the internal environment, metabolism, and energy balance.
2. They also help regulate muscular contraction, glandular secretion, and certain immune responses.
3. Hormones play a role in growth, development, and reproduction.

Chemistry of Hormones (p. 289)

1. Chemically, hormones are classified as lipid derivatives, amino acid derivatives, and peptides and proteins.
2. Lipid derivatives include steroid hormones and eicosanoids.

Mechanism of Hormonal Action (p. 291)

1. The amount of hormone released is determined by the body's need for the hormone.
2. Cells that respond to the effects of hormones are called target cells.
3. Receptors are found in the plasma membrane and in the nucleus of target cells.
4. The combination of hormone and receptor activates a chain of events in a target cell in which the physiological effects of the hormone are expressed.
5. One mechanism of hormonal action involves interaction of a hormone (lipid-soluble) with receptors inside the cell.
6. Another mechanism of hormonal action involves interaction of a hormone (water-soluble) with plasma membrane receptors; the second messenger may be cyclic AMP or other substances.

Control of Hormonal Secretions: Feedback Control (p. 292)

1. A negative feedback control mechanism prevents overproduction or underproduction of a hormone.
2. Hormone secretions are controlled by levels of circulating chemicals, nerve impulses, and releasing or inhibiting hormones from the hypothalamus.

Pituitary Gland (p. 293)

1. The pituitary gland is attached to the hypothalamus and consists of an anterior pituitary gland and a posterior pituitary gland.
2. Hormones of the pituitary gland are controlled by inhibiting and releasing hormones produced by the hypothalamus.
3. The blood vessels from the hypothalamus to the anterior pituitary gland transport hypothalamic releasing or inhibiting hormones.
4. The anterior pituitary gland consists of cells that produce human growth hormone (hGH), prolactin (PRL), thyroid-stimulating hormone (TSH), follicle-stimulating hormone (FSH), luteinizing hormone (LH), adrenocorticotropic hormone (ACTH), and melanocyte-stimulating hormone (MSH).
5. hGH stimulates body growth and is controlled by hypothalamic releasing and inhibiting hormones.
6. TSH regulates thyroid gland activities and is controlled by a hypothalamic releasing hormone.
7. FSH and LH stimulate secretion of estrogens and progesterone and maturation of oocytes by the ovaries and secretion of testosterone and production of sperm by the testes. FSH and LH are controlled by a hypothalamic releasing hormone.
8. PRL helps initiate milk secretion and is controlled by hypothalamic releasing and inhibiting hormones.
9. MSH increases skin pigmentation and is controlled by hypothalamic releasing and inhibiting hormones.
10. ACTH regulates the activities of the adrenal cortex and is controlled by a hypothalamic releasing hormone.
11. The posterior pituitary contains axon terminals of neurosecretory cells whose cell bodies are in the hypothalamus.
12. Hormones made by the hypothalamus and stored in the posterior pituitary gland are oxytocin or OT (stimulates contraction of uterus and ejection of milk) and antidiuretic hormone or ADH (stimulates water reabsorption by the kidneys and arteriole constriction).
13. OT secretion is controlled by uterine distension and sucking during nursing; ADH is controlled primarily by water concentration.

Thyroid Gland (p. 299)

1. The thyroid gland is located below the larynx.
2. The thyroid gland consists of thyroid follicles composed of follicular cells, which secrete the thyroid hormones thyroxine (T_4) and triiodothyronine (T_3), and parafollicular cells, which secrete calcitonin (CT).
3. Thyroid gland hormones regulate oxygen use and metabolic rate, cellular metabolism, and growth and development. Secretion is controlled by a hypothalamic releasing hormone.
4. Calcitonin (CT) lowers the blood level of calcium. Secretion is controlled by its own level in blood.

Parathyroid Glands (p. 301)

1. The parathyroid glands are embedded on the posterior surfaces of the thyroid.
2. The parathyroid glands consist of principal (chief) and oxyphil cells.
3. Parathyroid hormone (PTH) regulates the homeostasis of calcium and phosphate by increasing blood calcium level and decreasing blood phosphate level. Secretion is controlled by its own level in blood.

Adrenal Glands (p. 302)

1. The adrenal glands are located above the kidneys. They consist of an outer cortex and inner medulla.
2. The adrenal cortex is divided into three zones and each zone secretes different groups of hormones.
3. The outer zone secretes mineralocorticoids; the middle zone secretes glucocorticoids; and the inner zone secretes androgens.
4. Mineralocorticoids (for example, aldosterone) increase sodium and water reabsorption and decrease potassium reabsorption. Secretion is controlled by the renin–angiotensin pathway and blood level of potassium.
5. Glucocorticoids (for example, cortisol) promote normal metabolism, help resist stress, and serve as anti-inflammatories. Secretion is controlled by a hypothalamic releasing hormone.
6. Androgens secreted by the adrenal cortex have minimal effects.
7. The adrenal medulla secretes epinephrine and norepinephrine (NE), which produce effects similar to sympathetic responses. They are released under stress.

Pancreas (p. 308)

1. The pancreas is behind and slightly below the stomach. It is both an endocrine and an exocrine gland.
2. The endocrine portion consists of pancreatic islets or islets of Langerhans, made up of four types of cells—alpha, beta, delta, and F-cells.
3. Alpha cells secrete glucagon, beta cells secrete insulin, delta cells secrete somatostatin, and F-cells secrete pancreatic polypeptide.
4. Glucagon increases blood sugar level. Secretion is controlled by its own level in the blood.
5. Insulin decreases blood sugar level. Secretion is controlled by its own level in the blood.

Ovaries and Testes (p. 308)

1. The ovaries are located in the pelvic cavity and produce sex hormones related to the development of female sexual characteristics, menstrual cycle, pregnancy, and lactation.
2. The testes lie inside the scrotum and produce sex hormones related to the development of male sexual characteristics.

Pineal Gland (p. 310)

1. The pineal gland is located near the thalamus.
2. It secretes melatonin, which possibly regulates reproductive activities.

Thymus Gland (p. 310)

1. The thymus gland secretes several hormones related to immunity.
2. Thymosin, thymic humoral factor (THF), thymic factor (TF), and thymopoietin promote the development of T cells.

Other Endocrine Tissues (p. 310)

1. Body tissues other than those normally classified as endocrine glands contain endocrine tissue and secrete hormones.
2. These include the gastrointestinal tract, placenta, kidneys, skin, and heart. (See Exhibit 13.4.)

Stress and the General Adaptation Syndrome (GAS) (p. 311)

1. If stress is extreme or unusual, it triggers the general adaptation syndrome (GAS).
2. Unlike the homeostatic mechanisms, this syndrome does not maintain a constant internal environment. In fact, it does the opposite, preparing the body to meet an emergency.

Stressors (p. 311)

1. The stimuli that produce the GAS are called stressors.
2. A stressor can be almost any disturbance, from surgical operations and poisons to strong emotional responses.

Alarm Reaction (p. 312)

1. The alarm reaction is initiated by nerve impulses from the hypothalamus to the sympathetic division of the autonomic nervous system and adrenal medulla.
2. Responses are the immediate and short-lived, fight-or-flight responses that increase circulation, promote catabolism for energy production, and decrease nonessential activities.

Resistance Reaction (p. 312)

1. The resistance reaction is initiated by hypothalamic releasing hormones.
2. Glucocorticoids are produced in especially high concentrations during stress, with many distinct physiological effects.
3. Resistance reactions are long term and accelerate catabolism to provide energy to counteract stress.

Exhaustion (p. 312)

1. The stage of exhaustion results from dramatic changes during alarm and resistance reactions.
2. Exhaustion is caused mainly by loss of potassium, depletion of adrenal glucocorticoids, and weakened organs. If stress is too great, it may lead to death.

Stress and Disease (p. 312)

1. It appears that stress can lead to certain diseases.
2. Stress-related conditions include gastritis, peptic ulcers, hypertension, and migraine headaches.

Self-Quiz

1. Which of the following is NOT true concerning hormones?

   a. Responses by hormones are generally slower and longer-lasting than responses by the nervous system. b. Hormones are generally controlled by negative feedback systems. c. The hypothalamus inhibits the release of some hormones. d. Most hormones are released steadily throughout the day. e. Hormone secretion is determined by the body's needs in order to maintain homeostasis.

2. Which of the following statements concerning hormone action is NOT true?

   a. Hormones bring about changes in metabolic activities of cells. b. Target cells must have receptor sites for a hormone. c. Lipid-soluble hormones may directly enter target cells and activate the genes. d. A hormone that attaches to a membrane receptor is considered the first messenger. e. ATP is a common second messenger in target cells.

3. Which of the following statements is false?

   a. The secretion of hormones by the anterior pituitary gland is controlled by releasing hormones. b. The pituitary gland is attached to the hypothalamus by the infundibulum. c. Special blood vessels connect the posterior pituitary gland to the hypothalamus. d. The anterior pituitary gland comprises the majority of the pituitary gland. e. The posterior pituitary gland stores hormones produced by neurosecretory cells of the hypothalamus.

4. The hormone that promotes milk release from the mammary glands and that stimulates the uterus to contract is

   a. oxytocin (OT) b. prolactin (PRL) c. relaxin d. calcitonin e. follicle-stimulating hormone (FSH)

5. The parathyroid glands lie in contact with the

   a. thalamus b. thymus c. hypothalamus d. pancreas e. thyroid

6. The gland that prepares the body to react to stress by releasing epinephrine is the

   a. posterior pituitary gland b. anterior pituitary gland c. adrenal cortex d. adrenal medulla e. pancreas

7. The hormone that acts antagonistically to insulin is

   a. norepinephrine b. aldosterone c. glucagon d. parathyroid hormone e. thymosin

8. Glucocorticoid hormones are secreted by the

   a. adrenal medulla b. parafollicular cells c. pancreatic islets (islets of Langerhans) d. adrenal cortex e. kidneys

9. Pancreatic cells that secrete insulin are called

   a. chief cells b. beta cells c. F-cells d. alpha cells e. delta cells

10. Which of the following is NOT true concerning human growth hormone (hGH)?

    a. It stimulates protein synthesis. b. It has one primary target tissue in the body. c. It stimulates skeletal muscle growth. d. Hyposecretion in childhood results in dwarfism. e. Hypoglycemia can stimulate the release of hGH from the pituitary gland.

11. Follicle-stimulating hormone (FSH) acts on the _____ and luteinizing hormone (LH) acts on the _____.

    a. the ovaries; the testes b. the testes; the ovaries c. the ovaries and testes; ovaries and testes d. the ovaries; the mammary glands e. the ovaries and uterus; the testes

12. An injection of adrenocorticotropic hormone (ACTH) would

    a. stimulate the ovaries b. influence thyroid gland activity c. stimulate release of adrenal cortex hormones d. cause uterine contractions e. decrease urine output

13. Which of the following is NOT true concerning glucocorticoids?

    a. They help to control electrolyte balance. b. They help provide resistance to stress. c. They help promote normal metabolism. d. They are anti-inflammatory hormones. e. They provide the body with energy through gluconeogenesis.

14. Mineralocorticoids

    a. help prevent the loss of potassium from the body. b. are secreted based on the renin–angiotensin pathway. c. increase the rate of sodium loss in the urine. d. are involved in raising the body's blood pressure. e. increase water loss from the body by increasing urine production.

15. Match the following:

____ a. diabetes insipidus	A. hypersecretion of glucocorticoids
____ b. diabetes mellitus	B. hyposecretion of antidiuretic hormone (ADH)
____ c. myxedema	C. hyposecretion of insulin
____ d. Cushing's syndrome	D. hyposecretion of glucocorticoids
____ e. Addison's disease	E. hyposecretion of thyroid hormone
____ f. acromegaly	F. hyposecretion of parathyroid hormone
____ g. tetany	G. hypersecretion of parathyroid hormone

16. The hormone that is controlled by a positive feedback system is

    a. oxytocin b. melatonin c. insulin d. aldosterone e. testosterone

17. Melatonin is secreted by the

    a. anterior pituitary gland b. posterior pituitary gland c. pineal gland d. pancreas e. thymus gland

18. The hormone that controls the metabolic rate and cellular metabolism is

    a. relaxin b. thyroxine c. insulin d. aldosterone e. parathyroid hormone

19. Which of the following antagonistic hormones are correctly paired?

    a. parathyroid hormone/thyroid hormones b. parathyroid hormone/calcitonin c. oxytocin/glucocorticoids d. aldosterone/ oxytocin e. thyroid hormones/thymosin

Critical Thinking Applications

1. Rand was in a 50-mile bike-a-thon on a hot summer day. He was sweating profusely and he lost his water bottle. How will his hormones respond to this situation?

2. As Kelly walked into the A & P class (5 minutes late) she was handed an exam. She had missed class for a week when she decided to stay at the beach. Predict how her body will react to this stress.

3. Goiter (an enlarged thyroid gland) was fairly common in regions of the Midwest before iodized salt was in common use. Explain the relationship of iodine to the thyroid gland.

4. Beatrice had a large bag of chocolate chip cookies and a 16 oz cola for lunch. Draw a negative feedback system that illustrates the control of her blood sugar level.

5. Some bodybuilders use anabolic steroids (a form of testosterone) to enhance muscle growth. Using your understanding of feedback control of hormones, explain the effect of anabolic steroids on the body's testosterone production.

6. Melatonin may have a role in regulating sleep cycles. It has also been suggested as a possible "jet-lag" medication. Explain the relationship of melatonin secretion to light.

Answers to Figure Questions

13.1 Secretions of endocrine glands diffuse into extracellular spaces and then into the blood; exocrine secretions flow into ducts that lead into body cavities or to the body surface.

13.2 Liver or kidneys.

13.3 It brings the message of the first messenger, the water-soluble hormone, into the cell.

13.4 Anterior pituitary gland.

13.5 Thyroid gland, adrenal cortex, ovaries, testes.

13.6 Skeletal and muscular.

13.7 Oxytocin and antidiuretic hormone.

13.8 Uterine contraction and milk ejection.

13.9 Pain, stress, trauma, nicotine, high Na^+, morphine, and some anesthetics.

13.10 Follicular cells secrete T_3 and T_4, also known as thyroid hormones. Parafollicular cells secrete calcitonin.

13.11 Increases it.

13.12 Thyroid gland.

13.13 It increases their number and activity.

13.14 Outer zone secretes mineralocorticoids, middle zone secretes glucocorticoids, and inner zone secretes small amounts of androgens.

13.15 It constricts blood vessels and stimulates aldosterone secretion.

13.16 Low, due to negative feedback suppression.

13.17 Insulin and glucagon.

13.18 It has several effects that are opposite to those of insulin.

13.19 The GAS does not maintain a normal internal environment; homeostasis does.

THE CARDIOVASCULAR SYSTEM: BLOOD

a look ahead

Functions of Blood

Blood is a liquid connective tissue that has three general functions: transportation, regulation, and protection. *transports waste from cell to sites of excretion*

1. **Transportation.** Blood transports oxygen from the lungs to the cells of the body and carbon dioxide from the cells to the lungs. It also carries nutrients from the gastrointestinal tract to the cells, heat and waste products away from the cells, and hormones from endocrine glands to other body cells. *of dissolved gases O2 + CO2 + distribution of nutrients*

2. **Regulation.** Blood regulates pH through buffers. It also adjusts body temperature through the heat-absorbing and coolant properties of its water content and its variable rate of flow through the skin, where excess heat can be lost to the environment. Blood also influences the water content of cells, principally through dissolved ions and proteins.

3. **Protection.** The clotting mechanism protects against blood loss, and certain phagocytic white blood cells or specialized plasma proteins such as antibodies protect against foreign microbes and toxins.

Physical Characteristics of Blood

Blood is a viscous (sticky) fluid: it is heavier, thicker, and more viscous than water. Its temperature is about 38°C (100.4°F), its pH range is 7.35 to 7.45 (slightly alkaline), and its salt (NaCl) concentration is 0.90 percent. The blood volume of an average-sized male is 5 to 6 liters (5 to 6 qt); an average-sized female has 4 to 5 liters. Blood constitutes about 8 percent of the total body weight.

Components of Blood — *elements*

| objective: | *Discuss the formation, structure, and function of the components of blood.*

Blood is composed of two portions: (1) 45 percent of its volume is composed of formed elements (cells and cell fragments) and (2) 55 percent is plasma (liquid containing dissolved substances) (Figure 14.1).

Formed Elements — *Blood cells*

The *formed elements* of the blood are (Figure 14.2 on page 324):

I. **Erythrocytes** (red blood cells) *or corpuscles – 99%*

II. **Leukocytes** (white blood cells) *+ platelets = other 1%*

  A. **Granular leukocytes (granulocytes)**

    1. Neutrophils

    2. Eosinophils

    3. Basophils

  B. **Agranular leukocytes (agranulocytes)**

    1. Lymphocytes

    2. Monocytes

III. **Platelets (thrombocytes)**

Formation of Blood Cells

The process by which blood cells are formed is called **hemopoiesis** (hē-mō-poy-Ē-sis; *hemo* = blood; *poiem* = to make). During embryonic and fetal life, there are several centers for blood cell production: the yolk sac, liver, spleen, thymus gland, lymph nodes, and red bone marrow. After birth, however, hemopoiesis takes place in red bone marrow in long bones such as the humerus and femur; flat bones such as the sternum, ribs, and cranial bones; the vertebrae; and pelvis.

All blood cells originate from **hemopoietic stem cells,** red bone marrow cells derived from mesenchyme. These cells undergo differentiation into five types of cells from which the major types of blood cells develop (Figure 14.2, p. 324).

Erythrocytes (*Red Blood Cells—RBCs*)

RBC STRUCTURE **Erythrocytes** (e-RITH-rō-sīts; *erythros* = red; *cyte* = cell) or **red blood cells** (**RBCs**) are biconcave discs averaging about 8 μm in diameter* (Figure 14.3a, p. 325). Mature red blood cells are quite simple in structure. They have no nucleus or other organelles and cannot divide or carry on extensive metabolic activities. Essentially, they consist of a selectively permeable plasma membrane, cytosol, and a red *iron* pigment called **hemoglobin.** Hemoglobin carries oxygen to body cells and is responsible for the red color of blood. As you will see later, certain proteins (antigens) on the surfaces of red blood cells are responsible for the various blood groups such as ABO and Rh groups.

RBC FUNCTIONS The hemoglobin in erythrocytes combines with oxygen and with carbon dioxide and transports them through blood vessels. The hemoglobin molecule consists of four protein portions called **globin** and four nonprotein pigments called **hemes,** each of which is attached to a protein and contains iron (Figure 14.3b). As erythrocytes pass through the lungs, each iron atom in the hemoglobin molecules combines with a molecule of oxygen. The oxygen is transported to other tissues of the body. In the tissues, the iron–oxygen reaction reverses, and the oxygen is released to diffuse into the interstitial fluid and cells. Although some carbon dioxide is transported by hemoglobin, the greater portion is transported in blood plasma (see Chapter 18). Because erythrocytes lack mitochondria and generate ATP anaerobically, they do not consume any of the oxygen that they transport.

---

*1 μm = 1/25,000 of an inch or 1/10,000 of a centimeter (cm) or 1/1000 of a millimeter (mm).

Figure 14.1   Components of blood in a normal adult.

Blood is a connective tissue that consists of plasma (liquid) plus formed elements (erythrocytes, leukocytes, and platelets).

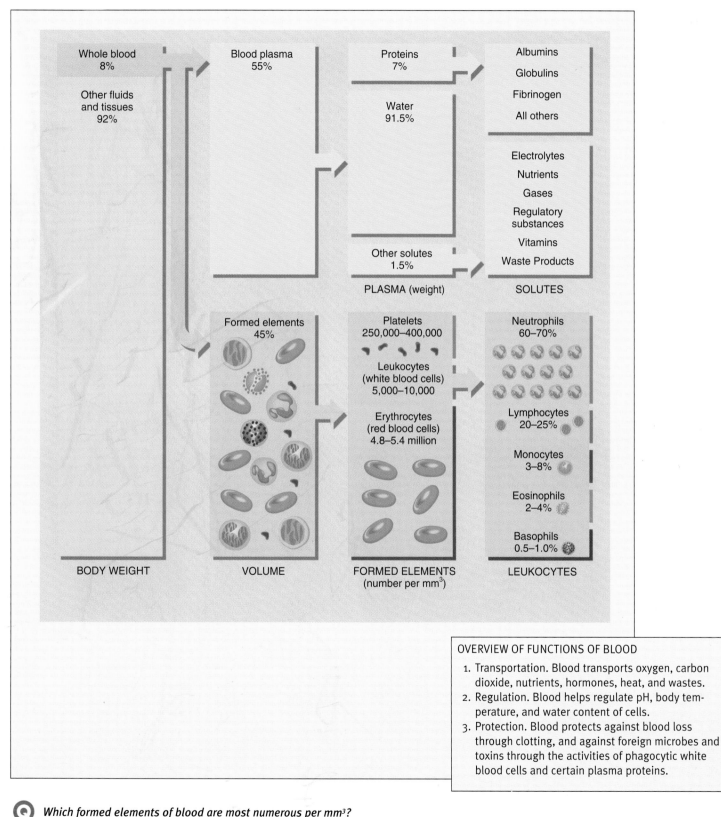

| BODY WEIGHT | VOLUME | FORMED ELEMENTS (number per mm³) | LEUKOCYTES |

Whole blood 8%

Other fluids and tissues 92%

Blood plasma 55%

Formed elements 45%

Proteins 7%

Water 91.5%

Other solutes 1.5%

PLASMA (weight)

Albumins

Globulins

Fibrinogen

All others

Electrolytes

Nutrients

Gases

Regulatory substances

Vitamins

Waste Products

SOLUTES

Platelets 250,000–400,000

Leukocytes (white blood cells) 5,000–10,000

Erythrocytes (red blood cells) 4.8–5.4 million

Neutrophils 60–70%

Lymphocytes 20–25%

Monocytes 3–8%

Eosinophils 2–4%

Basophils 0.5–1.0%

OVERVIEW OF FUNCTIONS OF BLOOD

1. Transportation. Blood transports oxygen, carbon dioxide, nutrients, hormones, heat, and wastes.
2. Regulation. Blood helps regulate pH, body temperature, and water content of cells.
3. Protection. Blood protects against blood loss through clotting, and against foreign microbes and toxins through the activities of phagocytic white blood cells and certain plasma proteins.

Q   *Which formed elements of blood are most numerous per mm³?*

Figure 14.2 Blood cells.

🔑 *Blood cell production is called hemopoiesis. After birth, it occurs only in red bone marrow.*

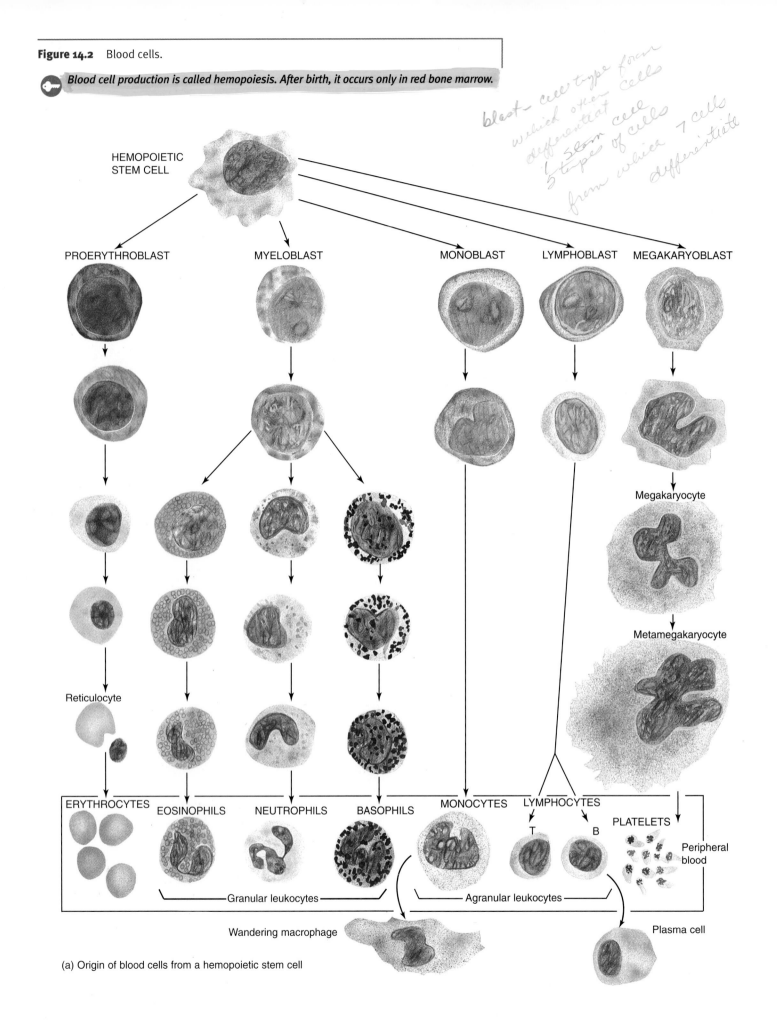

HEMOPOIETIC STEM CELL

PROERYTHROBLAST   MYELOBLAST   MONOBLAST   LYMPHOBLAST   MEGAKARYOBLAST

Megakaryocyte

Reticulocyte

Metamegakaryocyte

ERYTHROCYTES   EOSINOPHILS   NEUTROPHILS   BASOPHILS   MONOCYTES   LYMPHOCYTES   PLATELETS

T   B

Peripheral blood

———— Granular leukocytes ————   ———— Agranular leukocytes ————

Wandering macrophage   Plasma cell

(a) Origin of blood cells from a hemopoietic stem cell

Figure 14,2   (*Continued*)

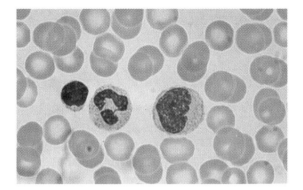

BLOOD SMEAR

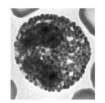

NEUTROPHIL

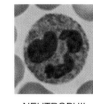

EOSINOPHIL

BASOPHIL

RED BLOOD
CELLS AND
PLATELET

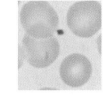

MONOCYTE

LARGE
LYMPHOCYTE

(b) Photomicrographs

List the following physical characteristics of blood: temperature, pH, and percentage of total body weight.

Figure 14.3   Shape of red blood cells (RBCs) and hemoglobin molecule. In (b), the four protein (globin) portions of the molecule are shown in blue; the heme (iron-containing) portions are shown in red. (Modified from R.E. Dickerson and I. Geis.)

Oxygen binds to the heme portion of hemoglobin for transportation.

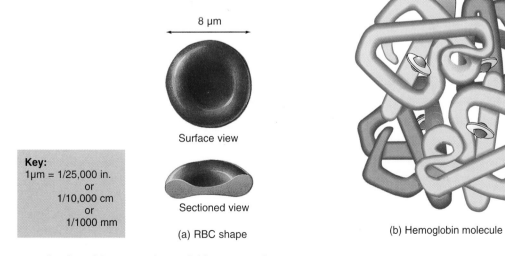

8 μm

Surface view

Sectioned view

Key:
1μm = 1/25,000 in.
or
1/10,000 cm
or
1/1000 mm

(a) RBC shape

(b) Hemoglobin molecule

How many molecules of O_2 can one hemoglobin transport?

RBC Life Span and Number   Red blood cells live only about 120 days because of wear and tear on their fragile plasma membranes as they squeeze through blood capillaries. Worn-out red blood cells are phagocytized by macrophages in the spleen, liver, and red bone marrow. The red blood cell's hemoglobin is subsequently recycled (Figure 14.4). The globin is split from the heme portions and broken down into amino acids that may be reused by other cells for protein synthesis. The heme is broken down into iron and biliverdin. The iron is stored in the liver until it is released from storage and transported to red bone marrow to be reused for hemoglobin synthesis. The noniron portion of heme is converted into *biliverdin,* a greenish pigment, and then into *bilirubin,* which is yellow in color. Bilirubin is excreted by the liver into the small intestine. As bilirubin passes into the large intestine, most of it is converted by bacteria into *urobilinogen.* Most is converted into a brown pigment (*stercobilin*), giving feces its characteristic color. Some urobilinogen is excreted in urine.

Jaundice is a yellowish coloration of the "white" of the eye (sclera), the skin, and mucous membranes due to a buildup of bilirubin in the body. Because the liver of a newborn functions poorly for the first week or so, many babies experience a mild form of jaundice called *neonatal* (*physiological*) *jaundice* that disappears as the liver matures. It is usually treated by exposing the infant to blue light, which converts bilirubin into substances the kidneys can excrete.

Production of RBCs   The process by which erythrocytes are formed is called *erythropoiesis* (e-rith′-rō-poy-Ē-sis). It takes place in red bone marrow (see Figure 14.2). Normally, erythropoiesis and red blood cell destruction proceed at the same pace. If the oxygen-carrying capacity of the blood falls because erythropoiesis is not keeping up with RBC destruction, a negative feedback system steps up erythrocyte production (Figure 14.5). The controlled condition is the amount of oxygen delivered to body tissues. Oxygen delivery may fall due to *anemia,* a lower than normal number of RBCs or quantity of hemoglobin, circulatory problems that reduce blood flow to tissues, chronic lung diseases, and high altitude (discussed in detail under Common Disorders). Cellular oxygen deficiency is called *hypoxia* (hī-POKS-ē-a). A person with prolonged hypoxia may develop *cyanosis* (sī-a-NŌ-sis), a bluish-purple skin coloration, most easily seen in the nails and mucous membranes. Whatever the cause, hypoxia stimulates the kidneys to step up release of a hormone called *erythropoietin* (e-rith′-rō-POY-ē-tin), or *EPO.* This hormone circulates through the blood to the red bone marrow, where it speeds the production of RBCs (erythropoiesis).

Figure 14.4   Formation and destruction of red blood cells and recycling of hemoglobin components.

The rate of RBC formation by red bone marrow normally equals the rate of RBC destruction by macrophages.

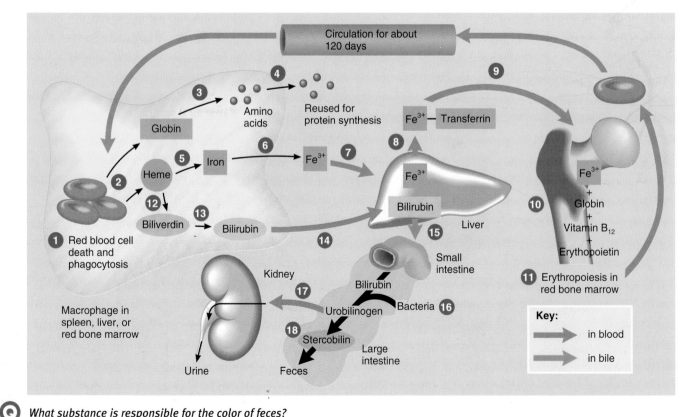

Q   *What substance is responsible for the color of feces?*

Figure 14.5   Negative feedback regulation of erythropoiesis (red blood cell formation).

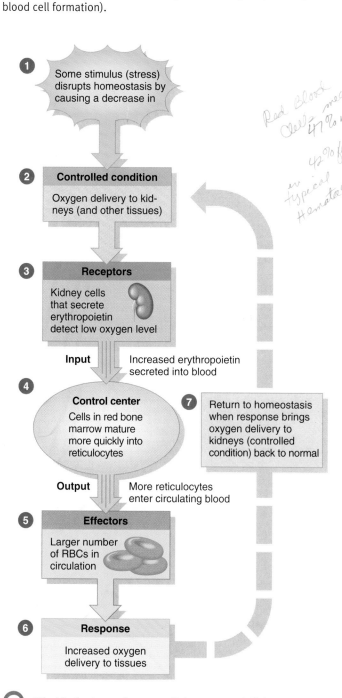

1 Some stimulus (stress) disrupts homeostasis by causing a decrease in

2 **Controlled condition**
Oxygen delivery to kidneys (and other tissues)

3 **Receptors**
Kidney cells that secrete erythropoietin detect low oxygen level

Input   Increased erythropoietin secreted into blood

4 **Control center**
Cells in red bone marrow mature more quickly into reticulocytes

7 Return to homeostasis when response brings oxygen delivery to kidneys (controlled condition) back to normal

Output   More reticulocytes enter circulating blood

5 **Effectors**
Larger number of RBCs in circulation

6 **Response**
Increased oxygen delivery to tissues

Q *What is the term given to cellular oxygen deficiency?*

One medical use of genetically engineered erythropoietin is to increase the amount of blood that can be produced and collected by individuals who choose to donate their own blood before surgery. Some athletes, seeking to improve their performance, use erythropoietin for blood "doping," a practice that is dangerous because it increases blood viscosity and thus the heart's workload. Blood doping is also considered dishonest by the International Olympic Committee.

A test that measures the rate of erythropoiesis is called a reticulocyte count. This and several other tests related to red blood cells are shown in Exhibit 14.1.

A healthy male has about 5.4 million red blood cells per cubic millimeter (mm^3) of blood, and a healthy female about 4.8 million. There are about 50 mm^3 in a drop of blood. The higher value in the male is caused by higher levels of testosterone, which stimulate the production of more red blood cells. To maintain normal quantities of erythrocytes, the body must produce new mature cells at the astonishing rate of 2 million per second. In order for red bone marrow to produce adequate numbers of healthy red blood cells, individuals must consume adequate amounts of iron, protein, folic acid, and vitamin B_{12} and produce enough *intrinsic factor* in the stomach to absorb the vitamin B_{12}. Intrinsic factor is necessary for the movement of vitamin B_{12} from the small intestine into the blood.

Leukocytes (White Blood Cells—WBCs)

WBC Structure and Types   Unlike red blood cells, *leukocytes* (LOO-kō-sīts; *leukos* = white; *cyte* = cell) or *white blood cells* (*WBCs*) have nuclei and do not contain hemoglobin (see Figure 14.2). Leukocytes fall into two major groups. The first is *granular leukocytes*. They have large characteristic granules in their cytoplasm that can be seen under a light microscope and possess lobed nuclei. There are three kinds of granular leukocytes, which are identified on the basis of how their specific granules stain: (1) *neutrophils* (NOO-trō-fils), which are 10 to 12 μm in diameter, have granules that stain pale lilac; (2) *eosinophils* (ē-ō-SIN-ō-fils), which are 10 to 14 μm in diameter, have granules that stain red or orange; and (3) *basophils* (BĀ-sō-fils), which are 8 to 10 μm in diameter, have granules that stain blue-purple.

The second group is called *agranular leukocytes* because their cytoplasmic granules cannot be seen with an ordinary light microscope due to their small size and poor staining qualities. *Lymphocytes* (LIM-fō-sīts) may be small, between 6 and 9 μm in diameter, or large, which are 10 to 15 μm in diameter, and monocytes (MON-ō-sīts) are 12 to 20 μm in diameter.

WBC Functions   The skin and mucous membranes of the body are continuously exposed to microbes, such as bacteria, and their toxins. Some of these microbes are capable of invading deeper tissues and causing disease. Once they enter the body, leukocytes combat microbes by phagocytosis or antibody production. Neutrophils and monocytes are actively *phagocytic*: they can ingest bacteria and dispose of dead cells (see Figure 3.6). Neutrophils respond first to bacterial invasion, carrying on phagocytosis and releasing the enzymes such as lysozyme, which destroys certain bacteria. Monocytes take longer to reach the site of infection than do neutrophils, but once they arrive, they do so in larger numbers and destroy more microbes. Monocytes that have migrated to infected tissues are called *wandering* (*tissue*) *macrophages* (*macro* = large; *phagen* = to eat). They also clean up cellular debris following an infection. Most leukocytes possess, to some degree, the ability to crawl through minute spaces between the cells that form the walls of capillaries and through connective and

---

Exhibit 14.1   Obtaining Blood Samples and Common Medical Tests Involving Blood

---

A. Obtaining Blood Samples

1. **Venipuncture.** Most frequently used procedure, involves withdrawal of blood from a vein. (Veins are used instead of arteries because they are closer to surface and more readily accessible and they contain blood at a much lower pressure.) Commonly used vein is the median cubital vein in front of the elbow (see Figure 16.8b). A tourniquet is wrapped around arm to stop blood flow through the veins. This make the veins below tourniquet stand out. Opening and closing the fist has the same effect.

---

2. **Fingerstick.** Using a sterile needle or lancet, a drop or two of capillary blood is taken from a finger, earlobe, or heel of the foot for evaluation.

---

3. **Arterial stick.** May be used to withdraw blood. Sample is most frequently taken from radial artery in the wrist or femoral artery in the thigh.

B. Testing Blood Samples

Once blood has been obtained, there are a number of tests the sample may be put through. These tests have various diagnostic values. Here are some examples of common tests:

---

1. Erythrocyte Sedimentation Rate (ESR) or Sed Rate

Diagnostic Value: Used as a screening test for a wide variety of infections, inflammations, and cancers. Test does not distinguish between specific diseases but is used to monitor the status of persons with rheumatoid arthritis and certain other rheumatic conditions.

Procedure: A sample of blood is placed vertically in a tube to allow the red blood cells, by gravity, to form a sediment at the bottom of the tube. The distance, in millimeters (mm), that the cells fall in 1 hour can be measured and is called the erythrocyte sedimentation rate. With certain diseases, the sed rate is increased.

Normal Values: Males: Under 50 years, less than 15 mm in 1 hour
                       Over 50 years, less than 20 mm in 1 hour
       Females: Under 50 years, less than 20 mm in 1 hour
                       Over 50 years, less than 30 mm in 1 hour

---

2. Reticulocyte (re-TIK-yoo-lō-sīt) Count

Diagnostic Value: To measure the rate of erythropoiesis and thus evaluate red bone marrow's response to anemia (low hemoglobin) or to monitor treatment for anemia.

Procedure: A blood sample is stained and examined to determine the percentage of reticulocytes in the total number of red blood cells.

Normal Values: 0.5—1.5 percent.

A high reticulocyte count might indicate the response to bleeding, hemolysis (rapid breakdown of erythrocytes), or the response to iron therapy in someone who is iron deficient. Low reticulocyte count in the presence of anemia might indicate inability of the red bone marrow to respond, owing to a nutritional deficiency, pernicious anemia, or leukemia.

---

3. Hematocrit (he-MAT-ō-krit), or Hct

Diagnostic Value: Hematocrit is the percentage of red blood cells in blood. A hematocrit of 40 means that 40 percent of the volume of blood is composed of red blood cells. The test is used to diagnose anemia and polycythemia (an increase in the percentage of the red blood cells) and abnormal states of hydration.

Procedure: A sample of venous blood is centrifuged (spun at a high speed, causing erythrocytes and plasma to separate), and the ratio of red blood cells to whole blood is measured.

Normal Values: Females: 38–46 percent (average 42)
                    Males: 40–54 percent (average 47)

A significant drop in hematocrit constitutes anemia, which may vary from mild (hematocrit of 35) to severe (hematocrit of less than 15). Polycythemic blood may have a hematocrit of 65 percent or higher. Athletes not uncommonly have a higher-than-average hematocrit, and the average hematocrit of persons living in mountainous terrain is greater than that of persons living at sea level.

(continued)

Exhibit 14.1   Obtaining Blood Samples and Common Medical Tests Involving Blood (Continued)

4. Differential White Blood Cell Count

Diagnostic Value: A routine part of the complete blood count (CBC) that may be helpful in evaluating infection or inflammation, determining the effects of possible poisoning by chemicals or drugs, monitoring blood disorders (for example, leukemia) and effects of chemotherapy, or detecting allergic reactions and parasitic infections.

Procedure: A sample of blood is spread on a glass slide and stained. Then the percentage of each type of white blood cell in a sample of 100 white blood cells is determined.

Normal Values:

	%
Neutrophils	60–70
Eosinophils	2–4
Basophils	0.5–1
Lymphocytes	20–25
Monocytes	3–8
	100

A high neutrophil count might result from bacterial infections, burns, stress, or inflammation; a low count might be caused by radiation, certain drugs, vitamin B_{12} deficiency, or systemic lupus erythematosus (SLE). A high eosinophil count could indicate allergic reactions, parasitic infections, autoimmune disease, or adrenal insufficiency; a low count could be caused by certain drugs, stress, or Cushing's syndrome. Basophils could be elevated in some types of allergic responses, leukemias, cancers, and hypothyroidism; decreases could occur during pregnancy, ovulation, stress, and hyperthyroidism. High lymphocyte counts could indicate viral infections, immune diseases, and some leukemias; low counts might occur as a result of prolonged severe illness, high steroid levels, and immunosuppression. A high monocyte count could result from certain viral or fungal infections, tuberculosis (TB), some leukemias, and chronic diseases; below normal monocyte levels rarely occur.

5. Complete Blood Count (CBC)

Diagnostic Value: The most commonly ordered blood test. It screens for anemia and various infections by assessing the white blood cell count, red blood cell count, and platelet count.

Procedure: It is performed on a blood sample and consists of cell counting by a Coulter counter and blood smears for evaluating the white blood cells and red blood cells. Usually included are a determination of red blood cell count, hemoglobin, hematocrit, white blood cell count, and platelet count.

---

epithelial tissue. This movement is called *emigration* (em′-i-GRĀ-shun; *e* = out; *migrare* = to move). See Figure 12.6. This process was formerly referred to as diapedesis.

Eosinophils are believed to release enzymes, such as histaminase, that combat the effects of inflammation in allergic reactions and are effective against certain parasitic worms. Thus a high eosinophil count frequently indicates an allergic condition or a parasitic infection. Eosinophils also phagocytize antigen–antibody complexes (to be explained in Chapter 17).

Basophils are involved in inflammation and allergic reactions. Once they leave the capillaries and enter the tissues, they are known as mast cells and function to liberate heparin, histamine, and serotonin, substances that are involved in allergic reactions and that intensify the inflammatory reaction (see Chapter 17).

The major types of lymphocytes are B cells, T cells, and natural killer cells (described in Chapter 17). B cells are particularly effective in destroying bacteria and inactivating their toxins. T cells attack viruses, fungi, transplanted cells, and cancer cells. Natural killer cells attack a wide variety of infectious microbes and certain spontaneously arising tumor cells.

Just as red blood cells have surface proteins, so do white blood cells and all other nucleated cells in the body. These proteins, called *major histocompatibility* (MHC) *antigens,* are unique for each person (except for identical twins) and can be used to identify a tissue. If an incompatible tissue is transplanted, it is rejected by the recipient as foreign, due, in part, to differences in donor and recipient MHC antigens. The MHC antigens are used to type tissues to help prevent rejection.

An increase in the number of white blood cells present in the blood usually indicates an inflammation or infection. Because each type of white blood cell plays a different role, determining the percentage of each type in the blood assists in diagnosing the condition. This test is called a *differential white blood cell count* (see Exhibit 14.1) and measures the number of each kind of white cell in a sample of 100 white blood cells.

WBC LIFE SPAN AND NUMBER   Bacteria exist everywhere in the environment and have continuous access to the body through the mouth, nose, and pores of the skin. Furthermore, many cells, especially those of epithelial tissue, age and die daily, and their remains must be removed by phagocytic WBCs. However, a WBC can phagocytize only a certain amount of substances before these substances interfere with the WBC's own metabolic activities and bring on its death. Consequently, the life span of most WBCs is only a few days, and during a period of infection they may live only a few hours. However, there are some WBCs, called B and T cells, that circulate in the body for years.

Leukocytes are far less numerous than red blood cells, averaging from 5000 to 10,000 cells per cubic millimeter (mm³) of blood. Red blood cells therefore outnumber white blood cells about 700 to 1. The term *leukocytosis* (loo'-kō-sī-TŌ-sis) refers to an increase in the number of white blood cells. If the increase exceeds 10,000/mm³, an infection is usually indicated. An abnormally low level of white blood cells (below 5000/mm³) is termed *leukopenia* (loo-kō-PĒ-nē-a). It often occurs in an individual who has an abnormality of red bone marrow or a viral infection.

PRODUCTION OF WBCs   Leukocytes develop in red bone marrow. The production of the five types of WBCs is shown in Figure 14.2a.

Platelets

PLATELET STRUCTURE   One of the cell types that hemopoietic stem cells differentiate into is called a megakaryoblast (see Figure 14.2a). Megakaryoblasts develop into megakaryocytes, large cells that break up into fragments of cytoplasm. Each fragment becomes enclosed by a piece of the cell membrane and is then called a *platelet*. Platelets are disc-shaped structures that do not contain a nucleus but do contain many granules that play a role in blood clotting. They range from 2 to 4 μm in diameter.

PLATELET FUNCTION   Platelets prevent blood loss by initiating a chain of reactions involving various plasma proteins that results in blood clotting. This mechanism is described shortly.

PLATELET LIFE SPAN AND NUMBER   Platelets have a short life span, only 5 to 9 days. Between 250,000 and 400,000 platelets appear in each cubic millimeter (mm³) of blood.

PLATELET PRODUCTION   Platelets are produced in red bone marrow according to the developmental sequence shown in Figure 14.2a.

A summary of the formed elements in blood is presented in Exhibit 14.2.

Plasma   THE MATRIX OF Blood

objective: *List the components and functions of blood plasma.*

When the formed elements are removed from blood, a straw-colored liquid called *plasma* is left. Plasma consists of 91.5 percent water and 8.5 percent solutes. The solutes include plasma proteins, nutrients, gases, electrolytes, waste products of metabolism, enzymes, and hormones. Some of the proteins in plasma are found elsewhere in the body, but those confined to blood are called *plasma proteins*. *Albumins*, which are synthesized by the liver and constitute 55 percent of plasma proteins, are largely responsible for maintaining water balance between blood and tissues and thus maintaining blood volume. *Globulins* comprise 38 percent of the plasma proteins. One group (antibodies) is produced by lymphatic tissue and functions in immunity. *Fibrinogen* makes up about 7 percent of plasma proteins and functions in the blood-clotting mechanism along with platelets. It is also produced by the liver.

Hemostasis

Hemostasis (hē-mō-STĀ-sis) refers to the stoppage of bleeding. When blood vessels are damaged, three basic mechanisms help prevent blood loss: (1) vascular spasm, (2) platelet plug formation, and (3) blood coagulation (clotting). These mechanisms are effective in preventing hemorrhage in smaller blood vessels, but extensive hemorrhage requires medical treatment.

Vascular Spasm

When a blood vessel is damaged, the smooth muscle in its wall contracts immediately. Such a *vascular spasm* reduces blood loss for several minutes to several hours, during which time the other hemostatic mechanisms go into operation. The spasm is probably caused by reflexes involving pain receptors due to damage to the vessel wall.

Platelet Plug Formation

When platelets come into contact with parts of a damaged blood vessel, their characteristics change drastically. They begin to enlarge and their shapes become even more irregular. They also become sticky and begin to adhere to collagen fibers in the wound. They produce substances that activate more platelets, causing them to stick to the original platelets. The accumulation and attachment of large numbers of platelets form a mass called a *platelet plug* (Figure 14.6 on page 332). Initially, it is loose, but it becomes quite tight when reinforced by fibrin threads formed during coagulation. A platelet plug is very effective in a small vessel and can stop blood loss completely if the hole is small.

Coagulation

Normally, blood maintains its liquid state as long as it remains in the vessels. If it is drawn from the body and not treated, however, it thickens and forms a gel. Eventually, the gel separates from the liquid. The straw-colored liquid, called *serum*, is plasma minus its clotting proteins. The gel is called a *clot* (*thrombus*) and consists of a network of insoluble protein fibers in which the blood cells are trapped (Figure 14.7 on page 332).

The process of clotting is called *coagulation* (kō-ag-yoo-LĀ-shun). If the blood clots too easily, the result can be *thrombosis*—clotting in an unbroken blood vessel. If the blood takes too long to clot, excessive bleeding can result.

Clotting involves various chemicals known as *coagulation factors* (calcium ions, enzymes, and molecules associated with platelets or damaged tissues). Clotting is a complex process in which coagulation factors activate each other. That is, the first coagulation factor activates the second, the second activates the third, and so on. We will describe clotting in three basic stages:

Stage 1. Formation of *prothrombinase* (*prothrombin activator*).

Stage 2. *Conversion of prothrombin* (a plasma protein formed by the liver) into the enzyme *thrombin,* by prothrombinase.

Exhibit 14.2 | Summary of the Formed Elements in Blood   *Never let monkeys eat Bananas*

NAME AND APPEARANCE	NUMBER	CHARACTERISTICS	FUNCTIONS
Red blood cells (RBCs) or erythrocytes	4.8 million/mm^3 in females; 5.4 million/mm^3 in males	7–8 μm diameter; nucleus has 2–5 lobes connected by thin strands of chromatin; cytoplasm has very fine, pale lilac granules.	Hemoglobin within RBCs transports most of the oxygen and part of the carbon dioxide in the blood.
White blood cells (WBCs) or leukocytes ⟨Granulocytes⟩	5000–10,000/mm^3	Most live for a few hours to a few days.	Combat pathogens and other foreign substances that enter the body. White blood cells leave the bloodstream by emigration.
Neutrophils Chromatin	60–70% of all WBCs	10–12 μm diameter; nucleus usually has 2–5 lobes connected by thin strands of chromatin; cytoplasm has very fine, pale lilac granules.	Phagocytosis. Destruction of bacteria with lysozyme.
Eosinophils	2–4% of all WBCs	10–12 μm diameter; nucleus usually has 2 lobes; large, red-orange granules fill the cytoplasm.	Combat the effects of histamine in allergic reactions, phagocytize antigen—antibody complexes, and destroy certain parasitic worms.
Basophils	½–1% of all WBCs	8–10 μm diameter; nucleus is bilobed or irregular in shape; large cytoplasmic granules appear deep blue-purple.	Liberate heparin, histamine, and serotonin in allergic reactions that intensify the overall inflammatory response.
⟨**Agranulocytes**⟩			
Lymphocytes (T cells, B cells, and natural killer cells)	20–25% of all WBCs	Small lymphocytes are 6–9 μm in diameter; large lymphocytes are 10–14 μm in diameter; nucleus is round or slightly indented; cytoplasm forms a rim around the nucleus that looks sky blue; the larger the cell, the more cytoplasm is visible.	Mediate immune responses, including antigen–antibody reactions. B cells develop into plasma cells, which secrete antibodies. T cells attack invading viruses, cancer cells, and transplanted tissue cells. Natural killer cells attack a wide variety of infectious microbes and certain spontaneously arising tumor cells.
Monocytes	3–8% of all WBCs	12–20 μm diameter; nucleus is oval, kidney-shaped, or horseshoe-shaped; cytoplasm is blue-gray and has foamy appearance.	Phagocytosis (after transforming into fixed or wandering macrophages).
Platelets (thrombocytes)	250,000–400,000/mm^3 *per cu. mm blood*	2–4 μm diameter cell fragments that live for 5–9 days; contain many granules but no nucleus.	Form platelet plug in hemostasis (stoppage of bleeding); release chemicals that promote vascular spasm and blood clotting.

Colors are those seen when using Wright's stain.

Some lymphocytes, called T and B memory cells, can live for many years once they are established. Most white blood cells, however, have life spans ranging from a few hours to a few days.

Figure 14.6　Platelet plug formation.

🔑 *A platelet plug can stop blood loss completely if the hole in a blood vessel is small.*

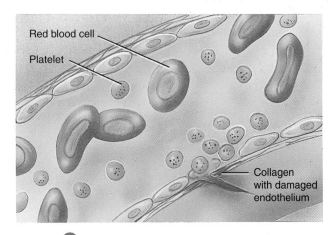

Red blood cell
Platelet
Collagen with damaged endothelium

1 Platelets stick to wound inside blood vessel

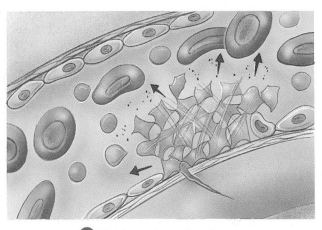

2 Platelets activate other platelets

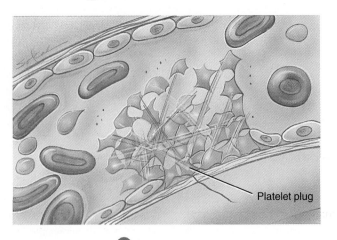

Platelet plug

3 Platelets form plug

❓ *What are the three mechanisms involved in hemostasis?*

Stage 3. *Conversion of soluble fibrinogen* (another plasma protein formed by the liver) into *insoluble fibrin* by thrombin. Fibrin forms the threads of the

Figure 14.7　Scanning electron micrograph (SEM) of fibrin threads, platelets, and red blood cells in a blood clot.

🔑 *A blood clot is a gel that contains formed elements of the blood entangled in fibrin threads.*

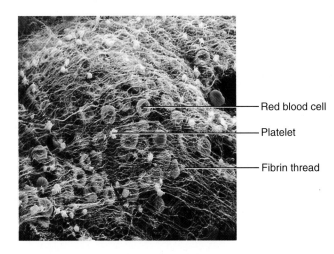

Red blood cell
Platelet
Fibrin thread

Scanning electron micrograph (550x)

❓ *What is serum?*

clot. (Cigarette smoke contains at least two substances that interfere with fibrin formation.)

Prothrombinase is formed in both the extrinsic and intrinsic pathways of blood clotting (Figure 14.8).

Extrinsic Pathway

The *extrinsic pathway* of blood clotting occurs rapidly, within seconds if trauma is severe. It is so named because the formation of prothrombinase is initiated by a tissue protein called *tissue factor* (TF), also called *thromboplastin,* which is found on the surfaces of cells *outside* the cardiovascular system. Damaged tissues release tissue factor (Figure 14.8a). Following several additional reactions that require calcium ions (Ca^{2+}) and several coagulation factors, tissue factor is eventually converted into prothrombinase. This completes the extrinsic pathway and stage 1 of clotting. In stage 2, prothrombinase and Ca^{2+} convert prothrombin into thrombin. In stage 3, thrombin, in the presence of Ca^{2+}, converts fibrinogen, which is soluble, to fibrin, which is insoluble.

Intrinsic Pathway

The *intrinsic pathway* of blood clotting is more complex than the extrinsic pathway and it operates more slowly, usually requiring several minutes. The intrinsic pathway is so named because the formation of prothrombinase starts on the surfaces of endothelial cells that line blood vessels, cells *within* the cardiovascular system. The intrinsic pathway is triggered when blood comes into contact with the damaged endothelial cells (Figure 14.8b). This also damages the platelets, causing them to

Figure 14.8   Blood clotting.

In clotting, coagulation factors activate each other, resulting in a cascade of reactions that include positive feedback cycles.

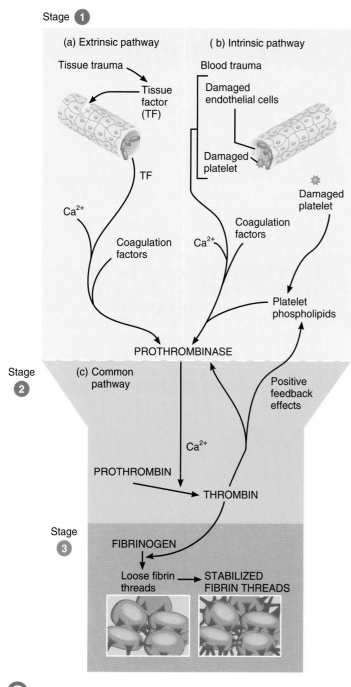

Stage **1**

(a) Extrinsic pathway

(b) Intrinsic pathway

Tissue trauma

Tissue factor (TF)

Blood trauma

Damaged endothelial cells

TF

Damaged platelet

Ca^{2+}

Damaged platelet

Coagulation factors

Coagulation factors

Ca^{2+}

Coagulation factors

Platelet phospholipids

PROTHROMBINASE

Stage **2**

(c) Common pathway

Positive feedback effects

Ca^{2+}

PROTHROMBIN

THROMBIN

Stage **3**

FIBRINOGEN

Loose fibrin threads → STABILIZED FIBRIN THREADS

Q   What is the outcome of stage 1 of clotting?

release phospholipids. Then, following several additional reactions that require the presence of Ca^{2+} and several coagulation factors, prothrombinase is formed. This completes the intrinsic pathway and stage 1 of clotting. From this point on, the reactions for stages 2 and 3 are similar to those of the extrinsic pathway. Once thrombin is formed, it causes more platelets to adhere to each other, resulting in the release of more platelet phospholipids. This is another example of a positive feedback cycle.

Once the clot is formed, it plugs the ruptured area of the blood vessel and thus prevents bleeding. Permanent repair of the blood vessel can then take place. In time, fibroblasts from connective tissue in the ruptured area and new endothelial cells repair the lining of the blood vessel.

Hemophilia (*hemo* = blood; *philein* = to love) refers to several different hereditary deficiencies of coagulation. The effects of all the forms of the disorder are so similar that they are hardly distinguishable from one another, but each is a deficiency of a different blood-clotting factor. Hemophilia is characterized by subcutaneous and intramuscular hemorrhaging, nosebleeds, blood in the urine, and joint pain and damage due to joint hemorrhaging. Treatment includes applying pressure to accessible bleeding sites and transfusing fresh plasma or the appropriate deficient clotting factor to relieve the bleeding tendency. The inheritance of hemophilia is discussed in Chapter 24.

Clot Retraction and Fibrinolysis

Normal coagulation involves two additional events after clot formation: clot retraction and fibrinolysis. **Clot retraction** is the tightening of the fibrin clot. The fibrin threads attached to the damaged surfaces of the blood vessel gradually contract because of platelets pulling on them. As the clot retracts, it pulls the edges of the damaged vessel closer together. Thus the risk of hemorrhage is further decreased.

The second event following clot formation is **fibrinolysis** (fī-brin-OL-i-sis), breaking up of the blood clot. When a clot is formed, an inactive plasma enzyme called *plasminogen* is incorporated into the clot. Body tissues and blood contain substances that can convert plasminogen to an active enzyme called *plasmin*. This enzyme can dissolve the clot by digesting fibrin threads. In addition to dissolving large clots in tissues, plasmin also removes very small clots in intact blood vessels before the clot can grow and impair blood flow to tissues.

Clot formation is a vital mechanism that prevents excessive loss of blood from the body. To form clots, the body needs calcium and vitamin K. Vitamin K is not involved in actual clot formation but is required for the synthesis of prothrombin and certain coagulation factors. The vitamin is normally produced by bacteria that live in the large intestine. Applying a thrombin or fibrin spray on a rough surface such as gauze may also promote clotting.

Hemostatic Control Mechanisms

Even though thrombin has a positive feedback effect on producing a blood clot, clot formation occurs locally at the site of damage; it does not extend beyond the wound site into general circulation. One reason for this is that some of the coagulation factors are carried away by blood flow so that their concentrations are not high enough to bring about widespread clotting. In addition, fibrin itself has the ability to absorb and inactivate up to nearly 90 percent of thrombin formed from prothrombin. This helps stop the spread of thrombin into the blood and thus inhibits clotting except at the wound.

A number of substances that inhibit coagulation are present in blood. Such substances are called *anticoagulants*. *Heparin,* for example, is an anticoagulant produced by mast cells, basophils, and endothelial cells lining blood vessels. It inhibits the conversion of prothrombin to thrombin, thereby preventing blood clot formation. Heparin prepared from animals is often used in open heart surgery and during hemodialysis.

Clotting in Blood Vessels

Even though the body has anticoagulating mechanisms, blood clots sometimes form in blood vessels. Such clots may be initiated by roughened endothelial surfaces of a blood vessel as a result of atherosclerosis (accumulation of fatty substances on arterial walls), trauma, or infection. These conditions induce platelets to stick together. Clots in blood vessels may also form when blood flows too slowly, allowing coagulation factors in local areas to increase in concentration and initiate coagulation. Clotting in an unbroken blood vessel (usually a vein) is called *thrombosis* (*thrombo* = clot). A thrombus may dissolve spontaneously, but if not, there is the possibility it will become dislodged and carried in the blood. If the clot occurs in an artery, the clot may block the circulation to a vital organ. A blood clot, bubble of air, fat from broken bones, or a piece of debris transported by the bloodstream is called an *embolus* (*em* = in; *bolus* = a mass). When an embolus becomes lodged in the lungs, the condition is called *pulmonary embolism.*

Clot-dissolving (thrombolytic) agents are chemical substances injected into the body that dissolve blood clots in order to restore circulation. They activate plasminogen. Examples of clot-dissolving agents are streptokinase and tissue plasminogen activators.

Grouping (Typing) of Blood

objective: *Explain the classification of blood into ABO and Rh groupings.*

The surfaces of red blood cells contain genetically determined antigens called *isoantigens* or *agglutinogens* (ag'-loo-TIN-ō-jens). There are at least 24 blood group systems based on isoantigens, but the two major ones are ABO and Rh.

ABO

ABO blood grouping is based on two isoantigens called A and B (Figure 14.9). Individuals whose red blood cells manufacture only antigen A are said to have blood type A. Those who manufacture only antigen B are type B. Individuals who manufacture both antigens A and B are type AB. Those who manufacture neither antigen are type O.

These four blood types are not equally distributed. They occur in different percentages in different ethnic groups (Exhibit 14.3).

The blood plasma of people who are type A, B, or O contains antibodies referred to as *isoantibodies* or *agglutinins* (a-GLOO-ti-nins). These are *anti-A antibody,* which reacts with antigen A, and *anti-B antibody,* which reacts with antigen B.

The antibodies formed by each of the four blood types are shown in Figure 14.9. You do not produce antibodies that react with the antigens of your own red blood cells, but you do have an antibody for any antigen you do not have. A blood

---

Figure 14.9 Antigens and antibodies involved in the ABO blood grouping system.

🔑 *The antibodies in your plasma do not react with the antigens on your red blood cells.*

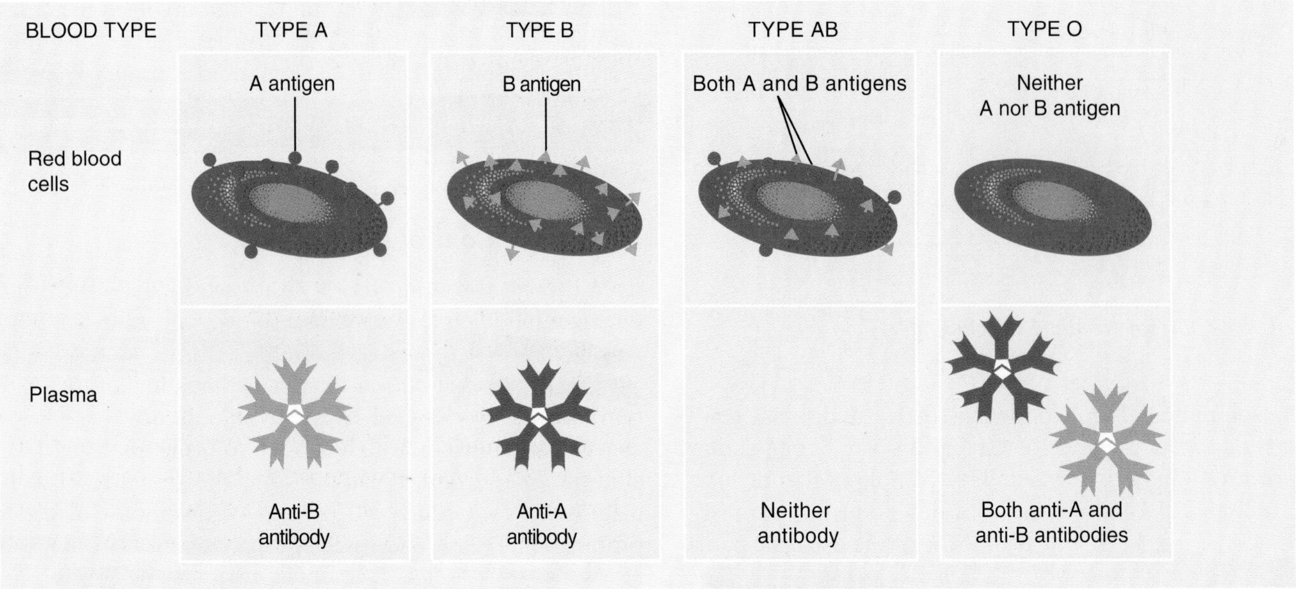

BLOOD TYPE	TYPE A	TYPE B	TYPE AB	TYPE O
Red blood cells	A antigen	B antigen	Both A and B antigens	Neither A nor B antigen
Plasma	Anti-B antibody	Anti-A antibody	Neither antibody	Both anti-A and anti-B antibodies

❓ *Which antibodies are found in type O blood?*

Exhibit 14-3	Incidence of Blood Types in the United States				
	BLOOD TYPE (PERCENTAGE)				
POPULATION	O	A	B	AB	Rh⁺
White	45	40	11	4	85
Black	49	27	20	4	95
Korean	32	28	30	10	100
Japanese	31	38	21	10	100
Chinese	42	27	25	6	100
Native American	79	16	4	1	100

transfusion (trans-FYOO-zhun), which is the transfer of whole blood or blood components directly into the bloodstream (red bone marrow can also be transfused into the bloodstream), is most frequently given when blood volume is low, for example, during shock. When blood is transfused, care must be taken to avoid incompatibilities. In an incompatible blood transfusion, the donated red blood cells bind to the recipient's antibodies, causing the red blood cells to *agglutinate* or clump (Figure 14.10). Agglutinated cells become lodged in small capillaries throughout the body and, over a period of hours, the cells swell, rupture, and release hemoglobin into the blood. Such a reaction is called *hemolysis* (*lysis* = dissolve). The degree of *agglutination* depends on the amount of antibody in the blood.

Individuals with type AB blood do not have any anti-A or anti-B antibodies in their plasma and are sometimes called universal recipients because they can *theoretically* receive blood from donors of all four blood types. They have no antibodies to attack antigens on donated red blood cells (see Figure 14.9). Individuals with type O blood have no A or B antigens on their red blood cells and are sometimes referred to as universal donors because they can *theoretically* donate blood to recipients of all four blood types. Type O persons requiring blood may receive only type O blood, though in practice, use of the terms universal recipient and universal donor is misleading and dangerous because there are other antigens and antibodies in blood besides those associated with the ABO system that can cause transfusion problems. Thus blood should be carefully matched before transfusion.

Knowledge of blood types is also used in paternity suits, linking suspects to crimes, and as part of studies to establish a relationship among ethnic groups. In about 80 percent of the population, soluble antigens of the ABO type appear in saliva and other bodily fluids. In criminal investigations it is possible to type such fluids from saliva residues on a cigarette or from semen in cases of rape.

Rh

The **Rh system** of blood classification is so named because it was first worked out in the blood of the *Rhesus* monkey. Like the ABO grouping, the Rh system is based on isoantigens on the surfaces of red blood cells. Individuals whose red blood cells have the Rh antigens are designated **Rh⁺**. Those who lack Rh antigens are designated **Rh⁻**. The percentages of Rh⁺ and Rh⁻ individuals in various populations are shown in Exhibit 14.3.

Under normal circumstances, human plasma does not contain anti-Rh antibodies. Unlike the ABO system in which antibodies develop spontaneously in blood, in the Rh system, antibodies develop only upon exposure to Rh antigens. For example, if an Rh⁻ person receives Rh⁺ blood, the body starts to make anti-Rh antibodies that will remain in the blood. If a second transfusion of Rh⁻ blood is given later, the previously formed anti-Rh antibodies will react against the donated blood and a severe reaction may occur.

The most common problem with Rh incompatibility may arise during pregnancy when a small amount of the fetus's blood may leak from the placenta into the mother's bloodstream, with the greatest possibility of transfer occurring at delivery. If the fetus is Rh⁺ and the mother is Rh⁻, she, upon exposure to the Rh⁺ fetal cells or cellular fragments, will make anti-Rh antibodies. If she becomes pregnant again, her anti-Rh antibodies will cross the placenta and make their way into the bloodstream of the baby. If the fetus is Rh⁻, no problem will occur, because Rh⁻ blood does not have the Rh antigen. If the fetus is Rh⁺, hemolysis may occur in the fetal blood. The hemolysis brought on by fetal–maternal incompatibility is called *hemolytic disease of the newborn* (HDN) or *erythroblastosis fetalis*.

When a baby is born with this condition, blood is slowly removed and replaced a little at a time, with Rh⁻ blood. Recall that Rh⁻ blood does not contain the Rh antigen. It is even possible to transfuse blood into the unborn child if the disease is diagnosed before birth. More important, though, the disorder can be prevented with an injection of anti-Rh antibodies called anti-Rh gamma globulin (RhoGAM) administered to Rh⁻ mothers at 26 to 28 weeks of gestation and right after delivery, miscarriage, or abortion. These antibodies tie up the fetal antigens, if present, so the mother cannot respond to the foreign antigens by producing antibodies. Thus, the fetus of the next pregnancy is protected. In the case of an Rh⁺ mother, there are no complications, because she cannot make anti-Rh antibodies.

Figure 14.10 Comparison of normal and agglutinated red blood cells.

 An incompatible blood transfusion can cause red blood cells to agglutinate.

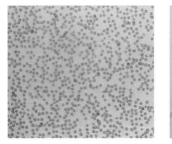 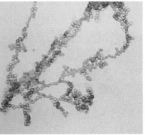

| (a) Normal | (b) Agglutinated |

 What is the name of the process in which RBCs rupture and release hemoglobin?

Many people fear cholesterol, that evil substance that silently accumulates along artery walls, year after year, until it eventually kills its victim by occluding blood flow to an important organ such as the heart or brain. But cholesterol is not the only villian in this atherosclerosis melodrama. While cholesterol contributes to the formation of arterial plaque, the antagonist delivering the final blow is often a blood clot that forms in a blood vessel and subsequently blocks a narrowed artery, cutting off circulation to the tissues downstream.

Health professionals have been promoting a "heart-healthy" lifestyle for years, to prevent or at least significantly slow the process of atherosclerosis by preventing hypertension, obesity, high blood lipid levels, and NIDDM. A heart-healthy lifestyle also reduces the tendency of the blood to clot when it encounters injured endothelial tissue.

Atherosclerosis, Hypertension, and Risk of Clotting in Blood Vessels

Atherosclerosis increases one's risk of clot formation, because arterial plaque activates platelets, which then adhere to the injured blood vessel wall, initiating a process known as platelet aggregation. Simply preventing atherosclerosis and thus, platelet aggregation, through a heart-healthy lifestyle and/or drug therapy helps to prevent clotting. A heart-healthy lifestyle also helps stop clotting by preventing hypertension; even borderline hypertension is associated with increased platelet aggregation and diminished fibrinolytic activity.

Smoking

If you need yet another reason to quit smoking, here it is. Smoking increases blood fibrinogen levels. Increased fibrinogen levels are associated with increased clotting risk. High fibrinogen levels increase platelet aggregation and fibrin deposition, contributing to both clotting and plaque deposition.

Physical Activity

Regular physical activity increases plasma volume. An increase in plasma volume means that the blood is more dilute, or "thinner," with a lower hematocrit and fibrinogen level, and consequently a reduced risk of blood clotting. Several studies have shown that vigorous exercise also reduces platelet aggregability and enhances fibrinolytic activity. These effects may help to explain why active people are at lower risk for heart disease and stroke. Like most of the beneficial effects of exercise, these changes are short-lived, disappearing after two or three days. Regular, lifelong physical activity is the type of exercise that reduces clotting risk. A sedentary lifestyle, on the other hand, leads to increased clotting risk; blood thickens as plasma volume decreases. Sedentary people have stickier blood platelets, which in conjunction with higher levels of fibrinogen, are more likely to form blood clots.

Stress

In cave people days, fighting and fleeing were often associated with bleeding, so it makes sense that enhanced clotting speed is part of the fight-or-flight response, even though clotting is not so useful nowadays. Prolonged mental stress impairs fibrinolysis by decreasing the activity of an enzyme called tissue plasminogen activator (tPA), which helps break down fibrinogen.

Heart-healthy Diet: Reducing Atherogenesis and Thrombogenesis

People with high serum cholesterol levels exhibit disturbances in coagulation, fibrinolysis, and platelet behavior. Lowering blood lipid levels by diet or drug therapy seems to reverse these disturbances, and may be one mechanism whereby a heart-healthy lifestyle reduces heart disease risk, in addition to its effects on atherogenesis. An interesting study from Denmark found that volunteers who stuck to a low-fat, high-fiber diet showed increased fibrinolytic activity, and thus a reduced risk of blood clot formation.

A moderate alcohol intake (one to two drinks per day) has been associated with a reduced heart disease risk. This risk reduction may be due in part to the increase in tPA levels observed in moderate drinkers.

critical thinking

Why are people at risk for clot formation told to avoid sitting for extended periods of time, such as on long airplane flights or car rides?

Why could a "heart-healthy" lifestyle also be called a "clot-prevention" lifestyle?

Common Disorders

Anemia

Anemia is a condition in which the oxygen-carrying capacity of the blood is reduced. There are several types of anemia (described next) and all are characterized by reduced numbers of RBCs or below-normal amounts of hemoglobin. All types of anemia lead to fatigue and intolerance to cold, both of which are related to lack of oxygen needed for energy and heat production, and to paleness, which is due to low hemoglobin content.

Iron-deficiency anemia is the most prevalent anemia in the world. It is caused by inadequate absorption or excessive loss of iron. It occurs most frequently in females, young children, and the elderly.

Pernicious anemia is due to insufficient production of erythrocytes because of lack of vitamin B_{12}, which results from an inability of the stomach to produce intrinsic factor. Intrinsic factor allows absorption of vitamin B_{12} in the small intestine.

An excessive loss of erythrocytes through bleeding (wounds, ulcers, heavy menstruation) is called *hemorrhagic anemia.*

Hemolytic anemia is characterized by distortion in the shape of erythrocytes that are progressing toward hemolysis. The premature destruction of red blood cells may result from parasites, toxins, and antibodies from incompatible blood (Rh^- mother and Rh^+ fetus, for instance).

Destruction or inhibition of red bone marrow results in *aplastic anemia.* Toxins, radiation, and certain medications are causes.

The erythrocytes of a person with *sickle-cell anemia* (*SCA*) manufacture an abnormal kind of hemoglobin. When an erythrocyte gives up its oxygen to the interstitial fluid, its hemoglobin tends to form long, stiff, rodlike structures that bend the erythrocyte into a sickle shape. The sickled cells rupture easily. Prolonged oxygen reduction may eventually cause extensive tissue damage. Furthermore, they tend to get stuck in blood vessels and can cut off blood supply to an organ altogether.

Polycythemia

The term *polycythemia* (pol'-ē-sī-THĒ-mē-a) refers to an abnormal increase in the number of red blood cells in which hematocrit is above 55, the upper limit of normal.

Infectious Mononucleosis (IM)

Infectious mononucleosis (*IM*) is a contagious disease primarily affecting lymphatic tissue throughout the body but also affecting the blood. It is caused by the Epstein–Barr virus (EBV). It occurs mainly in children and young adults. The virus most commonly enters the body through intimate oral contact. Symptoms include fatigue, headache, dizziness, sore throat, swollen and tender lymph nodes, and fever.

Leukemia

Clinically, *leukemia* is classified on the basis of the duration and character of the disease, that is, acute or chronic. Acute leukemia is a malignant disease of blood-forming tissues characterized by uncontrolled production and accumulation of immature leukocytes. In chronic leukemia, there is an accumulation of mature leukocytes in the bloodstream because they do not die at the end of their normal life span. The *human T-cell leukemia-lymphoma virus-1* (*HTLV-1*) is strongly associated with some types of leukemia.

In acute leukemia, anemia and bleeding problems result from the crowding out of normal bone marrow cells by the overproduction of immature cells, preventing normal production of red blood cells and platelets. The abnormal accumulation of immature leukocytes may be reduced by using x-rays and anti-leukemic drugs. ∎

Medical Terminology and Conditions

Acute normovolemic (nor'mō-vō-LĒ-mik) *hemodilution* Removal of blood immediately before surgery and replacing it with a cell-free solution to maintain normal blood volume for adequate circulation. At the end of surgery, when bleeding has been controlled, the collected blood is returned to the body.

Autologous (aw-TOL-o-gus; *auto* = self) *preoperative transfusion* (trans-FYOO-zhun) Donating one's own blood for up to six weeks before elective surgery. Also called *predonation.*

Autologous intraoperative transfusion (*AIT*) Procedure in which blood lost during surgery is suctioned from the patient, treated with an anticoagulant, and reinfused into the patient.

Blood bank A stored supply of blood for future use by the donor or others. Because blood banks have now assumed additional and diverse functions (immunohematology reference work, continuing medical education, bone and tissue storage, and clinical consultation), they are more appropriately referred to as *centers of transfusion medicine.*

Cyanosis (sī'-a-NŌ-sis; *cyano* = blue) Slightly bluish, dark purple skin coloration most easily seen in the nail beds and mucous membranes, due to increased quantity of reduced hemoglobin (hemoglobin not combined with oxygen) in systemic blood.

Exchange transfusion Removing blood from the recipient while alternately replacing it with donor blood. This method is used for treating hemolytic disease of the newborn (HDN) and poisoning.

Gamma globulin (GLOB-yoo-lin) A pooled sample of antibodies from many donors (or developed in a lab against specific diseases) that is injected into a person to temporarily boost the immune system. Such antibodies provide immunity to diseases such as measles and hepatitis A. Specific antibodies can also be manufactured by introducing particular antigens into an animal, such as a horse, and collecting the antibodies from the animal serum.

Hemochromatosis (hē-mō-krō′-ma-TŌ-sis; *heme* = iron; *chroma* = color) Disorder of iron metabolism characterized by excess deposits of iron in tissues, especially the liver and pancreas, that result in bronze coloration of the skin, cirrhosis, diabetes mellitus, and bone and joint abnormalities.

Hemorrhage (HEM-or-ij; *rrhage* = bursting forth) Bleeding, either internal (from blood vessels into tissues) or external (from blood vessels directly to the surface of the body).

Multiple myeloma (mī′-e-LŌ-ma) Malignant disorder of plasma cells in red bone marrow; symptoms (pain, osteoporosis, hypercalcemia, thrombocytopenia, kidney damage) are caused by the growing tumor cell mass or antibodies produced by malignant cells.

Platelet concentrates A preparation of platelets obtained from freshly drawn whole blood and used for transfusions in platelet-deficiency disorders such as hemophilia.

Porphyria (por-FĒ-rē-a; *porphyria* = purple) Any group of inherited disorders caused by the accumulation in the body of substances called porphyrins (molecules formed during the synthesis of hemoglobin and other important molecules). The buildup is due to inherited enzyme deficiencies. Symptoms include a rash or skin blistering brought on by sunlight, abdominal pain, and nervous system disturbances from certain drugs, such as barbiturates and alcohol.

Septicemia (sep′-ti-SĒ-mē-a; *sep* = decay; *emia* = condition of blood) Toxins or disease-causing bacteria growing in the blood. Also called *blood poisoning.*

Thrombocytopenia (throm′-bō-sī′-tō-PĒ-nē-a; *thrombo* = clot; *penia* = poverty) Very low platelet count that results in a tendency to bleed from capillaries.

Transfusion (trans-FYOO-zhun) Transfer of whole blood, blood components (red blood cells only or plasma only), or bone marrow directly into the bloodstream.

Venesection (vēn′-e-SEK-shun; *veno* = vein) Opening of a vein for withdrawal of blood. Although *phlebotomy* (fle-BOT-ō-mē; *phlebo* = vein; *tome* = to cut) is a synonym for venesection, in clinical practice, phlebotomy refers to therapeutic bloodletting, such as removing some blood to lower the viscosity of blood of a patient with polycythemia.

Whole blood Blood containing all formed elements, plasma, and plasma solutes in natural concentration.

Study Outline

Functions of Blood (p. 322)
1. Blood transports oxygen, carbon dioxide, nutrients, wastes, and hormones.
2. It helps to regulate pH, body temperature, and water content of cells.
3. It prevents blood loss through clotting and combats microbes and toxins through certain phagocytic white blood cells or specialized plasma proteins.

Physical Characteristics of Blood (p. 322)
1. Physical characteristics of blood include a viscosity greater than that of water, a temperature of 38°C (100.4°F), and a pH of 7.35 to 7.45.
2. Blood constitutes about 8 percent of body weight in an adult.

Components of Blood (p.322)
1. Blood consists of 55 percent plasma and 45 percent formed elements.
2. The formed elements in blood include erythrocytes (red blood cells), leukocytes (white blood cells), and platelets.

Formation of Blood Cells (p. 322)
1. Blood cells are formed from hemopoietic stem cells in red bone marrow.
2. The process is called hemopoiesis.

Erythrocytes (Red Blood Cells—RBCs) (p. 322)
1. Erythrocytes are biconcave discs without nuclei that contain hemoglobin.
2. The function of the hemoglobin in red blood cells is to transport oxygen and some carbon dioxide.
3. Red blood cells live about 120 days. A healthy male has about 5.4 million/mm^3 of blood; a healthy female about 4.8 million/mm^3.
4. After phagocytosis of aged red blood cells by macrophages, hemoglobin is recycled.
5. Erythrocyte formation, called erythropoiesis, occurs in adult red bone marrow of certain bones. It is stimulated by hypoxia, which stimulates release of erythropoietin by the kidneys.
6. A reticulocyte count is a diagnostic test that indicates the rate of erythropoiesis.
7. A hematocrit (Hct) measures the percentage of red blood cells in whole blood.

Leukocytes (White Blood Cells—WBCs) (p. 327)
1. Leukocytes are nucleated cells. The two principal types are granular (neutrophils, eosinophils, basophils) and agranular (lymphocytes and monocytes).
2. The general function of leukocytes is to combat inflammation and infection. Neutrophils and macrophages (which develop from monocytes) do so through phagocytosis.

3. Eosinophils combat the effects of histamine in allergic reactions, phagocytize antigen–antibody complexes, and combat parasitic worms; basophils develop into mast cells that liberate *release* heparin, histamine, and serotonin in allergic reactions that intensify the inflammatory response.

4. B cells (lymphocytes) are effective against bacteria and other toxins. T cells (lymphocytes) are effective against viruses, fungi, and cancer cells. Natural killer cells attack microbes and tumor cells.

5. White blood cells usually live for only a few hours or a few days. Normal blood contains 5000 to 10,000/mm³.

Platelets (p. 330) Thrombocytes

1. Platelets are disc-shaped structures without nuclei.

2. They are formed from megakaryocytes and are involved in clotting.

3. Normal blood contains 250,000 to 400,000/mm³.

Plasma (p. 330)

1. Plasma contains 91.5 percent water and 8.5 percent solutes.

2. Principal solutes include proteins (albumins, globulins, fibrinogens), nutrients, hormones, respiratory gases, electrolytes, and waste products.

Hemostasis (p. 330)

1. Hemostasis refers to the stoppage of bleeding.

2. It involves vascular spasm, platelet plug formation, and blood coagulation (clotting).

3. In vascular spasm, the smooth muscle of a blood vessel wall contracts to slow blood loss.

4. Platelet plug formation is the clumping of platelets to stop bleeding.

5. A clot is a network of insoluble protein fibers (fibrin) in which formed elements of blood are trapped.

6. The chemicals involved in clotting are known as coagulation (clotting) factors.

7. Blood clotting involves a series of reactions that may be divided into three stages: formation of prothrombinase (prothrombin activator), conversion of prothrombin into thrombin, and conversion of soluble fibrinogen into insoluble fibrin.

8. Stage 1 of clotting is initiated by the interplay of the extrinsic and intrinsic pathways of blood clotting.

9. Normal coagulation requires vitamin K and also involves clot retraction (tightening of the clot) and fibrinolysis (dissolution of the clot).

10. Anticoagulants (for example, heparin) prevent clotting.

11. Clotting in an unbroken blood vessel is called thrombosis. A thrombus that moves from its site of origin is called an embolus.

Grouping (Typing) of Blood (p. 334)

1. In the ABO system, isoantigens A and B determine blood type. Plasma contains isoantibodies—antibodies, designated anti-A antibody and anti-B antibody, that react with antigens that are foreign to the individual.

2. In the Rh system, individuals whose erythrocytes have Rh antigens are classified as Rh⁺. Those who lack the antigen are Rh⁻.

Self-Quiz

1. Which of the following statements is NOT true concerning red blood cells?

   a. The production of red blood cells is known as erythropoiesis.   b. Red blood cells originate from hemopoietic stem cells.   c. Red blood cells possess a lobed nucleus and granular cytoplasm.   d. The liver is involved in the destruction and recycling of red blood cell components.   e. Hypoxia causes an increase in the production of red blood cells.

2. A primary function of erythrocytes is to

   a. maintain blood volume   b. help blood clot   c. provide immunity against some diseases   d. clean up debris following infection   e. deliver oxygen to the cells of the body

3. If a differential white blood cell count indicated higher than normal numbers of basophils, what may be occurring in the body?

   a. chronic infection   b. allergic reaction   c. leukopenia   d. initial response to invading bacteria   e. hemostasis

4. In a person with blood type A, the isoantibodies that would normally be present in the plasma are (is)

   a. anti-B antibody   b. anti-A antibody   c. both anti-A and anti-B antibodies   d. neither anti-A nor anti-B antibodies   e. anti-O antibodies

5. Hemolytic disease of the newborn (erythroblastosis fetalis) may result in the fetus of a second pregnancy if

   a. the mother is Rh⁺ and the baby Rh⁻   b. the mother is Rh⁺ and the baby Rh⁺   c. the mother is Rh⁻ and the baby Rh⁻   d. the mother is Rh⁻ and the baby Rh⁺   e. the father is Rh⁻ and the mother is Rh⁺

6. Blood's normal platelet count per cubic millimeter would be

   a. 5,000–10,000   b. 4.8–5.2 million   c. 250,000–400,000   d. 50,000–200,000   e. 60–70

7. If a person has a vitamin K deficiency,

   a. his blood vessels will undergo vascular spasms   b. thrombosis will be stimulated   c. coagulation will be inhibited   d. hemoglobin cannot be produced   e. he will develop nutritional anemia

8. Which of the following is true concerning blood clotting?

   a. The intrinsic pathway of blood clotting utilizes tissue factor to form prothrombinase.   b. Heparin inhibits the conversion of thrombin to prothrombin.   c. Fibrin threads reinforce platelet plugs.   d. Blood flowing too quickly through vessels may initiate coagulation.   e. Plasmin converts fibrinogen into fibrin.

9. Match the following white blood cells with their functions.:

___E___ a. wandering macrophages develop from these cells

___D___ b. produce antibodies

___C___ c. involved in allergic reactions

___A___ d. first to respond to bacterial invasion

___B___ e. destroy antigen–antibody complexes; combat inflammation

   A. neutrophils

   B. eosinophils

   C. basophils

   D. lymphocytes

   E. monocytes

10. Concerning hemostasis, place the following steps in the correct order.

    I. Clot retraction

    II. Prothrombinase formed

    III. Fibrinolysis by plasmin

    IV. Vascular spasm

    V. Conversion of prothrombin into thrombin

    VI. Platelet plug formation

    VII. Conversion of fibrinogen into fibrin

_IV_, _VI_, _II_, _V_, _VII_, _I_, _III_

11. Blood is characterized as having a(n)

a. pH range of 7.0–7.4   b. average volume in females of 6.5 liters   c. temperature of 38°C   d. viscosity equal to that of water   e. volume of plasma equal to 45 percent

12. The substance that causes the yellowish color of jaundice is

a. albumin   b. leukocytes   c. biliverdin   d. bilirubin   e. hemoglobin

13. A red blood cell in a person with A⁻ blood will have _____ on its membrane surface.

a. isoantigen A   b. isoantigen B   c. major histocompatibility antigen A   d. isoantigen A and isoantigen Rh   e. isoantigen B and isoantigen Rh

14. The agranular leukocytes

a. are monocytes and lymphocytes   b. are neutrophils and lymphocytes   c. turn into granular leukocytes following emigration   d. are not counted in a differential white blood cell count   e. are involved in the production of platelets

15. Lack of adequate clotting would most likely be caused by a problem with

a. erythropoietin   b. globulins   c. fibrinogen   d. agglutination   e. intrinsic factor

Critical Thinking Applications

1. Biliary atresia is a condition in which the ducts that transport bile out of the liver do not function. A baby with this condition will have yellow eyes (yellow sclera) and other problems. What is causing the yellow eyes?

2. Taylor cut her finger while slicing a bagel. List the steps involved in hemostasis.

3. Compare the colors seen while a "black and blue" heals to the pigments released and when red blood cells are recycled in the liver. Try to correlate the bruise colors in sequence with the conversion of the red blood cell pigments.

4. A college student complained of lack of energy and constant fatigue. The nurse noticed that her skin appeared pale and colorless. A complete blood count (CBC) was ordered. What is the purpose of a CBC?

5. Cigarette smoking may promote clotting in an unbroken blood vessel. What is the name of this condition and what types of problems may it cause in the body?

6. A woman with blood type A⁺ married a man with blood type B⁻. She is pregnant with their second child. Is there a likelihood of hemolytic disease of the newborn (erythroblastosis fetalis) in this pregnancy? Explain.

Answers to Figure Questions

14.1   Erythrocytes.

14.2   Temperature, 38°C (100.4°F); pH 7.35 to 7.45; body weight, 8 percent.

14.3   Four—one attached to each heme group.

14.4   Stercobilin.

14.5   Hypoxia.

14.6   Vascular spasm, platelet plug formation, and blood coagulation.

14.7   Blood plasma minus the clotting proteins.

14.8   Formation of prothrombinase.

14.9   Anti-A and anti-B antibodies.

14.10 Hemolysis.

THE CARDIOVASCULAR SYSTEM: HEART

a look ahead

*T*he *heart* is the center of the cardiovascular system. Whereas the term *cardio* refers to the heart, the term *vascular* refers to blood vessels (or abundant blood supply). The heart propels blood through thousands of miles of blood vessels, and it is magnificently designed for this task. Although we ignore its activity most of the time, the heart's capacity for work is remarkable. Even at rest, your heart pumps 30 times its own weight each minute, about 5 liters (5.3 quarts) of blood to the body. At this rate, the heart would pump more than 7000 liters (1800 gallons) of blood in a day and 5 million liters (1.3 million gallons) in a year. However, because you don't spend all your time "resting" and because your heart pumps more vigorously when you are active, the actual flow is much larger.

The cardiovascular system provides the "pump" for circulating constantly refreshed blood through an estimated 100,000 kilometers (60,000 miles) of blood vessels. As blood flows through body tissues, nutrients and oxygen move from the blood into the interstitial fluid and then into cells. At the same time the blood picks up wastes, carbon dioxide, and heat. This chapter explores the structure and function of the heart, an organ that endures a lifetime of pumping with never a minute's rest.

The study of the heart and diseases associated with it is known as *cardiology* (kar-dē-OL-ō-jē; *cardio* = heart).

Location of the Heart

The heart is situated between the two lungs, in the mediastinum, the mass of tissues between the lungs from the sternum to the vetebral column (see Figure 1.8). About two-thirds of the heart lies to the left of the body's midline (Figure 15.1). The heart is shaped like a blunt cone about the size of your closed fist. Its pointed end, the *apex,* is formed by the tip of the left ventricle, a lower chamber of the heart, and rests on the diaphragm. The major blood vessels enter and exit at the *base* of the heart, which is formed by the atria (upper chambers of the heart), mostly the left atrium.

Pericardium

objective: *Describe the structure and functions of the pericardium.*

The heart is enclosed and held in place by the *pericardium* (*peri* = around). It consists of two portions, the fibrous pericardium and the serous pericardium (Figure 15.2, p. 344). The outer *fibrous pericardium* is very dense irregular connective tissue that prevents overstretching of the heart, provides a tough protective membrane around the heart, and anchors the heart in the mediastinum. The inner *serous pericardium* is a thinner, more delicate membrane that forms a double layer around the heart: (1) the outer *parietal layer* of the serous pericardium is directly beneath the fibrous pericardium; (2) the inner *visceral layer* of the serous pericardium, also called the *epicardium,* is beneath the parietal layer, attached to the myocardium (muscle) of the heart. Between the parietal and visceral layers of the serous pericardium is a thin film of serous fluid. This fluid, known as *pericardial fluid,* prevents friction between the membranes as the heart moves. The space occupied by the pericardial fluid is called the *pericardial cavity.*

An inflammation of the pericardium is known as *pericarditis.* Pericarditis, if untreated, is a life-threatening condition. Because the pericardium cannot stretch to accommodate excessive fluid or blood buildup, the heart is subjected to compression. This compression is known as *cardiac tamponade* (tam′-pon-ĀD) and can result in cardiac failure.

Heart Wall

The wall of the heart (Figure 15.2a) is composed of three layers: (1) epicardium (external layer), (2) myocardium (middle layer), and (3) endocardium (inner layer). The *epicardium* (*epi* = on top), which is also the visceral layer of serous pericardium, is the thin, transparent outer layer of the wall. It is composed of mesothelium and connective tissue.

The *myocardium* (*myo* = muscle) consists of cardiac muscle tissue, which constitutes the bulk of the heart. This tissue is found only in the heart and is specialized in structure and function. The myocardium is responsible for the pumping action of the heart. Cardiac muscle fibers (cells) are involuntary, striated, and branched, and the tissue is arranged in interlacing bundles of fibers (see Figure 25.2b).

Cardiac muscle fibers form two separate networks—one atrial and one ventricular. The unique structure of cardiac muscle fibers is responsible for the unique functions of the heart. Each fiber connects with other fibers in the networks by thickenings of the sarcolemma called *intercalated discs.* Within the discs are *gap junctions* that aid in the conduction of muscle action potentials between cardiac muscle fibers. The gap junctions provide bridges for the rapid spread of excitation (muscle action potentials) from one fiber to another. The intercalated discs also link cardiac muscle fibers to one another so they do not pull apart. Each network contracts as a functional unit. The atria contract as one unit and the ventricles as another.

The *endocardium* (*endo* = within) is a thin layer of simple squamous epithelium (endothelium) that lines the inside of the myocardium and covers the valves of the heart and the tendons attached to the valves. It is continuous with the epithelial lining of the large blood vessels.

Inflammations of the epicardium, myocardium, and endocardium are referred to as *epicarditis, myocarditis,* and *endocarditis.*

Figure 15.1   Position of the heart and associated blood vessels in the thoracic cavity. In this and subsequent illustrations, vessels that carry oxygenated blood are colored red; vessels that carry deoxygenated blood are colored blue.

The heart is located in the mediastinum with about two-thirds of its mass to the left of the midline.

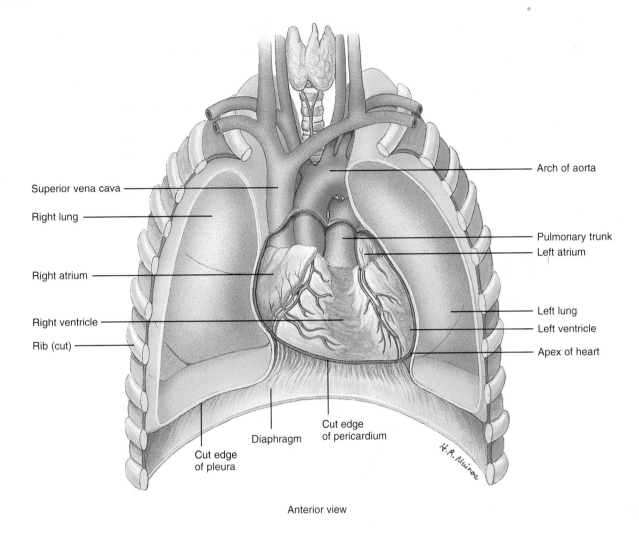

Anterior view

What is the mediastinum?

Chambers of the Heart

The interior of the heart is divided into four cavities called *chambers* that receive and pump blood (Figure 15.3, p. 345). The two upper chambers are called the right and left *atria* (singular is *atrium; atrium* = court or hall). The atria are separated by a partition called the **interatrial septum** (*septum* = partition). The two lower chambers are the right and left *ventricles* (*ventricle* = little belly). They are separated from each other by an **interventricular septum.** Externally, a groove known as the **coronary sulcus** (SUL-kus) separates the atria from the ventricles. It encircles the heart and contains fat and coronary blood vessels.

The thickness of the myocardium of the chambers varies according to the amount of work they have to perform. The walls of the atria are thin compared to those of the ventricles because the atria need only enough cardiac muscle tissue to deliver blood into the ventricles (Figure 15.3c). The right ventricle pumps blood only to the lungs (pulmonary circulation), the left ventricle pumps blood to all other parts of the body (systemic circulation). Thus the left ventricle must work harder than the right ventricle to maintain the same rate of blood flow. The structure of the two ventricles is related to this functional difference: the muscular wall of the left ventricle is considerably thicker than the wall of the right ventricle.

Figure 15.2   Pericardium and heart wall.

🔑 *The pericardium is a triple-layered sac that surrounds and protects the heart.*

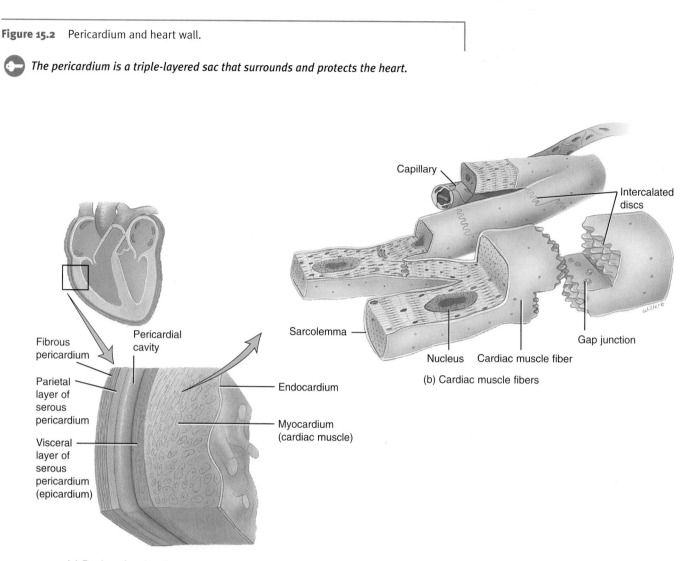

Fibrous pericardium

Pericardial cavity

Parietal layer of serous pericardium

Visceral layer of serous pericardium (epicardium)

Endocardium

Myocardium (cardiac muscle)

Capillary

Intercalated discs

Sarcolemma

Nucleus   Cardiac muscle fiber

Gap junction

(b) Cardiac muscle fibers

(a) Portion of pericardium and right ventricular heart wall

❓ *Which structure is both a part of the pericardium and part of the heart wall?*

Great Vessels of the Heart

The right atrium receives **deoxygenated blood** (blood that has given up its oxygen to cells) through three veins. Veins are blood vessels that *return blood to the heart*. The **superior vena cava** (*vena* = vein; *caria* = space) or **SVC** brings blood mainly from parts of the body above the heart; the **inferior vena cava** (**IVC**) brings blood mostly from parts of the body below the heart; and the **coronary sinus** drains blood from most of the vessels supplying the wall of the heart (Figure 15.3b,c). The right atrium then delivers the deoxygenated blood into the right ventricle, which pumps it into the

pulmonary trunk. The pulmonary trunk divides into a **right** and **left pulmonary artery,** each of which carries blood to the corresponding lung. Arteries are blood vessels that *carry blood away from the heart*. In the lungs, the deoxygenated blood releases its carbon dioxide and takes on oxygen. **Oxygenated blood** (blood that has not given up its oxygen to cells) is transported to the left atrium via four **pulmonary veins.** The blood then passes into the left ventricle, which pumps the blood into the **ascending aorta.** From here the oxygenated blood is passed into the **coronary arteries, arch of the aorta, thoracic aorta,** and **abdominal aorta.** These blood vessels and their branches transport the blood to all body parts.

Figure 15.3   Structure of the heart.

🔑 *The four chambers of the heart are the two upper atria and two lower ventricles.*

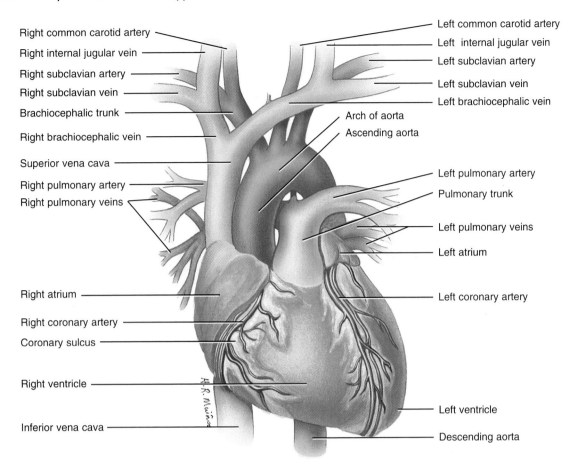

Right common carotid artery
Right internal jugular vein
Right subclavian artery
Right subclavian vein
Brachiocephalic trunk
Right brachiocephalic vein
Superior vena cava
Right pulmonary artery
Right pulmonary veins
Right atrium
Right coronary artery
Coronary sulcus
Right ventricle
Inferior vena cava

Left common carotid artery
Left internal jugular vein
Left subclavian artery
Left subclavian vein
Left brachiocephalic vein
Arch of aorta
Ascending aorta
Left pulmonary artery
Pulmonary trunk
Left pulmonary veins
Left atrium
Left coronary artery
Left ventricle
Descending aorta

(a) Anterior external view

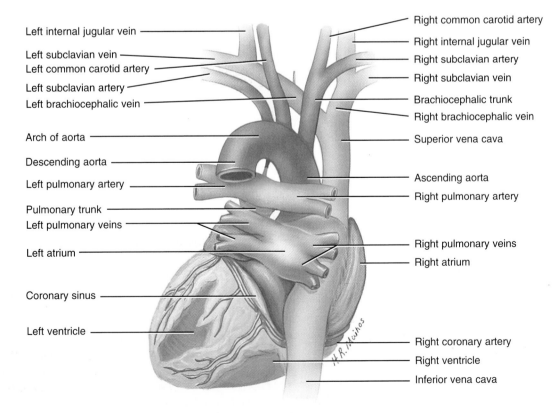

Left internal jugular vein
Left subclavian vein
Left common carotid artery
Left subclavian artery
Left brachiocephalic vein
Arch of aorta
Descending aorta
Left pulmonary artery
Pulmonary trunk
Left pulmonary veins
Left atrium
Coronary sinus
Left ventricle

Right common carotid artery
Right internal jugular vein
Right subclavian artery
Right subclavian vein
Brachiocephalic trunk
Right brachiocephalic vein
Superior vena cava
Ascending aorta
Right pulmonary artery
Right pulmonary veins
Right atrium
Right coronary artery
Right ventricle
Inferior vena cava

(b) Posterior external view

Figure 15.3   *(Continued)*

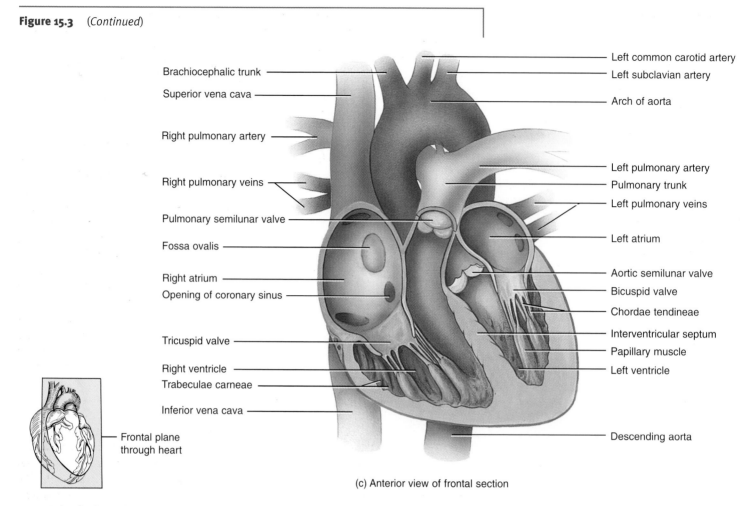

Brachiocephalic trunk

Superior vena cava

Right pulmonary artery

Right pulmonary veins

Pulmonary semilunar valve

Fossa ovalis

Right atrium

Opening of coronary sinus

Tricuspid valve

Right ventricle

Trabeculae carneae

Inferior vena cava

Frontal plane through heart

Left common carotid artery

Left subclavian artery

Arch of aorta

Left pulmonary artery

Pulmonary trunk

Left pulmonary veins

Left atrium

Aortic semilunar valve

Bicuspid valve

Chordae tendineae

Interventricular septum

Papillary muscle

Left ventricle

Descending aorta

(c) Anterior view of frontal section

Q *What is the path of blood flow from the heart to the lungs, back to the heart, from the heart to the rest of the body, and back to the heart again?*

Valves of the Heart

objective: *Describe the structure and functions of the valves of the heart.*

As each chamber of the heart contracts, it pushes a portion of blood into a ventricle or out of the heart through an artery. To keep the blood from flowing backward, the heart has four *valves* composed of dense connective tissue. These valves open and close in response to the pressure changes as the heart contracts and relaxes.

Atrioventricular (AV) Valves

Atrioventricular (*AV*) valves lie between the atria and ventricles (Figure 15.3c). The atrioventricular valve between the right atrium and right ventricle is also called the *tricuspid valve* because it consists of three cusps (flaps). These cusps are fibrous tissues covered with endocardium. The pointed ends of the cusps project into the ventricle. Tendonlike cords called *chordae tendineae*

(KOR-dē TEN-di-nē-ē; *corda* = cord; *tendo* = tendon) connect the pointed ends to cardiac muscle projections located on the inner surface of the ventricles called *papillary* (*papilla* = nipple) *muscles.* The chordae tendineae keep the valves from pushing up into the atria when the ventricles contract.

The atrioventricular valve between the left atrium and left ventricle is called the *bicuspid* (*mitral*) *valve*. It has two cusps that work in the same way as the cusps of the tricuspid valve. For blood to pass from an atrium to a ventricle, an atrioventricular (AV) valve must open. As you will see later, the opening and closing of the valves are due to pressure differences across the valves. When blood moves from an atrium to a ventricle, the valve is pushed open, the papillary muscles relax, and the chordae tendineae slacken (Figure 15.4a). When a ventricle contracts, the pressure of the ventricular blood drives the cusps upward until their edges meet and close the opening (Figure 15.4b). At the same time, contraction of the papillary muscles and tightening of the chordae tendineae help prevent the valve from swinging upward into the atrium.

Semilunar Valves

Near the origin of the pulmonary trunk and aorta are valves designed to prevent blood from flowing back into the heart. These valves are referred to as **semilunar valves** (see Figure 15.3c). The **pulmonary semilunar valve** lies in the opening where the pulmonary trunk leaves the right ventricle. The **aortic semilunar valve** is situated at the opening between the left ventricle and the aorta.

Each valve consists of three semilunar (half-moon or crescent-shaped) cusps. Each cusp is attached to the artery wall. Like the atrioventricular valves, the semilunar valves permit blood to flow in only one direction—in this case, from the ventricles into the arteries.

Valvular heart disease refers to any condition in which one or more of the heart valves operates improperly. Such incompetent valves could result in **regurgitation,** the backflow of blood through an incompletely closed valve. This could lead to congestive heart failure in which the heart is no longer capable of supplying the oxygen demands of the body.

A summary of the flow of blood through the heart is presented in Figure 15.5.

Blood Supply of the Heart

objective: *Describe the clinical importance of the blood supply of the heart.*

The wall of the heart, like any other tissue, has its own blood vessels. The flow of blood through the numerous vessels in the myocardium is called **coronary (cardiac) circulation.**

The principal coronary vessels are the **left** and **right coronary arteries,** which originate as branches of the ascending aorta (see Figure 15.3a). Each artery branches and subbranches to deliver oxygen and nutrients. Most of the deoxygenated blood, which carries the carbon dioxide and wastes, is collected by a large vein on the back of the heart, the **coronary sinus** (see Figure 15.3b), which empties into the right atrium.

Most heart problems result from faulty coronary circulation. If a reduced oxygen supply weakens cells but does not actually kill them, the condition is called **ischemia** (is-KĒ-mē-a; *ischein* = to suppress; *haima* = blood). **Angina pectoris** (an-JĪ-na, or AN-ji-na, PEK-to-ris), meaning "chest pain," results from ischemia of the myocardium. Common causes include stress, strenuous exertion after a heavy meal, narrowing of coronary arteries, high blood pressure, and fever. Symptoms include chest pain, accompanied by tightness or pressure, and difficult breathing. Sometimes weakness, dizziness, and perspiration occur.

A much more serious problem is **myocardial infarction** (in-FARK-shun), or **MI,** commonly called a heart attack. **Infarction** is the death of an area of tissue because of an interrupted blood supply. Myocardial infarction may result from a blood clot in one of the coronary arteries. Tissue past the blood clot dies and is replaced by noncontractile scar tissue, which causes the heart muscle to lose some of its strength. The aftereffects depend partly on the size and location of the infarct (dead) area. In addition to killing normal heart tissue, an infarction may disturb the flow of action potentials through the heart that cause the heart to beat. This can cause sudden death (**ventricular fibrillation**; see Common Disorders at the end of the chapter) but may be reversed by timely cardiopulmonary resuscitation (CPR). Individuals who survive an MI may develop new blood vessels that carry blood through secondary channels following obstruction of the principal channel.

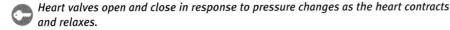

Figure 15.4   Atrioventricular (AV) valves. The bicuspid and tricuspid valves operate in a similar manner.

🔑 *Heart valves open and close in response to pressure changes as the heart contracts and relaxes.*

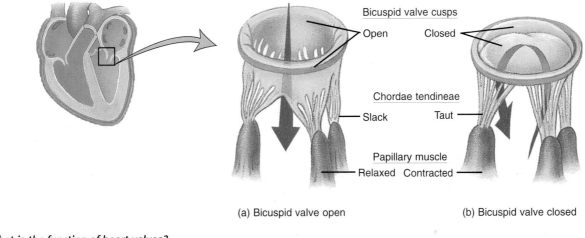

Bicuspid valve cusps — Open   Closed

Chordae tendineae — Slack   Taut

Papillary muscle — Relaxed   Contracted

(a) Bicuspid valve open   (b) Bicuspid valve closed

❓ *What is the function of heart valves?*

Figure 15.5   Blood flow through the heart.

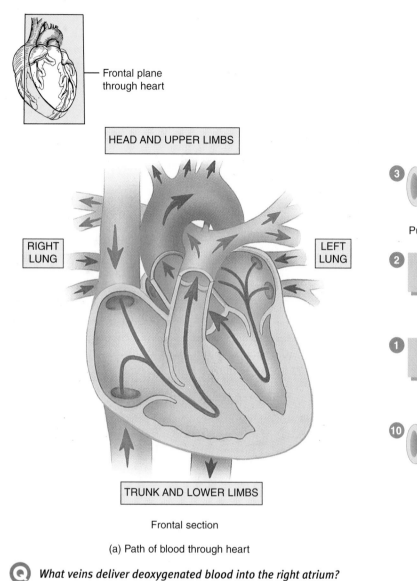

Frontal plane through heart

HEAD AND UPPER LIMBS

RIGHT LUNG

LEFT LUNG

TRUNK AND LOWER LIMBS

Frontal section

(a) Path of blood through heart

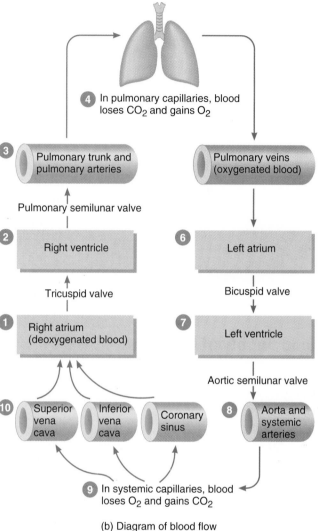

4 In pulmonary capillaries, blood loses CO_2 and gains O_2

3 Pulmonary trunk and pulmonary arteries

Pulmonary veins (oxygenated blood)

Pulmonary semilunar valve

2 Right ventricle

6 Left atrium

Tricuspid valve

Bicuspid valve

1 Right atrium (deoxygenated blood)

7 Left ventricle

Aortic semilunar valve

10 Superior vena cava   Inferior vena cava   Coronary sinus

8 Aorta and systemic arteries

9 In systemic capillaries, blood loses O_2 and gains CO_2

(b) Diagram of blood flow

Q *What veins deliver deoxygenated blood into the right atrium?*

Conduction System of the Heart

objective: *Explain how each heartbeat is initiated and maintained.*

The heart is innervated by the autonomic nervous system (ANS; see Chapter 11), which increases or decreases heart rate but does not initiate contraction. The heart can go on beating without any direct stimulus from the nervous system because it has an intrinsic regulating system called the ***conduction system.*** The conduction system is composed of specialized muscle tissue that generates and distributes action potentials that stimulate the cardiac muscle fibers to contract. All cardiac muscle is capable of ***self-excitation***; that is, it spontaneously and rhythmically generates action potentials that result in contraction of the muscle.

The components of the conduction system are the sinoatrial node, atrioventricular node, atrioventricular bundle, bundle branches, and conduction myofibers. The ***sinoatrial (SA) node*** is located in the right atrial wall just below the opening to the superior vena cava (Figure 15.6a). The SA node initiates each heartbeat and thereby sets the basic pace for the heart rate. For this reason, it is called the "pacemaker." The normal rate of the self-excitation SA node is about 75 times per minute in adults while at rest. Because the SA node spontaneously generates action potentials faster than other components of the conduction system, nerve impulses from the SA node spread to the other areas and stimulate them so frequently they are not able to generate action potentials at their own inherent rates. Thus the faster SA node sets the rhythm for the rest of the heart. The rate set by the SA node may be altered by several factors.

Figure 15.6   Conduction system of the heart. The arrows in (a) indicate the flow of impulses through the atria. Recall from Chapter 9 that a millivolt (mV) is equal to one-thousandth of a volt; a 1.5-volt battery will power an ordinary flashlight.

🔑 *The conduction system ensures that the cardiac chambers contract in a coordinated manner.*

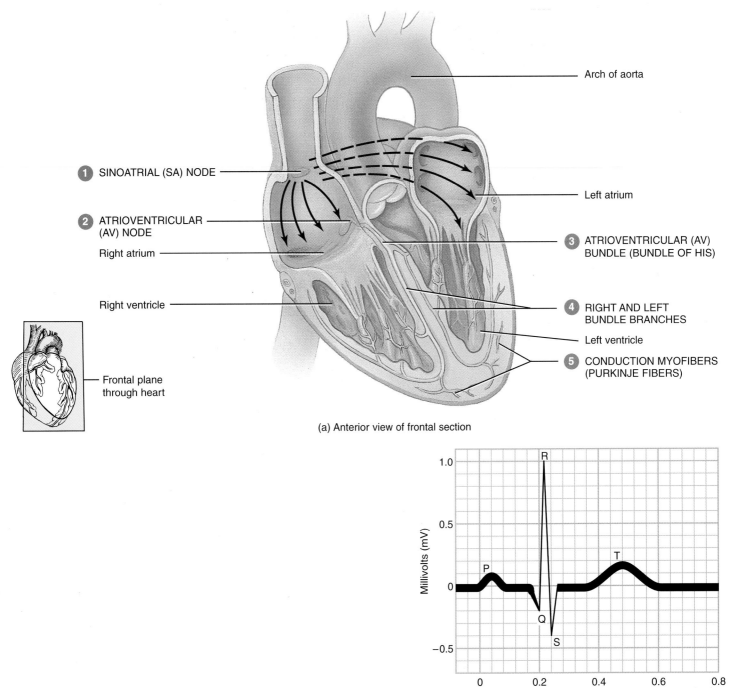

(a) Anterior view of frontal section

🔍 *Which component of the conduction system provides the only electrical connection between the atria and ventricles?*

(b) Normal electrocardiogram of a single heartbeat

When an action potential is initiated by the SA node, it spreads out over both atria, causing them to contract and depolarizing the *atrioventricular (AV) node.* The AV node is located in the interatrial septum. The action potential travels quite rapidly from the SA node, through the atria, and down to the AV node. At the AV node, the action potential slows con-

siderably. This delay allows time for the atria to empty their blood into the ventricles and finish their contraction before the ventricles begin theirs.

From the AV node, the action potential passes to a tract of conducting fibers at the top of the interventricular septum called the *atrioventricular (AV) bundle* or *bundle of His*

(HISS). From here the action potential passes toward the heart's apex through the *right* and *left bundle branches* in the interventricular septum. Actual contraction of the ventricles is stimulated by the *conduction myofibers* (*Purkinje fibers*) that emerge from the bundle branches and distribute the action potential to all the ventricular myocardial cells at about the same time. The action potential travels rapidly from the AV node through the ventricular myocardium.

A summary of the conduction system is presented in Exhibit 15.1.

Electrocardiogram (ECG or EKG)

| objective: | *Describe the meaning and diagnostic value of an electrocardiogram.*

Transmission of action potentials through the conduction system generates electric currents that can be detected on the body's surface. A recording of the electrical changes that accompany the heartbeat is called an *electrocardiogram* (e-lek′-trō-KAR-dē-ō-gram), which is abbreviated as either ECG or EKG.

Each portion of a heartbeat produces a different action potential. These action potentials are graphed as a series of up-and-down waves during an ECG. Three clearly recognizable waves normally accompany each cardiac cycle (Figure 15.6b). The first, called the *P wave*, indicates atrial depolarization—the spread of an action potential from the SA node through the two atria. A fraction of a second after the P wave begins, the atria contract. The second wave, called the *QRS wave*, represents ventricular depolarization, that is, the spread of the action potential through the ventricles. The third wave, the *T wave*, indicates ventricular repolarization. There is no wave to show atrial repolarization because the stronger QRS wave masks this event.

Variations in the size and duration of deflection waves of an ECG are useful in diagnosing abnormal cardiac rhythms and conduction patterns and in following the course of recovery from a heart attack. It can also detect the presence of fetal life.

In major disruptions of the conduction system, an irregular heart rhythm may occur. Normal heart rhythm can be restored and maintained with an *artificial pacemaker*, a device that sends out small electrical charges to the right atrium, right ventricle, or both in order to stimulate the heart.

Blood Flow Through the Heart

| objective: | *Explain how blood flows through the heart.*

The movement of blood through the heart is directly related to changes in blood pressure, which is caused by changes in the size of the chambers, and which brings about the opening and closing of the valves. Blood flows through the heart from areas of higher blood pressure to areas of lower blood pressure. When the walls of the atria are stimulated to contract by the SA node, the size of the atrial chambers is decreased, which thereby increases blood pressure within them. This increased blood pressure forces the AV valves open and atrial blood flows into the ventricles. The ventricle walls stay relaxed until after the atria are finished contracting. When a chamber's wall is relaxed, the blood pressure in that chamber is decreased.

When the walls of the ventricles contract, ventricular blood pressure increases to a higher level than that in the arteries so ventricular blood pushes the semilunar valves open and blood flows into the arteries. Also, at the same time, the shape of the AV valve cusps causes them to be pushed shut, to prevent backflow of ventricular blood into the atria where the pressure is now lower because the walls there have relaxed.

Cardiac Cycle (Heartbeat)

| objective: | *Describe the phases of a cardiac cycle (heartbeat).*

The term *systole* (SIS-tō-lē) refers to the phase of contraction of a chamber of the heart; *diastole* (dī′-AS-tō-lē) is the phase of relaxation.

Exhibit 15.1	Summary of Conduction System's Structures, Locations, and Functions	
STRUCTURE	LOCATION	FUNCTION
Sinoatrial (SA) node	Right atrial wall	Initiates each heartbeat and sets basic pace for heart rate. Sends action potential to both artria, causing them to contract.
Atrioventricular (AV) node	Interatrial septum	Picks up action potential from SA node and passes it to atrioventricular (AV) bundle.
Atrioventricular (AV) bundle (bundle of His)	Top of interventricular septum	Picks up action potential from AV node and passes it to bundle branches.
Bundle branches	Interventricular septum	Picks up action potential and passes it to conduction myofibers.
Conduction myofibers (Purkinje fibers)	Ventricular myocardium	Picks up action potential from bundle branches and passes it to ventricular myocardial cells.

Phases

For the purposes of our discussion, we will divide the *cardiac cycle* into the following phases (as you read the description, refer to Figure 15.7):

1 **Relaxation period.** At the end of a cardiac cycle when the ventricles start to relax, all four chambers are in diastole. This is the beginning of the *relaxation*. Repolarization of the ventricular muscle fibers (T wave in the ECG) initiates relaxation. As the ventricles relax, pressure within the chambers drops, and blood starts to flow from the pulmonary trunk and aorta back toward the ventricles. As this blood becomes trapped in the semilunar cusps, however, the semilunar valves close. As the ventricles continue to relax, the space inside expands, and the pressure falls. When ventricular pressure drops below atrial pressure, the AV valves open and ventricular filling begins.

2 **Ventricular filling.** The major part of ventricular filling (75 percent) occurs just after the AV valves open. This occurs without atrial contraction. Firing of the SA node results in atrial depolarization, noted as the P wave in the ECG. Atrial contraction follows the P wave, which also marks the end of the relaxation period. Atrial contraction accounts for the remaining 25 percent of the blood that fills the ventricles. Throughout the period of ventricular filling, the AV valves are open and the semilunar valves are closed.

3 **Ventricular systole (contraction).** Firing of the AV node causes ventricular depolarization, which is represented by the QRS complex in the ECG. Then, ventricular contraction begins, and blood is pushed up against the AV valves, forcing them shut. For a very brief period of time, all four valves are closed again. As ventricular contraction continues, pressure inside the chambers rises sharply. When left ventricular pressure rises above the pressure in the arteries, both semilunar valves open, and ejection of blood from the heart begins. This lasts until the ventricles start to relax. Then, the semilunar valves close and another relaxation period begins.

Timing

If the average heart rate is 75 beats per minute, then each cardiac cycle requires about 0.8 sec. In a complete cycle, the atria are in systole 0.1 sec and in diastole 0.7 sec. The ventricles are in systole 0.3 sec and in diastole 0.5 sec. The last 0.4 sec of the cycle is the *relaxation period* and all chambers are in diastole. When the heart beats faster than normal, the relaxation period is shortened accordingly.

Sounds

The sound of the heartbeat comes primarily from turbulence in blood flow created by the closure of the valves, not from the contraction of the heart muscle. The first sound (S1), *lubb,* is a long, booming sound from the AV valves closing after ventricular systole begins. The second sound, a short, sharp sound (S2), *dupp,* is from the semilunar valves closing at the end of ventricular systole. There is a pause between cycles. Thus, the cardiac cycle is heard as lubb, dupp, pause; lubb, dupp, pause; lubb, dupp, pause.

Heart sounds provide valuable information about the mechanical operation of the heart. A *heart murmur* is an abnormal sound that consists of a flow noise that is heard before, between, or after the lubb–dupp or that may mask normal heart sounds. Although some heart murmurs are "innocent," meaning they do not suggest a heart problem, most often a heart murmur indicates a valve disorder. One example of a valve disorder is *mitral stenosis* (narrowing of the mitral valve by scar formation or congenital defect). Another cause of a heart murmur is *mitral valve prolapse* (MVP), an inherited disorder in which a portion of a mitral valve is pushed back too far (prolapsed) during ventricular contraction and the cusps do not close properly. As a result, a small volume of blood may flow back into the left atrium during ventricular systole. Even so, mitral valve prolapse often does not pose a serious threat. In fact, it is found in up to 10 percent of otherwise healthy young persons.

Cardiac Output (CO)

Although the heart can beat independently, it is regulated by events occurring in the rest of the body. All body cells must receive a certain amount of oxygenated blood each minute to maintain health and life. When cells are very active, as during exercise, they need even more blood.

The amount of blood ejected from the left ventricle into the aorta (or right ventricle into the pulmonary trunk) per minute is called the *cardiac output* (CO). Cardiac output is determined by (1) the amount of blood pumped by the left (or right) ventricle during each beat and (2) the number of heartbeats per minute. The amount of blood ejected by a ventricle during each contraction is called the *stroke volume* (SV). In a resting adult, stroke volume averages 70 ml and heart rate is about 75 beats per minute. Thus the average cardiac output in a resting adult is

$$\text{Cardiac output} = \text{stroke volume} \times \text{beats per minute}$$

$$= 70 \text{ ml} \times 75/\text{min}$$

$$= 5250 \text{ ml/min or } 5.25 \text{ liters/min}$$

Factors that increase stroke volume or heart rate tend to increase cardiac output. Factors that decrease stroke volume or heart rate tend to decrease cardiac output.

The force of ventricular contraction is partly determined by the length of cardiac muscle fibers. The more cardiac fibers are stretched by the filling of a chamber with blood, the stronger the walls will contract to eject the blood. This relationship is referred to as *Starling's law of the heart*. The situation is somewhat like stretching a rubber band; the more you stretch it, the harder it contracts. The operation of Starling's law of the heart is important in maintaining equal blood output from both ventricles.

Figure 15.7 Cardiac cycle. (a) ECG. (b) Left atrial, left ventricular, and aortic pressure changes along with the opening and closing of valves. (c) Left ventricular volume changes. (d) Heart sounds. (e) Phases of the cardiac cycle.

🔑 *A cardiac cycle is composed of all the events associated with one heartbeat.*

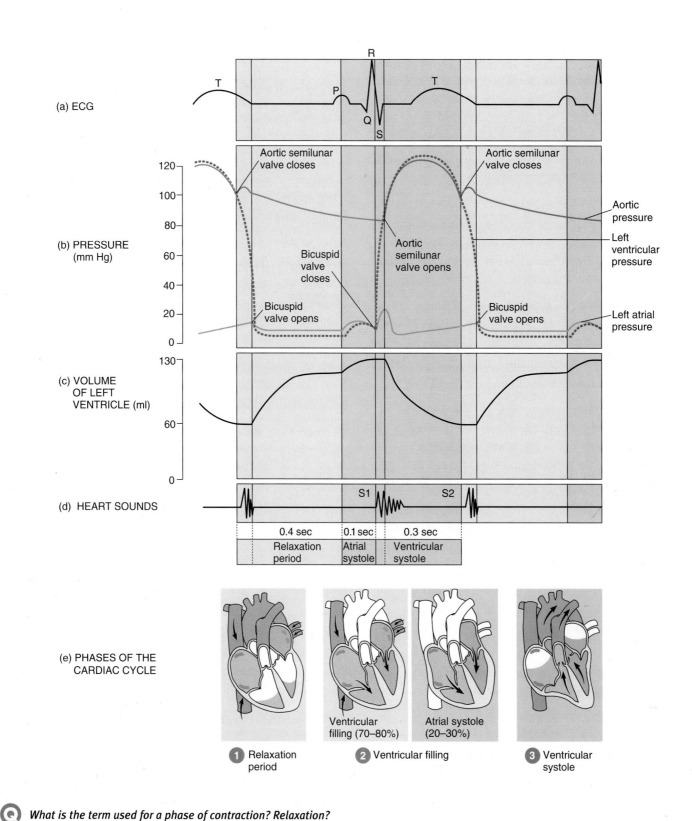

(a) ECG

(b) PRESSURE (mm Hg)

Aortic semilunar valve closes

Aortic semilunar valve closes

Aortic semilunar valve opens

Bicuspid valve closes

Bicuspid valve opens

Bicuspid valve opens

Aortic pressure

Left ventricular pressure

Left atrial pressure

(c) VOLUME OF LEFT VENTRICLE (ml)

(d) HEART SOUNDS

S1   S2

0.4 sec | 0.1 sec | 0.3 sec

Relaxation period | Atrial systole | Ventricular systole

(e) PHASES OF THE CARDIAC CYCLE

Ventricular filling (70–80%)

Atrial systole (20–30%)

1 Relaxation period

2 Ventricular filling

3 Ventricular systole

❓ *What is the term used for a phase of contraction? Relaxation?*

Heart Rate

objective: *Identify the factors that affect heart rate.*

Cardiac output depends on heart rate as well as stroke volume. In fact, changing heart rate is the body's principal mechanism of short-term control over cardiac output and blood pressure. The sinoatrial (SA) node that initiates contraction would, if left to itself, set an unvarying heart rate. However, the body's need for blood supply varies under different conditions, so several regulatory mechanisms exist. They are stimulated by such factors as chemicals present in the body, temperature, emotional state, and age.

In certain pathological states, stroke volume may fall dangerously low. If the ventricular myocardium is weak or damaged by an infarction, it cannot contract strongly. Or blood volume may be reduced by excessive bleeding, causing stroke volume to fall because the cardiac fibers are not sufficiently stretched. In these cases, the body attempts to maintain a safe cardiac output by increasing the rate and strength of contraction.

The heart rate is regulated by several factors, but the most important control of rate and strength of contraction is the autonomic nervous system.

Autonomic Control

Within the medulla oblongata of the brain is a group of neurons called the **cardiovascular center.** Arising from this center are sympathetic fibers that travel down a tract in the spinal cord and then pass outward to **cardiac accelerator nerves** that innervate the conduction system, atria, and ventricles (Figure 15.8). When this part of the cardiovascular center is stimulated, nerve impulses travel along the sympathetic fibers. This causes them to release norepinephrine (NE), which increases the rate of heartbeat and the strength of contraction.

Also arising from the cardiovascular center are parasympathetic fibers that reach the heart via the **vagus (X) nerves.** When this part of the cardiovascular center is stimulated, nerve impulses are transmitted along the parasympathetic fibers to the conduction system and atria that cause the release of acetylcholine (ACh). This decreases the rate of heartbeat and strength of contraction by slowing the SA and AV nodes.

The autonomic control of the heart is therefore the result of opposing sympathetic (stimulatory) and parasympathetic (inhibitory) influences. Receptors in the cardiovascular system inform the cardiovascular center so that a balance between stimulation and inhibition is maintained. For example, **baroreceptors,** neurons sensitive to blood pressure changes, are strategically located in the arch of the aorta, internal carotid arteries (arteries in the neck that supply blood to the brain), and right atrium. If there is an increase in blood pressure, the baroreceptors send nerve impulses along the glossopharyngeal (IX) nerves that stimulate the **cardioinhibitory** center and inhibit the cardiovascular center (Figure 15.8). The cardiovascular center responds by putting out more parasympathetic impulses along the vagus (X) nerves to the heart. The resulting decreases in heart rate and force of contraction lower cardiac output and thus lower blood pressure. Also, the cardiovascular center sends out decreased sympathetic impulses along nerves that normally cause vasoconstriction. The result is vasodilation, which also lowers blood pressure back to normal.

If, on the other hand, blood pressure falls, baroreceptors do not stimulate the cardiovascular center. As a result, heart rate and force of contraction increase, cardiac output increases, vasoconstriction occurs, and blood pressure increases to normal (Figure 15.9 on p. 355).

Chemicals

Certain chemicals in the body have an effect on heart rate. For example, epinephrine, produced by the adrenal medulla in response to sympathetic stimulation, increases the excitability of the SA node, which, in turn, increases the rate and strength of contraction. Elevated levels of potassium (K^+) or sodium (Na^+) decrease the heart rate and strength of contraction. Excess K^+ apparently interferes with the generation of nerve impulses and excess Na^+ interferes with calcium (Ca^{2+}) participation in muscular contraction. An excess of calcium increases heart rate and strength of contraction.

When oxygen demands of the body are low, heart rate is decreased. In response to increased oxygen demands, heart rate increases.

Temperature

Increased body temperature, from fever or strenuous exercise, for example, causes the AV node to discharge impulses faster and thereby increases heart rate. Decreased body temperature, from exposure to cold or deliberately cooling the body prior to surgery, decreases heart rate and strength of contraction.

Emotions

Strong emotions such as fear, anger, and anxiety, along with a multitude of physiological stressors, increase heart rate through the general adaptation syndrome (see Figure 13.19). Mental states such as depression and grief tend to stimulate the cardioinhibitory center and decrease heart rate.

Gender and Age

Gender is another factor: the heartbeat is somewhat faster in normal females than normal males. Age is yet another factor: the heartbeat is fastest at birth, moderately fast in youth, average in adulthood, and below average in old age.

Risk Factors in Heart Disease

objective: *List and explain the risk factors involved in heart disease.*

About 1.5 million people suffer a myocardial infarction every year in the United States and of these more than 500,000 die suddenly before reaching a hospital. However, the prevalence of heart disease has diminished in recent years, due in part to

Figure 15.8 Autonomic nervous system regulation of heart rate.

🔑 *The cardiovascular center in the medulla oblongata controls both sympathetic and parasympathetic nerves that innervate the heart.*

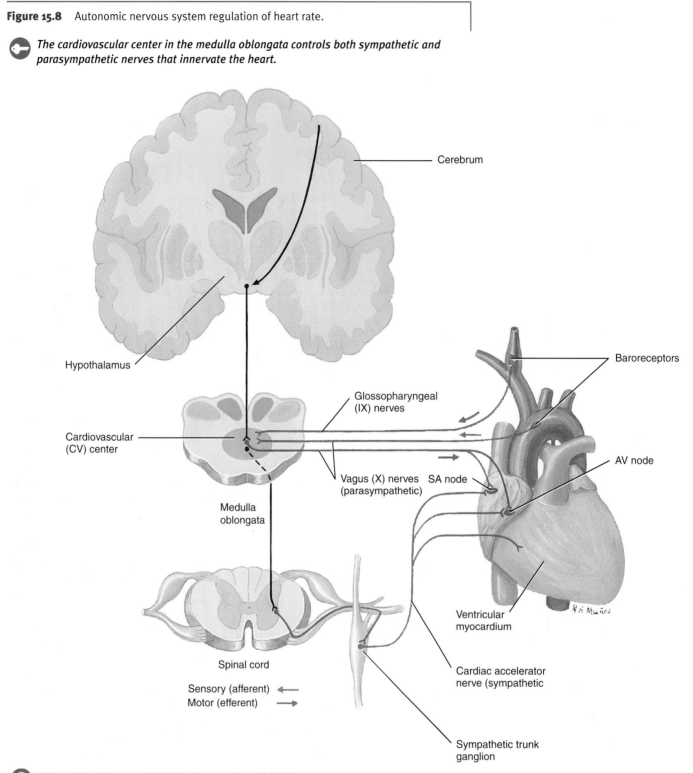

Cerebrum

Hypothalamus

Glossopharyngeal (IX) nerves

Baroreceptors

Cardiovascular (CV) center

AV node

Vagus (X) nerves (parasympathetic)   SA node

Medulla oblongata

Ventricular myocardium

Spinal cord

Sensory (afferent) ⟵
Motor (efferent) ⟶

Cardiac accelerator nerve (sympathetic

Sympathetic trunk ganglion

❓ *What effect does acetylcholine have on heart rate?*

changes in lifestyle. Some of the causes of heart disease can be foreseen and prevented. People who develop combinations of certain risk factors are more likely to have heart attacks. ***Risk factors*** are characteristics, symptoms, or signs present in a person free of disease that are statistically associated with a greater chance of developing a disease. The major risk factors in heart disease are:

1. High blood cholesterol level.
2. High blood pressure.

Figure 15.9   Negative feedback control of blood pressure involving baroreceptors.

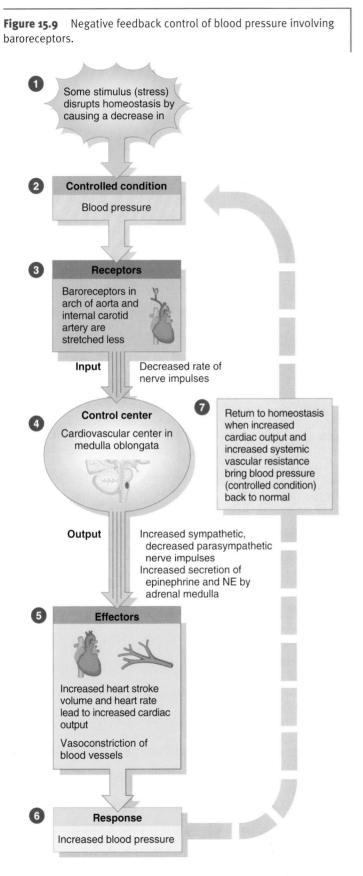

1 Some stimulus (stress) disrupts homeostasis by causing a decrease in

2 **Controlled condition**

Blood pressure

3 **Receptors**

Baroreceptors in arch of aorta and internal carotid artery are stretched less

Input Decreased rate of nerve impulses

4 **Control center**

Cardiovascular center in medulla oblongata

7 Return to homeostasis when increased cardiac output and increased systemic vascular resistance bring blood pressure (controlled condition) back to normal

Output Increased sympathetic, decreased parasympathetic nerve impulses
Increased secretion of epinephrine and NE by adrenal medulla

5 **Effectors**

Increased heart stroke volume and heart rate lead to increased cardiac output

Vasoconstriction of blood vessels

6 **Response**

Increased blood pressure

Q *Does vasoconstriction or vasodilation lower blood pressure?*

3. Cigarette smoking.

4. Obesity.

5. Lack of regular exercise.

6. Diabetes mellitus.

7. Genetic predisposition (family history of heart disease at an early age).

8. Male gender (after age 70, the risk of heart attack is similar in males and females).

The first five risk factors can all be modified to reduce one's risk; for example, losing weight, becoming more physically active, or quitting smoking. High blood cholesterol is discussed shortly, and high blood pressure is discussed in the next chapter. Nicotine in cigarette smoke enters the bloodstream and constricts small blood vessels. It also stimulates the adrenal gland to oversecrete epinephrine and norepinephrine, which elevate heart rate and blood pressure. Overweight people develop extra capillaries to nourish fat tissue. The heart has to work harder to pump the blood through more vessels.

Without exercise, the return of blood through the heart gets less help from contracting skeletal muscles. In addition, regular exercise strengthens the smooth muscle of blood vessels and enables them to assist general circulation. Exercise also increases cardiac efficiency and output. This is best accomplished by aerobics, any activity—such as brisk walking, running, or swimming—that works large body muscles for at least 20 minutes. Three 20-minute aerobic workouts per week seem to be sufficient for maintaining basic cardiovascular efficiency. A well-trained athlete can achieve a cardiac output up to about six times that of a sedentary person ("couch potato") during activity because training causes enlargement of the heart. The heart rate of a trained athlete is about 40 to 60 beats per minute. Other benefits of physical conditioning are an increase in high-density lipoprotein ("good cholesterol"), a decrease in triglyceride (fat) levels, and improved lung function. Exercise also helps reduce blood pressure, anxiety, and depression; control weight; and increase the body's ability to dissolve blood clots.

In diabetes mellitus, fat breakdown dominates glucose breakdown. As a result, cholesterol levels get progressively higher and result in fatty masses in blood vessels, a situation that may lead to high blood pressure. High blood pressure drives fat into the vessel wall, which encourages more fatty masses and clot formation.

Up to age 70, there is a 10- to 15-year lag in the extent of heart disease in females compared with males; after that, the rates of disease in both sexes are similar. This lower incidence in females is due to estrogens secreted up to menopause and estrogen replacement therapy after menopause. Regardless of gender, the incidence of heart disease increases with age.

Other factors may also contribute to the development of heart disease. Alcoholism (which damages cardiac and skeletal muscle), high blood level of fibrinogen (which promotes blood clot formation), and renin (which increases blood pressure) all increase the risk of a heart attack.

*a*nswer the following question: What disease kills the most women in North America? The correct answer is *heart disease*. Cardiovascular diseases (diseases of the heart and blood vessels, including heart attack and stroke) are the leading cause of death for both men and women. Each year 2.5 million women in the United States are hospitalized for and 500,000 women die from cardiovascular disease.

If you answered *breast cancer*, you are not alone. Many people have the impression that breast cancer is the leading cause of death for women, because the incidence of breast cancer has risen dramatically over the past twenty years. But of 2,000 postmenopausal women in the United States, 20 women in a given year will get heart disease, and 12 will die from it. In comparison, 6 women in the same year will get breast cancer, and 2 will die from it.

Heart Disease: An Equal Opportunity Killer

Until recently, most of the research on the prevention and treatment of cardiovascular disease has focused primarily on men, and therefore many of our ideas about heart disease have evolved from a male model of these disorders. Fortunately, the research focus continues to widen, and many studies now include women, as well as men and women from underrepresented groups. But many people still believe that heart disease is something that happens to men, not women. These beliefs have implications for both the prevention and treatment of cardiovascular disease in women.

Where do these beliefs come from? Young women do have an advantage over young men in terms of risk for heart disease. Women tend to develop arterial disease about ten years later than men. Men aged 25–64 years old have twice the incidence of heart disease of women in that age group.

But that is where the advantage ends. For the 70+ age group, heart disease rates for men and women are similar. What's worse, women who develop arterial disease tend to fare more poorly than men with similar diagnoses.

Why Do Women Have Poorer Prognoses?

Many reasons have been proposed to explain why women with heart disease fare so poorly, but we need more research before jumping to any conclusions. Some evidence suggests that physicians prescribe more surgical interventions for men than women. (Others have suggested that since surgery is not always beneficial, maybe women get better care in this regard!) While surgical and medical outcomes for women have improved over the years, they still lag behind, perhaps because surgical instruments, techniques, and drug therapies were designed for male arteries and bodies, which tend to be larger than women's. More fundamentally, however, women are not taught and thus may not recognize the warning signs of heart disease. Family members may not take Mom's chest pain and discomfort as seriously as Dad's, and may not urge her to get to the doctor. Thus, women typically appear in their doctor's office later in the course of the disease than men do. Also, a sedentary lifestyle may delay diagnosis, and women who are 65 and older are more likely than their male peers to be sedentary. Exertion often brings on chest pain in people with poor coronary circulation; sedentary people may not get chest pain until the disease is more advanced.

Research shows that by the time women begin treatment for arterial disease, they tend to be older and have more coexisting health problems than men. These factors contribute to women's poorer prognosis following heart disease diagnosis and treatment.

critical thinking

A female friend tells you that since cardiovascular disease occurs primarily in postmenopausal women, women need not be concerned about risk factors until after menopause. When should women be concerned about controlling risk factors?

Why might a cardiologist treat a patient with coexisting health problems less aggressively (less surgical intervention) than a patient in better health?

Common Disorders

Coronary Artery Disease (CAD)

Coronary artery disease (*CAD*) refers to conditions that cause narrowing of the coronary arteries so that blood flow to the heart is reduced. This results in *coronary heart disease* (*CHD*), a condition in which the heart muscle is damaged due to an inadequate amount of blood. The principal causes of coronary artery disease are atherosclerosis, coronary artery spasm, and a clot in a coronary artery.

Atherosclerosis

Thickening of the walls of arteries and loss of elasticity are the main characteristics of a group of diseases called *arteriosclerosis* (ar-tē′-rē-ō′-skle-RŌ-sis; *sclerosis* = hardening). One form of arteriosclerosis is *atherosclerosis* (ath′-er-ō-skle-RŌ-sis), a process in which smooth muscle cells of arterial walls proliferate and fatty substances, especially cholesterol and triglycerides (ingested fats), are deposited in the walls of arteries in response to certain stimuli. The fatty substances, cholesterol, and smooth muscle fibers form a mass called an *atherosclerotic plaque,* which obstructs blood flow in vessels (Figure 15.10). Contributing factors in atherosclerosis include high blood pressure, carbon monoxide in cigarettes, diabetes mellitus, and high cholesterol level. It has been found that a vegetarian diet and mild exercise can not only stop the progression of atherosclerosis, but actually reverse the process.

Figure 15.10   Photomicrograph of an artery partially obstructed by an atherosclerotic plaque.

Atherosclerosis is one form of arteriosclerosis (thickening of arterial walls with loss of elasticity).

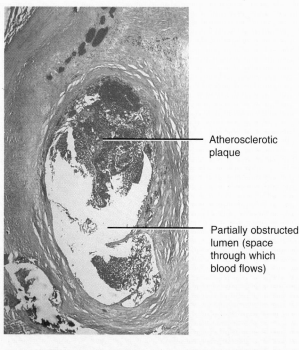

Atherosclerotic plaque

Partially obstructed lumen (space through which blood flows)

Transverse section

Q *What substances are part of an atherosclerotic plaque?*

Treatment options for coronary artery disease include drugs (nitroglycerine, beta blockers, and clot-dissolving agents) and various surgical and nonsurgical procedures designed to increase the blood supply to the heart.

Coronary artery bypass grafting (*CABG*) is a surgical procedure in which a blood vessel from another part of the body is used ("grafted") to bypass the blocked region of a coronary artery. A piece of the grafted blood vessel is sutured between the aorta and the unblocked portion of the coronary artery.

A nonsurgical procedure used to treat CAD is termed *percutaneous transluminal coronary angioplasty* (*PTCA*) (*percutaneous* = through the skin; *trans* = across; *lumen* = channel in a tube; *angio* = blood vessel; *plasty* = to mold or to shape). In this procedure, a balloon catheter (plastic tube) is inserted into an artery of an arm or leg and gently guided up into a coronary artery. Then, while dye is released, angiograms (x-rays of blood vessels) are taken to identify the plaques. Next, the catheter is advanced to the point of obstruction and a balloonlike device is inflated with air to squash the plaque against the blood vessel wall.

Coronary Artery Spasm

Atherosclerosis results in an obstruction to blood flow. Obstruction can also be caused by *coronary artery spasm,* in which the smooth muscle of a coronary artery undergoes a sudden contraction, resulting in vasoconstriction. Coronary artery spasm typically occurs in individuals with atherosclerosis and may result in chest pain, heart attacks, and sudden death. Factors related to coronary artery spasm include smoking, stress, and a vasoconstrictor chemical released by platelets.

Congenital Defects

A defect that exists at birth is called a *congenital defect.*

In *patent ductus arteriosus,* the ductus arteriosus (temporary blood vessel) between the aorta and the pulmonary trunk, which normally closes shortly after birth, remains open.

Interventricular septal defect is caused by an incomplete closure of the interventricular septum.

Valvular stenosis is a narrowing of one of the valves regulating blood flow in the heart.

Tetralogy of Fallot (tet-RAL-ō-jē; FAL-ō) is a combination of four defects: an interventricular septal defect, an aorta that emerges from both ventricles instead of from the left ventricle only, a stenosed pulmonary semilunar valve, and an enlarged right ventricle.

Arrhythmias

Arrhythmia (a-RITH-mē-a) or *dysrhythmia* is a general term referring to an irregularity in heart rhythm. It results when there is a disturbance in the conduction system of the heart. Arrhythmias are caused by factors such as caffeine, nicotine, alcohol, anxiety, certain drugs, hyperthyroidism, potassium deficiency, and certain heart diseases. One serious arrhythmia is called a *heart block.* The most common blockage is in the atrioventricular (AV) node, which conducts impulses from the atria to the ventricles. This disturbance is called *atrioventricular* (*AV*) *block.*

In *atrial flutter,* the atrial rhythm averages between 240 and 360 beats per minute. The condition is essentially rapid atrial contractions accompanied by AV block. *Atrial fibrillation* is an uncoordinated contraction of the atrial muscles that causes the atria to contract irregularly and still faster. When the muscle fibrillates, the muscle fibers of the atrium quiver individually instead of contracting together. The quivering cancels out the pumping of the atrium. *Ventricular fibrillation* (VF) is characterized by asynchronous, haphazard, ventricular muscle contractions. Ventricular contraction becomes ineffective and circulatory failure and death occur.

Congestive Heart Failure (CHF)

Congestive heart failure (CHF) is a chronic or acute state that results when the heart is not capable of supplying the oxygen demands of the body. Causes include coronary artery disease,

congenital defects, long-term high blood pressure, myocardial infarctions, and valve disorders. Often, one side of the heart starts to fail before the other. If the left ventricle fails first, it can't pump out all the blood it receives. As a result, blood backs up in the lungs. The result is pulmonary edema, fluid accumulation in the lungs which can suffocate an untreated person. If the right ventricle fails first, blood backs up in the systemic vessels. In this case, the resulting peripheral edema is usually most noticeable in the feet and ankles.

Cor Pulmonale (CP)

Cor pulmonale (kor pul-mōn-ALE; *cor* = heart; *pulmon* = lung), *or CP,* refers to enlargement of the right ventricle from disorders that bring about hypertension (high blood pressure) in the circulation of blood in the lungs. ■

Medical Terminology and Conditions

Cardiac angiography (an'-jē-OG-ra-fē; *angio* = vessel; *cardio* = heart; *graph* = writing) Procedure in which a cardiac catheter (plastic tube) is used to inject a radiopaque contrast medium into blood vessels or heart chambers. The procedure may be used to visualize blood vessels and the ventricles to assess structural abnormalities. Angiography can also be used to inject clot-dissolving drugs, such as streptokinase or tissue plasminogen activator (t-PA), into a coronary artery to dissolve an obstructing thrombus.

Cardiac arrest (KAR-dē-ak a-REST) A clinical term meaning cessation of an effective heartbeat. The heart may be completely stopped or quivering ineffectively (ventricular fibrillation).

Cardiac catheterization (kath'-e-ter-i-ZĀ-shun) Procedure that is used to visualize the heart's coronary arteries, chambers, valves, and great vessels. It may also be used to measure pressure in the heart and blood vessels; to assess cardiac output; and to measure the flow of blood through the heart and blood vessels, the oxygen content of blood, and the status of heart valves and conduction system. The basic procedure

involves inserting a catheter (plastic tube) into a peripheral vein (for right heart catheterization) or artery (for left heart catheterization) and guiding it under fluoroscopy (x-ray observation).

Cardiomegaly (kar'-dē-ō-MEG-a-lē; *mega* = large) Heart enlargement.

Incompetent valve (in-KOM-pe-tent VALV) Any valve that does not close properly, thus permitting a backflow of blood; also called *valvular insufficiency.*

Palpitation (pal'-pi-TĀ-shun) A fluttering of the heart or abnormal rate or rhythm of the heart.

Paroxysmal tachycardia (par'-ok-SIZ-mal tak'-e-KAR-dē-ā) A period of rapid heartbeats that begins and ends suddenly.

Sudden cardiac death The unexpected cessation of circulation and breathing due to an underlying heart disease such as ischemia, myocardial infarction, or a disturbance in cardiac rhythm.

Study Outline

Location of the Heart (p. 342)

1. The heart is situated between the lungs in the mediastinum.
2. About two-thirds of the heart's mass is to the left of the midline.

Pericardium (p. 342)

1. The pericardium consists of an outer fibrous layer and an inner serous pericardium.
2. The serous pericardium is composed of a parietal and visceral layer.
3. Between the parietal and visceral layers of the serous pericardium is the pericardial cavity, a space filled with pericardial fluid that prevents friction between the two membranes.

Wall; Chambers; Vessels; and Valves (pp. 342–346)

1. The wall of the heart has three layers: epicardium, myocardium, and endocardium.
2. The chambers include two upper atria and two lower ventricles.
3. The blood flows through the heart from the superior and inferior venae cavae (plural for vena cava) and the coronary sinus to the right atrium, through the tricuspid valve to the right ventricle, through the pulmonary trunk to the lungs, through the pulmonary veins into the left atrium, through the bicuspid valve to the left ventricle, and out through the aorta.
4. Four valves prevent backflow of blood in the heart.
5. Atrioventricular (AV) valves, between the atria and their ventricles, are the tricuspid valve on the right side of the heart and the bicuspid (mitral) valve on the left.
6. The atrioventricular valves, chordae tendineae, and their papillary muscles stop blood from backflowing into the atria.
7. The two arteries that leave the heart both have a semilunar valve.

Blood Supply of the Heart (p. 347)

1. Coronary circulation delivers oxygenated blood to the myocardium and removes carbon dioxide from it.
2. Deoxygenated blood returns to the right atrium via the coronary sinus.
3. Complications of this system are angina pectoris and myocardial infarction (MI).

Conduction System of the Heart (p. 348)

1. The conduction system consists of tissue specialized for action potential conduction.
2. Components of this system are the sinoatrial (SA) node (pacemaker), atrioventricular (AV) node, atrioventricular (AV) bundle (bundle of His), bundle branches, and conduction myofibers (Purkinje fibers).

Electrocardiogram (ECG or EKG) (p. 350)

1. The record of electrical changes during each cardiac cycle is referred to as an electrocardiogram (ECG).

2. A normal ECG consists of a P wave (spread of action potential from SA node over atria), QRS wave (spread of action potential through ventricles), and T wave (ventricular repolarization).
3. The ECG is used to diagnose abnormal cardiac rhythms and conduction patterns, detect the presence of fetal life, and follow the course of recovery from a heart attack.

Blood Flow Through the Heart (p. 350)

1. Blood flows through the heart from areas of higher to lower pressure.
2. The pressure developed is related to the size and volume of a chamber.
3. The movement of blood through the heart is controlled by the opening and closing of the valves and the contraction and relaxation of the myocardium.

Cardiac Cycle (Heartbeat) (p. 350)

1. A cardiac cycle consists of systole (contraction) and diastole (relaxation) of the chambers of the heart.
2. The phases of the cardiac cycle are (a) the relaxation period, (b) ventricular filling, and (c) ventricular systole.
3. With an average heartbeat of 75/min, a complete cardiac cycle requires 0.8 sec.
4. The first heart sound (lubb) represents the closing of the atrioventricular valves. The second sound (dupp) represents the closing of semilunar valves.

Cardiac Output (CO) (p. 351)

1. Cardiac output (CO) is the amount of blood ejected by the left ventricle into the aorta per minute. It is calculated as follows: CO = stroke volume × beats per minute.
2. Stroke volume (SV) is the amount of blood ejected by a ventricle during each systole.

Heart Rate (p. 353)

1. Heart rate and strength of contraction are controlled by the cardiovascular center in the medulla oblongata and may be increased by sympathetic stimulation and decreased by parasympathetic stimulation.
2. Baroreceptors are nerve cell receptors that respond to changes in blood pressure. They act on the cardiac centers in the medulla.
3. Other influences on heart rate include chemicals (epinephrine, sodium, potassium), temperature, emotion, gender, and age.

Risk Factors in Heart Disease (p. 353)

1. Risk factors include high blood cholesterol, high blood pressure, cigarette smoking, obesity, lack of regular exercise, diabetes mellitus, genetic predisposition, gender, and age.
2. Among the benefits of aerobic exercise are increased cardiac output, increased high-density lipoprotein, decreased triglycerides, improved lung function, decreased blood pressure, and weight control.

Self-Quiz

1. Which chamber of the heart has the thickest layer of myocardium?

   a. right ventricle   b. right atrium   c. left ventricle
   d. left atrium   e. coronary sinus

2. The valve between the left atrium and left ventricle is the

   a. bicuspid (mitral) valve   b. aortic semilunar valve
   c. tricuspid valve   d. pulmonary semilunar valve
   e. sinoatrial valve

3. Which part of the heart pumps blood to the lungs?

   a. left atrium   b. left ventricle   c. pulmonary trunk
   d. right atrium   e. right ventricle

4. Which of the following statements describes the pericardium?

   a. It is a layer of simple squamous epithelium.   b. It lines the inside of the myocardium.   c. It is continuous with the epithelial lining of the large blood vessels.   d. It is responsible for the contraction of the heart.   e. It anchors the heart in place in the mediastinum.

5. Which part of the heart first receives oxygenated blood returning from the lungs?

   a. left ventricle   b. left atrium   c. right atrium
   d. right ventricle   e. right auricle

6. The "pacemaker" of the heart is the

   a. sinoatrial (SA) node   b. atrioventricular (AV) node
   c. conduction myofibers (Purkinje fibers)   d. atrioventricular (AV) bundle   e. intercalated disc

7. In normal heart action

   a. the right atrium and ventricle contract, followed by the contraction of the left atrium and ventricle   b. the order of contraction is: right atrium then right ventricle then left atrium then left ventricle   c. the two atria contract together, then the two ventricles contract together   d. the right atrium and left ventricle contract, followed by the contraction of the left atrium and right ventricle   e. all four chambers of the heart contract and then relax simultaneously

8. Heart rate and strength of contraction are controlled by the cardiovascular center, which is located in the

   a. cerebrum   b. pons   c. right atrium   d. medulla oblongata   e. AV node

9. Heart sounds are produced by

   a. contraction of the myocardium   b. closure of the heart valves   c. the flow of blood in the coronary arteries
   d. the flow of blood in the ventricles   e. the transmission of action potentials through the conduction system

10. Using the situations below, indicate if the heart rate would speed up or slow down.

    _H_ a. sympathetic stimulation   A. speed up
         of the sinoatrial (SA) node   B. slow down

    _A_ b. decrease in blood pressure

    _A_ c. female gender

    _B_ d. parasympathetic
         stimulation of heart
         conduction system

    _A_ e. release of epinephrine

    _B_ f. elevated K+ levels

    _B_ g. release of acetylcholine

    _A_ h. strenuous exercise

    _B_ i. stimulation by vagus
         nerve

    _A_ j. fear, anger, stress

    _B_ k. cooling the body

    _B_ l. old age

11. The portion of the ECG that corresponds to atrial depolarization is the

    a. R peak   b. space between the T wave and P wave
    c. T wave   d. P wave   e. QRS wave

12. Cardiac output

    a. equals stroke volume (SV) × blood pressure (BP)
    b. equals stroke volume (SV) × beats per minute   c. is calculated using the formula for Starling's law of the heart
    d. is about 70 ml in an average adult male   e. is (the volume of blood in the ventricles) − (the volume of blood in the atria)

13. The opening of the semilunar valves is due to

    a. the pressure in the ventricles exceeding the pressure in the aorta and pulmonary trunk   b. the pressure in the ventricles exceeding the pressure in the atria   c. the pressure in the atria exceeding the pressure in the ventricles   d. the pressure in the atria exceeding the pressure in the aorta and pulmonary trunk   e. the pressure in the aorta and pulmonary trunk exceeding the pressure in the ventricles

14. Match the following:

    _B_ a. may cause heart   A. pericarditis
         murmur   B. mitral valve prolapse
    _G_ b. heart compression   C. myocardial infarction
    _A_ c. inflammation of   D. angina pectoris
         heart covering   E. diastole
    _F_ d. heart chamber   F. systole
         contraction   G. cardiac tamponade
    _D_ e. chest pain from
         ischemia
    _C_ f. heart attack
    _E_ g. heart chamber
         relaxation

Critical Thinking Applications

1. Compare the structure and function of the atrioventricular (AV) heart valves to that of a parachute or come up with your own comparison.

2. Your uncle had an artificial pacemaker put in after his last bout with heart trouble. What is the function of a pacemaker? For what heart structure does the pacemaker substitute?

3. Kathryn was strolling across a four-lane highway when a car suddenly appeared out of nowhere. As she finished sprinting across the road, she felt her heart racing. Trace the route of the signal from the brain to the heart.

4. Compare the causes and effects of angina pectoris to a myocardial infarction.

5. During some types of surgery, the body and blood of the patient are cooled. What effect will the lowered temperature have on heart rate, strength of contraction, and cardiac output? Why are these effects desirable during surgery?

6. Think about your lifestyle. Do you have any of the risk factors for heart disease? List two risk factors that can be influenced by changes in lifestyle and explain how.

Answers to Figure Questions

15.1 A mass of tissue in the middle of the thoracic cavity, between the lungs and between the sternum and backbone.

15.2 Visceral layer of the serous pericardium (epicardium).

15.3 Right atrium, tricuspid valve, right ventricle, pulmonary semilunar valve, pulmonary trunk, pulmonary arteries, lungs, pulmonary veins, left atrium, bicuspid valve, left ventricle, aortic semilunar valve, aorta, rest of body, and superior vena cava, inferior vena cava, and coronary sinus into right atrium.

15.4 To prevent the backflow of blood.

15.5 Superior vena cava, inferior vena cava, and coronary sinus.

15.6 Atrioventricular (AV) bundle.

15.7 Systole and diastole.

15.8 It decreases heart rate and strength of contraction.

15.9 Vasodilation.

15.10 Fatty substances, cholesterol, and smooth muscle fibers.

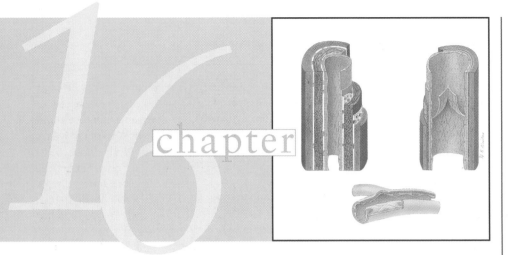

16 chapter

THE CARDIOVASCULAR SYSTEM: BLOOD VESSELS

lood vessels form a network of tubes that carry blood away from the heart to the tissues of the body and then return it to the heart. *Arteries* (AR-ter-ēs) are the vessels that carry blood *away from the heart* to the tissues. Large arteries leave the heart and divide into medium-sized arteries that branch out into the various regions of the body. These medium-sized arteries then divide into small arteries, which, in turn, divide into still smaller arteries called *arterioles* (ar-TER-ē-ōls). Arterioles within a tissue or organ branch into countless microscopic vessels called *capillaries* (KAP-i-lar′-ēs). Through the walls of the capillaries, substances are exchanged between the blood and body tissues. Before leaving the tissue, groups of capillaries reunite to form small veins called *venules* (VEN-yools). These, in turn, merge to form progressively larger tubes called veins. *Veins* (VĀNZ) are blood vessels that convey blood from the tissues *back to the heart.*

Arteries

Arteries have walls constructed of three layers of tissue that surround a hollow space, called a *lumen,* through which the blood flows (Figure 16.1a). The inner layer is composed of simple squamous epithelium called *endothelium* and elastic tissue. The middle layer consists of smooth muscle and elastic fibers. The outer layer is composed principally of elastic and collagen fibers.

As a result of the structure of the middle coat especially, arteries have two major properties: elasticity and contractility. With respect to elasticity, when the ventricles of the heart contract and eject blood into the large arteries, the arteries expand to accommodate the extra blood. Then, as the ventricles relax, the elastic recoil of the elastic fibers in the arteries forces the blood onward. The contractility of an artery comes from its smooth muscle, which is supplied by the sympathetic division of the autonomic nervous system. When sympathetic stimulation increases, the smooth muscle contractions increase, thereby narrowing or constricting the lumen, a process called *vasoconstriction. Vasodilation,* or an increase in lumen size, results from a decrease in sympathetic stimulation and consequent relaxation of smooth muscle. Some vasoconstriction is necessary to maintain circulation (more detail on this later).

Arterioles

An *arteriole* (*arteriola* = small artery) is a very small (almost microscopic) artery that delivers blood to capillaries. As arterioles get smaller in size, they eventually consist of little more than a layer of endothelium covered by a few smooth muscle fibers (see Figure 16.2).

Arterioles play a key role in regulating blood flow from arteries into capillaries. During vasoconstriction, blood flow to capillaries is restricted; during vasodilation, the flow is significantly increased.

Capillaries

Capillaries (*capillaris* = hairlike) are microscopic vessels that connect arterioles to venules (Figure 16.1c). The primary function of capillaries is to permit the exchange of nutrients and wastes between the blood and tissue cells. Capillaries are found near almost every cell in the body, but their activity varies with the activity of the tissue. Body tissues with high metabolic activity (nervous, muscular, liver, kidneys) require more oxygen and nutrients so they need more capillaries to supply blood. The epidermis, cornea and lens of the eye, and cartilage do not contain capillaries.

The structure of the capillaries is admirably suited to their purpose. Because capillary walls are composed of a single layer of endothelial cells and a basement membrane, substances in the blood pass easily through them to reach tissue cells and vice versa. Depending on how tightly the endothelial cells are joined, different types of capillaries permit varying degrees of permeability. The structure of capillaries is vital for the tissues' homeostasis because the walls of all other vessels are too thick to permit the exchange of substances between blood and tissue cells.

In some regions, capillaries pass directly from arterioles to venules; in other places, they form extensive branching networks between the two vessels. These networks increase the surface area for diffusion and thereby allow a rapid exchange of large quantities of materials. Blood normally flows through only a small portion of the capillary network when metabolic needs are low. But when a tissue becomes active, the entire capillary network fills with blood. The flow of blood in capillaries is regulated by smooth muscle fibers scattered along arterioles and *precapillary sphincters,* rings of smooth muscle at their origin. Smooth muscle fibers and precapillary sphincters contract (vasoconstriction) and relax (vasodilation) to regulate blood flow through them (Figure 16.2 on page 365).

Venules

When several capillaries unite, they form small veins called *venules* (*venula* = little vein). Venules collect blood from capillaries and drain it into veins. Venules are similar in structure to arterioles; their walls are thinner near the capillaries and thicker as they progress toward the heart.

Veins

Veins are structurally similar to arteries but their middle and inner layers are thinner. The outer layer of veins is the thickest layer. The inner layer may fold inward to form valves (see Figure 16.1b). Despite these differences, veins are still flexible enough to accommodate variations in the volume and pressure of blood passing through them. The lumen of a vein is wider than that of a corresponding artery.

By the time blood leaves the capillaries and moves into veins, it has lost a great deal of pressure. This can be observed in the blood leaving a cut vessel. Blood from a vein flows slow-

Figure 16.1   Comparative structure of blood vessels. The relative size of the capillary in (c) is enlarged for emphasis. Note the valve in the vein.

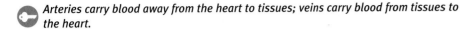

🔑 *Arteries carry blood away from the heart to tissues; veins carry blood from tissues to the heart.*

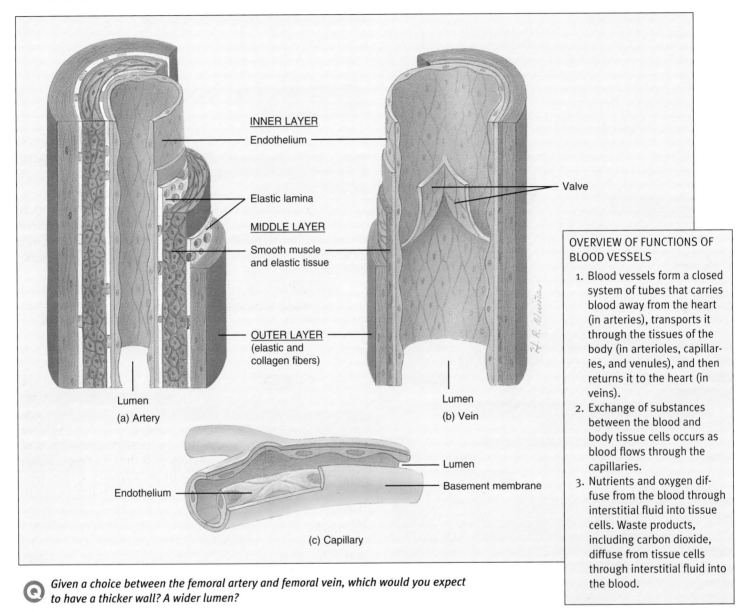

INNER LAYER
Endothelium

Elastic lamina

Valve

MIDDLE LAYER

Smooth muscle
and elastic tissue

OUTER LAYER
(elastic and
collagen fibers)

Lumen
(a) Artery

Lumen
(b) Vein

Lumen

Basement membrane

Endothelium

(c) Capillary

OVERVIEW OF FUNCTIONS OF BLOOD VESSELS

1. Blood vessels form a closed system of tubes that carries blood away from the heart (in arteries), transports it through the tissues of the body (in arterioles, capillaries, and venules), and then returns it to the heart (in veins).
2. Exchange of substances between the blood and body tissue cells occurs as blood flows through the capillaries.
3. Nutrients and oxygen diffuse from the blood through interstitial fluid into tissue cells. Waste products, including carbon dioxide, diffuse from tissue cells through interstitial fluid into the blood.

❓ *Given a choice between the femoral artery and femoral vein, which would you expect to have a thicker wall? A wider lumen?*

ly and evenly. Blood from an artery gushes in rapid spurts. The structural differences between arteries and veins reflect this pressure difference. For example, the walls of veins are not as strong as those of arteries. The low pressure in veins, however, has its disadvantages. When you stand, the pressure pushing blood up the veins in your lower limbs is barely enough to balance the force of gravity pushing it back down. For this reason, many veins, especially those in the limbs, have valves that prevent backflow (see Figure 16.6).

In people with weak venous valves, gravity forces blood back down into the vein. This pressure overloads the vein and pushes its wall outward. After repeated overloading, the walls lose their elasticity and become stretched and flabby. Such dilated and tortuous veins caused by leaky valves are called

varicose veins. They may be due to heredity, mechanical factors (prolonged standing and pregnancy), or aging. Veins close to the surface of the legs are highly susceptible to varicosities. Veins that lie deeper are not as vulnerable because surrounding skeletal muscles prevent their walls from overstretching.

See Exhibit 16.1 for a summary of arteries, capillaries, and veins.

Blood Reservoirs

The volume of blood in various parts of the cardiovascular system varies considerably. Veins and venules contain about 60 percent of the blood in the system, arteries and arterioles about

Figure 16.2   Capillaries.

🔑 *Arterioles regulate blood flow into capillaries where nutrients, gases, and wastes are exchanged between blood and tissue cells.*

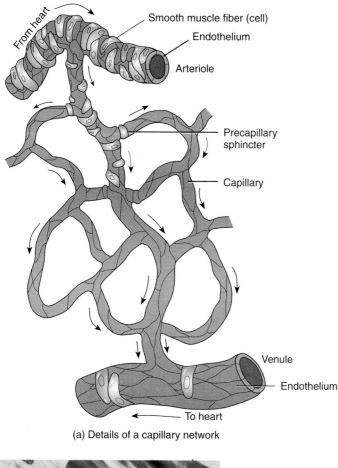

(a) Details of a capillary network

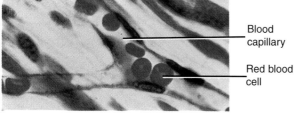

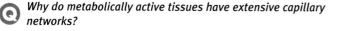

(b) Photomicrograph showing red blood cells squeezing through capillaries

❓ *Why do metabolically active tissues have extensive capillary networks?*

15 percent, pulmonary vessels about 12 percent, the heart about 8 percent, and capillaries about 5 percent. Because veins contain so much of the blood, they are referred to as *blood reservoirs.* Blood can be moved quickly to other parts of the body if the need arises, for example, to skeletal muscles when there is increased muscular activity. The principal blood reservoirs are the veins of the abdominal organs (especially the liver and spleen) and the skin.

Deoxiginated Blood

Physiology of Circulation

Factors that Determine Blood Flow

Blood flow refers to the amount of blood that passes through a blood vessel in a given period of time. Blood flows or circulates through two major sets of blood vessels, to the lungs where blood gets rid of carbon dioxide and picks up oxygen (pulmonary circulation) and to the rest of the body where blood delivers oxygen and removes carbon dioxide (systemic circulation). In the discussion that follows, we will concentrate on systemic circulation.

Blood flow is determined by two main factors: (1) blood pressure and (2) resistance (opposition), the force of friction as blood travels through blood vessels.

Blood Pressure

Blood pressure (BP) is the pressure exerted by blood on the wall of a blood vessel. In clinical use, the term refers to pressure in arteries. BP is influenced by cardiac output, blood volume, and resistance. Blood flows through its system of closed vessels because of different blood pressures in various parts of the cardiovascular system. Blood flow is directly proportional to blood pressure; that is, as pressure increases, flow increases. Blood always flows from regions of higher blood pressure to regions of lower blood pressure. The average pressure in the aorta is about 100 millimeters of mercury (mm Hg). As blood leaves the aorta and flows through systemic circulation, its pressure falls progressively to 0 mm Hg by the time it reaches the right atrium (Figure 16.3).

Resistance

Resistance refers to the opposition to blood flow that results from friction between blood and blood vessel walls. Resistance

Exhibit 16.1	Comparison of Arteries, Capillaries, and Veins	
	STRUCTURE	FUNCTION
Arteries	Composed of three layers around a lumen.	Carry blood away from heart.
Capillaries	Composed of a layer of endothelium and basement membrane around a lumen.	Permit exchange of nutrients and wastes between blood and tissue cells.
Veins	Composed of same three layers as arteries around a lumen but have a thicker outer layer and a thinner middle and inner layer. Also contain valves.	Carry blood back to heart.

Figure 16.3   Blood pressures in various parts of the cardiovascular system. The dashed line is average blood pressure.

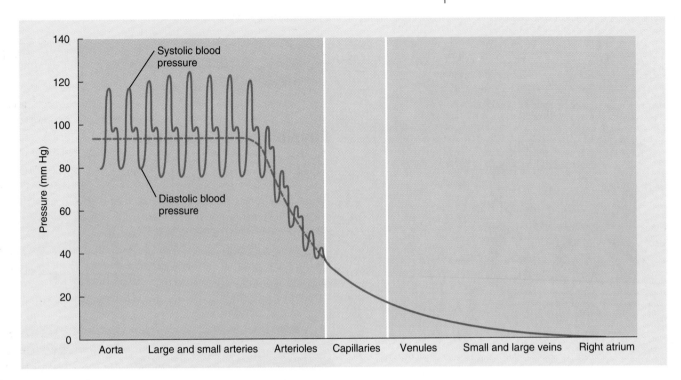

Q   *What is the relationship between blood pressure and blood flow?*

is related to (1) blood viscosity, (2) blood vessel length, and (3) blood vessel radius.

The viscosity (adhesiveness or stickiness) of blood is related to the ratio of red blood cells and solutes, especially plasma proteins, to plasma (fluid). Any condition that increases the viscosity of blood, such as dehydration or an unusually high number of red blood cells (polycythemia), increases blood pressure. A depletion of plasma proteins or red blood cells, as a result of anemia or hemorrhage, decreases blood viscosity and blood pressure. The longer a blood vessel, the greater the resistance as blood flows through it because of more contact between blood and the vessel wall. The smaller the radius of the blood vessel, the greater resistance it offers to blood flow.

Factors that Affect Arterial Blood Pressure

objective: *Describe the factors that affect blood pressure.*

Three factors influence arterial blood pressure: (1) cardiac output, (2) blood volume, and (3) peripheral resistance.

Cardiac Output (CO)

Cardiac output (CO), the amount of blood ejected by the left ventricle into the aorta each minute, is the principal determinant of blood pressure. As noted in Chapter 15, CO is calculated by multiplying stroke volume (the amount ejected by a ventricle during each contraction) by heart rate. In a normal, resting adult, it is about 5.25 liters/min (70 ml × 75 beats/min).

Also recall from Chapter 15 that CO is partially regulated by the cardiovascular center and that certain chemicals (epinephrine, potassium, sodium, and calcium), temperature, emotions, gender, and age affect heart rate. By their effect on CO, these factors also affect blood pressure. Blood pressure varies directly with CO. Any increase in CO increases blood pressure. Conversely, any decrease in CO lowers blood pressure.

Blood Volume

Blood pressure is directly related to the **volume of blood** in the cardiovascular system. The normal volume of blood in a human body is about 5 liters (5 qt). Any decrease in this volume, as from hemorrhage, decreases the amount of blood that is circulated through the arteries each minute. As a result, blood pressure drops. Conversely, anything that increases blood volume, such as high salt intake and therefore water retention, increases blood pressure.

Peripheral Resistance

Peripheral resistance refers to the resistance to blood flow in peripheral circulation, that is, away from the heart. Most resistance occurs in arterioles, capillaries, and venules. Resistance in arteries and veins is less because of their larger diameter. A major function of arterioles is to control peripheral resistance—and therefore blood pressure and flow—by changing their diameters. This regulation is governed by the vasomotor center in the medulla oblongata.

Homeostasis of Blood Pressure Regulation

objective: *Describe how blood pressure is regulated.*

In order to maintain homeostasis, blood pressure must be kept within a normal range. Whereas high blood pressure can do a great deal of damage to the heart, brain, and kidneys, low blood pressure can result in the delivery of inadequate amounts of oxygen and nutrients to body cells to meet their metabolic needs. In order to maintain a normal homeostatic range, there is a regulating center in the brain that receives input from receptors throughout the body, higher brain centers, and chemicals.

Vasomotor Center

In the cardiovascular center in the medulla oblongata is a cluster of sympathetic neurons referred to as the ***vasomotor*** (*vas* = vessel; *motor* = movement) ***center.*** This center controls the diameter of blood vessels, especially arterioles of the skin and abdominal viscera. It is the integrating center for blood pressure control. It continually sends impulses to the smooth muscle in arteriole walls that result in a moderate state of vasoconstriction at all times, which helps maintain peripheral resistance and blood pressure. By increasing the number of sympathetic impulses, the vasomotor center brings about increased vasoconstriction and raises blood pressure. By decreasing the number of sympathetic impulses, it causes vasodilation and lowers blood pressure. In other words, the sympathetic division of the autonomic nervous system can bring about either vasoconstriction or vasodilation by varying the frequency of impulses.

The vasomotor center is modified by any number of inputs from baroreceptors, chemoreceptors, higher brain centers, and various hormones, all of which influence blood pressure.

Baroreceptors

Baroreceptors are neurons sensitive to blood pressure that are located in the aorta, internal carotid arteries (arteries in the neck that supply blood to the brain), and other large arteries in the neck and chest. They send impulses to the cardiovascular center to increase or decrease cardiac output and thus help regulate blood pressure (see Figure 15.8). This mechanism acts not only on the heart but also on the arterioles. Recall that sympathetic stimulation via cardiac accelerator nerves increases heart rate and cardiac output, whereas parasympathetic stimulation via the vagus (X) nerves decreases heart rate and cardiac output. If there is an increase in blood pressure, the baroreceptors send more nerve impulses to the cardiovascular center and parasympathetic stimulation causes a decrease in cardiac output, with a consequent decrease in blood pressure. Or, if there is a decrease in blood pressure, the baroreceptors send fewer nerve impulses to the cardiovascular center and sympathetic stimulation increases cardiac output, thus increasing blood pressure. The baroreceptors also send impulses to the vasomotor center in the cardiovascular center. In response, the vasomotor center either decreases sympathetic stimulation to arterioles and veins, resulting in vasodilation and a decrease in blood pressure, or increases sympathetic stimulation, resulting in vasoconstriction and an increase in blood pressure (Figure 16.4). Note that although autonomic control of the heart is the result of opposing sympathetic and parasympathetic stimulation, autonomic control of blood vessels is exclusively the result of sympathetic stimulation.

Chemoreceptors

Neurons that are sensitive to chemicals in the blood are called ***chemoreceptors.*** They are located in the two ***carotid bodies,*** which are found in the common carotid arteries, and in the ***aortic bodies,*** which are located in the arch of the aorta. Chemoreceptors are sensitive to lower than normal levels of oxygen (O_2) and even more so to higher than normal levels of carbon dioxide (CO_2) and hydrogen ions (H^+) and send impulses to the vasomotor center in the cardiovascular center when these conditions exist. In response, the vasomotor center increases sympathetic stimulation to arterioles to bring about vasoconstriction and an increase in blood pressure.

Regulation by Higher Brain Centers

Higher brain centers, such as the cerebral cortex, influence blood pressure in response to strong emotions. During periods of intense anger or sexual excitement, for example, the cerebral cortex relays impulses to the hypothalamus, and then on to the vasomotor center. From here, impulses to arterioles cause increased vasoconstriction and an increase in blood pressure. Also, sympathetic impulses to the adrenal medulla cause the release of epinephrine and norepinephrine, which prolong many sympathetic responses, including vasoconstriction and higher blood pressure. When a person is grieving, impulses from higher brain centers decrease vasomotor center stimulation, producing vasodilation and a decrease in blood pressure. A frequent result is fainting because blood flow to the brain is diminished.

Hormones

Several ***hormones*** affect blood pressure. Epinephrine and norepinephrine (NE), produced by the adrenal medulla, increase the rate and force of heart contractions and bring about vasoconstriction of arterioles in the skin and abdomen (see Figure 13.19). Alcohol inhibits release of ADH, depresses the vasomotor center, and brings about vasodilation, which lowers blood pressure. When arterial blood pressure decreases, certain kidney cells secrete renin, which raises blood pressure by causing vasoconstriction and increasing sodium ion and water reabsorption (see Figure 13.15). Histamine, produced by mast cells, is a vasodilator in the inflammatory response (see Chapter 17). ***Atrial natriuretic peptide*** (ANP), released by cells in the atria of the heart, lowers blood pressure by causing vasodilation and by promoting loss of salt and water in the urine, which reduces blood volume.

Autoregulation

Autoregulation refers to a local, automatic adjustment of blood flow in a given region of the body in response to the particular needs of that tissue. It is independent of vasomotor control. Some stimuli that cause autoregulation are physical. For example, warming promotes vasodilation whereas cooling causes vasoconstriction. Other stimuli are chemical. In response to low oxygen supplies, the cells in the immediate area produce ***vasoactive factors,*** chemicals that alter blood vessel diameter. Some, such as potassium ions (K^+), hydrogen ions (H^+), carbon

Figure 16.4   The cardiovascular center receives input from baroreceptors, chemoreceptors, and higher brain centers. It provides output to both the sympathetic and parasympathetic divisions of the autonomic nervous system.

The cardiovascular (CV) center in the medulla oblongata is the main region for nervous system regulation of the heart and blood vessels.

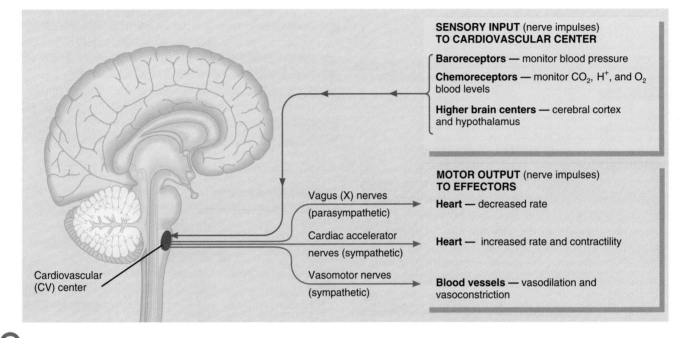

SENSORY INPUT (nerve impulses)
TO CARDIOVASCULAR CENTER

Baroreceptors — monitor blood pressure

Chemoreceptors — monitor CO_2, H^+, and O_2 blood levels

Higher brain centers — cerebral cortex and hypothalamus

MOTOR OUTPUT (nerve impulses)
TO EFFECTORS

Vagus (X) nerves (parasympathetic)
Heart — decreased rate

Cardiac accelerator nerves (sympathetic)
Heart — increased rate and contractility

Vasomotor nerves (sympathetic)
Blood vessels — vasodilation and vasoconstriction

Cardiovascular (CV) center

What types of effector tissues are regulated by the cardiovascular center?

dioxide (CO_2), and lactic acid, produce vasodilation. These substances cause local arterioles to dilate and precapillary sphincters to relax. The result is an increased flow of blood to the tissue, which restores oxygen levels to normal. Other vasoactive factors, such as eicosanoids (prostaglandins), produce vasoconstriction. These substances cause local arterioles to constrict and precapillary sphincters to contract, thus decreasing blood flow to a tissue. Autoregulation is important for meeting the nutritional demands of active tissues, such as muscle tissue.

Capillary Exchange

objective: *Discuss how materials are exchanged between blood and body cells.*

The velocity of blood flow in capillaries is the slowest in the cardiovascular system. This is important because it allows time for the exchange of materials between blood and body tissues.

The movement of water and dissolved substances, except proteins, through capillary walls depends on opposing forces or pressures. One pressure involved is called *hydrostatic pressure.* It is the pressure of blood within capillaries. The other pressure is called *osmotic pressure.* It is the pressure of a fluid due to its solute concentration. The higher the solute concentration, the greater its osmotic pressure. At the arterial end of a blood capillary, hydrostatic pressure is greater than osmotic pressure and this tends to move fluid out of blood capillaries into the surrounding interstitial (tissue) fluid, a process called *filtration* (Figure 16.5).

At the venous end of a blood capillary, osmotic pressure is greater than hydrostatic pressure and this tends to move fluid from interstitial fluid into the venous end of blood capillaries. (Blood has a higher osmotic pressure than tissue fluid because of its large number of proteins.) Fluid movement from interstitial fluid into blood capillaries is called *reabsorption.* This prevents fluid from moving in one direction only and accumulating in interstitial spaces.

About 85 percent of the fluid filtered at the arterial end of the capillary is reabsorbed at the venous end. The rest of the filtered fluid and any proteins that escape from blood into interstitial fluid are returned by the lymphatic system to the cardiovascular system.

Factors that Aid Venous Return

objective: *Describe how blood returns to the heart.*

A number of factors help venous return, the flow of blood back to the heart through veins: (1) pumping action of the heart, (2) velocity of blood flow, (3) skeletal muscle contractions, (4) valves in veins, and (5) breathing.

Pumping Action of the Heart

The pressure difference from venules (averaging about 16 mm Hg) to the right atrium (0 mm Hg), although small, is sufficient to aid venous return to the heart. In addition, during ventricular contraction the size of the atria increases. This action sucks

Figure 16.5 Capillary exchange.

Hydrostatic pressure moves fluid out of capillaries (filtration) whereas osmotic pressure moves fluid into capillaries (reabsorption).

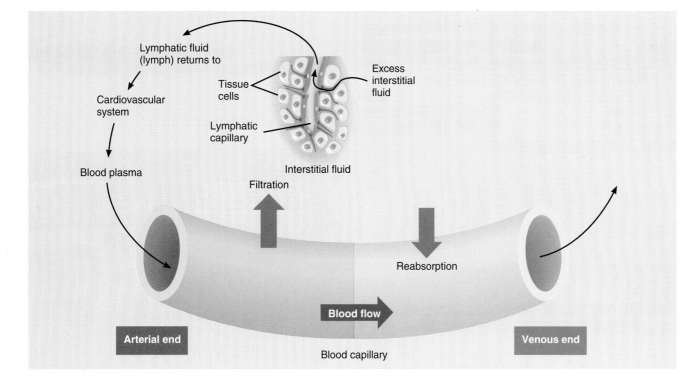

Lymphatic fluid (lymph) returns to

Cardiovascular system

Blood plasma

Tissue cells

Lymphatic capillary

Excess interstitial fluid

Interstitial fluid

Filtration

Reabsorption

Blood flow

Arterial end

Venous end

Blood capillary

What happens to fluid and proteins that are not reabsorbed at the arterial end of a capillary?

blood into the atria from the large veins and contributes to venous return, especially when heartbeat is fast.

Velocity of Blood Flow

Velocity of blood flow depends on the cross-sectional area of the blood vessel. Each time an artery branches, the total cross-sectional area of all the branches is greater than that of the original vessel. Blood flows slowest where the cross-sectional area is greatest. Thus the velocity of blood decreases as it flows from the aorta to arteries to arterioles to capillaries. Velocity in the capillaries is the slowest in the cardiovascular system. Decreased velocity allows adequate exchange (diffusion) time between capillaries and tissues. As blood vessels leave capillaries and approach the heart, their cross-sectional area decreases. Therefore the velocity of blood *increases* as it flows from capillaries to venules to veins to the heart.

Skeletal Muscle Contractions and Venous Valves

Skeletal muscle contractions and valves in veins work in combination to aid venous blood return. Many veins, especially in the limbs, contain valves. When skeletal muscles contract, they tighten around the veins passing through them and the valves open. This pressure drives the blood toward the heart—the action is called *milking* (Figure 16.6). When the muscles relax, the valves close to prevent backflow. When people are immobilized through injury or disease, they

Figure 16.6 Role of skeletal muscle contractions and venous valves in returning blood to the heart.

Venous return depends on the pumping action of the heart, velocity of blood flow, skeletal muscle contractions, valves in veins, and breathing.

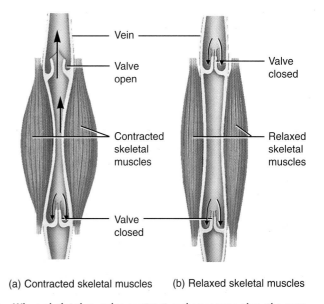

Vein

Valve open

Contracted skeletal muscles

Valve closed

Valve closed

Relaxed skeletal muscles

(a) Contracted skeletal muscles

(b) Relaxed skeletal muscles

When skeletal muscles contract and squeeze veins, the pressure drives blood toward the heart. What is the action called?

don't have the advantage of these contractions, which means venous return is slower and the heart has to work harder. Periodic leg massage helps.

Breathing

Breathing is important in venous circulation because during inspiration (breathing in), the diaphragm moves downward. This causes a decrease in pressure in the thoracic (chest) cavity and an increase in pressure in the abdominal cavity. As a result, blood moves from the abdominal veins (higher pressure) into the thoracic veins (lower pressure). When the pressures reverse during expiration (breathing out), blood in the veins is prevented from backflowing by the valves.

Checking Circulation

objective: *Define pulse and blood pressure and describe how they are measured.*

Pulse

The alternate expansion and elastic recoil of an artery with each contraction of the left ventricle is called *pulse*. Pulse is strongest in the arteries closest to the heart. It becomes weaker as it passes through the arterial system, and it disappears altogether in the capillaries. The radial artery at the wrist is most commonly used to feel the pulse. Others include the brachial artery along the medial side of the biceps brachii muscle; the common carotid artery, next to the voice box, which is frequently monitored in cardiopulmonary resuscitation; the popliteal artery behind the knee; and the dorsalis pedis artery above the instep of the foot.

The pulse rate is the same as the heart rate, about 75 beats per minute in the resting state. The term *tachycardia* (tak′-i-KAR-dē-a; *tachy* = fast) refers to a rapid heart or pulse rate (over 100/min). *Bradycardia* (brād′-i-KAR-dē-a; *brady* = slow) indicates a slow heart or pulse rate (under 60/min).

Measurement of Blood Pressure (BP)

In clinical use, the term *blood pressure* (BP) refers to the pressure in arteries exerted by the left ventricle when it undergoes systole (contraction) and the pressure remaining in the arteries when the ventricle is in diastole (relaxation). Blood pressure is usually taken in the left brachial artery on the arm (see Figure 16.8a), and it is measured by a *sphygmomanometer* (sfig′-mō-ma-NOM-e-ter; *sphygmo* = pulse). After the pressure cuff is inflated, the artery is compressed so that blood flow stops. As the cuff is deflated, the artery opens, and a spurt of blood passes through. This results in the first sound heard through the stethoscope. This sound corresponds to *systolic blood pressure* (SBP)—the force with which blood is pushing against arterial walls during ventricular contraction. The pressure recorded when the sounds suddenly become faint is called *diastolic blood pressure* (DBP). It measures the force of blood remaining in arteries during ventricular relaxation. Whereas systolic pressure indicates the force of

the left ventricular contraction, diastolic pressure provides information about the resistance of blood vessels.

The normal blood pressure of a young adult male is 120 mm Hg systolic and 80 mm Hg diastolic, reported as "120 over 80" and written as 120/80. In young adult females, the pressures are 8 to 10 mm Hg less. People who exercise regularly and are in good physical condition tend to have lower blood pressures. Thus blood pressure slightly lower than 120/80 may be a sign of good health and fitness.

Shock and Homeostasis

Shock occurs when the cardiovascular system cannot deliver sufficient oxygen and nutrients to meet the needs of body cells. The underlying cause is inadequate cardiac output. As a result, cellular metabolism is abnormal, and cellular death may occur. Shock may be due to hemorrhage, dehydration, burns, and excessive vomiting, diarrhea, or sweating.

Symptoms of shock vary with the severity of the condition, including the following:

1. Low blood pressure in which the systolic blood pressure is lower than 90 mm Hg as a result of generalized vasodilation and decreased cardiac output.

2. Clammy, cool, pale skin due to vasoconstriction of skin blood vessels.

3. Sweating due to sympathetic stimulation.

4. Reduced urine formation due to low blood pressure and increased levels of aldosterone and antidiuretic hormone (ADH).

5. Altered mental state due to reduced oxygen supply to the brain.

6. Tachycardia (rapid heart rate and pulse) due to sympathetic stimulation and increased levels of epinephrine.

7. Weak, rapid pulse due to generalized vasodilation and reduced cardiac output.

8. Thirst due to loss of extracellular fluid.

9. Nausea due to impaired circulation to digestive organs.

Circulatory Routes

objective: *Compare the major routes that blood takes through various regions of the body.*

Blood vessels are organized into routes that circulate blood throughout the body.

Figure 16.7 shows the basic *circulatory routes*. As noted earlier, the two basic circulatory routes are systemic and pulmonary. *Systemic circulation* includes all the arteries and arterioles that carry oxygenated blood from the left ventricle to capillaries plus the veins and venules that carry deoxygenated blood to the right atrium. Blood leaving the aorta and traveling

Figure 16.7 Circulatory routes. The overall pattern of circulation. Heavy black arrows indicate systemic circulation, thin black arrows pulmonary circulation, and thin red arrows hepatic portal circulation. The details of pulmonary circulation are shown in Figure 16.9 and the details of hepatic portal circulation are shown in Figure 16.11.

Blood vessels are organized into routes that deliver blood to various tissues of the body.

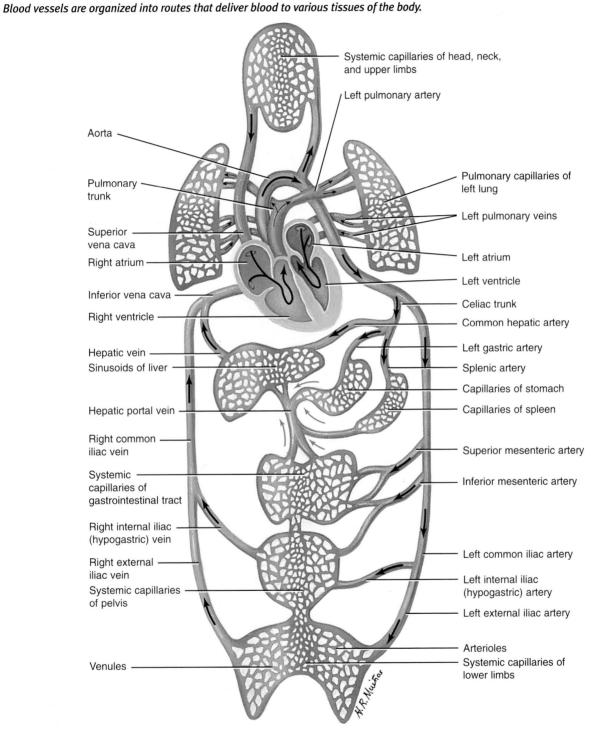

What are the two principal circulatory routes?

through the systemic arteries is a bright red color. As it moves through the capillaries, it loses its oxygen and takes on carbon dioxide, so that the blood in the systemic veins is a dark red color.

When blood returns to the heart from the systemic route, it is pumped out of the right ventricle through the *pulmonary circulation* to the lungs. In the lungs, it loses its carbon dioxide and takes on oxygen. It is now bright red again. It returns to the left atrium of the heart and reenters systemic circulation.

Another major route—*fetal circulation*—exists only in the fetus and contains special structures that allow the developing fetus to exchange materials with its mother (see Figure 24.8).

Systemic Circulation

The flow of oxygenated blood from the left ventricle to all parts of the body and back to the right atrium is called *systemic circulation.* Its function is to carry oxygen and nutrients to body tissues and to remove carbon dioxide and other wastes. All systemic arteries branch from the *aorta,* which arises from the left ventricle of the heart.

As the aorta emerges from the left ventricle, it passes upward and behind the pulmonary trunk (see Figure 16.9). At this point, it is called the *ascending aorta.* The ascending aorta gives off two coronary arteries to the heart muscle. Then the ascending aorta turns to the left, forming the *arch of the aorta,* before descending as the *descending aorta.* The descending aorta lies close to the vertebral bodies, passes through the diaphragm, and divides at the level of the fourth lumbar vertebra into two *common iliac arteries,* which carry blood to the lower limbs. The section of the descending aorta between the arch of the aorta and the diaphragm is referred to as the *thoracic aorta.* The section between the diaphragm and the common iliac arteries is termed the *abdominal aorta.* Each section of the aorta gives off arteries that continue to branch into arteries leading to organs and finally into the arterioles and capillaries that service all the tissues of the body, except the air sacs of the lungs.

Deoxygenated blood is returned to the heart through the systemic veins. All the veins of the systemic circulation flow into either the *superior* or *inferior vena cava* or the *coronary sinus.* They, in turn, empty into the right atrium. The principal blood vessels of systemic circulation are shown in Figure 16.8.

Pulmonary Circulation

The flow of deoxygenated blood from the right ventricle to the air sacs of the lungs and the return of oxygenated blood from the air sacs to the left atrium is called *pulmonary* (*pulmo* = lung) *circulation* (Figure 16.9 on p. 375). The *pulmonary trunk* emerges from the right ventricle and then divides into two branches. The *right pulmonary artery* runs to the right lung; the *left pulmonary artery* goes to the left lung. The pulmonary arteries are the only postnatal (after birth) arteries that carry deoxygenated blood. On entering the lungs, the branches divide and subdivide until ultimately they form capillaries around the air sacs in the lungs. Carbon dioxide passes from the blood into the air sacs to

be breathed out of the lungs. Oxygen breathed in passes from the air sacs into the blood. The capillaries unite, venules and veins are formed, and, eventually, two *pulmonary veins* from each lung transport the oxygenated blood to the left atrium. The pulmonary veins are the only postnatal veins that carry oxygenated blood. Contractions of the left ventricle then send the blood into systemic circulation.

Cerebral Circulation

Inside the cranium is an arrangement of blood vessels at the base of the brain called the *cerebral arterial circle* (*circle of Willis*) (Figure 16.10 on p. 375). From this circle arises arteries supplying most of the brain. This flow of blood is called *cerebral circulation* and is a subdivision of systemic circulation. The circle is formed by the union of the *anterior cerebral arteries* (branches of the internal carotid arteries) and the *posterior cerebral arteries* (branches of the basilar arteries). Communicating arteries connect the anterior and posterior cerebral arteries. The circle equalizes blood pressure to the brain and provides alternate routes, should any arteries become damaged.

Hepatic Portal Circulation

A vein that begins and ends in capillaries is called a portal vein. Such a vein associated with the liver is called the hepatic portal vein. It begins in the capillaries of digestive organs and ends in capillary-like structures in the liver called sinusoids. In *hepatic* (*hepato* = liver) *portal circulation,* another subdivision of systemic circulation, venous blood from the gastrointestinal organs and spleen is delivered to the hepatic portal vein and enters the liver before returning to the heart. This blood is rich with substances absorbed from the gastrointestinal tract. The liver monitors these substances before they pass into the general circulation. It may remove surplus glucose and store it. It modifies certain digested substances so they may be used by cells. It also detoxifies harmful substances absorbed by the gastrointestinal tract and destroys bacteria by phagocytosis.

The hepatic portal vein is formed by the union of the splenic and superior mesenteric veins (Figure 16.11 on page 376). Deoxygenated blood from the hepatic portal vein enters sinusoids. At the same time the liver receives and acts on deoxygenated blood, it receives oxygenated blood from the systemic circulation via the hepatic artery to nourish its tissues. The oxygenated blood mixes with the deoxygenated blood in sinusoids. Ultimately, all blood leaves the sinusoids of the liver through the hepatic veins, which enter the inferior vena cava.

Fetal Circulation

The circulatory system of a fetus, called *fetal circulation,* differs from an adult's because the lungs, kidneys, and digestive organs of a fetus are nonfunctional. The fetus derives its oxygen and nutrients from the maternal blood and eliminates its carbon dioxide and wastes into the maternal blood. The details of fetal circulation are discussed in Chapter 24.

Figure 16.8   Major blood vessels of systemic circulation.

All systemic arteries branch from the aorta; systemic veins drain their blood into the inferior and superior venae cavae and coronary sinus.

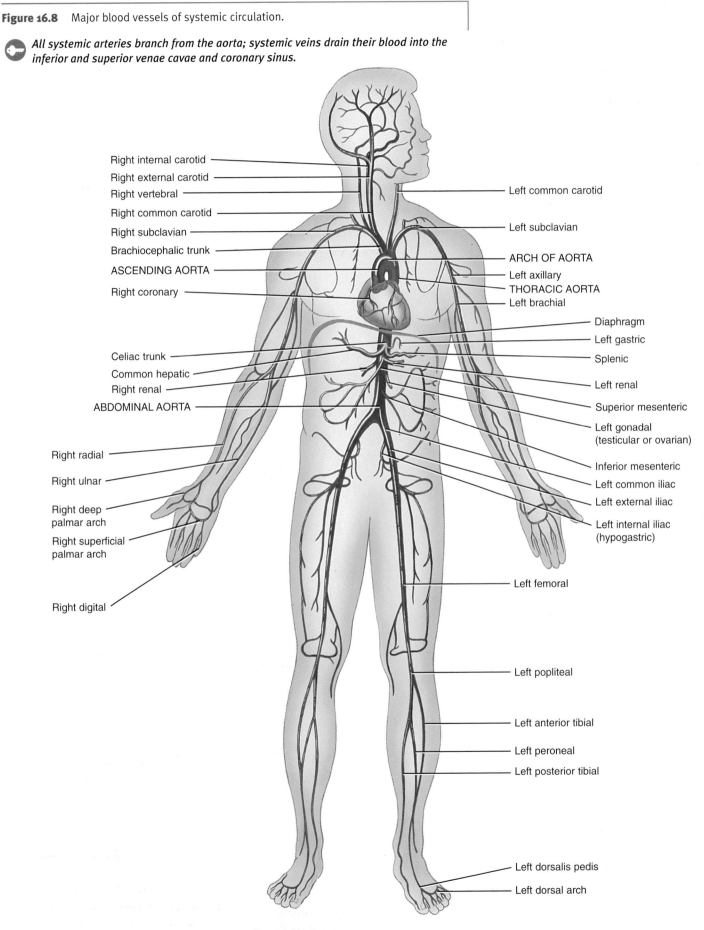

Right internal carotid

Right external carotid

Right vertebral

Right common carotid

Right subclavian

Brachiocephalic trunk

ASCENDING AORTA

Right coronary

Celiac trunk

Common hepatic

Right renal

ABDOMINAL AORTA

Right radial

Right ulnar

Right deep palmar arch

Right superficial palmar arch

Right digital

Left common carotid

Left subclavian

ARCH OF AORTA

Left axillary

THORACIC AORTA

Left brachial

Diaphragm

Left gastric

Splenic

Left renal

Superior mesenteric

Left gonadal (testicular or ovarian)

Inferior mesenteric

Left common iliac

Left external iliac

Left internal iliac (hypogastric)

Left femoral

Left popliteal

Left anterior tibial

Left peroneal

Left posterior tibial

Left dorsalis pedis

Left dorsal arch

(a) Overall anterior view

Figure 16.8   *(Continued)*

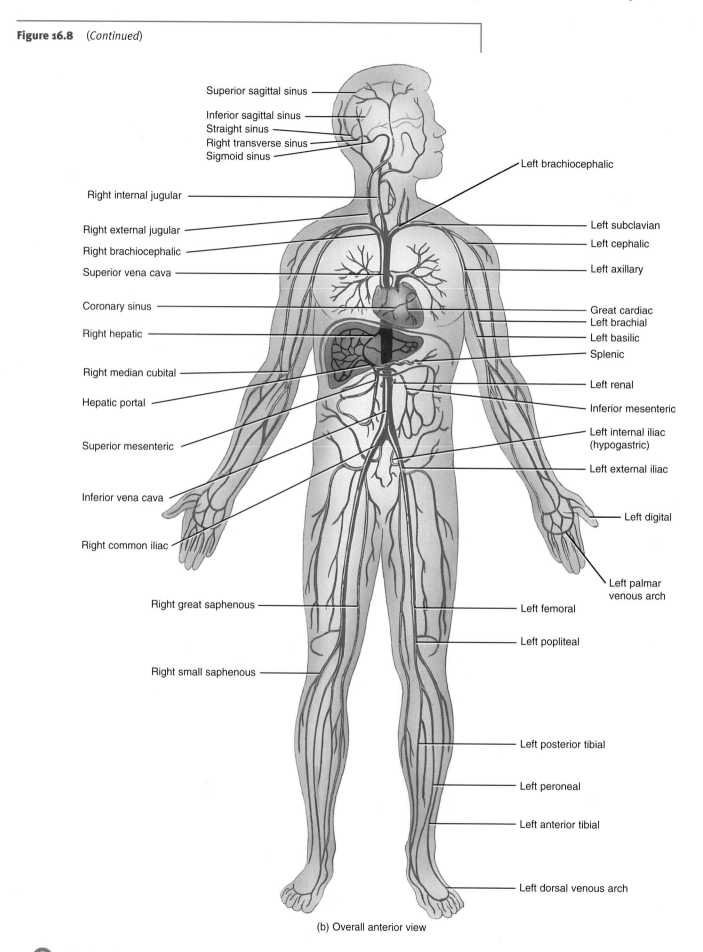

Superior sagittal sinus

Inferior sagittal sinus

Straight sinus

Right transverse sinus

Sigmoid sinus

Left brachiocephalic

Right internal jugular

Left subclavian

Right external jugular

Left cephalic

Right brachiocephalic

Left axillary

Superior vena cava

Coronary sinus

Great cardiac

Left brachial

Right hepatic

Left basilic

Splenic

Right median cubital

Left renal

Hepatic portal

Inferior mesenteric

Superior mesenteric

Left internal iliac (hypogastric)

Left external iliac

Inferior vena cava

Left digital

Right common iliac

Left palmar venous arch

Right great saphenous

Left femoral

Left popliteal

Right small saphenous

Left posterior tibial

Left peroneal

Left anterior tibial

Left dorsal venous arch

(b) Overall anterior view

Q   *All veins of systemic circulation deliver their blood into which heart chamber?*

Figure 16.9 Pulmonary circulation.

Pulmonary circulation brings deoxygenated blood from the right ventricle to the lungs and returns oxygenated blood from the lungs to the left atrium.

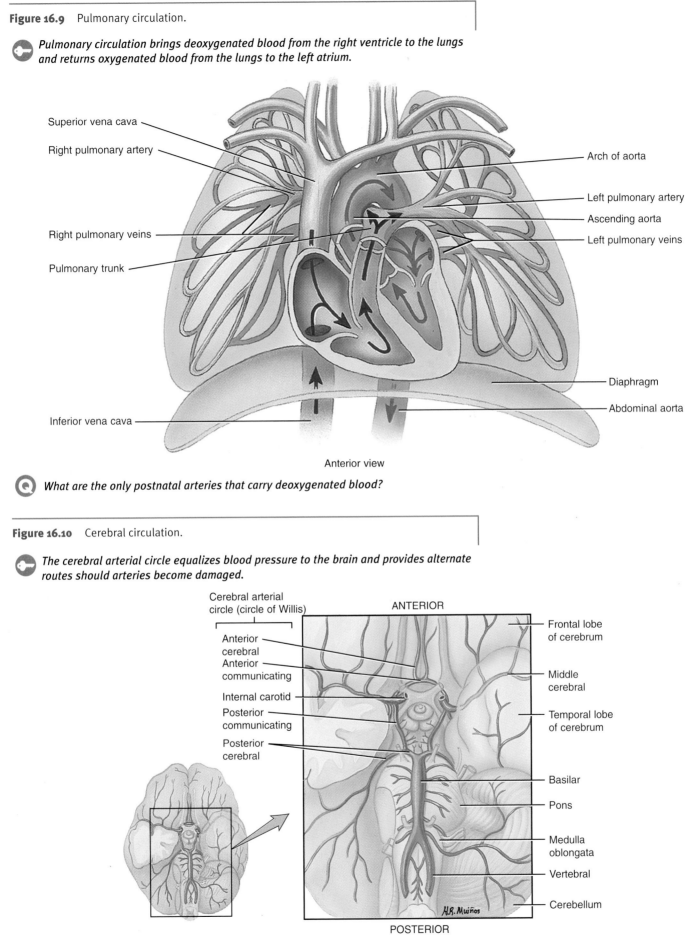

Superior vena cava

Right pulmonary artery

Right pulmonary veins

Pulmonary trunk

Inferior vena cava

Arch of aorta

Left pulmonary artery

Ascending aorta

Left pulmonary veins

Diaphragm

Abdominal aorta

Anterior view

Q *What are the only postnatal arteries that carry deoxygenated blood?*

Figure 16.10 Cerebral circulation.

The cerebral arterial circle equalizes blood pressure to the brain and provides alternate routes should arteries become damaged.

Cerebral arterial circle (circle of Willis)

ANTERIOR

Anterior cerebral

Anterior communicating

Internal carotid

Posterior communicating

Posterior cerebral

Frontal lobe of cerebrum

Middle cerebral

Temporal lobe of cerebrum

Basilar

Pons

Medulla oblongata

Vertebral

Cerebellum

H.R. Muños

POSTERIOR

Inferior view of base of brain

Q *What are the two main arteries that form the cerebral arterial circle?*

Figure 16.11   Hepatic portal circulation.

The hepatic portal circulation delivers venous blood from the gastrointestinal organs and spleen to the liver.

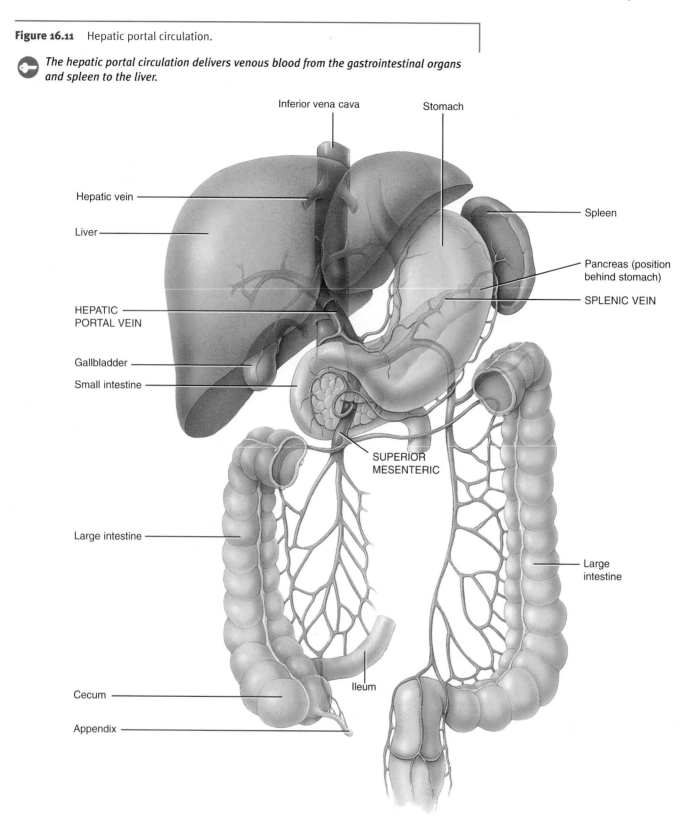

Anterior view of veins draining into the hepatic portal vein

Q   *Which veins carry blood away from the liver?*

*n*ot so long ago scientists believed that once plaque formed in an artery, it never went away. Medical researchers thought that appropriate lifestyle and pharmacological treatments could slow the process of atherosclerosis, but could not undo damage already done. In recent years, however, researchers have discovered that, given the right conditions, the body's own healing processes can sometimes reverse arterial plaque buildup. This news has provided hope for the millions of Americans who have been diagnosed with arterial disease.

Exercise Control

The first step in preventing, slowing, and hopefully reversing arterial disease is to control the risk factors associated with its progression. Of course, some risk factors are out of our control: family history, race, gender, and age. But many factors can be modified by lifestyle change, and sometimes by lifestyle change combined with medication, if lifestyle change alone does not produce adequate control.

Who is not familiar by now with public health recommendations for a heart-healthy lifestyle? No smoking; regular exercise (at least 30 minutes of moderate-intensity exercise per day); stress management; and a heart-healthy diet. Heart-healthy diet recommendations include limiting fat intake to 30 percent of daily caloric intake, and increasing consumption of plant foods, such as grains, fruits, and vegetables. These recommendations help prevent arterial disease by reducing obesity, blood lipids, platelet stickiness, and blood pressure, and by improving blood sugar control in people at risk for type II diabetes.

Are Heart-healthy Recommendations Enough?

People formulating public health guidelines such as "consume less than 30 percent of your daily caloric intake as fat" have a tough job. Their goal is to weigh all available scientific evidence and propose recommendations that are simple and practical enough for everyone to follow. Some researchers believe that while current public health recommendations are fine for many North Americans, they are not stringent enough for people at high risk for developing arterial disease and those already diagnosed with atherosclerosis. These researchers believe that some people are willing and able to make greater lifestyle modifications that offer a significant health advantage over the general public health recommendations, and should not be limited by these general recommendations.

Several studies have compared participants who either (1) received conventional medical care, which included advice to quit smoking, manage stress, reduce fat intake to less than 30 percent of daily caloric intake, and exercise regularly; or (2) followed more stringent lifestyle modifications. In these studies, the latter groups experienced significantly greater degrees of plaque regression, or at least slower progression.

What's "stringent"? Most stringent are the lifestyles followed by patients who participated in Dean Ornish's Lifestyle Heart Trial. They consumed a 6.8 percent fat diet, exercised over four hours a day, quit smoking, and participated in over an hour of stress management and relaxation training per day. And while their lifestyles were the most stringent, their results were also the most striking. Eighty-two percent of the patients in this group showed a regression of plaque in their coronary arteries, compared to 10 percent of the patients in the conventional care group. Conclusion: lifestyle can make a difference.

Other studies have used less demanding lifestyle changes but have still found significant, albeit less striking, differences between treatment and control groups. Studies which have included drug therapy along with lifestyle change have also observed plaque regression in many of their subjects. The best predictors of plaque regression in all studies were a decrease in LDL cholesterol and an increased aerobic capacity.

critical thinking

Why do you think public health officials recommend a diet that supplies up to 30 percent of its calories from fat, when studies have found that such a diet does not lead to plaque regression? Do you think this recommendation should be lower? Discuss your reasons why or why not. (Remember, this guideline is supposed to apply to all North Americans, not just those at risk for arterial disease.)

Common Disorders

Hypertension

Hypertension, or persistently high blood pressure, is defined as systolic blood pressure of 140 mm Hg or greater and diastolic blood pressure of 90 mm Hg or greater. Recall that a blood pressure of 120/80 is normal in a healthy adult male. In industrialized societies, hypertension is the most common disorder affecting the heart and blood vessels; it is a major cause of heart failure, kidney disease, and stroke. Blood pressure values for adults may be classified as follows:

Normal	Systolic less than 130 mm Hg; diastolic less than 85 mm Hg
High-normal	Systolic 130–139; diastolic 85–89
Hypertension	Systolic 140 or greater; diastolic 90 or greater

High blood pressure can be controlled with diet, exercise, and/or medication.

Aneurysm

An *aneurysm* (AN-yoo-rizm) is a thin, weakened section of the wall of an artery or a vein that bulges outward, forming a balloon-like sac of the blood vessel. Common causes include atherosclerosis, syphilis, congenital blood vessel defects, and trauma.

Deep-Venous Thrombosis (DVT)

Venous thrombosis, the presence of a thrombus in a vein, typically occurs in deep veins of the lower limbs and is referred to as *deep-venous thrombosis* (*DVT*). A serious complication is *pulmonary embolism,* in which the thrombus dislodges and finds its way into the pulmonary arterial blood flow. ■

Medical Terminology and Conditions

Aortography (ā′-or-TOG-ra-fē) X-ray examination of the aorta and its main branches after injection of radiopaque dye.

Arteritis (ar′-te-RĪ-tis; *itis* = inflammation of) Inflammation of an artery, probably due to an autoimmune response.

Cardiac arrest (KAR-dē-ak a-REST) Complete stoppage of the heartbeat.

Carotid endarterectomy (ka-ROT-id end′-ar-ter-EK-tō-mē) The removal of atherosclerotic plaque from the carotid artery to restore greater blood flow to the brain.

Claudication (klaw′-di-KĀ-shun) Pain and lameness caused by defective circulation of the blood in vessels of the limbs.

Hypercholesterolemia (hī-per-kō-les′-ter-ōl-Ē-mē-a; *hyper* = over; *heme* = blood) An excess of cholesterol in the blood.

Hypotension (hī′-pō-TEN-shun; *hypo* = below; *tension* = pressure) Low blood pressure; most commonly used to describe an acute drop in blood pressure, which occurs in shock.

Occlusion (o-KLOO-shun) The closure or obstruction of the lumen of a structure such as a blood vessel.

Orthostatic (or′-thō-STAT-ik) *hypotension* (*ortho* = straight; *statikos* = causing to stand) An excessive lowering of systemic blood pressure on assuming an erect or semierect posture; it is usually a sign of disease. May be caused by excessive fluid loss, certain drugs (antihypertensives), and cardiovascular or neurogenic factors. Also called *postural hypotension.*

Phlebitis (fle-BĪ-tis; *phleb* = vein) Inflammation of a vein, often in a leg. The condition is often accompanied by pain and redness of the skin over the inflamed vein. It is frequently caused by trauma or bacterial infection.

Raynaud's (rā-NOZ) *disease* A vascular disorder, primarily of females, characterized by bilateral attacks of ischemia (lack of sufficient blood), usually of the fingers and toes, in which the skin becomes pale and exhibits burning and pain; it is brought on by cold or emotional stimuli.

Syncope (SIN-kō-pē) A temporary cessation of consciousness; a faint. One cause might be insufficient blood supply to the brain.

Thrombectomy (throm-BEK-tō-mē; *thrombo* = clot) An operation to remove a blood clot from a blood vessel.

Thrombophlebitis (throm′-bō-fle-BĪ-tis) Inflammation of a vein with clot formation. Superficial thrombophlebitis occurs in veins under the skin, especially the calf.

White coat (*office*) *hypertension* A syndrome found in patients who have elevated blood pressures while being examined by health care personnel but otherwise have normal blood pressure.

Study Outline

Arteries (p. 363)

1. Arteries carry blood away from the heart. Their walls consist of three coats.
2. The structure of the middle coat gives arteries their two major properties, elasticity and contractility.

Arterioles (p. 363)

1. Arterioles are small arteries that deliver blood to capillaries.
2. Through constriction and dilation, they assume a key role in regulating blood flow from arteries into capillaries and thus arterial blood pressure.

Capillaries (p. 363)

1. Capillaries are microscopic blood vessels through which materials are exchanged between blood and tissue cells.
2. Capillaries branch to form an extensive network throughout a tissue that increases the surface area and thus allows a rapid exchange of large quantities of material.
3. Precapillary sphincters regulate blood flow through capillaries.

Venules (p. 363)

1. Venules are small vessels that continue from capillaries and merge to form veins.
2. They drain blood from capillaries into veins.

Veins (p. 363)

1. Veins consist of the same three coats as arteries but have less elastic tissue and smooth muscle.
2. They contain valves to prevent backflow of blood.
3. Weak valves can lead to varicose veins.

Blood Reservoirs (p. 364)

1. Systemic veins are collectively called blood reservoirs.
2. They store blood that, through vasoconstriction, can move to other parts of the body if the need arises.
3. The principal reservoirs are the veins of the abdominal organs (liver and spleen) and skin.

Physiology of Circulation (p. 365)

1. Blood flow is determined by blood pressure and resistance.
2. Blood flows from regions of higher to lower pressure. The pressure gradient for systemic circulation is from the aorta to large arteries to small arteries to arterioles to capillaries to venules to veins to venae cavae to the right atrium.
3. Resistance refers to the opposition to blood flow caused by friction between blood and the walls of blood vessels.
4. Resistance is determined by blood viscosity, blood vessel length, and blood vessel radius.
5. Factors that affect arterial blood pressure include cardiac output (CO), blood volume, and peripheral resistance.
6. Any factor that increases cardiac output increases blood pressure.

7. As blood volume increases, blood pressure increases.
8. Peripheral resistance refers to the opposition to blood flow in peripheral circulation (away from the heart).
9. In order to maintain a normal homeostatic range of blood pressure, there is a regulating cardiovascular center in the medulla oblongata that receives input from receptors throughout the body (baroreceptors, chemoreceptors), higher brain centers, and chemicals.
10. The vasomotor center in the cardiovascular center controls the diameter of arterioles to regulate blood pressure.
11. Baroreceptors, neurons sensitive to pressure, send impulses to the cardiovascular center and vasomotor center to regulate blood pressure.
12. Chemoreceptors—neurons sensitive to concentrations of oxygen, carbon dioxide, and hydrogen ions—send impulses to the vasomotor center to regulate blood pressure.
13. The cerebral cortex can influence the vasomotor center and this affects blood pressure.
14. Hormones such as epinephrine and norepinephrine raise blood pressure, whereas atrial natriuretic peptide lowers it.
15. Autoregulation refers to local, autonomic adjustments of blood flow in response to tissue needs.
16. The movement of water and dissolved substances (except proteins) through capillaries depends on pressure differences between the capillaries and interstitial spaces.
17. Blood return to the heart is maintained by the pumping action of the heart, velocity of blood flow, skeletal muscle contractions, valves in veins (especially in the limbs), and breathing.

Checking Circulation (p. 370)

Pulse (p. 370)

1. Pulse is the alternate expansion and elastic recoil of an artery with each heartbeat. It may be felt in any artery that lies near the surface or over a hard tissue.
2. A normal rate is about 75 beats per minute.

Measurement of Blood Pressure (BP) (p. 370)

1. Blood pressure is the pressure exerted by blood on the wall of an artery when the left ventricle undergoes systole and then diastole. It is measured by a sphygmomanometer.
2. Systolic blood pressure (SBP) is the force of blood recorded during ventricular contraction. Diastolic blood pressure (DBP) is the force of blood recorded during ventricular relaxation. The average blood pressure is 120/80 mm Hg.

Shock and Homeostasis (p. 370)

1. Shock is a failure of the cardiovascular system to deliver adequate amounts of oxygen and nutrients to meet the metabolic needs of cells.
2. Symptoms include low blood pressure; clammy, cool, pale skin; sweating; decreased urinary output; altered mental state; tachycardia; weak, rapid pulse; thirst; and nausea.

Circulatory Routes (p. 370)

1. The two major circulatory routes are systemic circulation and pulmonary circulation.

2. Two of the subdivisions of systemic circulation are hepatic portal circulation and cerebral circulation.

Systemic Circulation (p. 372)

1. Systemic circulation takes oxygenated blood from the left ventricle through the aorta to all parts of the body and returns deoxygenated blood to the right atrium.

2. The aorta is divided into the ascending aorta, the arch of the aorta, and the descending aorta. Each section gives off arteries that branch to supply the whole body.

3. Deoxygenated blood is returned to the heart through the systemic veins. All the veins of systemic circulation flow into either the superior or inferior vena cava or the coronary sinus. They, in turn, empty into the right atrium.

Pulmonary Circulation (p. 372)

1. Pulmonary circulation takes deoxygenated blood from the right ventricle to the air sacs of the lungs and returns oxygenated blood from the air sacs to the left atrium.

2. It allows blood to be oxygenated for systemic circulation.

Cerebral Circulation (p. 372)

1. Cerebral circulation arises from an arrangement of arteries at the base of the brain called the cerebral arterial circle (circle of Willis).

2. The circle equalizes blood pressure to the brain and provides alternate routes for blood.

Hepatic Portal Circulation (p. 372)

1. Hepatic portal circulation collects deoxygenated blood from the veins of the gastrointestinal tract and spleen and directs it into the hepatic portal vein of the liver.

2. This routing allows the liver to extract and modify nutrients and detoxify harmful substances in the blood.

3. The liver also receives oxygenated blood from the hepatic artery.

Self-Quiz

1. Which of the following statements about blood vessels is true?

   a. Capillaries contain valves.   b. Arterial walls are generally thicker and contain more elastic tissue than venous walls.   c. Veins carry blood away from the heart.   d. Blood flows most rapidly through veins.   e. Blood pressure in arteries is always lower than in veins.

2. Which of the following statements is NOT true?

   a. Regulation of blood vessel diameter originates from the vasomotor center of the pons.   b. The cerebral cortex can play a role in regulating blood pressure.   c. Baroreceptors may stimulate the vasomotor center.   d. Chemoreceptors are particularly sensitive to blood CO_2 levels.   e. Histamine is considered a vasodilator.

3. Blood flows through the blood vessels because of the

   a. establishment of a concentration gradient   b. elastic recoil of the veins   c. establishment of a pressure gradient   d. viscosity (stickiness) of the blood   e. suction created by the heart

4. Which of the following represents pulmonary circulation as the blood flows from the right ventricle?

   a. pulmonary trunk ⟶ pulmonary veins ⟶ pulmonary capillaries ⟶ pulmonary arteries   b. pulmonary arteries ⟶ pulmonary capillaries ⟶ pulmonary trunk ⟶ pulmonary veins   c. pulmonary capillaries ⟶ pulmonary trunk ⟶ pulmonary arteries ⟶ pulmonary veins   d. pulmonary trunk ⟶ pulmonary arteries ⟶ pulmonary capillaries ⟶ pulmonary veins   e. pulmonary veins ⟶ pulmonary capillaries ⟶ pulmonary arteries ⟶ pulmonary trunk

5. The blood vessels that allow the exchange of nutrients, wastes, oxygen, and carbon dioxide between the blood and tissues are the

   a. arteries   b. arterioles   c. veins   d. venules   e. capillaries

6. Match the following descriptions to the appropriate blood vessel:

   _E_ a. composed of single layer of endothelial cells and a basement membrane

   _D_ b. formed by reuniting capillaries

   _B_ c. regulate blood flow to capillaries

   _C_ d. may contain valves

   _A_ e. carry blood away from heart

   A. arteries
   B. arterioles
   C. veins
   D. venules
   E. capillaries

7. The blood vessels that serve as the main reservoirs of blood for the body are

   a. arteries   b. arterioles   c. veins   d. venules   e. capillaries

8. Match the following:

   _H_ a. source of all systemic arteries

   _D_ b. supplies lower limbs

   _E_ c. heart's blood system

   _F_ d. returns blood from lower limbs

   _A_ e. transports blood to liver

   _B_ f. leads to lungs

   _C_ g. returns blood from lungs

   _I_ h. supplies brain

   _G_ i. returns blood from head and upper body

   A. hepatic portal vein
   B. pulmonary trunk
   C. pulmonary veins
   D. common iliac arteries
   E. coronary circulation
   F. inferior vena cava
   G. superior vena cava
   H. aorta
   I. cerebral arterial circle

9. If a blood pressure reading is 120/80,

   a. 120 is the diastolic pressure   b. 80 is the pressure of the blood against the arteries during ventricular relaxation   c. 120 is the blood pressure and 80 is the beats per minute   d. 80 is the reading taken when the first sound is heard   e. the patient has a severe problem with hypertension

10. Blood pressure would be decreased by

    a. epinephrine and norepinephrine   b. atrial natriuretic peptide   c. a decrease in peripheral blood vessel radius   d. an increase in blood viscosity   e. an increase in cardiac output

11. Venous return to the heart is enhanced by all of the following EXCEPT

    a. skeletal muscle "milking"   b. valves in veins   c. increased velocity of flow in veins versus capillaries   d. pressure gradient produced by cardiac pumping   e. expiration during breathing

Critical Thinking Applications

1. The injection of a local anesthetics by a dentist often contains epinephrine. What would epinephrine do to the blood vessels in the vicinity of the dental work? Why might this effect be desired?

2. Chester stood straight and very still during a long, crowded, hot and stuffy funeral service. By the end of the service, he felt light-headed and two people standing nearby fainted. How does standing still affect blood flow?

3. Jerry's favorite snack is a big bowl of popcorn with LOTS of butter and salt. He'll eat an entire bowl during his favorite show along with at least a liter of soda. How will his snack attack affect his blood volume and blood pressure?

4. When a blood donor starts to feel faint at the blood drive, Chris places a handful of ice on the donor's neck. The donor's blood pressure (and his blood flow out the vein) recover promptly. Which division of the nervous system is responsible for the donor's quick recovery and how did this occur?

5. Emergency medical personnel were taking the blood pressure of a passenger at the scene of a train accident. The blood pressure was 90/60, her skin was cool and clammy, and her pulse was 120 beats per minute. The passenger had been seen wandering about the accident scene and seemed to be confused about how she got there. Explain the passenger's condition.

6. You've read about varicose veins. Why aren't there varicose arteries?

Answers to Figure Questions

16.1 Artery; vein.
16.2 They use oxygen and produce wastes more rapidly than inactive tissues.
16.3 As pressure increases, flow increases.
16.4 Cardiac muscle in the heart and smooth muscle in blood vessel walls.
16.5 They are returned by the lymphatic system to the cardiovascular system.
16.6 Milking.
16.7 Systemic and pulmonary.
16.8 Right atrium.
16.9 Pulmonary arteries.
16.10 Anterior and posterior cerebral arteries.
16.11 Hepatic veins.

17 chapter

THE LYMPHATIC SYSTEM, NONSPECIFIC RESISTANCE, AND IMMUNITY

a look ahead

student learning objectives

The *lymphatic* (lim-FAT-ik) *system* consists of a fluid called lymph, vessels called lymphatic vessels (lymphatics) that transport lymph, a number of structures and organs, all of which contain lymphatic (lymphoid) tissue, and red bone marrow, which houses cells that develop into white blood cells called lymphocytes (Figure 17.1). *Lymphatic tissue* is a specialized form of reticular connective tissue (see Exhibit 3.2) that contains large numbers of lymphocytes. Recall from Chapter 14 that there are two types of lymphocytes that participate in immune responses: (1) *B cells* and (2) *T cells*. B cells develop into plasma cells that protect us against disease by producing antibodies. T cells protect us against disease by attacking and destroying foreign cells and microbes.

Lymphatic nodules are oval-shaped masses of lymphatic tissue that are also not enclosed by a capsule. Some lymphatic nodules are found singly in mucous membranes of the gastrointestinal tract, respiratory passageways, urinary tract, and reproductive tract. This lymphatic tissue is referred to as *mucosa associated lymphoid tissue* (MALT). Other lymphatic nodules occur in clusters in specific parts of the body, such as the tonsils in the throat, aggregated lymphatic follicles (Peyer's patches) in the small intestine, and in the appendix. The *lymphatic organs* of the body—the lymph nodes, spleen, and thymus gland—all contain lymphatic tissue surrounded by a capsule.

Functions

The lymphatic system has several functions.

1. **Draining interstitial fluid.** Lymphatic vessels drain tissue spaces of protein-containing fluid (interstitial fluid). This fluid is formed by filtration at the arterial end of a capillary (see Figure 16.5). The lymphatic vessels return the proteins, along with any fluid not reabsorbed, to the cardiovascular system.

2. **Transporting lipids and lipid-soluble vitamins.** Lymphatic vessels transport triglycerides and some vitamins (A, D, E, and K) from the gastrointestinal tract to the blood (see Chapter 19).

3. **Protection against invasion.** Lymphatic tissue functions in surveillance and defense. That is, lymphocytes, with the aid of macrophages, protect the body from foreign cells, microbes, and cancer cells.

Lymph and Interstitial Fluid

objective: *Explain how lymph and interstitial fluid are related.*

Interstitial fluid and lymph are basically the same fluid. The major difference between the two is location. When the fluid bathes the cells, it is called *interstitial fluid* or *intercellular fluid.* When it flows through the lymphatic vessels, it is called *lymph* (lympha = clear water). Both fluids are similar in com-

position to plasma. The principal chemical difference is that they contain less protein than plasma because most plasma protein molecules cannot filter through the capillary wall.

Each day, about 20 liters of fluid seep from blood into tissue spaces. This fluid, as well as the plasma proteins it contains, must be returned to the cardiovascular system to maintain normal blood volume and functions. About 85 percent of the protein-containing fluid filtered from the arterial end of a blood capillary is reabsorbed at the venous end of the capillary. This fluid is returned to the blood directly. The remaining 15 percent of fluid is returned to the blood indirectly. It first passes into the lymphatic system and then into blood via the subclavian veins (see Figure 16.5).

Lymphatic Capillaries and Lymphatic Vessels

Lymphatic vessels originate as *lymphatic capillaries*, microscopic vessels in spaces between cells (Figure 17.2a on page 385). They are found throughout the body, with the exception of avascular tissue, the central nervous system, and red bone marrow. They are slightly larger than blood capillaries and have a unique structure that permits interstitial fluid to flow into them, but not out. The endothelial cells that make up the wall of a lymphatic capillary are not attached end to end, but rather, the ends overlap (Figure 17.2b). When pressure is greater in the interstitial fluid than lymph, the cells separate slightly, like a one-way swinging door, and fluid enters the lymphatic capillary. When pressure is greater inside the lymph capillary, the cells adhere more closely and fluid cannot escape back into interstitial fluid. Whereas blood capillaries link two larger blood vessels that form part of a circuit, lymphatic capillaries begin in the tissues and carry the lymph that forms there toward a larger lymphatic vessel.

Lymphatic vessels (*lymphatics*) resemble veins in structure but have thinner walls, more valves, and contain lymph nodes at various intervals. Just as blood capillaries converge to form venules and veins, lymphatic capillaries unite to form larger and larger lymph vessels called lymphatic vessels (see Figure 17.2). Ultimately, lymphatic vessels deliver lymph into two main channels: the thoracic duct and the right lymphatic duct. These are described shortly.

Lymphatic Tissue

objective: *Compare the structure and functions of the various types of lymphatic tissue.*

Lymph Nodes

Lymph nodes (Figure 17.3 on page 386) are oval or bean-shaped organs located along lymphatic vessels. They are scattered throughout the body, usually in groups (see Figure 17.1).

Figure 17.1 The lymphatic system.

🔑 *The lymphatic system consists of lymph, lymphatic vessels, lymphatic tissue, and red bone marrow.*

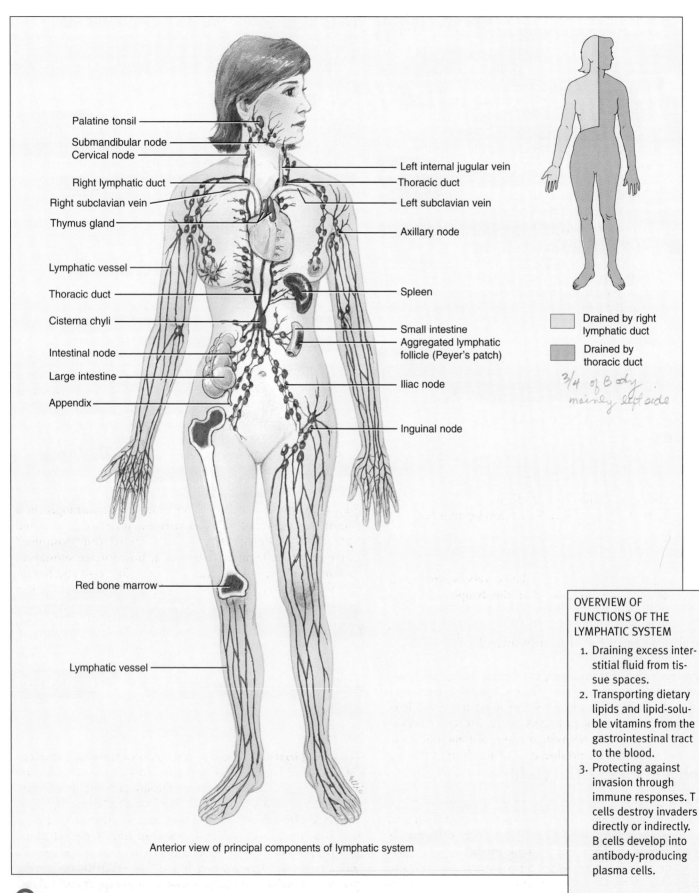

Palatine tonsil
Submandibular node
Cervical node
Right lymphatic duct
Right subclavian vein
Thymus gland
Lymphatic vessel
Thoracic duct
Cisterna chyli
Intestinal node
Large intestine
Appendix
Red bone marrow
Lymphatic vessel

Left internal jugular vein
Thoracic duct
Left subclavian vein
Axillary node
Spleen
Small intestine
Aggregated lymphatic follicle (Peyer's patch)
Iliac node
Inguinal node

Drained by right lymphatic duct
Drained by thoracic duct

3/4 of Body
mainly left side

Anterior view of principal components of lymphatic system

OVERVIEW OF FUNCTIONS OF THE LYMPHATIC SYSTEM

1. Draining excess interstitial fluid from tissue spaces.
2. Transporting dietary lipids and lipid-soluble vitamins from the gastrointestinal tract to the blood.
3. Protecting against invasion through immune responses. T cells destroy invaders directly or indirectly. B cells develop into antibody-producing plasma cells.

Q *What is lymphatic tissue?*

Figure 17.2   Lymphatic capillaries.

Lymphatic capillaries are found throughout the body except in vascular tissues, the central nervous system, portions of the spleen, and red bone marrow.

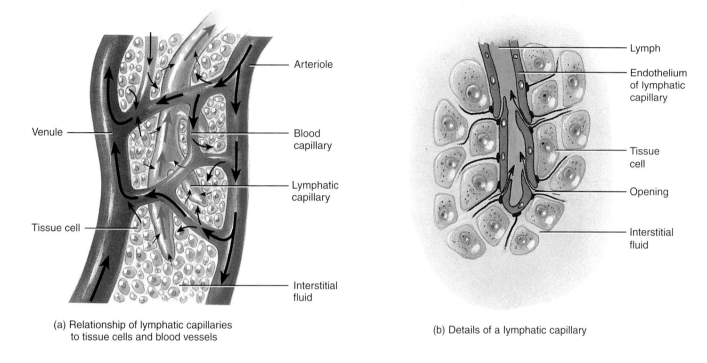

Arteriole

Venule

Blood capillary

Lymphatic capillary

Tissue cell

Interstitial fluid

(a) Relationship of lymphatic capillaries to tissue cells and blood vessels

Lymph

Endothelium of lymphatic capillary

Tissue cell

Opening

Interstitial fluid

(b) Details of a lymphatic capillary

Is lymph more similar to blood plasma or interstitial fluid?

Each node is covered by a *capsule* of dense connective tissue. Internally, nodes are divided into *follicles*, which are regions of T cells, macrophages, and B cells that develop into antibody-producing plasma cells. Throughout the lymph node are channels called *lymphatic sinuses* and reticular fibers.

Lymph enters a node through several *afferent* (*ad* = to; *ferre* = to carry) *lymphatic vessels*. These vessels have valves that open toward the node so that the lymph is directed *inward*. Lymph flows through the lymphatic sinuses and leaves a node through one or two *efferent* (*ef* = away) *lymphatic vessels*, which contain valves that open away from the node to convey lymph *outward*.

Lymph nodes filter the lymph passing from tissue spaces through lymphatic vessels on its way back to the cardiovascular system. Reticular fibers trap foreign substances, which are then destroyed through one or more means: macrophages by phagocytosis, T cells by releasing various antimicrobial substances, and plasma cells, produced from B cells that produce antibodies. Plasma cells and T cells produced in lymph nodes can circulate to other parts of the body.

The location of lymph nodes and the direction of lymph flow are important in the diagnosis and prognosis of the spread of cancer by *metastasis* (me-TAS-ta-sis). Cancer cells may be spread by way of the lymphatic system, producing aggregates of tumor cells wherever they lodge. Such secondary tumor sites are predictable by the direction of lymph flow from the organ primarily involved.

Tonsils

Tonsils are groups of large lymphatic nodules arranged in a ring at the junction of the oral cavity and pharynx (throat) and nasal cavity and pharynx. Tonsils are strategically positioned to participate in immune responses against foreign substances that are swallowed or inhaled. They contain T cells and B cells. The single *pharyngeal* (fa-RIN-jē-al) *tonsil* or *adenoid* (when inflamed) is embedded in the posterior wall of the upper part of the throat (see Figure 18.2). The paired *palatine* (PAL-a-tīn) *tonsils* are situated in the back of the mouth (see Figure 19.4). These are the ones commonly removed by a *tonsillectomy*. The paired *lingual* (LIN-gwal) *tonsils* are located at the base of the tongue and may also have to be removed by a tonsillectomy (see Figure 12.3a).

Spleen

The oval *spleen* is the largest single mass of lymphatic tissue in the body (see Figure 17.1). It is covered by a capsule of dense connective tissue and lies between the stomach and diaphragm. The spleen contains plasma cells, red blood cells, macrophages, and leukocytes. Because the spleen has no afferent lymphatic vessels or lymphatic sinuses, it does not filter lymph. It does, however, contain spaces for blood storage, one of the spleen's main functions. During severe blood loss, sympathetic impulses cause the release of stored blood to maintain blood volume and pressure.

Figure 17.3   Structure of a lymph node. Arrows indicate direction of lymph flow.

🔑 *Lymph nodes are present throughout the body, usually clustered in groups.*

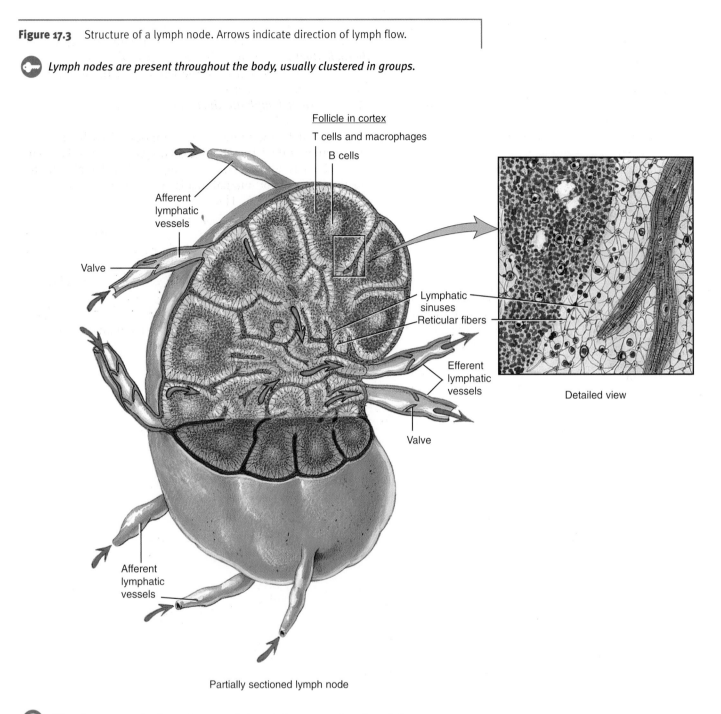

Follicle in cortex
T cells and macrophages
B cells

Afferent lymphatic vessels

Valve

Lymphatic sinuses
Reticular fibers

Efferent lymphatic vessels

Valve

Detailed view

Afferent lymphatic vessels

Partially sectioned lymph node

❓ *What happens to foreign substances in lymph that enter a lymph node?*

The spleen is the site of B cell proliferation into antibody-producing plasma cells. Also, cells within the spleen phagocytize bacteria and worn-out and damaged red blood cells and platelets. During early fetal development, the spleen participates in blood cell formation.

The spleen is the most frequently damaged organ in cases of abdominal trauma, particularly those involving severe blows to the lower left chest or upper abdomen that fracture the protecting ribs. Such a crushing injury may *rupture the spleen,* which causes severe hemorrhage and shock. Prompt removal of the spleen, called a *splenectomy,* is need-ed to prevent the patient from bleeding to death. The functions of the spleen are then assumed by other structures, particularly red bone marrow and the liver.

Thymus Gland

The **thymus gland** is a two-lobed organ located behind the sternum and between the lungs (see Figure 17.1). Each lobe is covered by a connective tissue **capsule.** Internally, the thymus gland consists of T cells, macrophages, and epithelial cells that produce thymic hormones.

The thymus gland is large in the infant but does not reach its maximum size until puberty. After puberty, much of the thymic tissue is replaced by fat and connective tissue but the gland continues to be functional, although to a lesser degree.

Its role in immunity is to help produce and distribute T cells to other lymphoid organs. These cells destroy invading microbes by producing various substances. Recall from Chapter 13 that hormones produced by the thymus gland, such as thymosin and thymopoietin, promote the proliferation and maturation of T cells. This is discussed in detail shortly.

Lymph Circulation

objective: *Describe how lymph circulates throughout the body.*

Lymph from lymphatic capillaries is passed to lymphatic vessels and through lymph nodes. Efferent vessels from the nodes pass on to other nodes. The efferent vessels from the last nodes in a chain unite to form *lymph trunks*.

The principal trunks pass their lymph into two main channels: the thoracic duct and the right lymphatic duct. The ***thoracic duct*** is the main collecting duct of the lymphatic system and receives lymph from the left side of the head, neck, and chest, the left upper limb, and the entire body below the ribs. The ***right lymphatic duct*** drains lymph from the upper right side of the body (see Figure 17.3).

Ultimately, the thoracic duct empties all its lymph into the junction of the left internal jugular vein and left subclavian vein, and the right lymphatic duct empties all its lymph into the junction of the right internal jugular vein and right subclavian vein (see Figure 17.3). Thus lymph is drained back into the blood, and the cycle repeats itself continuously. The sequence of fluid flow is: arteries (blood plasma) $\longrightarrow$ blood capillaries (blood plasma) $\longrightarrow$ interstitial spaces (interstitial fluid) $\longrightarrow$ lymphatic capillaries (lymph) $\longrightarrow$ lymphatic vessels (lymph) $\longrightarrow$ lymphatic ducts (lymph) $\longrightarrow$ subclavian veins (blood plasma) (Figure 17.4).

The flow of lymph from tissue spaces to the large lymphatic ducts to the subclavian veins is maintained primarily by the "milking action" of skeletal muscles (see Figure 16.6).

Know for Boards

Figure 17.4   Relationship of lymphatic system to the cardiovascular system.

The flow of fluid is from arteries (blood plasma) to blood capillaries (blood plasma) to interstitial spaces (interstitial fluid) to lymphatic capillaries (lymph) to lymphatic vessels (lymph) to lymphatic ducts (lymph) to subclavian veins (blood plasma).

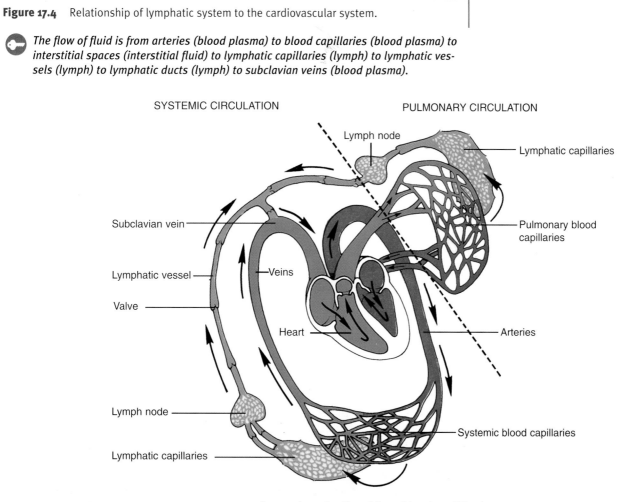

Arrows show direction of flow of lymph and blood

Which vessels of the cardiovascular system (arteries, veins, or capillaries) produce lymph?

Skeletal muscle contractions compress lymphatic vessels and force lymph toward the subclavian veins. Lymphatic vessels, like veins, contain valves, and the valves ensure the one-way movement of lymph toward the subclavian veins (see Figures 17.1 and 17.4).

Lymph flow is also maintained by respiratory movements, which create a pressure difference between the two ends of the lymphatic system. Lymph flows from the abdominal region, where the pressure is higher, toward the thoracic region, where it is lower.

Edema, an excessive accumulation of interstitial fluid in tissue spaces, may be caused by an obstruction, such as an infected node, or by a blockage of lymphatic vessels. Another cause is excessive lymph formation and increased permeability of blood capillary walls. Edema may also result from a rise in capillary blood pressure, in which interstitial fluid is formed faster than it can pass into lymphatic vessels.

Nonspecific Resistance to Disease

The human body continually attempts to maintain homeostasis by counteracting the activities of disease-producing organisms, called *pathogens* (PATH-ō-jens), or their toxins. The ability to ward off disease is called *resistance.* Vulnerability or lack of resistance is called *susceptibility.* Defenses against disease may be grouped into two broad areas: nonspecific resistance and specific resistance. *Nonspecific resistance* represents a wide variety of body reactions that provide a general response against invasion by a wide range of pathogens. *Specific resistance* or *immunity* is the production of a specific antibody against a specific pathogen or its toxin. We will first consider mechanisms of nonspecific resistance.

Skin and Mucous Membranes

objective: *Explain how the skin and mucous membranes protect the body against disease.*

The skin and mucous membranes of the body are the first line of defense against disease-causing microorganisms. They possess certain mechanical and chemical factors that combat the initial attempt of a microbe to cause disease.

Mechanical Factors

The *intact skin,* as noted in Chapter 5, consists of two distinct portions called the epidermis and dermis. The closely packed cells of the epidermis, its many layers, and the presence of keratin provide a formidable physical barrier to the entrance of microbes. In addition, periodic shedding of epidermal cells helps to remove microbes on the skin surface. The intact surface of healthy epidermis seems to be rarely penetrated by bacteria. But when the epithelial surface is broken, from cuts, burns, and so on, a subcutaneous infection often develops. The bacteria most likely to cause such an infection are staphy-lococci, which normally inhabit the hair follicles and sweat glands of the skin.

Mucous membranes, like the skin, also consist of two layers: an epithelial layer and an underlying connective tissue layer. Mucous membranes line body cavities that open to the exterior, such as the gastrointestinal, respiratory, urinary, and reproductive tracts. The epithelial layer of a mucous membrane secretes a fluid called *mucus,* which prevents the cavities from drying out. Because mucus is slightly viscous, it traps many microbes that enter the respiratory and gastrointestinal tracts. The mucous membrane of the nose has mucus-coated *hairs* that trap and filter air containing microbes, dust, and pollutants. The mucous membrane of the upper respiratory tract contains *cilia,* microscopic hairlike projections of the epithelial cells (see Figure 18.4), that move in such a manner that they pass inhaled dust and microbes that have become trapped in mucus toward the throat. This so-called ciliary escalator keeps the mucus moving toward the throat at a rate of 1 to 3 cm per hour. Coughing and sneezing speed up the escalator.

Microbes are also prevented from entering the lower respiratory tract by a small lid of cartilage called the *epiglottis* that covers the voice box during swallowing (see Figure 18.2).

The cleansing of the urethra by the *flow of urine* is another mechanical factor that prevents microbial colonization in the urinary system.

Defecation and *vomiting* might also be considered mechanical processes that expel microbes. For example, in response to microbes, such as those that might irritate the lining of the gastrointestinal tract, the muscles of the tract contract vigorously to attempt to expel the microbes (diarrhea).

A mechanism that protects the eyes is the *lacrimal* (LAK-rimal) *apparatus* (see Figure 12.4), a group of structures that manufactures and drains away tears, which are spread over the surface of the eyeball by blinking. Normally, the tears are carried away by evaporation or pass into the nose as fast as they are produced. The continual washing action of tears helps to keep microbes from settling on the surface of the eye. If an irritating substance makes contact with the eye, the lacrimal glands start to secrete heavily to dilute and wash away the irritant.

Saliva, produced by the *salivary glands,* washes microbes from the surface of the teeth and the mucous membrane of the mouth, much as tears wash the eyes. This helps to prevent colonization by microbes.

Chemical Factors

Certain chemical factors also account for the high degree of resistance of skin and mucous membranes to microbial invasion.

Oil glands of the skin secrete an oily substance called *sebum* that forms a protective film over the surface of the skin. The skin's acidity, with a pH of about 4, probably discourages the growth of many microorganisms. Sweat glands of the skin produce *perspiration,* which flushes some microorganisms from the skin surface. Perspiration also contains *lysozyme,* an enzyme capable of breaking down cell walls of certain bacteria. Lysozyme is also found in tears, saliva, nasal secretions, and tissue fluids.

Gastric juice is produced by the glands of the stomach. The very high acidity of gastric juice (pH about 2) destroys bacteria and most bacterial toxins.

Vaginal secretions are also highly acidic, which discourages bacterial growth.

Antimicrobial Substances

objective: *Describe how various antimicrobial substances protect the body against disease.*

In addition to the mechanical and chemical barriers of the skin and mucous membranes, the body also produces certain antimicrobial substances. They provide a second line of defense should microbes penetrate the skin and mucous membranes.

Interferons (IFNs)

Body cells infected with viruses produce proteins called **interferons** (in'-ter-FĒR-ons) or **IFNs**. IFNs diffuse to uninfected neighboring cells and bind to surface receptors. This enables uninfected cells to inhibit viral replication. IFNs appear to protect cells from invasion against many different viruses. Some IFNs have been incorporated into a nasal spray that may prevent the spread of colds.

In clinical trials, IFNs have exhibited limited effects against some tumors, such as malignant melanoma. One type of IFN, called alpha IFN (Intron A®), is approved in the United States for treating several virus-associated disorders, including Kaposi's (kap'-ō-SĒS) sarcoma, a cancer that often occurs in patients infected with HIV, the virus that causes AIDS (acquired immunodeficiency syndrome; see p. 400); genital herpes, caused by herpesvirus; and hepatitis B and C, caused by the hepatitis B and C viruses. Alpha IFN is also being tested to see if it can slow the development of AIDS in HIV-infected people. A form of beta IFN slows the progression of multiple sclerosis (MS).

Complement

Complement is actually a group of at least 20 proteins found in blood plasma and on plasma membranes. When activated, these proteins "complement" or enhance certain immune, allergic, and inflammatory reactions. Some complement proteins create holes in the plasma membrane of the microbe, causing the contents of the microbe to leak out, a process called **cytolysis.** Some attract phagocytes to the site of inflammation, called **chemotaxis,** and increase blood flow to the area by causing vasodilation of arterioles. Others bind to the surface of a microbe and promote phagocytosis. This is called **opsonization** (op-sō-ni-ZĀ-shun).

Natural Killer (NK) Cells

When microbes penetrate the skin and mucous membranes or bypass the antimicrobial substances in blood, the next, nonspecific line of defense consists of natural killer cells and phagocytes. We will first consider natural killer cells.

In addition to B cells and T cells, which participate in immune responses, there is another group of lymphocytes that destroys intruders. These lymphocytes, called **natural killer** (**NK**) **cells,** have the ability to kill a wide variety of infectious microbes plus certain tumor cells. NK cells are present in the spleen, lymph nodes, red bone marrow, and blood. NK cells may damage or cause cytolysis of a microbe. NK cells are defective or decreased in number in some cancer patients and in patients with AIDS.

Phagocytosis

objective: *Define phagocytosis and explain how it occurs.*

Phagocytosis (*phagein* = to eat; *cyto* = cell) is the ingestion of microbes or any foreign particles or cell debris by cells called phagocytes.

Kinds of Phagocytes

Phagocytes fall into two broad categories: granular leukocytes (microphages) and macrophages. The **granular leukocytes** of blood (see Figure 14.2) vary in the degree to which they function as phagocytes, neutrophils being the most phagocytic.

When an infection occurs, both neutrophils and monocytes migrate to the infected area. During this migration, the monocytes enlarge and develop into active phagocytic cells called **macrophages** (MAK-rō-fā-jez) (see Figure 14.2). Because these cells leave the blood and migrate to infected areas, they are called **wandering macrophages.** Some macrophages, called **fixed macrophages,** remain in certain tissues and organs of the body. Fixed macrophages are found in the skin and subcutaneous layer, liver, lungs, brain, spleen, lymph nodes, and red bone marrow.

Mechanism

For convenience of study, phagocytosis is divided into three phases: chemotaxis, adherence, and ingestion (Figure 17.5).

CHEMOTAXIS   **Chemotaxis** (kē'-mo-TAK-sis; *chemo* = chemicals; *taxis* = arrangement) is the chemical attraction of phagocytes to microorganisms. Chemicals that attract phagocytes might come from the microbes themselves, white blood cells, damaged tissue cells, or activated complement.

ADHERENCE   In **adherence,** the phagocyte attaches itself to the microorganism or other foreign material. Sometimes, adherence occurs easily and the microorganism is readily phagocytized (see Figure 3.9a,b). Microorganisms can more readily be phagocytized if they are first coated with certain plasma proteins (complement) that promote attachment (opsonization).

INGESTION   Following adherence, **ingestion** occurs. In the process, the plasma membrane of the phagocyte extends projections, called **pseudopods,** that engulf the microorganism. When the pseudopods meet, they fuse, surrounding the microorganism with a sac called a **phagocytic vesicle.**

After phagocytosis has occurred, the phagocytic vesicle and lysosome fuse to form a single, larger structure called a

Figure 17.5   Phagocytosis.

The stages of phagocytosis are chemotaxis, adherence, and ingestion.

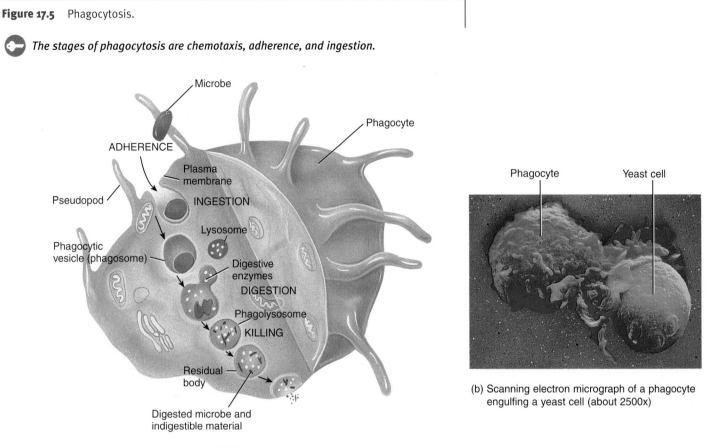

(a) Phases of phagocytosis

(b) Scanning electron micrograph of a phagocyte engulfing a yeast cell (about 2500x)

Which cells are most actively phagocytic?

phagolysosome. Some bacteria are killed in only 10 to 30 minutes. Indigestible materials remain in structures called *residual bodies,* which the cell disposes of by exocytosis.

Inflammation

objective: *Describe how inflammation occurs and why it is important.*

Inflammation is a nonspecific, defensive response of the body to stress due to tissue damage. Among the conditions that may produce inflammation are pathogens (disease-producing microbes), abrasions, chemical irritations, distortion or disturbances of cells, and extreme temperatures. Thus, not all inflammations are caused by infections. Regardless of the cause, the injury may be viewed as a form of stress.

Symptoms

Inflamation is characterized by four symptoms: *redness, pain, heat,* and *swelling.* A fifth symptom can be *loss of function,* depending on the site and extent of the injury. Inflammation is an attempt to dispose of microbes, toxins, or foreign material at the site of injury, to prevent their spread to other organs, and

to prepare the site for tissue repair. Thus, the inflammatory response is an attempt to restore tissue homeostasis.

Stages

There are three basic stages of the inflammatory response: (1) vasodilation and increased permeability of blood vessels, (2) phagocyte migration, and (3) repair.

VASODILATION AND INCREASED PERMEABILITY OF BLOOD VESSELS   Immediately following tissue damage, there is vasodilation and increased permeability of blood vessels in the area of the injury (Figure 17.6). *Vasodilation* is an increase in diameter of the blood vessels. *Increased permeability* means that substances normally retained in blood are permitted to pass from the blood vessels. Vasodilation allows more blood to go to the damaged area, and increased permeability permits defensive substances in the blood such as antibodies, phagocytes, and clot-forming chemicals to enter the injured area.

One of the principal substances that contribute to vasodilation, increased permeability, and other aspects of the inflammatory response is *histamine* (Figure 17.6). It is found in many body cells, especially mast cells in connective tissue, basophils, platelets, and phagocytes. Prostaglandins intensify the effects of histamine.

Figure 17.6   Inflammatory response. Several substances stimulate vasodilation, increased permeability of blood vessels, chemotaxis (attracting phagocytes to the area), emigration, and phagocytosis.

The three stages of inflammation are (1) vasodilation and increased permeability of blood vessels, (2) phagocyte migration, and (3) tissue repair.

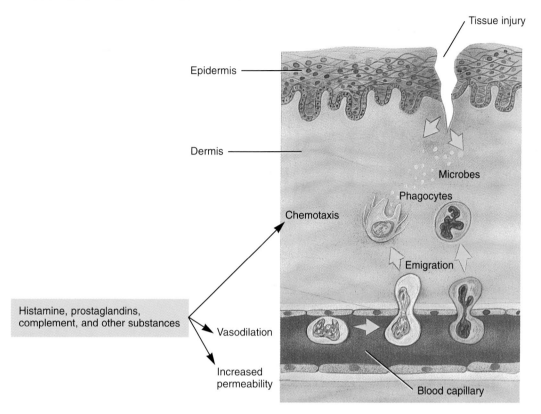

Phagocytes migrate from blood to site of tissue injury

What causes each of the following symptoms of inflammation: redness, pain, heat, and swelling

Dilation of arterioles and increased permeability of capillaries produce heat, redness, and swelling (edema). Heat and redness result from the large amount of blood that accumulates in the area.

Pain results from injury to nerve fibers or from toxic chemicals released by microorganisms. Some of the substances that promote vasodilation and increased dilation, especially prostaglandins, also affect nerve endings and cause pain.

The increased permeability of capillaries causes leakage of fibrinogen into tissues. Fibrinogen is then converted to an insoluble, thick network called fibrin, which traps the invading organisms, preventing their spread. This network eventually forms a clot (see Figure 14.8) that isolates the invading microbes or their toxins.

PHAGOCYTE MIGRATION   Generally, within an hour after the inflammatory process starts, phagocytes appear on the scene. As large amounts of blood accumulate, neutrophils begin to stick to the inner surface of the endothelium (lining) of blood vessels (see Figure 17.6). Then the neutrophils begin to squeeze through the wall of the blood vessel to reach the damaged area.

This is called *emigration* and it depends on chemotaxis. Neutrophils attempt to destroy the invading microbes by phagocytosis.

As the inflammatory response continues, monocytes follow the neutrophils into the infected area. Once in the tissue, monocytes become transformed into wandering macrophages that add to the phagocytic activity of fixed macrophages. The neutrophils predominate in the early stages of infection but tend to die off rapidly. Macrophages then enter the picture. Macrophages are several times more phagocytic than neutrophils and large enough to engulf tissue that has been destroyed, neutrophils that have been destroyed, and invading microbes.

REPAIR   In all but very mild inflammations, pus is produced. *Pus* is a thick fluid that contains living, as well as nonliving, white blood cells and debris from other dead tissue. Pus formation continues until the infection subsides. At times, the pus pushes to the surface of the body or into an internal cavity for dispersal. If the pus cannot drain out of the body, an abscess develops. An *abscess* is an excessive accumulation of pus in a confined space. Common examples are pimples and boils.

Fever

The most frequent cause of *fever,* an abnormally high body temperature, is infection from bacteria (and their toxins) and viruses. The high body temperature inhibits some microbial growth and speeds up body reactions that aid repair. (Fever is discussed in more detail on pp. 478–480).

The factors that contribute to nonspecific resistance are summarized in Exhibit 17.1.

Immunity (Specific Resistance to Disease)

objective: *Define immunity and compare it with nonspecific resistance to disease.*

Despite the various kinds of nonspecific resistance, they all have one thing in common. They are designed to protect the body

Exhibit 17.1	Summary of Nonspecific Resistance
COMPONENT	**FUNCTIONS**
SKIN AND MUCOUS MEMBRANES	
MECHANICAL FACTORS	
Epidermis of skin	Forms a physical barrier to the entrance of microbes.
Mucous membranes	Inhibit the entrance of many microbes, but not as effective as intact skin.
Mucus	Traps microbes in respiratory and gastrointestinal tracts.
Hairs	Filter out microbes and dust in nose.
Cilia	Together with mucus, trap and remove microbes and dust from upper respiratory tract.
Epiglottis	Prevents microbes and dust from entering lower respiratory tract.
Lacrimal apparatus	Tears dilute and wash away irritating substances and microbes.
Saliva	Washes microbes from surfaces of teeth and mucous membranes of mouth.
Urine	Washes microbes from urethra.
Defecation and vomiting	Expel microbes from body.
CHEMICAL FACTORS	
Sebum	Forms protective oily film on skin.
Acid pH of skin	Discourages growth of many microbes.
Lysozyme	Antimicrobial substance in perspiration, tears, saliva, nasal secretions, and tissue fluids.
Vaginal secretions	Highly acidic secretions that discourage bacterial growth.
Gastric juice	Destroys bacteria and most toxins in stomach.
ANTIMICROBIAL SUBSTANCES	
Interferons (IFNs)	Protect uninfected host cells from viral infection.
Complement	Causes cytolysis of microbes, promotes phagocytosis, and contributes to inflammation.
NATURAL KILLER (NK) CELLS	Kill a wide variety of microbes and certain tumor cells.
PHAGOCYTOSIS	Ingestion of foreign matter by neutrophils and macrophages.
INFLAMMATION	Confines and destroys microbes and initiates tissue repair.
FEVER	Inhibits growth of some microbes and speeds up body reactions that aid repair.

from *any* kind of pathogen. They are not specifically directed against a particular microbe. Specific resistance to disease, called *immunity,* involves the production of a specific type of cell or specific molecule (antibody) to destroy a particular antigen. An *antigen* is any substance—such as microbes, foods, drugs, pollen, or tissue—that brings about an immune response once the immune system has recognized it as foreign. The branch of science that deals with the responses of the body when challenged by antigens is referred to as *immunology* (im′-yoo-NOL-ō-jē; *immunis* = free).

Acquired immunity refers to immunity gained as a result of contact with an antigen. It may be obtained either actively or passively, or by natural or artificial means. Following are the types of acquired immunity: [*specifically talking about Bacteria or Viruses*]

1. *Naturally acquired active immunity* is obtained when a person's immune system comes into contact with microbes in the course of daily living and the body responds by producing antibodies and/or T cells (described shortly). An example is immunity to chickenpox. This type of immunity may be lifelong or last only a few years.

2. *Naturally acquired passive immunity* involves the natural transfer of antibodies from an immunized donor to a nonimmunized recipient. For example, an expectant mother can pass some of her antibodies to the fetus across the placenta to provide immunity to diseases such as diphtheria, rubella, or polio. Certain antibodies are also passed from the mother to her nursing infant in breast milk. Naturally acquired passive immunity generally lasts only a few weeks or months.

3. *Artificially acquired active immunity* results from vaccination. The individual is given a *vaccine* (vak-SĒN), a preparation containing an antigen from a specific microbe that stimulates antibody production. Vaccines consist of (1) *microbes* that have been killed or weakened so that they do not produce symptoms of the disease and/or (2) *toxoids,* infectious bacterial toxins (poisons) that have been chemically inactivated. Vaccines provide immunity while sparing a person the discomfort of the symptoms of a particular disease. Vaccines are used to prevent bacterial diseases such as diphtheria, pneumococcal pneumonia, pertussis (whooping cough), tetanus, and some types of meningitis and viral diseases such as influenza, measles, mumps, rubella, poliomyelitis, rabies, and hepatitis B. Some vaccines, such as those for poliomyelitis and measles, provide lifelong immunity; others, such as those for pneumococcal pneumonia and meningococcal meningitis, last only a few years.

4. *Artificially acquired passive immunity* is gained by an injection of antibodies from an outside source. For [*HcP.A*] example, a person might be given gamma globulin (antibody-rich serum). Antibodies produced by one person can also be transferred to another person.

[*or Rabies*]

We will examine the mechanism of immunity by first discussing the nature of antigens and antibodies.

Antigens (Ags)

> **objective:** *Explain the relationship between an antigen and an antibody.*

An *antigen* (*immunogen*), or *Ag,* is any chemical substance that, when introduced into the body, is recognized as nonself by the immune system. It causes the body to produce specific antibodies and/or specific T cells that react with the antigen. Antigens have two important characteristics. The first is the ability to stimulate the formation of specific antibodies. The second is the ability of the antigen to react specifically with the produced antibodies.

The entire microbe, such as a bacterium or virus, or just parts of microbes may act as an antigen, for example, bacterial structures such as flagella, capsules, and cell walls; bacterial toxins (poisons) are highly antigenic. Nonmicrobial examples of antigens include foods, drugs, pollen, cancer cells, transplanted tissues or organs, or serum from other humans, animals, or insects.

As a rule, antigens are foreign substances. They are not usually part of the chemistry of the body. The body's own substances, recognized as self, do not normally act as antigens; substances identified as nonself, however, stimulate antibody production. There are certain conditions, though, in which the distinction between self and nonself breaks down, and antibodies that attack the body are produced. These autoimmune diseases are described on p. 402.

Now let us look at some of the characteristics of antibodies.

Antibodies (Abs)

An *antibody* (*Ab*) is a protein produced by plasma cells in response to the presence of an antigen and is capable of combining specifically with the antigen.

The combination of antibody and antigen depends on the size and shape of their combining sites (Figure 17.7). The combination is very much like the fit between a lock and key. The portion of an antibody that recognizes and combines with an antigen is called an *antigen binding site.*

Antibodies belong to a group of plasma proteins called *globulins,* and for this reason antibodies are also known as *immunoglobulins* (im′-yoo-nō-GLOB-yoo-lins), or *Igs.* There are five different classes of immunoglobulins, designated as IgG, IgA, IgM, IgD, and IgE. Each has a distinct chemical structure and a specific biological role (Exhibit 17.2 on page 395).

Cell-Mediated and Antibody-Mediated Immunity

> **objective:** *Compare the functions of cell-mediated and antibody-mediated immunity.*

The body defends itself against invading agents by means of its immune system, which consists of two closely allied components.

Figure 17.7   Relationship of an antigen to an antibody.

An antigen induces the body to produce specific antibodies that combine with the antigen.

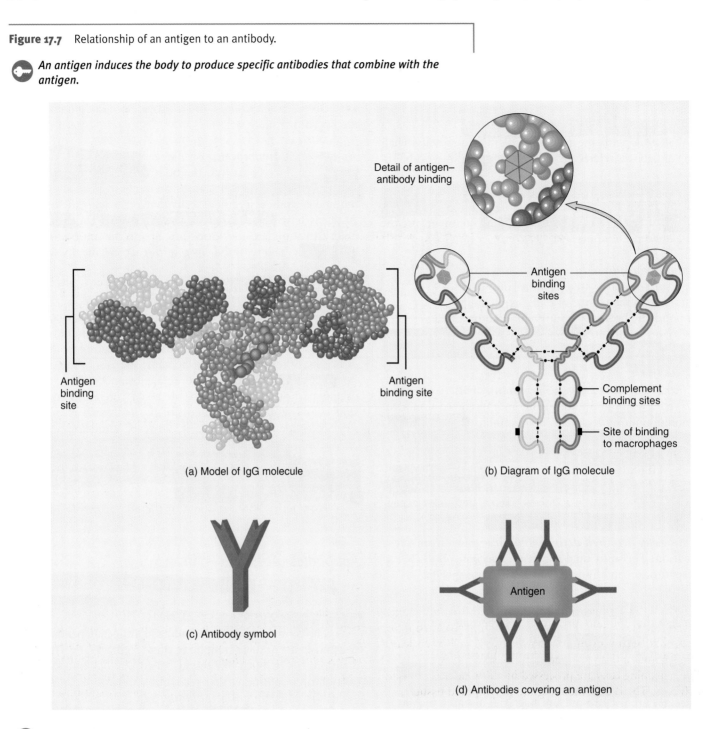

(a) Model of IgG molecule

(b) Diagram of IgG molecule

Detail of antigen–antibody binding

Antigen binding sites

Complement binding sites

Site of binding to macrophages

Antigen binding site

Antigen binding site

(c) Antibody symbol

Antigen

(d) Antibodies covering an antigen

What is the chemical nature of antibodies?

One component involves the formation of T cells that directly attack foreign agents and destroy them. This is *cell-mediated* (*cellular*) *immunity* and is particularly effective against fungi, parasites, intracellular viral infections, cancer cells, and foreign tissue transplants. *Antibody-mediated* (*humoral*) *immunity* refers to antibodies that are dissolved in blood plasma and lymph, the body's fluids or "humors." In antibody-mediated immunity, B cells develop into plasma cells that produce antibodies. This type of immunity is particularly effective against bacterial and viral infections.

The cells involved in both types of immunity originate in lymphoid tissue located in the lymph nodes, spleen, gastrointestinal tract, and red bone marrow. Lymphoid tissue is strategically located to intercept an invading agent before it can spread too extensively into general circulation.

Formation of T Cells and B Cells

Both T cells and B cells develop from cells in red bone marrow (stem cells), which they leave for two different areas of the

Exhibit 17.2 | Classes of Immunoglobulins (Igs)

NAME AND STRUCTURE	CHARACTERISTICS AND FUNCTIONS
IgG	Most abundant, about 75% of all antibodies in the blood, found in blood, lymph and the intestines. Protect against bacteria and viruses by enhancing phagocytosis, neutralizing toxins, and triggering the complement system. They are the only class of antibody to pass the placenta from mother to fetus and thereby confer considerable immune protection in newborns.
IGA	Make up about 15% of all antibodies in the blood. Found mainly in sweat, tears, saliva, mucus, milk, and gastrointestinal secretions. Smaller quantities present in blood and lymph. Levels decrease during stress, lowering resistance to infection. Provide localized protection on mucous membranes against bacteria and viruses.
IgM	About 5–10% of all antibodies in the blood, first antibodies to be secreted by plasma cells after an initial exposure to any antigen; found in blood and lymph. Activate complement and cause agglutination and lysis of microbes. Also present on the surfaces of B cells, where they serve as antigen receptors. In blood plasma, the anti-A and anti-B antibodies of the ABO blood group, which bind to A and B antigens during incompatible blood transfusions, are also IgM antibodies.
IgD	Less than 1% of all antibodies in the blood; found in blood, in lymph, and on the surfaces of B cells as antigen receptors. Involved in activation of B cells.
IgE	Less than 0.1% of all antibodies in the blood; located on mast cells and basophils. Involved in allergic and hypersensitivity reactions; provide protection against parasitic worms.

body. About half of them first migrate to the thymus gland, where they become T cells. The name T cell is derived from the processing that occurs in the thymus gland. T cells then leave the thymus gland and enter lymphoid tissue.

The remaining cells are processed in red bone marrow and become B cells. The B cells then migrate to lymphoid tissue.

T Cells and Cell-Mediated Immunity   See pg 331

Before T cells can attack foreign invaders (antigens), the T cells must become *sensitized* (*activated*); that is, they must be able to recognize the foreign invaders. *Sensitization* (sen'-si-ti-ZĀ-shun) involves macrophages, which process and present the antigens to T cells (Figure 17.8). In the processing phase, the macrophage phagocytizes the antigen; in the presentation phase, the partially digested antigen is displayed on the surface of the macrophage, along with self-antigens found on the surface of white blood cells called *human leukocyte associated* (*HLA*) *antigens*. These HLA antigens are unique for each person. Macrophages also participate in the immune response by secreting powerful protein growth factors, which stimulate the division of sensitized T cells to form a very large population of sensitized T cells.

There are literally millions of different kinds of T cells, because each must be able to recognize and attack a different specific antigen. Once sensitized, T cells increase in size and number to produce a *clone* (group of genetically identical cells). The cells in the clone then differentiate into various cell types, all of which play an important role in the immune response.

1. *Helper T cells* play a central role in the immune response. Their major function is to stimulate other cells of the immune system to fight off intruders. Helper T cells release proteins into the blood that cause cytotoxic T cells (described next) to grow and divide, attract phagocytes, and enhance the phagocytic ability of macrophages. Helper T cells also stimulate the development of B cells into antibody-producing plasma cells.

2. *Cytotoxic T cells* (*killer T cells*) destroy target cells on contact. Cytotoxic T cells provide protection against viral infections, slowly developing bacterial infections, cancer cells, and transplanted cells. Cytotoxic T cells release chemicals that make holes in the plasma membrane of the antigen, which allows extracellular fluid to

Figure 17.8   Role of T cells in cell-mediated immunity.

Before T cells can recognize foreign invaders, the T cells must be able to recognize them, a process called sensitization.

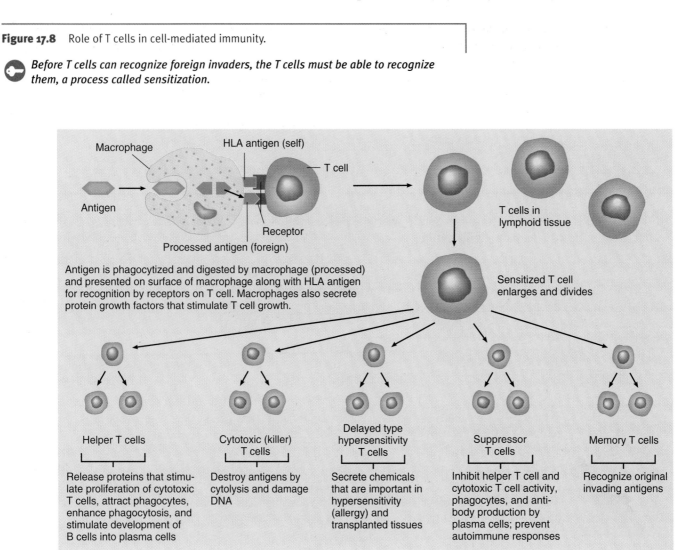

Antigen is phagocytized and digested by macrophage (processed) and presented on surface of macrophage along with HLA antigen for recognition by receptors on T cell. Macrophages also secrete protein growth factors that stimulate T cell growth.

Sensitized T cell enlarges and divides

Helper T cells	Cytotoxic (killer) T cells	Delayed type hypersensitivity T cells	Suppressor T cells	Memory T cells
Release proteins that stimulate proliferation of cytotoxic T cells, attract phagocytes, enhance phagocytosis, and stimulate development of B cells into plasma cells	Destroy antigens by cytolysis and damage DNA	Secrete chemicals that are important in hypersensitivity (allergy) and transplanted tissues	Inhibit helper T cell and cytotoxic T cell activity, phagocytes, and antibody production by plasma cells; prevent autoimmune responses	Recognize original invading antigens

Q   *What is the role of macrophages in the activation of T cells?*

flow in and the cell bursts (cytolysis). Cytotoxic T cells also secrete a substance that damages DNA in target cells, causing the cells to die and attract macrophages, enhance their phagocytic activity, and prevent their migration from the site of action.

3. *Delayed hypersensitivity cells* secrete chemicals that recruit defensive cells such as macrophages. Delayed hypersensitivity cells are activated in response to allergic reactions (such as poison ivy) and transplanted tissues.

4. *Suppressor T cells* are believed to depress parts of the immune response by inhibiting helper T cell and cytotoxic T cell activity, inhibiting phagocytes, and reducing antibody production by plasma cells. Suppressor T cells are also important in preventing autoimmune responses, where the immune system begins to attack the body's own tissues, as occurs, for example, in rheumatoid arthritis.

5. *Memory T cells* are descendants of other T cells that remain in lymphoid tissue and are programmed to recognize the original invading antigen. Should the pathogen invade the body at a later date, the memory cells initiate a far swifter reaction than during the first invasion. In fact, the second response is so swift that the pathogens are usually destroyed before any signs or symptoms of the disease occur. Memory cells permit sensitized persons to retain immunity for years.

Cell-mediated immunity augments our natural resistance and plays a crucial role in the initiation of antibody-mediated immunity.

B Cells and Antibody-Mediated Immunity

The body contains not only millions of different T cells but also millions of different B cells, each capable of responding to a specific antigen. Whereas killer T cells leave lymphoid tissue to confront a foreign antigen, B cells do not. B cells in the lymph nodes, spleen, or lymphoid tissue of the gastrointestinal tract can respond to an unprocessed antigen, although their response is much more intense if the antigen is processed. B cells specific for the antigen are sensitized. Some of the B cells enlarge and divide and develop into a population of *plasma cells* (Figure 17.9). Plasma cells secrete antibodies. The division and development of B cells into plasma cells are influenced by substances secreted by macrophages and helper T cells. These include small protein hormones called interleukins.

The sensitized B cells that do not differentiate into plasma cells remain as *memory B cells,* ready to respond more rapidly and forcefully should the same antigen appear at a future time.

Antibodies produced by plasma cells travel in lymph and blood to the invasion site, where they may attack antigens in several ways (described shortly).

A summary of the various cells involved in immune responses is presented in Exhibit 17.3.

The immune response is much more rapid and intense after a second or subsequent exposure to an antigen than after the initial exposure. After an initial contact with an antigen, there is a period of several days during which no antibody is present. Then there is a slow rise in the antibody level, first IgM and then IgG, followed by a gradual decline (Figure 17.10). This process is called the *primary response.* Memory cells may remain for decades. Every time the antigen is contacted again, there is an immediate proliferation of memory cells, and the plasma level of antibodies rises to higher levels (mainly IgG antibodies) than were initially produced. This accelerated, more intense response, is called the *secondary response.*

The secondary response provides the basis for immunization against certain diseases. When you receive the initial immunization, your body is sensitized. Should you encounter the pathogen again as an infecting microbe or "booster dose," your body responds by boosting the antibody to a higher level. However, booster doses of an immunizing agent must be given periodically, to maintain high antibody levels adequate for protection against the pathogenic organism.

Actions of Antibodies

The five classes of antibodies have functions that differ somewhat (see Exhibit 17.2), but all attack antigens in several ways:

1. **Neutralizing antigen.** The reaction of antibody with antigen blocks or neutralizes some bacterial toxins and prevents attachment of some viruses to body cells.

Figure 17.9   Role of B cells in antibody-mediated immunity.

🔑 *B cells develop into antibody-producing plasma cells.*

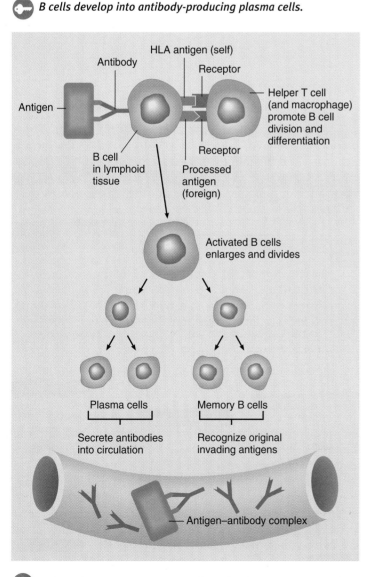

❓ *What substances stimulate B cells to develop into plasma cells?*

Figure 17.10   Secretion of antibodies in the primary (after first exposure) and secondary (after second exposure) responses to the same antigen.

🔑 *The secondary response is the basis for successful immunization by vaccination.*

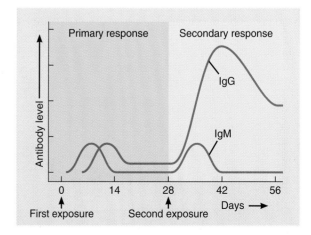

❓ *Which antibody is at a higher level during the secondary response?*

Exhibit 17.3	Summary of Cells that Are Important in Immune Responses
CELL	**FUNCTIONS**
Macrophage	Phagocytosis; processing and presentation of foreign antigens to T cells; secretion of protein growth factors that stimulate T cell growth and division of T cells to form a large population of sensitized T cells.
Helper T cell	Activates cytotoxic T cells, attracts phagocytes, enhances phagocytosis; and stimulates development of B cells into plasma cells.
Cytotoxic (killer) T cell	Releases chemicals that cause cytolysis of target cells and damage target cell DNA; attracts macrophages, enhances their phagocytic activity and prevents them from leaving the site of actions.
Delayed type hypersensitivity T cell	Attracts macrophages in response related to hypersensitivity (allergy) and tissue transplantation.
Suppressor T cell	Inhibits activity of helper T cells and cytotoxic T cells, inhibits phagocytosis, reduces antibody protection by plasma cells, and prevents autoimmune responses.
Memory T cell	Remains in lymphoid tissue and recognizes original invading antigens, even years after infection.
Natural killer (NK) cell	Destroys foreign cells by cytolysis.
B cell	Differentiates into antibody-producing plasma cell.
Plasma cell	Descendant of B cell; the plasma cell that produces antibodies.
Memory B cell	Ready to respond more rapidly and forcefully than initially should the same antigen challenge the body in the future.

2. **Immobilization of bacteria.** Some antibodies may cause bacteria to lose their motility, which limits their spread into nearby tissues.

3. **Agglutination of antigen.** The antigen–antibody reaction may connect pathogens to one another, causing agglutination (clumping together). Phagocytic cells more readily ingest agglutinated microbes.

4. **Activation of complement.** Antigen–antibody complexes activate complement proteins, which then work to remove microbes as previously described.

5. **Enhancing phagocytosis.** Antibodies enhance the activity of phagocytes by causing agglutination and precipitation and by coating microbes so that they are more susceptible to phagocytosis, a process known as *opsonization.*

6. **Providing fetal and newborn immunity.** Resistance of the fetus and newborn baby to infection stems mainly from maternal IgG antibodies that pass across the placenta before birth and IgA antibodies that are absorbed from breast milk after birth.

The Skin and Immunity

The skin not only provides nonspecific resistance to disease, but it is also an active part of the immune system. Langerhans cells in the epidermis assume a role in immunity according to the following mechanism.

If an antigen penetrates the epidermis, it binds to Langerhans cells, which present the antigen to helper T cells, which causes T cells to proliferate. The T cells then enter the lymphatic system and spread throughout the body.

Immunology and Cancer

When a normal cell transforms into a cancer cell, it may display cell surface components called *tumor antigens*. These are molecules that are rarely, if ever, displayed on the surface of normal cells. If the immune system can recognize tumor antigens as nonself, it can destroy the cancer cells carrying them. Such an immune response is called *immunologic surveillance* and is carried out by cytotoxic T cells, macrophages, and natural killer cells. It appears to be most effective in eliminating tumor cells that arise due to a cancer-causing virus.

Despite immunologic surveillance, some cancer cells escape destruction, a phenomenon called *immunologic escape.* One possible explanation is that tumor cells shed their tumor antigens, thus evading recognition. Another theory is that antibodies produced by plasma cells bind to tumor antigens, preventing recognition by cytotoxic T cells.

Immunologic techniques using monoclonal antibodies may be employed in the future to treat cancer. A *monoclonal antibody* (**MAb**) is a pure antibody produced from a single clone of identical cells. It is obtained by joining a specific B cell with a cultured cell that is capable of dividing endlessly. The resulting cell, called a *hybridoma* (hī-bri-DŌ-ma), is a long-term source of pure antibodies.

Clinical uses of monoclonal antibodies include the diagnosis of strep throat, pregnancy, allergies, and diseases such as hepatitis, rabies, and some sexually transmitted diseases. MAbs have also been used to detect cancer at an early stage and to determine the extent of metastasis. They may also be useful in preparing vaccines to counteract the rejection associated with transplants, to treat autoimmune diseases, and perhaps to treat AIDS.

Lifestyle, Immune Function, and Resistance to Disease

*i*f you want to observe the relationship between lifestyle and immune function, visit a college campus. You will probably see that as the semester progresses and the workload accumulates, an increasing number of students can be found in the waiting rooms of student health services.

Is Stress the Culprit . . . ?

Stress has been implicated as hazardous to immune function. Researchers in the field of *psychoneuroimmunology (PNI)* have found many communication pathways that link the nervous, endocrine, and immune systems. Chronic stress affects the immune system in several ways. For example, cortisol, a hormone secreted by the adrenal cortex in association with the stress response, inhibits immune system activity, perhaps one of its energy conservation effects. PNI research appears to justify what many people have observed since the beginning of time: your thoughts, feelings, moods, and beliefs influence your level of health and the course of disease. Especially toxic to the immune system are feelings of helplessness, hopelessness, fear, and social isolation.

People resistant to the negative health effects of stress are more likely to experience a sense of control over the future, a commitment to their work, expectations of generally positive outcomes for themselves, and feelings of social support. To increase your stress resistance, cultivate an optimistic outlook, get involved in your work, and build good relationships with others.

. . . or Lifestyle?

When the work piles up and you're feeling stressed, health habits tend to deteriorate. Just when you need a healthful lifestyle the most to buffer the negative effects of stress, you can hardly find the time to brush your teeth. Shopping for healthful food? Preparing homecooked meals? Working out? Who has the time?!

Sleep on It

Adequate sleep and relaxation are essential for a healthy immune system. But when there aren't enough hours in the day, you may be tempted to steal some from the night. While skipping sleep may give you a few more hours of productive time in the short run, in the long run you end up even farther behind, especially if getting sick keeps you out of commission for several days, blurs your concentration, and blocks your creativity.

Even if you make time to get your eight hours, stress can cause insomnia. If you find yourself mentally writing papers and solving problem sets as you lie in bed trying to fall asleep, it's time to improve your stress management and relaxation skills! Be sure to "change the channel" and unwind from the day before going to bed.

Bad Habits

Midterm and end-of-semester stress tend to fuel nervous habits. Some of these habits decrease our resistance to disease. Smoking interferes with the action of the upper respiratory tract cilia that work to remove infectious agents from the respiratory system. Smokers are much more likely to develop chronic bronchitis from a simple cold than nonsmokers. Like sleep deprivation, drug and alcohol abuse may reduce stress for the moment, but ultimately damage your health.

An Apple a Day?

The folks who recommended an apple a day were on the right track. What you eat can help keep the doctor away by keeping your immune system functioning at optimal levels. Consuming a variety of foods, including plenty of fruits and vegetables, will generally supply the necessary nutritive ingredients.

Resistance Training

Research suggests that people who exercise regularly come down with fewer colds and flus than their sedentary friends. While moderately vigorous exercise appears to be helpful, beware: overtraining can decrease immunity. And don't wait until you are sick to exercise. Exercising while sick can further deplete your already exhausted energy stores.

critical thinking

Have you ever observed a connection between stress and illness in your own life? Do you feel it was caused by stress or other lifestyle factors?

Common Disorders

Acquired Immunodeficiency Syndrome (AIDS)

Never before have humans been confronted with an epidemic in which the primary disease lowers the victim's immunity, and then one or more unrelated diseases produce the fatal symptoms. *Acquired immunodeficiency syndrome* (*AIDS*) is caused by *human immunodeficiency virus* (*HIV*). The initial response to HIV invasion is a modest decline in the number of circulating helper T cells. Infected people experience a brief flu-like illness, with chills and fever, but the immune system fights back by making antibodies against HIV and the number of helper T cells recovers nearly to normal. Although infected people test positive for HIV antibodies, they typically have few clinical signs or symptoms and do not yet have AIDS. Over the next several years, the virus slowly destroys the helper T cell population. As immune responses weaken, people develop certain *indicator diseases* (diseases that are rare in the general population but are common in AIDS patients). At this point, the diagnosis of AIDS is made. A simplified definition of AIDS includes anyone infected with HIV and having a helper T cell count under 200/mm³. (A normal helper T cell count is 1200 mm³.)

Two indicator diseases provided the original clues that a new disorder had appeared. AIDS was first recognized in June 1981 as a result of reports from the Los Angeles area to the Centers for Disease Control and Prevention (CDC) of several cases of a very rare type of pneumonia caused by a fungus. The pneumonia, called *Pneumocystis carinii* (noo-mō-SIS-tis kar-RIN-ē-ī) pneumonia (PCP) occurred among homosexual males. At about the same time, the CDC also received reports from New York and Los Angeles concerning an increase in the incidence of Kaposi's sarcoma (KS) among homosexual males. It had been a rare, generally benign skin cancer usually found in elderly Jewish or Italian men. In AIDS patients it is aggressive and rapidly fatal. KS arises from endothelial cells of blood vessels and produces painless purple or brownish skin lesions that resemble bruises.

In the U.S. AIDS is present in all 50 states. The primary victims are homosexual men, intravenous drug users, and patients who received blood products before 1985, when testing of all donated blood for HIV antibodies began. Others at high risk are the heterosexual partners of HIV-infected individuals. Worldwide, however, an estimated 75 percent of those who have AIDS were infected through heterosexual contacts. At present, the average incubation period (time interval from infection with HIV to full-blown AIDS) is about 10 years. About 5 to 8 percent of those infected with HIV have now survived 13 to 16 years, however, without developing AIDS. It is hoped that studies of these people, termed *nonprogressives,* will reveal the mechanism of their successful resistance to destruction of the immune system.

The World Health Organization (WHO) projects that 40 million people will be infected with HIV throughout the world by the year 2000. More than half will be women and a quarter will be children. The current number of infected Americans is estimated to be 1 to 1½ million, and most do not even know they carry the virus. All carriers of the virus are assumed to be infected for life and are capable of transmitting the virus to others.

HIV: Structure and Infection

Viruses consist of a core of nucleic acid (DNA or RNA) covered by a protein coat. Some viruses also have an envelope (outer layer) composed of a phospholipid bilayer penetrated by proteins (Figure 17.11). Outside a living host cell, a virus has no life functions and is unable to replicate. However, once a virus makes contact with a host cell, the viral nucleic acid enters the host cell. Once inside, the viral nucleic acid uses the host cell's resources to make copies of itself and new protein coats and envelopes. These components then assemble into viruses that eventually leave the infected cell to infect other cells.

As viruses replicate, they damage or kill their host cells. Moreover, the body's own defenses can kill the infected cell as well as the viruses it harbors. Not only can the AIDS virus damage and kill host cells, it can also lie dormant for years. Moreover, it has more complex means of destroying the immune system than suspected.

Infection with HIV begins when the virus binds to a protein receptor on the surface of helper T cells and other susceptible cells. Helper T cells are killed, and as the process progresses, there is a decline in immune functioning. Recall that helper T cells cooperate with B cells to amplify antibody production. The reduction in the number of T cells inhibits antibody production by HIV. Because helper T cells orchestrate a large portion of our immune defense, their destruction leads to a progressive collapse of the immune system and susceptibility to opportunistic infections. *Opportunistic infections* are ones that take advantage of a person's compromised immunity. For about 60 percent of patients, *Pneumocystis carinii* pneumonia (PCP) is the opportunistic infection that marks the progression from HIV infection to AIDS.

Figure 17.11   Human immunodeficiency virus (HIV), the causative agent of AIDS.

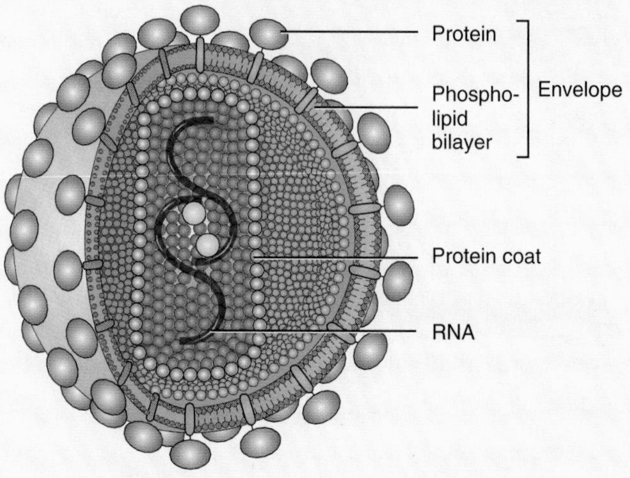

Structure of HIV

Which cells of the immune system are primarily depressed by AIDS?

The AIDS virus also attacks macrophages, brain cells (where it causes AIDS dementia complex), blood vessels, red bone marrow cells, colon cells, and skin cells. Heart and liver cells are also target cells. The virus might kill these cells or multiply in them, thus spreading the virus.

Present research suggests that macrophages may actually be the first cells invaded by the AIDS virus. It seems that once the virus invades macrophages, it then spreads to helper T cells. Macrophages serve as reservoirs for the virus, and even though the viruses multiply within the macrophages, they are not themselves killed. A very important consequence of this is that commonly used screening tests that detect antibodies to the AIDS virus in the blood may be negative even though the person has the AIDS virus hidden in macrophages. While stored in macrophages, the virus does not trigger the production of antibodies that would appear in blood. A test is available to detect the presence of AIDS viruses in macrophages, rather than AIDS antibodies in blood.

After infection with the AIDS virus, the host develops antibodies against several proteins in the virus within 6 weeks to 6 months after exposure. The presence of these antibodies in blood is used as a basis for diagnosing AIDS. Normally, antibodies so developed are protective, but in the case of the AIDS virus, this is not necessarily so because HIV can remain hidden inside body cells.

Symptoms of AIDS

The various stages through which an HIV-infected patient passes are correlated with decreasing helper T cell counts. After infection by HIV, there are usually mild mononucleosis-like symptoms at first that may be overlooked. These include fatigue, fever, swollen glands, and headache.

In the next stage, which lasts for several years, chronically swollen lymph nodes develop in the neck, armpits, and groin. Then, helper T cell counts decline even more, and the patients fail to respond to most skin tests that measure delayed type hypersensitivity, a measure of the individual's ability to produce a cellular immune response against specific proteins injected under the skin. Now, the HIV-infected person has AIDS. Opportunistic infections develop and many patients suffer from AIDS dementia complex, which is characterized by a progressive loss of function in motor activities, cognition, and behavior. Most persons reaching this stage die within two years.

HIV Outside the Body

Outside the body, HIV is fragile and can easily be eliminated. For example, dishwashing and clotheswashing by exposing the virus to 135°F (56°C) for 10 minutes will kill HIV. Chemicals such as hydrogen peroxide (H_2O_2), rubbing alcohol, Lysol, household bleach, and germicidal skin cleaner (such as Betadine and Hibiclens) are very effective, as is standard chlorination in swimming pools and hot tubs.

Transmission

Although HIV has been isolated from several body fluids, the only documented transmissions are by way of breast milk from a nursing mother, blood, semen, and vaginal secretions. The virus is found both in macrophages and free in these fluids.

HIV is transmitted by sexual contact between males and females through vaginal, oral, or anal intercourse. HIV is also effectively transmitted through exchanges of blood, for example, by contaminated hypodermic needles, contact with open wounds, or using the same razor blade for shaving. Infected mothers may transmit the virus to their infants before or during birth. It does not appear that people become infected as a result of routine, nonsexual contacts.

No evidence exists that AIDS can be spread through kissing, although "deep kissing" with exchange of saliva poses a theoretical danger. There is no known case of transmission from a mosquito bite. It also appears that health-care personnel who take proper routine barrier precautions when dealing with body fluids (gloves, masks, safety glasses) are not at risk unless the barriers fail.

Drugs Against HIV

Medical scientists are engaged in what is probably the greatest concentrated effort ever to find a cure for a single disease. One of the problems in treating HIV is that HIV can lie dormant in body cells. In addition, HIV can infect a variety of cells, including those in the central nervous system that are protected by the blood–brain barrier. Added to this is the problem of opportunistic infections, which may be very difficult to treat. And any therapy must overcome the problem that antiviral agents may also harm host cells. Thus scientists are trying to devise strategies for disrupting specific viral activities.

To date, several drugs are used to inhibit HIV replication and slow the progression of AIDS. They prevent HIV from making DNA from RNA, an activity needed for viral replication. The first and still most commonly used drug to treat AIDS is AZT (azidothymidine) or Retrovir. Among patients taking AZT, there is a slowing in the progression of symptoms. The main side effects are red bone marrow damage and anemia. Eventually, the virus develops resistance to the drug. Other drugs are DDI (dideoxyinosine), DDC (dideoxycytidine), D4T (stavudine) or Zerit, saquinavir, ritonavir, and indinavir which may be used in patients who do not respond to AZT or have become resistant to it. Although patients taking drugs show improvement in immunologic functions, for example, increased helper T cell counts, there are serious side effects such as pancreatitis and inflammation of peripheral nerves.

Alpha-interferon is believed to inhibit the final stage of virus production. It reduces the spread of Kaposi's sarcoma and is being tried both alone and in combination with other drugs. A host of other drugs are also being tested. It is quite possible that AIDS, like cancer, will have to be treated with a variety of drugs.

Vaccines Against HIV

Considerable effort has been expended to develop a vaccine against HIV, with little success. A vaccine would stimulate the production of antibodies to block the virus before it could infect body cells. Development of an effective vaccine has been impeded by the ability of HIV to mutate so quickly and the lack of a suitable experimental animal model. Most animals are not susceptible to HIV infection. And although chimpanzees can be infected, they are in short supply and very expensive to maintain. Moreover, there may be a shortage of volunteers when a potential vaccine is ready for testing in humans. The goal of any vaccination is to provide protection to healthy people so they do not succumb to a disease.

Besides HIV, What Else?

Several aspects of HIV continue to puzzle scientists. Why do some people live for several years after infection without symptoms while others rapidly progress to death? Why did it originally spread in the United States mainly by homosexual contacts whereas heterosexual transmission predominates in Africa? Why doesn't the body's immune system wipe out the invader before it causes massive destruction of helper T cells? One possibility is that a second infective agent (a *cofactor*) acts together with HIV to cause the damage. A candidate for such a cofactor is mycoplasma. Classified as bacteria, mycoplasmas are the smallest and simplest organisms that can live without a host. Experiments published in 1991 showed that coinfection of cultured helper T cells with HIV and a mycoplasma enhanced the ability of HIV to cause cell death.

Prevention of Transmission

At present, the only means of preventing AIDS is to block transmission of HIV. Sexual transmission of HIV can be prevented by abstinence from vaginal, oral, and anal intercourse with infected persons. Sexual transmission can be reduced by the use of effective barrier methods (condoms and spermicides such as nonoxynol 9) during intercourse. Infection from donated blood and blood products is now very rare. It is important to note that no one can acquire AIDS by donating blood. In the United States all blood has been tested for HIV antibodies since 1985. HIV transmission via contaminated hypodermic needles could be avoided by sterilization of the needles with bleach before each use. In 1994 researchers found that AZT taken during pregnancy dramatically decreases the risk of transmitting the virus to a fetus. Until there is effective drug therapy or an effective vaccine, blocking the spread of HIV depends on education and safer sexual practices.

Autoimmune Diseases

Under normal conditions, the body's immune mechanism is able to recognize its own tissues and chemicals and it does not produce T cells or B cells against its own substances. Such recognition of self is called *immunologic tolerance.*

At times, however, immunologic tolerance breaks down, and this leads to an *autoimmune disease* (*autoimmunity*). For reasons still not understood, certain tissues undergo changes that cause the immune system to recognize them as foreign antigens and attack them. Among human autoimmune diseases are rheumatoid arthritis (RA), systemic lupus erythematosus (SLE), rheumatic fever, glomerulonephritis, hemolytic and pernicious anemias, Addison's disease, Graves' disease, type I diabetes mellitus, myasthenia gravis, multiple sclerosis (MS), and ulcerative colitis.

Systemic Lupus Erythematosus

Systemic lupus erythematosus (er-e′-thēm-a-TŌ-sus), *SLE,* or *lupus* (*lupus* = wolf) is an autoimmune, noncontagious, inflammatory disease of connective tissue, occurring mostly in young women. In SLE, damage to blood vessel walls results in the release of chemicals that mediate inflammation. The blood vessel damage can be associated with virtually every body system. The disease may be triggered by drugs, such as penicillin, sulfa, or tetracycline, exposure to excessive sunlight, injury, emotional upset, infection, or other stress.

Symptoms of SLE include painful joints, slight fever, fatigue, mouth ulcers, weight loss, enlarged lymph nodes and spleen, photosensitivity, rapid loss of large amounts of scalp hair, and sometimes an eruption across the bridge of the nose and cheeks called a "butterfly rash." The most serious complications of the disease involve inflammation of the kidneys, liver, spleen, lungs, heart, and the central nervous system.

Severe Combined Immunodeficiency (SCID)

Severe combined immunodeficiency (*SCID*) is a rare disease in which both B cells and T cells are missing or inactive in providing immunity. Perhaps the most famous patient with SCID was David, the "bubble boy," who lived in a sterile plastic chamber for all but 15 days of his life; he died on February 22, 1984, at age 12. He was the oldest untreated survivor of SCID.

Hypersensitivity (Allergy)

A person who is overly reactive to an antigen is said to be *hypersensitive* (*allergic*). Whenever an allergic reaction occurs, there is tissue injury. The antigens that induce an allergic reaction are called *allergens.* Almost any substance can be an allergen for some individuals. Common allergens include certain foods (milk, peanuts, shellfish, eggs), antibiotics (penicillin, tetracycline), vitamins (thiamine, folic acid), protein drugs (insulin, ACTH, estradiol), vaccines (pertussis, typhoid), venoms (honeybee, wasp, snake), cosmetics, chemicals in plants such as poison ivy, pollens, dust, molds, iodine-containing dyes used in certain x-ray procedures, and even microbes.

Some allergic reactions, such as hives, eczema, swelling of the lips or tongue, abdominal cramps, and diarrhea, are referred to as *localized* (affecting one part or a limited area). Others are considered *systemic* (affecting several parts or the entire body). An example is acute anaphylaxis (anaphylactic shock), in which the person develops respiratory symptoms (wheezing and shortness of breath) as bronchioles constrict, usually accompanied by cardiovascular failure and collapse due to vasodilation and fluid loss from blood.

Tissue Rejection

Transplantation is the replacement of an injured or diseased tissue or organ. Usually, the body recognizes the proteins in the transplanted tissue or organ as foreign and produces antibodies against them. This is known as *tissue rejection.* Rejection can be somewhat reduced by matching donor and recipient HLA antigens and by administering drugs that inhibit the body's ability to form antibodies. Until recently, *immunosuppressive drugs* suppressed not only the recipient's immune rejection of the donor organ but also the immune response to all antigens as well. This makes patients susceptible to infectious diseases. A drug called *cyclosporine,* derived from a fungus, has largely overcome this problem for kidney, heart, and liver transplants. It has very little effect on B cells.

Hodgkin's Disease (HD)

Hodgkin's disease (*HD*) is a form of cancer, usually arising in lymph nodes. It may arise from a combination of genetic predisposition, disturbance of the immune system, and an infectious agent (Epstein–Barr virus). ∎

Medical Terminology and Conditions

Adenitis (ad′-e-NĪ-tis; *adeno* = gland; *itis* = inflammation of) Enlarged, tender, and inflamed lymph nodes resulting from an infection.

Hypersplenism (hī′-per-SPLĒN-izm; *hyper* = over) Abnormal splenic activity due to splenic enlargement and associated with an increased rate of destruction of normal blood cells.

Lymphadenectomy (lim-fad′-e-NEK-tō-mē; *ectomy* = removal) Removal of a lymph node.

Lymphadenopathy (lim-fad′-e-NOP-a-thē; *patho* = disease) Enlarged, sometimes tender lymph glands.

Lymphangioma (lim-fan′-jē-Ō-ma; *angio* = vessel; *oma* = tumor). A benign tumor of the lymphatic vessels.

Lymphangitis (lim′-fan-JĪ-tis) Inflammation of lymphatic vessels.

Lymphedema (lim′-fe-DĒ-ma; *edema* = swelling) Accumulation of lymph fluid producing subcutaneous tissue swelling.

Lymphoma (lim′-FŌ-ma) Any tumor composed of lymphatic tissue.

Splenomegaly (splē′-nō-MEG-a-lē; *mega* = large) Enlarged spleen.

Study Outline

Functions (p. 383)

1. The lymphatic system consists of lymph, lymphatic vessels, and structures and organs that contain lymphatic tissue (specialized reticular tissue containing large numbers of lymphocytes).
2. The lymphatic-tissue-containing components of the lymphatic system are diffuse lymphatic tissue, lymphatic nodules, and lymphatic organs (lymph nodes, spleen, and thymus gland).
3. The lymphatic system drains tissue spaces of excess fluid and returns proteins that have escaped from blood to the cardiovascular system, transports lipids and some vitamins from the gastrointestinal tract to the blood, and provides immunity.

Lymph and Interstitial Fluid (p. 383)

1. Interstitial fluid and lymph are basically the same. When the fluid bathes body cells, it is called interstitial fluid; when it is found in lymph vessels, it is called lymph.
2. These fluids differ chemically from plasma in that both contain less protein.

Lymphatic Capillaries and Lymphatic Vessels (p. 383)

1. Lymphatic vessels begin as lymphatic capillaries in tissue spaces between cells.
2. Lymphatic capillaries merge to form larger vessels, called lymphatic vessels, which ultimately converge into the thoracic duct or right lymphatic duct.
3. Lymphatic vessels have thinner walls and more valves than veins.

Lymphatic Tissue (p. 383)

1. Lymph nodes are oval structures located along lymphatic vessels.
2. Lymph enters nodes through afferent lymphatic vessels and exits through efferent lymphatic vessels.
3. Lymph passing through the nodes is filtered. Lymph nodes also produce plasma cells and T cells.
4. A tonsil is a group of large lymphatic nodules embedded in mucous membranes. The tonsils are the pharyngeal, palatine, and lingual.
5. The spleen is the single largest mass of lymphatic tissue in the body. It produces B cells, phagocytizes bacteria and worn-out red blood cells, and stores blood. During early fetal life, the spleen produces blood cells.
6. The thymus gland functions in immunity by producing T cells.

Lymph Circulation (p. 387)

1. The passage of lymph is from interstitial fluid, to lymphatic capillaries, to lymphatic vessels, to lymph trunks, to the thoracic duct or right lymphatic duct, to the subclavian veins.
2. Lymph flows as a result of skeletal muscle contractions, respiratory movements, and valves in the lymphatics.

Nonspecific Resistance to Disease (p. 388)

1. The ability to ward off disease using a number of defenses is called resistance. Lack of resistance is called susceptibility.
2. Nonspecific resistance refers to a wide variety of body responses against a wide range of pathogens.
3. Nonspecific resistance includes mechanical factors (skin, mucous membranes, lacrimal apparatus, saliva, mucus, hairs, cilia, epiglottis, flow of urine, defecation, and vomiting), chemical factors (lysozyme, gastric juice, acid pH of skin, vaginal secretions, sebum, and perspiration), antimicrobial substances (interferons and complement), natural killer (NK) cells, phagocytosis, inflammation, and fever.

Immunity (Specific Resistance to Disease) (p. 392)

1. Specific resistance to disease involves the production of a specific lymphocyte or antibody against a specific antigen and is called immunity.
2. Acquired immunity is gained during life as a result of contact with an antigen. It may be obtained actively or passively or naturally or artificially.
3. Antigens (Ags) are chemical substances that, when introduced into the body, stimulate the production of antibodies that react with the antigen.

4. Examples of antigens are microbes (such as bacteria or viruses), pollen, incompatible blood cells, and transplants.

5. Antibodies (Abs) are proteins produced in response to antigens.

6. Based on chemistry and structure, antibodies are distinguished into five principal classes, each with specific biological roles (IgG, IgA, IgM, IgD, and IgE). See Exhibit 17.2.

7. Cell-mediated immunity refers to destruction of antigens by T cells.

8. Antibody-mediated immunity refers to destruction of antigens by antibodies, which are produced by descendants of B cells called plasma cells.

9. T cells and B cells develop from cells in red bone marrow.

10. Macrophages process and present antigens to T cells and B cells and secrete substances that induce production of T cells and B cells.

11. There are five kinds of T cells: helper T cells, which stimulate growth and division of cytotoxic T cells, attract phagocytes, enhance phagocytosis by macrophages, and stimulate development of B cells; cytotoxic (killer) T cells, which destroy antigens on contact by causing cytolysis and DNA damage; delayed hypersensitivity cells recruit macrophages in response to allergic reactions and transplanted tissue;

suppressor T cells inhibit cytotoxic T cell helper T cell activity, inhibit phagocytes, depress antibody production by plasma cells, and prevent autoimmune responses; and memory T cells recognize antigens at a later date.

12. Natural killer (NK) cells are lymphocytes that resemble cytotoxic T cells that kill many infectious microbes and tumor cells by cytolysis. NK cells do not have to interact with lymphocytes or antigens.

13. B cells develop into antibody-producing plasma cells under the influence of thymic hormones and substances secreted by macrophages and T cells. Memory B cells recognize the original, invading antigen.

14. The secondary response provides the basis for immunization against certain diseases.

15. Functionally, antibodies neutralize antigens, immobilize bacteria, agglutinate antigens, activate complement, and enhance phagocytosis.

16. The immunologic properties of the skin are due to keratinocytes and Langerhans cells.

17. Cancer cells contain tumor-specific antigens and are usually destroyed by the body's immune system (immunologic surveillance); some cancer cells escape detection and destruction, a phenomenon called immunologic escape.

Self-Quiz

1. Which of the following is NOT true concerning the lymphatic system?

   a. Lymphatic vessels transport lipids from the gastrointestinal tract to the blood.   b. Lymph is more similar to interstitial fluid than to blood.   c. Lymphatic tissue is found in only a few isolated organs in the body.   d. The unique structure of lymphatic capillaries allows blood to flow into them but not out of them.   e. The lymphatic vessels closely resemble veins in structure.

2. Which of the following best represents lymph flow from the interstitial spaces back to the blood?

   a. lymphatic capillaries ⟶ lymphatic ducts ⟶ lymphatic vessels ⟶ subclavian veins   b. subclavian veins ⟶ lymphatic capillaries ⟶ lymphatic vessels ⟶ lymphatic ducts   c. lymphatic capillaries ⟶ lymphatic vessels ⟶ lymphatic ducts ⟶ subclavian veins   d. lymphatic ducts ⟶ lymphatic vessels ⟶ lymphatic capillaries ⟶ subclavian veins   e. lymphatic capillaries ⟶ lymphatic vessels ⟶ subclavian veins ⟶ lymphatic ducts

3. Which of the following substances is (are) produced by viral-infected cells to protect uninfected cells from viral invasion?

   a. complement   b. prostaglandins   c. fibrin   d. interferons   e. histamine

4. The transfer of antibodies from a pregnant mother to her fetus is a type of

   a. naturally acquired passive immunity   b. artificially acquired active immunity   c. naturally acquired active immunity   d. artificially acquired passive immunity   e. nonspecific resistance

5. Cells involved in attacking and destroying foreign agents such as fungi, parasites, cancer cells, and foreign tissues are

   a. T cells   b. plasma cells   c. B cells   d. natural killer cells   e. viruses

6. Which of the following statements about B cells is (are) true?

   a. They are incapable of leaving lymphatic tissue to directly confront a foreign antigen.   b. Some may develop into plasma cells that secrete antibodies.   c. Nonactivated B cells are called memory B cells.   d. Cytotoxic (killer) B cells travel in lymph and blood to react with the foreign antigens.   e. All of the above are true.

7. The secondary response in antibody-mediated immunity

   a. is characterized by a slow rise in antibody levels and then a gradual decline   b. is an intense response by memory cells to produce antibodies when an antigen is contacted again   c. produces fewer but more responsive antibodies than occur during the primary response   d. occurs when you first receive a vaccination against some disease   e. is rarely seen except in autoimmune disorders

8. Which of the following is NOT true concerning HIV and AIDS?

   a. HIV can lie dormant for years.   b. HIV attacks helper T cells.   c. Initially, HIV may be hidden in macrophages and antibodies may not develop.   d. HIV is easily transmitted by casual contact such as handholding, coughing, and sharing a meal.   e. Most death from AIDS occurs because of secondary infections due to the lack of a fully functioning immune response.

9. The ability of the body's immune system to recognize its own tissues is known as

   a. immunologic escape   b. immunologic tolerance   c. autoimmunity   d. nonspecific resistance   e. hypersensitivity

10. Place the following steps involved in the inflammatory process in the correct order:

    I. Migration of neutrophils

    II. Vasodilation and increased permeability

    III. Formation of pus

    IV. Migration of monocytes

    V. Formation of fibrin network to form a clot

    VI. Transformation of monocytes into wandering macrophages

    II , V , I , IV , VI , III

11. Match the following:

    B a. destroy antigens by cytolysis

    C b. stimulates other cells of the immune system

    D c. involved in allergic reactions and rejection of transplanted tissue

    E d. programmed to recognize the original invading antigen; allow immunity to last for years

    A e. depress parts of the immune response; help prevent autoimmune responses

    G f. function in nonspecific resistance

    F g. develop into plasma cells

    A. suppressor T cells

    B. cytotoxic T cells

    C. helper T cells

    D. delayed hypersensitivity T cells

    E. memory T cells

    F. B cells

    G. Natural killer cells

12. The lymphatic organ(s) that functions in the production of T cells and hormones that promote the maturation of T cells is the

    a. tonsils   b. spleen   c. red bone marrow   d. thymus   e. lymph trunks

13. Nonspecific resistance includes all of the following chemicals and/or cell types EXCEPT

    a. complement   b. immunoglobulin   c. natural killer cells   d. lysozyme   e. interferon

14. Place the phases of phagocytosis in the proper order:

    1. Adherence to foreign material

    2. Chemotaxis of phagocytes

    3. Exocytosis of indigestible materials

    4. Ingestion of foreign material

       a. 1,2,3,4   b. 2,1,3,4   c. 2,1,4,3   d. 1,4,3,2   e. 4,3,2,1

15. Inflammation is NOT characterized by

    a. pain   b. redness   c. swelling   d. putrid   odor   e. heat

16. Humoral immunity involves

    a. antibody production   b. cytotoxic T cell proliferation   c. memory T cell production   d. cheering up the patient (psychoneuroimmunology)   e. production of pus

17. Skin functions in specific immunity by helping to increase T cell numbers following

    a. sebum adhering to a microorganism   b. antigen binding to Langerhans cells   c. UV light stimulation of macrophages   d. lysozyme killing bacteria   e. mucus trapping microbes in the ciliary escalator

18. A skin rash, itching, and swelling following exposure to poison ivy or poison oak is an example of

    a. nonspecific resistance   b. hypersensitivity   c. autoimmunity   d. chemotaxis   e. acute anaphylaxis

Critical Thinking Applications

1. Irene got a splinter in her foot at the beach. She neglected to clean it properly and it became infected. Trace the route of the microbes through the lymphatic system.

2. Bob got a nasty cut in his hand while cleaning the house gutters. Describe his body's response to the invasion of microbes through the broken skin.

3. Before blood screening for HIV was started, many hemophiliacs received blood products contaminated with HIV and contracted AIDS. What is the likelihood of catching HIV from casual, everyday contact with someone with AIDS?

4. Kathryn is miserable every spring. She has itchy, watery eyes and nasal congestion. What is her problem and why is her body reacting this way?

5. Baby Carlos is due for his MMR "shot." The MMR vaccine is an immunization for measles, mumps, and rubella. How does immunization protect children against disease?

6. Following an infection with chickenpox, the virus can lie dormant in spinal nerves for many years. If the virus becomes active again the result is called shingles. Will shingles still be possible when chickenpox is prevented by vaccination? Explain.

Answers to Figure Questions

17.1   Reticular connective tissue that contains large numbers of lymphocytes.

17.2   Interstitial fluid, because the protein content is low.

17.3   Macrophages may phagocytize them, T cells may destroy them by releasing antimicrobial substances, or antibodies produced by plasma cells may destroy them.

17.4   Capillaries.

17.5   Neutrophils and macrophages.

17.6   Redness, increased blood flow due to vasodilation; pain, injury of nerve fibers, irritation by toxins from microbes, kinins, prostaglandins, pressure from edema; heat, increased blood flow, heat release by locally increased metabolic reactions; swelling, leakage of fluid from capillaries due to increased permeability.

17.7   Proteins.

17.8   Processing and presentation of antigens.

17.9   Interleukins.

17.10 IgG.

17.11 Helper T cells.

THE RESPIRATORY SYSTEM

a look ahead

student learning objectives

ells continually use oxygen (O_2) for the metabolic reactions that release energy from nutrient molecules and produce ATP. At the same time, these reactions release carbon dioxide (CO_2). Oxygen consumption and carbon dioxide production occur in mitochondria due to the cellular respiration that occurs there. Because an excessive amount of CO_2 produces acidity that is toxic to cells, the excess CO_2 must be eliminated quickly and efficiently. The two systems that supply oxygen and eliminate carbon dioxide are the cardiovascular system and the respiratory system. They participate equally in respiration. The respiratory system provides for gas exchange, intake of O_2 and elimination of CO_2, whereas the cardiovascular system transports the gases in the blood between the lungs and the cells. Failure of either system has the same effect on the body: disruption of homeostasis and rapid death of cells from oxygen starvation and buildup of waste products. In addition to functioning in gas exchange, the respiratory system also contains receptors for the sense of smell, filters inspired air, and produces sounds.

The exchange of gases between the atmosphere, blood, and cells is called **respiration.** Three basic processes are involved. The first process, **pulmonary** (*pulmo* = lung) **ventilation,** or breathing, is the inspiration (inflow) and expiration (outflow) of air between

the atmosphere and the lungs. The second process, **external (pulmonary) respiration,** is the exchange of gases between the lungs and blood; blood gains O_2 and gives up CO_2. The third process, **internal (tissue) respiration,** is the exchange of gases between the blood and cells; blood gives up O_2 and gains CO_2.

The **respiratory system** consists of the nose, pharynx (throat), larynx (voice box), trachea (windpipe), bronchi, and lungs (Figure 18.1). Structurally, the respiratory system consists of two portions. (1) The term **upper respiratory system** refers to the nose, pharynx, and associated structures. (2) The **lower respiratory system** refers to the larynx, trachea, bronchi, and lungs. Functionally, the respiratory system also consists of two portions. (1) The **conducting portion** consists of a series of interconnecting cavities and tubes—nose, pharynx, larynx, trachea, bronchi, bronchioles, and terminal bronchioles—that conduct air into the lungs. (2) The **respiratory portion** consists of those portions of the respiratory system where the exchange of gases occurs—respiratory bronchioles, alveolar ducts, alveolar sacs, and alveoli.

The branch of medicine that deals with the diagnosis and treatment of diseases of the ears, nose, and throat is called **otorhinolaryngology** (ō′-tō-rī′-nō-lar′-in-GOL-ō-jē; *oto* = ear, *rhino* = nose).

Figure 18.1   Organs of the respiratory system in relation to surrounding structures.

🔑 *The upper respiratory system includes the nose, pharynx, and associated structures. The lower respiratory system includes the larynx, trachea, bronchi, and lungs.*

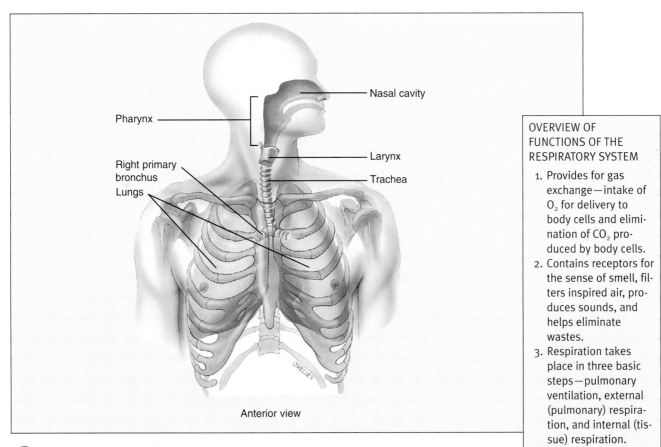

Pharynx

Right primary bronchus

Lungs

Nasal cavity

Larynx

Trachea

Anterior view

OVERVIEW OF FUNCTIONS OF THE RESPIRATORY SYSTEM

1. Provides for gas exchange—intake of O_2 for delivery to body cells and elimination of CO_2 produced by body cells.
2. Contains receptors for the sense of smell, filters inspired air, produces sounds, and helps eliminate wastes.
3. Respiration takes place in three basic steps—pulmonary ventilation, external (pulmonary) respiration, and internal (tissue) respiration.

❓ *Which structures are part of the conducting portion of the respiratory system?*

Organs

Nose

objective: *Describe the structure of the nose and how it functions in breathing.*

Anatomy

The **nose** has a visible external portion and an internal portion inside the skull (Figure 18.2). The external portion consists of bone and cartilage covered with skin and lined with mucous membrane. It has two openings called the **external nares** (NA-rēz; singular is **naris**) or **nostrils.**

The internal portion of the nose is a large cavity in the skull that lies below the cranium and above the mouth. The internal nose is connected to the throat (pharynx) through two openings called the **internal nares.** Four paranasal sinuses (frontal, sphenoidal, maxillary, and ethmoidal) and the nasolacrimal ducts also connect to the internal nose. The cavity inside the external and internal portions of the nose is called the **nasal cavity** and is divided into right and left sides by a partition called the **nasal septum.** The septum consists of the perpendicular plate of the ethmoid bone, vomer, and cartilage (see Figure 6.11).

Functions

The interior structures of the nose are specialized for three basic functions: it (1) warms, moistens, and filters incoming air; (2) receives olfactory stimuli; and (3) provides a resonating chamber for speech sounds.

When air enters the nostrils, it passes through coarse hairs that filter out large dust particles. The air then passes by three shelves formed by the superior, middle, and inferior **nasal conchae** (**turbinates**), which extend out of the wall of the cavity. Mucous membrane lines the cavity and its shelves. The conchae increase the surface area for warming, moistening, and filtering incoming air. Above the superior nasal conchae is the **olfactory epithelium,** which receives olfactory stimuli.

The mucous membrane of the nasal cavity contains pseudostratified ciliated columnar epithelium with many goblet cells and an extensive blood supply in the underlying connective tissue. As the air whirls around the conchae, it is warmed by the capillaries in the connective tissue. Mucus secreted by the goblet cells moistens the air and traps dust particles. The cilia move the mucus and dust packages toward the pharynx, where they can be swallowed or eliminated from the body. Substances in cigarette smoke inhibit movement of cilia. When this happens, only coughing can remove mucus–dust packages from the airways. This is one reason that smokers cough often.

Figure 18.2   Respiratory organs in the head and neck.

🗝 *As air passes through the nose, it is warmed, filtered, and moistened.*

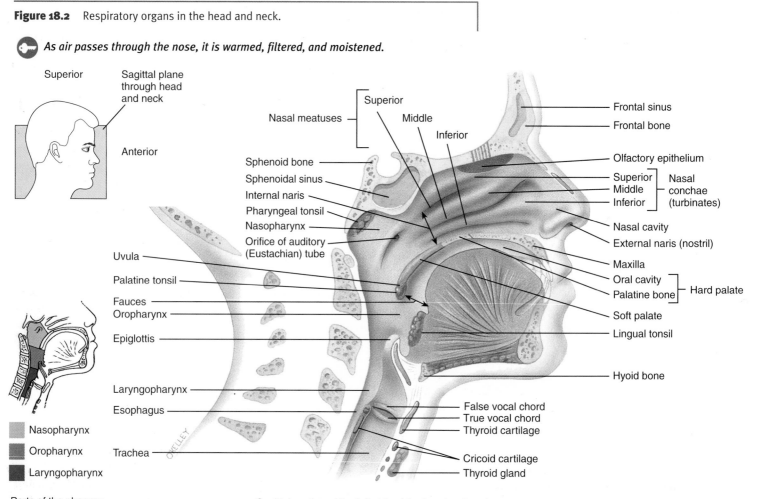

Parts of the pharynx   Sagittal section of the left side of the head and neck

🅠 *What is the path taken by air molecules into and through the nose?*

Pharynx

objective: *Describe the structure of the pharynx and how it functions in breathing.*

The *pharynx* (FAIR-inks), or throat, is a tube that starts at the internal nares and extends part way down the neck (see Figure 18.2). It lies just behind the nasal and oral cavities and just in front of the cervical (neck) vertebrae. Its wall is composed of skeletal muscle and lined with mucous membrane. The pharynx functions as a passageway for air and food and provides a resonating chamber for speech sounds.

The uppermost portion of the pharynx, called the *nasopharynx,* connects with the two internal nares, two openings that lead into the auditory (Eustachian) tubes, and the single opening into the oropharynx. The posterior wall contains the pharyngeal tonsil (adenoid when it becomes inflamed). Through the internal nares, the nasopharynx exchanges air with the nasal cavities and receives dust-laden mucus. It is lined with pseudostratified ciliated columnar epithelium, and the cilia move the mucus down toward the mouth. The nasopharynx also exchanges small amounts of air with the auditory (Eustachian) tubes to equalize air pressure between the pharynx and middle ear.

The middle portion of the pharynx, the *oropharynx,* has an opening from the mouth, called the *fauces* (FAW-sēz; *fauces* = throat), and serves as a common passageway for air, food, and drink. It is lined with nonkeratinized stratified squamous epithelium since it is subject to abrasion from coarse food particles. Two pairs of tonsils, the palatine and lingual, are found in the oropharynx (see also Figure 12.3b).

The lowest portion of the pharynx, the *laryngopharynx* (la-rin′-gō-FAIR-inks), extends downward from the hyoid bone and connects with the esophagus (food tube) in back and the larynx (voice box) in front. It is also lined with nonkeratinized stratified squamous epithelium. Like the oropharynx, the laryngopharynx is both a respiratory and a digestive pathway.

Larynx

objective: *Describe the structure of the larynx (voice box) and explain how it functions in breathing and voice production.*

The *larynx* (LAR-inks), or voice box, is a short passageway that connects the pharynx with the trachea. It lies in the midline of the neck in front of the fourth, fifth, and sixth cervical vertebrae (C4–C6). ∝ C7

Anatomy

The wall of the larynx is composed of nine pieces of cartilage (Figure 18.3). Three are single and three are paired. The three single pieces are the thyroid cartilage, epiglottis, and cricoid cartilage. Of the paired cartilages, the arytenoid cartilages are the most important.

The *thyroid cartilage* (*Adam's apple*) consists of hyaline cartilage and forms the front wall of the larynx and gives it its triangular shape. It is often larger in males than in females due to the influence of male sex hormones during puberty.

The *epiglottis* (*epi* = above; *glotta* = tongue) is a large, leaf-shaped piece of elastic cartilage covered by epithelium lying on top of the larynx (see also Figure 18.2). The "stem" of the epiglottis is attached to the thyroid cartilage, but the "leaf" portion is unattached and free to move up and down like a trap door. During swallowing, the larynx elevates, causing the epiglottis to cover the glottis, like a lid, closing it off. The *glottis* consists of the true vocal cords (discussed shortly) and the space between them. The epiglottis closes off the larynx so liquids and foods are routed into the esophagus, part of the digestive system, and kept out of the larynx and air passageways below it. When anything but air passes into the larynx, a cough reflex attempts to expel the material.

The *cricoid* (KRĪ-koyd) *cartilage* is a ring of hyaline cartilage attached to the first ring of cartilage of the trachea.

The paired *arytenoid* (ar′-i-TĒ-noyd) *cartilages* are mostly hyaline cartilage located above the cricoid cartilage. They attach to the true vocal cords and pharyngeal muscles and function in voice production.

Voice Production

The mucous membrane of the larynx forms two pairs of folds: an upper pair called the *false vocal cords* and a lower pair called the *true vocal cords* (see Figure 18.2). The false vocal cords function to hold the breath against pressure in the thoracic cavity such as might occur when a person strains to lift a heavy object. They do not produce sound.

It is the true vocal cords that produce sound. They contain elastic ligaments stretched between pieces of rigid cartilage like the strings on a guitar. Muscles attach to both the cartilage and to the true vocal cords themselves. When the muscles contract, they pull the elastic ligaments tight, which moves the vocal cords out into the air passageway. The air pushed against the vocal cords causes them to vibrate and sets up sound waves in the air in the pharynx, nose, and mouth. The greater the pressure of air, the louder the sound.

Pitch is controlled by the tension of the true vocal cords. If they are pulled taut, they vibrate more rapidly and a higher pitch results. Lower sounds are produced by decreasing the muscular tension. Due to the influence of male sex hormones, vocal cords are usually thicker and longer in males than in females. They therefore vibrate more slowly, giving men a generally lower range of pitch than women.

Laryngitis is an inflammation of the larynx, most often caused by a respiratory infection or irritants and excessive shouting or coughing and results in hoarseness or loss of voice. Inflammation interferes with the contraction of the cords or causes them to swell to the point where they cannot vibrate freely. Long-term smokers often acquire a permanent hoarseness from the damage done by chronic inflammation.

Trachea

objective: *Describe the structure and functions of the trachea and bronchi.*

The *trachea* (TRĀ-kē-a), or windpipe, is a passageway for air that is located anterior to the esophagus. It extends from the larynx to the upper part of the fifth thoracic vertebra (T5), where it divides into right and left primary bronchi (see Figure 18.1).

The wall of the trachea is lined with mucous membrane and is supported by cartilage. The mucous membrane is

Figure 18.3　Larynx. The arrow in the inset in (c) indicates the direction from which the vocal cords are viewed (superior).

 The larynx is composed of nine pieces of cartilage.

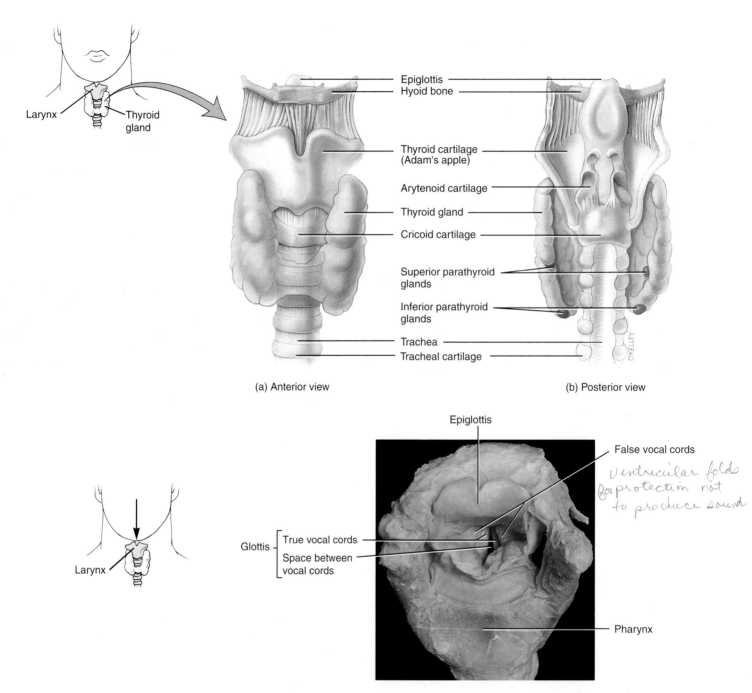

(a) Anterior view

(b) Posterior view

(c) Superior view

How does the epiglottis prevent foods and liquids from entering the larynx?

pseudostratified ciliated columnar epithelium, consisting of ciliated columnar cells, goblet cells, and basal cells (Figure 18.4), and provides the same protection against dust as the membrane lining the nasal cavity and larynx. Whereas the cilia in the upper respiratory tract move mucus and trapped particles *down* toward the pharynx, the cilia in the lower respiratory tract move mucus and trapped particles *up* toward the pharynx. The cartilage layer consists of about 16 to 20 incomplete rings of hyaline cartilage that look like letter "C's" stacked one on top of another. The open part of each **C**-shaped cartilage ring faces the esophagus and permits it to expand slightly into the trachea during swallowing. The solid parts of the C-shaped cartilage rings provide a rigid support so the tracheal wall does not collapse inward and obstruct the air passageway. The rings of cartilage may be felt under the skin below the larynx.

Respiratory passageways may become obstructed in several ways. The rings of cartilage may be accidentally crushed; the mucous membrane may become inflamed and swell so much that it closes off the passageway; inflamed membranes secrete a great deal of mucus that may clog the lower respiratory passages; or a large object may be breathed in (aspirated). The passageways must be cleared quickly. If the obstruction is above the level of the chest, a *tracheostomy* (trā-kē-OS-tō-mē) may be performed, in which an incision is made in the trachea below the cricoid cartilage and a tracheal tube is inserted to create an emergency air passageway. Another method is *intubation.* A tube is inserted into the mouth or nose and passed down through the larynx and trachea. The tube pushes back any flexible obstruction and provides a passageway for air. If mucus is clogging the trachea, it can be suctioned out through the tube.

Bronchi

at the carina

The trachea divides into a *right primary bronchus* (BRONkus; *bronchos* = windpipe), which goes to the right lung, and a *left primary bronchus,* which goes to the left lung (Figure 18.5). The right primary bronchus is more vertical, shorter, and wider than the left. As a result, foreign objects in the air passageways are more likely to enter and lodge in the right

than the left. Like the trachea, the primary bronchi (BRONGkē) contain incomplete rings of cartilage and are lined by pseudostratified ciliated columnar epithelium.

On entering the lungs, the primary bronchi divide to form smaller bronchi—the *secondary bronchi,* one for each lobe of the lung (the right lung has three lobes; the left lung has two). The secondary bronchi continue to branch, forming still smaller bronchi, called *tertiary bronchi,* that divide into *bronchioles.* Bronchioles, in turn, branch into even smaller tubes called *terminal bronchioles.* This continuous branching from the trachea resembles a tree trunk with its branches and is commonly referred to as the *bronchial tree.*

As the branching becomes more extensive in the bronchial tree, several structural changes occur. First, rings of cartilage are replaced by strips of cartilage that finally disappear in the bronchioles. Second, as the cartilage decreases, the amount of smooth muscle increases. Smooth muscle encircles the lumen in spiral bands and its contraction is affected by both the autonomic nervous system and various chemicals. Third, the mucous membrane changes from pseudostratified ciliated columnar epithelium to simple cuboidal (nonciliated) epithelium (and later simple squamous epithelium) in the terminal bronchioles.

The parasympathetic division of the autonomic nervous system and mediators of allergic reactions, such as histamine, cause bronchiole constriction, while the sympathetic division and epinephrine cause bronchiole dilation. The fact that bronchiole walls contain smooth muscle and no cartilage is clinically significant. During an *asthma attack* the muscles go into spasm. Because there is no supporting cartilage, the spasms can

---

Figure 18.4   Histology of the trachea.

The trachea is anterior to the esophagus and extends from the larynx to the top border of the fifth thoracic vertebra.

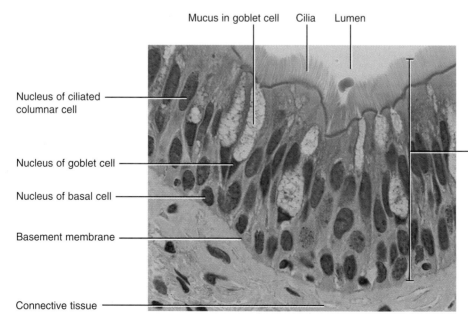

Mucus in goblet cell   Cilia   Lumen

Nucleus of ciliated columnar cell

Nucleus of goblet cell

Nucleus of basal cell

Basement membrane

Connective tissue

Epithelium (pseudostratified ciliated columnar epithelium)

Photomicrograph of tracheal epithelium (850x)

What is the benefit of not having cartilage between the trachea and esophagus?

Figure 18.5   Air passageways to the lungs. Note the bronchial tree.

The bronchial tree begins at the trachea and ends at the terminal bronchioles.

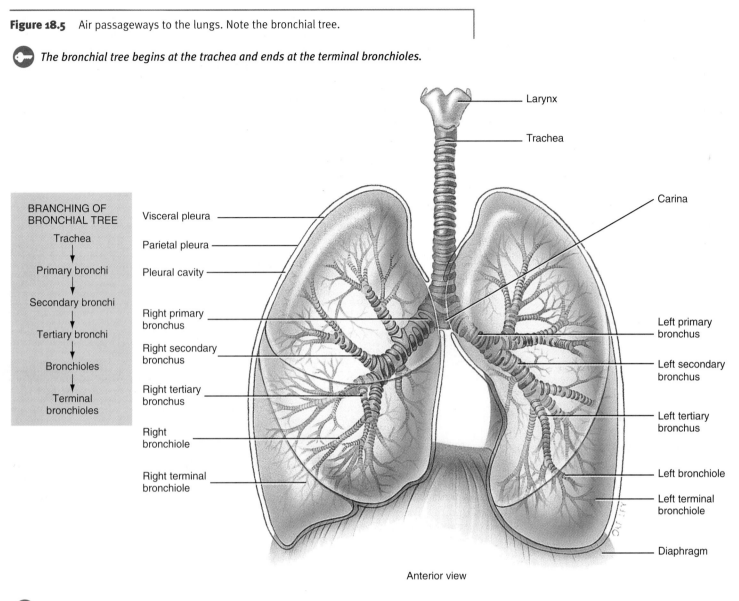

Anterior view

How many lobes and secondary bronchi are present in each lung?

close off the air passageways. Movement of air through constricted bronchial tubes causes breathing to sound louder.

Bronchoscopy is the visual examination of the bronchi through a *bronchoscope,* an illuminated, tubular instrument that is passed through the trachea into the bronchi.

Lungs

objective: *Describe the structure of the lungs and their role in breathing.*

The *lungs* (*lunge* = lightweight, because the lungs float) are paired, cone-shaped organs lying in the thoracic cavity. They are separated from each other by the heart and other structures in the mediastinum (see Figure 15.1). The *pleural membrane* is a double-layered serous membrane that encloses and protects each lung (Figure 18.5). The outer layer is attached to the wall of the thoracic cavity and diaphragm and is called the *parietal pleura.* The inner layer, the *visceral pleura,* covers the lungs themselves. Between the visceral and parietal

pleurae is a small space, the *pleural cavity,* which contains a lubricating fluid secreted by the membranes. This fluid reduces friction between the membranes and allows them to move easily during breathing.

Pleurisy, or inflammation of the pleural membrane, causes friction during breathing that can be quite painful when the swollen membranes rub against each other.

Gross Anatomy

The lungs extend from the diaphragm to just slightly above the clavicles and lie against the ribs. The broad bottom portion of the lung is the *base* (Figure 18.6). The narrow top portion of the lung is the *apex.* The *hilus* is an area on the medial side through which bronchi, pulmonary vessels, lymphatic vessels, and nerves enter and exit. The left lung has a concavity, the *cardiac notch,* in which the heart lies.

The right lung is thicker and broader than the left. It is also somewhat shorter because the diaphragm is higher on the right side to allow for the size of the liver that lies below it.

Lobes and Fissures

Each lung is divided into lobes by one or more deep grooves called fissures. The left lung has one fissure (oblique) and two lobes; the right lung has two fissures (oblique and horizontal) and three lobes (Figure 18.6). Each lobe receives its own secondary bronchus. Thus the right primary bronchus branches into three secondary bronchi called the *superior, middle,* and *inferior secondary bronchi.* The left primary bronchus branches into a *superior* and an *inferior secondary bronchus.*

Lobules

Each lobe of the lungs is divided into regions called *bronchopulmonary segments,* each supplied by its own tertiary bronchus. The segments, in turn, are divided into many small compartments called *lobules* (Figure 18.7). Each lobule contains a lymphatic vessel, an arteriole, a venule, and a branch from a terminal bronchiole wrapped in elastic connective tissue. Terminal bronchioles subdivide into microscopic branches called *respiratory bronchioles.* Respiratory bronchioles, in turn, subdivide into several (2–11) *alveolar ducts.* From the trachea to the alveolar ducts, there are about 25 levels of branching of the respiratory passageways. That is, the trachea divides into primary bronchi (first level), the primary bronchi divide into secondary bronchi (second level), and so on.

Around the circumference of alveolar ducts are numerous alveoli and alveolar sacs. An *alveolus* (al-VĒ-ō-lus) is a cup-shaped projection lined with epithelium. *Alveolar sacs* are two or more alveoli that share a common opening (Figure 18.7). The wall of an alveolus consists of two types of epithelial cells (Figure 18.8 on page 415): (1) *Squamous pulmonary epithelial cells* consist of simple squamous epithelium and are the predominant cells that form a continuous lining of the alveolar wall and under which is a basement membrane. Through these walls gaseous exchange takes place. (2) *Septal cells* are rounded or cuboidal epithelial cells that produce a phospholipid substance called *surfactant* (sur-FAK-tant), which is described shortly. Also associated with the wall of an alveolus are *alveolar macrophages,* phagocytic cells that remove dust particles and debris from alveolar spaces. Around an alveolus are a pulmonary arteriole and venule that form a capillary network. The capillaries consist of a single layer of endothelial cells and a basement membrane.

Alveolar–Capillary (Respiratory) Membrane

The exchange of respiratory gases between the lungs and blood takes place by diffusion across the alveolar and capillary walls. The respiratory gases move across what is termed the *alveolar–capillary (respiratory) membrane* (Figure 18.8b). It consists of the wall of the alveolus, a basement membrane underneath the alveolar wall, a capillary basement membrane that is often fused to the alveolar basement membrane, and endothelial cells of the capillary. Despite having several layers, the alveolar–capillary membrane averages only 0.5 μm* in thickness, which allows efficient diffusion. Moreover, it is estimated that the lungs contain 300 million alveoli, providing an immense surface area (70 m^2 or 750 ft^2, about the area of a handball court) for the exchange of gases.

Blood Supply

There is a double blood supply to the lungs. Deoxygenated blood from the heart passes through the pulmonary trunk, which divides into a left pulmonary artery that enters the left

---

*1 μm (micrometer) = 1/25,000 of an inch or 1/1,000,000 of a meter.

Figure 18.6   Lungs. The arrows in the inset indicate the direction from which the lungs are viewed (lateral).

The subdivisions of the lungs are lobes → bronchopulmonary segments → lobules.

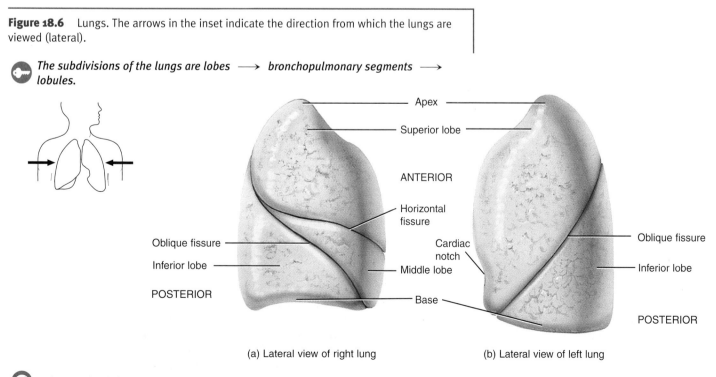

(a) Lateral view of right lung   (b) Lateral view of left lung

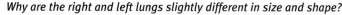

Q *Why are the right and left lungs slightly different in size and shape?*

Figure 18.7   Lobule of the lung.

🔑 *Alveolar sacs are two or more alveoli that share a common opening.*

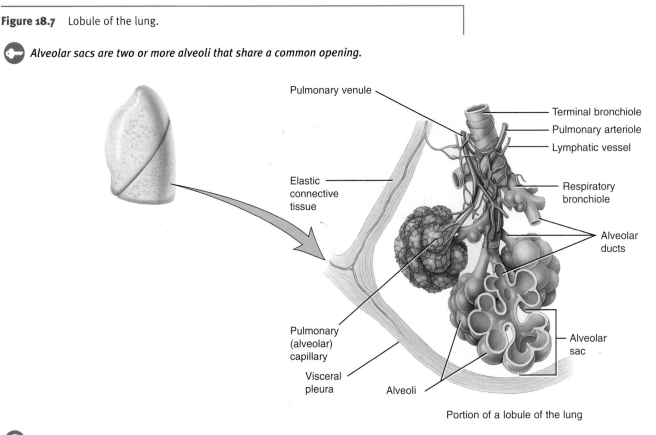

Pulmonary venule

Terminal bronchiole
Pulmonary arteriole
Lymphatic vessel

Elastic connective tissue

Respiratory bronchiole

Alveolar ducts

Pulmonary (alveolar) capillary

Alveolar sac

Visceral pleura   Alveoli

Portion of a lobule of the lung

❓ *What are the major parts of a lobule of a lung?*

lung and a right pulmonary artery that enters the right lung. The return of the oxygenated blood is by way of the pulmonary veins, which drain into the left atrium (see Figure 16.9).

Oxygenated blood for the lungs' own tissues is delivered through bronchial arteries, branches of the aorta. Venous blood is drained from the tissues of the lungs by the bronchial veins.

Respiration

The principal purpose of *respiration* is to supply the cells of the body with oxygen and remove the carbon dioxide produced by cellular activities. The three basic processes of respiration are pulmonary ventilation, external (pulmonary) respiration, and internal (tissue) respiration.

Pulmonary Ventilation

objective: *Explain how inspiration (breathing in) and expiration (breathing out) take place.*

Pulmonary ventilation (**breathing**) is the process by which gases are exchanged between the atmosphere and lung alveoli. Air flows between the atmosphere and lungs because a pressure difference exists between them. We breathe in when the pressure inside the lungs is less than the air pressure in the atmosphere. We breathe out when the pressure inside the lungs is greater than the pressure in the atmosphere. Pulmonary ventilation involves two phases: inspiration and expiration.

Inspiration

Breathing in is called *inspiration* (*inhalation*). Just before each inspiration, the air pressure inside the lungs equals the pressure of the atmosphere, which is about 760 mm Hg (millimeters of mercury), or 1 atmosphere (atm), at sea level. For air to flow into the lungs, the pressure inside the lungs must become lower than the pressure in the atmosphere. This is achieved by increasing the volume (size) of the lungs.

The pressure of a volume of gas is created by the number of molecules of a gas present and the size of the area they occupy. If the size of the container is decreased, then the pressure inside the container increases. If the size of a container is increased, the pressure of the air inside it decreases. This may be demonstrated by placing a gas in a container that has a movable piston (Figure 18.9). The initial pressure is created by the gas molecules striking the wall of the container. If the piston is pushed down, the gas is concentrated in a smaller volume. This means that the same number of gas molecules is striking less wall space. The gauge shows that the pressure doubles as the gas is compressed to half its volume. In other words, the same number of molecules in half the space produces twice the pressure. If the piston is raised to increase the volume, the pressure decreases.

In order for inspiration to occur, the lungs' size must be expanded. This increases lung volume and thus decreases the pressure in the lungs. The first step in increasing lung volume is contraction of the principal inspiratory muscles—the diaphragm and external intercostals (Figure 18.10a on page 416). The diaphragm is a dome-shaped skeletal muscle that forms the floor of the thoracic cavity. Contraction of the diaphragm means it

Figure 18.8   Structure of an alveolus.

The exchange of respiratory gases occurs by diffusion across the alveolar–capillary (respiratory) membrane.

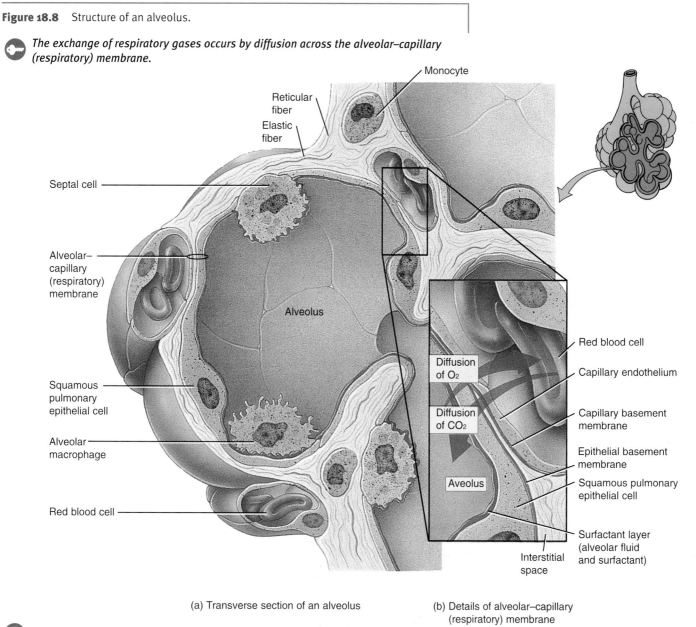

(a) Transverse section of an alveolus

(b) Details of alveolar–capillary (respiratory) membrane

What is the function of septal cells?

flattens (it normally curves up), which increases the size of the thoracic cavity from top to bottom. This movement accounts for about 75 percent of the air that enters the lungs during inspiration. At the same time the diaphragm contracts, the external intercostals contract. This pulls the ribs up and moves the sternum forward, increasing the size of the thoracic cavity from front to back (Figure 18.10b). Advanced pregnancy, obesity, confining clothing, or eating a large meal can prevent a complete descent of the diaphragm and may cause shortness of breath.

The term applied to normal quiet breathing is *eupnea* (yoop-NĒ-a; *eu* = normal). During deep, labored inspiration, accessory muscles of inspiration (sternocleidomastoid, scalenes, and pectoralis minors) also participate to increase the size of the thoracic cavity (Figure 18.10a). Inspiration is an active process because it requires contraction of one or more muscles. Following expansion of the thoracic cavity, expansion of the lungs is aided by movement of the pleural membrane. Normally, the parietal and visceral pleurae are strongly attached to each

Figure 18.9   The volume of gas varies inversely with the pressure.

The pressure of a gas is due to the number of molecules of the gas present and the size of the area they occupy.

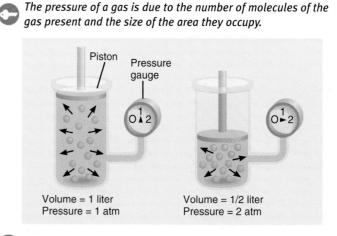

Volume = 1 liter
Pressure = 1 atm

Volume = 1/2 liter
Pressure = 2 atm

If the volume is decreased to ¼ liter, how does the pressure change?

Figure 18.10   Pulmonary ventilation: muscles of inspiration and expiration.

During deep, labored inspiration, accessory muscles of inspiration (sternocleidomas-toids, scalenes, and pectoralis minors) participate.

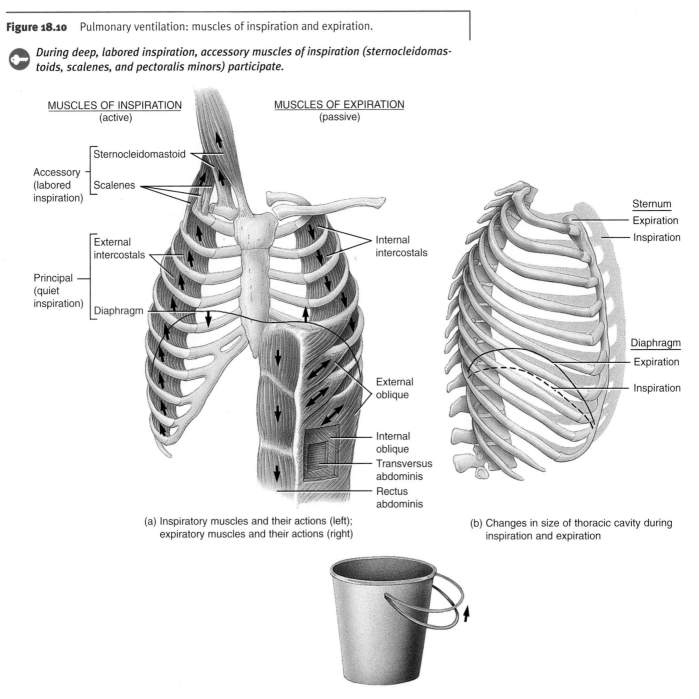

MUSCLES OF INSPIRATION
(active)

MUSCLES OF EXPIRATION
(passive)

Sternocleidomastoid

Accessory
(labored
inspiration)

Scalenes

Internal
intercostals

External
intercostals

Principal
(quiet
inspiration)

Sternum
Expiration
Inspiration

Diaphragm

External
oblique

Diaphragm
Expiration
Inspiration

Internal
oblique
Transversus
abdominis
Rectus
abdominis

(a) Inspiratory muscles and their actions (left);
expiratory muscles and their actions (right)

(b) Changes in size of thoracic cavity during
inspiration and expiration

(c) During inspiration, the ribs move upward
and outward like the handle on a bucket

Right now, what muscle is mainly responsible for your breathing?

other due to surface tension created by their moist adjoining sur-faces. As the diaphragm and external intercostals contract and the thoracic cavity expands, the parietal pleura lining the cavity is pulled in all directions, and the visceral pleura is pulled along with it. This increases the volume of the lungs.

When the volume of the lungs increases, the pressure inside the lungs, called the *alveolar pressure*, drops from 760 to 758 mm Hg. A pressure difference is thus established between the atmosphere and the alveoli. Air rushes from the atmosphere into the lungs, and an inspiration takes place. Air continues to move into the lungs until alveolar pressure equals atmospheric pressure.

A summary of inspiration is presented in Figure 18.11.

Expiration

Breathing out, called *expiration* (*exhalation*), is also achieved by a pressure difference, but in this case the difference is reversed so that the pressure in the lungs is greater than the pressure of the atmosphere. Normal expiration, unlike inspira-tion, is a passive process because no muscular contractions are involved. It depends partly on the elasticity of the lungs. Expiration starts when the inspiratory muscles relax. As the external intercostals relax, the ribs move downward, and as the diaphragm relaxes, it curves upward (Figure 18.10a). These movements decrease the size of the thoracic cavity.

During expiration, as the size of the thoracic cavity decreases, elastic fibers in the lungs that were stretched by inspiration now recoil, and lung volume decreases. Alveolar pressure increases to 762 mm Hg, and air moves from the area of higher pressure in the alveoli to the area of lower pressure in the atmosphere.

Expiration becomes active during higher levels of ventilation and when air movement out of the lungs is inhibited. During these times, muscles of expiration—abdominal and internal intercostals—contract to move the lower ribs downward and compress the abdominal viscera, thus forcing the diaphragm upward (Figure 18.10a).

A summary of expiration is presented in Figure 18.11.

Collapsed Lung

Intrapleural pressure, the pressure between the visceral and parietal pleurae, is normally subatmospheric, that is, below atmospheric pressure (760 mm Hg). Maintenance of a low intrapleural pressure is vital to the functioning of the lungs because it keeps the alveoli slightly inflated. The alveoli are so elastic that at the end of an expiration they attempt to recoil inward and collapse on themselves like the walls of a deflated balloon. A collapsed lung is prevented by the slightly lower pressure in the pleural cavities. A collapsed lung may be caused by air entering the pleural cavities from a surgical incision or chest wound, an airway obstruction, or lack of surfactant.

Surfactant is a phospholipid produced by the septal cells of the walls of alveoli that coats the alveoli and prevents them from sticking together following expiration. This allows the alveoli to

---

Figure 18.11 Pulmonary ventilation: pressure changes. At the beginning of inspiration, the diaphragm contracts, the chest expands, the lungs are pulled outward, and alveolar pressure decreases. As the diaphragm relaxes, the lungs recoil inward. Alveolar pressure rises, forcing air out until alveolar pressure equals atmospheric pressure.

 Air moves into the lungs when alveolar pressure is less than atmospheric pressure and out of the lungs when alveolar pressure is greater than atmospheric pressure.

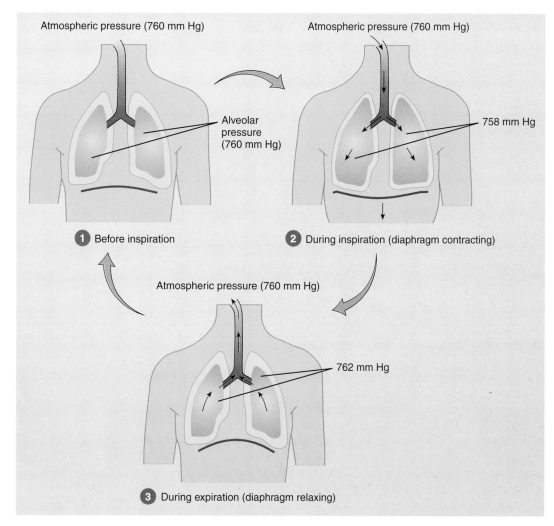

Atmospheric pressure (760 mm Hg)

Atmospheric pressure (760 mm Hg)

Alveolar pressure (760 mm Hg)

758 mm Hg

1 Before inspiration

2 During inspiration (diaphragm contracting)

Atmospheric pressure (760 mm Hg)

762 mm Hg

3 During expiration (diaphragm relaxing)

What is alveolar pressure during inspiration? Expiration?

be reinflated during the next inspiration and prevents alveolar collapse. The presence of surfactant and elastic fibers in the lungs provides them with a characteristic referred to as **compliance** (kom-PLĪ-ans), the ability to expand or inflate.

Pulmonary Air Volumes and Capacities

In clinical practice, the word **respiration** (**ventilation**) means one inspiration plus one expiration. The healthy adult averages about 12 respirations a minute while at rest. During each respiration, the lungs exchange various amounts of air with the atmosphere. A lower-than-normal amount of exchange is usually a sign of pulmonary malfunction.

Pulmonary Volumes

The record of pulmonary volumes and capacities is called a **spirogram** (Figure 18.12). Following are examples of average adult male pulmonary volumes. During the process of normal quiet breathing, about 500 ml of air move into the respiratory passageways with each inspiration. The same amount moves out with each expiration. This volume of air inspired (or expired) is called **tidal volume** (**TV**). Only about 350 ml (70 percent) of the tidal volume actually reach the respiratory portion of the respiratory system (respiratory bronchioles, alveolar ducts, alveolar sacs, and alveoli). The other 150 ml (30 percent) remain in air spaces of the conducting portion of the respiratory system (nose, pharynx, larynx, trachea, bronchi,

and bronchioles). The spaces are known as the **anatomic dead space.** The total air taken in during 1 minute is called the **minute volume of respiration** (**MVR**). It is calculated by multiplying the tidal volume by the normal breathing rate per minute. An average volume would be 500 ml times 12 respirations per minute, or 6000 ml/min.

By taking a very deep breath, we can inspire a good deal more than 500 ml. This excess inhaled air, called the **inspiratory reserve volume** (**IRV**), averages 3100 ml above the 500 ml of tidal volume. Thus the respiratory system can pull in 3600 ml of air.

If we inhale normally and then exhale as forcibly as possible, we should be able to push out 1200 ml of air in addition to the 500-ml tidal volume. These extra 1200 ml are called the **expiratory reserve volume** (**ERV**).

Even after the expiratory reserve volume is expelled, a good deal of air remains in the lungs because the lower intrapleural pressure keeps the alveoli slightly inflated and some air also remains in the noncollapsible air passageways. This air, the **residual volume** (**RV**), amounts to about 1200 ml.

Opening the thoracic cavity allows the intrapleural pressure to equal the atmospheric pressure, forcing out some of the residual volume. The air still remaining is called the **minimal volume.** Minimal volume provides a medical and legal tool for determining whether a baby was born dead or died after birth. The presence of minimal volume can be demonstrated by placing a piece of lung in water and watching it float. Fetal lungs contain no air, and so the lung of a stillborn will not float in water.

Figure 18.12   Spirogram of pulmonary volumes and capacities.

🔑 *Lung capacities are combinations of various lung volumes.*

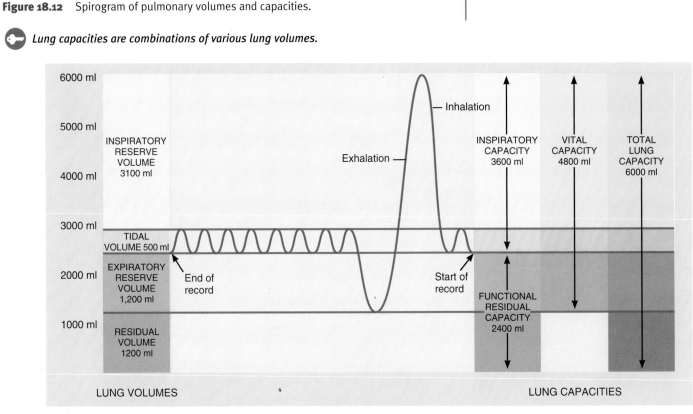

🔍 *Breathe in as deeply as possible and then exhale as much air as you can. Which pulmonary capacity have you demonstrated?*

Pulmonary Capacities

Lung capacity can be calculated by combining various lung volumes (Figure 18.12). Following are examples of average adult male pulmonary capacities. *Inspiratory capacity* (*IC*) is the total inspiratory ability of the lungs following a normal tidal volume expiration. It is the sum of tidal volume plus inspiratory reserve volume (3600 ml). *Functional residual capacity* (*FRC*) is the volume of air remaining in the lungs after a normal tidal volume expiration. It is the sum of residual volume plus expiratory reserve volume (2400 ml). *Vital capacity* (*VC*) is the maximum amount of air that can be expired following maximum inspiration. It is the sum of tidal volume, inspiratory reserve volume, and expiratory reserve volume (4800 ml). Finally, *total lung capacity* (*TLC*) is the maximum amount of air in the lungs following maximum inspiration and equals the sum of all volumes (6000 ml).

A summary of pulmonary volumes and capacities is given in Exhibit 18.1.

Exchange of Respiratory Gases

objective: *Explain how oxygen and carbon dioxide are exchanged between the lungs and blood and between blood and body cells.*

To understand how respiratory gases (oxygen and carbon dioxide) are exchanged in the body, you need to know something about pressures of gases.

Each gas in a mixture of gases has its own pressure and behaves as if no other gases are present. This *partial pressure* is denoted as *p*. The total pressure of the mixture is calculated by adding all the partial pressures.

Atmospheric air is a mixture of several gases—oxygen, carbon dioxide, nitrogen, water vapor, and a number of other gases that appear in such small quantities that we ignore them. Atmospheric pressure is the sum of the partial pressures of all these gases:

$$\text{Atmospheric pressure (760 mm Hg)} = pO_2 + pCO_2 + pN_2 + pH_2O$$

We can determine the partial pressure of each gas in the mixture by multiplying the percentage of the gas in the mixture by the total pressure of the mixture. For example, because oxygen comprises 21 percent of atmospheric air, this would be our calculation:

$$\text{Atmospheric } pO_2 = 21\% \times 760 \text{ Hg} = 160 \text{ mm Hg}$$

The partial pressures of the respiratory gases in the atmosphere, alveoli, blood, and tissue cells are shown in Exhibit 18.2. Partial pressures are important in determining the movement of oxygen and carbon dioxide in the body.

External (Pulmonary) Respiration

External (*pulmonary*) *respiration* is the movement of oxygen and carbon dioxide between the alveoli of the lungs and pulmonary capillaries across the alveolar–capillary (respiratory) membrane (Figure 18.13a). This exchange converts *deoxygenated blood* (depleted of some O_2) into *oxygenated*

Exhibit 18.1	Summary of Pulmonary Volumes and Capacities	
VOLUME OR CAPACITY	VALUE	DEFINITION
Pulmonary volumes		
Tidal volume (TV)[a]	500 ml	Amount of air inhaled or exhaled with each breath during normal quiet breathing.
Inspiratory reserve volume (IRV)	3100 ml	Amount of air that can be forcefully inhaled over and above tidal volume.
Expiratory reserve volume (ERV)	1200 ml	Amount of air that can be forcefully exhaled over and above tidal volume.
Residual volume (RV)	1200 ml	Amount of air remaining in the lungs after a forced exhalation.
Pulmonary capacities		
Inspiratory capacity (IC)	3600 ml	Maximum inspiratory capacity of the lungs following a normal tidal volume expiration (IC = TV + IRV).
Functional residual capacity (FRC)	2400 ml	Volume of air remaining in lungs after a normal tidal volume expiration (FRC = ERV + RV).
Vital capacity (VC)	4800 ml	Maximum amount of air that can be expired following maximum inspiration (VC = TV + IRV + ERV).
Total lung capacity (TLC)	6000 ml	Maximum amount of air in lungs following maximum inspiration (TLC = TV + IRV + ERV + RV).

[a] Only 350 ml of tidal volume reach the alveoli; 150 ml remain in *anatomic dead space*.

Exhibit 18.2	Partial Pressures (mm Hg) of Respiratory Gases in Atmospheric Air, Alveolar Air, Blood, and Tissue Cells				
	ATMO-SPHERIC AIR (SEA LEVEL)	ALVEOLAR AIR	DEOXY-GENATED BLOOD	OXY-GENATED BLOOD	TISSUE CELLS
pO_2	160	105	40	105	40
pCO_2	0.3	40	45	40	45

blood (saturated with O_2). When blood circulates through the pulmonary capillaries, it receives O_2 from the alveoli and loses CO_2 to the alveoli. Although this is referred to as an exchange of gases, in fact each gas diffuses from the area where its partial pressure is higher to the area where its partial pressure is lower. Each gas moves independently of the other.

Figure 18.13   Changes in partial pressures (in mm Hg) during external and internal respiration.

Gases diffuse from areas of higher partial pressure to areas of lower partial pressure.

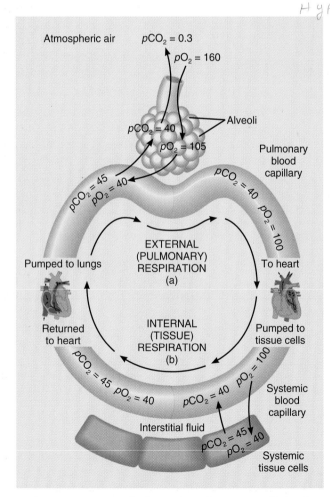

As Figure 18.13a shows, O_2 diffuses from the alveoli, where its partial pressure is 105 mm Hg, into the capillary blood, where pO_2 is only 40 mm Hg. The process continues until the pO_2 values in both areas are equal and blood returns to the heart with a pO_2 of 105 mm Hg. In the meantime, the pCO_2 of capillary blood is 45 mm Hg when it arrives at the alveoli. CO_2 diffuses into the alveoli, where the pCO_2 is only 40 mm Hg, until the pCO_2 values in both areas are equal. (Expiration constantly keeps alveolar pCO_2 at 40 mm Hg.) Blood returning to the heart also has a pCO_2 of 40 mm Hg.

External respiration is related to several structural features of the lungs. The alveolar–capillary (respiratory) membrane is thin enough to allow diffusion. The surface area of the alveoli over which diffusion may occur is large, many more times the total surface area of the skin. Lying over the alveoli are countless capillaries—so many that 900 ml of blood are able to participate in gas exchange at any time. Finally, the capillaries are so narrow that the red blood cells must flow through them in single file, giving each red blood cell maximum exposure to the available oxygen.

The efficiency of external respiration depends on several factors. One is altitude. Alveolar pO_2 must be higher than blood pO_2 for oxygen to diffuse from the alveoli into the blood. As a person ascends in altitude, atmospheric pO_2 decreases, alveolar pO_2 decreases correspondingly, and less oxygen diffuses into the blood. The common symptoms of **high altitude sickness**—shortness of breath, nausea, dizziness—are due to low concentrations of oxygen in the blood.

Another factor that affects external respiration is the total surface area available for O_2–CO_2 exchange. Any pulmonary disorder that decreases the functional surface area of the alveolar–capillary membranes decreases the efficiency of external respiration.

A third factor is the minute volume of respiration. Certain drugs, such as morphine, slow down the respiration rate, thereby decreasing the amount of O_2 and CO_2 that can be exchanged between the alveoli and the blood.

A final factor is the amount of O_2 that reaches the alveoli. This depends on keeping the air passageways intact and clear and having enough oxygen available to breathe.

Internal (Tissue) Respiration

When external respiration is completed, oxygenated blood returns to the heart to be pumped to tissue cells. The movement of O_2 and CO_2 between tissue capillaries and tissue cells is called **internal (tissue) respiration** (Figure 18.13b). It converts oxygenated blood into deoxygenated blood.

The pO_2 of blood pumped to the tissues is higher (105 mm Hg) than the pO_2 in the cells (40 mm Hg) because they are constantly using up O_2 to produce energy (Figure 18.13b). Oxygen diffuses out of the capillaries into tissue cells and blood pO_2 drops to 40 mm Hg by the time the blood reaches the venules.

Because cells are constantly producing CO_2, the pCO_2 of tissue cells (45 mm Hg) is higher than that of capillary blood (40 mm Hg). Carbon dioxide continuously diffuses into the capillaries so that the pCO_2 of blood returning to the heart rises to 45 mm Hg. The deoxygenated blood now returns to the heart to be pumped to the lungs for another cycle of external respiration.

What causes oxygen (O_2) to enter pulmonary capillaries from alveoli and to enter tissue cells from systemic capillaries?

Transport of Respiratory Gases

> **objective:** *Describe how the blood transports oxygen and carbon dioxide.*

The transport of respiratory gases between the lungs and body tissues is a function of the blood. When O_2 and CO_2 enter the blood, certain physical and chemical changes occur that aid in gas transport and exchange.

Oxygen

Oxygen does not dissolve well in water and therefore only about 1.5 percent of O_2 is carried in plasma, which is mostly water. About 98.5 percent of O_2 is carried in chemical combination with hemoglobin in red blood cells (Figure 18.14).

The heme portion of hemoglobin contains four atoms of iron, each capable of combining with a molecule of O_2 (see Figure 14.3). Oxygen and hemoglobin combine in an easily reversible reaction to form *oxyhemoglobin* as follows:

Figure 18.14   Transport of oxygen and carbon dioxide in the blood.

🔑 **Most O_2 is transported by hemoglobin as oxyhemoglobin within red blood cells; most CO_2 is transported in blood plasma as bicarbonate ions.**

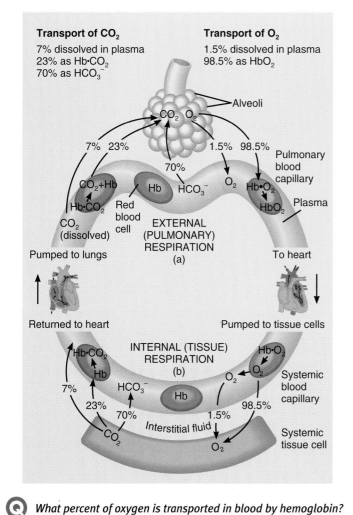

🅠 *What percent of oxygen is transported in blood by hemoglobin?*

$$Hb \quad + \quad O_2 \quad \rightleftharpoons \quad HbO_2$$

Reduced hemoglobin   Oxygen   Oxyhemoglobin hemoglobin
(uncombined hemoglobin)   (combined hemoglobin)

HEMOGLOBIN AND pO_2

The most important factor that determines how much O_2 combines with hemoglobin is pO_2. When blood pO_2 is high, hemoglobin binds with large amounts of O_2 and is *fully saturated*; that is, every available iron atom has combined with a molecule of O_2. When blood pO_2 is low, O_2 is released from hemoglobin. Therefore, in pulmonary capillaries, where the pO_2 is lower, hemoglobin does not hold as much O_2 and the O_2 is released for diffusion into the tissue cells.

HEMOGLOBIN AND OTHER FACTORS

The amount of O_2 released from hemoglobin is determined by several other factors. One factor is the relationship between hemoglobin and pH. In an acid environment, O_2 splits more readily from hemoglobin. Another factor is temperature. Within limits, as temperature increases, so does the amount of O_2 released from hemoglobin.

CARBON MONOXIDE (CO) POISONING

Carbon monoxide (CO) is a colorless and odorless gas found in exhaust fumes from automobiles and tobacco smoke. One of its interesting features is that it combines with hemoglobin very much like O_2 does, except that the combination of carbon monoxide and hemoglobin is over 200 times stronger. In addition, in concentrations as small as 0.1 percent, carbon monoxide will combine with half the hemoglobin molecules, dangerously reducing the oxygen-carrying capacity of the blood. Increased levels of carbon monoxide lead to hypoxia and result in *carbon monoxide poisoning*. The condition may be treated by administering pure O_2, which hastens the separation of carbon monoxide from hemoglobin.

HYPOXIA

Hypoxia (hī-POK-sē-a; *hypo* = below or under) refers to a deficiency of O_2 at the tissue level. It may be caused by a low pO_2 in arterial blood, as from high altitudes; too little functioning hemoglobin in the blood, as in anemia; inability of the blood to carry O_2 to tissues fast enough to sustain their needs, as in heart failure; or inability of tissues to use O_2 properly, as in cyanide poisoning.

Carbon Dioxide

Carbon dioxide is carried by the blood in several forms (Figure 18.14). About 7 percent is dissolved in plasma. About 23 percent combines with the globin portion of hemoglobin to form *carbaminohemoglobin* ($Hb \cdot CO_2$). CO_2 and globin combine in tissue capillaries, where pCO_2 is relatively high. In pulmonary capillaries, where pCO_2 is relatively low, the CO_2 splits apart from globin and enters the alveoli by diffusion.

About 70 percent of CO_2 is transported in plasma as bicarbonate ions (HCO_3^-):

$$CO_2 + H_2O \rightleftharpoons H_2CO_3 \rightleftharpoons H^+ + HCO_3^-$$

As CO_2 diffuses into tissue capillaries and enters the red blood cells, it combines with water to form carbonic acid (H_2CO_3). The carbonic acid breaks down into hydrogen ions (H^+) and bicarbonate ions (HCO_3^-). HCO_3^- diffuses out of the red blood cells to the plasma. The net effect is that CO_2

is carried from tissue cells as bicarbonate ions in plasma. Deoxygenated blood returning to the lungs, then, contains CO_2 dissolved in plasma, CO_2 combined with globin, and CO_2 incorporated in bicarbonate ions in plasma.

In the pulmonary capillaries, the events reverse. The CO_2 dissolved in plasma diffuses into the alveoli. The CO_2 combined with globin splits and diffuses into the alveoli. The bicarbonate ions (HCO_3^-) reenter the red blood cells and recombine with H^+ to form H_2CO_3, which splits into CO_2 and H_2O. The CO_2 leaves the red blood cells and diffuses into the alveoli.

Control of Respiration

objective: *Explain how the nervous system controls breathing and list the factors that can alter the rate of breathing.*

Although breathing can be controlled voluntarily to some extent, the nervous system usually controls respirations automatically to meet the body's demands without our conscious concern.

Nervous Control

The size of the thorax changes by action of the respiratory muscles. These muscles contract and relax in response to nerve impulses transmitted from centers in the brain. The area from which nerve impulses are sent is located in the brain stem and is referred to as the **respiratory center.** The respiratory center consists of groups of neurons in the brain stem that are functionally divided into three areas: (1) the medullary rhythmicity (rith-MIS-i-tē) area in the medulla oblongata, (2) the pneumotaxic (noo-mō-TAK-sik) area in the pons, and (3) the apneustic (ap-NOO-stik) area in the pons (Figure 18.15).

Medullary Rhythmicity Area

The **medullary rhythmicity area** controls the basic rhythm of respiration. In the normal resting state, inspiration usually lasts for about 2 seconds and expiration for about 3 seconds. This is the basic rhythm of respiration. In the medullary rhythmicity area are found both inspiratory and expiratory areas. Let us first consider the role of the inspiratory neurons.

Figure 18.15 Approximate location of areas of the respiratory center.

🔑 *The respiratory center is located in the medulla oblongata and pons of the brain stem.*

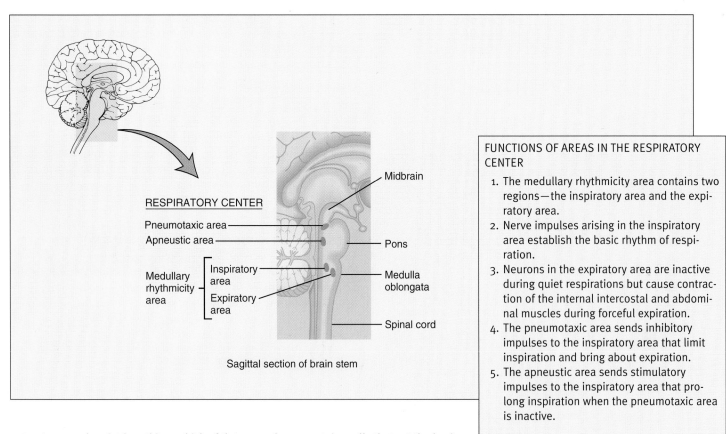

RESPIRATORY CENTER
Pneumotaxic area
Apneustic area
Medullary rhythmicity area
Inspiratory area
Expiratory area

Midbrain
Pons
Medulla oblongata
Spinal cord

Sagittal section of brain stem

FUNCTIONS OF AREAS IN THE RESPIRATORY CENTER

1. The medullary rhythmicity area contains two regions—the inspiratory area and the expiratory area.
2. Nerve impulses arising in the inspiratory area establish the basic rhythm of respiration.
3. Neurons in the expiratory area are inactive during quiet respirations but cause contraction of the internal intercostal and abdominal muscles during forceful expiration.
4. The pneumotaxic area sends inhibitory impulses to the inspiratory area that limit inspiration and bring about expiration.
5. The apneustic area sends stimulatory impulses to the inspiratory area that prolong inspiration when the pneumotaxic area is inactive.

❓ *In normal, quiet breathing, which of the areas shown contains cells that set the basic rhythm of respiration?*

The basic rhythm of respiration is determined by nerve impulses generated in the inspiratory area. Nerve impulses from the active inspiratory area travel via the phrenic and intercostal nerves to the muscles of inspiration (diaphragm and external intercostal muscles), causing them to contract and inspiration occurs. When the inspiratory neurons become inactive again, the muscles relax, expiration occurs, and the cycle repeats itself over and over.

The expiratory neurons remain inactive during normal quiet respiration. However, during high levels of ventilation, impulses discharged from the expiratory area cause contraction of the internal intercostals and abdominal muscles that decrease the size of the thoracic cavity to cause forced (labored) breathing.

Pneumotaxic Area

Although the medullary rhythmicity area controls the basic rhythm of respiration, the other areas help coordinate the transition between inspiration and expiration. The **pneumotaxic area** in the upper pons (Figure 18.15) transmits inhibitory impulses to the inspiratory area. In effect, these impulses help turn off the inspiratory area before the lungs become too full

of air. By limiting inspiration, the pneumotaxic area brings about expiration.

Apneustic Area

The **apneustic area** in the lower pons (Figure 18.15) also helps coordinate breathing. It stimulates the inspiratory area to prolong inspiration, thus inhibiting expiration. When the pneumotaxic area is active, it overrides the apneustic area.

A summary of the nervous control of respiration is presented in Figure 18.16.

Regulation of Respiratory Center Activity

Although the basic rhythm of respiration is set and coordinated by the respiratory center, the rhythm can be modified in response to demands of the body.

Cortical Influences

The respiratory center has connections with the cerebral cortex, which means we can voluntarily alter our pattern of breathing (Figure 18.16). We can even refuse to breathe at all

---

Figure 18.16   Summary of the nervous control of respiration.

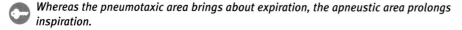

Whereas the pneumotaxic area brings about expiration, the apneustic area prolongs inspiration.

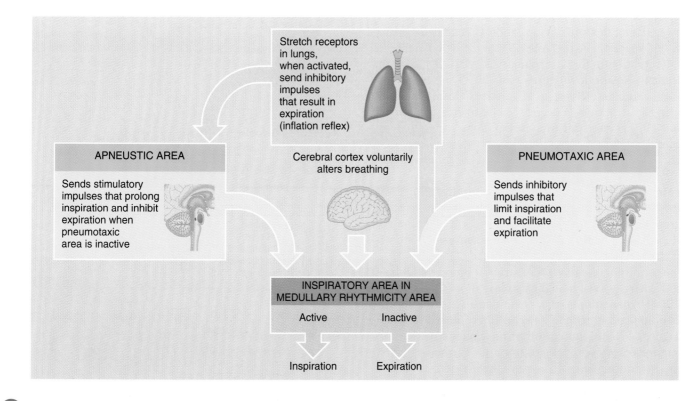

How do the durations of inspiration and expiration compare during periods of rest?

for a short time. Voluntary control is protective because it enables us to prevent water or irritating gases from entering the lungs. However, holding your breath causes CO_2 and H^+ to build up in the blood. At a certain blood level, the inspiratory area is stimulated, impulses are sent along the phrenic and intercostal nerves to inspiratory muscles, and breathing resumes whether or not the person wishes. It is impossible to kill yourself by holding your breath. Even if you faint, breathing resumes when consciousness is lost.

Inflation Reflex

Located in the walls of bronchi and bronchioles throughout the lungs are nerve cells called **stretch receptors.** When the receptors become stretched during overinflation of the lungs, impulses are sent along the vagus (X) nerves to the inspiratory and apneustic areas to cause inhibition of further inspiration. The result is that expiration follows. As the lungs deflate, the stretch receptors are no longer stimulated. Because the inspiratory and apneustic areas are no longer inhibited, a new inspiration begins. This process is referred to as the **inflation reflex,** a protective mechanism that prevents overinflation of the lungs (Figure 18.16).

Chemical Stimuli

Certain chemical stimuli determine how fast and deeply we breathe. The ultimate goal of the respiratory system is to maintain proper levels of CO_2 and O_2 and the system is highly responsive to changes in the blood levels of either. Although it is convenient to speak of carbon dioxide as the chemical stimulus for regulating the rate of respiration, it is actually hydrogen ions (H^+) that assume this role. CO_2 in the blood combines with water (H_2O) to form carbonic acid (H_2CO_3). But the H_2CO_3 quickly breaks down into H^+ and HCO_3^-. Any increase in CO_2 will cause an increase in H^+ (decrease in pH) and any decrease in CO_2 will cause a decrease in H^+ (increase in pH). In effect, it is H^+ that alters the rate of respiration rather than the CO_2 molecules. *Although the following discussion refers to levels of CO_2 and their effect on respiration, keep in mind that it is really the H^+ that causes the effects.*

Within the medulla oblongata is a group of neurons called the *chemosensitive area* that is highly sensitive to blood concentrations of CO_2. In the arch of the aorta and common carotid arteries are neurons called *chemoreceptors* that are sensitive to changes in CO_2 and O_2 levels in the blood.

Under normal circumstances, arterial blood pCO_2 is 40 mm Hg. If there is even a slight increase in pCO_2—a condition called *hypercapnia*—the chemosensitive area in the medulla oblongata and the CO_2 chemoreceptors are stimulated (Figure 18.17). Stimulation of the chemosensitive area and chemoreceptors sends nerve impulses to the brain that cause the inspiratory area to become highly active, and the rate and depth of respiration increase. This increased rate and depth, *hyperventilation,* allows the body to expel more CO_2 until the pCO_2 is lowered to normal. If arterial pCO_2 is

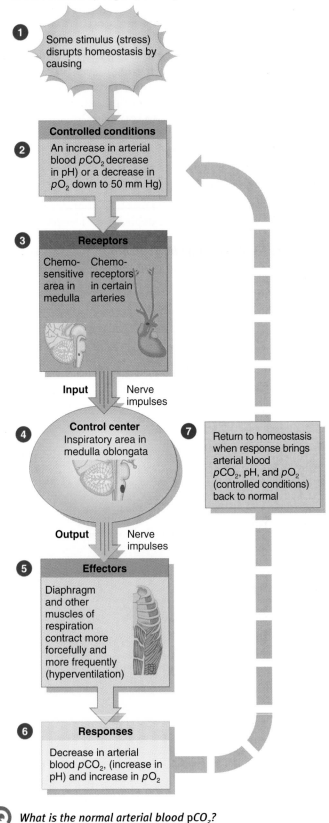

Figure 18.17   Negative feedback control of breathing by changes in blood pCO_2 (H^+ concentration) and pO_2.

The chemosensitive area in the medulla oblongata is sensitive to blood levels of CO_2 (H^+) and O_2.

1 Some stimulus (stress) disrupts homeostasis by causing

2 **Controlled conditions**
An increase in arterial blood pCO_2 decrease in pH) or a decrease in pO_2 down to 50 mm Hg)

3 **Receptors**
Chemosensitive area in medulla   Chemoreceptors in certain arteries

Input   Nerve impulses

4 **Control center**
Inspiratory area in medulla oblongata

Output   Nerve impulses

5 **Effectors**
Diaphragm and other muscles of respiration contract more forcefully and more frequently (hyperventilation)

6 **Responses**
Decrease in arterial blood pCO_2, (increase in pH) and increase in pO_2

7 Return to homeostasis when response brings arterial blood pCO_2, pH, and pO_2 (controlled conditions) back to normal

What is the normal arterial blood pCO_2?

lower than 40 mm Hg, the chemosensitive area and chemoreceptors are not stimulated and stimulatory impulses are not sent to the inspiratory area. Consequently, the area sets its own moderate pace until CO_2 accumulates and the pCO_2 rises to 40 mm Hg. A slow rate and depth of respiration is called *hypoventilation*.

The oxygen chemoreceptors are sensitive only to large decreases in the pO_2 because hemoglobin remains about 85 percent or more saturated at pO_2 values all the way down to about 50 mm Hg. If arterial pO_2 falls from a normal of 105 mm Hg to about 50 mm Hg, the chemoreceptors become stimulated and send impulses to the inspiratory area and respiration increases. But if the pO_2 falls much below 50 mm Hg, the cells of the inspiratory area suffer oxygen starvation and do not respond well to any chemical receptors. They send fewer impulses to the inspiratory muscles, and the respiration rate decreases or breathing ceases altogether.

Other Influences

Certain arteries also contain **baroreceptors** (pressure receptors) that are stimulated by a rise in blood pressure. Although these baroreceptors are concerned mainly with the control of circulation (see Figure 16.4), they also help control respiration. For example, a sudden rise in blood pressure decreases the rate of respiration, and a drop in blood pressure increases the respiratory rate.

Other factors that control respiration are:

1. *Temperature.* An increase in body temperature, as during a fever or vigorous muscular exercise, increases the rate of respiration. A decrease in body temperature decreases respiratory rate. A sudden cold stimulus such as plunging into cold water causes *apnea* (AP-nē-a), a temporary cessation of breathing.

2. *Pain.* A sudden, severe pain brings about apnea, but a prolonged pain triggers the general adaptation syndrome (see Figure 13.19) and increases respiratory rate.

3. *Stretching the anal sphincter muscle.* This increases the respiratory rate and is sometimes employed to stimulate respiration in a person who has stopped breathing.

4. *Irritation of air passages.* Mechanical or chemical irritation of the pharynx or larynx brings about an immediate cessation of breathing followed by coughing or sneezing.

A summary of stimuli that affect ventilation is presented in Exhibit 18.3.

Exhibit 18.3 Summary of Stimuli that Affect Ventilation Rate and Depth

STIMULI THAT INCREASE RATE AND DEPTH OF VENTILATION	STIMULI THAT DECREASE (OR INHIBIT) RATE AND DEPTH OF VENTILATION
Increase in arterial blood pCO_2 above 40 mm Hg (causes an increase in H^+ and a decrease in pH).	Decrease in arterial blood pCO_2 below 40 mm Hg (causes a decrease in H^+ and an increase in pH).
Decrease in arterial blood pO_2 from 105 mm Hg to 50 mm Hg.	Decrease in arterial blood pO_2 below 50 mm Hg.
Decrease in blood pressure.	Increase in blood pressure.
Increase in body temperature.	Decrease in body temperature decreases rate of respiration, and sudden cold stimulus causes apnea.
Prolonged pain.	Severe pain causes apnea.
Stretching anal sphincter.	Irritation of pharynx or larynx by touch or chemicals causes apnea.

Smoking: A Breathtaking Experience

Cigarette smoking has been called the single most preventable cause of death and disability in North America. In fact, all forms of tobacco use challenge the body's ability to maintain homeostasis and health. Now that you have a good understanding of the complexity of the cardiovascular and respiratory systems, it is easy to understand why. Here are a few of smoking's most deadly effects.

Where There's Smoke . . .

. . . there is injury. Consider the structure of the alveoli, a miraculous design that allows you to extract life-giving oxygen and cleanse your body of metabolic waste. Chronic exposure to smoke gradually destroys the elasticity of these sensitive structures. The result is *emphysema*, a progressive degeneration of the alveoli that causes breathing and oxygen assimilation to become increasing difficult. The exact mechanism whereby smoke destroys the structure and function of the alveoli is unknown. One hypothesis suggests that as the macrophages and leukocytes in the alveoli produce enzymes to destroy the foreign material introduced by the smoke, these enzymes may inadvertently destroy lung tissue as well.

Bronchitis results from inflammation of the upper respiratory tract; it is characterized by excessive mucus production. In smokers, chronic bronchitis may result from the irritation of cigarette smoke. "Smokers' cough" is a symptom of bronchitis. If left untreated, bronchitis increases the risk of more serious infections and permanent airway damage.

Calling for Cancer

Lung cancer was a rare disease in the early 1900s. Now it is the leading cancer killer for both men and women, thanks primarily to smoking. Cigarette smoke contains a number of known carcinogens, including benzoapyrene, N-nitrosoamines, and radioactive particles such as radon and polonium, which may initiate and promote the cellular changes leading to cancer.

Smoking increases risk of cancers at other sites as well. Cancers of the oral cavity (mouth, tongue, lip, cheek, and throat) and larynx occur in pipe and cigar smokers as well as cigarette smokers, and in people who use smokeless tobacco. When smoke is inhaled, carcinogens are absorbed from the airways into the bloodstream, and can thus contribute to cancers at nonrespiratory sites. Cancers of the esophagus, stomach, kidney, pancreas, and bladder are more common in smokers than non-smokers. Women who smoke have higher rates of cervical and breast cancer. Recent studies have also suggested links between smoking and colon cancer and some types of leukemia.

A Heart-breaking Habit

Because it has several negative effects on the cardiovascular system, smoking is one of the leading causes of death from heart attacks and strokes. Smoking is associated with injury to the endothelial lining of the arteries. Remember that this injury is the first step in the process of atherosclerosis. Chemicals in cigarette smoke may activate LDL cholesterol to perform its atherogenic work of depositing cholesterol in the artery wall. Smoking also decreases cardioprotective HDL cholesterol levels. Smoking increases blood fibrinogen levels, platelet aggregation, and thus clotting risk.

Nicotine, an addictive chemical found in tobacco products, is a *sympathomimetic* substance, which means it mimics sympathetic nervous system stimulation. It stimulates the release of epinephrine and norepinephrine, raises blood sugar and blood fat levels, and increases heart rate and blood pressure. This extra work for the heart requires more oxygen; however, the heart actually receives less, because oxygen concentration in the blood is lower in a smoker due partial replacement of oxygen by carbon monoxide (present in cigarette smoke). Oxygen also has a harder time reaching the heart muscle because nicotine constricts the coronary arteries.

critical thinking

Smoking reduces the amount of oxygen reaching body tissues, which interferes with the body's ability to make collagen. How does this help explain why smokers heal more slowly from ulcers, surgery, and fractures?

Why do you think smokers experience premature aging of the skin?

Common Disorders

Lung Cancer

A common lung cancer, *bronchogenic carcinoma,* starts in the walls of the bronchi. The constant irritation by inhaled smoke and pollutants causes the goblet cells of the bronchial epithelium to enlarge. They respond by secreting excessive mucus. The basal cells also respond to the stress by undergoing cell division so fast that they push into the area occupied by the goblet and columnar cells.

If the stress still continues, the basal cells of the bronchial tubes continue to divide and break through the basement membrane; columnar and goblet cells disappear and may be replaced with squamous cancer cells. If this happens, the malignant growth spreads through the lung and may block a bronchial tube. The occurrence of bronchogenic carcinoma is over 20 times higher in cigarette smokers than it is in nonsmokers.

Asthma

Diseases such as asthma, bronchitis, and emphysema have in common some degree of obstruction of the air passageways. The term *chronic obstructive pulmonary disease* (*COPD*) is used to refer to these disorders. Among the symptoms that might indicate significant airflow obstruction are cough, wheezing, and dyspnea (painful or labored breathing).

Asthma is a reaction, usually allergic, characterized by attacks of wheezing and difficult breathing. Attacks are brought on by spasms of the smooth muscles in the walls of the smaller bronchi and bronchioles, causing the passageways to close partially.

Bronchitis

Bronchitis is inflammation of the bronchi characterized by enlargement of glands and goblet cells lining the bronchial airways. Cigarette smoking is the leading cause of chronic bronchitis, that is, bronchitis that lasts for at least 3 months of the year for two successive years.

Emphysema

In *emphysema* (em'-fi-SĒ-ma), the alveolar walls lose their elasticity and remain filled with air during expiration. As increasing numbers of alveoli are damaged, the lungs become permanently inflated because they have lost elasticity. To adjust to the increased lung size, the size of the chest cage increases, resulting in a "barrel chest." Emphysema is generally caused by a long-term irritation. Cigarette smoke, air pollution, and occupational exposure to industrial dust are the most common irritants.

Pneumonia

Pneumonia refers to an acute infection or inflammation of the alveoli. The alveolar sacs fill up with fluid and dead white blood cells, reducing the amount of air space in the lungs.

Tuberculosis (TB)

The bacterium *Mycobacterium tuberculosis* produces an infectious, communicable disease called *tuberculosis* (*TB*). TB most often affects the lungs and the pleurae. The bacteria destroy parts of the lung tissue, and the tissue is replaced by fibrous connective tissue.

Respiratory Distress Syndrome (RDS) of the Newborn

Respiratory distress syndrome (*RDS*) *of the newborn* is also called *glassy lung disease* or *hyaline membrane disease* (*HMD*). At birth, the fluid-filled respiratory passageway must become an air-filled passageway, and the collapsed alveoli must expand and function in gaseous exchange. The success of this transition depends largely on surfactant, the substance produced by alveolar cells. In the newborn whose lungs are deficient in surfactant, the alveoli collapse almost to their original uninflated state.

Respiratory Failure

Respiratory failure refers to a condition in which the respiratory system cannot supply sufficient oxygen to maintain metabolism or cannot eliminate enough carbon dioxide to prevent respiratory acidosis (a lower than normal pH). Respiratory failure always causes dysfunction in other organs as well.

Sudden Infant Death Syndrome (SIDS)

Sudden infant death syndrome (*SIDS*), also called crib death, kills more infants between the ages of 1 week and 12 months than any other disease. SIDS occurs without warning. Although about half its victims had an upper respiratory infection within 2 weeks of death, the babies tended to be otherwise healthy. Many hypotheses have been proposed as explanations for SIDS but the precise causes are not clear. A main cause is thought to be hypoxia in infants who sleep in a prone position (on their stomach) and rebreathe exhaled air trapped in a depression of the mattress. For this reason, it is suggested that normal newborns be placed on their backs for sleeping. The seasonal distribution of SIDS incidence, being lowest during the summer months and highest in the late fall and winter, suggests an infectious agent, viruses being the most likely possibility. Most cases occur at a time in life when antibody levels are low and the infant is at a critical period of susceptibility. Alternatively, infants may be more likely to die from overheating due to being dressed too warmly in the winter months.

Common Cold and Influenza (Flu)

Hundreds of viruses are responsible for the *common cold*. A group of viruses, called rhinoviruses, account for about 40 percent of all colds in adults. Typical symptoms include sneezing, excessive nasal secretion, and congestion. Generally, there is no fever.

Influenza (*flu*) is also caused by a virus. Its symptoms include chills, fever (usually higher than 101°F), headache, and muscular aches.

Pulmonary Embolism

Pulmonary embolism refers to the presence of a blood clot or other foreign substance in a pulmonary arterial vessel that obstructs circulation to lung tissue.

Pulmonary Edema

Pulmonary edema refers to an abnormal accumulation of interstitial fluid in the interstitial spaces and alveoli of the lungs. ■

Medical Terminology and Conditions

Asphyxia (as-FIK-sē-a; *sphyxis* = pulse) Oxygen starvation due to low atmospheric oxygen or interference with ventilation, external respiration, or internal respiration.

Aspiration (as′-pi-RA-shun; *spirare* = breathe) Inhalation of a foreign substance such as water, food, or foreign body into the bronchial tree; drawing of a substance in or out by suction.

Cardiopulmonary resuscitation (kar′-dē-ō-PUL-mo-ner-ē resus′-i-TA-shun) *(CPR)* The artificial establishment of normal or near-normal respiration and circulation.

Cheyne–Stokes respiration (CHAN STOKS res′-pi-RA-shun) A repeated cycle of irregular breathing beginning with shallow breaths that increase in depth and rapidity, then decrease and cease altogether for 15 to 20 seconds. Cheyne–Stokes is normal in infants. It is also often seen just before death from pulmonary, cerebral, cardiac, and kidney disease.

Diphtheria (dif-THE-rē-a; *diphthera* = membrane) An acute bacterial infection that causes the mucous membranes of the oropharynx, nasopharynx, and larynx to enlarge and become leathery. Enlarged membranes obstruct airways and may cause death from asphyxiation.

Dyspnea (DISP-nē-a; *dys* = painful, difficult; *pnoia* = breath) Painful or labored breathing.

Epistaxis (ep′-i-STAK-sis) Loss of blood from the nose due to trauma, infection, allergy, neoplasms, and bleeding disorders. It can be arrested by cautery with silver nitrate, electrocautery, and firm packing. Also called **nosebleed.**

Heimlich (abdominal thrust) maneuver (HIM-lik ma-NOO-ver) First-aid procedure designed to clear the air passageways of obstructing objects. It is performed by applying a quick upward thrust that causes sudden elevation of the diaphragm and forceful, rapid expulsion of air in the lungs; this action forces air out the trachea to eject the obstructing object. The Heimlich maneuver is also used to expel water from the lungs of near-drowning victims before resuscitation is begun.

Hemoptysis (hē-MOP-ti-sis; *hemo* = blood; *ptein* = to spit) Spitting of blood from the respiratory tract.

Orthopnea (or′-THOP-nē-a; *ortho* = straight) Difficult breathing that occurs in the horizontal position.

Pneumonectomy (noo′-mō-NEK-tō-mē; *pneumo* = lung; *tome* = cutting) Surgical removal of a lung.

Rales (RALS) Sounds sometimes heard in the lungs that resemble bubbling, rattling, or crackling. Rales are to lungs what murmurs are to the heart. They are due to the presence of an abnormal type or amount of fluid or mucus in the terminal bronchi and alveoli; indicative of pneumonia.

Respirator (RES-pi-rā′-tor) An apparatus fitted to a mask over the nose and mouth, or hooked directly to an endotracheal or tracheotomy tube, that is used to assist or support ventilation or to provide medication to the air passages under pressure.

Rhinitis (rī-NI-tis; *rhino* = nose) Chronic or acute inflammation of the mucous membrane of the nose.

Tachypnea (tak′-ip-NE-a; *tachy* = rapid; *pnoia* = breath) Rapid breathing.

Wheeze (WEZ) A whistling sound made in breathing, usually during expiration; resulting from constriction or obstruction of a respiratory passageway. It occurs in conditions such as asthma, congestive heart failure, edema, and tumors.

Study Outline

Organs (p. 408)
1. Respiratory organs include the nose, pharynx, larynx, trachea, bronchi, and lungs.
2. They act with the cardiovascular system to supply oxygen and remove carbon dioxide from the blood.

Nose (p. 408)
1. The external portion of the nose is made of cartilage and skin and is lined with mucous membrane. Openings to the exterior are the external nares.
2. The internal portion of the nose communicates with the paranasal sinuses and nasopharynx through the internal nares.
3. The internal and external nose is divided by a septum.
4. The nose is adapted for warming, moistening, filtering air; olfaction; and serving as a resonating chamber for special sounds.

Pharynx (p. 409)
1. The pharynx (throat) is a muscular tube lined by a mucous membrane.
2. The anatomic regions are nasopharynx, oropharynx, and laryngopharynx.
3. The nasopharynx functions in respiration. The oropharynx and laryngopharynx function both in digestion and in respiration.

Larynx (p. 409)
1. The larynx (voice box) is a passageway that connects the pharynx with the trachea.
2. It contains the thyroid cartilage (Adam's apple); the epiglottis, which prevents food from entering the larynx; the cricoid cartilage, which connects the larynx and trachea; and three paired cartilages, the most important of

which are the arytenoid cartilages that assist in movement of the true vocal cords.

3. The larynx contains true vocal cords that produce sound and false vocal cords that function in holding the breath against pressure in the thoracic cavity, as when straining to lift a heavy weight.

4. Taut true vocal cords produce high pitches; relaxed ones produce low pitches.

Trachea (p. 409)

1. The trachea (windpipe) extends from the larynx to the primary bronchi.

2. It is composed of smooth muscle and C-shaped rings of cartilage and is lined with pseudostratified ciliated columnar epithelium.

Bronchi (p. 411)

1. The bronchial tree consists of the trachea, primary bronchi, secondary bronchi, tertiary bronchi, bronchioles, and terminal bronchioles.

2. Two walls of bronchi contain rings of cartilage; walls of bronchioles contain smooth muscle.

Lungs (p. 412)

1. Lungs are paired organs in the thoracic cavity. They are enclosed by the pleural membrane. The parietal pleura is the outer layer; the visceral pleura is the inner layer.

2. The right lung has three lobes separated by two fissures; the left lung has two lobes separated by one fissure plus a depression, the cardiac notch.

3. Each lobe consists of lobules, which contain lymphatic vessels, arterioles, venules, terminal bronchioles, respiratory bronchioles, alveolar ducts, alveolar sacs, and alveoli.

4. Gas exchange occurs across the alveolar–capillary (respiratory) membrane.

Respiration (p. 414)

Pulmonary Ventilation (p. 414)

1. Pulmonary ventilation or breathing consists of inspiration and expiration.

2. The movement of air into and out of the lungs depends on pressure changes.

3. Inspiration occurs when alveolar pressure falls below atmospheric pressure. Contraction of the diaphragm and external intercostal muscles increases the size of the thorax so that the lungs expand. Expansion of the lungs decreases alveolar pressure, so that air moves along the pressure gradient from the atmosphere into the lungs.

4. Expiration occurs when alveolar pressure is higher than atmospheric pressure. Relaxation of the diaphragm and external intercostal muscles decreases the size of the thorax and lung volume, and alveolar pressure increases so that air moves from the lungs to the atmosphere.

5. During forced inspiration, accessory muscles of inspiration are also used.

6. Forced expiration involves contraction of the internal intercostals and abdominal muscles.

7. A collapsed lung may be caused by air in the pleural cavities, airway obstruction, or lack of surfactant.

Pulmonary Air Volumes and Capacities (p. 418)

1. Among the pulmonary air volumes exchanged in ventilation are tidal volume, inspiratory reserve, expiratory reserve, residual volume, and minimal volume.

2. Pulmonary lung capacities, the sum of two or more volumes, include inspiratory, functional residual, vital, and total.

3. The minute volume of respiration is the total air taken in during 1 minute (tidal volume times 12 respirations per minute).

Exchange of Respiratory Gases (p. 419)

1. The partial pressure of a gas is the pressure exerted by that gas in a mixture of gases. It is symbolized by p.

2. Each gas in a mixture of gases exerts its own pressure and behaves as if no other gases are present.

External (Pulmonary) Respiration; Internal (Tissue) Respiration (pp. 419–420)

1. In external and internal respiration, O_2 and CO_2 move from areas of their higher partial pressure to areas of their lower partial pressure.

2. External (pulmonary) respiration is the exchange of gases between alveoli and pulmonary blood capillaries. It is aided by a thin alveolar–capillary (respiratory) membrane, a large alveolar surface area, and a rich blood supply.

3. Internal (tissue) respiration is the exchange of gases between tissue capillaries and tissue cells.

Transport of Respiratory Gases (p. 421)

1. Most oxygen, 98.5 percent, is carried by the iron atoms of the heme in hemoglobin; the rest is carried in plasma.

2. The association of O_2 and hemoglobin is affected by pO_2, pH, temperature, and pCO_2.

3. Hypoxia refers to O_2 deficiency at the tissue level.

4. Carbon dioxide is transported in three ways. About 7 percent is dissolved in plasma, 23 percent combines with the globin of hemoglobin, and 70 percent is converted to bicarbonate ions (HCO_3^-) that travel in plasma.

Control of Respiration (p. 422)

Nervous Control (p. 422)

1. The respiratory center consists of a medullary rhythmicity area (inspiratory and expiratory area), pneumotaxic area, and apneustic area.

2. The inspiratory area sets the basic rhythm of respiration.

3. The pneumotaxic and apneustic areas coordinate inspiration and expiration.

Regulation of Respiratory Center Activity (p. 423)

1. Respirations may be modified by a number of factors, both in the brain and outside.

2. Among the modifying factors are cortical influences, the inflation reflex, chemical stimuli (CO_2 and O_2 levels), blood pressure, temperature, pain, stretching the anal sphincter, and irritation to the respiratory mucosa.

Self-Quiz

1. Which of the following is NOT true concerning the pharynx?

   a. Food, drink, and air pass through the oropharynx and laryngopharynx. b. The auditory (Eustachian) tubes have openings in the nasopharynx. c. The pseudostratified ciliated epithelium of the nasopharynx helps move dust-laden mucus toward the mouth. d. The palatine and lingual tonsils are located in the laryngopharynx. e. The wall of the pharynx is composed of skeletal muscle lined with mucous membranes.

2. During speaking, you raise your voice's pitch. This is possible because

   a. the epiglottis vibrates rapidly b. you have increased the air pressure pushing against the vocal cords c. you have increased the tension on the true vocal cords d. your true vocal cords have become thicker and longer e. the true vocal cords begin to vibrate more slowly

3. Which of the following statements is NOT true concerning the lungs?

   a. Pulmonary blood vessels enter the lungs at the hilus. b. The left lung is thicker and broader because the liver lies below it. c. The right lung is composed of three lobes. d. The top portion of the lung is the apex. e. Bronchi are lined with ciliated pseudostratified columnar epithelium.

4. Which sequence of events best describes inspiration?

   a. contraction of diaphragm and external intercostals ⟶ increase in size of thoracic cavity ⟶ decrease in thoracic and alveolar pressure b. relaxation of diaphragm and external intercostals ⟶ decrease in size of thoracic cavity ⟶ increase in thoracic and alveolar pressure c. contraction of diaphragm and external intercostals ⟶ decrease in size of thoracic cavity ⟶ decrease in thoracic and alveolar pressure d. relaxation of diaphragm and external intercostals ⟶ increase in size of thoracic cavity ⟶ increase in thoracic and alveolar pressure e. contraction of diaphragm and external intercostals ⟶ decrease in size of thoracic cavity ⟶ increase in thoracic and alveolar pressure

5. Expiration

   a. occurs when alveolar pressure reaches 758 mm Hg b. is normally considered an active process requiring muscle contraction c. requires the alveolar pressure to be greater than the atmospheric pressure d. involves the expansion of the pleural membranes e. occurs when atmospheric pressure is equal to the pressure in the lungs

6. The amount of air that enters the respiratory passages during normal quiet breathing is the

   a. tidal volume b. inspiratory capacity c. vital capacity d. total lung capacity e. minimal volume

7. If the total pressure of a mixture of gases is 760 mm Hg and gas Z makes up 20 percent of the total mixture, then the partial pressure of gas Z would be

   a. 20 mm Hg b. 38 mm Hg c. 152 mm Hg d. 380 mm Hg e. 760 mm Hg

8. When carbon dioxide binds to hemoglobin, which compound is formed?

   a. carbonic acid b. bicarbonate ions c. oxyhemoglobin d. carbon monoxide e. carbaminohemoglobin

9. How does hypercapnia affect respiration?

   a. It increases the rate and depth of respiration. b. It decreases the rate and depth of respiration. c. It causes hypoventilation. d. It does not change the rate of respiration. e. It activates stretch receptors in the lungs.

10. Air would flow INTO the lungs along the following route:

    5 1. bronchioles

    2 2. primary bronchi

    3 3. secondary bronchi

    6 4. terminal bronchioles

    4 5. tertiary bronchi

    1 6. trachea

    a. 6, 1, 2, 3, 5, 4 b. 6, 5, 3, 4, 2, 1 c. 6, 2, 3, 5, 4, 1 d. 6, 2, 3, 5, 1, 4 e. 6, 1, 4, 5, 3, 1

11. Oxygen splits MORE readily from hemoglobin when

    a. temperature increases (within normal body limits) b. acid decreases c. blood pH increases d. pO_2 is high e. CO_2 is low

12. The alveolar–capillary membrane does NOT contain

    a. capillary endothelial cells b. septal cells c. alveolar macrophages d. partial rings of hyaline cartilage e. squamous epithelial cells

13. Normal rate and depth of breathing is called

    a. apnea b. dyspnea c. eupnea d. pneumonia e. hypercapnia

14. The exchange of gases between the lungs and blood is

    a. internal respiration b. external respiration c. exhalation d. inspiration e. pulmonary ventilation

15. Surfactant

    a. is the type of mucus present in the nasal cavity b. is a product of squamous epithelial cells in the alveoli c. lubricates the true vocal cords d. is found between the pleural membranes e. is produced by the septal cells.

Critical Thinking Applications

1. A high school ski club from the coast of Massachusetts took a trip to Colorado for some real mountain skiing. After a day of skiing, several students complained of shortness of breath, dizziness, and nausea. Diagnose their condition and explain what is causing the problem.

2. A community college student drove home after a hard night of studying, pulled into the garage and fell asleep in the car with the motor running. At the hospital the next day, a nurse told him he was lucky to be alive. Why was he in danger from car motor exhaust?

3. Your three-year-old niece Pattee likes to get her own way all the time! Pattee wants to eat 20 chocolate kisses (1 for each finger and toe), but you will give her only 3. She is at this moment "holding her breath until she turns blue." Is she in danger of death?

4. The respiratory system and the cardiovascular system work together to supply the body with oxygen and to remove carbon dioxide. Smoking has many negative effects on the functioning of both these systems. List at least three effects of smoking on the respiratory and/or the cardiovascular system.

5. A child is brought into the emergency room with great difficulty breathing after inhaling a collapsed balloon. What instruments can be used to locate and remove the balloon? If the balloon has reached the end of the trachea, where is it likely to lodge?

6. Hyperventilating before holding your breath when swimming under water may cause confusion in the signals going to your respiratory center; it may even result in passing out from lack of oxygen while still under water. Explain how hyperventilation would effect your O_2 and CO_2 levels and how the levels of these gases affect the respiratory center.

Answers to Figure Questions

18.1 Nose, pharynx, larynx, trachea, bronchi, and bronchioles (except the respiratory bronchioles).

18.2 External nares $\longrightarrow$ nasal cavity $\longrightarrow$ internal nares.

18.3 During swallowing, the epiglottis closes over the larynx.

18.4 Because the tissues between the esophagus and trachea are soft, the esophagus can bulge into the trachea when you swallow food.

18.5 Left lung, two of each; right lung, three of each.

18.6 Because two-thirds of the heart lies to the left of the midline, the left lung contains a cardiac notch to accommodate the position of the heart. The right lung is thicker and broader to accommodate the liver lying below it.

18.7 Lymphatic vessel, arteriole, venule, and branch of a terminal bronchiole wrapped in elastic connective tissue.

18.8 They secrete surfactant.

18.9 The pressure would increase to 4 atm.

18.10 If you are at rest while reading, your diaphragm.

18.11 Inspiration = 758 mm Hg; expiration = 762 mm Hg.

18.12 Vital capacity.

18.13 A difference in pO_2 that promotes diffusion in the indicated direction.

18.14 98.5 percent.

18.15 The medullary rhythmicity area.

18.16 Inspiration takes about 2 seconds; expiration takes about 3 seconds.

18.17 40 mm Hg.

19 chapter

THE DIGESTIVE SYSTEM

*F*ood is vital for life because it is the source of energy that drives the chemical reactions occurring in every cell. Energy is needed for activities such as muscle contraction, conduction of nerve impulses, and secretory and absorptive activities of many cells. Food is also needed for the nutrients (molecules) it contains that are used in the body's chemical reactions. Food as it is consumed, however, is not in a state suitable for use. It must be broken down into molecules that can pass through the wall of the gastrointestinal tract. The breaking down of food molecules for use by cells is called *digestion,* and the organs that collectively perform this function comprise the *digestive system.*

The medical specialty that deals with the structure, function, diagnosis, and treatment of diseases of the stomach and intestines is called *gastroenterology* (gas′-trō-en′-ter-OL-ō-jē; *gastro* = stomach; *enteron* = intestines).

Digestive Processes

The digestive system prepares food for use by the cells through five basic activities.

1. **Ingestion.** Taking food into the mouth (eating).

2. **Mixing and movement of food.** Muscular contractions mix food and secretions and move food along the gastrointestinal tract.

3. **Digestion.** The breakdown of food by both chemical and mechanical processes. *Chemical digestion* is a series of reactions that break down the large, complex carbohydrate, lipid, and protein molecules that we eat into simple molecules small enough to pass through the walls of the digestive organs and eventually into the body's cells. *Mechanical digestion* consists of various movements that aid chemical digestion. For example, food is chewed by the teeth before it is swallowed; the smooth muscles of the stomach and small intestine churn the food so it is thoroughly mixed with enzymes that digest foods. Waves of muscular contraction called *peristalsis* (per′-i-STAL-sis; *peri* = around; *stalsis* = contraction) move materials along the gastrointestinal tract.

4. **Absorption.** The passage of digested food from the gastrointestinal tract into the cardiovascular and lymphatic systems for distribution to cells.

5. **Defecation.** The elimination of indigestible substances from the gastrointestinal tract.

Organization

objective: *Classify the organs of the digestive system and describe the histology of the gastrointestinal tract.*

The organs of digestion may be divided into two main groups. First is the *gastrointestinal* (**GI**) *tract* or *alimentary canal,* a continuous tube that begins at the mouth and ends at the anus

(Figure 19.1). Organs of the gastrointestinal tract include the mouth, pharynx, esophagus, stomach, small intestine, and large intestine. The GI tract contains the food from the time it is eaten until it is digested and prepared for elimination.

The second group of organs composing the digestive system are the *accessory structures*: the teeth, tongue, salivary glands, liver, gallbladder, and pancreas. Teeth protrude into the GI tract and aid in the physical breakdown of food. The other accessory structures, except for the tongue, lie totally outside the tract and produce or store secretions necessary for chemical digestion. These secretions are released into the tract through ducts.

General Histology

The wall of the GI tract consists of four layers of tissue. From the inside out, they are the mucosa, submucosa, muscularis, and serosa (Figure 19.2 on page 435).

Mucosa

The *mucosa,* or inner lining of the tract, is a mucous membrane. It is composed of a layer of epithelium in direct contact with the contents of the GI tract, areolar connective tissue, and a thin layer of smooth muscle (muscularis mucosae). This thin, smooth muscle layer causes folds in the mucosa to increase surface area for digestive activities.

Submucosa

The *submucosa* consists of areolar connective tissue that binds the mucosa to the muscle layer. It contains numerous blood and lymphatic vessels for the absorption of the breakdown products of digestion and nerves that control GI tract secretions.

Muscularis

The *muscularis* is a thick layer of muscle. In the mouth, pharynx, and upper esophagus, it consists in part of skeletal muscle that produces voluntary swallowing. In the rest of the tract, the muscularis consists of smooth muscle usually arranged as an inner sheet of circular fibers and an outer sheet of longitudinal fibers. Involuntary contractions of these smooth muscles help break down food physically, mix it with digestive secretions, and propel it through the tract. The muscularis also contains nerves that control GI tract movements (motility).

Serosa and Peritoneum

The *serosa* is the outermost layer of the GI tract. It is a serous membrane composed of simple squamous epithelium and connective tissue. The serosa secretes serous fluid that allows the tract to glide easily against other organs.

Recall from Chapter 4 that the *peritoneum* (per′-i-tō-NĒ-um; *peri* = around; *tonos* = tension) is the largest serous membrane of the body. Whereas the parietal peritoneum lines the wall of the abdominal cavity, the visceral peritoneum covers some of the organs in the cavity and is their serosa.

The peritoneum contains large folds that weave between the viscera. The folds bind the organs to each other and to the

Figure 19.1   Organs of the digestive system and related structures.

Organs of the gastrointestinal (GI) tract are the mouth, pharynx, esophagus, stomach, small intestine, and large intestine. Accessory structures of the digestive system are the teeth, tongue, salivary glands (parotid, sublingual, and submandibular), liver, gallbladder, and pancreas.

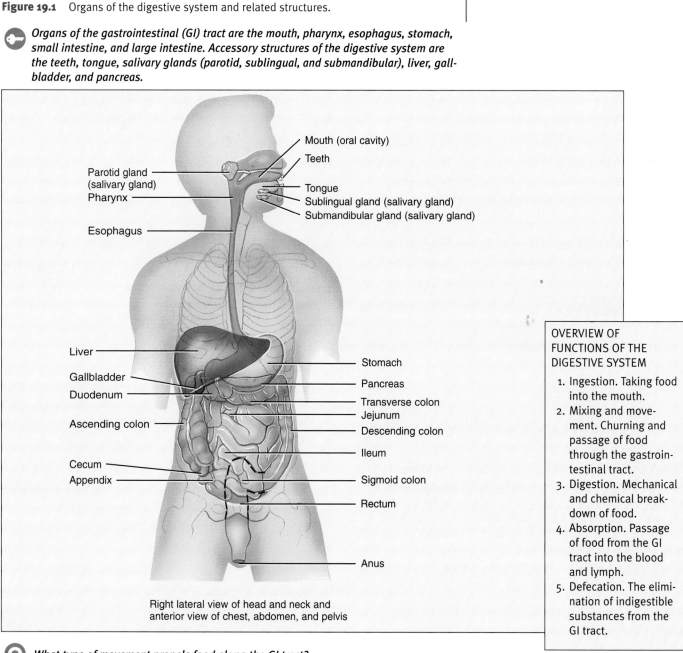

Right lateral view of head and neck and anterior view of chest, abdomen, and pelvis

OVERVIEW OF FUNCTIONS OF THE DIGESTIVE SYSTEM

1. Ingestion. Taking food into the mouth.
2. Mixing and movement. Churning and passage of food through the gastrointestinal tract.
3. Digestion. Mechanical and chemical breakdown of food.
4. Absorption. Passage of food from the GI tract into the blood and lymph.
5. Defecation. The elimination of indigestible substances from the GI tract.

Q *What type of movement propels food along the GI tract?*

walls of the abdominal cavity and contain blood and lymphatic vessels and nerves that supply the abdominal organs. One part of the peritoneum, the ***mesentery*** (MEZ-en-ter′-ē; *meso* = middle; *enteron* = intestine), binds the small intestine to the posterior abdominal wall (Figure 19.3b). The *falciform* (FAL-si-form; *falx* = sickle-shaped) *ligament* attaches the liver to the anterior abdominal wall and diaphragm (Figure 19.3a). The ***greater omentum*** (*omentum* = fat skin) drapes over the transverse colon and coils of the small intestine (Figure 19.3a). Because the greater omentum contains large quantities of adipose tissue, it looks like a "fatty apron" draped over the organs. The greater omentum contains many lymph nodes. If an infection occurs in the intestine, the lymph nodes combat the infection and help prevent it from spreading to the abdominal and pelvic organs.

Mouth (Oral Cavity)

objective: *Describe the structure of the mouth and explain its role in digestion.*

The ***mouth,*** also referred to as the ***oral*** or ***buccal*** (BUK-al; *bucca* = cheeks) *cavity,* is formed by the cheeks, hard and soft palates, and tongue (Figure 19.4 on page 436).

The ***lips*** are fleshy folds around the opening of the mouth. They are covered on the outside by skin and on the inside by a mucous membrane. During chewing, the lips and cheeks help keep food between the upper and lower teeth. They also assist in speech.

The hard palate consisting of the maxillae and palatine bones form most of the roof of the mouth. The rest is formed by the muscular soft palate. Hanging from the soft palate is a

Figure 19.2 Three-dimensional drawing depicting the various layers and related structures of the gastrointestinal tract.

🔑 *The four layers of the GI tract from inside to outside are the mucosa, submucosa, muscularis, and serosa.*

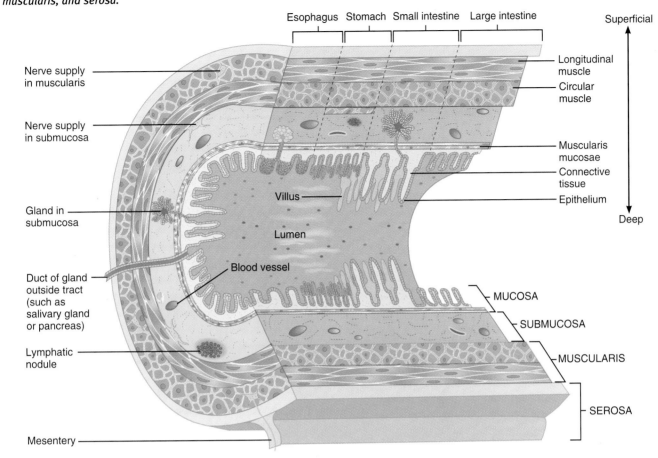

Sectional views of layers of the GI tract

❓ *What is the function of the nerves in the wall of the gastrointestinal tract?*

Figure 19.3 Views of the abdomen and pelvis indicating the relationship of the parts of the peritoneum to each other and to organs of the digestive system.

🔑 *The peritoneum is the largest serous membrane in the body.*

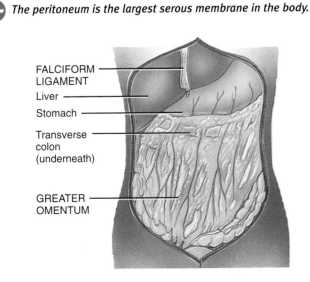

(a) Anterior view

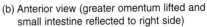

(b) Anterior view (greater omentum lifted and small intestine reflected to right side)

❓ *Which part of the peritoneum binds the small intestine to the posterior abdominal wall?*

Figure 19.4   Structures of the mouth (oral cavity).

🔑 *The mouth is formed by the cheeks, hard and soft palates, and tongue.*

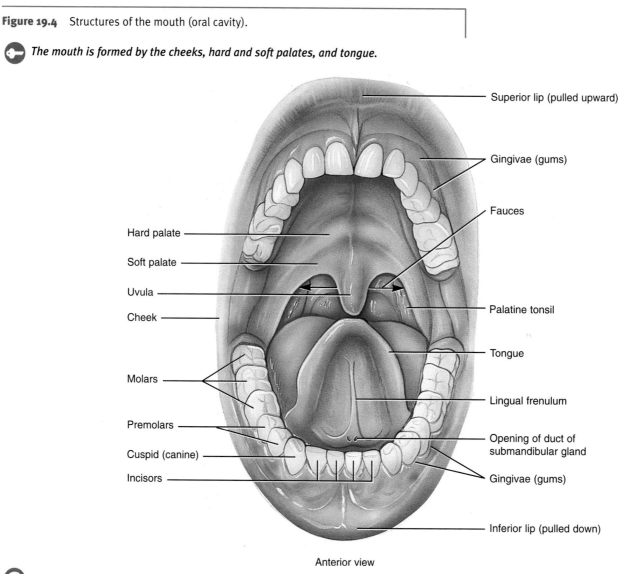

Superior lip (pulled upward)

Gingivae (gums)

Fauces

Hard palate

Soft palate

Uvula

Cheek

Palatine tonsil

Tongue

Molars

Lingual frenulum

Premolars

Cuspid (canine)

Opening of duct of
submandibular gland

Incisors

Gingivae (gums)

Inferior lip (pulled down)

Anterior view

❓ *What are the functions of the muscles of the tongue?*

projection called the *uvula* (YOU-vyoo-la). At the back of the soft palate, the mouth opens into the oropharynx through an opening called the *fauces* (FAW-sēs). Just behind this is the palatine tonsils.

Tongue

The *tongue* forms the floor of the oral cavity. It is an accessory structure of the digestive system composed of skeletal muscle covered with mucous membrane (see Figure 12.3).

The muscles of the tongue maneuver food for chewing, shape the food into a rounded mass, called a *bolus,* force the food to the back of the mouth for swallowing, and alter the shape and size of the tongue for swallowing and speech. The *lingual frenulum,* a fold of mucous membrane in the midline of the undersurface of the tongue, limits the movement of the tongue posteriorly (Figure 19.4). The lingual tonsil lies at the base of the tongue.

The upper surface and sides of the tongue are covered with projections called *papillae* (pa-PIL-ē). Some contain taste buds. Note the taste zones of the tongue in Figure 12.3a.

Salivary Glands

The fluid called saliva is secreted by three pairs of *salivary glands,* accessory structures that lie outside the mouth and pour their contents into ducts emptying into the oral cavity. Their names are parotid, submandibular, and sublingual (see Figure 19.1).

The *parotid glands* are located under and in front of the ears between the skin and the masseter muscle. The *submandibular glands* lie beneath the base of the tongue in the floor of the mouth. The *sublingual glands* lie in front of the submandibular glands.

Mumps is an inflammation and enlargement of the parotid glands accompanied by fever, malaise, a sore throat, and swelling on one or both sides of the face. In about 20 to 35 percent of males past puberty, the testes may also become inflamed, and although it rarely occurs, it may result in sterility.

Composition of Saliva

Chemically, *saliva* is 99.5 percent water and 0.5 percent solutes. One is the digestive enzyme salivary amylase, which starts the digestion of starch in the mouth. The water in saliva provides a medium for dissolving foods so they can be tasted and for starting digestive reactions. Mucus in saliva lubricates food so it can easily be swallowed. The enzyme lysozyme destroys bacteria to protect the mucous membrane from infection and the teeth from decay.

Secretion of Saliva

Salivation (sal-i-VĀ-shun), the secretion of saliva, is entirely under nervous control. Parasympathetic stimulation of the facial (VII) and glossopharyngeal (IX) nerves promotes continuous secretion of saliva to keep the tongue and lips moist. Sympathetic stimulation dominates during stress, resulting in dryness of the mouth. During dehydration, the salivary glands stop secreting to conserve water. The resulting dryness in the mouth contributes to the sensation of thirst. Drinking will then not only moisten the mouth but also restore the homeostasis of body water.

Food stimulates the glands to secrete heavily. Whenever food is tasted, smelled, touched by the tongue, or even thought about, parasympathetic impulses increase secretion of saliva. Saliva continues to be secreted heavily some time after food is swallowed. This flow washes out the mouth and dilutes and buffers chemical remnants of any irritating substances.

Teeth

The *teeth* (*dentes*) are accessory structures of the digestive system located in bony sockets of the mandible and maxillae. The sockets are covered by the *gingivae* (jin-JI-vē; singular is *gingiva*) or *gums* and lined with the *periodontal* (*peri* = around; *odous* = tooth) *ligament*, dense fibrous connective tissue that anchors the teeth to bone, keeps them in position, and acts as a shock absorber during chewing (Figure 19.5).

A typical tooth has three principal parts. The *crown* is the exposed portion above the level of the gums. The *root* consists of one to three projections embedded in the socket.

Figure 19.5   Parts of a typical tooth.

There are 20 teeth in a complete deciduous set and 32 teeth in a complete permanent set.

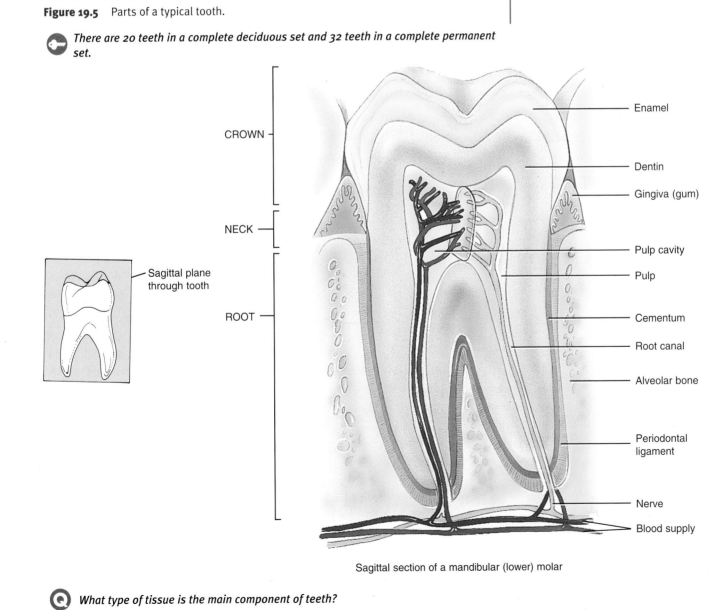

CROWN

NECK

Sagittal plane through tooth

ROOT

Enamel

Dentin

Gingiva (gum)

Pulp cavity

Pulp

Cementum

Root canal

Alveolar bone

Periodontal ligament

Nerve

Blood supply

Sagittal section of a mandibular (lower) molar

What type of tissue is the main component of teeth?

The **neck** is the junction line of the crown and root, near the gum line.

Teeth are composed primarily of **dentin,** a bonelike substance (calcified connective tissue) that gives the tooth its basic shape and rigidity. The dentin encloses the **pulp cavity,** a space in the crown filled with **pulp,** a connective tissue containing blood vessels, nerves, and lymphatic vessels. Narrow extensions of the pulp cavity run through the root of the tooth and are called **root canals.** Each root canal has an opening at its base through which enter blood vessels bearing nourishment, lymphatic vessels affording protection, and nerves providing sensation. The dentin of the crown is covered by **enamel,** a bonelike substance that consists primarily of densely packed calcium phosphate and calcium carbonate. Enamel is the hardest substance in the body. It protects the tooth from the wear of chewing and is a barrier against acids that easily dissolve the dentin. The dentin of the root is covered by **cementum,** another bonelike substance that attaches the root to the periodontal ligament.

The branch of dentistry that is concerned with the prevention, diagnosis, and treatment of diseases that affect the pulp, root, periodontal ligament, and alveolar bone is known as **endodontics** (en′-dō-DON-tiks; *endo* = within). **Orthodontics** (or′-thō-DON-tiks; *ortho* = straight), by contrast, is a branch of dentistry that is concerned with the prevention and correction of abnormally aligned teeth.

Humans have two sets of teeth. The **deciduous teeth** begin to erupt at about 6 months of age, with the incisors, and one pair appears about each month thereafter until all 20 are present. They are generally lost in the same sequence between 6 and 12 years of age. The **permanent teeth** appear between age 6 and adulthood. There are 32 teeth in a complete permanent set.

Humans also have different teeth for different functions (see Figure 19.4). **Incisors** are closest to the midline, are chisel-shaped, and are adapted for cutting into food; **cuspids** (canines) are next to the **incisors** and have one pointed surface (cusp) to tear and shred food; **premolars** have two cusps to crush and grind food; and **molars** have more than two blunt cusps, to crush and grind food.

Digestion in the Mouth

Mechanical

In chewing, or **mastication** (mas′-ti-KĀ-shun; *masticare* = to chew), the tongue moves food, the teeth grind it, and the saliva mixes with it. As a result, the food is reduced to a soft, flexible mass called a **bolus** that is easily swallowed.

Chemical

The enzyme **salivary amylase** (AM-i-lās) starts the breakdown of starch in the mouth. Carbohydrates are either monosaccharide and disaccharide sugars or polysaccharide starches (see Chapter 2). Most of the carbohydrates we eat are polysaccharides. Because only monosaccharides can be absorbed into the bloodstream, ingested disaccharides and polysaccharides must be broken down. Salivary amylase breaks the chemical bonds in starches to reduce the long-chain polysaccharide to the di-

saccharide maltose. Food usually is swallowed too quickly for all the starches to be reduced to disaccharides in the mouth, but salivary amylase in the swallowed food continues to act on starches in the stomach for about another hour until the stomach acids eventually inactivate it.

Pharynx

The **pharynx** (FAIR-inks) is a tube that extends from the internal nares to the esophagus in back and the larynx in front (see Figure 19.6). The pharynx is composed of skeletal muscle and lined by mucous membrane. Food that is swallowed passes from the mouth into the oropharynx and laryngopharynx before passing into the esophagus. Muscular contractions of the oropharynx and laryngopharynx help propel food into the esophagus.

Esophagus

The **esophagus** (e-SOF-a-gus; *oisein* = to carry; *phagema* = food) is a muscular tube that lies behind the trachea (see Figure 18.2). It begins at the end of the laryngopharynx, passes through the mediastinum, pierces the diaphragm, and ends at the top of the stomach. It transports food to the stomach and secretes mucus, which aids transport.

Swallowing

objective: *Describe the stages involved in swallowing.*

Swallowing is a mechanism that moves food from the mouth to the stomach. It is helped by saliva and mucus and involves the mouth, pharynx, and esophagus. Swallowing is divided into three stages: (1) the voluntary stage, (2) the pharyngeal stage, and (3) the esophageal stage.

Swallowing starts when the bolus is forced to the back of the mouth cavity and into the oropharynx by the movement of the tongue upward and backward against the palate. This is the **voluntary stage.**

With the passage of the bolus into the oropharynx, the involuntary **pharyngeal stage** of swallowing begins (Figure 19.6b). During this time, breathing is temporarily interrupted. This occurs when the soft palate and uvula move upward to close off the nasopharynx, the epiglottis seals off the larynx, and the vocal cords come together. After the bolus passes through the oropharynx, the respiratory passageways reopen and breathing resumes.

In the **esophageal stage,** food is pushed through the esophagus by peristalsis. It occurs as follows (Figure 19.6c): In the section of esophagus above the bolus, the circular muscle fibers contract. This constricts the wall of the esophagus and squeezes the bolus downward. Meanwhile, longitudinal fibers around the bottom of the bolus contract. This shortens the lower section, pushing its walls outward so it can receive the bolus. The contractions move down the esophagus, pushing the food into the stomach.

Figure 19.6   Swallowing. During the pharyngeal stage of swallowing, the tongue rises against the palate, the nasopharynx is closed off, the larynx rises, the epiglottis seals off the larynx, and the bolus is passed into the esophagus.

🔑 *Swallowing is a mechanism that moves food from the mouth into the stomach.*

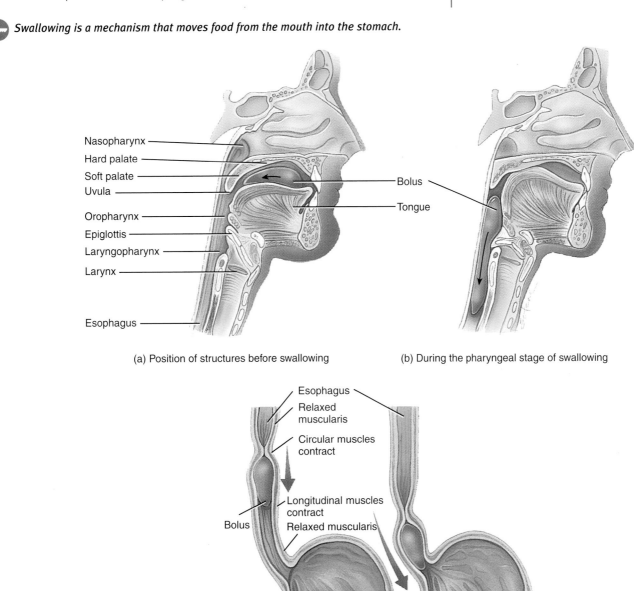

(a) Position of structures before swallowing

(b) During the pharyngeal stage of swallowing

(c) Esophageal stage of swallowing

❓ *Is swallowing a voluntary or involuntary action?*

Sometimes, after food has entered the stomach, the stomach contents can back up into the lower esophagus. Hydrochloric acid (HCl) from the contents can irritate the esophageal wall, resulting in a burning sensation known as ***heartburn,*** because it is experienced in the region near the heart, although it is not related to any cardiac problem. Heartburn can be treated by taking antacids that neutralize the hydrochloric acid and lessen the discomfort.

Stomach

objective: *Explain the structure and functions of the stomach in digestion.*

The ***stomach*** is a J-shaped region of the GI tract directly under the diaphragm. The superior portion of the stomach is connected to the esophagus. The inferior portion empties into the duodenum, the first part of the small intestine.

Anatomy

The stomach is divided into four main areas: cardia, fundus, body, and pylorus (Figure 19.7). The *cardia* (CAR-dē-a) surrounds the superior opening of the stomach. The portion above and to the left of the cardia is the *fundus* (FUN-dus). Below the fundus is the large central portion of the stomach, called the *body.* The narrow, inferior region is the *pylorus* (pī-LOR-us; *pyel* = gate; *ouros* = guard). Between the pylorus and duodenum of the small intestine is the *pyloric sphincter* or *valve* (a sphincter is a thick circle of muscle around an opening).

The stomach wall is composed of the same four basic layers as the rest of the GI tract, with certain modifications. When the stomach is empty, the mucosa lies in large folds, called *rugae* (ROO-gē = wrinkles), that can be seen with the naked eye. The surface of the mucosa is a layer of nonciliated simple columnar epithelial cells called *mucous surface cells* (Figure 19.8). Epithelial cells also extend downward, forming many narrow channels called *gastric pits* which are lined with secretory cells called *gastric glands.* Secretions from the gastric glands flow into the gastric pits and then into the lumen of the stomach. The glands contain three types of *exocrine gland* cells that secrete their products into the stomach: mucous neck cells, chief cells, and parietal cells. In addition, gastric glands include one type of hormone-producing cell (G cell).

Both mucous surface cells and *mucous neck cells* secrete mucus. The *chief cells* secrete an inactive gastric enzyme called *pepsinogen. Parietal cells* produce hydrochloric acid, which helps convert pepsinogen to the active enzyme pepsin, and intrinsic factor, involved in the absorption of vitamin B_{12}. Inadequate production of intrinsic factor can result in pernicious anemia because vitamin B_{12} is needed for red blood cell production. The secretions of the mucous, chief, and parietal cells are collectively called *gastric juice.* The *G cells* secrete the hormone gastrin into the bloodstream. Gastrin stimulates secretion of gastric juice, increases motility (movement) of the GI tract, and relaxes the pyloric sphincter.

The submucosa of the stomach is composed of areolar connective tissue, which connects the mucosa to the muscularis. The muscularis has three rather than two layers of smooth muscle: an outer longitudinal layer, a middle circular layer, and an inner oblique layer. The muscularis breaks food into small particles, mixes it with gastric juice, and passes it to the duodenum. The serosa (simple squamous epithelium and areolar connective tissue) covering the stomach is part of the visceral peritoneum.

Figure 19.7 External and internal anatomy of the stomach.

🔑 *The four regions of the stomach are the cardia, fundus, body, and pylorus.*

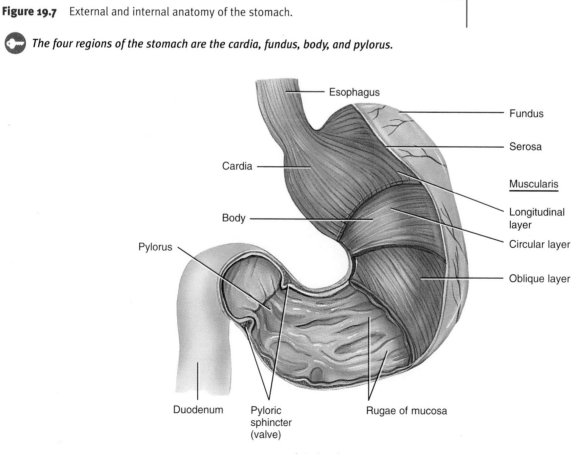

Anterior view

❓ *After a very large meal, does your stomach have rugae?*

Figure 19.8 Histology of the stomach.

The secretions of the mucous, chief, and parietal cells are referred to as gastric juice.

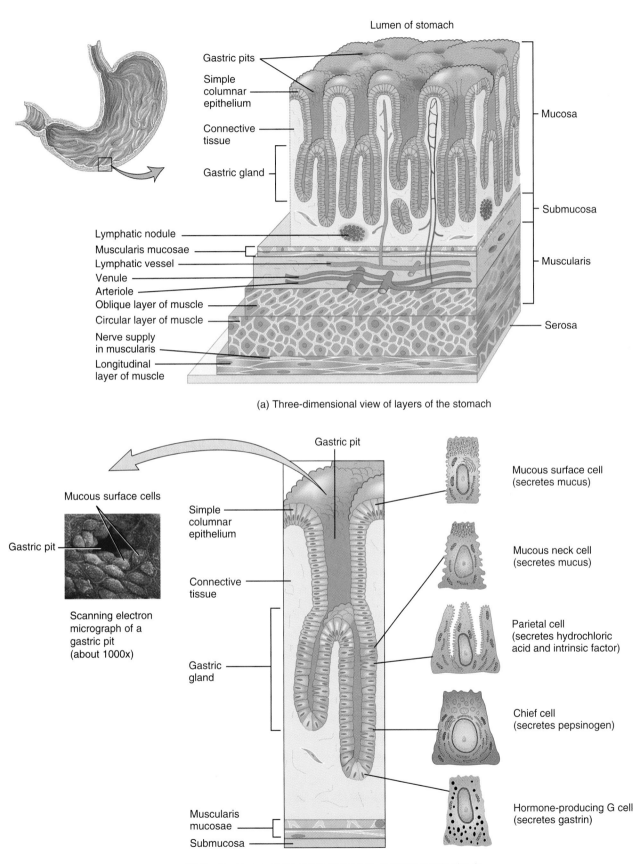

(a) Three-dimensional view of layers of the stomach

(b) Sectional view of the stomach mucosa showing gastric glands

What are the functions of gastrin?

Digestion in the Stomach

Mechanical

Several minutes after food enters the stomach, gentle, rippling, peristaltic movements called *mixing waves* pass over the stomach about every 20 seconds. These waves macerate food, mix it with the secretions of the gastric glands, and change it to a thin liquid called *chyme* (kīm; *chymos* = juice). Each mixing wave forces a small amount of the gastric contents into the duodenum through the pyloric sphincter. Most of the food is forced back into the body of the stomach where it is subjected to further mixing. The next wave pushes it forward again and forces a little more into the duodenum. The forward and backward movements of the gastric contents are responsible for almost all the mixing in the stomach.

Chemical

The principal chemical activity of the stomach is to begin the digestion of proteins by *pepsin*. Pepsin breaks certain peptide bonds between the amino acids making up proteins. Thus a protein chain of many amino acids is broken down into smaller fragments called *peptides*. Pepsin is most effective in the very acidic environment of the stomach (pH 2).

What keeps pepsin from digesting the protein in stomach cells along with the food? First, pepsin is secreted in an inactive form (pepsinogen) so it cannot digest the proteins in the chief cells that produce it. It is not converted into active pepsin until it comes in contact with the hydrochloric acid secreted by the parietal cells. Second, the stomach cells are protected by mucus secreted by mucous cells. The mucus coats the mucosa to form a barrier between it and the gastric juices.

Regulation of Gastric Secretion

Stimulation

The secretion of gastric juice and contraction of smooth muscle in the stomach wall are regulated by both nervous and hormonal mechanisms. Parasympathetic impulses from the medulla oblongata are transmitted via the vagus (X) nerves.

The impulses promote peristalsis and stimulate the gastric glands to secrete.

The sight, smell, taste, or thought of food initiates a reflex that stimulates gastric secretion before food even enters the stomach. The gastric glands are stimulated by parasympathetic impulses that originate in the cerebral cortex.

Once food reaches the stomach, both nervous and hormonal mechanisms ensure that gastric secretion continues. Food of any kind causes stretching, which stimulates receptors in the wall of the stomach to send impulses to the medulla and back to the gastric glands. Partially digested proteins, caffeine, and a high pH of stomach chyme also stimulate the secretion of gastric juice by triggering the release of stomach gastrin.

Inhibition

The presence of food in the small intestine ultimately inhibits parasympathetic stimulation and stimulates sympathetic activity. This inhibits gastric secretion. Negative emotions, such as anger, may also slow down digestion by stimulating the sympathetic nervous system.

Several small intestinal hormones are involved in inhibiting gastric secretion. One is *gastric inhibitory peptide* (GIP), which inhibits gastric secretion and mobility. Another is *secretin* (se-KRĒ-tin), which decreases gastric secretion. Finally, there is *cholecystokinin* (kō-lē-sis′-tō-KĪN-in) or **CCK** (Exhibit 19.1), which inhibits gastric emptying.

Regulation of Gastric Emptying

Gastric emptying is the release of chyme from the stomach into the duodenum. The stomach empties all its contents into the duodenum within 2 to 6 hours after ingestion. Foods rich in carbohydrate spend the least time in the stomach. Protein foods are somewhat slower, and emptying is slowest after a meal containing large amounts of triglycerides (fats).

Stomach emptying is regulated by both nervous and hormonal controls. The presence of chyme in the duodenum triggers a reflex that slows gastric emptying. It is also inhibited by cholecystokinin. These controls keep the rate of stomach emp-

Exhibit 19.1	Hormonal Control of Gastric Secretion, Pancreatic Secretion, and Secretion and Release of Bile			
HORMONE	WHERE PRODUCED	STIMULANT		ACTION
Gastrin	Pyloric mucosa	Stretching of stomach, partially digested proteins and caffeine in stomach, and high pH of stomach chyme		Stimulates secretion of gastric juice, increases movement of GI tract, and relaxes pyloric sphincter and ileocecal sphincter.
Gastric inhibitory peptide (GIP)	Intestinal mucosa	Fatty acids and glucose in small intestine		Stimulates release of insulin by pancreas, inhibits gastric secretion, and slows gastric emptying.
Secretin	Intestinal mucosa	Acid chyme that enters the small intestine		Inhibits secretion of gastric juice and stimulates secretion of pancreatic juice rich in bicarbonate ions.
Cholecystokinin (CCK)	Intestinal mucosa	Partially digested proteins (amino acids) and triglycerides (fatty acids)		Inhibits gastric emptying, stimulates the secretion of pancreatic juice rich in digestive enzymes, causes ejection of bile from the gallbladder, and induces a feeling of satiety (feeling full to satisfaction).

tying at the same rate the intestine can handle its final digestion of the chyme.

Vomiting is the forcible expulsion of the contents of the upper GI tract (stomach and sometimes duodenum) through the mouth. The strongest stimuli for vomiting are irritation and stretching of the stomach. Other stimuli include unpleasant sights, dizziness, and certain drugs such as morphine. Impulses are transmitted to the vomiting center in the medulla oblongata. Impulses to the upper GI tract organs, diaphragm, and abdominal muscles bring about the vomiting act. Basically, vomiting involves squeezing the stomach between the diaphragm and abdominal muscles and expelling the contents into the esophagus and through the mouth. Prolonged vomiting, especially in infants and elderly people, can be serious because the loss of gastric juice and fluids leads to disturbances in fluid and acid–base balance.

Absorption

The stomach wall is impermeable to the passage of most materials into the blood, so most substances are not absorbed until they reach the small intestine. However, the stomach does absorb some water, electrolytes, certain drugs (especially aspirin), and alcohol. The absorption of alcohol by the stomach of females is faster than that in males. The difference is attributed to smaller amounts of the enzyme alcohol dehydrogenase in the stomachs of females. The enzyme breaks down alcohol in the stomach, reducing the amount of alcohol that enters the blood.

After the stomach, the next organ of the GI tract involved in the breakdown of food is the small intestine. Chemical digestion in the small intestine depends not only on its own secretions but also on activities of three accessory structures: the pancreas, liver, and gallbladder.

Pancreas

| objective: | *Describe the structure of the pancreas, liver, and gallbladder and explain their functions in digestion.* |

Anatomy

The *pancreas* (*pan* = all; *kreas* = flesh) lies behind the stomach (Figure 19.9). Secretions pass from the pancreas to the duodenum by the *pancreatic duct,* which unites with the common bile duct from the liver and gallbladder to enter the duodenum as a single duct.

The pancreas is made up of two types of cells. Small clusters of glandular epithelial cells, the *pancreatic islets* (*islets of Langerhans*), constitute the endocrine portion of the pancreas, which produces the hormones glucagon and insulin (see Figure 13.17b). The *acini* (AS-i-nē) are exocrine glands that secrete a mixture of digestive enzymes called *pancreatic juice* (see Figure 13.17b).

Pancreatic Juice

Pancreatic juice is a clear, colorless liquid that consists mostly of water, some salts, sodium bicarbonate, and enzymes. The sodium bicarbonate gives pancreatic juice a slightly alkaline pH (7.1 to 8.2) that stops the action of pepsin from the stomach and creates the proper environment for the enzymes in the small intestine. The enzymes in pancreatic juice include a carbohydrate-digesting enzyme called *pancreatic amylase;* several protein-digesting enzymes called *trypsin* (TRIP-sin), *chymotrypsin* (KĪ-mō-trip-sin), and *carboxypeptidase* (kar-bok′-sē-PEP-ti-dās); the principal triglyceride-digesting enzyme in adults called *pancreatic lipase;* and nucleic-acid-digesting enzymes called *ribonuclease* and *deoxyribonuclease.*

The protein-digesting enzymes are produced in inactive form to prevent them from digesting the pancreas itself. The inactive form is activated in the small intestine.

Regulation of Pancreatic Secretions

Pancreatic secretion, like gastric secretion, is also regulated by both nervous and hormonal mechanisms. The secretion of pancreatic enzymes is also stimulated by parasympathetic fibers from the vagus (X) nerves and by the hormones secretin and cholecystokinin (see Exhibit 19.1).

Liver

The *liver* is the heaviest gland of the body, weighing about 1.4 kg (about 3 lb) in the average adult. It is located under the diaphragm, mostly on the right side of the body. It is covered by a connective tissue capsule, which, in turn, is covered by peritoneum, the serous membrane that covers all the viscera.

Anatomy

The liver is divided by the falciform ligament into two principal lobes: the *right lobe* and the *left lobe* (Figure 19.9).

The lobes are made up of functional units called *lobules* (Figure 19.10 on page 445). A lobule consists of rows of *hepatocytes* (*hepato* = liver) arranged radially around a *central vein.* Hepatocytes produce bile, the function of which is described shortly. Between the rows of hepatocytes are large capillaries called *sinusoids,* through which blood passes. The sinusoids are partly lined with phagocytic *stellate reticuloendothelial* (*Kupffer's*) *cells* that destroy worn-out white and red blood cells, bacteria, and toxic substances.

Bile produced by the hepatocytes enters the *right* and *left hepatic ducts,* which unite to leave the liver as the *common hepatic duct* (see Figure 19.9). The common hepatic duct then joins the *cystic duct* from the gallbladder and the two tubes become the *common bile duct.* The common bile duct and pancreatic duct then enter the duodenum in a single duct.

Blood Supply

The liver receives blood from two sources. From the hepatic artery it obtains oxygenated blood, and from the hepatic portal vein (see Figure 16.11) it receives deoxygenated blood containing newly absorbed nutrients. Branches of both the hepatic artery and the hepatic portal vein carry blood to the sinusoids of the lobules, where oxygen, most of the nutrients, and toxins

Figure 19.9   Relation of the pancreas to the liver, gallbladder, and duodenum. The inset shows details of the common bile duct and pancreatic duct forming the common duct that empties into the duodenum.

🔑 *A lobule consists of hepatocytes arranged around a central vein.*

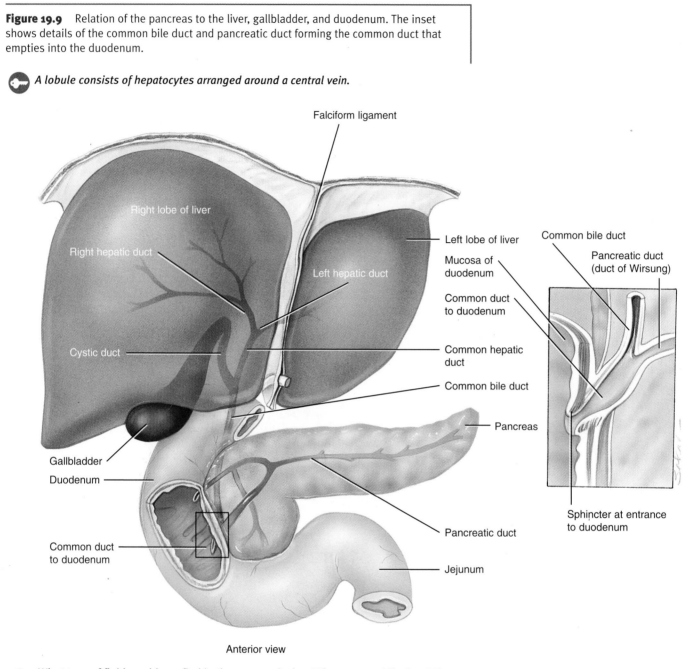

Anterior view

❓ *What type of fluid would you find in the pancreatic duct? The common bile duct? The common duct to the duodenum?*

are extracted by the hepatocytes. Reticuloendothelial (Kupffer's) cells remove microbes and foreign or dead matter from the blood. Nutrients are stored or used to make new materials. The poisons are stored or detoxified. Products manufactured by the hepatic cells and nutrients needed by other cells are secreted back into the blood. The blood then drains into the central vein, the hepatic vein, the inferior vena cava, and finally to the heart.

Bile 

Hepatocytes secrete *bile,* a yellow, brownish, or olive-green liquid. It has a pH of 7.6 to 8.6. Bile consists mostly of water and

bile salts, cholesterol, a phospholipid called lecithin, bile pigments, and several ions. It is produced in the liver but stored in the gallbladder.

Bile is partially an excretory product and partially a digestive secretion. Bile salts aid in *emulsification,* the conversion of triglyceride (fat) globules into a suspension of triglyceride droplets, and absorption of triglycerides following their digestion. The tiny triglyceride droplets present a very large surface area for the action of pancreatic and intestinal lipase to digest them rapidly. The principal bile pigment is *bilirubin.* When worn-out red blood cells are broken down, iron, globin, and bilirubin (derived from heme) are released. The iron and globin are recycled, but some of the bilirubin is excreted in bile.

Figure 19.10 Histology of the liver.

🔑 *Pancreatic enzymes are involved in digestion of starches (polysaccharides), proteins, triglycerides, and nucleic acids.*

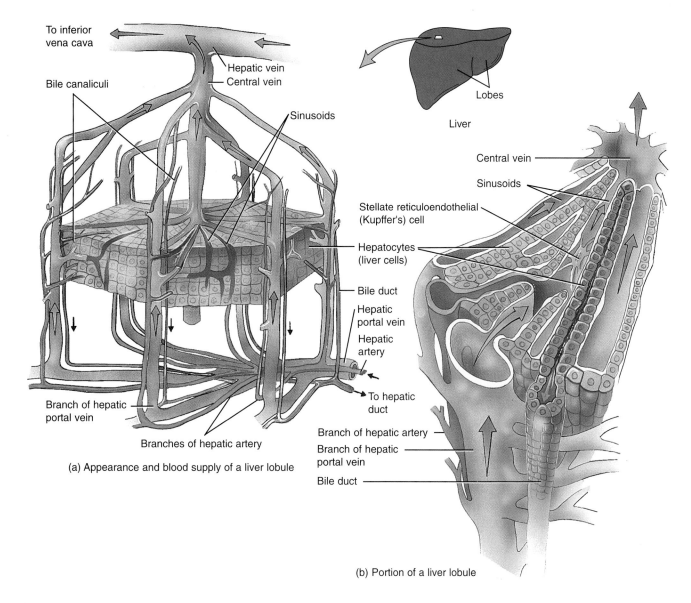

(a) Appearance and blood supply of a liver lobule

(b) Portion of a liver lobule

❓ *From the hepatic portal vein, what is the path of blood flow through the liver?*

Bilirubin eventually is broken down in the intestine, and one of its breakdown products (stercobilin) gives feces their normal color (see Figure 14.4).

Regulation of Bile Secretion

The rate at which bile is secreted is determined by both nervous and hormonal factors. Parasympathetic stimulation by way of the vagus (X) nerves increases the production of bile. Secretin also stimulates the secretion of bile (see Exhibit 19.1). Cholecystokinin (CCK) causes ejection of bile from the gallbladder.

Functions of the Liver

Many of the liver's vital functions are related to metabolism and are discussed in Chapter 20. Briefly, however, the main functions include the following:

1. **Carbohydrate metabolism.** In carbohydrate metabolism, the liver is especially important in maintaining a normal blood glucose level. For example, the liver can convert glucose to glycogen when blood sugar level is high and break down glycogen to glucose when blood sugar level is low. The liver can also convert certain amino acids and

lactic acid to glucose when blood sugar level is low; convert other sugars, such as fructose and galactose, into glucose; and convert glucose to triglycerides for storage.

2. **Lipid metabolism.** With respect to triglyceride metabolism, the liver stores some triglycerides and breaks down fatty acids.

3. **Protein metabolism.** Hepatocytes synthesize most plasma proteins, such as globulins, albumin, prothrombin, and fibrinogen. Liver enzymes can convert one amino acid into another. The liver alters amino acids so that they can be used for ATP production or converted to carbohydrates or triglycerides. It converts ammonia (NH_3), a toxic substance produced by amino acid alteration, into the much less toxic urea for excretion in urine.

4. **Removal of drugs and hormones.** The liver can detoxify substances such as alcohol or excrete into bile drugs such as penicillin, erythromycin, and sulfonamides. It can also chemically alter or excrete thyroid hormones and steroid hormones, such as estrogens and aldosterone.

5. **Excretion of bile.** As noted earlier, bilirubin, derived from the heme of worn-out red blood cells, is absorbed by the liver from the blood and secreted into bile. Most of the bilirubin in bile is metabolized in the large intestine by bacteria and eliminated in feces.

6. **Synthesis of bile salts.** Bile salts are used in the small intestine for the emulsification and absorption of triglycerides, cholesterol, phospholipids, and lipoproteins.

7. **Storage.** In addition to glycogen, the liver stores vitamins (A, B_{12}, D, E, and K) and minerals (iron and copper).

8. **Phagocytosis.** The stellate reticuloendothelial (Kupffer's) cells phagocytize worn-out red and white blood cells and some bacteria.

9. **Activation of vitamin D.** The skin, liver, and kidneys participate in the activation of vitamin D.

Gallbladder

The *gallbladder* (*galla* = bile) is a pear-shaped sac located in a depression under the liver (see Figure 19.9). The muscularis, the middle coat of the wall, consists of smooth muscle fibers that contract following hormonal stimulation to eject the contents of the gallbladder into the *cystic* (*kysitis* = bladder) *duct.*

Functions

The gallbladder concentrates and stores bile until it is needed in the small intestine. Bile enters the small intestine through the common bile duct. When the small intestine is empty, a valve around the common duct closes and bile backs up into the cystic duct to the gallbladder for storage.

Emptying of the Gallbladder

When triglycerides enter the small intestine, cholecystokinin (CCK) is released to stimulate contraction of the gallbladder's

muscular layer. Bile is then emptied into the common bile duct to flow into the small intestine (see Exhibit 19.1).

Small Intestine

objective: *Explain how the small intestine is adapted for digestion and absorption.*

Most digestion and absorption occur in the *small intestine.* It is a tube that averages about 2.5 cm (1 inch) in diameter and about 6.50 m (21 feet) in length in a cadaver and about 3 m (10 feet) in length in a living person (due to the difference in muscle tone).

Anatomy

The small intestine is divided into three segments (Figure 19.11). The *duodenum* (doo'-ō-DĒ-num), the shortest part (about 25 cm or 10 inches), originates at the pyloric sphincter of the stomach and merges with the jejunum. *Duodenum* means "twelve"; the structure is twelve fingers' breadth in length. The *jejunum* (jē-JOO-num) is about 1 m (3 feet) long and extends to the ileum. *Jejunum* means "empty," because at death it is found empty. The final portion of the small intestine, the *ileum* (IL-ē-um), measures about 2 m (6 feet) and joins the large intestine at the *ileocecal* (il'-ē-ō-SĒ-kal) *sphincter* (*valve*).

Because almost all the absorption of nutrients occurs in the small intestine, its structure is specially adapted for this function. Its length provides a large surface area for absorption and that area is further increased by modifications in the structure of its wall.

Figure 19.11 Divisions of the small intestine.

Most digestion and absorption occur in the small intestine.

DUODENUM

Stomach

Large intestine

JEJUNUM

ILEUM

Anterior view

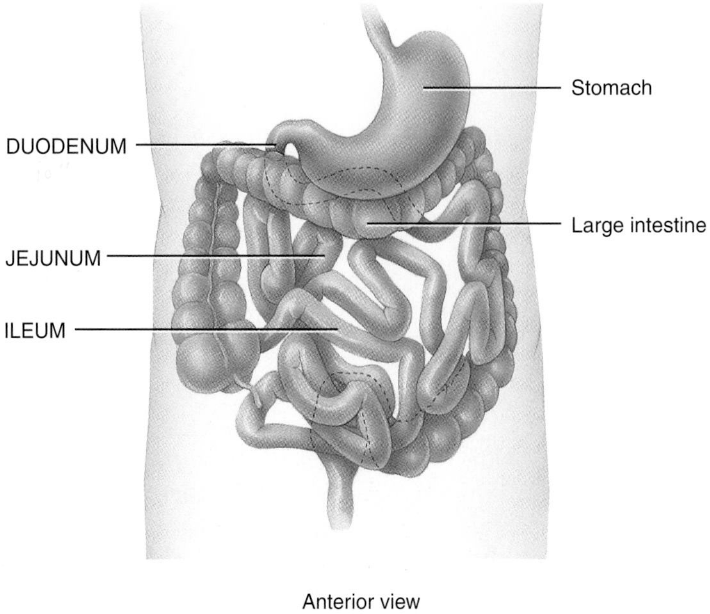

Which portion of the small intestine is the longest?

The wall of the small intestine is composed of the same four layers that make up most of the GI tract. However, special features of both the mucosa and the submucosa allow the small intestine to complete the processes of digestion and absorption (Figure 19.12). The mucosa forms a series of fingerlike *villi* (= tuft of hair; singular is *villus*). These projections are 1/2 to 1 mm long and give the intestinal mucosa a velvety appearance (Figure 19.12). The large number of villi (20–40 per square millimeter) vastly increases the surface area of the epithelium available for absorption and digestion. Each villus has a core of connective tissue that contains an arteriole, a venule, a capillary network, and a *lacteal* (LAK-tē-al), which is a lymphatic capillary. Nutrients absorbed by the epithelial cells covering the villus pass through the wall of a capillary or a lacteal to enter blood or lymph, respectively.

The epithelium of the mucosa consists of simple columnar epithelium that contains absorptive cells, goblet cells (secrete mucus), hormone-producing cells, and Paneth cells. The membrane of the absorptive cells features *microvilli* (mī´-krō-VIL-i), microscopic, cytoplasmic projections that greatly increase the surface area and allow larger amounts of digested nutrients to be digested and absorbed. It is estimated that 1 square millimeter of small intestine contains about 200 million microvilli.

In addition to the microvilli and villi, a third set of projections called *circular folds* further increases the surface area for absorption and digestion. The folds are permanent ridges in the mucosa. They enhance absorption by causing the chyme to spi-ral, rather than to move in a straight line, as it passes through the small intestine. The folds and villi decrease in size in the ileum, so most absorption occurs in the duodenum and jejunum. The absorptive surface area of the small intestine is increased about 600 times by the villi, microvilli, and circular folds.

The mucosa contains many cavities lined with glandular epithelium. Cells lining the cavities form the *intestinal glands* and secrete intestinal juice. *Paneth cells* are found in the deepest parts of the intestinal glands. They secrete lysozyme, an enzyme that kills bacteria, and are also capable of phagocytosis. They may have a role in regulating the microbial population in the intestines. Hormone-producing cells, also in the deepest part of the intestinal glands, secrete three hormones: secretin, cholecystokinin, and gastric inhibitory peptide. Their functions are described in Exhibit 19.1.

The connective tissue of the mucosa of the small intestine has an abundance of mucosa-associated lymphoid tissue (MALT). *Solitary lymphatic nodules* are most numerous in the ileum. Groups of lymphatic nodules, referred to as *aggregated lymphatic follicles* (*Peyer's patches*), are also numerous in the ileum. The submucosa of the duodenum contains *duodenal glands.* They secrete an alkaline mucus that helps neutralize gastric acid in the chyme.

The muscularis of the small intestine consists of two layers of smooth muscle. The outer, thinner layer contains longitudinally arranged fibers. The inner, thicker layer contains circularly

---

Figure 19.12   Small intestine. Shown are various structures that adapt the small intestine for digestion and absorption.

 Villi, microvilli, and circular folds increase the surface area of the small intestine for digestion and absorption.

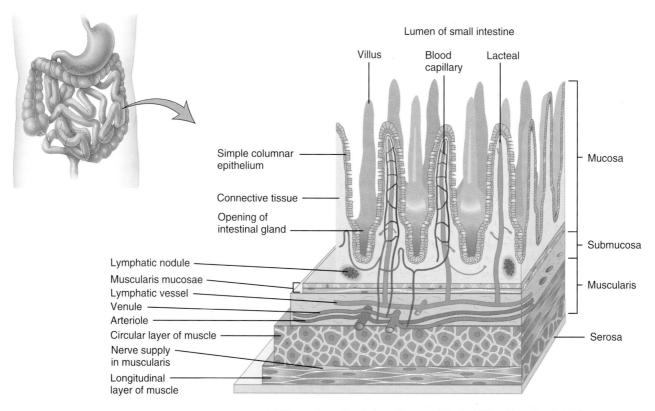

(a) Three-dimensional view of layers of the small intestine showing villi

Figure 19.12   *(Continued)*

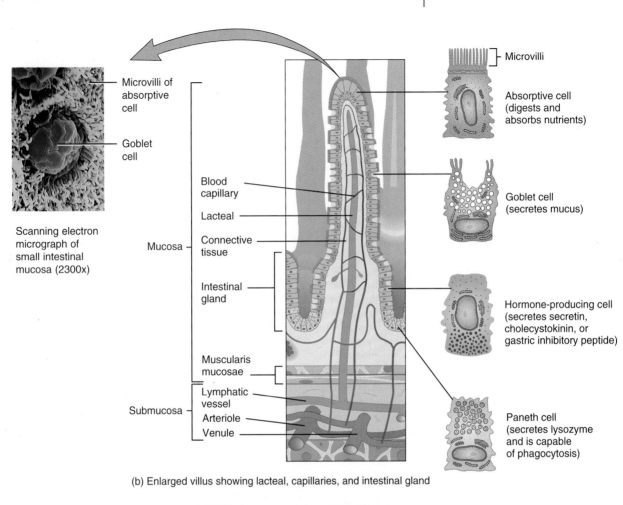

Microvilli of absorptive cell

Goblet cell

Scanning electron micrograph of small intestinal mucosa (2300x)

Mucosa

Blood capillary

Lacteal

Connective tissue

Intestinal gland

Muscularis mucosae

Lymphatic vessel

Arteriole

Venule

Submucosa

Microvilli

Absorptive cell (digests and absorbs nutrients)

Goblet cell (secretes mucus)

Hormone-producing cell (secretes secretin, cholecystokinin, or gastric inhibitory peptide)

Paneth cell (secretes lysozyme and is capable of phagocytosis)

(b) Enlarged villus showing lacteal, capillaries, and intestinal gland

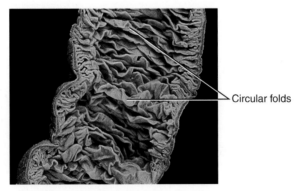

Circular folds

(c) Photograph of jejunum cut open to expose the circular folds

Q *How many villi and microvilli are present in a square millimeter of small intestinal mucosa?*

arranged fibers. Except for a major portion of the duodenum, the serosa completely surrounds the small intestine.

Intestinal Juice

Intestinal juice, secreted by the intestinal glands, is a clear yellow fluid that has a pH of 7.6 (slightly alkaline) and contains water and mucus. The juice is rapidly reabsorbed by the villi and provides a vehicle for the absorption of substances from chyme as they come in contact with the villi.

Intestinal enzymes are synthesized in the epithelial cells that line the villi and most digestion by enzymes of the small intestine occurs in or on the epithelial cells that line the villi.

Digestion in the Small Intestine

Mechanical

The movements of the small intestine are of two types: segmentation and peristalsis. **Segmentation** is the major movement. It is strictly a localized contraction in areas containing food that sloshes chyme back and forth, mixing it with digestive juices and bringing food particles into contact with the mucosa for absorption. It does not push the intestinal contents along the tract. The movement is similar to alternately squeezing opposite ends of a tube of toothpaste. Segmentation depends mainly on parasympathetic impulses. Sympathetic impulses decrease intestinal motility.

Peristalsis propels the chyme through the intestinal tract. Peristaltic contractions in the small intestine are normally very weak compared to those in the esophagus or stomach. Peristalsis, like segmentation, is initiated by stretching and controlled by the autonomic nervous system.

Chemical

In the mouth, salivary amylase converts starch (polysaccharide) to maltose (disaccharide). In the stomach, pepsin converts proteins to peptides (small proteins). Thus chyme entering the small intestine contains partially digested carbohydrates and proteins. The completion of digestion is a collective effort of pancreatic juice, bile, and intestinal juice in the small intestine.

CARBOHYDRATES　Starches not reduced to maltose by the time chyme leaves the stomach are broken down by *pancreatic amylase,* an enzyme in pancreatic juice that acts in the small intestine. Although amylase acts on both glycogen and starches, it does not act on the polysaccharide cellulose, an indigestible plant fiber.

Three enzymes in intestinal juice digest the disaccharides into monosaccharides: *maltase* splits maltose into two molecules of glucose; *sucrase* breaks sucrose into a molecule of glucose and a molecule of fructose; and *lactase* digests lactose into a molecule of glucose and a molecule of galactose. This completes the digestion of carbohydrates because monosaccharides are small enough to be absorbed.

PROTEINS　Protein digestion starts in the stomach, where *pepsin* fragments proteins into peptides. Enzymes found in pancreatic juice (trypsin, chymotrypsin, and carboxypeptidase) continue the digestion, though their actions differ somewhat because each splits peptide bonds between different amino acids. Protein digestion is completed by *peptidases,* enzymes produced by epithelial cells that line the villi. The final products of protein digestion are amino acids.

LIPIDS　In an adult, most lipid digestion occurs in the small intestine. In the first step of lipid digestion, bile salts emulsify globules of triglycerides (fats) into small triglyceride droplets. These droplets consist of a molecule of glycerol and three molecules of fatty acid (see Figure 2.9). Now the triglyceride-splitting enzyme can get at the lipid molecules. In the second step, *pancreatic lipase,* found in pancreatic juice, breaks down each triglyceride molecule by removing two of the three fatty acids from glycerol; the third remains attached to the glycerol, thus forming fatty acids and monoglycerides, the end products of triglyceride digestion.

NUCLEIC ACIDS　Both intestinal juice and pancreatic juice contain **nucleases** that digest nucleotides into their constituent pentoses and nitrogenous bases. **Ribonuclease** acts on ribonucleic acid nucleotides, and **deoxyribonuclease** acts on deoxyribonucleic acid nucleotides.

A summary of digestive enzymes is presented in Exhibit 19.2.

Regulation of Intestinal Secretion and Motility

The most important means for regulating small intestinal secretion and motility are local reflexes in response to the presence of chyme. Segmentation movements depend mainly on intestinal stretching, which initiates nerve impulses to the central nervous system. Local reflexes and returning parasympathetic impulses from the CNS increase motility. Sympathetic impulses decrease intestinal motility. The first remnants of a meal reach the beginning of the large intestine in about 4 hours.

Absorption

All the chemical and mechanical phases of digestion from the mouth down through the small intestine are directed toward changing food into molecules that can pass through the epithelial cells of the mucosa into the underlying blood and lymphatic vessels. These are monosaccharides, amino acids, fatty acids, and monoglycerides. Passage of these digested nutrients from the gastrointestinal tract into the blood or lymph is called *absorption.*

About 90 percent of all absorption takes place in the small intestine. The other 10 percent occurs in the stomach and large intestine. Any undigested or unabsorbed material left in the small intestine is passed on to the large intestine. Absorption in the small intestine occurs by diffusion, facilitated diffusion, osmosis, and active transport.

Carbohydrate Absorption

Essentially all carbohydrates are absorbed as monosaccharides. Glucose and galactose are transported into epithelial cells of the villi by active transport. Fructose is transported by facilitated diffusion (Figure 19.13a on page 451). Transported monosaccharides then move out of the epithelial cells by facilitated diffusion and enter the capillaries of the villi. From here they are transported to the liver via the hepatic portal system, then through the heart and to general circulation (Figure 19.13b).

Protein Absorption

Most proteins are absorbed as amino acids and the process occurs mostly in the duodenum and jejunum. Amino acid transport into epithelial cells of the villi occurs by active transport and from there by diffusion into the bloodstream (Figure 19.12a). They follow the same route as monosaccharides (Figure 19.13b).

Lipid Absorption

As a result of emulsification and triglyceride digestion, triglycerides are broken down into monoglycerides and fatty acids. Short-chain fatty acids (fewer than 10 to 12 carbon atoms)

Exhibit 19.2	Summary of Digestive Enzymes		
ENZYME	SOURCE	SUBSTRATE	PRODUCT
Carbohydrate-digesting			
Salivary amylase	Salivary glands	Starches (polysaccharides)	Maltose (disaccharide)
Pancreatic amylase	Pancreas	Starches (polysaccharides)	Maltose (disaccharide)
Maltase	Small intestine	Maltose	Glucose
Sucrase	Small intestine	Sucrose	Glucose and fructose
Lactase	Small intestine	Lactose	Glucose and galactose
Protein-digesting			
Pepsin (activated from pepsinogen by hydrochloric acid)	Stomach (peptic cells)	Proteins	Peptides
Trypsin (activated from trypsinogen by enterokinase)	Pancreas	Proteins	Peptides
Chymotrypsin (activated from chymotrypsinogen by trypsin)	Pancreas	Proteins	Peptides
Carboxypeptidase (activated from procarboxypeptidase by trypsin)	Pancreas	Terminal amino acid at carboxyl (acid) end of peptides	Peptides and amino acids
Peptidases	Small intestine	Terminal amino acids at amino end of peptides and dipeptides	Amino acids
Lipid-digesting			
Pancreatic lipase	Pancreas	Triglycerides (fats) that have been emulsified by bile salts	Fatty acids and monoglycerides
Nucleases			
Ribonuclease	Pancreas and small intestine	Ribonucleic acid nucleotides	Pentoses and nitrogenous bases
Deoxyribonuclease	Pancreas and small intestine	Deoxyribonucleic acid nucleotides	Pentoses and nitrogenous bases

pass into the epithelial cells by diffusion and follow the same route taken by monosaccharides and amino acids (Figure 19.13a,b). Most fatty acids, however, are long-chain fatty acids. They and the monoglycerides are transported with the help of bile salts. The bile salts form tiny spheres called *micelles* (mī-SELZ). During triglyceride digestion, fatty acids and monoglycerides dissolve in the micelles, and it is in this form that they reach the epithelial cells of the villi.

Within the epithelial cells, many monoglycerides, glycerol, and fatty acids are recombined into triglycerides by the epithelial cell. Here, the triglycerides come together and are coated with proteins to form large spheres called *chylomicrons* (Figure 19.13a). The protein coat keeps the globules from sticking to each other. They leave the epithelial cell to enter the lacteal of a villus. From here, they move into the lymphatic system and, eventually, empty into the bloodstream by way of the subclavian veins (Figure 19.13b).

The plasma lipids—fatty acids, triglycerides, cholesterol—are insoluble in water and body fluids. In order to be transported in blood and utilized by body cells, they must be combined with protein transporters to make them soluble. The combination of lipid and protein is referred to as a *lipoprotein.* Chylomicrons are one example. Other lipoproteins include *low-density lipoproteins (LDLs)* and *high-density lipoproteins (HDLs).*

LDLs are rich in cholesterol and also contain some phospholipids. The function of LDLs is to transport cholesterol from adipose and muscle tissue to other body tissues. In these tissues LDL is used for activities such as steroid hormone synthesis and the manufacture of cell membranes. High LDL levels are associated with the development of atherosclerosis because some of the cholesterol may be deposited in arteries ("bad" cholesterol). HDLs are rich in phospholipids and cholesterol. Their function is to transport cholesterol from body tissues to the liver. In the liver some of the HDL is catabolized to become a component of bile (bile salts). High HDL levels are associated with a decreased risk of cardiovascular disease because they may remove cholesterol from arterial walls ("good" cholesterol). The American Heart Association recommends that the total blood level of cholesterol should be less than 200 mg/dl and that the ratio of total cholesterol to HDL should be less than 3:1.

Figure 19.13 Absorption of digested nutrients in the small intestine. For simplicity, all digested foods are shown in the lumen of the small intestine, even though some nutrients are digested in or on epithelial cells of the villi.

Long-chain fatty acids and monoglycerides are absorbed into lacteals; other products of digestion enter blood capillaries.

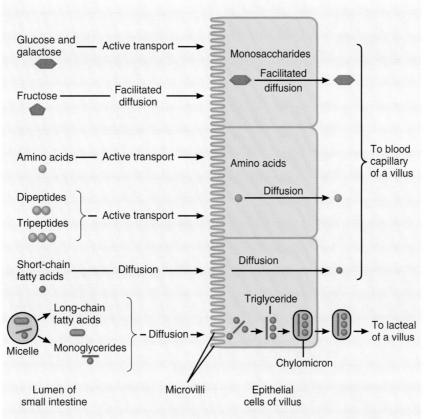

(a) Mechanisms for movement of nutrients through epithelial cells of the villi

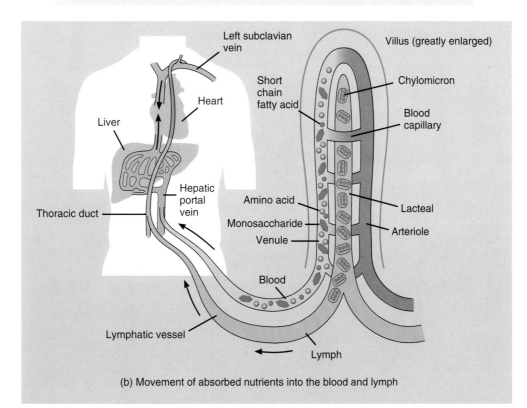

(b) Movement of absorbed nutrients into the blood and lymph

How are fat-soluble vitamins (A, D, E, K) absorbed?

Water Absorption

The total volume of fluid that enters the small intestine each day is about 9.3 liters (about 9.8 qt). Nearly all this fluid is absorbed in the small intestine; the remainder, about 0.1 liter (100 ml), passes into the large intestine, where most of it is also absorbed. Water absorption occurs by osmosis.

Electrolyte Absorption

Many of the electrolytes absorbed by the small intestine come from gastrointestinal secretions and some are part of ingested foods and liquids. Sodium ions (Na^+) are actively transported out of intestinal epithelial cells after they have moved into epithelial cells by diffusion and active transport. Thus most of the Na^+ in gastrointestinal secretions is reclaimed and not lost in the feces. Chloride, iodide, and nitrate ions can passively follow sodium ions or be actively transported. Calcium ions are also actively absorbed, and their movement depends on parathyroid hormone (PTH) and vitamin D. Other electrolytes such as iron, potassium magnesium, and phosphate are also absorbed by active transport.

Vitamin Absorption

Fat-soluble vitamins (A, D, E, and K) are absorbed along with ingested dietary triglycerides in micelles by diffusion. In fact, they cannot be absorbed unless they are ingested with some triglycerides. Most water-soluble vitamins, such as the B vitamins and C, are absorbed by diffusion. Vitamin B_{12} must be combined with intrinsic factor produced by the stomach for its absorption in the small intestine by active transport.

Large Intestine

objective: *Describe the structure of the large intestine and explain its function in digestion, feces formation, and defecation.*

The overall functions of the large intestine are the completion of absorption, the manufacture of certain vitamins, the formation of feces, and the expulsion of feces from the body.

Anatomy

The *large intestine* averages about 6.5 cm (2.5 inches) in diameter and about 1.5 m (5 feet) in length. It extends from the ileum to the anus and is attached to the posterior abdominal wall by its mesocolon (see Figure 19.3b). The large intestine is divided into four principal regions: cecum, colon, rectum, and anal canal (Figure 19.14).

At the opening of the ileum into the large intestine is a fold of mucous membrane called the *ileocecal sphincter.* It allows materials from the small intestine to pass into the large intestine. Hanging below the ileocecal sphincter is a pouch called the *cecum.* Attached to the cecum is the *appendix.*

The open end of the cecum merges with a long tube called the *colon* (*kolon* = food passage). The colon is divided into ascending, transverse, descending, and sigmoid portions. The *ascending colon* ascends on the right side of the abdomen, reaches the undersurface of the liver, and turns to the left. The colon continues across the abdomen to the left side as the *transverse colon.* It curves beneath the lower end of the spleen

---

Figure 19.14   Anatomy of the large intestine.

🔑 *The subdivisions of the large intestine are the cecum, colon, rectum, and anal canal.*

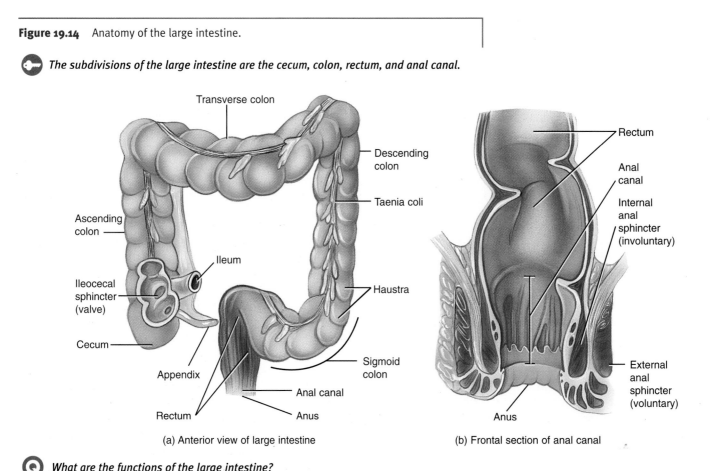

(a) Anterior view of large intestine   (b) Frontal section of anal canal

❓ *What are the functions of the large intestine?*

on the left side and passes downward as the *descending colon.* The *sigmoid colon* begins near the left iliac crest and terminates as the *rectum.*

The terminal 2 to 3 cm (1 inch) of the rectum is called the *anal canal.* The mucous membrane of the anal canal is arranged in longitudinal folds containing arteries and veins. The opening of the anal canal to the exterior is called the *anus.* It has an internal sphincter of smooth (involuntary) muscle and an external sphincter of skeletal (voluntary) muscle. Normally, the anus is closed except during the elimination of the wastes (feces) of digestion.

The medical specialty that deals with the diagnosis and treatment of disorders of the rectum and anus is called *proctology* (prok-TOL-ō-jē; *proct* = rectum; *logos* = study of).

The wall of the large intestine differs from that of the small intestine in several respects. No villi or permanent circular folds are found in the mucosa. The epithelium of the mucosa is simple columnar epithelium that contains mostly absorptive and numerous goblet cells (Figure 19.15). The cells form long tubes called intestinal glands. The absorptive cells function primarily in water absorption. The goblet cells secrete mucus to lubricate the colonic contents. Solitary lymphatic nodules also are found in the mucosa. The muscularis consists of an external layer of longitudinal muscles and an internal layer of circular muscles. Unlike other parts of the gastrointestinal tract, portions of the longitudinal bands, the *taeniae coli* (TĒ-ni-ēKŌ-lī; *taenia* = flat band), run the length of most of the large intestine. Tonic contractions of the bands gather the colon into a series of pouches called *haustra* (HAWS-tra), which give the colon its puckered appearance. *Haustrum* is singular.

Varicosities in any veins involve inflammation and enlargement. Varicosities of the rectal veins are known as *hemorrhoids* (*piles*). Initially contained within the anus (first degree), they gradually enlarge until they prolapse or extend outward on defecation (second degree) and finally remain prolapsed through the anal orifice (third degree). Hemorrhoids may be caused by constipation and pregnancy.

Digestion in the Large Intestine

Mechanical

The passage of chyme from the ileum into the cecum is regulated by the ileocecal sphincter. The sphincter normally remains mildly contracted so the passage of chyme is usually a slow process, but immediately following a meal, there is a reflex action that intensifies peristalsis and any chyme in the ileum is forced into the cecum.

One movement characteristic of the large intestine is *haustral churning.* In this process, the haustra remain relaxed and distended (stretched) while they fill up. When the distention reaches a certain point, the walls contract and squeeze the contents into the next haustrum. *Peristalsis* also occurs, although at a slower rate than in other portions of the tract. The third type of movement is *mass peristalsis,* a strong peristaltic wave that begins in about the middle of the transverse colon and drives the colonic contents into the rectum. Food in the stomach initiates this reflex action. Thus mass peristalsis usually takes place three or four times a day, during a meal or immediately after.

Chemical

The last stage of digestion occurs through bacterial, not enzymatic, action because no enzymes are secreted. Up to 40 percent of fecal mass is bacteria. Bacteria ferment any remaining carbohydrates and release hydrogen, carbon dioxide, and methane gas, which contribute to flatus (gas) in the colon. They also convert remaining proteins to amino acids and break down the amino acids into simpler substances: indole, skatole, hydrogen sulfide, and fatty acids. Some of these are carried off in the feces and contribute to their odor. The rest is absorbed and transported to the liver, converted to less toxic compounds, and excreted in the urine. Bacteria also decompose bilirubin to simpler pigments (urobilinogen), which give feces their brown color. Several vitamins needed for normal metabolism, including some B vitamins and vitamin K, are synthesized by bacterial action and absorbed.

When bacteria of the large intestine leave the intestine because of a perforation or ruptured appendix, they can cause *peritonitis,* an acute inflammation of the peritoneum. Complications of peritonitis include shock, renal failure, respiratory insufficiency, and liver failure.

Absorption and Feces Formation

By the time the chyme has remained in the large intestine 3 to 10 hours, it has become solid or semisolid as a result of absorption and is now known as *feces.* Chemically, feces consist of inorganic salts, sloughed off epithelial cells from the mucosa of the GI tract, bacteria, products of bacterial decomposition, undigested parts of food, and water.

Defecation

Mass peristalsis from the sigmoid colon initiates the defecation reflex. Filling of the rectum stretches its wall and stimulates stretch receptors, which send sensory impulses to the spinal cord. Motor impulses return along parasympathetic fibers to the muscles of the lower colon, rectum, and anus. *Defecation,* or emptying of feces via the anus, results when the internal anal sphincter opens due to pressure caused by involuntary contraction of longitudinal rectal muscles and voluntary contractions of the diaphragm and abdominal muscles. The external sphincter is voluntarily controlled. If it is voluntarily relaxed, defecation occurs; if it is voluntarily constricted, defecation can be postponed after about age 3. If defecation does not occur, the feces move back into the sigmoid colon until the next wave of mass peristalsis again stimulates the stretch-sensitive receptors, creating the desire to defecate.

Diarrhea refers to defecation of liquid feces caused by increased movement of the intestines. Because chyme passes too quickly through the small intestine and feces pass too quickly through the large intestine, there is not enough time for absorption. Like vomiting, diarrhea can result in dehydration and electrolyte imbalances. Diarrhea may be caused by stress and microbes that irritate the gastrointestinal mucosa.

Constipation refers to infrequent or difficult defecation or dry feces. It is caused by decreased motility of the intestines, in which feces remain in the colon for prolonged periods of time. Water absorption continues and feces become dry and hard. Constipation may be caused by improper bowel habits, spasms of the colon, insufficient bulk in the diet, lack of exercise, and stress.

Figure 19.15   Histology of the large intestine.

Intestinal glands formed by simple columnar and goblet cells extend the full thickness of the mucosa.

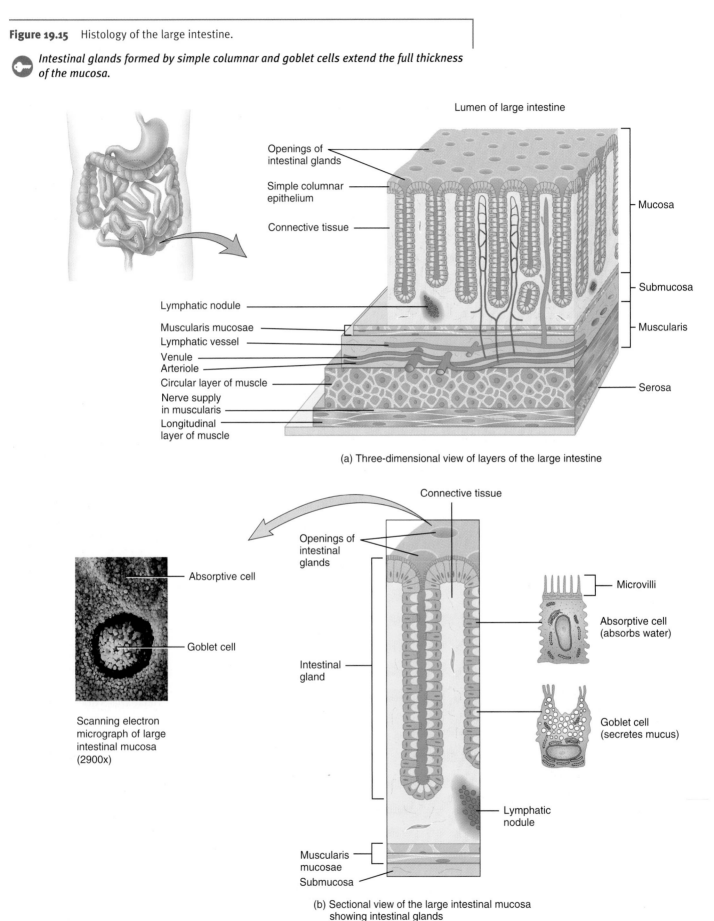

Lumen of large intestine

Openings of intestinal glands

Simple columnar epithelium

Connective tissue

Mucosa

Submucosa

Muscularis

Serosa

Lymphatic nodule

Muscularis mucosae

Lymphatic vessel

Venule

Arteriole

Circular layer of muscle

Nerve supply in muscularis

Longitudinal layer of muscle

(a) Three-dimensional view of layers of the large intestine

Connective tissue

Absorptive cell

Goblet cell

Scanning electron micrograph of large intestinal mucosa (2900x)

Openings of intestinal glands

Intestinal gland

Muscularis mucosae

Submucosa

Microvilli

Absorptive cell (absorbs water)

Goblet cell (secretes mucus)

Lymphatic nodule

(b) Sectional view of the large intestinal mucosa showing intestinal glands

How does the muscularis of the large intestine differ from other portions of the GI tract?

Our culture has devised the perfect recipe for the development of eating problems: Combine an abundant food supply with sedentary occupations, and then establish an unrealistic standard of thinness by which to judge self-worth. Add a dash of stress and low self-esteem, and voilà: problems with body image and eating behavior. No wonder the rate of eating disorders is reaching epidemic proportions in North America, especially among college women.

Eating Disorders and Disordered Eating

Most readers are by now familiar with clinical eating disorders. People with *anorexia nervosa* (usually young girls and women) severely restrict food intake and lose 15 percent or more of their body weight. Though emaciated, people with anorexia still believe they are too fat, and try to lose even more weight, sometimes to the point of hospitalization and even death. People with *bulimia* engage in frequent episodes of binge eating, in which they consume large volumes of food within a discrete period of time. These binges are characterized by feelings of being out of control, and are followed by feelings of self-depre-cation and low self-esteem. People with bulimia then attempt to purge the extra calories with self-induced vomiting, fasting, laxatives, diuretics, or excessive exercise.

The concept of eating disorders can lead the reader to the erroneous conclusion that problems with food

and weight are either something you have or don't have. In reality, problems with food and body image run along a continuum, from occasional fasting and cookie binges to clinical eating disorders. Many people have some degree of disordered eating but fall short of the clinical diagnosis for an eating disorder. Disordered eating behaviors include chronic dieting, following very-low-calorie diets, and obsessing about food and body weight.

People with disordered eating may also indulge in pathogenic weight control methods, such as those used by bulimics for purging. They may become addicted to drugs, such as diet pills, cocaine, and anabolic steroids, or take up cigarette smoking and other unhealthy behaviors in misguided efforts to control food intake and body weight.

The dangers of disordered eating range from mild effects, such as wasting too much time thinking about food to the development of life-threatening eating disorders. Unfortunately, many people with disordered eating behaviors develop more serious problems over time if the psychological problems prompting the behaviors are not addressed.

Dangerous Diets

Eating disorders involve more than eating behavior; they are symptomatic of serious psychological problems best addressed by qualified specialists. Eating disorders are difficult to treat, which makes the prevention of eating problems especially

important. One line of prevention is to encourage young people to maintain an appropriate weight by developing lifelong healthful eating and exercise habits rather than to expect to lose weight quickly on short-term restrictive diets (which don't work very well anyway).

While eating disorders are not just about eating, all eating disorders begin with a desire to lose weight and a decision to diet. It is easy for dieting behavior to develop into problems with food, especially if the diet is quite restricted. In such cases, dieters develop uncontrollable cravings, a normal response to starvation. When this happens, dieters frequently lose control and the fragile structure of restrained eating falls apart, leading to binge eating, a sense of disgust with oneself, and a resolve once again to regain control of one's eating behavior. Obviously, not every dieter develops an eating disorder. But encouragement of dieting behavior promotes disordered eating in people prone to such a problem.

critical thinking

Why do you think women are more likely than men to develop eating disorders?

Eating behavior is triggered by many factors other than physiological hunger. What are some of these factors?

Common Disorders

Dietary Fiber and the Digestive System

A dietary deficiency that has received recent attention is lack of adequate *dietary fiber* (bulk or roughage). Dietary fiber consists of indigestible plant substances, such as cellulose, lignin, and pectin, found in fruits, vegetables, grains, and beans. *Insoluble fiber,* which does not dissolve in water, includes the woody or structural parts of plants such as fruit and vegetable skins and the bran coating around wheat and corn kernels. Insoluble fiber passes through the GI tract largely unchanged and speeds up the passage of material through the tract. *Soluble fiber,* which does dissolve in water, is found in abundance in beans, oats, barley, broccoli, prunes, apples, and citrus fruits. It tends to slow the passage of material through the tract.

People who choose a fiber-rich, unrefined diet may reduce their risk of developing obesity, diabetes, atherosclerosis, gallstones, hemorrhoids, diverticulitis, appendicitis, and colon cancer. Each of these conditions is directly related to the digestion and metabolism of food and the operation of the digestive system. There is also evidence that insoluble fiber may help protect against colon cancer and that soluble fiber may help lower blood cholesterol level.

Dental Caries

Dental caries, or tooth decay, involve a gradual demineralization (softening) of the enamel and dentin. If untreated, various microorganisms may invade the pulp, causing inflammation and infection with subsequent death of the pulp and abscess of the bone surrounding the root. Such teeth are treated by root canal therapy. Dental caries begin when bacteria, acting on sugars, produce acids that demineralize the enamel.

Periodontal Disease

Periodontal disease refers to a variety of conditions characterized by inflammation and degeneration of the gums, bone, periodontal ligament, and cementum.

Peptic Ulcers

An *ulcer* is a craterlike lesion in a membrane. Ulcers that develop in areas of the gastrointestinal tract exposed to acidic gastric juice (stomach and duodenum) are called *peptic ulcers*. A major cause is a bacterium (*Helicobacteria pylori*).

Appendicitis

Appendicitis is an inflammation of the appendix. Appendectomy (surgical removal of the appendix) is recommended in all suspected cases because it is safer to operate than to risk gangrene, rupture, and peritonitis.

Tumors

Both benign and malignant *tumors* can occur in all parts of the gastrointestinal tract. *Colorectal cancer* is one of the most deadly and common malignant diseases, ranking second to lung cancer in males and breast cancer in females. Development of colorectal cancer involves both environmental and genetic factors. Dietary fiber, retinoids, calcium, and selenium may be protective, whereas intake of animal fat and protein may cause an increase in the disease. Genetics plays a very important role in that an inherited predisposition contributes to more than half of all cases of colorectal cancer. Signs and symptoms of colorectal cancer include changes in the normal pattern of bowel habits (diarrhea, constipation), cramping, abdominal pain, and rectal bleeding. The only definitive treatment for gastrointestinal tumors, if they cannot be removed endoscopically, is surgery.

Diverticulitis

Diverticula are saclike outpouchings of the wall of the colon in places where the muscularis has become weak. The development of diverticula is called *diverticulosis,* and an inflammation in diverticula is known as *diverticulitis.*

Hepatitis

Hepatitis is an inflammation of the liver caused by viruses, drugs, or chemicals, including alcohol.

Hepatitis A (*infectious hepatitis*) is caused by the hepatitis A virus and is spread by fecal contamination of food, clothing, toys, eating utensils, and so forth (fecal–oral route). It does not cause lasting liver damage.

Hepatitis B (*serum hepatitis*) is caused by the hepatitis B virus and is spread primarily by sexual contact and contaminated syringes and transfusion equipment. It can also be spread by any secretion of fluid by the body (tears, saliva, semen). Hepatitis B can produce chronic liver inflammation. Vaccines are available for hepatitis B and are required by certain individuals, such as health-care providers.

Hepatitis C is a form of hepatitis that cannot be traced to either the hepatitis A or hepatitis B viruses. It is clinically similar to hepatitis B and is often spread by blood transfusions. The hepatitis C virus can cause cirrhosis and possibly liver cancer.

Hepatitis D is caused by hepatitis D virus. It is transmitted like hepatitis B and, in fact, a person must be coinfected with hepatitis B before contracting hepatitis D. Hepatitis D results in severe liver damage and has a fatality rate higher than that of people infected with hepatitis B virus alone.

Hepatitis E is caused by hepatitis E virus and is spread like hepatitis A. Although it does not cause chronic liver disease, hepatitis E virus is responsible for a very high mortality rate in pregnant women.

Cirrhosis

Cirrhosis refers to a distorted or scarred liver as a result of chronic inflammation. Cirrhosis may be caused by hepatitis, certain chemicals, parasites that infect the liver, and alcoholism.

Gallstones

If there are insufficient bile salts or lecithin in bile, or if there is excessive cholesterol, the cholesterol may crystallize to form *gallstones*. Once formed, gallstones grow in size and number and may cause minimal, intermittent, or complete obstruction to the flow of bile from the gallbladder into the duodenum.

Anorexia Nervosa

Anorexia nervosa is a chronic disorder characterized by self-induced weight loss, negative perception of body-image, and physiological changes that result from nutritional depletion. Patients with anorexia nervosa fixate on their weight and, often, insist on having a bowel movement every day despite the lack of adequate food intake. They abuse laxatives, which worsens the fluid/electrolyte/nutrient deficiencies. The disorder is found predominantly in young, single females and may be inherited.

Bulimia

A disorder that typically affects single, middle-class, young, white females is known as *bulimia* (*bous* = ox; *limos* = hunger), or *binge–purge syndrome*. It is characterized by overeating at least twice a week followed by purging by self-induced vomiting, strict dieting or fasting, vigorous exercise, or use of laxatives or diuretics. This binge–purge cycle occurs in response to fears of being overweight, stress, depression, and physiological disorders such as hypothalamic tumors. ∎

Medical Terminology and Conditions

Botulism (BOCH-yoo-lism; *botulus* = sausage) A type of food poisoning caused by a toxin produced by *Clostridium botulinum*. The bacterium is ingested when improperly cooked or preserved foods are eaten. The toxin inhibits nerve impulse transmission at synapses by inhibiting the release of acetylcholine. Symptoms include paralysis, nausea, vomiting, blurred or double vision, difficulty in speech and swallowing, dryness of the mouth, and general weakness.

Canker (KANG-ker) *sore* Painful ulcer on the mucous membrane of the mouth that affects females more frequently than males and usually occurs between ages 10 and 40; may be an autoimmune reaction.

Colitis (ko-LĪ-tis) Inflammation of the mucosa of the colon and rectum in which absorption of water and salts is reduced, producing watery, bloody feces and, in severe cases, dehydration and salt depletion. Spasms of the irritated muscularis produce cramps.

Colostomy (ko-LOS-tō-mē; *stomoun* = provide an opening) The diversion of the fecal stream through an opening in the colon, creating a surgical "stoma" (artificial opening) that is affixed to the exterior of the abdominal wall. This opening serves as a substitute anus through which feces are eliminated.

Dysphagia (dis-FĀ-jē-a; *dys* = abnormal; *phagein* = to eat) Difficulty in swallowing that may be caused by inflammation, paralysis, obstruction, or trauma.

Enteritis (en'-ter-Ī-tis; *enteron* = intestine) An inflammation of the intestine, particularly the small intestine.

Flatus (FLA-tus) Air (gas) in the stomach or intestine, usually expelled through the anus. If the gas is expelled through the mouth, it is called *eructation* or *belching* (burping). Flatus may result from gas released during the breakdown of foods in the stomach or from swallowing air or gas-containing substances such as carbonated drinks.

Hernia (HER-nē-a) Protrusion of an organ or part of an organ through a membrane or cavity wall, usually the abdominal cavity.

Irritable bowel (IR-i-ta-bul BOW-el) *syndrome* (*IBS*) Disease of the entire gastrointestinal tract in which persons with this condition may react to stress by developing symptoms such as cramping and abdominal pain associated with alternating patterns of diarrhea and constipation. Excessive amounts of mucus may appear in the stools, and other symptoms include flatulence, nausea, and loss of appetite. The condition is also known as *irritable colon* or *spastic colitis*.

Nausea (NAW-sē-a; *nausia* = seasickness) Discomfort characterized by a loss of appetite and the sensation of impending vomiting. Its causes include local irritation of the gastrointestinal tract, a systemic disease, brain disease or injury, overexertion, or the effects of medication or drug overdosage.

Traveler's diarrhea Infectious disease of the gastrointestinal tract that results in loose, urgent bowel movements, cramping, abdominal pain, malaise, nausea, and occasionally fever and dehydration. It is acquired through ingestion of food or water that has become contaminated with fecal material containing mostly bacteria.

Study Outline

Digestive Processes (p. 433)

1. Food is prepared for use by cells by five basic activities: ingestion, mixing and movement, mechanical and chemical digestion, absorption, and defecation.
2. Chemical digestion is a series of reactions that break down the large carbohydrate, lipid, and protein molecules of food into molecules that are usable by body cells.
3. Mechanical digestion consists of movements that aid chemical digestion.
4. Absorption is the passage of nutrients from digested food in the digestive tract into blood or lymph for distribution to cells.
5. Defecation is emptying of the rectum.

Organization (p. 433)

1. The organs of digestion are divided into two main groups: those composing the gastrointestinal (GI) tract and accessory structures.
2. The GI tract is a continuous tube running from the mouth to the anus.
3. The accessory structures are the teeth, tongue, salivary glands, liver, gallbladder, and pancreas.
4. The basic arrangement of tissues in the GI tract from the inside outward is the mucosa, submucosa, muscularis, and serosa (peritoneum).
5. Parts of the peritoneum include the mesentery, falciform ligament, and greater omentum.

Mouth (Oral Cavity) (p. 434)

1. The mouth is formed by the cheeks, palates, lips, and tongue, which aid mechanical digestion.
2. The opening from the mouth to the throat is the fauces.

Tongue (p. 436)

1. The tongue forms the floor of the oral cavity. It is composed of skeletal muscle covered with mucous membrane.
2. The upper surface and sides of the tongue are covered with papillae. Some papillae contain taste buds.

Salivary Glands (p. 436)

1. The major portion of saliva is secreted by the salivary glands, which lie outside the mouth and pour their contents into ducts that empty into the oral cavity.
2. There are three pairs of salivary glands: the parotid, submandibular, and sublingual.
3. Saliva lubricates food and starts the chemical digestion of carbohydrates.
4. Salivation is entirely under nervous control.

Teeth (p. 437)

1. The teeth, or dentes, project into the mouth and are adapted for mechanical digestion.
2. A typical tooth consists of three principal portions: crown, root, and neck.
3. Teeth are composed primarily of dentin and are covered by enamel, the hardest substance in the body.

4. There are two dentitions: deciduous and permanent.

Digestion in the Mouth (p. 438)

1. Through mastication food is mixed with saliva and shaped into a bolus.
2. Salivary amylase converts polysaccharides (starches) to disaccharides (maltose).

Pharynx (p. 438)

1. Food that is swallowed passes from the mouth into the oropharynx.
2. From the oropharynx, food passes into the laryngopharynx.

Esophagus (p. 438)

1. The esophagus is a muscular tube that connects the pharynx to the stomach.
2. It passes a bolus into the stomach by peristalsis.

Swallowing (p. 438)

1. Swallowing moves a bolus from the mouth to the stomach.
2. It consists of a voluntary stage, pharyngeal stage (involuntary), and esophageal stage (involuntary).

Stomach (p. 439)

Anatomy (p. 440)

1. The stomach begins at the bottom of the esophagus and ends at the pyloric sphincter.
2. The anatomic subdivisions of the stomach are the cardia, fundus, body, and pylorus.
3. Adaptations of the stomach for digestion include rugae; glands that produce mucus, hydrochloric acid, a protein-digesting enzyme (pepsin), intrinsic factor, and gastrin; and a three-layered muscularis for efficient mechanical movement.

Digestion in the Stomach (p. 442)

1. Mechanical digestion consists of mixing waves.
2. Chemical digestion consists of the conversion of proteins into peptides by pepsin.
3. Mixing waves and gastric secretions reduce food to chyme.

Regulation of Gastric Secretion (p. 442)

1. Gastric secretion and motility are regulated by nervous and hormonal mechanisms.
2. Parasympathetic impulses and gastrin cause secretion of gastric juices.
3. The presence of food in the small intestine, gastric inhibitory peptide, secretin, and cholecystokinin inhibit gastric secretion.

Regulation of Gastric Emptying (p. 442)

1. Gastric emptying is stimulated in response to stretching and gastrin released in response to the presence of certain types of food.

2. Gastric emptying is inhibited by reflex action and hormones (secretin and cholecystokinin).

Absorption (p. 443)

1. The stomach wall is impermeable to most substances.
2. Among the substances absorbed are some water, certain electrolytes and drugs, and alcohol.

Pancreas (p. 443)

1. The pancreas is connected to the duodenum by the pancreatic duct.
2. Pancreatic islets (islets of Langerhans) secrete hormones and constitute the endocrine portion of the pancreas.
3. Acini cells secrete pancreatic juice; they constitute the exocrine portion of the pancreas.
4. Pancreatic juice contains enzymes that digest starch (pancreatic amylase), proteins (trypsin, chymotrypsin, and carboxypeptidase), triglycerides (pancreatic lipase), and nucleic acids (nucleases).
5. Pancreatic secretion is regulated by nervous control (parasympathetic fibers from the vagus nerves) and hormonal mechanisms (secretin and cholecystokinin).

Liver (p. 443)

1. The liver has left and right lobes.
2. The lobes are made up of lobules that contain hepatocytes (liver cells), sinusoids, stellate reticuloendothelial (Kupffer's) cells, and a central vein.
3. Hepatocytes of the liver produce bile that is transported by a duct system to the gallbladder for storage.
4. Bile emulsifies triglycerides.
5. Bile secretion is regulated by nervous control (parasympathetic stimulation by way of the vagus nerves) and hormonal mechanisms (secretin and cholecystokinin).
6. The liver also functions in carbohydrate, triglyceride, and protein metabolism; removal of drugs and hormones; excretion of bile; synthesis of bile salts; storage of vitamins and minerals; phagocytosis; and activation of vitamin D.

Gallbladder (p. 446)

1. The gallbladder is a sac located in a depression under the liver.
2. The gallbladder stores and concentrates bile.
3. Bile is ejected into the common bile duct under the influence of cholecystokinin.

Small Intestine (p. 446)

Anatomy (p. 446)

1. The small intestine extends from the pyloric sphincter to the ileocecal sphincter.
2. It is divided into duodenum, jejunum, and ileum.
3. It is highly adapted for digestion and absorption. Its glands produce enzymes and mucus, and the microvilli, villi, and circular folds of its wall provide a large surface area for digestion and absorption.
4. Intestinal enzymes break down foods in or on epithelial cells of the mucosa.

Intestinal Juice; Digestion in the Small Intestine (p. 448–449)

1. Mechanical digestion in the small intestine involves segmentation and peristalsis.
2. Intestinal enzymes in intestinal juice, pancreatic juice, and bile break down disaccharides to monosaccharides; protein digestion is completed by peptidase enzymes; triglycerides are broken down into fatty acids and monoglycerides by pancreatic lipase; and nucleases break down nucleic acids to pentoses and nitrogen bases.

Regulation of Intestinal Secretion and Motility (p. 449)

1. The most important mechanism is local reflexes.
2. They are initiated in the presence of chyme.
3. Vasoactive intestinal polypeptide (VIP) stimulates production of intestinal juice.

Absorption (p. 449)

1. Absorption is the passage of nutrients from digested food in the gastrointestinal tract into the blood or lymph.
2. Monosaccharides, amino acids, and short-chain fatty acids pass into the blood capillaries.
3. Long-chain fatty acids and monoglycerides are absorbed as part of micelles, resynthesized to triglycerides, and transported as chylomicrons.
4. Chylomicrons are taken up by the lacteal of a villus.
5. The small intestine also absorbs water, electrolytes, and vitamins.

Large Intestine (p. 452)

Anatomy (p. 452)

1. The large intestine extends from the ileocecal sphincter to the anus.
2. Its subdivisions include the cecum, colon, rectum, and anal canal.
3. The mucosa contains numerous absorptive and goblet cells and the muscularis contains taeniae coli.

Digestion in the Large Intestine (p. 453)

1. Mechanical movements of the large intestine include haustral churning, peristalsis, and mass peristalsis.
2. The last stages of chemical digestion occur in the large intestine through bacterial, rather than enzymatic, action. Substances are further broken down and some vitamins are synthesized.

Absorption and Feces Formation (p. 453)

1. The large intestine absorbs water, electrolytes, and vitamins.
2. Feces consist of water, inorganic salts, epithelial cells, bacteria, and undigested foods.

Defecation (p. 453)

1. The elimination of feces from the rectum is called defecation.
2. Defecation is a reflex action aided by voluntary contractions of the diaphragm and abdominal muscles.

Self-Quiz

1. Food is prevented from going into the trachea during swallowing by the
   a. esophagus   b. uvula   c. epiglottis   d. tongue   e. pharynx

2. The accessory structures associated with the mouth do NOT include the
   a. tongue   b. teeth   c. parotid glands   d. sublingual glands   e. gastric glands

3. Protein digestion first occurs in which structure?
   a. mouth   b. stomach   c. small intestine   d. liver   e. large intestine

4. Which of the following is NOT produced in the stomach?
   a. sodium bicarbonate (NaHCO₃)   b. gastrin   c. pepsinogen   d. mucus   e. hydrochloric acid (HCl)

5. Which of the following would correctly describe the esophagus?
   a. Food enters the esophagus from the pyloric region of the stomach.   b. It extends from the larynx to the bronchi and contains cartilage rings.   c. Food moves through the entire esophagus under voluntary control.   d. It allows the passage of chyme.   e. It is a collapsible muscular tube extending from pharynx to stomach.

6. The gastrointestinal tract is protected from the acidic chyme mixture by
   a. HCl   b. pepsin   c. trypsin   d. mucus and NaHCO₃   e. cilia

7. After leaving the stomach chyme passes next through which of the following?
   a. pancreas   b. large intestine   c. duodenum   d. stomach   e. esophagus

8. Most chemical digestion occurs in the
   a. liver   b. stomach   c. duodenum   d. large intestine   e. pancreas

9. The end products of protein digestion are
   a. monosaccharides   b. polypeptides   c. amino acids   d. phospholipids   e. fatty acids

10. If you have just digested triglyceride (fat) molecules, you would expect to find increased amounts of
    a. glucose   b. amino acids   c. fatty acids and glycerol   d. nucleic acids   e. amylase

11. Absorption is defined as
    a. the elimination of solid wastes from the digestive system   b. a reflex action controlled by the autonomic nervous system   c. the breakdown of foods by enzymatic action   d. the passage of nutrients from the gastrointestinal tract into the bloodstream   e. the mechanical breakdown of triglycerides

12. Which of the following structures enhance absorption of nutrients by the small intestine?
    a. gastric glands   b. rugae   c. haustra   d. villi and microvilli   e. mesentery

13. If an incision was made into the stomach, the tissue layers would be penetrated in the order
    a. mucosa, muscularis, serosa, submucosa   b. mucosa, muscularis, submucosa, serosa   c. serosa, muscularis, mucosa, submucosa   d. muscularis, submucosa, mucosa, serosa   e. serosa, muscularis, submucosa, mucosa

14. The hormone that stimulates gastric secretion and emptying is
    a. gastrin   b. cholecystokinin (CCK)   c. secretin   d. gastric inhibitory peptide (GIP)   e. bile

15. Which of the following are NOT correctly paired?
    a. esophagus—peristalsis   b. mouth—mastication   c. large intestine—mass peristalsis   d. small intestine—segmentation   e. stomach—deglutition

16. An enzyme that digests proteins is
    a. lipase   b. maltase   c. trypsin   d. ribonuclease   e. amylase

Critical Thinking Applications

1. Amelia enjoys eating fresh lemons. She holds a large lemon wedge with her front teeth and sucks on the juice. Her dentist is concerned that the frequent exposure to acid will affect her teeth. Why?

2. In a classic experiment, Pavlov conditioned dogs to salivate at the sound of a bell in anticipation of being fed. People may likewise salivate in anticipation of a delicious meal. Explain why.

3. A couple enjoying a meal in a restaurant were talking, laughing, and eating heartily. The man suddenly grabbed at his throat and made strangled noises. The woman called for help saying, "he can't breathe!" What happened?

4. Imagine that a nasal spray to administer cholecystokinin (CCK) becomes available. How would CCK treatments affect the digestive system and the appetite?

5. Howie ate two large bowls of vegetarian chili at The Mexican Smorgasbord and Cappuccino Stand. Chili contains one or more types of beans and many vegetables providing a large amount of dietary fiber. What effect will this meal have on Howie's large intestine?

6. After eating the chili, a large helping of strudel, and a cappuccino with extra cream, Howie experienced a burning feeling near the center of his chest. He loosened his belt and felt uncomfortably full. Explain the cause of Howie's burning feeling?

Answers to Figure Questions

19.1   Peristalsis.

19.2   They help regulate secretions and movements of the tract.

19.3   Mesentery.

19.4   They maneuver food for chewing, shape food into a bolus, force food to the back of the mouth for swallowing, and alter the shape of the tongue for swallowing and speech production.

19.5   Connective tissue, specifically dentin.

19.6   Both. Initiation of swallowing is voluntary and the action is carried out by skeletal muscles. Completion of swallowing—moving a bolus along the esophagus and into the stomach—involves peristalsis of smooth muscle and is involuntary.

19.7   Probably not, because as the stomach fills, the rugae stretch out.

19.8   Stimulates secretion of gastric juice, increases motility of the GI tract, and relaxes the pyloric sphincter.

19.9   Pancreatic juice (fluid and digestive enzymes); bile; pancreatic juice plus bile.

19.10 Hepatic portal vein $\longrightarrow$ branch of hepatic portal vein $\longrightarrow$ sinusoid $\longrightarrow$ central vein $\longrightarrow$ hepatic vein $\longrightarrow$ inferior vena cava.

19.11 Ileum.

19.12 About 20–40 villi and 200 million microvilli.

19.13 In micelles.

19.14 Completion of absorption, synthesis of certain vitamins, formation of feces, and elimination of feces.

19.15 It contains taeniae coli that form haustra.

20 chapter

NUTRITION AND METABOLISM

a look ahead

*I*n this chapter we will discuss the major groups of nutrients; how food intake is regulated; guidelines for healthy eating; how each group of nutrients is used for energy, growth, and repair of the body; and how various factors affect the body's metabolic rate.

Nutrients

Nutrients are chemical substances in food that provide energy, form new body components, or assist in the functioning of various body processes. The six principal classes of nutrients are carbohydrates, lipids, proteins, minerals, vitamins, and water. Carbohydrates, proteins, and lipids are digested by enzymes in the gastrointestinal tract. Most of the end products of digestion are used to produce energy to sustain life processes. Some are used to synthesize new structural molecules in cells or new regulatory molecules, such as hormones and enzymes.

Some minerals and many vitamins are part of enzyme systems that catalyze metabolic reactions involving carbohydrates, proteins, and lipids.

Water has five major functions. It is an excellent solvent and suspending medium; it participates in hydrolysis reactions; it acts as a coolant; it lubricates; and it helps to maintain a constant body temperature.

Regulation of Food Intake

In the hypothalamus of the brain are two groups of neurons related to food intake. When the *feeding (hunger) center* is stimulated in animals, they begin to eat heartily, even if they are already full. When the *satiety center* is stimulated in animals, they stop eating, even if they have been starved for days. Other parts of the brain that function in hunger and satiety are the cerebral cortex, brain stem, and limbic system.

Food intake is affected by blood glucose levels. When blood glucose levels are low, feeding increases. When blood glucose levels are high, the activity of the satiety center is sufficiently high to inhibit the feeding center, and feeding is depressed. To a lesser extent, low levels of amino acids in the blood also enhance feeding, whereas high levels depress eating.

As the amount of adipose (fat) tissue increases in the body, the rate of feeding usually decreases. It is theorized that substances, perhaps fatty acids, are released from stored triglycerides (fats) in adipose tissue and these substances then activate the satiety center and inhibit the feeding center.

Food intake is also affected by body temperature. Whereas a cold environment enhances eating, a warm environment depresses it. Food intake is also regulated by stretching of the gastrointestinal tract, particularly the stomach and duodenum. When these organs are stretched, a reflex is initiated that activates the satiety center and depresses the feeding center.

The hormone cholecystokinin (CCK), secreted when triglycerides enter the small intestine, also inhibits eating. Psychological factors may override the usual intake mechanisms, for example, in obesity, anorexia nervosa, and bulimia.

Guidelines for Healthy Eating

Ideas about what comprises healthy eating behavior vary from one culture to another, and studies of different populations have provided important clues about the impact of nutrition on health. For several generations, the United States has been a "meat and milk" country; the daily consumption of dairy products and beef was viewed as a healthy luxury not shared by many others throughout the world. In Asia, on the other hand, meals traditionally have included rice, a lot of vegetables, some fish, and little meat. The rate of heart disease in Japan, where meals tend to have less fat and more carbohydrate, is far lower than in the U.S. Several studies suggest that one reason is the higher consumption of saturated fats and cholesterol in the U.S.

Experts now suggest the following distribution of calories: (1) 50 to 60 percent from carbohydrates, with less than 15 percent from simple sugars, (2) less than 30 percent from fats (triglycerides are the main type of dietary fat), with no more than 10 percent as saturated fats, and (3) about 12 to 15 percent from proteins. The guidelines for healthy eating are:

1. Eat a variety of foods.
2. Maintain healthy weight.
3. Choose foods low in fat, saturated fat, and cholesterol.
4. Eat plenty of vegetables, fruits, and grain products.
5. Use sugars only in moderation.
6. Use salt and sodium only in moderation (less than 6 grams daily).
7. If you drink alcoholic beverages, do so in moderation (less than 1 ounce of the equivalent of pure alcohol per day).

To help people achieve a good balance of vitamins, minerals, carbohydrates, fats, and proteins in their food, the U.S. Department of Agriculture developed the Food Guide Pyramid (Figure 20.1). The sections of the pyramid show how many servings of five major food groups to eat per day. The smallest number of servings corresponds to a 1600 Cal/day diet, whereas the largest number of servings corresponds to a 2800 Cal/day diet. Because they should be consumed in largest quantity, food rich in complex carbohydrates—the bread, cereal, rice, and pasta food group—form the base of the pyramid. Vegetables and fruits form the next level. The health benefits of eating generous amounts of these foods are well documented. To be eaten in smaller quantities are foods on the next level up—the milk, yogurt, and cheese group and the meat, poultry,

Figure 20.1   The Food Guide Pyramid. The smallest number of servings corresponds to 1600 Calories per day whereas the largest number of servings corresponds to 2800 Calories per day. Each example given equals one serving.

🔑 *The sections of the pyramid show how many servings of five major food groups to eat each day.*

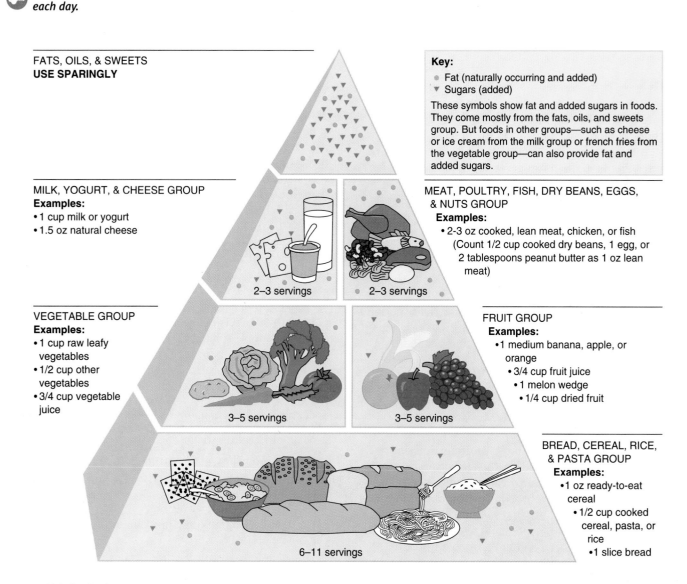

FATS, OILS, & SWEETS
USE SPARINGLY

Key:
- Fat (naturally occurring and added)
- ▼ Sugars (added)

These symbols show fat and added sugars in foods. They come mostly from the fats, oils, and sweets group. But foods in other groups—such as cheese or ice cream from the milk group or french fries from the vegetable group—can also provide fat and added sugars.

MILK, YOGURT, & CHEESE GROUP
Examples:
- 1 cup milk or yogurt
- 1.5 oz natural cheese

2–3 servings

MEAT, POULTRY, FISH, DRY BEANS, EGGS, & NUTS GROUP
Examples:
- 2-3 oz cooked, lean meat, chicken, or fish (Count 1/2 cup cooked dry beans, 1 egg, or 2 tablespoons peanut butter as 1 oz lean meat)

2–3 servings

VEGETABLE GROUP
Examples:
- 1 cup raw leafy vegetables
- 1/2 cup other vegetables
- 3/4 cup vegetable juice

3–5 servings

FRUIT GROUP
Examples:
- 1 medium banana, apple, or orange
- 3/4 cup fruit juice
- 1 melon wedge
- 1/4 cup dried fruit

3–5 servings

BREAD, CEREAL, RICE, & PASTA GROUP
Examples:
- 1 oz ready-to-eat cereal
- 1/2 cup cooked cereal, pasta, or rice
- 1 slice bread

6–11 servings

❓ *Which foods shown contain cholesterol and most of the saturated fatty acids in the diet?*

fish, dry beans, eggs, and nuts group. These two food groups have higher fat and protein content than the food groups below them. To lower your daily intake of fats, choose low-fat foods from these groups—nonfat milk and yogurt, low-fat cheese, fish, and poultry (remove the skin). The top of the pyramid is not a food group but rather a caution to use fats, oils, and sweets sparingly.

We still do not know with certainty what levels and types of carbohydrate, fat, and protein are optimal in the diet. Around the world, different populations eat radically different diets adapted to their particular lifestyle. Two or three servings of meat per day may be too much. Animal protein increases loss of calcium in the urine and red meat consumption is associated with increased risk of colon, breast, and prostate cancers. The food guide pyramid does not distinguish among the different types of fatty acids—saturated, polyunsaturated, and monoun-saturated—in dietary fats. Atherosclerosis and coronary artery disease are prevalent in populations that consume large amounts of saturated fats and cholesterol. But populations living around the Mediterranean Sea have low rates of coronary artery disease despite eating a diet that contains up to 40 percent of the calories as fats. Most of their fat comes from olive oil, however, which is rich in monounsaturated fatty acids and has no cholesterol. Thus the type of fat we eat may be more critical than the absolute amount. Besides olive oil, canola oil, avocados, nuts, and peanut oil are rich in monounsaturated fatty acids.

Metabolism

objective: *Define metabolism and describe its importance in homeostasis.*

Metabolism (me-TAB-ō-lizm; *metabole* = change) refers to all the chemical reactions of the body. The body's metabolism may be thought of as an energy-balancing act between anabolic (synthesis) and catabolic (decomposition) reactions.

Anabolism

Chemical reactions that combine simple substances into more complex molecules are collectively known as *anabolism* (a-NAB-ō-lizm; *ana* = upward). Overall, anabolic reactions require energy, which is supplied by catabolic reactions (Figure 20.2). One example of an anabolic process is the formation of peptide bonds between amino acids, thereby building the amino acids into proteins. Fatty acids and glycerol are built into triglycerides (fats) and simple sugars into polysaccharides through anabolism.

Catabolism

The chemical reactions that break down complex organic compounds into simple ones are collectively known as *catabolism* (ka-TAB-ō-lizm; *cata* = downward; *ballein* = to throw). Catabolic reactions release the chemical energy in organic molecules. This energy is then stored in ATP until it is needed for anabolic reactions. The role of ATP in catabolic and anabolic reactions is shown in Figure 20.2.

Chemical digestion is a catabolic process in which the breaking of bonds of food molecules releases energy. Another example is oxidation (cellular respiration), to be described shortly.

Metabolism and Enzymes

Chemical reactions occur when chemical bonds between substances are made or broken. Enzymes are proteins that serve as

Figure 20.2   Catabolism, anabolism, and ATP. When simple compounds are combined to form complex compounds (anabolism), ATP provides energy for synthesis. When large compounds are split apart (catabolism), some of the energy is transferred to and trapped in ATP and then utilized to drive anabolic reactions. Most of the energy is given off as heat.

The body's metabolism is an energy-balancing act between anabolism and catabolism.

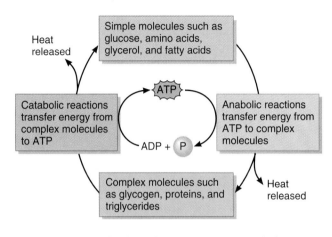

Is the process of making digestive enzymes anabolic or catabolic?

catalysts to speed up chemical reactions. The characteristics and functions of enzymes may be reviewed in Chapter 2.

Most enzymes work together with ions such as calcium, iron, and zinc that hold the enzyme to the substrate or with substances called *coenzymes* that accept or donate atoms to a substrate. Many coenzymes are derived from vitamins. For example, the coenzyme *NAD$^+$* is derived from the B vitamin niacin and the coenzyme *FAD* is derived from vitamin B_2 (riboflavin).

Oxidation–Reduction Reactions

Oxidation refers to the *removal* of electrons (e^-) and hydrogen ions (H^+) from a molecule. This is equivalent to the removal of a hydrogen atom ($e^- + H^+ \longrightarrow H$). An example of an oxidation reaction is the conversion of lactic acid into pyruvic acid:

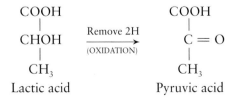

The two freed hydrogen atoms are transferred to another compound by coenzymes.

Reduction refers to the *addition* of electrons and hydrogen ions (hydrogen atoms) to a molecule. Reduction is the opposite of oxidation. An example of a reduction reaction is the conversion of pyruvic acid into lactic acid:

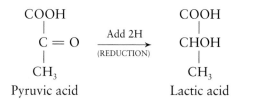

Oxidation and reduction reactions are always coupled; that is, whenever a substance is oxidized, another is reduced at the same time. This coupling of reactions is referred to as **oxidation–reduction.**

Oxidation is usually an energy-releasing reaction. Cells take nutrients (potential energy sources) and break them down, thereby releasing energy. For example, when a cell oxidizes a molecule of **glucose** ($C_6H_{12}O_6$), the energy in the chemical bonds of the glucose molecule is removed and trapped by adenosine triphosphate (ATP). ATP is thus an energy source for energy-requiring reactions.

Oxidation–reduction reactions play important roles in carbohydrate, lipid, and protein metabolism. Also, most body heat is a result of oxidation of the food we eat.

Carbohydrate Metabolism

objective: *Explain how the body uses carbohydrates.*

During digestion, polysaccharides and disaccharides are digested to monosaccharides—glucose, fructose, and galactose—which are absorbed in the small intestine and carried to the liver, where fructose and galactose are then converted to glucose (see Figure 19.13a). The liver is the only organ that has the necessary enzymes to make this conversion. Thus the story of carbohydrate metabolism is really the story of glucose metabolism.

Fate of Carbohydrates

Because glucose is the body's preferred source of energy, the fate of absorbed glucose depends on the body cells' energy needs. If the cells require immediate energy, the glucose is oxidized by the cells. Each gram of carbohydrate produces about 4.0 kilocalories (kcal). (The kilocalorie content of a food is a measure of the heat it releases upon oxidation. Caloric value is described later in the chapter.) Glucose not needed for immediate use may be converted to glycogen by the liver (glycogenesis) and stored there or in skeletal muscle fibers (cells). If these glycogen storage areas are filled up, the liver cells can transform the glucose to triglycerides (lipogenesis) for storage in adipose tissue. When the cells need more energy, the glycogen and triglycerides can be converted back to glucose.

Before glucose can be used by body cells, it must pass through the plasma membrane and enter the cytosol. This occurs through facilitated diffusion. The rate of glucose transport is greatly increased by insulin.

Glucose Catabolism

The **oxidation** of glucose is also known as **cellular respiration** (Figure 20.3). It involves

1. glycolysis
2. the Krebs cycle, and
3. the electron transport chain.

Glycolysis

Glycolysis (glī-KOL-i-sis; *glyco* = sugar; *lysis* = breakdown) refers to the oxidation of glucose to two molecules of pyruvic acid (see Figure 20.3). It takes place in the cytosol of most body cells. Glycolysis is also called **anaerobic cellular respiration** because it occurs without oxygen. In glycolysis, several compounds are formed on the way to producing pyruvic acid. Overall, for each molecule of glucose that undergoes glycolysis, there is a net gain of two molecules of ATP (although four molecules of ATP are produced, two are used in the process). In addition, two molecules of the energy-containing coenzyme NADH + H$^+$ are produced (NAD$^+$ is the oxidized form, NADH + H$^+$ is the reduced form).

The fate of pyruvic acid depends on the availability of oxygen. If conditions remain anaerobic, as during strenuous exercise, pyruvic acid is converted into lactic acid. The lactic acid may be transported to the liver, where it is converted back to pyruvic acid. Or, it may remain in the cells until aerobic (with oxygen) conditions are restored (recall the mechanisms for the repayment of the oxygen debt described on p. 163) and it is then converted to pyruvic acid in the cells.

Under anaerobic conditions, glucose is only partially oxidized. Complete oxidation of glucose requires the presence of oxygen and is referred to as **aerobic cellular respiration.** During aerobic cellular respiration, pyruvic acid enters mitochondria, where it is oxidized to produce a great deal more energy (ATP) than anaerobic cellular respiration. Aerobic cellular respiration consists of a number of chemical reactions known as the Krebs cycle and electron transport chain.

Krebs Cycle

The **Krebs cycle** is a series of oxidation–reduction reactions that occur in the mitochondria of cells. In order for pyruvic acid to enter the Krebs cycle, however, it must first be changed into an acetyl group and combined with coenzyme A (CoA) to form a substance called **acetyl coenzyme A** (see Figure 20.3). In this step, the energy-containing coenzyme NADH + H$^+$ plus carbon dioxide (CO_2) are produced.

During the Krebs cycle, a series of oxidation–reduction reactions breaks down acetyl coenzyme A, releases CO_2, produces a substance called guanosine triphosphate (GTP), and reduces various energy-containing coenzymes (NADH + H$^+$ and FADH$_2$). (FAD is the oxidized form, FADH$_2$ is the reduced

Figure 20.3   Oxidation of glucose: cellular respiration.

The oxidation of glucose involves glycolysis, the Krebs cycle, and the electron transport chain.

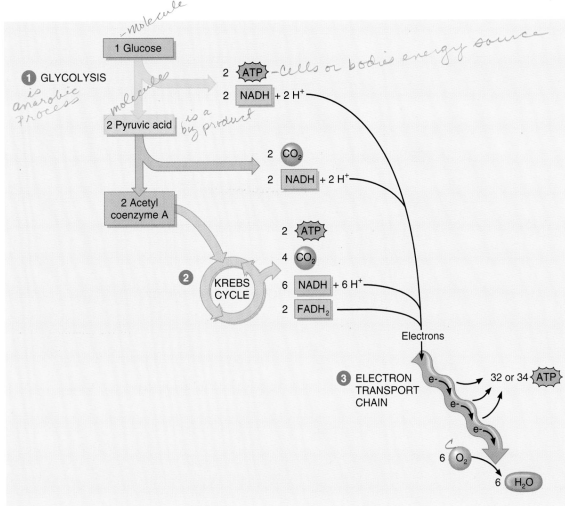

Handwritten margin notes:
anaerobic
– molecule
is anaerobic process
molecules
is a by product
– cells or bodies energy source
lactic acid builds up in muscle – leads to cell death – shakyness – burning in muscle tissue
– need more oxygen lactic acid lactic acid is reconverted to pyruvic acid quite quickly

Q **Why is the production of reduced coenzymes in the Krebs cycle important?**

form.) The CO_2 is transported by the blood to the lungs, where it is exhaled. The GTP is a high-energy compound that is used to change ADP into ATP (GTP is the energy equivalent of ATP). The reduced coenzymes now contain the stored (potential) energy that was in glucose, pyruvic acid, and acetyl coenzyme A. But the reduced coenzymes must first go through the electron transport chain before the energy will be available for cellular activities.

Electron Transport Chain

The *electron transport chain* (see Figure 20.3) is also a series of oxidation–reduction reactions that take place in mitochondria and that transfer the energy that was stored in the reduced coenzymes (NADH + H^+ and $FADH_2$) to 32 or 34 molecules of ATP (depending on which electron carriers are used in the electron transport chain). As NADH + H^+ and $FADH_2$ are oxi-

dized, they pass their electrons to a series of electron carriers and this transfer generates the ATP. The energy used to produce ATP represents only 43 percent of the energy originally stored in glucose because the rest of the energy is released as heat. In the electron transport chain, oxygen is required as the final electron acceptor and water is produced.

The complete catabolism of a molecule of glucose can be summarized as follows:

$$\text{Glucose} + 6 \text{ Oxygen} \longrightarrow \begin{array}{c} 36 \text{ or } 38 \\ \text{ATP} \end{array} + \begin{array}{c} 6 \text{ Carbon} \\ \text{dioxide} \end{array} + 6 \text{ Water}$$

A summary of the sites of the principal events of the various stages of the complete oxidation of glucose is shown in Figure 20.4.

Glycolysis, the Krebs cycle, and the electron transport chain provide all the ATP for cellular activities. And because

Figure 20.4 Sites of the principal events in the complete oxidation of glucose.

🔑 *Except for glycolysis, which occurs in cytosol, all other reactions involving the complete oxidation of glucose occur in mitochondria.*

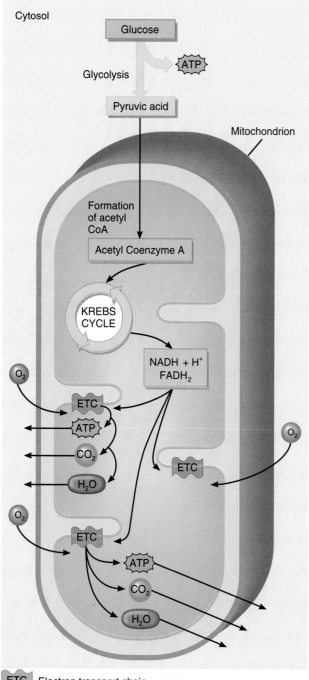

ETC -Electron transport chain

❓ *How many molecules of ATP are produced during the complete oxidation of glucose?*

the Krebs cycle and electron transport chain are aerobic processes, the cells cannot carry on their activities for long without sufficient oxygen.

Glucose Anabolism

Most of the glucose in the body is catabolized to supply energy (ATP). However, some glucose participates in anabolic reactions. For example, some is used in the synthesis of glycogen.

Glucose Storage and Release

If glucose is not needed immediately for energy, it combines with many other molecules of glucose to form a long-chain molecule called *glycogen* (Figure 20.5). This process is called *glycogenesis* (glī-kō-JEN-e-sis; *glyco* = sugar; *genesis* = origin). The body can store about 500 g (about 1.1 lb) of glycogen, roughly 75 percent in the skeletal muscles and the rest in the liver. Liver and skeletal muscle fibers (cells) are very active in glycogenesis. Glycogenesis is stimulated by insulin from the pancreas.

Glycogenesis decreases blood sugar level. When blood sugar level drops too much, glucagon is released from the pancreas and it stimulates the liver to convert glycogen back into glucose, a process called *glycogenolysis* (glī-kō-je-NOL-i-sis; *lysis* = breakdown) (Figure 20.5). The glucose is released into the blood and transported to cells for catabolism. Glycogenolysis usually occurs between meals.

Converting non-carbo-to a carbohydrate

Formation of Glucose from Proteins and Fats: Gluconeogenesis

When your liver runs low on glycogen, it is time to eat. If you don't, your body starts catabolizing triglycerides (fats) and proteins. Actually, the body normally catabolizes some of its triglycerides and a few of its proteins, but large-scale triglyceride and protein catabolism does not happen unless you are starving, eating very few carbohydrates, or suffering from an endocrine disorder.

Triglyceride and amino acid molecules are converted in the liver to glucose (Figure 20.5). The process by which glucose is formed from noncarbohydrate sources is called *gluconeogenesis* (gloo'-kō-nē'-o-JEN-e-sis). Gluconeogenesis occurs when the liver is stimulated by cortisol from the adrenal cortex, glucagon from the pancreas, and thyroxine from the thyroid gland. *Converting fat to carbohydr.*

✗ Lipid Metabolism

objective: *Explain how the body uses lipids.*

Lipids are second to carbohydrates as a source of energy. Triglycerides are ultimately digested into fatty acids and monoglycerides.

Figure 20.5   Glycogenesis, glycogenolysis, and gluconeogenesis. The glycogenesis pathway, the conversion of glucose into glycogen, is stimulated by insulin. The glycogenolysis pathway, the conversion of glycogen into glucose, is stimulated by glucagon. Gluconeogenesis involves the conversion of noncarbohydrate molecules (amino acids and glycerol) into glucose and is stimulated by cortisol, glucagon, and thyroxine.

 *About 500 g (about 1.1 lb) of glycogen are stored in skeletal muscles and the liver.*

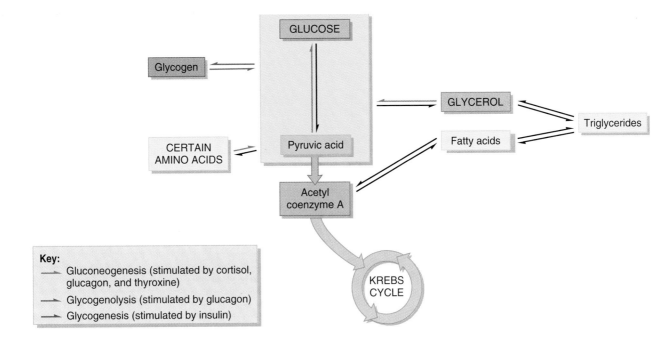

What cells can carry out gluconeogenesis and glycogenesis?

Fate of Lipids

Like carbohydrates, lipids may be oxidized to produce ATP. Each gram of triglyceride produces about 9.0 kilocalories. If the body has no immediate need to utilize triglycerides this way, they are stored in adipose tissue (fat depots) and in the liver.

Other lipids are used as structural molecules or to synthesize other essential substances. For example, phospholipids are constituents of plasma membranes; myelin sheaths speed up nerve impulse conduction; and cholesterol is used to synthesize bile salts and steroid hormones.

Triglyceride Storage

The major function of adipose tissue is to store triglycerides until they are needed for energy in other parts of the body. It also insulates and protects. About 50 percent of stored triglycerides are deposited in subcutaneous tissue.

Lipid Catabolism: Lipolysis

Triglycerides stored in adipose tissue constitute 98 percent of all body energy reserves. The body can store much more triglyceride than it can store glycogen. Moreover, the energy yield of triglycerides is more than twice that of carbohydrates. Triglycerides are nevertheless the body's second-favorite source of energy because they are more difficult to catabolize than carbohydrates.

Glycerol

Before triglyceride molecules can be metabolized for energy, they must be split into glycerol and fatty acids, a process called *lipolysis* (lip-OL-i-sis). Then, the glycerol and fatty acids are catabolized separately (Figure 20.6).

Glycerol is converted to glyceraldehyde-3-phosphate, a compound also formed during the catabolism of glucose. It then continues the catabolic sequence to pyruvic acid or goes on to become glucose. This is one example of gluconeogenesis.

Fatty Acids

Fatty acids are too complex to be catabolized directly by body cells. They must first undergo a series of reactions in the liver called *beta oxidation,* which result in a number of molecules of *acetyl coenzyme A (CoA).* The acetyl CoA may then enter the Krebs cycle in cells (Figure 20.6).

Figure 20.6   Metabolism of lipids. The breakdown of triglycerides into glycerol and fatty acids is called lipolysis. Glycerol may be converted to glyceraldehyde-3-phosphate, which can then be converted to glucose or enter the Krebs cycle for oxidation. Fatty acids undergo beta oxidation and enter the Krebs cycle via acetyl coenzyme A. Fatty acids also can be converted into ketone bodies (ketogenesis). Lipogenesis is the synthesis of lipids from glucose or amino acids.

🔑 *Glycerol and fatty acids are catabolized in separate pathways.*

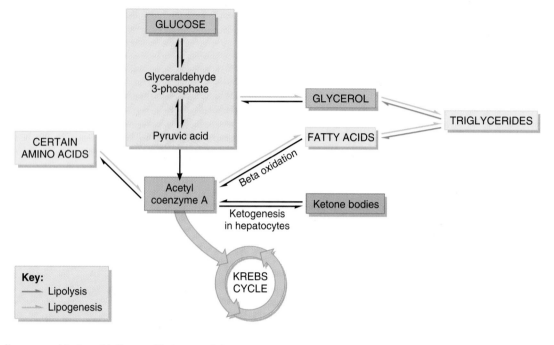

ⓠ *What cells carry out beta oxidation and ketogenesis?*

As part of normal fatty acid catabolism, the liver converts some acetyl CoA molecules into substances known as **ketone bodies,** a process known as **ketogenesis** (kē′-tō-JEN-e-sis) (Figure 20.6). They then leave the liver to enter body cells, where they are broken down into acetyl CoA, which enters the Krebs cycle for oxidation.

Because the body prefers glucose as a source of energy, ketone bodies are generally produced in very small quantities. When the number of ketone bodies in the blood rises above normal—a condition called **ketosis**—the ketone bodies, most of which are acids, must be buffered by the body. If too many accumulate, they use up the body's buffers and the blood pH falls, which leads to acidosis, or abnormally low blood pH, and possibly death.

Lipid Anabolism: Lipogenesis

Hepatocytes and fat cells can synthesize lipids from glucose or amino acids through a process called **lipogenesis** (lip′-ō-JEN-e-sis). Lipogenesis occurs when a greater quantity of carbohydrate enters the body than can be used for energy or stored as glycogen. The excess carbohydrates are synthesized into triglycerides. The conversion of glucose to lipids involves the formation of glycerol and acetyl CoA, which can be converted to fatty acids (see Figure 20.5). The process is enhanced by insulin.

Some amino acids can be converted into acetyl CoA, which can then be converted into triglycerides (Figure 20.6). When there is more protein in the diet than can be utilized as such, the excess is converted to and stored as triglycerides.

Protein Metabolism

| objective: | *Explain how the body uses proteins.*

During digestion, proteins are broken down into their constituent amino acids. Although proteins may be catabolized for energy, amino acids are mainly used to synthesize new proteins for body growth and repair.

Fate of Proteins

The amino acids that enter body cells are almost immediately synthesized into proteins. Proteins are used for structural components of the body (collagen, elastin, and keratin) or for functional purposes (actin and myosin for contraction, fibrinogen for clotting, hemoglobin for carrying oxygen, antibodies for fighting infection, and enzymes and hormones to alter the rate of metabolic processes). Excess amino acids may be converted to triglycerides or glycogen or glucose for energy. Each gram of protein produces about 4.0 kilocalories.

Protein Catabolism

A certain amount of protein catabolism occurs in the body each day, though much of this is only partial catabolism. Proteins are extracted from worn-out cells, such as red blood cells, and broken down into free amino acids. Some amino acids are converted into other amino acids, peptide bonds are re-formed, and new proteins are made as part of the constant turnover in all cells.

    If other energy sources are used up or inadequate and protein intake is high, the liver can convert protein to triglycerides or glucose or oxidize it to carbon dioxide and water. Like fatty acids, the liver must process amino acids before they can be used by the cells for energy. The liver removes the nitrogen from amino acids in a process called *deamination* (dē-am′-i-NĀ-shun). The resulting keto acids can then be converted to acetyl CoA and used for energy in the cells' Krebs cycle. Amino acids can be altered in various ways to enter the Krebs cycle at various locations. The gluconeogenesis of amino acids into glucose may be reviewed in Figure 20.5. The conversion of amino acids into fatty acids (lipogenesis) or ketone bodies (ketogenesis) is shown in Figure 20.6.

Protein Anabolism

Protein anabolism involves the formation of peptide bonds between amino acids to produce new proteins. Protein anabolism, or synthesis, is carried out on the ribosomes of almost every cell in the body, directed by the cells' DNA and RNA (see Figure 3.15). The synthesized proteins function as enzymes, hormones, structural components of cells, and so forth. To balance protein catabolism and protein anabolism, dietary guidelines recommend 60 to 75 g (0.8 g/kg of body weight) of protein per day. However, this should be increased in times of growth or tissue repair such as during pregnancy or following surgery.

    Of the 20 amino acids in your body, 10 are referred to as *essential amino acids* because they cannot be synthesized by the body. They are synthesized by plants or bacteria and so foods containing these amino acids are "essential" for human growth and must be part of the diet. *Nonessential amino acids* can be synthesized by body cells and the appropriate essential and nonessential amino acids must be present in cells before protein synthesis can occur.

    A summary of carbohydrate, lipid, and protein metabolism is presented in Figure 20.7 and Exhibit 20.1

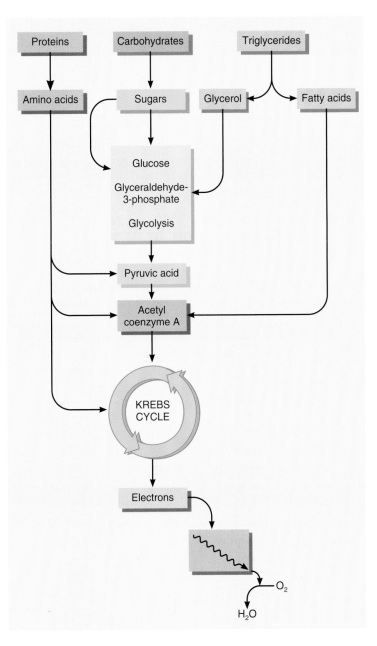

Figure 20.7 Summary of metabolism. Carbohydrates, triglycerides, and proteins enter glycolysis or the Krebs cycle at various points. Glycolysis and the Krebs cycle are catabolic funnels through which high-energy electrons from all kinds of organic molecules flow on their energy-releasing pathways.

Regulation of Metabolism

Absorbed nutrients may be oxidized for energy, stored, or converted into other molecules based on the body's needs. The pathway taken by a particular nutrient is controlled by enzymes and regulated by hormones. Hormones and enzymes are ineffective, however, without the proper minerals and vitamins. Some minerals and many vitamins are components of the enzymes that catalyze the metabolic reactions.

Exhibit 20.1 | Summary of Metabolism

PROCESS	COMMENT
Carbohydrate metabolism	
Glucose catabolism	Complete oxidation of glucose, also called aerobic cellular respiration, is the chief source of energy (ATP) in most cells. It requires glycolysis, Krebs cycle, and electron transport chain. The complete oxidation of one molecule of glucose yields 36 or 38 molecules of ATP.
Glycolysis	Conversion of glucose into pyruvic acid produces some ATP. Reactions do not require oxygen and are thus termed anaerobic cellular respiration.
Krebs cycle	Series of oxidation–reduction reactions in which coenzymes (NAD⁺ and FAD) pick up hydrogen atoms from oxidized organic acids; some ATP is produced. CO_2, H_2O, and heat are by-products. Reactions are aerobic.
Electron transport chain	Third set of reactions in glucose catabolism is a series of oxidation–reduction reactions, in which most of the ATP is produced. Reactions are aerobic.
Glucose anabolism	Some glucose is converted into glycogen (glycogenesis) for storage if not needed immediately for energy. Glycogen can be converted back to glucose (glycogenolysis) for energy (ATP). The conversion of proteins, glycerol, and lactic acid into glucose is called gluconeogenesis.
Lipid metabolism	
Triglyceride catabolism	Triglycerides are broken down into glycerol and fatty acids. Glycerol may be converted into glucose (gluconeogenesis) or catabolized via glycolysis. Fatty acids are catabolized via beta oxidation into acetyl CoA that is catabolized in the Krebs cycle. Acetyl CoA can also be converted into ketone bodies (ketogenesis).
Triglyceride anabolism	Synthesis of triglycerides from glucose and amino acids is lipogenesis. Triglycerides are stored in adipose tissue.
Protein metabolism	
Catabolism	Amino acids are deaminated to enter the Krebs cycle and then are oxidized. Amino acids may be converted into glucose (gluconeogenesis) and fatty acids or ketone bodies.
Anabolism	Protein synthesis is directed by DNA and uses the cell's RNA and ribosomes.

Minerals

objective: *Explain how the body uses minerals.*

Minerals are inorganic substances. They constitute about 4 percent of the total body weight and are concentrated most heavily in the skeleton. Minerals known to perform functions essential to life include calcium, phosphorus, sodium, chlorine, potassium, magnesium, iron, sulfur, iodine, manganese, cobalt, copper, zinc, selenium, and chromium. Other minerals—aluminum, silicon, arsenic, and nickel—are present in the body, but their functions have not yet been determined.

Some minerals, such as calcium, phosphorus, sodium, potassium, chlorine, magnesium, and sulfur, are referred to as *macrominerals* because they are needed in the diet at levels of 100 mg per day or more. Other minerals, such as iron, iodine, copper, zinc, and fluorine, are called *microminerals* or *trace elements* because levels of under 100 mg per day are sufficient.

Exhibit 20.2 describes some vital minerals.

Vitamins

objective: *Explain how the body uses vitamins.*

Vitamins are organic nutrients required in minute amounts to maintain growth and normal metabolism. Unlike carbohydrates, triglycerides, or proteins, vitamins do not provide energy or serve as building materials. They regulate physiological processes, primarily by serving as coenzymes.

Most vitamins cannot be synthesized by the body; they must be ingested in foods or pills. (Vitamin D is an example of a vitamin made by the skin.) Other vitamins, such as vitamin K, are produced by bacteria in the gastrointestinal tract. The body can assemble some vitamins if the raw materials called *provitamins* are provided. Vitamin A is produced by the body from the provitamin carotene, a chemical present in spinach, carrots, liver, and milk. No single food contains all the required vitamins—one of the best reasons for eating a balanced diet.

Exhibit 20.2	Minerals Vital to the Body	
MINERAL	**COMMENTS**	**IMPORTANCE**
Macrominerals		
Calcium	Most abundant cation in body. Appears in combination with phosphorus. About 99 percent is stored in bone and teeth. Remainder stored in muscle, other soft tissues, and blood plasma. Absorption occurs only in the presence of vitamin D. Sources are milk, egg yolk, shellfish, green leafy vegetables.	Formation of bones and teeth, blood clotting, muscle contraction and nerve activity, glycogen metabolism, synthesis and release of neurotransmitters, and functions as a second messenger.
Phosphorus	About 80 percent found in bones and teeth. Remainder distributed in muscle, brain cells, and blood. More functions than any other mineral. Sources are dairy products, meat, fish, poultry, and nuts.	Formation of bones and teeth. Constitutes a major buffer system of blood. Plays important role in muscle contraction and nerve activity. Component of many enzymes. Involved in transfer and storage of energy (ATP). Component of DNA and RNA.
Sulfur	Constituent of many proteins (such as collagen and insulin) and some vitamins (thiamine and biotin). Sources include beef, liver, lamb, fish, poultry, eggs, cheese, and beans.	As component of hormones and vitamins, regulates various body activities. Needed for energy (ATP) production in aerobic cellular respiration.
Sodium	Most abundant cation found in extracellular fluids, some in bones. Normal intake of NaCl (table salt) supplies required amounts.	Strongly affects distribution of water through osmosis. Part of bicarbonate buffer system. Functions in nerve impulse conduction and muscle contraction.
Potassium	Principal cation in intracellular fluid. Normal food intake supplies required amount.	Functions in transmission of nerve impulses and muscle contraction.
Chlorine	Found in extracellular and intracellular fluids. Principal anion of extracellular fluid. Normal intake of NaCl supplies required amounts.	Assumes roles in acid–base balance of blood, water balance, and formation of HCl in stomach.
Magnesium	Component of soft tissues and bone. Widespread in various foods, such as green leafy vegetables, seafood, and whole-grain cereals.	Required for normal functioning of muscle and nervous tissue. Participates in bone formation. Constituent of many coenzymes. Deficiency linked to high blood cholesterol, high blood pressure, pregnancy problems, and spasms of blood vessels.

(continued)

On the basis of solubility, vitamins are divided into two principal groups: fat-soluble and water-soluble. *Fat-soluble* vitamins are absorbed along with ingested dietary lipids by the small intestine. In fact, they cannot be absorbed unless they are ingested with some lipids. Fat-soluble vitamins are generally stored in cells, particularly hepatocytes, so reserves can be built up. The fat-soluble vitamins are A, D, E, and K. *Water-soluble* vitamins, such as vitamins C and B, are absorbed with water in the gastrointestinal tract and dissolve in the body fluids. Excess quantities are excreted in the urine, not stored.

Exhibit 20.3 on page 475 lists the principal vitamins, their sources, functions, and related disorders.

Exhibit 20.2 | Minerals Vital to the Body (Continued)

MINERAL	COMMENTS	IMPORTANCE
Microminerals (trace elements)		
Iron	About 66 percent found in hemoglobin of blood. Remainder distributed in skeletal muscles, liver, spleen, and enzymes. Normal losses of iron occur by shedding of hair, epithelial cells, and mucosal cells, and in sweat, urine, feces, and bile. Sources are meat, liver, shellfish, egg yolk, bean, legumes, dried fruits, nuts, and cereals.	Found in electron transport chain. As component of hemoglobin, carries O_2 to body cells. Involved in formation of ATP from catabolism. Large amounts of stored iron are associated with an increased risk of cancer. This may be related to the ability of iron to catalyze the production of oxygen radicals and serve as a nutrient for cancer cells.
Iodine	Essential component of thyroid hormones. Sources are seafood, cod-liver oil, iodized salt, and vegetables grown in iodine-rich soil.	Required by thyroid gland to synthesize thyroid hormones, hormones that regulate metabolic rate.
Copper	Some stored in liver and spleen. Sources include eggs, whole-wheat flour, beans, beets, liver, fish, spinach, and asparagus.	Required with iron for synthesis of hemoglobin. Found in electron transport chain. Component of enzyme necessary for melanin pigment formation.
Zinc	Important component of certain enzymes. Widespread in many foods, especially meats.	Important in carbon dioxide metabolism. Necessary for normal growth and wound healing, normal taste sensations and appetite, and normal sperm counts in males. As a component of peptidases, it is involved in protein digestion. Deficiency may inhibit immunity and slow learning process. Excess may raise cholesterol level.
Fluorine	Component of bones, teeth, and other tissues.	Appears to improve tooth structure and inhibit tooth decay. Inhibits bone resorption.
Manganese	Some stored in liver and spleen.	Activates several enzymes. Needed for hemoglobin synthesis, urea formation, growth, reproduction, lactation, bone formation, and possibly production and release of insulin, and inhibiting cell damage.
Cobalt	Constituent of vitamin B_{12}.	As part of B_{12}, required for red blood cell formation.
Chromium	Found in high concentrations in brewer's yeast. Also found in wine and some brands of beer.	Necessary for the proper utilization of dietary sugars and other carbohydrates by optimizing the production and effects of insulin. Helps increase blood levels of HDL, while decreasing levels of LDL.
Selenium	Found in seafood, meat, chicken, grain cereals, egg yolk, milk, mushrooms, and garlic.	An antioxidant. Prevents chromosome breakage and may assume a role in preventing certain birth defects and certain types of cancer (esophagus).

Exhibit 20.3	The Principal Vitamins		
VITAMIN	**COMMENT AND SOURCE**	**FUNCTIONS**	**DEFICIENCY SYMPTOMS AND DISORDERS**
Fat-Soluble	All require bile salts and some dietary lipids for adequate absorption.		
A	Formed from provitamin beta-carotene (and other provitamins) in GI tract. Stored in liver. Sources of carotene and other provitamins include yellow and green vegetables; sources of vitamin A include liver and milk.	Maintains general health and vigor of epithelial cells. Beta-carotene acts as an antioxidant to inactivate free radicals. Its potential role in cancer prevention is currently under investigation.	Deficiency results in atrophy and keratinization of epithelium, leading to dry skin and hair, increased incidence of ear, sinus, respiratory, urinary, and digestive system infections, inability to gain weight, drying of cornea and ulceration (**xerophthalmia**), nervous disorders, and skin sores.
		Essential for formation of photopigments, light-sensitive chemicals in photoreceptors of retina.	**Night blindness** or decreased ability for dark adaptation.
		Aids in growth of bones and teeth apparently by helping to regulate activity of osteoblasts and osteoclasts.	Slow and faulty development of bones and teeth.
D	In the presence of sunlight, the skin, liver, and kidneys form vitamin D. Stored in tissues to slight extent. Most excreted via bile. Dietary sources include fish-liver oils, egg yolk, fortified milk.	Essential for absorption and utilization of calcium and phosphorus from GI tract. Works with parathyroid hormone (PTH) to maintain Ca^{2+} homeostasis.	Defective utilization of calcium by bones leads to **rickets** in children and **osteomalacia** in adults. Possible loss of muscle tone.
E (tocopherols)	Stored in liver, adipose tissue, and muscles. Sources include fresh nuts and wheat germ, seed oils, green leafy vegetables.	Believed to inhibit catabolism of certain fatty acids that help form cell structures, especially membranes. Involved in formation of DNA, RNA, and red blood cells. May promote wound healing, contribute to the normal structure and functioning of the nervous system, and prevent scarring. Believed to help protect liver from toxic chemicals like carbon tetrachloride. Acts as an antioxidant to inactivate free radicals.	May cause the oxidation of monounsaturated fats, resulting in abnormal structure and function of mitochondria, lysosomes, and plasma membranes. A possible consequence is hemolytic anemia. Deficiency also causes muscular dystrophy in monkeys and sterility in rats.
K	Produced by intestinal bacteria. Stored in liver and spleen. Dietary sources include spinach, cauliflower, cabbage, liver.	Coenzyme essential for synthesis of several clotting factors by liver, including prothrombin.	Delayed clotting time results in excessive bleeding.
Water-Soluble	Absorbed along with water in GI tract and dissolved in body fluids.		
B$_1$ (thiamine)	Rapidly destroyed by heat. Not stored in body. Excessive intake eliminated in urine. Sources include whole-grain products, eggs, pork, nuts, liver, yeast.	Acts as coenzyme for many different enzymes that are involved in carbohydrate metabolism. Essential for synthesis of acetylcholine.	Improper carbohydrate metabolism leads to buildup of pyruvic and lactic acids and insufficient production of ATP for muscle and nerve cells. Deficiency leads to: (1) **beriberi**—partial paralysis of smooth muscle of GI tract, causing digestive disturbances; skeletal muscle paralysis; atrophy of limbs; (2) **polyneuritis**—due to degeneration of myelin sheaths; impaired reflexes related to kinesthesia, impaired sense of touch, stunted growth in children, and poor appetite.

Exhibit 20.3 | The Principal Vitamins (Continued)

VITAMIN	COMMENT AND SOURCE	FUNCTIONS	DEFICIENCY SYMPTOMS AND DISORDERS
B_2 (riboflavin)	Not stored in large amounts in tissues. Most is excreted in urine. Small amounts supplied by bacteria of GI tract. Dietary sources include yeast, liver, beef, veal, lamb, eggs, whole-grain products, asparagus, peas, beets, peanuts.	Component of certain coenzymes (for example, FAD) in carbohydrate and protein metabolism.	Deficiency may lead to improper utilization of oxygen resulting in blurred vision, cataracts, and corneal ulcerations. Also dermatitis and cracking of skin, lesions of intestinal mucosa, and development of one type of anemia.
Niacin (nicotinamide)	Derived from amino acid tryptophan. Sources include yeast, meats, liver, fish, whole-grain products, peas, beans, nuts.	Essential component of coenzymes (such as NAD^+) in oxidation—reduction reactions. In lipid metabolism, inhibits production of cholesterol and assists in triglyceride breakdown.	Principal deficiency is **pellagra**, characterized by dermatitis, diarrhea, and psychological disturbances.
B_6 (pyridoxine)	Synthesized by bacteria of GI tract. Stored in liver, muscle, brain. Other sources include salmon, yeast, tomatoes, yellow corn, spinach, whole-grain products, liver, yogurt.	Essential coenzyme for normal amino acid metabolism. Assists production of circulating antibodies. May function as coenzyme in triglyceride metabolism.	Most common deficiency symptom is dermatitis of eyes, nose, and mouth. Other symptoms are retarded growth and nausea.
B_{12} (cyanocobalamin)	Only B vitamin not found in vegetables; only vitamin containing cobalt. Absorption from GI tract depends on HCl and intrinsic factor secreted by gastric mucosa. Sources include liver, kidney, milk, eggs, cheese, meat.	Coenzyme necessary for red blood cell formation, formation of amino acid methionine, entrance of some amino acids into Krebs cycle, and manufacture of choline (used to synthesize acetylcholine).	Pernicious anemia, neuropsychiatric abnormalities (ataxia, memory loss, weakness, personality and mood changes, and abnormal sensations), and impaired osteoblast activity.
Pantothenic acid	Stored primarily in liver and kidneys. Some produced by bacteria of GI tract. Other sources include kidney, liver, yeast, green vegetables, cereal.	Constituent of coenzyme A essential for transfer of pyruvic acid into Krebs cycle, conversion of lipids and amino acids into glucose, and synthesis of cholesterol and steroid hormones.	Experimental deficiency tests indicate fatigue, muscle spasms, neuromuscular degeneration, insufficient production of adrenal steroid hormones.
Folic acid (folate, folacin)	Synthesized by bacteria of GI tract. Dietary sources include green leafy vegetables and liver.	Component of enzyme systems synthesizing purines and pyrimidines built into DNA and RNA. Essential for normal production of red and white blood cells.	Production of abnormally large red blood cells (macrocytic anemia). Higher risk of neural tube defects in babies born to folate-deficient mothers.
Biotin	Synthesized by bacteria of GI tract. Dietary sources include yeast, liver, egg yolk, kidneys.	Essential coenzyme for conversion of pyruvic acid to other acids and synthesis of fatty acids.	Mental depression, muscular pain, dermatitis, fatigue, nausea.
C (ascorbic acid)	Rapidly destroyed by heat. Some stored in glandular tissue and plasma. Sources include citrus fruits, tomatoes, green vegetables.	Exact role not understood. Promotes many metabolic reactions, particularly protein metabolism, including laying down of collagen in formation of connective tissue. As coenzyme, may combine with poisons, rendering them harmless until excreted. Works with antibodies, promotes wound healing, and functions as an antioxidant. Role in cancer prevention is under investigation.	Scurvy; anemia; many symptoms related to poor connective tissue growth and repair including tender swollen gums, loosening of teeth (alveolar processes also deteriorate), poor wound healing, bleeding (vessel walls fragile because of connective tissue degeneration), and retardation of growth.

[handwritten: 1 gram of carb is 4 Kilo Calories]
[handwritten: 1 gram protein = 4 calories]
[handwritten: 1 gram fat = 9 Calories]

Metabolism and Body Heat

objective: *Explain how metabolism and body heat are related.*

We will now consider the relationship of foods to body heat, heat gain and loss, and the regulation of body temperature.

Measuring Heat

Heat is a form of energy that can be measured as **temperature** and expressed in units called calories. A *kilocalorie* (*kcal*) is equal to 1000 calories and is defined as the amount of heat required to raise the temperature of 1000 g of water 1°C from 14°C to 15°C. The kilocalorie is the unit we use to express the caloric value of foods and to measure the body's metabolic rate. Thus, when we say that a particular food item contains 500 "calories," we are actually referring to kilocalories. Knowing the caloric value of foods is important; if we know the amount of energy the body uses for various activities, we can adjust our food intake. In this way we can control body weight by taking in only enough kilocalories to sustain our activities.

Production of Body Heat

Most of the heat produced by the body comes from oxidation of the food we eat. The rate at which this heat is produced—the *metabolic rate*—is measured in kilocalories. The following factors affect metabolic rate:

1. *Exercise.* During strenuous exercise, the metabolic rate increases. It may increase to as much as 15 to 20 times the normal rate.

2. *Nervous system.* In a stress situation, the sympathetic nervous system causes the release of norepinephrine, which increases metabolic rate.

3. *Hormones.* In addition to norepinephrine, several other hormones affect metabolic rate. Epinephrine is also secreted under stress. Increased secretions of thyroid hormones increase metabolic rate. Testosterone and human growth hormone also increase metabolic rate.

4. *Body temperature.* The higher the body temperature, the higher the metabolic rate. Metabolic rate may be substantially increased during fever and strenuous exercise.

5. *Ingestion of food.* The ingestion of food, especially proteins, can raise metabolic rate by as much as 10 to 20 percent.

6. *Age.* The metabolic rate of a child, in relation to its size, is about double that of an elderly person because of the high rates of reactions related to growth. Metabolic rate decreases with age.

7. *Others.* Other factors that affect metabolic rate are gender (lower in females, except during pregnancy and lac-tation), climate (higher in tropical regions), sleep (lower), and malnutrition (lower).

Basal Metabolic Rate (BMR)

Because many factors affect metabolic rate, it is measured under standard conditions designed to reduce or eliminate those factors as much as possible. These conditions of the body are called the *basal state* and the measurement obtained is the *basal metabolic rate* (**BMR**). Basal metabolic rate is a measure of the rate at which the body breaks down foods (and therefore releases heat). BMR is also a measure of how much thyroxine the thyroid gland is producing, because thyroxine regulates the basal state.

Loss of Body Heat

Body heat is produced by the oxidation of foods we eat. This heat must be removed continuously or body temperature would rise steadily. The principal routes of heat loss are radiation, evaporation, convection, and conduction.

Radiation

Radiation is the transfer of heat from a warmer object to a cooler one without physical contact. Your body loses heat by the radiation of heat waves to cooler objects nearby such as ceilings, floors, and walls. If these objects are at a higher temperature, you absorb heat by radiation. At rest, about 60 percent of body heat is lost by radiation in a room at 21°C (70°F).

Evaporation

Evaporation is the conversion of a liquid to a vapor. Every gram of water evaporating from the skin takes heat with it. About 22 percent of heat loss occurs through evaporation. Under extreme conditions, about 4 liters (1 gal) of perspiration is produced each hour and this volume can remove 2000 kilocalories of heat from the body. The higher the relative humidity, however, the lower the rate of evaporation.

Convection

Convection is the transfer of heat by the movement of a liquid or gas between areas of different temperature. When cool air makes contact with the body, it becomes warmed and is carried away by convection currents. Then, more cool air makes contact and is carried away. The faster the air moves, the faster the rate of convection. About 15 percent of body heat is lost to the air by convection.

Conduction

In *conduction,* body heat is transferred to a substance or object in contact with the body, such as chairs, clothing, or jewelry. About 3 percent of body heat is lost via conduction.

Homeostasis of Body Temperature Regulation

If the amount of heat production equals the amount of heat loss, you maintain a constant body temperature near 37°C (98.6°F). If your heat-producing mechanisms generate more heat than is lost by your heat-losing mechanisms, your body temperature rises. If your heat-losing mechanisms give off more heat than is generated by heat-producing mechanisms, your temperature falls. A too high temperature destroys body proteins, while a too low temperature causes cardiac arrhythmias; both can result in death.

Hypothalamic Thermostat

Body temperature is regulated by mechanisms that attempt to keep heat production and heat loss in balance. The control center for these mechanisms is a group of neurons in the hypothalamus called the **preoptic area.** If blood temperature rises, these neurons fire impulses more frequently. If blood temperature goes down, the neurons fire impulses less frequently.

The impulses are sent to two other portions of the hypothalamus: the **heat-losing center,** which initiates a series of responses that lower body temperature, and the **heat-promoting center,** which initiates a series of responses that raise body temperature. The heat-losing center is mainly parasympathetic in function; the heat-promoting center is primarily sympathetic.

Mechanisms of Heat Production

Suppose the environmental temperature is low or that some factor causes blood temperature to decrease (Figure 20.8). These cause body temperature (controlled condition) to fall below normal. Both stresses stimulate thermoreceptors in the skin and hypothalamus (receptors) that send nerve impulses (input) to the preoptic area (control center). The preoptic area, in turn, activates the heat-promoting center (control center). In response, the heat-promoting center discharges nerve impulses (output) to effectors that automatically set into operation several responses designed to increase and retain body heat and bring body temperature back up to normal. This cycle is a negative feedback system that attempts to raise body temperature to normal.

VASOCONSTRICTION   Sympathetic nerves stimulate blood vessels of the skin to constrict. This decreases the flow of warm blood from the internal organs to the skin, thus decreasing the transfer of heat from the internal organs to the skin and raising internal body temperature.

SYMPATHETIC STIMULATION OF METABOLISM   Sympathetic nerves stimulate the adrenal medulla to secrete epinephrine and norepinephrine into the blood. The hormones increase cellular metabolism, which, in turn, increases heat production.

THYROID HORMONES   As a result of a series of steps initiated by the hypothalamic control center, the thyroid gland releases thyroid hormones into the blood. Because increased levels of thyroid hormones increase the metabolic rate, body temperature is increased.

SKELETAL MUSCLES   Stimulation of the heat-promoting center causes stimulation of parts of the brain that increase muscle tone. **Shivering** increases the rate of heat production. During maximal shivering, body heat production can rise to about four times the normal rate.

Mechanisms of Heat Loss

Now suppose some stress raises body temperature (controlled condition) above normal (Figure 20.9 on page 480). The stress or higher temperature of the blood stimulates thermoreceptors (receptors) in the skin and hypothalamus that send nerve impulses (input) to the preoptic area (control center). The preoptic area, in turn, stimulates the heat-losing center (control center) and inhibits the heat-promoting center. The heat-losing center discharges nerve impulses (output) to blood vessels (effectors) in the skin, causing them to dilate. The skin becomes warm, and the excess heat is lost to the environment as an increased volume of blood flows from the core of the body into the skin. At the same time, metabolic rate and shivering are decreased. The high temperature of the blood stimulates sweat glands of the skin (effectors) to produce perspiration. As the water of the perspiration evaporates from the surface of the skin, the skin is cooled. All these responses reverse the heat-promoting effects and decrease body temperature (response). This brings body temperature down to normal.

Fever

A **fever** is an abnormally high body temperature. The most frequent cause of fever is infection from bacteria and viruses. Other causes are heart attacks, tumors, tissue destruction by x-rays, surgery or trauma, and reactions to vaccines. The mechanism of fever production is believed to occur as follows. When phagocytes ingest certain bacteria, they secrete a chemical that circulates to the hypothalamus and induces neurons of the preoptic area to secrete prostaglandins. Prostaglandins reset the hypothalamic thermostat at a higher temperature, and temperature-regulating reflex mechanisms will then act to bring the core body temperature up to this new setting. Aspirin, acetaminophen (Tylenol®), and ibuprofen (Advil®) reduce fever by inhibiting synthesis of prostaglandins. (Fever may also be reduced by peripheral cooling in which a cooling blanket is used.)

Suppose that as a result of fever the body thermostat is set at 39.4°C (103°F). Now the heat-promoting mechanisms (vasoconstriction, increased metabolism, shivering) are operating at full force. Thus, even though body temperature is climbing higher than normal—say, 38.3°C (101°F)—the skin remains cold and shivering occurs. This condition, called a **chill,** is a definite sign that body temperature is rising. After several hours, body temperature reaches the setting of the thermostat and the chills disappear. But the body will continue to regulate temperature at 39.4°C (103°F) until the stress is removed. When the stress is removed, the thermostat is reset at normal—37°C (98.6°F). The heat-losing mechanisms (vasodilation and sweating) go into operation to decrease body tem-

Figure 20.8   Negative feedback mechanisms that increase heat production.

When stimulated, the heat-promoting center in the hypothalamus raises body temperature.

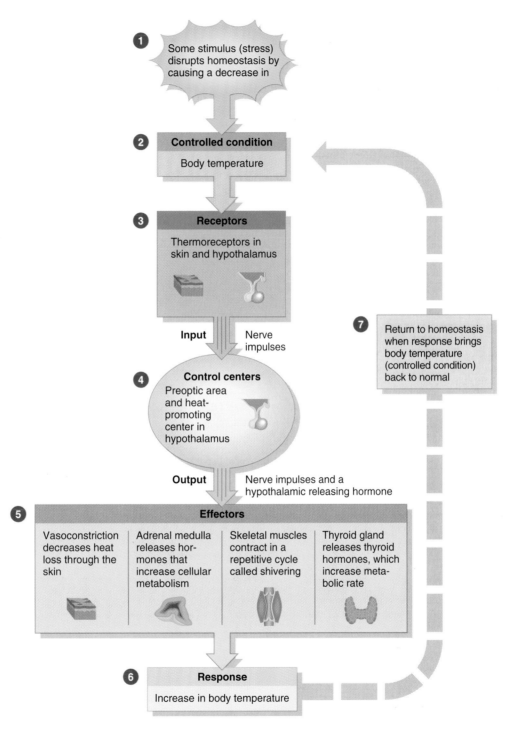

① Some stimulus (stress) disrupts homeostasis by causing a decrease in

② **Controlled condition**
Body temperature

③ **Receptors**
Thermoreceptors in skin and hypothalamus

Input   Nerve impulses

④ **Control centers**
Preoptic area and heat-promoting center in hypothalamus

Output   Nerve impulses and a hypothalamic releasing hormone

⑦ Return to homeostasis when response brings body temperature (controlled condition) back to normal

⑤ **Effectors**

| Vasoconstriction decreases heat loss through the skin | Adrenal medulla releases hormones that increase cellular metabolism | Skeletal muscles contract in a repetitive cycle called shivering | Thyroid gland releases thyroid hormones, which increase metabolic rate |

⑥ **Response**
Increase in body temperature

What factors can increase your metabolic rate and thus increase your rate of heat production?

perature. The skin becomes warm and the person begins to sweat. This phase of the fever is called the *crisis* and indicates that body temperature is falling.

Up to a certain point, fever has a beneficial effect on the body. High body temperature is believed to inhibit the growth of some bacteria. Fever also increases heart rate so that white

Figure 20.9 Negative feedback mechanisms that promote heat loss.

When stimulated, the heat-losing center in the hypothalamus lowers body temperature.

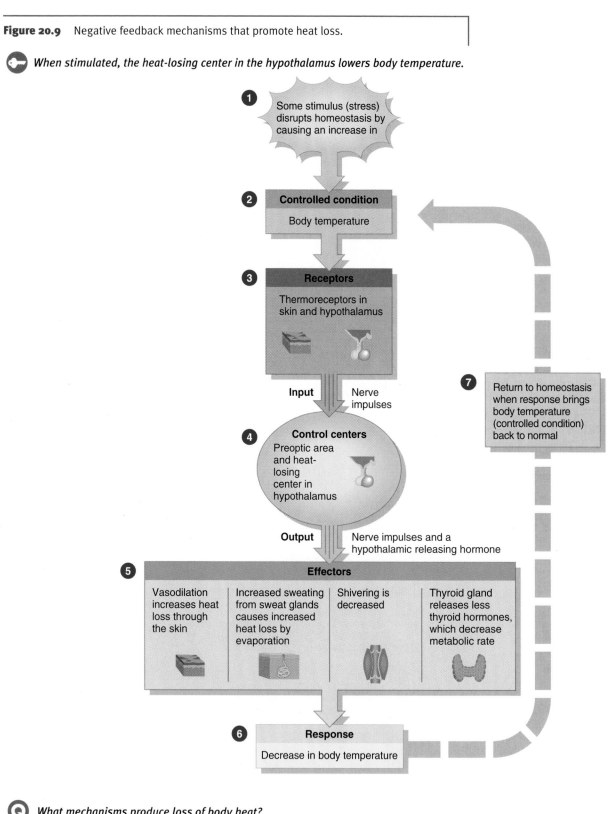

1 Some stimulus (stress) disrupts homeostasis by causing an increase in

2 **Controlled condition**
Body temperature

3 **Receptors**
Thermoreceptors in skin and hypothalamus

Input — Nerve impulses

4 **Control centers**
Preoptic area and heat-losing center in hypothalamus

Output — Nerve impulses and a hypothalamic releasing hormone

5 **Effectors**

| Vasodilation increases heat loss through the skin | Increased sweating from sweat glands causes increased heat loss by evaporation | Shivering is decreased | Thyroid gland releases less thyroid hormones, which decrease metabolic rate |

6 **Response**
Decrease in body temperature

7 Return to homeostasis when response brings body temperature (controlled condition) back to normal

What mechanisms produce loss of body heat?

blood cells are delivered to sites of infection more rapidly and their secretions are increased. In addition, antibody production and T cell proliferation are increased. Also, heat speeds up the rate of chemical reactions that may help body cells repair themselves more quickly during a disease. Among the compli-

cations of fever are dehydration, acidosis, and permanent brain damage. As a rule, death results if body temperature rises to 44 to 46°C (112 to 114°F). At the other end of the scale, death usually results when body temperature falls to 21 to 24°C (70 to 75°F).

wellness focus

How Fat Is Too Fat?

North Americans reap many benefits living in the land of plenty. One of the side-effects, however, is obesity. According to the U.S. National Center for Health Statistics, one-third of the adult population of the United States is obese. Assessment of obesity is based on standard weight-for-height tables. Women are considered obese if they weigh 120 percent or more of their desirable weight, men 124 percent. The one-third statistic is alarming because of obesity's association with many leading causes of mortality, including cardiovascular disease and cancer. Another cause for concern arises from weight trends in recent years. Just 15 years ago, "only" 25 percent of U.S. adults weighed in as obese, which means there has been a 32 percent increase in the incidence obesity in 15 years. On average, U.S. adults weigh about 8 pounds more than they did ten years ago. Obesity rates are rising in children as well.

Is Obesity Really a Problem?

Many studies have found associations between excess body weight and health problems for both men and women. Weight-associated problems include hypertension, high cholesterol levels, type II diabetes, atherosclerosis, certain cancers, arthritis, gallstones, and lower-back disorders. Risk increases in a dose-response fashion: the greater the excess weight, the greater the risk. The typical middle-aged weight gain experienced by many North Americans—20 to 30 pounds—substantially increases health risks. For example, one study found that women who added 22–40 pounds after age 18 had a 70 percent greater risk of death from heart disease and a 20 percent greater risk of cancer than women who had maintained their weight.

How Fat Is Too Fat?

While studies of large groups of people support the relationship between obesity and health risk, it is hard to say how fat is too fat for any given individual. The only general guidelines available are the height-weight tables. In addition to consulting a height-weight table, the following issues should be considered.

Body Composition

More important than weight is body composition: How much of your weight is adipose tissue? Very muscular people may find themselves "too heavy" according to the tables, when in reality they are very healthy.

Location of Fat Stores

People who carry extra fat on the torso are at greater risk for hypertension, type II diabetes, and arterial disease than people whose extra fat resides in the hips and thighs. However, lower-body obesity is still a health problem; people with "pear shapes" are still at higher risk for obesity-related disorders than people who are not overweight.

Medical History and Family Medical History

Weight loss is especially important for people who have obesity-related health problems or a family history of these disorders.

Age

What if you are over 70 years old, a bit overweight, but apparently healthy, with normal blood pressure, blood lipids, and blood sugar? Weight loss in this case is probably not too important. In fact, many nutritionists recommend an extra 10 or 15 pounds for people over 70 to help them resist wasting if they should become ill. People undergoing chemotherapy, for example, have a better prognosis if somewhat overweight.

Setting a Realistic Weight Goal

If you are way over your desirable weight, it may not be realistic to get down to a weight on the chart. The good news is that the first 10 to 15 pounds of weight loss has the greatest health benefits, and can result in a significant reduction of blood pressure, cholesterol, and sugar.

A focus on the scale can be frustrating. The best way to maintain good health or improve health status is to focus on fitness rather than fatness. Cultivate a healthful lifestyle, and health benefits will accrue whether or not weight loss occurs.

critical thinking

Why do you think obesity rates are rising so quickly in North America?

Common Disorders

Obesity

Obesity is defined as a body weight 10 to 20 percent above the desirable standard due to accumulation of fat. Even moderate obesity is hazardous to health. It is a risk factor in cardiovascular disease, hypertension, pulmonary disease, diabetes mellitus (type II), arthritis, certain cancers (uterine and colon), varicose veins, and gallbladder disease.

Starvation

Starvation is characterized by the loss of energy stores in the form of glycogen, triglycerides, and proteins. It may arise from inadequate intake of nutrients or the inability to digest, absorb, or metabolize ingested nutrients. Starvation may be voluntary, as in fasting or anorexia nervosa, or involuntary, as in deprivation or disease (such as diabetes mellitus and cancer).

Phenylketonuria (PKU)

Phenylketonuria (fen′-il-kē′-tō-NOO-rē-a) or *PKU* is a genetic error of metabolism characterized by elevated levels of the amino acid phenylalanine in the blood. It is frequently associated with mental retardation. The artificial sweetener aspartame (Nutrasweet®) contains phenylalanine; its consumption should be restricted in children with PKU because they cannot metabolize it and it may produce neuronal damage.

Cystic Fibrosis (CF)

Cystic fibrosis (*CF*) is an inherited disease of the exocrine glands that affects the pancreas, respiratory passageways, and salivary and sweat glands. It is the most common lethal genetic disease of Caucasians: 5 percent of the population are thought to be genetic carriers. One of the prominent features of CF is blockage of the pancreatic ducts so that the digestive enzymes cannot reach the small intestine. Because pancreatic juice contains the main fat-digesting enzyme, the person fails to absorb fats or fat-soluble vitamins and thus suffers from vitamin A, D, and K deficiency diseases. ■

Study Outline

Nutrients (p. 463)

1. Nutrients are chemical substances in food that provide energy, act as building blocks in forming new body components, serve as storage molecules, or assist in the functioning of various body processes.
2. There are six major classes of nutrients: carbohydrates, lipids, proteins, minerals, vitamins, and water.

Regulation of Food Intake (p. 463)

1. Two centers in the hypothalamus relate to regulation of food intake, the feeding center and satiety center; the feeding center is constantly active but may be inhibited by the satiety center.
2. Among the stimuli that affect the feeding and satiety centers are glucose, amino acids, lipids, body temperature, distention, and cholecystokinin (CCK).

Guidelines for Healthy Eating (p. 463)

1. Nutrition experts suggest dietary calories be 50 to 60 percent from carbohydrates, 30 percent or less from fats, and 12 to 15 percent from proteins, although the optimal levels of these nutrients are not known for sure.
2. The Food Guide Pyramid shows how many servings of five food groups to eat each day to attain the number of calories and variety of nutrients needed for wellness.

Metabolism (p. 465)

1. Metabolism refers to all chemical reactions of the body and has two phases: catabolism and anabolism.
2. Anabolism consists of a series of synthesis reactions whereby small molecules are built up into larger ones that form the body's structural and functional components. Anabolic reactions use energy.
3. Catabolism refers to decomposition reactions that provide energy.
4. Anabolic reactions require energy, which is supplied by catabolic reactions.
5. Metabolic reactions are catalyzed by enzymes, proteins that speed up chemical reactions without themselves being changed.
6. Oxidation refers to the removal of electrons and hydrogen ions from a molecule; oxidation releases energy.
7. Reduction is the opposite of oxidation; oxidation and reduction reactions are always coupled.

Carbohydrate Metabolism (p. 466)

1. During digestion, polysaccharides and disaccharides are converted to monosaccharides, which are transported to the liver.
2. Carbohydrate metabolism is primarily concerned with glucose metabolism.

Fate of Carbohydrates (p. 466)

1. Some glucose is oxidized by cells to provide energy; it moves into cells by facilitated diffusion; insulin stimulates glucose movement into cells.
2. Excess glucose can be stored by the liver and skeletal muscles as glycogen or converted to fat.

Glucose Catabolism (p. 466)

1. Glucose oxidation is also called cellular respiration.

2. The complete oxidation of glucose to CO_2 and H_2O involves glycolysis, the Krebs cycle, and the electron transport chain.

Glycolysis (p. 466)

1. Glycolysis is also called anaerobic respiration because it occurs without oxygen.

2. Glycolysis refers to the breakdown of glucose into two molecules of pyruvic acid. It occurs in the cytosol.

3. When oxygen is in short supply, pyruvic acid is converted to lactic acid; under aerobic conditions, pyruvic acid enters the Krebs cycle.

4. Glycolysis yields a net of two molecules of ATP and two molecules of $NADH + H^+$.

Krebs Cycle (p. 466)

1. The Krebs cycle occurs in mitochondria and begins when pyruvic acid is converted to acetyl coenzyme A.

2. Then a series of oxidations and reductions of various organic acids take place; coenzymes ($NADH + H^+$ and $FADH_2$) are reduced.

3. The energy originally in glucose and then pyruvic acid is now in the reduced coenzymes.

Electron Transport Chain (p. 467)

1. The electron transport chain is a series of oxidation–reduction reactions that occur in mitochondria in which the energy in the coenzymes is liberated and transferred to ATP for storage.

2. The complete oxidation of glucose can be represented as follows:

$$\text{Glucose} + 6\ \text{Oxygen} \rightarrow 36\ \text{or}\ 38 + 6\ \text{Carbon} + 6\ \text{Water}$$
$$\text{ATP} \qquad \text{dioxide}$$

Glucose Anabolism (p. 468)

1. The conversion of glucose to glycogen for storage in the liver and skeletal muscle is called glycogenesis. It occurs extensively in liver and skeletal muscle fibers (cells) and is stimulated by insulin.

2. The body can store about 500 g of glycogen.

3. The conversion of glycogen back to glucose is called glycogenolysis; it occurs between meals.

4. Gluconeogenesis is the conversion of fat and protein molecules into glucose.

Lipid Metabolism (p. 468)

1. Lipids are second to carbohydrates as a source of energy.

2. During digestion, triglycerides are ultimately broken down into fatty acids and glycerol.

Fate of Lipids (p. 469)

1. Some triglycerides may be oxidized to produce ATP.

2. Some triglycerides are stored in adipose tissue.

3. Other lipids are used as structural molecules or to synthesize essential molecules. Examples include phospholipids of plasma membranes, lipoproteins that transport cholesterol, thromboplastin for blood clotting, and cholesterol used to synthesize bile salts and steroid hormones.

Triglyceride Storage (p. 469)

1. Triglycerides are stored in adipose tissue.

2. Most storage occurs in the subcutaneous layer.

Lipid Catabolism: Lipolysis (p. 469)

1. Triglycerides must be split into fatty acids and glycerol before they can be catabolized.

2. Glycerol can be converted into glucose by conversion into glyceraldehyde-3-phosphate.

3. Fatty acids are catabolized through beta oxidation, yielding acetyl coenzyme A, which can enter the Krebs cycle.

4. The formation of ketone bodies by the liver is a normal phase of fatty acid catabolism, but an excess of ketone bodies, called ketosis, may cause acidosis.

Lipid Anabolism: Lipogenesis (p. 470)

1. The conversion of glucose or amino acids into lipids is called lipogenesis.

2. The process is stimulated by insulin.

Protein Metabolism (p. 470)

1. During digestion, proteins are broken down into amino acids.

2. Protein anabolism and catabolism must be balanced through daily dietary intake to prevent protein depletion.

Fate of Proteins (p. 471)

1. Amino acids that enter cells are almost immediately synthesized into proteins.

2. Proteins function as enzymes, hormones, structural elements, and so forth; are stored as fat or glycogen; or are used for energy.

Protein Catabolism (p. 471)

1. Before amino acids can be catabolized, they must be converted to substances that can enter the Krebs cycle.

2. Amino acids may also be converted into glucose, fatty acids, and ketone bodies.

Protein Anabolism (p. 471)

1. Protein synthesis is directed by DNA and RNA and carried out on the ribosomes of cells.

2. Before protein synthesis can occur, all the essential and nonessential amino acids must be present.

Regulation of Metabolism (p. 471)

1. Absorbed nutrients may be oxidized, stored, or converted, based on the needs of the body.

2. The pathway taken by a particular nutrient is controlled by enzymes and regulated by hormones.

Minerals (p. 472)

1. Minerals are inorganic substances that help regulate body processes. They are classified as macrominerals and microminerals.

2. Minerals known to perform essential functions are calcium, phosphorus, sodium, chlorine, potassium, magnesium, iron, sulfur, iodine, manganese, cobalt, copper, zinc,

selenium, and chromium. Their functions are summarized in Exhibit 20.2.

Vitamins (p. 472)

1. Vitamins are organic nutrients that maintain growth and normal metabolism. Many function in enzyme systems.

2. Fat-soluble vitamins (A, D, E, K) are absorbed with lipids.

3. Water-soluble vitamins (B, C) are absorbed with water.

4. The functions and deficiency disorders of the principal vitamins are summarized in Exhibit 20.3.

Metabolism and Body Heat (p. 477)

1. A kilocalorie (kcal) is the amount of energy required to raise the temperature of 1000 g of water 1°C from 14 to 15°C.

2. The kilocalorie is the unit of heat used to express the caloric value of foods and to measure the body's metabolic rate.

Production of Body Heat (p. 477)

1. Most body heat is a result of oxidation of the food we eat. The rate at which this heat is produced is known as the metabolic rate.

2. Metabolic rate is affected by exercise, the nervous system, hormones, body temperature, ingestion of food, age, sex, climate, sleep, and malnutrition.

3. Measurement of the metabolic rate under conditions designed to minimize influential factors is called the basal metabolic rate (BMR).

Loss of Body Heat (p. 477)

1. Radiation is the transfer of heat from one object to another without physical contact.

2. Evaporation is the conversion of a liquid to a vapor, as in perspiration.

3. Convection is the transfer of body heat by the movement of air that has been warmed by the body.

4. Conduction is the transfer of body heat to a substance or object in contact with the body.

Homeostasis of Body Temperature Regulation (p. 478)

1. A normal body temperature is maintained by a delicate balance between heat-producing and heat-losing mechanisms.

2. The hypothalamic thermostat exists as a group of neurons (preoptic area) that stimulate the heat-losing and heat-promoting centers, also in the hypothalamus.

3. Mechanisms that produce heat are vasoconstriction, sympathetic stimulation, skeletal muscle contraction, and thyroxine production.

4. Mechanisms of heat loss include vasodilation, decreased metabolic rate, decreased skeletal muscle contraction, and perspiration.

5. Fever is an abnormally high body temperature.

6. Fever is caused by bacteria, viruses, heart attacks, tumors, tissue destruction by x-rays and trauma, and reactions to vaccines.

Self-Quiz

1. The hunger and satiety centers are found in the
   a. hippocampus b. hypothalamus c. thalamus d. cerebral cortex e. pituitary gland

2. A protein that speeds up the rate of a chemical reaction without being changed is called a(n)
   a. substrate b. complex c. oxidizer d. enzyme e. reducer

3. Match the following:
   D a. removal of electrons and hydrogen ions (H⁺) from a molecule
   E b. the oxidation of glucose
   C c. the conversion of glucose to pyruvic acid
   B d. building simple molecules into more complex ones
   F e. addition of electrons and hydrogen ions (H⁺) to a molecule
   A f. the breakdown of organic compounds

   A. catabolism
   B. anabolism
   C. glycolysis
   D. oxidation
   E. cellular respiration
   F. reduction

4. The process by which glucose is formed from noncarbohydrate compounds is called
   a. Krebs cycle b. glycogenesis c. glycogenolysis d. glycolysis e. gluconeogenesis

5. Which of the following statements summarizes the complete breakdown of glucose?
   a. glucose + 6 water ⟶ 36 or 38 ATP + 6 CO₂ + 6 O₂
   b. glucose + 6 oxygen ⟶ 36 or 38 ATP + 6 CO₂ + 6 water c. glucose + ATP ⟶ 31 or 38 CO₂ + 6 water
   d. glucose + pyruvic acid ⟶ 36 or 38 ATP + 6 O₂
   e. glucose + citric acid ⟶ 31 or 38 ATP + 6 CO₂

6. Organic nutrients that regulate physiological processes by serving as coenzymes are
   a. vitamins b. minerals c. trace elements d. amino acids e. essential amino acids

7. Those amino acids that cannot be synthesized by the body and must be derived through the diet are known as
   a. coenzymes b. ketones c. essential amino acids d. nonessential amino acids e. polypeptides

8. Body temperature is controlled by the
   a. pons b. thyroid gland c. hypothalamus d. medulla oblongata e. autonomic nervous system

9. Loss of body heat is greatest due to
   a. radiation b. conduction c. convection d. evaporation e. vasodilation

10. Which of the following fuel sources is second to carbohydrates as a source of energy for the body?

    a. sugars   b. fats   c. proteins   d. minerals   e. vitamins

11. Cellular respiration includes the following steps in order:
    a. Krebs cycle, glycolysis, electron transport   b. Krebs cycle, electron transport, glycolysis   c. glycolysis, electron transport, Krebs cycle   d. electron transport, Krebs cycle, glycolysis   e. glycolysis, Krebs cycle, electron transport

12. Ketosis may result following

    a. excessive lipolysis   b. excessive lipogenesis   c. excessive glycolysis   d. excessive deamination   e. slowed glycogenesis

13. One example of a fat-soluble vitamin is

    a. vitamin B_{12} b. vitamin B_2 c. niacin   d. vitamin K   e. pantothenic acid

Critical Thinking Applications

1. Hilda is taking an aerobic exercise class with a new instructor who tells the class to keep going until they "feel the burn" in their muscles. Hilda wonders if the "burn" is caused by lactic acid, is it really aerobic exercise? Explain Hilda's concern.

2. Doreen ingests a considerable amount of vitamin C each day. What are the functions of vitamin C in the body? Does Doreen need to worry about her body storing toxic levels of vitamin C? Why? Why not?

3. During a hot summer day, a group of sunbathers roasts on the beach. What mechanism causes their body temperature to increase? Several of the sunbathers eventually get up and jump into the cool water. What mechanisms will decrease their body temperatures?

4. You've probably heard the saying "feed a cold but starve a fever." Does this make sense physiologically? Discuss the effects of body temperature and ingestion of food on metabolic rate.

5. Garth has been sitting in the bleachers during a blustery fall football game. He's feeling chilled to the bone. List several actions that can increase his body temperature.

6. Many people take a multivitamin with minerals once a day. Some people think if one is good, more is better. What problems may be caused by an overdose of fat-soluble vitamins?

Answers to Figure Questions

20.1   Milk, yogurt, cheeses, and meats.

20.2   Anabolic because it involves the synthesis of complex molecules.

20.3   The reduced coenzymes will later yield ATP in the electron transport chain.

20.4   36 or 38, depending on which carriers are used on the electron transport chain.

20.5   Liver cells; liver and skeletal muscle cells.

20.6   Liver cells.

20.8   Exercise, sympathetic nervous system, hormones (epinephrine, thyroid hormones, testosterone, human growth hormone), body temperature, and ingestion of food.

20.9   Vasodilation, decreased metabolic rate, decreased skeletal muscle contraction, and perspiration.

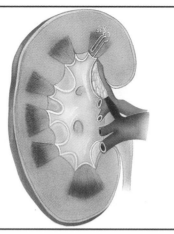

21 chapter

THE URINARY SYSTEM

*I*n metabolizing nutrients, body cells produce wastes—carbon dioxide, excess water, and heat. In addition, protein catabolism produces toxic nitrogenous wastes such as ammonia and urea. Also, essential ions such as sodium (Na^+), chloride (Cl^-), sulfate (SO_4^{2-}), phosphate (HPO_4^{2-}), and hydrogen (H^+) tend to build up in excess quantity. All the toxic materials and the excess essential materials must be excreted (eliminated) from the body in order for homeostasis to be maintained.

Several organs contribute to the job of waste elimination from the body.

1. **Kidneys.** Excrete water, nitrogenous wastes from protein catabolism, some bacterial toxins, H^+, and inorganic salts (electrolytes) plus some heat and carbon dioxide.

2. **Lungs.** Excrete carbon dioxide, heat, and a little water.

3. **Skin (sweat glands).** Excrete heat, water, and carbon dioxide plus small quantities of salts and urea.

4. **Gastrointestinal tract.** Eliminates solid, undigested foods and excretes carbon dioxide, water, salts, and heat.

The prime function of the ***urinary system*** is to help maintain homeostasis by controlling the composition and volume of blood. It does so by removing and restoring selected amounts of water and solutes. Two kidneys, two ureters, one urinary bladder, and a single urethra make up the system (Figure 21.1).

The kidneys have several functions:

1. **Regulation of blood volume and composition.** They regulate the composition and volume of the blood and remove wastes from the blood. In the process, urine is formed. They also excrete selected amounts of various wastes including excess H^+, which helps control blood pH.

2. **Regulation of blood pressure.** They help regulate blood pressure by secreting the enzyme renin, which activates the renin–angiotensin pathway. This results in an increase in blood pressure.

3. **Contribution to metabolism.** The kidneys contribute to metabolism by (1) performing gluconeogenesis (synthesis of new glucose molecules) during periods of fasting or starvation, (2) secreting erythropoietin, a hormone that stimulates production of red blood cells, and (3) participating in synthesis of vitamin D.

Urine is excreted from each kidney through its ureter and is stored in the urinary bladder until it is expelled from the body through the urethra. When the kidneys do not function properly to continually remove waste products, one result is

---

Figure 21.1   Organs of the female urinary system in relation to surrounding structures.

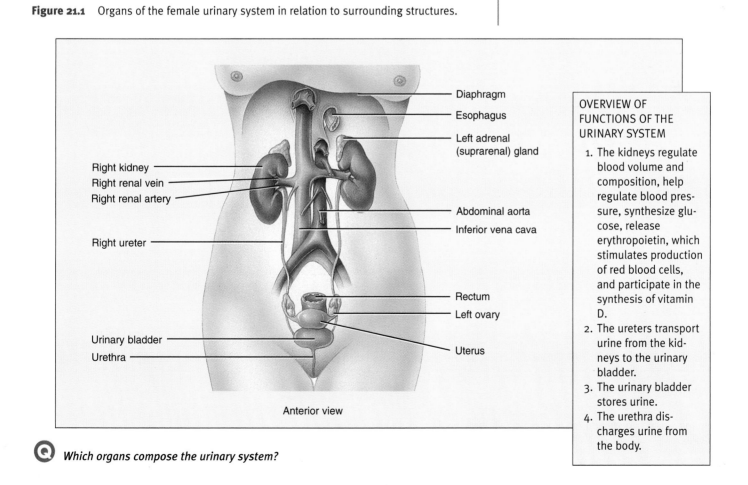

Right kidney
Right renal vein
Right renal artery

Right ureter

Urinary bladder
Urethra

Diaphragm
Esophagus
Left adrenal (suprarenal) gland

Abdominal aorta
Inferior vena cava

Rectum
Left ovary

Uterus

Anterior view

OVERVIEW OF FUNCTIONS OF THE URINARY SYSTEM

1. The kidneys regulate blood volume and composition, help regulate blood pressure, synthesize glucose, release erythropoietin, which stimulates production of red blood cells, and participate in the synthesis of vitamin D.
2. The ureters transport urine from the kidneys to the urinary bladder.
3. The urinary bladder stores urine.
4. The urethra discharges urine from the body.

Q *Which organs compose the urinary system?*

uremia (yoo-RĒ-mē-a; *emia* = condition of blood), a toxic level of urea or other nitrogenous wastes in the blood.

The specialized branch of medicine that deals with the structure, function, and diseases of the male and female urinary systems and the male reproductive system is known as *nephrology* (nef-ROL-ō-jē; *nephros* = kidney; *logos* = study of). The branch of surgery related to male and female urinary systems and the male reproductive system is *urology* (yoo-ROL-ō-jē; *urina* = urine or urinary tract).

Kidneys

> **objective:** *Describe the structure and blood supply of the kidneys.*

The paired *kidneys* are reddish organs shaped like kidney beans. They lie just above the waist against the back wall of the abdominal cavity. Because they are outside the peritoneal lin-

ing of the abdominal cavity, their position is described as *retroperitoneal* (re′-trō-per-i-tō-NĒ-al; *retro* = behind). Other retroperitoneal structures are the ureters and adrenal glands. The kidneys are partially protected by the eleventh and twelfth pairs of ribs and the right kidney is slightly lower than the left because the liver occupies a larger area on the right side.

External Anatomy

The average adult kidney measures about 10 to 12 cm (4 to 5 inches) long, 5.0 to 7.5 cm (2 to 3 inches) wide, and 2.5 cm (1 inch) thick. Near the center of the concave border is an indentation called the *renal* (*renalis* = kidney) *hilus* (HĪ-lus), through which the ureter leaves the kidney and blood and lymphatic vessels and nerves enter and exit.

Each kidney is enclosed in a *renal capsule,* a smooth, transparent, fibrous membrane that serves as a barrier against trauma and infection (Figure 21.2). A mass of fatty tissue surrounds the renal capsule, cushions the kidney, and, along with

---

Figure 21.2   Internal anatomy of the kidney.

The kidneys are covered by a renal capsule, fatty tissue, and a thin layer of dense irregular connective tissue.

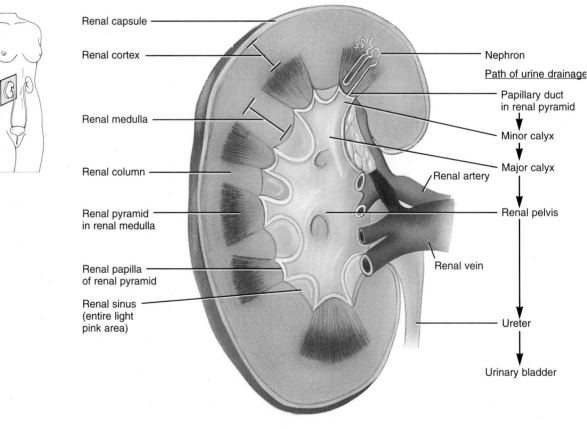

Frontal section of right kidney

Q *Through which structure do blood vessels, lymphatic vessels, and nerves enter the kidneys?*

a thin layer of dense irregular connective tissue, anchors the kidney to the back abdominal wall.

Internal Anatomy

The outer granular appearing part of the kidney is a reddish area called the *renal cortex* (*cortex* = rind or back) and the inner striated (striped) appearing part is a red-brown region called the *renal medulla* (*medulla* = inner portion) (Figure 21.2). Within the renal medulla are 8 to 18 striated, triangular structures, the *renal pyramids.* The striations are straight tubules and blood vessels. The bases of the renal pyramids face the renal cortex, and their tips, called *renal papillae,* face the center of the kidney. The renal cortex extends into the areas between the renal pyramids to form the *renal columns.*

The functioning portion of each kidney consists of approximately 1 million microscopic units called nephrons (described shortly). Urine formed by the nephrons drains into collecting ducts and then larger papillary ducts. Near the renal hilus is a large cavity called the *renal pelvis.* The edge of the renal pelvis has cuplike extensions called *major* and *minor calyces* (KĀ-li-sēz; *calyx* = cup; singular is *calyx*). Each minor calyx collects urine from papillary ducts in the renal pyramids, which moves to the major calyces, drains into the pelvis, and leaves through the ureter.

Blood Supply

The nephrons are largely responsible for removing wastes from the blood and regulating its fluid and electrolyte content. Thus they are abundantly supplied with blood vessels. The right and left *renal arteries* transport about one-fourth the total cardiac output to the kidneys (Figure 21.3). At rest approximately 1200 ml pass through the kidneys every minute.

Before or immediately after entering through the renal hilus, the renal artery divides into several branches that enter the kidney. The arteries divide into smaller and smaller vessels (segmental, interlobar, arcuate, interlobular arteries) that eventually terminate as *afferent* (*ad* = toward; *ferre* = to carry) *arterioles* (see Figure 21.4).

One afferent arteriole is distributed to each nephron, where the arteriole divides into the tangled capillary network called the *glomerulus* (glō-MER-yoo-lus; *glomus* = ball; *ulus* = small). The plural is *glomeruli.* The capillaries of the glomerulus permit the passage of some substances but restrict the passage of others. Once blood is filtered by the glomeruli, they unite to form an *efferent* (*efferens* = to bring out) *arteriole,* which is smaller in diameter than the afferent arteriole. This variation in diameter helps raise the blood pressure in the glomerulus. The afferent–efferent arteriole situation is unique because blood usually flows out of capillaries into venules and not into other arterioles.

In the predominant type of nephron, each efferent arteriole divides again to form a network of capillaries, called the *peritubular* (*peri* = around) *capillaries,* which reclaim useful substances from filtered blood and eliminate other substances into it. These capillaries eventually reunite to form peritubular venules and then interlobular and segmen-

tal veins. Ultimately, all the veins deliver their blood to the *renal veins.*

Nephron

The functional unit of the kidney is the *nephron* (NEF-ron) (Figure 21.4 on page 491). Nephrons filter blood; that is, they permit the passage of some substances out of the blood but restrict the passage of others. As the filtered liquid (filtrate) moves through nephrons, it is further processed by the addition of wastes and excess substances and the return of useful materials back to the blood. As a result of these activities of nephrons, urine is formed.

A nephron consists of two portions: a *renal corpuscle* (KŌR-pus-sul; *corpus* = body; *cle* = tiny) where blood plasma is filtered and a *renal tubule* into which the filtrate passes. Renal corpuscles all lie in the renal cortex. Each corpuscle has two parts—the *glomerulus* (capillary network) and the *glomerular* (*Bowman's*) *capsule,* a double-walled epithelial cup that surrounds the glomerulus. Their arrangement is analogous to a fist (glomerulus) punched into a limp balloon (glomerular capsule) until the fist is covered by two layers of balloon with a space in between. The cells that make up the inner wall of the glomerular capsule adhere closely to the endothelial cells of the glomerular capillaries. Together, they form a *filtration* (*endothelial–capsular*) *membrane* that acts as a filter. This membrane permits the passage of fluid and solutes from blood in glomeruli into renal tubules but restricts the passage of blood cells and large protein molecules.

From the renal corpuscle, filtered fluid passes into the renal tubule. The first portion of the renal tubule is called the *proximal convoluted tubule.* The walls of proximal convoluted tubules contain numerous microvilli to provide a large surface area for the exchange of substances between renal tubules and blood vessels. From the proximal convoluted tubules, fluid and solutes pass into the *loop of Henle (nephron loop).* The first portion of this loop is called the *descending limb of the loop of Henle.* In some nephrons, the descending limb dips down into the renal medulla of the kidney. Then, the fluid and solute pass into the second portion of the loop of Henle, called the *ascending limb of the loop of Henle.* Back in the cortex of the kidney, the fluid passes into a *distal convoluted tubule.* The distal convoluted tubules of several nephrons empty into a single *collecting duct.* Collecting ducts then unite and converge until eventually there are only several hundred large *papillary ducts* at the apices of the renal pyramids, which drain into the minor calyces. The collecting ducts and papillary ducts extend from the renal cortex through the renal medulla to the renal pelvis. On average, there are about 30 papillary ducts per renal papilla.

Juxtaglomerular Apparatus (JGA)

The smooth muscle cells in the wall of the afferent arteriole have round nuclei (instead of long), and their cytoplasm contains granules (instead of fibers). They are called *juxtaglomerular cells.* The cells of the ascending limb of the loop of Henle, which are next to the afferent arteriole, are considerably narrower and taller than the other cells. They are

Figure 21.3   Blood supply of the right kidney.

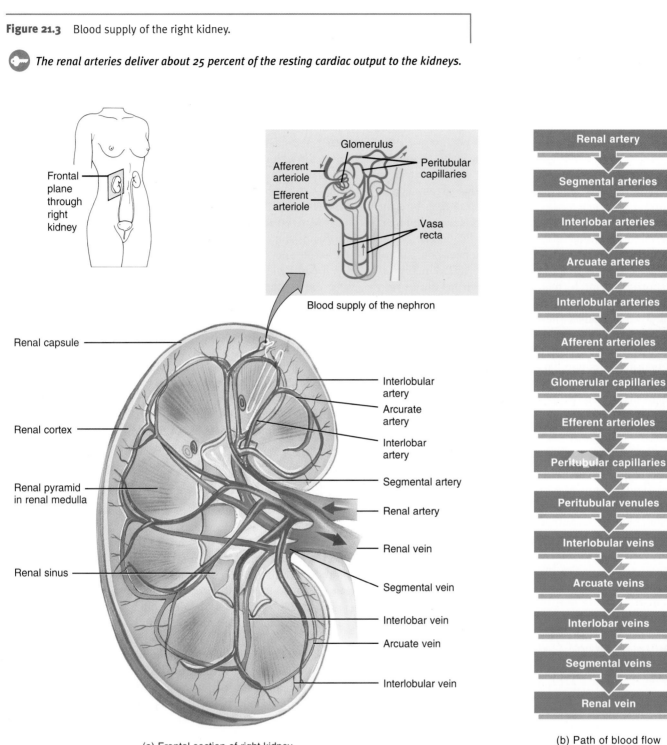

The renal arteries deliver about 25 percent of the resting cardiac output to the kidneys.

(a) Frontal section of right kidney

(b) Path of blood flow

How much blood enters the renal arteries each minute?

Figure 21.4   Parts of a nephron and its blood supply.

Nephrons are the functional units of the kidneys.

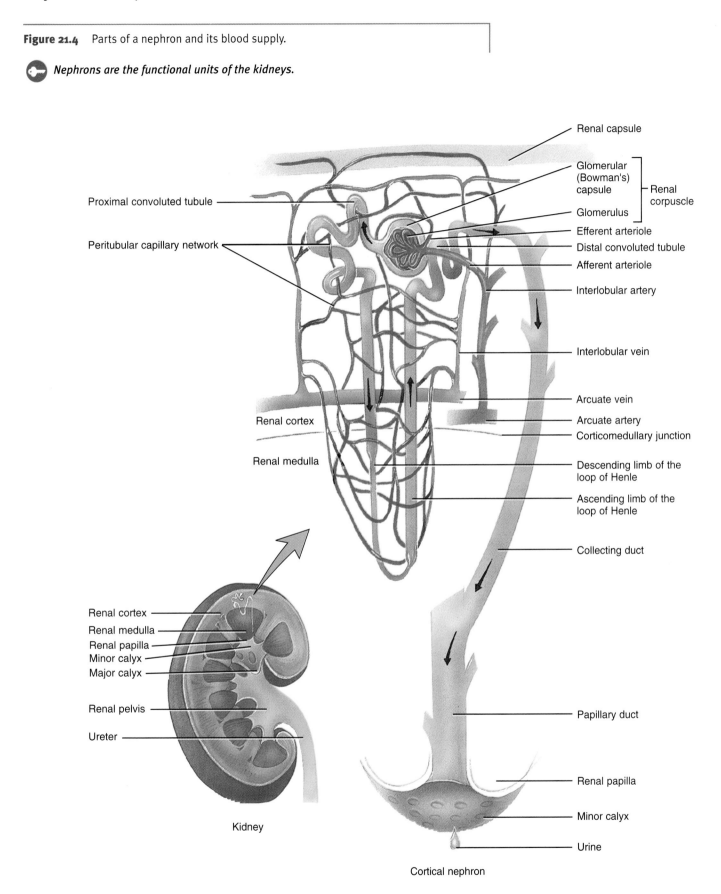

Cortical nephron

Imagine you are a water molecule that has just entered the proximal convoluted tubule of the nephron and will eventually be part of the urine. What parts of the nephron will you travel through (in order) before you reach the renal pelvis?

known as the *macula densa.* Together with the modified cells of the afferent arteriole they constitute the *juxtaglomerular apparatus* or *JGA* (Figure 21.5), which helps regulate renal blood pressure. This is described shortly.

Functions

objective: *Describe how the kidneys filter the blood and regulate its volume, chemical composition, and pH.*

The major work of the urinary system is done by the nephrons. The other parts of the system are primarily passageways and storage areas. Nephrons carry out three important functions: (1) they control blood concentration and volume by removing selected amounts of water and solutes; (2) they help regulate blood pH; and (3) they remove toxic wastes from the blood.

As nephrons go about these activities, they remove many materials from the blood, return the ones the body requires, and eliminate the remainder. The eliminated materials are collectively called *urine.* The entire volume of blood in the body is filtered by the kidneys approximately 60 times a day. Urine

Figure 21.5   Juxtaglomerular apparatus (JGA).

The juxtaglomerular apparatus consists of the juxtaglomerular cells and the macula densa.

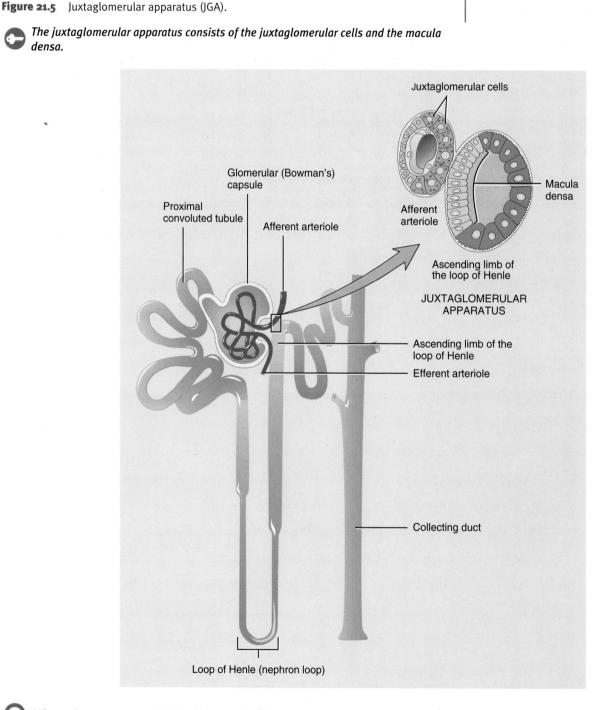

What substance secreted by juxtaglomerular cells helps to regulate blood pressure?

formation involves three processes: (1) glomerular filtration, (2) tubular reabsorption, and (3) tubular secretion.

Glomerular Filtration

Production of Filtrate

The first step in the regulation of blood composition and volume by the kidneys is *glomerular filtration.* Filtration—the forcing of fluids and dissolved substances smaller than a certain size through a membrane by pressure—begins at the filtering membrane when blood enters the glomerulus. Blood pressure in the kidneys, which is about 18 mm Hg, forces water and dissolved blood components (plasma) through the membrane. The resulting fluid is called the *filtrate.* Filtrate consists of all the materials present in the blood except for the blood cells and most proteins, which are too large to pass through the filtering membrane. Exhibit 21.1 compares the constituents of plasma, glomerular filtrate, and urine during a 24-hour period. Although the values shown are typical, they vary considerably according to diet.

Nephrons are especially structured for filtering blood. First, each glomerular capsule contains a tremendous length of highly coiled glomerular capillaries, presenting a vast surface area for filtration. Second, the filtering membrane is very thin and selectively permeable and thus is structurally adapted for filtration. It permits the passage only of smaller molecules. Therefore water, glucose, vitamins, amino acids, small proteins, nitrogenous wastes, and ions pass into the glomerular capsule. Large proteins and the cells in blood do not normally

pass through. Third, the efferent arteriole is smaller in diameter than the afferent arteriole, so there is usually high resistance to the outflow of blood from the glomerulus. Consequently, blood pressure is higher in the glomerular capillaries than in the capsule so fluid flows from an area of higher pressure to one of lower pressure.

Glomerular Filtration Rate (GFR)

The amount of filtrate that forms in both kidneys every minute is called the *glomerular filtration rate* (*GFR*). In the normal adult, this rate is about 125 ml/min—about 180 liters (48 gal) a day. It is very important for the kidneys to maintain a constant GFR in order to maintain the homeostasis of glomerular filtration. If the GFR is too high, needed substances pass so quickly through nephrons that they are unable to be reabsorbed and pass out of the body as part of urine. On the other hand, if the GFR is too low, nearly all the filtrate is reabsorbed and the kidneys would not eliminate the appropriate waste products.

GFR is directly related to several factors. For example, changes in the size of afferent and efferent arterioles can change GFR. Constriction of the afferent arterioles decreases blood flow into the glomerulus, which decreases GFR. Constriction of the efferent arteriole, which takes blood out of the glomerulus, increases GFR.

In some kidney diseases, such as glomerulonephritis (inflammation of glomeruli), glomerular capillaries become so permeable that plasma proteins pass from blood into filtrate. As a result, water is drawn from blood into filtrate by osmosis

---

Exhibit 21.1 | Chemicals in Plasma, Filtrate, and Urine During 24-Hour Period[a]

SUBSTANCE	PLASMA[b] (TOTAL AMOUNT)	FILTRATE IMMEDIATELY AFTER PASSING INTO GLOMERULAR CAPSULE[c]	REABSORBED FROM FILTRATE[d]	URINE
Water	3000 ml	180,000 ml	178,500 ml	1500 ml
Proteins	200	2	1.9	0.1[e]
Chloride ions (Cl⁻)	10.7	639	633.7	6.3
Sodium ions (Na⁺)	9.7	579.6	575	4.6
Bicarbonate ions (HCO$_3$⁻)	4.6	274.5	274.5	0
Glucose	3	180	180	0
Urea	48	53	28	25
Potassium ions (K⁺)	0.5	29.6	27.6	2
Uric acid	0.15	8.5	7.7	0.8
Creatinine	0.03	1.6	0	1.6

[a] All values, except for water, are expressed in grams. The chemicals are arranged in sequence from highest to lowest concentration in plasma.

[b] These substances are present in glomerular blood plasma before filtration.

[c] These substances pass from glomerular blood plasma through the filtering membrane before reabsorption.

[d] These substances have been filtered.

[e] Although trace amounts of protein normally appear in urine, we assume for purposes of discussion that all of it is reabsorbed from filtrate.

and GFR increases. A kidney stone that blocks the ureter or an enlarged prostate gland decreases GFR.

Regulation of GFR

GFR is regulated by three principal mechanisms: (1) renal autoregulation, (2) hormonal regulation, and (3) neural regulation.

RENAL AUTOREGULATION   The ability of the kidneys to maintain a constant blood pressure and GFR despite changes in systemic arterial blood pressure is called *renal autoregulation.* It operates by a negative feedback system that involves the macula densa cells of the juxtaglomerular apparatus or JGA (Figure 21.6). When GFR (controlled condition) is low due to low blood pressure, the filtrate with low sodium ion (Na⁺), chloride ion (Cl⁻), and water content flows past the macula densa (receptor). The macula densa detects the low Na⁺, Cl⁻, and water content of the filtrate and cells of the JGA (control center) decrease their secretion of a vasoconstrictor substance (output). This causes vasodilation of the afferent arterioles (effectors). This allows more blood to flow into the glomerular capillaries, which increases GFR (response) and brings about a return to homeostasis.

HORMONAL REGULATION   Two hormones contribute to the regulation of GFR—angiotensin II and atrial natriuretic peptide (ANP). Figure 21.7 illustrates the renin–angiotensin system. When blood pressure and therefore GFR (controlled conditions) decrease, juxtaglomerular cells and macula densa cells of the JGA (receptors) detect decreased stretch and decreased delivery of Na⁺, Cl⁻, and water, respectively. The juxtaglomerular cells then secrete an enzyme called *renin* (input) into the blood. In the blood, renin acts on a large plasma protein produced by the liver called *angiotensinogen* (control center) and converts it to angiotensin I. As angiotensin I passes through the lungs, it is converted to *angiotensin II,* which is an active hormone.

Angiotensin II (output) is transported in the blood and has important actions on several effectors:

1. **Vasoconstriction of arterioles.** Angiotensin II constricts efferent arterioles (effectors), which increases glomerular blood pressure (response) and raises GFR back to normal.

2. **Stimulation of aldosterone secretion by the adrenal cortex.** Angiotensin II stimulates the adrenal cortex (effector) to secrete aldosterone, which increases retention of Na⁺, Cl⁻, and water by the kidneys. Water retention increases blood volume (response), which restores blood pressure and GFR to normal.

3. **Stimulation of thirst center in the hypothalamus.** Angiotensin II acts on the thirst center in the hypothalamus (effector) to increase water intake. This results in an increase in blood volume (response), which restores blood pressure and GFR to normal.

4. **Stimulation of ADH secretion from the posterior pituitary gland.** Angiotensin II stimulates the release of

Figure 21.6   Negative feedback regulation of glomerular filtration rate by the juxtaglomerular apparatus (JGA).

🔑 *The ability of the kidneys to maintain a constant blood pressure despite changes in systemic arterial pressures is called renal autoregulation.*

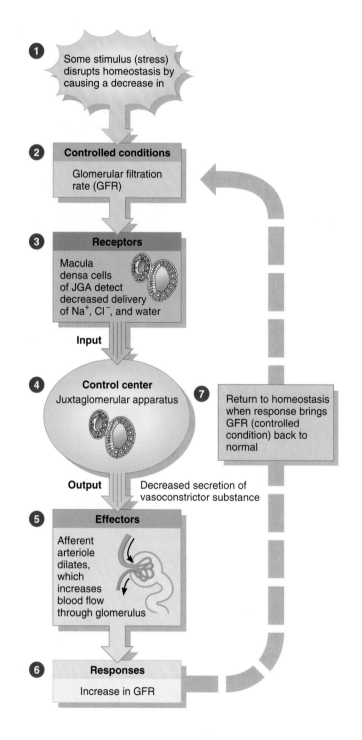

❓ *What happens to GFR if blood pressure in the kidneys increases?*

Figure 21.7   Renin–angiotensin system in the regulation of blood pressure and glomerular filtration rate.

Angiotensin II is the active hormone that acts on the receptors to bring about the appropriate responses.

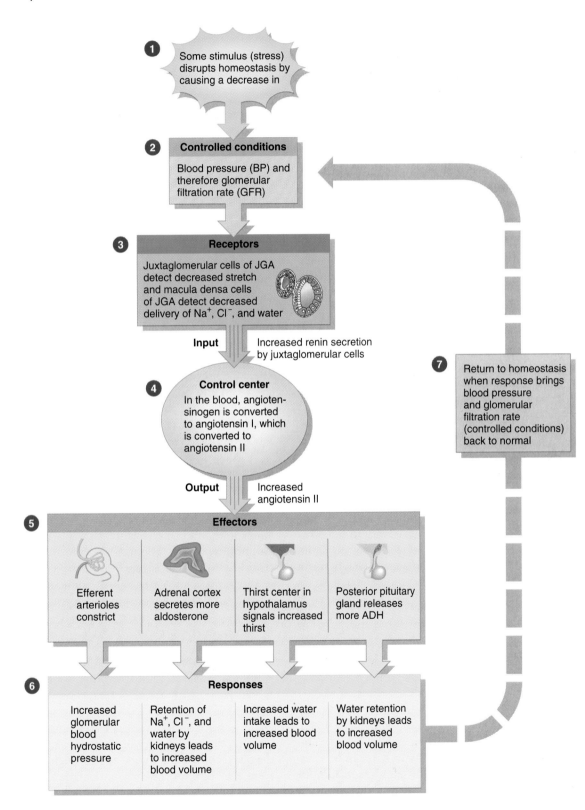

What enzymes and hormones are involved in this system?

antidiuretic hormone (ADH) from the posterior pituitary gland (effector). ADH promotes water retention by the kidneys and increases blood volume (response), which restores blood pressure and GFR to normal.

In all its actions, angiotensin II helps restore normal blood pressure, which normalizes GFR and brings a return to homeostasis.

A second hormone that influences glomerular filtration and other renal processes is *atrial natriuretic peptide* (*ANP*). As its name suggests, it is secreted by cells in the atria (superior chambers) of the heart. ANP promotes both excretion of water (diuresis) and excretion of sodium (natriuresis). Secretion of ANP is stimulated by increased stretching of the atria, as occurs when blood volume increases. ANP increases GFR, perhaps by increasing the permeability of the filtering membrane or by dilating the afferent arterioles. It also suppresses secretion of ADH, aldosterone, and renin; stimulates peripheral vasodilation; and reduces thirst.

NEURAL REGULATION   Like most blood vessels of the body, those of the kidneys are supplied by vasoconstrictor fibers from the sympathetic division of the autonomic nervous system. At rest, sympathetic stimulation is minimal and renal blood vessels are maximally dilated. With moderate sympathetic stimulation, both afferent and efferent arterioles constrict to the same degree. Blood flow into and out of the glomerulus is inhibited to the same extent, which decreases GFR only slightly. With greater sympathetic stimulation, as might occur during exercise, hemorrhage, or a fight-or-flight response, vasoconstriction of the afferent arterioles predominates, which, in turn, greatly decreases glomerular blood flow and GFR. Strong sympathetic stimulation also causes the adrenal medulla to secrete epinephrine, which also decreases GFR by bringing about afferent arteriolar vasoconstriction.

Tubular Reabsorption

As the filtrate passes through the renal tubules and collecting ducts, about 99 percent of it is reabsorbed back into the blood of the peritubular capillaries. Thus only about 1 percent of the filtrate actually leaves the body—about 2 liters a day. The movement of the filtrate back into the blood of the peritubular capillaries is called *tubular reabsorption.* It is the second step in the regulation of blood composition and volume by the kidneys.

Tubular reabsorption is carried out by epithelial cells throughout the renal tubules and collecting ducts, but mostly in the proximal convoluted tubules. It involves processes such as osmosis, diffusion, and active transport. Tubular reabsorption is a very discriminating process. Only specific amounts of certain substances are reabsorbed, depending on the body's needs at the time. Materials that are reabsorbed include water, glucose, amino acids, urea, and ions such as Na^+, K^+, Cl^-, and HCO_3^-. The maximum amount of a substance that can be reabsorbed under any condition is called the substance's *tubular maximum* (*Tm*) or *renal threshold.* Tubular reabsorption allows the body to retain most of its nutrients. Wastes such as urea are only partially reabsorbed. Exhibit 21.1 compares the values for the chemicals in the filtrate immediately after passing into the glomerular (Bowman's) capsule with those reabsorbed from the filtrate. It will give you an idea of how much of the various substances the kidneys reabsorb.

Following are a few examples of how tubular reabsorption occurs. Glucose is reabsorbed by active transport. Normally, all the glucose filtered by the glomeruli (125 mg/100 ml of filtrate/min) is reabsorbed by the proximal convoluted tubules. However, if the plasma concentration of glucose is above normal, the glucose transport mechanism cannot reabsorb it all because it exceeds the tubular maximum and so the excess remains in the urine. This is called *glycosuria,* a symptom of untreated diabetes mellitus.

Most sodium ions are reabsorbed by the proximal convoluted tubules by active transport. Na^+ reabsorption varies with its concentration in extracellular fluid. When the Na^+ concentration of the blood is low, there is a drop in blood pressure, and the renin–angiotensin pathways goes into operation (see Figure 21.7). In one part of the pathway, angiotensin II stimulates aldosterone secretion by the adrenal glands. This, in turn, increases Na^+ and water reabsorption. Thus extracellular fluid volume increases and blood pressure is restored to normal. In the absence of aldosterone, the Na^+ is not reabsorbed and passes into urine for excretion.

When Na^+ moves out of the proximal convoluted tubules into the peritubular blood, chloride ions (Cl^-), negatively charged ions, follow the positive Na^+ out of the tubules (unlike charges attract each other). Thus, the movement of Na^+ influences the movement of Cl^- and other anions into the blood.

Cl^-, and not Na^+, is actively transported from other portions of the renal tubules. Cl^- moves actively and Na^+ follows passively. Whether Na^+ moves actively and Cl^- follows passively or vice versa, the end result is the same.

Almost 90 percent of the water reabsorption occurs together with reabsorption of solutes such as Na^+, Cl^-, and glucose. Of the 180 liters of water that pass into the proximal convoluted tubules every day, this type of water reabsorption reclaims about 160 liters. Passage of most of the remaining water in the filtrate can be regulated. This type of water reabsorption represents about 10 percent of the water reabsorbed from the filtrate. It is regulated mainly by ADH via negative feedback to control the water content of the blood. In the absence of ADH, portions of the nephron are virtually impermeable to water, and urine contains a large volume of water. However, in the presence of ADH, portions of the nephrons become quite permeable to water.

When the water concentration of blood is low (controlled condition), osmoreceptors in the hypothalamus (receptors) detect the change (Figure 21.8). They send nerve impulses (input) that cause the hypothalamus and posterior pituitary gland (control center) to release ADH into the blood. ADH then acts on cells (effectors), which are present in the final portion of the distal convoluted tubule and throughout the collecting duct. Here, ADH makes the cells more permeable to water and water moves through them into the blood. As more water is reabsorbed, the blood's water concentration increases (response). Thus the controlled condition is brought back to normal.

Figure 21.8 Negative feedback regulation of water reabsorption by ADH.

🔑 *Most water reabsorption occurs together with the reabsorption of solutes such as Na⁺, Cl⁻, and glucose.*

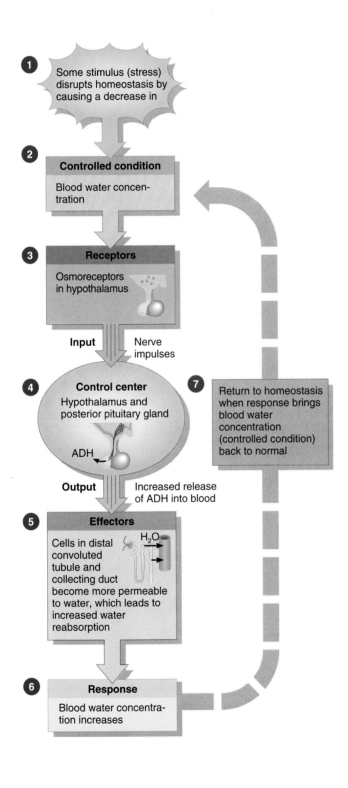

1 Some stimulus (stress) disrupts homeostasis by causing a decrease in

2 **Controlled condition**

Blood water concentration

3 **Receptors**

Osmoreceptors in hypothalamus

Input | Nerve impulses

4 **Control center**

Hypothalamus and posterior pituitary gland

ADH

7 Return to homeostasis when response brings blood water concentration (controlled condition) back to normal

Output | Increased release of ADH into blood

5 **Effectors**

Cells in distal convoluted tubule and collecting duct become more permeable to water, which leads to increased water reabsorption

H_2O

6 **Response**

Blood water concentration increases

❓ *Besides ADH, which other hormones contribute to the regulation of water reabsorption?*

Examine Exhibit 21.2 to see where various substances are absorbed as they pass through different parts of the nephron.

Tubular Secretion

The third process involved in the regulation of blood composition and volume by the kidneys is *tubular secretion.* Whereas tubular reabsorption removes substances from the filtrate into the blood, tubular secretion adds materials to the filtrate from the blood. Tubular secretion occurs in epithelial cells throughout the renal tubules and collecting ducts. The secreted substances include potassium ions (K^+) and hydrogen ions (H^+), ammonium ions (NH_4^+), creatinine, urea, and certain drugs, such as penicillin. Tubular secretion has two principal functions, to rid the body of certain materials and to help control blood pH.

Secretion of K^+ is very important. If the K^+ concentration in plasma nearly doubles, disturbances in cardiac rhythm may develop. At even higher concentrations, cardiac arrest may occur.

The body must maintain normal blood pH (7.35 to 7.45) even though a normal diet provides more acid-producing foods than alkali-producing foods. To raise blood pH, that is, to make it more alkaline, the renal tubules secrete H^+ into the filtrate, which is also what normally makes urine acidic. As a result of H^+ secretion, urine normally has a pH of 6.

Examine Exhibit 21.2 to see where various substances are secreted as they pass through different parts of a nephron.

Exhibit 21.2	Summary of Filtration, Reabsorption, and Secretion
PROCESS	REGION AND SUBSTANCE INVOLVED
Glomerular filtration	Renal corpuscle (filtration membrane): Blood in glomerular capillaries under pressure results in the formation of filtrate that contains water, glucose, some amino acids, Na⁺, Cl⁻, HCO₃⁻, K⁺, urea, uric acid, creatinine, and other solutes in the same concentration as in blood plasma. Plasma proteins and cellular elements of blood normally do not pass through the filtering membranes and are not found in filtrate.
Tubular reabsorption *99%*	Renal tubule (mostly proximal convoluted tubule) and collecting duct: reabsorbed materials include water, glucose, amino acids, urea, Na⁺, K⁺, Cl⁻, and HCO₃⁻.
Tubular secretion	Renal tubule and collecting duct: Secreted materials include K⁺, H⁺, NH₄⁺, creatinine, urea, and some drugs (for example, penicillin).

Exhibit 21.2 and Figure 21.9 summarize glomerular filtration, tubular reabsorption, and tubular secretion in the nephrons.

Hemodialysis Therapy

objective: *Describe the principle and importance of hemodialysis.*

If the kidneys are so impaired by disease or injury that they are unable to excrete wastes and regulate pH and electrolyte and water concentration of the plasma, then the blood must be filtered by artificial means. The process is called *hemodialysis* (*hemo* = blood) and an artificial kidney machine is used (Figure 21.10).

The patient's blood flows through tubing made of selectively permeable dialysis membrane. As blood flows through the tubing, waste products diffuse from the blood into the dialysis solution or *dialysate* (dī-AL-i-sāt) surrounding the dialysis membrane. Also, if nutrients are provided in the solution, they can diffuse from the dialysate into the blood. The dialysate is con-

tinuously replaced to maintain favorable concentration gradients between the solution and the blood. After passing through the dialysis tubing, the blood flows back to the body. In removing wastes from the blood, the dialysis membrane performs one of the kidney's principal functions. Hemodialysis typically is performed three times a week, each session lasting for several hours. There are serious drawbacks to hemodialysis. Blood cells can be damaged and anticoagulants must be added, which can lead to bleeding problems. The slow rate at which the blood can be processed makes the treatment time-consuming.

Ureters

objective: *Describe the structure and functions of the ureters.*

Once urine is formed by the nephrons and passed into collecting ducts, it drains into the calyces. The minor calyces join to become the major calyces that unite to become the renal pelvis. From the renal pelvis, the urine drains into the ureters and is carried to the urinary bladder, where it is discharged from the

Figure 21.9   Summary of functions of a nephron. (1) As blood flows through the glomerulus, it is filtered by the filtering membrane. As the filtrate passes through the renal tubule and collecting duct, certain components are selectively reabsorbed (2) into blood and other substances are secreted (3) into the filtrate for elimination in urine. Both reabsorption and secretion occur at various points along the length of the renal tubule and collecting duct.

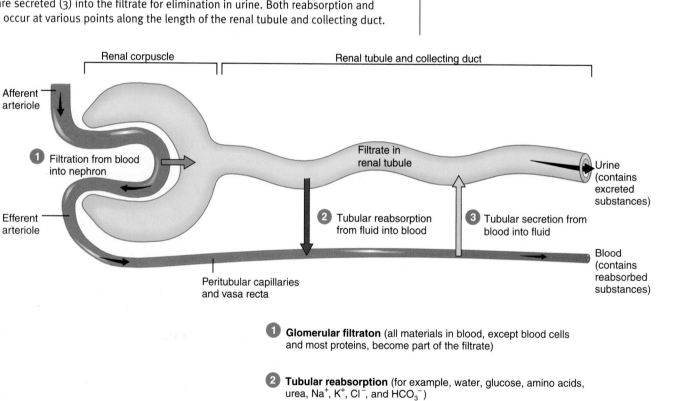

1 **Glomerular filtraton** (all materials in blood, except blood cells and most proteins, become part of the filtrate)

2 **Tubular reabsorption** (for example, water, glucose, amino acids, urea, Na^+, K^+, Cl^-, and HCO_3^-)

3 **Tubular secretion** (for example, K^+, H^+, urea, and creatinine)

Q *Which structures reabsorb and/or secrete ions?*

marily by peristaltic contractions of the muscular walls of the ureters, but hydrostatic pressure and gravity also contribute.

Figure 21.10   Operation of an artificial kidney. The blood route is indicated in red and blue. The route of the dialysate is indicated in gold.

 Dialysis is the separation of large particles from smaller ones through a selectively permeable membrane.

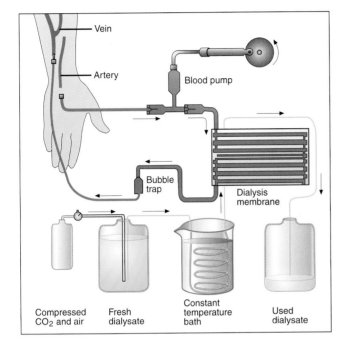

Compressed CO_2 and air   Fresh dialysate   Constant temperature bath   Used dialysate

Q *What happens to plasma proteins during hemodialysis?*

marily by peristaltic contractions of the muscular walls of the ureters, but hydrostatic pressure and gravity also contribute.

Urinary Bladder

objective: *Describe the structure and functions of the urinary bladder.*

The ***urinary bladder*** is a hollow muscular organ situated in the pelvic cavity behind the pubic symphysis. In the male, it is directly in front of the rectum. In the female, it is in front of the vagina and below the uterus. It is a freely movable organ held in position by folds of the peritoneum. The shape of the urinary bladder depends on how much urine it contains. When empty, it looks like a deflated balloon. It becomes spherical when slightly stretched and, as urine volume increases, becomes pear-shaped and rises into the abdominal cavity.

Structure

At the base of the urinary bladder is a small triangular area, the ***trigone*** (TRĪ-gōn; *trigonium* = triangle) (Figure 21.11). The opening into the urethra (*internal urethral orifice*) is in the apex of this triangle. At the two points of the base of the triangle, the ureters drain into the urinary bladder.

The mucous membrane of the urinary bladder contains transitional epithelium that is able to stretch—a marked advantage for an organ that must continually inflate and deflate. Rugae (folds in the mucosa) are also present. The muscular layer of the wall consists of three layers of smooth muscle. Around the opening to the urethra is an ***internal urethral sphincter*** composed of smooth muscle. The opening and closing of this sphincter is involuntary. Below the internal sphincter is the ***external urethral sphincter,*** which is composed of skeletal muscle and is a modification of the urogenital diaphragm muscle. Within limits, the opening and closing of the sphincter are voluntary. The outermost coat, the serous coat, is formed by the peritoneum, which covers the superior surface of the organ; the rest has a fibrous covering.

Functions

Urine is expelled from the urinary bladder by an act called ***micturition*** (mik'-too-RISH-un; *micturere* = to urinate), commonly known as urination or voiding. It is brought about by a combination of involuntary and voluntary nerve impulses similar to defecation.

The average capacity of the urinary bladder is 700 to 800 ml. When the amount of urine in the urinary bladder exceeds 200 to 400 ml, stretch receptors in the urinary bladder wall transmit impulses to the lower portion of the spinal cord. These impulses, by way of sensory tracts to the cerebral cortex, initiate both a conscious desire to void urine and a subconscious reflex that causes the parasympathetic impulses to relax the internal urethral sphincter. Then the conscious portion of the brain sends impulses to the external urethral sphincter to

body through the single urethra. Beyond the minor calyces, urine is in no way modified in either volume or composition.

Structure

The body has two ***ureters*** (YOO-re-ters)—one for each kidney. Each ureter is an extension of the renal pelvis of the kidney and extends to the urinary bladder (see Figure 21.1). Like the kidneys, the ureters are retroperitoneal.

The ureters pass under the urinary bladder for several centimeters, causing the bladder to compress the ureters and thus prevent backflow of urine when pressure builds up in the bladder when it is full, prior to urination. If this physiological valve is not operating, cystitis (urinary bladder inflammation) may develop into a kidney infection.

The inner layer of the wall of the ureter is a mucous membrane. Mucus prevents the cells from coming in contact with urine, which has a drastically different pH and solute concentration from the cell's internal environment. The middle layer of the wall of the ureter consists of smooth muscle.

Functions

The principal function of the ureters is to transport urine from the renal pelvis into the urinary bladder. Urine is conveyed pri-

relax and urination takes place. Although emptying of the urinary bladder is controlled by reflex, it may be initiated voluntarily and stopped at will because of cerebral control of the external urethral sphincter.

A lack of voluntary control over micturition is referred to as *incontinence*. Under about 2 years of age, incontinence is normal because neurons to the external urethral sphincter muscle are not completely developed. Infants void whenever the urinary bladder is sufficiently distended to arouse a reflex stimulus. Proper training overcomes incontinence if the latter is not caused by emotional stress or irritation of the urinary bladder.

urethra to the exterior, the *external urethral orifice,* lies between the clitoris and vaginal opening. In males, the urethra passes vertically through the prostate gland, the urogenital diaphragm, and finally the penis (see Figures 23.1 and 23.6).

Functions

The urethra is the terminal portion of the urinary system. It is the passageway for discharging urine from the body. The male urethra also serves as a duct through which reproductive fluid (semen) is discharged from the body.

Urethra

objective: *Describe the structure and functions of the urethra.*

The *urethra* is a small tube leading from the floor of the urinary bladder to the exterior of the body (see Figure 21.11). In females, it lies directly behind the pubic symphysis and is embedded in the front wall of the vagina. The opening of the

Urine

objective: *Describe the normal and abnormal components of urine.*

The by-product of the kidneys' activities is *urine,* named for one of its constituents—uric acid. The volume, pH, and solute concentration of urine vary with the state of the internal envi-

---

Figure 21.11   Ureters, urinary bladder, and urethra (female).

🔑 *Urine is stored in the urinary bladder until it is expelled by an act called micturition.*

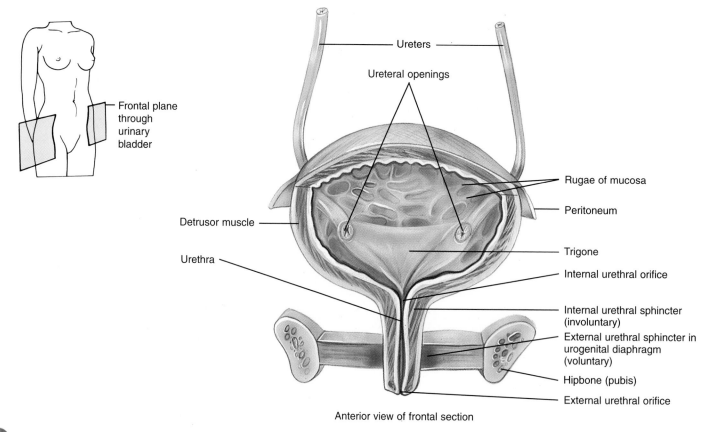

Anterior view of frontal section

🅠 *What is a lack of voluntary control over micturition called?*

ronment. An analysis of the volume and physical and chemical properties of urine, called ***urinalysis*** *(UA)*, tells us much about the state of the body.

Volume

The volume of urine eliminated per day in the normal adult varies between 1000 and 2000 ml (1 to 2 qt). Urine volume is influenced by blood pressure (low pressure triggers the renin–angiotensin pathway which decreases urine volume), blood concentration (low water concentration induces secretion of antidiuretic hormone, which decreases urine volume), temperature (high temperatures decrease urine volume), diuretics (increase urine production), mental state (nervousness can increase urine flow), and general health.

Physical Characteristics

The principal physical characteristics of normal urine are summarized in Exhibit 21.3. Normal urine is sterile; that is, it contains no microbes.

Exhibit 21.3	Physical Characteristics of Normal Urine
CHARACTERISTIC	**DESCRIPTION**
Volume	About two liters in 24 hours but varies considerably.
Color	Yellow or amber but varies with urine concentration and diet. Color is due to urochrome (pigment produced from breakdown of bile). Concentrated urine is darker in color. Diet (reddish colored urine from beets), medications, and certain diseases affect color. Kidney stones may produce blood in urine.
Turbidity	Transparent when freshly voided but becomes turbid (cloudy) upon standing.
Odor	Mildly aromatic but becomes ammonia-like upon standing. Some people inherit the ability to form methylmercaptan from digested asparagus that gives urine a characteristic odor. Urine of diabetics has a fruity odor due to presence of ketone bodies.
pH	Ranges between 4.6 and 8.0; average 6.0; varies considerably with diet. High-protein diets increase acidity; vegetarian diets increase alkalinity.
Specific gravity	Specific gravity (density) is the ratio of the weight of a volume of a substance to the weight of an equal volume of distilled water. It ranges from 1.001 to 1.035. The higher the concentration of solutes, the higher the specific gravity.

Chemical Composition

Water accounts for about 95 percent of the total volume of urine. The remaining 5 percent consists of solutes derived from cellular metabolism and outside sources such as drugs. The solutes are described in Exhibit 21.4.

Exhibit 21.4	Principal Solutes in Normal Urine
SOLUTE	**COMMENTS**
ORGANIC	
Urea	Composes 60–90% of all nitrogen-containing material in urine; derived from ammonia produced by deamination of amino acids, which combines with carbon dioxide to form urea; amount excreted increases with increased dietary protein intake.
Creatinine	Normal constituent of blood. Derived primarily from breakdown of creatine phosphate in muscle tissue.
Uric acid	Product of catabolism of nucleic acids (DNA and RNA) derived from food or cellular destruction. Because of its insolubility, uric acid tends to crystallize and is a common component of kidney stones.
Urobilinogen	Bile pigment derived from breakdown of hemoglobin.
Other substances	May be present in small quantities, depending on diet and general health. Include carbohydrates, pigments, fatty acids, mucin, enzymes, and hormones.
INORGANIC	
Na^+, K^+	Amount excreted varies with dietary intake and level of aldosterone.
Cl^-, Mg^{2+}	Amount excreted varies with dietary intake.
SO_4^{2-}	Derived from amino acids. Amount excreted varies with dietary protein intake.
$H_2PO_4^-$ HPO_4, PO_4^{3-}	Serve as buffers in blood and urine. Parathyroid hormone increases urinary excretion.
NH_4^+	Derived from protein catabolism and from deamination of the amino acid glutamine in kidneys. Amount produced by kidneys may vary with need to produce HCO_3^- to offset acidity of blood and tissue fluids.
Ca^{2+}	Amount excreted varies with dietary intake. Parathyroid hormone increases urinary excretion.

Abnormal Constituents

In a pathological state, traces of substances not normally present may appear in the urine, or normal constituents may appear in abnormal amounts. A urinalysis provides information that aids diagnosis.

A summary of some abnormal constituents of urine is presented in Exhibit 21.5.

Exhibit 21.5	Summary of Abnormal Constituents in Urine
ABNORMAL CONSTITUENT	**COMMENTS**
Albumin	Normal constituent of plasma, but it usually appears in only very small amounts in urine because it is too large to pass through the pores in capillary walls. The presence of excessive albumin in the urine—**albuminuria** (al′-byoo-mi-NOO-rē-a)—indicates an increase in the permeability of filtering membranes due to injury or disease, increased blood pressure, or irritation of kidney cells by substances such as bacterial toxins, ether, or heavy metals.
Glucose	The presence of glucose in the urine is called **glucosuria** (gloo-kō-soo-rē-a) and usually indicates diabetes mellitus. Occasionally, it may be caused by stress, which can cause excessive amounts of epinephrine to be secreted. Epinephrine stimulates the breakdown of glycogen and liberation of glucose from the liver.
Red blood cells (erythrocytes)	The presence of red blood cells in the urine is called **hematuria** (hēm-a-TOO-rē-a) and generally indicates a pathological condition. One cause is acute inflammation of the urinary organs as a result of disease or irritation from kidney stones. Other causes include tumors, trauma, and kidney disease. One should make sure the urine sample was not contaminated with menstrual blood from the vagina.
White blood cells (leukocytes)	The presence of white blood cells and other components of pus in the urine, referred to as **pyuria** (pī-YOO-rē-a), indicates infection in the kidney or other urinary organs.
Ketone bodies	High levels of ketone bodies, called **ketosis** (kē-TŌ-sis), may indicate diabetes mellitus, anorexia, starvation, or simply too little carbohydrate in the diet.
Bilirubin	When red blood cells are destroyed by macrophages, the globin portion of hemoglobin is split off and the heme is converted to biliverdin. Most of the biliverdin is converted to bilirubin, which gives bile its major pigmentation. An above-normal level of bilirubin in urine is called **bilirubinuria** (bil′-ē-roo-bi-NOO-rē-a).
Urobilinogen	The presence of urobilinogen (breakdown product of hemoglobin) in urine is called **urobilinogenuria** (yoo′-rō-bi-lin′-o-je-NOO-rē-a). Traces are normal, but increased urobilinogen may be due to hemolytic or pernicious anemia, infectious hepatitis, biliary obstruction, jaundice, cirrhosis, congestive heart failure, or infectious mononucleosis.
Casts	Casts are tiny masses of material that have hardened and assumed the shape of the lumen of a tubule in which they formed. They are then flushed out of the tubule when filtrate builds up behind them. Casts are named after the cells or substances that compose them or on the basis of their appearance. For example, there are white blood cells casts, red blood cell casts, and epithelial cell casts that contain cells from the walls of the tubules.
Kidney stones	Occasionally, the crystals of salts found in urine solidify into insoluble stones called ***kidney stones***. Conditions leading to stone formation include the ingestion of excessive mineral salts, too low water intake, abnormally alkaline or acidic urine, and overactive parathyroid glands. Kidney stones usually form in the pelvis of the kidney, where they cause pain, hematuria, and pyuria.
Microbes	The number and type of bacteria vary with specific infections in the urinary tract. The most common fungus to appear in urine is *Candida albicans*, a cause of vaginitis. The most frequent protozoan seen is *Trichomonas vaginalis*, a cause of vaginitis in females and urethritis in males.

wellness focus
Infection Prevention for Recurrent UTIs

*U*rinary tract infections (UTIs) are the most common bacterial infections for women, and the second most common illness (after colds) for them. Men get UTIs, too, but much less frequently. The female's shorter urethra allows bacteria to enter the urinary bladder more easily, where they set up housekeeping and multiply, feeding on the urine stored there. In addition, the urethral and anal openings are closer in women. Eighty-five percent of first-time UTIs are caused by *Escherichia coli* (*E. coli*) bacteria that have migrated to the urethra from the anal area. *E. coli* bacteria are necessary for proper digestion and are welcome in the intestinal tract, but they cause much pain and suffering if they infect the urinary system.

About 10 to 15 percent of women develop UTIs several times a month. The patient sometimes feels she is living on antibiotics and can't seem to stay healthy without them. These women must examine their lifestyles very closely to track down and eliminate possible causes for repeated infections.

Infection Prevention
Personal hygiene is the first line of prevention. Care must be taken to avoid transporting bacteria from the anal area to the urethra. Girls should be taught to wipe from front to back and to wash hands thoroughly after using the toilet. When bathing, women and girls should wash from front to back as well.

Menstrual blood provides an excellent growth medium for bacteria. Sanitary napkins and tampons should be changed frequently. Some women find that switching from tampons to napkins or from napkins to tampons reduces the frequency of UTIs. Deodorant tampons and napkins and superabsorbent tampons can increase irritation. Intercourse during days of menstrual flow should be avoided, since blood can get pushed up into the urethra.

People who are prone to UTIs should drink at least 2 to 2.5 liters of fluid daily. Caffeinated beverages should be avoided, since these can be irritating. Cranberry and blueberry juice increase the acidity of the urine and such an acid environment is not favorable for the growth of *E. Coli*. These juices also contain a compound which prevents *E. coli* from adhering to urinary bladder cells, which may help to decrease bacterial growth. Voiding frequently, every 2 to 3 hours, helps to prevent recurrent UTIs, since this expels bacteria and eliminates the urine needed for their growth. Most people find that it is natural to urinate this often when they drink 2 liters of fluid daily!

One study found that the urinary bladders of women with recurrent UTIs were stretched from urinating infrequently. A full bladder stretches the bladder wall and compresses the blood vessels located here. A decreased blood supply means fewer infection-fighting immune cells. More frequent urination prevents this problem, so women are advised not to "hold it."

Partners in Health
Sexual intercourse is frequently associated with the onset of UTIs in women. Women who find that sex brings on UTIs learn to develop stringent personal hygiene and teach their partners as well. Women should drink plenty of water before and after sex, and urinate as soon afterward as possible. This flushes out bacteria that may have entered the urethra.

At times a woman's partner may be the source of bacterial transmission, and when UTIs continue to recur, he should be tested for asymptomatic urethritis, which is the term for any bacterial infection of the urethra other than gonorrhea. Sometimes treating the partner with antibiotics cures both parties.

critical thinking

One of the basic tenets of the wellness philosophy is that the health-care system works best when patients work as partners with their providers to understand, treat, and prevent illness. Explain why treatment of recurrent UTIs is a good illustration of this belief.

Common Disorders

Glomerulonephritis (Bright's Disease)

Glomerulonephritis (*Bright's disease*) is an inflammation of the kidneys that involves the glomeruli. A common cause is toxins given off by streptococci bacteria that have recently infected another part of the body (strep throat, boils, or impetigo). The glomeruli may become so inflamed that the permeability of the filtering membranes increases to the point where blood cells and proteins enter the filtrate and pass into the urine. Blood in the urine is a common first symptom.

Pyelitis and Pyelonephritis

Pyelitis (*pyelos* = pelvis) is an inflammation of the renal pelvis and its calyces. *Pyelonephritis,* an inflammation of one or both kidneys, involves the nephrons and the renal pelvis. The disease is generally a complication of infection elsewhere in the body.

Cystitis

Cystitis is an inflammation of the urinary bladder. Symptoms include burning on urination or painful urination, urgency and frequent urination, and low back pain.

Nephrotic Syndrome

Nephrotic syndrome refers to protein in the urine, primarily albumin, that results in low blood level of albumin, edema, and high blood levels of cholesterol, phospholipids, and triglycerides. It is caused by increased permeability of the filtering membrane.

Renal Failure

Renal failure is a decrease or cessation of glomerular filtration. In *acute renal failure* (**ARF**), the kidneys abruptly stop working entirely or almost entirely. The main feature of ARF is suppression of urine flow. *Chronic renal failure* (**CRF**) refers to the progressive and generally irreversible decline in glomerular filtration rate (GFR).

Renal failure causes edema from salt and water retention; acidosis due to inability of the kidneys to excrete acidic substances; increased levels of urea due to impaired renal excretion of metabolic waste products; elevated potassium levels that can lead to cardiac arrest; anemia, because the kidneys no longer produce renal erythropoietic factor required for red blood cell production; and osteomalacia, because the kidneys are no longer able to convert vitamin D to its active form for calcium absorption from the small intestine.

Urinary Tract Infections (UTIs)

The term *urinary tract infection* (*UTI*) is used to describe either an infection of a part of the urinary system or the presence of large numbers of microbes in urine. UTIs include urethritis (inflammation of the urethra), cystitis, pyelitis, and pyelonephritis. Symptoms associated with UTIs include burning on urination or painful urination, urinary urgency and frequency, pubic and back pain, passage of cloudy or blood-tinged urine, chills, fever, nausea, vomiting, and urethral discharge, usually in males. UTIs are more common in females due to their shorter urethra and closeness of the urethral opening to the anus. ∎

Medical Terminology and Conditions

Anuria (a-NOO-rē-a) Daily urine output of less than 50 ml.

Azotemia (az-ō-TĒ-mē-a; *azo* = nitrogen-containing; *emia* = condition of blood) Presence of urea or other nitrogenous elements in the blood.

Cystocele (SIS-tō-sēl; *cyst* = bladder; *cele* = cyst) Hernia of the urinary bladder.

Diuresis (dī-yoo-RĒ-sis; *dia* = through; *urina* = urine). Increased excretion of urine.

Dysuria (dis-YOO-rē-a; *dys* = painful; *uria* = urine) Painful urination.

Enuresis (en'-yoo-RĒ-sis; *enourein* = to void urine) Bedwetting; a very common pediatric problem (15 to 22 percent of five-year-old children suffer from it) that occurs more frequently in boys than girls; in most cases, the disorder resolves itself spontaneously without medical treatment; suspected causes include deep sleep patterns, nocturnal urine production, and urinary bladder capacity abnormalities. Also referred to as *nocturia.*

Intravenous pyelogram (in'-tra-VĒ-nus PĪ-e-lō-gram') or *IVP* (*intra* = within; *veno* = vein; *pyelo* = pelvis of kidney; *gram* = written or recorded) x-ray film of the kidneys after injection of a dye.

Polyuria (pol'-ē-YOO-rē-a; *poly* = much) Excessive urine.

Stricture (STRIK-chur) Narrowing of the lumen of a canal or hollow organ, as the ureter or urethra.

Uremia (yoo-RĒ-mē-a; *emia* = condition of blood) Toxic levels of urea in the blood resulting from severe malfunction of the kidneys.

Study Outline

1. The primary function of the urinary system is to regulate the concentration and volume of blood by removing and restoring selected amounts of water and solutes. It also excretes wastes.

2. The organs of the urinary system are the kidneys, ureters, urinary bladder, and urethra.

Kidneys (p. 488)

External Anatomy; Internal Anatomy (p. 488–489)

1. The kidneys are retroperitoneal organs attached to the posterior abdominal wall.

2. Each kidney is enclosed in a renal capsule, which is surrounded by adipose tissue.

3. Internally, the kidneys consist of a renal cortex, renal medulla, renal pyramids, renal papillae, renal columns, calyces, and a renal pelvis.

4. The nephron is the functional unit of the kidneys.

5. A nephron consists of a renal corpuscle (glomerulus and glomerular capsule) and renal tubule (proximal convoluted tubule, descending limb of the loop of Henle, ascending limb of the loop of Henle, and distal convoluted tubule).

6. The filtering unit of a nephron is the filtering (endothelial–capsular) membrane.

7. Blood enters the kidney through the renal artery and leaves through the renal vein.

8. The juxtaglomerular apparatus (JGA) consists of the juxtaglomerular cells and the macula densa of the distal convoluted tubule.

Functions (p. 492)

1. Nephrons regulate the composition and volume of blood and in the process form urine.

2. Nephrons regulate the composition and volume of blood by glomerular filtration, tubular reabsorption, and tubular secretion.

3. The primary force behind glomerular filtration is hydrostatic pressure.

4. Most substances in plasma are filtered by the glomerular (Bowman's) capsule. Normally, blood cells and most proteins are not filtered.

5. The amount of filtrate that forms in both kidneys every minute is glomerular filtration rate (GFR). It is regulated by renal autoregulation, hormonal regulation, and neural regulation.

6. The renin–angiotensin pathway refers to the release of renin (by the juxtaglomerular cells) and conversion of angiotensin I to angiotensin II that is initiated by too low levels of Na^+ and Cl^-. Angiotensin II restores blood pressure and GFR to normal.

7. Tubular reabsorption retains substances needed by the body, including water, glucose, amino acids, and ions (Na^+, K^+, Ca^{2+}, Cl^-, HCO_3^-, and HPO_4^{2-}). The maximum amount of a substance that can be absorbed is called tubular maximum (Tm).

8. About 90 percent of the reabsorbed water is reabsorbed with sodium and glucose, the rest by ADH.

9. Tubular secretion discharges chemicals not needed by the body into the urine. Included are excess ions, nitrogenous wastes, and certain drugs.

10. The kidneys help maintain blood pH by excreting H^+ ions, which make urine acidic and the blood more alkaline.

11. Tubular secretion also helps maintain proper levels of potassium in the blood.

Hemodialysis Therapy (p. 498)

1. Filtering blood through an artificial device is called hemodialysis.

2. Hemodialysis is normally performed three times a week with each session lasting 4 to 5 hours.

Ureters (p. 499)

1. The ureters are retroperitoneal.

2. The ureters transport urine from the renal pelvis to the urinary bladder, primarily by peristalsis.

Urinary Bladder (p. 499)

1. The urinary bladder is posterior to the pubic symphysis. Its function is to store urine prior to micturition.

2. Micturition is the act by which urine is expelled from the urinary bladder.

Urethra (p. 500)

1. The urethra is a tube leading from the floor of the urinary bladder to the exterior.

2. Its function is to discharge urine from the body.

Urine (p. 500)

1. Urine volume is influenced by blood pressure, blood concentration, temperature, diuretics, and emotions.

2. The physical characteristics of urine evaluated in a urinalysis (UA) are color, odor, turbidity, pH, and specific gravity.

3. Chemically, normal urine contains about 95 percent water and 5 percent solutes. The solutes include urea, creatine, uric acid, ketone bodies, salts, and ions.

4. Abnormal constituents diagnosed through urinalysis include albumin, glucose, erythrocytes, leukocytes, ketone bodies, bilirubin, urobilinogen, casts, kidney stones, and microbes.

Self-Quiz

1. Which does NOT fit the description of the kidneys?
   a. They are protected by the 11th and 12th pairs of ribs. b. The average adult kidney is 10 to 12 cm (4 to 5 inches) long and 5.0 to 7.5 cm (2 to 3 inches) wide. c. The left kidney is lower than the right in order to accommodate the large size of the liver. d. Each kidney is surrounded by fat and connective tissue. e. The kidneys are retroperitoneal.

2. Place the following structures in the correct order for the flow of urine:
   1. Renal pyramids
   2. Minor calyx
   3. Renal pelvis
   4. Major calyx
   5. Collecting ducts
   6. Ureters
      a. 1,2,4,3,6,5   b. 5,1,4,2,3,6   c. 5,1,2,4,3,6
      d. 3,5,1,2,4,6   e. 1,5,2,4,3,6

3. Oxygenated blood enters the kidney through the
   a. renal vein   b. renal portal system   c. nephronal arteries   d. renal artery   e. hilus artery

4. The functional unit of the kidney where the urine is actually produced is the
   a. juxtaglomerular apparatus   b. glomerulus   c. calyx   d. nephron   e. pyramid

5. What forces the plasma to be filtered across the glomerulus?
   a. a full urinary bladder   b. control by the nervous system   c. water retention   d. the pressure of the blood   e. the pressure of urine in the glomerulus

6. Glomerular filtration rate (GFR) is the
   a. rate of bladder filling   b. amount of filtrate formed in both kidneys each minute   c. amount of filtrate reabsorbed at the collecting ducts   d. amount of blood delivered to the kidney each minute   e. amount of urine formed per hour

7. Tubular maximum (Tm) refers to
   a. how much filtrate the glomerular (Bowman's) capsule can hold   b. how much urine the collecting ducts can hold   c. how much filtrate is secreted into the distal convoluted tubule   d. amount of a substance that can be reabsorbed from the tubules   e. amount of a substance that is secreted by the tubules

8. Which of the following is secreted into the urine from the blood?
   a. hydrogen ions   b. amino acids   c. glucose   d. a and b   e. all of the above

9. Which of the following does NOT influence the volume of urine produced by the body?
   a. blood pressure   b. respiratory rate   c. blood water concentration   d. temperature   e. emotions

10. Which of the following is the stimulus for the renin–angiotensin system?
    a. increased blood sugar   b. increased sodium retention   c. overhydration   d. lowered blood pressure   e. high Cl^- concentration in the blood

11. In the nephron, filtrate is reabsorbed from the renal tubules back into the
    a. glomerulus   b. peritubular capillaries   c. efferent arteriole   d. afferent arteriole   e. renal vein

12. Which of the following is NOT a function of the nephrons?
    a. They are storage areas for urine.   b. They control the chemical composition of blood.   c. They regulate the pH of blood.   d. They remove toxins from blood.   e. They control the volume of blood.

13. Aldosterone enhances the reabsorption of which substance in the collecting ducts?
    a. K^+ b. Ca^{2+} c. Cl^- d. Na^+ e. H^+

14. The hormone that promotes excretion of water and sodium is
    a. ADH (antidiurectic hormone)   b. angiotensin II   c. atrial natriuretic peptide (ANP)   d. aldosterone   e. renin

15. The urinary bladder
    a. functions in micturition   b. is retroperitoneal in position   c. expels urine through the trigone   d. is connected to the kidney by two urethras   e. prevents backflow of urine into the kidneys by the action of the internal sphincter

Critical Thinking Applications

1. You are fascinated by glomerular filtration and would like to explain it to a friend who knows nothing about human anatomy and physiology. Use the process of making coffee with an automatic coffeemaker to describe glomerular filtration to your friend.

2. Yesterday, you attended a large, outdoor party where beer was the only beverage available. You remember having to urinate many, many times yesterday, and today you feel dehydrated. What hormone is affected by alcohol and how does this affect your kidney function?

3. Urine itself is normally sterile (no microbes present), so what could be the reason for the appearance of microbes in a urine sample? What could be the source of the microbes?

4. Sarah is a precocious 1-year-old whose parents would like her to be the first toilet-trained child in preschool. So far they have been unsuccessful in their efforts. Should they be concerned about Sarah's incontinence? Why? Why not?

5. Bette was recently diagnosed with diabetes mellitus after her urine tested positive for glucose. What normally happens to glucose as it passes through the kidney? Is glucose considered a normal constituent of urine?

6. One warm spring day, Brian went on a long hike with only one small water bottle. After a bad tumble down a cliff, Brian called a park ranger on his cellular phone, and the ranger took Brian to a hospital, where Brian supplied a very scanty urine sample that was cloudy and orange in color. Explain the characteristics of the urine sample.

Answers to Figure Questions

21.1   Kidneys, ureters, urinary bladder, and urethra.

21.2   Hilus.

21.3   About 1200 ml.

21.4   Proximal convoluted tubule ⟶ descending limb of the loop of Henle ⟶ ascending limb of the loop of Henle ⟶ distal convoluted tubule ⟶ collecting duct ⟶ papillary duct ⟶ minor calyx ⟶ major calyx ⟶ renal pelvis.

21.5   Renin.

21.6   GFR also increases.

21.7   Enzymes—renin; hormones—angiotensin II and aldosterone.

21.8   Aldosterone and atrial natriuretic peptide.

21.9   Proximal convoluted tubule, ascending and descending limbs of the loop of Henle, distal convoluted tubule, and collecting duct.

21.10 They remain in the blood because they are too large to pass through the pores in the dialysis membrane.

21.11 Incontinence.

chapter 22

FLUID, ELECTROLYTE, AND ACID–BASE BALANCE

a look ahead

student learning objectives

*T*he term ***body fluid*** refers to body water and its dissolved substances. (There is no pure water in the body; all body water contains solutes.) Fluids comprise an average of 55 to 60 percent of total body weight.

Fluid Compartments and Fluid Balance

About two-thirds of body fluid is within cells and is termed ***intracellular*** (*intra* = within) ***fluid*** (**ICF**). The other third, called ***extracellular*** (*extra* = outside) ***fluid*** (**ECF**), includes all other body fluids (Figure 22.1). About 80 percent of the ECF is ***interstitial*** (*inter* = between) ***fluid*** and 20 percent is ***blood plasma,*** the liquid portion of blood. Some of the extracellular fluid is localized in specific places such as lymph in lymphatic vessels; cerebrospinal fluid in the brain; synovial fluid in joints; aqueous humor and vitreous body in the eyes; endolymph and perilymph in the ears; pleural, pericardial, and peritoneal fluids between serous membranes; and glomerular filtrate in the kidneys.

Selectively permeable membranes separate body fluids into distinct compartments. A compartment may be as small as the interior of a single cell or as large as the combined interiors of the heart and blood vessels. Plasma membranes of individual cells separate intracellular fluid from interstitial fluid. Although fluids are in constant motion from one compartment to another, the volume of fluid in each compartment remains fairly stable—another example of homeostasis.

Water is the main component of all body fluids. When we say that the body is in ***fluid balance,*** we mean that the required amount of water is present and proportioned among the various compartments according to their needs. Osmosis is the primary method of water movement in and out of body compartments. The concentration of solutes in the fluids is therefore a major determinant of fluid balance. Most solutes in body fluids are electrolytes—substances that dissociate into ions. Fluid balance, then, means water balance, but it also implies electrolyte balance. The two are inseparable.

Water

Water is by far the largest single component of the body. Infants have the highest proportion of water, up to 75 percent of body

---

Figure 22.1   Body fluid compartments.

🔑 *Fluids comprise an average of 55 to 60 percent body weight.*

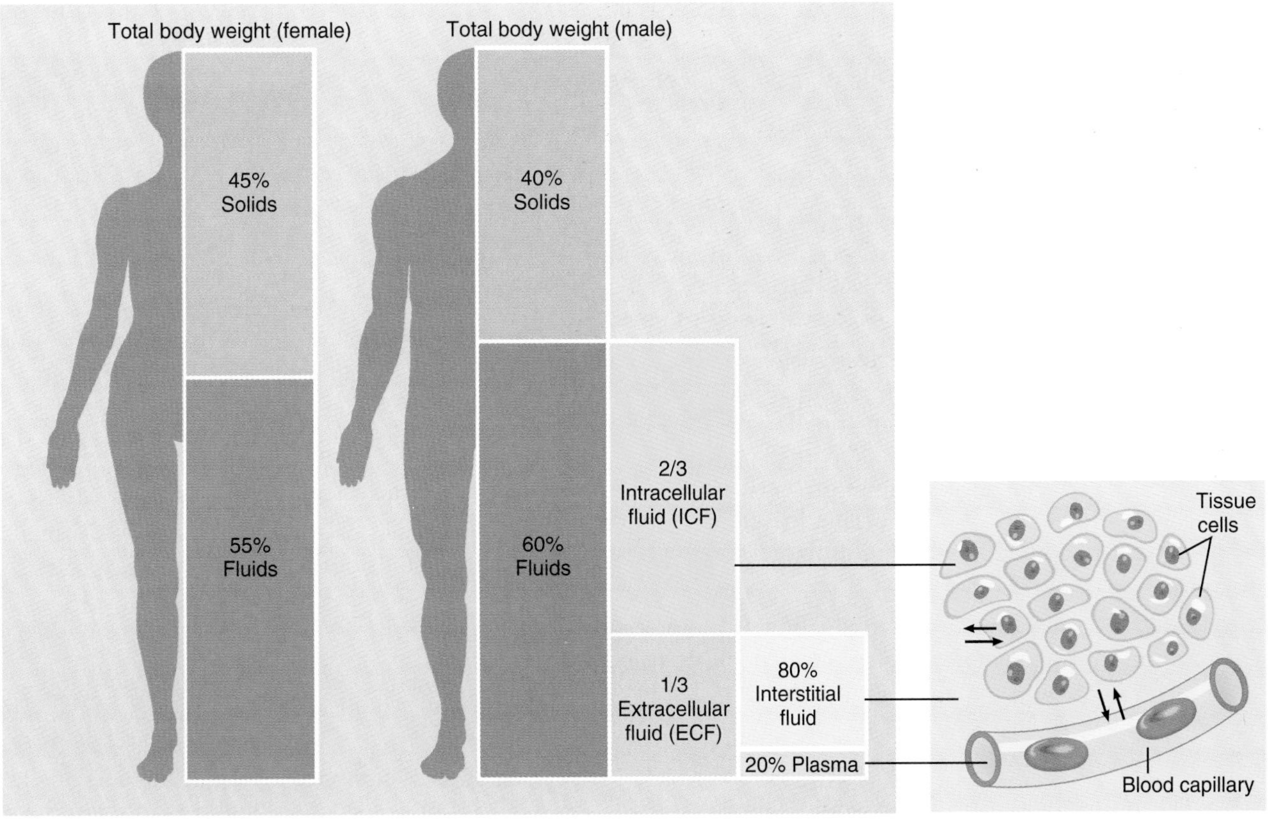

Total body weight (female)

Total body weight (male)

45% Solids

40% Solids

55% Fluids

60% Fluids

2/3 Intracellular fluid (ICF)

1/3 Extracellular fluid (ECF)

80% Interstitial fluid

20% Plasma

Tissue cells

Blood capillary

(a) Distribution of body water in an average lean, adult female and male

(b) Exchange of water among body fluid compartments

 What is body fluid?

weight. The percentage decreases with age. Because fat is basically water-free, lean people have a greater proportion of water than fat people. In a normal adult male, water accounts for about 60 percent of body weight. Because females have more subcutaneous fat, on average, their total body water is lower, accounting for about 55 percent of body weight.

Fluid Intake (Gain) and Output (Loss)

objective: *Explain the routes of fluid intake and output and explain how intake and output are regulated.*

The primary source of body fluid is derived from liquids (1600 ml) and foods (700 ml) taken into the body by mouth and then absorbed from the gastrointestinal tract. This is called *ingested* *(preformed)* *water* and amounts to about 2300 ml/day. Another source is *metabolic water,* the water produced through anabolism, which amounts to about 200 ml/day. Thus total fluid input averages about 2500 ml/day (Figure 22.2).

Under normal conditions, fluid intake equals fluid output, so the body maintains a constant volume. The approximate amounts of water lost from each avenue of fluid output are 1500 ml/day by the kidneys as urine, 600 ml/day by the skin, mostly through evaporation (400 ml/day) and some through perspiration (200 ml/day), 300 ml/day by the lungs as water vapor, and 100 ml/day by the gastrointestinal tract as feces. Therefore total fluid output is about 2500 ml/day (Figure 22.2).

Regulation of Fluid Intake

Thirst is a powerful regulator of fluid intake and operates in the following way (Figure 22.3):

1. When water loss is greater than water gain, the result is *dehydration,* which may be mild or severe.

2. Dehydration stimulates thirst in at least three ways: (1) it decreases production of saliva, (2) it increases blood osmotic pressure, and (3) it decreases blood volume.

3. When production of saliva decreases, it causes dryness of the mucosa of the mouth and pharynx. When blood osmotic pressure increases, it stimulates receptors, called osmoreceptors, in the hypothalamus. When blood volume decreases, blood pressure drops. This change stimulates release of renin by juxtaglomerular cells of the kidneys and renin promotes the synthesis of angiotensin II.

4. A dry mouth and pharynx, stimulation of osmoreceptors in the hypothalamus, and increased angiotensin II in the blood all stimulate the thirst center in the hypothalamus (see Figure 21.8).

5. As a result, the sensation of thirst is increased.

6. This normally leads to increased fluid intake, if fluids are available.

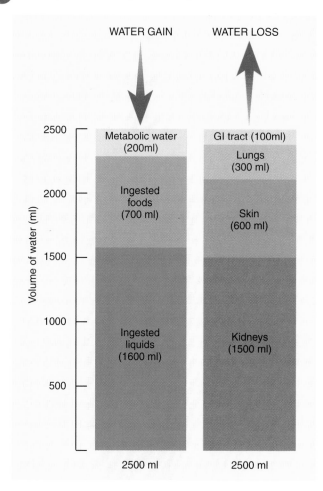

Figure 22.2 Summary of fluid intake (gain) and output (loss) per day under normal conditions.

🔑 *Normally, water loss equals water gain.*

WATER GAIN — WATER LOSS

Volume of water (ml)

Water Gain: Metabolic water (200ml), Ingested foods (700 ml), Ingested liquids (1600 ml) — 2500 ml

Water Loss: GI tract (100ml), Lungs (300 ml), Skin (600 ml), Kidneys (1500 ml) — 2500 ml

❓ *How would each of the following affect fluid balance: hyperventilation, vomiting, fever, and diuretics?*

7. As a result, normal fluid volume is restored and this relieves dehydration. The net effect of the cycle is that fluid intake balances the fluid output.

Initially, quenching of thirst results from wetting the mucosa of the mouth and pharynx, but the major inhibition of thirst is believed to occur as a result of a decrease in osmotic pressure (increased blood volume ⟶ increased blood pressure) in fluids of the hypothalamus.

Regulation of Fluid Output

The primary regulator of fluid output is urine formation. Normally, fluid loss is adjusted by antidiuretic hormone (ADH), atrial natriuretic peptide (ANP), and aldosterone. ADH and aldosterone slow fluid loss in the urine, whereas ANP causes increased fluid loss in the urine (see Chapter 21). Under abnormal conditions, other factors may influence fluid

Figure 22.3   Pathways for stimulation of thirst by dehydration.

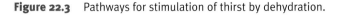

🔑 *Dehydration occurs when water loss is greater than water gain.*

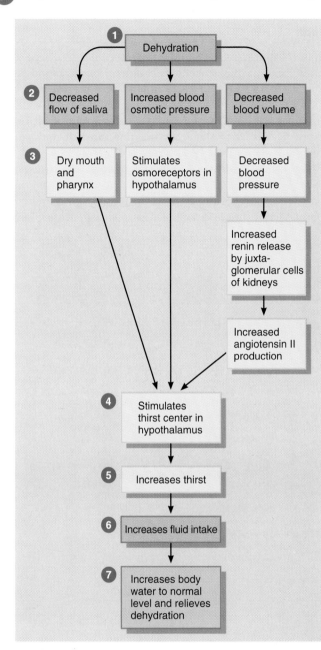

❓ *Is this a negative feedback system or a positive feedback system? Why?*

loss. If the body is dehydrated, blood pressure falls, the glomerular filtration rate decreases, and water is conserved. Conversely, excessive fluid in blood results in an increase of blood pressure, glomerular filtration rate, and fluid output. Hypertension (high blood pressure) produces the same effect. Hyperventilation leads to increased fluid loss through the loss of water vapor to the lungs. Vomiting and diarrhea result in fluid loss from the gastrointestinal tract. Finally, fever, heavy perspiration, and destruction of extensive areas of the skin from burns bring about excessive water loss through the skin.

Electrolytes

objective: *Describe the general functions of electrolytes and how they are distributed.*

Body fluids contain a variety of dissolved chemicals. Some are compounds with covalent bonds; that is, the atoms that compose the molecule share electrons, do not form ions, and do not conduct an electric current. They are called ***nonelectrolytes*** and include most organic compounds, such as glucose, urea, and creatine.

Other compounds, called ***electrolytes*** (e-LEK-trō-līts), have at least one ionic bond. They are so named because they do conduct an electric current. When they dissolve in a fluid, they dissociate into positive ions (***cations***) and negative ions (***anions***). Acids, bases, and salts are electrolytes. Most electrolytes are inorganic compounds, but a few are organic (for example, some proteins).

Electrolytes serve four general functions in the body. (1) Many are essential minerals. (2) Because they are more numerous than nonelectrolytes, electrolytes control the osmosis of water between body compartments. (3) They help maintain the acid–base balance required for normal cellular activities. (4) They carry electrical current, which allows production of action potentials and controls secretion of some hormones and neurotransmitters. Electrical currents also are important during development.

Concentration of Electrolytes

In osmosis, water moves from an area with fewer particles in solution (lower osmotic pressure) to an area with more particles in solution (higher osmotic pressure). Osmotic pressure is related to the *number* of particles in a solution rather than the size of the particles. An electrolyte exerts a far greater effect on osmosis than a nonelectrolyte because an electrolyte breaks apart into at least two particles, both of them charged. Suppose the nonelectrolyte glucose and an electrolyte (NaCl) are placed in solution:

$$C_6H_{12}O_6 \xrightarrow{H_2O} C_6H_{12}O_6 = \text{one particle}$$

$$\underset{\substack{\text{Sodium} \\ \text{chloride}}}{NaCl} \xrightarrow{H_2O} Na^+ + Cl^- = \text{two particles}$$

Because glucose does not break apart when dissolved in water, a molecule of glucose contributes only one particle to the solution. Sodium chloride, on the other hand, contributes two particles, or ions, and has a two times greater effect on solute concentration as glucose.

To determine how much effect an electrolyte has on concentration, it is necessary to know its concentration in a solution. The concentration of an ion in a certain volume of solution is commonly expressed in ***milliequivalents per liter*** (***mEq/liter***)—the total number of cations and anions (electrical charges) in each liter of solution.

Distribution of Electrolytes

Figure 22.4 compares the principal chemical constituents of plasma, interstitial fluid, and intracellular fluid. The chief difference between plasma and interstitial fluid is that plasma contains many protein anions, whereas interstitial fluid has hardly any. Because normal capillary membranes are practically impermeable to protein, the protein stays in the plasma and does not move out of the blood into the interstitial fluid. Plasma also contains fewer Na^+ ions and Cl^- ions than interstitial fluid. In other respects the two fluids are similar.

Intracellular fluid differs considerably from extracellular fluid, however. In extracellular fluid, the most abundant cation is sodium (Na^+) and the most abundant anion is chloride (Cl^-). In intracellular fluid, the most abundant cation is potassium (K^+) and the most abundant anion is phosphate (HPO_4^{2-}). Also, there are many more protein anions in intracellular fluid than in extracellular fluid.

Functions and Regulation

objective: *Discuss the functions and regulation of sodium, potassium, calcium, magnesium, chloride, and phosphate.*

Sodium

Sodium (Na^+), the most abundant extracellular ion, represents about 90 percent of extracellular cations. Normal blood Na^+ is about 136 to 142 mEq/liter. Sodium is necessary for the transmission of action potentials in nervous and muscle tissue. It is a component of buffer systems and it plays a significant role in fluid and electrolyte balance by creating most of the osmotic pressure of extracellular fluid (ECF).

The kidneys normally excrete excess Na^+ and conserve it during periods of sodium restriction. The Na^+ level in the blood is controlled primarily by aldosterone, antidiuretic hormone (ADH), and atrial natriuretic peptide (ANP).

Figure 22.4   Comparison of electrolyte concentrations in plasma, interstitial fluid, and intracellular fluid. The height of each column represents the total electrolyte concentration.

🔑 *An electrolyte has a far greater effect on osmosis than a nonelectrolyte because an electrolyte breaks apart into at least two particles.*

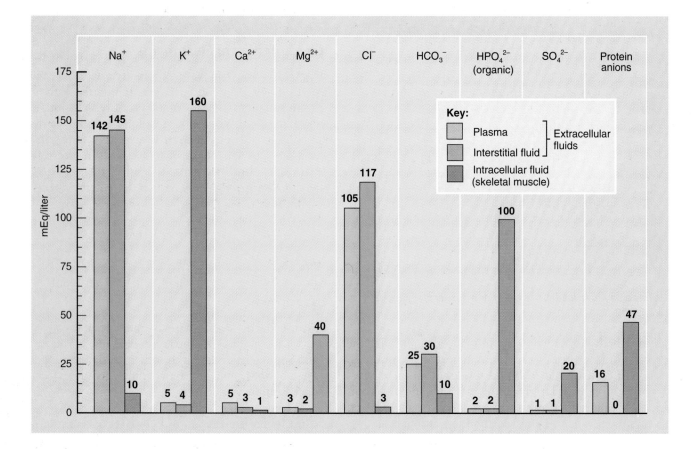

Ⓠ *What is the major cation and the major anion in ECF?*

Aldosterone, which is secreted by the adrenal cortex, acts on the distal convoluted tubules and collecting ducts of the nephron of the kidneys, causing them to increase their reabsorption of Na^+ (see Figure 13.15). The Na^+ thus moves from the filtrate back into the blood and establishes an osmotic gradient that causes water to follow from the filtrate back into the blood as well. Aldosterone is secreted in response to reduced blood volume or cardiac output, decreased extracellular Na^+, increased extracellular potassium, and physical stress. A decrease in Na^+ concentration inhibits the release of antidiuretic hormone (ADH) by the pituitary gland, which, in turn, permits more water to be excreted in urine and the Na^+ level to be restored to normal. The atria of the heart produce the hormone atrial natriuretic peptide (ANP) that also increases Na^+ and water excretion by the kidneys.

Sodium may be lost from the body through excessive perspiration, vomiting, diarrhea, taking certain diuretics, and burns. A lower-than-normal blood Na^+ level can result in *hyponatremia* (hī'-pō-na-TRĒ-mē-a; *natrium* = sodium). It is characterized by muscular weakness, dizziness, headache, hypotension, tachycardia, and shock. Severe sodium loss can result in mental confusion, stupor, and coma. A higher-than-normal blood Na^+ level, called *hypernatremia,* may occur because of excess water loss, water deprivation, or sodium intake. The average daily intake of Na^+ far exceeds the body's normal daily requirements. Because Na^+ is the major determinant of the osmotic pressure of ECF, too high Na^+ levels cause water to move out of body cells into ECF, resulting in cellular dehydration. Symptoms include intense thirst, fatigue, restlessness, agitation, and coma.

Potassium

Potassium (K^+) is the most abundant cation in intracellular fluid. Normal blood potassium is about 3.8 to 5.0 mEq/liter. Potassium assumes a key role in the functioning of nervous and muscle tissue. Abnormal serum K^+ levels adversely affect neuromuscular and cardiac function. Potassium helps maintain fluid volume in cells and regulates extracellular fluid pH. When K^+ moves out of the cell, it is replaced by Na^+ and hydrogen ions; it is the shift of hydrogen ions that affects pH.

The blood level of K^+ is under the control of mineralocorticoids, mainly aldosterone. When K^+ concentration is high, more aldosterone is secreted and more K^+ is excreted. This process occurs in the distal convoluted tubules and collecting ducts of the kidneys by way of tubular secretion. When K^+ concentration is low, aldosterone secretion decreases and less K^+ is secreted in urine.

A lower-than-normal level of K^+, called *hypokalemia,* (hī'-pō-ka-LĒ-mē-a; *kalium* = potassium), may result from vomiting, diarrhea, high sodium intake, kidney disease, or taking certain diuretics. Symptoms include cramps, fatigue, flaccid paralysis, nausea, vomiting, mental confusion, increased urine output, shallow respirations, and changes in the electrocardiogram.

A higher-than-normal blood K^+ level, called *hyperkalemia,* is characterized by irritability, anxiety, abdominal cramping, diarrhea, weakness (especially of the lower limbs), and paresthesia (abnormal sensation, such as burning or prickling). Hyperkalemia can also cause death from fibrillation of the heart.

Calcium

Calcium is the most abundant mineral in the body. Most of it, about 98 percent, is not in ionic form and is combined with phosphate (HPO_4^{2-}) in the skeleton and teeth as calcium and phosphorus salts. The remaining calcium is found in ionic form as Ca^{2+} in extracellular fluid and cells of all tissues, especially skeletal muscle. Normal blood Ca^{2+} is about 4.6 to 5.5 mEq/liter. Calcium is not only a structural component of bones and teeth and a second messenger, it is also involved in blood clotting, neurotransmitter release, maintenance of muscle tone, and excitability of nervous and muscle tissue.

Calcium blood levels are regulated principally by parathyroid hormone (PTH) and calcitonin (CT) (see Figure 13.12). PTH is released when Ca^{2+} blood level is low. PTH stimulates osteoclasts to release Ca^{2+} (and phosphate) into the blood. PTH also helps to increase absorption of Ca^{2+} from the gastrointestinal tract and increases the rate at which Ca^{2+} is reabsorbed in the kidneys and returned to the blood. CT, from the thyroid gland, is released when blood Ca^{2+} levels are high. It decreases the blood level of Ca^{2+} by stimulating osteoblasts and inhibiting osteoclasts. In the presence of CT, osteoblasts remove Ca^{2+} (and phosphate) from blood and deposit them in bone.

An abnormally low level of Ca^{2+} is called *hypocalcemia* (hī'-pō-kal-SĒ-mē-a). It may be due to increased Ca^{2+} loss, reduced Ca^{2+} intake, elevated levels of phosphate (as one goes up, the other goes down), or altered regulation as might occur in hypoparathyroidism. It is characterized by numbness and tingling of the fingers, hyperactive reflexes, muscle cramps, tetany, and convulsions; it may also lead to bone fractures. Hypocalcemia may also cause spasms of laryngeal muscles that can cause death by asphyxiation.

Hypercalcemia, an abnormally high level of Ca^{2+}, is characterized by lethargy, weakness, anorexia, nausea, vomiting, polyuria, itching, bone pain, depression, confusion, paresthesia, stupor, and coma.

Magnesium

Magnesium (Mg^{2+}) is primarily an intracellular electrolyte. Normal blood Mg^{2+} is about 1.3 to 2.1 mEq/liter. In an adult, about 50 percent of the body's Mg^{2+} is in bone; about 45 percent is in intracellular fluid; and about 5 percent is in extracellular fluid. Magnesium activates enzymes for the metabolism of carbohydrates and proteins, triggers the sodium–potassium pump, and preserves the structure of DNA, RNA, and ribosomes. It is also important in neuromuscular activity, neural transmission in the central nervous system, and myocardial functioning.

Magnesium level is regulated by aldosterone. When Mg^{2+} concentration is low, increased aldosterone secretion acts on the kidneys so that more Mg^{2+} is reabsorbed.

Magnesium deficiency, called *hypomagnesemia* (hī'-pō-mag'-ne-SĒ-mē-a), may be caused by malabsorption, diarrhea, alcoholism, malnutrition, excessive lactation, diabetes mellitus, or diuretics. Symptoms include weakness, irritability, tetany, delirium, convulsions, confusion, anorexia, nausea, vomiting, paresthesia, and cardiac arrhythmias.

Hypermagnesemia, or Mg^{2+} excess, occurs almost exclusively in persons with renal failure who take medications containing magnesium. Other causes include Addison's disease, acute diabetic acidosis, severe dehydration, and hypothermia. Symptoms include flaccidity, hypotension, muscle weakness or paralysis, nausea, vomiting, and altered mental functioning.

Chloride

Chloride (Cl^-) is the major extracellular anion. However, it easily diffuses between extracellular and intracellular compartments. This makes Cl^- important in regulating osmotic pressure differences between the compartments. Also, in the gastric mucosal glands, Cl^- combines with hydrogen to form hydrochloric acid (HCl). Normal blood Cl^- is about 95 to 103 mEq/liter.

Part of Cl^- regulation is indirectly under the control of aldosterone. Aldosterone regulates Na^+ reabsorption, and Cl^- follows Na^+ passively because the negatively charged Cl^- follows the positively charged Na^+.

An abnormally low level of Cl^- in the blood, called *hypochloremia* (hī'-pō-klō-RĒ-mē-a), may be caused by excessive vomiting, dehydration, and certain diuretics. Symptoms include muscle spasms, alkalosis, depressed respirations, and even coma.

Phosphate

Phosphate (HPO_4^{2-}) is principally an intracellular electrolyte. Normal blood HPO_4^{2-} is about 1.7 to 2.6 mEq/liter. About 85 percent of the HPO_4^{2-} in an adult is in bones and teeth as calcium phosphate salts. The remainder is mostly combined with lipids (phospholipids), proteins, carbohydrates, and other organic molecules to form cellular membranes and to synthesize nucleic acids (DNA and RNA) and high-energy compounds such as adenosine triphosphate (ATP). In addition, HPO_4^{2-} plays an important role in buffering reactions (the phosphate buffer system is discussed later in the chapter).

Phosphate blood levels are regulated by PTH and CT. PTH stimulates osteoclasts to release HPO_4^{2-} from mineral salts of bone matrix and causes renal tubular cells to excrete HPO_4^{2-}. CT lowers high HPO_4^{2-} levels by stimulating osteoblasts and inhibiting osteoclasts. In the presence of CT, osteoblasts remove HPO_4^{2-} from blood, combine it with calcium, and deposit it in bone.

An abnormally low level of HPO_4^{2-}, called *hypophosphatemia* (hī'-pō-fos'-fa-TĒ-mē-a), may occur from polyuria, decreased intestinal absorption, increased utilization of phosphate, or alcoholism. Symptoms include confusion, seizures, coma, chest and muscle pain, numbness and tingling of the fingers, lack of coordination, memory loss, and lethargy.

Hyperphosphatemia occurs most often in response to renal insufficiency because the kidneys increase the intake of HPO_4^{2-} or fail to excrete excess HPO_4^{2-}. The primary complication is the accumulation of calcium phosphate in soft tissues, joints, and arteries. Symptoms include anorexia, nausea, vomiting, muscular weakness, hyperactive reflexes, tetany, and tachycardia.

Movement of Body Fluids

objective: *Describe how fluids move between compartments.*

Between Plasma and Interstitial Compartments

The movement of fluid between plasma and interstitial compartments occurs across capillary membranes. The movement of water and dissolved substances, except proteins, through capillary walls is mostly by diffusion and bulk flow. *Bulk flow* is a passive process that involves the movement of large numbers of dissolved substances and fluid as a result of hydrostatic (water) pressure. Fluid movement between plasma and interstitial fluid is due to several opposing forces, or pressures. Some forces push fluid out of capillaries into the surrounding interstitial (tissue) spaces. Opposing forces push fluid from interstitial spaces into blood capillaries, which prevents fluid from accumulating in interstitial spaces.

Whether fluids enter or leave capillaries depends on how various pressures relate to each other. The difference between the forces that move fluid out of plasma and the forces that push it into plasma is the *net filtration pressure* (**NFP**). The NFP at the arterial end of a capillary is slightly greater than that at the venous end. Thus, at the arterial end of a capillary, fluid moves out (filtered) from plasma into the interstitial compartment; at the venous end of a capillary, fluid moves in (reabsorbed) from the interstitial to the plasma compartment (see Figure 16.5). Not all the fluid filtered at one end of the capillary is reabsorbed at the other. The fluid not reabsorbed and any proteins that escape from capillaries pass into lymphatic capillaries. From here, the fluid (lymph) moves through lymphatic vessels to the thoracic duct or right lymphatic duct to enter the cardiovascular system (see Figure 17.1). Under normal conditions, there is a state of balance at the arterial and venous ends of a capillary in which filtered fluid and absorbed fluid, as well as fluid picked up by the lymphatic system, are nearly equal.

Between Interstitial and Intracellular Compartments

The principal cation inside the cell is K^+, whereas the principal cation outside is Na^+ (see Figure 22.4). When a fluid imbalance between these two compartments occurs, it is usually caused by a change in the Na^+ or K^+ concentration.

As previously discussed, sodium balance in the body is controlled by aldosterone, ANP, and ADH. Certain conditions, however, can result in a decrease in the Na^+ concentration in interstitial fluid (Figure 22.5).

Figure 22.5   Interrelations between fluid imbalance and Na⁺ imbalance.

Na⁺ loss results in water loss.

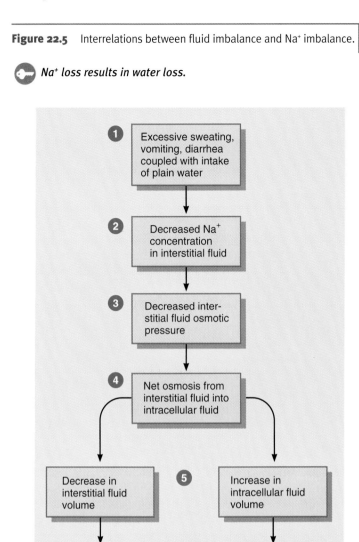

What are the effects of water intoxication?

① Both body water and Na⁺ are lost during excessive sweating, vomiting, or diarrhea. If the lost fluid is replaced by drinking plain water, body fluids become more dilute.

② This dilution can cause the Na⁺ concentration of plasma and then of interstitial fluid to fall below the normal range (hyponatremia).

③ When its Na⁺ concentration decreases, the interstitial fluid's osmotic pressure also falls.

④ The result is net osmosis of water from interstitial fluid into intracellular fluid.

⑤ This osmosis-driven water movement has two serious consequences. The first effect is to decrease interstitial fluid volume. The second effect is to increase intracellular fluid volume.

⑥ Decreased interstitial fluid volume causes water to move from plasma into interstitial fluid. The increase in intracellular fluid volume causes cells to swell, a condition called *water intoxication.*

⑦ As water moves out of plasma, blood volume decreases, which may lead to circulatory shock. Among the symptoms of water intoxication are disoriented behavior, convulsions, coma, and possible death.

To prevent this sequence of events in cases of severe electrolyte and water loss, solutions given for intravenous or oral rehydration therapy (ORT) include a small amount of table salt (NaCl).

Acid–Base Balance

objective: *Explain how buffers, respiration, and kidney excretion help maintain pH.*

Before reading this section, you might want to review the discussion of acids, bases, and pH in Chapter 2. In addition to controlling water movement, electrolytes help regulate the body's acid–base balance. The overall acid–base balance depends on the hydrogen ion (H^+) concentration of body fluids, particularly extracellular fluid. Recall that acids ionize to produce H^+. The majority of H^+ are produced as a result of the metabolic activities of cells. For example, during the aerobic respiration of glucose, carbon dioxide (CO_2) is produced. It combines with water in extracellular fluid to form carbonic acid (H_2CO_3), which breaks down into H^+ and bicarbonate ions (HCO_3^-):

$$\underset{\substack{\text{Carbon}\\\text{dioxide}}}{CO_2} + \underset{\text{Water}}{H_2O} \rightleftharpoons \underset{\substack{\text{Carbonic}\\\text{acid}}}{H_2CO_3} \rightleftharpoons \underset{\substack{\text{Hydrogen}\\\text{ion}}}{H^+} + \underset{\substack{\text{Bicarbonate}\\\text{ion}}}{HCO_3^-}$$

In a healthy person, the pH of the extracellular fluid is between 7.35 and 7.45. Homeostasis of this narrow range is essential to survival and depends on three major mechanisms: buffer systems, respiration, and kidney excretion.

Buffer Systems

Most human *buffer systems* consist of a weak acid and a weak base that function to prevent drastic changes in the pH of a body fluid by rapidly changing strong acids and bases into weak acids and bases. A strong acid dissociates into H^+ more easily than a weak acid. Strong acids therefore lower pH more than weak ones because strong acids contribute more H^+. Similarly, strong bases raise pH more than weak ones because strong bases dissociate more easily into OH^-. The principal buffer systems are the carbonic acid–bicarbonate system, the phosphate system, the hemoglobin system, and the protein system.

Carbonic Acid–Bicarbonate Buffer System

As a result of metabolic processes, the body normally produces more acids than bases. A strong acid like HCl can be buffered by a weak base, sodium bicarbonate ($NaHCO_3$), to produce a weaker acid, carbonic acid (H_2CO_3), and a salt (NaCl):

$$HCl + NaHCO_3 \rightleftharpoons H_2CO_3 + NaCl$$

For example, this reaction occurs when alkaline secretions from the pancreas mix with acidic chyme from the stomach.

Although the body needs more bicarbonate to neutralize acids, strong bases must also be buffered to prevent the blood pH from rising too high. In this case weak acids such as carbonic acid can neutralize strong bases such as sodium hydroxide (NaOH) to weaker bases like sodium bicarbonate:

$$NaOH + H_2CO_3 \rightleftharpoons NaHCO_3 + H_2O$$

Therefore the *carbonic acid–bicarbonate buffer system* is an important regulator of blood pH.

Phosphate Buffer System

The *phosphate buffer system* helps regulate pH in red blood cells but is especially important in kidney tubular fluids. With the use of this system, the kidneys help maintain normal blood pH by the acidification of urine. The two phosphate buffers act in the same manner as the bicarbonate buffer system.

Hemoglobin Buffer System

The *hemoglobin buffer system* buffers carbonic acid in the blood. When blood moves from the arterial end of a capillary to the venous end, the carbon dioxide given up by body cells enters the erythrocytes and combines with water to form carbonic acid (Figure 22.6). Simultaneously, oxyhemoglobin gives up its oxygen to the body cells, and some becomes reduced hemoglobin, which carries a negative charge. The hemoglobin anion attracts the H+ from the carbonic acid and becomes an acid that is even weaker than carbonic acid.

Protein Buffer System

The *protein buffer system* is the most abundant buffer in body cells and plasma. The amino acids in proteins contain at least one carboxyl group (COOH) and one amino group (NH_2). In solution, a carboxyl group may become ionized (COOH $\longrightarrow$ COO$^-$ + H+) and buffers bases by combining its H+ with excess hydroxide ions (OH$^-$) to form water (H+ + OH$^-$ $\longrightarrow$ H_2O). In the presence of excess H+, free COO$^-$ will recombine with H+ to form COOH, thereby raising the pH of the fluids again. Thus the carboxyl group can buffer both acids and bases.

The amino group (NH_2) can also act as an acid or a base. By accepting free H+ (NH_3^+), it removes them from a body

Figure 22.6   Hemoglobin buffer system. When CO_2 enters red blood cells, it combines with water to form carbonic acid (H_2CO_3), which dissociates into bicarbonate ions (HCO_3^-) and hydrogen ions (H+). Oxyhemoglobin gives up its oxygen and combines with excess H+ to form HbH (a weak acid). The bicarbonate may move out of the cell and combine with sodium ions (Na+) in the plasma. In this way, much of the carbon dioxide is carried back to the lungs in the form of sodium bicarbonate ($NaHCO_3$).

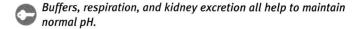

Buffers, respiration, and kidney excretion all help to maintain normal pH.

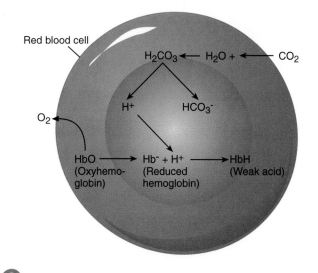

What do all buffers have in common?

fluid. By releasing H+ from NH_3^+ in the presence of excess OH$^-$ to form water, it lowers the pH again.

Respiration

Respiration also plays a role in maintaining blood pH. An increase in the carbon dioxide concentration in body fluids as a result of cellular respiration lowers pH (makes the fluid more acid). This is illustrated by the following equation:

$$\underset{\substack{\text{Carbon} \\ \text{dioxide}}}{CO_2} + \underset{\text{Water}}{H_2O} \rightleftharpoons \underset{\substack{\text{Carbonic} \\ \text{acid}}}{H_2CO_3} \rightleftharpoons \underset{\substack{\text{Hydrogen} \\ \text{ion}}}{H^+} + \underset{\substack{\text{Bicarbonate} \\ \text{ion}}}{HCO_3^-}$$

Conversely, a decrease in the carbon dioxide concentration of body fluids raises the pH.

The pH of body fluids may be adjusted by a change in the rate and depth of breathing, an adjustment that usually takes from 1 to 3 minutes. If the rate and depth of breathing are increased, more carbon dioxide is exhaled and blood pH rises. A slowed respiratory rate means less carbon dioxide is exhaled, causing blood pH to fall. Breathing rate can be altered up to eight times the normal rate; thus respiration can greatly influence the pH of body fluids.

The pH of body fluids, in turn, affects the rate of breathing (Figure 22.7). If, for example, the blood becomes

Figure 22.7   Negative feedback regulation of blood pH by the respiratory system.

The pH of body fluids affects the rate of respiration.

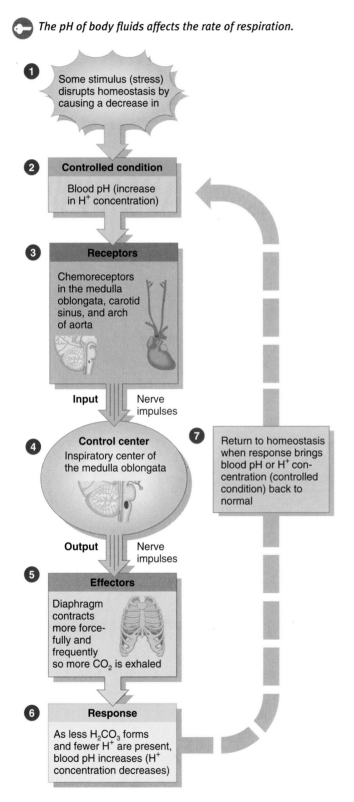

1 Some stimulus (stress) disrupts homeostasis by causing a decrease in

2 Controlled condition

Blood pH (increase in H$^+$ concentration)

3 Receptors

Chemoreceptors in the medulla oblongata, carotid sinus, and arch of aorta

Input   Nerve impulses

4 Control center

Inspiratory center of the medulla oblongata

7 Return to homeostasis when response brings blood pH or H$^+$ concentration (controlled condition) back to normal

Output   Nerve impulses

5 Effectors

Diaphragm contracts more forcefully and frequently so more CO$_2$ is exhaled

6 Response

As less H$_2$CO$_3$ forms and fewer H$^+$ are present, blood pH increases (H$^+$ concentration decreases)

If you hold your breath for 30 seconds, what will happen to your blood pH?

more acidic, the increase in H$^+$ (controlled condition) stimulates chemoreceptors in the respiratory center in the medulla oblongata and chemoreceptors in the arch of the aorta and carotid sinus (receptors). This stimulates the inspiratory area in the medulla oblongata (control center). As a result, the diaphragm and other muscles of respiration (effectors) contract more forcefully and frequently and respirations increase in rate and depth so more CO$_2$ is exhaled. As less H$_2$CO$_3$ forms and fewer H$^+$ are present, blood pH increases (response). The same effect occurs if the blood concentration of carbon dioxide increases. Increased respiration removes more carbon dioxide from blood to reduce the H$^+$ concentration. On the other hand, if the pH of the blood increases, or CO$_2$ concentration decreases (responses), the respiratory center is inhibited and respirations decrease. The decreased respirations allow CO$_2$ to accumulate in blood and the H$^+$ concentration to increase.

The respiratory mechanism normally can eliminate more acid or base than all the buffers combined.

Kidney Excretion

The role of the kidneys in maintaining pH was discussed on p. 497. This is done principally by the secretion of H$^+$ into filtrate.

Acid–Base Imbalances

In *acidosis,* blood pH decreases below 7.35. In *alkalosis,* pH increases above 7.45. A change in blood pH that leads to acidosis or alkalosis can be compensated to return pH to normal. *Compensation* refers to the physiological response to an acid–base imbalance that attempts to normalize blood pH. If a person has an altered pH due to metabolic causes, respiratory mechanisms (hyperventilation or hypoventilation) can help compensate for the alteration. Respiratory compensation occurs within minutes and is maximized within hours. Conversely, if a person has an altered pH due to respiratory causes, metabolic mechanisms (kidney excretion of H$^+$ or absorption of HCO$_3^-$) can compensate for the alteration. Metabolic compensation may begin in minutes but takes days to be maximized.

Acidosis and alkalosis affect the body in different ways. The principal physiological effect of acidosis is depression of the central nervous system through depression of synaptic transmission. Blood pH below 7 causes disorientation and coma. In fact, patients with severe acidosis usually die in a state of coma. On the other hand, the major physiological effect of alkalosis is overexcitability of the nervous system through facilitation of synaptic transmission. The overexcitability occurs both in the central nervous system and in peripheral nerves. Nerves conduct impulses repetitively even when not stimulated, causing nervousness, muscle spasms, and even convulsions and death.

wellness focus

Prolonged Physical Activity—A Challenge to Fluid and Electrolyte Balance

heavy or prolonged physical activity can lead to dehydration and disrupt fluid and electrolyte homeostasis. During physical activity, muscle contraction generates a great deal of heat, up to 100 times more than when you are at rest. The body can get rid of this extra heat by increasing blood flow to the skin, where heat is given off by radiation and convection, and by activating the sweat glands to increase heat loss by evaporation. In very hot weather, radiation and convection do not work, so the body must rely primarily on evaporation to cool itself. Strenuous exercise in hot weather may cause the loss of over 2 liters (about 2 quarts) of water per hour from the skin and lungs. Such losses can lead to dehydration and hyperthermia if fluids are not replaced.

Don't Sweat It

Dehydration refers to a loss of body fluid that amounts to 1 percent or more of total body weight. It is most common during exercise in high temperatures but can also occur during very low levels of physical activity in a hot environment or during strenuous exercise in a thermally neutral environment. Fluid deficits of 5 percent are common in athletic events such as football, soccer, tennis, and distance running. Symptoms include irritability, fatigue, and loss of appetite. Dehydration levels greater than 7 percent may cause heat exhaustion.

With dehydration, water is lost from all body compartments. The decrease in blood volume has deleterious effects on physical performance, since it decreases the amount of blood the heart can pump per beat. Muscles need oxygen to work, so performance decreases as cardiac output is reduced. The body tries to maintain blood volume to the muscles by constricting vessels in the skin, so less heat is lost and body temperature rises. Intracellular electrolyte changes may also occur with dehydration and interfere with optimal performance.

Thirst is the body's signal that its water level is getting too low, and it motivates a person to drink. Unfortunately, thirst is not a reliable indicator of fluid needs. People tend to drink just enough to relieve their parched throats. The thirst mechanism is especially unreliable in children and older adults. Aging decreases the kidneys' ability to retain water when the body needs fluids, which increases the susceptibility of older people to dehydration.

Fluid Replacement

People experiencing severe dehydration may find that plain water is not the optimal solution. Studies have shown that when a dehydrated person consumes water, the water dilutes the blood as plasma volume is replenished. This removes the feeling of thirst, which protects against low plasma electrolyte levels. In other words, as electrolyte levels drop, the sensation of thirst goes away so the blood will not become any more dilute. The kidneys sense the increase in fluids in renal tubules and begin to excrete water. In laboratory studies, subjects consuming plain water after dehydration needed to urinate long before hydration was complete. In other words, although plasma volume increased to some extent when water was consumed, it did not return to its desirable level, and subjects did not rehydrate all body cells and extracellular compartments.

Enter Gatorade and other sports drinks. When sodium is taken along with water, dehydrated subjects rehydrate to a greater level than subjects taking only water. Sodium helps to restore plasma volume and to retain water in the blood without inhibiting thirst. Restoring intracellular and extracellular electrolyte levels restores fluid balance in these compartments as well. Sports drinks certainly aren't necessary for the recreational athlete who plays a leisurely game of tennis doubles or a person who walks briskly for half an hour, but may be helpful if you exercise to the point of dehydration.

critical thinking

You are a 150-pound athlete, and you work out in a hot weight room for two hours. You weigh yourself after your workout and see you have lost 5 pounds. Describe what has happened in your body to make you lose those 5 pounds.

Study Outline

Fluid Compartments and Fluid Balance (p. 509)

1. Body fluid refers to water and its dissolved substances.
2. About two-thirds of the body's fluid is located inside cells and is called intracellular fluid (ICF).
3. The other third is called extracellular fluid (ECF). It includes interstitial fluid, plasma, lymph, cerebrospinal fluid, synovial fluid, fluids of the eyes and ears, glomerular filtrate, and pleural, pericardial, and peritoneal fluids.
4. Fluid balance means that the proportion of water in each fluid compartment is kept in homeostasis.
5. Fluid balance and electrolyte balance are inseparable.

Water (p. 509)

1. Water is the largest single constituent in the body, making up about 60 percent of body weight depending on the amount of fat present and age.
2. Primary sources of fluid intake are ingested liquids and foods and water produced by catabolism.
3. Avenues of fluid output are the kidneys, skin, lungs, and gastrointestinal tract.
4. The stimulus for fluid intake is dehydration, which causes thirst sensations. Under normal conditions, fluid output is adjusted by aldosterone, antidiuretic hormone (ADH), and atrial natriuretic peptide (ANP).

Electrolytes (p. 511)

1. Electrolytes are chemicals that dissolve in body fluids and dissociate into either cations (positive ions) or anions (negative ions) because of their ionic bonds.
2. Electrolytes serve four functions in the body: they serve as essential minerals for normal metabolism, control proper fluid movement between compartments, help regulate pH, and carry an electric current.
3. Electrolyte concentration is expressed in milliequivalents per liter (mEq/liter).
4. Electrolytes have a greater effect on osmosis than nonelectrolytes.
5. Plasma, interstitial fluid, and intracellular fluid contain varying kinds and amounts of electrolytes.
6. Sodium (Na^+) is the most abundant extracellular ion. It is involved in nerve impulse transmission and muscle contraction and creates most of the osmotic pressure of ECF. Its level is controlled by aldosterone.
7. Potassium (K^+) is the most abundant cation in intracellular fluid. It is involved in maintaining fluid volume, nerve impulse conduction, muscle contraction, and regulating pH. Its level is controlled by aldosterone, ADH, and ANP.

8. Calcium (Ca^{2+}), is the most abundant mineral in the body. Most is nonionic and is a structural component of bones and teeth. Some is ionic and is in extracellular fluid and cells of all tissues, especially skeletal muscle. Calcium also functions in blood clotting, neurotransmitter release, muscle contraction, secretion, and heartbeat. Its level is controlled by parathyroid hormone (PTH) and calcitonin (CT).
9. Magnesium (Mg^{2+}) is primarily an intracellular electrolyte that activates several enzyme systems. Its level is controlled by aldosterone.
10. Chloride (Cl^-) is the major extracellular anion. It assumes a role in regulating osmotic pressure and forming HCl. Its level is controlled indirectly by aldosterone.
11. Phosphate (HPO_4^{2-}) is principally an intracellular ion that is a structural component of bones and teeth. It is also required for the synthesis of nucleic acids and ATP and for buffer reactions. Its level is controlled by PTH and CT.

Movement of Body Fluids (p. 514)

1. At the arterial end of a capillary, fluid moves from plasma into interstitial fluid (filtration). At the venous end, fluid moves in the opposite direction (reabsorption). This movement occurs as a result of diffusion and bulk flow.
2. There is a state of near equilibrium at the arterial and venous ends of a capillary between filtered fluid and absorbed fluid plus that picked up by the lymphatic system.
3. Fluid movement between interstitial and intracellular compartments occurs by osmosis and depends on the movement of sodium.
4. Fluid imbalance between interstitial and intracellular compartments may lead to water intoxication and circulatory shock.

Acid–Base Balance (p. 515)

1. The overall acid–base balance of the body is maintained by controlling the H^+ concentration of body fluids, especially extracellular fluid.
2. The normal pH of extracellular fluid is 7.35 to 7.45.
3. Homeostasis of pH is maintained by buffers, respirations, and kidney excretion.
4. The important buffer systems include carbonic acid–bicarbonate, phosphate, hemoglobin, and protein buffer systems.
5. An increase in rate of respiration increases pH; a decrease in rate decreases pH.

Acid–Base Imbalances (p. 517)

1. Acidosis is a blood pH below 7.35. Its principal effect is depression of the central nervous system (CNS).
2. Alkalosis is a blood pH above 7.45. Its principal effect is overexcitability of the CNS.

Self-Quiz

1. The largest single component of the body is
   a. fat   b. muscle   c. connective tissue   d. water   e. blood

2. Thirst is a primary stimulus for the intake and maintenance of body fluids. The center for thirst is located in the
   a. kidneys   b. adrenal cortex   c. hypothalamus   d. cerebral cortex   e. liver

3. Substances that dissociate into ions when dissolved in body fluids are known as
   a. neurotransmitters   b. enzymes   c. acids   d. non-electrolytes   e. electrolytes

4. Which of the following is NOT one of the functions of electrolytes in the body?
   a. control of fluid movement between compartments   b. regulation of pH   c. source of minerals for metabolism   d. energy source   e. carrier of electric current

5. Parathyroid hormone (PTH) and calcitonin (CT) control blood levels of which of the following ions?
   a. magnesium   b. sodium   c. calcium   d. potassium   e. chloride

6. Fluid movement between intracellular fluid and extracellular fluid depends primarily on the movement of
   a. sodium   b. chloride   c. calcium   d. phosphate   e. magnesium

7. Aldosterone is secreted in response to
   a. increased blood pressure   b. decreased blood volume or cardiac output   c. increased calcium levels   d. increased sodium levels   e. increased water levels

8. Which of the following statements concerning acid–base balance in the body is false?
   a. An increase in respiration rate increases pH.   b. Normal pH of extracellular fluid is 7.35 to 7.45.   c. Buffers are an important mechanism in the maintenance of pH balance.   d. A blood pH of 7.2 is called alkalosis.   e. Respiratory acidosis is characterized by high CO_2 levels.

9. Fluid moves from the plasma into the interstitial fluid
   a. at the venous end of a capillary   b. at the arterial end of a capillary   c. when blood pressure equals net filtration pressure   d. when the net filtration pressure is −9 mm Hg   e. only in cases of overhydration

10. The principal effect of acidosis is
    a. excitability of the central nervous system   b. hyperventilation   c. increased bicarbonate levels   d. depression of the central nervous system   e. a decrease in kidney function

11. A lower-than-normal blood sodium level is termed
    a. hyperchloremia   b. hypocalcemia   c. hyponatremia   d. hypokalemia   e. hypernatremia

12. Match the following:

    E a. Anion of the intracellular fluid, it is a structural component of bones and teeth.
    A b. The most abundant ion in the body, it is controlled by parathyroid hormone (PTH) and calcitonin (CT).
    D c. The most abundant extracellular cation; creates an osmotic pressure difference.
    F d. Controlled by aldosterone, this intracellular ion activates enzymes.
    C e. The most abundant intracellular fluid ion; is involved in nerve and muscle homeostasis.
    B f. Found in the intracellular fluid, ion assists Na^+ in regulating osmotic pressure.

    A. calcium
    B. chloride
    C. potassium
    D. sodium
    E. phosphate
    F. magnesium

13. In a normal adult male, water accounts for about _____ of total body weight.
    a. 80%   b. 75%   c. 60%   d. 45%   e. 33% (1/3)

14. Most human buffer systems consist of
    a. a weak acid and a weak base   b. a strong acid and a strong base   c. a strong acid such as HCl   d. an electrolyte and a nonelectrolyte   e. a weak base and a gas

15. The most abundant buffer in body cells and plasma is the _____ buffer system.
    a. hemoglobin   b. carbonic acid   c. protein   d. bicarbonate   e. phosphate

Critical Thinking Applications

1. When alcohol is ingested in an alcoholic beverage, it is transported in body water. Females generally have a higher percentage of body fat and a lower percentage of body water than males of the same weight. Use these facts to explain why blood alcohol levels rise higher in a female than in a male after drinking the same amount of alcohol.

2. Sports drinks are supposed to replenish both water and electrolytes lost during strenuous exercise. If you were going to create a recipe for your own special sports drink, what would you include in it to help you recover from strenuous exercise?

3. Seawater contains over three times as much salt (NaCl) as blood plasma. Imagine that you are stranded in a lifeboat without food or water. Should you drink the seawater to prevent dehydration? Why? Why not?

4. Two-year-old Timon had a busy morning at the "mom and tot" swim program. Today's lesson included lots of

underwater exercises in blowing bubbles. After the lesson, Timon was definitely not himself and seemed disoriented. What could be the cause of Timon's behavior?

5. Ed had a very busy schedule at work today. He was able to bolt down some junk food between meetings, but now he is suffering from acid indigestion (excess secretion of stomach acid). He plans to drink a "bicarb" as soon as he gets home. How will the bicarbonate solution help his stomach problem?

6. A patient was hospitalized with renal failure due to a complication during pregnancy. How will her body attempt to compensate for the increase in the pH of her blood?

Answers to Figure Questions

22.1   Water plus dissolved substances.

22.2   All would increase fluid loss.

22.3   Negative, because the result (an increase in fluid intake) is opposite to the initiating stimulus (dehydration).

22.4   Major cation = Na^+; major anion = Cl^-.

22.5   Disoriented behavior, convulsions, coma, and even death.

22.6   They prevent drastic changes in pH by changing strong acids and bases into weak ones.

22.7   It will decrease slightly as CO_2 and H^+ accumulate.

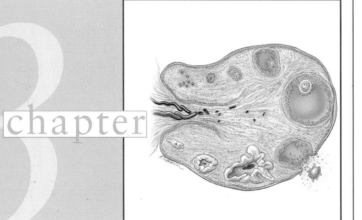

23-chapter

THE REPRODUCTIVE SYSTEMS

a look ahead

*R*eproduction is the process by which new individuals of a species are produced and the genetic material is passed from generation to generation. This maintains continuation of the species. Cell division in a multicellular organism is necessary for growth and repair and it involves passing of genetic material from parent cells to daughter cells.

The organs of the male and female reproductive systems may be grouped by function. (1) The testes and ovaries, also called *gonads* (*gonos* = seed), produce gametes—sperm cells and oocytes, respectively. The gonads also secrete hormones. (2) The *ducts* of the reproductive systems transport and store gametes. (3) *Accessory sex glands* produce materials that support gametes. (4) Finally, several *supporting structures* have various roles in reproduction.

Male Reproductive System

objective: *Describe the structure and functions of the male reproductive organs.*

The organs of the male reproductive system are the testes (male gonads), which produce sperm and hormones; a number of ducts that either store or transport sperm to the exterior; accessory sex glands that secrete semen; and several supporting structures, including the penis (Figure 23.1).

Scrotum

The *scrotum* (SKRŌ-tum; *scrotum* = bag) is a pouch that supports the testes; it consists of loose skin, superficial fascia, and smooth muscle (Figure 23.1). Internally, it is divided by a septum into two sacs, each containing a single testis.

The location of the scrotum and contraction of its muscle fibers regulate the temperature of the testes. The production and survival of sperm require a lower than normal blood temperature. Because the scrotum is outside the body cavities, it supplies an environment about 3°C below body temperature. On exposure to cold and during sexual arousal, skeletal muscles contract to elevate the testes, moving them closer to the pelvic cavity, where they can absorb body heat. Exposure to warmth reverses the process.

Testes

The *testes* (TES-tēz; singular is *testis*), or *testicles,* are paired oval glands that develop on the embryo's posterior abdominal wall and usually begin their descent into the scrotum in the seventh month of fetal development. When the testes do not descend, the condition is referred to as *cryptorchidism* (krip-TOR-ki-dizm).

The testes are covered by a dense white fibrous capsule that extends inward and divides each testis into internal compartments called *lobules* (Figure 23.2a on page 525). Each of the 200 to 300 lobules contains one to three tightly coiled *seminiferous* (*semen* = seed; *ferre* = to carry) *tubules* that produce sperm by a process called spermatogenesis. This process is considered shortly.

Seminiferous tubules are lined with spermatogenic (sperm-forming) cells in various stages of development (Figure 23.2b). The most immature spermatogenic cells, the *spermatogonia* (sper′-ma-tō-GŌ-nē-a; *sperm* = seed; *gonium* = generation or offspring), lie against the basement membrane toward the outside of the tubules. Toward the lumen of the tubule are layers of progressively more mature cells in order of advancing maturity: primary spermatocytes, secondary spermatocytes, spermatids, and sperm. By the time a *sperm cell,* or *spermatozoon* (sper′-ma-tō-ZŌ-on; *zoon* = life), has nearly reached maturity, it is released into the lumen of the tubule and begins to be moved through a series of ducts.

Between the developing sperm cells in the tubules are *sustentacular* (sus-ten-TAK-yoo-lar; *sustentare* = to support) or *Sertoli cells.* These cells support, protect, and nourish developing spermatogenic cells; phagocytize degenerating spermatogenic cells; and secrete the hormone inhibin that helps regulate sperm production by inhibiting the secretion of follicle-stimulating hormone (FSH). Between the seminiferous tubules are clusters of *interstitial endocrinocytes* (*interstitial cells of Leydig*). These cells secrete the male hormone testosterone, the most important androgen (AN-drō-jen), a substance controlling development of male characteristics.

Spermatogenesis

objective: *Describe how sperm cells are produced.*

The process by which the seminiferous tubules of the testes produce sperm is called *spermatogenesis* (sper′-ma-tō-JEN-e-sis). It consists of three stages: reduction division (meiosis I), equatorial division (meiosis II), and spermiogenesis.

MEIOSIS: OVERVIEW In sexual reproduction, a new organism is produced by the union and fusion of two different cells, one produced by each parent. These cells, called *gametes* (*gameto* = too many), are the secondary oocyte (potential mature ovum) produced in the female gonads (ovaries) and the sperm produced in the male gonads (testes). The union and fusion of gametes is called *fertilization* and the cell thus produced is known as a *zygote* (ZĪ-gōt; *zygosis* = a joining). The zygote contains a mixture of chromosomes (DNA) from the two parents and, through its repeated mitotic division, develops into a new organism.

Gametes differ from all other body cells (somatic cells) in that they contain the *haploid* (one-half) *chromosome number,* 23, a single set of chromosomes. It is symbolized by *n*. Somatic cells, such as brain, stomach, kidney, and so on, contain 46 chromosomes in their nuclei. Of the 46 chromosomes, 23 are a complete set from one parent that contains one copy of all the genes necessary for carrying out the activities of the cell. The other 23 chromosomes are another set from the other parent, which codes for the same traits. Because somatic cells contain two sets of chromosomes, they are referred to as *diploid* (DIP-loyd; *di* = two) *cells,* symbolized as *2n*. In a diploid cell, two chromosomes that belong to a pair are called *homologous* (hō-MOL-ō-gus; *homo* = same) *chromosomes,* or *homologues.*

Figure 23.1 Male organs of reproduction and surrounding structures.

Reproductive organs are adapted to produce new individuals and pass on genetic material from one generation to the next.

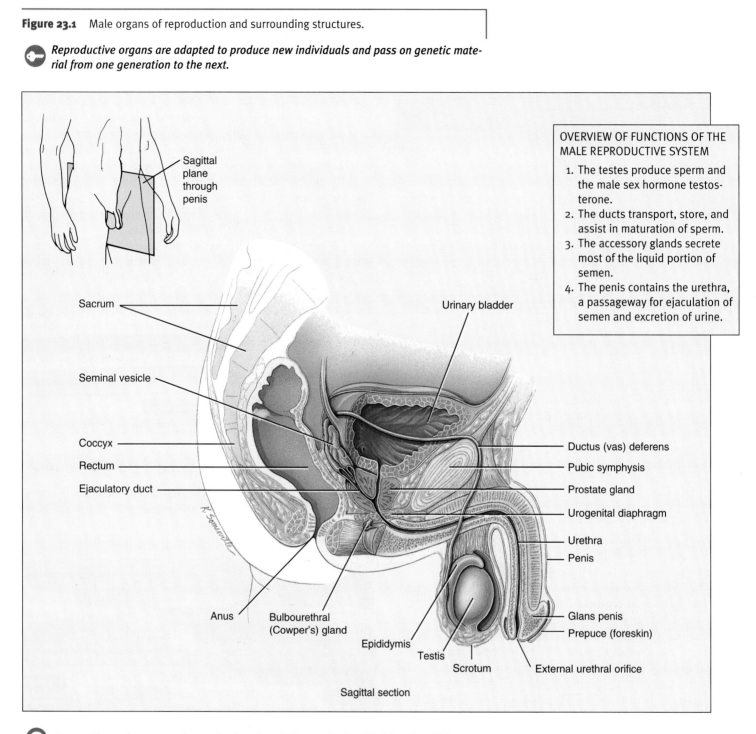

OVERVIEW OF FUNCTIONS OF THE MALE REPRODUCTIVE SYSTEM

1. The testes produce sperm and the male sex hormone testosterone.
2. The ducts transport, store, and assist in maturation of sperm.
3. The accessory glands secrete most of the liquid portion of semen.
4. The penis contains the urethra, a passageway for ejaculation of semen and excretion of urine.

Sagittal plane through penis

Sacrum

Seminal vesicle

Coccyx

Rectum

Ejaculatory duct

Anus

Bulbourethral (Cowper's) gland

Epididymis

Testis

Scrotum

Urinary bladder

Ductus (vas) deferens

Pubic symphysis

Prostate gland

Urogenital diaphragm

Urethra

Penis

Glans penis

Prepuce (foreskin)

External urethral orifice

Sagittal section

Among the male organs of reproduction, how is the penis classified functionally?

If gametes had the same number of chromosomes as somatic cells, the zygote formed from their fusion would have double the diploid number, or 92, and with every succeeding generation, the number of chromosomes would continue to double and normal development would not occur. The chromosome number does not double with each generation because of a special nuclear division called *meiosis* that occurs only in the production of gametes. In meiosis, a developing sperm cell or secondary oocyte gives up its duplicate set of chromosomes so that the mature gamete has only 23.

Review the process of mitosis in Chapter 3. A basic difference between mitosis and meiosis is that in mitosis, a parent cell divides into two daughter cells, each of which receives the same chromosome number as the parent cell. In meiosis, the daughter cells have only half as many chromosomes as the parent cell. Let us see how this happens.

Stages of Spermatogenesis Spermatogenesis begins during puberty and continues through life. It takes about 74 days. The spermatogonia or sperm stem cells (Figure 23.3 on page 526)

Figure 23.2   Anatomy of the testes. The stages of spermatogenesis are shown in (b). Arrows in (b) indicate the progression from least to most mature spermatogenic cells. The (*n*) and (2*n*) refer to haploid and diploid chromosome number, to be described shortly.

The testes are the male gonads, which produce haploid sperm.

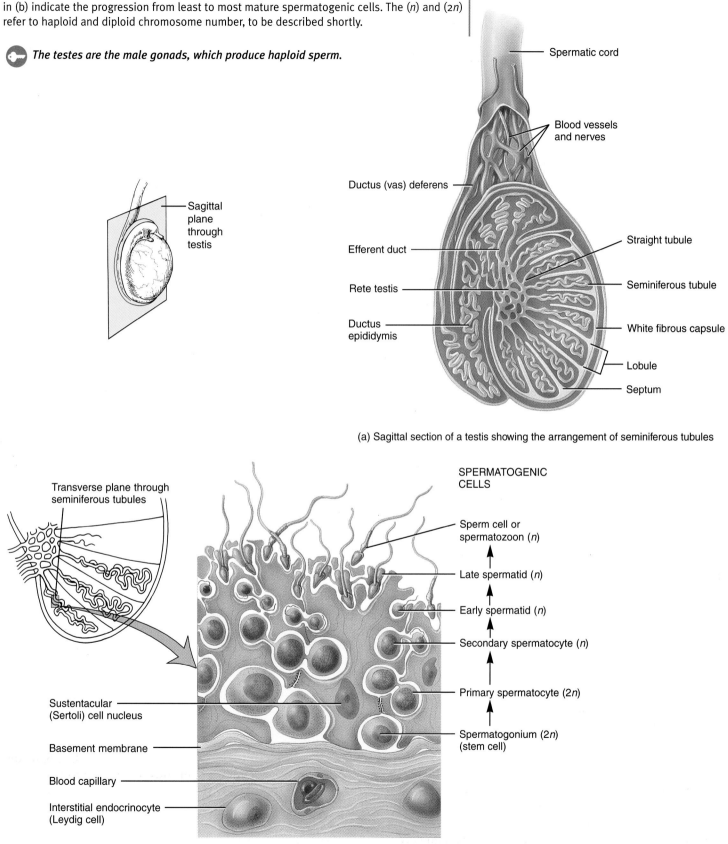

(a) Sagittal section of a testis showing the arrangement of seminiferous tubules

(b) Transverse section of a portion of a seminiferous tubule

Which spermatogenic cells in a seminiferous tubule are most mature and least mature?

Figure 23.3   Spermatogenesis. The designation *2n* means diploid (46 chromosomes); *n* means haploid (23 chromosomes).

🔑 *Spermiogenesis involves the maturation of spermatids into sperm.*

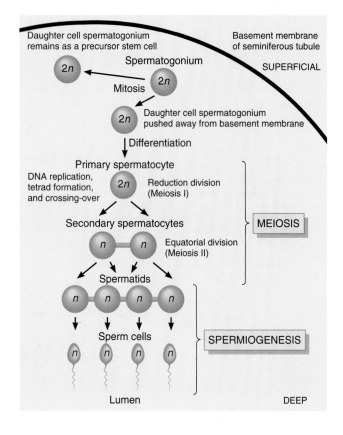

Q *Why is meiosis I also called reduction division?*

that line the seminiferous tubules contain the diploid chromosome number (46). Following division, some spermatogonia remain near the basement membrane to prevent depletion of the cell population. Other spermatogonia undergo developmental changes and become ***primary spermatocytes*** (SPER-ma-tō-sītz′). Like spermatogonia, they are diploid (*2n*); that is, they have 46 chromosomes.

1. **Reduction division (meiosis I).** Each primary spermatocyte enlarges before dividing. Then, two nuclear divisions take place as part of meiosis. In the first, DNA is replicated and the 46 chromosomes (each made up of two identical chromatids, come together as 23 homologous pairs of chromosomes. The four chromatids of each homologous pair then twist around each other and portions of one chromatid may be exchanged with portions of another (***crossing-over***), which permits recombination of genes. Thus the sperm eventually produced will be genetically unlike each other and unlike the cell that produced them.

    Next, the homologous pairs separate and one member of each pair migrates to opposite poles of the dividing

nucleus. The cells formed by this first nuclear division (reduction division) are called ***secondary spermatocytes.*** Each cell has 23 chromosomes—the haploid number. Each chromosome of the secondary spermatocytes, however, is made up of two identical chromatids, but the genes may be rearranged as a result of crossing-over.

2. **Equatorial division (meiosis II).** The second nuclear division of meiosis is equatorial division. There is no replication of DNA. The chromosomes (each composed of two identical chromatids) line up near the center of the nucleus and the chromatids of each chromosome separate from each other. The cells formed from the equatorial division are called ***spermatids.*** Each contains half the original chromosome number, 23, and is haploid. Each primary spermatocyte therefore produces four spermatids by meiosis through two rounds of cell division (reduction division and equatorial division).

3. **Spermiogenesis.** In the final stage of spermatogenesis, called ***spermiogenesis*** (sper′-mē-ō-JEN-e-sis), each spermatid develops into a single ***sperm cell*** (***spermatozoon***). Spermatids develop a head and a tail and are then nourished by sustentacular (Sertoli) cells (see Figure 23.2b).

Sperm

Sperm are produced at the rate of about 300 million per day and, once ejaculated, have a life expectancy of about 48 hours in the female reproductive tract. A sperm cell is composed of a head, a midpiece, and a tail (Figure 23.4). In the *head* are the

Figure 23.4   Parts of a sperm cell.

🔑 *About 300 million sperm mature each day.*

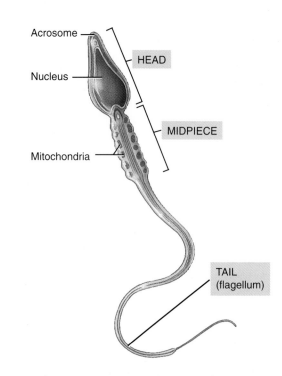

Q *What are the functions of each part of a sperm cell?*

nuclear material and a dense granule called the ***acrosome*** (*acro* = atop), which contains enzymes that help the sperm cell penetrate a secondary oocyte. Mitochondria in the ***midpiece*** carry on the metabolism that provides energy for locomotion. The ***tail***, a typical flagellum, propels the sperm cell.

Testosterone and Inhibin

objective: *Explain the functions of the male reproductive hormones.*

At the onset of puberty, the anterior pituitary gland starts to secrete two hormones that have profound effects on male reproductive organs: LH (luteinizing hormone) and FSH (follicle-stimulating hormone). Their release is controlled by a releasing hormone from the hypothalamus called gonadotropin releasing hormone (GnRH). LH stimulates interstitial endocrinocytes (interstitial cells of Leydig) to secrete the hormone testosterone. FSH and testosterone act on the seminiferous tubules to initiate spermatogenesis.

Testosterone (tes-TOS-te-rōn) is synthesized from cholesterol in the testes. It has a number of effects on the male body:

1. During prenatal development, it facilitates the development of male internal genitals.

2. It stimulates the descent of the testes just prior to birth.

3. At puberty, it brings about development and enlargement of the male sex organs and the development of male secondary sex characteristics. These include pubic, axillary, facial, and chest hair (within hereditary limits); temporal hairline recession; thickening of the skin; increased sebaceous (oil) gland secretion; growth of skeletal muscles and bones; and enlargement of the larynx and deepening of the voice.

4. It stimulates metabolic rate and is the basis for sex drive.

Testosterone production is controlled by a negative feedback system with both the hypothalamus and the anterior pituitary gland (Figure 23.5). LH stimulates the production of testosterone. When testosterone concentration in the blood (controlled condition) increases to a certain level, it inhibits the release of GnRH by cells in the hypothalamus (receptors). As a result, there is less GnRH that flows from the hypothalamus to the anterior pituitary gland (input). Cells in the anterior pituitary gland (control center) then release less LH so the concentration of LH in systemic blood falls (output). With less stimulation, the interstitial endocrinocytes (Leydig cells) in the testes (effectors) secrete less testosterone (response) and there is a return to homeostasis. However, if the testosterone concentration in the blood falls too low, GnRH is again released by the hypothalamus and stimulates secretion of LH by the anterior pituitary gland, which stimulates testosterone production by the testes.

Inhibin, a hormone secreted by sustentacular (Sertoli) cells, inhibits the secretion of FSH. Once the degree of spermatogenesis required for male reproductive functions has been achieved, sustentacular cells secrete inhibin. Inhibin feeds back negatively to the anterior pituitary gland to inhibit FSH and thus decrease spermatogenesis. If spermatogenesis is proceed-

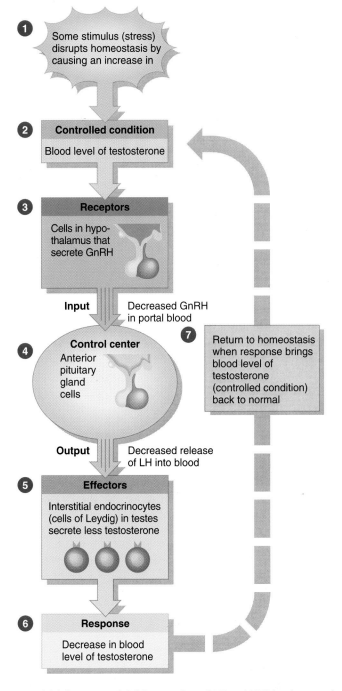

Figure 23.5 Negative feedback control of blood level of testosterone. Testosterone production is controlled by a negative feedback system that involves the hypothalamus and anterior pituitary gland.

1 Some stimulus (stress) disrupts homeostasis by causing an increase in

2 **Controlled condition**
Blood level of testosterone

3 **Receptors**
Cells in hypothalamus that secrete GnRH

Input   Decreased GnRH in portal blood

4 **Control center**
Anterior pituitary gland cells

7 Return to homeostasis when response brings blood level of testosterone (controlled condition) back to normal

Output   Decreased release of LH into blood

5 **Effectors**
Interstitial endocrinocytes (cells of Leydig) in testes secrete less testosterone

6 **Response**
Decrease in blood level of testosterone

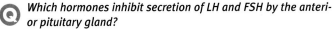

Q *Which hormones inhibit secretion of LH and FSH by the anterior pituitary gland?*

ing too slowly, lack of inhibin production permits FSH secretion and an increased rate of spermatogenesis.

Male Puberty

Puberty (PŪ-ber-tē; *puber* = marriageable age) refers to the period of time when secondary sex characteristics begin to develop and the potential for sexual reproduction is reached. Male puberty begins at an average age of 10 to 11 and ends at an average age of 15 to 17. The factors that determine the onset of puberty are poorly understood, but the sequence of events is well established. During prepubertal years, levels of LH, FSH, and testosterone are low. At around age six or seven, a prepubertal growth spurt occurs that is probably related to secretion of adrenal androgens and human growth hormone (hGH).

The onset of puberty is signaled by sleep-associated surges in LH and, to a lesser extent, FSH secretion. As puberty advances, elevated LH and FSH levels are present throughout the day and are accompanied by increased levels of testosterone. The rises in LH and FSH are believed to result from increased GnRH secretion and enhanced responsiveness of the anterior pituitary gland to GnRH. With sexual maturity, the hypothalamic–pituitary gland system becomes less sensitive to the feedback inhibition of testosterone on LH and FSH secretion.

The changes in the testes that occur during puberty include maturation of sustentacular cells and initiation of spermatogenesis. The anatomical and functional changes associated with puberty are the result of increased testosterone secretion. Usually, the first sign is enlargement of the testes. About a year later, the penis increases in size. The internal reproductive structures increase in size over a period of several years. Development of the secondary sex characteristics occurs and a growth spurt takes place as elevated testosterone levels increase both bone and muscle growth.

Ducts

Ducts of the Testis

Following their production, sperm are moved through the seminiferous tubules to the **straight tubules** (see Figure 23.2a). The straight tubules lead to a series of ducts in the testis called the **rete** (RĒ-tē; *rete* = network) **testis**. Some of the cells lining the rete testis possess cilia that help move the sperm along. The sperm are next transported out of the testis into the epididymis (see Figure 23.2a).

Epididymis

The **epididymis** (ep′-i-DID-i-mis; *epi* = above; *didymos* = testis) is a comma-shaped organ that lies along the posterior border of the testis (see Figures 23.1 and 23.2a) and consists mostly of a tightly coiled tube, the **ductus** (*ducere* = to lead) **epididymis**. The ductus epididymis is the site of sperm maturation. Over a 10 to 14 day period, their motility increases. The ductus epididymis also stores sperm for up to a month or more, after which they are expelled or reabsorbed.

Ductus (Vas) Deferens

As the ductus epididymis becomes less convoluted and its diameter increases, it is referred to as the **ductus (vas) deferens** (see Figure 23.2a). The ductus (vas) deferens ascends along the back border of the testis, penetrates the inguinal canal (a passageway in the front abdominal wall), and enters the pelvic cavity, where it loops over the side and down the back surface of the urinary bladder (see Figure 23.1). The ductus (vas) deferens has a heavy coat of three layers of muscle. It stores sperm

for up to several months and propels them toward the urethra during ejaculation by peristaltic contractions of its muscular coat. Sperm cells that are not ejaculated are reabsorbed.

Traveling with the ductus (vas) deferens as it ascends in the scrotum are blood vessels, autonomic nerves, and lymphatic vessels that together make up the **spermatic cord,** a supporting structure of the male reproductive system.

One method of sterilization of males is called **vasectomy** (vas-EK-tō-mē; *tome* = incision). It is a relatively uncomplicated procedure typically performed under local anesthesia in which a portion of each ductus (vas) deferens is removed. In the procedure, an incision is made in the scrotum, the ducts are tied in two places, and the portion between the ties is removed. Although sperm production continues in the testes, the sperm cannot reach the exterior because the ducts are cut; the sperm degenerate and are destroyed by phagocytosis. Vasectomy has no effect on sexual desire and performance and, if performed correctly, is virtually 100 percent effective. However, it has recently been linked to a higher incidence of prostate cancer. Vasectomy is reversible, with a 45 to 60 percent chance of regaining fertility.

Ejaculatory Duct

Behind the urinary bladder are the **ejaculatory** (e-JAK-yoo-la-tō′-rē; *ejectus* = to throw out) **ducts** (Figure 23.6), formed by the union of the duct from the seminal vesicle (to be described shortly) and ductus (vas) deferens. The ejaculatory ducts eject sperm into the urethra.

Urethra

The **urethra** is the terminal duct of the system, serving as a passageway for sperm or urine. In the male, the urethra passes through the prostate gland, urogenital diaphragm, and penis (see Figure 23.1). The opening of the urethra to the exterior is called the **external urethral orifice.**

Accessory Sex Glands

Whereas the ducts of the male reproductive system store and transport sperm cells, the **accessory sex glands** secrete most of the liquid portion of semen (the combined secretions from the accessory sex glands and testes). The paired **seminal** (*seminalis* = pertaining to seed) **vesicles** (VES-i-kuls) are pouchlike structures, lying at the base of the urinary bladder in front of the rectum (Figure 23.6). They secrete an alkaline, viscous fluid that contains the sugar fructose, prostaglandins, and clotting proteins (unlike those found in blood). The alkaline nature of the fluid helps neutralize acid in the female reproductive tract. This acid would inactivate and kill sperm if not neutralized. The fructose is used for ATP production by sperm. Prostaglandins contribute to sperm motility and viability and may also stimulate muscular contraction within the female reproductive tract. Clotting proteins help semen coagulate after ejaculation. Fluid secreted by the seminal vesicles constitutes about 60 percent of the volume of semen.

The **prostate** (PROS-tāt) **gland** is a single, doughnut-shaped gland about the size of a chestnut (Figure 23.6). It is below the urinary bladder and surrounds the upper portion of the urethra. The prostate gland slowly increases in size from

Figure 23.6   Male reproductive organs in relation to surrounding structures.

🔑 *The accessory sex glands secrete most of the liquid portion of semen.*

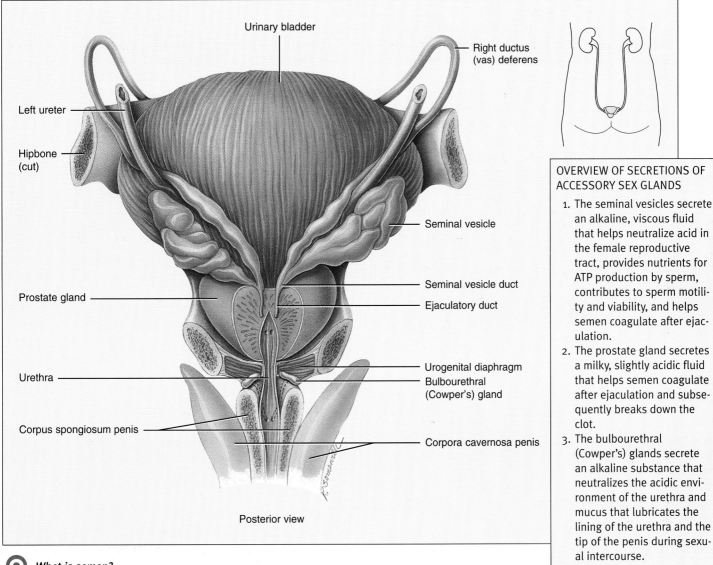

OVERVIEW OF SECRETIONS OF ACCESSORY SEX GLANDS

1. The seminal vesicles secrete an alkaline, viscous fluid that helps neutralize acid in the female reproductive tract, provides nutrients for ATP production by sperm, contributes to sperm motility and viability, and helps semen coagulate after ejaculation.
2. The prostate gland secretes a milky, slightly acidic fluid that helps semen coagulate after ejaculation and subsequently breaks down the clot.
3. The bulbourethral (Cowper's) glands secrete an alkaline substance that neutralizes the acidic environment of the urethra and mucus that lubricates the lining of the urethra and the tip of the penis during sexual intercourse.

Posterior view

❓ *What is semen?*

birth to puberty, and then a rapid growth spurt occurs. The size attained by the third decade remains stable until about age 45, when further enlargement may occur. (See prostate disorders on p. 544.) The prostate secretes a milky, slightly acidic fluid that contains nutrients (used by sperm for ATP production) and several enzymes, such as prostate specific antigen (PSA) and lysozyme, and clotting enzymes. PSA and clotting enzymes liquefy coagulated semen. Secretions of the prostate gland enter the urethra through many prostatic ducts. They make up about 25 percent of the volume of semen and contribute to sperm motility and viability.

The paired ***bulbourethral*** (bul′-bō-yoo-RĒ-thral) or *Cowper's glands* are about the size of peas. They are located beneath the prostate gland on either side of the urethra and within the urogenital diaphragm (Figure 23.6). The bulbourethral glands secrete an alkaline substance into the urethra

that protects sperm by neutralizing the acid environment of the urethra. They also produce mucus that lubricates the end of the penis during sexual intercourse.

Semen

Semen (*semen* = sew) is a mixture of sperm and the secretions of the seminal vesicles, prostate gland, and bulbourethral glands. The average volume of semen for each ejaculation is 2.5 to 5 ml, and the average range of spermatozoa ejaculated is 50 to 150 million/ml. When the number of sperm falls below 20 million/ml, the male is likely to be infertile. The very large number is required because only a small percentage eventually reach the secondary oocyte. Also, though only a single sperm cell fertilizes a secondary oocyte, fertilization requires the combined action of a larger number of sperm to digest the material covering the secondary oocyte. The acrosome of a sperm produces enzymes that dissolve the barrier.

Semen has a slightly alkaline pH of 7.20 to 7.70. The prostatic secretion gives semen a milky appearance, and fluids from the seminal vesicles and bulbourethral glands give it a sticky consistency. Semen provides sperm with a transportation medium and nutrients. It neutralizes the acid environment of the male urethra and the female vagina. It also contains enzymes that activate sperm after ejaculation, and an antibiotic that kills bacteria in semen and the female reproductive tract.

Once ejaculated into the vagina, liquid semen coagulates rapidly because of clotting proteins produced by the seminal vesicles. The clot liquefies in about 10 to 20 minutes because of enzymes produced by the prostate gland that break down the clot. Abnormal or delayed liquefaction of coagulated semen may cause complete or partial immobilization of sperm, thus inhibiting their movement through the cervix of the uterus.

Penis

The **penis** is used to introduce sperm into the vagina (Figure 23.7). The penis is a cylinder-shaped organ that consists of a root, a body, and glans penis. It is composed of three masses of erectile tissue. The two dorsal masses are called the **corpora cavernosa penis** (*corpus* = body; *caverna* = hollow). The ventral mass, the **corpus spongiosum penis**, contains the urethra.

Figure 23.7   Internal structure of the penis. The inset in (b) shows details of the skin and fascia.

🔑 *The penis contains a pathway for ejaculation of semen and excretion of urine.*

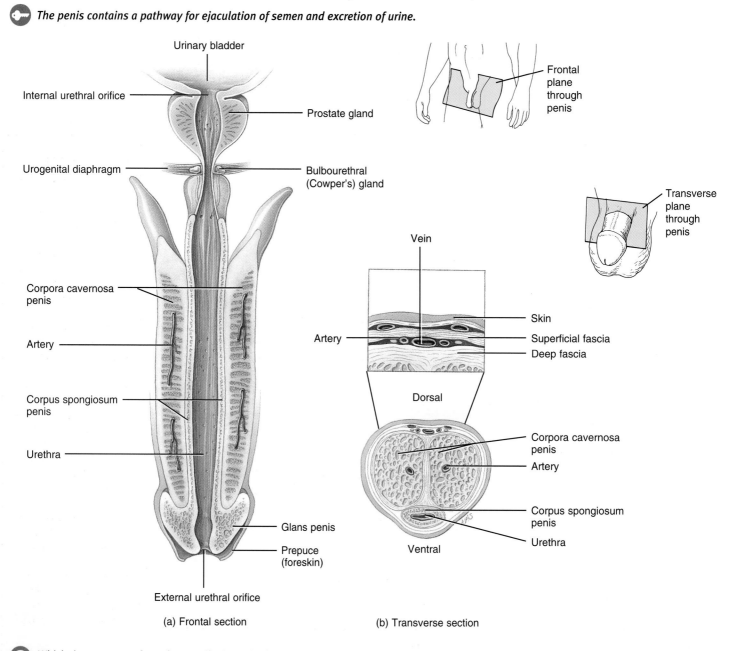

(a) Frontal section

(b) Transverse section

🔍 *Which tissue masses form the erectile tissue in the penis and why do they become rigid?*

All three masses are enclosed by fascia and skin and consist of erectile tissue containing blood sinuses. Under the influence of sexual stimulation, the arteries supplying the penis dilate, and large quantities of blood enter the sinuses, resulting in an *erection,* a parasympathetic reflex. Details of male sexual responses are presented in the next chapter. A smooth muscle sphincter at the base of the urinary bladder closes during ejaculation to prevent the mixing of acidic urine with semen in the urethra, which could immobilize sperm. The sphincter also keeps semen from entering the urinary bladder.

The distal end of the corpus spongiosum penis is a slightly enlarged region called the *glans penis,* which means shaped like an acorn. In the glans penis is the opening of the urethra (the *external urethral orifice*) to the exterior. Covering the glans penis is the loosely fitting *prepuce* (PRĒ-pyoos), or *foreskin.*

Circumcision (*circumcido* = to cut around) is a surgical procedure in which part or all of the prepuce is removed. It is usually performed just after delivery, on the third or fourth day after birth, or on the eighth day as part of a Jewish religious rite.

Female Reproductive System

objective: *Describe the structure and functions of the female reproductive organs.*

The female organs of reproduction include the ovaries (female gonads), which produce secondary oocytes (cells that develop into mature ova, or eggs, following fertilization) and the female sex hormones progesterone, estrogens, inhibin, and relaxin; the uterine (Fallopian) tubes that transport ova to the uterus (womb); the vagina; and external organs—the vulva, or pudendum (Figure 23.8). The mammary glands also are considered part of the female reproductive system.

Figure 23.8   Female organs of reproduction and surrounding structures.

The female organs of reproduction include the ovaries, uterine (Fallopian) tubes, uterus, vagina, vulva, and mammary glands.

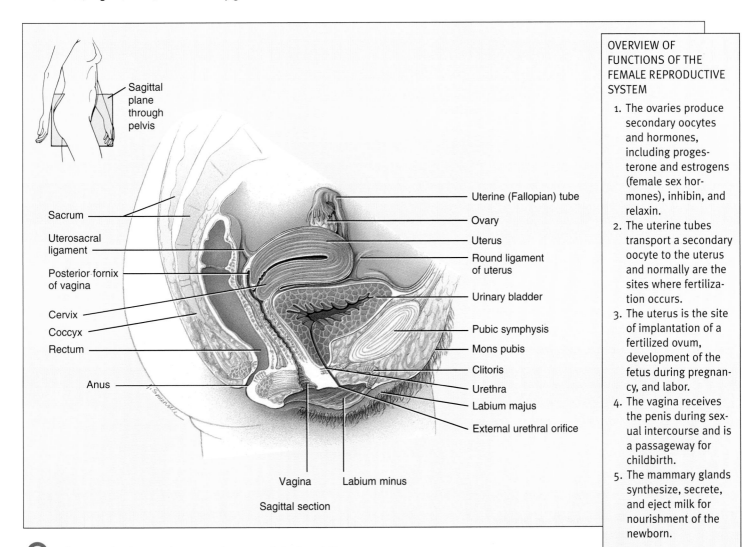

Sagittal plane through pelvis

Sacrum
Uterosacral ligament
Posterior fornix of vagina
Cervix
Coccyx
Rectum
Anus

Uterine (Fallopian) tube
Ovary
Uterus
Round ligament of uterus
Urinary bladder
Pubic symphysis
Mons pubis
Clitoris
Urethra
Labium majus
External urethral orifice

Vagina   Labium minus

Sagittal section

OVERVIEW OF FUNCTIONS OF THE FEMALE REPRODUCTIVE SYSTEM

1. The ovaries produce secondary oocytes and hormones, including progesterone and estrogens (female sex hormones), inhibin, and relaxin.
2. The uterine tubes transport a secondary oocyte to the uterus and normally are the sites where fertilization occurs.
3. The uterus is the site of implantation of a fertilized ovum, development of the fetus during pregnancy, and labor.
4. The vagina receives the penis during sexual intercourse and is a passageway for childbirth.
5. The mammary glands synthesize, secrete, and eject milk for nourishment of the newborn.

What term is given to the external organs of the female?

The specialized branch of medicine that deals with the diagnosis and treatment of diseases of the female reproductive system is **gynecology** (gī′-ne-KOL-ō-jē; *gyneco* = woman).

Ovaries

The **ovaries** (*ovarium* = egg receptacle) are paired organs that in fetal life arise from the same embryonic tissue as the testes. They are the size and shape of unshelled almonds. One lies on each side of the pelvic cavity, held in place by broad, ovarian, and suspensory ligaments (Figure 23.9). Each contains a hilus where nerves, blood, and lymphatic vessels enter. They consist of the following parts (Figure 23.10):

1. **Germinal epithelium.** A surface layer of simple cuboidal epithelium.

2. **Stroma.** A region of connective tissue composed of an outer layer (the *cortex*) that contains ovarian follicles and an inner layer (the *medulla*) that contains the nerve, blood, and lymph supply.

3. **Ovarian follicles** (*folliculus* = little bag) consist of *oocytes* in various stages of development and their surrounding cells. When the surrounding cells form a single layer, they are called **follicular cells.** Later in development, when they form several layers, they are referred to as **granulosa cells.** The surrounding cells nourish the

---

Figure 23.9   Uterus and associated structures. The left side of the figure has been sectioned to show internal structures.

 The uterus is the site of menstruation, implantation of a fertilized ovum, development of the fetus, and labor.

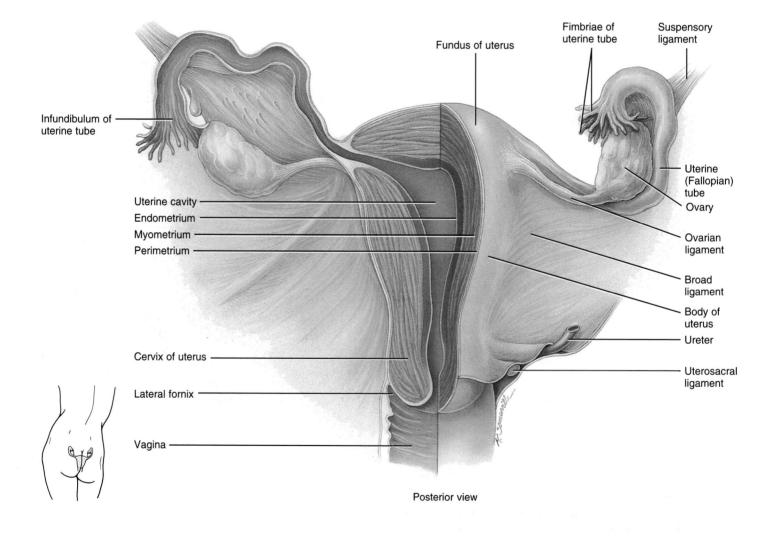

Posterior view

 What ligaments hold the ovaries in position?

Figure 23.10   Histology of the ovary. The arrows indicate the sequence of developmental stages that occur as part of the ovarian cycle.

The ovaries are the female gonads; they produce haploid oocytes.

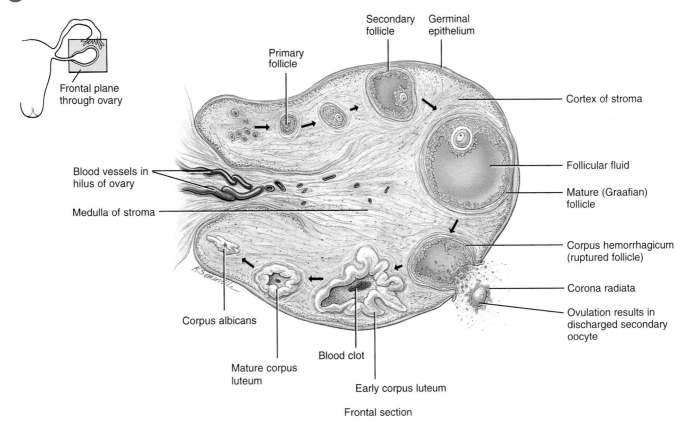

Frontal section

What structures in the ovary contain endocrine tissue and what hormones do they secrete?

developing oocyte and begin to secrete estrogens as the follicle grows larger.

4. A *mature (Graafian) follicle* is a large, fluid-filled follicle that soon will rupture and expel a secondary oocyte, a process called *ovulation.*

5. A *corpus luteum* (= yellow body) contains the remnants of an ovulated mature follicle. The corpus luteum produces progesterone, estrogens, relaxin, and inhibin until it degenerates and turns into white fibrous tissue called a *corpus albicans* (= white body).

Oogenesis

objective: *Describe how ova are produced.*

The formation of a haploid (*n*) ovum in the ovary is called *oogenesis* (ō'-ō-JEN-e-sis). Whereas spermatogenesis begins in males at puberty, oogenesis begins in females before they are even born. Oogenesis occurs in essentially the same manner as spermatogenesis. It involves meiosis and maturation.

1. **Reduction division (meiosis I).** During early fetal development, germ cells in the ovaries differentiate into *oogonia* (ō-ō-GŌ-nē-a; *oo* = egg), which can give rise to cells that develop into secondary oocytes (Figure 23.11). Oogonia are diploid (*2n*) cells that divide mitotically. At about the third month of prenatal development, some oogonia develop into larger diploid (*2n*) cells called *primary* (*primus* = first) *oocytes* (Ō-ō-sītz). Although these cells start reduction division (meiosis I) before birth, they do not complete it until after puberty. Some primary oocytes develop into *primary follicles.* These consist of a primary oocyte surrounded first by one layer of follicular cells and then by six to seven layers of *granulosa cells.* As a follicle grows, it forms a clear glycoprotein layer, called the *zona pellucida* (pe-LOO-si-da) between the primary oocyte and the granulosa cells. The innermost layer of granulosa cells becomes firmly attached to the zona pellucida and is called the *corona radiata* (*corona* = crown; *radiata* = radiation). The granulosa cells begin to secrete follicular fluid, which builds

Figure 23.11 Oogenesis. The designation *2n* means diploid; *n* means haploid.

🔑 *In an oocyte, equatorial division is completed only if fertilization occurs.*

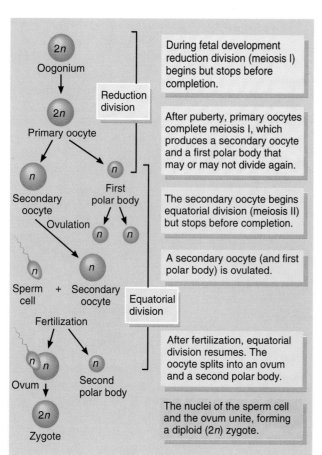

During fetal development reduction division (meiosis I) begins but stops before completion.

After puberty, primary oocytes complete meiosis I, which produces a secondary oocyte and a first polar body that may or may not divide again.

The secondary oocyte begins equatorial division (meiosis II) but stops before completion.

A secondary oocyte (and first polar body) is ovulated.

After fertilization, equatorial division resumes. The oocyte splits into an ovum and a second polar body.

The nuclei of the sperm cell and the ovum unite, forming a diploid (2n) zygote.

❓ *How does the age of a primary oocyte in a female compare with the age of a primary spermatocyte in a male?*

up in the cavity of the follicle. The follicle is now termed a *secondary follicle.*

After puberty, under the influence of the hormones secreted by the anterior pituitary gland, each month meiosis resumes in one secondary follicle. The diploid primary oocyte completes reduction division (meiosis I), and two haploid cells of unequal size, both with 23 chromosomes (*n*) of two chromatids each, are produced. The smaller cell produced by reduction division, called the *first polar body,* is essentially a packet of discarded nuclear materials. The larger cell, known as the *secondary oocyte,* receives most of the cytoplasm. Once a secondary oocyte is formed, it starts the equatorial division (meiosis II) and then stops at this stage. The follicle in which these events are taking place, termed the *mature (Graafian) follicle* (also called a *vesicular ovarian follicle*) will soon rupture and release its secondary oocyte. This process is called *ovulation.*

2. **Equatorial division (meiosis II).** At ovulation, usually one secondary oocyte (with the first polar body and corona radiata) is expelled into the pelvic cavity. Normally, the cells are swept into the uterine (Fallopian) tube. If fertilization does not occur, the secondary oocyte degenerates. If sperm are present in the uterine tube and one penetrates the secondary oocyte (fertilization), however, equatorial division (meiosis II) resumes. The secondary oocyte splits into two haploid (*n*) cells of unequal size. The larger cell is the *ovum,* or mature egg; the smaller one is the *second polar body.* The nuclei of the sperm cell and the ovum then unite, forming a diploid (*2n*) *zygote.* The first polar body may also undergo another division to produce two polar bodies. If it does, the primary oocyte ultimately gives rise to a single haploid (*n*) ovum and three haploid (*n*) polar bodies, which all degenerate. Thus one oogonium gives rise to a single gamete (ovum), whereas one spermatogonium produces four gametes (sperm).

Spermatogenesis and oogenesis differ in other ways as well. Spermatogenesis is a continuous process that begins in puberty and continues throughout life; oogenesis that results in the production of a secondary oocyte begins at the first menstruation and ends at menopause. Also, sperm cells are quite small, have flagella for locomotion, and contain few nutrients; a secondary oocyte is larger, lacks flagella, and contains more nutrients for nourishment until implantation occurs in the uterus.

Uterine (Fallopian) Tubes

The female body contains two *uterine (Fallopian) tubes* that extend laterally from the uterus and transport the secondary oocytes from the ovaries to the uterus (see Figure 23.9). The open funnel-shaped end of each tube, the *infundibulum,* lies close to the ovary and is surrounded by fingerlike projections called *fimbriae* (FIM-brē-ē; *fimbrae* = fringe), which help gather secondary oocytes into the tube following ovulation. From the infundibulum the uterine tube extends across and down to the upper outer corners of the uterus.

About once a month a mature (Graafian) follicle ruptures, releasing a secondary oocyte, a process called *ovulation.* The secondary oocyte is swept into the uterine tube by the ciliary action of the epithelium of the infundibulum. The oocyte is then moved along the tube by the cilia of the tube's mucous lining and the peristaltic contractions of the muscle layer.

If the secondary oocyte is fertilized by a sperm cell, it usually occurs in the uterine tube. Fertilization may occur any time up to about 24 hours following ovulation. The fertilized ovum (zygote) descends into the uterus within 7 days. An unfertilized secondary oocyte disintegrates.

Ectopic (*ektopos* = displaced) *pregnancy* (*EP*) refers to the implantation of an embryo or fetus outside the inner lining (endometrium) of the uterine cavity. Most occur in the uterine tube. Some occur in the ovaries, abdomen, uterine cervix, and broad ligaments. The basic cause of a tubal pregnancy is impaired passage of the fertilized ovum through the uterine tube, which might be due to conditions such as pelvic inflammatory disease (PID), previous uterine tube surgery, previous ectopic pregnancy, pelvic tumors, or developmental abnormalities. Ectopic pregnancy is characterized by one or two missed menses (periods), followed by vaginal bleeding and acute pelvic pain.

Uterus

The *uterus* (womb) is the site of menstruation, implantation of a fertilized ovum, development of the fetus during pregnancy, and labor. It is situated between the urinary bladder and the rectum and is shaped like an inverted pear (see Figures 23.8 and 23.9).

Parts of the uterus include the dome-shaped portion above the uterine tubes called the *fundus,* the tapering central portion called the *body,* and the narrow portion opening into the vagina called the *cervix.* The interior of the body is called the *uterine cavity.*

The uterus is supported and held in position by several ligaments. These are the broad, uterosacral, cardinal, and round ligaments (see Figures 23.8 and 23.9). The broad liga-

ment also forms part of the outer layer of the uterus, the *perimetrium* (*peri* = around).

The middle muscular layer of the uterus, the *myometrium* (*myo* = muscle), forms the bulk of the uterine wall. It consists of smooth muscle. During childbirth, coordinated contractions of the muscles help expel the fetus.

The innermost part of the uterine wall, the *endometrium* (*endo* = within), is a mucous membrane composed of two layers. The *stratum basalis* is a permanent layer lying next to the myometrium. The *stratum functionalis* surrounds the uterine cavity. It nourishes a growing fetus or is shed each month during menstruation if fertilization does not occur. Following menstruation, it is replaced by the stratum basalis. The endometrium contains many glands whose secretions nourish sperm and the zygote.

Blood is supplied to the uterus by the *uterine arteries* (Figure 23.12). Branches of the uterine arteries penetrate deeply into the myometrium. Just before the branches enter the endometrium, they divide into two kinds of arterioles. The *straight arterioles* terminate in the stratum basalis and supply it with the materials necessary to regenerate the stratum func-

---

Figure 23.12   Blood supply of the uterus. The inset shows the details of the blood vessels of the endometrium.

🔑 *Straight arterioles supply the necessary materials for regeneration of the stratum functionalis.*

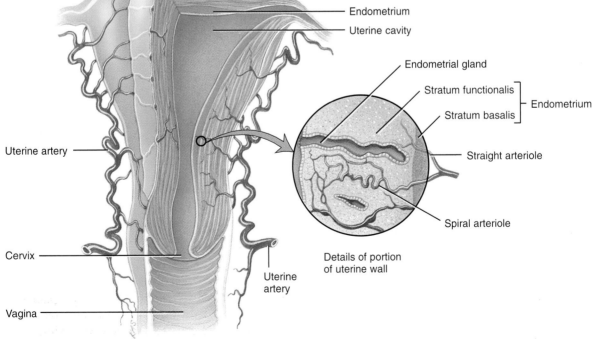

Anterior view with left side partially sectioned

❓ *What is the significance of the two layers of the endometrium?*

tionalis. The *spiral arterioles* penetrate the functionalis and change markedly during the menstrual cycle. Blood leaves the uterus via the *uterine veins.*

Early stages of *cancer of the uterus* can be detected through the *Papanicolaou* (pap-a-NIK-ō-la-oo) *test,* or *Pap smear.* In this generally painless procedure, a few cells from the part of the vagina surrounding the cervix and from the cervix are removed with a swab and examined microscopically. Malignant cells have a characteristic appearance that allows diagnosis even before symptoms occur. Estimates indicate that the Pap smear is more than 90 percent reliable in detecting cancer of the cervix.

Vagina

The *vagina* (*vagina* = sheath) serves as a passageway for menstrual flow and childbirth. It is also the receptacle for the penis during sexual intercourse (see Figures 23.8 and 23.9). It is situated between the urinary bladder and the rectum. A recess, called the *fornix* (*fornix* = arch or vault), surrounds the cervix. The fornix makes it possible for a woman to use contraceptive diaphragms.

The mucosa of the vagina is continuous with that of the uterus and cervix and lies in a series of transverse folds, the *rugae* (ROO-jē). The mucosa of the vagina has an acid environment that retards microbial growth. The muscular layer is composed of smooth muscle that can stretch to receive the penis during intercourse and allow for birth of the fetus. At the vaginal opening, the *vaginal orifice,* there may be a thin fold of mucous membrane called the *hymen* (*hymen* = membrane), which partially covers it (see Figure 23.13).

Vulva

The term *vulva* (VUL-va; *volvere* = to wrap around), or *pudendum* (pyoo-DEN-dum), refers to the external genitalia of the female (Figure 23.13). Its components are as follows.

The *mons pubis* (MONZ PYOO-bis; *mons* = mountain) is an elevation of adipose tissue covered by coarse pubic hair, which cushions the pubic symphysis. From the mons pubis, two longitudinal folds of skin, the *labia majora* (LĀ-bē-a ma-JŌ-ra; *labium* = lip), extend down and back. The labia majora and scrotum are equivalent structures in terms of fetal origin. The labia majora contain adipose tissue and sebaceous (oil) and sudoriferous (sweat) glands; they are covered by pubic hair. Inside the labia majora are two folds of skin called the *labia minora* (MĪ-nō-ra). The labia minora do not contain pubic hair or fat and have few sudoriferous (sweat)

Figure 23.13   Components of the vulva.

🔑 *Like the penis, the clitoris is capable of enlargement upon tactile stimulation.*

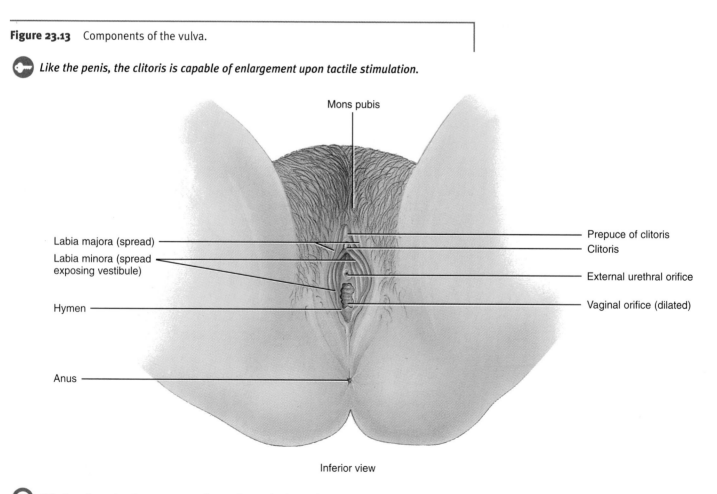

Mons pubis

Labia majora (spread)

Labia minora (spread exposing vestibule)

Hymen

Anus

Prepuce of clitoris

Clitoris

External urethral orifice

Vaginal orifice (dilated)

Inferior view

❓ *What surface structures are anterior to the vaginal opening? Lateral to it?*

glands; they do, however, contain numerous sebaceous (oil) glands.

The *clitoris* (KLI-to-ris) is a small, cylindrical mass of erectile tissue and nerves. It is located at the front junction of the labia minora. A layer of skin called the *prepuce* (foreskin) is formed at a point where the labia minora unite and cover the body of the clitoris. The exposed portion of the clitoris is the *glans.* The clitoris and glans penis of the male are equivalent structures. Like the penis, the clitoris is capable of enlargement upon tactile stimulation and assumes a role in sexual excitement.

The region between the labia minora is called the *vestibule.* In the vestibule are the hymen (if present); *vaginal orifice,* the opening of the vagina to the exterior; *external urethral orifice,* the opening of the urethra to the exterior; and on either side of the vaginal orifice, the openings of the ducts of the *paraurethral (Skene's) glands.* These glands are in the wall of the urethra and secrete mucus. The paraurethral glands and the male prostate are equivalent structures. On either side of the vaginal orifice itself are the *greater vestibular (Bartholin's) glands,* which produce a mucus for lubrication during sexual intercourse. The greater vestibular glands and the male bulbourethral glands are equivalent structures.

Perineum

The *perineum* (per'-i-NĒ-um) is the diamond-shaped area between the thighs and buttocks of both males and females (Figure 23.13). It contains the external genitals and anus.

In the female, the region between the vagina and anus is known as the *clinical perineum.* If the vagina is too small to accommodate the head of an emerging fetus, the skin, vaginal epithelium, subcutaneous fat, and muscle of the clinical perineum may tear. Moreover, the tissues of the rectum may be damaged. To avoid this, a small incision called an *episiotomy* (e-piz'-ē-OT-ō-mē; *epision* = pubic region; *tome* = incision) is made in the perineal skin and underlying tissues just prior to delivery. After delivery the episiotomy is sutured in layers.

Mammary Glands

The *mammary* (*mamma* = breast) *glands* are modified sudoriferous (sweat) glands that lie over the pectoralis major and serratus anterior muscles and are attached to them by a layer of connective tissue (Figure 23.14). Internally, each mammary gland consists of 15 to 20 *lobes* arranged radially and separat-

---

Figure 23.14   Mammary glands.

The mammary glands function in the synthesis, secretion, and ejection of milk (lactation).

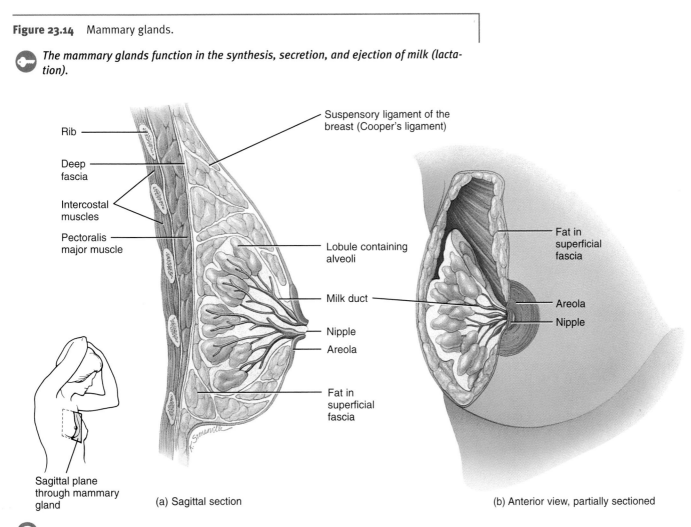

(a) Sagittal section

(b) Anterior view, partially sectioned

What hormones regulate synthesis and release of milk?

ed by adipose tissue and strands of connective tissue called *suspensory ligaments of the breast (Cooper's ligaments),* which support the breast. In each lobe are smaller *lobules,* in which milk-secreting cells referred to as *alveoli (alveolus = small cavity)* are found. Breast size has nothing to do with the amount of milk produced. Milk secretion by alveoli is transported through ducts that terminate in the *nipple.* The circular pigmented area of skin surrounding the nipple is called the *areola* (a-RĒ-ō-la; *areolar* = small space).

At birth, both male and female mammary glands are undeveloped and appear as slight elevations on the chest. With the onset of puberty under the influence of estrogens and progesterone, the female breasts begin to develop; the duct system matures, and fat is deposited, which increases breast size to varying degrees. The areola and nipple also grow and become pigmented.

The function of the mammary glands is milk secretion and ejection, together called *lactation.* The secretion of milk following delivery is due largely to the hormone prolactin (PRL), with help from progesterone and estrogens. The ejection of milk occurs in the presence of oxytocin (OT), which is released by the posterior pituitary gland in response to sucking by the infant. Lactation is considered in detail in the next chapter.

Female Puberty

The factors that determine the onset of female puberty are no better understood than those that determine male puberty. Prepubertal levels of LH, FSH, and estrogens are low. At around age seven or eight, the secretion of adrenal androgens begins to increase. The onset of puberty is signaled by sleep-associated surges in LH and FSH. As puberty progresses, LH and FSH levels increase throughout the day. The rising levels stimulate the ovaries to secrete estrogens, which bring about development of the secondary sex characteristics. These include fat distribution to the breasts, abdomen, mons pubis, and hips; increased vascularization of the skin; voice pitch; broad pelvis; and hair pattern. (Budding of the breasts is the first observable sign of puberty.) Estrogens also stimulate the growth of the uterine (Fallopian) tubes, uterus, and vagina and bring on *menarche* (me-NAR-kē), the onset of menstruation, at an average age of 12. But, the first ovulation does not take place until 6 to 9 months after menarche because the positive feedback of estrogens on LH and FSH is the last step in the maturation of the hypothalamic–pituitary gland–ovarian loop, which is discussed next.

Female Reproductive Cycle

objective: *Explain the functions of the female reproductive hormones and then define the menstrual and ovarian cycles and explain how they are related.*

The general term *female reproductive cycle* refers to the menstrual and ovarian cycles and the hormonal cycles that regulate them. It also includes other cyclic changes in female reproductive organs. At this point, we will consider the menstrual and ovarian cycles and their hormonal regulation.

The *menstrual* (*mens* = monthly) cycle is a series of changes in the endometrium of a nonpregnant female. Each month the endometrium is prepared to receive a fertilized ovum. If no fertilization occurs, the stratum functionalis of the endometrium is shed. The *ovarian cycle* is a monthly series of events associated with the maturation of an oocyte.

Hormonal Regulation

The menstrual cycle and ovarian cycle are controlled by gonadotropin releasing hormone (GnRH) from the hypothalamus (Figure 23.15). GnRH stimulates the release of follicle-stimulating hormone (FSH) and luteinizing hormone (LH) from the anterior pituitary gland. FSH stimulates the initial secretion of estrogens by growing follicles. LH stimulates the further development of ovarian follicles and their secretion of estrogens, brings about ovulation, promotes formation of the corpus luteum, and stimulates the production of estrogens, progesterone, relaxin, and inhibin by the corpus luteum.

Estrogens, which are secreted by follicular cells, promote development and maintenance of female reproductive structures, especially the endometrial lining of the uterus, and secondary sex characteristics, including the breasts. Estrogens also control fluid and electrolyte balance and, with human growth hormone (hGH), increase protein anabolism. Moderate levels of estrogens in the blood inhibit the release of GnRH by the hypothalamus, which, in turn, inhibits the secretion of FSH by the anterior pituitary gland. This inhibition provides the basis for action of one kind of contraceptive (birth control) pill.

Progesterone, secreted mainly by the corpus luteum, works with estrogens to prepare the endometrium for implantation of a fertilized ovum and the mammary glands for milk secretion. High levels of progesterone also inhibit GnRH and luteinizing hormone (LH).

Inhibin is secreted by growing follicles and the corpus luteum. It inhibits secretion of FSH and, to a lesser extent, LH.

Relaxin is produced by the corpus luteum and placenta during the late stages of pregnancy. It facilitates delivery by relaxing the pubic symphysis and helping to dilate the cervix.

Exhibit 23.1 presents a summary of hormones that regulate the female reproductive cycle.

Menstrual Phase (Menstruation)

The menstrual cycle ranges from 24 to 35 days, with an average of 28 days. The menstrual cycle may be divided into three phases: the menstrual phase, the preovulatory phase, and the postovulatory phase (Figure 23.16 on page 540).

The *menstrual* (MEN-stroo-al) *phase,* also called *menstruation* (men'-stroo-Ā-shun) or *menses* (*mensis* = month), lasts for roughly the first 5 days of the cycle. (By convention, the first day of menstruation marks the first day of a new cycle.)

Figure 23.15   Secretion and physiological effects of estrogens, progesterone, relaxin, and inhibin.

The uterine and ovarian cycles are controlled by GnRH and ovarian hormones.

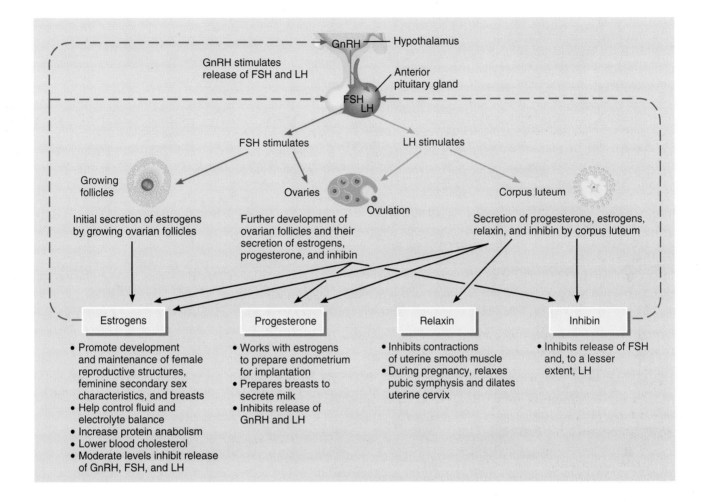

Which hormone helps facilitate delivery?

Exhibit 23.1	Summary of Hormones that Regulate the Female Reproductive Cycle	
HORMONE	WHERE PRODUCED	FUNCTION
Estrogens	Ovaries and placenta	Development and maintenance of female reproductive structures, especially the endometrium of the uterus; development of female secondary sex characteristics; development of breasts; control of fluid and electrolyte balance; increase protein anabolism.
Progesterone	Ovaries and placenta	Works with estrogens to prepare endometrium for implantation; prepares mammary glands to secrete milk.
Inhibin	Ovaries (and testes)	Inhibits secretion of FSH (and LH to a lesser extent).
Relaxin	Ovaries and placenta	Relaxes pubic symphysis and dilates uterine cervix.

Figure 23.16   Correlation of ovarian and uterine cycles with the hypothalamic and anterior pituitary gland hormones. In the cycle shown, fertilization and implantation have not occurred.

The preovulatory phase is more variable in length than the other phases.

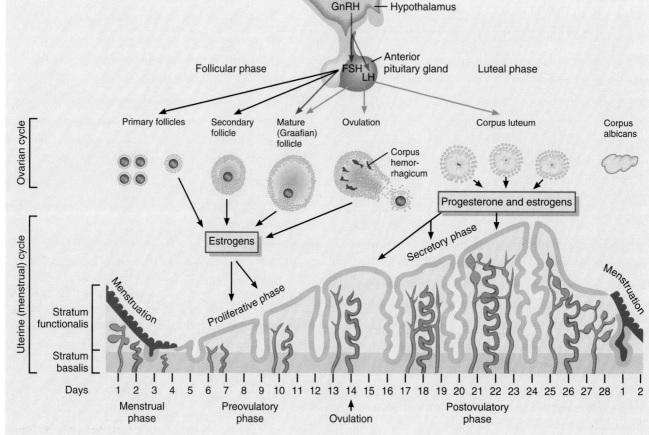

Which hormones directly stimulate proliferation of the endometrium?

EVENTS IN THE OVARIES   During the menstrual phase, about 20 small secondary follicles, some in each ovary, begin to enlarge. Follicular fluid, secreted by the granulosa cells, accumulates in the enlarging cavity in the follicle while the oocyte remains near the edge of the follicle.

EVENTS IN THE UTERUS   Menstrual flow from the uterus consists of 50 to 150 ml of blood and tissue cells from the endometrium. This discharge occurs because the declining level of estrogens and progesterone causes the uterine spiral arteries to constrict. As a result the cells they supply become ischemic (deficient in blood) and start to die. Eventually, the entire stratum functionalis sloughs off. At this time the endometrium is very thin because only the stratum basalis remains. The menstrual flow passes from the uterine cavity to the cervix and through the vagina to the exterior.

Preovulatory Phase

The **preovulatory phase,** the second phase of the female reproductive cycle, is the time between menstruation and ovulation. The preovulatory phase of the cycle is more variable in length than the other phases and accounts for most of the difference when cycles are shorter or longer than 28 days. It lasts from days 6 to 13 in a 28-day cycle.

EVENTS IN THE OVARIES   Under the influence of FSH, about 20 secondary follicles continue to grow and begin to secrete estrogens and inhibin. By about day 6, one follicle in one ovary has outgrown all the others and is thus the dominant follicle. Estrogens and inhibin secreted by the dominant follicle decrease the secretion of FSH, which causes the other less well-developed follicles to stop growing and die.

The one dominant follicle becomes the *mature (Graafian) follicle* that continues to enlarge until it is ready for ovulation. This follicle forms a blisterlike bulge on the surface of the ovary. During the final maturation process, the dominant follicle continues to increase its production of estrogens under the influence of an increasing level of LH (Figure 23.17). Estrogens are the primary ovarian hormones before ovulation, but small amounts of progesterone are produced by the mature follicle (Figure 23.17).

With reference to the ovaries, the menstrual phase and preovulatory phase together are termed the *follicular* (fō-LIK-yoo-lar) *phase* because ovarian follicles are growing and developing.

EVENTS IN THE UTERUS   Estrogens liberated into the blood by growing ovarian follicles (described next) stimulate the growth of the endometrium. Cells of the stratum basalis undergo mitosis and produce a new stratum functionalis. As the endometrium thickens, the endometrial glands develop and the arterioles lengthen as they penetrate the stratum functionalis. The thickness of the endometrium approximately doubles. With reference to the uterus, the preovulatory phase is also termed the *proliferative phase* because the endometrium is proliferating.

Ovulation

Ovulation, the rupture of the mature (Graafian) follicle with release of the secondary oocyte into the pelvic cavity, usually occurs on day 14 in a 28-day cycle. The *high* levels of estrogens during the last part of the preovulatory phase exert a *positive feedback* effect on both LH and GnRH and cause ovulation as follows: When estrogens are present in high enough concentration, they stimulate the hypothalamus to release more GnRH and the anterior pituitary gland to produce more LH. GnRH promotes release of FSH and more LH by the anterior pituitary gland. The LH surge brings about rupture of the mature follicle and expulsion of a secondary oocyte. The ovulated oocyte and its zona pellucida and corona radiata cells are usually swept into the uterine tube (any lost in the pelvic cavity disintegrate).

After ovulation, the mature follicle collapses and blood within it forms a clot due to minor bleeding during rupture of the follicle to become the *corpus hemorrhagicum* (*hemo* = blood; *rhegnynai* = to burst forth). See Figure 23.10. The clot is absorbed by the remaining follicular cells, which enlarge, and form the corpus luteum under the influence of LH.

An over-the-counter home test that detects the LH surge associated with ovulation predicts ovulation a day in advance.

Postovulatory Phase

The *postovulatory phase* of the female reproductive cycle is the most constant in duration and lasts for 14 days, from days 15 to 28 in a 28-day cycle. It represents the time between ovulation and the onset of the next menses. After ovulation, LH secretion stimulates the remnants of the mature follicle to develop into the corpus luteum. During its 2-week lifespan, the corpus luteum secretes increasing quantities of progesterone and some estrogens, relaxin, and inhibin (Figure 23.17).

EVENTS IN ONE OVARY   If the secondary oocyte is fertilized and begins to divide, the corpus luteum persists past its normal 2-week lifespan. It is maintained by *human chorionic* (kō-rē-ON-ik) *gonadotropin* (*hCG*), a hormone produced by the chorion of the embryo as early as 8 to 12 days after fertilization. The presence of hCG in maternal blood or urine is an indication of pregnancy. The chorion eventually develops into part of the placenta. As the pregnancy progresses, the placenta itself begins to secrete estrogens to support pregnancy and progesterone to support pregnancy and breast development for lactation. Once the placenta begins its secretion, the role of the corpus luteum becomes minor. With reference to the ovaries, this phase of the cycle is also called the *luteal phase.*

If hCG does not rescue the corpus luteum, after 2 weeks its secretions decline and it degenerates into a scar called the *corpus albicans*. The lack of progesterone and estrogens due to degeneration of the corpus luteum then causes menstruation. In addition, the decreased levels of progesterone, estrogens, and inhibin promote the release of GnRH, FSH, and LH, which stimulate follicular growth and a new ovarian cycle begins. A summary of these hormonal interactions is presented in Figure 23.18.

EVENTS IN THE UTERUS   Progesterone and estrogens produced by the corpus luteum promote growth of the endometrial glands, which begin to secrete glycogen, and vascularization and thickening of the endometrium. These preparatory

Figure 23.17   Relative concentrations of anterior pituitary gland hormones (FSH and LH) and ovarian hormones (estrogens and progesterone) during a normal menstrual cycle.

🔑 *Estrogens are the primary ovarian hormones before ovulation; after ovulation, both progesterone and estrogens are secreted by the corpus luteum.*

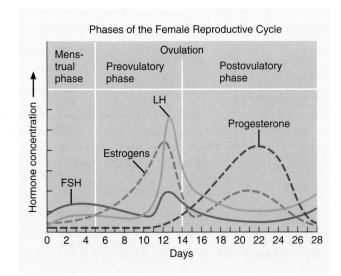

❓ *An over-the-counter home test that predicts ovulation detects the presence of which hormone?*

changes are maximal about one week after ovulation, corresponding to the time of possible arrival of a fertilized ovum. With reference to the uterus, this phase of the cycle is called the **secretory phase** because of the secretory activity of the endometrial glands.

Menopause

The menstrual cycle normally occurs once each month from menarche to **menopause** (*mens* = monthly; *pause* = to stop), the last menses. The advent of menopause is signaled by the **climacteric** (klī-MAK-ter-ik): menstrual cycles become less frequent. The climacteric typically begins between ages 40 and 50. Some women experience hot flashes, copious sweating, headache, insomnia, and emotional instability. In post-menopausal women there is some atrophy of the ovaries, uterine (Fallopian) tubes, uterus, vagina, external genitalia, and breasts.

Menopause occurs because of degeneration of the aging ovaries. Over the years of menstrual cycling, the number of primary follicles diminishes, causing the production of ova, estrogens, and progesterone by the ovary to decrease.

Figure 23.18 Summary of hormonal interactions of the uterine and ovarian cycles.

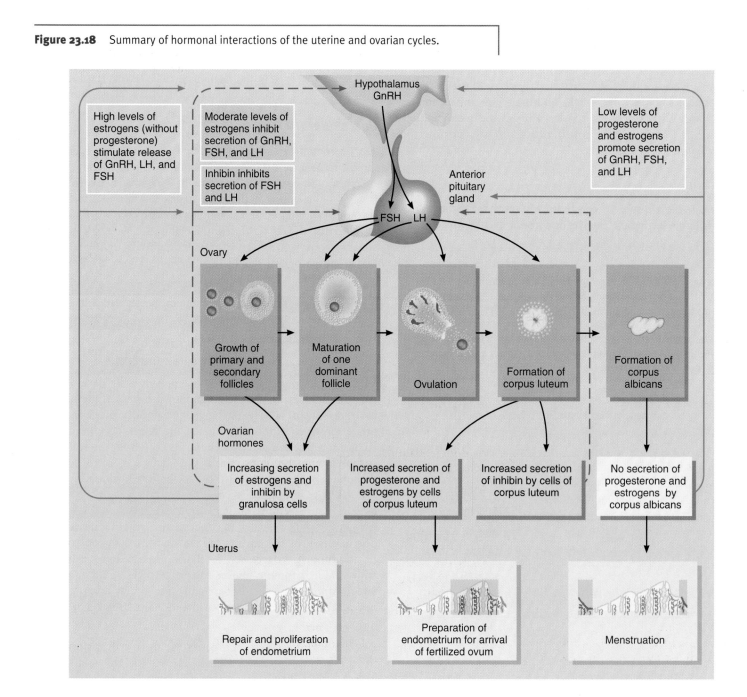

When declining levels of estrogens and progesterone stimulate secretion of GnRH, is this a positive or negative feedback effect? Why?

wellness focus

*t*he female reproductive cycle can be disrupted by many factors, including weight loss, low body weight, disordered eating, and vigorous physical activity. Many athletes experience intense pressure from coaches, parents, peers, and themselves to lose weight in order to improve performance. Consequently, many develop disordered eating behaviors and take on other harmful weight-loss practices in a struggle to maintain a very low body weight. Combine this drive for thinness with several hours of exercise every day, and you can see why female athletes are more likely to experience menstrual irregularity than their sedentary peers. This is particularly true for athletes who participate in sports that emphasize large amounts of vigorous physical activity, a lean physique, or minimum weight requirements. The athletes with the highest rates of menstrual irregularity include runners, gymnasts, dancers, figure skaters, and divers.

Many female athletes do not regard menstrual irregularity as a serious symptom, but believe it is a natural result of exercise training. Some feel amenorrhea, the absence of menstrual periods, offers a competitive advantage, since the athlete need not deal with symptoms of premenstrual syndrome (PMS) or dysmenorrhea. This is unfortunate, because menstrual irregularity may be caused by a serious underlying disorder for which the athlete should receive prompt medical treatment.

The Female Athlete Triad: Disordered Eating, Amenorrhea, and Osteoporosis

When menstrual irregularity is apparently caused by exercise training and not associated with another physical disorder, it is still a cause for concern. If the absence of menses is accompanied by a lack of ovulation, then a significant source of production of estrogen is missing. Remember that the ovarian follicles produce estrogen when stimulated by FSH and LH. If ovulation is not occurring, then the ovarian follicles, and later the corpus luteum, are not producing estrogen. Chronically low levels of estrogens may have several important health consequences in addition to those related to reproductive functions. One of these is a loss of bone minerals, since estrogens help bones retain calcium. The loss of the protective effect of estrogens explains why many women experience a decline in bone density that may develop into osteoporosis after menopause, when levels of estrogens drop. Amenorrheic runners have been shown to experience a similar effect. While short periods of menstrual irregularity in young athletes, may cause no lasting harm, long-term cessation of the menstrual cycle may be accompanied by a loss of bone mass, or in adolescent athletes, a failure to achieve an adequate bone mass, both of which can lead to premature osteoporosis and irreversible bone damage. In fact, in one study amenorrheic runners in their twenties were found to have bone densities similar to those of postmenopausal women 50 to 70 years old. A few of these runners had already sustained vertebral crush fractures, the "caving in" of the vertebrae responsible for the kyphosis (dowager's hump) associated with osteoporosis. More common in amenorrheic runners were recurring stress fractures in weakened bones.

Such studies led researchers to coin the term "the female athlete triad" for this syndrome, comprised of disordered eating, amenorrhea, and osteoporosis.

Follow-up studies on these runners found that with supplementation, of estrogens, bone density increased somewhat, but never reached age-appropriate levels, even after several years. These studies underscore the importance of preventing the triad in the first place, since current treatment strategies appear to be inadequate.

Sticks and Stones and . . . Exercise?

It is somewhat ironic that dedicated athletes should experience premature osteoporosis, since physical activity in general is associated with a *reduced* risk of osteoporosis. Exercise has been shown to increase bone density, especially if the exercise involves bone stress, such as running and aerobic dance. However, in the presence of disordered eating and overtraining, exercise may simply add insult to injury.

critical thinking

Do you think the existence of the "female athlete triad" suggests that girls and women should be discouraged from participating in athletics and other forms of vigorous physical activity? Why? Why not?

Common Disorders

Sexually Transmitted Disease (STD)

The general term *sexually transmitted disease* (STD) applies to any of the large groups of diseases that are contracted sexually or may be contracted otherwise but are then transmitted to a sexual partner. The group includes conditions traditionally specified as *venereal disease* (VD) (from Venus, goddess of love in Roman mythology), such as gonorrhea, syphilis, and genital herpes, as well as other conditions, such as AIDS (p. 400) and hepatitis B (p. 456).

Gonorrhea

Gonorrhea (gon'-ō-RĒ-a; *gonos* = seed; *rhein* = to flow) or *"clap"* is an infectious sexually transmitted disease that primarily affects the mucous membrane of the urogenital tract, the rectum, and occasionally the eyes. It is caused by the bacterium *Neisseria gonorrhoeae*. Discharges from the infected mucous membranes transmit the bacteria by direct contact, usually sexual. During passage of a newborn through the birth canal, the infant's eyes may become infected. Males usually suffer inflammation of the urethra with pus and painful urination. In females, infection may occur in the urethra, vagina, and cervix, often with a discharge of pus. Frequently, however, there are no noticeable symptoms in males or females. Treatment is with antibiotics, mainly ceftriaxone. Untreated gonorrhea in females may lead to sterility as a result of damage to the uterine (Fallopian) tubes.

Syphilis

Syphilis is a sexually transmitted disease caused by the bacterium *Treponema pallidum*. It is transmitted through sexual contact or exchange of blood or through the placenta to a fetus. The disease progresses through several stages: primary, secondary, latent, and sometimes tertiary. During the *primary stage*, the chief symptom is a painless open sore, called a *chancre* (SHANGK-ker), at the point of contact. The chancre heals in 1 to 5 weeks. From 6 to 24 weeks later, symptoms such as a skin rash, fever, and aches in the joints and muscles usher in the *secondary stage*. These symptoms disappear in about 4 to 12 weeks and the disease ceases to be infectious, but a blood test for the presence of the bacteria generally remains positive. During this "symptomless" period, called the *latent stage*, the bacteria may invade body structures. When other structures become infected (usually the brain, heart valves, and aorta), the disease is said to be in the *tertiary stage*. Treatment is with penicillin.

Genital Herpes

The sexually transmitted disease *genital herpes* is common in North America and, unlike syphilis and gonorrhea, genital herpes is incurable. Type I herpes simplex virus causes most infections above the waist, such as cold sores. Type II herpes simplex virus causes most infections below the waist, such as painful genital blisters on the prepuce, glans penis, and penile shaft in males and on the vulva or sometimes high up in the vagina in females. The blisters disappear and reappear in most patients, but the virus itself remains in the body and causes recurrence of symptoms such as fever, enlarged and sometimes tender lymph nodes, and numerous clusters of genital blisters.

Trichomoniasis

The microorganism *Trichomonas vaginalis*, a flagellated protozoan, causes *trichomoniasis* (trik'-ō-mō-NĪ-a-sis), an inflammation of the mucous membrane of the vagina in females and the urethra in males. Symptoms include a yellow vaginal discharge and severe vaginal itch in women. Metronidazole taken orally is used for treatment.

Chlamydia

Chlamydia (kla-MID-ē-a), a sexually transmitted disease caused by the bacterium *Chlamydia trachomatis*, currently is one of the most prevalent and damaging of the sexually transmitted diseases. In males, it is characterized by frequent, painful or burning urination and low back pain. In females, urethritis may spread through the reproductive tract and develop into inflammation of the uterine (Fallopian) tubes, which increases the risk of ectopic pregnancy and sterility. As in gonorrhea, infants' eyes may become infected during birth.

Genital Warts

Warts are an infectious disease caused by viruses. Sexual transmission of *genital warts* is common and is caused by the *human papillomavirus* (HPV). It is estimated that nearly one million persons a year develop genital warts in the United States. Patients with a history of genital warts may be at increased risk for certain types of cancer (cervical, vaginal, anal, vulval, and penile). There is no cure for genital warts. Treatment consists of cryotherapy (freezing) with liquid nitrogen, electrocautery (burning), excision, laser surgery, and topical application of certain drugs. Alpha interferon is also injected to treat genital warts.

Male Disorders

Testicular Cancer

Testicular cancer is one of the most common cancers seen in males and is associated with males with a history of late-descended or undescended testes. Most testicular cancers arise from the sperm-producing cells. Early signs of testicular cancer are a mass in the testis, often with pain or discomfort. Treatment involves removal of the diseased testis.

Prostate Disorders

The prostate gland is susceptible to infection, enlargement, and benign and malignant tumors. Because the prostate surrounds the urethra, any of these disorders can obstruct the flow of urine and result in serious changes in the urinary bladder, ureters, and kidneys and may perpetuate urinary tract infections. Therefore, if the obstruction cannot be relieved by other means, surgical removal (prostatectomy) of part of or the entire gland is necessary.

Prostate cancer is the leading cause of death from cancer in men in the United States. Its incidence is related to age, race, occupation, geography, and ethnic origin. Both benign and malignant growths are common in elderly men. Both types of tumors put pressure on the urethra, making urination painful and difficult. At times, excessive pressure due to backed-up urine destroys kidney tissue and gives rise to an increased susceptibility to infection.

Sexual Functional Abnormalities

Impotence (*impotenia* = lack of strength) is the inability of an adult male to attain or hold an erection long enough for sexual intercourse. Impotence may be the result of diabetes mellitus, physical abnormalities of the penis, systemic disorders such as syphilis, vascular disturbances (arterial or venous obstructions), neurological disorders, testosterone deficiency, drugs (alcohol, antidepressants, antihistamines, antihypertensives, narcotics, nicotine, and tranquilizers), or psychic factors.

Male infertility (*sterility*) is an inability to fertilize the ovum. It does not imply impotence. Male fertility requires production of adequate amounts of viable, normal sperm by the testes, unobstructed transportation of sperm through the seminal tract, and satisfactory deposition in the vagina.

Female Disorders

Menstrual Abnormalities

Amenorrhea (ā-men-ō-RĒ-a; *a* = without; *men* = month; *rhein* = to flow) is the absence of menstruation. It can be caused by endocrine disorders, congenitally abnormal development of the ovaries or uterus, significant decreases in body weight, or continuous, rigorous athletic training.

Dysmenorrhea (dis-men-ō-RĒ-a; *dys* = difficult) is painful menstruation caused by forceful contraction of the uterus. It is often accompanied by nausea, vomiting, diarrhea, headache, fatigue, and nervousness. Some cases are caused by pathological conditions such as uterine tumors, ovarian cysts, endometriosis, and pelvic inflammatory disease (PID). However, other cases of dysmenorrhea are not related to any pathologies.

Abnormal uterine bleeding includes menstruation of excessive duration or excessive amount, too frequent menstruation, intermenstrual bleeding, and postmenopausal bleeding. These abnormalities may be caused by improper hormonal regulation, emotional factors, fibroid tumors of the uterus, or systemic diseases.

Premenstrual syndrome (PMS) refers to severe physical and emotional distress occurring late in the postovulatory phase of the menstrual cycle and sometimes overlapping with menstruation. Symptoms usually increase in severity until the onset of menstruation and then dramatically disappear. Among the symptoms are edema, weight gain, breast swelling and tenderness, abdominal distention, backache, joint pain, constipation, skin eruptions, fatigue and lethargy, greater need for sleep, depression or anxiety, irritability, mood swings, headache, poor coordination and clumsiness, and cravings for sweet or salty foods. The basic cause of PMS is unknown.

Toxic Shock Syndrome (TSS)

Toxic shock syndrome (TSS) is primarily a disease of previously healthy young, menstruating females who use tampons, but it is also recognized in males, children, and nonmenstruating females. Clinically, TSS is characterized by high fever up to 40.6°C (105°F), sore throat or very tender mouth, headache, fatigue, irritability, muscle soreness and tenderness, conjunctivitis, diarrhea and vomiting, abdominal pain, vaginal irritation, and rash. Other symptoms include lethargy, unresponsiveness, memory loss, hypotension, peripheral vasoconstriction, respiratory distress syndrome, intravascular coagulation, decreased platelet count, renal failure, circulatory shock, and liver involvement.

Toxin-producing strains of the bacterium *Staphylococcus aureus* are necessary for development of the disease. The risk is greatest in females who use highly absorbent tampons. These tampons absorb magnesium that is normally present in the vagina. When the amount of magnesium in the vagina is reduced, *S. aureus* produces large amounts of toxin that cause the disease. TSS can also occur as a complication of influenza and influenza-like illness.

Ovarian Cysts

Ovarian cysts are fluid-containing sacs within the ovary. Follicular cysts have thin walls and contain a serous albuminous material. Cysts may also arise from the corpus luteum or the endometrium.

Endometriosis

Endometriosis (en′-dō-mē-trē-Ō-sis; *endo* = within; *metri* = uterus; *osis* = condition) is a benign condition characterized by the growth of endometrial tissue outside the uterus. The tissue enters the pelvic cavity via the open uterine (Fallopian) tubes and may be found on the ovaries, surface of the uterus, sigmoid colon, pelvic and abdominal lymph nodes, cervix, abdominal wall, kidneys, and urinary bladder. Symptoms include premenstrual or unusual menstrual pain. The unusual pain is caused by the displaced tissue sloughing off at the same time the normal uterine endometrium is being shed during menstruation. Infertility can be a consequence. Endometriosis disappears at menopause or when the ovaries are removed.

Female Infertility

Female infertility, or the inability to conceive, occurs in about 10 percent of married females in the United States. Once it is established that ovulation occurs regularly, the reproductive tract is examined for functional and anatomical disorders to determine the possibility of union of the sperm and the ovum in the uterine tube. Female infertility may be caused by obstruction in the uterine (Fallopian) tubes, ovarian disease, and certain conditions of the uterus. An upset in hormone balance, so that the endometrium is not adequately prepared to receive the fertilized ovum, may also be the problem.

Disorders Involving the Breasts

The breasts of females are highly susceptible to cysts and tumors. In females, *fibrocystic disease* is the most common cause of a breast lump in which one or more cysts (fluid-filled sacs) and thickening of alveoli (clusters of milk-secreting cells) develop. The condition occurs mainly in females between the ages of 30 and 50 and is probably due to a hormonal imbalance; a relative excess of estrogens or deficiency of progesterone in the postovulatory (luteal) phase of the reproductive cycle may be responsible. Fibrocystic disease usually causes one or both breasts to become lumpy, swollen, and tender about a week or so before menstruation begins.

Benign *fibroadenoma* is a common tumor of the breast. It occurs most often in young women. Fibroadenomas have a firm rubbery consistency and are easily moved about within the mammary tissue. The usual treatment is excision of the growth. The breast itself is not removed.

One in nine American women faces the prospect of *breast cancer.* After lung cancer, it is the second-leading cause of death from cancer in U.S. women but seldom occurs in men. In females, breast cancer is rarely seen before age 30, and its occurrence rises rapidly after menopause. An estimated 5 percent of the 180,000 cases diagnosed each year in the United States, particularly those that arise in younger women, stem from inherited genetic mutations (changes in the DNA). Two genes that increase susceptibility to breast cancer now have been identified. Breast cancer is generally not painful until it becomes quite advanced, so often it is not discovered early or, if noted, is ignored. Any lump, no matter how small, should be reported to a physician at once. Early detection—by breast self-examination and mammograms—is the best way to increase the chance of survival.

It is estimated that 95 percent of breast cancers are first detected by women themselves. Each month after the menstrual period the breasts should be thoroughly examined for lumps, puckering of the skin, or discharge.

The most effective technique for detecting tumors less than 1 cm (about $\frac{1}{2}$ in.) in diameter is *mammography* (mam-OG-ra-fē; *graphein* = to record). It is a type of radiography using very sensitive x-ray film. The image of the breast, called a *mammogram,* is obtained by placing the breasts, one at a time, on a flat surface and using a flat plate to compress the breast for better imaging. A supplementary procedure for evaluating breast abnormalities is *ultrasound.* Although ultrasound cannot detect tumors less than 1 cm in diameter, it can be used to determine whether a lump is a benign, fluid-filled cyst or a solid and therefore possibly malignant tumor.

Among the factors that increase the risk of breast cancer development are (1) a family history of breast cancer, especially in a mother or sister; (2) never having a child or having a first child after age 34; (3) previous cancer in one breast; (4) exposure to ionizing radiation, such as x-rays; (5) excessive fat and alcohol intake; and (6) cigarette smoking. Recent studies in the United States show that modern, low-dose birth control pills do not increase a woman's risk of developing breast cancer.

The American Cancer Society recommends the following steps to help diagnose breast cancer as early as possible:

1. A mammogram should be taken between the ages of 35 and 39, to be used later for comparison (baseline mammogram).

2. A physician should examine the breasts every three years when a female is between the ages of 20 and 40, and every year after 40.

3. Females with no symptoms should have a mammogram every year or two between ages 40 and 49, and every year after 50.

4. Females of any age with a history of breast cancer, a strong family history of the disease, or other risk factors should consult a physician to determine a schedule for mammography.

5. All females over 20 should develop the habit of monthly breast self-examination.

Treatment for breast cancer may involve hormone therapy, chemotherapy, radiation therapy, *lumpectomy* (removal of just the tumor and immediate surrounding tissue), a modified or radical mastectomy, or a combination of these. A *radical mastectomy* (*mastos* = Greek for breast; *ektome* = excision) involves removal of the affected breast along with the underlying pectoral muscles and the axillary lymph nodes. Lymph nodes are removed because metastasis of cancerous cells is usually through lymphatic or blood vessels. Radiation treatment and chemotherapy may follow the surgery to ensure the destruction of any stray cancer cells. By using artificial implants, skin, fat, and muscles from other parts of the body, the breast can be reconstructed after a radical mastectomy. Using these techniques, it is possible to reconstruct a natural-looking breast.

Cervical Cancer

Another common disorder of the female reproductive tract is *cervical cancer,* carcinoma of the cervix of the uterus. The condition starts with *cervical dysplasia* (dis-PLĀ-sē-a), a change in the shape, growth, and number of the cervical cells. If the condition is minimal, the cells may regress to normal. If it is severe, it may progress to cancer. Cervical cancer may be detected in most cases in its earliest stages by a Pap smear. There is some evidence linking cervical cancer to penile virus (papillomavirus) infections of male sexual partners. Depending on the progress of the disease, treatment may consist of excision of lesions, radiotherapy, chemotherapy, and hysterectomy (removal of the uterus).

Ovarian Cancer

Ovarian cancer ranks fifth among all malignancies of females and usually occurs in females in their 50s. Typically, the disease is not detected in an early stage because there are no symptoms. Also there are no effective methods for prevention or screening. In the majority of patients, diagnosis is not made until the disease has spread within or beyond the pelvis. For patients with the most common type of ovarian cancer, long-term survival is observed in only about one-third of all patients.

Pelvic Inflammatory Disease (PID)

Pelvic inflammatory disease (PID) is a collective term for any extensive bacterial infection of the pelvic organs, especially the uterus, uterine (Fallopian) tubes, or ovaries. PID is most commonly caused by the bacterium that causes gonorrhea, but any bacterium can trigger infection. Often the early symptoms of PID, which include increased vaginal discharge and pelvic pain, occur just after menstruation. As infection spreads, fever may develop in advanced cases along with painful abscesses of the reproductive organs. Early treatment with antibiotics (tetracycline or penicillin) can stop the spread of PID. ■

Medical Terminology and Conditions

Colposcopy (kol-POS-kō-pē) A procedure used to directly examine the vaginal and cervical mucosa with a low-power binocular microscope called a colposcope, which magnifies the mucous membrane from about 6 to 40 times its actual size. This is often the first test done after an abnormal Pap smear.

Culdoscopy (kul-DOS-kō-pē; *skopein* = to examine) A procedure in which a culdoscope (endoscope) is used to view the pelvic cavity. The approach is through the posterior wall of the vagina.

Endocervical curettage (ku-re-TAZH; *curette* = scraper) The cervix is dilated and the endometrium of the uterus is scraped with a spoon-shaped instrument called a curette. This procedure is commonly called a *D and C* (*dilation and curettage*).

Hermaphroditism (her-MAF-rō-di-tizm′) Presence of both male and female sex organs in one individual.

Hypospadias (hī′-pō-SPĀ-dē-as; *hypo* = below; *span* = to draw) A displaced urethral opening. In the male, the opening may be on the underside of the penis, at the penoscrotal junction, between the scrotal folds, or in the perineum. In the female, the urethra opens into the vagina.

Leukorrhea (loo′-kō-RĒ-a; *leuco* = white; *rrhea* = discharge) A nonbloody vaginal discharge that may occur at any age and affects most women at some time.

Smegma (SMEG-ma; *smegma* = soap) A secretion, consisting principally of sloughed off epithelial cells, found chiefly about the external genitalia and especially under the foreskin of the male.

Vaginitis (vaj′-i′NĪ-tis) Inflammation of the vagina.

Study Outline

Male Reproductive System (p. 523)

1. Reproduction is the process by which new individuals of a species are produced and the genetic material is passed on from one generation to the next.
2. The organs of reproduction are grouped as gonads (produce gametes), ducts (transport and store gametes), accessory sex glands (produce materials that support gametes), and supporting structures.
3. The male structures of reproduction include the testes, ductus epididymis, ductus (vas) deferens, ejaculatory duct, urethra, seminal vesicles, prostate gland, bulbourethral (Cowper's) glands, and penis.

Scrotum (p. 523)

1. The scrotum supports the testes.
2. It regulates the temperature of the testes by elevating them closer to the pelvic cavity.

Testes (p. 523)

1. The testes are oval-shaped glands (gonads) in the scrotum containing seminiferous tubules, in which sperm cells are made; sustentacular (Sertoli) cells, which nourish sperm cells and produce inhibin; and interstitial endocrinocytes (cells of Leydig), which produce the male sex hormone testosterone.
2. Secondary oocytes and sperm are collectively called gametes, or sex cells, and are produced in gonads.
3. Somatic cells divide by mitosis, the process in which each daughter cell receives the full complement of 23 chromosome pairs (46 chromosomes). Somatic cells are therefore diploid.
4. Immature gametes divide by meiosis in which the pairs of chromosomes are split so that the mature gamete has only 23 chromosomes. Sperm are therefore haploid.

5. Spermatogenesis occurs in the testes. It results in the formation of four haploid sperm cells.
6. Spermatogenesis consists of reduction division, equatorial division, and spermiogenesis.
7. Mature sperm consist of a head, midpiece, and tail. Their function is to fertilize a secondary oocyte.
8. At puberty, gonadotropin releasing hormone (GnRH) stimulates anterior pituitary gland secretion of LH and FSH. LH stimulates interstitial endocrinocytes to produce testosterone. FSH and testosterone initiate spermatogenesis.
9. Testosterone controls the growth, development, and maintenance of sex organs; stimulates bone growth, protein anabolism, and sperm maturation; and stimulates development of male secondary sex characteristics.
10. Inhibin is produced by sustentacular (Sertoli) cells. Its inhibition of FSH helps regulate the rate of spermatogenesis.

Male Puberty (p. 527)

1. Puberty refers to the period of time when secondary sex characteristics begin to develop and the potential for sexual reproduction is reached.
2. The onset of male puberty is signaled by increased levels of LH, FSH, and testosterone.

Ducts (p. 528)

1. The duct system of the testes includes the seminiferous tubules, straight tubules, and rete testis.
2. Sperm are transported out of the testes into an adjacent organ, the epididymis.
3. The ductus epididymis is the site of sperm maturation and storage.
4. The ductus (vas) deferens stores sperm and propels them toward the urethra during ejaculation.

5. Alteration of the ductus (vas) deferens to prevent fertilization is called vasectomy.

6. The ejaculatory ducts are formed by the union of the ducts from the seminal vesicles and ductus (vas) deferens and eject sperm into the urethra.

7. The male urethra passes through the prostate gland, urogenital diaphragm, and penis.

Accessory Sex Glands (p. 528)

1. The seminal vesicles secrete an alkaline, viscous fluid that constitutes most of the volume of semen and contributes to sperm viability. It also contains proteins that help semen coagulate after ejaculation.

2. The prostate gland secretes a slightly acidic fluid that is used by sperm for ATP production and liquifies coagulated semen.

3. The bulbourethral (Cowper's) glands secrete mucus for lubrication and an alkaline substance that neutralizes acid.

Semen (p. 529)

1. Semen is a mixture of sperm and accessory gland secretions.

2. It provides the fluid in which sperm are transported, provides nutrients, and neutralizes the acidity of the male urethra and female vagina.

Penis (p. 530)

1. The penis is the male organ of copulation.

2. Expansion of its blood sinuses under the influence of sexual excitation is called erection.

Female Reproductive System (p. 531)

1. The female organs of reproduction include the ovaries (gonads), uterine (Fallopian) tubes, uterus, vagina, and vulva.

2. The mammary glands are considered part of the reproductive system.

Ovaries (p. 532)

1. The ovaries are female gonads located in the upper pelvic cavity on either side of the uterus.

2. They produce secondary oocytes, discharge secondary oocytes (ovulation), and secrete estrogens, progesterone, inhibin, and relaxin.

3. Oogenesis occurs in the ovaries. It results in the formation of a single haploid ovum.

4. Oogenesis consists of reduction division and equatorial division.

Uterine (Fallopian) Tubes (p. 534)

1. The uterine (Fallopian) tubes transport secondary oocytes from the ovaries to the uterus and are the normal sites of fertilization.

2. Implantation outside the uterus is called an ectopic pregnancy.

Uterus (p. 535)

1. The uterus is an inverted, pear-shaped organ that functions in menstruation, implantation of a fertilized ovum, development of a fetus during pregnancy, and labor.

2. The innermost layer of the uterine wall is the endometrium, which undergoes marked changes during the menstrual cycle.

Vagina (p. 536)

1. The vagina is a passageway for the menstrual flow, the receptacle for the penis during sexual intercourse, and the lower portion of the birth canal.

2. It is capable of considerable stretching to accomplish its functions.

Vulva (p. 536)

1. The vulva is a collective term for the external genitals of the female.

2. It consists of the mons pubis, labia majora, labia minora, clitoris, vestibule, vaginal and urethral orifices, paraurethral (Skene's), and greater vestibular (Bartholin's) glands.

Perineum (p. 537)

1. The perineum is a diamond-shaped area at the inferior end of the trunk between the thighs and buttocks.

2. An incision in the perineal skin prior to delivery is called an episiotomy.

Mammary Glands (p. 537)

1. The mammary glands are modified sweat glands located over the pectoralis major muscles. Their function is to secrete and eject milk (lactation).

2. Mammary gland development depends on estrogens and progesterone.

3. Milk secretion is due mainly to the hormone prolactin (PRL).

Female Puberty (p. 538)

1. The onset of female puberty is signaled by increased levels of LH, FSH, and estrogens.

2. The hormones bring about development of the secondary sex characteristics.

Female Reproductive Cycle (p. 538)

1. The purpose of the menstrual cycle is to prepare the endometrium each month for the reception of a fertilized egg. The ovarian cycle is associated with the maturation of an ovum each month.

2. The menstrual and ovarian cycles are regulated by GnRH, which stimulates the release of FSH and LH.

3. FSH stimulates the initial secretion of estrogens by the ovaries. LH stimulates further development of ovarian follicles, ovulation, and the secretion of estrogens and progesterone by the ovaries.

4. Estrogens stimulate the growth, development, and maintenance of female reproductive structures; stimulate the development of secondary sex characteristics; regulate fluid and electrolyte balance; and stimulate protein anabolism.

5. Progesterone works with estrogens to prepare the endometrium for implantation and the mammary glands for milk secretion.

6. Inhibin inhibits secretion of FSH.

7. Relaxin relaxes the pubic symphysis and helps dilate the uterine cervix to facilitate delivery.

8. During the menstrual phase (days 1 to 5), the stratum functionalis of the endometrium is shed, discharging blood, tissue fluid, mucus, and epithelial cells.

9. During the preovulatory phase, a group of follicles in the ovaries begin to undergo final maturation. One follicle outgrows the others and becomes dominant while the others degenerate. At the same time endometrial repair occurs in the uterus. Estrogens are the dominant ovarian hormones during the preovulatory phase.

10. Ovulation is the rupture of the dominant mature (Graafian) follicle and the release of a secondary oocyte into the pelvic cavity. It is brought about by a surge of LH.

11. During the postovulatory phase, both progesterone and estrogens are secreted in large quantity by the corpus luteum of the ovary and the uterine endometrium thickens in readiness for implantation.

12. If fertilization and implantation do not occur, the corpus luteum degenerates, and the resulting low levels of estrogens and progesterone allow discharge of the endometrium followed by initiation of another uterine and ovarian cycle.

13. If fertilization and implantation do occur, the corpus luteum is maintained by placental hCG, and the corpus luteum and later the placenta secrete progesterone and estrogens to support pregnancy and breast development for lactation.

14. Climacteric is a period of infrequent menstrual cycles; menopause is the cessation of the reproductive cycles.

Self-Quiz

1. During development, sperm are nourished by

   a. fructose   b. mitochondria   c. interstitial endocrinocytes (cells of Leydig)   d. sustentacular (Sertoli) cells   e. the acrosome

2. Which of the following statements is true concerning meiosis?

   a. Meiosis is the process by which somatic cells divide.   b. The haploid chromosome number is symbolized by $2n$.   c. Gametes are formed after one reduction division.   d. Gametes contain the haploid chromosome number.   e. Gametes contain 46 chromosomes in their nuclei.

3. Which of the following is NOT true regarding the uterus?

   a. It is the site of menstruation.   b. It is the site of implantation of a fertilized ovum.   c. It is the site of ovulation.   d. It is the site of labor.   e. It is the site of development of the fetus.

4. Menses is triggered by a

   a. rapid rise in luteinizing hormone (LH)   b. rapid fall of luteinizing hormone (LH)   c. drop in estrogens and progesterone   d. rise in estrogens and progesterone   e. rise in inhibin

5. During the postovulatory phase, the hormone that is most dominant is

   a. progesterone   b. relaxin   c. luteinizing hormone (LH)   d. follicle-stimulating hormone (FSH)   e. inhibin

6. Place the following in the correct order for the passage of sperm from the testes to the outside of the body.

   5 1. ductus (vas) deferens
   3 2. rete testes
   1 3. seminiferous tubules
   4 4. epididymis
   6 5. urethra
   2 6. straight tubules

   a. 6, 3, 2, 4, 1, 5   b. 3, 2, 6, 4, 1, 5   c. 3, 6, 2, 4, 1, 5   d. 3, 6, 2, 4, 5, 1   e. 2, 4, 6, 1, 3, 5

7. All but which of the following are true concerning semen?

   a. The average volume is 2.5 to 5 ml.   b. The average sperm count is just below 20 million.   c. pH is about 7.40.   d. Coagulation occurs rapidly following ejaculation.   e. Semen contains an antibiotic.

8. Match the following:

   _B_ a. regulates ovarian cycle

   _F_ b. stimulates the initial secretion of estrogens by growing follicles

   _A_ c. stimulates ovulation to occur

   _E_ d. stimulates the growth, development, and maintenance of the female reproductive system

   _D_ e. works with estrogens to prepare the uterus for implantation of a fertilized ovum

   _C_ f. assists with labor by helping to dilate the cervix and relax the pubic symphysis

   A. luteinizing hormone (LH)

   B. gonadotropic releasing hormone (GnRH)

   C. relaxin

   D. progesterone

   E. estrogens

   F. follicle-stimulating hormone (FSH)

9. Sperm production occurs in the

   a. epididymis   b. seminiferous tubules   c. seminal vesicles   d. rete testis   e. ejaculatory duct

10. In the male, the gland that surrounds the urethra at the base of the bladder is the

    a. prostate gland   b. glans penis   c. seminal vesicle   d. bulbourethral gland   e. greater vestibular gland

11. Following ovulation, the secondary oocyte enters the

    a. cervix   b. fornix   c. infundibulum   d. fundus   e. vagina

12. The time of maturation of the follicles corresponds best to

    a. ovulation   b. menstrual phase   c. preovulatory phase   d. secretory phase   e. postovulatory phase

13. The cessation of menstrual cycles is termed

    a. menarche   b. menopause   c. menstruation   d. menses   e. postovulatory phase

14. Circumcision is the removal of the

    a. prepuce   b. hymen   c. perineum   d. glans penis   e. vas deferens

Critical Thinking Applications

1. Trace the route of the vas deferens from its origin to its junction with the ducts of the seminal vesicles, naming all the structures through which it passes. Where along this route would a vasectomy be performed?

2. Mittelschmerz is the name for abdominal pain that lasts for a few hours and that occurs at about the midpoint of a woman's menstrual cycle. Using your knowledge of the ovarian cycle, offer an explanation for the cause of the pain.

3. A woman with a history of pelvic inflammatory disease was admitted to the hospital with severe pain in the pelvic region. She had missed one period (menses) and a pregnancy test was positive. What could be the medical problem with this pregnancy?

4. Thirty-five-year-old Janelle has been advised to have a hysterectomy due to medical problems. She is worried that the procedure will cause menopause. Is this a valid concern? Explain what is involved in this procedure.

5. Imagine that you are a sperm cell preparing for your journey toward the ovulated secondary oocyte. List your requirements for the long journey through the female reproductive tract and the eventual fertilization of the oocyte. How does the sperm's structure and the components of the semen help meet these requirements?

6. Phil has promised his wife that he will get a vasectomy after the birth of their next child. However, he is concerned about possible negative effects on his virility. What would you tell Phil about the procedure to calm his concerns?

Answers to Figure Questions

23.1   Supporting structure.

23.2   Spermatogonia (stem cells) are least mature; sperm cells (spermatozoa) are most mature.

23.3   Because the number of chromosomes in each cell is reduced by half.

23.4   Head contains enzymes for penetration of secondary oocyte and DNA; midpiece contains mitochondria for ATP production; tail is a flagellum that provides propulsion.

23.5   Testosterone inhibits secretion of LH through the inhibition of GnRH and inhibin inhibits secretion of FSH.

23.6   The combined secretions from the accessory sex glands and testes.

23.7   Two corpora cavernosa penis and one corpus spongiosum penis contain blood sinuses that fill with blood. This blood cannot flow out of the penis as quickly as it flows in and the trapped blood stiffens the tissue.

23.8   Vulva or pudendum.

23.9   Broad, ovarian, and suspensory ligaments.

23.10 Ovarian follicles secrete estrogens and the corpus luteum secretes estrogens, progesterone, relaxin, and inhibin.

23.11 Primary oocytes are present in the ovary at birth, so they are as old as the woman is. In males, primary spermatocytes are continually being formed from spermatogonia and thus are only a few days old.

23.12 The stratum basalis provides cells to replace those shed (stratum functionalis) during each menstruation.

23.13 Anterior: mons pubis, clitoris, prepuce, and external urethral orifice; lateral: labia minora and labia majora.

23.14 Synthesis: prolactin, estrogens, progesterone; release: oxytocin.

23.15 Relaxin.

23.16 Estrogens.

23.17 Luteinizing hormone (LH).

23.18 Negative, because the response is opposite to the stimulus. Decreasing estrogens and progesterone stimulate release of GnRH, which, in turn, increases production and release of estrogens.

chapter 24

DEVELOPMENT AND INHERITANCE

a look ahead

Now that you have learned about the structure and function of the female and male reproductive systems, we will examine the physiology of sexual intercourse, the principal events of pregnancy and lactation, methods of birth control, and the basic principles of inheritance.

Sexual Intercourse

objective: *Describe the roles of the male and female in sexual intercourse.*

Sexual intercourse, or *copulation* (in humans, called *coitus;* KŌ-itus), is the process by which sperm are deposited in the vagina.

Male Sexual Act

Erection

The male role in the sexual act starts with *erection,* the enlargement and stiffening of the penis. An erection may be initiated in the cerebrum by stimuli such as anticipation, memory, and visual sensation, or it may be a reflex brought on by stimulation of the touch receptors in the penis, especially in the glans. In either case, parasympathetic impulses from the sacral portion of the spinal cord to the penis cause dilation of the arteries to the penis, filling spaces in the corpora cavernosa and corpus spongiosum with blood, resulting in erection.

Lubrication

Parasympathetic impulses from the sacral portion of the spinal cord also cause the bulbourethral (Cowper's) glands to secrete mucus through the urethra, which affords some *lubrication* for intercourse. However, fluids lubricating the walls of the vagina and a small amount of cervical mucus provide most of this lubrication.

Orgasm (*Climax*)

Tactile stimulation of the penis brings about emission and ejaculation. When sexual stimulation becomes intense, rhythmic sympathetic impulses leave the lumbar spinal cord and stimulate peristaltic contractions of the ducts of the testes, epididymides, and ductus (vas) deferens that propel sperm into the urethra—a process called *emission.* Simultaneously, peristaltic contractions of the seminal vesicles and prostate gland expel seminal and prostatic fluid along with the sperm. All these mix with the mucus of the bulbourethral glands to produce semen. Other rhythmic parasympathetic impulses sent from the sacral spinal cord stimulate skeletal muscles at the base of the penis, which then expels the semen. The propulsion of semen from the urethra to the exterior constitutes an *ejaculation.*

A number of sensory and motor responses accompany ejaculation, including rapid heart rate, increased blood pressure, increased respiration, and pleasurable sensations. These responses, together with the muscular events involved in ejaculation, constitute an *orgasm (climax).*

Female Sexual Act

Erection

The female role in the sex act involves erection, lubrication, and orgasm. Stimulation of the female depends on both psychic and tactile responses. Stimulation of the female genitalia results in clitoral erection and widespread sexual arousal. This response is controlled by parasympathetic impulses from the sacral spinal cord.

Lubrication

Parasympathetic impulses from the sacral spinal cord stimulate *lubrication* of the walls of the vagina. The cervical mucosa and greater vestibular (Bartholin's) glands produce a small amount of mucus.

Orgasm (*Climax*)

When genital stimulation reaches maximum intensity, it initiates reflexes that produce an *orgasm (climax).* Female orgasm causes increased respirations, pulse, and blood pressure and muscular contractions of the reproductive organs that may aid fertilization of an ovum (to be described shortly). However, unlike males who experience a single climax with ejaculation, females may have multiple orgasms of varying intensity with accompanying pleasurable sensations, but no ejaculation.

About 15 minutes after semen is deposited in the vagina, it liquefies to allow sperm to swim into the cervix. Prostaglandins in semen and oxytocin (OT) released from the posterior pituitary gland during female orgasm may stimulate uterine contractions to further aid the movement of sperm toward the uterine (Fallopian) tubes.

Development During Pregnancy

Once sperm and a secondary oocyte are developed through meiosis and maturation and the sperm are deposited in the vagina, pregnancy can occur. *Pregnancy* is the sequence of events that normally includes fertilization, implantation, and embryonic and fetal growth that terminates in birth about 38 weeks later. We will now examine these events in detail.

Fertilization

objective: *Explain how a secondary oocyte is fertilized and implanted in the uterus.*

The term *fertilization* (fer-til-i-ZĀ-shun; *fertilis* = fruitful) refers to the union of the genetic material from a sperm cell and secondary oocyte into a single nucleus. Of the 300 to 500 million sperm cells introduced into the vagina, less than 1 percent reach the secondary oocyte. Fertilization normally occurs in

Note:
Sperm to be
need in Vagina
usually no less
than 10 hours
to fertilize

the upper portion of the uterine (Fallopian) tube about 12 to 24 hours after ovulation.

In order for sperm to fertilize a secondary oocyte, they must shed the protective covering of the acrosome during their ascent through the female reproductive tract, in a process called *capacitation* (ka′-pas′-i-TĀ-shun). This refers to the functional changes that sperm undergo in the female reproductive tract to allow fertilization to occur. This permits enzymes from the acrosome to penetrate the *corona radiata,* several layers of cells around the oocyte, and *zona pellucida* (pe-LOO-si-da), a clear glycoprotein layer under the corona radiata (Figure 24.1). The enzymes from many sperm are needed before one sperm cell's head can finally break through. Once this occurs, the zona pellucida undergoes chemical changes that block the entry of other sperm.

When a sperm cell penetrates a secondary oocyte, it triggers the completion of equatorial division (meiosis II) and oogenesis. This produces a second polar body (which disintegrates) and a larger ovum with 23 chromosomes. The nuclear membranes of the ovum and sperm cell disintegrate: the haploid chromosomes of each cell then combine to form a diploid cell called the fertilized ovum or *zygote* (ZĪ-gōt; *zygosis* = a joining).

Fraternal twins are produced from the independent release of two secondary oocytes and the subsequent fertilization of each by different sperm cells. They are the same age and are in the uterus at the same time, but they are genetically as dissimilar as any other siblings. They may or may not be the same sex. *Identical twins* develop from a single fertilized ovum that splits at an early stage in development. They contain exactly the same genetic material and are always the same sex.

Formation of the Morula

Approximately 24 to 30 hours after fertilization, rapid cell division of the zygote takes place. This early division of the zygote is called *cleavage.* Although cleavage increases the number of cells, it does not result in an increase in the size of the zygote. Successive cleavages produce a solid mass of tiny cells, the *morula* (MOR-yoo-la) or mulberry, within 3 to 4 days after fertilization.

Development of the Blastocyst

As the number of cells in the morula increases, it continues to move down through the uterine (Fallopian) tube and enters the uterine cavity. By this time, which is 4½ to 5 days after fertilization, the morula forms a hollow ball of cells, now referred to as a *blastocyst* (Figure 24.2).

The blastocyst consists of an outer layer, the *trophoblast* (TRŌF-ō-blast; *troph* = nourish), which surrounds a fluid-filled cavity (*blastocele*), and an *inner cell mass,* which develops into the *embryo* by the end of the second week. The trophoblast and parts of the inner cell mass will form the chorion and placenta, which protect and nourish the growing organism (described shortly).

Implantation

During the next stages of development, the zona pellucida disintegrates and the blastocyst is nourished by "uterine milk," secretions from endometrial (uterine) glands. The attachment

---

Figure 24.1   Fertilization. Sperm cell penetrating the corona radiata and zona pellucida around a secondary oocyte.

🔑 *During fertilization, genetic material from a sperm cell merges with that of a secondary oocyte to form a single nucleus.*

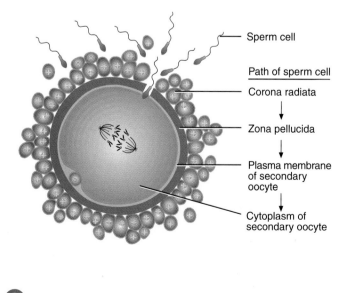

Sperm cell

Path of sperm cell
Corona radiata

Zona pellucida

Plasma membrane of secondary oocyte

Cytoplasm of secondary oocyte

❓ *What is capacitation?*

Figure 24.2   Blastocyst.

🔑 *Cleavage refers to the early, rapid mitotic divisions in a zygote.*

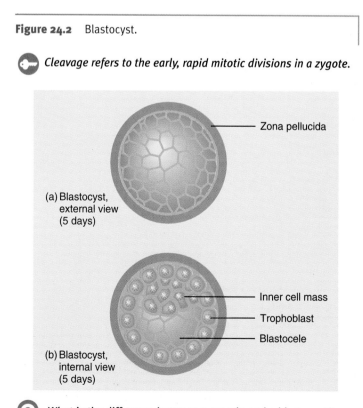

Zona pellucida

(a) Blastocyst, external view (5 days)

Inner cell mass
Trophoblast
Blastocele

(b) Blastocyst, internal view (5 days)

❓ *What is the difference between a morula and a blastocyst?*

of the blastocyst to the endometrium about 6 days after fertilization is called *implantation* (Figure 24.3). In the process, the trophoblast secretes enzymes that digest the uterine lining and allow the blastocyst to burrow into the endometrium during the second week after fertilization. The placenta will develop between the inner cell mass and the endometrial wall to provide nutrients for the growth of the embryo (more detail on this shortly). By now, the trophoblast has begun to secrete *human chorionic gonadotropin* (*hCG*), a hormone that maintains the corpus luteum and is the basis of positive pregnancy tests. The hormone may cause the woman to feel nauseated (often called *morning sickness*). Blood levels of hCG increase to a maximum during the ninth week of pregnancy, then the level decreases.

A summary of the principal events associated with fertilization and implantation is presented in Figure 24.4.

Figure 24.3   Implantation. Shown is the blastocyst in relation to the endometrium of the uterus at various time intervals after fertilization.

🔑 *Implantation refers to the attachment of a blastocyst to the endometrium, which occurs about 6 days after fertilization.*

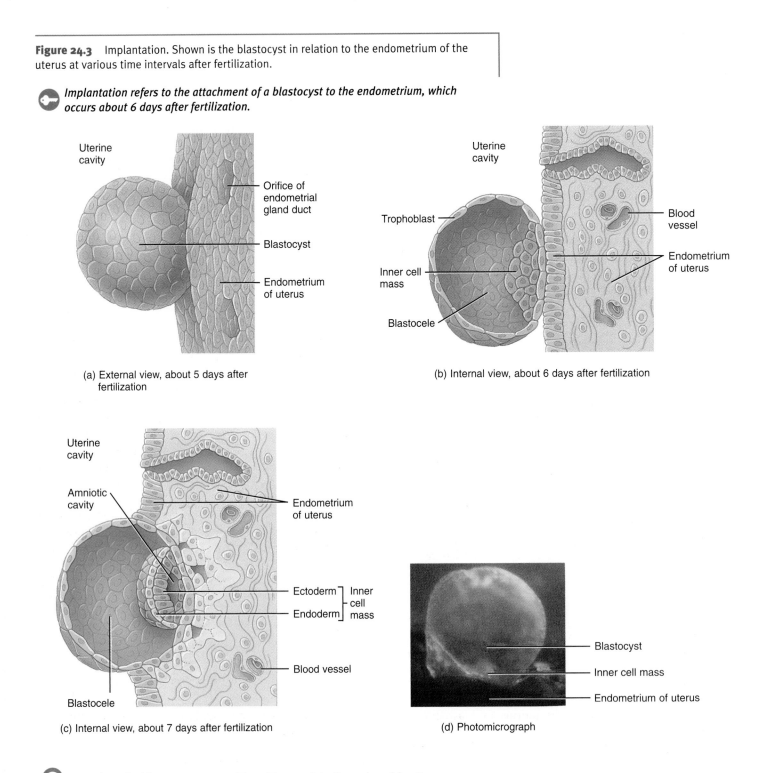

(a) External view, about 5 days after fertilization

(b) Internal view, about 6 days after fertilization

(c) Internal view, about 7 days after fertilization

(d) Photomicrograph

❓ *How does the blastocyst merge with and burrow into the endometrium?*

Figure 24.4   Summary of events associated with fertilization and implantation.

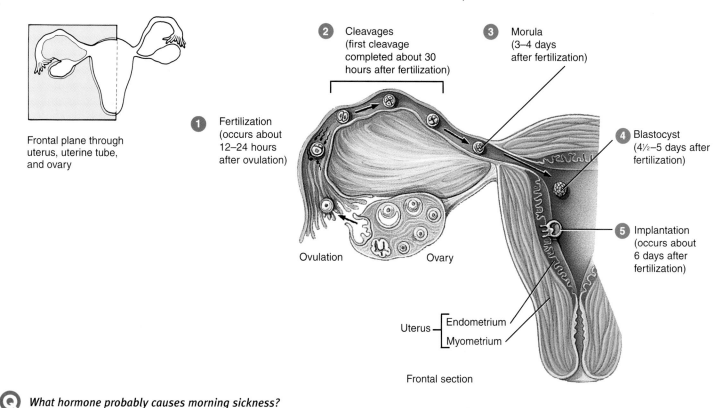

Frontal plane through
uterus, uterine tube,
and ovary

1 Fertilization
(occurs about
12–24 hours
after ovulation)

2 Cleavages
(first cleavage
completed about 30
hours after fertilization)

3 Morula
(3–4 days
after fertilization)

4 Blastocyst
(4½–5 days after
fertilization)

5 Implantation
(occurs about
6 days after
fertilization)

Ovulation   Ovary

Uterus — Endometrium
         — Myometrium

Frontal section

Q   *What hormone probably causes morning sickness?*

In Vitro Fertilization (IVF)

objective: *Discuss alternative procedures to natural fertilization and implantation.*

In vitro fertilization (**IVF**) literally means fertilization in a laboratory dish. In the IVF procedure, the female is given follicle-stimulating hormone (FSH) soon after menstruation, so that several secondary oocytes, rather than the typical single one, will be produced (superovulation). Administration of luteinizing hormone (LH) may also ensure the maturation of the secondary oocytes. Next, a small incision is made near the navel, and the secondary oocytes are removed from the follicles with a laparoscope fitted with a suction device and then transferred to a solution of the male's sperm. Once fertilization has taken place, the fertilized ovum is put into another medium and observed for cleavage. When the fertilized ovum reaches the 8-cell or 16-cell stage, it is introduced into the uterus for implantation and subsequent growth. It is also possible to freeze embryos to permit parents a successive pregnancy several years later or allow a second attempt at implantation if the first attempt is unsuccessful.

Embryo transfer is a procedure in which a husband's semen is used to artificially inseminate a fertile secondary oocyte donor. After fertilization, the morula or blastocyst is transferred from the donor to the infertile wife who carries it to term. Embryo transfer is indicated for females who are infertile or who do not want to pass on their own genes because they are carriers of a serious genetic disorder. In the embryo

transfer procedure, the donor is monitored to determine the time of ovulation by checking her blood levels of luteinizing hormone (LH) and by ultrasound. The wife is also monitored to make sure that her ovarian cycle is synchronized with that of the donor. Once ovulation occurs in the donor, she is artificially inseminated with the husband's (or another male's) semen. Four days later, a morula or blastocyst is flushed from the donor's uterus through a soft plastic catheter and transferred to the wife's uterus, where it grows and develops until the time of birth. Embryo transfer is an office procedure that requires no anesthesia.

In *gamete intrafallopian transfer* (**GIFT**) the goal is to mimic the normal process of conception by uniting sperm and a secondary oocyte in the prospective mother's uterine tubes. In the procedure, the female is given FSH and LH to stimulate the production of several secondary oocytes. The secondary oocytes are removed with a laparoscope, mixed with a solution of the male's sperm outside the body, and then immediately inserted into the uterine (Fallopian) tubes.

Read this

Embryonic Development

objective: *Describe the principal events associated with embryonic development.*

The first two months of development are considered the *embryonic period.* During this time, the developing human is called an *embryo* (*bryein* = grow). The study of development from the fertilized egg through the eighth week in utero is termed *embryology* (embrē-OL-ō-jē). The months of development after the second month are considered the *fetal period,* and during this time the developing human is called a *fetus* (*feo* = to bring forth). By the end of the embryonic period, the beginnings of all principal adult organs are present, the embryonic membranes are developed, and by the end of the third month, the placenta is functioning.

Beginnings of Organ Systems

After implantation, the inner cell mass of the blastocyst begins to differentiate into the three *primary germ layers*: ectoderm, endoderm, and mesoderm. These are the embryonic tissues from which all tissues and organs of the body will develop.

In the human, the germ layers form so quickly that it is difficult to determine the exact sequence of events. Within eight days after fertilization, a layer of cells of the inner cell mass forms the *ectoderm* and separates to form a fluid-filled space, the *amniotic cavity* (Figure 24.5a). The layer of the inner cell mass near the blastocele develops into the *endoderm,* which also separates to form an amniotic fluid-filled space, the *yolk sac* (Figure 24.5a).

About the 12th day after fertilization, the cells of the inner cell mass are called the *embryonic disc.* They will form the embryo. At this stage, the embryonic disc contains ectodermal and endodermal cells. About the 14th day, the cells of the embryonic disc differentiate into three distinct layers: the upper ectoderm, the middle *mesoderm,* and the lower endoderm (Figure 24.5b). The ectoderm, mesoderm, and endoderm are attached to the trophoblast (now called the *chorion*) by a structure called the *body stalk* (future umbilical cord). See Figure 24.5c.

The endoderm becomes the epithelial lining of the gastrointestinal tract, respiratory tract, and a number of other organs. The mesoderm forms the peritoneum, muscle, bone, and other connective tissue. The ectoderm develops into the skin and nervous system. See Exhibit 24.1 on page 558 for more detail.

Embryonic Membranes

A second major event that occurs during the embryonic period is the formation of the *embryonic membranes.* These membranes lie outside the embryo and protect and nourish the embryo and, later, the fetus. The membranes are the yolk sac, amnion, chorion, and allantois.

In many species such as birds whose young develop inside a shelled egg, the *yolk sac* is a membrane that is the primary source of nourishment for the embryo (Figure 24.5c). However, a human embryo receives nutrients from the endometrium. The yolk sac remains small and functions as an early site of blood formation.

The *amnion* is a thin, protective membrane that forms by the eighth day after fertilization and initially overlies the embryonic disc (Figure 24.5). As the embryo grows, the amnion entirely surrounds the embryo, creating a cavity that becomes filled with *amniotic fluid* (Figure 24.6 on page 558). Most amniotic fluid is initially produced from maternal blood. Later, the fetus makes daily contributions to the fluid by excreting urine into the amniotic cavity. Amniotic fluid serves as a shock absorber for the fetus, helps regulate fetal body temperature, and prevents attachments between the skin of the fetus and surrounding tissues. Embryonic cells are sloughed off into amniotic fluid; they can be examined in the procedure called *amniocentesis* (am'-nē-ō-sen-TĒ-sis), which is described on p. 563. The amnion usually ruptures just before birth and with its fluid constitutes the "bag of waters."

Figure 24.5 Formation of the primary germ layers and associated structures.

The primary germ layers (ectoderm, mesoderm, and endoderm) are the embryonic tissues from which all tissues and organs develop.

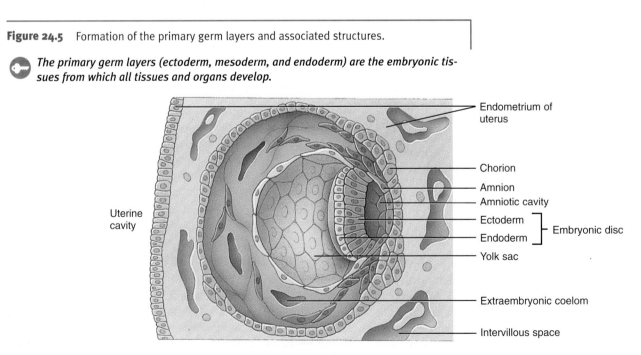

(a) Internal view, about 12 days after fertilization

Figure 24.5 (*Continued*)

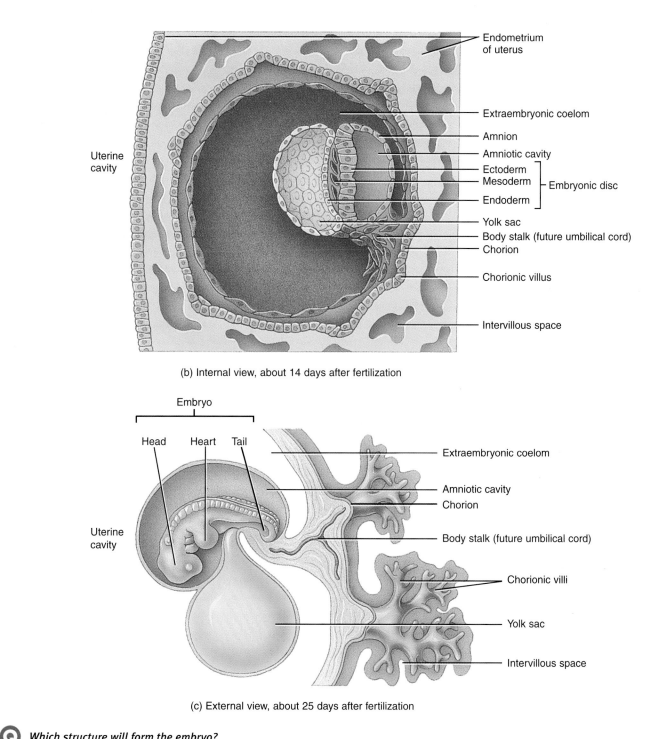

(b) Internal view, about 14 days after fertilization

(c) External view, about 25 days after fertilization

Q *Which structure will form the embryo?*

The *chorion* (KŌR-ē-on) is derived from the trophoblast and surrounds the embryo and, later, the fetus. The chorion becomes the principal embryonic part of the placenta, the structure for exchange of materials between mother and fetus. It also produces human chorionic gonadotropin (hCG).

The *allantois* (a-LAN-tō-is; *allas* = sausage) is a small projection of the yolk sac that contains blood vessels. It serves as an early site of blood formation. Later its blood vessels serve as the umbilical connection in the placenta between mother and fetus. This connection is the umbilical cord.

Exhibit 24.1 | Structures Developed from the Three Primary Germ Layers

ENDODERM	MESODERM	ECTODERM
Epithelial lining of gastrointestinal tract (except the oral cavity and anal canal) and the epithelium of its glands	All skeletal, most smooth, and all cardiac muscle	All nervous tissue
Epithelial lining of urinary bladder, gallbladder, and liver	Cartilage, bone, and other connective tissues	Epidermis of skin
	Blood, bone marrow, and lymphatic tissue	Hair follicles, arrector pili muscles, nails, and epithelium of skin glands (sebaceous and sudoriferous)
Epithelial lining of pharynx, auditory (Eustachian) tubes, tonsils, larynx, trachea, bronchi, and lungs	Endothelium of blood vessels and lymphatic vessels	Lens, cornea, and internal eye muscles
Epithelium of thyroid, parathyroid, pancreas, and thymus glands	Dermis of skin	Internal and external ear
	Fibrous tunic and vascular tunic of eye	Neuroepithelium of sense organs
Epithelial lining of prostrate and bulbourethral (Cowper's) glands, vagina, vestibule, urethra, and associated glands such as the greater (Bartholin's) vestibular and lesser vestibular glands	Middle ear	Epithelium of oral cavity, nasal cavity, paranasal sinuses, salivary glands, and anal canal
	Mesothelium of ventral body cavity	Epithelium of pineal gland, pituitary gland, and adrenal medulla
	Epithelium of kidneys and ureters	
	Epithelium of adrenal cortex	
	Epithelium of gonads and genital ducts	

Figure 24.6 Embryonic membranes.

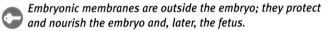

🔑 *Embryonic membranes are outside the embryo; they protect and nourish the embryo and, later, the fetus.*

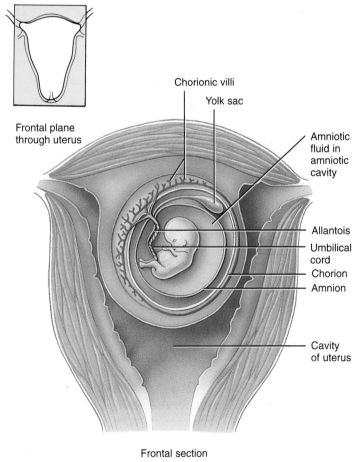

Frontal plane through uterus

Chorionic villi
Yolk sac
Amniotic fluid in amniotic cavity
Allantois
Umbilical cord
Chorion
Amnion
Cavity of uterus

Frontal section

❓ *How do the amnion and chorion differ in function?*

Placenta and Umbilical Cord

The *placenta* (pla-SEN-ta) is fully developed by the third month of pregnancy. It is shaped like a flat cake and is formed by the inner portion of the chorion of the embryo and a portion of the mother's endometrium (Figure 24.7). It allows the fetus and mother to exchange nutrients and wastes and secretes hormones necessary to maintain pregnancy.

During embryonic life, fingerlike projections of the chorion, called *chorionic villi* (kō′-rē-ON-ik VIL-ē), grow into the endometrium of the uterus (Figure 24.7). They contain fetal blood vessels of the body stalk. They are bathed in maternal blood sinuses called *intervillous* (in-ter-VIL-us) *spaces.* Thus maternal and fetal blood vessels are brought close together, but maternal and fetal blood do not normally mix. Oxygen and nutrients from the mother's blood diffuse into capillaries of the villi. From the capillaries the nutrients circulate into the *umbilical vein.* Wastes leave the fetus through the *umbilical arteries,* pass into the capillaries of the villi, and diffuse into the maternal blood. The *umbilical* (um-BIL-i-kul) *cord* consists of the umbilical arteries, umbilical vein, and supporting connective tissue called mucous connective tissue (Figure 24.7).

After the birth of the baby, the placenta detaches from the uterus and is termed the *afterbirth.* At this time, the umbilical cord is severed, leaving the baby on its own. The small portion (about an inch) of the cord that remains still attached to the infant begins to wither and falls off, usually within 12 to 15 days after birth. The area where the cord was attached becomes covered by a thin layer of skin and scar tissue forms. The scar is the *umbilicus (navel).*

Fetal Circulation

The cardiovascular system of a fetus, called *fetal circulation,* differs from an adult's because the lungs, kidneys, and gas-

Figure 24.7   Placenta and umbilical cord.

The placenta is formed by the chorion of the embryo and part of the endometrium of the mother.

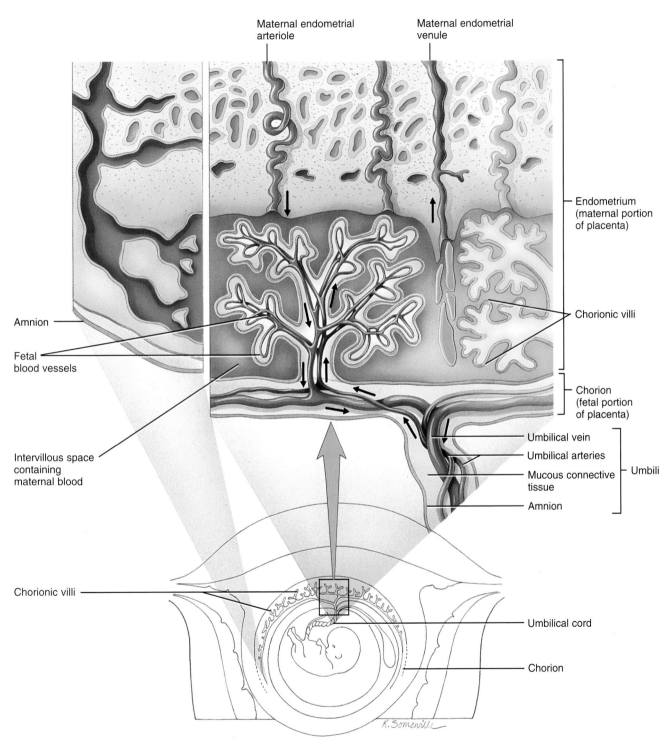

Overall structure

Q   *What is the function of the placenta?*

trointestinal tract of a fetus are not functioning. The fetus derives its oxygen and nutrients from the maternal blood and eliminates its carbon dioxide and wastes into maternal blood.

The exchange of materials between fetal and maternal circulation occurs through the placenta. Blood passes from the fetus to the placenta via two *umbilical arteries* in the umbilical cord

(Figure 24.8a). At the placenta, the blood picks up oxygen and nutrients and eliminates carbon dioxide and wastes. The oxygenated blood returns from the placenta via a single *umbilical vein*, which ascends to the liver of the fetus, where it divides into two branches. Some blood flows through the branch that joins the hepatic portal vein and enters the liver. The fetal liver manufac-

Figure 24.8 Fetal circulation and changes at birth. The boxed areas indicate the fate of certain fetal structures once postnatal circulation is established.

🔑 *The lungs, kidneys, and gastrointestinal organs begin to function at birth.*

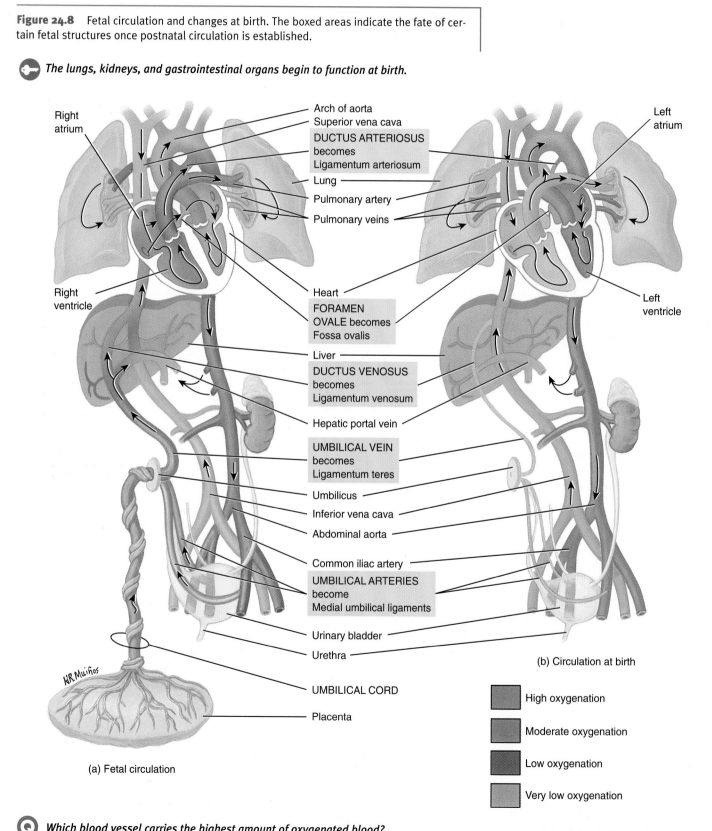

(a) Fetal circulation

(b) Circulation at birth

High oxygenation

Moderate oxygenation

Low oxygenation

Very low oxygenation

🔍 *Which blood vessel carries the highest amount of oxygenated blood?*

tures red blood cells but does not function in digestion. Therefore most of the blood flows into the second branch, the *ductus venosus* (DUK-tus ve-NŌ-sus). The ductus venosus eventually passes its blood to the inferior vena cava, bypassing the liver.

In general, circulation through other portions of the fetus is similar to postnatal circulation. Deoxygenated blood returning from the lower regions is mingled with oxygenated blood from the ductus venosus in the inferior vena cava. This mixed blood then enters the right atrium. The circulation of blood through the upper portion of the fetus is also similar to postnatal flow. Deoxygenated blood returning from the upper regions of the fetus is collected by the superior vena cava, and it also passes into the right atrium.

Most of the blood does not pass through the right ventricle to the lungs, as it does in postnatal circulation, because the fetal lungs do not operate. In the fetus, an opening called the *foramen ovale* (fō-RĀ-men ō-VAL-ē) exists in the septum between the right and left atria. A valve in the inferior vena cava directs about a third of the blood through the foramen ovale so that it may be sent directly into systemic circulation. The blood that does descend into the right ventricle is pumped into the pulmonary trunk, but little of this blood actually reaches the lungs. Most is sent through the *ductus arteriosus* (ar-tē-rē-Ō-sus), a small vessel connecting the pulmonary trunk with the aorta that allows most blood to bypass the fetal lungs. The blood in the aorta is carried to all parts of the fetus through its systemic branches. When the common iliac arteries branch into the external and internal iliacs, part of the blood flows into the internal iliacs. It then goes to the umbilical arteries and back to the placenta for another exchange of materials. The only fetal vessel that carries fully oxygenated blood is the umbilical vein.

At birth, when lung, renal, digestive, and liver functions are established, the special structures of fetal circulation are no longer needed. For example, the placenta is delivered by the mother as the "*afterbirth*"; the foramen ovale normally closes shortly after birth to become the *fossa ovalis,* a depression in the interatrial septum; the ductus venosus, ductus arteriosus, and umbilical vessels constrict and eventually turn into ligaments (Figure 24.8b).

Fetal Growth   *Read this*

objective: *Describe the principal events associated with fetal growth.*

During the *fetal period,* organs established by the primary germ layers grow rapidly and the organism takes on a human appearance. A summary of changes associated with the embryonic and fetal period is presented in Exhibit 24.2.

Hormones of Pregnancy   *Read this*

objective: *Explain the functions of the hormones secreted during pregnancy.*

The corpus luteum is maintained for at least the first three or four months of pregnancy, during which time it continues to secrete *progesterone* and *estrogens* (Figure 24.9a). These hor-

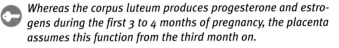

Figure 24.9   Hormones of pregnancy.

Whereas the corpus luteum produces progesterone and estrogens during the first 3 to 4 months of pregnancy, the placenta assumes this function from the third month on.

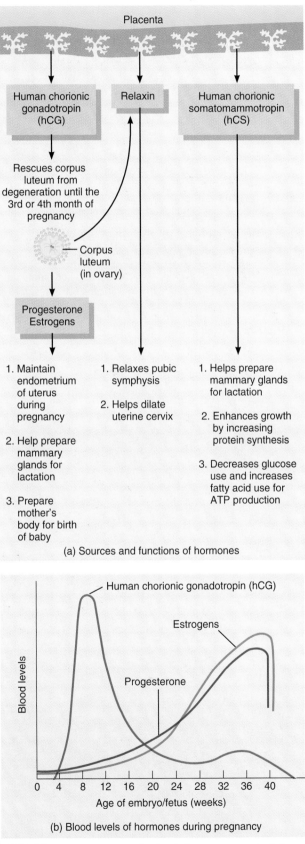

(a) Sources and functions of hormones

(b) Blood levels of hormones during pregnancy

Which hormone serves as the basis for pregnancy tests?

Exhibit 24.2 | Changes Associated with Embryonic and Fetal Growth

END OF MONTH	APPROXIMATE SIZE AND WEIGHT	REPRESENTATIVE CHANGES
1	0.6 cm ($^3/_{16}$ in.)	Eyes, nose, and ears not yet visible. Vertebral column and vertebral canal form. Small buds that will develop into limbs form. Heart forms and starts beating. Body systems begin to form. The central nervous system appears at the start of the third week.
2	3 cm (1$^1/_4$ in.) 1 g ($^1/_{30}$ oz)	Eyes far apart, eyelids fused, nose flat. Ossification begins. Limbs become distinct and digits are well formed. Major blood vessels form. Many internal organs continue to develop.
3	7$^1/_2$ cm (3 in.) 30 g (1 oz)	Eyes almost fully developed but eyelids still fused, nose develops a bridge, and external ears are present. Ossification continues. Limbs are fully formed and nails develop. Heartbeat can be detected. Urine starts to form. Fetus begins to move, but it cannot be felt by mother. Body systems continue to develop.
4	18 cm (6$^1/_2$–7 in.) 100 g (4 oz)	Head large in proportion to rest of body. Face takes on human features and hair appears on head. Many bones ossified, and joints begin to form. Rapid development of body systems.
5	25–30 cm (10–12 in.) 200–450 g ($^1/_2$–1 lb)	Head less disproportionate to rest of body. Fine hair (lanugo) covers body. Brown fat forms and is the site of heat production. Fetal movements commonly felt by mother (quickening). Rapid development of body systems.
6	27–35 cm (11–14 in.) 550–800 g (1$^1/_4$–1$^1/_2$ lb)	Head becomes even less disproportionate to rest of body. Eyelids separate and eyelashes form. Substantial weight gain. Skin wrinkled. Type II alveolar cells begin to produce surfactant.
7	32–42 cm (13–17 in.) 1100–1350 g (2$^1/_2$–3 lb)	Head and body more proportionate. Skin wrinkled. Seven-month fetus (premature baby) is capable of survival. Fetus assumes an upside-down position. Testes start to descend into scrotum.
8	41–45 cm (16$^1/_2$–18 in.) 2000–2300 g (4$^1/_2$–5 lb)	Subcutaneous fat deposited. Skin less wrinkled. Chances of survival much greater at end of eighth month.
9	50 cm (20 in.) 3200–3400 g (7–7$^1/_2$ lb)	Additional subcutaneous fat accumulates. Lanugo shed. Nails extend to tips of fingers and maybe even beyond.

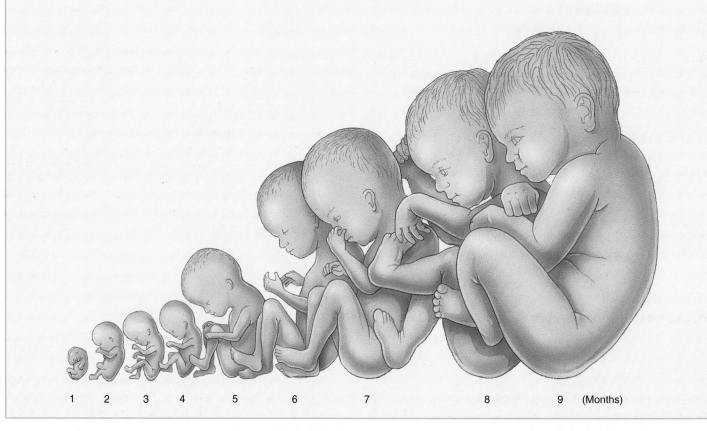

    1   2   3   4   5   6   7   8   9   (Months)

mones maintain the endometrium (lining) of the uterus during pregnancy and prepare the mammary glands for lactation (secretion and ejection of milk).

The chorion of the placenta secretes **human chorionic gonadotropin (hCG)**, which stimulates continued production of estrogens and progesterone by the corpus luteum to maintain the attachment of the embryo or fetus to the endometrium of the uterus. Excretion of hCG from blood into the urine of pregnant women occurs from about the eighth day of pregnancy, peaking about the ninth week (Figure 24.9b). Recall that excretion of hCG in the urine serves as the basis for most home pregnancy tests. hCG can be detected in blood even before menstruation is missed.

The placenta begins to secrete estrogens after the third or fourth week and progesterone by the sixth week of pregnancy. The secretion of hCG is then greatly reduced because the secretions of the corpus luteum are no longer essential. From the third to the ninth month, estrogens and progesterone are secreted by the placenta to prepare the mother's body for birth of the baby and lactation. After delivery, levels of estrogens and progesterone in the blood decrease to normal.

Relaxin, produced by the placenta and corpus luteum, relaxes the pubic symphysis and helps dilate the uterine cervix toward the end of pregnancy to ease delivery.

The chorion also produces **human chorionic somatomammotropin (hCS)**. This hormone is believed to stimulate development of breast tissue for lactation. It also enhances growth by promoting protein anabolism in tissues. Therefore, hCS causes decreased use of glucose by the mother, thus making more available for fetal metabolism, while promoting the release of fatty acids from fat deposits as an alternative energy source for the mother.

Gestation

Read this

The period of time a zygote, embryo, and fetus is carried in the female reproductive tract is called **gestation** (jes-TĀ-shun; *gestare* = to bear). The human gestation period is about 38 weeks, counted from the estimated day of fertilization. The specialized branch of medicine that deals with pregnancy, labor, and the period of time immediately following delivery (about 42 days) is called **obstetrics** (ob-STET-riks; *obstetrix* = midwife).

By about the end of the third month of gestation, the uterus occupies most of the pelvic cavity, and as the fetus continues to grow, the uterus extends higher and higher into the abdominal cavity. By the end of a full-term pregnancy, the uterus practically fills all the abdominal cavity, causing displacement of the maternal intestines, liver, and stomach upward, elevation of the diaphragm, and widening of the thoracic cavity. In the pelvic cavity, there is compression of the ureters and urinary bladder.

In addition to the anatomical changes associated with pregnancy, there are also certain pregnancy-induced physiological changes. For example, cardiac output (CO) rises by 20 to 30 percent by the 27th week due to increased maternal blood flow to the placenta and increased metabolism.

Pulmonary function is altered in several ways. Tidal volume increases, expiratory reserve decreases, and shortness of breath occurs during the last few months.

A general decrease in gastrointestinal tract motility can result in constipation and a delay in gastric emptying time. Nausea, vomiting, and heartburn also occur.

Pressure on the urinary bladder by the enlarging uterus can produce urinary symptoms, such as frequency, urgency, and stress incontinence.

Changes in the skin during pregnancy include increased pigmentation around the eyes and cheekbones in a mask-like pattern, in the areolae of the breasts, and in the linea alba of the lower abdomen. Striae (stretch marks) over the abdomen occur as the uterus enlarges, and hair loss also increases.

Prenatal Diagnostic Techniques

Amniocentesis

In **amniocentesis** (am′-nē-ō-sen-TĒ-sis; *amnio* = amnion; *centesis* = puncture), a sample of amniotic fluid is withdrawn and analyzed to diagnose genetic disorders or determine fetal maturity or well-being. By using ultrasound and palpation, the positions of the fetus and placenta are first determined. About 10 ml of fluid containing sloughed-off fetal cells are removed by hypodermic needle puncture of the uterus, usually 14 to 16 weeks after conception. The fluid and cells are then examined. Close to 300 chromosomal disorders and over 50 inheritable biochemical defects can be detected through amniocentesis, including Down syndrome, hemophilia, certain muscular dystrophies, Tay–Sachs disease, sickle-cell anemia, and cystic fibrosis. When both parents are known or suspected to be genetic carriers of any one of these disorders, or when maternal age is 35 or over, amniocentesis is advisable.

Chorionic Villi Sampling (CVS)

Chorionic (ko-rē-ON-ik) **villi sampling (CVS)** is a test that picks up the same defects as amniocentesis but can be performed during the first trimester of pregnancy, as early as 8 weeks of gestation, and results are available within 24 hours. Moreover, the procedure does not require penetration of the abdominal wall, uterine wall, or amniotic cavity.

In CVS, a catheter (tube) is placed through the vagina into the uterus and then to the chorionic villi under ultrasound guidance. About 30 mg of tissue are suctioned out and subjected to chromosomal analysis. Chorion cells and fetal cells contain identical genetic information. The procedure is believed to be as safe as amniocentesis.

Labor

| **objective:** | *Describe the stages of labor.* |

Labor is the process by which uterine contractions expel the fetus through the vagina to the outside.

The onset of labor appears to be hormonally directed. Just prior to birth, the muscles of the uterus contract rhythmically and forcefully. Both placental and ovarian hormones seem to play a role in these contractions. Because progesterone inhibits uterine contractions, labor cannot take place until its effects are diminished. At the end of gestation, progesterone level falls. Now, the level of estrogens in the mother's blood is sufficient to overcome the inhibiting effects of progesterone, and labor begins. Oxytocin (OT) also stimulates uterine contractions (see Figure 13.8). Relaxin assists by relaxing the pubic symphysis and helping to dilate the uterine cervix. Prostaglandins may also play a role in labor.

Uterine contractions occur in waves that start at the top of the uterus and move downward. These waves expel the fetus.

Labor can be divided into three stages (Figure 24.10).

❶ The *stage of dilation* is the time (6 to 12 hours) from the onset of labor to the complete dilation of the cervix. During this stage there are regular contractions of the uterus and complete dilation (10 cm) of the cervix. If the amniotic sac does not rupture spontaneously, it is done artificially.

❷ The *stage of expulsion* is the time (10 minutes to several hours) from complete cervical dilation to delivery.

❸ The *placental stage* is the time (5 to 30 minutes or more) after delivery until the placenta or "afterbirth" is expelled by powerful uterine contractions. These contractions also constrict blood vessels that were torn during delivery to prevent hemorrhage.

During the six week period of time following delivery of the baby and placenta, called the *puerperium* (pyoo′-er-PĒ-rē-um), the reproductive organs and maternal physiology return to the pre-pregnancy state.

Lactation

objective: *Discuss how lactation occurs and how it is controlled.*

The term *lactation* (lak′-TĀ-shun) refers to the secretion and ejection of milk by the mammary glands. The major hormone in promoting lactation is *prolactin* (PRL) from the anterior pituitary gland. It is released in response to a hypothalamic releasing hormone. Even though PRL levels increase as the pregnancy progresses, there is no milk secretion because estrogens and progesterone inhibit the PRL. Following delivery, the levels of estrogens and progesterone in the mother's blood decrease following delivery of the placenta and the inhibition is removed.

The principal stimulus in maintaining prolactin secretion during lactation is the sucking action (suckling) of the infant. Sucking initiates impulses from the receptors in the nipples to the hypothalamus and PRL is released. The sucking action also initiates impulses to the hypothalamus that stimulate the release of *oxytocin* (*OT*) by the posterior pituitary gland. Oxytocin promotes *milk ejection* (*milk let-down*), the process by which milk is moved from the alveoli of the mammary glands into the ducts where it can be sucked.

During late pregnancy and the first few days after birth, the mammary glands secrete a cloudy fluid called *colostrum* (first milk). It is not as nutritious as true milk because it contains

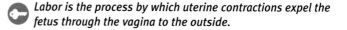

Figure 24.10   Stages of labor.

🔑 *Labor is the process by which uterine contractions expel the fetus through the vagina to the outside.*

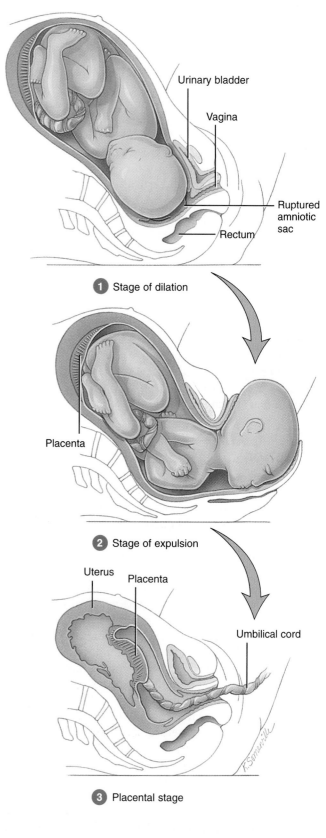

Urinary bladder

Vagina

Ruptured amniotic sac

Rectum

❶ Stage of dilation

Placenta

❷ Stage of expulsion

Uterus   Placenta

Umbilical cord

❸ Placental stage

🅠 *What event marks the beginning of the stage of expulsion?*

less lactose and virtually no fat. True milk appears on about the fourth day. Colostrum and maternal milk contain antibodies that protect the infant during the first few months of life.

Following birth, the PRL level starts to diminish. However, each time the mother nurses the infant, nerve impulses from the nipples to the hypothalamus cause the release of PRH and a tenfold increase in PRL that lasts about an hour, which provides milk for the next nursing period. If this surge of PRL is blocked by injury or disease or if nursing is discontinued, the mammary glands lose their ability to secrete milk in a few days.

Birth Control

objective: *Name and compare the effectiveness of birth control methods.*

Several types of **birth control** (prevention of conception or birth) are available, each with its own advantages and disadvantages. The methods discussed here are sterilization, hormonal, intrauterine, barrier, chemical, physiological (natural), coitus interruptus (withdrawal), and induced abortion. See Exhibit 24.3 for a summary of birth control methods.

Sterilization

One means of **sterilization** of males is **vasectomy**. Sterilization in females is usually by **tubal ligation** (lī-GĀ-shun), in which the uterine tubes are tied closed and then cut. The secondary oocyte is thus prevented from passing to the uterus, and the sperm cannot reach the oocyte.

Hormonal Methods

By adjusting hormone levels, it is possible to interfere with production of gametes (sperm and ova) or implantation of a fertilized ovum in the uterus. This may be accomplished by use of an **oral contraceptive** or **the pill,** which is used by about 28 percent of American women. Although several types of pills are available, the ones most commonly used contain a high concentration of progesterone and a low concentration of estrogens (combination pill). These two hormones act via negative feedback on the anterior pituitary gland to decrease the secretion of FSH and LH and on the hypothalamus to inhibit secretion of GnRH. The low levels of FSH and LH usually prevent both follicular development and ovulation; thus pregnancy cannot occur. Even if ovulation does occur, as it does in some cases, oral contraceptives also alter cervical mucus so that it is more hostile to sperm and may make the endometrium less receptive to implantation.

Among the noncontraceptive benefits of oral contraceptives are regulation of the length of menstrual cycles, decreased menstrual flow (and therefore decreased risk of anemia), and prevention of ovarian cysts. The pill also provides protection against endometrial and ovarian cancers. Oral contraceptives may not be advised for women with a history of thromboembolic disorders (predisposition to blood clotting), hypertension, heart disease, cerebral blood vessel damage, or liver malfunction. Women who take the pill and smoke face far higher odds of having a heart attack or stroke than do non-

smoking pill users. Smokers should quit or use an alternative birth control method. Oral contraceptives do not provide any protection against sexually transmitted diseases.

If daily pill-taking is not desired, a woman may opt for **Norplant** or **Depo-provera,** two other hormonal methods of contraception. Norplant is six slender hormone-containing capsules that are surgically implanted under the skin of the arm using local anesthesia. They slowly and continually release progestin, which inhibits ovulation and thickens the cervical mucus. The effects last for 5 years, and Norplant is as reliable as sterilization (less than 1 percent failure rate). Removing the Norplant capsules restores fertility. Over a 5-year period Norplant is less expensive than most birth control pills. Depo-provera is given as an intramuscular injection once every 3 months. It contains a hormone similar to progesterone that prevents maturation of the ovum and causes changes in the uterine lining that make it less likely for pregnancy to occur.

The quest for an efficient male oral contraceptive has been disappointing. The challenge is to find substances that will block production of functional sperm without disrupting the ability to have an erection.

Intrauterine Devices (IUDs)

An **intrauterine device (IUD)** is a small object made of plastic, copper, or stainless steel that is inserted into the cavity of the uterus. IUDs cause changes in the uterine lining that block implantation of a fertilized ovum. They are more easily tolerated by women who have had a child. The dangers associated with the use of IUDs in some females include pelvic inflammatory disease (PID), infertility, and excess menstrual bleeding and pain. Females in monogamous relationships are at a lesser risk of developing PID, and recent research discounts IUDs as a cause of PID.

Barrier Methods

Barrier methods are designed to prevent sperm from gaining access to the uterine cavity and uterine tubes. Among the barrier methods are use of a condom, vaginal pouch, diaphragm, or cervical cap.

The **condom** is a nonporous, elastic (latex or similar material) covering placed over the penis that prevents sperm from being deposited in the female reproductive tract. The **vaginal pouch,** sometimes called a **female condom,** is made of two flexible rings connected by a polyurethane sheath. One ring lies inside the sheath and is inserted into the vagina to fit over the cervix. The other ring remains outside the vagina and covers the female external genitals. Proper use of condoms with each act of sexual intercourse, especially when used with a spermicide (sperm-killing chemical), is a fairly reliable method of birth control. Condom use also reduces, but does not eliminate, the risk of acquiring a sexually transmitted disease (STD) such as AIDS or syphilis. Even when used properly, condoms fail to protect against pregnancy and disease transmission 10 to 20 percent of the time.

The **diaphragm** is a rubber dome-shaped structure that fits over the cervix and is generally used together with a spermicide. The diaphragm stops the sperm from passing into the cervix. The chemical kills the sperm cells. The diaphragm must be initially fit-

Exhibit 24.3	Summary of Birth Control Methods

METHOD	COMMENTS
Sterilization	In males, *vasectomy*, in which each ductus (vas) deferens is cut and tied. In females, *tubal ligation* (li-GĀ-shun), in which both uterine (Fallopian) tubes are cut and tied. Failure rate: 0.1% (0.4%).[a]
Hormonal	Except for total abstinence or surgical sterilization, hormonal methods are the most effective means of birth control. *Oral contraceptives*, commonly known as "the pill," usually include both an estrogen and a progestin. Side effects include nausea, occasional light bleeding between periods, breast tenderness or enlargement, fluid retention, and weight gain. Pill users may have an increased risk of infertility. Failure rate: 0.1% (3%). *Norplant* is a set of implantable, progestin-loaded cylinders that slowly release hormone for 5 years. Fertility is restored by removing the cylinders. Failure rate: 0.3% (0.3%). *Depo-provera* is given as an intramuscular injection once every 3 months. It contains a hormone similar to progesterone that prevents maturation of the ovum and causes changes in the uterine lining that make it less likely for pregnancy to occur. Failure rate: 0.3% (0.3%).
Intrauterine device (IUD)	Small object made of plastic, copper, or stainless steel and inserted into uterus by physician. May be left in place for long periods of time. (Copper T is approved for 8 years of use.) Some women cannot use them because of expulsion, bleeding, or discomfort. Not recommended for women who have not had children because uterus is too small and cervical canal too narrow. Failure rate: 0.8% (3%).
Barrier	A *condom* is a thin, strong sheath of latex or similar material worn by male to prevent sperm from entering the vagina. A similar device for use by a woman is called a *vaginal pouch*. Failures are caused by the condom or pouch tearing or slipping off after climax or not putting it on soon enough. Failure rate: 2% (12%). A *diaphragm* is a flexible rubber dome inserted into the vagina to cover the cervix, providing a barrier to sperm. Usually used with spermicidal cream or jelly. Must be left in place at least 6 hours after intercourse and may be left in place as long as 24 hours. Must be fitted by a health-care professional and refitted every two years and after each pregnancy. Offers high level of protection if used with spermicide. Occasional failures are caused by improper insertion or displacement during sexual intercourse. Diaphragm alone failure rate: 6% (18%). A *cervical cap* is a thimble-shaped latex device that fits snugly over the cervix of the uterus. Used with a spermicide, must be fitted by a health-care professional. May be left in place for up to 48 hours, and it is not necessary to reintroduce spermicide before sexual intercourse. Failure rate: 6% (18%).
Chemical	Sperm-killed chemicals inserted into vagina to coat vaginal surfaces and cervical opening. Provide protection for about 1 hour. Effective when used alone but significantly more effective when used with diaphragm or condom. Spermicide alone failure rate: 3% (21%).
Physiological	In the *rhythm method*, sexual intercourse is avoided just before and just after ovulation (about 7 days). Failure rate is about 20% even in females with regular menses. In the *sympto-thermal method*, signs of ovulation are noted (increased basal body temperature, clear and stretchy cervical mucus, opening of the external os, elevation and softening of the cervix, abundant cervical mucus, and pain associated with ovulation), and sexual intercourse is avoided. Failure rate: 1–9% (20%).
Coitus interruptus	Withdrawal of penis from vagina before ejaculation occurs. Failure rate: 4% (18%).
Induced abortion	Surgical or drug-induced removal of products of conception at an early stage from uterus or from uterine tube in cases of tubal pregnancy. Surgical removal from the uterus may involve vacuum aspiration (suction), saline solution, or surgical evacuation (scraping). Drug-induced abortions make use of RU 486, which blocks the action of progesterone.

[a] Failure rates: The first number is the expected percentage of women experiencing an unintended pregnancy in the first year of continuous and proper use. In parenthesis is the typical percentage of women who become pregnant, including those who forgot to use their birth control. With no method of birth control, the failure rate is 85%.

ted by a physician. Toxic shock syndrome (TSS) and recurrent urinary tract infections are associated with diaphragm use in some females, and a diaphragm does not protect against STDs.

The *cervical cap* is a thimble-shaped contraceptive device made of latex or plastic that measures about 4 cm (1.5 in.) in diameter. It fits over the cervix of the uterus and is held in posi-

tion by suction. Like the diaphragm, the cervical cap is used with a spermicide. Also, like the diaphragm, it must be fitted initially by a physician or other trained personnel. Advantages of the cervical cap in comparison to the diaphragm are (1) the cap can be worn up to 48 hours versus 24 hours for the diaphragm, and (2) because the cap fits tightly and rarely leaks, it is not necessary to reintroduce spermicide before intercourse. The cervical cap is not recommended for females with known or suspected cervical or uterine malignancies and current vaginal or cervical infections. Like the diaphragm, the cervical cap does not protect against STDs.

Chemical Methods

Chemical methods of contraception are spermicidal agents. Various foams, creams, jellies, suppositories, and douches that contain spermicidal agents make the vagina and cervix unfavorable for sperm survival and are available without prescription. The most widely used spermicide is nonoxynol-9. Its action is to disrupt plasma membranes of sperm, thus killing them. They are most effective when used with a diaphragm or condom.

Physiological (Natural) Methods

Physiological (**natural**) **methods** are based on knowledge of certain physiological changes that occur during the menstrual cycle. In females with normal menstrual cycles, especially, these events help to predict on which day ovulation is likely to occur.

The first physiological method, developed in the 1930s, is known as the ***rhythm method.*** It takes advantage of the fact that a secondary oocyte is fertilizable for only 24 hours and is available only during a period of three to five days in each menstrual cycle. During this time, the couple refrains from intercourse (three days before ovulation, the day of ovulation, and three days after ovulation). The effectiveness of the rhythm method is poor because few women have absolutely regular cycles.

Another natural family planning system, developed during the 1950s and 1960s, is the ***sympto-thermal method.*** According to this method, couples are instructed to know and understand certain signs of fertility and infertility. Recall that the signs of ovulation include increased basal body temperature; the production of clear, stretchy cervical mucus; abundant cervical mucus; and pain associated with ovulation. If the couple refrains from sexual intercourse when the signs of ovulation are present, the chance of pregnancy is decreased.

Coitus Interruptus (Withdrawal)

Coitus (KŌ-i-tus; *coitio* = coming together) ***interruptus*** refers to withdrawal of the penis from the vagina just prior to ejaculation. Failures with this method are related to either failure to withdraw before ejaculation or pre-ejaculatory escape of sperm-containing fluid from the urethra. This method offers no protection against STDs.

Induced Abortion

Abortion refers to the premature expulsion from the uterus of the products of conception, usually before the 20th week of pregnancy. An abortion may be spontaneous (naturally occurring), sometimes called a miscarriage, or induced (intentionally performed). When birth control methods fail to prevent an unwanted pregnancy, ***induced abortion*** may be performed. Induced abortions may involve vacuum aspiration (suction), infusion of a saline solution, or surgical evacuation (scraping).

Certain drugs, most notably the drug ***RU 486 (mifepristone),*** can induce abortion, a so-called nonsurgical abortion. RU 486 blocks the action of progesterone. Progesterone prepares the uterine endometrium for implantation and then maintains the uterine lining after implantation. If progesterone levels fall during pregnancy or if the action of the hormone is blocked, menstruation occurs, and the embryo is sloughed off along with the uterine lining. Within 12 hours after taking RU 486, the endometrium starts to degenerate and then begins to slough off within 72 hours. Prostaglandin, which stimulates uterine contractions, is given after RU 486 to aid in expulsion of the endometrium. RU 486 can be taken up to 5 weeks after conception. One side effect of the drug is uterine bleeding. RU 486 is being tested in clinical trials in the U.S. and has been used for several years in France, Sweden, the United Kingdom, and China.

Inheritance

| objective: | *Describe the basic principles of inheritance.* |

Inheritance is the passage of hereditary traits from one generation to another. The branch of biology that studies inheritance is *genetics* (je-NET-iks).

Genotype and Phenotype

The nuclei of all human cells except gametes contain 23 pairs of chromosomes. One chromosome from each pair comes from the mother, and the other comes from the father. *Homologous chromosomes,* the two chromosomes of a pair, contain genes that control the same traits. If a chromosome contains a gene for height, its homologous chromosome will also contain a gene for height. Genes that control the same inherited trait, for example, height, eye color, or hair color, and that occupy the same position on homologous (paired) chromosomes are known as *alleles.*

The relationship of genes to heredity is illustrated by the inheritance of a disorder called *phenylketonuria.* Individuals with this disorder lack the gene that produces an enzyme necessary to convert an amino acid called phenylalanine into an amino acid called tyrosine. High levels of phenylalanine are toxic to the brain during the early years of life when the brain is developing and result in mental retardation. The normal gene for conversion of phenylalanine into tyrosine is symbolized as *P*, whereas the abnormal gene is symbolized as *p* (Figure 24.11). The chromosome concerned with normal conversion of phenylalanine into tyrosine has either *p* or *P* on it. Its homologous chromosome also has *p* or *P*. Thus every individual will have one of the following genetic makeups, or *genotypes* (JĒ-nō-tīps): *PP, Pp,* or *pp*. Although people with genotypes of *Pp* have the abnormal gene, only those with genotype *pp* have phenylketonuria because the normal gene, when present, dominates over the abnormal one. A gene that dominates is called the *dominant gene,* and the trait expressed is said to be a dominant trait. The gene that is not expressed is called the *recessive gene,* and the trait it controls is called the recessive trait.

By tradition, we symbolize the dominant gene with a capital letter and the recessive one with a lowercase letter. An individual with the same genes on homologous chromosomes (for example, *PP* or *pp*) is said to be homozygous for the trait. An individual with different genes on homologous chromosomes (for

Figure 24.11 Inheritance of phenylketonuria (PKU).

🔑 *Whereas genotype refers to genetic makeup, phenotype refers to the physical or outward expression of a gene.*

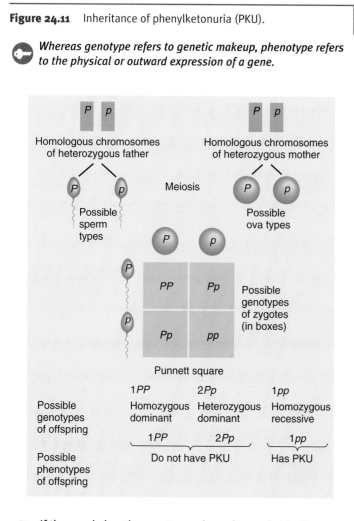

Punnett square

Possible genotypes of offspring	1 *PP* Homozygous dominant	2 *Pp* Heterozygous dominant	1 *pp* Homozygous recessive
	1 *PP*	2 *Pp*	1 *pp*
Possible phenotypes of offspring	Do not have PKU		Has PKU

❓ *If the couple has the genotypes shown here, what is the percent chance that their first child will have PKU?*

Exhibit 24.4	Selected Hereditary Traits in Humans
DOMINANT	**RECESSIVE**
Coarse body hair	Fine body hair
Male pattern baldness	Baldness
Normal skin pigmentation	Albinism
Freckles	Absence of freckles
Astigmatism	Normal vision
Near- or farsightedness	Normal vision
Normal hearing	Deafness
Broad lips	Thin lips
Tongue roller	Inability to roll tongue into a U shape
PTC taster	PTC nontaster
Large eyes	Small eyes
Polydactylism (extra digits)	Normal digits
Brachydactylism (short digits)	Normal digits
Syndactylism (webbed digits)	Normal digits
Feet with normal arches	Flat feet
Hypertension	Normal blood pressure
Diabetes insipidus	Normal excretion
Huntington's disease	Normal nervous system
Normal mentality	Schizophrenia
Migraine headaches	Normal
Widow's peak	Straight hairline
Curved (hyperextended) thumb	Straight thumb
Normal Cl⁻ transport	Cystic fibrosis
Hypercholesterolemia (familial)	Normal cholesterol level

example, *Pp*) is said to be **heterozygous** for the trait. **Phenotype** (FĒ-nō-tīp; *pheno* = showing) refers to how the genetic makeup is expressed in the body. A person with *Pp* has a different genotype from one with *PP*, but both have the same phenotype—which in this case is the normal conversion of phenylalanine into tyrosine. It should be noted that individuals who carry a recessive gene but do not express it (*Pp*) can pass the gene on to their offspring. Such individuals are called **carriers.**

To determine how gametes containing haploid chromosomes unite to form diploid fertilized eggs, special charts called **Punnett squares** are used. Usually, possible genes in the male gametes (sperm cells) are placed at the side of the chart and possible genes in the female gametes (secondary oocytes) are at the top (Figure 24.11). The four spaces on the chart represent the possible combinations (genotypes) of male and female gametes that could form fertilized eggs.

Exhibit 24.4 lists some of the inherited structural and functional traits in humans.

Genes for severe disorders are more frequently recessive than dominant, but this is not always true. For example, the neurological disorder **Huntington's disease** is caused by a dominant gene. Huntington's disease is characterized by degeneration of nervous tissue, usually leading to mental disturbances and death. The first signs of Huntington's disease do not occur until adulthood, very often after the person has already produced offspring. A genetic test is now available to determine whether or not an individual carries the gene for the disorder.

Variations on Dominant–Recessive Inheritance

Most patterns of inheritance are not simple **dominant–recessive inheritance** in which only dominant and recessive genes interact. In fact, most inherited traits are influenced by more than one gene and most genes can influence more than a single trait. Following are some examples.

In **incomplete dominance,** neither member of a pair of genes (alleles) is dominant over the other and the heterozygote has a phenotype intermediate between the homozygous dominant and homozygous recessive. An example of incomplete dominance in humans is the inheritance of *sickle-cell anemia* (**SCA**). Individuals with the homozygous dominant genotype $Hb^A Hb^A$ form normal hemoglobin. Those with the homozygous recessive genotype $Hb^S Hb^S$ have the disease called sickle-cell anemia and have severe anemia. Although they are usually healthy, those with the heterozygous genotype $Hb^A Hb^S$ have minor problems with anemia since they produce both normal and sickle-cell hemoglobin. People in this last category are carriers; they are sometimes referred to as having the "sickle-cell" trait.

Although a single individual inherits only two alleles for each gene, some genes may have more than two alternate forms and this is the basis for *multiple-allele inheritance.* One example of multiple-allele inheritance in humans is the inheritance of the ABO blood group. The four blood types (phenotypes) of the ABO group—A, B, AB, and O—result from the inheritance of six combinations of three different forms of a single gene that code for the same trait called the *I* gene: (1) I^A produces the A antigen, (2) I^B produces the B antigen, and (3) *i* produces neither A nor B antigen. Each person inherits two *I* genes, one from each parent, that give rise to the various phenotypes. The six possible genotypes produce four blood types as follows:

Genotype	Blood type
$I^A I^A$ or $I^A i$	A
$I^B I^B$ or $I^B i$	B
$I^A I^B$	AB
ii	O

Note that both I^A and I^B are inherited as dominant traits but *i* is inherited as a recessive trait. Since an individual with type AB blood has characteristics of both type A and type B red blood cells expressed in the phenotype, genes I^A and I^B are said to be *codominant.* In other words, both genes are expressed in the heterozygote equally. Depending on the parental blood types, different offspring may have one, two, three, or four different blood types.

Most inherited traits are not controlled by one gene but rather by the combined effects of many genes. This is referred to as *polygenic* (*poly* = many) *inheritance.* A polygenic trait shows a continuous gradation of small differences between extremes among individuals. Examples of polygenic traits include skin color, hair color, eye color, height, and body build. Suppose that skin color is controlled by three separate genes, each having two alleles: *A, a; B, b;* and *C, c.* Whereas a person with the genotype *AABBCC* is very dark skinned, an individual with the genotype *aabbcc* is very light skinned; and a person with the genotype *AaBbCc* has an intermediate skin color. Parents having an intermediate skin color may have children with very light, very dark, or intermediate skin color.

Genes and the Environment

The phenotypic expression of a particular gene is influenced not only by the alleles of the gene present but also by the environment. Put another way, a given phenotype is the result of both genotype and environment. For example, height, along with body build and hair color, is controlled by the combined effects of many genes. Even though a person inherits genes for tallness, full potential may not be reached due to environmental factors, such as disease or malnutrition during the growth years.

Exposure of a developing embryo or fetus to certain environmental factors can damage the developing organism or even cause death. A *teratogen* (*terato* = monster) is any agent or influence that causes physical defects in the developing embryo. Examples are pesticides, defoliants, industrial chemicals, oral anticoagulants, anticonvulsants, antitumor agents, thyroid drugs, thalidomide, diethylstilbestrol (DES), LSD, marijuana, cocaine, alcohol, nicotine, and ionizing radiation. Thalidomide, a sedative in use in the 1960s, affected limb development in fetuses. A pregnant female who uses cocaine subjects the fetus to risks of underweight, retarded growth, attention and orientation problems, alteration in visual processing, hyperirritability, a tendency to stop breathing, sudden infant death syndrome (SIDS), malformed or missing organs, strokes, and seizures. The risks of spontaneous abortion (miscarriage), premature birth, and stillbirth also increase from fetal exposure to cocaine.

Alcohol is the number one fetal teratogen. Intrauterine exposure to alcohol results in *fetal alcohol syndrome* (**FAS**), affecting 1 per 1000 live births. FAS is one of the major causes of mental retardation in the United States. Other symptoms include slow growth before and after birth, poor eyesight, small head, facial irregularities such as narrow eye slits and sunken nasal bridge, defective heart and other organs, malformed arms and legs, genital abnormalities, damage to the central nervous system, mental retardation, and learning disabilities, plus behavioral problems, such as hyperactivity, extreme nervousness, and poor attention span.

The latest evidence indicates a causal relationship between cigarette smoking during pregnancy and low infant birth weight and also an association between smoking and cardiac abnormalities, anencephaly (the absence of a cerebrum in the cranium), and higher fetal and infant mortality rate. Maternal smoking also appears to be a significant factor in the development of cleft lip and palate and has been tentatively linked with sudden infant death syndrome (SIDS). Infants nursing from smoking mothers have been found to have an increased incidence of gastrointestinal disturbances. Finally, infants of smoking mothers have an increased incidence of respiratory problems (pneumonia and bronchitis) during the first year of life.

Ionizing radiations are potent teratogens. Treatment of pregnant mothers with large doses of x-rays and radium during the embryo's susceptible period of development may cause microcephaly (small size of head in relation to the rest of the body), mental retardation, and skeletal malformations. Caution is advised for diagnostic x-rays during the first trimester of pregnancy.

Inheritance of Gender (Sex)

One pair of chromosomes in cells differs in males and females (Figure 24.12a). In females, the pair consists of two chromosomes designated as X chromosomes. One X chromosome is also present in males, but its mate is a chromosome called a Y chromosome. The XX pair in the female and the XY pair in the male are called the *sex chromosomes.* All other chromosomes are called *autosomes.*

The sex chromosomes are responsible for the gender (sex) of the individual (Figure 24.12b). When a spermatocyte undergoes meiosis to reduce its chromosome number, one daughter cell will contain the X chromosome and the other will contain the Y chromosome. Secondary oocytes produce only

Figure 24.12   Inheritance of gender (sex). In (a) the sex chromosomes, pair 23, are indicated in the colored box.

🔑 *Gender is determined at the time of fertilization.*

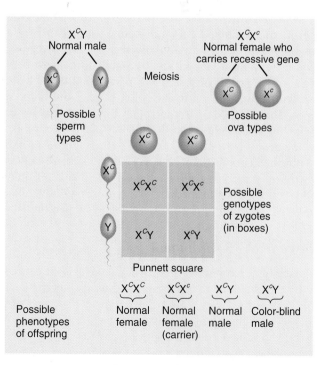

(a) Normal human male chromosomes

(b) Sex determination

❓ *What are chromosomes other than sex chromosomes called?*

X-containing ova. If the secondary oocyte is subsequently fertilized by an X-bearing sperm cell, the offspring will be female (XX). Fertilization by a Y sperm cell produces a male (XY). Thus gender is determined at fertilization.

Red–Green Color Blindness and Sex-Linked Inheritance

The sex chromosomes also are responsible for the transmission of a number of nonsexual traits. Genes for these traits appear on X chromosomes but are absent from Y chromosomes. This produces a pattern of heredity that is different from the pattern described earlier. Let us consider the most common type of color blindness, called red–green color blindness. The gene for *red–green color blindness* is a recessive one designated *c*. Normal color vision, designated *C*, dominates. The *C/c* genes are located on the X chromosome. The Y chromosome does not contain the segment of DNA that programs this aspect of

vision. Thus the ability to see colors depends entirely on the X chromosomes. The genetic possibilities are

$X^C X^C$ Normal vision female

$X^C X^c$ Normal vision female carrying the recessive gene

$X^c X^c$ Red-green color-blind female

$X^C Y$ Normal vision male

$X^c Y$ Red-green color-blind male

Only females who have two X^c chromosomes are red–green color-blind. In $X^C X^c$ females, the trait is dominated by the normal, dominant gene. Males, on the other hand, do not have a second X chromosome that would dominate the trait. Therefore all males with an X^c chromosome will be red-green color-blind. The inheritance of color blindness is illustrated in Figure 24.13.

Traits inherited in the manner just described are called *sex-linked traits.* Another sex-linked trait is *hemophilia*—a condition in which the blood fails to clot or clots very slowly after an injury (see Chapter 14). Like the trait for red–green color blindness, hemophilia is caused by a recessive gene. If *H* represents normal clotting and *h* represents abnormal clotting, then $X^h X^h$ females will have the disorder. Males with $X^H Y$ will be normal; males with $X^h Y$ will be hemophiliac. Actually, clotting time varies somewhat among hemophiliacs, so the condition may be affected by other genes as well.

A few other sex-linked traits in humans are certain forms of diabetes, night blindness, one form of cataract, white forelocks, juvenile glaucoma, and juvenile muscular dystrophy.

Figure 24.13   Inheritance of red–green color blindness.

🔑 *Red–green color blindness and hemophilia are examples of sex-linked traits.*

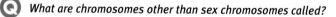

❓ *What would be the genotype of a red–green color-blind female?*

wellness focus

Breast Milk: Mother Nature's Approach to Infection Prevention

*U*NICEF and the World Health Organization both recommend that women breast-feed infants and children for "two years and beyond." Yet in the United States, only 56 percent of mothers of newborns leave the hospital breast-feeding their infants, and by the time these babies are 3 months old, only 20 percent of their mothers continue to breast-feed. It seems that parents in the United States regard formula as the nutritional equivalent of breast milk, and a lot more convenient. Continuing research on the composition of breast milk suggests this is not the case. Breast-feeding offers several health advantages, most important of which is protection from infection.

More Than Clean

Physicians in both industrialized and developing countries have long observed that breast-fed babies contract fewer infections than do babies fed formula. This observation was originally attributed to the fact that formula must often be mixed with water and placed in bottles, and so could easily become contaminated. We now know that breast milk contains a panoply of ingredients that enhance an infant's ability to fight disease, including antibodies and some immune cells.

Keeping the Tract Intact When Under Attack

Several substances in breast milk enhance both immunity and nonspecific resistance to disease in the baby's gastrointestinal (GI) tract. One family of these is the *secretory immunoglo-bin A (IgA)* antibodies. When the baby's mother encounters pathogens, she manufactures antibodies specific to each agent. These pass into the breast milk and escape breakdown in the baby's GI tract, protected by a so-called secretory component. Once in the baby's GI tract, these antibodies bind with the targeted infectious agents and prevent them from passing through the lining of the GI tract. This protection is especially important in the earliest days of life, because the infant does not even begin to make secretory IgA until several weeks or months after birth. The protection provided by the antibodies in breast milk is helpful for the first few years, because a child's own immune response is not fully mature until age 5.

The secretory IgA antibodies perform their work neatly and discretely. They disable pathogens without harming helpful gut flora or causing inflammation. This is important, because while inflammation helps fight infection, sometimes the process overwhelms the GI tract, and infants suffer more from the inflammatory process than the infection itself when inflammation destroys healthy tissue. The large quantities of the immune system molecule *interleukin-10* found in breast milk also help inhibit inflammation. And a substance called *fibronectin* enhances the phagocytic activity of macrophages, inhibits inflammation, and helps to repair tissues damaged by inflammation.

Several other breast milk molecules help disable harmful microbes. *Mucins* and certain *oligosaccharides* (sugar chains) and *glycoproteins* (carbohydrate-protein compounds) bind to microbes and prevent them from gaining a foothold on the gut lining.

Discouraging Unwelcome Visitors

Breast milk components make the baby's digestive tract an unfriendly environment for bacteria in several other ways. Many of breast milk's immune cells, including *T* lymphocytes and macrophages, attack invading microbes directly. Some breast milk compounds decrease the supply of nutrients such as iron and vitamin B12 needed by harmful bacteria to survive. A substance called *bifidus factor* promotes the growth of helpful gut flora, which help crowd out pathogens. *Retinoic acids*, a group of vitamin A precursors, reduce the ability of viruses to replicate. And some of the hormones and growth factors present in breast milk stimulate the baby's GI tract to mature more quickly, making it less vulnerable to dangerous invaders.

critical thinking

Describe why preventing infection through breast-feeding is preferable to giving babies antibiotics.

What are some of the factors that deter women in North America from breast-feeding their babies?

Medical Terminology and Conditions

Abortion (a-BOR-shun) Premature expulsion from the uterus of the products of conception—embryo or nonviable fetus. May be caused by abnormal development of the embryo, placental abnormalities, endocrine disturbances, certain diseases, trauma, and stress. *Induced (nonspontaneous) abortions* are brought on intentionally by methods such as vacuum aspiration (suction curettage), up to the 12th week of pregnancy; dilation and evacuation (D&E), commonly used between the 13th and 15th weeks of pregnancy and sometimes up to 20 weeks; and use of saline or prostaglandin preparations to induce labor and delivery, usually after the 15th week of pregnancy. *Spontaneous abortions (miscarriages)* occur without apparent cause. Recent evidence suggests that as many as one-third of all successful fertilizations end in spontaneous abortions, most of which occur even before a female or her physician is aware that she is pregnant.

Breech presentation A malpresentation in which the fetal buttocks or lower limbs, instead of the head, enter the birth canal first; most common cause is prematurity.

Mutation (myoo-TĀ-shun; *mutare* = change) A permanent heritable change in a gene that may cause it to have a different function than it had previously.

Preeclampsia (prē′-e-KLAMP-sē-a) A syndrome characterized by sudden hypertension, large amounts of protein in urine, and generalized edema; possibly related to an autoimmune or allergic reaction to the presence of a fetus; when the condition is also associated with convulsions and coma, it is referred to as *eclampsia*.

Puerperal (pyoo-ER-per-al; *puer* = child; *parere* = to bring forth) *fever* Infectious disease of childbirth that originates in the birth canal and affects the endometrium. It may spread to other pelvic structures and lead to septicemia.

Study Outline

Sexual Intercourse (p. 552)

1. The role of the male in the sex act involves erection, lubrication, and orgasm (climax).
2. The female role also involves erection, lubrication, and orgasm (climax).

Development During Pregnancy (p. 552)

1. Pregnancy is a sequence of events that normally includes fertilization, implantation, embryonic growth, fetal growth, and birth.
2. Its various events are hormonally controlled.

Fertilization; Implantation (pp. 552–553)

1. Fertilization refers to the penetration of a secondary oocyte by a sperm cell and the subsequent union of the genetic material of the sperm cell and secondary oocyte chromosomes to form a single nucleus.
2. Penetration is facilitated by enzymes produced by sperm.
3. Normally only one sperm cell fertilizes a secondary oocyte.
4. Early rapid cell division of a zygote is called cleavage.
5. The solid mass of tiny cells produced by cleavage is a morula.
6. The morula develops into a blastocyst, a hollow ball of cells differentiated into a trophoblast (future embryonic membranes) and inner cell mass (future embryo).
7. The attachment of a blastocyst to the endometrium is called implantation; it occurs by enzymatic degradation of the endometrium.

Embryonic Development (p. 555)

1. During embryonic growth, the primary germ layers and embryonic membranes are formed and the placenta is functioning.
2. The primary germ layers—ectoderm, mesoderm, and endoderm—form all tissues of the developing organism.
3. Embryonic membranes include the yolk sac, amnion, chorion, and allantois.
4. Fetal and maternal materials are exchanged through the placenta.
5. Fetal circulation involves the exchange of materials between fetus and mother.
6. The fetus derives its oxygen and nutrients and eliminates its carbon dioxide and wastes through the maternal blood supply by means of the placenta.
7. At birth, when lung, digestive, and liver functions are established, the special structures of fetal circulation are no longer needed.

Fetal Growth (p. 561)

1. During the fetal period, organs established by the primary germ layers grow rapidly.
2. The principal changes associated with embryonic and fetal growth are summarized in Exhibit 24.2.

Hormones of Pregnancy (p. 561)

1. Pregnancy is maintained by human chorionic gonadotropin (hCG), estrogens, and progesterone.
2. Human chorionic somatomammotropin (hCS) facilitates breast development, decreases glucose catabolism, and

promotes the release of fatty acids as an alternate energy source.

3. Relaxin relaxes the pubic symphysis and helps dilate the uterine cervix toward the end of pregnancy.

4. Inhibin inhibits secretion of FSH and might regulate secretion of hGH.

Gestation (p. 563)

1. The time an embryo or fetus is carried in the uterus is called gestation.

2. Human gestation lasts about 38 weeks from the estimated day of fertilization.

3. During gestation, the mother undergoes anatomical and physiological changes in preparation for birth.

Prenatal Diagnostic Techniques (p. 563)

1. Amniocentesis is the withdrawal of amniotic fluid to diagnose biochemical defects and chromosomal disorders, such as hemophilia, Tay–Sachs disease, sickle-cell anemia, and Down syndrome.

2. Chorionic villi sampling (CVS) involves withdrawal of chorionic villi for chromosomal analysis.

3. CVS can be done earlier than amniocentesis and the results are available in a shorter time.

Labor (p. 563)

1. Labor is the process by which uterine contractions expel the fetus through the vagina to the outside.

2. The birth of a baby involves dilation of the cervix, expulsion of the fetus, and delivery of the placenta.

Lactation (p. 564)

1. Lactation refers to the secretion and ejection of milk by the mammary glands.

2. Secretion is influenced by prolactin (PRL), estrogens, and progesterone.

3. Ejection is influenced by oxytocin (OT).

Birth Control (p. 565)

1. Methods include sterilization (vasectomy, tubal ligation), hormonal, intrauterine devices, barriers (condom, vaginal pouch, diaphragm, cervical cap), chemical (spermicides), physiological (rhythm, sympto-thermal method), coitus interruptus, and induced abortion.

2. See Exhibit 24.3 for a summary of birth control methods.

Inheritance (p. 567)

1. Inheritance is the passage of hereditary traits from one generation to another.

2. The genetic makeup of an organism is its genotype. The traits expressed are its phenotype.

3. Dominant genes control a particular trait; expression of recessive genes is inhibited by dominant genes.

4. Most patterns of inheritance do not conform to the simple dominant–recessive patterns. Examples are incomplete dominance (sickle-cell anemia, multiple-allele inheritance of the ABO blood group) and polygenic inheritance (skin color).

5. A given phenotype is the result of the interactions of genotype and the environment.

6. Teratogens are agents that cause physical defects in developing embryos. They include thalidomide, drugs, alcohol, nicotine, and ionizing radiation.

7. Gender is determined by the Y chromosome of the male at fertilization.

8. Red–green color blindness and hemophilia primarily affect males because there are no counterbalancing dominant genes on the Y chromosomes.

Self-Quiz

1. Match the following:

___F___ a. early division of the zygote that increases cell number but not size

___C___ b. solid mass of cells 3–4 days following fertilization

___A___ c. a hollow ball of cells found in the uterine cavity about 5 days after fertilization

___D___ d. portion of the blastocyst that develops into the embryo

___B___ e. portion of the blastocyst that will form the chorion and placenta

___E___ f. results from fertilization of the sperm and egg

A. blastocyst
B. trophoblast
C. morula
D. inner cell mass
E. zygote
F. cleavage

2. Which of the following is the embryonic membrane that most closely surrounds the developing fetus?

(a) amnion   b. umbilicus   c. chorion   d. placenta   e. zona pellucida

3. Fetal circulation differs from that of an adult in that

a. the fetus transports oxygenated blood in the umbilical artery   (b) the gastrointestinal tract, kidneys, and lungs are not functioning   c. the heart is not functioning   d. the fetus utilizes CO_2 and the adult utilizes O_2 e. the blood flow is reversed in the arteries and veins

4. The foramen ovale is located

(a) between the atria   b. between the ventricles   c. between the right atrium and right ventricle   d. between the right ventricle and the pulmonary artery   e. between the pulmonary artery and pulmonary vein

5. The high levels of estrogens and progesterone needed to maintain pregnancy are provided by the placenta from

a. the 1st–3rd month   b. the 1st–9th month   c. the 1st–6th month   d. the 3rd–6th month   (e) the 3rd–9th month

6. The hormone that causes a home pregnancy test to show a positive result for pregnancy is

   a. FSH   b. progesterone   c. hCG   d. hGH   e. LH

7. The period of time from conception of the zygote to delivery of the fetus is called

   a. gestation   b. fertilization   c. placentation   d. gastrulation   e. implantation

8. Which of the following is NOT associated with maternal physiological changes during pregnancy?

   a. increased cardiac output   b. decreased pulmonary expiratory reserve volume   c. decreased gastrointestinal tract motility   d. increased frequency and urgency of urination   e. decreased production of estrogen

9. Homologous chromosomes

   a. contain genes that control the same trait   b. contain genes that control different traits   c. are inherited only from the mother   d. are inherited only from the father   e. contain all identical alleles

10. Sex-linked traits are carried on the

    a. autosomes   b. X and Y chromosomes   c. X chromosomes only   d. Y chromosomes only   e. Y chromosome in males and X chromosomes in females

11. Birth control pills are a combination of ovarian hormones that prevent pregnancy by

    a. neutralizing the pH of the vagina   b. inhibiting motility of the sperm   c. causing changes in the uterine lining   d. inhibiting the secretion of LH and FSH from the pituitary gland   e. chemically killing sperm

12. The afterbirth is expelled from the uterus during the _____ stage of labor.

    a. parturition   b. dilation   c. puerperium   d. placental   e. expulsion

13. The linings of the gastrointestinal tract, urinary bladder, and respiratory system develop from which of the following primary germ layers?

    a. amnion   b. chorion   c. ectoderm   d. endoderm   e. mesoderm

14. A person who is homozygous for a dominant trait on an autosome would have the genotype

    a. AA   b. Aa   c. aa   d. $X^A X^A$ e. $X^a X^a$

15. The genotype of a normal female is

    a. XX and 44 autosomes   b. XY and 44 autosomes   c. 46 autosomes   d. 46 X chromosomes   e. XX and 46 pairs of autosomes

Critical Thinking Applications

1. Your neighbors put up a sign to announce the birth of their twins, a girl and a boy. Another neighbor said "Oh, how sweet! I wonder if they're identical." Without even seeing the twins, what can you tell her?

2. Vanessa has been gaining weight despite trying to watch her diet. She has a history of irregular menstrual cycles and her best estimate of the time of her last menses is 3 or 4 months ago. Her doctor suggests that she may be pregnant and that a home pregnancy test may not be reliable at this point. Why not?

3. You are at the gas station when an obviously pregnant woman asks the attendant to call for help because her "waters just broke." What's happening to her?

4. A friend's child was called a "blue baby" when the hole in his heart did not close properly after his birth. The hole was later closed surgically and the baby's color is normal. Identify the hole in the heart in the fetus. Why does it need to close after birth?

5. Toward the end of pregnancy, a woman will often walk with a distinct "wobble" in her movement. Which hormone is most responsible for this wobbly gait and how does it produce this effect?

6. A woman with normal vision has a daughter with red–green color blindness. The mother claims that the daughter is the offspring of a man who has normal color vision. Is it possible for him to be the father of this child? Why? Why not?

Answers to Figure Questions

24.1   Functional changes experienced by sperm after they have been deposited in the female reproductive tract that allow them to fertilize a secondary oocyte.

24.2   Morula is a solid ball of cells; blastocyst is the rim of cells surrounding a cavity (blastocele) plus an inner cell mass.

24.3   It secretes digestive enzymes that eat away the endometrial lining at the site of implantation.

24.4   Human chorionic gonadotropin (hCG).

24.5   Embryonic disc.

24.6   The amnion functions as a shock absorber, whereas the chorion forms the fetal portion of the placenta.

24.7   Exchange of materials between fetus and mother.

24.8   Umbilical vein.

24.9   Human chorionic gonadotropin (hCG).

24.10 Complete dilation of the cervix.

24.11 Twenty-five percent.

24.12 Autosomes.

24.13 $X^c X^c$.

Answers to self-quizzes

Chapter 1

1. c 2. a. A b. B c. E d. C e. F f. G g. D 3. b.
4. a. skin and associated structures b. regulates body temperature, protects, eliminates wastes, makes vitamin D, receives stimuli c. lymphatic, immune d. protects against disease, filters body fluid, transports fats, returns proteins and plasma to the cardiovascular system e. respiratory f. nose, throat, larynx, trachea, lungs g. testes, ovaries, vagina, uterine tubes, uterus, penis h. reproduces the organism i. urinary j. regulates the composition of blood; eliminates wastes; regulates fluid and electrolyte balance and volume; helps maintain acid–base balance; helps control red blood cell production 5. D 6. d 7. a 8. c 9. d 10. a 11. c 12. D
13. b 14. a. D b. A c. H d. F e. G f. E g. C h. B 15. e
16. b

Chapter 2

1. d 2. c 3. b 4. b 5. a 6. d 7. e 8. c 9. d 10. a
11. e 12. c 13. d 14. b 15. e 16. c 17. a. R,D b. D
c. D d. R e. R f. D g. D h. R i. R,D j. R,D 18. a. F b.
D c. A d. E e. C f. B

Chapter 3

1. b 2. a 3. c 4. c 5. c 6. b 7. a. B b. F c. G d. H
e. C f. E g. D h. A 8. a 9. d 10. b 11. b 12. a. C b. E
c. F d. B e. D f. A 13. d 14. a 15. e

Chapter 4

1. c 2. d 3. e 4. a 5. c 6. a 7. a. C b. G c. E d. D
e. H f. F g. B h. A i. I 8. b 9. d 10. b

Chapter 5

1. d 2. e 3. a 4. e 5. e 6. b 7. c 8. a 9. a 10. a
11. b 12. a 13. c 14. b 15. d 16. b 17. a. D b. F c.
E d. G e. H f. A g. C h. B i. I

Chapter 6

1. b 2. d 3. d 4. b 5. d 6. c 7. a 8. a 9. a 10. a.
E b. D c. A d. C e. B 11. e 12. a. AX b. AP c. AP d.
AX e. AP f. AP g. AP h. AX i. AP j. AP k. AX l. AP m.
AX n. AP o. AX p. AP q. AX r. AX s. AP t. AX u. AX v.
AP w. AX x. AX y. AX z. AP aa. AP bb. AP cc. AX dd. AX
13. a. C b. E c. B d. D e. A

Chapter 7

1. d 2. b 3. e 4. d 5. e 6. c 7. b 8. c 9. b 10. c
11. c. 12. a 13. a. A b. G c. B d. C e. D f. E g. F h. J
i. I j. H

Chapter 8

1. c 2. c 3. e 4. b 5. b 6. d 7. a 8. c 9. c 10. a
11. d 12. b 13. e 14. b 15. c 16. a 17. b 18. d 19.
a. B b. C c. E d. G e. N f. H g. O h. P i. T j. I k. Y l. D
m. Q n. W o. X p. M q. V r. K s. A t. L u. S v. U w. R
x. J y. F

Chapter 9

1. c 2. b 3. e 4. d 5. d 6. c 7. d 8. c 9. a. F b. T
c. F d. T 10. a. H b. A c. D d. E e. I f. G g. J h. F i.
C j. B k. L l. N m. K n. M 11. d 12. e

Chapter 10

1. c 2. a 3. e 4. d 5. a 6. c 7. c 8. b 9. a 10. a
11. e 12. a 13. e 14. b 15. a and e 16. c 17. b

Chapter 11

1. d 2. e 3. a 4. c 5. e 6. d 7. a. F b. T c. T d. F
8. e 9. d

Chapter 12

1. a. J b. E c. F d. I e. C f. G g. H h. A i. B j. D 2. e 3. b 4. b 5. a 6. a 7. b 8. d 9. a 10. D 11. b 12. b 13. e 14. a 15. c 16. c 17. b

Chapter 13

1. d 2. e 3. c 4. a 5. e 6. d 7. c 8. d 9. b 10. b 11. c 12. c 13. a 14. b 15. a. B b. C c. E d. A e. D f. G g. F 16. a 17. c 18. b 19. a

Chapter 14

1. c 2. e 3. b 4. b 5. d 6. c 7. c 8. c 9. a. A b. D c. C d. A e. B 10. IV, VI, II, V, VII, I, III 11. c 12. D 13. A 14. A 15. c

Chapter 15

1. c 2. a 3. e 4. e 5. b 6. a 7. c 8. d 9. b 10. a. A b. A c. A d. B e. A f. B g. B h. A i. B j. A k. B l. B 11. d. 12. b 13. a 14. a. B b. G c. A d. F e. D f. C g. E

Chapter 16

1. b 2. a 3. c 4. d 5. e 6. a. E b. D c. B d. C e. A 7. c 8. a. H b. D c. E d. F e. A f. B g. C h. I i. G 9. b 10. b 11. e

Chapter 17

1. c 2. c 3. d 4. a 5. a 6. b 7. b 8. d 9. b 10. II, V, I, IV, VI, III 11. a. B b. C c. D d. E e. A f. G g. F 12. d 13. b 14. b 15. d 16. a 17. b 18. b

Chapter 18

1. d 2. c 3. b 4. a 5. c 6. a 7. c 8. e 9. a 10. d 11. a 12. d. 13. c 14. b 15. e

Chapter 19

1. c 2. e 3. b 4. A 5. a 6. d 7. c 8. c 9. c 10. c 11. d 12. D 13. e 14. a. 15. e 16. c

Chapter 20

1. b 2. d 3. a. D b. E c. C d. B e. F f. A 4. e 5. b 6. a 7. c 8. c 9. a 10. b 11. 12. a 14. d

Chapter 21

1. c 2. c 3. d 4. d 5. d 6. b 7. d 8. a 9. b 10. d 11. b 12. a 13. d 14. c 15. a

Chapter 22

1. e 2. c 3. e 4. d 5. c 6. a 7. b 8. d 9. b 10. d 11. c 12. a. E b. A c. D d. F e. C f. B 13. c 14. a 15. c

Chapter 23

1. d 2. d 3. c 4. c 5. a 6. c 7. b 8. a. B b. F c. A d. E e. D f. C 9. b 10. a 11. c 12. c 13. b 14. a

Chapter 24

1. a. F b. C c. A d. D e. B f. E g. E 2. a 3. b 4. A 5. e 6. c 7. a 8. e 9. a 10. c 11. d 12. D 13. D 14. A 15. a.

Glossary of terms

Pronunciation Key

1. The strongest accented syllable appears in capital letters, for example, bilateral (bī-LAT-er-al) and diagnosis (dī-ag-NŌ-sis).

2. If there is a secondary accent, it is noted by a prime symbol ('), for example, constitution (kon'-sti-TOO-shun) and physiology (fiz'-ē-OL-ō-jē). Any additional secondary accents are also noted by a prime symbol, for example, decarboxylation (dē'-kar-bok'-si-LĀ-shun).

3. Vowels marked with a line above the letter are pronounced with the long sound, as in the following common words:
ā as in māke
ē as in bē
ī as in īvy
ō as in pōle

4. Vowels not so marked are pronounced with the short sound, as in the following words:
e as in *bet*
i as in *sip*
o as in *not*
u as in *bud*

5. Other phonetic symbols are used to indicate the following sounds:
a as in *above*
oo as in *sue*
yoo as in *cute*
oy as in *oil*

Abdomen (ab-DŌ-men) The area between the diaphragm and the pelvis.

Abdominal (ab-DOM-i-nal) **cavity** Superior portion of the abdominopelvic cavity that contains the stomach, spleen, liver, gallbladder, pancreas, small intestine, and most of the large intestine.

Abdominal thrust maneuver A first-aid procedure for choking. Employs a quick, upward thrust against the diaphragm that forces air out of the lungs with sufficient force to get any lodged material. Also called the *Heimlich* (HĪM-lik) *maneuver.*

Abdominopelvic (ab-dom'-i-nō-PEL-vic) **cavity** Inferior component of the ventral body cavity that is subdivided into an upper abdominal cavity and a lower pelvic cavity.

Abduction (ab-DUK-shun) Movement away from the axis or midline of the body or one of its parts.

Abortion (a-BOR-shun) The premature loss (spontaneous) or removal (induced) of the embryo or nonviable fetus; any failure in the normal process of developing or maturing.

Abrasion (a-BRĀ-zhun) A portion of skin that has been scraped away.

Abscess (AB-ses) A localized collection of pus and liquefied tissue in a cavity.

Absorption (ab-SORP-shun) The taking up of liquids by solids or of gases by solids or liquids; intake of fluids or other substances by cells of the skin or mucous membranes; the passage of digested foods from the gastrointestinal tract into blood or lymph.

Accessory duct A duct of the pancreas that empties into the duodenum about 2.5 cm (1 in.) superior to the ampulla of Vater (hepatopancreatic ampulla). Also called the *duct of Santorini* (san'-tō-RĒ-ne).

Accommodation (a-kom-ō-DĀ-shun) A change in the curvature of the eye lens to adjust for vision at various distances; focusing.

Acetabulum (as'-e-TAB-yoo-lum) The rounded cavity on the external surface of the hipbone that receives the head of the femur.

Acetylcholine (as'-ē-til-KŌ-lēn) **(ACh)** A neurotransmitter, liberated at synapses in the central nervous system, that stimulates skeletal muscle contraction.

Achille's tendon *See Calcaneal tendon.*

Acid (AS-id) A proton donor, or substance that dissociates into hydrogen ions (H⁺) and anions, characterized by an excess of hydrogen ions and a pH less than 7.

Acidosis (as-i-DŌ-sis) A condition in which blood pH ranges from 7.35 to 6.80 or lower.

Acini (AS-i-nē) Masses of cells in the pancreas that secrete digestive enzymes.

Acquired immunodeficiency syndrome (AIDS) Caused by a virus called human immunodeficiency virus (HIV). A disorder characterized by a positive HIV-antibody test, low T4 (helper) cell count, and certain indicator diseases (for example Kaposi's sarcoma, *Pneumocystis carinii* pneumonia, tuberculosis, fungus diseases). Other symptoms include fever or night sweats, coughing, sore throat, fatigue, body aches, weight loss, and enlarged lymph nodes.

Acromegaly (ak'-rō-MEG-a-lē) Condition caused by hypersecretion of human growth hormone (hGH) during adulthood, characterized by thickened bones and enlargement of other tissues.

Actine (AK-tin) The contractile protein that makes up thin myofilaments in a muscle fiber (cell).

Action potential A wave of negativity that self-propagates along the outside surface of the membrane of a neuron or muscle fiber (cell); a rapid change in membrane potential that involves a depolarization following a repolarization. Also called a *nerve action potential nerve impulse* as it relates to a neuron and a *muscle action potential* as it relates to a muscle fiber (cell).

Active transport The movement of substances, usually ions, across cell membranes, against a concentration gradient, requiring the expenditure of energy (ATP).

Acupuncture (AK-yoo-punk'-chur) The insertion of a needle into a tissue for the purpose of drawing fluid or relieving pain. It is also an ancient Chinese practice employed to cure illnesses by inserting needles into specific locations of the skin.

Acute (a-KYOOT) Having rapid onset, severe symptoms, and a short course; not chronic.

Adam's apple *See Thyroid cartilage.*

Adaptation (ad'-ap-TĀ-shun) The adjustment of the pupil of the eye to light variations. The property by which a neuron relays a decreased frequency of action potentials from a receptor even though the strength of the stimulus remains constant; the decrease in perception of a sensation over time while the stimulus is still present.

Adduction (ad-DUK-shun) Movement toward the axis or midline of the body or one of its parts.

Adenohypophysis (ad'-e-nō-hī-POF-i-sis) The anterior portion of the pituitary gland.

Adenoids (AD-e-noyds) Inflamed and enlarged pharyngeal tonsils.

Adenosine triphosphate (a-DEN-ō-sēn-tri-FOS-fāt) (*ATP*) The universal energy-carrying molecule manufactured in all living cells as a means of capturing and storing energy. It consists of the purine base *adenine* and the five-carbon sugar *ribose,* to which are added, in linear array, three *phosphate* molecules.

Adenylate cyclase (a-DEN-i-lāt SĪ-Klās) An enzyme in the postsynaptic membrane that is activated when certain neurotransmitters bind to their receptors; the enzyme converts ATP into cyclic AMP.

Adipocyte (AD-i-pō-sīt) Fat cell, derived from a fibroblast.

Adrenal cortex (a-DRĒ-nal KOR-teks) The outer portion of an adrenal gland, divided into three zones, each of which has a different cellular arrangement and secretes different hormones.

Adrenal glands Two glands located superior to each kidney. Also called the *suprarenal* (soo'-pra-RĒ-nal) *glands.*

Adrenal medulla (me-DUL-a) The inner portion of an adrenal gland, consisting of cells that secrete epinephrine and norepinephrine (NE) in response to the stimula-tion of preganglionic sympathetic neurons.

Adrenergic (ad'-ren-ER-jik) **fiber** A nerve fiber that when stimulated releases norepinephrine (noradrenaline) at a synapse.

Adrenocorticotropic (ad-rē'-nōkor-ti-kŌ-TRŌ-pik) **hormone** (**ACTH**) A hormone produced by the anterior pituitary gland that influences the production and secretion of certain hormones of the adrenal cortex.

Adrenoglomerulotropin (a-drē'-nō-glō-mer'-yoo-lō-TRŌ-pin) A hormone secreted by the pineal gland that may stimulate aldosterone secretion.

Adventitia (ad-ven-TISH-ya) The outermost covering of a structure or organ.

Aerobic (air-Ō-bik) Requiring molecular oxygen.

Afferent arteriole (AF-er-ent ar-TĒ-rē-ōl) A blood vessel of a kidney that breaks up into a capillary network called a glomerulus, there is one afferent arteriole for each glomerulus.

Agglutination (a-gloo'-ti-NĀ-shun) Clumping of microorganisms or blood corpuscles; typically an antigen–antibody reaction.

Aggregated lymphatic follicles Aggregated lymphatic nodule that are most numerous in the ilium. Also called *Peyer's* (PĪ-erz) *patches.*

Aging Normal process accompanied by a progessive alteration of the body's homeostatic adaptive responses.

AIDS See *Acquired immunodeficiency syndrome.*

Albinism (AL-bin-izm) Abnormal, nonpathological, partial, or total absence of pigment in skin, hair, and eyes.

Albumin (al-BYOO-min) The most abundant (60 percent) and smallest of the plasma proteins, which functions primarily to regulate osmotic pressure of plasma.

Aldosterone (al-do-STĒR-ōn) A mineralocorticoid produced by the adrenal cortex that brings about sodium and water reabsorption and potassium excretion.

Alimentary (al-i-MEN-ta-rē) Pertaining to nutrition.

Alkaline (Al-ka-līn) Containing more hydroxyl ions (OH$^-$) than hydrogen ions (H$^+$) to produce a pH higher than 7.

Alkalosis (al-ka-LŌ-sis) A condition in which blood pH ranges from 7.45 to 8.00 or higher.

Alleles (a-LĒLS) Genes that control the same inherited trait (such as height or eye color) and are located on the same position (locus) on homologous chromosomes.

Allergen (Al-er-jen) An antigen that evokes a hypersensitivity reaction.

Allergic (a-LER-jik) Pertaining to or sensitive to an allergen.

All-or-none principle In muscle physiology, muscle fibers (cells) of a motor unit contract to their fullest extent or not at all. In neuron physiology, if a stimulus is strong enough to initiate an action potential, a nerve impulse is transmitted along the entire neuron at a constant and minimum strength.

Alveolar (al-VĒ-ō-lar)–**capillary membrane** Thin membrane formed by the alveolar wall, capillary endothelial cells, and their basement membranes, through which respiratory gases are exchanged between alevoli and pulmonary blood (see *external respiration*).

Alveolar (al-V-Ē-ō-lar) **duct** Branch of a respiratory bronchiole around which alveoli and alveolar sacs are arranged.

Alveolar macrophage (MAK-rō-fāj) Cell found in the alveolar walls of the lungs that is highly phagocytic. Also called a *dust cell.*

Alveolar sac A collection or cluster of alveoli that share a common opening.

Alveolus (al-VĒ-ō-lus) A small hollow or cavity; an air sac in the lungs; milk-secreting portion of a mammary gland. *Plural, alveoli* (al-VĒ-ō-lī).

Alzheimer's (ALTZ-hī-merz) **disease** (**AD**) Disabling neurological disorder characterized by dysfunction and death of specific cerebral neurons, resulting in widespread intellectual impairment, personality changes, and fluctuations in alertness.

Amenorrhea (ā-men-ō-RĒ-a) Absence of menstruation.

Amino acid An organic acid, containing a carboxyl group (COOH) and an amino group (NH$_2$), that is the building unit from which proteins are formed.

Amnesia (am-NĒ-zē-a) A lack or loss of memory.

Amniocentesis (am'-nē-ō-sen-TĒ-sis) Removal of amniotic fluid by inserting a needle transabdominally into the amniotic cavity.

Amnion (AM-nē-on) The innermost fetal membrane; a thin transparent sac that holds the fetus suspended in amniotic fluid. Also called the "*bag of waters.*"

Amniotic (am′-nē-OT-ik) **fluid** Fluid in the amniotic cavity, the space between the developing embryo (or fetus) and amnion; the fluid is initially produced as a filtrate from maternal blood and later from fetal urine.

Amphiarthrosis (am′-fē-ar-THRŌ-sis) Articulation midway between a diarthrosis and synarthrosis, in which the articulating bony surfaces are separated by an elastic substance to which both are attached, so that the mobility is slight.

Ampulla of Vater *See Hepatopancreatic ampulla.*

Anabolism (a-NAB-ō-lizm) Synthetic energy-requiring reactions whereby small molecules are built up into larger ones.

Anaerobic (an-AIR-ō-bik) Not requiring molecular oxygen.

Analgesia (an-al-JĒ-zē-a) Pain relief.

Anamnestic (an′-am-NES-tik) **response** Accelerated, more intense production of antibodies on a subsequent exposure to an antigen after the initial exposure.

Anaphase (AN-a-fāz) The third stage of mitosis in which the chromatids that have separated at the centromeres move to opposite poles of the cell.

Anaphylaxis (an′-a-fi-LAK-sis) Against protection; a hypersensitivity (allergic) reaction in which IgE antibodies attach to mast cells and basophils, causing them to produce mediators of anaphylaxis (histamine, leukotrienes, kinins, and prostaglandins) that bring about increased blood permeability, increased smooth muscle contraction and increased mucus production. Examples are hayfever, hives, and anaphylactic shock.

Anatomical (an′-a-TOM-i-kal) **position** A position of the body universally used in anatomical descriptions in which the body is erect, facing the observer, the upper limbs are at the sides, the palms of the hands are facing forward, and the feet are on the floor.

Anatomic dead space Spaces of the nose, pharynx, larynx, trachea, bronchi, and bronchioles that contain 150 ml of tidal volume; the air does not reach the alveoli to participate in gas exchange.

Anatomy (a-NAT-ō-mē) The structure or study of structure of the body and the relation of its parts to each other.

Androgen (AN-drō-jen) Substance producing or stimulating male characteristics, such as the male hormone testosterone.

Anemia (a-NĒ-mē-a) Condition of the blood in which the number of functional red blood cells or their hemoglobin content is below normal.

Anesthesia (an′-es-THĒ-zha) A total or partial loss of feeling or sensation, usually defined with respect to loss of pain sensation; may be general or local.

Aneurysm (AN-yoo-rizm) A saclike enlargement of a blood vessel caused by a weakening of its wall.

Angina pectoris (an-JĪ-na or AN-ji-na PEK-tō-ris) A pain in the chest related to reduced coronary circulation that may or may not involve heart or artery disease.

Angiotensin (an-jē-ō-TEN-sin) Either of two forms of a protein associated with regulation of blood pressure. Angiotensin I is produced by the action of renin on angiotensinogen and is converted by the action of a plasma enzyme into angiotensin II, which stimulates aldosterone secretion by the adrenal cortex.

Anomaly (a-NOM-a-lē) An abnormality that may be a developmental (congenital) defect; a variant from the usual standard.

Anorexia nervosa (an-ō-REK-sē-a ner-VŌ-sa) A chronic disorder characterized by self-induced weight loss, body-image and other perceptual disturbances, and physiologic changes that result from nutritional depletion.

Antagonist (an-TAG-ō-nist) A muscle that has an action opposite that of the prime mover (agonist) and yields to the movement of the prime mover.

Anterior (an-TĒR-ē-or) Nearer to or at the front of the body. Also called *ventral.*

Anterolateral (an′-ter-ō-LAT-er-al) **pathway** Sensory pathway that conveys information related to pain, temperature, crude touch, pressure, tickle, and itch.

Antibiotic (an′-ti-bī-OT-ik) Literally, "anti-life"; a chemical produced by a microorganism that is able to inhibit the growth of or kill other microorganisms.

Antibody (AN-ti-bod′-ē) A protein produced by certain cells in response to a specific antigen; the antibody combines with that antigen to neutralize, inhibit, or destroy it. Also called an *immunoglobulin* (im-yoo-nō-GLOB-yoo-lin) or *Ig.*

Antibody-mediated immunity That component of immunity in which lymphocytes (B cells) develop into plasma cells that produce antibodies that destroy antigens. Also called *humoral* (YOO-mor-al) *immunity.*

Anticoagulant (an-ti-cō-AG-yoo-lant) A substance that is able to delay, suppress, or prevent the clotting of blood.

Antidiuretic (an′-ti-dī-yoo-RET-ik) Substance that inhibits urine formation.

Antidiuretic hormone (ADH) Hormone produced by neurosecretory cells in the paraventricular and supraoptic nuclei of the hypothalamus that stimulates water reabsorption from kidney cells into the blood and vasoconstriction of arterioles. Also called *vasopressin* (vāz-ō-PRESS-in).

Antigen (AN-ti-jen) Any substance that when introduced into the tissues or blood induces the formation of antibodies and reacts only with those specific antibodies.

Anuria (a-NOO-rē-a) A daily urine output of less than 50 ml.

Anus (Ā-nus) The distal end and outlet of the rectum.

Aorta (ā-OR-ta) The main systemic trunk of the arterial system of the body; emerges from the left ventricle.

Aphasia (a-FĀ-zē-a) Loss of ability to express oneself properly through speech or loss of verbal comprehension.

Apnea (AP-ne-a) Temporary cessation of breathing.

Aponeurosis (ap′-ō-noo-RŌ-sis) A sheetlike tendon joining one muscle with another or with bone.

Apoptosis (ap-ō-TŌ-sis) A normal type of cell death that removes unneeded cells during embryological development, regulates the number of cells in tissues, and eliminates many potentially dangerous cells such as cancer cells.

Appendicitis (a-pen-di-SĪ-tis) Inflammation of the vermiform appendix.

Aqueous humor (AK-wē-us HYOO-mor) The watery fluid, similar in composition to cerebrospinal fluid, that fills the anterior cavity of the eye.

Arachnoid (a-RAK-noyd) The middle of the three coverings (meninges) of the brain or spinal cord.

Areflexia (a′-rē′-FLEK-sē-a) Absence of reflexes.

Areola (a-RĒ-ō-la) Any tiny space in a tissue. The pigmented ring around the nipple of the breast.

Arm The portion of the upper limb from the shoulder to the elbow.

Arrector pili (a-REK-tor PI-lē) Smooth muscles attached to hairs; contraction pulls the hairs into a more vertical position, resulting in "goose bumps."

Arrhythmia (a-RITH-mē-a) Irregular heart rhythm. Also called a *dysrhythmia.*

Arteriole (ar-TĒ-rē-ōl) A small, almost microscopic artery that delivers blood to a capillary.

Arteriosclerosis (ar-tē-rē-ō-skle-RŌ-sis) Group of diseases characterized by thickening of the walls of arteries and loss of elasticity.

Artery (AR-ter-ē) A blood vessel that carries blood away from the heart.

Arthritis (ar-THRĪ-tis) Inflammation of a joint.

Arthroscopy (ar-THROS-co-pē) A procedure for examining the interior of a joint, usually the knee, by inserting an arthroscope into a small incision; used to determine extent of damage, remove torn cartilage, repair cruciate ligaments, and obtain samples for analysis.

Articular (ar-TIK-yoo-lar) **capsule** Sleevelike structure around a synovial joint composed of a fibrous capsule and a synovial membrane.

Articular cartilage (KAR-ti-lij) Hyaline cartilage attached to articular bone surfaces.

Articular disc Fibrocartilage pad between articular surfaces of bones of some synovial joints. Also called a *meniscus* (men-IS-cus).

Articulation (ar-tik'-yoo-LĀ-shun) A joint; a point of contact between bones, cartilage and bones, or teeth and bones.

Artificial pacemaker A device that generates and delivers electrical signals to the heart to maintain a regular heart rhythm.

Aseptic (ā-SEP-tik) Free from any infectious or septic material.

Asphyxia (as FIX-ē-a) Unconsciousness due to interference with the oxygen supply of the blood.

Aspiration (as'-pi-RĀ-shun) Inhalation of a foreign substance (water, food, or foreign body) into the bronchial tree; drainage of a substance in or out by suction.

Association area A portion of the cerebral cortex connected by many motor and sensory fibers to other parts of the cortex. The association areas are concerned with motor patterns, memory, concepts of word-hearing and word-seeing, reasoning, will, judgment, and personality traits.

Asthma (AZ-ma) Usually allergic reaction characterized by smooth muscle spasms in bronchi resulting in wheezing and difficult breathing. Also called *bronchial asthma.*

Astigmatism (a-STIG-ma-tizm) An irregularity of the lens or cornea of the eye causing the image to be out of focus and producing faulty vision.

Astrocyte (AS-trō-sīt) A neuroglial cell having a star shape that supports neurons in the brain and spinal cord and attaches the neurons to blood vessels.

Ataxia (a-TAK-sē-a) A lack of muscular coordination, lack of precision.

Atelectasis (at'-ē-LEK-ta-sis) A collapsed or airless state of all or part of the lung, which may be acute or chronic.

Atherosclerosis (ath'-er-ō-skle-RŌ-sis) A process in which fatty substances (cholesterol and triglycerides) are deposited in the walls of medium and large arteries in response to certain stimuli (hypertension, carbon monoxide, dietary cholesterol). Following endothelial damage, monocytes stick to the tunica interna, develop into macrophages, and take up cholesterol and low-density lipoproteins. Smooth muscle fibers (cells) in the tunica media ingest cholesterol. This results in the formation of an atherosclerotic plaque that decreases the size of the arterial lumen.

Atom Unit of matter that comprises a chemical element; consists of a nucleus and electrons.

Atomic number Number of protons in an atom.

Atrial fibrillation (Ā-trē-al fib-ri-LĀ-shun) Asynchronous contraction of the atria that results in the cessation of atrial pumping.

Atrial natriuretic (na'-trē-yoo-RET-ik) **peptide** (ANP) Peptide hormone produced by the atria of the heart in response to stretching that inhibits aldosterone production and thus lowers blood pressure.

Atrioventricular (AV) (ā'-trē-ō-ven-TRIK-yoo-lar) **bundle** The portion of the conduction system of the heart that begins at the atrioventricular (AV) node, passes through the cardiac skeleton separating the atria and the ventricles, then runs a short distance down the interventricular septum before splitting into right and left bundle branches. Also called the *bundle of His* (HISS).

Atrioventricular (AV) **node** The portion of the conduction system of the heart made up of a compact mass of conducting cells located near the orifice of the coronary sinus in the right atrial wall.

Atrioventricular (AV) **valve** A structure made up of membranous flaps or cusps

that allows blood to flow in one direction only, from an atrium into a ventricle.

Atrium (Ā-trē-um) A superior chamber of the heart.

Atrophy (AT-rō-fē) Wasting away or decrease in size of a part, due to a failure, abnormality of nutrition, or lack of use.

Auditory ossicle (AW-di-tō-rē- OS-si-kul) One of the three small bones of the middle ear called the malleus, incus, and stapes.

Auditory tube The tube that connects the middle ear with the nose and nasopharynx region of the throat. Also called the *Eustachian* (yoo-STĀ-kē-an) *tube.*

Auscultation (aws-kul-TĀ-shun) Examination by listening to sounds in the body.

Autoimmunity An immunologic response against a person's own tissue antigens.

Autologous preoperative transfusion (aw-TOL-ō-gus prē-OP-er-a-tiv trans-FYOO-zhun) Donating one's own blood up to six weeks before elective surgery to ensure an abundant supply and reduce transfusion complications such as those that may be associated with diseases such as AIDS and hepatitis. Also called *predonation.*

Autolysis (aw-TOL-i-sis) Destruction of a cell by its own lysosomal enzymes.

Autonomic nervous system (ANS) Visceral efferent neurons, both sympathetic and parasympathetic, that transmit nerve impulses from the central nervous system to smooth muscle, cardiac muscle, and glands; so named because this portion of the nervous system was thought to be self-governing or spontaneous.

Autophagy (aw-TOF-a-jē) Process by which worn-out organelles are digested within lysosomes.

Autopsy (AW-top-sē) The examination of the body after death.

Autoregulation (aw-tō-reg-yoo-LĀ-shun) A local, automatic adjustment of blood flow in a given region of the body in response to tissue needs.

Autosome (AW-tō-sōm) Any chromosome other than the pair of sex chromosomes.

Axilla (ak-SIL-a) The small hollow beneath the arm where it joins the body at the shoulders. Also called the *armpit.*

Axon (AK-son) The usually single, long process of a nerve cell that carries a nerve impulse away from the cell body.

Babinski (ba-BIN-skē) **sign** Extension of the great toe, with or without fanning of the other toes, in response to stimulation of the outer margin of the sole of the foot; normal up to 1½ years of age.

Ball-and-socket joint A synovial joint in which the rounded surface of one bone moves within a cup-shaped depression or fossa of another bone, as in the shoulder or hip joint. Also called a *spheroid* (SFĒR-oyd) *joint.*

Baroreceptor (bar′-ō-re-SEP-tor) Nerve cell capable of responding to changes in blood pressure. Also called a *pressoreceptor.*

Bartholin's glands See *Greater vestibular glands.*

Basal ganglia (BĀ-sal GANG-glē-a) Paired clusters of cell bodies that make up the central gray matter in each cerebral hemisphere, including the caudate nucleus, lentiform nucleus, claustrum, and amygdaloid body. Also called *cerebral nuclei* (SER-e-bral NOO-klē-ī).

Basal metabolic (met′-a-BOL-ik) **rate (BMR)** The rate of metabolism measured under standard or basal conditions.

Base The broadest part of a pyramidal structure. A nonacid or a proton acceptor, characterized by excess of hydroxide ions (OH^-) and a pH greater than 7. A ring-shaped, nitrogen-containing organic molecule that is one of the components of a nucleotide, for example, adenine, guanine, cytosine, thymine, and uracil.

Basement membrane Thin, extracellular layer consisting of basal lamina secreted by epithelial cells and reticular lamina secreted by connective tissue cells.

Basilar (BAS-i-lar) **membrane** A membrane in the cochlea of the inner ear that separates the cochlear duct from the scala tympani and on which the spiral organ (organ of Corti) rests.

Basophil (BĀ-sō-fil) A type of white blood cell characterized by a pale nucleus and large granules that stain blue-purple with basic dyes.

B cell A lymphocyte that develops into an antibody-producing plasma cell or a memory cell.

Belly The abdomen. The gaster or prominent, fleshy part of a skeletal muscle.

Benign (be-NĪN) Not malignant; favorable for recovery; a mild disease.

Bilirubin (bil-ē-ROO-bin) A red pigment that is one of the end products of hemoglobin breakdown in the liver cells and is excreted as a waste material in the bile.

Biliverdin (bil-ē-VER-din) A green pigment that is one of the first products of hemoglobin breakdown in the liver cells and is

converted to bilirubin or excreted as a waste material in bile.

Biofeedback Process by which an individual gets constant signals (feedback) about various visceral biological functions.

Biopsy (BĪ-op-sē) Removal of tissue or other material from the living body for examination, usually microscopic.

Blastocele (BLAS-tō-sēl) The fluid-filled cavity within the blastocyst.

Blastocyst (BLAS-tō-sist) In the development of an embryo, a hollow ball of cells that consists of a blastocele (the internal cavity), trophoblast (outer cells), and inner cell mass.

Blastomere (BLAS-tō-mer) One of the cells resulting from the cleavage of a fertilized ovum.

Blastula (BLAS-tyoo-la) An early stage in the development of a zygote.

Blind spot Area in the retina at the end of the optic (II) nerve in which there are no light receptor cells.

Blood The fluid that circulates through the heart, arteries, capillaries, and veins and that constitutes the chief means of transport within the body.

Blood–brain barrier (BBB) A special mechanism that prevents the passage of materials from the blood to the cerebrospinal fluid and brain.

Blood pressure (BP) Pressure exerted by blood as it presses against and attempts to stretch blood vessels, especially arteries; the force is generated by the rate and force of heartbeat; clinically, a measure of the pressure in arteries during ventricular systole and ventricular diastole.

Blood–testis barrier A barrier formed by sustentacular (Sertoli) cells that prevents an immune response against antigens produced by sperm cells and developing cells by isolating the cells from the blood.

Body cavity A space within the body that contains various internal organs.

Body fluid Body water and its dissolved substances; comprises about 60 percent of total body weight.

Bolus (BŌ-lus) A soft, rounded mass, usually food, that is swallowed.

Bowman's capsule See *Glomerular capsule.*

Brachial plexus (BRĀ-kē-al PLEK-sus) A network of nerve fibers of the anterior rami of spinal nerves C5, C6, C7, C8, and T1. The nerves that emerge from the brachial plexus supply the upper limbs.

Bradycardia (brād′-i-KAR-dē-a) A slow heartbeat or pulse rate.

Brain A mass of nervous tissue located in the cranial cavity.

Brain stem The portion of the brain immediately superior to the spinal cord, made up of the medulla oblongata, pons, and midbrain.

Brain wave Electrical activity produced as a result of action potentials of brain cells.

Broca's (BRŌ-kaz) **area** Motor area of the brain in the frontal lobe that translates thoughts into speech. Also called the *motor speech area.*

Bronchi (BRONG-kē) Branches of the respiratory passageways including primary bronchi (the two divisions of the trachea), secondary or lobar bronchi (divisions of the primary that are distributed to the lobes of the lung), and tertiary or segmental bronchi (divisions of the secondary that are distributed to bronchopulmonary segments of the lung).

Bronchial tree The trachea, bronchi, and their branching structures.

Bronchiole (BRONG-kē-ōl) Branch of a tertiary bronchus further dividing into terminal bronchioles (distributed to lobules of the lung), which divide into respiratory bronchioles (distributed to alveolar sacs).

Bronchitis (brong-KĪ-tis) Inflammation of the bronchi characterized by hypertrophy and hyperplasia of seromucous glands and goblet cells that line the bronchi and which results in a productive cough.

Bronchogenic carcinoma (brong′-kō-JEN-ik kar′-si-NŌ-ma) Cancer originating in the bronchi.

Bronchoscopy (brong-KOS-kō-pē) Visual examination of the interior of the trachea and bronchi with a bronchoscope to biopsy a tumor, clear an obstruction, take cultures, stop bleeding, or deliver drugs.

Bronchus (BRONG-kus) One of the two large branches of the trachea. *Plural, bronchi* (BRONG-kē).

Buccal (BUK-al) Pertaining to the cheek or mouth.

Buffer (BUF-er) **system** A pair of chemicals, one a weak acid and one the salt of the weak acid, which functions as a weak base, that resists changes in pH.

Bulbourethral (bul′-bō-yoo-RĒ-thral) **gland** One of a pair of glands located inferior to the prostate gland on either side of the urethra that secretes an alkaline fluid into the cavernous urethra. Also called a *Cowper's* (KOW-perz) *gland.*

Bulimia (boo-LIM-ē-a) A disorder characterized by overeating, at least twice a week, followed by purging by self-induced vomiting, strict dieting or fasting, vigorous exercise, or use of laxatives.

Bulk flow The movement of large numbers of ions, molecules, or particles in the same direction as a result of pressure differences (osmotic, hydrostatic, or air pressure).

Bundle of His *See Atrioventricular (AV) bundle.*

Bunion (BUN-yun) Lateral deviation of the great toe that produces inflammation and thickening of the bursa, bone spurs, and calluses.

Burn An injury in which proteins are destroyed (denatured) as a result of heat (fire, steam), chemicals, electricity, or the ultraviolet rays of the sun.

Bursa (BUR-sa) A sac or pouch of synovial fluid located at friction points, especially about joints.

Bursitis (bur-SĪ-tis) Inflammation of a bursa.

Buttocks (BUT-oks) The two fleshy masses on the posterior aspect of the lower trunk, formed by the gluteal muscles.

Cachexia (kah-KEK-sē-ah) A state of ill health, malnutrition, and wasting.

Calcaneal tendon The tendon of the soleus, gastrocnemius, and plantaris muscles at the back of the heel. Also called the *Achilles* (a-KIL-ēz) *tendon.*

Calcification (kal′-si-fi-KĀ-shun) Hardening by deposits of calcium salts.

Calcitonin (kal-si-TŌ-nin) **(CT)** A hormone produced by the thyroid gland that lowers the calcium and phosphate levels of the blood by inhibiting bone breakdown and accelerating calcium absorption by bones.

Callus (KAL-lus) A growth of new bone tissue in and around a fractured area, ultimately replaced by mature bone. An acquired, localized thickening.

Calorie (KAL-ō-rē) A unit of heat. A calorie (cal) is the standard unit and is the amount of heat necessary to raise 1 g of water 1°C from 14° to 15°C. The kilocalorie (kcal), used in metabolic and nutrition studies, is the amount of heat necessary to raise 1000 g of water 1°C and is equal to 1000 cal.

Calyx (KĀL-iks) Any cuplike division of the kidney pelvis. *Plural, calyces* (KĀ-li-sēz).

Canaliculus (kan′-a-LIK-yoo-lus) A small channel or canal, as in bones, where they connect lacunae. *Plural, canaliculi* (kan′-a-LIK-yoo-lī).

Canal of Schlemm *See Scleral venous sinus.*

Cancer (KAN-ser) A malignant tumor of epithelial origin tending to infiltrate and give rise to new growths or metastases. Also called *carcinoma* (kar′-si-NŌ-ma).

Capacitation (ka′-pas-i-TĀ-shun) The functional changes that sperm undergo in the female reproductive tract that allow them to fertilize a secondary oocyte.

Capillary (KAP-i-lar′-ē) A microscopic blood vessel located between an arteriole and venule through which materials are exchanged between blood and body cells.

Carbohydrate (kar′-bō-HĪ-drāt) An organic compound containing carbon, hydrogen, and oxygen in a particular amount and arrangement and comprised of sugar subunits; usually has the formula $(CH_2O)_n$.

Carbon monoxide (CO) poisoning Hypoxia due to increased levels of carbon monoxide as a result of its preferential and tenacious combination with hemoglobin rather than with oxygen.

Carcinogen (kar-SIN-ō-jen) Any substance that causes cancer.

Carcinoma (kar′-si-NŌ-ma) A malignant tumor consisting of epithelial cells.

Cardiac (KAR-dē-ak) **arrest** Cessation of an effective heartbeat in which the heart is completely stopped or in ventricular fibrillation.

Cardiac cycle A complete heartbeat consisting of systole (contraction) and diastole (relaxation) of both atria plus systole and diastole of both ventricles.

Cardiac muscle An organ specialized for contraction, composed of striated muscle fibers (cells), forming the wall of the heart, and stimulated by an intrinsic conduction system and visceral efferent neurons.

Cardiac output (CO) The volume of blood pumped from one ventricle of the heart (usually measured from the left ventricle) in 1 min; about 5.2 liters/min under normal resting conditions.

Cardioacceleratory (kar-dē-ō-ak-SEL-er-a-tō-rē) **center (CAC)** A group of neurons in the medulla oblongata from which cardiac nerves (sympathetic) arise; nerve impulses along the nerves release epinephrine that increases the rate and force of heartbeat.

Cardioinhibitory (kar-dē-ō-in-HIB-i-tō-rē) **center (CIC)** A group of neurons in the medulla oblongata from which parasympathetic fibers that reach the heart via

the vagus (X) nerve arise; nerve impulses along the nerves release acetylcholine that decreases the rate and force of heartbeat.

Cardiology (kar-dē-OL-ō-jē) The study of the heart and diseases associated with it.

Cardiopulmonary resuscitation (rē-sus-i-TĀ-shun) **(CPR)** A technique employed to restore life or consciousness to a person apparently dead or dying; includes external respiration (exhaled air respiration) and external cardiac massage.

Cardiovascular (kar-dē-ō-VAS-kyoo-lar) **center** Groups of neurons scattered within the medulla oblongata that regulate heart rate, force of contraction, and blood vessel diameter.

Carotene (KAR-o-tēn) Antioxidant vitamin; yellow-orange pigment present in the stratum corneum of the epidermis. Accounts for the yellowish coloration of skin. Also termed *beta carotene.*

Carotid sinus A dilated region of the internal carotid artery immediately above the branching of the common carotid artery that contains receptors that monitor blood pressure.

Carpus (KAR-pus) A collective term for the eight bones of the wrist.

Cartilage (KAR-ti-lij) A type of connective tissue consisting of chondrocytes in lacunae embedded in a dense network of collagen and elastic fibers and a matrix of chondroitin sulfate.

Cast A small mass of hardened material formed within a cavity in the body and then discharged from the body; can originate in different areas and be composed of various materials.

Castration (kas-TRĀ-shun) The removal of the testes.

Catabolism (ka-TAB-ō-lizm) Chemical reactions that break down complex organic compounds into simple ones with the release of energy.

Catalyst (KAT-a-list) A substance that can speed up a chemical reaction by increasing the frequency of collisions or by lowering activation energy without itself being altered.

Cataract (KAT-a-rakt) Loss of transparency of the lens of the eye or its capsule or both.

Catheter (KATH-i-ter) A tube that can be inserted into a body cavity through a canal or into a blood vessel; used to remove fluids, such as urine and blood, and to introduce diagnostic materials or medication.

Cecum (SĒ-kum) A blind pouch at the proximal end of the large intestine to which the ileum is attached.

Cell The basic structural and functional unit of all organisms; the smallest structure capable of performing all the activities vital to life.

Cell division Process by which a cell reproduces itself that consists of a nuclear division (mitosis) and a cytoplasmic division (cytokinesis); types include somatic and reproductive cell division.

Cell inclusion A lifeless, often temporary, constituent in the cytoplasm of a cell as opposed to an organelle.

Cell-mediated immunity That component of immunity in which specially sensitized lymphocytes (T cells) attach to antigens to destroy them. Also called *cellular immunity.*

Cellular respiration *See Oxidation.*

Cementum (se-MEN-tum) Calcified tissue covering the root of a tooth.

Center of ossification (os′-i-fi-KĀ-shun) An area in the cartilage model of a future bone where the cartilage cells hypertrophy and then secrete enzymes that result in the calcification of their matrix, resulting in the death of the cartilage cells, followed by the invasion of the area by osteoblasts that then lay down bone.

Central canal A circular channel running longitudinally in the center of an osteon (Haversian system) of mature compact bone, containing blood and lymphatic vessels and nerves. Also called a *Haversian* (ha-VĒR-shun) *canal.* A microscopic tube running the length of the spinal cord in the gray commissure.

Central nervous system (CNS) That portion of the nervous system that consists of the brain and spinal cord.

Centrioles (SEN-trē-ōlz) Paired, cylindrical structures within a centrosome, each consisting of a ring of microtubules and arranged at right angles to each other; function in the formation and regeneration of flagella and cilia.

Centromere (SEN-trō-mēr) The clear, constricted portion of a chromosome where the two chromatids are joined; serves as the point of attachment for the chromosomal microtubules.

Centrosome (SEN-trō-sōm) A rather dense area of cytoplasm, near the nucleus of a cell, containing a pair of centrioles; organizes the mitotic spindle during cell division.

Cephalic (se-FAL-ik) Pertaining to the head; superior in position.

Cerebellum (ser-e-BEL-um) The portion of the brain lying posterior to the medulla oblongata and pons, concerned with coordination of movements.

Cerebral aqueduct (SER-ē-bral AK-we-dukt) A channel through the midbrain connecting the third and fourth ventricles and containing cerebrospinal fluid.

Cerebral arterial circle A ring of arteries forming an anastomosis at the base of the brain between the internal carotid and basilar arteries and arteries supplying the brain. Also called the *circle of Willis.*

Cerebral cortex The surface of the cerebral hemispheres, 2–4 mm thick, consisting of six layers of nerve cell bodies (gray matter) in most areas.

Cerebral palsy (PAL-zē) A group of motor disorders resulting in muscular uncoordination and loss of muscle control and caused by damage to motor areas of the brain (cerebral cortex, basal ganglia, and cerebellum) during fetal life, birth, or infancy.

Cerebrospinal (se-rē′-brō-SPĪ-nal) **fluid (CSF)** A fluid produced in the choroid plexuses and ependymal cells of the ventricles of the brain that circulates in the ventricles and the subarachnoid space around the brain and spinal cord.

Cerebrovascular (se-rē′-brō-VAS-kyoo-lar) **accident (CVA)** Destruction of brain tissue (infarction) resulting from disorders of blood vessels that supply the brain. Also called a *stroke.*

Cerebrum (SER-ē-brum) The two hemispheres of the forebrain, making up the largest part of the brain.

Cerumen (se-ROO-men) Waxlike secretion produced by ceruminous glands in the external auditory meatus (ear canal).

Ceruminous (se-ROO-mi-nus) **gland** A modified sudoriferous (sweat) gland in the external auditory meatus that secretes cerumen (ear wax).

Cervical dysplasia (dis-PLĀ-sē-a) A change in the shape, growth, and number of cervical cells of the uterus that, if severe, may progress to cancer.

Cervical plexus (PLEK-sus) A network of neuron fibers formed by the anterior rami of the first four cervical nerves.

Cervix (SER-viks) Neck; any constricted portion of an organ, such as the lower cylindrical part of the uterus.

Cesarean (se-SA-rē-an) **section** Procedure in which a low, horizontal incision is made through the abdominal wall and uterus for removal of the baby and placenta. Also called a *C-section.*

Chemical element Unit of matter that cannot be decomposed into a simpler substance by ordinary chemical reactions. Examples include hydrogen (H), carbon (C), and oxygen (O).

Chemical reaction The combination or breaking apart of atoms in which chemical bonds are formed or broken and new products with different properties are produced.

Chemoreceptor (kē′-mō-rē-SEP-tor) Receptor outside the central nervous system on or near the carotid and aortic bodies that detects the presence of chemicals.

Chemotherapy (kē-mō-THER-a-pē) The treatment of illness or disease by chemicals.

Chiropractic (kī-rō-PRAK-tik) A system of treating disease by using one's hands to manipulate body parts, mostly the vertebral column.

Chlamydia (kla-MID-ē-a) Most prevalent sexually transmitted disease, characterized by frequent, painful, or burning urination and low back pain.

Cholecystectomy (kō′-lē-sis-TEK-tō-mē) Surgical removal of the gallbladder.

Cholesterol (kō-LES-te-rol) Classified as a lipid, the most abundant steroid in animal tissues; located in cell membranes and used for the synthesis of steroid hormones and bile salts.

Cholinergic (kō′-lin-ER-jik) **fiber** A nerve ending that liberates acetylcholine at a synapse.

Chondrocyte (KON-drō-sīt) Cell of mature cartilage.

Chordae tendineae (KOR-dē TEN-di-nē-ē) Tendonlike, fibrous cords that connect the heart valves with the papillary muscles.

Chorion (KŌ-rē-on) The outermost fetal membrane that becomes the principal embryonic portion of the placenta; serves a protective and nutritive function.

Chorionic villi (kō′-rē-ON-ik VIL-ī) Finger-like projections of the chorion that grow into the decidua basalis of the endometrium and contain fetal blood vessels.

Chorionic villi sampling (CVS) The removal of a sample of chorionic villi tissue by means of a catheter to analyze the tissue for prenatal genetic defects.

Choroid (KŌ-royd) One of the vascular coats of the eyeball.

Choroid plexus (PLEK-sus) A vascular structure located in the roof of each of the four ventricles of the brain; produces cerebrospinal fluid.

Chromatid (KRŌ-ma-tid) One of a pair of identical connected nucleoprotein strands that are joined at the centromere and separate during cell division, each becoming a chromosome of one of the two daughter cells.

Chromatin (KRŌ-ma-tin) The threadlike mass of the genetic material consisting principally of DNA, which is present in the nucleus of a nondividing or interphase cell.

Chromotophilic substance Rough endoplasmic reticulum in the cell bodies of neurons that function in protein synthesis. Also called *Nissl bodies.*

Chromosome (KRŌ-mō-sōm) One of the 46 small, dark-staining bodies that appear in the nucleus of a human diploid (2*n*) cell during cell division.

Chronic (KRON-ik) Long term or frequently recurring; applied to a disease that is not acute.

Chronic obstructive pulmonary disease (COPD) Any disease, such as asthma, bronchitis, or emphysema, in which there is some degree of obstruction of air passageways.

Chylomicron (kī-lō-MĪK-ron) Protein-coated spherical structure that contains triglycerides, phospholipids, and cholesterol and is absorbed into the lacteal of a villus.

Chyme (kīm) The semifluid mixture of partly digested food and digestive secretions found in the stomach and small intestine during digestion of a meal.

Ciliary (SIL-ē-ar'-ē) **body** One of the three portions of the vascular tunic of the eyeball, the others being the choroid and the iris; includes the ciliary muscle and the ciliary processes.

Cilium (SIL-ē-um) A hair or hairlike process projecting from a cell that may be used to move the entire cell or to move substances along the surface of the cell.

Circadian (ser-KĀ-dē-an) **rhythm** A cycle of active and nonactive periods in organisms determined by internal mechanisms and repeating about every 24 hours.

Circle of Willis *See Cerebral arterial circle.*

Circulation time Time required for blood to pass from the right atrium, through pulmonary circulation, back to the left ventricle, through systemic circulation to the foot, and back again to the right atrium; normally about 1 min.

Circumcision (ser'-kum-SIZH-un) Surgical removal of the foreskin (prepuce), the fold of skin over the glans penis.

Circumduction (ser'-kum-DUK-shun) A movement at a synovial joint in which the distal end of a bone moves in a circle while the proximal end remains relatively stable.

Cirrhosis (si-RŌ-sis) A liver disorder in which the parenchymal cells are destroyed and replaced by connective tissue.

Cleavage The rapid mitotic divisions following the fertilization of a secondary oocyte, resulting in an increased number of progressively smaller cells, called blastomeres, so that the overall size of the zygote remains the same.

Cleft palate Condition in which the palatine processes of the maxillae do not unite before birth; cleft lip, a split in the upper lip, is often associated with cleft palate.

Climacteric (klī-mak-TER-ik) Cessation of the reproductive function in the female or diminution of testicular activity in the male.

Climax (KLĪ-max) The peak period of moments of greatest intensity during sexual excitement.

Clitoris (KLI-to-ris) An erectile organ of the female located at the anterior junction of the labia minora that is homologous to the male penis.

Clot The end result of a series of biochemical reactions that changes liquid plasma into a gelatinous mass; specially, the conversion of fibrinogen into a tangle of polymerized fibrin molecules.

Clot retraction (rē-TRAK-shun) The consolidation of a fibrin clot to pull damaged tissue together.

Coagulation (cō-ag-yoo-LĀ-shun) Process by which a blood clot is formed.

Cochlea (KŌK-lē-a) A winding, cone-shaped tube forming a portion of the inner ear and containing the spiral organ (organ of Corti).

Coenzyme A type of cofactor; a nonprotein organic molecule that is associated with and activates an enzyme; many are derived from vitamins. An example is nicotinamide adenine dinucleotide (NAD), derived from the B vitamin niacin.

Coitus (KŌ-i-tus) Sexual intercourse. Also called *copulation* (cop-yoo-LĀ-shun).

Colitis (kō-LĪ-tis) Inflammation of the mucosa of the colon and rectum in which absorption of water and salts is reduced, producing watery, bloody feces, and, in severe cases, dehydration and salt depletion. Spasms of the irritated muscularis produce cramps.

Collagen (KOL-a-jen) A protein that is the main organic constituent of connective tissue.

Colon The division of the large intestine consisting of ascending, transverse, descending, and sigmoid portions.

Color blindness Any deviation in the normal perception of colors, resulting from the lack of usually one of the photopigments of the cones.

Colostomy (kō-LOS-tō-mē) The diversion of feces through an opening in the colon, creating a surgical opening at the exterior of the abdominal wall.

Colostrum (kō-LOS-trum) A thin, cloudy fluid secreted by the mammary glands a few days prior to or after delivery before true milk is secreted.

Coma (KŌ-ma) Final stage of brain failure that is characterized by total unresponsiveness to all external stimuli.

Common bile duct A tube formed by the union of the common hepatic duct and the cystic duct that empties bile into the duodenum at the hepatopancreatic ampulla (ampulla of Vater).

Compact (dense) bone tissue Bone tissue that contains few spaces between osteons (Haversian systems); forms the external portion of all bones and the bulk of the diaphysis (shaft) of long bones.

Complement (KOM-ple-ment) A group of at least 20 proteins found in serum that forms a component of nonspecific resistance and immunity by bringing about cytolysis, inflammation, and opsonization.

Complete blood count (CBC) Hematology test that usually includes hemoglobin determination, hematocrit, red and white blood cell count, differential white blood cell count, and platelet count.

Compound A chemical substance composed of two or more different elements.

Computed tomography (tō-MOG-ra-fē) **(CT)** X-ray technique that provides a cross-sectional image of any area of the body. Also called *computed axial tomography* (**CAT**).

Concha (KONG-ka) A scroll-like bone found in the skull. *Plural, conchae* (KONG-kē).

Concussion (kon-KUSH-un) Traumatic injury to the brain that produces no visible bruising but may result in abrupt, temporary loss of consciousness.

Conduction myofiber Muscle fiber (cell) in a ventricle of the heart specialized for conducting an action potential to the myocardium; part of the conduction system of the heart. Also called a *Purkinje* (pur-KIN-je) *fiber.*

Conduction system A series of autorhythmic cardiac muscle fibers that generates and distributes electrical impulses to stimulate coordinated contraction of the heart chambers; includes the sinoatrial (SA)node, the atrioventricular (AV)node, the atrioventricular (AV) bundle, the right and left bundle branches, and the conduction myofibers (Purkinje fibers).

Conductivity (kon′-duk-TIV-i-tē) The ability to carry the effect of a stimulus from one part of a cell to another; highly developed in nerve and muscle fibers (cells).

Condyloid (KON-di-loyd) **joint** A synovial joint structured so that an oval-shaped condyle of one bone fits into an elliptical cavity of another bone, permitting side-to-side and back-and-forth movements, as at the joint at the wrist between the radius and carpals. Also called an *ellipsoidal* (e-lip-SOY-dal) *joint.*

Cone The light-sensitive receptor in the retina concerned with color vision.

Congenital (kon-JEN-i-tal) Present at the time of birth.

Congestive heart failure (CHF) Chronic or acute state that results when the heart is not capable of supplying the oxygen demands of the body.

Conjunctiva (kon′-junk-TĪ-va) The delicate membrane covering the eyeball and lining the eyes.

Conjunctivitis (kon-junk′-ti-VĪ-tis) Inflammation of the conjunctiva, the delicate membrane covering the eyeball and lining the eyelids.

Connective tissue The most abundant of the four basic tissue types in the body, performing the functions of binding and supporting; consists of relatively few cells in a generous matrix (the ground substance and fibers between the cells).

Consciousness (KON-shus-nes) A state of wakefulness in which an individual is fully alert, aware, and oriented as a result of feedback between the cerebral cortex and reticular activating system.

Constipation (con-sti-PĀ-shun) Infrequent or difficult defecation caused by decreased motility of the intestines.

Contraception (kon′-tra-SEP-shun) The prevention of conception or impregnation without destroying fertility.

Contractility (kon′-trak-TIL-i-tē) The ability of cells or parts of cells to generate force actively to undergo shortening and change form for purposeful movements. Muscle fibers (cells) exhibit a high degree of contractility.

Contusion (kon-TOO-shun) Condition in which tissue below the skin is damaged, but the skin is not broken.

Convergence (con-VER-jens) An anatomical arrangement in which the synaptic end bulbs of several presynaptic neurons terminate on one postsynaptic neuron; the medial movement of the two eyeballs so that both are directed toward a near object being viewed in order to produce a single image.

Convulsion (con-VUL-shun) Violent, involuntary, tetanic contractions of an entire group of muscles.

Cornea (KOR-nē-a) The nonvascular, transparent fibrous coat through which the iris can be seen.

Coronal (kō-RŌ-nal) **plane** A plane that runs vertical to the ground and divides the body into anterior and posterior portions. Also called a *frontal plane.*

Coronary (KOR-ō-na-rē) **artery bypass grafting (CABG)** Surgical procedure in which a portion of a blood vessel is removed from another part of the body and grafted onto a coronary artery so as to bypass an obstruction in the coronary artery.

Coronary circulation The pathway followed by the blood from the ascending aorta through the blood vessels supplying the heart and returning to the right atrium. Also called *cardiac circulation.*

Coronary sinus (SĪ-nus) A wide venous channel on the posterior surface of the heart that collects the blood from the coronary circulation and returns it to the right atrium.

Cor pulmonale (kor pul-mōn-AL-ē) **(CP)** Right ventricular hypertrophy from disorders that bring about hypertension in pulmonary circulation.

Corpus albicans (KOR-pus AL-bi-kanz) A white fibrous tissue in the ovary that forms after the corpus luteum regresses.

Corpus callosum (kal-LŌ-sum) The great commissure of the brain between the cerebral hemispheres.

Corpuscle of touch The sensory receptor for the sensation of touch; found in the dermal papillae, especially in palms and soles. Also called a *Meissner's* (MĪS-nerz) *corpuscle.*

Corpus luteum (LOO-tē-um) A yellow endocrine gland in the ovary formed when a follicle has discharged its secondary oocyte; secretes estrogens, progesterone, relaxin, and inhibin.

Cortex (KOR-teks) An outer layer of an organ. The convoluted layer of gray matter covering each cerebral hemisphere.

Costal cartilage (KOS-tal KAR-ti-lij) Hyaline cartilage that attaches a rib to the sternum.

Cowper's gland *See Bulbourethral gland.*

Cramps A spasmodic, especially a tonic, contraction of one of many muscles, usually painful.

Cranial (KRĀ-nē-al) **cavity** A subdivision of the dorsal body cavity formed by the cranial bones and containing the brain.

Cranial nerve One of 12 pairs of nerves that leave the brain, pass through foramina in the skull, and supply the head, neck, and part of the trunk; each is designated by a Roman numeral and a name.

Creatine phosphate (KRĒ-a-tin FOS-Fāt) High-energy molecule in skeletal muscle fibers (cells) that is used to generate ATP rapidly; on decomposition, creatine phosphate breaks down into creatine, phosphate, and energy—the energy is used to generate ATP from ADP.

Crenation (krē-NĀ-shun) The shrinkage of red blood cells into knobbed, starry forms when placed in a hypertonic solution.

Crista (KRIS-ta) A crest or ridged structure. A small elevation in the ampulla of each semicircular duct that serves as a receptor for dynamic equilibrium.

Crossing-over The exchange of a portion of one chromatid with another in a tetrad during meiosis. It permits an exchange of genes among chromatids and is one factor that results in genetic variation.

Crypt of Lieberkühn *See Intestinal gland.*

Cryptorchidism (krip-TOR-ki-dizm) The condition of undescended testes.

Cupula (KUP-yoo-la) A mass of gelatinous material covering the hair cells of a crista, a receptor in the ampulla of a semicircular canal stimulated when the head moves.

Cushing's syndrome Condition caused by a hypersecretion of glucocorticoids characterized by spindly legs, "moon face," "buffalo hump," pendulous abdomen, flushed facial skin, poor wound healing, hyperglycemia, osteoporosis, weakness, hypertension, and increased susceptibility to disease.

Cutaneous (kyoo-TĀ-nē-us) Pertaining to the skin.

Cyanosis (sī-a-NŌ-sis) Reduced hemoglobin (unoxygenated) concentration of blood of more than 5 g/dl that results in a blue or dark purple discoloration that is most easily seen in nail beds and mucous membranes.

Cyclic AMP (cyclic adenosine-3′,5′-monophosphate) Molecule formed from ATP by the action of the enzyme adenylate cyclase; serves as an intracellular messenger (second messenger) for some hormones.

Cyst (SIST) A sac with a distinct connective tissue wall, containing a fluid or other material.

Cystic (SIS-tik) **duct** The duct that transports bile from the gallbladder to the common bile duct.

Cystic fibrosis (fī-BRŌ-sis) Inherited disease of secretory epithelia that affects the respiratory passageways, pancreas, salivary glands, and sweat glands; the most common lethal genetic disease among the white population.

Cystitis (sis-TĪ-tis) Inflammation of the urinary bladder.

Cystoscope (SIS-to-skōp) An instrument used to examine the inside of the urinary bladder.

Cytokinesis (sī-tō-ki-NĒ-sis) Distribution of the cytoplasm into two separate cells during cell division; coordinated with nuclear division (mitosis).

Cytology (sī-TOL-ō-jē) The study of cells.

Cytolysis (sī-TOL-i-sis) The rupture of living cells in which the contents leak out.

Cytoplasm (SĪ-tō-plazm) The substance comprised of cytosol, all organelles (except the nucleus), and inclusions that surrounds organelles and is located within a cell's plasma membrane and external to its nucleus. Also called *protoplasm.*

Cytosol (SĪ-tō-sol) Semifluid portion of cytoplasm in which organelles and inclusions are suspended and solutes are dissolved. Also called *intracellular fluid.*

Deafness Lack of the sense of hearing or a significant hearing loss.

Deciduous (dē-SID-yoo-us) Falling off or being shed seasonally or at a particular stage of development. In the body, referring to the first set of teeth.

Decompression sickness A condition characterized by joint pain and neurologic symptoms; follows from a too-rapid reduction of environmental pressure or decompression, so that nitrogen that dissolved in body fluid under pressure comes out of solution as bubbles that form air emboli and occlude blood vessels. Also called *caisson* (KĀ-son) *disease* or *bends.*

Deep Away from the surface of the body.

Deep-venous thrombosis (DVT) The presence of a thrombus in a vein, usually a deep vein of the lower limbs.

Defecation (def-e-KĀ-shun) The discharge of feces from the rectum.

Defibrillation (dē-fib-ri-LĀ-shun) Delivery of a very strong electrical current to the heart in an attempt to stop ventricular fibrillation.

Degeneration (dē-jen-er-Ā-shun) A change from a higher to a lower state; a breakdown in structure.

Dehydration (dē-hī-DRĀ-shun) Excessive loss of water from the body or its parts.

Delirium (de-LIR-ē-um) A transient disorder of abnormal cognition (perception, thinking, and memory) and disordered attention that is accompanied by disturbances of the sleep–wake cycle and psychomotor behavior (hyperactivity or hypoactivity of movements and speech). Also called *acute confusional state (ACS).*

Dementia (de-MEN-shē-a) An organic mental disorder that results in permanent or progressive general loss of intellectual abilities such as impairment of memory, judgment, and abstract thinking and changes in personality; most common cause is Alzheimer's disease.

Demineralization (de-min′-er-al-i-ZĀ-shun) Loss of calcium and phosphorus from bones.

Dendrite (DEN-drīt) A nerve cell process that carries a nerve impulse toward the cell body.

Dental caries (KA-rēz) Gradual demineralization of the enamel and dentin of a tooth that may invade the pulp and alveolar bone. Also called *tooth decay.*

Dentin (DEN-tin) The bony tissues of a tooth enclosing the pulp cavity.

Deoxyribonucleic (dē-ok′-sē-rī′-bō-nyoo-KLĒ-ik) **acid (DNA)** A nucleic acid in the shape of a double helix constructed of nucleotides consisting of one of four nitrogenous bases (adenine, cytosine, guanine, or thymine), deoxyribose, and a phosphate group; encoded in the nucleotides is genetic information.

Depolarization (dē-pō-lar-i-ZĀ-shun) Used in neurophysiology to describe the reduction of voltage across a cell membrane; expressed as a movement toward less negative (more positive) voltages on the interior side of the cell membrane.

Dermal papilla (pa-PILL-a) Fingerlike projection of the papillary (upper) region of the dermis that may contain blood capillaries or corpuscles of touch (Meissner's corpuscles).

Dermatology (der-ma-TOL-ō-jē) The medical specialty dealing with diseases of the skin.

Dermatome (DER-ma-tōm) The cutaneous area developed from one embryonic spinal cord segment and receiving most of its innervation from one spinal nerve. An instrument for incising the skin or cutting thin transplants of skin.

Dermis (DER-mis) A layer of dense connective tissue lying deep to the epidermis; the true skin or corium.

Developmental anatomy The study of development from the fertilized egg to the adult form. The branch of anatomy called embryology is generally restricted to the study of development from the fertilized egg through the eighth week in utero.

Diabetes insipidus (dī-a-BĒ-tēz in-SIP-i-dus) Condition caused by hyposecretion of antidiuretic hormone (ADH) and characterized by thirst and excretion of large amounts of urine.

Diabetes mellitus (MEL-i-tus) Hereditary condition caused by hyposecretion of insulin and characterized by hyperglycemia, increased urine production, excessive thirst, and excessive eating.

Diagnosis (dī-ag-NŌ-sis) Distinguishing one disease from another or determining the nature of a disease from signs and symptoms by inspection, palpation, laboratory tests, and other means.

Dialysis (dī-AL-i-sis) The process of separating small molecules from large by the difference in their rates of diffusion through a selectively permeable membrane.

Diaphragm (DĪ-a-fram) Any partition that separates one area from another, especially the dome-shaped skeletal muscle between the thoracic and abdominal cavities. Also a dome-shaped structure that fits over the cervix, usually with a spermicide, to prevent conception.

Diaphysis (dī-AF-i-sis) The shaft of a long bone.

Diarrhea (dī-a-RĒ-a) Frequent defecation of liquid feces caused by increased motility of the intestines.

Diarthrosis (dī-′ar-THRŌ-sis) Articulation in which opposing bones move freely, as in a hinge joint.

Diastole (dī-AS-tō-lē) In the cardiac cycle, the phase of relaxation or dilation of the heart muscle, especially of the ventricles.

Diastolic (dī-as-TOL-ik) **blood pressure** The force exerted by blood on arterial walls during ventricular relaxation; the lowest blood pressure measured in the large arteries, about 80 mm Hg under normal conditions for a young, adult male.

Diencephalon (dī-en-SEF-a-lon) A part of the brain consisting primarily of the thalamus and the hypothalamus.

Differential (dif-fer-EN-shal) **white blood cell count** Determination of the number of each kind of white blood cell in a sample of 100 cells for diagnostic purposes.

Differentiation (dif′-e-ren′-shē-Ā-shun) Acquisition of specific functions different from those of the original general type.

Digestion (dī-JES-chun) The mechanical and chemical breakdown of food to simple molecules that can be absorbed and used by body cells.

Dilation (dī-LĀ-shun) **and curettage** (ku-re-TAZH) Following dilation of the uterine cervix, the uterine endometrium is scraped with a curette (spoon-shaped instrument). Also called a *D and C.*

Diploid (DIP-loyd) Having the number of chromosomes characteristically found in the somatic cells of an organism. Symbolized 2*n.*

Direct (motor) pathways Collections of upper motor neurons with cell bodies in the motor cortex that project axons into the spinal cord, where they synapse with lower motor neurons or association neurons in the anterior horns. Also called the *pyramidal pathways.*

Disease Any change from a state of health.

Dislocation (dis-lō-KĀ-shun) Displacement of a bone from a joint with tearing of ligaments, tendons, and articular capsules. Also called *luxation* (luks-Ā-shun).

Distal (DIS-tal) Farther from the attachment of a limb to the trunk or a structure; farther from the point of origin.

Diuretic (dī-yoo-RET-ik) A chemical that inhibits sodium reabsorption, reduces antidiuretic hormone (ADH) concentration, and increases urine volume by inhibiting facultative reabsorption of water.

Divergence (di-VER-jens) An anatomical arrangement in which the synaptic end bulbs of one presynaptic neuron terminate on several postsynaptic neurons.

Diverticulitis (dī-ver-tik-yoo-LĪ-tis) Inflammation of diverticula, saclike outpouchings of the colonic wall, when the muscularis becomes weak.

Dominant gene A gene that is able to override the influence of the complementary gene on the homologous chromosome; the gene that is expressed.

Donor insemination (in-sem′-i-NĀ-shun) The deposition of semen within the vagina or cervix at a time during the menstrual cycle when pregnancy is most likely to occur. It may be homologous (using the husband's semen) or heterologous (using a donor's semen). Also called *artificial insemination.*

Dorsal body cavity Cavity near the dorsal surface of the body that consists of the cranial cavity and vertebral cavity.

Down syndrome (DS) An inherited defect due to an extra copy of chromosome 21. Symptoms include mental retardation; a small skull, flattened from front to back; a short, flat nose; short fingers; and a widened space between the first two digits of the hand and foot. Also called *trisomy 21.*

Duct of Santorini *See Accessory duct.*

Duct of Wirsung *See Pancreatic duct.*

Ductus arteriosus (DUK-tus ar-tē-rē-Ō-sus) A small vessel connecting the pulmonary trunk with the aorta; found only in the fetus.

Ductus (vas) deferens (DEF-er-ens) The duct that conducts sperm from the epididymis to the ejaculatory duct. Also called the *seminal duct.*

Ductus venosus (ve-NŌ-sus) A small vessel in the fetus that helps the circulation bypass the liver.

Duodenum (doo′-ō-DĒ-num) The first 25 cm (10 in.) of the small intestine.

Dura mater (DYOO-ra MĀ-ter) The outer membrane (meninx) covering the brain and spinal cord.

Dynamic equilibrium (ē-kwi-LIB-rē-um) The maintenance of body position, mainly the head, in response to sudden movements such as rotation.

Dysfunction (dis-FUNK-shun) Absence of complete normal function.

Dyslexia (dis-LEX-sē-a) Impairment of the brain's ability to translate images received from the eyes or ears into understandable language.

Dysmenorrhea (dis′-men-ō-RĒ-a) Painful menstruation.

Dysphagia (dis-FĀ-jē-a) Difficulty in swallowing.

Dysplasia (dis-PLĀ-zē-a) Change in the size, shape, and organization of cells due to chronic irritation or inflammation; may revert to normal if stress is removed or may progress to neoplasia.

Dyspnea (DISP-nē-a) Shortness of breath.

Dystocia (dis-TŌ-sē-a) Difficult labor due to factors such as pelvic deformities, malpositioned fetus, and premature rupture of fetal membranes.

Dysuria (dis-YOO-rē-a) Painful urination.

Eardrum A thin, semitransparent partition of fibrous connective tissue between the external auditory meatus and the middle ear. Also called the *tympanic* (tim-PAN-ik) *membrane.*

Ectoderm The outermost of the three primary germ layers that gives rise to the nervous system and the epidermis of skin and its derivatives.

Ectopic (ek-TOP-ik) Out of the normal location, as in ectopic pregnancy.

Edema (e-DĒ-ma) An abnormal accumulation of interstitial fluid.

Effective filtration pressure (Peff) Net pressure that expresses the relationship between the force that promotes glomerular filtration and the forces that oppose it.

Effector (e-FEK-tor) The organ of the body, either a muscle or a gland, that responds to a motor neuron impulse.

Efferent arteriole (EF-er-ent ar-TĒ-rē-ōl) A vessel of the renal vascular system that transports blood from the glomerulus to the peritubular capillary.

Ejaculation (e-jak-yoo-LĀ-shun) The reflex ejection or expulsion of semen from the penis.

Ejaculatory (e-JAK-yoo-la-tō′-rē) **duct** A tube that transports sperm from the ductus (vas) deferens to the prostatic urethra.

Elasticity (e-las-TIS-i-tē) The ability of tissue to return to its original shape after contraction or extension.

Electrocardiogram (e-lek′-trō-KAR-dē-ō-gram) **(ECG or EKG)** A recording of the electrical changes that accompany the cardiac cycle and can be recorded on the surface of the body; may be resting, stress, or ambulatory.

Electroencephalogram (e-lek′-trō-en-SEF-a-lō-gram) **(EEG)** A recording of the electrical impulses of the brain to diagnose certain diseases (such as epilepsy), furnish information regarding sleep and wakefulness, and confirm brain death.

Electrolyte (ē-LEK-trō-līt) Any compound that separates into ions when dissolved in water and is able to conduct electricity.

Electron transport chain A series of oxidation–reduction reactions in the catabolism of glucose that occur on the inner mitochondrial membrane and in which energy is released and transferred for storage to ATP.

Ellipsoidal joint *See Condyloid joint.*

Embolism (EM-bō-lizm) Obstruction or closure of a vessel by an embolus.

Embolus (EM-bō-lus) A blood clot, bubble of air, fat from broken bones, mass of bacteria, or other debris or foreign material transported by the blood.

Embryo (EM-brē-ō) The young of any organism in an early stage of development; in humans, the developing organism from fertilization to the end of the eighth week in utero.

Embryology (em′-brē-OL-ō-jē) The study of development from the fertilized egg to the end of the eighth week in utero.

Embryo transfer A procedure in which semen is used to inseminate artificially a fertile secondary oocyte donor and the morula or blastocyst is then transferred from the donor to the infertile woman, who then carries it to term.

Emigration (em′-e-GRĀ-shun) Process whereby white blood cells leave the bloodstream by slowing down, rolling along the endothelium, and squeezing between the endothelial cells. Adhesion molecules help WBCs stick to the endothelium. Also known as *migration* or *extravasation.* Formerly called *diapedesis.*

Emission (ē-MISH-un) Propulsion of sperm cells into the urethra in response to peristaltic contractions of the ducts of the testes, epididymides, and ductus (vas) deferens as a result of sympathetic stimulation.

Emphysema (em′-fi-SĒ-ma) A swelling or inflation of air passages due to loss of elasticity in the alveoli.

Emulsification (ē-mul′-si-fi-KĀ-shun) The dispersion of large fat globules to smaller uniformly distributed particles in the presence of bile.

Enamel (e-NAM-el) The hard, white substance covering the crown of a tooth.

Endocardium (en-dō-KAR-dē-um) The layer of the heart wall, composed of endothelium and smooth muscle, that lines the inside of the heart and covers the valves and tendons that hold the valves open.

Endochondral ossification (en′-dō-KON-dral os′-i-fi-KĀ-shun) The replacement of cartilage by bone. Also called *intracartilaginous* (in′-tra-kar′-ti-LAJ-i-nus) *ossification.*

Endocrine (EN-dō-krin) **gland** A gland that secretes hormones into the blood; a ductless gland.

Endocrinology (en′-dō-kri-NOL-ō-jē) The science concerned with the structure and functions of endocrine glands and the diagnosis and treatment of disorders of the endocrine system.

Endocytosis (en′-dō-sī-TŌ-sis) The uptake into a cell of large molecules and particles in which a segment of plasma membrane surrounds the substance, encloses it, and brings it in; includes phagocytosis, pinocytosis, and receptor-mediated endocytosis.

Endoderm The innermost of the three primary germ layers of the developing embryo that gives rise to the gastrointestinal tract, urinary bladder and urethra, and respiratory tract.

Endodontics (en′-dō-DON-tiks) The branch of dentistry concerned with the prevention, diagnosis, and treatment of diseases that affect the pulp, root, periodontal ligament, and alveolar bone.

Endometriosis (en′-dō-MĒ-trē-ō′-sis) The growth of endometrial tissue outside the uterus.

Endometrium (en′-dō-MĒ-trē-um) The mucous membrane lining the uterus.

Endoplasmic reticulum (en′-do-PLAZ-mik re-TIK-yoo-lum) **(ER)** A network of channels running through the cytoplasm of a cell that serves in intracellular transportation, support, storage, synthesis, and packaging of molecules. Portions of ER where ribosomes are attached to the outer surface are called **rough** or **granular reticulum;** portions that have no ribosomes are called **smooth** or **agranular reticulum.**

End organ of Ruffini *See Type II* **cutaneous** *mechanoreceptor.*

Endoscopy (en-DOS-kō-pē) The visual examination of any cavity of the body using an endoscope, an illuminated tube with lenses.

Endosteum (en-DOS-tē-um) The membrane that lines the medullary cavity of bones, consisting of osteoprogenitor cells and scattered osteoclasts.

Endothelium (en′-dō-THĒ-lē-um) The layer of simple squamous epithelium that lines the cavities of the heart and blood and lymphatic vessels.

Energy The capacity to do work.

Enterogastric (en-te-rō-GAS-trik) **reflex** A reflex that inhibits gastric secretion; initiated by food in the small intestine.

Enuresis (en′-yoo-RĒ-sis) Involuntary discharge of urine, complete or partial, after age 3.

Enzyme (EN-zīm) A substance that affects the speed of chemical changes; an organic catalyst, usually a protein.

Eosinophil (ē-ō-SIN-ō-fil) A type of white blood cell characterized by granules that stain red or pink with acid dyes.

Ependymal (e-PEN-de-ma) **cells** Neuroglial cells that cover choroid plexuses and produce cerebrospinal fluid (CSF); they also line ventricles of the brain and probably assist in the circulation of CSF.

Epicardium (ep′-i-KAR-dē-um) The thin outer layer of the heart wall, composed of serous tissue and mesothelium. Also called the *visceral pericardium.*

Epidemic (ep′-i-DEM-ik) A disease that occurs above the expected level among individuals in a population.

Epidermis (ep-i-DERM-is) The outermost, thinner layer of skin, composed of keratinized stratified squamous epithelium.

Epididymis (ep′-i-DID-i-mis) A comma-shaped organ that lies along the posterior border of the testis and contains the ductus epididymis, in which sperm undergo maturation. *Plural, epididymides* (ep′-i-DID-i-midēz).

Epidural (ep′-i-DOO-ral) **space** A space between the spinal dura mater and the vertebral canal, containing areolar connective tissue and a plexus of veins.

Epiglottis (ep′-i-GLOT-is) A large, leaf-shaped piece of cartilage lying on top of the larynx, with its "stem" attached to the thyroid cartilage and its "leaf" portion unattached and free to move up and down to cover the glottis (vocal folds and rima glottidis).

Epilepsy (EP-i-lep′-sē) Neurological disorder characterized by short, periodic attacks of motor, sensory, or psychological malfunction.

Epinephrine (ep′-i-NEF-rin) Hormone secreted by the adrenal medulla that produces actions similar to those that result from sympathetic stimulation. Also called *adrenaline* (a-DREN-a-lin).

Epiphyseal (ep′-i-FIZ-ē-al) **line** The remnant of the epiphyseal plate in a long bone.

Epiphyseal plate The hyaline cartilage plate between the epiphysis and diaphysis that is responsible for the lengthwise growth of long bones.

Epiphysis (ē-PIF-i-sis) The end of a long bone, usually larger in diameter than the shaft (diaphysis).

Episiotomy (e-piz′-ē-OT-ō-mē) A cut made with surgical scissors to avoid tearing of the perineum at the end of the second stage of labor.

Epistaxis (ep′-i-STAK-sis) Loss of blood from the nose due to trauma, infection, allergy, neoplasm, and bleeding disorders. Also called *nosebleed.*

Epithelial (ep′-i-THĒ-lē-al) **tissue** The tissue that forms glands or the outer part of the skin and lines blood vessels, hollow organs, and passages that lead externally from the body.

Equilibrium (ē′-kwi-LIB-rē-um) A state of balance; a condition in which opposing forces exactly counterbalance each other.

Erection (ē-REK-shun) The enlarged and stiff state of the penis (or clitoris) resulting from the engorgement of the spongy erectile tissue with blood.

Erythema (er′-e-THĒ-ma) Skin redness usually caused by engorgement of the capillaries in the lower layers of the skin.

Erythrocyte (e-RITH-rō-sīt) Red blood cell.

Erythropoiesis (e-rith′-rō-poy-Ē-sis) The process by which erythrocytes (red blood cells) are formed.

Erythropoietin (e-rith′-rō-POY-ē-tin) A hormone formed from a plasma protein that stimulates erythrocyte (red blood cell) production.

Esophagus (e-SOF-a-gus) A hollow muscular tube connecting the pharynx and the stomach.

Essential amino acids Those 10 amino acids that cannot be synthesized by the human body at an adequate rate to meet its needs and therefore must be obtained from the diet.

Estrogens (ES-tro-jens) Female sex hormones produced by the ovaries concerned with the development and maintenance of female reproductive structures and secondary sex characteristics, fluid and electrolyte balance, and protein anabolism. Examples are β-estradiol, estrone, and estriol. See pg. 135

Etiology (ē′-tē-OL-ō-jē) The study of the causes of disease, including theories of origin and the organisms, if any, involved.

Euphoria (yoo-FOR-ē-a) A subjectively pleasant feeling of well-being marked by confidence and assurance.

Eupnea (yoop-NĒ-a) Normal quiet breathing.

Eustachian tube See *Auditory tube.*

Euthanasia (yoo′-tha-NĀ-zē-a) The practice of ending a life in case of incurable diseases.

Eversion (ē-VER-zhun) The movement of the sole outward at the ankle joint.

Exacerbation (eg-zas′-er-BĀ-shun) An increase in the severity of symptoms or of disease.

Excitability (ek-sīt′-a-BIL-i-tē) The ability of muscle tissue to receive and respond to stimuli; the ability of nerve cells to respond to stimuli and convert them into nerve impulses.

Excrement (EKS-kre-ment) Material cast out from the body as waste, especially fecal matter.

Excretion (eks-KRĒ-shun) The process of eliminating waste products from a cell, tissue, or the entire body; or the products excreted.

Exocrine (EK-sō-krin) **gland** A gland that secretes substances into ducts that empty at covering or lining epithelium or directly onto a free surface.

Exocytosis (ex′-ō-si-TŌ-sis) A process of discharging cellular products too big to go through the membrane. Particles for export are enclosed by Golgi membranes when they are synthesized. Vesicles pinch off from the Golgi complex and carry the enclosed particles to the interior surface of the cell membrane, where the vesicle membrane and plasma membrane fuse and the contents of the vesicle are discharged.

Expiration (ek-spi-RĀ-shun) Breathing out; expelling air from the lungs into the atmosphere. Also called *exhalation.*

Extensibility (ek-sten′-si-BIL-i-tē) The ability of muscle tissue to be stretched when pulled.

Extension (ek-STEN-shun) An increase in the angle between two bones; restoring a body part to its anatomical position after flexion.

External ear The outer ear, consisting of the pinna, external auditory canal, and tympanic membrane or eardrum.

External nares (NA-rēz) The external nostrils, or the openings into the nasal cavity on the exterior of the body.

External respiration The exchange of respiratory gases between the lungs and blood. Also called *pulmonary respiration.*

Extracellular fluid (ECF) Fluid outside body cells, such as interstitial fluid and plasma.

Extrinsic (ek-STRIN-sik) Of external origin.

Exudate (EKS-yoo-dāt) Escaping fluid or semifluid material that oozes from a space that may contain serum, pus, and cellular debris.

Facilitated diffusion (fa-SIL-i-tā-ted dif-YOO-zhun) Diffusion in which a substance not soluble by itself in lipids is transported across a selectively permeable membrane with the help of a transporter (carrier) protein.

Facilitation (fa-sil-i-TĀ-shun) The process in which a nerve cell membrane is partially depolarized by a subliminal stimulus so that a subsequent subliminal stimulus can further depolarize the membrane to reach the threshold of nerve impulse initiation.

Fallopian tube See *Uterine tube.*

Fascia (FASH-ē-a) A fibrous membrane covering, supporting, and separating muscles.

Fascicle (FAS-i-kul) A small bundle or cluster, especially of nerve or muscle fibers (cells). Also called a *fasciculus* (fa-SIK-yoo-lus). *Plural, fasciculi* (fa-SIK-yoo-lī).

Fat See *Triglyceride.*

Fauces (FAW-sēz) The opening from the mouth into the pharynx.

Febrile (FĒ-bril) Feverish; pertaining to a fever.

Feces (FĒ-sēz) Material discharged from the rectum and made up of bacteria, excretions, and food residue. Also called *stool.*

Feedback system A circular sequence of events in which information about the status of a situation is continually reported (fed back) to a central control region.

Female reproductive cycle General term for the ovarian and uterine cycles, the hormonal changes that accompany them, and cyclic changes in the breasts and cervix; includes changes in the endometrium of a nonpregnant female that prepares the lining of the uterus to receive a fertilized ovum. Less correctly termed *menstrual cycle.*

Fertilization (fer′-ti-li-ZĀ-shun) Penetration of a secondary oocyte by a sperm cell and subsequent union of the nuclei of the cells.

Fetal (FĒ-tal) **alcohol syndrome (FAS)** Term applied to the effects of intrauterine exposure to alcohol, such as slow growth, defective organs, and mental retardation.

Fetal circulation The cardiovascular system of the fetus, including the placenta and special blood vessels involved in the exchange of materials between fetus and mother.

Fetus (FĒ-tus) The latter stages of the developing young of an animal; in humans, the developing organism in utero from the beginning of the third month to birth.

Fever An elevation in body temperature above its normal temperature of 37°C (98.6°F).

Fibrillation (fi-bri-LĀ-shun) Involuntary brief twitch of a muscle that is not visible under the skin and is not associated with movement of the affected muscle.

Fibrin (FĪ-brin) An insoluble protein that is essential to blood clotting; formed from fibrinogen by the action of thrombin.

Fibrinogen (fi-BRIN-ō-jen) A high-molecular-weight protein in the blood plasma that by the action of thrombin is converted to fibrin.

Fibrinolysis (fi-brin-OL-i-sis) Dissolution of a blood clot by the action of a proteolytic enzyme that converts insoluble fibrin into a soluble substance.

Fibroblast (FĪ-brō-blast) A large, flat cell that secretes most of the matrix (extracellular) material of areolar and dense connective tissues.

Fibrosis (fi-BRŌ-sis) Abnormal formation of fibrous tissue.

Fight-or-flight response The effect of the stimulation of the sympathetic division of the autonomic nervous system.

Filtrate (fil-TRĀT) The fluid produced when blood is filtered by the filtration membrane.

Filtration (fil-TRĀ-shun) The passage of a liquid through a filter or membrane that acts like a filter.

Filtration membrane A filtering membrane in a nephron of a kidney consisting of the glomerulus and inner wall of the glomerular (Bowman's) capsule.

Fissure (FISH-ur) A groove, fold, or slit that may be normal or abnormal.

Fixed macrophage (MAK-rō-fāj) Stationary phagocytic cell found in the liver, lungs, brain, spleen, lymph nodes, subcutaneous tissue, and bone marrow. Also called a *histiocyte* (HIS-tē-ō-sīt).

Flaccid (FLAS-sid) Relaxed, flabby, or soft; lacking muscle tone.

Flagellum (fla-JEL-um) A hairlike, motile process on the surface of a bacterium or protozoan. *Plural, flagella* (fla-JEL-a).

Flatfoot A condition in which the ligaments and tendons of the arches of the foot are

weakened and the height of the longitudinal arch decreases.

Flatus (FLĀ-tus) Air (gas) in the stomach or intestines, commonly used to denote passage of gas rectally.

Flexion (FLEK-shun) A folding movement in which there is a decrease in the angle between two bones.

Flexor reflex A protective reflex in which flexor muscles are stimulated while extensor muscles are inhibited.

Fluoroscope (FLOOR-ō-skōp) An instrument for visual observation of the body by means of x-ray.

Follicle-stimulating (FOL-i-kul) **hormone (FSH)** Hormone secreted by the anterior pituitary gland that initiates development of ova and stimulates the ovaries to secrete estrogens in females and initiates sperm production in males.

Fontanel (fon'-ta-NEL) A membrane-covered spot where bone formation is not yet complete, especially between the cranial bones of an infant's skull.

Foramen (fo-RĀ-men) A passage or opening; a communication between two cavities of an organ or a hole in a bone for passage of vessels or nerves. *Plural, foramina* (fo-RĀ-men-a).

Foramen ovale (ō-VAL-ē) An opening in the fetal heart in the septum between the right and left atria. A hole in the greater wing of the sphenoid bone that transmits the mandibular branch of the trigeminal (V) nerve.

Fossa (FOS-a) A furrow or shallow depression.

Fracture (FRAK-chur) Any break in a bone.

Fragile X syndrome Inherited disorder characterized by learning difficulties, mental retardation, and physical abnormalities; due to a defective gene on the X chromosome.

Frenulum (FREN-yoo-lum) A small fold of mucous membrane that connects two parts and limits movement.

Frontal plane A plane at a right angle to a midsagittal plane that divides the body or organs into anterior and posterior portions. Also called a *coronal* (kō-RŌ-nal) *plane.*

Gallbladder A small pouch that stores bile, located under the liver, which is emptied via the cystic duct.

Gallstone A concretion, usually consisting of cholesterol, formed anywhere between bile canaliculi in the liver and the hepatopancreatic ampulla (ampulla of Vater), where bile enters the duodenum. Also called a *biliary calculus.*

Gamete (GAM-ēt) A male or female reproductive cell; the sperm cell or ovum.

Gamete intrafallopian transfer (GIFT) A procedure in which aspirated secondary oocytes are combined with a solution containing sperm outside the body and the mixture is then immediately inserted into the uterine (Fallopian) tubes.

Ganglion (GANG-glē-on) A group of nerve cell bodies that lie outside the central nervous system. *Plural, ganglia* (GANG-glē-a).

Gangrene (GANG-rēn) Death and rotting of a considerable mass of tissue that usually is caused by interruption of blood supply followed by bacterial (*Clostridium*) invasion.

Gastric (GAS-trik) **glands** Glands in the mucosa of the stomach composed of cells that empty their secretions into narrow channels called gastric pits. Types of gastric cells include chief cells (secrete pepsinogen), parietal cells (secrete hydrochloric acid and intrinsic factor), mucous cells (secrete mucus), and G cells (secrete gastin).

Gastroenterology (gas'-trō-en'ter-OL-ō-jē) The medical specialty that deals with the structure, function, diagnosis, and treatment of diseases of the stomach and intestines.

Gastrointestinal (gas-trō-in-TES-ti-nal) **(GI) tract** A continuous tube running through the ventral body cavity extending from the mouth to the anus. Also called the *alimentary* (al'-i-MEN-tar-ē) *canal.*

Gene (JĒN) Biological unit of heredity; a segment of DNA located in a definite position on a particular chromosome; a sequence of DNA that codes for a particular mRNA, rRNA, or tRNA.

General adaptation syndrome (GAS) Wide-ranging set of bodily changes triggered by a stressor that gears the body to meet an emergency.

Generator potential The graded depolarization that results in a change in the resting membrane potential in a receptor (specialized neuronal ending).

Genetic engineering The manufacture and manipulation of genetic material.

Genetics The study of heredity.

Genital herpes (JEN-i-tal HER-pēz) A sexually transmitted disease caused by type II herpes simplex virus.

Genotype (JĒ-nō-tīp) The total hereditary information carried by an individual; the genetic makeup of an organism.

Geriatrics (jer'-ē-AT-riks) The branch of medicine devoted to the medical problems and care of elderly persons.

Gestation (jes-TĀ-shun) The period of intrauterine fetal development.

Giantism (GĪ-an-tizm) Condition caused by hypersecretion of human growth hormone (hGH) during childhood, characterized by excessive bone growth and body size. Also called *gigantism.*

Gingivae (jin-JI-vē) Gums. They cover the alveolar processes of the mandible and maxilla and extend slightly into each socket.

Gland Single or group of specialized epithelial cells that secrete substances.

Glaucoma (glaw-KŌ-ma) An eye disorder in which there is increased intraocular pressure due to an excess of aqueous humor.

Gliding joint A synovial joint having articulated surfaces that are usually flat, permitting only side-to-side and back-and-forth movements, as between carpal bones, tarsal bones, and the scapula and clavicle. Also called the *arthrodial* (ar-THRŌ-dē-al) *joint.*

Glomerular (glō-MER-yoo-lar) **capsule** A double-walled globe at the proximal end of a nephron that encloses the glomerular capillaries. Also called *Bowman's* (Bō-manz) *capsule.*

Glomerular (glō-MER-yoo-lar) **filtration** The first step in urine formation in which substances in blood are filtered at the filtering membrane and the filtrate enters the proximal convoluted tubule of a nephron.

Glomerular filtration rate (GFR) The total volume of fluid that enters all the glomerular (Bowman's) capsules of the kidneys in 1 min; about 125 ml/min.

Glomerulus (glō-MER-yoo-lus) A rounded mass of nerves or blood vessels, especially the microscopic tuft of capillaries that is surrounded by the glomerular (Bowman's) capsule of each kidney tubule.

Glottis (GLOT-is) The vocal folds (true vocal cords) in the larynx and the space between them (rima glottidis).

Glucagon (GLOO-ka-gon) A hormone produced by the alpha cells of the pancreas that increases the blood glucose level.

Glucocorticoids (gloo-kō-KOR-ti-koyds) A group of hormones of the adrenal cortex.

Gluconeogenesis (gloo'-kō-nē'-JEN-e-sis) The conversion of a substance other than carbohydrate into glucose.

Glucose (GLOO-kōs) A six-carbon sugar, $C_6H_{12}O_6$; the major energy source for every cell type in the body. Its metabolism is possible by every known living cell for the production of ATP.

Glucosuria (glū-kō-SOO-rē-a) The presence of glucose in the urine; may be temporary or pathological. Also called *glycosuria.*

Glycogen (GLĪ-kō-jen) A highly branched polymer of glucose containing thousands of subunits; functions as a compact store of glucose molecules in liver and muscle fibers (cells).

Glycogenesis (glī'-kō-JEN-e-sis) The process by which many molecules of glucose combine to form a molecule called glycogen.

Glycogenolysis (glī-kō-je-NOL-i-sis) The process of converting glycogen to glucose.

Glycolysis (glī-KŌL-i-sis) A series of chemical reactions that break down glucose into pyruvic acid, with a net gain of two molecules of ATP.

Goblet cell A goblet-shaped unicellular gland that secretes mucus. Also called a *mucous cell.*

Goiter (GOY-ter) An enlargement of the thyroid gland.

Golgi (GOL-jē) **complex** An organelle in the cytoplasm of cells consisting of four to eight flattened channels, stacked on one another, with expanded areas at their ends; functions in packaging secreted proteins, lipid secretion, and carbohydrate synthesis.

Golgi tendon organ *See Tendon organ.*

Gonad (GŌ-nad) A gland that produces gametes and hormones; the ovary in the female and the testis in the male.

Gonadotropic (gō'-nad-ō-TRŌ-pik) **hormone** A hormone that regulates the functions of the gonads.

Gonorrhea (gon'-ō-RĒ-a) Infectious, sexually transmitted disease caused by the bacterium *Neisseria gonorrhoeae* and characterized by inflammation of the urogenital mucosa, discharge of pus, and painful urination.

Graafian follicle *See Mature follicle.*

Gray matter Area in the central nervous system and ganglia consisting of nonmyelinated nerve tissue.

Greater omentum (ō-MEN-tum) A large fold in the serosa of the stomach that hangs down like an apron over the front of the intestines.

Greater vestibular (Ves-TIB-yoo-lar) **glands** A pair of glands on either side of the vaginal orifice that open by a duct into the space between the hymen and labia minora. Also called *Bartholin's* (BAR-to-lins *glands.*

Groin (GROYN) The depression between the thigh and the trunk, the inguinal region.

Gross anatomy The branch of anatomy that deals with structures that can be studied without using a microscope. Also called *macroscopic anatomy.*

Growth An increase in size due to an increase in the number of cells or an increase in the size of existing cells as internal components increase in size or an increase in the size of intercellular substances.

Gustatory (GUS-ta-tō-rē) Pertaining to taste.

Gynecology (gī'-ne-KOL-ō-jē) The branch of medicine dealing with the study and treatment of disorders of the female reproductive system.

Gynecomastia (gīn'-e-kō-MAS-tē-a) Excessive growth (benign) of the male mammary glands due to secretion of sufficient estrogens by an adrenal gland tumor (feminizing adenoma).

Gyrus (JĪ-rus) One of the folds of the cerebral cortex of he brain. *Plural, gyri* (JĪ-rī). Also called a *convolution.*

Hair A threadlike structure produced by hair follicles that develops in the dermis. Also called *pilus* (PI-lus).

Hair follicle (FOL-li-kul) Structure composed of epithelium surrounding the root of a hair from which hair develops.

Hallucination (ha-loo'-si-NĀ-shun) A sensory perception of something that does not really exist in the world, that is, a sensory experience created from within the brain.

Haploid (HAP-loyd) Having half the number of chromosomes characteristically found in the somatic cells of an organism; characteristic of mature gametes. Symbolized *n.*

Hard palate (PAL-at) The anterior portion of the roof of the mouth, formed by the maxillae and palatine bones and lined by mucous membrane.

Haustra (HAWS-tra) The sacculated elevations of the colon.

Haversian canal *See Central canal.*

Haversian system *See Osteon.*

Head The superior part of a human, cephalic to the neck. The superior or proximal part of a structure.

Heart A hollow muscular organ lying slightly to the left of the midline of the chest that pumps the blood through the cardiovascular system.

Heart block An arrhythmia (dysrhythmia) of the heart in which the atria and ventricles contract independently because of a blocking of electrical impulses through the heart at a critical point in the conduction system.

Heartburn Burning sensation in the esophagus due to reflux of hydrochloric acid (HCl) from the stomach.

Heart–lung machine A device that pumps blood, functioning as a heart, and removes carbon dioxide from blood and oxygenates it, functioning as lungs; used during heart transplantation, open-heart surgery, and coronary artery bypass grafting.

Heart murmur (MER-mer) An abnormal sound that consists of a flow noise that is heard before the normal lubb–dupp or that may mask normal heart sounds.

Heat exhaustion Condition characterized by cool, clammy skin, profuse perspiration, and fluid and electrolyte (especially salt) loss that results in muscle cramps, dizziness, vomiting, and fainting. Also called *heat prostration.*

Heatstroke Condition produced when the body cannot easily lose heat and characterized by reduced perspiration and elevated body temperature. Also called *sunstroke.*

Heimlich maneuver *See Abdominal thrust maneuver.*

Hematocrit (hē-MAT-ō-krit) **(Hct)** The percentage of blood made up of red blood cells. Usually calculated by centrifuging a blood sample in a graduated tube and then reading off the volume of red blood cells and total blood.

Hematology (hē′-ma-TOL-ō-jē) The study of blood.

Hematoma (hē′-ma-TŌ-ma) A tumor or swelling filled with blood.

Hematuria (hē′-ma-TOOR-ē-a) Blood in the urine.

Hemodialysis (hē′-mō-dī-AL-i-sis) Filtering of the blood by means of an artificial device so that certain substances are removed from the blood as a result of the difference in rates of their diffusion through a selectively permeable membrane while the blood is being circulated outside the body.

Hemoglobin (hē′-mō-GLŌ-bin) **(Hb)** A substance in erythrocytes (red blood cells) consisting of the protein globin and the iron-containing red pigment heme and constituting about 33 percent of the cell

volume; involved in the transport of oxygen and carbon dioxide.

Hemolysis (hē-MOL-i-sis) The escape of hemoglobin from the interior of the red blood cells into the surrounding medium; results from disruption of the integrity of the cell membrane by toxins or drugs, freezing or thawing, or hypotonic solutions.

Hemolytic disease of the newborn A hemolytic anemia of a newborn child that results from the destruction of the infant's erythrocytes (red blood cells) by antibodies produced by the mother; usually the antibodies are due to an Rh blood type incompatibility. Also called *erythroblastosis fetalis* (e-rith′-rō-blas-TŌ-sis fe-TAL-is).

Hemophilia (hē′-mō-FĒL-ē-a) A hereditary blood disorder where there is a deficient production of certain factors involved in blood clotting, resulting in excessive bleeding into joints, deep tissues, and elsewhere.

Hemopoiesis (hēm′-ō-poy-Ē-sis) Blood cell production occurring in the red bone marrow of bones. Also called *hematopoiesis* (hēm′-a-tō-poy-Ē-sis).

Hemopoietic (hē′-mō-poy-Ē-tic) **stem cell** Immature stem cell in red bone marrow that gives rise to precursors of all the different mature red blood cells. Previously called a *hemocytoblast* (hē-mō-SĪ-tō-blast) or pluripotent hematopoietic stem cell.

Hemorrhage (HEM-or-rij) Bleeding; the escape of blood from blood vessels, especially when it is profuse.

Hemorrhoids (HEM-ō-royds) Dilated or varicosed blood vessels (usually veins) in the anal region. Also called *piles.*

Hemostasis (hē-MŌS-tā-sis) The stoppage of bleeding.

Heparin (HEP-a-rin) An anticoagulant given to slow the conversion of prothrombin to thrombin, thus reducing the risk of blood clot formation; also found naturally in basophils and most cells.

Hepatic (he-PAT-ik) Refers to the liver.

Hepatic portal circulation The flow of blood from the gastrointestinal organs to the liver before returning to the heart.

Hepatitis (hep-a-TĪ-tis) Inflammation of the liver due to a virus, drugs, and chemicals.

Hepatocyte (he-PAT-ō-sīt) A liver cell.

Hepatopancreatic (hep′-a-tō-pan′-krē-A-tik) **ampulla** A small raised area in the duodenum where the combined common bile duct and main pancreatic duct empty

into the duodenum. Also called the *ampulla of vater* (VA-ter).

Hering–Breuer reflex *See Inflation reflex.*

Hernia (HER-nē-a) The protrusion or projection of an organ or part of an organ through a membrane or cavity wall, usually the abdominal cavity.

Herniated (her′-nē-A-ted) **disc** A rupture of an intervertebral disc so that the nucleus pulposus protrudes into the vertebral cavity. Also called a *slipped disc.*

Heterozygous (he-ter-ō-ZĪ-gus) Possessing a pair of different genes on homologous chromosomes for a particular hereditary characteristic.

High altitude sickness Disorder caused by decreased levels of alveolar PO_2 as altitude increases and characterized by headache, fatigue, insomnia, shortness of breath, nausea, and dizziness. Also called *acute mountain sickness.*

Hilus (HĪ-lus) An area, depression, or pit where blood vessels and nerves enter or leave an organ. Also called a *hilum.*

Hinge joint A synovial joint in which a convex surface of one bone fits into a concave surface of another bone, such as the elbow, knee, ankle, and interphalangeal joints. Also called a *ginglymus* (JIN-gli-mus) *joint.*

Hirsutism (HER-soot-izm) An excessive growth of hair in females and children, with a distribution similar to that in adult males, due to the conversion of vellus hairs into large terminal hairs in response to higher-than-normal levels of androgens.

Histamine (HISS-ta-mēn) Substance found in many cells, especially mast cells, basophils, and platelets, released when the cells are injured; results in vasodilation, increased permeability of blood vessels, and bronchiole constriction.

Histology (hiss-TOL-ō-jē) Microscopic study of the structure of tissues.

Hives (HĪVZ) Condition of the skin marked by reddened elevated patches that are often itchy; may be caused by infections, trauma, medications, emotional stress, food additives, and certain foods.

Hodgkin's disease (HD) A malignant disorder, usually arising in lymph nodes.

Homeostasis (hō′-mē-ō-STĀ-sis) The condition in which the body's internal environment remains relatively constant, within physiological limits.

Homologous chromosomes Two chromosomes that belong to a pair. Also called *homologues.*

Homozygous (hō-mō-ZĪ-gus) Possessing a pair of similar genes on homologous chromosomes for a particular hereditary characteristic.

Hormone (HOR-mōn) A secretion of endocrine tissue that alters the physiological activity of target cells of the body.

Human growth hormone (hGH) Hormone secreted by the anterior pituitary gland that stimulates growth of body tissues, especially skeletal and muscular. Also known as *somatotropin* and *somatotropic hormone* (*STH*).

Human leukocyte associated (HLA) antigens Surface proteins on white blood cells and other nucleated cells that are unique for each person (except for identical twins) and are used to type tissues and help prevent rejection.

Hunger center A cluster of neurons in the lateral nuclei of the hypothalamus that, when stimulated, brings about feeding.

Hymen (HĪ-men) A thin fold of vascularized mucous membrane at the vaginal orifice.

Hypercapnia (hī′-per-KAP-nē-a) An abnormal increase in the amount of carbon dioxide in the blood.

Hyperglycemia (hī′-per-glī-SĒ-mē-a) An elevated blood sugar level.

Hyperplasia (hī′-per-PLĀ-zē-a) An abnormal increase in the number of normal cells in a tissue or organ, increasing its size.

Hyperpolarization (hī′-per-PŌL-a-ri-zā′-shun) Increase in the internal negativity across a cell membrane, thus increasing the voltage and moving it farther away from the threshold value.

Hypersecretion (hī′-per-se-KRĒ-shun) Overactivity of glands resulting in excessive secretion.

Hypersensitivity (hī′-per-sen-si-TI-vi-tē) Overreaction to an allergen that results in pathological changes in tissues. Also called *allergy*.

Hypertension (hī′-per-TEN-shun) High blood pressure.

Hyperthermia (hī′-per-THERM-ē-a) An elevated body temperature.

Hypertonia (hī′-per-TŌ-nē-a) Increased muscle tone that is expressed as spasticity or rigidity.

Hypertonic (hī′-per-TON-ik) Having an osmotic pressure greater than that of a solution with which it is compared.

Hypertrophy (hī-PER-trō-fē) An excessive enlargement or overgrowth of tissue without cell division.

Hyperventilation (hī′-per-ven-ti-LĀ-shun) A rate of respiration higher than that required to maintain a normal level of plasma PCO_2.

Hypophysis (hī-POF-i-sis) Pituitary gland.

Hypoplasia (hī-pō-PLĀ-zē-a) Defective development of tissue.

Hyposecretion (hī-pō-se-KRĒ-shun) Underactivity of glands resulting in diminished secretion.

Hypospadias (hī′-pō-SPĀ-dē-as) A displaced urethral opening. In the male, the opening may be on the underside of the penis, at the penoscrotal junction, between the scrotal folds, or in the perineum. In the female, the urethra opens into the vagina.

Hypothalamus (hī′-pō-THAL-a-mus) A portion of the diencephalon, lying beneath the thalamus and forming the floor and part of the wall of the third ventricle.

Hypothermia (hī-pō-THER-mē-a) Lowering of body temperature below 35°C (95°F); in surgical procedures, it refers to deliberate cooling of the body to slow down metabolism and reduce oxygen needs of tissues.

Hypotonia (hī′-pō-TŌ-nē-a) Decreased or lost muscle tone in which muscles appear flaccid.

Hypotonic (hī′-pō-TON-ik) Having an osmotic pressure lower than that of a solution with which it is compared.

Hypoventilation (hī-pō-ven-ti-LĀ-shun) A rate of respiration lower than that required to maintain a normal level of plasma PCO_2.

Hypoxia (hī-POKS-ē-a) Lack of adequate oxygen at the tissue level.

Hysterectomy (his-te-REK-tō-mē) The surgical removal of the uterus.

Ileocecal (il′-ē-ō-SĒ-kal) **sphincter** A fold of mucous membrane that guards the opening from the ileum into the large intestine. Also called the *ileocecal valve*.

Ileum (IL-ē-um) The terminal portion of the small intestine.

Immunity (i-MYOON-i-tē) The state of being resistant to injury, particularly by poisons, foreign proteins, and invading parasites, due to the presence of antibodies.

Immunoglobulin (im-yoo-nō-GLOB-yoo-lin) (**Ig**) An antibody synthesized by plasma cells derived from B lymphocytes in response to the introduction of antigen. Immunoglobulins are divided into five kinds (IgG, IgM, IgA, IgD, IgE) based primarily on the larger protein component present in the immunoglobulin.

Immunology (im′-yoo-NOL-ō-jē) The branch of science that deals with the responses of the body when challenged by antigens.

Implantation (im-plan-TĀ-shun) The insertion of a tissue or a part into the body. The attachment of the blastocyst to the lining of the uterus 7–8 days after fertilization.

Impotence (IM-pō-tens) Weakness; inability to copulate; failure to maintain an erection long enough for sexual intercourse.

Incontinence (in-KON-ti-nens) Inability to retain urine, semen, or feces, through loss of sphincter control.

Indirect (motor) pathways Motor tracts that convey information from the brain down the spinal cord for somatic movements, coordination of body movements with visual stimuli, skeletal muscle tone and posture, and balance. Also known as *extrapyramidal pathways*.

Infant respiratory distress syndrome (RDS) A disease of newborn infants, especially premature ones, in which insufficient surfactant is produced and breathing is labored. Also called *hyaline* (HĪ-a-lin) *membrane disease* (*HMD*).

Infarction (in-FARK-shun) The presence of a localized area of necrotic tissue, produced by inadequate oxygenation of the tissue.

Infection (in-FEK-shun) Invasion and multiplication of microorganisms in body tissues, which may be inapparent or characterized by cellular injury.

Infectious mononucleosis (mon-ō-nook′-lē-Ō-sis) (**IM**) Contagious disease caused by the Epstein–Barr virus (EBV) and characterized by an elevated mononucleocyte and lymphocyte count, fever, sore throat, stiff neck, cough, and malaise.

Inferior (in-FĒR-ē-or) Away from the head or toward the lower part of a structure. Also called *caudad* (KAW-dad).

Inferior vena cava (VĒ-na CĀ-va) (**IVC**) Large vein that collects blood from parts of the body inferior to the heart and returns it to the right atrium.

Infertility Inability to conceive or to cause conception. Also called *sterility*.

Inflammation (in′-fla-MĀ-shun) Localized, protective response to tissue injury designed to destroy, dilute, or wall off the infecting agent or injured tissue; characterized by redness, pain, heat, swelling, and sometimes loss of function.

Inflation reflex Reflex that prevents overinflation of the lungs. Also called *Hering–Breuer reflex.*

Ingestion (in-JES-chun) The taking in of food, liquids, or drugs by mouth.

Inheritance The acquisition of body characteristics and qualities by transmission of genetic information from parents to offspring.

Inner cell mass A region of cells of a blastocyst that differentiates into the three primary germ layers—ectoderm, mesoderm, and endoderm—from which all tissues and organs develop; also called an *embryoblast.*

Inorganic (in′-or-GAN-ik) **compound** Compound that usually lacks carbon, usually is small, and contains ionic bonds. Examples include water and many acids, bases, and salts.

Insertion (in-SER-shun) The manner or place of attachment of a muscle to the bone that it moves.

Insomnia (in-SOM-nē-a) Difficulty in falling asleep and, usually, frequent awakening.

Inspiration (in-spi-RĀ-shun) The act of drawing air into the lungs.

Insula (IN-su-la) A triangular area of cerebral cortex that lies deep within the lateral cerebral fissure, under the parietal, frontal, and temporal lobes, and cannot be seen in an external view of the brain. Also called the *island* or *isle of Reil* (RĪL).

Insulin (IN-su-lin) A hormone produced by the beta cells of the pancreas that decreases the blood glucose level.

Integumentary (in-teg′-yoo-MEN-tar-ē) Relating to the skin.

Intercalated (in-TER-ka-lāt-ed) **disc** An irregular transverse thickening of sarcolemma that holds cardiac muscle fibers (cells) together and gap junctions that aid in conduction of muscle action potentials.

Intercostal (in′-ter-KOS-tal) **nerve** A nerve supplying a muscle located between the ribs.

Interferons (in′-ter-FĒR-ons) **(IFNs)** Three principal types of protein (alpha, beta, gamma) naturally produced by virus-infected host cells that induce uninfected cells to synthesize antiviral proteins that inhibit intracellular viral replication in uninfected host cells.

Internal ear The inner ear or labyrinth, lying inside the temporal bone, containing the organs of hearing and balance.

Internal nares (NA-rēz) The two openings posterior to the nasal cavities opening into the nasopharynx. Also called the *choanae* (kō-A-nē).

Internal respiration The exchange of respiratory gases between blood and body cells. Also called *tissue respiration.*

Interphase (IN-ter-fāz) The period during its life cycle when a cell is carrying on every life process except division; the stage between two mitotic divisions. Also called *metabolic phase.*

Interstitial cell of Leydig *See Interstitial endocrinocyte.*

Interstitial (in′-ter-STISH-al) **endocrinocyte** A cell located in the connective tissue between seminiferous tubules in a mature testis that secretes testosterone. Also called an *interstitial cell of Leydig* (LĪ-dig).

Interstitial (in′-ter-STISH-al) **fluid** The portion of extracellular fluid that fills the microscopic spaces between the cells of tissues; the internal environment of the body. Also called *intercellular* or *tissue fluid.*

Intervertebral (in′-ter-VER-te-bral) **disc** A pad of fibrocartilage located between the bodies of two vertebrae.

Intestinal gland Gland that opens onto the surface of the intestinal mucosa and secretes digestive enzymes. Also called a *crypt of Lieberkühn* (LĒ-ber-kyoon).

Intracellular (in′-tra-SEL-yoo-lar) **fluid (ICF)** Fluid located within cells.

Intramembranous ossification (in′-tra-MEM-bra-nus os′-i′-fi-KĀ-shun) The method of bone formation in which the bone is formed directly in membranous tissue.

Intrauterine device (IUD) A small metal or plastic object inserted into the uterus for the purpose of preventing pregnancy.

Intrinsic factor (IF) A glycoprotein synthesized and secreted by the parietal cells of the gastric mucosa that facilitates vitamin B_{12} absorption in the small intestine.

Intubation (in′-too-BĀ-shun) Insertion of a tube through the nose or mouth into the larynx and trachea for entrance of air or to dilate a stricture.

In utero (YOO-ter-ō) Within the uterus.

Inversion (in-VER-zhun) The movement of the sole inward at the ankle joint.

Ion (Ī-on) Any charged particle or group of particles; usually formed when a substance, such as a salt, dissolves and dissociates.

Ionization (ī-on-i-ZĀ-shun) Separation of inorganic acids, bases, and salts into ions when dissolved in water. Also called *dissociation* (dis′-sō-sē-Ā-shun).

Iris The colored portion of the eyeball seen through the cornea that consists of circular and radial smooth muscle; the black hole in the center of the iris is the pupil.

Ischemia (is-KĒ-mē-a) A lack of sufficient blood to a part due to obstruction of circulation.

Islet of Langerhans *See Pancreatic islet.*

Isoantibody (Ī-sō-AN-ti-bod′-ē) A specific antibody in blood serum capable of causing the clumping of bacteria, blood cells, or particles. Also called an *agglutinin.*

Isoantigen (Ī-sō-AN-ti-jen) A genetically determined antigen located on the surface of red blood cells; basis for the ABO grouping and Rh system of blood classification. Also called an *agglutinogen.*

Isometric contraction A muscle contraction in which tension on the muscle increases, but there is only minimal muscle shortening so that no movement is produced.

Isotonic (ī′-sō-TON-ik) Having equal tension or tone. Having equal osmotic pressure between two different solutions or between two elements in a solution.

Isotonic contraction A muscle contraction in which tension remains constant, but the muscle shortens and pulls on another structure to produce movement.

Isotope (Ī-sō-tōpe′) A chemical element that has the same atomic number as another but a different atomic weight. Radioactive isotopes change into other elements with the emission of certain radiations.

Jaundice (JAWN-dis) A condition characterized by yellowness of skin, white of eyes, mucous membranes, and body fluids because of a buildup of bilirubin.

Jejunum (jē-JOO-num) The middle portion of the small intestine.

Juxtaglomerular (juks-ta-glō-MER-yoo-lar) **apparatus (JGA)** Consists of the macula densa (cells of the distal convoluted tubule adjacent to the afferent and efferent arteriole) and juxtaglomerular cells (modified cells of the afferent and sometimes efferent arteriole); secretes renin when blood pressure starts to fall.

Karyotype (KAR-ē-ō-tip) An arrangement of chromosomes based on shape, size, and position of centromeres.

Keratinocyte (ker-A-tin′-ō-sīt) The most numerous of the epidermal cells that function in the production of keratin.

Ketone (KĒ-tōn) **bodies** Substances produced primarily during excessive triglyceride metabolism, such as acetone, acetoacetic acid, and β-hydroxybutyric acid.

Ketosis (kē-TŌ-sis) Abnormal condition marked by excessive production of ketone bodies.

Kidney (KID-nē) One of the paired reddish organs located in the lumbar region that regulates the composition and volume of blood and produces urine.

Kidney stone A concentration, usually consisting of calcium oxalate, uric acid, and calcium phosphate crystals, that may form in any portion of the urinary tract. Also called a *renal calculus.*

Kilocalorie (KIL-ō-kal′-ō-rē) **(kcal)** The amount of heat required to raise the temperature of 1000 g of water 1°C; the unit used to express the heating value of foods and to measure metabolic rate.

Korotkoff (kō-ROT-kof) **sounds** The various sounds that are heard while taking blood pressure.

Krebs cycle A series of biochemical reactions that occur in the matrix of mitochondria in which electrons are transferred to coenzymes and carbon dioxide is formed. The electrons carried by the coenzymes then enter the electron transport chain, which generates a large quantity of ATP. Also called the *citric acid cycle or tricarboxylic acid (TCA) cycle.*

Kupffer's cell *See Stellate reticuloendothelial cell.*

Kyphosis (kī-FŌ-sis) An exaggeration of the thoracic curve of the vertebral column, resulting in a "round-shouldered" or hunchback appearance.

Labia majora (LĀ-bē-a ma-JO-ra) Two longitudinal folds of skin extending downward and backward from the mons pubis of the female.

Labia minora (min-OR-a) Two small folds of mucous membrane lying medial to the labia majora of the female.

Labor The process by which the product of conception is expelled from the uterus by its contractions through the vagina to the outside of the body.

Labyrinth (LAB-i-rinth) Intricate communicating passageway, especially in the internal ear.

Laceration (las′-er-Ā-shun) Wound or irregular tear of the skin.

Lacrimal (LAK-ri-mal) Pertaining to tears.

Lacrimal (LAK-ri-mal) **canal** A duct, one on each eyelid, commencing at the medial margin of an eyelid and conveying tears medially into the nasolacrimal sac.

Lacrimal gland Secretory cells located at the superior anterolateral portion of each orbit that secrete tears into excretory ducts that open onto the surface of the conjunctiva.

Lactation (lak-TĀ-shun) The secretion and ejection of milk by the mammary glands.

Lacteal (LAK-tē-al) One of many intestinal lymphatic vessels in villi that absorb fat from digested food.

Lacuna (la-KOO-na) A small, hollow space, such as that found in bones in which the osteoblasts lie. *Plural, lacunae* (la-KOO-nē).

Lamellae (la-MEL-ē) Concentric rings located in compact bone.

Lamellated corpuscle Oval pressure receptor located in subcutaneous tissue and consisting of concentric layers of connective tissue wrapped around an afferent nerve fiber. Also called a *Pacinian* (pa-SIN-ē-an) *corpuscle.*

Lamina propria (PRŌ-prē-a) The connective tissue layer of a mucous membrane.

Large intestine The portion of the gastrointestinal tract extending from the ileum of the small intestine to the anus, divided structurally into the cecum, colon, rectum, and anal canal.

Laryngitis (la-rin-JĪ-tis) Inflammation of the mucous membrane lining the larynx.

Laryngopharynx (la-rin-gō-FAR-inks) The inferior portion of the pharynx, extending downward from the level of the hyoid bone to divide posteriorly into the esophagus and anteriorly into the larynx.

Larynx (LAR-inks) The voice box, a short passageway that connects the pharynx with the trachea.

Lateral (LAT-er-al) Farther from the midline of the body or a structure.

Lateral ventricle (VEN-tri-kul) A cavity within a cerebral hemisphere that communicates with the lateral ventricle in the other cerebral hemisphere and with the third ventricle by way of the interventricular foramen.

Learning The ability to acquire knowledge or a skill through instruction or experience.

Lens A transparent organ constructed of proteins (crystallins) lying posterior to the pupil and iris of the eyeball and anterior to the vitreous body.

Lesion (LĒ-zhun) Any localized, abnormal change in tissue formation.

Lethargy (LETH-ar-jē) A condition of drowsiness or indifference.

Leukemia (loo-KĒ-mē-a) A malignant disease of the blood-forming tissues characterized by either uncontrolled production and accumulation of immature leukocytes in which many cells fail to reach maturity (acute) or an accumulation of mature leukocytes in the blood because they do not die at the end of their normal life span (chronic).

Leukocyte (LOO-kō-sīt) A white blood cell. Also spelled *leucocyte.*

Leukocytosis (loo′-kō-sī-TŌ-sis) An increase in the number of white blood cells, characteristic of many infections and other disorders.

Leukopenia (loo-kō-PĒ-nē-a) A decrease of the number of white blood cells below 5000/mm³.

Leukoplakia (loo-kō-PLĀ-kē-a) A disorder in which there are white patches in the mucous membranes of the tongue, gums, and cheeks.

Leydig (LĪ-dig) **cell** *See Interstitial endocrinocyte.*

Libido (li-BĒ-dō) The sexual drive, conscious or unconscious.

Ligament (LIG-a-ment) Dense, regularly arranged connective tissue that attaches bone to bone.

Limbic system A portion of the forebrain, sometimes termed the visceral brain, concerned with various aspects of emotion and behavior, that includes the limbic lobe, dentate gyrus, amygdaloid body, septal nuclei, mamillary bodies, anterior thalamic nucleus, olfactory bulbs, and bundles of myelinated axons.

Lipase (LĪ-pās) A triglyceride-splitting enzyme.

Lipid An organic compound composed of carbon, hydrogen, and oxygen that is usually insoluble in water, but soluble in alcohol, ether, and chloroform; examples include triglycerides, phospholipids, steroids, and prostaglandins.

Lipogenesis (li-pō-GEN-e-sis) The synthesis of lipids from glucose or amino acids by liver cells.

Lipolysis (lip-OL-i-sis) The splitting of a triglyceride (fat) molecule into glycerol and fatty acids.

Lipoprotein (lip′-ō-PRŌ-tēn) Protein-containing lipid that is produced by the liver and combines with cholesterol and triglycerides (fats) to make it water-soluble for transportation by the cardiovascular system; high levels of low-density lipoproteins (LDLs) are associated with increased risk of atherosclerosis, while high levels of high-density lipoproteins (HDLs) are associated with decreased risk of atherosclerosis.

Lithotripsy (LITH-ō-trip′-sē) A noninvasive procedure in which shock waves generated by a lithotriptor are used to pulverize kidney stones or gallstones.

Liver Large gland under the diaphragm that occupies most of the right hypochondriac region and part of the epigastric region; functionally, it produces bile salts, heparin, and plasma proteins; converts one nutrient into another; detoxifies substances; stores glycogen, minerals, and vitamins; carries on phagocytosis of blood cells and bacteria; and helps activate vitamin D.

Lordosis (lor-DŌ-sis) An exaggeration of the lumbar curve of the vertebral column.

Lumbar (LUM-bar) Region of the back and side between the ribs and pelvis; loin.

Lumbar plexus (PLEK-sus) A network formed by the anterior branches of spinal nerves L1 through L4.

Lumen (LOO-men) The space within an artery, vein, intestine, or a tube.

Lung One of the two main organs of respiration, lying on either side of the heart in the thoracic cavity.

Lunula (LOO-nyoo-la) The moon-shaped white area at the base of a nail.

Luteinizing (LOO-tē-in′-ĭz-ing) **hormone (LH)** A hormone secreted by the anterior pituitary gland that stimulates ovulation, progesterone secretion by the corpus luteum, and readies the mammary glands for milk secretion in females and stimulates testosterone secretion by the testes in males.

Lyme (LĪM) **disease** Disease caused by a bacterium (*Borrelia burgdorferi*) and transmitted to humans by ticks (mainly deer ticks); may be characterized by a bull's eye rash. Symptoms include joint stiffness, fever and chills, headache, stiff neck, nausea, and low back pain. Later stages may involve cardiac and neurologic problems and arthritis.

Lymph (limf) Fluid confined in lymphatic vessels and flowing through the lymphatic system to be returned to the blood.

Lymphatic tissue A specialized form of reticular tissue that contains large numbers of lymphocytes.

Lymphatic vessel A large vessel that collects lymph from lymphatic capillaries and converges with other lymphatic vessels to form the thoracic and right lymphatic ducts.

Lymph node An oval or bean-shaped structure located along lymphatic vessels.

Lymphocyte (LIM-fō-sīt) A type of white blood cell, found in lymph nodes, associated with the immune system.

Lymphokines (LIM-fō-kīns) Powerful proteins secreted by T cells that endow T cells with their ability to assist in immunity.

Lymphoma (lim′-FŌ-ma) Any tumor composed of lymphatic tissue.

Lysosome (LĪ-sō-sōm) An organelle in the cytoplasm of a cell, enclosed by a single membrane and containing powerful digestive enzymes.

Lysozyme (LĪ-sō-zīm) A bactericidal enzyme found in tears, saliva, and perspiration.

Macrophage (MAK-rō-fāj) Phagocytic cell derived from a monocyte. May be fixed or wandering.

Macula (MAK-yoo-la) A discolored spot or a colored area. A small, thickened region on the wall of the utricle and saccule that serves as a receptor for static equilibrium.

Macula lutea (LOO-tē-a) The yellow spot in the center of the retina.

Magnetic resonance imaging (MRI) A diagnostic procedure that focuses on the nuclei of atoms of a single element in a tissue, usually hydrogen, to determine if they behave normally in the presence of an external magnetic force; used to indicate the biochemical activity of a tissue. Formerly called *nuclear magnetic resonance (NMR)*.

Malaise (ma-LĀYZ) Discomfort, uneasiness, and indisposition, often indicative of infection.

Malignant (ma-LIG-nant) Referring to diseases that tend to become worse and cause death; especially the invasion and spreading of cancer.

Malignant melanoma (mel′-a-NŌ-ma) A usually dark, malignant tumor of the skin containing melanin.

Malnutrition (mal′-noo-TRISH-un) State of bad or poor nutrition that may be due to inadequate food intake, imbalance of nutrients, malabsorption of nutrients, improper distribution of nutrients, increased nutrient requirements, increased nutrient losses, or overnutrition.

Mammary (MAM-ar-ē) **gland** Modified sudoriferous (sweat) gland of the female that secretes milk for the nourishment of the young.

Mammography (mam-OG-ra-fē) Procedure for imaging the breasts (xeromammography or film-screen mammography) to evaluate for breast disease or screen for breast cancer.

Marrow (MAR-ō) Soft, spongelike material in the cavities of bone. Red bone marrow produces blood cells; yellow bone marrow, formed mainly of fatty tissue, has no blood-producing function.

Mass number Total number of protons and neutrons in an atom.

Mast cell A cell found in areolar connective tissue along blood vessels that produces histamine, a dilator of small blood vessels during inflammation.

Mastectomy (mas-TEK-tō-mē) Surgical removal of breast tissues.

Mastication (mas′-ti-KĀ-shun) Chewing.

Matrix (MĀ-triks) The ground substance and fibers between cells in a connective tissue.

Matter Anything that occupies space and has mass.

Mature follicle A relatively large, fluid-filled follicle containing a secondary oocyte and surrounding granulosa cells that secrete estrogens. Also called a *Graafian* (GRAF-ē-an) *follicle.*

Meatus (mē-Ā-tus) A passage or opening, especially the external portion of a canal.

Mechanoreceptor (me-KAN-ō-rē′-sep-tor) Receptor that detects mechanical deformation of the receptor itself or adjacent cells; stimuli so detected include those related to touch, pressure, vibration, proprioception, hearing, equilibrium, and blood pressure.

Medial (MĒ-dē-al) Nearer the midline of the body or of a structure.

Mediastinum (mē′-dē-as-TĪ-num) A broad, median partition, actually a mass of tissue found between the pleurae of the lungs that extends from the sternum to the vertebral column.

Medulla (me-DULL-la) An inner layer of an organ, such as the medulla of the kidneys.

Medulla oblongata (ob′-long-GA-ta) The most inferior part of the brain stem.

Medullary (MED-yoo-lar′-ē) **cavity** The space within the diaphysis of a bone that contains yellow bone marrow. Also called the *marrow cavity.*

Medullary rhythmicity (rith-MIS-i-tē) *area* Portion of the respiratory center in the medulla that controls the basic rhythm of respiration.

Meibomian gland *See Tarsal gland.*

Meiosis (mē-Ō-sis) A type of cell division restricted to sex-cell production involving two successive nuclear divisions that result in daughter cells with the haploid (*n*) number of chromosomes.

Meissner's corpuscle *See Corpuscle of touch.*

Melanin (MEL-a-nin) A dark black, brown, or yellow pigment found in some parts of the body such as the skin.

Melanocyte (MEL-a-nō-sīt′) A pigmented cell located between or beneath cells of the deepest layer of the epidermis that synthesizes melanin.

Melanocyte-stimulating hormone (MSH) A hormone secreted by the anterior pituitary gland that stimulates the dispersion of melanin granules in melanocytes in amphibians; continued administration produces darkening of skin in humans.

Melatonin (mel-a-TŌN-in) A hormone secreted by the pineal gland that may inhibit reproductive activities.

Membrane A thin, flexible sheet of tissue composed of an epithelial layer and an underlying connective tissue layer, as in an epithelial membrane, or of areolar connective tissue only, as in a synovial membrane.

Membranous labyrinth (mem-BRA-nus LAB-i-rinth) The portion of the labyrinth of the inner ear that is located inside the bony labyrinth and separated from it by the perilymph; made up of the membranous semicircular canals, the saccule and utricle, and the cochlear duct.

Memory The ability to recall thoughts; commonly classified as short term (activated) and long term.

Menarche (me-NAR-kē) Beginning of the menstrual function.

Ménière's (men-YAIRZ) **syndrome** A type of labyrinthine disease characterized by fluctuating loss of hearing, vertigo, and tinnitus due to an increased amount of endolymph that enlarges the labyrinth.

Meninges (me-NIN-jēz) Three membranes covering the brain and spinal cord, called the dura mater, arachnoid, and pia mater. *Singular, meninx* (MEN-ninks).

Meningitis (men-in-JĪ-tis) Inflammation of the meninges, most commonly the pia mater and arachnoid.

Menopause (MEN-ō-pawz) The termination of the menstrual cycles.

Menstrual (MEN-stroo-al) **cycle** *See Female reproductive cycle.*

Menstruation (men′-stroo-Ā-shun) Periodic discharge of blood, tissue fluid, mucus, and epithelial cells that usually lasts for 5 days; caused by a sudden reduction in estrogens and progesterone. Also called the **menstrual phase** or **menses**.

Merkel (MER-kel) **cell** Cell found in the epidermis of hairless skin that makes contact with a tactile (Merkel) disc that functions in touch.

Merocrine (MER-ō-krin) **gland** A secretory cell that remains intact throughout the process of formation and the discharge of the secretory product, as in the salivary and pancreatic glands.

Mesenchyme (MEZ-en-kīm) An embryonic connective tissue from which all other connective tissues arise.

Mesentery (MEZ-en-ter′-ē) A fold of peritoneum attaching the small intestine to the posterior abdominal wall.

Mesoderm (MEZ-ō-derm) The middle of the three primary germ layers that gives rise to connective tissues, blood and blood vessels, and muscles.

Metabolism (me-TAB-ō-lizm) The sum of all the biochemical reactions that occur within an organism, including the synthetic (anabolic) reactions and decomposition (catabolic) reactions.

Metacarpus (met′-a-KAR-pus) A collective term for the five bones that make up the palm of the hand.

Metaphase (MET-a-phāz) The second stage of mitosis in which chromatid pairs line up on the equatorial plane of the cell.

Metaphysis (me-TAF-i-sis) Growing portion of a bone.

Metaplasia (met′-a-PLĀ-zē-a) The abnormal change of one type of adult, differentiated cell into another; for example, such as that which occurs in bronchogenic carcinoma.

Metastasis (me-TAS-ta-sis) The spread of cancer to surrounding tissues (local) or to other body sites (distant).

Metatarsus (met′-a-TAR-sus) A collective term for the five bones located in the foot between the tarsals and the phalanges.

Micelle (mī-SEL) A spherical aggregate of bile salts that dissolves fatty acids and monoglycerides so that they can be transported into small intestinal epithelial cells.

Microfilament (mī-krō-FIL-a-ment) Rodlike cytoplasmic structure about 6 nm in diameter; comprises contractile units in muscle fibers (cells) and provides support, shape, and movement in nonmuscle cells.

Microglia (mī-krō-GLĒ-a) Neuroglial cells that carry on phagocytosis. Also called **brain macrophages** (MAK-rō-fāj-ez).

Microphage (MĪK-rō-fāj) Granular leukocyte that carries on phagocytosis, especially neutrophils and eosinophils.

Microtubule (mī-krō-TOOB-yool′) Cylindrical cytoplasmic structure, ranging in diameter from 18 to 30 nm, consisting of the protein tubulin; provides support, structure, and transportation.

Microvilli (mī-krō-VIL-ē) Microscopic, fingerlike projections of the cell membranes of small intestinal cells that increase surface area for absorption.

Micturition (mik′-too-RISH-un) The act of expelling urine from the urinary bladder. Also called **urination** (yoo-ri-NĀ-shun).

Midbrain The part of the brain between the pons and the diencephalon. Also called the **mesencephalon** (mes′-en-SEF-a-lon).

Middle ear A small, epithelial-lined cavity hollowed out of the temporal bone, separated from the external ear by the eardrum and from the internal ear by a thin bony partition containing the oval and round windows; extending across the middle ear are the three auditory ossicles. Also called the **tympanic** (tim-PAN-ik) **cavity.**

Midsagittal plane A vertical plane through the midline of the body that divides the body or organs into *equal* right and left sides. Also called a **median plane.**

Milk let-down reflex Contraction of alveolar cells to force milk into ducts of mammary glands, stimulated by oxytocin (OT), which is released from the posterior pituitary gland in response to suckling action.

Mineral Inorganic, homogeneous solid substance that may perform a function vital to life; examples include calcium, sodium, potassium, iron, phosphorus, and chlorine.

Mineralocorticoids (min′-er-al-ō-KOR-ti-koyds) A group of hormones of the adrenal cortex.

Mitochondrion (mī′tō-KON-drē-on) A double-membraned organelle that plays a central role in the production of ATP; known as the "powerhouse" of the cell.

Mitosis (mī-TŌ-sis) The orderly division of the nucleus of a cell that ensures that each new daughter nucleus has the same number and kind of chromosomes as the original parent nucleus. The process includes the replication of chromosomes and the distribution of the two sets of chromosomes into two separate and equal nuclei.

Mitotic spindle Collective term for a football-shaped assembly of microtubules that is responsible for the movement of chromosomes during cell division.

Mitral (MĪ-tral) **stenosis** (ste-NŌ-sis) Narrowing of the mitral valve by scar formation or a congenital defect.

Mitral valve prolapse (PRŌ-laps) or **MVP** An inherited disorder in which a portion of a mitral valve is pushed back too far (prolapsed) during contraction due to expansion of the cusps and elongation of the chordae tendineae.

Molecule (MOL-e-kyool) The chemical combination of two or more atoms.

Monocyte (MON-ō-sīt′) A type of white blood cell characterized by agranular cytoplasm; the largest of the leukocytes.

Monosaturated fat A fat that contains one double covalent bond between its carbon atoms; it is not completely saturated with hydrogen atoms. Examples are olive oil and peanut oil.

Mons pubis (monz PYOO-bis) The rounded, fatty prominence over the pubic symphysis covered by coarse pubic hair.

Morbid (MOR-bid) Diseased; pertaining to disease.

Morning sickness Sensation of nausea during pregnancy that may be related to the hormone human chorionic gonadotropin (hCG).

Morula (MOR-yoo-la) A solid mass of cells produced by successive cleavages of a fertilized ovum a few days after fertilization.

Motor area The region of the cerebral cortex that governs muscular movement, particularly the precentral gyrus of the frontal lobe.

Motor end plate Portion of the sarcolemma of a muscle fiber (cell) in close approximation with an axon terminal.

Motor neuron (NOO-ron) A neuron that conducts nerve impulses from the brain and spinal cord to effectors that may be either muscles or glands. Also called an *efferent neuron.*

Motor unit A motor neuron together with the muscle fibers (cells) it stimulates.

Mucosa-associated lymphatic tissue (MALT) Lymphatic nodules scattered throughout the lamina propria (connective tissue) of mucous membranes lining the gastrointestinal tract, respiratory airways, urinary tract, and reproductive tract.

Mucous (MYOO-kus) **cell** A unicellular gland that secretes mucus. Also called a *goblet cell.*

Mucous membrane A membrane that lines a body cavity that opens to the exterior. Also called the *mucosa* (myoo-KŌ-sa).

Mucus The thick fluid secretion of mucous glands and mucous membranes.

Mumps Inflammation and enlargement of the parotid glands, accompanied by fever and extreme pain during swallowing.

Muscle An organ composed of one of three types of muscle tissue (skeletal, cardiac, or visceral), specialized for contraction to produce voluntary or involuntary movement of parts of the body.

Muscle action potential A stimulating impulse that travels along a sarcolemma and then into transverse tubules; it is generated by acetylcholine from synaptic vesicles that alters permeability of the sarcolemma to sodium ions (Na^+).

Muscle fatigue (fa-TĒG) Inability of a muscle to maintain its strength of contraction or tension; may be related to insufficient oxygen, depletion of glycogen, and/or lactic acid buildup.

Muscle tissue A tissue specialized to produce motion in response to muscle action potentials by its qualities of contractility, extensibility, elasticity, and excitability. Types include skeletal, cardiac, and smooth.

Muscle tone A sustained, partial contraction of portions of a skeletal muscle in response to activation of stretch receptors.

Muscular dystrophies (DIS-trō-fēz′) Inherited muscle-destroying diseases, characterized by degeneration of the individual muscle fibers (cells), which leads to progressive atrophy of the skeletal muscle.

Muscularis (MUS-kyoo-la′-ris) A muscular layer (coat or tunic) of an organ.

Mutation (myoo-TĀ-shun) Any change in the sequence of bases in the DNA molecule resulting in a permanent alteration in some inheritable characteristic.

Myasthenia (mī-as-THĒ-nē-a) **gravis** Weakness of skeletal muscles caused by antibodies directed against acetylcholine receptors that inhibit muscle contraction.

Myelin (MĪ-e-lin) **sheath** Multilayered lipid and protein covering, formed by neurolemmocytes (Schwann cells) and oligodendrocytes around axons of many peripheral and central nervous system neurons.

Myocardial infarction (mī′-ō-KAR-dē-al in-FARK-shun) (**MI**) Gross necrosis of myocardial tissue due to interrupted blood supply. Also called *heart attack.*

Myocardium (mī′-ō-KAR-dē-um) The middle layer of the heart wall, made up of cardiac muscle, comprising the bulk of the heart, and lying between the epicardium and the endocardium.

Myofibril (mī′-ō-FĪ-bril) A threadlike structure, running longitudinally through a muscle fiber (cell) consisting mainly of thick myofilaments (myosin) and thin myofilaments (actin).

Myoglobin (mī-ō-GLŌ-bin) The oxygen-binding, iron-containing conjugated protein complex present in the sarcoplasm of muscle fibers (cells); contributes the red color to muscle.

Myogram (MĪ-ō-gram) The record or tracing produced by the myograph, the apparatus that measures and records the effects of muscular contractions.

Myology (mī-OL-ō-jē) The study of the muscles.

Myometrium (mī′-ō-MĒ-trē-um) The smooth muscle layer of the uterus.

Myopathy (mī-OP-a-thē) Any disease of muscle tissue.

Myosin (MĪ-ō-sin) The contractile protein that makes up the thick myofilaments of muscle fibers (cells).

Nail A hard plate, composed largely of keratin, that develops from the epidermis of the skin to form a protective covering on the dorsal surface of the distal phalanges of the fingers and toes.

Nasal (NĀ-zal) **cavity** A mucosa-lined cavity on either side of the nasal septum that opens onto the face at an external naris and into the nasopharynx at an internal naris.

Nasal septum (SEP-tum) A vertical partition composed of bone (perpendicular plate of ethmoid and vomer) and cartilage, covered with a mucous membrane, separating the nasal cavity into left and right sides.

Nasolacrimal (nā′-zō-LAK-ri-mal) **duct** A canal that transports the lacrimal secretion (tears) from the nasolacrimal sac into the nose.

Nasopharynx (nā′-zō-FAR-inks) The uppermost portion of the pharynx, lying posterior to the nose and extending down to the soft palate.

Nausea (NAW-sē-a) Discomfort characterized by loss of appetite and sensation of impending vomiting.

Negative feedback The principle governing most control systems; a mechanism of response in which a stimulus initiates actions that reverse or reduce the stimulus.

Neonatal (nē′-ō-NĀ-tal) Pertaining to the first 4 weeks after birth.

Neoplasm (NĒ-ō-plazm) Any abnormal formation or growth; usually a malignant tumor.

Nephron (NEF-ron) The functional unit of the kidney.

Nephrotic (ne-FROT-ik) **syndrome** A condition in which the filtering membrane leaks, allowing large amounts of protein to escape into urine.

Nerve A cordlike bundle of nerve fibers (axons and/or dendrites) and their associated connective tissue coursing together outside the central nervous system.

Nerve fiber General term for any process (axon or dendrite) projecting from the cell body of a neuron.

Nerve impulse A wave of negativity (depolarization) that self-propagates along the outside surface of the plasma membrane of a neuron; also called a *nerve action potential.*

Nervous tissue Tissue that initiates and transmits nerve impulses to coordinate homeostasis.

Net filtration pressure (NFP) Net pressure that expresses the relationship between the forces that promote fluid movement at the arterial and venous ends of a capillary and that promote glomerular filtration in the kidneys.

Neuralgia (noo-RAL-jē-a) Attacks of pain along the entire course or branch of a peripheral sensory nerve.

Neuritis (noo-RĪ-tis) Inflammation of a single nerve, two or more nerves in separate areas, or many nerves simultaneously.

Neuroglia (noo-RŌG-lē-a) Cells of the nervous system that are specialized to perform the functions of connective tissue. The neuroglia of the central nervous system are the astrocytes, oligodendrocytes, microglia, and ependyma; neuroglia of the peripheral nervous system include the neurolemmocytes (Schwann cells) and the ganglion satellite cells. Also called *glial* (GLĒ-al) *cells.*

Neurolemma (noo-rō-LEM-ma) The peripheral, nucleated cytoplasmic layer of the neurolemmocyte (Schwann cell). Also called *sheath of Schwann* (SCHVON).

Neurolemmocyte A neuroglial cell of the peripheral nervous system that forms the myelin sheath and neurolemma of a nerve fiber by wrapping around a nerve fiber in a jelly-roll fashion. Also called a *Schwann* (SCHVON) *cell.*

Neurology (noo-ROL-ō-jē) The branch of science that deals with the normal functioning and disorders of the nervous system.

Neuromuscular (noo-rō-MUS-kyoo-lar) **junction** The area of contact between the axon terminal of a motor neuron and a portion of the sarcolemma of a muscle fiber (cell). Also called a *myoneural* (mī-ō-NOO-ral) *junction.*

Neuron (NOO-ron) A nerve cell, consisting of a cell body, dendrites, and an axon.

Neurotransmitter One of a variety of molecules synthesized within the nerve axon terminals, released into the synaptic cleft in response to a nerve impulse, and affecting the membrane potential of the postsynaptic neuron. Also called a *transmitter substance.*

Neutrophil (NOO-trō-fil) A type of white blood cell characterized by granules that stain pale lilac with a combination of acidic or basic dyes.

Nipple A pigmented, wrinkled projection on the surface of the mammary gland that is the location of the openings of the lactiferous ducts for milk release.

Nissl bodies *See Chromatophilic substance.*

Node of Ranvier *See Neurofibral node.*

Norepinephrine (nor′-ep-ē-NEF-rin) **(NE)** A hormone secreted by the adrenal medulla that produces actions similar to those that result from sympathetic stimulation. Also called *noradrenaline* (nor-a-DREN-a-lin).

Nuclear medicine The branch of medicine concerned with the use of radioisotopes in the diagnosis of disease and therapy.

Nuclease (NOO-klē-ās) An enzyme that breaks nucleotides into pentoses and nitrogenous bases; examples are ribonuclease and deoxyribonuclease.

Nucleic (noo-KLĒ-ic) **acid** An organic compound that is a long polymer of nucleotides, with each nucleotide containing a pentose sugar, a phosphate group, and one of four possible nitrogenous bases (adenine, cytosine, guanine, and thymine or uracil).

Nucleolus (noo-KLĒ-ō-lus) Nonmembranous spherical body within the nucleus composed of protein, DNA, and RNA that functions in the synthesis and storage of ribosomal RNA.

Nucleus (NOO-klē-us) A spherical or oval organelle of a cell that contains the hereditary factors of the cell, called genes. A cluster of unmyelinated nerve cell bodies in the central nervous system. The central portion of an atom made up of protons and neutrons.

Nucleus pulposus (pul-PŌ-sus) A soft, pulpy, highly elastic substance in the center of an intervertebral disc, a remnant of the notochord.

Nutrient A chemical substance in food that provides energy, forms new body components, or assists in the functioning of various body processes.

Nystagmus (nis-TAG-mus) Rapid, involuntary, rhythmic movement of the eyeballs; horizontal, rotary, or vertical.

Obesity (ō-BĒS-i-tē) Body weight 10 to 20 percent over a desirable standard as a result of excessive accumulation of fat. Types of obesity are hypertrophic (adult-onset) and hyperplastic (lifelong).

Oblique (ō-BLĒK) **plane** A plane that passes through the body or an organ at an angle between the transverse plane and either the midsagittal, parasagittal, or frontal plane.

Obstetrics (ob-STET-riks) The specialized branch of medicine that deals with pregnancy, labor, and the period of time (about 6 weeks) immediately following delivery.

Occlusion (ō-KLOO-zhun) The act of closure or state of being closed.

Occult (o-KULT) Obscure or hidden from view, as, for example, occult blood in stools or urine.

Olfactory (ōl-FAK-tō-rē) Pertaining to smell.

Oligodendrocyte (o-lig-ō-DEN-drō-sīt) A neuroglial cell that supports neurons and produces a phospholipid myelin sheath around axons of neurons of the central nervous system.

Oliguria (ol′-i-GYOO-rē-a) Daily urinary output usually less than 250 ml.

Oncogene (ONG-kō-jēn) Gene that has the ability to transform a normal cell into a cancerous cell.

Oncology (ong-KOL-ō-jē) The study of tumors.

Oogenesis (ō′-ō-JEN-e-sis) Formation and development of the ovum.

Opsonization (op-sō-ni-ZĀ-shun) The action of some antibodies that renders bacteria and other foreign cells more susceptible to phagocytosis. Also called *immune adherence.*

Optic (OP-tik) Refers to the eye, vision, or properties of light.

Optic chiasm (kī-AZ-m) A crossing point of the optic (II) nerves, anterior to the pituitary gland.

Optic disc A small area of the retina containing openings through which the fibers of the ganglion neurons emerge as the optic (II) nerve. Also called the *blind spot.*

Oral contraceptive (OC) A hormone compound, usually a high concentration of progesterone and a low concentration of estrogens, that is swallowed and prevents ovulation, and thus pregnancy. Also called *"the pill."*

Orbit (OR-bit) The bony, pyramid-shaped cavity of the skull that holds the eyeball.

Organ A structure composed of two or more different kinds of tissues with a specific function and usually a recognizable shape.

Organelle (or-gan-EL) A permanent structure within a cell with characteristic morphology that is specialized to serve a specific function in cellular activities.

Organic (or-GAN-ik) **compound** Compound that always contains carbon and hydrogen and the atoms are held together by covalent bonds. Examples include carbohydrates, lipids, protein, and nucleic acids (DNA and RNA).

Organism (OR-ga-nizm) A total living form; one individual.

Orgasm (OR-gazm) Sensory and motor events involved in ejaculation in the male and involuntary contraction of the perineal muscles in the female at the climax of sexual intercourse.

Orifice (OR-i-fis) Any aperture or opening.

Origin (OR-i-jin) The place of attachment of a muscle to the more stationary bone, or the end opposite the insertion.

Oropharynx (or'-ō-FAR-inks) The second portion of the pharynx, lying posterior to the mouth and extending from the soft palate down to the hyoid bone.

Orthopedics (or'-thō-PĒ-diks) The branch of medicine that deals with the preservation and restoration of the skeletal system, articulations, and associated structures.

Orthopnea (or'-THOP-nē-a) Dyspnea that occurs in the horizontal position.

Osmoreceptor (oz'-mō-re-CEP-tor) Receptor in the hypothalamus that is sensitive to changes in blood osmotic pressure and in response to high osmotic pressure (low water concentration) causes synthesis and release of antidiuretic hormone (ADH).

Osmosis (os-MŌ-sis) The net movement of water molecules through a selectively permeable membrane from an area of higher water concentration to an area of lower water concentration until an equilibrium is reached.

Osmotic pressure The pressure required to prevent the movement of pure water into a solution containing solutes when the two solutions are separated by a selectively permeable membrane.

Ossification (os'-i-fi-KĀ-shun) Formation of bone. Also called *osteogenesis* (os-tē-ō-JEN-e-sis).

Osteoblast (OS-tē-ō-blast') Cell formed from an osteoprogenitor cell that participates in bone formation by secreting some organic components and inorganic salts.

Osteoclast (OS-tē-ō-clast') A large multinuclear cell that develops from a monocyte and destroys or resorbs bone tissue.

Osteocyte (OS-tē-ō-sīt') A mature bone cell that maintains the daily activities of bone tissue.

Osteology (os'-tē-OL-ō-jē) The study of bones.

Osteon (OS-tē-on) The basic unit of structure in adult compact bone, consisting of a central (Haversian) canal with its concentrically arranged lamellae, lacunae, osteocytes, and canaliculi. Also called a *Haversian* (ha-VER-shun) *system*.

Osteoporosis (os'-tē-ō-pō-RŌ-sis) Age-related disorder characterized by decreased bone mass and increased susceptibility to fractures as a result of decreased levels of estrogens.

Otolith (Ō-tō-lith) A particle of calcium carbonate embedded in the otolithic membrane that functions in maintaining static equilibrium.

Otolithic (ō-tō-LITH-ik) **membrane** Thick, gelatinous, glycoprotein layer located directly over hair cells of the macula in the saccule and utricle of the inner ear.

Oval window A small opening between the middle ear and inner ear into which the footplate of the stapes fits. Also called the *fenestra vestibuli* (fe-NES-tra ves-TIB-yoo-lē).

Ovarian (ō-VAR-ē-an) **cycle** A monthly series of events in the ovary associated with the maturation of an ovum.

Ovarian follicle (FOL-i-kul) A general name for oocytes (immature ova) in any stage of development, along with their surrounding epithelial cells.

Ovary (Ō-var-ē) Female gonad that produces ova and the hormones estrogens, progesterone, and relaxin.

Ovulation (ō-vyoo-LĀ-shun) The rupture of a mature ovarian (Graafian) follicle with discharge of a secondary oocyte into the pelvic cavity.

Ovum (Ō-vum) The female reproductive or germ cell; an egg cell.

Oxidation (ok-si-DĀ-shun) The removal of electrons and hydrogen ions (hydrogen atoms) from a molecule or, less commonly, the addition of oxygen to a molecule that results in a decrease in the energy content of the molecule. The oxidation of glucose in the body is also called *cellular respiration.*

Oxygen debt The volume of oxygen required to oxidize the lactic acid produced by muscular exercise.

Oxyhemoglobin (ok'-sē-HĒ-mō-glō-bin) (HbO_2) Hemoglobin combined with oxygen.

Oxytocin (ok'-sē-TŌ-sin) (**OT**) A hormone secreted by neurosecretory cells in the paraventricular and supraoptic nuclei of the hypothalamus that stimulates contraction of the smooth muscle fibers (cells) in the pregnant uterus and contractile cells around the ducts of mammary glands.

Pacinian corpuscle *See Lamellated corpuscle.*

Palate (PAL-at) The horizontal structure separating the oral and the nasal cavities; the roof of the mouth.

Palpate (PAL-pāt) To examine by touch; to feel.

Palpitation (pal'-pi-TĀ-shun) A fluttering of the heart or abnormal rate or rhythm of the heart.

Pancreas (PAN-krē-as) A soft, oblong organ lying along the greater curvature of the stomach and connected by a duct to the duodenum. It is both exocrine (secreting pancreatic juice) and endocrine (secreting insulin, glucagon, and somatostatin).

Pancreatic (pan'-krē-AT-ik) **duct** A single, large tube that unites with the common bile duct from the liver and gallbladder and drains pancreatic juice into the duodenum at the hepatopancreatic ampulla (ampulla of Vater). Also called the *duct of Wirsung.*

Pancreatic islet A cluster of endocrine gland cells in the pancreas that secretes insulin, glucagon, and somatostatin. Also called *islet of Langerhans* (LANG-er-hanz).

Papanicolaou (pap'-a-NIK-ō-la-oo) **test** A cytological staining test for the detection and diagnosis of premalignant and malignant conditions of the female genital tract. Cells scraped from the genital epithelium are smeared, fixed, stained, and examined microscopically. Also called a *Pap smear.*

Papilla (pa-PIL-a) A small nipple-shaped projection or elevation.

Paralysis (pa-RAL-a-sis) Loss or impairment of motor function due to a lesion of nervous or muscular origin.

Paranasal sinus (par'-a-NĀ-zal SĪ-nus) A mucus-lined air cavity in a skull bone that communicates with the nasal cavity. Paranasal sinuses are located in the frontal, maxillary, ethmoid, and sphenoid bones.

Parasagittal plane A vertical plane that does not pass through the midline and divides the body or organs into *unequal* right and left sides.

Parasympathetic (par'-a-sim-pa-THET-ik) **division** One of the two subdivisions of the autonomic nervous system, having cell bodies of preganglionic neurons in nuclei in the brain stem and in the lateral gray matter of the sacral portion of the spinal cord; primarily concerned with activities that conserve and restore body energy. Also called the *craniosacral* (krā-nē-ō-SĀ-kral) *division.*

Parathyroid (par'-a-THĪ-royd) **gland** One of usually four small endocrine glands embedded on the posterior surfaces of the lateral lobes of the thyroid gland.

Parathyroid hormone (PTH) A hormone secreted by the parathyroid glands that decreases blood phosphate level and increases blood calcium level.

Parietal (pa-RĪ-e-tal) Pertaining to or forming the outer wall of a body cavity.

Parietal cell The secreting cell of a gastric gland that produces hydrochloric acid and intrinsic factor. Also called an *oxyntic cell.*

Parietal pleura (PLOO-ra) The outer layer of the serous pleural membrane that encloses and protects the lungs; the layer that is attached to the wall of the pleural cavity.

Parkinson's disease Progressive degeneration of the basal ganglia and substantia nigra of the cerebrum resulting in decreased production of dopamine (DA) that leads to tremor, slowing of voluntary movements, and muscle weakness.

Parotid (pa-ROT-id) **gland** One of the paired salivary glands located inferior and anterior to the ears connected to the oral cavity via a duct (Stensen's) that opens into the inside of the cheek opposite the upper second molar tooth.

Pathogen (PATH-ō-jen) A disease-producing organism.

Pectoral (PEK-tō-ral) Pertaining to the chest or breast.

Pediatrician (pē'-dē-a-TRISH-un) A physician who specializes in the care and treatment of children and their illnesses.

Pelvic (PEL-vik) **cavity** Inferior portion of the abdominopelvic cavity that contains the urinary bladder, sigmoid colon, rectum, and internal female and male reproductive structures.

Pelvic inflammatory disease (PID) Collective term for any extensive bacterial infection of the pelvic organs, especially the uterus, uterine (Fallopian) tubes, and ovaries.

Pelvis The basinlike structure formed by the two hipbones, sacrum, and coccyx; the expanded, proximal portion of the ureter, lying within the kidney and into which the major calyces open.

Penis (PĒ-nis) The male copulatory organ, used to introduce sperm into the female vagina.

Pepsin Protein-digesting enzyme secreted by zymogenic (chief) cells of the stomach as the inactive form pepsinogen, which is converted to active pepsin by hydrochloric acid.

Peptic ulcer An ulcer that develops in areas of the gastrointestinal tract exposed to hydrochloric acid; classified as a gastric ulcer if in the lesser curvature of the stomach and as a duodenal ulcer if in the first part of the duodenum. A major cause is the bacterium *Helicobacter pylori.*

Perforating canal A minute passageway through which blood vessels and nerves from the periosteum pass into compact bone. Also called *Volkmann's* (FŌLK-manz) *canal.*

Pericardial (per'-i-KAR-dē-al) **cavity** Small potential space between the visceral and parietal layers of the serous pericardium that contains pericardial fluid.

Pericarditis (per'-i-KAR-dī-tis) Inflammation of the pericardium.

Pericardium (per'-i-KAR-dē-um) A loose-fitting membrane that encloses the heart, consisting of an outer fibrous layer and an inner serous layer.

Perichondrium (per'-i-KON-drē-um) The membrane that covers cartilage.

Perilymph (PER-i-lymf) The fluid contained between the bony and membranous labyrinths of the inner ear.

Perineum (per'-i-NĒ-um) The pelvic floor; the space between the anus and the scrotum in the male and between the anus and the vulva in the female.

Periodontal (per-ē-ō-DON-tal) **disease** A collective term for conditions characterized by degeneration of gingivae, alveolar bone, periodontal ligament, and cementum.

Periodontal ligament The periosteum lining the alveoli (sockets) for the teeth in the alveolar processes of the mandible and maxillae.

Periosteum (per'-ē-OS-tē-um) The membrane that covers bone and consists of connective tissue, osteoprogenitor cells, and osteoblasts and that is essential for bone growth, repair, and nutrition.

Peripheral (pe-RIF-er-al) Located on the outer part or a surface of the body.

Peripheral nervous system (PNS) The part of the nervous system that lies outside the central nervous system—nerves and ganglia.

Peripheral resistance (re-ZIS-tans) Resistance (impedance) to blood flow as a result of the force of friction between blood and the walls of blood vessels that is related to viscosity of blood and blood vessel length and diameter.

Peristalsis (per'-i-STAL-sis) Successive muscular contractions along the wall of a hollow muscular structure.

Peritoneum (per'-i-tō-NĒ-um) The largest serous membrane of the body that lines the abdominal cavity and covers the viscera.

Peroxisome (pe-ROKS-i-sōm) Organelle similar in structure to a lysosome that contains enzymes related to hydrogen peroxide metabolism; abundant in liver cells.

Perspiration Substance produced by sudoriferous (sweat) glands containing water, salts, urea, uric acid, amino acids, ammonia, sugar, lactic acid, and ascorbic acid; helps maintain body temperature and eliminate wastes.

Peyer's patches *See Aggregated lymphatic follicles.*

pH A symbol of the measure of the concentration of hydrogen ions (H+) in a solution. The pH scale extends from 0 to 14, with a value of 7 expressing neutrality, values lower than 7 expressing increasing acidity, and values higher than 7 expressing increasing alkalinity.

Phagocytosis (fag'-ō-sī-TŌ-sis) The process by which cells (phagocytes) ingest particulate matter; especially the ingestion and destruction of microbes, cell debris, and other foreign matter.

Phalanx (FĀ-lanks) The bone of a finger or toe. *Plural, phalanges* (fa-LAN-jēz).

Phantom pain A sensation of pain as originating in a limb that has been amputated.

Pharmacology (far'-ma-KOL-ō-jē) The science that deals with the effects and uses of drugs in the treatment of disease.

Pharynx (FAR-inks) The throat; a tube that starts at the internal nares and runs partway down the neck where it opens into the esophagus posteriorly and the larynx anteriorly.

Phenotype (FĒ-nō-tīp) The observable expression of genotype; physical characteristics of an organism determined by genetic makeup and influenced by interaction between genes and internal and external environmental factors.

Phlebitis (fle-BĪ-tis) Inflammation of a vein, usually in the lower limbs.

Phospholipid (fos'-fō-LIP-id) **bilayer** Arrangement of phospholipid molecules in two parallel rows in which the hydrophilic "heads" face outward and the hydrophobic "tails" face inward.

Photoreceptor Receptor that detects light on the retina of the eye.

Physiology (fiz'-ē-OL-ō-jē) Science that deals with the functions of an organism or its parts.

Pia mater (PĪ-a MĀ-ter) The inner membrane (meninx) covering the brain and spinal cord.

Pineal (PĪN-ē-al) **gland** The cone-shaped gland located in the roof of the third ventricle. Also called the *epiphysis cerebri* (ē-PIF-i-sis se-RĒ-brē).

Pinocytosis (pi'-nō-sī-TŌ-sis) The process by which cells ingest liquid.

Pituitary (pi-TOO-i-tar'-ē) **dwarfism** Condition caused by hyposecretion of human growth hormone (hGH) during the growth years and characterized by childlike physical traits in an adult.

Pituitary gland A small endocrine gland lying in the sella turcica of the sphenoid bone and attached to the hypothalamus by the infundibulum; nicknamed the "master gland." Also called the *hypophysis* (hī-POF-i-sis).

Pivot joint A synovial joint in which a rounded, pointed, or conical surface of one bone articulates with a ring formed partly by another bone and partly by a ligament, as in the joint between the atlas and axis and between the proximal ends of the radius and ulna. Also called a *trochoid* (TRŌ-koyd) *joint.*

Placenta (pla-SEN-ta) The special structure through which the exchange of materials between fetal and maternal circulations occurs. Also called the *afterbirth.*

Plantar flexion (PLAN-tar FLEK-shun) Bending the foot in the direction of the plantar surface (sole).

Plaque (plak) A cholesterol-containing mass in the tunica media of arteries. A mass of bacterial cells, dextran (polysaccharide), and other debris that adheres to teeth.

Plasma (PLAZ-ma) The extracellular fluid found in blood vessels; blood minus the formed elements.

Plasma cell Cell that produces antibodies and develops from a B cell (lymphocyte).

Plasma (cell) membrane Outer, limiting membrane that separates the cell's internal parts from extracellular fluid and the external environment.

Platelet (PLĀT-let) A fragment of cytoplasm enclosed in a cell membrane and lacking a nucleus; found in the circulating blood; plays a role in blood clotting.

Platelet plug Aggregation of platelets at a damaged blood vessel to prevent blood loss.

Pleura (PLOOR-a) The serous membrane that covers the lungs and lines the walls of the chest and diaphragm.

Pleural cavity Small potential space between the visceral and parietal pleurae.

Plexus (PLEK-sus) A network of nerves, veins, or lymphatic vessels.

Pneumonia (noo-MŌ-nē-a) Acute infection or inflammation of the alveoli of the lungs.

Podiatry (pō-DĪ-a-trē) The diagnosis and treatment of foot disorders.

Polar body The smaller cell resulting from the unequal division of cytoplasm during the meiotic divisions of an oocyte. The polar body has no function and is resorbed.

Polarized A condition in which opposite effects or states exist at the same time. In electrical contexts, having one portion negative and another positive; for example, a polarized nerve cell membrane has the outer surface positively charged and the inner surface negatively charged.

Poliomyelitis (pō'-lē-ō-mī-e-LĪ-tis) Viral infection marked by fever, headache, stiff neck and back, deep muscle pain and weakness, and loss of certain somatic reflexes; a serious form of the disease, *bulbar polio,* results in destruction of motor neurons in anterior horns of spinal nerves that leads to paralysis.

Polycythemia (pol'-ē-si-THĒ-mē-a) Disorder characterized by a hematocrit above the normal level of 55 in which hypertension, thrombosis, and hemorrhage occur.

Polyp (POL-ip) A tumor on a stem found especially on a mucous membrane.

Polysaccharide (pol'-ē-SAK-a-rīd) A carbohydrate in which tens or hundreds of monosaccharides are joined chemically.

Polyunsaturated fat A fat that contains more than one double covalent bond between its carbon atoms; examples are corn oil, safflower oil, and cottonseed oil.

Polyuria (pol'-ē-YOO-rē-a) An excessive production of urine.

Pons (ponz) The portion of the brain stem that forms a "bridge" between the medulla oblongata and the midbrain, anterior to the cerebellum.

Positive feedback A mechanism in which the response enhances the original stimulus.

Positron emission tomography (PET) A type of radioactive scanning based on the release of gamma rays when positrons collide with negatively charged electrons in body tissues; it indicates where radioisotopes are used in the body.

Posterior (pos-TĒR-ē-or) Nearer to or at the back of the body. Also called *dorsal.*

Posterior column–medial lemniscus (lem-NIS-kus) **pathway** Sensory pathway that conveys information related to proprioception, discriminative touch, stereognosis, weight discrimination, and vibrations.

Posterior root The structure composed of sensory fibers lying between a spinal nerve and the dorsolateral aspect of the spinal cord. Also called the *dorsal (sensory) root.*

Posterior root ganglion A group of cell bodies of sensory neurons and their supporting cells located along the posterior root of a spinal nerve. Also called a *dorsal (sensory) root ganglion* (GANG-glē-on).

Postganglionic neuron (pōst'-gang-lē-ON-ik NOO-ron) The second visceral motor neuron in an autonomic pathway, having its cell body and dendrites located in an autonomic ganglion and its unmyelinated axon ending at cardiac muscle, smooth muscle, or a gland.

Preeclampsia (prē'-e-KLAMP-sē-a) A syndrome characterized by sudden hypertension, large amounts of protein in urine, and generalized edema; it might be related to an autoimmune or allergic reaction due to the presence of a fetus.

Preganglionic (prē'-gang-lē-ON-ik) **neuron** The first visceral motor neuron in an autonomic pathway, with its cell body and dendrites in the brain or spinal cord and its myelinated axon ending at an autonomic ganglion, where it synapses with a postganglionic neuron.

Pregnancy Sequence of events that normally includes fertilization, implantation, embryonic growth, and fetal growth that terminates in birth.

Premenstrual syndrome (PMS) Severe physical and emotional stress occurring late in the postovulatory phase of the menstrual cycle and sometimes overlapping with menstruation.

Prepuce (PRE-pyoos) The loose-fitting skin covering the glans of the penis and clitoris. Also called the *foreskin.*

Pressure sore Tissue destruction due to a constant deficiency of blood to tissues overlying a bony projection that has been subjected to prolonged pressure against an object such as a bed, cast, or splint. Also called *bedsore, decutitus* (dē-KYOO-bi-tus) *ulcer,* or *trophic ulcer.*

Primary germ layer One of three layers of embryonic tissue, called ectoderm, mesoderm, and endoderm, that give rise to all tissues and organs of the organism.

Prime mover The muscle directly responsible for producing the desired motion. Also called an *agonist* (AG-ō-nist).

Proctology (prok-TOL-ō-jē) The branch of medicine that treats the rectum and its disorders.

Progeny (PROJ-e-nē) Refers to offspring or descendants.

Progesterone (prō-JES-te-rōn) **(PROG)** A female sex hormone produced by the ovaries that helps prepare the endometrium for implantation of a fertilized ovum and the mammary glands for milk secretion.

Prognosis (prog-NŌ-sis) A forecast of the probable results of a disorder; the outlook for recovery.

Projection (prō-JEK-shun) The process by which the brain refers sensations to their point of stimulation.

Prolactin (prō-LAK-tin) **(PRL)** A hormone secreted by the anterior pituitary gland that initiates and maintains milk secretion by the mammary glands.

Proliferation (prō-lif′-er-Ā-shun) Rapid and repeated reproduction of new parts, especially cells.

Pronation (prō-NĀ-shun) A movement of the forearm in which the palm is turned posteriorly or inferiorly.

Prophase (PRŌ-fāz) The first stage of mitosis during which chromatid pairs are formed and aggregate around the equatorial plane region of the cell.

Proprioception (prō-prē-ō-SEP-shun) The receipt of information from muscles, tendons, and the labyrinth that enables the brain to determine movements and position of the body and its parts. Also called *kinesthesia* (kin′-es-THĒ-zē-a).

Proprioceptor (prō′-prē-ō-SEP-tor) A receptor located in muscles, tendons, or joints that provides information about body position and movements.

Prostaglandin (pros′-ta-GLAN-din) **(PG)** A membrane-associated lipid composed of 20-carbon fatty acids with 5 carbon atoms joined to form a cyclopentane ring; synthesized in small quantities and basically mimics the activities of hormones.

Prostate (PROS-tāt) **gland** A doughnut-shaped gland inferior to the urinary bladder that surrounds the superior portion of the male urethra and secretes a slightly acid solution that contributes to sperm motility and viability.

Prosthesis (pros-THĒ-sis) An artificial device to replace a missing body part.

Protein An organic compound consisting of carbon, hydrogen, oxygen, nitrogen, and sometimes sulfur and phosphorus, and made up of amino acids linked by peptide bonds.

Prothrombin (prō-THROM-bin) An inactive protein synthesized by the liver, released into the blood, and converted to active thrombin in the process of blood clotting.

Proto-oncogene (prō′-tō-ONG-kō-jēn) Gene responsible for some aspect of normal growth and development; it may transform into an oncogene, a gene capable of causing cancer.

Proximal (PROK-si-mal) Nearer the attachment of a limb to the trunk or a structure; nearer to the point of origin.

Psychosomatic (sī′-kō-sō-MAT-ik) Pertaining to the relation between mind and body. Commonly used to refer to those physiological disorders thought to be caused entirely or partly by emotional disturbances.

Ptosis (TŌ-sis) Drooping, as of the eyelid or the kidney (nephrotosis).

Puberty (PYOO-ber-tē) The time of life during which the secondary sex characteristics begin to appear and the capability for sexual reproduction is possible; usually between the ages of 10 and 17.

Pubic symphysis (PYOO-bik SIM-fi-sis) A slightly movable cartilaginous joint between the anterior surfaces of the hipbones.

Pulmonary (PUL-mo-ner′-ē) Concerning or affected by the lungs.

Pulmonary circulation The flow of deoxygenated blood from the right ventricle to the lungs and the return of oxygenated blood from the lungs to the left atrium.

Pulmonary edema (e-DĒ-ma) An abnormal accumulation of interstitial fluid in the tissue spaces and alveoli of the lungs due to increased pulmonary capillary permeability or increased pulmonary capillary pressure.

Pulmonary embolism (EM-bō-lizm) **(PE)** The presence of a blood clot or other foreign substance in a pulmonary arterial blood vessel that obstructs circulation to lung tissue.

Pulmonary ventilation The inflow (inspiration) and outflow (expiration) of air between the atmosphere and the lungs. Also called *breathing*.

Pulp cavity A cavity within the crown and neck of a tooth, filled with pulp, a connective tissue containing blood vessels, nerves, and lymphatic vessels.

Pulse The rhythmic expansion and elastic recoil of an artery with each contraction of the left ventricle.

Pulse pressure The difference between the maximum (systolic) and minimum (diastolic) pressures; normally a value of about 40 mm Hg.

Pupil The hole in the center of the iris, the area through which light enters the posterior cavity of the eyeball.

Purkinje fiber *See Conduction myofiber.*

Pus The liquid product of inflammation containing leukocytes or their remains and debris of dead cells.

Pyemia (pī-Ē-mēa) Infection of the blood, with multiple abscesses, caused by pus-forming microorganisms.

Pyloric (pī-LOR-ik) **sphincter** A thickened ring of smooth muscle through which the pylorus of the stomach communicates with the duodenum. Also called the *pyloric valve.*

Pyuria (pī-YOO-rē-a) The presence of leukocytes and other components of pus in the urine.

Reactivity (rē-ak-TI-vi-tē) Ability of an antigen to react specifically with the antibody whose formation it induced.

Receptor A specialized cell or a nerve cell terminal modified to respond to some specific sensory modality, such as touch, pressure, cold, light, or sound, and convert it to a nerve impulse. A specific molecule or arrangement of molecules organized to accept only molecules with a complementary shape.

Receptor potential Depolarization of the plasma membrane of a receptor cell, which stimulates release of neurotransmitter from the cell; if the neuron connected to the receptor cell becomes depolarized to threshold, a nerve action potential (nerve impulse) is triggered.

Recessive gene A gene that is not expressed in the presence of a dominant gene on the homologous chromosome.

Recombinant DNA Synthetic DNA, formed by joining a fragment of DNA from one source to a portion of DNA from another.

Recruitment (rē-KROOT-ment) The process of increasing the number of active motor units. Also called *motor unit summation.*

Rectum (REK-tum) The last 20 cm (7 in.) of the gastrointestinal tract, from the sigmoid colon to the anus.

Reduction The addition of electrons and hydrogen ions (hydrogen atoms) to a molecule or, less commonly, the removal of oxygen from a molecule that results in an increase in the energy content of the molecule.

Referred pain Pain that is felt at a site remote from the place of origin.

Reflex Fast response to a change (stimulus) in the internal or external environment that attempts to restore homeostasis; passes over a reflex arc.

Reflex arc The most basic conduction pathway through the nervous system, connecting a receptor and an effector and consisting of a receptor, a sensory neuron, a center in the central nervous system for a synapse, a motor neuron, and an effector.

Refraction (rē-FRAK-shun) The bending of light as it passes from one medium to another.

Regeneration (rē-jen′-er-Ā-shun) The natural renewal of a structure.

Regurgitation (rē-gur′-ji-TĀ-shun) Return of solids or fluids to the mouth from the stomach; flowing backward of blood through incompletely closed heart valves.

Relapse (RĒ-laps) The return of a disease weeks or months after its apparent cessation.

Relaxin A female hormone produced by the ovaries that relaxes the pubic symphysis and helps dilate the uterine cervix to facilitate delivery.

Releasing hormone Chemical secretion of the hypothalamus that can stimulate secretion of hormones of the anterior pituitary gland.

Remodeling Replacement of old bone by new bone tissue.

Renal (RĒ-nal) Pertaining to the kidney.

Renal corpuscle (KOR-pus′-l) A glomerular (Bowman's) capsule and its enclosed glomerulus.

Renal erythropoietic (ē-rith′-rō-poy-Ē-tik) **factor** An enzyme released by the kidneys and liver in response to hypoxia that acts on a plasma protein to bring about the production of erythropoietin, a hormone that stimulates red blood cell production.

Renal failure Inability of the kidneys to function properly, due to abrupt failure (acute) or progressive failure (chronic).

Renal pelvis A cavity in the center of the kidney formed by the expanded, proximal portion of the ureter, lying within the kidney, and into which the major calyces open.

Renal pyramid A triangular structure in the renal medulla composed of the straight segments of renal tubules.

Renin (RĒ-nin) An enzyme released by the kidney into the plasma, where it converts angiotensinogen into angiotensin I.

Renin–angiotensin (an′-jē-ō-TEN-sin) **pathway** A mechanism for the control of aldosterone secretion by angiotensin II, initiated by the secretion of renin by the kidney in response to low blood pressure.

Repolarization (rē-pō-lar-i-ZĀ-shun) Restoration of a resting membrane potential following depolarization.

Reproduction (rē′-prō-DUK-shun) Either the formation of new cells for growth, repair, or replacement, or the production of a new individual.

Resistance Ability to ward off disease. The hindrance encountered by an electrical charge as it moves through a substance from one point to another. The hindrance encountered by blood as it flows through the vascular system or by air through respiratory passageways.

Respiration (res-pi-RĀ-shun) Overall exchange of gases between the atmosphere, blood, and body cells consisting of pulmonary ventilation, external respiration, and internal respiration.

Respiratory center Neurons in the reticular formation of the brain stem that regulate the rate of respiration.

Respiratory distress syndrome (RDS) of the newborn A disease of newborn infants, especially premature ones, in which insufficient amounts of surfactant are produced and breathing is labored. Also called *hyaline* (HĪ-a-lin) *membrane disease* (HMD).

Respiratory failure Condition in which the respiratory system cannot supply sufficient oxygen to maintain metabolism or eliminate enough carbon dioxide to prevent respiratory acidosis.

Resting membrane potential The voltage that exists between the inside and outside of a cell membrane when the cell is not responding to a stimulus; about – 70 to – 90 mV, with the inside of the cell negative.

Resuscitation (rē-sus′-i-TĀ-shun) Act of bringing a person back to full consciousness.

Retention (rē-TEN-shun) A failure to void urine due to obstruction, nervous contraction of the urethra, or absence of sensation or desire to urinate.

Reticular (re-TIK-yoo-lar) **activating system (RAS)** An extensive network of branched nerve cells running through the core of the brain stem. When these cells are activated, a generalized alert or arousal behavior results.

Reticular formation A network of small groups of nerve cells scattered among bundles of fibers beginning in the medulla oblongata as a continuation of the spinal cord and extending upward through the central part of the brain stem.

Retina (RET-i-na) The inner coat of the eyeball, lying only in the posterior portion of the eye and consisting of nervous tissue and a pigmented layer comprised of epithelial cells lying in contact with the choroid. Also called the *nervous tunic* (TOO-nik).

Retinal (RE′-ti-nal) A derivative of vitamin A that functions as the light-absorbing portion of the photopigment rhodopsin.

Retraction (rē-TRAK-shun) The movement of a protracted part of the body backward on a plane parallel to the ground, as in pulling the lower jaw back in line with the upper jaw.

Retroflexion (re-trō-FLEK-shun) A malposition of the uterus in which it is tilted posteriorly.

Retroperitoneal (re′-trō-per-i-tō-NĒ-al) External to the peritoneal lining of the abdominal cavity.

Reye (RĪ) **syndrome** Disease, primarily of children and teenagers, that is characterized by vomiting and brain dysfunction and may progress to coma and death; seems to occur following a viral infection, particularly chickenpox or influenza, and aspirin is believed to be a risk factor.

Rh factor An inherited agglutinogen (antigen) on the surface of red blood cells.

Rhodopsin (rō-DOP-sin) The photopigment in rods of the retina, consisting of a glycoprotein called opsin and a derivative of vitamin A called retinal.

Ribonucleic (rī′-bō-nyoo-KLĒ-ik) **acid (RNA)** A single-stranded nucleic acid constructed of nucleotides consisting of one of four possible nitrogenous bases (adenine, cytosine, guanine, or uracil), ribose, and a phosphate group; three types are messenger RNA (mRNA), transfer RNA (tRNA), and ribosomal RNA (rRNA), each of which cooperates with DNA for protein synthesis.

Ribosome (RĪ-bō-sōm) An organelle in the cytoplasm of cells, composed of ribosomal RNA and ribosomal proteins, that synthesizes proteins; nicknamed the "protein factory."

Rigor mortis State of partial contraction of muscles following death due to lack of ATP that causes cross bridges of thick myofilaments to remain attached to thin myofilaments, thus preventing relaxation.

Rod A visual receptor in the retina of the eye that is specialized for vision in dim light.

Root canal A narrow extension of the pulp cavity lying within the root of a tooth.

Rotation (rō-TĀ-shun) Moving a bone around its own axis, with no other movement.

Round window A small opening between the middle and inner ear, directly below the oval window, covered by the secondary tympanic membrane. Also called the *fenestra cochlea* (fe-NES-tra KŌK-lē-a).

Rugae (ROO-jē) Large folds in the mucosa of an empty hollow organ, such as the stomach and vagina.

Sacral plexus (PLEK-sus) A network formed by the anterior branches of spinal nerves L4 through S3.

Saddle joint A synovial joint in which the articular surface of one bone is saddle-shaped and the articular surface is shaped like the legs of the rider sitting in the saddle, as in the joint between the trapezium and the metacarpal of the thumb.

Sagittal (SAJ-i-tal) **plane** A vertical plane that divides the body or organs into left and right portions. Such a plane may be *midsagittal* (*median*), in which the divisions are equal, or *parasagittal*, in which the divisions are unequal.

Saliva (sa-LĪ-va) A clear, alkaline, somewhat viscous secretion produced by the three pairs of salivary glands; contains various salts, mucin, lysozyme, and salivary amylase.

Salivary gland One of three pairs of glands that lie outside the mouth and pour their secretory product (called saliva) into ducts that empty into the oral cavity; the parotid, submandibular, and sublingual glands.

Salt A substance that, when dissolved in water, ionizes into cations and anions, neither of which is hydrogen ions (H^+) nor hydroxide ions (OH^-).

Saltatory (sal-ta-TŌ-rē) **conduction** The propagation of an action potential (nerve impulse) along the exposed portions of a myelinated nerve fiber. The action potential appears at successive neurofibral nodes (nodes of Ranvier) and therefore seems to jump or leap from node to node.

Sarcoma (sar-KŌ-ma) A connective tissue tumor, often highly malignant.

Sarcomere (SAR-kō-mēr) A contractile unit in a striated muscle fiber (cell) extending from one Z disc to the next Z disc.

Sarcoplasmic reticulum (sar′-kō-PLAZ-mik re-TIK-yoo-lum) A network of saccules and tubes surrounding myofibrils of a muscle fiber (cell), comparable to endo-plasmic reticulum; functions to reabsorb calcium ions during relaxation and to release them to cause contraction.

Satiety center A collection of nerve cells located in the ventromedial nuclei of the hypothalamus that, when stimulated, brings about the cessation of eating.

Saturated fat A fat that contains no double bonds between any of its carbon atoms; all are single bonds and all carbon atoms are bonded to the maximum number of hydrogen atoms; found naturally in animal foods such as meat, milk, milk products, and eggs.

Schwann cell *See Neurolemmocyte.*

Sciatica (sī-AT-i-ka) Inflammation and pain along the sciatic nerve; felt at the back of the thigh running down the inside of the leg.

Sclera (SKLE-ra) The white coat of fibrous tissue that forms the outer protective covering over the eyeball except in the most anterior portion; the posterior portion of the fibrous tunic.

Scleral venous sinus A circular venous sinus located at the junction of the sclera and the cornea through which aqueous humor drains from the anterior chamber of the eyeball into the blood. Also called the *canal of Schlemm* (SHLEM).

Scoliosis (skō′-lē-Ō-sis) An abnormal lateral curvature from the normal vertical line of the backbone.

Scrotum (SKRŌ-tum) A skin-covered pouch that contains the testes and their accessory structures.

Sebaceous (se-BĀ-shus) **gland** An exocrine gland in the dermis of the skin, almost always associated with a hair follicle, that secretes sebum. Also called an *oil gland*.

Sebum (SĒ-bum) Secretion of sebaceous (oil) glands.

Secondary sex characteristic A feature characteristic of the male or female body that develops at puberty under the stimulation of sex hormones but is not directly involved in sexual reproduction, such as distribution of body hair, voice pitch, body shape, and muscle development.

Secretion (se-KRĒ-shun) Production and release from a gland cell of a fluid, especially a functionally useful product as opposed to a waste product.

Selective permeability (per′-mē-a-BIL-i-tē) The property of a membrane by which it permits the passage of certain substances but restricts the passage of others.

Sella turcica (SEL-a TUR-si-ka) A depression on the superior surface of the sphenoid bone that houses the pituitary gland.

Semen (SĒ-men) A fluid discharged at ejaculation by a male that consists of a mixture of sperm and the secretions of the seminal vesicles, prostate gland, and bulbourethral (Cowper's) glands.

Semicircular canals Three bony channels (anterior, posterior, lateral), filled with perilymph, in which lie the membranous semicircular canals filled with endolymph. They contain receptors for equilibrium.

Semilunar (sem′-ē-LOO-nar) **valve** A valve guarding the entrance into the aorta or the pulmonary trunk from a ventricle of the heart.

Seminal vesicle (SEM-i-nal VES-i-kul) One of a pair of convoluted, pouchlike structures, lying posterior and inferior to the urinary bladder and anterior to the rectum, that secrete a component of semen into the ejaculatory ducts.

Seminiferous tubule (sem′-i-NI-fer-us TOO-byool) A tightly coiled duct, located in a lobule of the testis, where sperm are produced.

Senescence (se-NES-ens) The process of growing old; the period of old age.

Senility (se-NIL-i-tē) A loss of mental or physical ability due to old age.

Sensation A state of awareness of external or internal conditions of the body.

Sensitization (sen′-si-ti-ZĀ-shun) Process by which T cells are able to recognize antigens; process of making a person susceptible to a substance by repeated injections of it.

Sensory area A region of the cerebral cortex concerned with the interpretation of sensory impulses.

Sensory neuron (NOO-ron) A neuron that conducts nerve impulses into the central nervous system. Also called an *afferent neuron*.

Septal defect An opening in the septum (interatrial or interventricular) between the left and right sides of the heart.

Septicemia (sep′-ti-SĒ-mē-a) Toxins or disease-causing bacteria in blood. Also called *"blood poisoning."*

Serosa (ser-Ō-sa) Any serous membrane; the outermost layer of an organ formed by a serous membrane; the membrane that lines the pleural, pericardial, and peritoneal cavities.

Serous (SIR-us) **membrane** A membrane that lines a body cavity that does not open to the exterior. Also called the *serosa* (se-RŌ-sa).

Serum Plasma minus its clotting proteins.

Sesamoid bones (SES-a-moyd) Small bones usually found in tendons.

Sex chromosomes The twenty-third pair of chromosomes, designated X and Y, that determine the genetic sex of an individual; in males, the pair is XY; in females, XX.

Sexual intercourse The insertion of the erect penis of a male into the vagina of a female. Also called *coitus* (KŌ-i-tus) or *copulation*.

Sexually transmitted disease (STD) General term for any of a large number of diseases spread by sexual contact. Also called a *venereal disease* (VD).

Sheath of Schwann *See Neurolemma.*

Shingles Acute infection of the peripheral nervous system caused by a virus.

Shinsplints Soreness or pain along the tibia probably caused by inflammation of the periosteum brought on by repeated tugging of the muscles and tendons attached to the periosteum. Also called *tibia stress syndrome.*

Shivering Involuntary contraction of a muscle that generates heat.

Shock Failure of the cardiovascular system to deliver adequate amounts of oxygen and nutrients to meet the metabolic needs of the body due to inadequate cardiac output. It is characterized by hypotension; clammy, cool, and pale skin; sweating; reduced urine formation; altered mental state; acidosis; tachycardia; weak, rapid pulse; and thirst. Types include hypovolemic, cardiogenic, obstructive, neurogenic, and septic.

Simple diffusion (dif-YOO-zhun) A passive process in which there is a net or greater movement of molecules or ions from a region of high concentration to a region of low concentration until equilibrium is reached.

Sinoatrial (si-nō-Ā-trē-al) **(SA) node** A compact mass of cardiac muscle fibers (cells) specialized for conduction, located in the right atrium beneath the opening of the superior vena cava. Also called the *sinuatrial node* or *pacemaker.*

Sinus (SĪ-nus) A hollow in a bone (paranasal sinus) or other tissue; a channel for blood (vascular sinus); any cavity having a narrow opening.

Sinusitis (sīn-yoo-SĪT-is) Inflammation of the mucous membrane of a paranasal sinus.

Skeletal muscle An organ specialized for contraction, composed of striated muscle fibers (cells), supported by connective tissue, attached to a bone by a tendon or an aponeurosis, and stimulated by somatic motor neurons.

Skin The outer covering of the body that consists of an outer, thinner epidermis (epithelial tissue) and an inner, thicker dermis (connective tissue) and that is anchored to the subcutaneous layer.

Skin cancer Any one of several types of malignant tumors that arise from epidermal cells. Types include basal cell carcinoma (arises from epidermis and rarely spreads), squamous cell carcinoma (arises from epidermis and has a variable tendency to spread), and malignant melanoma (arises from melanocytes and spreads rapidly).

Sleep A state of partial unconsciousness from which an individual can be aroused, caused by inactivation of the reticular activating system.

Sliding-filament theory The most commonly accepted explanation for muscle contraction in which actin and myosin myofilaments move into interdigitation with each other, decreasing the length of the sarcomeres.

Small intestine A long tube of the gastrointestinal tract that begins at the pyloric sphincter of the stomach, coils through the central and lower part of the abdominal cavity, and ends at the large intestine, divided into three segments: duodenum, jejunum, and ileum.

Smooth muscle An organ specialized for contraction, composed of smooth muscle fibers (cells), located in the walls of hollow internal structures, and innervated by a visceral efferent (motor) neuron.

Sodium–potassium pump An active transport system located in the cell membrane that transports sodium ions out of the cell and potassium ions into the cell at the expense of cellular ATP. It functions to keep the ionic concentrations of these elements at physiological levels.

Solution A homogeneous molecular or ionic dispersion of one or more substances (solutes) in a usually liquid-dissolving medium (solvent).

Somatic (sō-MAT-ik) **cell division** Type of cell division in which a single starting cell (parent cell) duplicates itself to produce two identical cells (daughter cells); consists of mitosis and cytokinesis.

Somatic nervous system (SNS) The portion of the peripheral nervous system made up of the somatic motor fibers that run between the central nervous system and the skeletal muscles and skin.

Spasm (spazm) A sudden, involuntary contraction of large groups of muscles.

Spastic (SPAS-tik) An increase in muscle tone (stiffness) associated with an increase in tendon reflexes and abnormal reflexes (Babinski sign).

Spermatogenesis (sper′-ma-tō-JEN-e-sis) The formation and development of sperm in the seminiferous tubules of the testes.

Spermatozoon (sper′-ma-tō-ZŌ-on). *See Sperm cell.*

Sperm cell A mature male gamete.

Spermicide (SPER-mi-sīd′) An agent that kills sperm.

Sphincter (SFINGK-ter) A circular muscle constricting an orifice.

Sphygmomanometer (sfig′-mō-ma-NOM-e-ter) An instrument for measuring arterial blood pressure.

Spina bifida (SPĪ-na BIF-i-da) A congenital defect of the vertebral column in which the halves of the neural arch of a vertebra fail to fuse in the midline.

Spinal (SPĪ-nal) **cord** A mass of nerve tissue located in the vertebral canal from which 31 pairs of spinal nerves originate.

Spinal nerve One of the 31 pairs of nerves that originate on the spinal cord from posterior and anterior roots.

Spinal tap Withdrawal of some of the cerebrospinal fluid from the subarachnoid space in the lumbar region for diagnostic purposes, introduction of various substances, and evaluation of the effects of treatment. Also called a *lumbar puncture.*

Spinothalamic (spi-nō-THAL-a-mik) **tracts** Sensory tracts that convey information up the spinal cord to the brain, related to pain, temperature, crude touch, and deep pressure.

Spiral organ The organ of hearing, consisting of supporting cells and hair cells that rest on the basilar membrane and extend into the endolymph of the cochlear duct. Also called the *organ of Corti* (KOR-tē).

Spirometer (spī-ROM-e-ter) An apparatus used to measure air capacity of the lungs.

Spleen (SPLĒN) Large mass of lymphatic tissue between the fundus of the stomach and the diaphragm that functions in phagocytosis, production of lymphocytes, and blood storage.

Spongy (cancellous) bone tissue Bone tissue that consists of an irregular latticework of thin plates of bone called trabeculae; spaces between trabeculae of some bones are filled with red bone marrow; found inside short, flat, and irregular bones and in the epiphyses (ends) of long bones.

Sprain Forcible wrenching or twisting of a joint with partial rupture or other injury to its attachments without dislocation.

Squamous (SKWĀ-mus) Flat.

Starling's law of the capillaries The movement of fluid between plasma and interstitial fluid is in a state of near equilibrium at the arterial and venous ends of a capillary; that is, filtered fluid and absorbed fluid plus that returned to the lymphatic system are nearly equal.

Starvation (star-VĀ-shun) The loss of energy stores in the form of glycogen, fats, and proteins due to inadequate intake of nutrients or inability to digest, absorb, or metabolize ingested nutrients.

Static equilibrium (ē-kwi-LIB-rē-um) The maintenance of posture in response to changes in the orientation of the body, mainly the head, relative to the ground.

Stellate reticuloendothelial (STEL-āte re-tik′-yoo-lō-en′-dō-THĒ-lē-al) **cell** Phagocytic cell that lines a sinusoid of the liver. Also called a *Kupffer's* (KOOP-ferz) *cell.*

Stenosis (sten-Ō-sis) An abnormal narrowing or constriction of a duct or opening.

Sterilization (ster′-i-li-ZĀ-shun) Elimination of all living microorganisms. The rendering of an individual incapable of reproduction (for example, castration, vasectomy, hysterectomy).

Stimulus Any stress that changes a controlled condition; any change in the internal or external environment that excites a receptor, a neuron, or a muscle fiber.

Stomach The J-shaped enlargement of the gastrointestinal tract directly under the diaphragm in the epigastric, umbilical, and left hypochondriac regions of the abdomen, between the esophagus and small intestine.

Strabismus (stra-BIZ-mus) A condition in which the visual axes of the two eyes differ, so that they do not fix on the same object.

Stratum basalis (STRA-tum ba-SAL-is) The outer layer of the endometrium, next to the myometrium, that is maintained during menstruation and gestation and produces a new functionalis following menstruation or parturition.

Stratum functionalis (funk′-shun-AL-is) The inner layer of the endometrium, the layer next to the uterine cavity, that is shed during menstruation and that forms the maternal portion of the placenta during gestation.

Stress Any stimulus that creates an imbalance in the internal environment.

Stressor A stress that is extreme, unusual, or long lasting and triggers the general adaptation syndrome.

Stroke volume The volume of blood ejected by either ventricle in one systole; about 70 ml.

Subarachnoid (sub′-a-RAK-noyd) **space** A space between the arachnoid and the pia mater that surrounds the brain and spinal cord and through which cerebrospinal fluid circulates.

Subcutaneous (sub′-kyoo-TĀ-nē-us) Beneath the skin. Also called *hypodermic* (hī-pō-DER-mik).

Subcutaneous layer A continuous sheet of areolar connective tissue and adipose tissue between the dermis of the skin and the deep fascia of the muscles. Also called the *superficial fascia* (FASH-ē-a).

Sublingual (sub-LING-gwal) **gland** One of a pair of salivary glands situated in the floor of the mouth under the mucous membrane and to the side of the lingual frenulum, with a duct (Rivinus's) that opens into the floor of the mouth.

Submandibular (sub′-man-DIB-yoo-lar) **gland** One of a pair of salivary glands found beneath the base of the tongue under the mucous membrane in the posterior part of the floor of the mouth, posterior to the sublingual glands, with a duct (Wharton's) situated to the side of the lingual frenulum. Also called the *submaxillary* (sub′-MAK-si-ler-ē) *gland.*

Submucosa (sub-myoo-KŌ-sa) A layer of connective tissue located beneath a mucous membrane, as in the gastrointestinal tract or the urinary bladder; the submucosa connects the mucosa to the muscularis layer.

Substrate A substance with which an enzyme reacts.

Subthreshold stimulus A stimulus of such weak intensity that it cannot initiate an action potential (nerve impulse). Also called a *subliminal stimulus.*

Sudden infant death syndrome (SIDS) Completely unexpected and unexplained death of an apparently well, or virtually well, infant; death usually occurs during sleep.

Sudoriferous (soo′-dor-IF-er-us) **gland** An apocrine or eccrine exocrine gland in the dermis or subcutaneous layer that produces perspiration. Also called a *sweat gland.*

Sulcus (SUL-kus) A groove or depression between parts, especially between the convolutions of the brain. *Plural, sulci* (SUL-sē).

Summation (sum-MĀ-shun) The algebraic addition of the excitatory and inhibitory effects of many stimuli applied to a nerve cell body. The increased strength of muscle contraction that results when stimuli follow in rapid succession.

Superficial (soo′-per-FISH-al) Located on or near the surface of the body.

Superficial fascia (FASH-ē-a) A continuous sheet of fibrous connective tissue between the dermis of the skin and the deep fascia of the muscles. Also called *subcutaneous* (sub′-kyoo-TĀ-nē-us) *layer.*

Superior (soo-PĒR-ē-or) Toward the head or upper part of a structure. Also called *cephalad* (SEF-a-lad) or *craniad.*

Superior vena cava (VĒ-na CĀ-va) **(SVC)** Large vein that collects blood from parts of the body superior to the heart and returns it to the right atrium.

Supination (soo-pi-NĀ-shun) A movement of the forearm in which the palm is turned anteriorly or superiorly.

Surfactant (sur-FAK-tant) A phospholipid substance produced by the lungs that decreases surface tension.

Susceptibility (sus-sep′-ti-BIL-i-tē) Lack of resistance of a body to the deleterious or other effects of an agent such as pathogenic microorganisms.

Sustentacular (sus′-ten-TAK-yoo-lar) **cell** A supporting cell of seminiferous tubules that produces secretions for supplying nutrients to sperm and the hormone inhibin. Also called a *Sertoli* (ser-TŌ-lē) *cell.*

Suture (SOO-cher) An immovable fibrous joint in the skull where bone surfaces are closely united.

Sympathetic (sim′-pa-THET-ik) **division** One of the two subdivisions of the autonomic nervous system, having cell bodies of preganglionic neurons in the lateral gray columns of the thoracic segment and first two or three lumbar segments of the spinal cord; primarily concerned with processes involving the expenditure of energy. Also called the *thoracolumbar* (thō′-ra-kō-LUM-bar) division.

Sympathomimetic (sim′-pa-thō-mi-MET-ik) Producing effects that mimic those brought about by the sympathetic division of the autonomic nervous system.

Symphysis (SIM-fi-sis) A line of union. A slightly movable cartilaginous joint such as the pubic symphysis between the anterior surfaces of the hipbones.

Symptom (SIMP-tum) A subjective change in body function not apparent to an observer, such as fever or nausea, that indicates the presence of a disease or disorder of the body.

Synapse (SIN-aps) The junction between the process of two adjacent neurons, the place where the activity of one neuron affects the activity of another; may be electrical or chemical.

Synaptic end bulb Expanded distal end of an axon terminal that contains synaptic vesicles. Also called *synaptic knob* or *end foot*.

Synaptic vesicle Membrane-enclosed sac in a synaptic end bulb that stores neurotransmitters.

Synarthrosis (sin′-ar-THRŌ-sis) An immovable joint.

Syncope (SIN-kō-pē) Faint; a sudden temporary loss of consciousness associated with loss of postural tone and followed by spontaneous recovery; most commonly caused by cerebral ischemia.

Syndrome (SIN-drōm) A group of signs and symptoms that occur together in a pattern that is characteristic of a particular disease or abnormal condition.

Synergist (SIN-er-jist) A muscle that assists the prime mover by reducing undesired action or unnecessary movement.

Synergistic (syn-er-GIS-tik) **effect** A hormonal interaction in which the effects of two or more hormones complement each other so that the target cell responds to the sum of the hormones involved. An example is the combined actions of estrogens, progesterone (PROG), prolactin (PRL), and oxytocin (OT) necessary for lactation.

Synovial (si-NŌ-vē-al) **cavity** The space between the articulating bones of a diarthrotic joint, filled with synovial fluid. Also called a *joint cavity.*

Synovial fluid Secretion of synovial membranes that lubricates joints and nourishes articular cartilage.

Synovial joint A fully movable or diarthrotic joint in which a synovial (joint) cavity is present between the two articulating bones.

Synovial membrane The inner of the two layers of the articular capsule of a synovial joint, composed of areolar connective tissue that secretes synovial fluid into the synovial (joint) cavity.

Syphilis (SIF-i-lis) A sexually transmitted disease caused by the bacterium *Treponema pallidum.*

System An association of organs that have a common function.

Systemic (sis-TEM-ik) Affecting the whole body; generalized.

Systemic circulation The routes through which oxygenated blood flows from the left ventricle through the aorta to all the organs of the body and deoxygenated blood returns to the right atrium.

Systemic lupus erythematosus (er-i-them-a-TŌ-sus) **(SLE)** An autoimmune, inflammatory disease that may affect every tissue of the body.

Systole (SIS-tō-lē) In the cardiac cycle, the phase of contraction of the heart muscle, especially of the ventricles.

Systolic (sis-TO-lik) **blood pressure** The force exerted by blood on arterial walls during ventricular contraction; the highest pressure measured in the large arteries, about 120 mm Hg under normal conditions for a young, adult male.

Tachycardia (tak′-i-KAR-dē-a) A rapid heartbeat or pulse rate.

Tactile (TAK-til) Pertaining to the sense of touch.

Target cell A cell whose activity is affected by a particular hormone.

Tarsal gland Sebaceous (oil) gland that opens on the edge of each eyelid. Also called *Meibomian* (mī-BŌ-mē-an) *gland.*

Tarsus (TAR-sus) A collective term for the seven bones of the ankle.

Tay–Sachs (TĀ-SAKS) **disease** Inherited, progressive neuronal degeneration of the central nervous system due to a deficient lysosomal enzyme that causes excessive accumulations of a lipid called ganglioside.

T cell A lymphocyte that becomes mature in the thymus gland and differentiates into one of several kinds of cells that function in cell-mediated immunity.

Tectorial (tek-TŌ-rē-al) **membrane** A gelatinous membrane projecting over and in contact with the hair cells of the spiral organ (organ of Corti) in the cochlear duct.

Teeth Accessory structures of digestion composed of calcified connective tissue and embedded in bony sockets of the mandible and maxilla that cut, shred, crush, and grind food. Also called *dentes* (DEN-tēz).

Telophase (TEL-ō-fāz) The final stage of mitosis in which the daughter nuclei become established.

Temporomandibular joint (TMJ) syndrome A disorder of the temporomandibular joint (TMJ) characterized by dull pain around the ear, tenderness of jaw muscles, a clicking or popping noise when opening or closing the mouth, limited or abnormal opening of the mouth, headache, tooth sensitivity, and abnormal wearing of the teeth.

Tendon (TEN-don) A white fibrous cord of dense, regularly arranged connective tissue that attaches muscle to bone.

Tendon organ A proprioceptive receptor, sensitive to changes in muscle tension and force of contraction, found chiefly near the function of tendons and muscles. Also called *Golgi* (GOL-jē) *tendon organ.*

Teratogen (TER-a-tō-jen) Any agent or factor that causes physical defects in a developing embryo.

Testis (TES-tis) Male gonad that produces sperm and the hormones testosterone and inhibin. Also called a *testicle.*

Testosterone (tes-TOS-te-rōn) A male sex hormone (androgen) secreted by interstitial endocrinocytes (cells of Leydig) of a mature testis; controls the growth and development of male sex organs, secondary sex characteristics, sperm, and body growth.

Tetanus (TET-a-nus) An infectious disease caused by the toxin of *Clostridium tetani*, characterized by tonic muscle spasms and exaggerated reflexes, lockjaw, and arching of the back; a smooth, sustained contraction produced by a series of very rapid stimuli to a muscle.

Tetany (TET-a-nē) A nervous condition caused by hypoparathyroidism and characterized by intermittent or continuous tonic muscular contractions of the limbs.

Tetralogy of Fallot (tet-RAL-ō-jē of fal-Ō) A combination of four congenital heart defects: (1) constricted pulmonary semilunar valve, (2) interventricular septal opening, (3) emergence of aorta from both ventricles instead of from the left only, and (4) enlarged right ventricle.

Thalamus (THAL-a-mus) A large, oval structure located above the midbrain, consisting of two masses of gray matter covered by a thin layer of white matter.

Therapy (THER-a-pē) The treatment of a disease or disorder.

Thermoreceptor (THER-mō-rē-sep-tor) Receptor that detects changes in temperature.

Thirst center A cluster of neurons in the hypothalamus that is sensitive to the osmotic pressure of extracellular fluid and brings about the sensation of thirst.

Thoracic (thō-RAS-ik) **cavity** Superior component of the ventral body cavity that contains two pleural cavities, the mediastinum, and the pericardial cavity.

Thoracic duct A lymphatic vessel that begins as a dilation called the cisterna chyli, receives lymph from the left side of the head, neck, and chest, the left arm, and the entire body below the ribs, and empties into the left subclavian vein. Also called the *left lymphatic* (lim-FAT-ik) *duct.*

Threshold potential The membrane voltage that must be reached in order to trigger an action potential (nerve impulse).

Threshold stimulus Any stimulus strong enough to initiate an action potential (nerve impulse). Also called a *liminal* (LIM-i-nal) *stimulus.*

Thrombin (THROM-bin) The active enzyme formed from prothrombin that acts to convert fibrinogen to fibrin.

Thrombosis (throm-BŌ-sis) The formation of a clot in an unbroken blood vessel, usually a vein.

Thrombus A clot formed in an unbroken blood vessel, usually a vein.

Thymus (THĪ-mus) **gland** A bilobed organ, located in the upper mediastinum posterior to the sternum and between the lungs, that plays a role in the immune responses of the body.

Thyroid cartilage (THĪ-royd KAR-ti-lij) The largest single cartilage of the larynx, consisting of two fused plates that form the anterior wall of the larynx. Also called the *Adam's apple.*

Thyroid gland An endocrine gland with right and left lateral lobes on either side of the trachea connected by an isthmus located in front of the trachea just below the cricoid cartilage.

Thyroid-stimulating hormone (TSH) A hormone secreted by the anterior pituitary gland that stimulates the synthesis and secretion of hormones produced by the thyroid gland.

Thyroxine (thī-ROK-sēn) **(T_4)** A hormone secreted by the thyroid that regulates organic metabolism, growth and development, and the activity of the nervous system.

Tic Spasmodic twitching made involuntarily by muscles that are ordinarily under voluntary control.

Tissue A group of similar cells and their intercellular substance joined together to perform a specific function.

Tissue factor (TF) A factor, or collection of factors, whose appearance initiates the blood clotting process. Also called *thromboplastin* (throm-bō-PLAS-tin).

Tissue rejection Phenomenon by which the body recognizes the protein (HLA antigens) in transplanted tissues or organs as foreign and produces antibodies against them.

Tonsil (TON-sil) A multiple aggregation of large lymphatic nodules embedded in mucous membrane.

Topical (TOP-i-kal) Applied to the surface rather than ingested or injected.

Torn cartilage A tearing of an articular disc in the knee.

Toxic (TOK-sik) Pertaining to poison; poisonous.

Toxic shock syndrome (TSS) A disease caused by the bacterium *Staphylococcus aureus*, occurring mainly among menstruating females who use tampons and characterized by high fever, sore throat, headache, fatigue, irritability, and abdominal pain.

Trabecula (tra-BEK-yoo-la) Irregular lattice work of thin plates of spongy bone. Fibrous cord of connective tissue serving as supporting fiber by forming a septum extending into an organ from its wall or capsule. *Plural, trabeculae* (tra-BEK-yoo-lē).

Trachea (TRĀ-kē-a) Tubular air passageway extending from the larynx to the fifth thoracic vertebra. Also called the *windpipe.*

Tracheostomy (trā-kē-OS-tō-mē) Creation of an opening into the trachea through the neck (below the cricoid cartilage), with insertion of a tube to facilitate passage of air or evacuation of secretions.

Tract A bundle of nerve fibers in the central nervous system.

Transcription (trans-KRIP-shun) The first step in the transfer of genetic information in which a single strand of a DNA molecule serves as a template for the formation of an RNA molecule.

Transfusion (trans-FYOO-shun) Transfer of whole blood, blood components, or red bone marrow directly into the bloodstream.

Transient ischemic (is-KĒ-mik) **attack (TIA)** Episode of temporary cerebral dysfunction caused by interference of the blood supply to the brain.

Transplantation (trans-plan-TĀ-shun) The replacement of injured or diseased tissues or organs with natural ones.

Translation The process by which information in the nitrogenous bases of mRNA directs the arrangement of amino acids on a protein.

Transverse plane A plane that runs parallel to the ground and divides the body or organs into superior and inferior portions. Also called a *horizontal plane.*

Transverse tubules (TOO-byools) **(T tubules)** Minute, cylindrical invaginations of the muscle fiber (cell) membrane that carry muscle action potentials deep into the muscle fiber.

Trauma (TRAW-ma) An injury, either a physical wound or psychic disorder, caused by an external agent or force, such as a physical blow or emotional shock; the agent or force that causes the injury.

Tremor (TREM-or) Rhythmic, involuntary, purposeless contraction of opposing muscle groups.

Tricuspid (trī-KUS-pid) **valve** Atrioventricular (AV) valve on the right side of the heart.

Trigeminal neuralgia (trī-JEM-i-nal noo-RAL-jē-a) Pain in one or more of the branches of the trigeminal (V) nerve. Also called *tic douloureux* (doo-loo-ROO).

Triglyceride (fat) A lipid compound formed from one molecule of glycerol and three molecules of fatty acids; the body's most highly concentrated source of energy. Adipose tissue, composed of adipocytes specialized for triglyceride storage and present in the form of soft pads between various organs for support, protection, and insulation.

Triiodothyronine (trī-ī-ōd-ō-THĪ-rō-nēn) **(T_3)** A hormone produced by the thyroid gland that regulates organic metabolism, growth and development, and the activity of the nervous system.

Trochlea (TROK-lē-a) A pulleylike surface.

Trophoblast (TRŌF-ō-blast) The outer covering of cells of the blastocyst.

Tropic (TRŌ-pik) **hormone** A hormone whose target is another endocrine gland.

True vocal cords Pair of mucous membrane folds below the ventricular folds that function in voice production. Also called *vocal folds.*

Tubal ligation (lī-GĀ-shun) A sterilization procedure in which the uterine (Fallopian) tubes are tied and cut.

Tuberculosis (too-berk-yoo-LŌ-sis) An infection of the lungs and pleurae caused by *Mycobacterium tuberculosis* resulting in destruction of lung tissue and its replacement by fibrous connective tissue.

Tubular reabsorption The movement of filtrate from renal tubules back into blood in response to the body's specific needs.

Tubular secretion The movement of substances in blood back into filtrate in response to the body's specific needs.

Tumor (TOO-mor) A growth of excess tissue due to an unusually rapid division of cells.

Tunica externa (eks-TER-na) The outer coat of an artery or vein, composed mostly of elastic and collagen fibers. Also called the *adventitia.*

Tunica interna (TOO-ni-ka in-TER-na) The inner coat of an artery or vein, consisting of a lining of endothelium, basement membrane, and internal elastic lamina. Also called the *tunica intima* (IN-ti-ma).

Tunica media (MĒ-dē-a) The middle coat of an artery or vein, composed of smooth muscle and elastic fibers.

Twitch contraction Rapid, jerky contraction of a muscle in response to a single stimulus.

Tympanic membrane See *Eardrum.*

Type II cutaneous mechanoreceptor A receptor embedded deeply in the dermis and deeper tissue that detects heavy and continuous touch sensations. Also called *end organ of Ruffini.*

Ulcer (UL-ser) An open lesion of the skin or a mucous membrane of the body with loss of substance and necrosis of the tissue.

Umbilical (um-BIL-i-kal) **cord** The long, ropelike structure containing the umbilical arteries and vein that connect the fetus to the placenta.

Umbilicus (um-BIL-i-kus or um-bil-Ī-kus) A small scar on the abdomen that marks the former attachment of the umbilical cord to the fetus. Also called the *navel.*

Uremia (yoo-RĒ-mē-a) Accumulation of toxic levels of urea and other nitrogenous waste products in the blood, usually resulting from severe kidney malfunction.

Ureter (YOO-re-ter) One of two tubes that connect the kidney with the urinary bladder.

Urethra (yoo-RĒ-thra) The duct from the urinary bladder to the exterior of the body that conveys urine in females and urine and semen in males.

Urinalysis The physical, chemical, and microscopic analysis or examination of urine.

Urinary (YOO-ri-ner-ē) **bladder** A hollow, muscular organ situated in the pelvic cavity posterior to the pubic symphysis.

Urinary tract infection (UTI) An infection of a part of the urinary tract or the presence of large numbers of microbes in urine.

Urine The fluid produced by the kidneys that contains wastes or excess materials and is excreted from the body through the urethra.

Urology (yoo-ROL-ō-jē) The specialized branch of medicine that deals with the structure, function, and diseases of the male and female urinary systems and the male reproductive system.

Uterine (YOO-ter-in) **tube** Duct that transports ova from the ovary to the uterus. Also called the *Fallopian* (fal-LŌ-pē-an) *tube* or *oviduct.*

Uterus (YOO-te-rus) The hollow, muscular organ in females that is the site of menstruation, implantation, development of the fetus, and labor. Also called the *womb.*

Utricle (YOO-tri-kul) The larger of the two divisions of the membranous labyrinth located inside the vestibule of the inner ear, containing a receptor organ for static equilibrium.

Uvula (YOO-vyoo-la) A soft, fleshy mass, especially the V-shaped pendant part, descending from the soft palate.

Vagina (va-JĪ-na) A muscular, tubular organ that leads from the uterus to the vestibule, situated between the urinary bladder and the rectum of the female.

Valence (VA-lens) The combining capacity of an atom; the number of deficit or extra electrons in the outermost electron shell of an atom.

Varicose (VAR-i-kōs) Pertaining to an unnatural swelling, as in the case of a varicose vein.

Vascular (VAS-kyoo-lar) Pertaining to or containing many blood vessels.

Vasectomy (va-SEK-tō-mē) A means of sterilization of males in which a portion of each ductus (vas) deferens is removed.

Vasoconstriction (vāz-ō-kon-STRIK-shun) A decrease in the size of the lumen of a blood vessel caused by contraction of the smooth muscle in the wall of the vessel.

Vasodilation (vās′-ō-DĪ-lā-shun) An increase in the size of the lumen of a blood vessel caused by relaxation of the smooth muscle in the wall of the vessel.

Vasomotor (vā-sō-MŌ-tor) **center** A cluster of neurons in the medulla oblongata that controls the diameter of blood vessels, especially arteries.

Vein A blood vessel that conveys blood from tissues back to the heart.

Vena cava (VĒ-na KĀ-va) One of two large veins that open into the right atrium, returning to the heart all of the deoxygenated blood from the systemic circulation except from the coronary circulation.

Venereal (ve-NĒ-rē-al) **disease (VD)** See *Sexually transmitted disease.*

Venesection (vēn′-e-SEK-shun) Opening of a vein for withdrawal of blood.

Ventral (VEN-tral) Pertaining to the anterior or front side of the body; opposite of dorsal.

Ventricle (VEN-tri-kul) A cavity in the brain or an inferior chamber of the heart.

Ventricular fibrillation (ven-TRIK-yoo-lar fib-ri-LĀ-shun) Asynchronous ventricular contractions that result in cardiovascular failure.

Vermiform appendix (VER-mi-form a-PEN-diks) A twisted, coiled tube attached to the cecum.

Vertebral (VER-te-bral) **canal** A cavity within the vertebral column formed by the vertebral foramina of all the vertebrae and containing the spinal cord. Also called the *spinal canal.*

Vertebral column The 26 vertebrae; encloses and protects the spinal cord and serves as a point of attachment for the ribs and back muscles. Also called the *spine, spinal column,* or *backbone.*

Vertigo (VER-ti-go) Sensation of spinning or movement.

Villus (VIL-lus) A projection of the intestinal mucosal cells containing connective tissue, blood vessels, and a lymphatic vessel; functions in the absorption of the end products of digestion. *Plural, villi* (VIL-Ī).

Viscera (VIS-er-a) The organs inside the ventral body cavity. *Singular, viscus* (VIS-kus).

Visceral (VIS-er-al) Pertaining to the organs or to the covering of an organ.

Visceral effector (e-FEK-tor) Cardiac muscle, smooth muscle, and glandular epithelium.

Visceral muscle An organ specialized for contraction, composed of smooth muscle fibers (cells), located in the walls of hollow internal structures, and stimulated by visceral efferent neurons.

Visceral pleura (PLOO-ra) The inner layer of the serous membrane that covers the lungs.

Vital signs Signs necessary to life that include temperature (T), pulse (P), respiratory rate (RR), and blood pressure (BP).

Vitamin An organic molecule necessary in trace amounts that acts as a catalyst in normal metabolic processes in the body.

Vitiligo (vit-i-LĪ-go) Patchy, white spots on the skin due to partial or complete loss of melanocytes.

Vitreous (VIT-rē-us) **body** A soft, jellylike substance that fills the vitreous chambers of the eyeball, lying between the lens and the retina.

Volkmann's canal See *Perforating canal.*

Vomiting Forcible expulsion of the contents of the upper gastrointestinal tract through the mouth.

Vulva (VUL-va) Collective designation for the external genitalia of the female. Also called the *pudendum* (poo-DEN-dum).

Wandering macrophage (MAK-rō-fāj) Phagocytic cell that develops from a monocyte, leaves the blood, and migrates to infected tissues.

White matter Aggregations or bundles of myelinated axons located in the brain and spinal cord.

White matter tract The treelike appearance of the white matter of the cerebellum when seen in midsagittal section. Also called *arbor vitae* (AR-bor VĒ-te). A series of branching ridges within the cervix of the uterus.

Yolk sac An extraembryonic membrane that connects with the midgut during early embryonic development, but is nonfunctional in humans.

Zona pellucida (pe-LOO-si-da) Clear glycoprotein layer between the primary oocyte and granulosa cells

Zygote (ZĪ-gōt) The single cell resulting from the union of a male and female gamete; the fertilized ovum.

Zymogenic (zī-mō-JEN-ik) **cell** One of the cells of a gastric gland that secretes the principal gastric enzyme precursor, pepsinogen. Also called a *peptic cell.*

Credits

photos

Chapter 1

Fig. 1.06a, transverse plane of brain: "Stephen A. Kieffer and E. Robert Heitzman, An Atlas of Cross-Sectional Anatomy. Harper & Row, Publishers, Inc. New York, 1979."
Fig. 1.06b, frontal plane of brain: Lester Bergman & Associates
Fig. 1.06c, midsagittal plane of brain: "©1988, Martin Rotker"

Chapter 3

Fig. 3.17a, late anaphase: CABISCO/Phototake NY
Fig. 3.17b, early prophase: CABISCO/Phototake NY
Fig. 3.17b, late prophase: CABISCO/Phototake NY
Fig. 3.17c, metaphase: CABISCO/Phototake NY
Fig. 3.17d, early anaphase: CABISCO/Phototake NY
Fig. 3.17d, late anaphase: CABISCO/Phototake NY
Fig. 3.17e, telophase: CABISCO/Phototake NY
Fig. 3.17f, daughter cells in early interphase: CABISCO/Phototake NY

Chapter 4

exhibit 4.01 A, EM- mesothelial lining: Biophoto/Photo Researchers
exhibit 4.01 B, EM- intestinal serosa: Douglas Merrill
exhibit 4.01 D, EM- kidney tubules: © Ed Reschke
exhibit 4.01 F, EM- epithelium of vilus from lining of small intestine: © Ed Reschke
exhibit 4.01 H, EM- Fallopian tube: Douglas Merrill
exhibit 4.01 J, EM- vagina: Biophoto/ Photo Researchers
exhibit 4.01 L, EM- duct of sweat gland: Biophoto/Photo Researchers
exhibit 4.01 N, EM- duct of submandibular salivary gland: Biophoto/ Photo Researchers
exhibit 4.01 P, EM- urinary bladder in relaxed state: Andrew J. Kuntzman
exhibit 4.01 R, EM- trachea: © Ed Reschke
exhibit 4.01 T, EM- secretory portion of sweat gland: Bruce Iverson
exhibit 4.01 V, EM- thyroid gland: Lester Bergman & Associates
exhibit 4.02 A, EM- mesenchyme of developing fetus: © R. Kessel/ VU

exhibit 4.02 C, EM- umbilical cord: Lester Bergman & Associates
exhibit 4.02 E, EM- subcutaneous tissue: Biophoto/Photo Researchers
exhibit 4.02 G, adipocytes of white fat of pancreas: © Ed Reschke
exhibit 4.02 I, EM- lymph node: Biophoto/Photo Researchers
exhibit 4.02 K, EM- tendon: Andrew J. Kuntzman
exhibit 4.02 M, EM- dermis of skin: © Ed Reschke
exhibit 4.02 O, EM- aorta: Biophoto/Photo Researchers
exhibit 4.02 Q, hyaline cartilage from trachea: Biophoto/Photo Researchers
exhibit 4.02 S, EM- fibrocartilage from patella: Biophoto/Photo Researchers
exhibit 4.02 U, EM- elastic cartilage from auricle of ear: Biophoto/ Photo Researchers

Chapter 5

Fig. 5.02a, EM- skin: Lester Bergman & Associates

Chapter 6

Fig. 6.01b, partially sectioned femur: Lester Bergman & Associates

Chapter 7

Fig. 7.03a joints between atlas and occipital bone and between backbone: ©1994 Evan J. Collela
Fig. 7.03b shoulder joint-hyperextension: ©1994 Evan J. Collela
Fig. 7.03b shoulder joint-extension: ©1994 Evan J. Collela
Fig. 7.03c elbow joint: ©1994 Evan J. Collela
Fig. 7.03d wrist joint: ©1994 Evan J. Collela
Fig. 7.03e hip joint: © 1983 by Gerard J. Tortora. Courtesy of Lynn and James Borghesi
Fig. 7.03f knee joint: © 1983 by Gerard J. Tortora. Courtesy of Lynn and James Borghesi
Fig. 7.03g hip joint: ©1994 Evan J. Collela
Fig. 7.03h joints between metacarpals and phlanges: © 1983 by Gerard J. Tortora. Courtesy of Lynn and James Borghesi
Fig. 7.03i shoulder joint: ©1994 Evan J. Collela
Fig. 7.03j joint between atlas and axis: Evan J. Collela
Fig. 7.03k shoulder joint: ©1994 Evan J. Collela

Fig. 7.04a-b, temporomandibular joint: © 1983 by Gerard J. Tortora. Courtesy of Lynne and James Borghesi

Fig. 7.04c temporomandibular joint: "© 1992 Scott, Foresman photo by John Moore"

Fig. 7.04d temporomandibular joint: "© 1992 Scott, Foresman photo by John Moore"

Fig. 7.04e intertarsal joint: © 1983 by Gerard J. Tortora. Courtesy of Lynne and James Borghesi

Fig. 7.04f intertarsal joint: © 1983 by Gerard J. Tortora. Courtesy of Lynne and James Borghesi

Fig. 7.04g ankle joint: Evan J. Collela

Fig. 7.04h proximal radioulnar joint: © 1983 by Gerard J. Tortora. Courtesy of Lynne and James Borghesi

Chapter 8

Fig. 8.02f, EM- muscle fibers: © Ed Reschke

Fig. 8.10b, EM- cardiac muscle fiber: © Ed Reschke

Fig. 8.11b, EM- longitudinal section of smooth muscle: Andrew J. Kuntzman

Chapter 13

Fig. 13.13b, EM- parathyroid gland: Biophoto/Photo Researchers

Fig. 13.14c, EM- subdivisions of adrenal gland: Andrew J. Kuntzman

Fig. 13.17b, EM- pancreatic islet: James Sheetz

Fig. 13.20a, acromegaly: "Hall/Evered, A Colour Atlas of Endocrinology, Mosby-Year Book/Wolfe Publishing"

Fig. 13.20b, cretinism: Lester Bergman & Associates

Fig. 13.20c, goiter: Kay/Peter Arnold

Fig. 13.20d, exophthalmos: Lester Bergman & Associates

Fig. 13.20e, Cushing's Syndrome: "Beverly M. K. Biller, Massachusetts General Hospital"

Fig. 13.20f, virilism: Lester Bergman & Associates

Chapter 14

Fig. 14.02b 'a', EM- neutrophil: © Ed Reschke

Fig. 14.02b 'b', EM- eosinophil: Biophoto/Photo Researchers

Fig. 14.02b 'c', EM- basophil: Biophoto/Photo Researchers

Fig. 14.02b 'd', photo- blood smear: Douglas Merrill

Fig. 14.02b 'e', EM- rbc's and platelet: Lester Bergman & Associates

Fig. 14.02b 'f', EM- monocyte: Biophoto/Photo Researchers

Fig. 14.02b 'g', EM- large lymphocyte: Lester Bergman & Associates

Fig. 14.07, EM- fibrin threads: "Lennart Nilsson, Our Body Victorious, Dell Publishing Company, © Boehringer Ingelheim International GmbH"

Fig. 14.10a, normal rbc: Lester Bergman & Associates

Fig. 14.10b, agglutinated rbc: Lester Bergman & Associates

Chapter 15

Fig. 15.10, EM- artery w/ atherosclerotic plaque: Sklar/Photo Researchers

Chapter 16

Fig. 16.02b, EM- rbcs squeezing through capillaries: "Lennart Nilsson. Behold Man. Dell Publishing Company, © Boehringer Ingelheim International GmbH"

Chapter 17

Fig. 17.05b, EM- phagocyte engulfing a yeast cell: © National Cancer Institute/Photo Researchers

Chapter 18

Fig. 18.03c, photo- thyroid- superior view + inset: "Mark Nielsen, University of Utah."

Chapter 19

Fig. 19.08b, EM- gastric pit: Hessler/VU

Fig. 19.12b, EM- small intestinal mucosa: "P. Motta and A. Familiari/ University La Sapienza, Rome/SPL/Photo Researchers"

Fig. 19.12c, color 'cadaver' photo of jejunum: "Mark Nielsen, University of Utah."

Fig. 19.15b, EM- large intestinal mucosa: "P. Motta/University La Sapienza, Rome/SPL/Photo Researchers"

Chapter 24

Fig. 24.03d, EM- blastocyst during implantation: "Roberts Rugh, Landrum B. Shettles, with Richard Einhorn: Conception to Birth: The Drama of Life's Beginnings. Copyright © 1971 by Roberts Rugh and Landrum B. Shettles. By permission of Harper & Row, Publishers, Inc."

Fig. 24.03d, EM- blastocyst during implantation: "Roberts Rugh, Landrum B. Shettles, with Richard Einhorn: Conception to Birth: The Drama of Life's Beginnings. Copyright ©1971 by Roberts Rugh and Landrum B. Shettles. By permission of Harper & Row, Publishers, Inc."

line art

Biaggio John Melloni, Ph.D.

12.09a, 12.09b, 12.10a, 12.10b, 12.10c, 12.11, 12.12a, 12.12b, 12.13a, 12.13b

Beth Willert

8.10a, 8.11a, 15.02a

Hilda Muinos

3.04, 3.05, 3.08, 8.03, 8.04, 8.05a, 8.05b, 8.05c, 8.06a-b, 8.07, 8.08, 9.01, 9.04a-e, exhibit 10.02b, exhibit 10.02c, exhibit 10.02d, exhibit 10.02e, exhibit 10.02e2, exhibit 10.02f, exhibit 10.02g, exhibit 10.02h, exhibit 10.02i, exhibit 10.02j, exhibit 10.02k, 11.02, 15.01, 15.02a, 15.03a, 15.03b, 15.03c and inset, 15.04 inset, 15.04a-b, 15.05 inset, 15.05a, 15.05b, 15.06a and inset, 15.07, 15.08, 16.01a-c, 16.07, 16.08a, 16.08b, 16.09, 16.10, 17.07a-d, 17.08, 17.09, 19.02, 19.08a, 19.08b, 19.12a, 19.12b, 19.15a, 19.15b, 24.08a-b

Jean Jackson

14.09, exhibit 17.02a, exhibit 17.02b, exhibit 17.02c, exhibit 17.02d, exhibit 17.02e

Jared Schneidman Design

1.02, 2.01, 2.02, 2.03a-c, 2.04, 2.05a-c, 2.06, 2.07a-b, 2.08a-c, 2.09a-b, 2.09b, 2.10, 2.11a-b, 2.12, 3.06a-b, 3.07a-c, 3.09, 3.11b, 3.14, 3.15, 3.16, 5.05, 6.05, 9.03, 9.05, 11.01a, 11.01b, 12.06a-c, 13.03a, 13.03b, 13.06, 13.08, 13.09, 13.11, 13.12, 13.15, 13.16, 13.18, 13.19, exhibit 14.02a, exhibit 14.02b, exhibit 14.02c, exhibit 14.02d, exhibit 14.02e, exhibit 14.02f, exhibit 14.02g, 14.01, 14.04, 14.05, 14.08, 15.09, 16.03, 16.04, 16.05, 18.09, 18.12, 18.13, 18.14, 18.15, 18.16, 18.17, 19.13a, 19.13b, 20.07, 20.08, 20.09, 21.03a inset, 21.03b, 21.06, 21.07, 21.08, 22.01a-b, 22.03, 22.04, 22.05, 22.07, 23.03, 23.05, 23.11, 23.15, 23.16, 23.17, 23.18, 24.02, 24.09a, 24.09b, 24.11, 24.12a, 24.12b, 24.13

Kevin Somerville

1.01, 1.03a, 1.03b, 1.04, 1.06a1, 1.06b1, 1.06c1, 1.07a-b, 1.08, 1.08 inset, 1.09a, 1.09b, 1.10, exhibit 4.01 C, exhibit 4.01 E, exhibit 4.01 G, exhibit 4.01 I, exhibit 4.01 K, exhibit 4.01 M, exhibit 4.01 O, exhibit 4.01 Q, exhibit 4.01 S, exhibit 4.01 U, exhibit 4.01 W, exhibit 4.02 B, exhibit 4.02 D, exhibit 4.02 H, exhibit 4.02 J, exhibit 4.02 L, exhibit 4.02 N, exhibit 4.02 P, exhibit 4.02 R, exhibit 4.02 T, exhibit 4.02 V, 5.02b, 5.03 inset, 5.04 inset, 5.04a-b, 6.32a inset, 6.32b inset, 7.01 inset, 7.05c inset, exhibit 9.01a, exhibit 9.01b, exhibit 9.01c, exhibit 9.01d, exhibit 9.01e, exhibit 9.01f, 10.02b inset, 10.03 inset, 10.06a inset, 10.07a inset, 10.09 inset, 10.10 inset, 12.04 inset, 16.11, 18.07, 18.08 inset, 18.08a-b, 18.10a-c, 19.08a inset, 19.11, 19.12a inset, 19.15a inset, 21.01, 21.02 inset, 21.11 inset, 23.01 and inset, 23.02a and inset, 23.02b and inset, 23.04, 23.06, 23.07a inset, 23.07a-b, 23.07b inset, 23.08 and inset, 23.09 and inset, 23.10 and inset, 23.12, 23.13, 23.14 inset, 23.14a-b, 24.03a, 24.03b, 24.03c, 24.03d, 24.04 and inset, 24.05a, 24.05b, 24.06 and inset, 24.07, 24.10, exhibit 24.02

Leonard Dank

6.01a, 6.02b, 6.03, 6.04, 6.06, 6.07a, 6.07b, 6.07c, 6.08 and inset, 6.09 and inset, 6.10 inset, 6.10a, 6.10b, 6.11 inset, 6.11a, 6.11b, 6.12 inset, 6.12a, 6.12b, 6.13a-b, 6.14a, 6.15 inset, 6.15a-b, 6.16 inset, 6.16a, 6.16b, 6.16c, 6.17, 6.17 inset, 6.18 inset, 6.18a-b, 6.19, 6.20, 6.21, 6.21a-b, 6.22, 6.23a, 6.23b, 6.24, 6.25a-b, 6.26, 6.26 inset, 6.27, 6.28, 6.29, 6.30, 6.31, 6.32a, 6.32b, exhibit 6.03a, exhibit 6.03b, exhibit 6.03c, exhibit 6.03d, 7.01, 7.02, 7.05a, 7.05b, 7.05c, 8.12a-b, 8.13a, 8.13b, 8.14a-b, 8.15, 8.16, 8.17a-b, 8.18, 8.19a-b, 8.19c-d, 8.20, 8.21, 8.22a-b, 8.23, 8.24a, 8.24b, 8.25a-b, 8.25c-d, 16.06a-b

Lauren Keswick

3.01, 3.02 and inset, 3.04 inset, 3.05 inset, 3.08 inset, 3.10 and inset, 3.11a and inset, 3.12 and inset, 3.13 and inset, 3.17a-f, exhibit 3.2, exhibit 4.02 F, 5.01, 5.03, 6.02a, exhibit 8.01b, exhibit 8.01c, exhibit 8.01d

Lynn O'Kelley

1.05, 12.01, 12.02a-b, 12.03a, 12.03b, 12.03c, 12.05a inset, 12.05b, 12.05b1, 12.09b inset, 12.10a inset, 12.12a inset, 12.13a inset, exhibit 12.02 g, exhibit 12.02b, exhibit 12.02c, exhibit 12.02d, exhibit 12.02e, exhibit 12.02f, exhibit 12.03b, exhibit 12.03c, exhibit 12.03d, exhibit 12.03e, 13.01, 13.04, 13.04 inset 1, 13.04 inset 2, 13.07, 13.07 inset 1, 13.07 inset 2, 13.10a and inset, 13.10c, 13.13a and inset, 13.13c, 13.14 a, 13.14 inset, 13.14b, 13.17a and inset, 13.17b, 18.01, 18.02, 18.02 inset 2, 18.03a-b, 18.05, 18.06 inset, 18.06a-b, 18.07 inset, 18.3a inset

Nadine Sokol

3.03, 6.14b, 6.19 inset, 6.21 inset, 6.22 inset, 6.23a inset, 6.23b inset, 6.24 inset, 6.25a inset, 6.25b inset, 6.28 inset, 6.29 inset, 6.30 inset, exhibit 6.02a, exhibit 6.02b, 8.01, 8.02a-e, exhibit 10.01b, exhibit 10.01c, exhibit 10.01d, exhibit 10.01e, exhibit 10.01f, exhibit 10.01g, 10.08 inset, 12.07a-e, 12.08, 13.02, 13.05, 14.02a, 14.03a, 14.03b, 14.06, 16.02a, 17.05a, 17.06, 17.11, 18.02 inset 1, 18.11, 18.15 inset, 19.01, 19.03a, 19.03b, 19.04, 19.05 and inset, 19.06c, 19.07, 19.09, 19.10 inset, 19.10a-b, 19.14a-b, 21.02, 21.03 inset, 21.03a, 21.04, 21.05, 21.09, 21.10, 21.11, 22.02, 22.06, 23.06 inset, 24.01, 24.05c

Page Two

20.01, 20.02, 20.03, 20.04, 20.05, 20.06

Sharon Ellis

9.02a-b, 10.01, 10.02a, 10.02b, 10.03, 10.04b, 10.05, 10.06a, 10.07a, 10.07b, 10.08, 10.09, 10.10, 10.11a, 10.11b, 10.12, 11.03, 12.04, 12.09a inset, 12.12c1, 12.12c2, 12.13 inset 1, 12.13 inset 2, 17.01 and inset, 17.02a, 17.02b, 17.03

Index

Note: Page numbers followed by the letter E indicate terms to be found in exhibits and corresponding art. Page numbers followed by the letter *f* indicate terms to be found in figures and corresponding captions.

Glossary of combining forms, word roots, prefixes, and suffixes

(continued from front endpapers)

Ambi- both sides Ambidextrous (am′-bi-DEK-strus), able to use either hand.

Ambly- dull Amblyaphia (am-blē-A-fē-a), dull sense of touch.

Andro- male, masculine Androgen (AN-drō-jen), male sex hormone.

Ankyl(o)- bent, fusion Ankylosed (ANG-ki-lōsd), fused joint.

Ante- before Antepartum (ant-ē-PAR-tum), before delivery of a baby.

Anti- against Anticoagulant (an-ti-kō-AG-yoo-lant), a substance that prevents coagulation (clotting) of blood.

Basi- base, foundation Basal (BĀ-sal), located near the base.

Bi- two, double, both Biceps (BĪ-seps), a muscle with two heads of origin.

Bili- bile, gall Biliary (BIL-ē-er-ē), pertaining to bile, bile ducts, or gallbladder.

Brachy- short Brachyesophagus (brā-kē-e-SOF-a-gus), short esophagus.

Brady- slow Bradycardia (brād′-ē-KARD-ē-a), abnormally slow resting heart rate

Cata- down, lower, under, against Catabolism (ka-TAB-a-lizm), metabolic breakdown into simpler substances.

Circum- around Circumrenal (ser-kum-RĒN-al), around the kidney.

Cirrh- yellow Cirrhosis (si-RŌ-sis), liver disorder that causes yellowing of skin.

Co-, Con-, Com- with, together Congenital (kon-JEN-i-tal), existing at birth.

Contra- against, opposite Contraception (kon-tra-SEP-shun), the prevention of conception.

Crypt- hidden, concealed Cryptorchidism (krip-TOR-ka-dizm′), undescended or hidden testes.

Cyano- blue Cyanosis (sī-a-NŌ-sis), bluish discoloration due to inadequate oxygen.

De- down, from Decay (de-KA), waste away from normal.

Demi-, hemi- half Hemiplegia (hem′-ē-PLE-jē-a), paralysis on one side of the body.

Di-, Diplo- two Diploid (DIP-loyd), having double the haploid number of chromosomes.

Dis- separation, apart, away from Disarticulate (dis′-ar-TIK-yoo-lāt′), to separate at a joint.

Dys- painful, difficult Dyspnea (disp-NĒ-a), difficult breathing.

E-, Ec-, Ex- out from, out of Eccentric (ek-SEN-trik), not located at the center.

Ecto-, Exo- outside Ectopic (ek-TOP-ik) pregnancy, gestation outside the uterine cavity.

Em-, En- in, on Empyema (em′-pi-Ē-ma), pus in a body cavity.

End-, Endo- inside Endocardium (en′-dō-KARD-ē-um), membrane lining the inner surface of the heart.

Epi- upon, on, above Epidermis (ep′-i-DER-mis), outermost layer of skin.

Eu- well Eupnea (YOOP-nē-a), normal breathing.

Ex-, Exo- out, away from Exocrine (EK-sō-krin), excreting outwardly or away from.

Extra- outside, beyond, in addition to Extracellular (ek′-stra-SEL-yoo-lar), outside the cell.

Fore- before, in front of Forehead (FOR-hed), anterior part of head.

Gen- originate, produce, form Genetics (gen-ET-iks), the study of heredity.

Gingiv- gum Gingivitis (jin′-je-VĪ-tus), inflammation of the gums.

Hemi- half Hemiplegia (hem-ē-PLĒ-jē-a), paralysis of only half of the body.

Heter-, Hetero- other, different Heterogeneous (het′-e-rō-JEN-ē-us), composed of different substances.

Homeo-, Homo- unchanging, the same, steady Homeostasis (hō′-mē-ō-STA-sis), achievement of a steady state.

Hyper- beyond, excessive Hyperglycemia (hī-per-gli-SĒ-mē-a), excessive amount of glucose in the blood.

Hypo- under, below, deficient Hypodermic (hī-pō-DER-mik), below the skin or dermis.

Idio- self, one's own, separate Idiopathic (id′-ē-o-PATH-ik), a disease without recognizable cause.

In-, Im- in, inside, not Incontinent (in-KON-ti-nent), not able to retain urine or feces.

Infra- beneath Infraorbital (in′-fra-OR-bi-tal), beneath the orbit.

Inter- among, between Intercostal (int′-er-KOS-tal), between the ribs.

Intra- within, inside Intracellular (in′-tra-SEL-yoo-lar), inside the cell.

Iso- equal, like Isotonic (ī-sō-TON-ik), equal tension or tone.

Later- side Lateral (LAT-er-al), pertaining to a side or farther from the midline.

Lepto- small, slender, thin Leptodermic (lep′-tō-DER-mik), having thin skin.

Macro- large, great Macrophage (MAK-rō-fāj), large phagocytic cell.

Mal- bad, abnormal Malnutrition (mal′-noo-TRISH-un), lack of necessary food substances.

Medi-, Meso- middle Medial (MĒD-ē-al), nearer to midline.

Mega-, Megalo- great, large Megakaryocyte (meg′-a-KAR-ē-ō-sit), giant cell of bone marrow.

Melan- black Melanin (MEL-a-nin), black or dark brown pigment found in skin and hair.

Meta- after, beyond Metacarpus (met-a-KAR-pus), the part of the hand between the wrist and fingers.

Micro- small Microtome (MĪ-krō-tōm), instrument for preparing very thin slices of tissue for microscopic examination.

Mono- one Monorchid (mon-OR-kid), having one testicle.

Neo- new Neonatal (nē-ō-NĀT-al), pertaining to the first weeks after birth.

Noct(i)- night Nocturia (nok-TOO-rē-a), involuntary urination occurring at night during sleep.

Null(i)- none Nullipara (nu-LIP-a-ra), woman with no children.

Nyct- night Nyctalopia (nik′-ta-LŌ-pē-a), night blindness.

Oligo- small, deficient Oliguria (ol-ig-YOO-rē-a), abnormally small amount of urine.

Ortho- straight, normal Orthopnea (or-THOP-nē-a), inability to breathe in any position except when straight or erect.

Pan- all Pancarditis (pan-kar-DĪ-tis), inflammation of the entire heart.

Para- near, beyond, apart from, beside Paranasal (par-a-NĀ-zal), near the nose.

Per- through Percutaneous (per′-kyoo-TĀ-nē-us), through the skin.

Peri- around Pericardium (per′-i-KARD-ē-um), membrane or sac around the heart.

Poly- much, many Polycythemia (pol′-i-sī-THĒ-mē-a), an excess of red blood cells.

Post- after, beyond Postnatal (pōst-NĀT-al), after birth.

Pre-, Pro- before, in front of Prenatal (prē-NĀ-T-al), before birth.

Prim- first Primary (PRĪ-me-rē), first in time or order.

Proto- first Protocol (PRŌ-to-kol), clinical report made from first notes taken.

Pseud-, Pseudo- false Pseudoangina (soo′-dō-an-JĪ-na), false angina.

Retro- backward, located behind Retroperitoneal (re′-trō-per′-it-on-Ē-al), located behind the peritoneum.

Schizo- split, divide Schizophrenia (skiz′-ō-FRE-nē-a), split personality mental disorder.

Semi- half Semicircular (sem′-i-SER-kyoo-lar) canals, canals in the shape of a half circle in the ears.

Sub- under, beneath, below Submucosa (sub′-myoo-KŌ-sa), tissue layer under a mucous membrane.

Super- above, beyond Superficial (soo-per-FISH-al), confined to the surface.

Supra- above, over Suprarenal (soo-pra-RĒN-al), adrenal gland above the kidney.